Comprehensive Desk Reference of
Polymer Characterization and Analysis

Comprehensive Desk Reference of
Polymer Characterization and Analysis

Robert F. Brady, Jr.
Editor

American Chemical Society
Washington, D.C.

2003

OXFORD
UNIVERSITY PRESS

Oxford   New York
Auckland   Bangkok   Buenos Aires   Cape Town   Chennai
Dar es Salaam   Delhi   Hong Kong   Istanbul   Karachi   Kolkata
Kuala Lumpur   Madrid   Melbourne   Mexico City   Mumbai   Nairobi
São Paulo   Shanghai   Taipei   Tokyo   Toronto

and an associated company in Berlin

Developed and distributed in partnership by
American Chemical Society and Oxford University Press

Published by Oxford University Press, Inc.
198 Madison Avenue, New York, New York 10016

www.oup.com

Oxford is a registered trademark of Oxford University Press.

Library of Congress Cataloging-in-Publication Data
Comprehensive desk reference of polymer characterization and analysis / Robert F.
Brady, Jr., editor.
    p.   cm.
  Includes bibliographic references and index.
  ISBN 0-8412-3665-8
    1. Polymers—Analysis—Handbooks, manuals, etc.   I. Brady, Robert F., 1942–
QD139.P6   C66   2002
547'.7046—dc21   2001056650

9 8 7 6 5 4 3 2 1

Printed in the United States of America
on acid-free paper

To Sharon

Preface

A large fraction of chemists are employed in the polymer industry. Many are new graduates, and many others come from other segments of the industry or from academia. Frequently these people enter the industry in some phase of polymer characterization and find the challenges daunting. This book is intended to be the first book they will turn to for an understanding of the work they do, how to do it, and how to interpret it. This book is intended to be a practical manual for the entry-level polymer scientist that, although it omits step-by-step instructions, gives a firm scientific foundation of basic principles and teaches how the measurements made relate to molecular structure and thus to physical properties. The book is not a research manual but, rather, a guide to performing and understanding polymer characterization, as well as an entree to the specialized literature of the analytical chemistry of polymers.

Techniques now being used to characterize polymers fall into three broad categories. Many established techniques have a solid theoretical basis, and it is clear how the macroscopic properties of polymers arise from the molecular structure and properties of the polymers themselves. But many other test methods used daily throughout the polymer industry are highly empirical and the value measured, rather than being determined entirely by the sample, is significantly influenced by the design and operation of the test apparatus. Lastly, there are emerging analytical techniques that, although they are not yet widely used in the industry, offer the opportunity to understand polymer structure and performance in greater depth and are beginning to influence the design and use of polymeric materials. This book endeavors to cover the most important techniques in each of these categories. An objective of this book is to set forth the scientific foundations (where one exists) of polymeric materials and the measurements made on them, in that way highlighting the strengths and limitations of each method.

We envision that the reader will pick up this book, quickly locate the information of interest, absorb what is needed, and return the book to the shelf until its next use. Thus each self-contained chapter presents easily accessible information using appropriate mathematics, clear illustrations, and unambiguous language. Each chapter provides information on when to select the technique, what types of information it provides, how to obtain this information, and how to interpret it. For the reader who needs to explore a topic in depth, each chapter has a comprehensive list of references to advanced topics.

Although this book focuses on synthetic polymers, biological polymers are included whenever appropriate so that those interested in biopolymers such as polysaccharides, nucleic acids, and proteins will find the book directly applicable to their needs. Methods used only for the characterization of biopolymers are beyond the scope of this book.

I thank the authors for their valuable contributions to this volume. I am also indebted to the staff at ACS Books, particularly to Anne Wilson for her valuable insights on the concept and contents of this book, and Margaret Brown for her expert assistance in producing this book.

Contents

Contributors

David F. Alliet
One Post Side Lane
Pittsford, NY 14534

David W. Ball
Department of Chemistry
Cleveland State University
1983 East 24th Street
Cleveland, OH 44115-2440

Howard G. Barth
Corporate Center for Analytical Sciences
DuPont Company Experimental Station
P.O. Box 80228
Wilmington, DE 19880-0228

Guy C. Berry
Department of Chemistry
Carnegie Mellon University
4400 Fifth Avenue
Pittsburgh, PA 15213-3890

Robert F. Brady, Jr.
706 Hope Lane
Gaithersburg, MD 20878-1883

John Cahoon
Department of Chemistry
Cleveland State University
6325 Aldenham Drive
Cleveland, OH 44143-3331

Donald A. Chernoff
Advanced Surface Microscopy
6009 Knyghton Road
Indianapolis, IN 46220

Anthony R. Dolan
Department of Chemistry
Natural Sciences Complex
State University of New York at Buffalo
Buffalo, NY 14260-3000

Barbara Foster
Microscopy/Marketing & Education
125 Paridon Street, Suite 102
Springfield, MA 01118-2130

Joseph A. Gardella Jr.
Department of Chemistry
Natural Sciences Complex
State University of New York at Buffalo
Buffalo, NY 14260-3000

Carin A. Helfer
The Maurice Morton Institute of Polymer Science
The University of Akron
260 South Forge Street
Akron, OH 44325-3909

Marcelo M. Hirschler
GBH International
2 Friar's Lane
Mill Valley, CA 94941

Brian K. Kochanowski
Solutia Inc.
730 Worcester Street
Springfield, MA 01151

William J. Koros
School of Chemical Engineering
Georgia Institute of Technology
Atlanta, GA 30332-0100

Ilario Losito
Department of Chemistry
Natural Sciences Complex
State University of New York at Buffalo
Buffalo, NY 14260-3000

Sergei Magonov
Digital Instruments
112 Robin Hill Road
Santa Barbara, CA 93117

John Marshall
Research and Commercial Development
Glasgow Caledonian University
70 Cowcaddens Road
Glasgow G4 0BA
United Kingdom

Wayne L. Mattice
The Maurice Morton Institute of Polymer Science
The University of Akron
260 South Forge Street
Akron, OH 44325-3909

E. Peter Maziarz III
Bausch and Lomb Healthcare
1400 North Goodman Street
Rochester, NY 14292-0450

Francisco Menduti
Departmento de Química Física
Universidad de Alcalá
Alcalá de Henares
Madrid, Spain

Peter A. Mirau
Lucent Technologies
600 Mountain Avenue
Murray Hill, NJ 07974-0636

Stephen L. Morgan
Department of Chemistry and Biochemistry
The University of South Carolina
Columbia, SC 29208

Christopher R. Mubarak
Department of Chemistry and Biochemistry
The University of South Carolina
Columbia, SC 29208

Marcus N. Myers
FFFractionation LLC
4797 South West Ridge Boulevard
Salt Lake City, UT 84118

Michael J. Owen
Dow Corning Corporation
200 West Salzburg Road
Midland, MI 48686-0994

Wei-Ping Pan
Department of Chemistry
Cleveland State University
6325 Aldenham Drive
Cleveland, OH 44143-3331

Jung Ok Park
School of Textile and Fiber Engineering
Georgia Institute of Technology
Atlanta, GA 30332-0295

Edward N. Peters
GE Plastics
1 Noryl Avenue
Selkirk, NY 12158

Lee N. Polite
Axion Analytical Laboratories, Inc.
14 N. Peoria Street, Suite 100
Chicago, IL 60607

Steven K. Pollack
Naval Research Laboratory
Center for Bio/Molecular Science and Engineering
4555 Overlook Avenue, SW
Washington, DC 20375

Thomas H. Richardson
Department of Chemistry
The Citadel
Charleston, SC 29409

Alan T. Riga
Department of Chemistry
Cleveland State University
6325 Aldenham Drive
Cleveland, OH 44143-3331

Ryong-Joon Roe
Department of Materials Science and Engineering
University of Cincinnati
498 Rhodes Hall
Cincinnati, OH 45221-0012

Luisa Sabbatini
Department of Chemistry
University of Bari
Bari, Italy

Mohan Srinivasarao
School of Textile and Fiber Engineering
Georgia Institute of Technology
Atlanta, GA 30332-0295

Laurie L. Switzer
Xerox Corporation
800 Phillips Road
Webster, NY 14580

William Stephen Tait
Pair O Docs Professionals, L.L.C.
510 Charmany Drive, Suite 55
Madison, WI 53719

Thomas P. White
Department of Chemistry
Natural Sciences & Mathematics Complex
State University of New York at Buffalo
Buffalo, NY 14260-3000

Garth L. Wilkes
Polymer Materials and Interfaces Laboratory and
Department of Chemical Engineering
Virginia Polytechnic Institute and State University
Blacksburg, VA 24061-0211

Troy D. Wood
Department of Chemistry
Natural Sciences & Mathematics Complex
State University of New York at Buffalo
Buffalo, NY 14260-3000

Catherine M. Zimmerman
150 West Warrenville Road
Napierville, IL 60563

Comprehensive Desk Reference of
Polymer Characterization and Analysis

1

Introduction to Polymer Characterization

EDWARD N. PETERS

What Are Polymers?

Polymers, which are also known as macromolecules, are ubiquitous. All forms of life depend on polymer molecules. Natural polymers such as carbohydrates, proteins, and nucleic acids are fundamental to the biological sustenance of life. From the earliest times, natural polymeric materials have been employed to satisfy human needs. These are exemplified by wood, hide, natural resins, rubbers, gums, and fibers like cotton, wool, and silk. Through human necessity and ingenuity, natural polymers have been modified to improve their utility and synthetic polymers have been developed. Synthetic polymers in the form of plastics, fibers, rubbers, adhesives, and coatings have come on the scene as the result of a continual search for man-made substances that can perform better or can be produced at a lower cost than natural materials (such as wood, glass, and metal), which require mining, refining, processing, milling, and machining.[1,2] Polymers can also increase productivity by producing finished parts and consolidating parts. For example, an item made of several metal parts that require separate fabrication and assembly can often be consolidated into one or two plastic parts. Such increases in productivity have led to fantastic growth in industries based on macromolecules.

The term *polymer* is derived from the Greek words *poly* and *meros* meaning "many parts". A *monomer* (*mono* means one) is a small molecule of low molecular weight, which combines many times with other monomers of the same or different type to form a polymer. Thus, hundreds to thousands of monomers link together to form very long, chain-like molecules, or polymers.[3] A simple model of a polymer may be represented by a string of beads, where each single bead represents a monomer unit. This simple model is shown in Figure 1.1.

An *oligomer* is a molecule with low molecular weight and is formed by the combination of a few to several monomer units. The term stems from the Greek word *oligo* meaning "a few". Oligomers do not have sufficient molecular weight to be practically useful. Oligomers are formed during the making of synthetic polymers and may be present in low amounts in polymers.

A repeating unit of a linear polymer is a portion of the polymer such that the complete molecule (except for end groups) might be produced by linking together to form sufficiently large numbers of the units. Polymer structures are normally drawn by showing only one repeating unit rather than trying to depict the long polymer chains.[4] For example, polystyrene would be represented by either of the structures shown in Figure 1.2. The subscript n represents the number of repeating units in the polymer molecule. This number is often not known exactly in synthetic polymers, and sometimes the subscript is not used. This representation of polymer structures implies that the whole molecule is made up of a sequence of such repeating units by linking the left-side atom to the right-side atom. The polymer structure is usually written only in terms of the repeating unit, and the end groups are not always known. In most cases, end groups in high molecular weight polymers are present in very low amounts and have a negligible effect on polymer properties. For example, a linear polymer with high molecular weight and with 1000 to 2000 repeating units would have only two end groups. A polymer with only 100 to 200 repeating units would still have only two units, but the concentration of end groups is much higher and the end groups could have a minor effect on properties.[4]

Most properties of polymers depend on the large molecular chain length. With low molecular weight chains, the molecular chains slide over each other and separate from each other when mechanical stress is applied. The simple model shown in Figure 1.3 characterizes this.

With increasing chain length, the molecules bend, twist, turn, tangle, and collapse in on themselves and their neighbors. It is largely this structural phenomenon of chain en-

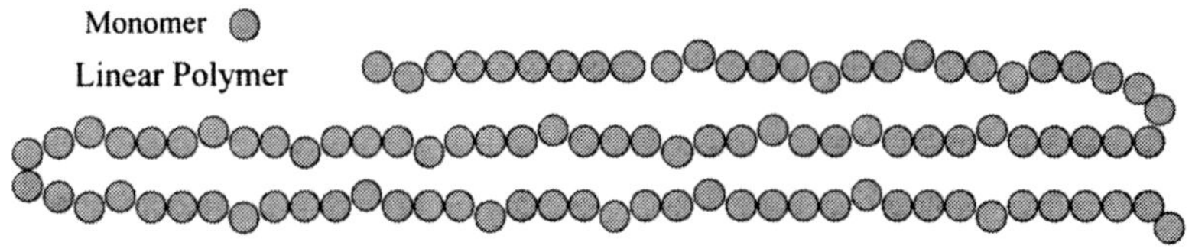

Monomer ⊚

Linear Polymer

Figure 1.1 Model of a linear polymer.

tanglement and the attractive forces between molecules that gives polymers their unique characteristics.[5,6] Indeed, when mechanical stress is applied to chains with high molecular weight, the chains stretch but do not separate from each other, as shown in the simple model in Figure 1.4. Moreover, the sum of the entanglement and intermolecular forces becomes greater than the strength of the individual covalent bonds in the polymer backbone. Hence the covalent bonds in the polymer backbone will break when enough stress is applied.

Some properties of polymers depend on chain entanglement and will increase as the molecular size increases—for example, strength and impact resistance. These properties will tend to level off gradually, as molecular size becomes very large. Other properties depend on chain disentanglement (chemical/solvent resistance, viscosity in the molten state), and these properties will increase continuously with molecular size. This relationship of molecular entanglement and disentanglement appears in Figure 1.5.

Not all polymers are alike. Although they all share certain unique characteristics, the individual properties of each depend on the chemical nature of the repeating unit(s), the shape and length of the polymer chains, the alignment of the molecular chains, and any structural isomerism.[7,8] In addition to carbon in the polymer backbone, there can also be heteroatoms such as oxygen or nitrogen. Moreover, the backbone of polymer chains can be inorganic with and without pendant organic groups.[9,10]

Departures from linear sequence leads to structures of increasing geometric complexity. A branched structure has short or long chain branches attached along the main chain or polymer backbone, as shown by the simple model in Figure 1.6. For branching to occur, some of the units—that is, the branch points—must possess bonding characteristics that are different from the predominant repeating unit.

Linear chains that are covalently linked together can represent network structures. Thus all the polymer chains are bonded to one another to form a three-dimensional or infinite network. Such structures are referred to as being *crosslinked* and are depicted by the simple structure in Figure 1.7.

Homopolymers consist of only one type of repeating monomer. When a chemically different repeating monomer is introduced into the polymer chain, such polymers are called *copolymers*. The properties of copolymers depend not only on the chemical nature and the relative proportions of the monomers but also on how the monomers are distributed along the chain. If there is a strong propensity for a given monomer to be followed by one of the other monomers, then the polymer will have an alternating pattern of repeat monomers. Figure 1.8 depicts a simple model of an alternating copolymer.

At the opposite extreme is the sequence of a block copolymer. In these materials, there is an overwhelming tendency for a monomer to be followed by another monomer of the same kind. There are several types of block copolymers, depending on the number of ordered repeat units per molecule. The simplest case is when each polymer chain has one sequence of one monomer followed by one sequence of the comonomer. This molecular geometry is called an A-B block, or a diblock, copolymer. A simple model appears as in Figure 1.9.

A triblock, or A-B-A block, copolymer has long sequences of one unit at the ends of the molecule and a long sequence of the co-unit in the middle. A triblock is depicted by the simple structure in Figure 1.10.

An alternating block copolymer, as depicted by the simple structure in Figure 1.11, has shorter sequences of one type of monomer alternating with shorter sequences of the comonomer.

A third classification is a random copolymer in which the different monomers are randomly distributed along the chain. A simple model appears in Figure 1.12. Neither monomer has any strong likelihood of being followed by any particular monomer.

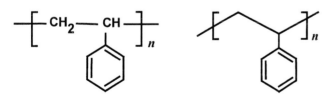

Figure 1.2 Two ways to represent the structure of polystyrene. Key: n = number of repeating units in the polymer molecules.

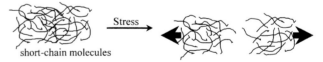

Figure 1.3 Chain disentanglement in short polymer (low molecular weight) chains.

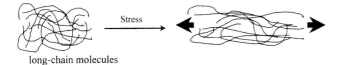

long-chain molecules

Figure 1.4 Chain entanglement in long polymer (high molecular weight) chains.

In addition to linear copolymers, there are branched, or graft, copolymers. Here the backbone of the polymer is composed of one type of monomer, and the side chains or grafts are composed of the comonomer. This is depicted by the simple structure in Figure 1.13.

Classification of Polymers

Polymers can be classified in several ways.[11,12] The two major classifications are thermosetting polymers and thermoplastic polymers.[13] As the name implies, thermosetting plastics, or thermosets, are set, cured, or hardened into a permanent shape.[14] The curing, which usually occurs rapidly under heat or ultraviolet (UV) light, leads to an irreversible cross linking of the polymer. Thermoplastics differ from thermosetting materials in that they do not set or cure.[15] When they are heated, thermoplastics merely soften to a mobile, flowable state where they can be shaped into useful objects. Upon cooling, the thermoplastics harden and hold their shape. Thermoplastics can be repeatedly softened by heat and shaped.

Amorphous Polymers

Polymers can also be classified as amorphous, meaning without structure, or semicrystalline materials. Most polymers are either completely amorphous or are semicrystalline with an amorphous component. Amorphous polymers are hard, rigid glasses below a fairly sharply defined tem-

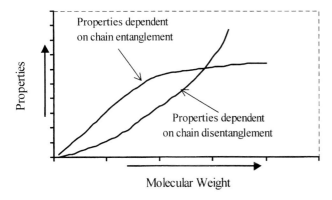

Figure 1.5 Properties versus molecular weight: entanglement and disentanglement.

perature, which is known as the glass-transition temperature, or T_g. Above the T_g the amorphous polymer becomes soft, rubbery, and flexible and can be shaped. The glass-transition temperature depends on the chemical nature of the repeating unit(s). Polymers can have widely different T_gs.

Amorphous polymers are composed of random entanglements of polymer chains that are in a state of motion. Polymer chains in the solid state are moving more slowly. Applying energy in the form of heat to the polymer makes the polymer chains move faster. When enough energy is applied to the polymer and the temperature goes above the T_g, the molecular motion increases substantially, and the chains start to disentangle. This is depicted in the simple model in Figure 1.14.

Mechanical properties show profound changes near the glass-transition temperature. Figure 1.15 illustrates how the stiffness (or modulus) of an amorphous polymer decreases as it goes above the T_g. The vertical axis indicates the stiffness of the polymer. The horizontal axis indicates the temperature (or the amount of heat energy) being added to the polymer. The stiffness remains constant below the T_g and then drops rapidly upon reaching the T_g. Thus a polymer goes from a stiff, glassy solid to a soft rubbery material when heated above its T_g. Further increases in temperature lead to chain disentanglement and flow. Cooling below the T_g reverses the process.

Elastomers, or rubbers, are polymers with glass-transition temperatures well below room temperature. Thus elastomeric polymers have rubberlike properties at room temperature—that is, they can be stretched substantially beyond their original length and will retract. Below their T_g, elastomers become rigid and lose their rubbery characteristics. Elastomeric polymers are typically cured or cross-linked into a permanent shape to prevent flow above the polymer T_g.

Crystalline Polymers

In crystalline and semicrystalline thermoplastics, the chemical structure allows the polymer chains to fold on themselves and pack together in an organized manner. The resulting organized regions exhibit the characteristics of crystals. The formation of very large ordered regions is rendered difficult because of the length and flexibility of the chains, as well as their unavoidable entanglements and imperfections. Therefore, crystallinity can never be complete, and there is always a portion of amorphous material present. For this reason, crystalline polymers are often called semicrystalline. Most semicrystalline polymers are not more than 90% crystalline, and the degree of crystallinity is often considerably less.

Every semicrystalline polymer has a crystalline melting point, T_m, that is above the glass transition temperature. Below the melting point there is not enough energy available to force these low-energy crystalline regions to break apart and flow. Above the T_m these crystals melt and the

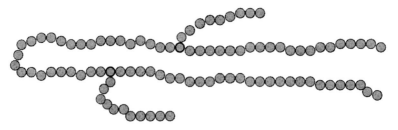

Figure 1.6 Model of a branched polymer.

Figure 1.7 Model of a network (or cross-linked) polymer.

Figure 1.8 Model of an alternating copolymer.

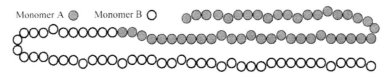

Figure 1.9 Model of a diblock (A-B) copolymer.

Figure 1.10 Model of a triblock (A-B-A) copolymer.

Figure 1.11 Model of an alternating block copolymer.

Figure 1.12 Model of a random copolymer.

polymer flows. The crystalline melting temperature depends on the chemical nature of the repeating unit(s). Hence polymers can have widely different T_ms.

The molecular model in Figure 1.16 depicts a semicrystalline polymer with crystalline regions, or spherulites, surrounded by an amorphous mass. As energy is applied to the semicrystalline polymer and the temperature rises above the T_g, there is an increased motion of the amorphous sections of the polymer chains. The crystalline regions remain intact. When sufficient energy is added to raise the temperature above the T_m, the crystalline regions melt, the chains disentangle, and the polymer will flow. The crystalline regions, or spherulites, can be seen in the photomicrograph of a semicrystalline polyester in Figure 1.17.

The degree of crystallinity and the morphology of the crystalline phase have an important effect on mechanical properties. Figure 1.18 illustrates the stiffness, or modulus, of a semicrystalline polymer. Semicrystalline plastics will become less rigid above their glass-transition temperature but will not flow because the polymer chains are held together by the crystalline regions. Above the crystalline melting point, the crystalline regions melt, there is a major decrease in stiffness, and the polymer flows.

Liquid crystalline polymers (LCPs) are best thought of as being a unique class of semicrystalline polymers. The name refers to the characteristic of the polymer to maintain a high level of crystalline order in the melt phase. Thus LCPs have stiff, crystalline, rodlike segments that are organized in large parallel arrays or domains in both the melt and solid states. These ordered domains provide liquid crystalline polymers with unique characteristics, compared to those of semicrystalline and amorphous polymers.

LCPs have a crystalline melting point, T_m, that is above the glass transition temperature. The T_m in an LCP is where the polymer makes a transition from a semicrystalline solid to a liquid crystalline state where the ordered rodlike domains in the melt. These domains are liquid crystalline. As more energy is applied, a second melting transition is reached, and the liquid crystalline state goes to an isotropic melt, T_i. Thus above the T_i there is no liquid crystalline behavior.

LCPs exhibit anisotropic behavior above the T_m and below the T_i, and the rodlike domains orient readily in the direction of flow. This orientation is maintained as the polymer solidifies. The molecular shape and crystalline structure of LCPs result in mechanical and electrical properties that are anisotropic with flow direction. In general, properties are higher along the direction of flow. The degree of order ultimately depends on the chemical structure of the polymer and the conditions under which it was processed.

Many of the mechanical and physical property differences between polymers can be attributed to their structure. In general, the ordering of semicrystalline and liquid crystalline thermoplastics makes them stiffer, stronger, harder, denser, and less resistant to impact than amorphous polymers. In addition, semicrystalline and liquid crystalline polymers have greater rigidity, density, lubricity, creep resistance, and solvent resistance than their amorphous counterparts. Moreover, semicrystalline polymers tend to shrink and warp more than amorphous polymers do.

Synthesis of Polymers

The process of forming polymers from monomers is called *polymerization*. A monomer must be capable of forming covalent bonds to two or more molecules—that is, its functionality must be ≥2. Three routes of achieving the difunctionality and bonding to two molecules are (1) opening a carbon-carbon double bond, (2) using molecules with two reactive functional groups, and (3) opening a cyclic molecule.[6,16]

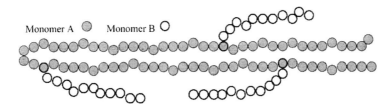

Figure 1.13 Model of a graft copolymer.

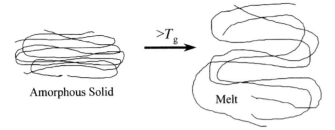

Figure 1.14 Molecular change of an amorphous polymer: random entanglements. Key: T_g = glass-transition temperature.

There are two types of synthetic polymerization processes: chain polymerization and step-growth polymerization.[17] These processes are sometimes referred to as *addition polymerization* and *condensation polymerization*, respectively.

Step-growth Polymerization

Step-growth polymerization is merely an extension of classical organic condensation reactions in which monofunctional reactants are used to prepare esters, amides, acetals, and the like.[18] The key difference in polymer formation is in the use of molecules with two or more reactive functional groups. Examples would be the formation of polyesters and polyamides. The use of a mixture of difunctional and multifunctional monomers will produce a branch or network polymer.

In the formation of polyesters, a dicarboxylic acid and dialcohol (or diol) are present in equimolar amounts. Each of the monomers has two functional groups that are capable of reacting with one another. The carboxyl group of a dicarboxylic acid will react with the hydroxyl group of a diol to form a 1:1 reaction product with an ester group and a water molecule. This is shown in Reaction 1 in Figure 1.19.

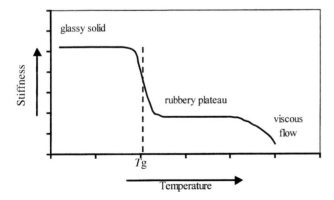

Figure 1.15 Amorphous polymer: stiffness versus temperature profile. Key: T_g = glass-transition temperature.

The product from this reaction is also difunctional. Hence further reaction can take place with concomitant increase in chain length. The new species formed is still difunctional and could react further. This is shown in Reaction 2 in Figure 1.19. After numerous reactions a polyester molecule is formed as shown in Reaction 3 in Figure 1.19.

Here again, the difunctionality is maintained in the product. The size of the molecules increases with each successive step. Thus in a step-wise manner, very long chain molecules are obtained. In step-growth polymerization, the molecular weight increases slowly in a step-wise manner, and it is only after high conversion rates that high molecular weights are obtained—that is, at the late stages of the reaction.

Thermoplastics prepared by step-growth polymerization include polyesters, polyamides, polyurethanes, and polycarbonates. Figure 1.20 shows reactions for the formation of some step-growth or condensation polymers. Polyesters can be prepared from the condensation reaction of diacids with diols, diacid chlorides with diols, or transesterification of diesters with diols.[19] In addition, polyesters can be prepared from a single monomer that contains both an OH and an acid group. Polyamides result from the condensation reaction of a diacid with a diamine. In addition, polyamides can be prepared from a monomer that contains both an amine and an acid group.[20,21] Polyurethanes are prepared from the reaction of diols with diisocyanates.[22] Polycarbonates are synthesized from the reaction of diphenols (bisphenols) with phosgene or by transesterification of bisphenols with diphenyl carbonate.[19,23]

Chain Polymerization

In contrast to slow step-growth polymerization, a chain polymerization reaction is very rapid and involves opening a double bond.[24] In chain polymerization, an active site at the chain end is able to react with an unsaturated (or heterocyclic) monomer molecule in such a way that, after the addition of the latter, the active site is preserved at the new chain end. Chain polymerization involves formation of a high molecular weight polymer at early stages of the reaction—that is, at low degrees of conversion. As soon as an active site is created, it rapidly adds monomer molecules in a sequence of chain reactions.

The most common types of chain polymerizations are the following:

- Free-radical polymerizations with free-radical active sites[25,26]
- Metallocene, single-site, or Ziegler-Natta catalyzed polymerizations in which active sites on the monomer form coordination bonds with a transition metal[27]
- Anionic polymerization with carbanion active sites[28,29]
- Cationic polymerization with carbocation ion active sites.[30–33]

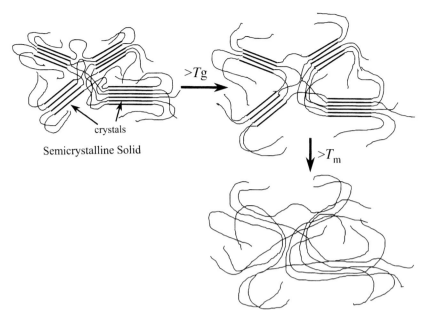

Figure 1.16 Molecular change of a semicrystalline polymer: spherulites and amorphous mass. Key: T_g = glass-transition temperature; T_m = crystalline melting point.

There are several steps in a chain polymerization processes. Free-radical polymerizations have been extensively studied and will be used to exemplify the phases of chain polymerization. These phases are depicted in Figure 1.21.

- In the initiation phase, an active site is formed on a monomer molecule. In free-radical polymerization,

the homolytic decomposition of an initiator forms radicals.

- Addition occurs when the double bond adds to the active site on the free radical.
- In the propagation or chain-growth phase, a double bond adds to the active site. The monomer is incorporated into the polymer chain, and the active site is restored at the chain end.
- The termination phase involves destruction of the active site on a polymer chain, whereby the growth reaction is stopped.
- Chain-transfer reactions involve removal of the active site from a polymer chain, which stops growing, and

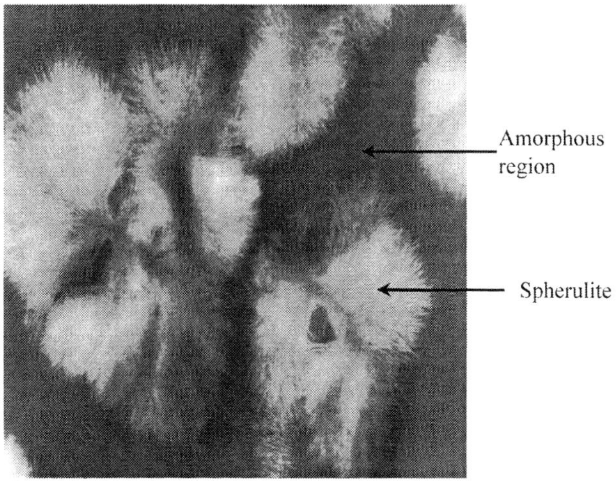

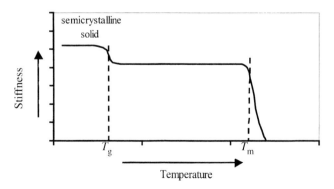

Figure 1.17 Photomicrograph of spherulites in a semicrystalline polymer. (Transmission electron microscopy courtesy of G. A. Hutchins of Corporate R&D, General Electric Company.)

Figure 1.18 Semicrystalline polymer: stiffness versus temperature profile. Key: T_g = glass-transition temperature; T_m = crystalline melting point.

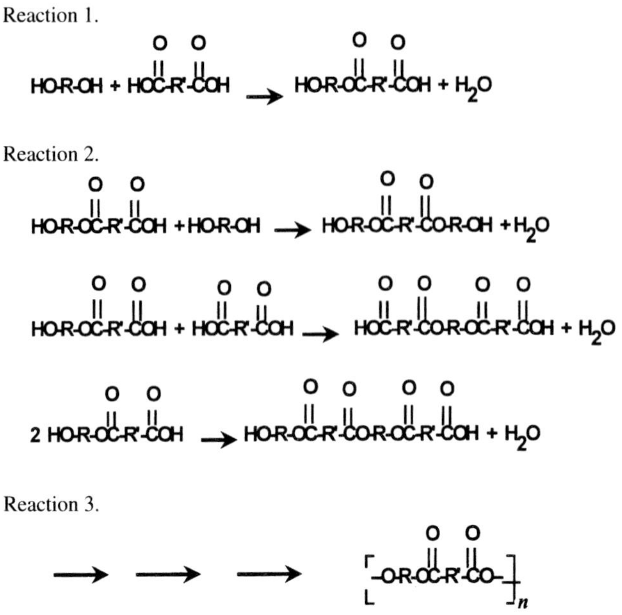

Reaction 1.

Reaction 2.

Reaction 3.

Figure 1.19 Step-growth polymerization.

transfer of the active site to another molecule where it can start a new polymer chain. Thus one active site can give rise to several chain molecules, instead of a single one if there were no transfers. Chain-transfer reactions exert a strong influence on the molecular weight of the polymers formed.

Comparisons of the key polymerization characteristics for chain- and step-growth polymerizations are summarized in Table 1.1.

Thermoplastics prepared by chain polymerization of olefins include polyethylene, polypropylene, polystyrene, poly-(vinyl chloride), polyacrylonitrile, and poly(methyl methacrylate). In addition to polymerization of olefinic monomers, cyclic monomers can undergo chain polymerization. This is exemplified by the polymerization of ethylene oxide to poly-(ethylene oxide). Figure 1.22 shows reactions for preparing some of these chain, or addition, polymers.

In chain polymerization, the addition of a monomer to a growing chain can occur in either of two orientations: either head-to-tail or head-to-head (or tail-to-tail) orientations. This is depicted by the model shown in Figure 1.23. The substituted carbon in the monomer is defined as the head, and the methylene is the tail. For most vinyl polymers, head-to-tail addition is the dominant mode of addition. The predominance of head-to-tail additions is the result of a combination of resonance and steric effects. Generally, the ionic or free-radical center occurs at the substituted carbon due to the possibility of resonance stabilization or electron delocalization through the substituent group. Head-to-tail addition is also sterically favored be-

cause a methylene group separates the substituent groups on successive repeat units.

However, head-to-head additions can occur at high polymerization temperatures where higher amounts of thermal energy make the less-favored head-to-head more accessible. Moreover, head-to-head additions are more abundant with halogen-containing monomers such as vinyl chloride and vinyl fluoride. Indeed, poly(vinyl fluoride) contains about 16% head-to-head units.[34]

Thermoset Polymerization

Thermoset resins can be produced by chain- or step-growth polymerization.[14] Figure 1.24 shows the structure of various monomers and oligomers added in small amounts to make the cross-links in chain-polymerized thermoset polymers. Sometimes thermoset polymers can be prepared directly from di-, tetra-, or multifunctional unsaturated monomers. For example, a mixture of styrene and divinyl benzene will undergo chain-growth polymerization to give cross-linked polystyrene.

Other times multifunctional oligomers are prepared first, for ease of handling, storage, and formulation. The oligomers are shaped and then cross-linked into final form. For example, unsaturated esters are prepared from the polymerization of various diols and maleic anhydride to give a viscous, unsaturated polyester oligomer. The multifunctional oligomer is dissolved in styrene to lower the viscosity. Free-radical chain polymerization of this mixture gives a cross-linked resin.

Other thermosetting monomers that are cured through radical reactions are diallyl phthalate, drying oils, and alkyd resins. Diallyl phthalate is prepared from the reaction of allyl alcohol with phthalic anhydride. Drying oils are unsaturated fatty acid esters of glycerol. Alkyd resins are oligomers prepared from unsaturated fatty acids, glycerol, and phthalic anhydride.

Examples of step-growth thermoset resins are phenolic, melamine, urea-based, epoxy, and polyurethane, resins.[14] Figure 1.25 shows the chemical reactions for the formation of some step-growth polymerized thermoset polymers.

Phenolic resins are prepared from phenol and formaldehyde. The trifunctional phenol reacts in the 2, 4, and 6 positions with formaldehyde to give a highly cross-linked structure. Normally, oligomers are prepared by step-growth polymerization because of the volatility of formaldehyde. Oligomers formed by condensation of phenol and formaldehyde in the presence of a catalyst and alkali are called *resole resins*. Resoles will cure and harden when they are heated. Oligomers formed under acidic conditions from phenol and formaldehyde are called *novolak resins*. Novolak resins are fusible and soluble but will not cure without the addition of a hardening agent.

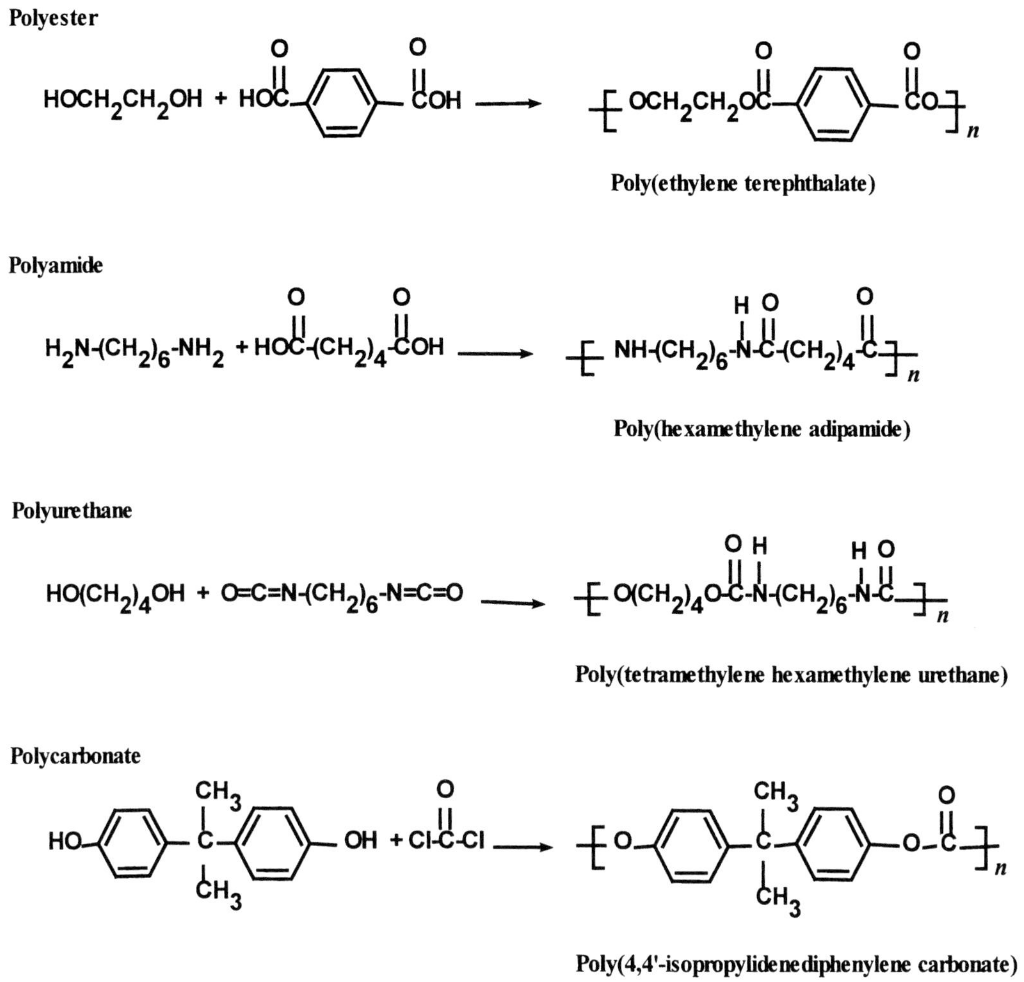

Polyester

Poly(ethylene terephthalate)

Polyamide

Poly(hexamethylene adipamide)

Polyurethane

Poly(tetramethylene hexamethylene urethane)

Polycarbonate

Poly(4,4'-isopropylidenediphenylene carbonate)

Figure 1.20 Chemical reactions for preparing step-growth polymers.

Melamine and urea-based resins (amino resins) are prepared by the reaction of melamine and urea, respectively, with formaldehyde. As with phenolic resins, oligomers are prepared first; then, after formulating and shaping, the oligomers are cured by heat to a highly cross-linked structure.

Epoxy resins typically contain two or more epoxy groups. The most common epoxy resins are prepared from the reaction of bisphenol A and epichlorohydrin to give a low molecular weight material. Epoxies show high reactivity toward amines and anhydrides. Usually tri- or tetrafunctional amines and cyclic aliphatic carboxylic acid anhydrides are used.

Thermoset polyurethanes are similar to thermoplastic polyurethanes, except multifunction alcohols (polyols) and multifunctional isocyanates are used to produce a cross-linked polymer.

Cross-linked polymers may also be prepared from linear polymers. For example, polyethylene can be cross-linked through radical sources such as peroxides or ionizing radiation.[35] Thus, cross-linked polyethylene is no longer a thermoplastic and cannot be re-melted and reshaped.

Polymerization Conditions

Polymer syntheses are carried out by a remarkable variety of processes that can be classified as homogeneous or heterogeneous, depending on the number of phases in the system.[4] The simplest process is the conversion of monomer to polymer while retaining at all times a single homogeneous liquid phase. These would include melt, bulk, and solution polymerization.[36,37]

Heterogeneous polymerizations include various multiphase processes. The continuous phase can be a gaseous monomer, and the polymerization takes place on catalyst particles. The continuous phase can be a liquid monomer or a monomer-solvent mixture; in both of these, the resultant

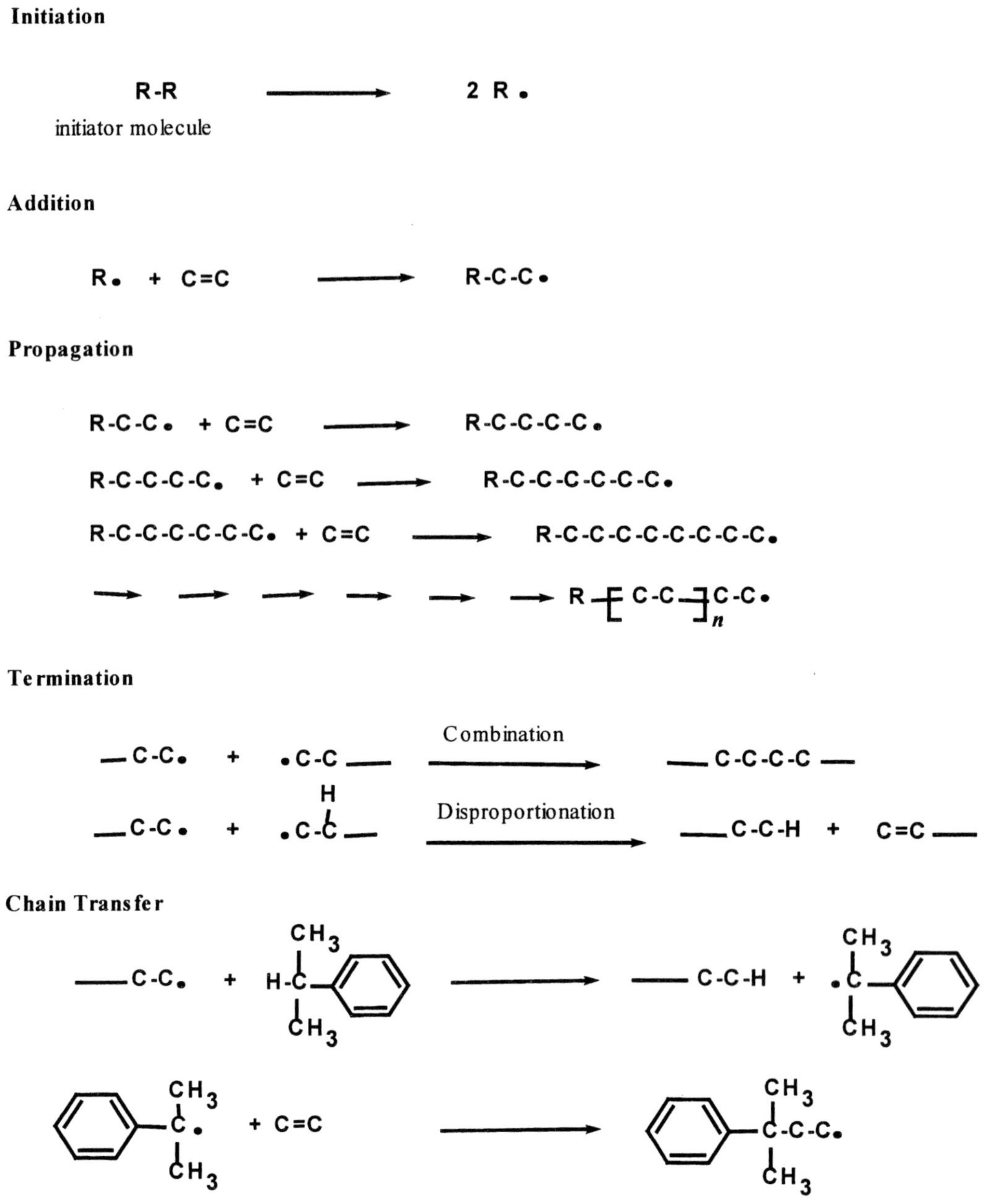

Initiation

$$R\text{-}R \longrightarrow 2\,R\bullet$$

initiator molecule

Addition

$$R\bullet \;+\; C=C \longrightarrow R\text{-}C\text{-}C\bullet$$

Propagation

$$R\text{-}C\text{-}C\bullet \;+\; C=C \longrightarrow R\text{-}C\text{-}C\text{-}C\text{-}C\bullet$$

$$R\text{-}C\text{-}C\text{-}C\text{-}C\bullet \;+\; C=C \longrightarrow R\text{-}C\text{-}C\text{-}C\text{-}C\text{-}C\text{-}C\bullet$$

$$R\text{-}C\text{-}C\text{-}C\text{-}C\text{-}C\text{-}C\bullet \;+\; C=C \longrightarrow R\text{-}C\text{-}C\text{-}C\text{-}C\text{-}C\text{-}C\text{-}C\text{-}C\bullet$$

$$\longrightarrow \longrightarrow \longrightarrow \longrightarrow \longrightarrow \longrightarrow R\!-\!\!\left[\!\! C\text{-}C \!\!\right]_n\!\! C\text{-}C\bullet$$

Termination

$$\underline{\quad}C\text{-}C\bullet \;+\; \bullet C\text{-}C\underline{\quad} \xrightarrow{\text{Combination}} \underline{\quad}C\text{-}C\text{-}C\text{-}C\underline{\quad}$$

$$\underline{\quad}C\text{-}C\bullet \;+\; \bullet C\text{-}\overset{H}{\underset{}{C}}\underline{\quad} \xrightarrow{\text{Disproportionation}} \underline{\quad}C\text{-}C\text{-}H \;+\; C=C\underline{\quad}$$

Chain Transfer

$$\underline{\quad}C\text{-}C\bullet \;+\; H\text{-}\overset{CH_3}{\underset{CH_3}{C}}\!\!-\!\!\bigcirc \longrightarrow \underline{\quad}C\text{-}C\text{-}H \;+\; \bullet\overset{CH_3}{\underset{CH_3}{C}}\!\!-\!\!\bigcirc$$

$$\bigcirc\!\!-\!\!\overset{CH_3}{\underset{CH_3}{C}}\bullet \;+\; C=C \longrightarrow \bigcirc\!\!-\!\!\overset{CH_3}{\underset{CH_3}{C}}\!\!-\!\!C\text{-}C\bullet$$

Figure 1.21 Steps in chain-growth polymerization.

Table 1.1 Comparison of chain- and step-growth polymerization

Characteristic	Chain growth	Step growth
Mechanism	Rapid addition of monomer to small number of active centers; chain reaction	Coupling of any two monomers or polymers; reaction occurs in steps
Monomer concentration	Decreases slowly during polymerization	Decreases rapidly before any high polymer is formed
Molecular weight increase	High polymer at low conversion	Increases continuously during polymerization; high polymer only after large conversions
Backbone	Usually a carbon chain	Functional groups; heteroatoms
Repeat unit	Same composition as monomer; different bonding	Different functional group than in either monomer; small molecule often eliminated during condensation reaction

Note: Based on data in references 4 and 11.

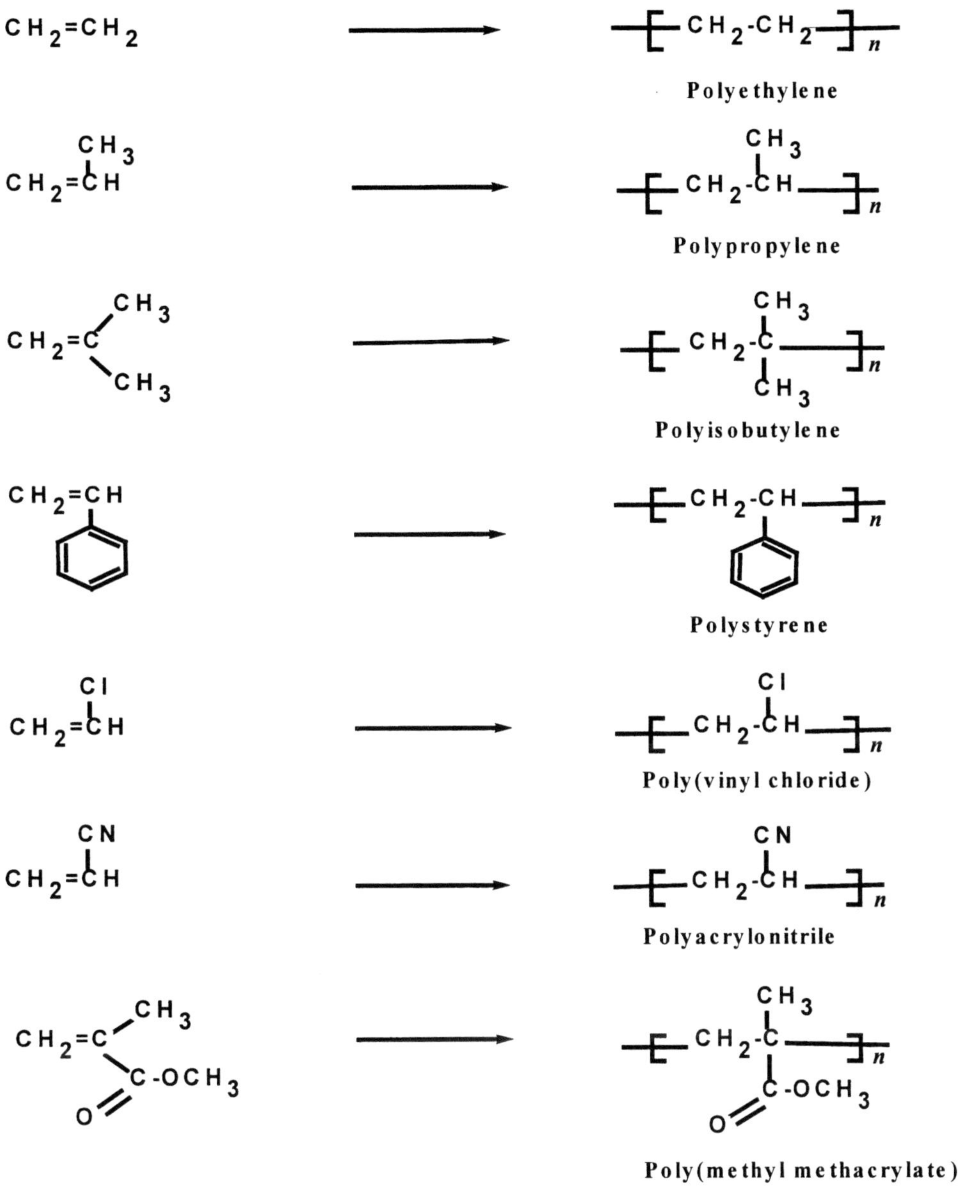

Figure 1.22 Chemical reactions for preparing addition polymers.

polymer is not soluble. In suspension and emulsion polymerization, the continuous phase is water, and the monomer is dispersed in the water in the form of droplets. In interfacial polymerization, there are two continuous phases—typically aqueous and organic—and the chemical reaction takes place at the interface of the two distinct phases.

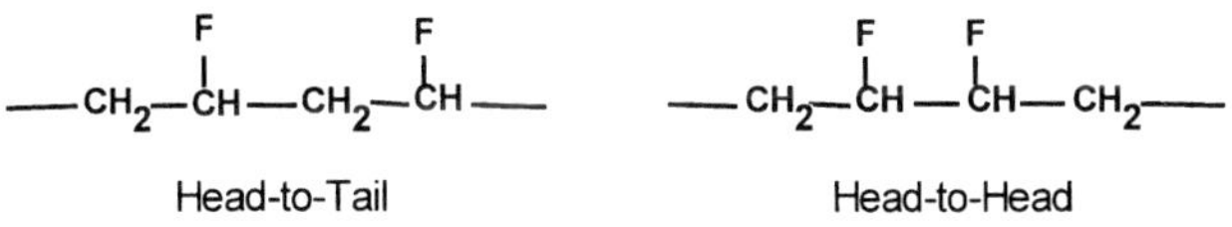

Figure 1.23 Head-to-tail and head-to-head isomers of poly(vinyl fluoride).

Bulk Polymerization

Bulk or mass polymerization is inherently the simplest and most direct method of converting monomer to polymer.[36] It involves adding a small amount of initiator and perhaps a chain-transfer agent to the monomer and heating to a temperature where the initiator breaks down to give free radicals. The viscosity increases drastically during polymerization.

The advantages of bulk or mass polymerizations are the following:

1. A highly pure polymer is obtained because only monomer, initiator, and perhaps chain-transfer agent are used.

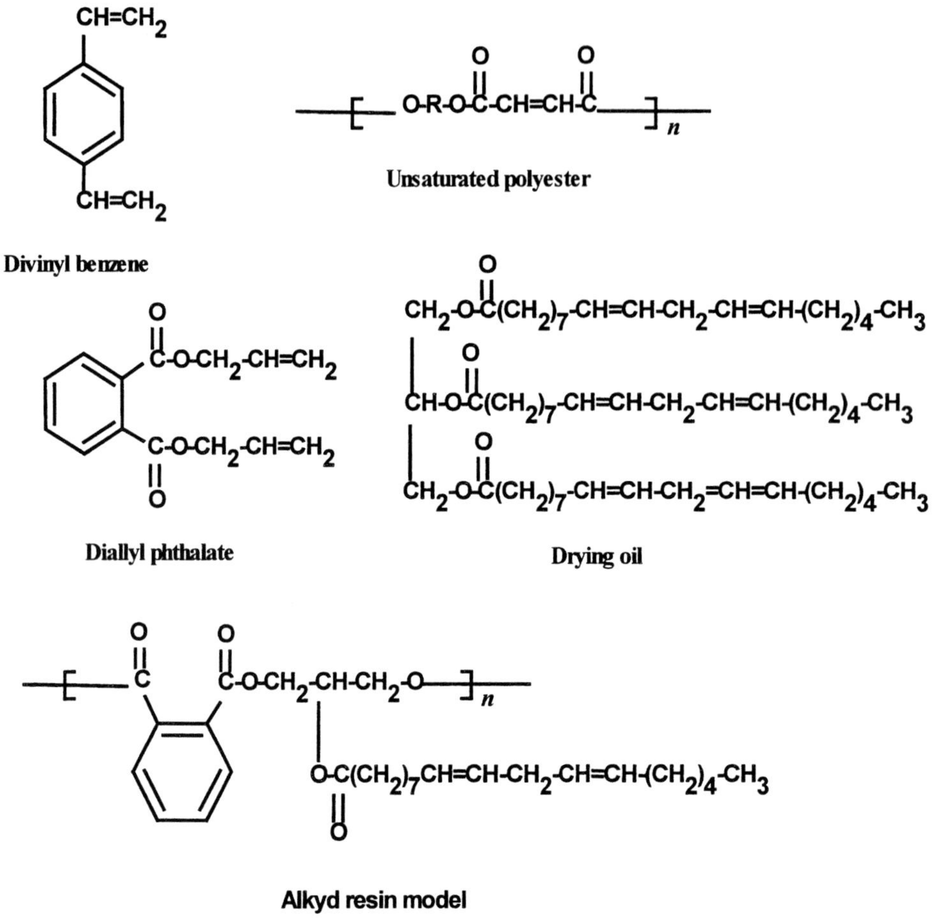

Figure 1.24 Thermoset monomers for chain-growth polymerization.

2. Objects may be conveniently cast to shape. For a polymer that is cross-linked in the synthesis reaction, casing to shape is the only way to obtain such objects short of machining from larger blocks.
3. Bulk polymerization provides the greatest possible polymer yield per reactor volume.

Some of the disadvantages are the following:

1. Bulk polymerization is often difficult to control due to the highly exothermic nature of vinyl polymerization, making thermal control of the reaction difficult.
2. The polymerization may have to run slowly to keep the reaction under control.
3. Auto acceleration can lead to an increase in rate and broader molecular weight distributions.
4. It can be difficult to remove the last traces of unreacted monomer.

Solution Polymerization

In solution polymerization, the monomer is diluted with an inert solvent in which both the monomer and the resultant polymer are soluble, in order to minimize many of the diffi-

culties encountered in bulk polymerization. The viscosity of the medium does increase, but not nearly as drastically as it does in bulk polymerization. Thermal control is much easier because of improved heat transfer.

The advantages of solution polymerization are the following:

1. Heat removal and control are easier.
2. For some applications—such as lacquers—the desired polymer solution can be obtained directly from the reactor.

Some of the disadvantages are the following:

1. Handling of solvent can be difficult.
2. Separation of the polymer and recovery of the solvent require additional operations.
3. Removal of the last traces of the solvent may be difficult.
4. The yield per reactor volume is lowered.

Suspension Polymerization

In a suspension polymerization, known also as suspension bead and pearl polymerization processes, a water-insoluble

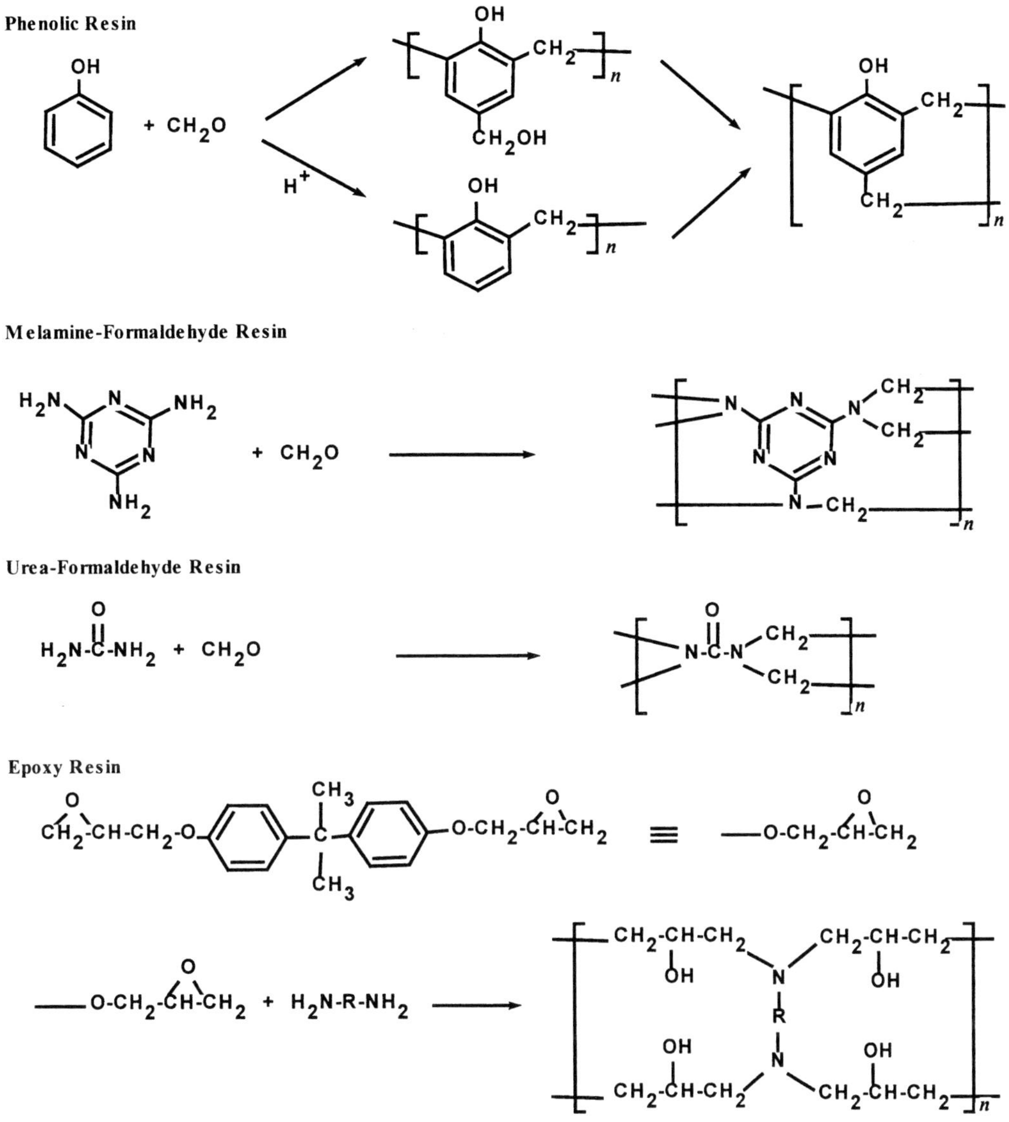

Phenolic Resin

Melamine-Formaldehyde Resin

Urea-Formaldehyde Resin

Epoxy Resin

Figure 1.25 Thermoset resins for step-growth polymerization.

monomer is dispersed in water in the form of droplets with diameters that vary from about 1 to 500 μm.[38] The suspension is stabilized by mechanical agitation and with a protective colloid or dispersant. A protective colloid is a water-soluble polymer such as poly(vinyl alcohol) whose function is to increase the viscosity of the continuous water phase and to hinder coalescence of monomer drops, but to be inert with regard to the polymerization. A finely divided insoluble inorganic salt such as magnesium carbonate may also be used. The insoluble salts collect at the droplet-water interface via surface tension and prevent the coalescence of the droplets at the stage when they are sticky. Polymerization is initiated by means of a monomer-soluble initiator so that the polymerization occurs in the individual droplets. The whole process may therefore be viewed as if each droplet is a single bulk reactor. Since water is the continuous

phase, the viscosity remains constant, and the large surface area of the individual droplets ensures good heat transfer. This method is widely used for the commercial production of many polymers and copolymers.

The major advantages of suspension polymerization are the following:

1. Heat removal and control are easy.
2. The polymer is obtained in a convenient, easily handled form.

The disadvantages include the following:

1. A somewhat less pure polymer than bulk polymerization results, since there are bound to be remnants of the suspending agent(s) absorbed on the particle surface.
2. There is a low yield per reactor volume.

Emulsion Polymerization

Emulsion polymerization, like suspension polymerization, takes place with water as the continuous phase.[39,40] In this case, however, the monomer droplets are dispersed in an aqueous phase by means of an emulsifying agent or surfactant to form a stable emulsion. In contrast to a suspension, a proper emulsion is thermodynamically stable and will remain so even if the agitation is stopped. The diameter of the monomer droplets is of the order of 0.1 μm, which is considerably less than the diameter of the particles from suspension polymerization. In addition, soap micelles contain some monomer molecules.

When surfactants such as fatty acid soaps are added to the water in low concentrations, they ionize and float around freely, much as sodium chloride ions would. The anions, however, consist of a highly polar hydrophilic (water-seeking) head and an organic, hydrophobic (water-fearing) tail. As the soap concentration is increased, a concentration is reached where anions begin to agglomerate in micelles rather than float around individually. These micelles have dimensions on the order of 0.1 μm. They consist of a tangle of the hydrophobic tails in the interior (getting away from the water as far as possible) with the hydrophilic heads on the outside. When an organic monomer is added to an aqueous micelle solution, it naturally prefers the organic environment at the interior of the micelles. Some of it congregates there, swelling the micelles until equilibrium is reached with the concentration force of surface tension. Most of the monomer, however, is distributed in the form of much larger monomer/soap droplets.

Free radicals are generated in the aqueous phase by the decomposition of water-soluble initiators, usually potassium persulfate or redox systems. The radicals thus generated in the aqueous phase bounce around until they encounter some monomer. Since the surface area presented by the monomer-swollen micelles is so much greater than that of the droplets, the probability is large that a radical will enter a monomer-swollen micelle rather than a droplet. As soon as the radical encounters the monomer within the micelle, it initiates polymerization. The conversion of monomer to polymer within the growing micelle lowers the monomer concentration therein, and the monomer begins to diffuse from uninitiated micelles and monomer droplets to the growing, polymer-containing micelles.

The product of an emulsion polymerization is latex-polymer particles on the order of 0.05 to 0.15 μm; these particles are stabilized by the soap. The latex is then coagulated either by adding an acid that converts the soap to an insoluble acid form or by adding an electrolyte salt, which builds up charges on the particles, causing them to agglomerate through electrostatic attraction.

The advantages of emulsion polymerization are the following:

1. Viscosity is low, and thermal control is easy.

2. The latex product may be directly usable in coatings and adhesives.
3. Soft or sticky polymers may be conveniently prepared and handled.
4. It is possible to obtain both high rates of polymerization and high average chain lengths through the use of high soap and low initiator concentrations.

Its disadvantages are the following:

1. It is extremely difficult to remove the various additives, such as emulsifiers and coagulants, from the polymer, and the high residual impurity level may degrade the polymer properties.
2. Considerable technology is required to recover the solid polymer.
3. There is a low yield per reactor volume.

Copolymers

Copolymerization is a common method for modifying the performance of polymers.[41] Here two monomers or comonomers are polymerized to form a polymer composed of two types of repeating units. As previously mentioned, copolymers can be alternating, random, or block. The copolymer can have different properties from those of the homopolymers made from either repeating unit.

The glass-transition temperature of both alternating and random copolymers varies with the composition. The T_g of a random copolymer can be approximated with the equation[42]

$$1/T_g = w_1/T_{g1} + w_2/T_{g2}$$

In this empirical equation w_1 and w_2 are the weight fraction of the two monomers, and T_{g1} and T_{g2} are the glass-transition temperatures (in degrees Kelvin) of the two homopolymers.

The formation of random or block copolymers is determined by the reactivity of the comonomers toward themselves and the other monomer.[43] In the propagation (or chain-growth) phase with two monomers, M_1 and M_2, there are four possible propagation reactions. Each reaction is characterized by a propagation rate constant, k, labeled by two-digit subscripts. These reactions and associated rate constants are depicted in Figure 1.26.

The reactivity ratios are designated r_1 and r_2 where

$$r_1 = k_{11}/k_{12} \quad \text{and} \quad r_2 = k_{22}/k_{21}$$

The r is the ratio of the two propagation rate constants—that is, the propagation rate constant for reacting with a like monomer divided by the propagation rate constant for reacting with a different monomer. Thus r_1 and r_2 represent the tendency for an active chain end $-M_1\bullet$ to react with M_1 versus M_2, and active chain $-M_2\bullet$ to react with M_2 versus M_1, respectively.

Propagation Reactions   Rate Constants

$$—M_1{\cdot} + M_1 \longrightarrow —M_1 - M_1{\cdot} \quad k_{11}$$

$$—M_1{\cdot} + M_2 \longrightarrow ——M_1 - M_2{\cdot} \quad k_{12}$$

$$——M_2{\cdot} + M_1 \longrightarrow ——M_2 - M_1{\cdot} \quad k_{21}$$

$$—M_2{\cdot} + M_2 \longrightarrow ——M_2 - M_2{\cdot} \quad k_{22}$$

Figure 1.26 Copolymerization propagation reactions.

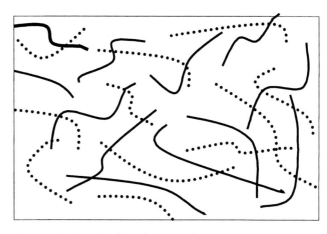

Figure 1.27 Miscible blend: single phase.

$r_1 > 1$ suggests that $-M_1{\bullet}$ prefers to add to M_1.
$r_1 < 1$ suggests that $-M_1{\bullet}$ prefers to add to M_2.

In terms of the types of copolymers discussed earlier, alternating copolymers are favored when the reactive species prefers to react predominantly with the other monomer. Thus both r_1 and r_2 are close to zero.

Block copolymers are favored when each reactive species prefers to react with its own monomer. Here both r_1 and r_2 are much greater than 1.

Random copolymers are favored when both reactive species have no strong preference toward either monomer. Hence r_1 and r_2 are similar in value and close to 1: that is, $r_1 \cong r_2 \cong 1$.

Blends and Alloys

Polymer blends and alloys are another method to modify and tailor performance in plastics.[44] In this instance, two different polymers are blended together to give a new product. It is sometimes more economically attractive to combine available polymers to produce desirable and novel blends than to prepare copolymers.[45] Thus new products can be developed rapidly. For example, one might look to blend a lower cost polymer that is deficient in toughness, chemical resistance, heat resistance, and flame resistance with a polymer that possesses these attributes. The available degrees of freedom make the opportunity challenging and provide almost infinite possibilities.

A polymer blend can have a single phase or multiple phases. The number of phases of the blend depends on the miscibility or solubility of the individual polymers with each other.

In a single-phase blend, the polymers dissolve in each other when they are mixed together, creating a single, continuous phase; this is an uncommon occurrence. The two different polymer chains are uniformly interdispersed, and this is depicted in the simple representation of the microstructure in Figure 1.27. Moreover, miscible blends exhibit a single T_g between the two polymers making up the blend.

Figure 1.28 depicts the T_gs of the individual polymers and a miscible blend.

Partially miscible blends have two discrete phases. However, the two individual polymers are partially miscible in each other. This miscibility is enough to provide adequate interfacial adhesion between the two discrete phases but not enough to form a single phase. Thus a small number of chains of each polymer are present in the other polymer phase; this is depicted in the simple microstructure in Figure 1.29. Since the individual polymers are partially soluble in each other, their T_gs are now slightly different from the individual polymers. Assume that the individual polymers have different T_gs: the lower T_g is increased slightly, and the higher T_g is decreased slightly. This is depicted in Figure 1.30.

The third alternative is immiscible blends in which neither polymer is soluble or slightly soluble in the other polymer. There are two discrete phases and essentially no molecules of one polymer in the domain of the other polymer. This is shown in the simple structure in Figure 1.31. The immiscible blends exhibit two T_gs that are the same as those of the individual polymers. This is depicted in Figure 1.32.

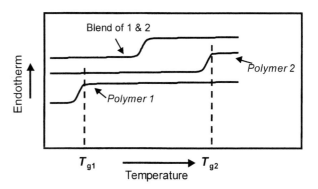

Figure 1.28 Thermogram showing the T_gs of a miscible blend.

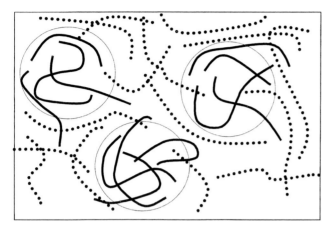

Figure 1.29 Partially miscible blend: two phases.

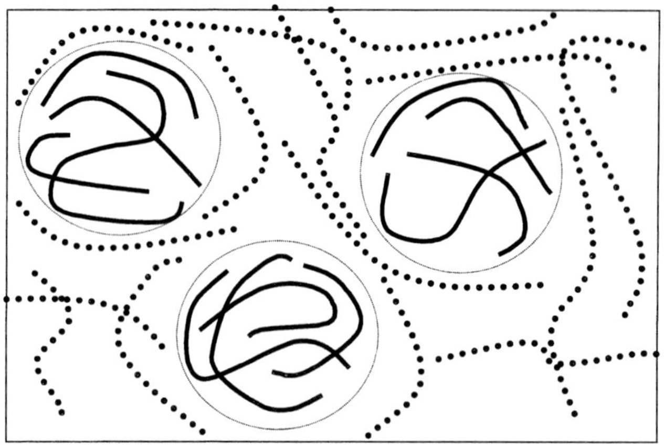

Figure 1.31 Immiscible blend: two phases.

In the two-phase systems, which are partially compatible and uncompatible, blends can be co-continuous, or one phase can be continuous while the other is dispersed in the continuous phase. The actual mixture depends on the amount of each polymer and the chemical nature of the polymers.

Properties of blends depend on the nature of the blend and its microstructure. Generally, properties of miscible blends are additive (linear behavior) based on the linear contribution from each polymer fraction. Some properties of partially miscible blends can achieve a synergistic combination (positive nonlinear behavior) where the properties are better than those predicted by linear behavior. In immiscible blends, there is very little interfacial adhesion between the two phases, and the properties are unpredictable and sometimes can have an antagonistic effect (negative nonlinear behavior) where the properties are worse than those predicted by linear behavior. Figure 1.33 shows a graphic representation of properties as a function of blend miscibility. In addition to having poor and unpredictable properties, the morphology (or microstructure) of immiscible blends is unstable. The discrete phases can coalesce into larger domains in the molten state. Moreover, there could be delamination in the solid state when a stress is applied to the material.

Compatibilization technology is used to circumvent these adverse characteristics of immiscible blends.[46,47] One method of compatibilization is the use of a graft or a block copolymer that contains segments of each of the individual polymers to improve the interfacial adhesion between the two phases and lead to enhanced properties. Since these graft and block copolymers contain portions of each polymer, they would tend to appear at the interface of the two phases and improve adhesion between the phases. This is depicted in Figure 1.34.

Structural Isomerism

Three possible types of structural isomerism are encountered in polymers:[12] positional isomerism, stereoisomerism, and geometrical isomerism. Positional isomerism (also called *orienticity*) can occur in addition polymers. Positional isomerism from either head-to-tail or head-to-head (or tail-to-

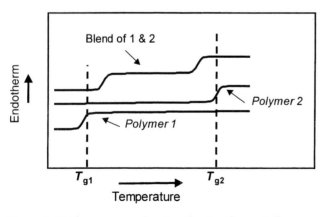

Figure 1.30 Thermogram showing the T_gs of a partially miscible blend.

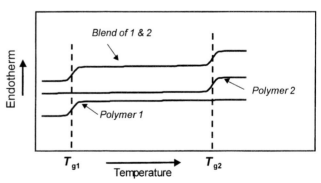

Figure 1.32 Thermogram showing the T_gs of an immiscible blend.

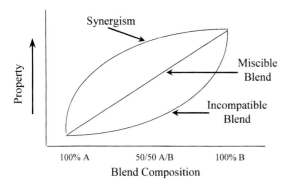

Figure 1.33 Properties versus blend composition.

tail) orientations was reviewed under the section on chain polymerization.

Stereoisomerism

Stereoisomerism or stereochemical configuration occurs in vinyl polymers in which two different groups are attached to the substituted carbon. These differences in stereoisomerism are also called *tacticity* from the Greek work *taktikos*, meaning "arranging" or "ordering". There are three different configurations in vinyl polymers. When all the substituents lie on the same side of the polymer backbone, the configuration is called *isotactic*. Figure 1.35 depicts a model of isotactic polystyrene. In Figure 1.35 the carbon-carbon bond, which forms the chain backbone, lies in the plane of the paper; the phenyl groups lie above the plane (heavy solid lines) or are on the same side, and the hydrogen atoms are below the plane (dashed line).

When the substituents lie on alternating sides of the polymer backbone, the configuration is called *syndiotactic*. A model of syndiotactic polystyrene appears in Figure 1.36.

The third possibility is that the substituents are distributed randomly along the polymer chain. This configuration is called *atactic*. Figure 1.37 shows a simple model of atactic polystyrene.

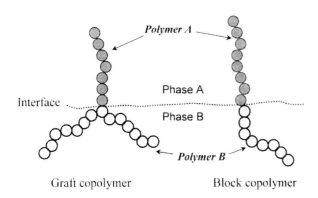

Graft copolymer   Block copolymer

Figure 1.34 Compatibilization with block and graft copolymers.

Differences in tacticity can play a major role in polymer properties. For example, there are major differences in properties and performance of atactic, isotactic, and syndiotactic polystyrene.[48,49] A comparison of some key properties appears in Table 1.2. Atactic polystyrene is an important commercial amorphous plastic. Syndiotactic polystyrene is a commercial semicrystalline plastic, which has superior chemical resistance vis-à-vis atactic polystyrene and, because of the high T_m, a much higher heat deflection temperature. Isotactic polystyrene is not important commercially because its very slow rate of crystallization requires long cycle times that make it uneconomical for most commercial applications.

Geometrical Isomerism

Geometrical isomerism is important in polymers made from diene monomers such as butadiene, isoprene, and chloroprene. There are several possibilities for the addition of a diene monomer during polymerization. If only one double bond is involved during polymerization, then 1,2-polymerization or 3,4-polymerization can occur. If both double bonds are involved together, then 1,4-addition occurs. In the case of 1,4-polymers, rotation is restricted because of the double bond in the chain and can give 1,4-*cis* or 1,4-*trans* repeating units with different geometrical arrangements about the double bond. The geometrical isomerism for polybutadiene is shown in Figure 1.38.

The *cis* isomer of 1,4-polybutadiene is a soft elastomer with a T_g of -108 °C. However, the *trans* isomer is harder with a T_g of -83 °C.[50]

The 1,2-addition products can exist in isotactic, syndiotactic, or atactic form. The stereoregularity in the isotactic and syndiotactic 1,2-polybutadienes gives rigid, crystalline polymers. The atactic form is elastomeric. The effect of tacticity on 1,2-polybutadiene properties is given in Table 1.3.

Polysaccharides do not contain any carbon-carbon double bonds, but they can have geometrical isomers.[51-53] Cellulose and starch are major examples of polysaccharides. The repeat unit in both polymers is D-glucose. However, the configuration of the glucoside linkage is different in the two polymers. Cellulose and starch have α-1,4 and β-1,4 linkages, respectively. This difference in geometrical isomerism results in cellulose being highly crystalline and not soluble in hot water, but starch being soluble in hot water and more easily hydrolyzed. The geometrical isomers of the D-glucose monomers in polysaccharides are depicted in Figure 1.39.

Polymer Molecular Weight

Pure chemical compounds are defined by their chemical composition, molecular structure, and a uniform molecular weight. However, polymers have heterogeneous molecular

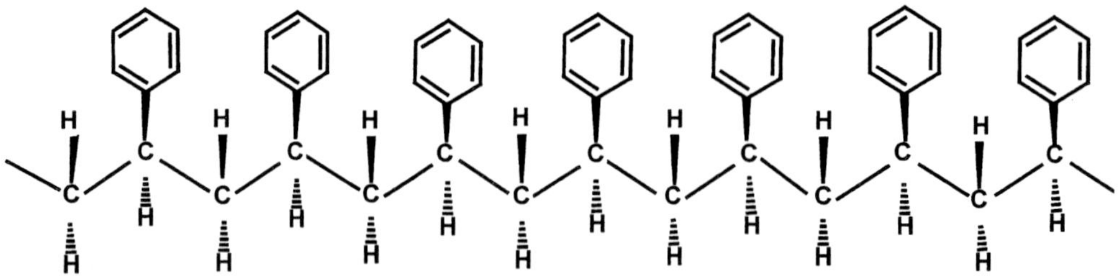

Figure 1.35 Configuration of isotactic polystyrene. Key: heavy solid lines = phenyl groups; dashed lines = hydrogen atoms; C—C = carbon bonds as chain backbone.

weights. Thus the molecular weight of the individual polymer chains can vary widely, although the composition remains the same for all molecules. It is merely the number of repeat units that varies from one molecule to another. Polymers that consist of molecules with different molecular weights are said to be *polydisperse*, and those with narrow molecular weight ranges are called *monodisperse*.

Therefore, instead of a molecular weight of a polymer (or degree of polymerization), the average molecular weight should be used. In addition, the type of average molecular weight should be specified, such as the number-average molecular weight and the weight-average molecular weight.[54]

The number-average molecular weight is based on the mole fraction of molecules and is expressed as

$$M_n = \Sigma \, x_i \, M_i$$

where i is the number of repeat units in a molecule, x_i is the mole fraction of molecules with i repeat units in the chain, and M_i is the molecular weight of the molecule with i repeat units. For a synthetic polymer, the value of i will vary over a range.

The weight-average molecular weight is based on the weight fraction of molecules and is expressed as

$$M_w = \Sigma \, w_i \, M_i$$

where w_i is the weight fraction of molecules with i repeat units in the chain and M_i is the molecular weight of the molecule with i repeat units. M_w is always larger than M_n.

The width of the molecular weight distribution (MWD) within a sample can be roughly characterized by the ratio of its two molecular weight averages. This ratio, the polydispersity index, is expressed as

$$MWD = M_w/M_n$$

The MWD is equal to 1 for a monodisperse polymer, in which all molecules would be strictly identical in number of units. With the exception of many biopolymers, no synthetic polymer is absolutely monodisperse.

The distribution of molecular weights can vary, and knowledge of the MWD can give insights to the polymer. The MWD can be narrow, broad, broad with a low molecular weight tail, or bimodal. Tails or broad MWD may suggest polymerization conditions, postblending of polymers of different molecular weights, or some degradation. Bimodal molecular weight distributions may suggest a polymerization process that has two different pathways. Models of molecular weight distribution are shown in Figures 1.40–1.42.

Plastic Applications, Uses, and Additives

The wide range of synthetic polymers offers a huge variety of specific performance attributes.[55-57] In addition, modification technology via copolymers, blends, and additives expands the use of plastics even more by increasing the variety of performance combinations that are available.

Synthetic and natural polymers are used extensively in commercial applications in the form of rigid plastics, rubbers, fibers, coatings, and adhesives. Additives are a key

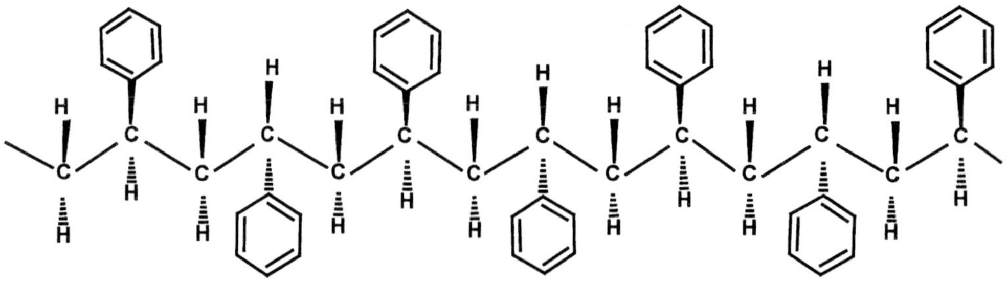

Figure 1.36 Configuration of syndiotactic polystyrene. Key: heavy solid lines = phenyl groups; dashed lines = hydrogen atoms; C—C = carbon bonds as chain backbone.

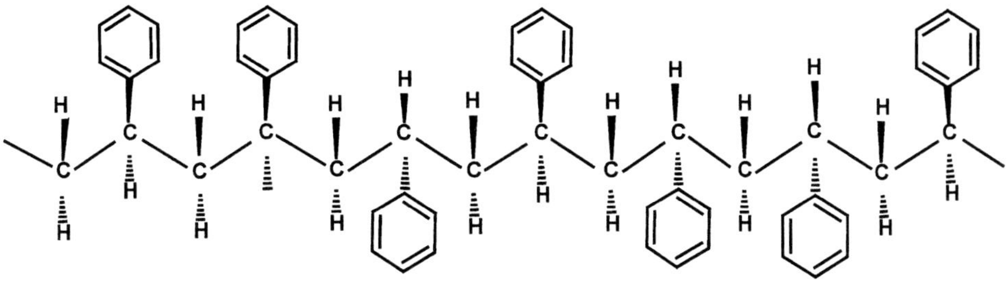

Figure 1.37 Configuration of atactic polystyrene. Key: heavy solid lines = phenyl groups; dashed lines = hydrogen atoms; C—C = carbon bonds as chain backbone.

part of modifying properties of polymers and transforming them into commercially viable products. Indeed, additives are found in most commercial materials and are a key part of the formulation in which performance is tailored for specific applications.[58–62]

Rigid Plastics

For rigid plastics, the end use temperature for the polymer must be below its T_g and/or below its T_m (if crystalline). Moreover, for thermosets there must be sufficient cross-linking to restrict molecular mobility.

Additives for rigid plastics can be classified in numerous ways. A functional classification includes curing agents, processing aids, stabilizers, performance-enhancing additives, aesthetic additives, and many others.

Curing agents are catalysts, monomers, or oligomers that are used in thermoset resins to polymerize and cross-link them. This would include diamines and acid anhydrides for curing epoxy resins, catalysts for curing unsaturated polyesters, and catalysts for curing polyurethanes.

Processing aids such as lubricants and mold releases facilitate the melt processing of plastics and fabrication of plastic parts. Examples are fatty acids, amides, esters and alcohols, metallic soaps, paraffin waxes, and silicones. Blowing agents, used in the processing of foamed plastics, decompose during processing to form a gas, leading to a cellular structure in the plastic.

Stabilizers protect plastics from degradation from energy sources such as heat and UV light. Antioxidants inhibit or reduce the rate of oxidative degradation of polymers at high temperatures, and they extend service life and increase the

stability during melt processing. Typically, antioxidants are hindered phenols, aromatic amines, thioesters, and phosphites. Heat stabilizers inhibit or retard the degradation of halogenated polymers and copolymers such as poly(vinyl chloride). Typical heat stabilizers are organotin and mercaptotin compounds, metal salts, and metal mercaptins. UV stabilizers such as benzophenones and benzotriazoles inhibit or reduce degradation of polymers from UV radiation.

Performance-enhancing additives increase key physical and mechanical properties of plastics. Reinforcing materials—such as glass fiber, carbon fiber, mica, talc, and clays—are inert solids that increase the stiffness and strength of the plastics. Some fillers such as calcium carbonate are added to lower the cost. Coupling agents are important when they are used in conjunction with fiber and filler additives. The use of silanes, titanates, and wetting agents improves the surface attraction between the polymer and the filler or fiber. Flame retardants are added to reduce the combustibility of a plastic.[63] Flame retardants can be halogenated when brominated and chlorinated organic additives are used in conjunction with a synergist such as antimony oxide. Nonhalogenated flame retardants include organophosphate esters, aluminum trihydroxide (hydrated alumina), and polyammoniumphosphate. Plasticizers such as phthalates, phosphates, and linear aliphatic esters impart flexibility to plastics. Impact modifiers are used to impart ductility and toughness to plastics. Typically, impact modifiers are rubbery polymeric molecules such as graft copolymers with a thermoplastic grafted onto the rubber or block copolymers where one of the blocks is a soft elastomeric block. Antistatic agents, which reduce the static charge on the surface of plastics, include quaternary ammonium compounds, sodium alkyl sulfonates, and anionic surface-active agents.

Aesthetic additives are colorants that impart hue (shade), brightness (value), and intensity (color strength) to plastics. Colorants include inorganic and organic pigments, dyes, and optical brighteners.

Elastomers

A rubber or elastomer is defined as a material that can be elongated to at least twice its original length and which will

Table 1.2 Effect of tacticity on polystyrene

Isomerism	T_m (°C)	Crystallization rate
Atactic	None	None
Isotactic	240	Very slow
Syndiotactic	270	Faster

Note: Based on data in references 48 and 49.

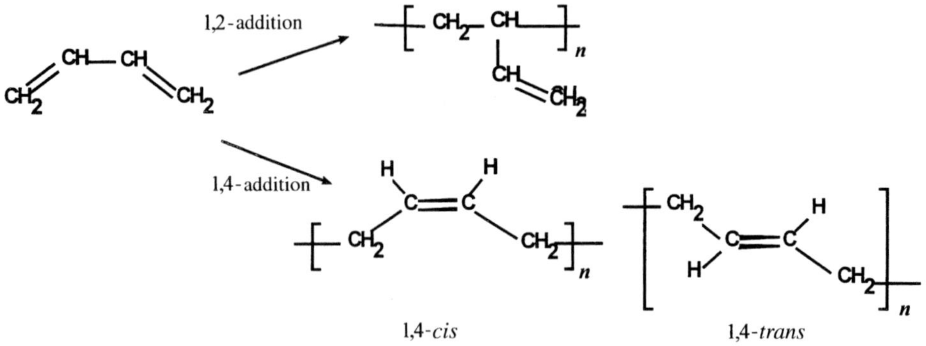

Figure 1.38 Geometrical isomerism of polybutadiene. Key: n = number of repeating units in the polymer molecules.

retract rapidly and forcibly to essentially its original length. A rubber should have high molecular chain length for optimal performance since the elasticity is due mainly to chain entanglement. For the molecules to move freely for rubber elasticity, the use temperature for the rubber must be above the T_g. The rubber can be a block copolymer with a hard segment with a T_g and/or T_m above the use temperature and a rubbery segment with a T_g below room temperature.[64,65] These block copolymers are termed *thermoplastic elastomers*. An example would be an A-B-A block copolymer with styrene-butadiene-styrene segments. The rubber can also be an amorphous polymer with a single T_g below room temperature, in which case it needs to be cross-linked to determine and maintain its shape. Various additives are used in rubber compounds. In unsaturated rubbers like natural rubber, polybutadiene, and polyisoprene, vulcanizing or curing additives such as sulfur compounds are used to cross-link the polymers through their double bonds. Accelerators, promoters, and activators are used to speed up the vulcanization process.[66,67]

For rubbers to develop reasonable properties, a reinforcing agent is typically used. In unsaturated rubbers, carbon black is typically used as a reinforcing agent. In addition, carbon black is an excellent absorber of light and stabilizes the rubber polymer from UV degradation. Fumed silica is used as a reinforcing agent in silicones. Extending fillers such as calcium carbonate and carbon black are used mainly to reduce the overall cost. Extending oils are hydrocarbons that are used in rubbers to plasticize the polymer, making it easier to process. In addition, extending oils are used to reduce the overall cost.

Antioxidants and stabilizers are important in rubbers and very important in unsaturated rubbers, which are more prone to degradation by oxygen and ozone. Colored rubbers can be obtained by use of pigments if carbon black is not used.

Silicone resins are characterized by an inorganic backbone with alternating silicon and oxygen atoms with organic groups attached to the silicon atoms.[68–70] Typical organic groups include methyl, phenyl, vinyl, and trifluoropropyl. Phenyl groups increase the heat resistance of the polymer.[71] Vinyl groups facilitate cross-linking via free radicals. Trifluoropropyl groups increase the chemical resistance.[72] These silicone structures are illustrated in Figure 1.43. The outstanding feature of silicone resins is their excellent resistance to high temperatures.

Silicone rubbers can be cured with peroxides or via room-temperature vulcanization technology. In room-temperature vulcanization technology, a silanol-terminated polysiloxane is mixed with a tri- or a tetrafunctional monomer such as trialkoxy- or triacetoxymethyl siloxane. Hydrolysis of a trialkoxy- or acetoxysilane bond produces a silanol that will condense with the silanol-terminated siloxane. Further hydrolysis and condensation leads to a silicone network, as shown in Figure 1.44. Fumed silica is used as a reinforcing filler to enhance properties of silicones.

Coatings

Coatings and surface finishes have been used for many thousands of years.[73–75] Coatings are typically applied to various surfaces to provide decoration and protection against corrosion or weathering. Surface coating compositions are usually viscous liquids consisting of a film-forming polymer called a "binder" and a volatile liquid. For colored coating, either a pigment or a combination of pigments is added. Film-forming polymers, or binders, can be fatty acid esters of glycerol, cellulose esters and ethers, vinyl polymers, acrylate polymers, alkyd resins, unsaturated polyesters, amino resins, phenolic resins, epoxy resins, and urethane polymers. Coatings can be classified as lacquers, varnishes, enamels, oil paints, and latex paints.

Table 1.3 Effect of tacticity on 1,2-polybutadiene

Isomerism	T_g (°C)	T_m (°C)
Atactic	−4	None
Isotactic	−15	126
Syndiotactic	−28	156

Note: Based on data in reference 50.

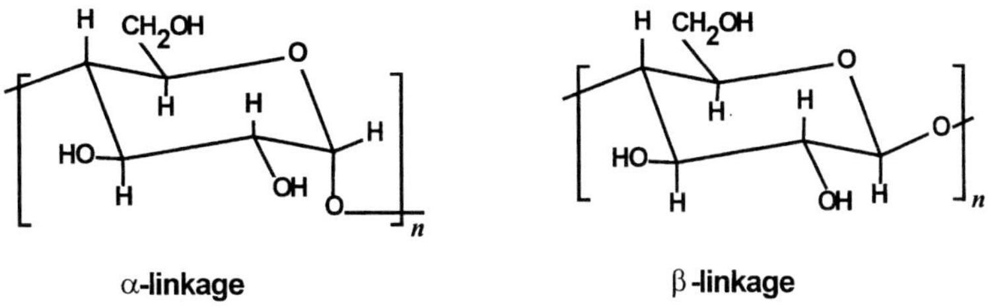

α-linkage   β-linkage

Figure 1.39 Geometrical isomerism of polysaccharides. Key: n = number of repeating units in the polymer molecules.

Lacquers are formulated with a high molecular weight thermoplastic and a solvent in which the polymer is soluble. The coating or polymer film is formed by evaporation from a solvent rather than by polymerization. Thus lacquer coatings remain thermoplastic. Various types of cellulose esters and ethers, vinyl, and acrylic resins are used in lacquers.

A varnish consists of a polymer dissolved in an inert solvent to adjust viscosity, a drying oil, and a catalyst to promote the cross-linking reaction of the drying oil with oxygen. A coating, or film, is formed by a reaction of oxygen that polymerizes and cross-links the drying oil through its double bonds. Drying oils are low molecular weight, liquid, unsaturated oils that readily react with oxygen from air and polymerize. The reaction with oxygen is catalyzed by cobalt and manganese compounds. Drying oils can be natural materials such as linseed, tung, soybean, and dehydrated castor oil. Synthetic drying oils are prepared by reacting unsaturated fatty acids with various synthetic resins. Enamel is typically a varnish with pigments that gives a high gloss coating.

Oil paints have film-formers based on unsaturated fatty acid esters of glycerol such as linseed oil and tung oil. The unsaturation provides sites for oxidation when the material is exposed to air. Cross-linking proceeds via a free-radical mechanism between the unsaturation on the fatty-acid chains, which results in an insoluble polymer network. The use of driers such as cobalt and manganese carboxylates accelerates the free-radical polymerization.

Oil paints also consist of a suspension of pigments in a drying oil and an inert solvent to control viscosity. The volatile solvent (such as mineral spirits) provides a means for lowering the viscosity into a useful range for formulation and application and does not become part of the final coating. After the coating has been applied to the surface, the volatile solvent evaporates, leaving the film-forming polymer and pigments.

Latex paints are water-based paints that have film-formers prepared by emulsion polymerization. Thus the polymer particles are dispersed in an aqueous carrier. The latex has low viscosity at a high level of solids, where the term *solids* refers to the nonaqueous content of the latex. The polymer emulsion or latex is typically based on polymers of acrylic and methacrylic esters and poly(vinyl acetate). Latex paints offer rapid drying, low odor, and ease of clean up.

The polymer emulsions are blended with pigments and additives. Pigment levels are typically 30% to 60%. Viscosity-control agents are used to thicken the aqueous phase and reduce settling of the pigments. Coalescing agents such as ethylene glycol monobutyl ether are used to coalesce the discrete particles of polymer during evaporation of the water into a uniform film.

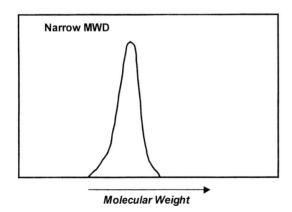

Figure 1.40 Narrow molecular weight distribution.

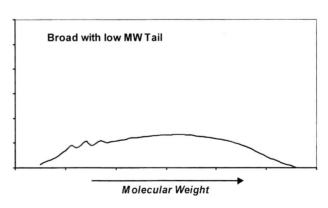

Figure 1.41 Broad molecular weight distribution.

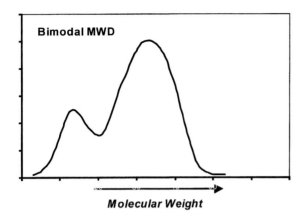

Figure 1.42 Bimodal molecular weight distribution.

Adhesives

An adhesive is defined as a substance capable of holding materials together by surface attachment.[76] Adhesives were first used thousands of years ago and were based on natural materials. Modern adhesives are primarily synthetic polymers. Adhesion results from an interaction between the adhesive and the substrate. This interaction can be mechanical (adhesive getting into surface voids), polar attraction between the adhesive and the substrate, and/or the formation of covalent bonds between the adhesive and substrate.

The two basic requirements for good adhesion are wetting and solidification. For optimal performance, an adhesive must contact the surface intimately. Thus while it is in the liquid state, the adhesive must flow into surface nooks and crannies, and it must wet the surfaces. Once good contact has been made, the adhesive must solidify or harden to provide adequate mechanical strength. The process of solidification can be accomplished by (1) drying via evaporation, (2) chemical reaction, and (3) cooling from a molten state to a solid. Adhesives can be classified as solvent-based, latex, pressure-sensitive, hot melt, and reactive adhesives.

Solvent-based adhesives are prepared by dissolving the adhesive polymer in an appropriate solvent. The solvent lowers the viscosity of the adhesive solution to facilitate wetting the surface. If the solvent can attack the surface, additional adhesion may be obtained. After application of the solvent-based adhesive and joining the surfaces, the solvent needs to evaporate so that maximum adhesion may develop.

Polymer latexes from emulsion polymerization have the adhesive polymer particles dispersed in a water emulsion. The latex facilitates flow and wetting the surface. As the water evaporates, the adhesive polymer particles must be able to coalesce. White glues are based on latex adhesives.

Pressure-sensitive adhesives are highly viscous polymers. When pressure is applied, the polymer will flow and contact the substrate surface. When pressure is released, there is enough adhesion between the adhesive polymer and substrate. This type of adhesive is used in pressure-sensitive tapes. The adhesive polymers used in pressure-sensitive adhesives are typically elastomers.

Hot melt adhesives are thermoplastic materials that are applied the to substrate in the molten state. Adhesion develops as the melt solidifies during cooling.

Reactive adhesives are monomeric compounds of low molecular weight, which are mixed with a hardener or curing agent. These two-component systems are applied to the substrate in the liquid state, react or cure, and form a solid by polymerization and cross-linking reactions. Added adhesion is obtained if the adhesive can chemically react with the surface of the substrate. Examples of reactive adhesive polymers are epoxies, cyanoacrylates, phenolics, and silicones.

Fibers

Synthetic and natural fibers are the building blocks of yarns and fabrics. Synthetic fibers are typically crystalline thermoplastics that were processed into fibers.[77] Modification of fibers comes from antistaining additives, flame retardants, and colorants. The flame retardant and/or dye and pigments used in fibers can be internal or external.

Characterization of Polymers

Polymer characterization is an essential step in working with polymers.[76-90] Polymer characterization is the bridge that links polymer performance to the chemical and physical structure of the polymer. The chemical characterization of polymers is not basically different from the analysis of organic compounds of low molecular weight. However, the generally low solubility, nonvolatile nature, chemical inertness, and complexity of some polymeric systems necessitate modifications of existing procedures.

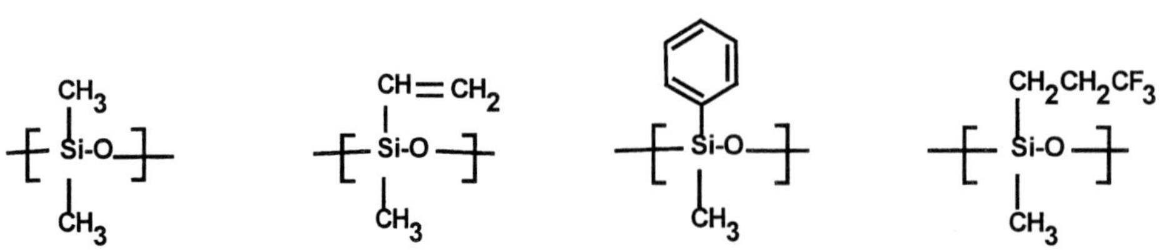

Figure 1.43 Structure of silicone resins.

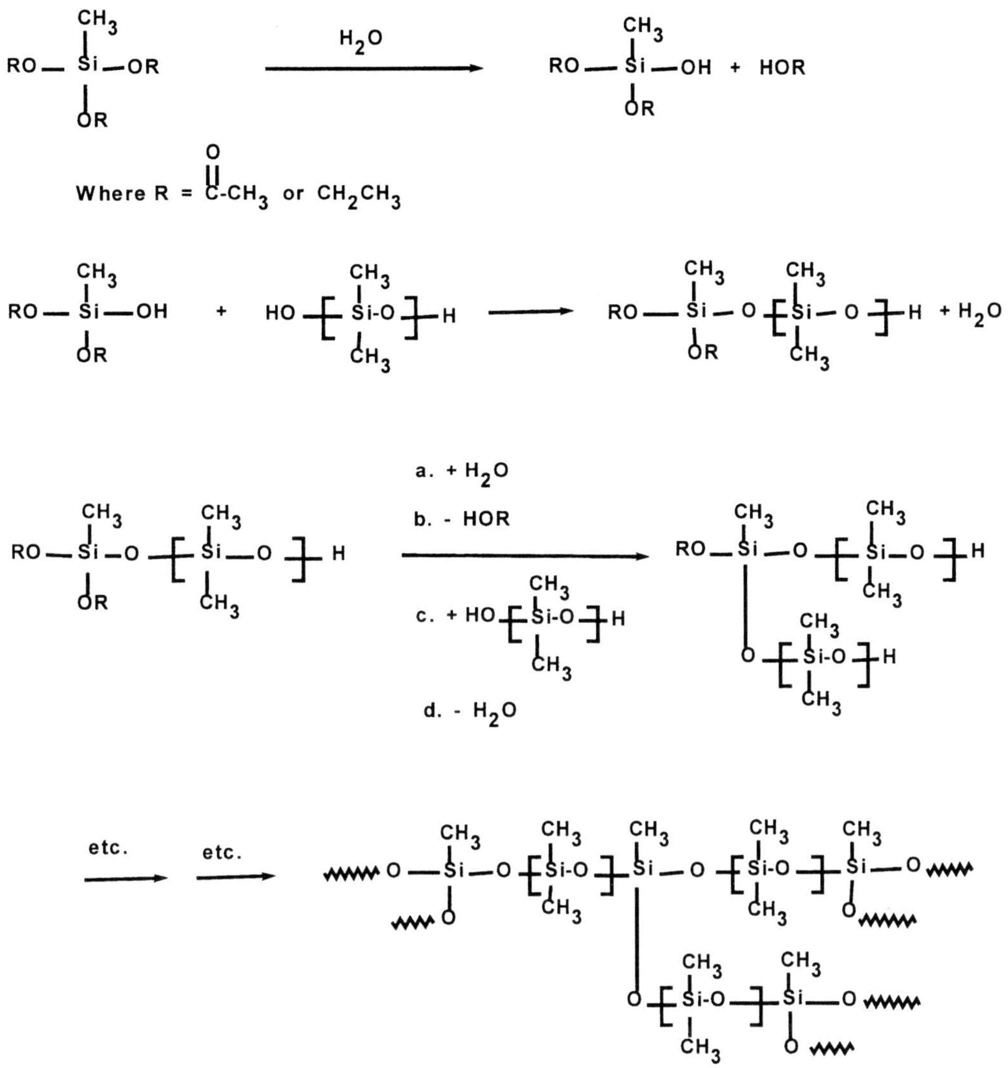

Figure 1.44 Chemistry of room temperature vulcanizing (RTV) silicone.

Polymer characterization focuses on two wide categories: polymer structure and polymer properties. Characterization of polymer structure examines the composition, molecular size, molecular structure, and molecular organization of polymer chains. A full characterization would include the following:

- The type of polymer
- Monomers used to make the polymer
- Residual catalyst, solvent, or monomer in polymer
- Heteroatoms in the polymer (halogen, P, Si, N, O, etc.)
- Functional groups in the polymer (ether, ester, amide, sulfone urethane, ketone, etc.)
- Molecular size (number- and weight-average molecular weight)
- Sequence distribution in copolymers (alternating, random, or block)
- Types and amount of end groups on the polymer

- Transition temperatures (T_g, T_m, T_c–crystallization temperature)
- Amount of crystallinity
- Crystal structure
- Structural isomerism (tacticity and positional and geometrical isomerism)
- Thermal stability (weight loss versus temperature)
- Heat capacity

The focus of polymer properties is on bulk or macroscopic properties of polymers such as characterization of polymer surfaces or performance properties of polymers. Such characterization would include the following:

- Surface energy, contact angle, critical surface tension, and interfacial tension
- Surface composition
- Surface roughness

- Polymer morphology of blends and alloys
- Adhesion of polymers to fillers and reinforcing agents
- Solubility and solvent resistance
- Swelling, gelation, and cross-link density
- Optical properties (birefringence, dichroism, optical activity, refractive index, and molecular refractivity).
- Rheology (the effect of shear and temperature on viscosity)
- Mechanical properties (tensile and flexural stress-strain, creep, and fracture toughness)
- Electrical properties (dielectric constant, conductivity, resistivity, and ferroelectric properties)
- Transport and barrier properties (permeability of liquids, gases, and vapors)
- Flammability (ignitability, combustion, and heat release)

Polymer characterization is directed toward a specific purpose. The scope of characterization can be relatively narrow or extremely broad, depending on the degree of scientific information desired and the commercial interest that develops. This would range from a research or manufacturing environment, where there is knowledge about the composition or targeted composition, to an unknown plastic with limited or no knowledge of composition.

In a newly synthesized polymer, characterization of structural details is facilitated by knowledge of the monomers used in the polymerization. The amount of each monomer incorporated into the polymer and the sequence distribution of the monomers could be determined by nuclear magnetic resonance (NMR) imaging. Stereoisomerisms in chain-growth polymers made with a stereospecific catalyst could be checked with NMR spectroscopy.

Gel permeation chromatography (GPC; also called size exclusion chromatography) or solution viscosity methods could determine molecular size. In solution viscosity, the polymer is dissolved in a solvent, and the viscosity of the solution is measured and compared to the viscosity of the solvent. The difference in viscosity gives a relative indication of molecular size versus standards. This simple method provides a single value as an indication of molecular size. GPC would give information on the number-average and weight-average molecular weight and molecular weight distribution (MWD). A broad, narrow, or bimodal (two polymerization pathways) MWD would imply something about the uniformity and quality of the polymer. Other classical techniques for measurement of molecular weights and MWD include membrane osmometry (MO) and vapor pressure osmometry (VPO).

Molecular size contributes to the viscosity of the melted polymer. Rheology looks at the melt viscosity as a function of shear rate and temperature. For a given polymer structure, larger molecules have higher viscosity in the melt and in solution.

An indication of thermal properties can be obtained by differential-scanning calorimetry (DSC). DSC gives a thermal fingerprint for identifying polymers and measures heat capacity and temperatures of transitions such as T_g, T_m, and T_c. The heat capacity or specific heat, C_p, is the amount of heat required to raise the temperature of 1 gram of plastic 1 degree centigrade. The C_p changes at the T_g of a polymer. DSC measures both exo- and endothermic transitions such as T_c (crystallization temperature) and T_m.

Transition temperatures can also be measured by dynamic mechanical analysis (DMA). DMA indicates how the modulus varies with temperature and frequency of loading. Thus DMA identifies mechanical transitions in solid polymers (T_g, T_c, and T_m). In addition, it defines the useful thermal range for property retention of a polymer.

Thermo-mechanical analysis (TMA) measures dimensional change in a polymer as a function of temperature: as dimensional changes at the T_g, T_c, and T_m. Other methods such as sonic modulus and torsional braid analysis (TBA) reveal transition temperatures.

Information about the thermal or thermo-oxidative stability of a polymer can be obtained from thermo-gravimetric analysis (TGA). In TGA a small sample of the polymer is heated at a constant rate, usually in an air or nitrogen environment. The temperature at the onset of weight loss suggests the limits of the thermal or thermo-oxidative stability of the polymer.

Information from structural characterization may suggest areas for further physical property characterization. For example, a high T_g would suggest a high heat deflection temperature of the new polymer.

In manufacturing operations, quality-control testing of polymers is important. Since details of the manufacturing process have usually been optimized, the focus is on quick methods for checking molecular size, composition, and key physical properties. Molecular size can be checked by solution viscosity or melt flow index. GPC and rheological measurements could also be used. Mechanical property characterization would include stress-strain and impact/toughness testing.

The characterization of an unknown plastic is more challenging and would first involve the estimation of the base resin(s) by infrared spectroscopy (IR), NMR, and UV/visible spectroscopy. This could be corroborated by elemental analysis for C, H, and heteroatoms by various methods such as Kjeldehl analysis, atomic absorption (AA), inductively coupled plasma spectroscopy (ICP), and X-ray fluorescence and energy-dispersive X-ray analysis (EDAX).

NMR could determine any stereoisomerism or the sequence distribution of monomers in copolymers. Molecular size would be determined by GPC.

Monomers used to make the polymer can be obtained by hydrolysis of the polymer if it is a condensation polymer or by thermolysis followed by various chromatographic and spectroscopic techniques and mass spectrometry. Residual monomer(s), solvents, and trace catalyst can also be characterized by gas chromatography (GC), high-performance liquid chromatography (HPLC), and gas chromatography/mass spectrometry (GC/MS). DSC would give a thermal fingerprint. If the plastic is a thermoplastic, determine the molecular weight. If it is a thermoset, then determine the cross-link density or swelling index.[91–94]

The presence of fillers and reinforcing additives can be identified by an EDAX. Elemental analysis would indicate the presence of flame retardants with halogen or nonhalogen (P, metal) additives. Residual inorganic catalysts can be characterized by AA, ICP, and EDAX.

The presence of various additives is an important part of polymer characterization. The additives can be organic, inorganic, or organometallic. The organic additives can be small molecules or polymeric. The amount of an additive can cover a wide range. Fillers, reinforcing materials, and plasticizers can be present in relatively high levels such as 10 to 40 weight percent. In contrast, stabilizers can be present at levels less than 1 weight percent. In addition, polymers can and usually do contain more than one additive to enhance performance. Moreover, some additives can react or decompose during processing, and this can considerably complicate the characterization.

Characterization of additives can involve the direct analysis of the polymer using nondestructive methods such as infrared and UV spectroscopy. A key criterion is that the additive has distinct absorption areas where the base polymer exhibits little or no absorption. EDAX would indicate the presence of heavier atoms and suggest the presence and type of inorganic additives such as fillers and reinforcing additives, pigments, and organometallic additives. Elemental analysis would indicate the presence of a flame retardant with halogen or nonhalogen (P, metal) additives. In addition, after burning away the polymer, inorganic or inorganic-containing additives can be characterized by weighing and analyzing the residue via AA and ICP. This requires that the inorganic material is nonvolatile under the preparative conditions.

Some additives can be isolated by extraction from the polymer. Typically, the polymer would be ground to a fine powder to increase the surface area, then extracted. The solvent should be a nonsolvent for the polymer, nonreactive with the additives or polymer, and a good solvent for the additives. Some additives such as antioxidants and UV stabilizers are reactive toward oxygen and UV light; hence, exposure to such environments may need to be minimized during extraction and characterization. The solvent extract can be concentrated and analyzed by HPLC or GC/MS methods. Infrared, UV, and NMR imaging can be used in the identification after removal and/or replacement of the solvent. The extract from the polymer may contain low molecular weight polymer, residual monomer, or solvent from the polymerization, which may complicate the analysis.

Some additives may have enough volatility that they may be released from the polymer during heating. Thus volatiles would be sent to a gas chromatograph for identification.

References

1. Wakeman, R. L. *The Chemistry of Commercial Plastics;* Reinhold: New York, 1947.
2. Mark, H. F. In *Proceedings of the Robert Welch Foundation, Conference on Chemical Research X. Polymers;* Milligan, W. O., Ed.; Welch Foundation: Houston, 1967; pp 19–55.
3. Staudinger, H. *Die hochmolekularen organischen Verbindungen Kautschuk und Cellulose;* Springer-Verlag: Berlin, 1932.
4. Rudin, A. *The Elements of Polymer Science and Engineering;* Academic Press: New York, 1982.
5. Deanin, R. D. *Polymer Structure, Properties and Applications;* Cahners: Boston, 1972.
6. Billingham, N. C.; Jenkins, A. D. In *Polymer Science;* Jenkins, A. D., Ed.; Elsevier: New York, 1972; Vol. 1, pp 1–117.
7. Peters, E. N. In *Mechanical Engineer's Handbook,* 2nd ed.; Kutz, M., Ed.; Wiley-Interscience: New York, 1998; pp 115–129.
8. Peters, E. N.; Arisman, R. K. In *Applied Polymer Science: 21st Century;* Lohse, D. J., Ed.; American Chemical Society: Washington, DC, 2000; pp 177–196.
9. Peters, E. N. In *Concise Encyclopedia of Chemical Technology;* Grayson, M., Ed.; John Wiley & Sons: New York, 1985; pp 651–652.
10. Peters, E. N. In *Kirk-Othmer Encyclopedia of Chemical Technology,* 3rd ed.; Grayson, M., Ed.; John Wiley & Sons: New York, 1981; Vol. 13, pp 398–413.
11. Hiemenz, P. C. *Polymer Chemistry: The Basic Concepts;* Marcel Dekker: New York, 1984.
12. Billmeyer, F. W. *Textbook of Polymer Science,* 3rd ed.; John Wiley & Sons: New York, 1984.
13. Bovey, F. A., Winslow, F. H., Eds. *Macromolecules: An Introduction to Polymer Science;* Academic Press: New York, 1979.
14. Goodman, S. H., Ed. *Handbook of Thermoset Plastics,* 2nd ed.; Plastics Design Library: Brookfield, CT, 1999.
15. Olabisi, O., Ed. *Handbook of Thermoplastics;* Marcel Dekker: New York, 1998.
16. Matyjaszewski, K., Ed. *Polymerizations: Mechanisms, Synthesis and Applications;* Marcel Dekker: New York, 1996.
17. Odian, G. *Principles of Polymerization,* 3rd ed.; John Wiley & Sons: New York, 1991.
18. Solomon, D. H., Ed. *Step Growth Polymerizations;* Marcel Dekker: New York, 1972.
19. Bottenbruch, L., Ed. *Engineering Thermoplastics: Polycarbonates, Polyacetals, Polyesters, Cellulose Esters;* Hanser Gardner: New York, 1996.
20. Kohan, M. I., Ed. *Nylon Plastics Handbook;* Hanser Gardner: New York, 1995.
21. Ahorani, S. M. *n-Nylons: Their Synthesis, Structure and Properties;* John Wiley & Sons: New York, 1997.
22. Uhlig, K. *Discovering Polyurethanes;* Hanser Gardner: New York, 1999.

23. LeGrand, D.G., Bendler, J. T., Eds. *Handbook of Polycarbonates: Science and Technology;* Marcel Dekker: New York, 1999.

24. Smith, D. A., Ed. *Addition Polymers: Formation and Characterization;* Plenum Press: New York, 1968.

25. Moad, G.; Solomon, D. H. *The Chemistry of Free Radical Polymerization;* Pergamon: Oxford, 1995.

26. Matyjaszewski, K., Ed. *Controlled Radical Polymerization;* American Chemical Society: Washington, DC, 1997.

27. Benedikt, G. M., Goodall, B. L., Eds. *Metallocene-Catalyzed Polymers: Materials, Properties, Processing and Markets;* Plastics Design Library: Brookfield, CT, 1998.

28. Morton, M. *Anionic Polymerization: Principles and Practice;* Academic Press: New York, 1983.

29. Szwarc, M. *Ionic Polymerization Fundamentals;* Hanser Gardner: Cincinnati, OH, 1996.

30. Kennedy, J. P.; Marechal, E. *Carbocationic Polymerization;* Krieger Publishing: Malabar, FL, 1982.

31. Plesch, P. H., Ed. *The Chemistry of Cationic Polymerizations;* Pergamon Press: New York, 1963.

32. Kennedy, J. P. *Cationic Polymerization of Olefins;* John Wiley & Sons: New York, 1975.

33. Kennedy, J. P.; Ivan, B. *Designed Polymers by Cationic Macromolecular Engineering;* Hanser Gardner: Cincinnati, OH, 1992.

34. Wilson, W. C.; Santee, E. R. *J. Polym. Sci., Part C* **1965,** *8,* 97.

35. Singh, A., Silverman, J., Eds. *Radiation Processing of Polymers;* Hanser Gardner: Cincinnati, OH, 1991.

36. Nauman, E. B. In *Encyclopedia of Polymer Science and Engineering;* Kroschwitz, J. L., Ed.; Wiley-Interscience: New York, 1985; Vol. 2, pp 500–514.

37. Swift, G.; Hughes, K. A. In *Encyclopedia of Polymer Science and Engineering;* Kroschwitz, J. L., Ed.; Wiley-Interscience: New York, 1989; Vol. 15, pp 402–418.

38. Grulke, E. A. In *Encyclopedia of Polymer Science and Engineering;* Kroschwitz, J. L., Ed.; Wiley-Interscience: New York, 1989; Vol. 16, pp 443–472.

39. Gilbert, R. G. *Emulsion Polymerization: A Mechanistic Approach;* Acedemic Press: Orlando, FL, 1995.

40. Bovey, F. A.; Kolthoff, I. M.; Medalia, A. I.; Meehan, E. J. In *High Polymers IX: Emulsion Polymerization;* Bovey, F. A., Ed.; John Wiley: New York, 1955.

41. Ham, G. E., Ed. *Copolymerization;* Wiley-Interscience: New York, 1964.

42. Gordon, M.; Taylor, J. S. *J. Appl. Chem.* **1952,** *2,* 493.

43. Greenley, R. Z. In *Polymer Handbook,* 3rd ed.; Brandrup, J., Immergut, E. H., Eds.; John Wiley & Sons: New York, 1989; pp 153–266.

44. Shonaike, G. O., Simon, G. P., Eds. *Polymer Blends and Alloys;* Society of Plastics Engineers: Brookfield, CT, 1999.

45. Olabisi, O.; Robeson, L. M.; Shaw, M. T. *Polymer-Polymer Miscibility;* Academic Press: New York, 1979.

46. Solc, K., Ed. *Polymer Compatibility and Incompatibility: Principles and Practice;* Harwood Academic Publishers: New York, 1982.

47. Majumdar, B.; Paul, D.R. In *Polymer Blends;* Paul, D. R., Bucknall, C. P., Eds.; John Wiley & Sons: New York, 1999; Vol. 2, pp 539–579.

48. Karasz, F. E.; Blair, H. E.; O'Reilly, J. M. *J. Phys. Chem.* **1965,** *69,* 2657.

49. Ishida, N.; Seimiya, T.; Kuramoto, M.; Uoi, M. *Macromolecules* **1986,** *19,* 2465.

50. Peyser, P. In *Polymer Handbook,* 3rd ed.; Brandrup, J., Immergut, E. H., Eds.; John Wiley & Sons: New York, 1989; p 213.

51. Pigman, W., Horton, D., Eds.; *The Carbohydrates,* 2nd ed., Academic Press: New York, 1970.

52. Aspinall, G. O., Ed. *The Polysaccharides;* Academic Press: New York, 1982; Vol. 1.

53. Gilbert, R., Ed. *Cellulosic Polymers;* Hanser Gardner: Cincinnati, OH, 1993.

54. Peebles, L. H. *Molecular Weight Distributions in Polymers;* Wiley-Interscience: New York, 1971.

55. Ulrich, H. *Introduction to Industrial Polymers,* 2nd ed.; Hanser Gardner: Cincinnati, OH, 1993.

56. Rubin, I. I. *Handbook of Plastic Materials and Technology;* John Wiley & Sons: New York, 1990.

57. Harper, C. A., Ed. *Handbook of Plastics and Elastomers;* McGraw-Hill: New York, 1975.

58. Gächter, R., Müller, H., Eds. *Plastics Additives,* 4th ed.; Hanser Gardner: Cincinnati, OH, 1993.

59. Stepek, J.; Daoust, H. *Additives for Plastics;* Springer-Verlag: New York, 1983.

60. Pritchard, G. *Plastics Additives: An A–Z Reference;* Kluwer Academic Publishers: Amsterdam, 1998.

61. Lutz, J. T., Ed. *Thermoplastic Polymer Additives: Theory and Practice;* Marcel Dekker: New York, 1989.

62. Wypych, G. *Handbook of Fillers: The Definitive User's Guide and Databook of Properties, Effects and Users,* 2nd ed.; Plastics Design Library: Brookfield, CT, 1999.

63. Peters, E. N. In *Flame Retardancy of Polymeric Materials;* Kuryler, W. C., Papa, A. J., Eds.; Marcel Dekker: New York, 1979; Vol. 5, pp 173–176.

64. Walker, B. M., Ed. *Handbook of Thermoplastics Elastomers;* Van Nostrand Reinhold: New York, 1979.

65. Holden, G.; Legge, N. R.; Quirk, R. *Thermoplastic Elastomers,* 2nd ed.; Hanser Gardner: Cincinnati, OH, 1996.

66. Ciullo, P. A.; Hewitt, N. *The Rubber Formulary;* Plastics Design Library: Brookfield, CT, 1999.

67. Morton, M., Ed. *Rubber Technology;* Van Nostrand Reinhold: New York, 1973.

68. Zeigler, J. M., Fearon, F. W. G., Eds. *Silicone-Based Polymer Science: A Comprehensive Resource;* American Chemical Society: Washington, DC, 1990.

69. Peters, E. N. *J. Macromol. Sci.—Rev. Macromol. Chem.* **1979,** *C17,* 173–200.

70. Peters, E. N. *Ind. Eng. Chem. Prod. Res. Dev.* **1984,** *23,* 28–32.

71. Peters, E. N.; Kawakami, J. H.; Kwiatkowski, G. T.; Joesten, L. B.; McNeil, D. W.; Owens, D. A.; Hedaya, E. *J. Polym. Sci., Polym. Phys. Ed.* **1977,** *15,* 723–732.

72. Peters, E. N.; Stewart, D. D.; Bohan, J. J.; Moffitt, R.; Beard, C. D.; Kwiatkowski, G. T.; Hedaya, E. *J. Polym. Sci., Polym. Chem. Ed.* **1977,** *15,* 973–981.

73. Wicks, Z. W., Jones, F. N., Pappas, S. P., Eds. *Organic Coatings: Science and Technology,* 2nd ed.; John Wiley & Sons: New York, 1999.

74. Freitag, W.; Stoye, D. *Paints, Coatings and Solvents,* 2nd ed.; John Wiley & Sons: New York, 1998.

75. Swaraj, P., Ed. *Surface Coatings: Science and Technology,* 2nd ed.; John Wiley & Sons: New York, 1996.

76. Pocius, A. V. *Adhesion and Adhesives Technology: An Introduction;* Hanser Gardner: Cincinnati, OH, 1996.
77. Fourne, F. *Synthetic Fibers;* Hanser Gardner: Cincinnati, OH, 1999.
78. Sibilia, J. P., Ed. *A Guide to Materials Characterization and Chemical Analysis;* VCH Publishers: New York, 1988.
79. Schröder, E.; Müller, G.; Arndt, K.-F. *Polymer Characterization;* Hanser Gardner: Cincinnati, OH, 1989.
80. Kroschwitz, J. I., Ed. *Polymers: Polymer Characterization and Analysis;* John Wiley & Sons: New York, 1990.
81. Crompton, T. R. *Practical Polymer Analysis;* Plenum Press: New York, 1993.
82. Hunt, B. J., James, M. I., Eds. *Polymer Characterization;* Blackie Academic & Professional: Glasgow, Scotland, 1993.
83. Campbell, D.; White, J. R. *Polymer Characterization: Physical Techniques;* Chapman & Hall: London, 1989.
84. Bath, H. G., Mays, J. W., Eds. *Modern Methods of Polymer Characterization;* John Wiley & Sons: New York, 1991.
85. Bath, H. G., Ed. *Polymer Analysis and Characterization, II;* Applied Polymer Symposium No. 45; John Wiley & Sons: New York, 1990.
86. Craver, C. D., Provder, T., Eds. *Polymer Characterization: Physical Property, Spectroscopic and Chromatographic Methods;* Advances in Chemistry Series No. 227; American Chemical Society: Washington, DC, 1990.
87. Haslam, J.; Willis, H. A. *Identification and Analysis of Plastics;* Van Nostrand Reinhold: Princeton, NJ, 1965.
88. Brown, R. P., Ed. *Handbook of Plastics Test Methods,* 2nd ed.; George Godwin: London, 1981.
89. Brown, R., Ed. *Handbook of Polymer Testing;* Society of Plastics Engineers: Brookfield, CT, 1999.
90. Kampf, G. *Characterization of Plastics by Physical Methods;* Hanser Gardner: Cincinnati, OH, 1986.
91. Labana, S. S. In *Encyclopedia of Polymer Science and Engineering;* Kroschwitz, J. L., Ed.; Wiley-Interscience: New York, 1986; Vol. 4, pp 350–395.
92. Priss, L. S., *J. Polym. Sci., Part C* **1975,** 53, 195.
93. *ASTM D3616;* American Society for Testing and Materials: Philadelphia, 1988.
94. *ASTM D2765;* American Society for Testing and Materials: Philadelphia, 1995.

General References

Campbell, I. M. *Introduction to Synthetic Polymers;* Oxford University Press: New York, 1994.
Carraher, C. E. *Seymour/Carraher's Polymer Chemistry,* 5th ed.; Marcel Dekker: New York, 2000.
Challa, G. *Polymer Chemistry: An Introduction;* Ellis Horwood: Chichester, England, 1993.
Cowie, J. M. G. *Polymers: Chemistry and Physics of Modern Materials,* 2nd ed.; Chapman and Hall: New York, 1991.
Ebdon, J. R., Ed. *Newer Methods of Polymer Synthesis;* Blackie Academic & Professional: Glasgow, Scotland, 1991.
Kroschwitz, J. I., Ed. *Concise Encyclopedia of Polymer Science and Engineering;* John Wiley & Sons: New York, 1998.
Mark, J. Ed. *Physical Properties of Polymers,* 2nd ed.; American Chemical Society: Washington, DC, 1993.
Munk, P. *Introduction to Macromolecular Science;* John Wiley & Sons: New York, 1989.
Painter, P. C.; Coleman, M. M. *Fundamentals of Polymer Science: An Introductory Text;* Technomic Publishing: Lancaster, PA, 1994.
Rosen, S. L. *Fundamental Principles of Polymeric Materials;* John Wiley & Sons: New York, 1993.
Salamone, J. C., Ed. *Concise Polymeric Materials Encyclopedia;* CRC Press: Boca Raton, FL, 1998.
Sperling, L. H., Ed. *Introduction to Physical Polymer Science,* 2nd ed.; John Wiley & Sons: New York, 1992.
Stevens, M. P. *Polymer Chemistry: An Introduction,* 2nd ed.; Oxford University Press: New York, 1990.
Stille, J. K., Campbell, T. W. *Condensation Monomers,* Krieger Publishing: Melbourne, FL, 1972.
Sun, S. F., *Physical Chemistry of Macromolecules: Basic Principles and Issues;* John Wiley & Sons: New York, 1994.
Van Krevelen, D. W. *Properties of Polymers,* 3rd ed.; Elsevier: Amsterdam, 1990.
Young, R. J., Lovell, P. A. *Introduction to Polymers,* 2nd ed.; Chapman & Hall: London, 1991.

2

Measurement of Molecular Weight and Molecular Weight Distribution

HOWARD G. BARTH

Polymers and polymeric materials play a pivotal role in advancing science and technology and in improving our standard of living. We are beginning to understand better how to control and tailor materials with unique properties and uses for all types of applications, from biomaterials to electronic devices. As a result, there is a greater need to apply and develop analytical methods to characterize these materials to measure fundamental properties for determining structural-property relationships, as well as for controlling and monitoring processing and quality assurance.

One critical property of a polymer is its molecular weight (MW) and molecular weight distribution (MWD), both of which greatly influence properties of the final product. In this chapter, definitions used in calculating average molecular weight values are reviewed, and the significance of these values is discussed. A survey of commonly used methods for measuring molecular weights is presented. More detailed treatment of some of these techniques is found in subsequent chapters, however. The main focus of this chapter is on synthetic polymers, although all of the techniques discussed here are applicable to biopolymers.

Polymer Terminology

Macromolecules are "large" molecules consisting of repeating units (referred to as *monomers*) that are joined together through two or more chemical bonds. If the macromolecule consists of 2 to about 20 monomers, the term *oligomer* is used; if the number of monomer units (n) or degree of polymerization (dp) is >20, we have a *polymer*. Two major categories of polymers are synthetic polymers (such as polyethylene and polystyrene) and polymers that occur naturally (such as biopolymers, which include proteins, nucleic acids, and polysaccharides). In proteins, the monomeric units are amino acids linked together by amide bridges. The monomeric units of nucleic acids are nucleosides covalently bon-

ded to one another through phosphodiester groups. In the case of polysaccharides, monomers are carbohydrates joined together by glycosidic linkages.

The molecular weight of polymers covers an immense range, from the hundreds for oligomers to ultrahigh molecular weight polymers extending into the low millions, to over 10^{10} g/mol for deoxyribonucleic acid (DNA). For synthetic polymers of commercial interest, however, the range is typically about 10^4 to 5×10^5 g/mol.

A homopolymer is defined as a polymer that consists of one type of monomeric unit, as in the case of polyethylene, that is, $[-CH_2CH_2-]_n$, or polycaprolactam (nylon 6), that is, $[-NH(CH_2)_6NHCO(CH_2)_4CO-]_n$. A copolymer consists of two different monomers that can be arranged in a random, alternating, or block configuration, as shown in Figure 2.1. In the case of a random or statistical copolymer, there is no preferential sequence distribution between the two different comonomers. As a result, the comonomers are arranged randomly along the chain. An alternating copolymer resembles a homopolymer; however, the term *homopolymer* is usually reserved for condensation types of polymerizations, as will be discussed later in this chapter. A block copolymer consists of runs of the same comonomer; polymerization chemistry and conditions control the length, order, and number of blocks. The arrangement of comonomers within a copolymer depends on the relative chemical reactivity of each comonomer, as well as polymerization conditions. The ordering of comonomers will also influence its physical, and often chemical, properties. Polymers consisting of three different types of monomers are called *terpolymers*, and if there are more than three monomers, the term *multicomponent copolymer* is sometimes employed.

One of the more interesting ways that the properties of a polymer can be controlled is through its topological, or three-dimensional, structure, as shown in Figure 2.2. For example, when short-chain branching is introduced along the backbone, melting behavior is greatly affected. Long-

Linear Polymeric Configurations

Homopolymer

   A-A-A-A-A-A-A-A-(A)ₙ

Copolymers

Random

   -A-A-B-A-B-B-A-B-B-B-

Alternating

   -A-B-A-B-A-B-A-B-A-B-

Block

   (A)ₙ-A-A-A-B-B-B-B-B-(B)ₘ

Figure 2.1 Configurations of homopolymer and copolymer chains composed of comonomers A and B. For copolymers, more than two different monomers may be used during polymerization.

chain branching, in contrast, consists of branches that can be as long as the polymer itself. As a result, long-chain branching would influence rheological (or viscoelastic) properties. Thermoset resins have three-dimensional cross-linked structures, which render the polymer insoluble; as such, the molecular weight is undefined. Other structures include star-shaped, graft or comb-like, dendritic copolymers, and cyclics, all of which have unique physical and chemical properties.

In addition to architectural differences, polymers in solution can differ in terms of molecular conformation or shape, depending on the chemical structure of the polymer, as well as the solvent and temperature. The majority of synthetic polymers assume a nearly random (Figure 2.2) or slightly extended conformation in solution. Branched polymers take on an almost spherical conformation, depending on the degree of branching. A classical example of a highly branched polymer is amylopectin starch. Globular proteins, although not branched, have compact spherical structures caused by intramolecular interactions that lead to chain folding. Polymers with hindered rotation along the chain have semirigid or extended conformation in solution, such as poly(*p*-phenylene terephthalamide), known as Kevlar, which produces high-strength fibers because of its highly ordered, crystalline structure.

Molecular Size versus Molecular Weight

Molecular weight is an intrinsic property of a polymer that can be calculated readily if the chemical structure and the degree of polymerization are known. Molecular size, however, is an extrinsic property of the polymer that depends on the molecular weight and the shape of a polymer, the

latter of which is a function of solvent and temperature. Although the focus of this chapter is on the determination of molecular weight and molecular weight distributions, often it is of interest to measure molecular size. Furthermore, several of the techniques for measuring molecular weight—including size exclusion chromatography and viscometry—actually measure molecular size, while light scattering measures both molecular size and molecular weight. In this section, I define size parameters and show the relationship between molecular weight and size.

Because a polymer chain in solution is in constant motion, one defines molecular size by taking into account the probability of finding the distance r between the two ends

Architectural Structures of Polymers

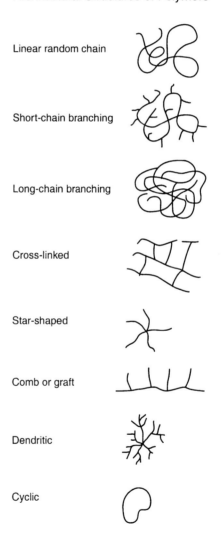

Linear random chain

Short-chain branching

Long-chain branching

Cross-linked

Star-shaped

Comb or graft

Dendritic

Cyclic

Figure 2.2 Different polymeric architectures of polymers. The majority of polymers are either linear, branched, or cross-linked. These architectural features impart unique properties to polymers. Cyclic polymers are usually oligomeric.

of a polymer chain. Based on random walk theory, the end-to-end distance of a random-chain polymer—a polymer chain that is perfectly flexible—consisting of n units of length l, is

$$<r^2> = nl^2 \qquad (1)$$

(The angular brackets indicate an average value—that is, distance r averaged over a long time period.)

Since the degree of polymerization n is directly proportional to molecular weight, higher molecular weight polymers have a larger size or end-to-end distance. If the polymer chain has hindered rotation or is less flexible, as in the case of poly(p-phenylene terephthalamide), its molecular size will be significantly greater than the size of a more flexible chain of the same molecular weight. Furthermore, the molecular size of a given polymer can be expanded or contracted, depending on the nature of the solvent and the temperature. For example, if a polymer is dissolved in a good solvent (that is, one in which the polymer chains are well solvated by solvent molecules), it will have a more extended conformation or have a larger size. Likewise, if a polymer is dissolved in a poorer solvent, in which solvent-polymer interactions are not as strong as polymer-intramolecular interactions, the chain will be less extended in solution.

A more useful size parameter than the end-to-end distance is the radius of gyration, which can be determined by using viscometry or light scattering. The radius of gyration R_g, defined as the root-mean-square distance R_i of the elements of a chain of mass m_i from its center of gravity, is:

$$<R_g^2>^{1/2} = (\Sigma_i m_i R_i^2 / \Sigma_i m_i)^{1/2} \qquad (2)$$

The relationship between the radius of gyration and end-to-end distance is simply

$$<R_g^2> = 1/6(<r^2>) \qquad (3)$$

These molecular size parameters typically are used for studying the physical chemistry of polymers in solution. Molecular size gives us insight into the extent and structure of polymer branching. The effect of polymer chemical structure on coil dimensions allows us to study chain flexibility, as does the influence of solvent type and temperature on coil expansion. These fundamental thermodynamic properties are needed to develop theoretical models to describe, understand, and predict polymer behavior. Furthermore, molecular shape can also be predicted from equations that relate molecular size to molecular weight, as will be discussed in the viscosity section that follows.

Knowledge of molecular size and shape is important for fundamental studies, especially when developing structure-property relationships and designing polymers of certain characteristics. In actual practice, however, determining size and shape parameters is best reserved for well-defined model polymers of known composition or for providing insight into structural differences among samples that behave differently. The discussion in this chapter is limited to molecular weight measurement, rather than size and shape, because of its practical importance in most industrial analytical laboratories.

Molecular Weight Averages and Distribution

Synthetic polymers have a distribution of polymer chains of different molecular weights; that is, they typically consist of hundreds to thousands of macromolecules of different molecular weights. The reason for this is that polymerization reactions depend on the probability of monomers and polymer-chain ends reacting with one another, changes in monomer and polymer concentrations during reaction, and the occurrence of side reactions that can terminate chain growth. Furthermore, it is often difficult to precisely control polymerization reactions in a plant setting. In the case of copolymers, the relative reactivity of each comonomer will also influence molecular weight distribution, as well as polymer chemical composition. In contrast, nucleic acids and nearly all proteins have well-defined molecular structures that are composed of a distinct molecular weight.

Figure 2.3 shows the molecular weight distributions of polymers produced from three common types of polymerization reactions. Figure 2.3A is typical of polystyrene produced by anionic polymerization. In this polymerization the active polymer end is an anionic group. If the reaction is conducted in the absence of impurities, the polymer will continue to grow until all of the monomer is reacted. This so-called living polymerization produces polymer with narrow molecular weight distribution. A condensation-polymerized polymer, as shown in Figure 2.3B, takes place between two bifunctional monomers—for example, a diacid (adipic acid) and a diamine (hexamethylenediamine) to produce nylon 6,6. As the reaction proceeds, the two monomers become depleted, and the end groups of the resulting polymer chains begin to react with each other, until all active polymer ends are terminated or used up. The molecular weight distribution, which is broader than in anionically polymerized polymer, depends on the extent of the reaction. That is, early in the polymerization, only species with low molecular weight are formed; as the reaction proceeds, higher molecular weight polymer is formed. Examples of condensation-polymerized polymers are polyesters, polyamides, and polyurethanes. The third example shown is free-radical addition, in which the growing end of the polymer chain is a free radical formed from a vinyl group. During polymerization, a vinyl monomer is added to the chain end. As shown in Figure 2.3C, the molecular weight distribution is quite broad, often contains branched structures, and may be multimodal. Examples of free-radical polymerized polymers include ethylene copolymers, polyacrylates, and polystyrene.

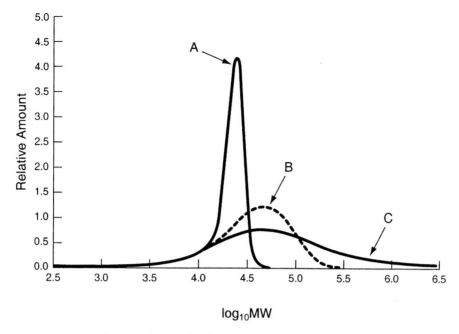

Figure 2.3 Typical molecular weight distributions of synthetic polymers obtained from (A) anionic polymerization, (B) condensation polymerization, and (C) free-radical polymerization. (Adapted with permission from Mori and Barth (1999). Copyright 1999 Springer-Verlag.)

Because of these molecular weight distributions (or populations of different molecular weights), the molecular weight of a polymer must be defined in terms of average molecular weights, or weighted averages of the distribution. To do this, we need to borrow statistical concepts to define moments or averages of these populations. Of the infinite number of possible averages that can be calculated, only the first three averages (discussed next) are the most useful in polymer characterization: number-average molecular weight (M_n); weight-average molecular weight (M_w); and z-average molecular weight (M_z). The reason is that these averages can also be measured independently using colligative properties, light scattering, and ultracentrifugation, respectively. Furthermore, in some cases, these averages can be related to the kinetics of the polymerization reaction.

The number-average molecular weight M_n is essentially the arithmetic mean in which the molecular weight of each chain is multiplied by the number of chains, then summed and divided by the total number of polymer chains:

$$M_n = (N_1M_1 + N_2M_2 + \ldots N_iM_i)/\Sigma N_i \qquad (4)$$

This value can be viewed as the average molecular weight per molecule.

The common representation of the number-average molecular weight is

$$M_n = \Sigma N_iM_i/\Sigma N_i \qquad (5)$$

Since $n_i = N_i/\Sigma N_i$, the mole fraction representation is simply

$$M_n = \Sigma n_iM_i \qquad (6)$$

With the exception of mass spectrometry, most analytical techniques detect and measure polymer weight or concentration, rather than numbers of chains. Thus, by substituting $N_i = W_i/M_i$, we obtain a more useful form of eq 5:

$$M_n = \Sigma W_i/\Sigma(W_i/M_i) \qquad (7a)$$

By examining eq 7a in more detail

$$M_n = \Sigma W_i/(W_1/M_1 + W_2/M_2 + \ldots W_i/M_i) \qquad (7b)$$

we can readily see that M_n values are more sensitive to the presence of low molecular weight molecules; conversely, high molecular weight molecules have less influence on M_n.

The weight-average molecular weight M_w is essentially a weight-average arithmetic mean, in which the molecular weight of each chain is multiplied by its weight, then summed and divided by the total weight of the sample:

$$M_w = (W_1M_1 + W_2M_2 + \ldots W_iM_i)/\Sigma W_i \qquad (8)$$

commonly represented as

$$M_w = \Sigma N_iM_i^2/\Sigma N_iM_i \qquad (9)$$

By the use of $N_i = W_i/M_i$, eq 9 can be written in terms of weight

$$M_w = \Sigma W_iM_i/\Sigma W_i \qquad (10a)$$

$$M_w = \Sigma(W_1M_1 + W_2M_2 + \ldots W_iM_i)/\Sigma W_i \qquad (10b)$$

or as a weight fraction

$$M_w = \Sigma W_i M_i \tag{11}$$

As can be seen from either eq 10b or eq 11, M_w values are sensitive to the presence of high molecular weight molecules; conversely, low molecular weight molecules have less influence on M_w.

Figure 2.4 depicts the molecular weight distribution of a condensation-polymerized polymer (similar to Figure 2.3B), also referred to as the Flory-Schulz distribution or the most-probable distribution. Figure 2.4A is a plot using a linear molecular weight scale, while Figure 2.4B is a plot based on a logarithmic molecular weight scale. Note that on the linear scale, M_n is at the peak maximum; whereas on the logarithmic scale, M_w is at the peak maximum.

In addition to M_n and M_w, higher molecular weight averages can be calculated, which take into account higher molecular weight components of the distribution.

The next two higher moments or averages of the molecular weight distribution are the z-average molecular weight M_z and the z + 1−average molecular weight M_{z+1}. The z designation originates from the German word *Zentrifuge* (centrifuge), which was used to determine M_z. The number and weight representations of these averages are as follows:

$$M_z = \Sigma N_i M_i^3 / \Sigma N_i M_i^2 \tag{12}$$

$$M_{z+1} = \Sigma N_i M_i^4 / \Sigma N_i M_i^3 \tag{13}$$

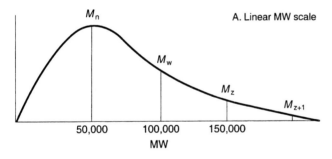

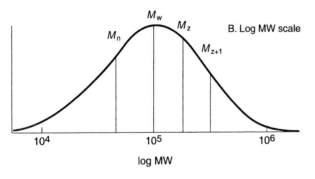

Figure 2.4 Flory-Schulz most-probable molecular weight distribution of a polymer synthesized using condensation polymerization. For this sample, $M_n = 50,000$, $M_w = 100,000$, $M_z = 150,000$, and $M_{z+1} = 200,000$. Note that the peak maximum occurs at M_n for a linear molecular weight scale (A) and at M_w for a logarithmic molecular weight scale (B).

$$M_z = \Sigma W_i M_i^2 / \Sigma W_i M_i \tag{14}$$

$$M_{z+1} = \Sigma W_i M_i^3 / \Sigma W_i M_i^2 \tag{15}$$

The general equations for molecular weight averages are

$$M_y = \Sigma N_i M_i^x / \Sigma N_i M_i^{x-1} \tag{16}$$

$$M_y = \Sigma W_i M_i^{x-1} / \Sigma W_i M_i^{x-2} \tag{17}$$

where x can be any positive integer, or even a fraction, as discussed later in the paragraph that follows.

Another average occasionally used is the viscosity-average molecular weight (M_v) calculated from intrinsic viscosity [η] measurements using the Mark-Houwink equation

$$[\eta] = KM^a \tag{18}$$

where K and a are coefficients that depend on the structure of the polymer, solvent, and temperature of the measurement. These coefficients are obtained empirically by plotting the log [η] versus log molecular weight of a series of monodisperse polymer standards. (Theoretical values of these coefficients are known, but seldom used in practice, since they are only valid for polymers dissolved in an ideal solvent and temperature.) This average is actually a fractional average of the distribution and is equal to

$$M_v = (\Sigma W_i M_i^a / \Sigma W_i)^{1/a} \tag{19}$$

Because M_v depends on experimental conditions (that is, the solvent and temperature), it is not a commonly used molecular weight average. Nevertheless, because of the simplicity of intrinsic viscosity measurements, M_v values are often used to compare relative molecular weight differences among the same types of samples or to estimate the molecular weight of a polymer when a and K values are known. Since a ranges from 0.5 to 0.8 for random chains, M_v values are close to M_w. (For semirigid or extended chains, in which $a \geq 1$, M_v will be equal to or greater than M_w.)

Most polymers have broad molecular weight distributions so that $M_{z+1} > M_z > M_w \geq M_v > M_n$, as shown in Figure 2.4, and these polymers are considered to be *polydisperse*. If the polymer consists of only a single molecular weight specie, the polymer is *monodisperse*; the molecular weight distribution is essentially one vertical line, and all averages are equal to one another. Nucleic acids and many proteins are monodisperse. Well-fractionated polymers or those produced using special polymerization reactions, such as anionic or group-transfer polymerization, can also lead to nearly monodisperse polymers.

The following example will demonstrate how average molecular weight data are calculated. Consider a solution prepared from three monodisperse polymer standards of molecular weights 50,000, 200,000, and 600,000 in which 5 mg of each are added to 10 mL of solvent. Since we are dealing with sample weight and not numbers of molecules, we can use eqs 7a, 10a, and 14 to compute M_n, M_w, and M_z, respectively. Since they are mixed in a $1:1:1$ weight ratio, the weight fraction of each is $1/3$, thus

$M_n = \Sigma W_i / \Sigma(W_i/M_i) = 1/(0.333/600,000 + 0.333/200,000$
$\quad + 0.333/50,000) = 112,000$

$M_w = \Sigma W_i M_i / \Sigma W_1 = (0.333 \times 600,000 + 0.333 \times 200,000$
$\quad + 0.333 \times 50,000)/1 = 283,000$

$M_z = \Sigma W_i M_i^2 / \Sigma W_i M_i = [0.333 \times (600,000)^2 + 0.333$
$\quad \times (200,000)^2 + 0.333 \times (50,000)^2]/(0.333 \times 600,000$
$\quad + 0.333 \times 200,000 + 0.333 \times 50,000) = 473,000$

with a polydispersity index of $M_w/M_n = 2.52$. Alternatively, similar results would have been obtained if we had used 5 mg for W, rather than the weight fraction 0.333.

The breadth of the molecular weight distribution can be quantified by taking the ratios of any two averages. Two commonly used ratios or polydispersity indices are M_w/M_n and M_z/M_w. For some types of polymers, the polydispersity of a polymer can be predicted from polymerization kinetics. In the case of condensation polymerization (Figure 2.3B), $M_w/M_n = 1 + p$, where p is the extent or completion of the reaction. As p becomes unity, M_w/M_n approaches 2, and M_z/M_w reaches a limit of 3. If the polydispersity of a given sample produced from a polycondensation reaction differs from these limits, then either side reactions have occurred or the reaction has not gone to completion. In the case of anionic polymerization reactions (Figure 2.3A), $M_w/M_n = 1 + 1/dp_n$, where dp_n is the number-average degree of polymerization. Here, nearly monodisperse polymers can be obtained, provided that impurities that cause termination reactions are absent. Free-radical polymerizations result in a broader molecular weight distribution (Figure 2.3C), in which M_w/M_n can range from 3 to 5. One of the main reasons for the higher polydispersity is that free-radical end groups are quite reactive and can either undergo different reactions or become "dead" or neutral, or both. Termination can be caused by chain transfer, in which the growth of a chain is terminated by abstracting a group and generating an active site on another chain. Reacting with an initiator molecule can also terminate an active chain end. Two growing chains can combine to form a "dead" chain, or a hydrogen atom can be transferred from one active chain to another to form a neutral end and an unsaturated (vinylic) end group; this is called a *disproportionation* reaction.

Multicomponent polymer systems, including polymer blends and those with different molecular structures (Figure 2.2) will have more complex distributions, depending on their composition. Naturally occurring polydisperse polymers, such as polysaccharides, cellulosics, natural rubber, and some types of proteins, like spider silk, can have widely varying polydispersities, depending on their source of origin and method of isolation.

In summary, the breadth and shape of the molecular weight distribution of polymers reflect the polymerization mechanism, kinetics, and process conditions for synthetic polymers or biological processes for naturally occurring

polymers. Deviations of these conditions can be detected by measuring average molecular weights or, better yet, the molecular weight distribution. These molecular weight parameters also play a major role in predicting and studying properties of polymeric materials, as discussed in the next section.

Influence of Molecular Weight on Polymer Properties

Solution, mechanical, and other physical properties of polymers are highly dependent on molecular weight averages and molecular weight distribution. (For an excellent treatment of this topic, see Van Krevelen's book listed at the end of this chapter.) Other factors that influence these and other properties are polymer chemical composition, molecular conformation (shape), and configuration about chemical bonds, molecular size, and structural features such as branching, sequence distribution of comonomers within a copolymer chain, end-group chemistry, and so on. However, knowledge of molecular weight averages and the molecular weight distribution can provide a great deal of insight into the properties of a polymer.

In general, the number-average molecular weight affects polymer properties most influenced by thermodynamic factors, as well as those properties that depend on the number of chain ends or molecules. Colligative properties, such as vapor pressure and osmotic pressure, fall into this latter category. Other properties that depend on M_n are the glass-transition temperature, density, specific heat capacity, and refractive index. Properties that are influenced more by chain ends are tackiness (adhesion) and mechanical properties such as tensile strength and impact resistance. In other words, low molecular weight components influence these properties to a greater extent than do higher molecular weight components, and as the molecular weight increases, the property reaches a limiting value.

A good example that illustrates the effect of M_n is tensile strength, which is the stress needed to break a material, like a fiber. As shown in Figure 2.5A, low molecular weight chains do not contribute to tensile strength; in fact, if these chains are present, the fiber will be weakened. As the molecular weight increases, tensile strength asymptotically levels off and reaches a limiting value. Given the molecular weight distribution of the two polymers shown in Figure 2.5B, sample 2 would have better tensile strength properties than sample 1. Since low molecular weight components have a dramatic effect on tensile strength, measurement of M_n values would be an appropriate average to examine.

The glass-transition temperature of a polymer (T_g) is another property that depends on M_n. Above T_g, because of the onset of segmental molecular motion, chains become flexible and the polymer becomes rubbery; below this temperature, the polymer behaves as a brittle, glassy material.

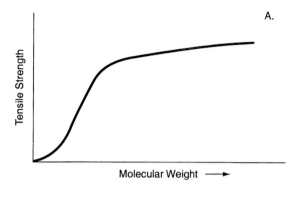

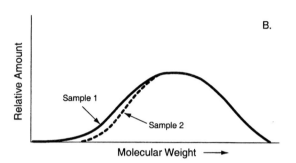

Figure 2.5 A. The influence of molecular weight on tensile strength of a polymer. (Adapted from Sperling (1992), by permission of John Wiley and Sons, Inc.) B. Molecular weight distributions of two samples, in which sample 2 would have better tensile strength than sample 1 because of the absence of low molecular weight components.

As shown in Figure 2.6A, low molecular weight polymers (essentially chain ends) strongly influence T_g values. With increasing molecular weight, a limiting T_g is reached. By examining the molecular weight distribution of two polymer samples, one can readily predict that sample 1 would have the lower T_g value (Figure 2.6B). As in the case of tensile strength, M_n values would be a sensitive indicator for this property.

Higher molecular weight averages, such as M_w and M_z, affect bulk properties of a polymer. A classical example is melt viscosity, which is a measure of the resistance to flow of a polymer above its melting point. As shown in Figure 2.7A, we see two discrete exponential regions when log of the melt viscosity is plotted against log molecular weight. The first region has a slope close to 1, whereas the second region has a greater dependency on molecular weight, and the slope is close to 3.5. The molecular weight at which the slope changes is called the *critical molecular weight*. At this point, polymer chains become entangled networks, and resistance to flow is greatly hindered. In the first region, we are dealing with a large deformation or movement of all chain segments of the polymer chain. Thus M_w would be a

good indicator that would reflect changes in melt viscosity in this region. In the entanglement region, viscoelastic (or rheological) properties become important, and these properties are greatly influenced by very high molecular weight chains. Thus, if we were selling polymers based on their viscoelastic behavior, quality-control measurements would include either measurement of M_z values or close examination of molecular weight distributions to find the high end of the distribution. In Figure 2.7B, we would predict that sample 2 would have the higher melt viscosity. In addition to molecular weight, other factors, such as the presence of cross-linked structures, which would generate large polymer networks, or long-chain branching, would also profoundly affect viscoelastic properties.

Solution viscosity, in which a polymer is dissolved in solvent, also is related to the movement or resistance of the entire polymer chain in a solvent. As in the case of melt viscosity, measurement of M_w would reflect differences of solution viscosity among samples. Recall that the viscosity-average molecular weight M_v (eq 19) is very close to M_w.

In actual practice, well-defined, nearly monodisperse model polymers of known chemical composition should be used to establish relationships between structure and prop-

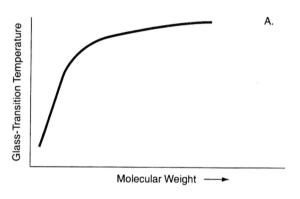

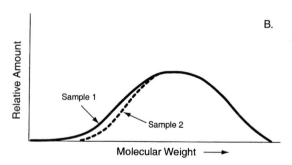

Figure 2.6 A. The influence of molecular weight on the glass-transition temperature (T_g) of a polymer. B. Molecular weight distributions of two samples, in which sample 2 would have a higher T_g than sample 1 because of the absence of low molecular weight components.

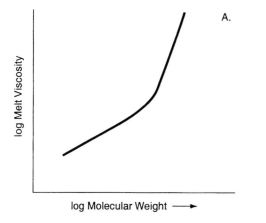

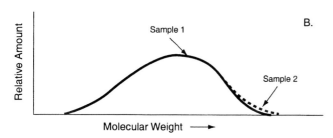

Figure 2.7 A. The influence of molecular weight on the melt viscosity of a polymer. At high molecular weights, there is a significant increase in melt viscosity brought about by entangled polymer chains. B. Molecular weight distributions of two samples, in which sample 2 would have a higher melt viscosity than sample 1 because of the presence of high molecular weight components.

erty. Based on these and other correlations, the molecular weights of products are then monitored to ensure that they fall within these specifications. In many instances, these correlations and specifications are obtained over time by using less well defined samples. In any event, because of potential variability during polymerization reactions and processing, determining molecular weight averages and distributions is important for process monitoring and control, quality assurance, troubleshooting, and basic research.

The most commonly used methods for making these measurements are outlined in the next section. It should be stressed, however, that relying solely on molecular weight averages could be misleading. For example, Figure 2.8 shows three computer-simulated molecular weight distributions that were constructed to have identical values of M_n, M_w, and M_z values. Based on the previous discussion here, and assuming that all three samples were different production lots of the same type of polymer, I would predict, for example, that polymer 3 would have the lowest tensile strength because of the high content of low molecular weight components. Furthermore, polymer 3 would have the highest melt viscosity because of the presence of a higher molec-

ular weight component. Consequently, depending on the polymer being analyzed and the problem being addressed, obtaining the complete molecular weight distribution, rather than the average values, is often preferable.

Molecular Weight Techniques

A list of techniques used for molecular weight measurements, along with approximate molecular weight ranges, is given in Table 2.1. Here I review only those methods that are commonly found in industrial analytical laboratories, as indicated with an asterisk in the table. Since our focus is on synthetic polymers, electrophoretic techniques commonly used for biopolymers are not included. For M_n measurements, ebulliometry (boiling point elevation) and cryoscopy (freezing point depression) are no longer used and will not be covered. Ultracentrifugation, the classical method for molecular weight measurements, will also not be treated since this method is not widely used for synthetic polymers and has essentially been replaced by size exclusion chromatography (SEC). Field-flow fractionation methods hold great promise for characterizing higher molecular weight polymers (and particles). Although the technique has several unique features not found in SEC, it is not commonly used at present.

Mass spectrometry (MS) is the most promising method for determining not only the molecular weight distribution but also the chemical composition. However, for electrospray ionization (ESI), the occurrence of multiple-charged species makes interpretation very difficult; the practical upper molecular weight range of ESI is about 5,000 to 10,000. At the time of this writing, most commercial instruments for matrix-assisted laser-desorption ionization (MALDI) can handle molecular weights of up to about 50,000 g/mol. At present, the major limitation of MALDI is that, depending on the instrument design, it does not give reliable molecular weight data for samples that have a polydispersity of greater than about 1.2. Because of this problem, SEC fractionation is required on samples before MALDI analysis can be done. Although MS is a very active and exciting area, especially when used in combination with SEC, coverage is beyond the scope of this chapter.

End-Group Analysis

If the terminal groups on a polymer chain are known and are chemically distinct from functional groups on the rest of the polymer, it may be possible to determine M_n by common analytical techniques, using chemical analysis or spectroscopy. However, in a given sample, the end group of interest must be present either on one end or on both ends of the chain. This information must be known and usually can be determined from the polymerization chemistry.

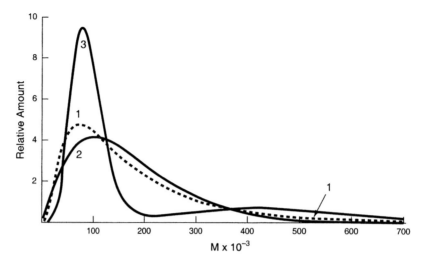

Figure 2.8 Computer-simulated molecular weight distributions of three polymers (1, 2, 3) that have the same M_n, M_w, and M_z averages. (Adapted with permission from Koningsveld (1970). Copyright 1970 Springer-Verlag.)

Since M_n is the average weight per molecule

$$M_n = \text{g of polymer/moles of polymer} \qquad (20)$$

and the number of moles of a polymer is equal to the number of moles of active chain ends divided by the number of active chain ends n, then

$$M_n = (\text{g} \cdot n)/\text{moles of active chain ends} \qquad (21)$$

Table 2.1 Techniques of determining molecular weight of synthetic polymers and approximate molecular weight ranges. Those methods followed by an asterisk are the most popular methods used.

Quantity measured	Technique	Approximate MW range
M_n	End group*	$<5 \times 10^4$
	Membrane osmometry*	10^4 to 5×10^5
	Vapor pressure osmometry*	$<10^5$
	Ebulliometry	$<10^4$
	Cryoscopy	$<10^4$
M_w	Light scattering*	10^3 to 10^7
	Ultracentrifugation	10^3 to 10^7
M_v	Viscometry*	$>10^2$
M_z	Ultracentrifugation	10^3 to 10^7
MWD	SEC*	10^2 to 10^7
	Ultracentrifugation	10^3 to 10^7
	Field-flow fractionation	
	Sedimentation	$>10^6$
	Thermal	$>10^4$
	Flow	$>10^4$
	Mass spectrometry	
	ESI	$<10^4$
	MALDI	500 to 10^5

Either titration, spectrophotometry, or nuclear magnetic resonance (NMR) spectroscopy is commonly used to measure the number of moles of active chain ends.

The upper M_n limit, which depends on the sensitivity of the analytical method used to measure end groups, is usually between 30,000 and 50,000. For high accuracy, impurities (such as either residual monomer or initiator) that have the same functional group as the end group being measured must be absent from the sample. For example, if a 10,000 molecular weight polymer contains 1% of a 100 g/mol impurity with the same functional group as the polymer end group, then M_n will be underestimated by about 40%. As previously stressed, the number of active chain ends per molecule must be known, otherwise reliable M_n values cannot be obtained. In view of this, M_n of branched polymers, which usually have unknown end-group content, cannot be determined by end-group analysis.

Colligative Properties

Colligative properties depend on the number of molecules (or particles) that are present rather than on their chemical structure. The three colligative properties are boiling point elevation (or vapor pressure depression), freezing point depression, and osmotic pressure. Since we are dealing with numbers of moles of polymer (eq 20), M_n values are directly related to colligative properties. Although freezing point depression and boiling point elevation can be used for measuring M_n, it is usually experimentally difficult to measure these values accurately and reproducibly, except for fairly low molecular weight polymers; the upper molecular limit is typically less than 10^4. Furthermore, when using cryos-

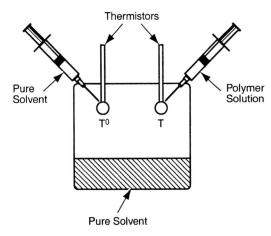

Figure 2.9 Schematic of a vapor pressure osmometer. With the use of a syringe, equal-sized droplets of a solvent and polymer solution are placed onto corresponding thermistors, both of which are in equilibrium with the solvent. Because of its higher vapor pressure, the pure solvent evaporates and cools its thermistor more than does the polymer solution. The difference in temperature between thermistors $(T - T^0)$ is used to calculate the number-average molecular weight of the polymer in solution (eq 23).

copy, polymer solubility at low temperatures may become a problem.

Vapor Pressure Osmometry

Because of the presence of dissolved polymer, the vapor pressure of a polymer solution P will be lower than that of the solvent alone P^0. The relationship between M_n and the decrease of vapor pressure is

$$\ln((P^0 - P)/P^0) = V^0((c/M_n) + A_2 c + A_3 c^2 + \dots) \quad (22)$$

where V^0 is the molar volume of the solvent, c is the concentration of polymer, and A_2 and A_3 are the vapor pressure second and third virial coefficients. As can be seen from this equation, the lower the molecular weight of a given concentration of polymer, the more molecules (particles) are present and the lower the vapor pressure. However, it is experimentally difficult to measure small changes in vapor pressure depression and to measure small changes in boiling point elevation directly. Instead, we measure the temperature difference caused by the evaporation of a droplet of solvent placed on one thermistor (T^0) and the temperature of condensation when solvent condenses onto a droplet of polymer solution placed on a second thermistor (T). As seen in Figure 2.9, the two thermistors are located within the same instrument and are exposed to an atmosphere of pure solvent. This technique is referred to as vapor pressure osmometry (VPO), although neither vapor pressure nor os-

motic pressure is measured. At steady state, this temperature difference is

$$T - T^0 = K_s(c/M_n) \quad (23)$$

where K_s is a calibration constant. In general, the upper M_n limit of this technique is usually less than 10^5.

Membrane Osmometry

Membrane osmometry (MO) can by explained by considering a polymer solution and pure solvent separated by a semipermeable membrane; that is, a membrane that will allow just solvent, and not polymer, to pass (Figure 2.10). Because of the polymer concentration gradient that exists between the compartments, solvent will diffuse across the membrane until both sides of the membrane have the same chemical potential. The liquid level of the polymer solution compartment will increase until the pressure that is developed, that is, the osmotic pressure, will prevent additional solvent from entering the polymer solution compartment. The osmotic pressure (π) is essentially the excess hydrostatic pressure that must be applied to the polymer solution so that both sides of the semipermeable membrane will have the same chemical potential. The relationship between osmotic pressure and M_n is

$$\pi = RT((c/M_n) + A'_2 c + A'_3 c^2 + \dots) \quad (24)$$

where R is the gas constant and A'_2 and A'_3 are the second and third osmotic pressure virial coefficients. At a given

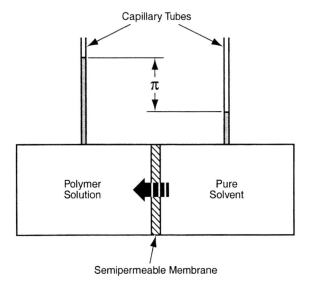

Figure 2.10 Schematic of a membrane osmometer. In an enclosed system, a polymer solution is separated from pure solvent by a semipermeable membrane that allows passage of only the solvent. Because of the polymer concentration gradient that exists on either side of the membrane, osmotic pressure π develops, which is related to the number-average molecular weight of the polymer in solution (eq 24).

polymer concentration, a low molecular weight polymer will develop a high osmotic pressure because it would contain more molecules than a higher molecular weight sample.

The upper molecular weight limit of MO is about 5×10^5. The lower limit, however, is dictated by the molecular weight cutoff of the membrane that is used, which is typically about 10,000.

Light Scattering

In this section, we will limit our discussion to Rayleigh light scattering in which a laser light source is used. When light interacts with the electron cloud in a macromolecule, it will induce a temporary dipole moment that oscillates in phase with the incident light source. Scattered light, which has the same wavelength as the incident beam, will radiate light in all directions. The intensity of scattered light (I_s) is proportional to the molecular mass or weight (M) and the concentration of the polymer (c) and inversely proportional to the fourth power of the wavelength,

$$I_s \propto Mc\alpha^2/\lambda^4 \qquad (25)$$

where α is the polarizability of the polymer. Thus for a given polymer concentration, high molecular weight samples will scatter more light than low molecular weight polymers. The complete Rayleigh equation that relates scattering intensity of a polymer solution to weight-average molecular weight and radius of gyration for a polydisperse sample is

$$i_s r^2/I_0 V = (4\pi^2 n_0^2/\lambda_0^4 N_A)(dn/dc)^2 (c([M_w P(\theta)]^{-1}$$
$$+ 2A_2 c)^{-1}) + i_{s,solvent} r^2/I_0 V \qquad (26)$$

where r is the distance from the scatterer to the detector, V is the scattering volume, i_s is the scattered light intensity of the polymer solution, $i_{s,solvent}$ is the scattered light intensity of the solvent, I_0 is the incident light intensity, n_0 is the refractive index of the medium, λ_0 is the wavelength of light, N_A is Avogadro's number, dn/dc is the specific refractive index of the polymer in solution, A_2 is the second virial coefficient, and $P(\theta)$ is the particle-scattering function, defined later in this section.

Equation 26 can be simplified by defining an optical constant K that is equal to all terms that are known or can be measured independent of the light-scattering experiment:

$$K = (4\pi^2 n_0^2/\lambda^4 N_A)(dn/dc)^2 \qquad (27)$$

The normalized scattering intensity is then set equal to R_θ, the Rayleigh ratio

$$R_{\theta,\,soln} = i_s r^2/I_0 V \qquad (28a)$$

$$R_{\theta,\,solvent} = i_{s,\,solvent} r^2/I_0 V \qquad (28b)$$

Since we are interested in the scattered intensity of the

polymer, the Rayleigh ratio of the polymer, referred to as the excess Rayleigh ratio, is simply

$$R_\theta = R_{\theta,\,soln} - R_{\theta,\,solvent} \qquad (29)$$

Substituting these equations into eq 25 and rearranging, we have a simplified and more useful form of the Rayleigh equation:

$$Kc/R_\theta = [M_w P(\theta)]^{-1} + 2A_2 c \qquad (30)$$

where

$$P(\theta)^{-1} = 1 + (16\pi^2 n_0^2 <R_g^2>_z)(3\lambda^2)^{-1}\sin^2(\theta/2) + \ldots \quad (31)$$

Please note that when light scattering is performed on a polydisperse polymer, the radius of gyration that is measured is the z-average. In other words, the higher molecular weight components are weighed more heavily than lower molecular weight species. In modern SEC, on-line light-scattering detectors are commonly employed to measure absolute molecular weights, as well as molecular size, as a function of the molecular weight distribution without the need for SEC column calibration. Because of the low concentrations that are employed in SEC, the $2A_2 c$ term in eq 30 is usually deleted.

In a classical light-scattering experiment, a series of polymer solutions of different concentrations are analyzed, and the scattered light intensity is measured as a function of angle with respect to the incident beam. As shown in Figure 2.11, one plots Kc/R versus $\sin^2(\theta/2) + kc$ for each polymer concentration, where k is an arbitrary constant used to spread out the data. For each concentration, the Kc/R_θ data are extrapolated to zero angle. Likewise, for each angle of measurement, we extrapolate to zero angle. The y-axis intercept is the reciprocal of M_w. The initial slope of the extrapolated zero-concentration data will give us the radius of gyration

$$\text{initial slope} = (16\pi^2 n_0^2 <R_g^2>_z)/3\lambda^2 M_w \qquad (32)$$

Although we have not discussed the significance of the second virial coefficient A_2, this value is equal to one-half of the initial slope of the extrapolated zero-angle data. Suffice it to say the second virial coefficient is a thermodynamic parameter related to the extent of polymer-solvent interaction.

Some light-scattering instruments can measure intensity of scattered light at an angle close to zero degrees with respect to the incident beam, and this is commonly referred to as "low-angle light scattering". In this region, there is negligible angle dependence, and M_w is the reciprocal of the intercept of a plot of Kc/R_θ versus kc. In this case, however, information regarding molecular size, that is, R_g, is lost.

In summary, light scattering is an absolute technique for measuring M_w, R_g, and A_2 values of polydisperse polymers. Depending on the specific refractive index of a polymer, the practical molecular weight range is about 10^3 to over 10^7. For accurate R_g measurements, the lower limit is approxi-

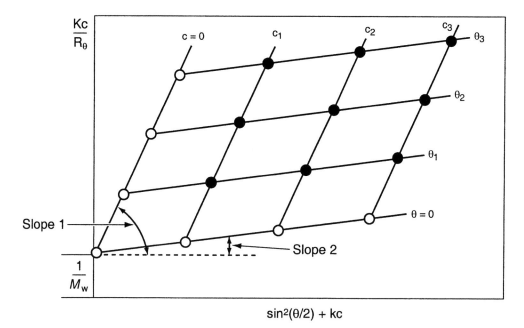

Figure 2.11 Double-extrapolation light-scattering (Zimm) plot in which both polymer concentration and angle of measurement are varied. The intercept is the reciprocal of M_w. The radius of gyration and the second virial coefficient can be obtained from slopes 1 and 2, respectively.

mately 10 nm, which corresponds to a molecular weight of about 10^5.

Viscometry

Viscometry, in particular measurements of intrinsic or inherent viscosity, is perhaps the most widely used method because of its low cost and simplicity. Historically, these measurements have been used since 1930 and were first used to help demonstrate the existence of polymer molecules.

To measure the intrinsic viscosity of a polymer, a series of polymer solutions of different concentrations c are prepared, and the time t it takes for a solution of a given concentration to flow through a capillary is measured with respect to the flow time of the solvent t_0. There are a number of capillary designs that are used, one of which is a Ubbelohde viscometer shown in Figure 2.12. Reservoir R is filled with polymer solution and then suctioned up to A. The time it takes for the meniscus of the polymer solution to fall from A to B is then measured; C is the capillary. For each of these measurements, the relative viscosity, specific viscosity, reduced viscosity, and inherent viscosity are calculated using eqs 33–36, respectively

Relative viscosity $\eta_r = t/t_0$ (33)

Specific viscosity $\eta_{sp} = \eta_r - 1$ (34)

Reduced viscosity $\eta_{red} = \eta_{sp}/c$ (35)

Inherent viscosity $\eta_{inh} = (\ln \eta_r)/c$ (36)

As shown in Figure 2.13, η_{sp}/c is plotted against polymer concentration, and the intercept at zero polymer concentration is the intrinsic viscosity

$$[\eta] = (\eta_{sp}/c)_{c \to 0}$$ (37)

Alternatively, η_{inh} can be plotted versus polymer concentration and extrapolated to zero polymer concentration to obtain the intrinsic viscosity:

$$[\eta] = [(\ln \eta_r)/c]_{c \to 0}$$ (38)

At a low polymer concentration, the "one-point method" also can be used in which

$$[\eta] = 1/c[2 \, (\eta_{sp} - \ln \eta_{rel})]^{1/2}$$ (39)

The importance of intrinsic viscosity is that it is a fundamental parameter related to molecular weight and molecular size. If we know both the molecular weight and the intrinsic viscosity of a linear polymer, we can readily determine the radius of gyration R_g using

$$[\eta] = \Phi_0 \, 6^{3/2} <R_g^2>^{3/2}/M$$ (40)

or its molecular hydrodynamic volume V_h

$$[\eta] = \Phi_0 \, 6^{3/2} V_h/M$$ (41)

where Φ_0 is the Flory constant.

Another important relationship is the Mark-Houwink equation, which relates intrinsic viscosity to molecular weight, previously defined in eq 18, $[\eta] = KM^a$, where a and K are coefficients that depend on polymer conformation, solvent, and temperature. The coefficient a is a conforma-

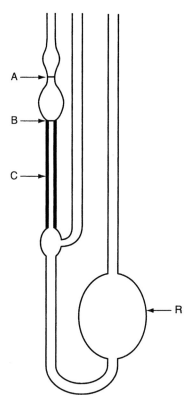

Figure 2.12 Schematic of a capillary Ubbelohde viscometer used for determining the intrinsic viscosity of polymer solutions. Key: A, B = points of measurement; C = capillary; R = reservoir.

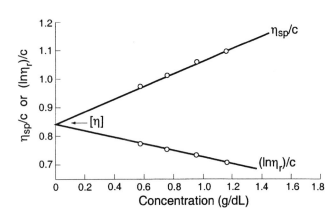

Figure 2.13 A typical reduced (η_{sp}/c) or inherent viscosity (($\ln \eta_r)/c$) viscosity plot of four polymer solutions of different concentrations. The common intercept at zero polymer concentration is the intrinsic viscosity.

tional parameter that ranges from 0 for spheres to 2 for rigid rods. For flexible linear chains, which includes most synthetic polymers dissolved in a good solvent, $0.5 < a < 0.8$. For branched or globular polymers, like proteins, a is less than 0.5. For those polymers that have an extended molecular conformation, a is greater than 0.8. The coefficients a and K are determined experimentally by plotting log [η] versus log molecular weight of a series of nearly monodisperse polymers dissolved in a given solvent at a specified temperature. The slope and intercept of such a plot are equal to a and K, respectively. These values are compiled in a number of sources, the most comprehensive being the *Polymer Handbook* (see selected references).

The Mark-Houwink equation can be used in several ways. First, if the coefficients are known, the molecular weight (actually the viscosity-average molecular weight) can be estimated. By knowing both the molecular weight and the intrinsic viscosity, the radius of gyration or hydrodynamic volume can be determined through eqs 40 and 41. Furthermore, the molecular conformation of a sample in a given solvent can be approximated from the coefficient a, by constructing a Mark-Houwink plot of a series of polymer samples of different molecular weights.

If a polymer is known to have long-chain branching (see Figure 2.2), the extent of branching can be quantified indirectly through hydrodynamic or molecular size parameters, using either intrinsic viscosity or light-scattering measurements. This concept is based on the fact that a branched polymer is smaller in size than a linear molecule of the same molecular weight. These branching parameters are expressed as a molecular size ratio of a branched to a linear polymer of the same molecular weight. In the case of intrinsic viscosity measurements, we can compute the ratio (g') of the intrinsic viscosity of the branched polymer to that of the linear polymer of the same molecular weight:

$$g' = ([\eta]_b/[\eta]_l)_M \qquad (42)$$

Thus, values less than unity signify branched structures. By the use of SEC with on-line viscometry, g' values can be readily determined across the molecular weight distribution.

If the radius of gyration and molecular weights are known from light-scattering measurements, the following ratio can be obtained:

$$g = [(R_g^2)_b/(R_g^2)_l]_M \qquad (43)$$

As with SEC and on-line viscometry, on-line light-scattering detection is also a powerful tool for determining g values across the distribution; however, the lower limit of R_g using light scattering is about 10 nm. For simple branching models, g can be theoretically related to the number of branches per molecule, provided that measurements are carried out under ideal (theta) solvent and temperature conditions. At theta conditions, interactions between polymer segments within a chain and among chains are equivalent to solvent-polymer interactions. The dimensions of a polymer chain under these ideal conditions can be calculated from bond angles, bond lengths, and hindered bond rotation. Theta conditions represent the thermodynamically unperturbed state of the chain in solution, in which molecular size can be calculated from first principles.

The relationship between intrinsic viscosity and light-scattering branching parameters is

$$g' = g^{\varepsilon} \tag{44}$$

where ε is a structure factor, not specified by theory, which ranges from 0.5 to 1.5. In practice, it is best not to convert g' to g^{ε} because of the uncertainty of ε and that most measurements are not under ideal (theta) solvent and temperature conditions.

To summarize, light scattering is an absolute technique for measuring M_w, the only polymer-dependent parameter that is needed is the specific refractive index dn/dc that is readily obtainable from simple experiments. With the advent of multiangle diode-array detectors, individual angular-dependent measurements are no longer needed. High sensitivity of these detectors allows one to make measurements at low polymer concentration, which precludes the concentration extrapolation procedure. Moreover, by coupling SEC to on-line light-scattering detection, only a single polymer solution needs to be analyzed to obtain complete molecular weight, size information, and branching information.

Size Exclusion Chromatography

Of all the methods that have been described, size exclusion chromatography (SEC), also referred to as gel permeation chromatography (GPC), is perhaps the most popular. The reason is that SEC analyses are relatively simple to perform using standard high-performance liquid chromatographic (HPLC) instrumentation. Essentially, all one needs is an isocratic HPLC pump, SEC columns, a detector, most commonly a differential refractometer, and a data system for calculating molecular weight averages. Like viscometry measurements, SEC is a relative technique that requires calibration or standardization using polymer samples of known molecular weight. However, once calibrated, SEC can give the molecular weight distribution and all molecular weight averages within 30 minutes.

Since SEC is already covered in chapter 4, here I examine the principles of this technique and review approaches used for molecular weight calibration.

Separation Mechanism

In SEC, macromolecules are separated in terms of hydrodynamic volume or size, and not necessarily by molecular weight. In an SEC experiment, a sample is injected into a chromatographic column that is packed with porous particles. As the sample passes through the column, larger molecules that are too big to penetrate the pores of the packing elute first, followed by smaller molecules, which can penetrate or diffuse into the pores and elute at a later time. The resulting SEC elution profile is essentially the molecular weight distribution. If the SEC system is calibrated with a series of polymer standards of known molecular weight, as shown in Figure 2.14, a relationship between log molecular weight and elution volume is obtained. This relationship or calibration curve can then be used to determine the molecular weight distribution and molecular weight averages of samples.

High molecular weight polymers that are too large to penetrate the pores of the packing elute within the interstitial or void volume V_0 of the column. Lower molecular weight polymers that approach the average pore size of the packing will penetrate or partition into the pores of the packing and elute later. When the molecular size of the polymer is relatively small with respect to the average pore size, it will freely diffuse into and out of the pores and will elute at V_t, the total mobile phase volume of the packed SEC column,

$$V_t = V_0 + V_i \tag{45}$$

where V_i is the pore volume of the packing. The elution volume of a polymer (V_R) can be described in terms of the SEC distribution coefficient K_{SEC}, which has defined limits of $0 \le K_{SEC} \le 1$

$$V_R = V_0 + K_{SEC}V_i \tag{46}$$

The most widely accepted theory, and perhaps the most useful, to help explain the nature of SEC is based on thermodynamic considerations. Thus K_{SEC} can be expressed in

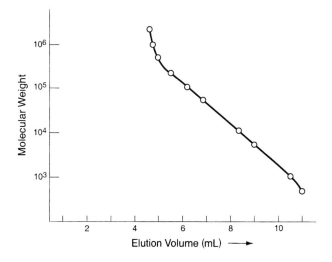

Figure 2.14 Typical size exclusion chromatography (SEC) calibration curve constructed by injecting a series of polymer standards of known molecular weight.

terms of entropy $\Delta S°$, which is the change in conformational entropy when polymer is transferred from the interstitial volume into the pores of the packing, and R is the gas constant:

$$K_{SEC} = \exp (\Delta S°/R) \qquad (47)$$

Since the conformational entropy of a polymer in solution is restricted inside the pores of the packing, as compared to being in the interstitial volume, $\Delta S°$ decreases during permeation into the pores; the driving force is the concentration gradient of the polymer between the interstitial volume and the pore volume. This conformational entropy loss governs SEC separation. As a result, the sign of $\Delta S°$ is negative and K_{SEC} values range from zero to unity, as previously stated.

Calibration

Over the years, a number of different column calibration approaches have been developed for SEC. (Details of these methods can be found in the appropriate texts listed at the end of this chapter.) The most common methods involve the injection of a series of nearly monodisperse polymer standards of defined molecular weight. If these standards are chemically and structurally similar to the sample, *absolute* average molecular weights are obtained; if not, *apparent* averages are calculated.

If monodisperse or nearly monodisperse standards are not available, a useful procedure is to employ a polydisperse sample in which the number- and weight-average molecular weights have been determined, by, for example, membrane osmometry and light scattering. In this procedure, these average molecular weights are input into the SEC data system, and the polydisperse standard is injected. By using an iterative procedure, the computer selects the best calibration curve so that the input molecular weight values are obtained. This broad molecular weight approach is very useful, provided that the molecular weight of the standard is sufficiently broad to encompass the anticipated range of samples and that the calibration curve is linear. The latter can be adjusted by the proper selection of SEC columns.

If a polymer has a known molecular weight distribution based on polymerization kinetics, and if either the M_n or M_w value is known, the integral-molecular weight distribution method can be employed. This method is useful only for polymer samples of known polymerization chemistry that can be defined by a kinetic model, such as for condensation polymerization reactions.

With the introduction of on-line viscometric detectors for SEC, there has been increased use of a rather interesting approach, called *universal calibration*. Universal calibration is based on the concept that the product of intrinsic viscosity and molecular weight is equal to the hydrodynamic volume of a polymer—that is, $[\eta]M = \Phi_0\, 6^{3/2}V_h$ (eq 41). If the product, log $[\eta]M$, rather than log molecular weight is plotted against elution volume, all polymers, irrespective of their chemical structure, will fall on this curve. The reason for this is that the SEC separation mechanism is based on molecular volume rather than on weight. In practice, a universal calibration curve is constructed by injecting polymer standards of known molecular weight and intrinsic viscosity. A polymer sample is then injected, and at each elution volume, a value of log $[\eta]M$ is acquired. To back out M from $[\eta]M$ values at each elution volume increment, we can use the Mark-Houwink equation $[\eta] = KM^a$ (eq 18), but only if coefficients a and K are known *accurately* for the polymer being analyzed. The equation used for this calculation is

$$\log M = (\log [\eta]M - \log K)/(a + 1) \qquad (48)$$

If the coefficients a and K are not known *accurately*, which is often indeed the case, an on-line viscometric detector can be used, whereby $[\eta]$ is determined directly at each elution volume increment.

The use of SEC with on-line viscometry and light-scattering detection systems has greatly increased in recent years. With these molecular weight–sensitive detectors, absolute measurements can be obtained across the molecular weight distribution of a sample. Moreover, by using these detectors in combination, molecular conformation, size, and branching characteristics can also be obtained.

To conclude, SEC can cover a broad molecular weight distribution, typically from 10^2 to 10^6; the upper molecular weight is limited by shear degradation of polymer chains, unavailability of large pore-size packings, and poor SEC resolution for very high molecular weight polymers. As we can see from eq 47, SEC is governed by conformational entropy; thus enthalpic forces, such as polymer-packing interactions, must be eliminated by judicious choice of packing material and mobile phase composition.

Summary

Molecular weight measurements of polymers is standard practice in almost all laboratories that deal with polymers and polymeric materials because of the profound effect these parameters can have on physical properties. The classical techniques of vapor pressure osmometry, membrane osmometry, light scattering, viscometry, and size exclusion chromatography are still the most important methods of choice. Although analytical ultracentrifugation is still used for characterizing biopolymers, especially their association phenomena, SEC has supplanted it for characterizing synthetic polymers because of its speed and low cost. Although they are not covered in this chapter, field-flow fractionation techniques have unique applicability, especially for high molecular weight polymers and particles. Mass spectroscopy using soft ionization techniques, such as matrix-assisted laser-desorption ionization (MALDI), is a rapidly developing area for synthetic polymers, especially when used in conjunction with SEC.

Selected Bibliography

Physical Chemistry and Properties of Polymers

Brandrup, J., Immergut, E. H. Eds. *Polymer Handbook;* John Wiley & Sons: New York, 1989.

Brydson, J. A. *Plastic Materials;* Butterworths: London, 1989.

Elias, H.-G. *Macromolecules: Structure and Properties;* Plenum Press: New York, 1977; Vol. 1.

Flory, P. J. *Principles of Polymer Chemistry;* Cornell University Press: Ithaca, 1953.

Grossberg, A. Yu; Khokhlov, A. R. *Giant Molecules;* Academic Press: San Diego, 1997.

Miller, M. L. *The Structure of Polymers;* Van Nostrand Reinhold: New York, 1966.

Morawetz, H. *Macromolecules in Solution;* Wiley: New York, 1975.

Peebles, L. H. *Molecular Weight Distributions in Polymers;* Interscience: New York, 1971.

Sperling, L. H. *Introduction to Physical Polymer Science;* Wiley-Interscience: New York, 1992.

Tager, A. *Physical Chemistry of Polymers;* MIR: Moscow, 1978.

Tanford, C. *Physical Chemistry of Macromolecules;* John Wiley: New York, 1966.

Van Krevelen, D. W. *Properties of Polymers;* Elsevier: Amsterdam, 1976.

Polymer Chemistry

Billmeyer, F. W. Jr. *Textbook of Polymer Science;* Wiley-Interscience: New York, 1984.

Hiemenz, P. C. *Polymer Chemistry;* Marcel Dekker: New York, 1984.

Munk, P. *Introduction to Macromolecular Science;* Wiley-Interscience: New York, 1989.

Seymour, R. B.; Carraher, C. E. Jr. *Polymer Chemistry;* Marcel Dekker: New York, 1988.

Young, R. J. *Introduction to Polymers;* Chapman & Hall, London, 1981.

Polymer Characterization

Barth, H. G. , Mays, J. W. Eds. *Modern Methods of Polymer Characterization;* Wiley: New York, 1991.

Cooper, A. R., Ed. *Determination of Molecular Weight;* Wiley: New York, 1989.

Pethrick, R. A.; Dawkins, J. V. *Modern Techniques for Polymer Characterization;* Wiley: Chichester, 1999.

Rabek, J. F. *Experimental Methods in Polymer Chemistry;* Wiley: Chichester, 1980.

Schröder, E.; Müller, G.; Arndt, K.-F. *Polymer Characterization;* Hanser: Munich, 1988.

Light Scattering

Huglin, M. B. Ed. *Light Scattering from Polymer Solutions;* Academic Press: New York, 1972.

Kratochvil, P. *Classical Light Scattering from Polymer Solutions;* Elsevier: Amsterdam, 1987.

Size Exclusion Chromatography and Separation Techniques

Balke, S. T. *Quantitative Column Liquid Chromatography;* Elsevier: Amsterdam, 1984.

Cantow, M. J. R. Ed. *Polymer Fractionation;* Academic Press: New York, 1967.

Dubin, P. L. Ed. *Aqueous Size Exclusion Chromatography;* Elsevier: Amsterdam, 1988.

Francuskiewicz, F. *Polymer Fractionation;* Springer-Verlag: Berlin, 1994.

Glöckner, G. *Polymer Characterization by Liquid Chromatography;* Elsevier: Amsterdam, 1987.

Glöckner, G. *Gradient HPLC of Copolymers and Chromatographic Cross-Fractionation;* Springer-Verlag: Berlin, 1991.

Hunt, B. J., Holding, S. B. Eds. *Size Exclusion Chromatography;* Chapman & Hall: New York, 1989.

Janca, J. Ed. *Steric Exclusion Liquid Chromatography;* Marcel Dekker: New York, 1984.

Koningsveld, R. *Adv. Polym. Sci.,* 1970, *7,* 1.

Mori, S.; Barth, H. G. *Size Exclusion Chromatography;* Springer-Verlag: Berlin, 1999.

Pash, H.; Trathnigg, B. *HPLC of Polymers;* Springer-Verlag: Berlin, 1997.

Patton, E. V. Size-Exclusion Chromatography. In *Physical Methods of Chemistry,* 2nd ed.; Rossiter, B. W., Hamilton, J. E., Eds.; Wiley Interscience, New York, 1989; Vol. IIIB, Chapter 7.

Schimpf, M., Caldwell, K., Giddings, J. C. Eds. *Field-Flow Fractionation Handbook;* Wiley-Interscience: New York, 2000.

Tung L. H. Ed. *Fractionation of Synthetic Polymers;* Marcel Dekker: New York, 1977.

Wu C.–S. Ed. *Handbook of Size Exclusion Chromatography;* Marcel Dekker: New York, 1995.

Yau, W. W.; Kirkland, J. J.; Bly, D. D. *Modern Size Exclusion Chromatography;* John Wiley: New York, 1979.

PART I

ATOMIC, FUNCTIONAL GROUP, AND MONOMERIC ANALYSIS OF POLYMERS

3

Instrumental Elemental Analysis of Polymeric Materials

JOHN MARSHALL

The elemental analysis of polymers, stated as such, would appear to be a relatively restricted field of activity. The implication of such a definition is that the requirement is for the quantification of major elemental constituents of the polymer such as carbon, hydrogen, nitrogen, sulfur, and oxygen. The determination of these elements does indeed provide some information about the average composition of the polymer in the bulk phase. Consequently, such an analysis is perhaps of most use in the relatively limited areas of assessing empirical formulas of the polymer in terms of repeat units or in the specific determination of end groups. However, polymers are very rarely used in their nascent state, and in practical applications polymers are employed in product formulations that result in materials with a much wider range of properties. This extension of material properties is achieved by the incorporation of a vast range of additives, which are designed to give particular effects in the end use of the product. Examples include the use of fiber reinforcement, mineral fillers, pigments, lubricants, flame retardant, antioxidants, plasticizers, impact modifiers, colorants, coupling agents, and heat and light stabilizers. Polymer systems may incorporate several such additives in a single compounding process. Other components such as catalyst residues survive from the production process, and metal contaminants may be picked up in processing and subsequent machining for end use. Within this context, the range of potential applications of elemental analysis techniques to the characterization of polymer systems is greatly enhanced in scope. This chapter is devoted to a discussion of the main techniques used for characterizing the elemental composition of polymer systems as encountered in real-world applications.

The requirements for analysis range from highly accurate, precise, and rapid bulk quantification, at the percentage level, to spatially resolved lateral and depth profiling for trace constituents. Consequently, no single analytical technique is likely to be suitable for all applications in this field.

The review here is restricted to instrumental methods of quantitative analysis, which have largely superseded "classical" elemental techniques such as volumetric and gravimetric analysis in the modern laboratory. Similarly, while neutron activation analysis is indeed widely applied to the elemental analysis of polymers, the requirement for a nuclear reactor to irradiate the sample places that technique sufficiently beyond the capabilities of most laboratories and is therefore considered to be outside the scope of this review. Atomic spectroscopy techniques are very widely used throughout the world for elemental analysis, whether at trace, minor or major component levels. The main spectrometric techniques of interest for the characterization of polymer systems are atomic absorption spectrometry (AAS), atomic emission spectrometry (AES), X-ray fluorescence spectrometry (XRF), and atomic mass spectrometry (e.g., inductively coupled plasma mass spectrometry; ICP-MS). The determination of nonmetals may be achieved using atomic spectrometry techniques, but alternative instrumental methods do exist to take advantage of the volatility of these elements for analytical purposes. The basis of these techniques is sample combustion, and the various approaches are grouped together in the Combustion Analysis section later in this chapter.

Review articles concerning the elemental analysis of polymeric materials have been published. The annual Atomic Spectrometry Updates on industrial analysis in the *Journal of Analytic Atomic Spectrometry* provide a good source of primary references concerning the elemental analysis of polymers, paints, and other synthetic materials.[1] The biennial "Coatings" reviews in *Analytical Chemistry* include references to atomic spectrometry and X-ray techniques as applied to the characterization of paints, coatings, and related materials.[2] A review of the major techniques used in the trace element analysis of plastics has been published.[3] The techniques discussed included atomic absorption spectrometry, atomic emission spectrometry, mass spectrometry, and

X-ray fluorescence spectrometry. The application of atomic spectrometry to the analysis of advanced materials has been discussed within the context of present capabilities and likely future needs, particularly in respect to the direct characterization of solids.[4] An overview of the use of atomic spectroscopy techniques in the analysis of paints (essentially polymer-filler-pigment composites) may also be of interest.[5]

Combustion Analysis

"Elemental analysis" is something of a catch-all phrase that is applied to techniques that can determine elements. State-of-the-art elemental analysis techniques such as X-ray fluorescence or plasma source mass spectrometry are unable to measure (at analytically useful levels) some of the most common elements on the planet. There are good theoretical reasons why this should be so, but it is often surprising to learn of the limitations of such multielement techniques, which claim jurisdiction over most of the periodic table. The quantitative determination of elements such as carbon, hydrogen, nitrogen, and oxygen, and, to a significant extent, sulfur, is the preserve of instrumental techniques based on combustion technology. These techniques consequently fulfill a vital role in the armory of the modern industrial analytical laboratory.

The determination of carbon, hydrogen, and nitrogen (CHN) in organic materials is carried out by thermal combustion using commercial instruments (e.g., Leco, Perkin Elmer, Carlo Erba) which are available worldwide. Ma[6] has reviewed developments in this field. The determination of CHN in organic samples is often referred to as *microanalysis* because of the size of the sample (micrograms) used. This can be seen as an advantage when the amount of sample is restricted—for example, in organic synthesis—or as a disadvantage in situations where better sensitivity or representative sampling is required. Indeed, the small sample size may be valuable in studies of homogeneity. The working range of such systems is typically restricted to CHN contents of greater than 0.1% w/w in normal operation. Instruments of this type operate by combusting the sample, usually in metallic capsules, and detecting the volatile gaseous products that result (e.g., carbon dioxide, water, and nitrogenous gases). A schematic diagram of such an instrument is shown in Figure 3.1. A number of methods of detecting these gases are used, usually involving thermal conductivity (for nitrogen) or infrared absorption (for carbon dioxide and water). It may be necessary to perform on-stream reduction (e.g., nitrogen oxides to molecular nitrogen) or to remove carbon dioxide (using sodium hydroxide) or water (via magnesium perchlorate) to allow accurate thermal conductivity measurements; it is also possible to perform a prior separation by gas chromatography.[7] These operations are carried out automatically. Oxygen may also be determined at percentage levels by high-temperature

(1000 to 1300 °C) volatilization from the sample, followed by combination with a carbon-rich medium to produce carbon monoxide. The latter is oxidized (e.g., over copper oxide in an oxygen-rich atmosphere) to generate carbon dioxide, which can then be detected as described above. The specific determination of total nitrogen in plastics can be carried out using the Kjeldahl method, as described in a standard test method.[8] Nitrogen may also be determined using oxidative combustion with chemiluminescence detection. The sample is combusted in an oxygen-rich high-temperature environment, and nitrogen species are converted to nitric oxide (NO). The NO is reacted with ozone and converted to excited nitrogen oxide (NO_2), which undergoes a radiative decay that can be detected by a photomultiplier tube. The quantity of the light emitted is proportional to the nitrogen content of the sample.[9] The chemiluminescence technique has been modified in our laboratory to allow the direct introduction of polymer samples for the determination of low levels (<1% w/w) of total nitrogen and requires a solid sampling boat.

The determination of sulfur and the halogens can be achieved by techniques such as X-ray fluorescence spectrometry and inductively coupled plasma atomic emission spectrometry (ICP-AES). However, these are not the most sensitive of methods, and they all can be subject to volatilization losses in sample preparation (e.g., by chemical digestion, fusion, or oxygen flask combustion). Instrumental combustion methods rely on the volatility of these elements to achieve quantitation and can be used to complement spectroscopic methods for total elemental analysis. For example, microcoulometry is one of the most sensitive techniques for the determination of nonmetals, such as sulfur and chlorine.[10] Microcoulometry has been applied with success in industrial laboratories to a wide range of sample types, including petrochemicals, monomers, and polymers. However, the technique, while absolute, depends on the reliability of the titration and can be subject to interference from other ions that may be present in the sample—for example, heavy metals and halogens. Within this constraint, the technique is very reliable.

Samples may be introduced in either solid liquid or gas forms, using syringe or boat techniques specific to the application. In the procedure for the determination of total sulfur, the sample is volatilized in a quartz tube furnace in an inert gas stream and then oxidized at high temperature in an oxygen-rich atmosphere, thus converting the free sulfur to sulfur dioxide.[11] The latter is passed to a titration cell that contains an electrolyte of potassium iodide and glacial acetic acid in distilled water. The sulfur dioxide thus retained is titrated with tri-iodide from the electrolyte. The tri-iodide consumed is coulometrically replaced, and the flowing current is proportional to the amount of sulfur in the sample.

A similar procedure is used for the determination of chlorine.[12] In this case, the combined chlorine from the sample

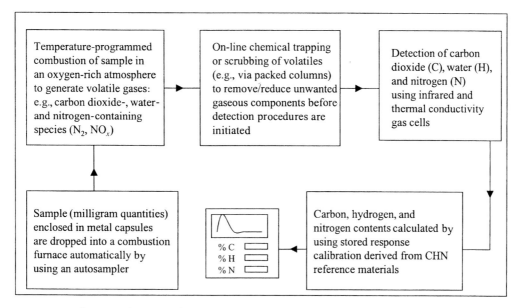

Figure 3.1 Schematic diagram of combustion apparatus for elemental (CHN) analysis based on a flowing stream absorbent–chemical treatment method.

is converted into hydrochloric acid by combustion, using the same furnace system as described for sulfur. The hydrochloric acid thus formed is subjected to an argentiometric titration, and the response to chloride (as total halogen) is measured coulometrically (replacement of silver titrant ions at the generator electrode). The combination of separate oxygen flask combustion and silver nitrate titration has been reported for the determination of total chlorine in isocyanates as part of a standard test method for polyurethane raw materials.[13] It was noted that erroneous results could be obtained when either bromine or iodine is present, as is the case for microcoulometry. The converse of this problem is that these elements can be determined using this method in the absence of the others. It is also possible to determine sulfur by oxygen combustion followed by infrared detection of the evolved sulfur dioxide in a flow-through gas cell. A method of this type has been applied to the analysis of rubber compounds.[14]

Atomic Absorption Spectrometry

Atomic absorption spectrometry (AAS) has been established as a commercially available technique for the determination of elemental composition for almost 40 years. It is beyond the scope of the present discussion to give an in-depth review of AAS, but the latest edition of the classic text by Welz and Sperling[15] provides a comprehensive overview of the field. A schematic diagram of an atomic absorption spectrometer is shown in Figure 3.2.

The technique destroys the sample integrity, however, because the sample is either fully or partially decomposed by the nature of the processes involved in the analysis.

Sample preparation for AAS generally involves the conversion of the solid materials by chemical means into solution form before instrumental analysis is undertaken. The sample is then introduced to an atomizer, such as a combustion flame or an electrically heated furnace. The atomizer is employed to thermally decompose the sample analyte compounds to allow a quantitative release of free atoms into the vapor phase in the optical path of the AAS instrument. Free atoms produced in this way may absorb atomic spectral line radiation that is produced by a dedicated source lamp (such as a hollow cathode; see Figure 3.3) containing the analyte of interest. The attenuation of source radiation by the analyte is detected by using an optical spectrometer and is related directly to concentration by calculation of the absorbance value from the incident and transmitted intensities observed. In most cases in AAS, analyte atoms are excited from the ground state to the first excited state by the absorption of the incident radiation, and the quantized energy associated with these transitions is related to specific wavelengths that are characteristic of each element. The function of a spectrometer in the measurement of atomic absorption is therefore to act as a filter to select only the wavelength of interest. A monochromator containing a diffraction grating is normally used to separate the various wavelengths, and spectral resolution is controlled via a variable slit function that allows the detector (usually a photomultiplier) to observe only a limited range of the spectrum (typically 0.2 to 1 nm bandpass). The great advantage of AAS is that a signal will only be observed if the atoms can absorb very narrow line radiation (of the order of a few picometers line width) emanating from the source. The source is designed to emit atomic line radiation characteristic of the analyte element, which forms the bulk of the ca-

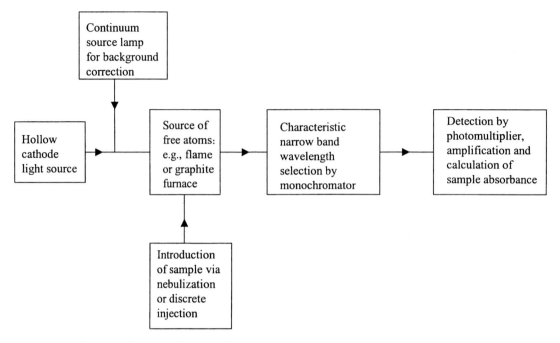

Figure 3.2 Schematic diagram of atomic absorption spectrometer.

thodic material of the lamp. Hence, the possibility for non-analyte elements to absorb this radiation is relatively small because the specific energies (wavelengths) of the allowable transitions between electronic states are defined by the atomic structure, and this will be different for every element. Spectral line overlaps from other elements in AAS are therefore comparatively rare; however, spectral interferences may arise when nonspecific attenuation of the incident radiation occurs. This type of interference may arise from scattering, either by particles or from molecular absorption, as a result of the incomplete breakdown of samples in the atomizer. Optical methods are available for correcting for interferences of this type when the purpose is to estimate the magnitude of the nonspecific signal in the proximity of the atomic line and to subtract it automatically during measurement. The use of a secondary deuterium lamp for nonspecific background measurement is perhaps the most common approach.[16] Alternatively, the source lamp can be pulsed to achieve a periodically broadened and self-reversed line profile in every half cycle, thus allowing the estimation of broadband attenuation of radiation displaced from the atomic line wavelength.[17] The advantage of this approach is that only one lamp is required. A disadvantage exists because full self-reversal of the line is required, and if this is not achieved, for example for less volatile elements, sensitivity will be reduced.

Zeeman-effect background correction relies on the application of an alternating current magnetic field to the atom cell (inverse mode) to achieve splitting of the energy level associated with atomic spectral line.[15] In the normal Zeeman effect, the two σ components of the split line that re-

sult are shifted in terms of energy—and, hence, wavelength—from the π component, which is unmoved. When the magnetic field is switched on, the nonspecific background signal can be measured at the original atomic line wavelength by instrumentally exploiting the polarized nature of the split line components. The total absorption signal is measured in the second half cycle, when the magnet is switched off, thus allowing a correction to be made. This approach is of particular importance in solid sampling applications where the amount of the sample entering the optical path of the spectrometer is considerably higher than when a prior chemical sample preparation step has been employed. Generally speaking, it is more important to use background correction methods with electrothermal atomization devices than with flame atom cells, because of the much higher levels of nonspecific attenuation encountered. That said, the application of a flame AAS spectrometer with Zeeman-effect background correction to the determination of calcium and magnesium in medical grade rubber has been described.[18]

The role of the atomization device in AAS is central to achieving accurate results. The output from an AAS measurement is a number rather than a spectrum, because for a given signal, data will only be collected at a fixed position in the wavelength range to an extent determined by the spectral bandpass of the instrument. Consequently, the technique is much more suited to quantitative rather than qualitative analysis. Physical and chemical interferences can have a profound effect on signal magnitude in AAS. These effects arise as a result of the presence of sample matrix species and how these behave in the atomizer (e.g., stable compound formation, influence in sample transport ef-

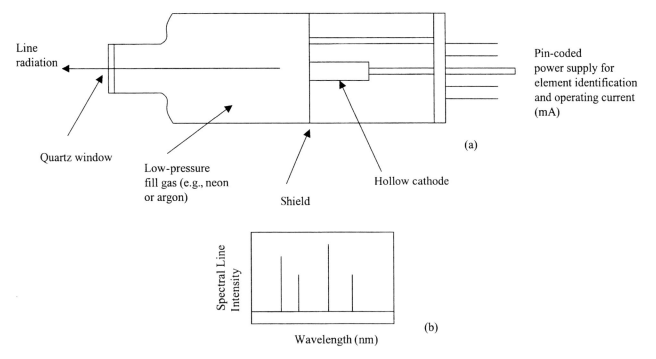

Figure 3.3 Schematic diagram of (a) a hollow cathode lamp, showing (b) spectral line characteristic output.

ficiency, and generation of nonspecific background signals). Developments in the field of AAS are primarily devoted to addressing the limitations in quantitative performance that are necessarily imposed by atomizer design. Flame and elec-trothermal atomizers are most widely used in AAS to determine the elemental content of polymers. The advantages and limitations of each atomizer in this type of application are described in the following sections.

Flame Atomic Absorption Spectrometry

The combustion flame is the most widely used atom cell for AAS. The air-acetylene flame (maximum temperature, 2200 °C) is probably most widely used for general-purpose determination of trace elements. The nitrous oxide–acety-lene flame operates at higher temperatures (maximum, 2900 °C) than the air-acetylene flame and tends to be re-served for use in applications in which the sample dissocia-tion is problematic. Flames can be run under a variety of fuels (acetylene) and oxidant conditions (e.g., air, nitrous oxide) to achieve the appropriate environment for analyte atomization. Fuel-lean conditions produce oxygen-rich flames, which exhibit higher temperatures. Fuel-rich flames, by contrast, are in relative terms cooler and have a high content of free radicals that provide a chemically re-ducing atmosphere for the breakdown—for example, of sta-ble oxides. The sample (usually a liquid or slurry) is contin-uously introduced at a rate of 4 to 5 mL/min via a nebulizer that produces a fine aerosol, which may be further frag-mented by an optionally fitted impact bead. The aerosol

thus generated is passed through a spray chamber fitted with a flow spoiler/baffles to discriminate against coarse droplets, which are discarded to a drain (see Figure 3.4). The fine droplets are included in only a few percent of the original sample mass and are swept through the spray chamber and into the slotted flame burner by the combus-tion gases. The flame itself is located directly in the absorp-tion path of the AAS. The flame desolvates the analyte spe-cies and then reduces the compounds present to free atoms, which then absorb radiation from the line-specific radiation source. The concentration of analyte present is directly pro-portional to the steady-state absorption signal generated.

Flame atomic absorption spectrometry (FAAS) is a well-established technique for elemental analysis. National and international standard methods based on FAAS for the analysis of synthetic and natural polymers, paints, and rub-bers have been published, and a number of these are cited for reference.[19–26] Although it is possible to adapt such methods for other purposes, with caution and attention to detail, it is important to note that a standard only claims to provide a valid means of determining specific elements in particular matrices. It is quite possible to have a method that is suitable for a particular analyte, but with a different polymer matrix (e.g., polyamide instead of polypropylene [PP]), or two different polyamides, one of which may con-tain glass fiber or pigment. The nature of the analyte com-pound, in terms of chemical composition and physical char-acteristics, may well differ from one situation to another, and this can have an influence on both the accuracy and the precision of the method. The method of preparation of

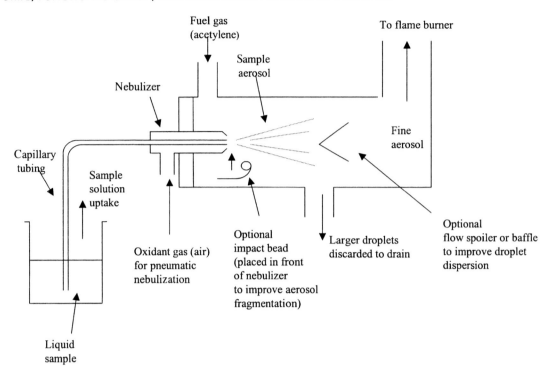

Figure 3.4 Concentric nebulizer and premix spray chamber for liquid sample introduction in flame atomic absorption spectrometry (AAS).

the polymer sample is therefore a critical step in ensuring the accuracy of the analysis. Flame AAS has now reached a degree of maturity in terms of instrumentation, and research efforts tend to be focused on overcoming the inherent limitations of the technique by the use of novel sample preparation methods.

The determination of trace metals in plastics by AAS and inductively coupled plasma atomic emission spectrometry using sample pretreatment methods has been the subject of a recent review.[27] The method of ashing the sample in a muffle furnace is often used to provide a gravimetric measure of the total inorganic content of a thermoplastic.[28] This approach has traditionally been used as a precursor for the determination of metal content of polymers by FAAS. As a method of sample preparation it has the advantage of removal of the organic components of the sample matrix by thermal oxidation and volatilization, and it can provide a means of significantly preconcentrating trace elements if enough sample is taken. The inorganic residue from ashing is taken into solution by using an appropriate mineral acid, the choice of which depends on the elements that are present. The disadvantages of this approach are that it is relatively slow and is subject to analyte losses by occlusion or volatilization. It is sometimes necessary to carry out a further sample preparation step based on fusion of the residue to break down analyte forms that are resistant to acidic attack. This type of method, employing potassium bisulfate for the fusion step and a further extraction into 6N sulfuric

acid, has been applied to the determination of chromium in polyolefins.[29]

Traditional wet oxidation is a well-established, if labor-intensive, method of sample preparation for the determination of the elemental content of polymeric materials. A protocol typical of this type of operation has been described for the determination of tin stabilizers in poly(vinyl chloride) (PVC) compositions.[30] The samples were decomposed by using sulfuric acid, followed by oxidation with hydrogen peroxide. An acetylene/nitrous oxide flame was used for the FAAS measurement to minimize the interference effects from mineral acids, which affect the response of tin in the cooler air/acetylene and air/hydrogen flames.

The potential for the occurrence of interferences that affect accuracy in flame AAS should never be underestimated, even when the methodology is relatively well established. An example of this was illustrated in a series of articles concerning the analysis of PVC when the widely employed method of sample treatment with sulfuric acid and hydrogen peroxide was found to be insufficient. It was found that the presence of carbon black, a reinforcing filler that affects the elastomeric properties of PVC, prevented the use of this method alone.[31] Further treatment of the sample with fuming nitric acid was necessary to destroy the carbon black, in order to permit the quantitative determination of cadmium, antimony, and tin in PVC. The precipitation of sulfates was found to be problematic when the sulfuric acid/hydrogen peroxide method was used for the pretreatment of PVC samples that contained high levels of alkaline earth elements (e.g., 10% calcium carbonate added to prevent the release of HCl during combustion) and lead as stabilizer.[32] It was reported that the precipitates thus formed could be dissolved in ethylenediaminetetraacetic acid (EDTA) in ammonia solution, thus allowing the quantitative determination of aluminum, antimony, and calcium. This method was further extended in a subsequent article to the determination of magnesium hardeners and basic lead stabilizers in commercially available PVC samples that contained high levels of calcium.[33] The approach was further modified for the determination of lead and aluminum in PVC samples that contained silicate filler.[34] Using the sulfuric acid/hydrogen peroxide decomposition, the precipitation of silica caused the adsorption of analytes from solution, thus giving rise to erroneously low results. The precipitate was dissolved in ammonia solution with EDTA added to prevent precipitation of metal hydroxides. The amount of ammonia solution added needed to be controlled to achieve stability in the determination of lead over a 24-hour period. It was recommended that the samples should be analyzed immediately after decomposition. It was also noted in all of the studies that care needed to be taken to avoid any interference effects that might arise from the presence of reagents or concomitants in the FAAS analysis of the resulting solutions. Selection of the appropriate flame conditions and sample dilution permitted calibration of the instrument with aqueous standard solutions. Potassium chloride was added as an ionization interference suppressant for the determination of aluminum.

Microwave digestion has increasingly been adopted in the last decade as a more effective way of achieving sample digestion for polymers.[35] Essentially, the sample is heated in a suitably inert vessel (either open to the atmosphere or sealed) in a microwave oven or focused via a waveguide in a digestion cavity. Both open and closed digestion systems are available commercially, as are systems for microwave ashing. My own experience, gathered over a period of 10 years, in application to the elemental analysis of polymers has involved work with pressurized digestion and ashing systems and with open digestion systems. Open systems are more suitable for the digestion of larger samples (up to several grams) and can be highly automated, with stepwise additions of reagents from reservoirs to multisample units under computer control. Simultaneous and sequential configurations for sample digestion are available. A further advantage of open systems is that there is no buildup of pressure when carrying out wet oxidation, as the acidic fumes can be vented to an alkaline scrubber unit, thus allowing the preparation of samples on the standard laboratory bench. Pressurized systems are restricted in the size of the sample that can be used (typically 100 to 250 mg, but dependent on the design of the vessel) because of the buildup of released gases during the digestion of the sample. Venting systems are available for each vessel to avoid the possi-

bility of explosive overpressurization. However, digestion under pressure can improve either the speed or the quality (or both) of the digestion. Pressurized systems tend to be operated on a batchwise basis of up to 12 samples at a time, with a single compromise set of microwave conditions, which are experienced by all the samples in the same oven. It is important to consider the effectiveness of the digestion because partially decomposed or solubilized polymer can affect signal intensity, particularly in FAAS. It is also worth noting that manufacturers tend to use "inert" polymers in the materials of construction of microwaveable vessels. Care should be taken to make sure that blank reagents are run regularly to check contamination and migration of trace elements, either from or to the vessel, as a result of previous experiments. Two essays on the stability of standard solutions in contact with acid-washed polymer surfaces are instructive in this context.[36,37] A comparison of dissolution procedures based on microwave digestion and oxygen flask combustion has been reported for the analysis of polymers.[38] Cadmium and lead were determined in polyamides, polyolefins, and poly(ethylene terephthalate) (PET). It was found that both methods gave acceptable agreement with a conventional wet ashing procedure and were considered to be time-saving and cost-effective pretreatments for FAAS. Microwave digestion has also been used as a sample preparation method for the FAAS determination of toxic elements in toy coatings.[39] Using this procedure, analytical recoveries of the order of 90% or better were obtained for antimony, arsenic, copper, lead, mercury, and selenium.

The analysis of PVC can be problematic because of the potential for loss of analyte species as volatile chlorides. However, the polymer matrix may be dissolved by heating in a solvent, and the resulting solution may be analyzed directly.[40] A recent example of this approach involved the dissolution of the PVC sample in dimethyl formamide and the addition of a small amount of acetic acid.[41] After dilution with acetone, both cadmium and lead at percent levels[42] and tin[43] in PVC were determined directly by FAAS. In a related approach involving sample dissolution, the determination of zinc in polystyrene has been carried out using a water and oil emulsion method.[44] The sample was dissolved with isobutyl methyl ketone (IBMK) and treated with an emulsifier (Brij-30 or Tween 80). The resulting stable emulsion was introduced directly to the flame atomizer, and zinc was determined in the range of 0.1 to 1 parts per million (ppm), by calibrating the system with IBMK-iodine/zinc-water or IBMK-iodine-potassium iodide/zinc-water standards.

The presence of polymers in solution can give rise to interference effects in FAAS, and this has been reported for the determination of trace metals (calcium, lithium, and magnesium) in polyacrylamide solutions.[45,46] The magnitude of the interference effect was greatest at a pH of 7, but the addition of acid or alkali resulted in partial precipitation of the polymer. The addition of an oxidizing agent (sodium chlorate) in an acid medium (acetic acid) was used to overcome the matrix effect, by precipitating the polyacrylamide molecules to form colloidal suspensions. The elimination of interference effects observed in the FAAS determination of calcium in rubber vulcanisates has been described.[47] Barium chloride was added to the test solution to eliminate suppression effects of sulfate on calcium to allow accurate determination. The optimization of FAAS for the determination of manganese in polymer-metal complexes may also be of some interest in this context.[48] Potential interference effects arising from the presence of 10 cationic metal species were assessed for the determination of manganese in the range of 0.1 to 10 ppm.

To identify the origin of a polymeric material, trace elements may be used as markers. For example, this approach has been used in the forensic examination of colored polyethylene bags.[49] A number of companies that manufacture either cross-linked polyacrylate adsorbents or products containing the material developed methods of detecting the dust in situ, based on the determination of the level of a sodium marker.[50] An interlaboratory trial was conducted to assess the potential of this method. Detection limits were established for this method in the range of 0.11 to 2.4 $\mu g\ m^{-3}$.

Antimony oxide is used as a flame retardant in many polymer materials, including acrylic fibers. The oxide contains undesirable levels of arsenic, and a method has been proposed for its determination in the presence of a large excess of antimony, based on a hydride generation AAS technique.[51] Antimony interferes in the generation of arsine, but the proposed method eliminated this effect by solvent extraction of arsenic (III) from a hydrochloric acid digest of the acrylic fiber. Titanium (III) chloride was used to ensure that the arsenic (V) produced by the digestion was converted to arsenic (III). After extraction with benzene, the arsenic (III) was back extracted quantitatively into aqueous solution prior to the analysis of the latter. Arsine was generated under controlled conditions by using a reduction with stannous chloride in the presence of potassium iodide in hydrochloric acid solution. The vapor was introduced to a nitrogen-hydrogen flame, which decomposed the arsine to free arsenic atoms before detection, by AAS. Analytical recoveries were reported for arsenic in the range of 96% to 104%. The method was considered suitable for the detection of arsenic in the range of 0.04 to 400 ppm in acrylic fibers containing antimony.

The analysis of solid and liquid paint samples has been carried out by FAAS using the Delves cup approach.[52] Originally developed for the rapid and direct determination of lead in whole blood, this approach is suitable for the determination of volatile elements only. In this method, the sample, either as a solid or as a solvent-suspended dried liquid film, was deposited in the cup and combusted directly in an air-acetylene flame. Cadmium and lead were determined in the paint samples with the same accuracy as when using conventional methods. The modern equivalent of the Delves cup is the electrothermal vaporization (ETV) device. The advantage of this procedure is that the sample container is heated electrically, and the temperature can be controlled in a more effective way to achieve sample pyrolysis before the analyte species are vaporized. An ETV device was coupled to a flame AAS system for the direct determination of rhodium in solid polymer and liquid polymer samples.[53] A modified graphite furnace was used to volatilize the sample after temperature-programmed ashing via an argon air carrier gas stream into the flame. Magnesium chloride was used as a matrix modifier in the ETV to assist analyte volatilization from the solid. A sample mass of up to 10 mg could be analyzed directly, or, alternatively, liquid microsamples in chlorobenzene could be introduced. Unfortunately, as with many solid sampling techniques, the precision obtained with the system was relatively poor and of the order of 10% to 20% relative standard deviation (RSD). This was attributed in part to uncertainty in the weighing procedure of relatively small masses. Further examples of solid sampling in AAS using electrothermal atomizers are discussed in the next section.

Electrothermal Atomic Absorption Spectrometry

Electrothermal atomic absorption spectrometry (ETAAS) was developed primarily to improve on the detection limits achievable by flame AAS. The graphite furnace was the first such "electrothermal atomizer", and the names are often used interchangeably in the literature, although other materials of construction can be used. As with the flame cell, the purpose of an electrothermal atomizer is to achieve complete atomization of the sample in order to quantify the presence of a specific analyte. A microsample (either solid or liquid) is introduced to the electrothermal atomizer (Figure 3.5), which is an electrically heated tubular (graphite) device that is capable of operating under programmable temperature control to a maximum of 3000 °C. The liquid sample is dried in stages at ca 110 °C for about 30 s, pyrolyzed or ashed at temperatures up to 1000 °C for typically 30 s to remove organic matter, and atomized at temperatures normally in the range of 2200 to 3000 °C for up to 5 s. The atomizer is situated in the optical path of the AAS instrument, the free atoms generated by the thermal action of the device absorb characteristic radiation, and the absorbance signal derived from this process obeys a linear relationship with the concentration of analyte present. Typically, ETAAS achieves a 100-fold improvement in sensitivity for most analytes, in comparison with that achieved by FAAS. This is achieved by total consumption of the sample and a much longer residence time for the free atoms in the observation volume of the spectrometer. The linear range for ETAAS, in terms of analyte concentration, extends about 2 to 3 orders of magnitude above the detection limit. Consequently, ETAAS is a trace technique, which is

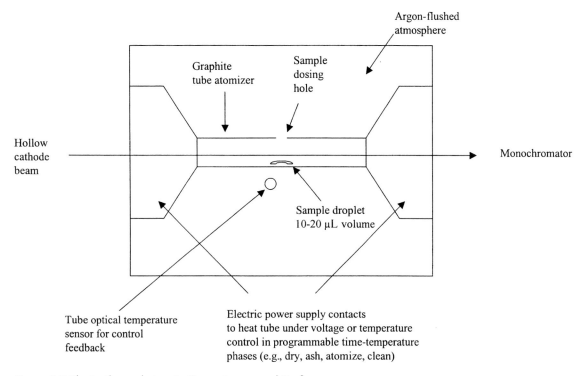

Figure 3.5 Electrothermal atomization using a graphite furnace.

largely complementary to FAAS in terms of the range of application. However, it should be noted that the sensitivity in ETAAS to nonspecific attenuation of source radiation (i.e., scatter and molecular absorption) and of chemical interference (stable compound formation) is also increased as a result of the same factors that improve analyte sensitivity. Therefore, the application of background correction techniques is much more important for ETAAS than for FAAS. Similarly, chemical interference effects that result from stable compound formation (e.g., oxides or carbides that do not decompose, or chlorides that volatilize prematurely below the temperature required for atomization) may appear severe. It is worth pointing out that, if they are compared on the basis of ratios of analyte to interferent, the performance of ETAAS in this respect is not much different from that of other trace metal techniques. However, the absolute amounts of interferent required to cause an analytically significant bias are small because the analyte concentrations determined by the technique are at the level of parts per billion. It is possible to avoid interference effects in some cases by sacrificing sensitivity and merely diluting the sample. A more appropriate response to problems of this type is to seek to develop sample preparation regimes that overcome the interference effect before the analysis is undertaken.

Consequently, efforts continue to be made to find more effective methods of sample preparation for ETAAS in application to the analysis of polymeric materials. For example, the determination of trace elements in fluorocarbon poly-

mers is problematic because analyte species may be volatilized as fluorides during sample preparation. This difficulty has been overcome by the development of a decomposition apparatus.[54] Samples were decomposed in a high purity stream of oxygen or air in a quartz boat at 600 °C for 30 min. The residual metals (calcium, copper, iron, and potassium) were dissolved in nitric acid and determined quantitatively by ETAAS. Acceptable analytical recoveries were reported, and blank values using the system were reported to be less than 1 ng.

A comparison has been made between sample dissolution and decomposition methods for the ETAAS determination of tin in plasticized PVC.[55] The performance of the conventional sulfuric acid/hydrogen peroxide decomposition method was evaluated for compatibility with ETAAS in comparison with metal extraction methods and a procedure involving solubilization of the polymer in tetrahydrofuran. The latter procedure was found to be the simplest and least time consuming. Diethyl ether and nitric acid extraction procedures did not achieve quantitative results. In a related approach, the determination of tin[43] and cadmium and lead[56] in PVC has been reported. In these cases, dimethyl formamide was used as solvent, and the dissolved sample was diluted in acetone before analysis by ETAAS was undertaken.

Antimony is used as a catalyst in the production of PET. A method for the determination of antimony has been described in which the sample was pretreated with 4% acetic acid and analyzed by ETAAS.[57] Using this approach, analytical recoveries for antimony were established in the range of 92%–97%, and precision was of the order of 3% relative. Results obtained using this method were validated by using a colorimetric method based on malachite green.

A method has been published for the determination of trace metal impurities in polyimides.[58] The polymer was dissolved in cyclohexane, dioxane and N-methylpyrrolidone, and the suitability of each solvent was assessed for use with ETAAS. The atomizer consisted of a transversely heated graphite tube fitted with an integrated platform. Zeeman-effect background correction was employed. The method of standard additions was used for quantification. The optimum temperatures set for drying, pyrolysis, and atomization steps were significantly higher than those normally used for the analysis of aqueous solutions. A range of elements—including calcium, copper, iron, potassium, silicon, and sodium—were determined using this approach, and the results obtained compared well with those generated when using the microwave nitric acid digestion method. It was noted that the solvent dissolution method yielded better detection limits, in comparison with the microwave digestion method, by a factor between 5- and 1200-fold, depending on the element. Aspects of reagent and tube contamination and atomizer tube lifetime were also evaluated. Carbonaceous residues built up on the atomizer tube surface during repeated analysis and affected the response for silicon but not for the other elements. The measurement of silicon by ETAAS is relatively insensitive and required the introduction of very concentrated sample solutions (150 mg/mL), which exacerbated the problem of matrix residue buildup. It was reported that the use of a palladium–magnesium nitrate matrix modifier enhanced the signal quality and reproducibility for the determination of silicon. It is worth noting in this context that refractory elements such as silicon are also subject to atomization interference that arises from carbide formation as a result of intimate contact with either the graphite tube itself or the carbon-based sample residues. It is necessary to use higher atomizer temperatures to decompose such compounds and to clean the tube between sample injections. This has a deleterious effect on tube lifetime. For this reason, attempts to employ solid sampling procedures for the analysis of polymers by ETAAS tend to be restricted to the determination of relatively volatile elements.

A novel combined separation-ETAAS technique has been proposed for the characterization of organo-tin-containing compounds.[59] Size exclusion chromatography (SEC) was used to separate the high and low molecular weight fractions of the polymer. Graphite furnace AAS was used to determine the tin content of these fractions, and this was related to the fractionated polymer molecular weights.

The development of solid sampling procedures for the analysis of plastics by AAS has been has the subject of a review.[60] There has been significant concern in recent years regarding the environmental impact of lead-based paint. A portable field-operable electrothermal AAS has been designed for this application for use in the field.[61] The atomizer unit consisted of a battery-powered tungsten coil atomizer. A relatively low light throughput optical spectrometer was employed for measurements, and this limited the signal-to-noise ratio that could be achieved. A limit of detection of 20 pg lead was reported, and the precision achieved in a typical analysis was 5% relative. The accuracy of the device was established by analysis of National Institute of Standards and Technology standard reference material paint (NIST SRM 1579a). An ETAAS method for the direct determination of lead in paint chips by means of slurry sample introduction has also been reported.[62] The use of ammonium hydrogen phosphate as a matrix modifier in ETAAS for the determination of lead in paint has also been investigated.[63]

The determination of lead stabilizer in PVC has been investigated using solid sampling ETAAS.[64] A pyrolytic graphite tube was used for the direct atomization of a 3-mg PVC microsample. Calibration of instrument response was achieved using aqueous standard solutions. It was stated that background correction was not required, as an alternative atomic line was used which was suitable for the determination of lead in the range of 0.1% to 1% m/m. The precision achieved using the method was in the range of 5% to 10% relative, which is fairly typical of the performance of solid sampling ETAAS methods. A solid sampling ETAAS method for the screening of antimony in PVC samples may also be of interest.[65]

Polycarbonate is used in the manufacture of optical fibers and compact disks. Transition metals present at ultra trace levels can affect the optical quality of such materials by inducing polymer degradation. A solid sampling ETAAS was reported for the detection of six transition metals and sodium in a polycarbonate matrix.[66] The accuracy of the method was established for the determination of cobalt, chromium, copper, iron, manganese, nickel, and sodium by reference to a comparative method. Antimony, cobalt, and manganese are present in PET as catalyst residues derived from the production process. Trace amounts of these elements were determined directly in PET films using solid sampling ETAAS.[67] It was found that optimization of the atomizer temperature program settings in the pyrolysis stage was essential to ensure that the film was completely charred before atomization occurred. Quantification was achieved by the method of standard additions, although in practice, it is often difficult to ensure that the spiked analyte will behave in exactly the same manner as analyte that is intrinsically bound within the solid sample. However, the results obtained, although exhibiting poorer overall precision, compared reasonably well with those derived from a flame AAS method based on a conventional digestion protocol.

The trace element fingerprint of an individual carpet fiber is potentially useful in tracing its origin. Single polymer fibers are an ideal size for direct analysis using solid sampling

ETAAS.[68] An investigation was conducted to examine the feasibility of using ETAAS and ETV-ICP-MS to link forensic evidence to crime scenes. It was reported that while ETAAS could be used for the analysis of a single fiber, the technique compared unfavorably to ETV-ICP-MS in terms of analytical linear range and restriction to the determination of a single element. However, it is worth noting that commercial AAS instruments are now available which can detect several elements simultaneously in a single atomization cycle; consequently, the latter limitation to the use of ETAAS in this application may now be overcome.

Zeeman-effect background correction has been used with some success in conjunction with solid sampling ETAAS. The solid sampling approach lends itself easily to homogeneity testing at the microscope. This ability has been used to assess homogeneity in the production of a polyethylene reference material. The determination of cadmium in polyethylene was carried out using Zeeman-effect ETAAS.[69,70] Four materials with cadmium contents in the range of 40 to 400 mg/kg were produced, and the homogeneity of the batch was tested by using solid sampling. An atomizer fitted with a solid sampling cup was described for use with Zeeman-effect ETAAS for the determination of chromium in photographic film[71] and of lead in PVC.[72] Zeeman-effect ETAAS also have been applied to the determination of tin in unplasticized PVC.[73] Samples were suspended in 4% acetic acid using calcium nitrate as a matrix modifier. The analytical recoveries found were generally acceptable, in the range of 95% to 106%. The precision of the method was estimated to be of the order of 3% to 4% relative, which is very respectable for a solid sampling method.

Polymers are often used in food contact applications, and the potential migration of chemical species from such materials is a matter of some concern to national and international regulatory authorities (e.g., the U.S. Food and Drug Administration and the European Union). Migration is assessed by the controlled exposure of the polymer to food simulants such as water, acetic acid, ethanol, and an oil, under protocols specified by regulatory authorities. Atomic absorption spectrometry is particularly suited to the detection of trace element migration, and the literature contains examples of this type of application. For example, PVC bottles are stabilized with organo-tin compounds to prevent degradation in processing. Consequently, the detection of organo-tin compounds in food simulants is of particular interest to consumers. An ETAAS method incorporating platform atomization with a matrix modifier has been proposed for the direct determination of methyl tin in corn oil leachates derived from PVC bottles.[74] It was shown that the method was capable of recovering tin at a spike level of 15 ng/g, and that the level of tin found in the extracts posed no safety problem to consumers. A similar study was carried out to determine the migration of barium and cobalt from polyethylene packaging into food.[75] The food simulants used to carry out the extraction were water, 3% acetic

acid, 10% ethanol, and sunflower oil. A flame instrument was used to detect barium without sample preconcentration, except the addition of ionization suppressant, in the aqueous simulants. Cobalt was preconcentrated from the simulants by using a chelating extraction procedure based on the conventional ammonium pyrrolidine dithiocarbamate (APDC)–isobutyl methyl ketone (IBMK) regime. Sunflower oil was ashed, and the residue was dissolved in hydrochloric acid before analysis was undertaken. Lower limits of detection were quoted for barium as 0.1 mg/kg in aqueous food simulants and 0.125 mg/kg in sunflower oil. Because of the preconcentration step used, the cobalt figures of merit were exactly 10-fold improved over those for barium.

The migration of chromium species from polyethylene samples used for packaging mineral water has been the subject of an investigation.[76] An attempt was made to separate chromium (III), by co-precipitation on an alumina carrier, from chromium (VI) species, before detection by AAS. It was reported that changes in chromium oxidation state (reduction of chromium (VI) to chromium (III)) during the analytical process prevented satisfactory speciation, although the total chromium content could be determined. Atomic absorption spectrometry has also been used for the determination of trace metals in Polish PET films designed for use in food packaging.[77] The analysis revealed that the metal content of five types of film and particularly the antimony levels were unacceptable for this type of application. Some 62 food packaging containers purchased commercially were assessed for the content of trace elements.[78] Migration studies on containers by AAS revealed that low levels of cadmium and mercury could be extracted in some cases and that these levels increased when UV irradiation and surface abrasion were applied. Concern about the level of toxic elements present in children's toys has led to the development of test methods designed to mimic the extraction potential of saliva.[79] In this study of aqueous extracts produced in accordance with the European Community (EC) Draft Standard EN 71 on the Safety of Toys (1982), cadmium was determined by AAS. Using this procedure, a limit of detection of 0.03 ppm was reported.

The extraction of metals from PVC pipes has been investigated in two studies. Atomic absorption spectrometry was used for the determination of calcium, magnesium, tin, and titanium in water extracts maintained at temperatures of 25 and 50 °C over a period of several weeks.[80] It was found that while titanium was not extracted, calcium magnesium and tin levels reached a maximum after 2 weeks, which remained when the sample was stored at 25 °C. However, magnesium levels were further enhanced at 50 °C to reach a further maximum of migration after 4 weeks. The examination of organotin leachates from chlorinated drinking water samples collected from PVC pipes has been described.[81] Organotins were isolated by solid phase extraction followed by GC-AAS detection. The levels of leachate were found to be dependent on the water temperature. Mono-alkyl tin was identified, and dialkyl tin levels of approximately 30 to 100 ng g^{-1} were quantified.

The physical characteristics of polymers can be monitored by using atomic spectroscopy techniques. For example, the effect of exposure of polyethylene (high density, low density, and linear low density) and polypropylene to a range of solvent conditions over a 5-month period was monitored using atomic absorption and emission spectrometry, among other techniques.[82] The samples were exposed to acetone, ammonia solution, isopropyl alcohol, hydrochloric acid, trichloroethane, and trifluoroethane, and interactions in terms of dissolution, degradation, and migration were monitored. A study of the permeation of metal ions though low-density polyethylene (LDPE) has been carried out.[83] The penetration of metals through the polymer was monitored by ETAAS as a function of temperature of the solvent and time. Diffusion and permeability coefficients were measured, and it was reported, perhaps unsurprisingly, that the extent of penetration of the inorganic salts was strongly dependent on the solvent employed. The transport properties of a cation exchange membrane composed of poly(styrene sulfonic acid) cross-linked with poly(vinyl alcohol) have been investigated by using a diaphragm cell.[84] The alkali metal ion content on either side of the membrane, which was maintained at 25 °C, was monitored by AAS.

The mechanism of evaporation of antimony-based flame retardants has been studied using AAS and thermal analysis.[85] Compositions containing antimony trioxide and halogenated additives in a range of molar ratios were used to explore the formation and evaporation of antimony halide species. The evolved species were detected as antimony by AAS in situ.

Atomic spectroscopy techniques can be used to study polymerization reactions. Thus in a study of the radical copolymerization of tributyl tin methacrylate, AAS was used to establish the reactive ratios of the products.[86] This method was compared with alternatives based on proton nuclear magnetic resonance (NMR) and Fourier transform infrared (FTIR) spectrometry. The polymerization of acrylamide using threonine/cerium salts and threonine/potassium permanganate initiators has been studied using AAS, among other techniques.[87] The dependence of the polymerization yields and molecular weights of the polymers produced was investigated as a function of reaction conditions and the mole ratio of monomer/initiator. The reaction of poly(propylene glycol) adipates and magnesium oxide has been monitored using AAS.[88] The incorporation of magnesium into the resin was investigated, and the levels of basic and neutral salts present in the mixture were calculated. The preparation and characterization of mercuriated copolymers has been the subject of a recent investigation.[89] A linear copolymer of methyl acrylate and N-vinyl carbazole was prepared in benzene by using azobis(isobutyronitrile)

(AIBN) as an initiator; then the copolymer was mercuriated by using mercuric trifluoroacetate. The tetrahydrofuran (THF) soluble polymers were purified several times by repeated precipitation using methanol and water as nonsolvents. Atomic absorption spectrometry was used in conjunction with FTIR and NMR to investigate the structure of the copolymers prepared. In a further example of the incorporation of metals in polymers, poly(selenyl thiophene) was synthesized electrochemically in acetonitrile, and its redox properties were examined.[90] The maximum oxidation state and doping level for the polymer was established by using chronocoulometry and AAS.

The mechanism of antifouling paints containing organo-lead and organo-tin compounds has been studied using AAS.[91] The systems investigated included elastomeric polymer dispersed in organic solvent and water-dispersible latexes. The action of the protective system was monitored, using AAS for the determination of metals and FTIR for the identification of the organic reaction products released. The combination of AAS and FTIR has also been used in the characterization of an electroconductive polymeric pigment.[92] The pigment, which exhibited microwave absorption properties, was produced by modifying polyacrylonitrile by treatment with copper sulfate and sodium thiosulfate in water. It was established that the resulting pigment was essentially the original polymer cross-linked by copper and sulfur atoms to the cyano-groups of the polyacrylonitrile. The incorporation of silver metal in polypyrrole films has been studied using AAS.[93] The films, which were synthesized in the presence of p-toluene sulfonate, were treated sequentially with 0.5 M sodium hydroxide to remove this functionality and 0.5 M nitric acid to reprotonate the material. The treated film was then exposed to nitric acid that contained levels of silver in ppm, and this resulted in the reduction of the ratio of silver ions to metal, and the metal was then concentrated in the polymer. The concentration of silver removed from solution was determined by AAS. The study of a semi-interpenetrating network consisting of sodium alginate and polyacrylic acid or polyacrylamide has been reported.[94] It was found that calcium ion concentration, as established by AAS, could be used to control the swelling of the polymer membrane produced. A spectral study of mechano-chemically synthesized polychelates may also be of some interest.[95] Techniques such as Mossbauer spectroscopy, electron spin resonance (ESR), and AAS were used to investigate the macromolecular compounds. It was established that the presence of metallic iron was primarily responsible for the observed magnetic properties of these materials.

A great deal of research is currently being conducted to determine the synthesis and application of polymer systems in the adsorption of metal ions. The main areas of use of this technology are in the industrial scale development of methods for metal removal (e.g., renewable ion exchange or chelation processes) and in speciation and preconcentration techniques for analytical measurement. Although a detailed treatment of this subject is beyond the scope of this review, a brief summary of the relevant literature citations is provided here. Polymeric materials are now widely used in analytical science for separation and preconcentration of trace metals. The emphasis of applications of this type is focused on the use of adsorbent to improve the performance of a particular analytical technique (e.g., AAS, ICP-AES, and ICP-MS), although to a large extent the technology will be transferable. The straightforward preconcentration of analytes from liquids on a polymeric column allows the elution of more concentrated solutions, which are often stripped of undesirable matrix species. The detection of the metals on the loaded polymer can be achieved destructively by both acid digestion and direct analysis, but it is more usual to use either acids or solvents as eluants before the sorbent is reused. Alternatively, the preconcentration approach can be used in reverse by allowing the analyte to pass through the columns unhindered while trapping interfering species. It is possible to carry out such liquid-solid separations based on the use of polymers functionalized with chelating groups,[96] and sorbent chelation extraction methods using hydrophobic solid supports.[97] A range of polymers has been used in analytical preconcentration studies. Polyurethane foams are a popular choice in this field of activity, and examples include the preconcentration of antimony, arsenic, bismuth, mercury, selenium, and tin on dithiocarbamate-loaded polyurethane[98] and platinum on polyurethane.[99] Other polymeric materials used include salicylic acid functionalized Amberlite XAD-2 for lead and zinc,[100] polydithiocarbamate resin for cadmium and lead,[101] and polyacrylamidoxime for cadmium, copper, iron (II), lead, and zinc (II).[102] The speciation of antimony has been achieved using a polymer that contains amidoxime and amine functional groups.[103] Quantitative sorption of total antimony (III) and antimony (V) was achieved at pH 2 and antimony (III) only at pH 10. After the preconcentration step (×50), the sorbent was injected into a graphite furnace as a suspension, for AAS determination of the adsorbed species.

The metal-binding properties of poly(N-vinylimidazole) hydrogels have been studied using a batch equilibrium process.[104] The effect of conditions of pH on the adsorption process was studied for a wide range of transition metal ions and binary mixtures using AAS. The influence of hydrogel cross-linking on the kinetics of sorption and overall capacity at equilibrium were also investigated. A study of the use of poly(methyl methacrylate) and poly(methyl acrylonitrile) for the preconcentration of copper from aqueous solutions has been published.[105] Metal ion adsorption on these polymers and related copolymers was investigated using AAS. The preparation of polyethylene glycol dimethacrylate-hydroxyethyl methacrylate microspheres for the adsorption of heavy metal ions from solution has been described.[106] The polymers, which became swollen after contact with

water, were functionalized for metal chelation by the incorporation of chemically bound Congo red ligands. Electrothermal AAS was used to examine the adsorption and desorption characteristics of cadmium on the microspheres under batch equilibrium conditions. The synthesis and chelating properties of N-aryl poly(acrylohydroxamic acid) resins have been described.[107] The extraction of divalent metal ions and the kinetics of the sorption process were studied. An assessment was also made of the potential of these resins in the chromatographic separation of metal ions. The preparation of porous metal chelating resins based on polyhydroxamic acid[108] and acrylonitrile-ethyl acrylate divinyl benzene[109] polymers may also be of interest in this context.

The use of polymeric materials in biological systems provides interesting opportunities for the study of trace-element interactions. For example, an evaluation of silicone and polyurethane biomaterials in urinary tract applications was undertaken using a model system, which simulated the probable physiological environment.[110] Using X-ray analysis and AAS, it was found that the silicone material was less prone to encrustation in the continuous flow model process. In a similar study, styrene-butadiene block copolymers were examined before and after immersion in a physiological serum for 6 months.[111] Using AAS, it was found that the material with higher styrene contents had a greater capability to absorb calcium, sodium, and potassium ions. Electrothermal AAS has been used in a study of the migration of silicone gel-filled implants into breast tissue.[112] The levels of silicon found ranged from 3 ppm to 6% (60,000 ppm) in dried tissue samples. A controlled release delivery system for the drug copper bis(acetylsalicylate) based on the use of poly(hydroxyethyl acrylate) has been described.[113] It was shown by AAS and UV-visible spectrometry that copper was released into solution, over a 7-day period, in the form of one or more complexes. Finally, an investigation of the effect of porosity of Biomer-segmented polyurethane on tissue growth and calcification in rats using X-ray analysis and AAS may be of some interest to workers in this field.[114]

Atomic Emission Spectrometry

Atomic emission spectrometry (AES) is the oldest of the techniques used for instrumental elemental analysis. The technique uses optical spectrometry, in the UV-visible and near-infrared regions of the spectrum (180–800 nm) for the detection of atomic (and ionic) emission generated by samples introduced to a high-temperature excitation source. An example of an atomic emission spectrometer is shown in Figure 3.6. A variety of sources have been developed over many decades for atomization and excitation of samples, including the flames, sparks, and arcs. In recent years, however, the inductively coupled plasma (ICP) has become well established as the primary source for atomic emission measurements. A detailed description of the ICP

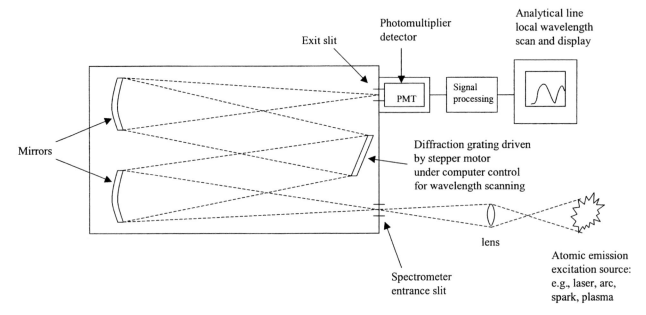

Figure 3.6 Schematic diagram of a sequential scanning atomic emission spectrometer (AES).

as a spectrometric source is provided in a recent book.[115] More recently, glow discharge and laser sources for AES have become popular because of their ability to directly characterize solids with spatial resolution. A general review of developments in atomic emission spectrometry is published annually.[116]

Laser Ablation AES

The prime advantage of laser ablation AES for the elemental analysis of solid plastic materials is that no sample preparation is required, obviating the need for time-consuming chemical pretreatment steps. In addition, laser sampling provides the opportunity to achieve a degree of spatial and depth resolution while at the same time providing bulk compositional measurement on a quantitative basis. Thus, laser AES has been applied to industrial on-line control analysis.[117] Rubber slabs made as a precursor to tire production were sampled on-line using a ruby laser. The light emitted from the interaction of the laser with the sample was observed spectroscopically by using a fiber optic link. A multichannel analyzer was used in the detection system to achieve good time resolution of the pulsed signal. The spatial element distributions of compound constituents were investigated by controlled laser scanning across the slab surface. In a related approach, a laser-induced breakdown spectroscopy system has been described which used a charged-coupled device array detector for spatial and temporal resolution of the emission signal.[118] This measurement approach allowed the removal of spectral interferences due to the presence of air in the analysis of poly(methyl methacrylate) materials. Laser-induced plasma emission spectrometry has been evaluated for the identifi-

cation of polymers using trace-element fingerprinting.[119] This technique was applied to the characterization of manufactured plastic articles, including car bumpers, dashboards, computer cases, and paints. Laser-induced breakdown spectroscopy has been applied to the identification of polymers.[120] A neodymium:yttrium aluminum garnet (Nd:YAG) laser operated at the fundamental wavelength was used to generate time-resolved emission spectra in the near-UV and visible regions. A neural network was used to classify the elemental spectra produced for a range of polymers, including PVC, PET, PE, and PP. The accuracy of the method was reported to be between 90% and 100%, depending on the system under investigation.

Laser-induced breakdown spectroscopy instruments have been designed for use in the identification of lead-based paints. For example, a portable system has been developed, which weighed 14.6 kg and operated on a 115 V supply.[121] Sampling of the paint in situ was achieved using a handheld laser probe, and the resultant emission was detected via a fiber optic link to the spectrometer. However, the sensitivity of the device was relatively poor (0.8% m/m) due to the use of the less sensitive 220.3 nm lead ion transition line.

Glow Discharge AES

The glow discharge source has been used in emission spectrometry since the late 1960s, but, until fairly recently, applications have been primarily in the field of metallurgical analysis. The principle of operation of the source is that an electrical discharge is created under low pressure in an inert gas atmosphere, and the plasma thus created interacts with the sample, which is usually held in, or is, or forms part of an electrode (typically the cathode; see Figure 3.7). The most common design of cell uses a direct current supply to sustain a glow discharge plasma. Ions in the gas phase are electrostatically attracted to the surface of the sample, and this bombardment results in a sputtering process that generates free atoms and ions from the bulk of the material to be examined. Emission spectrometry can then be used to observe the excited atoms and ions generated by this process. A brief review of the use of glow discharge spectroscopy in application to the broader materials field has been published.[122] A specific advantage offered by glow discharge (GD) spectrometry, in comparison with many of the other atomic spectroscopy techniques described in this chapter, is that because, over time, the sample material is gradually worn away at the surface as a result of the action of the sputtering process, the technique can be used for elemental depth profiling, usually at submicron resolution.

The use of the direct current glow discharge source has been relatively restricted in polymer applications, primarily because of the requirement for the sample to be conducting. However, Marcus[123] has reviewed the status of radiofrequency glow discharge sources that do not have the same limitations. The application of GD-AES to the depth profiling of transition metals in a silicate paint thin film on automotive glass was cited as an example of the potential of the technique for the analysis of nonconductors. A further application has been described in which GD-AES was applied to the depth profiling of painted steel materials of the type used in car bodies.[124] The spectral region interrogated by the detector was 11 nm wide, which allowed the simultaneous detection of iron, titanium, and zinc in 100-micron thick multilayers of paints and other corrosion resistance treatments, using a spectrograph fitted with a charge coupled device (CCD) array detector. It was suggested that the argon carrier gas signal could be used to normalize the source response for the purposes of quantification. Glow discharge AES has been applied to the analysis of acrylic and polyester coatings and to paint and lacquer thin films.[125] The spatial resolution achieved by continuous depth profiling for a number of elements simultaneously was in the nanometer to micron range. The use of GD-AES in depth profiling of polyaniline films grown using an electrodeposition process has been the subject of a further investigation.[126] Using X-ray spectrometry, the degree of incorporation of sulfate ions was also examined in this study. Glow discharges can also be used in plasma polymerization processes. An ultraviolet GD-AES spectrometer was devised to monitor the discharge polymerization of hexamethyldisilane and tetramethyl silane.[127] Glow discharge emission spectrometry has been used in conjunction with X-ray photoelectron spectroscopy to investigate a plasma polymerization process. Methane and three kinds of fluoromethanes were polymerized in a discharge, and the plasma polymerization mechanism was investigated by characterizing the polymerization behavior and the resultant polymer films.[128]

A new technique for the direct analysis of polymers by AES has been described; it involves a positionally stable sliding spark source.[129,130] A high voltage is applied to the sample until dielectric breakdown occurs and a current pulse (>100 A) is then applied to a secondary circuit for excitation. The spark is moved over the surface of the sample, and data are acquired by a multichannel spectrometer fitted with an array detector. The system was applied to the identification of polymers such as PVC and to nonchlorinated polymers such as polypropylene and polyethylene. The determination of such additives as flame retardants, stabiliz-

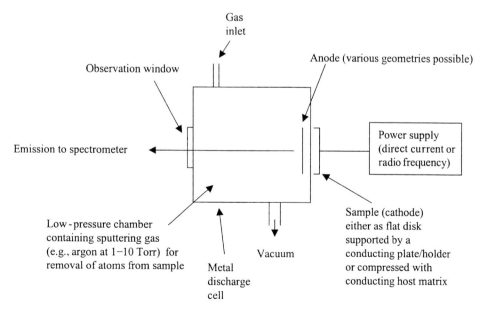

Figure 3.7 Schematic diagram of glow discharge (GD) atom–ion source for atomic emission spectrometry.

ers, fillers, and pigments at low or subpercent w/w levels in polymers and rubber was also described.

Inductively Coupled Plasma AES

Inductively coupled plasma atomic emission spectrometry is used in the polymer industry for the determination of both metals and nonmetals, and it is particularly suitable for the analysis of liquid samples. The sample in the liquid form is typically introduced to the ICP via a pneumatic nebulizer and spray chamber combination, which serves to act as a filter to allow only a fine aerosol to reach the discharges. The ICP itself is a radio frequency (2 kW) source sustained in argon.[116] Other gases may be used, but commercially the argon source predominates. A schematic diagram of an ICP source is shown in Figure 3.8. The discharge is initiated in a quartz torch by using a high voltage spark to provide a seed of electrons in the argon gas stream. Autoionization of the argon atoms takes place in a cascade process that is maintained by the application of radio frequency power via a load coil encircling the quartz torch. Consequently, the ICP is an electron rich, dynamic system in which the flowing gas (typically 15 liters/min) is ionized in the radiofrequency heat zone. The temperature at the center of the discharge is in the range of 6000 to 10,000 K, depending on the operating conditions selected and the area being viewed. The sample in the form of an aerosol is carried in an argon stream by an injector tube located in the center of the quartz torch, through the middle of the

discharge. The processes of desolvation, dissociation, atomization, and ionization take place as the analyte species traverse the source temperature gradient. Atomic and ionic emissions are observed optically by sequential scanning or using a simultaneous multichannel UV-visible spectrometer. The latter may be fitted (optionally) with a vacuum or nitrogen-purged optical path for measurements below 200 nm. The ICP is self-buffered from matrix effects due to ionization of matrix components, and analyses are relatively free of atomization interferences as a consequence of the high operating temperature. Spectral interferences, such as line overlap and background and scattered radiation from the plasma itself, may be more difficult to overcome; however, the multiplicity of atomic transitions afforded by the emission basis of the technique usually means that such problems may be resolved by appropriate line selection. Inductively coupled plasma atomic emission spectrometry is therefore both competitive with and complementary to AAS. Consequently, it is not surprising to find that similar sample preparation methods are used for both techniques.

As with AAS, the major drawback to the use of ICP-AES in the analysis of plastics is the time required for the preparation of samples before they can be measured. The conventional ashing method can be used as a precursor to dissolution of metal residues in nitric acid, as has been described for the ICP-AES determination of trace metals in polyolefins.[131]

High-temperature fusions are also generally applicable in the ICP-AES analysis of plastics and are particularly suit-

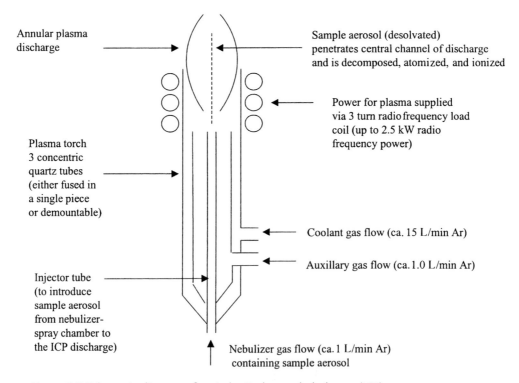

Figure 3.8 Schematic diagram of an inductively coupled plasma (ICP) source.

able for ensuring the complete solubilization of refractory elements. For example, the determination of aluminum and titanium (catalyst residues) in polypropylene has been carried out using a potassium bisulfate fusion approach after acid treatment.[132] The resulting melt was dissolved in dilute sulfuric acid and analyzed by ICP-AES. The precision of the method was about 8% relative. Fusion methods for ICP-AES often require other sample preparation steps before the either melt phase (e.g., ashing) or postmelt (acid dissolution). Consequently, such methods can be slow and laborious to carry out on a routine basis. This is exemplified by an ICP-AES study in which the high-temperature ashing and fusion-sample treatment of poly(arylene ether ketone) copolymers was reported to take up to 4 hours.[133] A wet oxidation method that gave comparable results in 30 minutes was preferred.

Wet oxidation of polymeric samples is widely used before AAS analysis is performed, and such protocols can be used to good effect in ICP-AES. Germanium is widely employed as a catalyst in the production of polyesters. An investigation of sample preparation methods for the determination of germanium in PET indicated that both dry ashing and wet oxidation methods employing mineral acids were suitable.[134] However, it was observed that dilute hydrochloric acid enhanced the ICP-AES signal response for germanium, whereas nitric and sulfuric acids did not. This was attributed to the improved transport efficiency for germanium in the hydrochloric acid medium as a result of the formation of more volatile chlorides. It is worth noting that variations in viscosity of samples as a result of either different acids or acid concentrations (or both) can give rise to such systematic biases. Spectral interferences are problematic in emission spectrometry, and the determination of germanium in PET has been examined from this point of view.[135] The analytical performance of the principal ICP-AES germanium lines at 265.1 and 303.9 nm were evaluated in terms of detection limit, background equivalent concentration, and interelement effects. Detection limits in the ng/mL range were reported, and it was concluded that matrix-matched standard calibration should be used to avoid systematic bias in the determination.

A wet oxidation method based on microwave digestion has been described for the determination of phosphorus in aromatic heterocyclic polymer films.[136] The samples were decomposed using sulfuric and nitric acids, and the ICP-AES methodology was validated using a spiking and reference standard protocol. Spectral interference on the measurement of phosphorus at 178.3 nm was observed, and this was attributed to sulfur emission arising from the presence of sulfuric acid. Alternative wavelengths at 213.6 and 214.9 nm were found, and they were free of interference. The use of mineral acids can be potentially hazardous to personnel, and options for the automation of such methods have been explored. In one example, a laboratory robot system has been devised for the automatic determination of

trace metal ions on the surface of epoxy laminates used in the production of printed circuit boards.[137] The system, which was modified to perform in a corrosive environment, was used to carry out mineral acid extractions by immersion of the circuit board in a mixture of hot nitric and hydrochloric acids. It was reported that analysis times were reduced, as was the potential for accidents such as acid spillage. A review of the application of ICP-AES to the analysis of paints and coatings may be of specific interest to workers in the surface coatings industry.[138] Protocols for the determination of lead in paints have been published.[139] The use of microwave digestion of this type of application has also been explored.[140]

Polymer dissolution methods are also practicable for ICP-AES analysis, although care has to be taken when selecting the organic solvents. This is more problematic in ICP-AES because solvent vapor loading can destabilize the discharge and give rise to undesirable spectral emissions, whereas in FAAS, an organic solvent will usually act as a fuel. An example of this approach to the determination of barium, calcium, phosphorus, and zinc in vinyl stabilizers by ICP-AES in organic media has been described.[141] The direct examination of waxes and other low melting point solids by ICP-AES has been described.[142] In this approach, the samples were introduced to the ICP-sample introduction system in the molten state and then nebulized at high temperatures. The technique was developed to assess the metal content of the wax during production.

Laser Ablation

Owing to the differences in the chemical and physical behavior of analytes, particularly in variable matrices, it is often difficult to achieve general applicability of conventional sample preparation methods for multielement techniques such as ICP-AES. Furthermore, the power and throughput of modern emission spectrometers is such that the actual time required for measurement in a well-defined application is only a small fraction of that required for sample preparation. Consequently, research effort has focused on developing techniques for the direct analysis of samples, thus obviating the need for extensive sample preparation regimes. Laser ablation has been explored as a method of achieving the direct analysis of solids by ICP-AES. For example, the direct analysis of poly(vinylidene fluoride), PVC, and polyethylene has been carried out using a Nd:YAG laser for solid sampling.[143] The laser was operated at three distinct wavelengths (266, 355, and 1064 nm), and the effect of the energy of the pulse on the sample topology was examined. It was reported that the optimum conditions required an input energy of 10 to 14 J/cm^2 at 266 nm and that the threshold level for ablation was a function of both the input energy and pulse wavelength. Relative standard deviations in the range of 2% to 5% were achieved in conjunction with a beam-masking technique.

Matrix effects are a significant problem in laser sampling of composite materials because of the effect of preferential vaporization of the more volatile analyte components, which results in analytical bias. The need for the direct use of reference materials can be overcome if the analytical technique is free from matrix interference. Backsurface ablation in LA-ICP-AES has been used in the analysis of poly-(vinyl alcohol) matrix standards for paint samples.[144] Samples were coated on a transparent quartz plate, and a laser beam was used to irradiate the sample through the substrate rather than at the sample surface as is done conventionally. The sample was completely removed from the substrate by a single laser pulse at a sample thickness of 35 microns. It was reported that this procedure yielded up to a six-fold increase in emission in comparison with front surface ablation, to an extent dependent on the physical characteristics of the metal oxides present. The absence of preferential vaporization effects with backsurface ablation was cited as an advantage of the technique and was shown to improve accuracy, albeit with the assistance of an internal standard. The method of coating the sample as a thin film on a substrate has also been applied to achieve analyte preconcentration.[145]

Metals were preconcentrated by ion exchange deposition on APDC-polystyene films, that were about 5 microns thick and cast on glass plates. The films were directly sampled using a Nd:YAG laser operating at 266 nm and ablated into the ICP before atomic emission detection was undertaken. It was observed that the film was removed completely from the glass substrate at the sampling spot. The advantage of the approach for isolating the analyte from the matrix was demonstrated in the determination of copper in a high salt matrix. Electrothermal vaporization has been used within an ablation cell to carry out a sample pretreatment function, before laser sampling of polymers and paints.[146] Emission signals from the ICP were detected using a multichannel spectrometer and were applied to qualitative analysis of the materials. However, because traceable reference materials are not widely available, it was not possible to test the quantitative performance of this system directly.

Electrothermal vaporization can also be used to introduce samples directly into the ICP, as either solid or liquid microsamples, as in AAS. The advantage of this approach is that the sample can be pyrolyzed in situ, thus allowing the phased release of analyte and matrix components under controlled temperature conditions, into the ICP. Since this technique involves the total consumption of the sample, unlike conventional nebulization, the sensitivity is extremely high. This can be useful in the determination of elements at trace levels or in the determination of elements that exhibit relatively poor sensitivity in the ICP, such as nonmetals. For example, a graphite furnace has been used to vaporize rubber samples directly into an ICP-AES instrument.[147] Sample weights of about 1 mg were used. The 182.0 nm line was employed for the analysis of rubber samples containing about 1% sulfur.

Chromatographic Separation

The use of chromatography as a "coupled" method of sample introduction for atomic spectrometry has been a growth area over the last decade. The benefits derived from the ability to achieve highly sensitive, selective, and quantitative measurement of separated chemical compounds and ionic species via atomic spectroscopic detection are now beginning to be realized. This field is generally referred to in the literature as "speciation" and is a subset of the wider "elemental analysis" activity. A number of examples of the use of coupled chromatography-AES for the characterization of polymer systems have been described. For example, the speciation and quantification of lower molecular weight fractions of poly(dimethylsiloxanes) by coupling directly high-performance liquid chromatography (HPLC) to ICP-AES has been investigated.[148] A reversed phase separation of low molecular mass silanols was carried out using acetonitrile as the mobile phase. Size exclusion chromatography was applied in conjunction with ICP-AES to the detection of nonpolar high molecular weight species using either tetrahydrofuran or xylene as eluant. In both cases, ICP-AES was used to detect silicon in the separated components. In a later conference report, the application of this technology to the detection of heteroatoms and metal-tagged functional groups in polymers, and in degradation studies were described.[149] The speciation of polyphosphate oligomers by gradient-elution HPLC coupled to a direct current plasma (DCP) atomic emission spectrometer has also been reported.[150]

Gas chromatography (GC) can also be coupled to plasma atomic emission spectrometers. The microwave plasma (MIP) is particularly suited as an AES detector for GC as it readily accepts gaseous and volatile samples and offers a high excitation potential that is particularly useful for the determination of metalloids and nonmetals (e.g., carbon, hydrogen, nitrogen, oxygen, sulfur, and silicon). Thus, a GC-MIP-AES system has been applied in the characterization of polymer pyrolysates.[151] Both silicone and poly(ethylene pyrolysate)s were examined, and it was found that the calculation of empirical formulas from element ratio data gave errors of less than 2% relative. Similarly, a GC-MIP-AES method for the measurement of low molecular weight silicones in poly(dimethylsiloxane) oil has been described.[152] The GC-MIP-AES approach has also been applied to the detection of acrylamide polymerized into a styrene butadiene polymer.[153] The samples were pyrolyzed, trapped, separated by GC, and detected by atomic emission. The performance of the technique was compared with conventional GC and GC-MS techniques for this application. The pyrolysis GC-MIP-AES technique has also been applied to the detection of fumaric acid and itaconic acid monomers as minor components of emulsion polymers.[154] It was necessary to derivatize the acids with primary amines to achieve detection as elemental nitrogen species by GC-MIP-AES. The data were combined with those from GC-MS to elucidate the structure of the pyrolysis products. The thermal degradation of a flame-retardant brominated polycarbonate has been studied using analytical pyrolysis techniques, including GC-MIP-AES.[155] The evolution of hydrogen bromide and antimony synergist species such as the tribromide was monitored as part of an investigation of the mechanism of polymer degradation and of the flame-retarding process.

There has been a significant growth in the design and use of polymers for the extraction and preconcentration of metals from solution, and ICP-AES has had a significant role to play in such studies, in both the development and application phases. The advantage of ICP-AES in such applications arises from the ability of the technique to carry out multielement analysis. Consequently, multielement preconcentration regimes have been explored.

The use of dithiocarbamate-loaded polyurethane foam has been described for solid phase column extraction of trace metals.[156] Arsenic, bismuth, mercury, antimony, selenium, and tin were quantitatively extracted from water and methanol elutes and determined by ICP-AES. A poly(dithiocarbamate) resin has been used with ICP-AES for the separation and preconcentration of chromium (III) and chromium (VI) species.[157] In a similar vein, a poly(dithiocarbamate) resin for the preconcentration of platinum group metals has been evaluated.[158] Macroporous poly(vinyl thioproprionamide) chelating resin has been used for the preconcentration of noble metals.[159] It was found that even after repeated usage, analytical recoveries over 92% were obtained. In another study, a macroporous poly(vinylamidine thiocyanate thiourea) was used for the preconcentration of beryllium, copper, chromium, lanthanum, titanium, vanadium, and yttrium.[160] The use of poly(ethylene oxide) (PEO) phases for extraction of metals has been explored.[161] The removal of barium and lead divalent ions from highly acidic media was achieved very effectively using a combination of PEO and trichloroacetic acid. Additional studies indicated that polystyrene beads coated with PEO could be used for the quantitative extraction of copper iron, indium gold, and thallium divalent ions from acid solutions. A sodium acrylate maleic anhydride copolymer doped with diethylenetriamine has been used for the adsorption of gold, ruthenium, bismuth, and mercury ions.[162] The capacity of the materials under variable conditions of temperature, acidity, and interference were studied using ICP-AES. The application of a polyacrylacylisothiourea-chelating fiber for the preconcentration of gold, platinum, and ruthenium ions has also been described.[163] An investigation of the capacity and reuse of the material as a function of parameters such as acidity and flow rate was conducted, using ICP-AES as a diagnostic tool. Similarly, polyacrylonitrile has been used in the preconcentration of gold.[164]

The synthesis and characterization of a chelating ion exchange resin based on carboxymethylated poly(ethylene imine)-poly(methylene) polyphenylene isocyanate for use in preconcentration and separation methods in ICP-AES has been described.[165] The polymer was synthesized from the reaction of a polyamine-polyurea polymer with chloroacetic acid. The system was applied to the preconcentration of trace metals from saltwater. A polyamine polyurea chelating fiber may be of interest in this context.[166] A polyacrylamidoxime resin has also been used for the separation and simultaneous preconcentration of iron (III), copper, cadmium, lead, and zinc (II) from waters.[167]

Multielement techniques such as ICP-AES are particularly useful in the study of specific migration of trace elements from food packaging material.[168] Acetic acid extracts (4% v/v) were prepared by exposing foodware samples according to an Association of Official Analytical Chemists (AOAC) protocol for 24 h at room temperature. The determination of 18 elements in these food simulant solutions was carried out by ICP-AES, although lead was not determined because of poor sensitivity for this element. However, ICP-MS offers the same levels of multielement capability as ICP-AES, but with significantly better sensitivity for such applications.

The area of metal-polymer composites has been the subject of research activity, and ICP-AES has been used in the characterization of heterogeneous materials of this type. For example, the metal cluster loading of microphase-separated films of block copolymers of methyltetracyclododecene and 2-norbornene-5,6-dicarboxylic acid has been described.[169] Thus quantities of manganese and terbium doped zinc sulfide nanoclusters were introduced into the carboxylic acid that contained microdomains of the polymer. The levels of impurity dopant ions in these optically active nanoclusters were determined by ICP-AES. The chemical deposition of palladium into polyaniline and polypyrrole has been studied using ICP-AES as a diagnostic technique.[170] Such materials can be used in catalytic hydrogenation processes. The rate of uptake of the metal by the polymers and the composition before and after its use in catalysis was investigated by ICP-AES in conjunction with surface analysis techniques. The investigation of the magnetic properties produced by the oxidative polymerization of m-phenylenediamine has been described.[171] Using ICP-AES, it was established that the synthesized polymers contained traces of transition metals, particularly iron and nickel. The materials exhibited ferromagnetic behavior.

The gas permeability of polymer membranes has been the subject of investigation.[172] Poly(dimethyl phenylene oxide) membranes containing metals were used to enhance olefin transport through the structure. The inorganic compositions of the membranes were characterized using ICP-AES and X-ray diffraction. The properties of polymer-bound mixed-ligand complexes of europium have been studied by ICP-AES, elemental analysis, and X-ray photoelectron spectroscopy.[173] It was found that the polymer/mixed-ligand complexes were strongly fluorescent at ambient temperatures.

The use of polyaniline and poly-o-phenetidine spin-coated films as surface coatings for metals has been explored.[174] The corrosion protection properties of these films on copper and silver surfaces were monitored by ICP-AES in a water-based electrolyte. It was found that corrosion could be either enhanced or reduced, depending on both the monomer and film-preparation method employed. The characterization of colloidal palladium catalysts on poly(vinylpyrrolidine) polymer supports by ICP-AES in conjunction with a range of surface and microscopy techniques may also be of interest.[175] A method has been described for the extraction of iron from poly(phenylene oxide) materials synthesized using a Pruitt-Baggett catalyst.[176] It was found by ICP-AES that the iron could be gradually removed on exposure to 3M hydrochloric acid and that the iron-extractable content was related to the molecular weight of the polymer.

The preparation of Amberlite XAD-2 resins solvent impregnated with functionalized phosphinic acid from aqueous solution has been described.[177] The phosphorus content of the aqueous phase was determined as a function of pH by ICP-AES to allow the solvent loading of the resin phase to be estimated.

X-Ray Fluorescence Spectrometry

X-ray fluorescence spectrometry (XRF) is widely used in the elemental analysis of polymer systems.[178–180] The principle of the technique is that an X-ray source (such as an X-ray tube or radioisotope) is used to excite inner electron transitions within analyte atoms.[181] The replacement of the inner electron (from K, L, or M shells of the atom) by another from a higher energy level results in the secondary emission (fluorescence) of an X-ray photon characteristic of the element to be determined. The fluorescence derived from this process is denoted by the atomic shell involved (K, L, M). The technique is inherently multielement, and with a broadband excitation source, such as an X-ray tube, most elements in the periodic table will be excited to some extent. The result of this process is the generation of an X-ray spectrum that contains lines corresponding to specific atomic transitions (e.g., K_α, K_β, L_α, etc). The induced secondary X-rays can be detected using instrumentation based on either wavelength-dispersive (WDXRF) or energy-dispersive (EDXRF) principles. In the wavelength-dispersive approach, the X-ray photons emanating from the sample are collimated to restrict the angle of incidence and separated, in terms of wavelength, by the use of a diffraction crystal. Simultaneous wavelength-dispersive instruments incorporate a number of fixed-channel monochromators, each dedicated to the isolation of a single X-ray line. Sequential spec-

trometers use a device known as a *goniometer* to rotate the crystal to produce the effect of a wavelength scan, as seen by a single X-ray detector situated at a fixed position (see Figure 3.9a). Both scintillation counters and gas proportional flow counters are commonly used for the detection of the isolated X-ray photons in wavelength-dispersive instruments. In the energy-dispersive mode (Figure 3.9b) there is no pre-separation of X-ray photons in terms of wavelength. Instead, a solid state detector [e.g., either Si(Li) or Ge(Li) lithium-doped semiconductor] is used to distinguish directly between the incoming X-ray photons from the sample on the basis of energy. The energy of the incoming secondary photon from the sample can be directly related to the characteristic electronic transition in the atom that gave rise to it. The resolution of such a system is therefore critically dependent on the ability of the detector to perform pulse-height energy discrimination. Both types of XRF instrument are widely used for the analysis of polymers. A typical XRF spectrum of a polymer is shown in Figure 3.10. Calibration is usually achieved by comparison of sample intensities for the analyte elements to the response for standard material used for calibration purposes. It is also possible to use XRF in a screening mode, to achieve a qualitative analysis or semiquantitative analysis based on the use of theoretical and instrumental response factors.[182] In a practical situation, this method of rapidly identifying the presence of unknowns is particularly valuable. For example, the elemental content of plastics used in food containers requires scrutiny both at the point of production and in the marketplace.[78] The presence of toxic elements such as lead, mercury, and selenium was established by XRF examination of plastic food containers purchased in supermarkets.

Polymer systems vary greatly in terms of composition. It is important to recognize the effect of the sample matrix on the observed X-ray intensity. The extent of penetration of X-rays through the sample depends on the mass absorption coefficient for the sample matrix. Critical depth calculations can be performed to establish the appropriate physical sampling range (the depth in the sample beyond which no secondary line radiation emerges) for a known matrix composition and for a given analyte. Light elements generally tend to be more dominated by fluorescence from the surface layers of a sample because the energy of the X-rays produced are relatively low and therefore easily attenuated by the bulk matrix. The heavier elements tend to generate more energetic X-rays that pass through the sample matrix more easily, and thus secondary X-rays from these elements are observed from a greater depth in the sample. Consequently, from an analytical standpoint, the sampling volume in XRF varies for the same sample from analyte to analyte. The extent of this effect will depend on the element itself and the potential of the matrix elements for absorbing and scattering radiation. Unadulterated polymers generally tend to

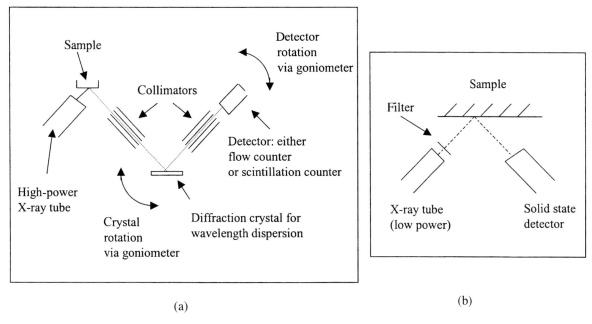

Figure 3.9 Schematic diagram of (a) wavelength-dispersive XRF spectrometer and (b) energy-dispersive XRF spectrometer.

be composed of light elements such as carbon, hydrogen, nitrogen, and oxygen, which do not strongly absorb X-rays. The effective critical sampling depth can be of the order of centimeters in such cases. Polymeric materials containing heavy fillers such as barium sulfate (often used in practical applications as X-ray shielding material) will significantly absorb X-rays, and the observed fluorescence intensity will be less than in an unfilled sample. An interference effect on the XRF signal response will therefore be observed. It may be necessary to use matrix-matched standards to correct for any changes in XRF response that may result from such effects.

However, polymer elemental reference materials are not widely available and are usually generated in-house for the purposes of quantitative XRF measurement. This is generally achieved by making use of variants in production-grade material to achieve a calibration range and then analyzing the homogenized material by a reference technique such as AAS, ICP-AES, or neutron activation analysis (NAA). It is possible to base the XRF calibration on these independently derived concentrations. Nevertheless, the mode of presentation of the sample to the XRF can result in a significant variation in analytical performance. Hot press molding of polymer discs is the preferred method to achieve both a homogenous material and consistency in the response of samples and standards. Alternative approaches, which can be viable in some circumstances, include the direct examination of the sample in granule or powder form.[183] It is obvious, but nevertheless worth pointing out, that the XRF response from a pressed disc, powder, or gran-

ulated material will be different and that the use of standards in one form (e.g., a disc) to calibrate the response from another form (e.g., granules) is inappropriate. The packing density of powder or granulated products is the critical parameter in controlling accuracy and precision of these approaches. Consideration has to be given to the changes in XRF response that arise from the void volume in such samples and to how the sample can be consistently presented to the instrument to avoid significant intensity variations that affect both accuracy and precision. A further factor to be taken into account concerns the orientation of the sample with respect to the excitation source. In some XRF instruments, the sample is excited from below the sample cup; in others, the sample is effectively inverted. In the case of pressed discs, the excitation geometry is relatively unimportant, all other things being equal, because the disc is physically impervious to changes in orientation and to externally imposed conditions (e.g., vacuum or helium path). However, both granules and powders have to be constrained in the cup by using a secondary medium such as polypropylene or poly(ethylene terephthalate) film. The thickness of this film will be of the order of a few microns, to reduce X-ray attenuation and scattered background, and to minimize "blank" signals arising from impurities such as catalyst residues present in the film, which contributes to the analytical signal. The film will be subject to the physical stress of containing the material, whether granulated or powdered, or even solutions or suspensions, and the film will tend to deform under the influence of gravity. This will have the effect of altering the distance between

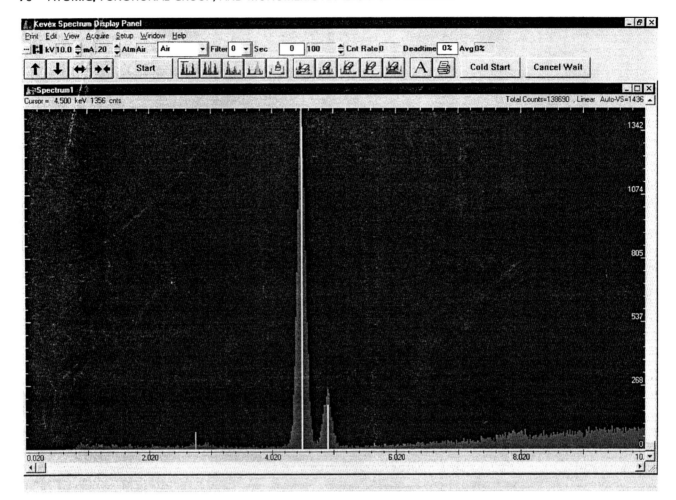

Figure 3.10 Energy-dispersive X-ray fluorescence spectrum of polypropylene, showing catalyst residue titanium K lines.

the excitation source and the sample and will give rise to variations in signal from sample to sample. Consequently, whether such methodology will be appropriate in any particular application will depend to a large extent on the variation that can be accepted in the result.

The preparation of standards for the XRF analysis of polymers has therefore been the subject of investigation. A general approach to the use of calibration standards in XRF has been outlined for a range of materials, including synthetic plastics.[184] More detailed practical instructions for the preparation of homogenous amino-plastic XRF standards have been described.[185] Before they were used, the samples were doped with light and heavy elements and homogenized. A standardization procedure for the XRF analysis of rubber samples has been proposed.[186] The quantification was achieved by using cadmium (^{109}Cd) and iron (^{55}Fe) radioisotope excitation sources. Calibration of the instrument response was based on the use of a minimum set of steps, using single-element standards and stable compounds, as well as reference materials. The calibrations thus derived were used in conjunction with matrix-correction procedures based on fundamental parameters to derive accurate

results on particular samples. The validity of this approach was established by reference to comparative results that were achieved through analysis of neutron activation. It is possible to create XRF calibration standards by the fabrication of synthetic thin films. The attraction of this approach is that the thickness of the standard is such that matrix absorption effects and background scatter can be minimized, thus allowing a broader range of application of the material. For example, the preparation of polymer film XRF standards for the determination of 18 elements in polymer film has been described.[187] The standard materials were prepared in sets of nine, containing pairs of elements chosen to avoid spectral line interference. The accuracy of the standards thus produced was in the range of 2%–9% for 15 of the elements investigated. It was reported that these standard films were stable for more than 1 year, if they were stored at ambient temperatures. In a related application, a fundamental parameters technique has been employed in the analysis of single- and multilayer thin films.[188] Pure element bulk standards were used as the basis for quantification. Matrix effects including primary and secondary fluorescence from the elements in all layers in the film were automatically corrected by using this approach. No previous information about the thickness of the unknown sample was required to employ this method. Results obtained in the analysis of the layered films using the method were reported to be in agreement with data obtained using AAS. The use of mathematical modeling software to achieve standardless calibration in the determination of cations in ion exchange resin-impregnated membranes may also be of interest here.[189]

It is worth noting that XRF can also be used directly in conjunction with a chemical sample preparation step, as in AAS or ICP-AES, thus avoiding the need for polymer standard materials. For example, a method has been proposed for the simultaneous determination of heavy metals in plastics by fusion of the sample with sodium hydroxide in the presence of sodium nitrate as an auxiliary oxidant.[190] Sodium diethyldithiocarbamate and sodium rhodizonate were used to precipitate antimony, arsenic, barium, cadmium, chromium, lead, and mercury at pH 8.5 using iron (III) as a carrier. The precipitate was collected on a filter paper and examined by XRF. This method was shown to be rapid and to give results that were comparable to those obtained for the recommended decomposition procedures for individual elements by AAS.

However, the destruction of the sample results in the loss of information about the distribution of elements present and the physical form of the sample. It should be noted that XRF may be used for physical measurements such as the thickness of coatings and estimation of color of plastics.[191] Examples of this include the nondestructive measurement of silicone coatings on paper and the determination of coating weight.[192] Such measurements can be achieved either by making use of the uniformity of the coating and a known or calculated depth of X-ray penetration, or by calibration of a standard set of materials of known coating weight and thickness using an external reference method. In either case, once the sample response is modeled, it is possible to obtain accurate information on physical parameters.

X-ray fluorescence spectrometry is a technique that is used in industrial laboratories for the purposes of quality and process control.[193,194] This is partly because of the advent of reliable and relatively inexpensive energy-dispersive instruments that can be fully automated. Thus, EDXRF instruments have found increasing application in areas such as the quality control of engineering plastic. Examples—including the determination of stabilizer, filler, pigment, and flame retardant—in polymer have been reported for a commercially available system.[195] It is claimed that such methods are more easily carried out by plant personnel than traditional chemical methods. A related article outlined the use of XRF for the on-plant determination of PVC polymer as chlorine, stabilizer as lead, and filler as calcium.[196]

Other examples of the use of commercial XRF instrumentation in industrial applications include the determination of aluminum, magnesium, and titanium, all of which are key catalyst elements in polypropylene, and levels of antiblock in polyethylene by estimation of silicon levels.[194] The use of XRF in the rubber industry is well established. The combination of X-ray diffraction and X-ray fluorescence spectrometry has been shown to provide an accurate and reliable approach to the determination of the filler composition of rubber formulations.[197] The methods were shown to offer particular advantages in the areas of positive identification and simplicity of sample preparation. X-ray fluorescence offers particular advantages for the determination of halogens where methods based on chemical sample preparation are laborious. For example, the identification and detection of chlorine in chlorine-containing rubbers by XRF has been described.[198]

It is often necessary to make use of wavelength-dispersive instrumentation to achieve the best sensitivity for light elements. Wavelength-dispersive spectrometers that employ multilayer crystals are increasingly being used to detect elements of low atomic number. Thus, analytical performance data have been published for the determination of fluorine in fluoropolymers.[199]

One of the principle advantages of XRF compared with other spectrometric techniques for the determination of elemental composition is that it can be used nondestructively. Consequently, it can be used for the direct analysis of components. For example, the determination of bromine-containing flame retardants in electronic components has been described using the technique.[200] Thus, plastics and circuit boards were examined directly using EDXRF, allowing the rapid identification and accurate quantification of halogen-containing components. In another example, XRF has been used to determine the gadolinium content of poly(methyl methacrylate) scintillators.[201] A limit of detection of 2 $\mu g\ g^{-1}$ gadolinium was reported. A general article on the benefits of controlling component quality using analytical techniques including XRF may be of interest in this context.[202] Warren[203] has compared XRF with other nondestructive techniques such as neutron absorption and scattering, gamma ray absorption, and ultrasonic methods for the determination of fiber and mineral content in engineering plastics and composites. The applications concerned the estimation of major components in materials formulation, and comments were made regarding the feasibility of EDXRF for operation in a production environment.

The advantages of the on-line application of XRF in polymer blending have been briefly described in a wider review of the subject.[204] The application of XRF for the on-line determination of a lubricant in polymer processing has been described.[205] An energy-dispersive instrument incorporating a radioisotope source was used for the measurement of calcium in the form of stearate in pelletized polymer via a slurry cell. Calcium was detected in a counting time of around 600s in the range of 80 to 150 ppm. A patent has described an on-line XRF system for the determination of sulfur in rubber.[206] The analyzer head enclosure was flushed with helium, and, using an immersion approach, the system was used to examine a moving sheet of rubber. There is an increasing demand for techniques for identifying materials from plastic wastes in connection with recycling. This sort of application requires that the analysis can be carried out very rapidly, and preferably nondestructively, with adequate sensitivity to provide a real-time decision as part of an automated process. X-ray fluorescence is a particularly useful technique for the identification of chlorine-containing plastics (e.g., PVC) in the presence of hydrocarbon-based polymers.[207] This type of approach may be used as part of a physical separation process such as bottle sorting, where the samples are passed down a chute past an X-ray analyzer head and then diverted, depending on the chlorine level found.

The last decade has seen the advent of a new mode of XRF operation based on total reflection geometry. The technique offers substantially improved detection limits in comparison to conventional XRF, due to the fact that the background signal generated is very small because the angle of incidence of source radiation is limited to a critical angle of a few degrees. Total reflection X-ray fluorescence spectrometry (TXRF) confines the X-rays to a relatively thin portion of the sample surface and thereby obviates many of the problems associated with matrix absorption and scattering of radiation that are encountered in the mainstream configuration. The application of TXRF in forensic science has been the subject of a review.[208] This article included an assessment of the technique for the examination of plastics, fibers, and inks. A further review, which included the use of TXRF for the analysis of polymers, may also make interesting reading.[209] Sensitivity limitations of XRF may be overcome in certain applications by the use of total reflection geometry; however, it may be less easy to obtain quantitative data because of the requirement for a flat sample and due to the sensitivity of the technique to surface effects. Nevertheless, TXRF has been used for the qualitative examination of acrylonitrile-butadiene-styrene (ABS) resins for catalyst residues and flame retardants, and of a range of first-row transition metals in pigmented materials.[210] Metal fillers containing chromium, iron, and nickel were detected in electro-conducting polymers, indicating the possible use of stainless steel filters. The analysis of thin metal films on acrylic substrates has been carried out using a TXRF instrument with a variable angle configuration.[211] It was demonstrated that this technique could also be used to estimate the thickness of the films, based on establishing the zero angle via an X-ray interference signal and calculation using a difference method.

The analysis of thin films of paint is often carried out in situ using portable XRF instruments.[212] A field method for the determination of Pb in paint using a portable XRF instrument has been described.[213] Excitation was provided using a cobalt radioisotope source and the energy dispersive spectrometer was fitted with a high resolution nitrogen-cooled Si(Li) detector. Results obtained using the system correlated well with those obtained using a low resolution instrument. The detection of lead in painted military structures has been the subject of investigation using XRF.[214] The influence of substrate matrix effects on analyte intensities was overcome by the use of a standardization approach. The equipment was able to detect lead-based paint that was covered by layers of lead-free paint. An improved method based on portable XRF has been described for the analysis of lead in paints.[215] This method relies on the use of lead L-shell fluorescence rather than the more conventional K-shell X-rays that have been employed previously to avoid matrix interferences. It was found that when using lead L-shell fluorescence, matrix substrate effects could be removed by ratioing mass absorption coefficients for the L and L lines, but with better sensitivity than the K-line method.

Polymeric materials are increasingly being used in all atomic spectrometry applications for preconcentration of trace elements. Since XRF is one of the most used techniques for polymer characterization, the use of sorbents and polymeric-chelating agents to improve sensitivity is a natural development of the technique. For example, uranium has been detected by XRF after it was preconcentrated on powdered polyurethane foam.[216] Uranium at part per billion (ppb) levels was sorbed as the salicylate complex on polyurethane foam at pH 4.0. The interference level of various ions and ligands was studied, and optimum conditions were developed to determine uranium in reference materials, wastewater, mine drainage, and seawater. In a similar vein, a method for the direct determination of gallium on polyurethane foam has been described.[217] Gallium chloride was extracted from 6M hydrochloric acid onto polyether-type polyurethane foam. The method was applied to the determination of gallium in aluminum bauxite and industrial residue samples. X-ray fluorescence spectrometry was also used to determine gold absorbed onto a polyurethane foam disc.[218] It was reported that traces of water affected the intensity of the Compton scatter peak. The limit of detection for the method was given as 0.4 g, and the linear range of response was 0 to 300 g. The use of poly(styrene)–azothiazone-2,4-dithione as a chelating sorbent for XRF has also been reported.[219] The interaction of the natural marine polymer chitin and its deacetylated derivative chitosan with chromium has been investigated using EDXRF and X-ray photoelectron spectroscopy.[220] The effects of parameters such as concentration, agitation, particle size, pH, and anion present on the uptake isotherms were also studied. The X-ray techniques were used to characterize the chitin and the deacylated derivative before and after equilibration with chromium.

X-ray fluorescence spectrometry may be of particular use in following the progress of polymerization reactions where the monomer contains functionality from atoms outside the traditional hydrocarbon constituency. The determination of comonomer content is one such area of application. Examples of this use of XRF in application to the determination of comonomer content of tetrafluoroethylene/perfluoro-(propylvinyl ether) copolymers[221] and the direct estimation of sulfur content in copolymer powders[222] have been published. The quantitation of sulfur-containing stabilizer groups built into PVC has been carried out by XRF.[223] It was reported that the adoption of this method avoided the need for labeling procedures and allowed the mechanism of PVC stabilization to be investigated. The incompleteness of the substitution of labile chlorines was reported to be a significant factor in the performance of tin stabilizers. The polymerization of poly(butyl cyanoacrylate) has been studied using XRF.[224] The reaction was initiated in tetrahydrofuran by triphenylphosphine and terminated by sulfur trioxide, and XRF was used to monitor the presence of sulfur and phosphorus in the polymer. It was found that these elements were in near equivalent proportions, but that some of the sulfur could be removed by anion exchange. It was inferred from the results of the investigation that all polymer chains initially had phosphonium groups, and that in the absence of any sulfuric acid formed by hydrolysis were macrozwitterions with terminal sulfite ions. The surface properties of a poly(vinyl acetate) (PVA) film grafted with a polyorganophosphazene that bore allylic functions were examined by using a number of techniques, including attenuated total reflection infrared spectrometry, XRF, and X-ray photoelectron spectroscopy.[225] It was found that the presence of covalently bound polyorganophosphazene on the surface of the film provided it with a hydrophobic character that enhanced its oxygen barrier properties. In a similar study, XRF was used in combination with infrared spectrometry in the study of exterior protective coatings for construction materials.[226] Water toluene extraction analysis and accelerated staining tests were developed and applied to the investigation of the performance of fluoro-resin-, acrylic-silicone-, and polyurethane-based coatings. The transport kinetics and diffusion of zinc chloride and lithium bromide into nylon-6 have been investigated by EDXRF.[227] Polymers were soaked in solutions of known concentration for varying lengths of time, and the concentration profiles of the metal ions were studied. A cross-sectional X-ray analysis of the materials revealed the extent of penetration of the salts into the polymer from solution. Plasma modification is often used to treat polymers and fibersto achieve different surface properties for particular applications. The combination of XRF, Fourier transform infrared (FTIR) spectrometry, and electron microscopy has been shown to be valuable in the investigation of the flame retardancy of fibers.[228] The deposition of fire-retardant monomers such as hexamethylenedisiloxane, ethyldichlorophosphate, and *tris*(butoxyethyl) phosphate was studied in a cold plasma polymerization process. Flame retardancy was examined using the oxygen index method.

The blister resistance of fiberglass-reinforced polyester composites used as marine laminates has been studied using XRF.[229] Materials that contain orthophthalic, isophthalic, and vinyl polyester resins and fiberglass were tested by immersion in water at 65 °C for up to 20 days. The blister initiation time, blister size, and blister density were estimated, and the levels of metal ions in the gel coats and laminated resins were determined by XRF, thus allowing a mechanism for short-term blistering to be developed.

In a similar, but broader, study, XRF, AAS, and gel permeation chromatography (GPC) were used to investigate disc crack formation and blistering in polyester and epoxy laminates.[230] Several hundred 4-ply chopped-strand mat laminates obtained by hand lay-up, using orthophthalic laminating resins and isophthalic gel coats, were immersed in distilled water at 45 °C and ambient temperature until blisters had developed. It was observed that cracks were more often encountered in polyester than in epoxy laminates.

The metal-polymer interface is of particular interest in many industrial applications. The bonding of natural rubber to brass-plated steel cord in steel-belted radial tires has been the subject of investigation using a combination of techniques, including FTIR, EDXRF, and AAS.[231] The inorganic techniques were used to examine the effectiveness of cobalt naphthenate to promote bonding and to enhance durability.

X-ray spectrometry has a niche application in the biological systems, because of its ability to provide spatially specific and nondestructive information about the level and distribution of elements in a complex environment. For example, EDXRF has been used in conjunction with scanning electron microscopy to study the surface encrustation of biomaterials in a simulated physiological environment.[110] It was reported that silicone was less susceptible to hydroxyapatite and struvite encrustation than were polyurethane biomaterials in a continuous flow model based on a Robbins device. X-ray analysis and FTIR have been used in combination to study the surface chemistry of microbially treated rubbers.[232] The microorganisms used were selected for biological activity toward sulfur, with a view to improve the compounding characteristics in virgin rubber stocks. It was established that the sulfur in the rubber was oxidized, as anticipated from a biodegradation model. This was confirmed by ion chromatography, which indicated the release of sulfur in the sulfate ion form. The effect of the porosity of Biomer-segmented polyurethane on tissue growth and calcination in rats has been investigated by X-ray analysis, scanning electron microscopy (SEM), and AAS.[114] It was found that tissue ingrowth into the material depended on porosity and that mineralization required an initial calcium site in the material.

Inorganic Mass Spectrometry

Inductively Coupled Plasma Mass Spectrometry

Inorganic mass spectrometry is a relatively recent addition to the armory of techniques used for the elemental characterization of polymers. The commercial introduction of inductively coupled plasma mass spectrometry (ICP-MS), in 1983, led to the widespread application of the technique for ultra trace level elemental analysis.[233] The coupling of the inductively coupled plasma (ICP) ion source, familiar from atomic emission spectrometry, to a quadrupole mass spectrometer, provided analytical advantages in terms of sensitivity (typically subparts per billion), speed of analysis, selectivity, and multielement capability over other approaches that involved optical atomic spectrometry. In practice, ICP-MS is generally used in applications where the sensitivity of ICP-AES is insufficient or where ETAAS may have been used in the past.

As with ICP-AES and AAS, it is necessary to chemically prepare solid polymer samples before they are analyzed by ICP-MS if using a conventional liquid sample introduction system such as a nebulizer. For example, palladium has been determined in a polymer matrix by ICP-MS using a sulfuric acid–hydrogen peroxide decomposition, not dissimilar to that which might be used in AAS or ICP-AES.[234] However, rhodium was added as an internal standard in order to compensate for changes in instrument response, and this is a practice both easily achieved and widely adopted in ICP-MS analysis. A detection limit of 5 ng/kg palladium was reported, giving some indication of the sensitivity of the technique. The combination of microwave digestion and ICP-MS is particularly attractive for the analysis of polymers.[4,235] This is because the better sensitivity of ICP-MS permits the analysis of smaller samples and the use of greater dilution factors, without impairing the quality of the analytical result.

However, ICP-MS offers specific advantages over competitive techniques such as AAS and ICP-AES. For example, ICP-MS can be used to detect halogens, which are difficult to detect at trace levels using optical techniques.[236] Samples were fused, using an excess of sodium carbonate, before bromine, chlorine, and iodine were measured. The resulting fluxes were dissolved and analyzed by ICP-MS. Recoveries consistently of the order of 80% were achieved for these analytes by using this approach to sample preparation.

Inductively coupled plasma mass spectrometry may also be used to determine elemental isotope ratios. For example, the manufacturer of paints that contained mixed-source lead pigments was identified on the basis of ICP-MS lead isotope data.[237] The isotope ratio facility of ICP-MS has also been used to identify disturbed paint surfaces in an investigation of the source of lead exposure in a works environment.[238] Isotope-dilution mass spectrometry has been used

as a high-accuracy reference technique in the certification of cadmium levels in polyethylene.[71,239] Results from the latter study by ID-MS indicated the presence of 40.9, 75.9, 198, and 407 mg/kg of cadmium in the test samples.

The high sensitivity offered by ICP-MS when coupled with sample introduction by electrothermal vaporization has been exploited in forensic examination of single fibers.[240] It was found that the rapid-scanning capabilities of quadrupole ICP-MS allowed the detection of up to six elements simultaneously. A direct injection nebulizer, which permits the total consumption of very small liquid samples, was also evaluated in this investigation but was found to offer insufficient sensitivity because of the large dilution factors that are involved.

Over the last decade, there has been growing interest in the use of lasers for direct solid sampling in ICP-MS. The scanning rate of quadrupole mass spectrometers is such that multielement analysis can be achieved on a single laser ablation event, which is important in the examination of "features" such as streaking in engineering plastics due to inhomogeneous dispersions or molding defects. For example, laser ablation ICP-MS has been used to analyze polymers taken from crime scenes.[241] It was noted that elemental associations could be identified by this technique in samples as small as 50 microns in size. However, for best sensitivity, multisite repetitive ablation or computer-controlled "rastering" can be used to generate a pseudocontinuous signal for bulk analysis. A detailed report has been published on the application of laser ablation ICP-MS to the analysis of a range of filled and unfilled polymers.[242] It was reported that semiquantitative analysis of propylene, nylon, PVC, polyethylene, and polyester could be carried out with an accuracy of a factor of 2 by ratioing to matrix carbon as an internal standard. To achieve reliable quantification, in-house polymer standards were used to calibrate instrument response. The technique generated data that agreed well with those obtained using WDXRF and neutron-activation analysis. The influence of instrumental parameters on the analytical performance that can be achieved by LA-ICP-MS was the subject of a further study.[243] The position of beam focus was found to be the most important parameter for controlling analyte intensity and reproducibility. Using the Nd:YAG laser at a focal point approximately 10 mm above the surface of the polymer sample yielded precision data consistently below 10% relative for analyte levels of the order of a few hundred parts per million. In another study, the determination of trace elements in polymers by ICP-MS was investigated using single-spot, multiple-spot, and raster laser ablation approaches.[244] It was reported that elements could be quantitatively detected in the range of 10 ppb to percent levels by using a combination of analogue and pulse-counting detector modes. Relative standard deviation data of under 5% were found to be achievable under optimum operating conditions. A brief investigation of the use of laser ablation ICP-MS, which was apart of a wider study of the semiquantitative and quantitative elemental analysis of polymers, also indicated that precision in the range of 3% to 10% was achievable.[235]

Glow Discharge Mass Spectrometry

Although glow discharge sources have been widely used in emission spectrometry for more than 30 years, their analytical utility in elemental mass spectrometry has only been exploited in the last decade or so. The glow discharge can be initiated between electrodes in a inert gas such as argon, using either a direct current[245] or a radio frequency power supply.[246] The sample is presented in one of the discharge electrodes and is sputtered into the gas phase by the action of the plasma gas; the ions thus produced are introduced to the inlet of the mass spectrometer. The direct current version of the source has not been widely used for polymer analysis, primarily because of the need for the sample to be conductive in order to sustain the discharge.

Nevertheless, a direct current glow discharge mass spectrometer has been used for the elemental characterization of polymers.[247] This was achieved by the application of a tantalum diaphragm acting as a secondary discharge cathode placed immediately in front of the nonconducting sample. The glow discharge cell was coupled to a double focusing mass spectrometer and was used to study poly(tetrafluoroethylene) (PTFE), polycarbonate, and PVC materials. It was found that discharge-operating conditions of 0.6 kV and 3 mA were satisfactory for the atomization of PTFE and polycarbonate, and signals for carbon, iron, and fluorine ions remained relatively stable over a period of 1 h. The characteristics of the discharge spectrum were found to be predominantly atomic, although some polyatomic cluster ions such as $C_2F_4^+$ were observed. This suggested that the source was potentially useful for quantitative elemental analysis. However, for PVC, discharge instability was observed under these conditions, and, after a time, the samples began to melt, despite the use of liquid nitrogen cooling of the cell. It was found to be necessary to use lower power conditions (1.2 kV, 0.6 mA) to successfully analyze PVC and to minimize the heating effects.

Marcus and co-workers[248] have described the development of a radio frequency glow discharge quadrupole mass spectrometer for the direct characterization of bulk polymers. A series of PTFE-based polymers were studied, using a glow discharge cell that could accommodate flat samples. The effect of discharge parameters on the intensities of polymeric fragment ions observed was studied. Samples were prepared by rinsing with methanol, then air-drying and loading into an appropriate flat sample holder. It was found that side chain structure in the polymers significantly affected the nature of the spectra observed, allowing the opportunity to differentiate between materials. Signal stability was reported to be better than 5% relative, and plasma stabilization times of less than 3 min were obtained. It was noted that the relative contributions by carbon-fluorine polyatomic ion fragments were governed by the relative bias potential of the direct current at the sample surface. Thus, when either high powers or low discharge pressures were applied, larger populations of high mass fractions were observed. It was also noted that since the direct current bias generated on the sample surface was inversely proportional to its thickness, the volume of the sample used could affect the proportion of low mass ions observed. There was no evidence for the presence of thermal volatilization processes, and surface examination of samples by scanning electron microscopy indicated that a sputtering mechanism predominated with this source. The opportunity for depth profiling was assessed by reference to the examination of a 1 micron layer of copper metal on a 1.5 mm thick PTFE substrate. A sputtering rate of 200 nm/min was established for this material.

In a further investigation,[249] a cryogenically cooled sample holder was constructed for use in the GD-MS analysis of polymeric materials. Samples were cooled by passing liquid nitrogen through the body of a brass sample backing plate. It was found that thermally sensitive polymers could be analyzed using this cooled radio frequency glow discharge source without evidence of degradation. The ion signals thus produced were found to exhibit greater temporal stability. The system was applied to the characterization of LDPE and PTFE samples.

Glow discharges are used in plasma polymerization processes, and mass spectrometry may be used in this context for diagnostic purposes. For example, a tetrafluoroethylene polymerization carried out in a glow discharge has been described.[250] Mass spectrometry was used to identify the presence of perfluoroalkane and perfluoroalkene intermediate compounds of various structures. The consumption rates for monomers and intermediate oligomers were obtained and were used to develop equations that described the kinetics of the polymerization process. In another study, kinetic principles were used to postulate the polymerization mechanism in a glow discharge using mass spectral measurement.[251] It was found that the rate equations derived from this process agreed well with previously determined rate constants for tetrafluoroethylene.

References

1. Fairman. B.; Hinds, M. W.; Nelms, S. M.; Penny, D. M.; Goodall, P. Industrial analysis: metals chemicals and advanced materials. *J. Anal. At. Spectrom.* **1999**, *14*, 1937.
2. Anderson, D. G. Coatings. *Anal. Chem.* **1997**, *69*, 15R.
3. Dorner, W. G. Analysis of plastics: VI. Methods of physical trace analysis. *Kunst. Plast.* **1985**, *32*, 23.

4. Marshall, J.; Franks, J., Application of atomic spectrometry to the analysis of advanced materials. *Anal. Proc.* **1990**, *27*, 240.

5. Norris, J. D. The use of selected spectrometric techniques for determining trace elements in paints and pigments. *Polym. Paint Colour J.* **1988**, *76*, 816.

6. Ma, T.S. Organic elemental analysis. *Anal. Chem.* **1988**, *60*, 175R.

7. *ASTM 5291–96*. Standard test method for instrumental determination of carbon hydrogen and nitrogen in petroleum products and lubricants; American Society for Testing and Materials: Philadelphia, 1996.

8. *ASTM D1013–1998*. Standard test method for determining total nitrogen in resins and plastics; American Society for Testing and Materials: Philadelphia, 1998.

9. *ASTM D4629–96*. Standard test method for trace nitrogen in liquid petroleum hydrocarbons by syringe/inlet oxidative combustion and chemiluminescence detection; American Society for Testing and Materials: Philadelphia, 1996.

10. White, D.C. Determination of low levels of sulfur in organics by combustion microcoulometry. *Anal. Chem.* **1977**, *49*, 1615.

11. *ASTM D3120*. Standard test method for trace quantities of sulfur in light liquid petroleum hydrocarbons by oxidative microcoulometry; American Society for Testing and Materials: Philadelphia, 1996.

12. Ladrach, W.; Van de Craats, F.; Gouverneur, P. Determination of traces of chlorine in organic liquids: a combustion–microcoulometric approach. *Anal. Chim. Acta* **1970**, *50*, 219.

13. *ASTM D4661–98*. Standard test methods for polyurethane raw materials: determination of total chlorine in isocyanates; American Society for Testing and Materials: Philadelphia, 1998.

14. Coz, D.; Baranwal, K.; Knowles, T. M. Rubber compound analysis and formula reconstruction; 150th ACS Rubber Division Meeting, fall 1996; Conference Preprints, Ed.; ACS, Rubber Div.: Louisville, Ky., 8–11 October 1996; Paper 68.

15. Welz, B.; Sperling, M. *Atomic Absorption Spectrometry*, 3rd ed.; Wiley-VCH: Weinheim, Germany, 1999.

16. Price, W. J. *Spectrochemical Analysis by Atomic Absorption*; Heyden: London, 1979.

17. Smith, S. B.; Hieftje, G. M. A new background correction method for atomic absorption spectrometry. *Appl. Spectrosc.* **1983**, *37*, 419.

18. Hong, A.; Dong, H.; Yang, X. Measurement of calcium and magnesium contents in medical grade silicone rubber using flame atomic absorption spectroscopy. *China Rubber Ind.* **1991**, *38*, 225.

19. *ASTM D3733*. Test method for silicon content of silicone polymers and silicone modified alkyds by atomic absorption spectroscopy; American Society for Testing and Materials: Philadelphia, 1993.

20. *ASTM D4004*. Test methods for rubber: determination of metal content by flame atomic absorption; American Society for Testing and Materials: Philadelphia, 1993.

21. *International Standard 6101/1-1981*. Rubber: determination of metal content. flame atomic absorption spectrometric method: Part 1. determination of zinc content; ISO: Geneva, 1981.

22. *ASTM D 4085-93* (reapproved 1997). Standard test method for metals in cellulose by atomic absorption spectrometry; American Society for Testing and Materials: Philadelphia, 1997.

23. *ASTM D 3335-85a* (reapproved 1994). Standard test method for low concentrations of lead, cadmium, and cobalt in paint by atomic absorption spectroscopy; American Society for Testing and Materials: Philadelphia, 1994.

24. *DIN 3599, Part 1*. Testing of rubber and elastomers: determination of the lead content, determination by atomic absorption spectroscopy for lead contents up to 0.1%; Deutches Institut für Normung: Berlin, Germany, 1978.

25. British Standards Institution. *Chemical Tests for Raw and Vulcanized Rubber: Part 28. Methods for the Determination of Copper Content*. Section 28.1 Atomic Absorption Spectrometry. British standard BS7164: Section 28.1: 1990 (ISO 6101–3:1988); 31 July 1990.

26. British Standards Institution. *Chemical Tests for Raw and Vulcanized Rubber: Part 26. Methods for the Determination of Manganese Content*. Section 26.1 Atomic Absorption Spectrometry. British standard BS7164: Section 26.1: 1990 (ISO 6101–4:1988); 31 July 1990.

27. Yamane, T. Determination of trace metals in plastics; *Idemitsu Giho* **1993**, *36*, 750.

28. *ASTM D5630*: Standard test method for ash content in thermoplastics; American Society for Testing and Materials: Philadelphia, 1994.

29. Tian, Y. Z.; Li, R. S. Determination of trace chromium in polyolefins by atomic absorption spectrometry; *Shiyou Huagong* **1986**, *15*, 701.

30. Mendiola, J. M.; Lopez, A. G. Atomic absorption spectrophotometry for analysis of tin in PVC compositions. *Rev. Plast. Mod.* **1981**, *41*, 550.

31. Belarra, M. A.; Azofra, C.; Anzano, J.M.; Castillo, J. Determination of metals in poly(vinyl chloride) by atomic absorption spectrometry: Part 3. Determination of cadmium, antimony, and tin in samples of poly(vinyl chloride) with carbon black. *J. Anal. At. Spectrom.* **1988**, *3*, 591.

32. Belarra, M. A.; Gallarta, F.; Anzano, J. M.; Castillo, J. R. Determination of metals in poly(vinyl chloride) by atomic absorption spectrometry: Part 1. Determination of calcium, aluminium and antimony in samples of poly(vinyl chloride) with a high content of alkaline earths. *J. Anal. At. Spectrom.* **1986**, *1*, 141.

33. Belarra, M. A.; Anzano, J. M.; Gallarta, F.; Castillo, J. R. Determination of metals in poly(vinyl chloride) by atomic absorption spectrometry: Part 2. Determination of lead and magnesium in samples of poly(vinyl chloride) with a high content of alkaline earths. *J. Anal. At. Spectrom.* **1987**, *2*, 77.

34. Belarra, M. A.; Azofra, C.; Anzano, J. M.; Castillo, J. R. Determination of metals in poly(vinyl chloride) by atomic absorption spectrometry: Part 4. Determination of lead and aluminium in samples of poly(vinyl chloride) with a high content of silicates. *J. Anal. At. Spectrom.* **1989**, *4*, 101.

35. Grillo, A.; Floyd, T. And they said it couldn't be done by microwave digestion. Presented at the 1990 Pittsburgh Conference and Exposition on Analytical Chemistry, New York, 5–9 March 1990.

36. Moody, J. R.; Lindstrom, R. M. Selection and cleaning of plastic containers for storage of trace element samples. *Anal. Chem.* **1977**, *49*, 2264.

37. Laxen, D. P. H.; Harrison, R. M. Cleaning methods for polythene containers prior to the determination of trace metals in freshwater samples. *Anal. Chem.* **1981**, *53*, 345.

38. Vollrath, A.; Otz, T.; Hohl, C.; Seiler, H. G. Comparison of dissolution procedures for the determination of cadmium and lead in plastics. *Fresenius Z. Anal. Chem.* **1992**, *344*, 269.

39. Alonso, M. F.; Rodriguez, A. G. Sample preparation for heavy element analysis in toy coatings. Presented at the 1990 Pittsburgh Conference and Exposition on Analytical Chemistry, New York, 5–9 March 1990.

40. Mendiola, J. M.; Lopez, A. G. Analysis of cadmium and zinc in PVC compositions by atomic absorption spectrophotometry. *Rev. Plast. Mod.* **1981**, *41*, 413.

41. Belarra, M.; Anzano, J. M.; Castillo, J. R. Determination of metallic additives in poly(vinyl chloride) by atomic absorption spectrometry. Presented at the 26th Colloquium on Spectroscopium Internationale, Sofia, Bulgaria, 2–9 July 1989.

42. Belarra, M. A.; Anzano, J. M.; Castillo, J. R. Determination of lead and cadmium in samples of poly(vinyl chloride) using organic solvents by flame atomic absorption spectrometry. *Fresenius Z. Anal. Chem.* **1989**, *334*, 118.

43. Belarra, M. A.; Anzano, J. M.; Sanchez, A.; Lavilla, I.; Castillo, J. R. Determination of tin in samples of poly(vinyl chloride) by flame and electrothermal atomic absorption spectrometry using organic solvents. *Microchem. J.* **1990**, *41*, 377.

44. Pedras Penalva, F.; Polo Diez, L.; Gil Galindo, C. Analytical applications of oil-water emulsions: determination of zinc in plastics by atomic absorption spectrometry. *Ann. Quim., Ser. B* **1988**, *84*, 252.

45. Lakatos, I. Direct FAES and FAAS analysis of macromolecular solutions. Presented at the Second Hungaro-Italian Symposium on Spectrochemistry, Budapest, 10–14 June 1985.

46. Lakatos, I.; Lakatos, J. Direct determination of trace elements in polyacrylamide solutions by flame atomic absorption spectrometry. *Magy. Kem. Foly.* **1986**, *92*, 343.

47. Yingfen, L.; Suiwei, X. Key to determination of calcium content in rubber vulcanizate by flame atomic absorption spectrometry. *China Rubber Ind.* **1988**, *35*, 36.

48. Zimakova, E. V.; Kareva, N. D. Atomic absorption determination of manganese in polymeric metal complexes and biological materials and organs. *Zavod Lab.* **1996**, *62*, 28.

49. Nir-El, Y. Forensic characteristics of colored polyethylene bags. *J. Forensic Sci.* **1994**, *39*, 758.

50. Forshey, P. A.; Turan, T. S.; Lemmo, J. S.; Cutie, S. S.; Pytynia, D. L. Analysis of sodium polyacrylate absorbent dust using ultra trace sodium analysis: a seven company collaborative study. *Anal. Chim. Acta* **1994**, *298*, 363.

51. Korenaga, T. Atomic absorption spectrophotometric determination of trace amounts of arsenic in acrylic fibers containing antimony oxide with solvent extraction and arsine generation. *Analyst* **1981**, *106*, 40.

52. Lau, O. W.; Li, K. L. Determination of lead and cadmium in paint by atomic absorption spectrophotometry utilizing the Delves microsampling technique. *Analyst* **1975**, *100*, 430.

53. Kannipoor, R.; Van Loon, J. C. Direct determination of rhodium in solid and liquid polymer samples. *Spectrosc. Lett.* **1987**, *20*, 871.

54. Takenaka, M.; Yamada, Y.; Hayashi, M.; Endo, H. Determination of ultra trace metallic impurities in fluorocarbon polymers by electrothermal atomic absorption spectrometry after decomposition with a combustion system. *Anal. Chim. Acta* **1996**, *336*, 151.

55. Sipos, M.; Adamis, T. Determination of tin in organo-tin stabilized poly(vinyl chloride) by graphite furnace atomic absorption spectrometry. *Acta Chim. Hung.* **1991**, *128*, 869.

56. Belarra, M. A.; Anzano, J. M.; Castillo, J. R. Determination of lead and cadmium in samples of poly(vinyl chloride) by electrothermal atomic absorption spectrometry using organic solvents. *Analyst* **1990**, *115*, 955.

57. Wang, L.; Bai, Y. Determination of Sb in the resin and products of PET by graphite furnace atomic absorption spectrometry. *Guangpuxue Yu Guangpu Fenxi* **1998**, *18*, 606–608

58. Krivan, V.; Koch, B. Determination of calcium, copper, iron, potassium sodium, and silicon in polyimides for microelectronics by electrothermal atomic absorption spectrometry involving sample dissolution in organic solvents. *Anal. Chem.* **1995**, *67*, 3148.

59. Parks, E. J.; Brinckman F. E. Characterization of bioactive organotin polymers: fractionation and determination of molecular weight by SEC-GFAA. In *Controlled Release of Pesticides and Pharmaceuticals*, Proceedings of the 7th International Symposium, Fort Lauderdale, 1980; D. H. Lewis, Ed. Plenum Publishing: New York, 1981; p 219.

60. Blinova, E. S.; Guzeev, I. D.; Miskar'yants, V. G. Atomic absorption analysis of solid samples. *Zavod. Lab.* **1988**, *54*, 27.

61. Sanford, C. L.; Thomas, S. E.; Jones, B. T. Portable battery powered tungsten coil atomic absorption spectrometry for lead determinations. *Appl. Spectrosc.* **1996**, *50* 174.

62. Bradshaw, D. K. Analysis of paint chips using slurry sample introduction for graphite furnace AAS. Presented at the Nineteenth Annual Meeting of the Federation of Analytical Chemistry and Spectroscopy Societies (FACSS), Philadelphia, 20–25 September, 1992.

63. Zhaon, D. Determination of lead in paint material of the inside wall by graphite furnace atomic absorption spectroscopy with matrix modifier. *Huanjing Wuran Yu Fangzhi* **1992**, *14*, 35.

64. Belarra, M.A.; Lavilla, I.; Anzano, J. M.; Castillo, J. R. Rapid determination of lead by analysis of solid samples using graphite furnace atomic absorption spectrometry. *J. Anal. At. Spectrom.* **1992**, *7*, 1075.

65. Belarra, M. A.; Belategui, I.; Lavilla, I.; Anzano, J. M.; Castillo, J. R. Screening of antimony in PVC by solid

sampling graphite furnace atomic absorption spectrometry. *Talanta* **1998**, *46*, 1265.

66. Satoh, E.; Yamamoto, Y.; Takyu, T. Determination of ultratrace elements in polycarbonate used as optical materials. *Anal. Sci.* **1991**, *7*, 1645.

67. Kawai, H.; Katayama, Y.; Ninomiya, Y.; Okuda, J. Determination of trace metals in poly(ethylene terephthalate) film by punched out film introduction and graphite furnace AAS. *Bunseki Kagaku* **1994**, *43*, 1193.

68. Koons, R. D. ICP-MS or ETAA: analysis of individual carpet fibers. Presented at the Twenty-first Annual Meeting of the Federation of Analytical Chemistry and Spectroscopy Societies (FACSS). St Louis, 2–7 October 1994.

69. Pauwels, J.; Hofmann, C.; Grobecker, K. H. Homogeneity determination of cadmium in plastic CRM's using solid sampling atomic absorption spectrometry. *Fresenius J. Anal. Chem.* **1993**, *345*, 475.

70. Pauwels, J.; Lamberty, A.; De Bievre, P.; Grobeker, K. H., Baupiess, C. Certified reference materials for the determination of cadmium in polyethylene. *Fresenius. J. Anal. Chem.* **1994**, *349*, 409.

71. Carnrick, G. R.; Lumas, B. K.; Barnett, W. B. The analysis of solid samples by graphite furnace atomic absorption using Zeeman background correction. Presented at the 1985 Pittsburgh Conference and Exposition on Analytical Chemistry, New Orleans, 25 February to 1 March 1985.

72. Carnrick, G. R.; Lumas, B. K.; Barnett, W. B.; Vollkopf, U. The analysis of solid samples by graphite furnace atomic absorption using Zeeman background correction. Presented at the Third Colloquium on Atomspektrometrische Spurenanalytik, Konstanz, Germany, 18–21 March 1985.

73. Guo, R.; Jiang, X. Determination of tin in unplasticized polyvinyl chloride pipes for drinking water supply by Zeeman-effect graphite furnace atomic absorption. *Weisheng Yanjiu* **1994**, *23*, 83.

74. Dominic, M. R.; Koch, C. P.; Mills, J. K. Improved method for detection of organotins in food simulants. *J. Vinyl Technol.* **1993**, *15*, 228.

75. Kaminska, E.; Rabiasz, B.; Wisniewska, A. Determination of barium and cobalt migration from plastics. *Polimery* **1999**, *44*, 288.

76. Kolasa, D.; Samsonowska, K.; Zorawska, K. Studies of chromium speciation using atomic absorption spectrometry in analysis of polyethylene plastics. *Chemik* **1998**, *51*, 100.

77. Maslowska, J.; Czerwinska, H. Studies of low contents of mineral components in PETP films. *Polim. Tworz. Wielk* **1985**, *30*, 31.

78. Meranger J. C.; Cunningham, H. M.; Giroux, A. Extraction of heavy metals from plastic food containers: an X-ray fluorescence and atomic absorption study. *Canad. J. Public Health* **1974**, *65*, 292.

79. Sans, R.; Garcia, J.; Puig, M. D.; Valero, A. Determination of cadmium by atomic and molecular absorption spectrometry: application to aqueous extracts from children's toys. *Rev. Plast. Mod.* **1986**, *51*, 767.

80. Dietz, G. R.; Banzer, J. D.; Millar, E. M. Water extraction of additives from PVC pipe. Presented at SPE, Safety and Health with Plastics, National Technology Conference, Denver, Colorado, November 1997; Confer 921, p 25.

81. Forsyth, D. S.; Meranger, J. C.; Jay, B.; Brule, D. Organotin leachates from chlorinated poly(vinyl chloride) pipe. Presented at the 5th COMTOX Symposium on Toxicology and Clinical Chemistry of Metals, Vancouver, BC, Canada, 10–13 July1995.

82. Iring, M.; Barabas, K.; Foldes, E.; Tudos, F.; Odor, L. Interactions of some liquid chemicals with polyolefins. *Angew. Makromol. Chem.* **1990**, *181*, 129.

83. Hampe, D.; Piringer, O. Studies on the permeation of inorganic salts through plastic films. *Food Addit. Contam.* **1998**, *15*, 209.

84. Uragami, T.; Nakamura, R.; Sugihara, M. Active and selective transport of alkali metal ions through crosslinked polystyrene sulfonic acid membranes. *Makromol. Chem. Rapid Commun.* **1982**, *3*, 467; Studies on Synthesis and Permeabilities of Special Polymer Membranes 45.

85. Simon, J.; Kantor, T.; Kozma, T.; Pungor, E. Thermal analysis of antimony trioxide/organohalide-based flame retardants including atomic absorption detection of the evolved species. *J. Thermal. Anal.* **1982**, *25*, 57.

86. Zhao, B.; Chen, W.; Yan, L.; Li, X.; Zhu, Q.; Lu, F. Comparison of reactive ratios measurement of PTBTM-CO-MMA copolymers. *Zhongguo Kexue Jishu Daxue Xuebao* **1995**, *25*, 65.

87. Ozeroglu, C.; Guney, O.; Mustafafaev, M. I. Oxidative polymerization of acrylamide in the presence of threonine. *Angew. Makromol. Chem.* **1997**, *249*, 1.

88. Rao, K. B.; Gandhi, K. S. Analysis of products of the thickening reaction between polyesters and magnesium oxide. *J. Polym. Sci. Polym. Chem.* **1985**, *23*, 2305.

89. Gereltu, B.; Thorpe, F. G. Mercuriation and iodination of linear copolymers of methyl acrylate and N-vinyl carbazole. *Macromolecular Rep.* **1994**, *31*, 127.

90. Peulon, V.; Barbey, G.; Outurquin, F.; Paulmier, C. Electrochemical synthesis and study of polyselenienylthiophene. *Synthetic Met.* **1993**, *53*, 115.

91. Kronstein, M. Controlled release of polymeric organometal toxicants. *IEC Prod. Res. Rev.* **1981**, *20*, 5.

92. Lekova, V.; Popov, C.; Ivanov, B.; Garwanska, R. Preparation, characterization and application of an electroconductive polymeric pigment with microwave absorption properties. *Fibres Text. East. Eur.* **1998**, *6*, 52.

93. Pickup, N. L.; Shapiro, J. S.; Wong, D. K. Y. Extraction of silver by polypyrrole films upon a base-acid treatment. *Anal. Chim. Acta* **1998**, *364*, 41.

94. Sue, R. K.; Soon, H. K.; Mu, S. J. Semi-interpenetrating network system for a polymer membrane. *Eur. Polym. J.* **1997**, *33*, 1009.

95. Oprea, C. V.; Popa, M. Polymeric materials with magnetic properties. *Polym. Plast. Technol. Engng.* **1993**, *32*, 375.

96. Singh, A. K.; Dhingra, S. K. Application of Dowex-2 loaded with sulfonephthalein dyes to the preconcentration of copper (II) and cadmium (II). *Analyst* **1992**, *117*, 889.

97. Lancaster, H. L.; Marshall, G. D.; Gonzalo, E. R.; Ruzicka, J.; Christian, G. D. Trace metal atomic absorption spectrometric analysis utilizing sorbent extraction on poly-

meric-based supports and renewable reagents. *Analyst* **1994**, *119*, 1459.

98. Arpadjan, S.; Vuchkova, L.; Kostadinova, E. Sorption of arsenic bismuth, mercury, antimony, selenium and tin on dithiocarbamate loaded polyurethane foam as a preconcentration method for their determination in water samples by simultaneous inductively coupled plasma atomic emission spectrometry and electrothermal atomic absorption spectrometry. *Analyst* **1997**, *122*, 243.

99. Brackenbury, K. F. G.; Jones, L.; Koch, K. R. Polyurethane foams as sorbents for noble metals: selective preconcentration of small amounts of platinum from hydrochloric acid containing tin (II) chloride followed by flame atomic absorption analysis. *Analyst* **1987**, *112*, 459.

100. Saxena, R.; Singh, A. K.; Rathore, D. P. S.; Salicylic acid functionalized PS sorbent amberlite XAD-2 synthesis and application as a preconcenrator in the determination of zinc and lead using atomic absorption spectrometry. *Analyst* **1995**, *120*, 403.

101. Yebra-Biurrun, M. C.; Bermejo-Barrera, A.; Bermejo-Barrera, M. P. Application of a poly(dithiocarbamate) resin with macroreticular support to the determination of trace amounts of cadmium and lead in non-saline water. *Analyst* **1991**, *106*, 1033.

102. Agrawal, Y. K.; Rao, K. V. Polyhydroxamic acids: Synthesis, ion exchange separation and AAs determination of divalent metal cations. *React. Polym.* **1995**, *25*, 79.

103. Garbos, S.; Bulska, E.; Hulanicki, A.; Shcherbinina, N. I.; Sedykh, E. M. Preconcentration of inorganic species of antimony by sorption on Polyorgs 31 followed by atomic absorption spectrometry. *Anal. Chim. Acta* **1997**, *342*, 167.

104. Rivas, B. L.; Maturana, H. A.; Molina, M. J.; Gomez-Anton, M. R.; Pierola, I. F. Metal ion binding properties of poly(N-vinylimidazole) hydrogels. *J. Appl. Polym. Sci.* **1998**, *67*, 1109.

105. Demirata-Ozturk, B.; Gumms, G.; Oncul-Koc, A.; Catagil-Giz, H. Preconcentration of copper ion in aqueous phase on methacrylate polymers. *J. Appl. Polym. Sci.* **1996**, *62*, 613.

106. Salih, B.; Denizli, A.; Engin, B.; Tuncel, A.; Piskin, E. Congo red attached poly(EGDMA-HEMA) microspheres as specific sorbents for removal of cadmium ions. *J. Appl. Polym. Sci.* **1996**, *60*, 871.

107. Agrawal, Y. K.; Rao, K. V. Polyhydroxamic acids: synthesis, ion exchange, separation and atomic absorption spectrophotometric determination of divalent metal ions. *React. Polym.* **1995**, *25*, 79.

108. Lee, T. S.; Hong, S. I. Synthesis and metal binding behaviour of hydroxamic acid resins from polyethyl acrylate crosslinked with divinylbenzene and hydrophilic cross-linking agent. *J. Polym. Sci. Polym. Chem.* **1995**, *33*, 203.

109. Lee, T. S.; Hong, S. I. Porous chelating resins from acrylonitrile–ethyl acrylate-divinylbenzene copolymers. *J. Macromol. Sci. A* **1995**, *A32*, 379.

110. Tunney, M. M.; Keane, P. F.; Gorman, S. P. Assessment of a urinary tract biomaterial encrustation using a modified Robbins device continuous flow model. *J. Biomed. Mater. Res. (Appl. Biomater.)* **1997**, *38*, 87.

111. Navarro, A.; Garcia-Garduno, M.; Vazquez-Polo, G.; Alvarez-Castillo, A.; Castano, V. M.; Martinez, E. Studies on the stability of a styrene-butadiene copolymer for use in biomedical applications. *J. Bioact. Compat. Pol.* **1995**, *10*, 258.

112. Wichems, D. N.; Calloway, C. P.; Fernando, R.; Jones, B. T.; Morykwas, M. J. Determination of silicone in breast tissue by graphite furnace continuum source atomic absorption spectrometry. *Appl. Spectrosc.* **1993**, *47*, 1577.

113. Allan, J. R.; Renton, A.; Smith, W. E.; Gerrard, D. L.; Birnie, J. Slow release of bis(acetylsalicylate) copper (II) from poly(hydroxyethyl acrylate) membranes. *Plast. Rubb. Comp. Process. Appln.* **1991**, *16*, 201.

114. Pollock, E.; Andrews, E. J.; Lentz, D.; Sheikh, K. Tissue ingrowth and porosity of biomer. *Trans. Am. Soc. Art. Int. Org.* **1981**, *27*, 405.

115. Hill, S. J., Ed. *Inductively Coupled Plasma Spectrometry and Its Applications*. Sheffield Academic Press: Sheffield, UK, 1999; Sheffield Analytical Chemistry, Vol. 1.

116. Evans, E. H.; Chenery, S.; Fisher, A.; Marshall, J.; Sutton, K. Atomic spectrometry update: atomic emission spectrometry. *J. Anal. At. Spectrom.* **1999**, *14*, 977.

117. Lorenzen, C. J.; Carlhoff, C. Industrial applications of laser-induced emission spectral analysis (LIESA) for process and quality control. Presented at the Winter Conference on Plasma Spectrochemistry, San Diego, 6–11 January 1992.

118. Milan, M.; Vadillo, J. M.; Laserna, J. J. Removal of air interference in laser induced breakdown spectrometry monitored by spatially and temporally resolved charge coupled device measurements. *J. Anal. At. Spectrom.* **1997**, *12*, 441.

119. Ediger, R. D. Plastic discrimination using laser induced plasma spectroscopy. Presented at the European Winter Conference on Plasma Spectrochemistry, Cambridge, UK, 8–13 January 1995.

120. Sattman, R.; Monch, I.; Krause, H.; Noll, R.; Couris, S.; Hatziapostolou, A.; Mavromanolakis, A.; Fotakis, C.; Larraui, E.; Miguel, R. Laser induced breakdown spectroscopy for polymer identification. *Appl. Spectrosc.* **1998**, *52*, 456.

121. Yamamoto, K. Y.; Cremers, D. A.; Ferris, M. J.; Foster, L. E. Detection of metals in the environment using a portable laser induced breakdown spectroscopy instrument. *Appl. Spectrosc.* **1996**, *50*, 222.

122. Marshall, K.; Valensi, D. Surface analysis: glow discharge spectroscopy. *Mater. World* **1995**, *3*, 471.

123. Marcus, R. K.; Harville, T. R.; Mei, Y.; Shick, C. R. R.f. powered glow discharges: elemental analysis across the solids spectrum. *Anal. Chem.* **1994**, *66*, 902A.

124. Fernandez, M.; Bordel, N.; Pereiro, R.; Sanz-Medel, A. Investigations on the use of radio frequency glow discharge optical emission spectrometry for in-depth profile analysis of painted coatings. *J. Anal. At. Spectrom.* **1997**, *12*, 1209.

125. Caroli, S.; Senofonte, O.; DelMonte Tamba, M. G.; Brenner, I. Applicability of low pressure discharges to the analysis of non-conducting solids. Presented at the 19th Annual Meeting of the Federation of Analytical Chemistry and Spectroscopy Societies (FACSS), Philadelphia, 20–25 September 1992.

126. Bouyssoux, G.; Gaillard, F.; Romand, M. Characterization of thin organic films on metallic substrates by X-ray emission and glow discharge optical spectrometries. Presented at the 27th Colloquium on Spectroscopium Internationale, Bergen, Norway, 9–14 June 1991.

127. Fonseca, J. L. C.; Tasker, S.; Apperley, D. C.; Badayal, J. P. S. Plasma-enhanced chemical vapour deposition of organosilicon materials:a comparison of hexamethyldisilane and tetramethylsilane precursors. *Macromolecules* **1996**, *29*, 1705.

128. Iriyama, Y.; Noda, M. Plasma polymerization of fluoromethanes. *J. Polym. Sci. Polym. Chem.* **1998**, *36*, 2043.

129. Seidel, T.; Golluch, A.; Beerwald, H.; Boehm, G. Sliding spark spectroscopy: characterization and application of a new radiation source in optical emission spectrometry for the identification of chlorine containing waste plastics and their inorganic additives. *Fresenius J. Anal. Chem.* **1993**, *347*, 92.

130. Golloch, A.; Siegmund, D. Sliding spark spectroscopy: rapid survey analysis of flame retardants and other additives in polymers. *Fresenius J. Anal. Chem.* **1997**, *358*, 804.

131. Elms, T. J.; Pierson, R. R. The determination of trace element residues in polyethylene by inductively coupled plasma atomic emission spectrometry. Presented at the 9th Australian Symposium on Analytical Chemistry, Sydney, 27 April–1 May 1987.

132. Oktavec, D.; Lehotay, J. Determination of titanium and aluminium in polypropylene by AAS and ICP-AES in the presence of sodium. *At. Spectrosc.* **1993**, *14*, 103.

133. Gapasin, E. C.; Austin, C. S. Analysis of residual catalyst in the production of a high performance polymer using ICP and IC. Presented at the Pittsburgh Conference and Exposition on Analytical Chemistry and Applied Spectroscopy, New York City, 5–9 March 1990.

134. Hezina, F.; Filipu, P. Determination of germanium in polyester matrix by inductively coupled plasma atomic emission spectrometry. Presented at the 9th Czechoslovak Spectroscopic Conference, Ceske Budejovice, Czechoslovakia, 22–24 June 1992.

135. Nakahara, T.; Wasa, T. Determination of low concentrations of germanium in polyethylene terephthalate by inductively coupled plasma atomic emission spectrometry with conventional solution nebulisation. *Bull. Univ. Osaka Prefect., Ser. A* **1992**, *41*, 13.

136. Burns, D. W. Microwave digestion of rigid-rod polymer films for phosphorus analysis by inductively coupled plasma atomic emission spectrometry. Presented at the 1992 Winter Conference on Plasma Spectrochemistry, San Diego, 6–11 January 1992.

137. Sinclair T.; Koch, S. A. Automation of trace dissolution from epoxy laminates with hot corrosive acids. *Adv. Lab. Autom. Rob.* 1989, 5, 185.

138. Benjamin, S. Use of inductively coupled plasma spectroscopy in a surface coatings laboratory. *Polym. Paint. Colour J.* **1991**, *181*, 248.

139. Williams, E. E.; Binstock, D. A.; Gutknecht, W. F. Preparation of lead-containing paint and dust method evaluation materials and verification of the preparation protocol by round-robin analysis. Report, EPA/600/R-93/235; Order no. PB94–141165. Springfield, VA: NTIS, 1993.

140. Paudyn, A. M.; Smith, R. G. Determination of elements in paints and paint scrapings by inductively coupled plasma atomic emission spectrometry using microwave assisted digestion. *Fresenius J. Anal. Chem.* **1993**, *345*, 695.

141. Nagourney, S. J.; Madan, R. K. Metals analysis in the plastics additives industry: quality assurance considerations. *J. Test. Eval.* **1991**, *19*, 77.

142. Fischer, J. L.; Rademeyer, C. J. High temperature nebulisation of waxes and other low melting point solids for ICPAES. Presented at the 3rd International Symposium on Analytical Chemistry in the Exploration, Mining and Processing of Materials, Sandton, South Africa, 2–7 August 1992.

143. Hemmerlin, M.; Mermet, J. M.; Bertucci, M.; Zydowicz, P. Determination of additives in PVC material by u.v. laser ablation inductively coupled plasma atomic emission spectrometry. *Spectrochim. Acta, Part B* **1997**, *52B*, 421.

144. Lam, K. K.; Chan, W. T. Novel laser sampling technique for inductively coupled plasma emission spectrometry. *J. Anal. At. Spectrom.* **1997**, *12*, 7.

145. Wing, T. C.; Yip, H. H. G. Sample preconcentration using ion exchange polymer film for laser ablation sampling inductively coupled plasma atomic emission spectrometry. *Anal. Chem.* **1997**, *69*, 4872.

146. Booth, P. K.; McLeod, C W. Rapid Survey Analysis of polymeric materials, by laser ablation inductively coupled plasma spectrometry. *Mikrochim. Acta* **1989**, *3*, 283.

147. Casetta, B.; Di Pasquale, G.; Soffientini, A. Setting up an ICP coupled with HGA for characterization of S in polymeric materials. *At. Spectrosc.* **1985**, *6*, 62.

148. Dorn, S. B.; Skelly-Frame, E. M. Development of a high performance liquid chromatograph inductively coupled plasma method for speciation and quantification of silicones; from silanols to polysiloxanes. *Analyst* **1994**, *119*, 1687.

149. Dorn, S. B.; Skelly-Frame, E. M. HPLC-ICP in the industrial research laboratory. Presented at the 1995 Pittsburgh Conference and Exposition on Analytical Chemistry and Applied Spectroscopy, New Orleans, 5–10 March 1995.

150. Biggs, W. R.; Gano, J. T.; Brown, R. J. Determination of polyphosphate distribution by LC separation and d.c. plasma AES detection. *Anal. Chem.* **1984**, *56*, 2653.

151. Perpall, H. J.; Uden, P. C.; Deming, R. L. Empirical and molecular formula determination of polymer pyrolysates by multi-referencing GC-MIP spectroscopy. *Spectrochim. Acta, Part B* **1987**, *42*, 243.

152. Kala, S.V.; Lykissa, E. D.; Lebovitz, R. M. Detection and characterization of polydimethylsiloxanes in biological tissues by gas chromatography atomic emission detection and mass spectrometry. *Anal. Chem.* **1997**, *69*, 1267.

153. Wang, F. C. Y.; Gerhart, B. Determination of acrylamide in emulsion polymers by pyrolysis–solvent trapping gas chromatography. *Anal. Chem.* **1996**, *68*, 3917.

154. Wang, F. C. Y.; Green, J. G.; Gerhart, B. Qualitative identification of fumaric acid and itaconic acid in emulsion polymers. *Anal. Chem.* **1996**, *68*, 2477.

155. Sato, H.; Kondo, K.; Tsuge, S.; Ohtani, H.; Sato, N. Mechanisms of thermal degradation of a polyester flame-retarded with antimony oxide/brominated polycarbonate studied by temperature programmed analytical pyrolysis. *Polym. Degradat. Stabil.* **1998**, *62*, 41.

156. Arpadjan, S.; Vuchkova, L.; Kostadinova, E. Sorption of arsenic bismuth, mercury, antimony, selenium and tin on dithiocarbamate loaded polyurethane foam as a preconcentration method for their determination in water samples by simultaneous inductively coupled plasma atomic emission spectrometry and electrothermal atomic absorption spectrometry. *Analyst* **1997**, *122*, 243

157. Miyazaki, A.; Barnes, R. M. Differential determination of chromium(VI) and chromium (III) with polydithiocarbamate chelating resin and inductively coupled plasma atomic emission spectrometry. *Anal. Chem.* **1981**, *53*, 364.

158. Amarasiriwardena, D.; Barnes, R. M. Optimization and evaluation of poly(dithiocarbamate) resin preconcentration for determination of platinum group metals in geological samples by ICP-AES and ICP-MS. Presented at the 1992 Winter Conference on Plasma Spectrochemistry, San Diego, 6–12 January 1992.

159. Su, Z.; Chang, X.; Xu, K.; Luo, X.; Zhan, G. Efficiency and mechanism of macroporous poly(vinylthiopriprionamide) chelating resin for adsorbing and separating noble metal ions and determination by atomic spectrometry. *Anal. Chim. Acta* **1992**, *268*, 323.

160. Su, Z.; Chang, X.; Xu, K.; Luo, X. Application of macroporous poly(vinyl amidine thiocyanate-thiourea)chelate resin for adsorbing and separating trace chromium, vanadium titanium beryllium yttrium lanthanum and copper: detection by inductively coupled plasma optical emission spectrometry. *Microchem. J.* **1991**, *44*, 78.

161. Lamb, J. D.; Nazarenko, A. Y.; Neilson, L. Novel extraction of metal ions from acidic media using poly(ethylene oxide) phases. *Anal. Comm.* **1998**, *35*, 145.

162. Zhixing, S.; Xun, L.; Xijun, C. Study of polymer-metal ions triple complex as materials for adsorption and separation. *J. Appl. Polym. Sci.* **1999**, *71*, 819.

163. Xijun C.; Zhixing, Su.; Guangyao, Z.; Xingyin, L.; Weyun G. Synthesis and efficiency of a polyacrylacylisothiourea chelating fibre for the preconcentration and separation of trace amounts of gold palladium and ruthenium from solution samples. *Analyst* **1994**, *119*, 1445.

164. Wang, Y. Atomic emission spectroscopic determination of traces of gold after its preconcentration on foamed plastics. *Lihua Jianyan Huaxue Fence* **1990**, *26* 280.

165. Horvath, Z.; Barnes, R. M. Carboxymethylated polyethyleneimine-polymethylenepolyphenylene isocyanate chelating ion exchange resin preconcentration for inductively coupled plasma spectrometry. *Anal. Chem.* **1986**, *58*, 1352.

166. Luo, X.; Chang, X.; Su, Z.; Zhang, G.; Lin, Q. Study on synthesis and performance of the polyamine polyurea chelate fibre: application in ICP spectrochemical analysis. *Fensi Huaxue* **1988**, *16*, 525.

167. Colella, M. B.; Siggia, S.; Barnes, R. M. Polyacrylamidoxime resin for determination of trace elements in natural waters. *Anal. Chem.* **1980**, *52*, 2347.

168. Hight, S. C. Inductively coupled plasma atomic emission spectrometric determination of lead, cadmium and other elements leached by acetic acid from foodware. Presented at the 27th Eastern Analytical Symposium, New York, 2–7 October 1988.

169. Kane, R. S.; Cohen, R. E.; Silbey, R. Synthesis of doped zinc sulfide nanoclusters within block copolymer nanoreactors. *Chem. Mat.* **1999**, *11*, 90.

170. Huang, S. W.; Neoh, K. G.; Shih, C. W.; Lim, D. S.; Kang, E. T.; Han, H. S.; Tan, K. L. Synthesis, characterization and catalytic properties of palladium-containing electroactive polymers. *Synthetic Met.* **1998**, *96*, 117.

171. Ichinohe, D.; Muranaka, T.; Kise, H. Oxidative polymerization of phenylenediamines by enzyme and magnetic properties of the products. *J. Appl. Polym. Sci.* **1988**, *70*, 717.

172. Bai, S.; Sridhar, S.; Khan, A. A. Metal-ion mediated separation of propylene from propane using PPG membranes. *J. Membrane Sci.* **1998**, *147*, 131.

173. Yun-Pu, W.; Yi, L.; Rong-Min, W.; Li, Y. Synthesis and fluorescence properties of the mixed complexes of Europium (III) with polymer ligand and thenoyl trifluoroacetone. *J. Appl. Polym. Sci.* **1997**, *66*, 755.

174. Brusic, V.; Angelopoulos, M.; Graham, T. Use of polyaniline and its derivatives in corrosion protection of copper and silver. Antec '95 Vol 2 Conference Proceedings, Boston, 7–11 May 1995; Society of Plastics Engineers: Boston, 1995; p 1397.

175. Huang, S.; He, B. One-step preparation of polymer-supported colloidal palladium catalysts and their catalytic properties: I. Synthesis and characterization. *React. Polym.* **1994**, *23*, 1.

176. Colak, N.; Gumgum, B. Determination of iron by inductively coupled plasma atomic emission spectrometry in stereoregular polypropylene oxide by Pruitt-Baggett catalyst. *Polym. Plast. Technol. Engng.* **1994**, *33*, 105.

177. Cortina, J. L.; Miralles, N.; Sastre, A.; Aguilar, M.; Profumo, A.; Pesavento, A. Solvent impregnated resins containing *d*-(2,4,-trimethylpentyl)phosphinic acid: I. Comparative study of di-(2,4,4-trimethylpentyl) phosphinic acid adsorbed into Amberlite XAD-2 and dissolved in toluene. *React. Polym.* **1993**, *21*, 89.

178. Bruna, J. M. New analytical techniques for polymers: II. X-ray fluorescence. *Rev. Plast. Mod.* **1995**, *69*, 550.

179. Ribar, B.; Skrbic, Z. Analysis of additives in PVC compounds by X-ray fluorescence spectroscopy. *Polimeri* **1986**, *7*, 357.

180. Dorner, W.G. Analysis of plastics: VI. Methods of physical trace analysis. *Kunst. Plast.* **1985**, *32*, 23.

181. Jenkins, R.; Gould, R. W.; Gedcke, D. *Quantitative X-Ray Spectrometry*; Marcel Dekker: New York, 1981.

182. Massonet, G. Comparison of X-ray fluorescence and X-ray diffraction techniques for the forensic analysis of automobile paint. In *Advances in Forensic Sciences*; Jacob, B., Bonte, W., Eds.; Berlin: Verlag Dr. Köster; Vol. 5; 1996; 321–326.

183. Warren, P. L.; Farges, O.; Horton, M.; Humber, J. Role of energy dispersive X-ray fluorescence in process analysis of plastic materials. *J. Anal. At. Spectrom.* **1987**, *2*, 245

184. Helan, V.; Ersepke, Z. Recalibration standards for X-ray fluorescence spectrometers. Comm. Eur. Communities,

[Rep.] EUR, 1992, EUR 14113, *Prog. Anal. Chem. Iron. Steel Ind.* 125.

185. Morgan, A. J.; Winters, C. Practical notes on the production of thin amino-plastic standards for quantitative X-ray microanalysis. *Micron Microsc. Acta* 1988, *19*, 209.

186. Kump, P.; Necemer, M.; Smodis, B.; Jacimovic, R. Multi-element analysis of rubber samples by X-ray fluorescence. *Appl. Spectrosc.* **1996**, *50*, 1373.

187. Dzubay, T. G.; Morosoff, N.; Whitaker, G. L.; Yasuda, H.; Bazan, F.; Bennet, R. L.; Cooper, J.; Courtney, W. J.; Frazier, C. A. Polymer film standards for X-ray fluorescence spectrometers. *J. Trace Microprobe Tech.* **1987**, *5*, 327

188. Huang, T. C., Thin film characterization by X-ray fluorescence. Presented at the 16th Annual Meeting of the Federation of Analytical Chemistry and Spectroscopy Societies, Chicago, 1–6 October 1989.

189. Bogert, J. R.; Kibler, J. M.; Shmotzer, J. K. Standardless EDXRF analysis for cations in ion exchange resin impregnated membranes, *Adv. X-ray Anal.* **1987**, *30*, 153.

190. Peris Martinez, V.; Bosch, Reig, F.; Gimeno Adelantado J. V.; Domenech Carbo, M. T. Multi-elemental determination of heavy elements in plastics using X-ray fluorescence after destruction of the polymer by molten sodium hydroxide. *Fresenius J. Anal. Chem.*, **1992**, *342*, 586.

191. Anon. IRIS: quality control of colour and coatings. *Plast'21* **1994**, No *39*, 54.

192. Lister, D. B.; Binns, J. W. X-raying silicone coatings provides accurate weight gauge. *Paper Film Foil Conv.* **1981**, *55*, 37.

193. Li, J. C. Determination of metal oxides in plastic by X-ray fluorescence spectrometry. *Guangpu Shiyanshi* **1993**, *10*, 22.

194. Hutchinson, B. H. Elemental analysis of polyolefins by X-ray fluorescence spectroscopy. *Annu. Tech. Conf. Soc. Plast. Eng.* **1992**, *1*, 1575.

195. Ellis, A. T. Oxford launches material test kits. *Plast. Rubb. Wkly.* **1989**, *1304*, 19.

196. Ellis, A. T. X-ray fluorescence analyses PVC compound constituents. *Brit. Plast. Rubb.* **1989**, Sept., 32.

197. Karmarchi, P. Applications of X-ray analytical methods in filled vulcanized elastomers. Presented at the ACS Spring Meeting, Dallas, 19–22 April 1988; Paper 71.

198. Shestakov, A. S.; Zomikov, I. P. X-ray fluorescent analysis of chloride in rubber. *Proizvod. Ispol'z. Elastomerov* **1997**, *4*, 8.

199. Hasany, S. M.; Rashid, F.; Raschid, A.; Rehman, H. Determination of fluorine in solids down to 120 ppm by wavelength-dispersive X-ray fluorescence spectrometry. *J. Radioanal. Nucl. Chem.* **1991**, *149*, 211.

200. Brodersen, K.; Danzer, B.; Miekisch, W.; Wolf, M. Identification and quantification of brominated flame retardants in electronic components. *GIT Fachz. Lab.* **1996**, *40*, 1132.

201. Blank, A. B. Vlasov, V. G.; Nartova, Z. M.; Shetsov, N. I. X-ray fluorescence control of gadolinium contents of plastic scintillators based on poly(methyl methacrylate). *Zh. Anal. Khim.* **1987**, *42*, 358.

202. Rosato, D. V. The complete molding operation: technology performance. In *Testing and Quality Control. Injection Molding Handbook: The Complete Molding Operation: Technology Performance, Economics;* Rosato, D. V., Ed.; Van Nostrand Reinhold: New York, 1986; p 683.

203. Warren, P. L. Comparison of non-destructive techniques for the measurement of fiber content in reinforced engineering plastics and composites. In *Non-destructive Characterisation of Materials, IV;* Ruud, C. O., Bussiere, J. F., Green, R. E., Eds.; Plenum Press: New York, 1991; p 307.

204. Marshall, J. Process analysis using atomic spectrometry. In *Spectroscopy in Process Analysis;* Chalmers, J. M., Ed.; Analytical Chemistry Vol. 4, Sheffield Academic Press: Sheffield, UK, 2000.

205. Kalnicky, D. J.; Ramanujam, R. S. On-line elemental analysis with X-ray fluorescence (XRF): technology, applications and case studies in the process industries. *Anal. Div.* **1991**, *25*, 263.

206. Hashizume, T.; Tokushige, K.; Yajima, S.; Kawasaki, Y.; Ogata, Y. Apparatus for the non-destructive on-line analysis on sulfur contents in rubbers. Japanese Patent Kokai Tokkyo Koho JP 07,294,458 [95,294,458] (Cl.G01N23/223) 10 Nov 1995, Appl. 94/88,743 25 Apr. 1994, 7pp.

207. Bledzki, A.; Nowaczek, W. Identification of plastics in waste materials and methods for their recycling. *Polim. Tworz. Wielk.* **1993**, *38*, 511.

208. Ninomiya, T.; Nomura, S.; Taniguchi, K. Probing into residual forensic evidence. Application of total reflection X-ray fluorescence spectrometry. *Hyomen Kagaku* **1990**, *11*, 189.

209. Taniguchi, K.; Ninomiya, T. Total reflection X-ray fluorescence spectrometry. *Tetsu to Hagane* **1990**, *76*, 1228.

210. Ninomiya, T.; Nomura, S.; Taniguchi, K. Elemental analysis of trace plastic residuals using total reflection X-ray fluorescence. *X-sen Bunseki no Shinpo* **1988**, *19*, 237.

211. Tsuji, K.; Wagatsuma, K.; Hirokawa, K. Take off angle-dependent X-ray fluorescence analysis of thin films on acrylic substrate. *J. Trace Microprobe Tech.* **1997**, *15*, 1.

212. McKnight, M. E. Field methods for measuring lead concentration in paint films. In *Waste Management Technologies.* Proceedings of the Annual Meeting of Air and Waste Management Association, Vancouver, BC, 1991; Air and Waste Management Association: Pittsburgh, 1991; Paper 91/134.5.

213. Driscoll, J. N.; Laliberte, R.; Wood, C. Field detection of lead in paint and soil by high resolution XRF. *Am. Lab.* **1995**, *27*, 34H.

214. Beitelman, A.; Drozdz, S.; Vogel, R. Evaluation of X-ray fluorescence unit for detecting lead in paint on military structures. Report, 1991, CERI-TR-M-91/16; Order no AD-A232 229, Gov. Rep. Announc. Index (US) **1991**, *91*, (15).

215. Grodzins, L.; Parsons, C.; Sackett, D.; Shefsky, S.; Tannian, B. Filed screening methods hazard. *Wastes Toxic Chem. Proc. Int. Symp.* **1995**, *2*, 1118.

216. Carvalho, M. S.; Dominguez, M. F.; Mantovano, J. L.; Filho, E. Q. S. Uranium determination at ppb levels by X-ray fluorescence after its preconcentration on polyurethane foam. *Spectrochim. Acta Part B* **1998**, *53B*, 1945.

217. Carvalho, M. S.; Medeiros, J. A.; Nobrega, A. W.; Mantovano, J. L.; Rocha, V. P. A. Direct determination of gallium on polyurethane foam by X-ray fluorescence. *Talanta* **1995**, *42*, 45.

218. Bao, S. X-ray fluorescence spectrophotometry for the direct determination of gold absorbed in polyurethane foam. *Fenxi Huaxue* **1995**, *23*, 410.

219. Semenova, E. V.; Bagasaro, K. N.; Blokhina, G. E. Use of polystyrene-azothianzane-2,4,-dithione for X-ray fluorescence determination of gold and silver in ore materials. *Zavod Lab.* **1988**, *54*, 33.

220. Maruca, R.; Suder, B. J.; Wightman, J. P. Interaction of heavy metals with chitin and chitosan. III. *Chromium P. Appl. Polym. Sci.* **1982**, *27*, 4827.

221. Yonemori, S.; Ichikura, E.; Sugizaki, M. Determination of the co-monomer content in tetrafluoroethylene/perfluoro(propylvinyl ether) copolymers by XRF analysis. *Bunseki Kagaku* **1992**, *41*, 593.

222. Rao, K. V.; Pandey, G. C. Estimation of sulfur content in copolymer powders by energy dispersive X-ray spectroscopy. *Polym. Test.* **1991**, *10*, 31.

223. Juirian, C.; van den Heuvel, M.; Hoentijen, G.; de Kreely, A. K. Stabilization of PVC by 2-(alkyltinthio)acetates: the use of X-ray fluorescence spectrometry to measure the incorporation of sulfur into PVC molecules. *Macromol. Chem. Rapid. Comm.* **1984**, 5, 235.

224. Costa, G.; Loonan, C.; Pepper, D. End group evidence of zwitterionic species in the anionic polymerization of cyanoacrylates by Lewis bases. *Macromol. Rapid Commun.* **1997**, *18*, 891.

225. Pemberton, L.; De Jaeger, R. Surface modification of polyvinyl alcohol by peroxide-initiated grafting of a polyorganophosphazene. *Chem. Mater.* **1996**, *8*, 1391.

226. Nakaya, T. Development of a staining preventive coating for architecture. *Prog. Org. Coatings* **1996**, *27*, 173.

227. Burford, R. P.; Harrauer, E. Diffusion of salt solutions into Nylon 6,6. *Polymer* **1983**, *24*, 1001.

228. Akovali, G.; Gundogan, G. Studies on flame retardancy of PAN fibre treated by flame-retardant monomers in cold plasma. *J. Appl. Polym. Sci.* **1990**, *41*, 2011.

229. Florio, T.; Rose, V. C.; Rockett, T. J. Effects of selected manufacturing procedures on water resistance of marine laminates. Presented at the 42nd Annual Conference and Expo '87; Proceedings Editors: SPI, Reinforced Plastics/Composites Institute, Cincinnati, OH, 1987, 2–6 February Session 15-F pp 4, 627.

230. Abeysinghe, H. P.; Ghotra, J. S.; Pritchard, G. Substances contributing to the generation of osmotic pressure in resins and laminates. *Composites* **1983**, *14*, 57.

231. Hamed, G. R.; Huang, J. Combining cobalt and resorcinolic bonding agents in brass-rubber adhesion. Presented at the ACS 138th Meeting, Fall 1990; Preprints, ACS Rubber Division, Washington DC, 9–12 October 1990; Paper 30.

232. Romine, R. A.; Romine, M. F.; Snowden-Swan, L. Microbial processing of waste tyre rubber. Presented at the 148th ACS Rubber Division Meeting, Fall 1995, Conference Preprints, ACS Rubber Division, Cleveland, OH, 17–20 Oct 1995; Paper 56, pp 13.

233. Jarvis, K. E.; Gray, A. L.; Houk, R. S. *Inductively Coupled Plasma Mass Spectrometry*; Blackie Academic & Professional, Chapman & Hall: London, 1992.

234. Klawers, G. H. C., v. Alphen, A. M. ICP-MS successfully applied in environmental and polymer analysis. Presented at the 3rd International Conference on Plasma Source Mass Spectrometry, University of Durham, Durham, UK, 13–18 September 1992.

235. Fordham, P. J.; Gramshaw, J. W.; Castle, L.; Crews, H. M.; Thompson, D.; Parry, S. J.; McCurdy, E. Determination of trace elements in food contact polymers by semiquantitative inductively coupled plasma mass spectrometry: performance evaluation using alternative multi-element techniques and in-house reference materials. *J. Anal. At. Spectrom.* **1995**, *10*, 303.

236. Marshall, J.; Franks, J. Determination of halogens in plastics by ICP-MS. Presented at the 16th FACSS Meeting, Chicago, 1–6 October 1989.

237. Hall, G. S.; Rabinowitz, M. Use of ICP-MS to characterize lead-based paints using stable lead isotopes and multi-element analyses. Presented at the Winter Conference on Plasma Spectrochemistry, Fort Lauderdale, FL, 8–13 January 1996.

238. Nixon, D. E.; Moyer, T.P. This old house: can lead isotope ratios be used to detect the exposure source? Presented at the 22nd FACSS, Cincinnati, 15–20 October 1995.

239. Gotz, A.; Lamberty, A.; De Bievre, P. CBNM certification of cadmium in polyethylene by isotope dilution mass spectrometry. *Int. J. Mass. Spectrom. Ion Processes* 1993, *123*, 1.

240. Koons, R. D. ICP-MS or ETA-AA: analysis of individual carpet fibers. Presented at the Federation of Analytical Chemistry and Spectroscopy Societies Meeting, St. Louis, 2–7 October 1994.

241. Watling, R. J.; Herbert, H. K.; Bruce, K. Use of laser ablation inductively coupled plasma mass spectrometry in the interpretation of scene of crime evidence. Presented at the Federation of Analytical Chemistry and Spectroscopy Societies Meeting, St. Louis, 2–7 October 1994.

242. Marshall, J.; Franks, J.; Abell, I. Determination of trace elements in solid plastic materials by laser ablation-inductively coupled plasma mass spectrometry. *J. Anal. At. Spectrom.* **1991**, *6*, 145.

243. Franks, J.; Marshall, J.; Abell, I. Optimization of laser ablation inductively coupled plasma mass spectrometry for the analysis of polymeric materials. Presented at the 5th Biennial National Atomic Spectroscopy Symposium, Loughborough, UK, 18–20 July 1990.

244. Lord, C. J. III; Nelson, E. W. Determination of trace elements in polymers by laser ablation inductively coupled plasma mass spectrometry. Presented at the 1992 Winter Conference on Plasma Spectrochemistry, San Diego, 6–11 January 1992.

245. Stuewer, D. Glow discharge mass spectrometry: aspects of a versatile analytical tool. In *Applications of Plasma Source Mass Spectrometry*, 2nd International Conference on Plasma Source Mass Spectrometry, Durham, 24–28 September 1990; Holland, G.; Eaton, A.N., Eds.; Royal Society of Chemistry: Cambridge, UK, 1991, p 1.

246. Marcus, R. K.; Harville, T. R.; Mei, Y.; Shick, C. R. RF-powered glow discharges: elemental analysis across the solids spectrum. *Anal. Chem.* **1994,** *66,* 902A.

247. Schelles, W.; Van Grieken, R. Direct current glow discharge mass spectrometry for elemental characterization of polymers. *Anal. Chem.* **1997,** *69,* 2931.

248. Shick, C. R.; DePalma, P. A.; Marcus, R. K. Radiofrequency glow discharge mass spectrometry for the characterization of bulk polymers. *Anal. Chem.* **1996,** *68,* 2113.

249. Gibeau, T. E.; Hartenstein, M. L.; Marcus, R. K. Cryogenically cooled sample holder for polymer sample analysis by radiofrequency glow discharge mass spectrometry. *J. Am. Soc. Mass. Spectrom.* **1997,** *8,* 1214.

250. Zyn, V. I.; Oparin, V. B.; Potapov, V. K. Kinetic mass spectrometry of the initial stages of polymerization of tetrafluoroethylene in a glow discharge. *Khim. Vys. Energ.* **1989,** *23,* 75.

251. Zen, V. I.; Parking, A. A.; Tkachuk, B. V. Kinetic principles in mass spectra during polymerization in a glow discharge. *Teor. Eksp. Khim.* **1981,** *17,* 605.

4

Chemical Analysis of Polymers

STEVEN K. POLLACK

This chapter covers chemical methods for analyzing polymeric materials, with major focus on methods for the quantitative determination of reactive functional groups present in polymers. These types of analyses typically have one of two purposes. The first is the determination of the number of end groups present in a given mass of polymer sample; this is needed to calculate the molar mass of the polymer. The second purpose is to determine the concentration of reactive groups; this is used in formulating thermosetting resin systems or for the synthesis of a linear copolymer from reactive oligomers. While many of the methods for the quantitative determination of reactive groups are the same for these two applications, an overview of how the data are used in both instances is given. In addition, chemical methods for identifying other functionalities that determine the chemical composition of polymers are described. The chapter focuses on the procedures that predominantly use spectrometric, wet chemical, and titrametric methods. Some of these methods are used in conjunction with gas chromatography. Assays based solely on chromatographic methods, including those for determining volatile additives such as plasticizers, are not covered here in any detail. The appropriate ASTM (American Society for Testing and Materials) procedures will be pointed out where relevant (Table 4.1).

End-group Analysis

For "conventional" molecules, the molecular weight (or molar mass) is defined as the mass of an Avogadro's number (N_A) of molecules of a well-defined molecular formula. However, for most polymers, there is no unique formula. A range of formulas exist, which depend on the molecular weight of each chain. As was shown in previous chapters, the number-average molecular weight of a polymer is defined by the expression

$$\overline{M}_n = \frac{\sum_{i=1}^{N} w_i}{\sum_{i=1}^{N} \frac{w_i}{M_i}} = \frac{\text{total mass of sample}}{\sum_{i=1}^{N} n_i}$$

$$= \frac{\text{total mass of sample}}{\text{total moles of polymer molecules}} \qquad (1)$$

where M_i is the molar mass of the ith fraction of a polymer sample, w_i is the mass of that fraction, n_i is the number of moles of that fraction present, and N is the total number of molar mass fractions the sample is divided into. Knowing the total mass of the polymer sample and the number of molecules, the number-average molecular weight can be determined.

In end-group analysis, the number of polymer molecules is determined by quantitating the number of polymer chain ends. The general formula for the average molecular weight determined in this fashion is

Table 4.1 American Society for Testing and Materials (ASTM) methods cited

Number	Name
D 1045	Method of sampling and testing plasticizers used in plastics
D 1638	Methods of testing urethane foam isocyanate raw materials
D 1652	Test methods for epoxy content of epoxy resins
D 2849	Methods of testing urethane foam polyol raw materials
D 4094	Test method for acid content of ethylene-acrylic acid copolymer
D 4273	Method for testing polyurethane raw materials: determination of primary hydroxyl contents of polyether polyols
D 4274	Method for testing polyurethane raw materials: determination of hydroxyl contents of polyols

Data are from *Annual Book of ASTM Standards, Section 8: Plastics*; American Society for Testing and Materials: Philadelphia, 2002.

$$\overline{M}_{end} = \frac{\text{mass of polymer}}{\text{moles of polymer}} = \frac{W}{n_{end}/q} = \frac{W \cdot q}{n_{end}} \quad (2)$$

where W is the mass of the polymer sample, n_{end} is the number of moles of chain ends in the sample, and q is the number of chain ends per molecule which contribute to the measured n_{end}. For example, for an oligomer of poly(ethylene glycol), there are two hydroxyl chain ends per chain, hence q is 2. These can be quantitated by a titrametric method (described below). But for polyethylene glycol monomethyl ether, q is 1, as the methoxy group will not contribute to the determination.

The limitations of this method of molecular weight determination are three-fold. First, as the molecular weight of the polymer increases, the molar concentration of measureable end groups decreases, thus increasing the uncertainty in the value of n_{end}. Secondly, there are some instances in which the value of q may not be uniquely defined for every chain. For example, in oligomers of the polyester derived from tetramethylene glycol and adipic acid, some chains may have both termini as hydroxyl groups, others have both as carboxylic acid groups, and still others have one of each. If the detection method used for the quantitation of end groups is sensitive only to one of these two functional groups, then the molecular weight determined will be in error. Similarly, if side reactions during the polymerization cause the loss of functional groups on some chains, such as acid-catalyzed dehydration of a glycol to a terminal olefin, then these chain ends will also be unavailable, thus rendering the assumed value for q too high and producing an overestimation of molecular weight. Finally, such contaminants as residual solvent and moisture in the sample affect the actual weight of the sample and may be capable of the same reaction as the chain ends. The former effect will cause an overestimate of molecular weight; the latter will cause an underestimate.

Reactive Group Analysis

Frequently, low molar mass polymers are used as building blocks in a thermosetting resin based on network polycondensation reactions. Here accurately knowing the number of functional groups is necessary for control of gel point and cross-link density of the material. For a thermosetting polymer, the number average functionality, f_{av}, is defined as

$$f_{av} = \frac{\sum f_i N_i}{\sum N_i} \quad (3)$$

where N_i is the initial number of moles of monomer i of functionality f_i. For example, glycerol ($HOCH_2CHOHCH_2OH$) has a functionality of 3, whereas pentaerythritol (1) has a functionality of 4. Based on Carother's theory of gelation, a network whose formulation has an average functionality of f_{av} will gel (convert from liquid to solid behavior) when the

1

extent of reaction, p (probability that the functional groups have reacted), has a value of

$$p_c = 2/f_{av} \quad (4)$$

This assumes that the components are formulated with stoichiometric balance.

Let us give two examples to show the effect of functionality on p_c. For a resin formulation of glycerol and phthalic anhydride (2) with mole ratio of $2:3$, the average functionality is

2

$$f_{av} = \frac{\sum f_i N_i}{\sum N_i} = \frac{2 \cdot 3 + 3 \cdot 2}{2 + 3} = \frac{12}{5} = 2.4 \quad (5)$$

and p_c is 0.83%. Thus, when 83.3% of the functional groups are converted to ester linkages, this material would gel. For a formulation of pentaerythritol and phthalic anhydride of $1:2$, the average functionality is 2.67 and the critical extent of reaction at gelation is 0.75. For formulations where one functional group is in excess, the analysis is far more complex. For example, if monomers A and B are polyols of differing functionality and C is a polyisocyanate added in excess, then the equation for f_{av} becomes

$$f_{av} = \frac{2rf_A f_B f_C}{f_A f_B + r\rho f_A f_C + r(1 - \rho)f_B f_C} \quad (6)$$

where the f_x is the functionality of monomer x and the constants r and ρ are given by

$$r = \frac{N_A f_A + N_B f_B}{N_C f_C} \quad (7)$$

and

$$\rho = \frac{N_B f_B}{N_A f_A + N_B f_B} \quad (8)$$

The critical conversion, p_c is still given by eq 4, but it is based on the conversion of the limiting functional group (in monomers A and B).

The preceding analysis highlights the need to have an accurate determination of the number of equivalents of functional groups in a component of a thermosetting resin.

When reactive oligomers with functionalities of 2 are used in combination with complementary low molar mass or oligomeric monomers, linear copolymers are produced. Again, Carothers has shown that the degree of polymerization in linear polycondensation is a function of the extent of reaction and stoichiometry. In general, the average degree of polymerization is defined as

$$\overline{DP} = \frac{1 + r}{r + 1 - 2rp} \quad (9)$$

where r, the stoichiometric imbalance, is defined as the ratio of the initial number of *limiting* functional groups to the initial number of functional groups in excess (and is of necessity less than or equal to 1). The extent of reaction, p, is the probability of complete consumption of the limiting functional group. When $r = 1$ (stoichiometric balance), the equation simplifies to

$$\overline{DP} = \frac{1}{1 - p} \quad (10)$$

Table 4.2 demonstrates the degree of polymerization as a function of the extent of reaction. Note that high degrees of polymerization are only reached when nearly all the functional groups have reacted. Using the example of the glycerol/phthalate resin given above, at a value for p equal to 0.833, the average degree of polymerization would only be 6 for a *linear* polycondensation reaction run to that extent of reaction.

Assuming complete reaction ($p = 1$), then the equation simplifies to

$$\overline{DP} = \frac{1 + r}{1 - r} \quad (11)$$

As shown in Table 4.3, small stoichiometric imbalances can lead to a dramatic decrease in the molecular weight of the final polymer.

Also, the intentional or accidental inclusion of monofunctional compounds further reduces r via the formula

$$r = \frac{N_A}{N_B + 2N_B'} \quad (12)$$

where N_A is the number of moles of functional group A (the

Table 4.2 Average degree of polymerization, $\overline{DP}$ versus the extent of reaction, p

p	$\overline{DP}$
0.900	10
0.990	100
0.999	1000

Table 4.3 Effect of stoichiometric imbalance, r, on the average degree of polymerization, $\overline{DP}$

Mole % of excess functional group	r	$\overline{DP}$
0.01	0. 9990	2,001
0.1	0.9901	201
1	0.9091	21

limiting functional group), N_B is the number of moles of functional group B, and N_B' is the number of moles of the monofunctional compound having a B-type functional group.

Functional Group Classes

A number of functional groups require quantitation. Table 4.4 gives a list of these, the polymers in which they are found, and a list of methods used to characterize them.

Hydroxyl Groups

Hydroxyl groups are found in polyglycols, polyesters, or polyurethanes synthesized with excess diol, and oligomeric epoxy resins. Hydroxyl groups are also present in polymeric polyols used as cross-linking agents in polyester and polyurethane resins. Partially hydrolyzed triglycerides are building blocks in alkyd resins. The *hydroxyl equivalent* is the number of equivalents of hydroxyl groups in 1 g of polymer. The more commonly used term is the *hydroxyl number*, which is the number of mg of potassium hydroxide (KOH) equivalent to the hydroxyl content in 1 g of polymer. For example, poly(ethylene glycol) with a molecular weight of 2 kg mol^{-1} has two OH end groups and has a hydroxyl equivalent of $2/2000 = 0.001$ and a hydroxyl number of

$$\frac{56.1 \text{ g/mol KOH} \times 2 \text{ OH/mol}}{2000 \text{g/mol PEG}} = 56.1$$

The basic procedure for determining hydroxyl groups is to react the polymer in a nonaqueous solvent with an excess of an acid anhydride (m moles), usually acetic anhydride or phthalic anhydride in the presence of a base catalyst. Each hydroxyl group is converted to n moles of ester and n moles of acid and leaves ($m - n$) moles of anhydride unreacted. The solution is then hydrolyzed with excess water to convert the remaining anhydride to acid giving a total of $2m - n$ moles of acid. A blank solution containing only the anhydride is also hydrolyzed to produce $2m$ moles of acid.

$$n \text{ ROH} + m \text{ R}'{-}\overset{\overset{\text{O}}{\|}}{\text{C}}{-}\text{O}{-}\overset{\overset{\text{O}}{\|}}{\text{C}}{-}\text{R}' \longrightarrow n \text{ ROCOR}'$$

$$+ n \text{ R}'\text{COOH} + (m - n) \text{ R}'{-}\overset{\overset{\text{O}}{\|}}{\text{C}}{-}\text{O}{-}\overset{\overset{\text{O}}{\|}}{\text{C}}{-}\text{R}'$$

Table 4.4 Functional groups found in typical polymers and the analytical methods used to quantitate them

Functional group	Classes of polymers	Analytical method
Aliphatic alcohols	Polyesters, polyethers, polyurethanes, epoxy resins, cellulosics	Acylation/titration
Aliphatic amines	Polyamides, polyureas, epoxy resins, amine terminated polybutadienes	Derivitization/UV
Alkoxides	Methacrylates and acrylates, cellulosics, polyvinylethers	Hydroiodic acid/GC
Carboxylic acids	Polyesters	Titration
Epoxy groups	Epoxy resins	HBr acidolysis, back-titration, derivitization/UV
Esters	Polyvinyl acetate, cellulosics, polyesters, alkyd resins	Saponification/GC
Isocyanates	Polyureas, polyurethanes	Urea formation/titration
Phenols	Polycarbonates, phenol-formaldehyde resins	Acylation/titration, bromination/titration
Si-H, Si-OH	Silicone rubber	IR
Unsaturation	Polybutadienes, silicone rubber, macromers	Bromination/titration, IR

$$\xrightarrow{\text{H}_2\text{O}} \quad n\,\text{ROCOR}' + \{n + 2(m - n)\}\,\text{R}'\text{COOH}$$
$$\text{or}$$
$$(2m - n)\,\text{R}'\text{COOH}$$

where ROH is the polymeric alcohol and R′ is either methyl (acetic anhydride) or phenyl (benzoic anhydride). Both solutions are titrated with standardized methanolic NaOH to a phenolphthalein endpoint. The volume of titrant used for the blank solution, B, provides a measure of the total amount of acid due to the anhydride ($2m$ moles). The volume of titrant used for the sample solution, S, provides of measure of the amount of acid not converted to ester ($2m - n$ moles). Using the equations

$$\text{hydroxyl equivalent} = \frac{(B - S)N}{W \cdot 1000}$$

$$\text{hydroxyl number} = \frac{(B - S) \cdot N \cdot 56.1}{W} \tag{13}$$

where B is the volume of titrant needed to neutralize the blank, S is the volume of titrant needed to neutralize the sample (both in milliliters), N is the normality of the titrant, and W is the weight of the sample (in grams). The details of the procedure are given in ASTM Standards D2849 and D4274. This includes corrections for any acid or base present in the initial alcohol.

A spectrophotometric assay of hydroxyl group concentration can be undertaken as well. Direct measurements of the hydroxyl group could be done using infrared spectrometry, but this approach suffers from interferences from water and other hydrogen-bonding species. A more useful assay is to derivatize the hydroxyl with a chromophore and determine the concentration via UV-visible spectrophotometry. To this end, a reactive naphthylsilane (3) is prepared from the re-

action of dimethyl amine and chlorodimethyl naphthylsilane. This is reacted with the hydroxyl functional polymer and the derivatized polymer is precipitated. The dried polymer is dissolved in a suitable solvent that is UV transparent at 282 nm. The molarity of derivatized end groups is given by the expression

$$n_{silyl} = \frac{A_{282.5}}{\varepsilon_{282.5}} = \frac{A_{282.5}}{7.33 \times 10^3} \tag{14}$$

assuming a 1-cm cuvette is used. Note that the molecular weight of the derivatizing agent is rather high (185 g mol^{-1}) requiring that polymer molecular weights derived from derivatives should be adjusted appropriately, especially for low molecular weight polyols. Phenolic hydroxyl can also be quantitated in this manner but it is not distinguishable from aliphatic hydroxyls. Phenols do react specifically with arylsulfonylchlorides, and spectrophotometric methods based on this have been developed.

Distinguishing between primary and secondary hydroxyl groups in polyols can be accomplished via either ^{13}C or ^{19}F nuclear magnetic resonance (NMR) spectrometry (ASTM D4273). In the former case, ^{13}C NMR can be used directly to quantitate the relative areas of primary [around $\delta(^{13}\text{C})$ 61 ppm] and secondary [around $\delta(^{13}\text{C})$ 65 ppm] hydroxyl carbons of simple polyglycols. Note that the method assumes equal intensity for methine (CH) and methylene (CH$_2$) carbons in proton-decoupled ^{13}C NMR and may give only semiquantitative data. However, it will be useful in a quality-control application. Alternatively, the polyol can be derivatized by acylation with trifluoroacetic anhydride to form trifluoroacetates and using ^{19}F NMR to quantitate the relative areas of the primary trifluoroacetates [around $\delta(^{19}\text{F})$ 75.6 ppm] and secondary trifluoroacetates [around $\delta(^{19}\text{F})$ 75.9 ppm]. Here quantitation is far more reliable, but it requires an NMR capable of operation in ^{19}F mode.

Carboxylic Acids

These groups are found in polyesters or polyamides synthesized with excess diacids or dianhydrides or with monomers that possess both an acid and a hydroxyl or an amine

3

group. Acids are also present in polyamic acids, the intermediate polymer formed in the synthesis of polyimides. Analogous to hydroxyl groups, the quantity of carboxylic acid is specified either as the *acid equivalent* (number of equivalents of acid groups in 1 g of polymer) or the *acid number* (number of mg of KOH equivalent to the acid content in 1 g of polymer). These are determined via direct nonaqueous titration, either with indicator or potentiometrically. Note that phenolic hydrogens are also acidic so that potentiometric titrations can be used to distinguish these two acid groups by their different potentiometric endpoints. Nonaqueous titration methods for carboxylic acids using indicators are given in ASTM D 1045 and D 4094.

Before leaving the determination of hydroxyl groups and acids, it is useful to discuss the end-group average molecular analysis of polyesters. In a typical polyester derived from diacids and diols, three classes of polymer chains may be present.

a. $HO_2C\wedge\!\wedge\!\wedge\!\wedge\!\wedge\!\wedge CO_2H$

b. $HO\wedge\!\wedge\!\wedge\!\wedge\!\wedge CO_2H$

c. $HO\wedge\!\wedge\!\wedge\!\wedge\!\wedge OH$

Since we require q, the number of detectable end groups, to calculate the molecular weight, this system presents a challenge. We first titrate a fixed weight of sample to a phenolphthalein endpoint with standardized KOH to determine the sample's *acid equivalent*. For the system above, we will have $(2a + b)$ equivalents of acid. We then independently react the sample with m equivalents of acetic anhydride as we would for a determination of the hydroxyl equivalent, obtaining the following products:

a. $HO_2C\wedge\!\wedge\!\wedge\!\wedge\!\wedge\!\wedge CO_2H$

b. $H_3COCO\wedge\!\wedge\!\wedge\!\wedge\!\wedge CO_2H + b\ CH_3CO_2H$

c. $H_3COCO\wedge\!\wedge\!\wedge\!\wedge\!\wedge OCOCH_3 + 2c\ CH_3CO_2H$

This produces $(2a + 2b + 2c)$ equivalents of acid (acetic acid and chain-end acids) and $[m - (b + 2c)]$ equivalents of unreacted anhydride. After hydrolyzing the anhydride, we get a total of $2m + 2a - 2c$ equivalents of acid. We perform a blank titration on the anhydride to obtain $2m$ equivalents of acid and subtract from that the number of equivalents of base needed in the sample titration, yielding the hydroxyl equivalents equal to $(-2a + 2c)$. Adding the hydroxyl equivalents to two times the acid equivalents, $(4a + 2b)$, yields $2(a + b + c)$. Thus the molecular weight of the polyester is calculated by the equation

$$\overline{M}_n = \frac{2W}{2 \times eq\ COOH + eq\ OH} \qquad (15)$$

where W is the mass of the sample (in grams).

Amines

Amine groups are found in polyamides, polyureas, and some amine-terminated polybutadienes. Polymeric poly-

amines and polyamidoamines are also used as cross-linking agents in epoxy and polyurea resins. Since amines are basic, they can be quantitated directly by nonaqueous titration. Useful solvents include benzyl alcohol (b.p. 175 °C) for homopolyamides and phenol:methanol (7:3). The polymer is dissolved at high temperature or at reflux and allowed to cool. The amino groups are titrated with standardized 0.1N HCl, either to a thymol blue endpoint or potentiometrically. A more sensitive approach is to derivatize the amine groups via aromatic nucleophilic substitution with 2,4-dinitrofluorobenzene in the presence of sodium bicarbonate. The yellow polyamide is precipitated, and the end group concentration is determined spectrophotometrically in 2-methylphenol(o-cresol) at 430 nm. Other derivatization reagents include 4-dimethylaminobenzaldehyde and ninhydrin (1,2,3-indanetrione hydrate). A calibration curve should be prepared by using a long-chain primary amine derivative as a reference sample. *Please note that o-cresol is a strong organic acid that can cause significant burns and therefore should be handled with appropriate safety equipment.*

Isocyanates

Isocyanates are the building blocks of polyurethanes and polyureas. They are quantitated (ASTM D 1638) by reaction with excess dibutyl amine to form the N,N-dibutylurea. Then excess unreacted amine is titrated potentiometrically (to a pH of 4.1 to 4.2 endpoint) with 1N hydrochloric acid to determine the amount of isocyanate. The *amine equivalent* of a polyisocyanate is the number of grams of polymer that will combine with 1 g of dibutylamine and is calculated as follows:

$$amine\ equivalent = 1000\ W/(B - S)N \qquad (16)$$

where W is the weight of sample (in grams), B is the volume of titrant needed for the blank (in milliliters), S is the volume of titrant needed for the sample (in milliliters), and N is the normality of the titrant.

Epoxy Groups

Epoxy groups are found in a variety of polymers. As glycidyl ethers

$$\left(\begin{array}{c} O \\ / \backslash \\ CH_2{-}CH{-}CH_2{-}O{-} \end{array}\right)$$

they are found in linear bis-phenol-A-based epoxy resins, epoxy-modified novolac resins, and reactive polyacylates and siloxanes. There are also a number of cyclohexene oxide-based epoxy systems. All are used in thermosetting systems cross-linked either by polyamines or via anhydrides. Epoxy groups are quantitated (ASTM D1652) by titration of the material with 0.1N HBr in glacial acetic acid (generated by bubbling anhydrous hydrogen bromide though glacial acetic acid and standardized against potassium acid phthalate) to a crystal violet endpoint. The HBr reacts with the epoxy groups to form a bromohydrin. A blank titration is also carried out. The *epoxy content* is the number of gram equivalents of epoxy groups per 100 g of resins and is calculated as follows:

$$\text{epoxy content} = N(S - B)/10W \qquad (17)$$

where N is the normality of the titrant, S is the volume of titrant used for the sample (in milliliters), B is the volume of titrant for the blank (in milliliters), and W is the weight of the sample (in grams). The weight per epoxy equivalent (WPE) is the number of grams of resin containing one gram equivalent of epoxy groups and is calculated as follows:

$$\text{WPE} = 1000S/N(S - B) \qquad (18)$$

Unsaturation

Unsaturation (carbon-carbon double bonds) is present in linear polymers based on the addition polymerization of diene monomers (butadiene, isoprene, chloroprene); ring-opening polymerization of cyclic olefins (polypentenomer from cyclopentene); and polycondensations involving unsaturated diacids (for example, maleic anhydride). These unsaturations are used in forming thermosetting resins and elastomers. The olefin content of these polymers is determined by the electrophilic addition of a halogen, usually bromine or iodine monochloride, to the double bonds and the redox titration of the remaining halogen with 0.1N sodium thiosulfate. Bromine is generally generated in situ using a mixture of potassium bromide and potassium bromate, whereas iodine monochloride is generated in situ by mixing iodine trichloride and molecular iodine. In the case of ICl, excess potassium iodide is added to convert it to molecular iodine before it is tritated. The degree of unsaturation is reported as an *iodine value*, which is defined as the number of grams of iodine that are chemically bonded by 100 g of polymer and is calculated as follows:

$$\text{iodine value} = 12.69(B - S)/W \qquad (19)$$

where B is the volume of titrant used for a blank (in milliliters), S is the volume of titrant used for the sample (in milliliters), and W is the weight of the sample (in grams).

Reactive Siloxane Polymers

Silicone fluids and gums (high molecular weight, linear polymers) often contain reactive groups, thus allowing for cross-linking reactions to produce elastomers. These groups include vinyl, alkoxysilyl (SiOR), hydrosilyl (Si-H), silanol (SiOH), and acetoxysilyl ($SiOCOCH_3$). The first can be quantitated by adding ICl followed by redox titration (see above). The presence of SiH is an interference, as it reduces ICl. The

reaction of mercuric acetate with terminal vinyl compounds can be used to quantitate vinyl silanes.

$$\equiv Si-CH=CH_2 + Hg(OAc)_2 \xrightarrow{\text{MeOH}}$$

$$\underset{\underset{OMe}{|}}{\equiv Si-CH-\overset{\overset{HgOAc}{|}}{CH_2}} + HOAc$$

The liberated acetic acid (1 equivalent per vinyl group) is titrated with potassium hydroxide after the excess mercuric acetate is reacted with calcium chloride or sodium bromide. SiH is an interference in this system as well. *It is worth noting that organomercury compounds are severely toxic, and appropriate precautions in their handling and disposal should be used.*

Alkoxysilyl groups can be quantitated by perchloric acid acetylation. A sample is added to a 1M to 2M acetic acid and a 0.06M to 0.15M perchloric acid solution in ethyl acetate or 1,2-dichloroethane. After 1 h, water, DMF, and pyridine are added to hydrolyze unreacted anhydride and acetoxysilanes formed. The reaction consumes 1 equivalent of acetic acid per alkoxysilyl group based on the following reaction scheme:

$$Ac_2O + H^+ \longrightarrow Ac_2OH^+$$

$$Ac_2OH^+ \longrightarrow Ac^+ + AcOH$$

$$Ac^+ + SiOR \longrightarrow \underset{\underset{Ac}{|}}{\overset{+}{Si-OR}}$$

$$\underset{\underset{Ac}{|}}{\overset{+}{Si-OR}} + Ac_2O \longrightarrow SiOAc + AcOR + Ac^+$$

Hydrolysis liberates 2 equivalents of acetic acid per remaining unreacted anhydride and 1 equivalent from the acetoxysilane. The net equation is:

$$SiOR + Ac_2O + H_2O \longrightarrow SiOH + AcOR + AcOH$$

The alkoxysilane equivalent is calculated as per the hydroxyl equivalents in eq 13.

Hydrosilyl groups can quantitated by measuring the volume of H_2 evolved when they react with an alcohol under base-catalyzed conditions.

$$Si-H + HOR \xrightarrow{\text{Base}} SiOR + H_2$$

The manometric procedure requires an appropriate mercury or digital manometer system and may provide difficult unless it is practiced routinely. The only interference is from disilanes (Si-Si). Analogous to the determination for vinylsilanes, SiH can be determined by the reduction of mercuric to mercurous acetate with the liberation of 2 equivalents of acetic acid per equivalent of SiH.

$$\text{—SiH} + 2\ Hg(OAc)_2 + CH_3OH \longrightarrow$$

$$\text{—Si—OCH}_3 + Hg_2(OAc)_2 + 2\ HOAc$$

Silanol (Si—OH) can be quantitated manometrically by reaction with metal hydrides (LiAlH$_4$) to liberate hydrogen. Free silanols readily condense to form siloxanes with the liberation of water. Acid-catalyzed condensation of silanols and the subsequent quantitation of liberated water, either volumetrically in a Dean-Stark apparatus or titrametrically with Karl-Fischer reagent, can be used to determine the amount of silanol groups. Direct titration of silanols is accomplished with the highly basic titrant lithium aluminum di-*n*-butylamide as per the reaction

$$LiAl[NBu_2]_4 + x\ SiOH \longrightarrow$$

$$x\ Bu_2NH + LiOSi + Al(OSi)_3$$

The titrant is air and moisture sensitive, so appropriate apparatus and conditions must be used.

Bibliography

Annual Book of ASTM Standards, Section 8: Plastics; American Society for Testing and Materials: Philadelphia, published annually.

Collins, E. A.; Barés, J.; Billmeyer, F. W. Jr. *Experiments in Polymer Science*; John Wiley & Sons: New York, 1973; pp 362–367.

Crompton, T. R. *Practical Polymer Analysis*; Plenum Press: New York, 1993; pp 241–255.

Crompton, T. R. *Manual of Plastics Analysis*; Plenum Press: New York, 1998.

Sandler, S. R.; Karo, W.; Bonesteel, J.-A.; Pierce, E. M. *Polymer Synthesis and Characterization: A Laboratory Manual*; Academic Press: New York, 1998; pp 163–171.

Schröder, E.; Müller, G.; Arndt, K.-F. *Polymer Characterization*; Hanser Publishers: New York, 1989; pp 18–31.

Smith, A. L. *The Analytical Chemistry of Silicones*; Wiley: New York, 1991.

PART II

CHROMATOGRAPHIC METHODS

5

Gel Permeation Chromatography

STEVEN K. POLLACK

Often, simply having the number average and/or weight average molecular weights, $\overline{M_n}$ and $\overline{M_w}$, of a polymer does not provide sufficient information to explain differences in the performance of two presumably identical batches of materials. Subtleties in the balance between small amounts of high or low molecular weight materials or in the blending of polymers of different molecular weights can have catastrophic consequences on the reproducibility of polymer processing and performance. Gel permeation chromatography (GPC), also known as size exclusion chromatography (SEC), is the most straightforward method for obtaining the complete molecular weight distribution (MWD) of a polymer sample. In addition to the information on the complete MWD, from the analytical data one can derive $\overline{M_n}$ and $\overline{M_w}$, as well as the viscosity and z-average quantities. As an additional benefit, the presence or absence of low molecular weight additives can often be determined simultaneously.

The formulas for the molecular weight averages that are cast in terms of sample weight are most useful for this analytical technique:

$$\overline{M_n} = \frac{\sum_{i=1}^{N} w_i}{\sum_{i=1}^{N} \frac{w_i}{M_i}} \qquad \overline{M_w} = \frac{\sum_{i=1}^{N} M_i w_i}{\sum_{i=1}^{N} w_i}$$

$$\overline{M_z} = \frac{\sum_{i=1}^{N} M_i^2 w_i}{\sum_{i=1}^{N} M_i w_i} \qquad \overline{M_\eta} = \left(\frac{\sum_{i=1}^{N} (M_i)^a w_i}{\sum_{i=1}^{N} w_i} \right)^{1/a} \qquad (1)$$

where w_i is the mass of the ith fraction of polymer in a sample, and M_i is the molar mass (molecular weight) of the molecules in the ith fraction. M_i is further defined as

$$M_i = x_i M_0 \qquad (2)$$

where x_i is the degree of polymerization of these chains, and M_0 is the molar mass of the structural repeat unit. The exponent "a" in the viscosity average molecular weight, $\overline{M_\eta}$, is the Mark-Houwink exponent. For GPC to be able to provide us with a measure of these averages, it must be able to obtain values of w_i and M_i for fractions of a polymer sample. The detection system provides for the former, while the separation process used either alone or in conjunction with specialized detectors provides the second.

These averages provide characteristic values that describe a given sample of polymer but do not completely describe its molecular weight distribution. The complete distribution can be of great importance because molecular weight has a strong effect on the physical properties of polymers. Figure 5.1 shows the typical behavior that most physical properties of a polymer exhibit as a function of molecular weight.

As Figure 5.1 indicates, until some critical molecular weight is reached, there is a strong dependence of molecular property on molecular weight. Above this molecular weight, the *limiting property* is reached. Trying to fabricate a part using polymers whose molecular weight falls in the former region runs the risk of highly irreproducible manufacturability and end-use behavior of the material. The logical solution is to make the polymers of as high a molecular weight as possible, avoiding the region of unstable properties. However, there is one property whose behavior is different—that of viscosity. Viscosity is a measure of a molten (or dissolved) polymer's resistance to flow. As we see from Figure 5.1, viscosity increases monotonically with molecular weight. Since many thermoplastic materials are fabricated by using the melt-flow properties of the polymer, changes in viscosity will affect the *processability*: as melt viscosity increases, the pressures and temperatures needed to fabricate a part increases. In addition to limitations on the machinery used in this fabrication, these conditions can lead to chemical breakdown of the polymers and can degrade the final physical properties. Thus we must keep molecular weight within a *manufacturing window*.

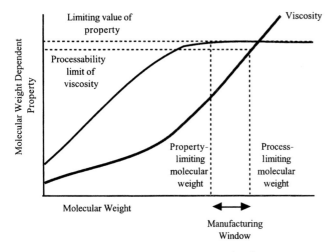

Figure 5.1 Molecular weight versus general polymer properties.

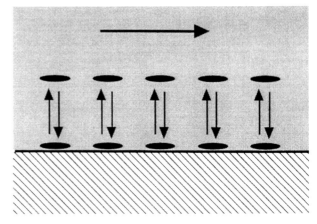

Figure 5.2 Absorption-based separation.

While the average molecular weight is an important material parameter, in many cases the breadth or shape (or both) of the molecular weight distribution can affect the bulk mechanical properties of a polymer. The presence of low molecular weight fractions can increase the toughness of a polymer (its ability to withstand mechanical stress before failure), but, in excess, low molecular weight can lower the strength. The presence of high molecular weight fractions can increase strength but can also increase brittleness and viscosity. GPC provides a determination of the complete molecular weight distribution. Before we describe the GPC experiment, a brief description of the more familiar adsorption chromatography is given to contrast it with the size exclusion based separation.

Adsorption Chromatography

In conventional adsorption chromatography, the separation of materials in a mixture is a consequence of their dynamic partitioning between the mobile phase and the stationary phase due to enthalpic interactions. There is an equilibrium between dissolved molecules and surface-adsorbed molecules that governs the relative elution time of each component (see Figure 5.2).

This equilibrium is determined by the specific interactions (such as hydrogen bonding, van der Waals interactions, Coulombic attractions, etc.) between the molecules and the surface versus their interactions with the solvent. Molecules with a larger number of sites for interaction will elute more slowly. Since van der Waals interactions are caused by the polarization of electrons in molecules, the larger the molecule, the larger the number of van der Waals interactions. However, chain molecules adsorbed on surfaces are subject to a decrease in their configurational entropy over their entropy in solution, and, as such, this process may be energet-

ically disfavored. Control of the selectivity of the process is dictated by changes in the surface chemistry of the stationary phase and the eluting power of the solvent.

Size Exclusion Chromatography

In the case of size exclusion chromatography, the mechanism of separation involves an equilibrium between molecules in the mobile phase and those molecules capable of entering (permeating) the internal volume of a porous stationary phase (see Figure 5.3).

Those molecules whose hydrodynamic volume (effective size in solution) makes their radius much smaller than the radius of the stationary phase's pores will freely exchange between the void volume, v_0, (the volume surrounding the

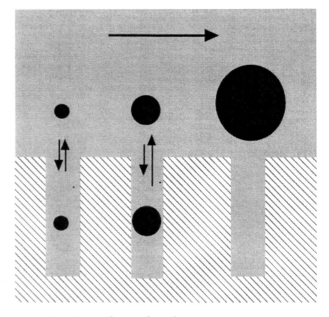

Figure 5.3 Size exclusion–based separation.

stationary phase), and the internal volume of the stationary phase, v_p. This increases the elution time for these molecules. Those molecules significantly larger than these pores are restricted to the void volume and elute rapidly. Molecules with sizes on the order of the diameter of the pores will establish an equilibrium distribution between the stationary phase and the mobile phase. The magnitude of the equilibrium constant will depend on how readily those molecules can deform to accommodate passage into the pore. As stated, any deformation of a polymer molecule from its equilibrium random coil conformation leads to a decrease in its entropy and thus will be disfavored. Therefore, the greater the deformation required, the less probable is permeation. This predicts a relationship between molecular size and elution time such that larger molecules elute first and smaller molecules elute later. This is opposite of what would be expected for adsorption processes, as the larger molecules would experience a greater number of interactions per molecule with the surface and thus be eluted later. In fact, it is critical to minimize all adsorption processes if the GPC measurement is to give useful data.

The typical relationship between molecular size (which for a given polymer is proportional to molecular weight) and elution volume can be seen best in Figure 5.4. There are three regions of behavior. In Regime 1, at low elution volume, all molecules above a given molecule weight elute at the same volume, which corresponds to v_0. This is the regime of *total exclusion*. Regime 3 corresponds to the largest elution volumes, and all molecules whose molecular weights are below some threshold will elute simultaneously, and this volume corresponds to $v_0 + v_p$. This is the region of *total permeation*. Regime 2 displays a monotonic increase in elution volume with decreasing molecular weight. The potential exists in this region of *selective permeation* to relate elution volume to molecular weight and ob-

tain quantitative data on the molecular weight distribution and on the average values already mentioned here.

To be useful, we must have some relation between the elution volume and molecular weight. The total volume of solvent that passes through the GPC column is the sum of the void volume v_0 (volume of solvent outside the porous beads) and the internal volume v_p (the volume of solvent in the pores). The volume of solvent needed to elute a particular species from the point of injection to the detector is known as the *elution volume* v_e and, based on separation by size exclusion mechanism only, is given by

$$v_e = v_0 + K_{SE}v_p \qquad (3)$$

where K_{SE} is the fraction of the internal pore volume penetrated by those particular polymer molecules. For very small molecules that can penetrate all the pores, $K_{SE} = 1$, and $v_e = v_0 + v_p$, whereas for very large molecules, $K_{SE} = 0$ and $v_e = v_0$. For molecules of intermediate size, K_{SE} and v_e lie between these limits.

Flow rates in the GPC experiment are typically around 1.0 cm^3 min^{-1}, are chosen to give enough time for the molecules to diffuse in and out of the pores. Under these circumstances, equilibrium is established between the concentration of polymer within the pores, c_p, and that of the polymer in the void volume, c_0. Thus, we can think of K_{SE} as an equilibrium constant defined as

$$K_{SE} = \frac{c_p}{c_0} \qquad (4)$$

The thermodynamics of this equilibrium and the Gibbs free energy for permeation is given by

$$\Delta G_p^0 = -RT \ln K_{SE} \qquad (5)$$

If the separation is exclusively by size exclusion, and involves no specific interactions of the polymer with the stationary phase, then the enthalpy change of the process is zero, and ΔG_p^0 must be controlled exclusively by entropic effects. Thus $\Delta G_p^0 = -T\Delta S_p^0$, where ΔS_p^0 is the change in entropy as a polymer chain enters a pore, and therefore

$$K_{SE} = \exp(\Delta S_p^0/R) \qquad (6)$$

If we confine a polymer coil into a pore, there will be a reduction in the number of available conformations, and thus a lowering of the entropy of the chain. It can be shown[1] for a gaussian (random) coil in a cylindrical pore that

$$\Delta S_p^0 = -RA_s(\bar{L}/2) \qquad (7)$$

where $\bar{L}$ is the mean diameter of the molecules and A_s is the surface area per unit pore volume. This leads to the expression

$$K_{SE} = \exp(-A_s\bar{L}/2) \qquad (8)$$

and the expression for the elution volume of

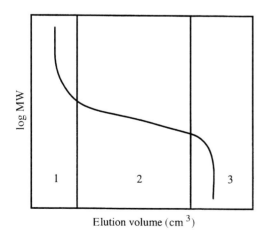

Figure 5.4 Typical relationship between elution volume and log(MW) in Regimes 1–3. Key: 1 = total exclusion; 2 = selective permeation; 3 = total permeation.

$$v_e = v_0 + v_p \exp(-A_s \bar{L}/2) \qquad (9)$$

This expression implies that v_e should vary exponentially with molecular size, or linearly with the logarithm of molecular size.

It is important to note here that molecular size or, put in formal terms, the *hydrodynamic volume* V_H, of the polymer chains in solution controls the relationship between molecular weight and elution volume. This volume is based on the volume of an equivalent sphere whose radius corresponds to the root mean square radius of gyration, $\langle s^2 \rangle^{1/2}$, of the polymer chain in solution. The theories that relate $\langle s^2 \rangle^{1/2}$ to molecular weight show that it is a strong function of molecular weight, polymer-solvent interactions and temperature.[1]

Figure 5.5 demonstrates how polymers of differing molecular weight can have the same value for V_H. The leftmost polymer is highly coiled and therefore many more monomer units can fit inside the equivalent sphere. The center polymer is less coiled and therefore fewer monomer units can be accommodated. Finally in the rightmost polymer, a rigid rod conformation is adopted and the smallest number of monomer units can fit inside the equivalent sphere. *All three of these polymers would have the same elution volume!* Taking the polymer on the left, if the solvent or temperature were changed to one that creates more favorable polymer-solvent interactions, this polymer would be less tightly coiled in solution and would, of necessity, require an equivalent sphere of larger diameter to contain it. In turn, this would decrease its elution time and, using the relationships between hydrodynamic volume and elution volume shown, have a larger *apparent* molecular weight.

The GPC Experiment

The typical GPC system consists of several subsystems. A schematic is given in Figure 5.6. We will describe each subsystem and discuss important operating procedures.

Solvents

Solvent is delivered to the GPC system from a solvent reservoir. The solvent reservoir can be as simple as the bottle the

Figure 5.5 Schematic representation of three different polymers (1, 2, and 3) of increasing chain stiffness possessing the same effective hydrodynamic volume.

solvent comes shipped in, or it can have a special design to maximize solvent use as well as fittings to degas the solvent continuously. Since most GPC solvent delivery systems are reciprocating piston pumps,[2] the filling stoke of the pump reduces the pressure on the solvent and can cause dissolved gases to come out of solution and form bubbles. These bubbles then lead to problems with pumping and detection and can potentially damage the GPC columns. To minimize this, solvents should be degassed. This is accomplished in several ways. At a minimum, solvents in their containers can be placed in an ultrasonic bath and sonicated for 2 to 5 min to remove dissolved gasses. The dissolved gasses can be displaced by purging with helium. Helium is sparingly soluble in most organic solvents and entrains the other dissolved gases from the solvent. This is typically done by having a gas line with a sintered metal fitting placed in the solvent reservoir. In some GPC systems, the control system of the GPC will intermittently turn the purge gas on and off. More sophisticated continuous degassing arrangements involve using a gas-porous membrane through which the solvent flows on one side and a vacuum is maintained on the opposite side. Generally, the solvent is transferred from the reservoir to the solvent delivery system via a teflon or polyethylene tube with a fritted steel filter at the end.

Solvents used for GPC need to be free of particulates that may plug the columns or increase their operating pressure, or both. Commercially prepared high-performance liquid chromatography (HPLC) grade solvents have been filtered and can be used as supplied. If non-HPLC grade solvents are to be used, if waste solvents re-used, or if modifiers have been dissolved in the solvent, then the solvent should be filtered through a pore size no greater than 0.45 µm before they are used. This is best done with a vacuum solvent filtration system. If a UV/vis detector is to be used, then the stated wavelength cutoff for the solvent should be noted, and the detector must be adjusted accordingly. Ethereal solvents such as tetrahydrofuran (THF) can accumulate peroxides over time unless they are stabilized by inhibitors such as butylated hydroxytoluene (BHT). They should be tested for peroxide buildup and discarded when excessive peroxides are detected. Other stabilizers are added to solvents for high-temperature GPC to minimize oxidation of the solvent and the polymers. Some manufacturers of GPC columns advise using unstabilized THF. *Be sure to read the literature supplied with the column before using any new solvent.*

Solvent Delivery System

Most modern GPC systems use a single- or multiple-head reciprocating pump to deliver solvent at the required pressures and flow rates. Systems with sophisticated controllers and multiple solvents are rarely needed. The pump must be capable of delivering flow rates from 0.1 to 10.0 mL/min with less than 1% variation in flow rate at pressures up

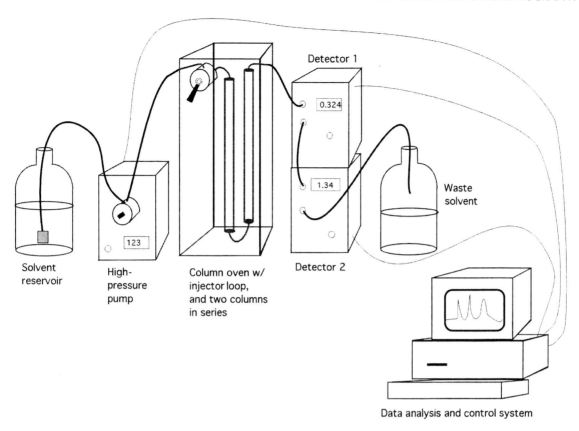

Figure 5.6 Setup of typical gel permeation chromatography (GPC) system.

4000 psi (276 bar, or 27.6 MPa). The flow must also be pulsation-free to minimize detector artifacts. Solvent delivery systems require routine replacement of the high-pressure seals and check valves, especially when they are used continuously with aggressive solvents. Also, failure to remove particulates from solvents will shorten the life of the seals. Generally, the pump has a capability (either electronic or manual) to temporarily redirect the solvent flow to an auxiliary outlet to allow for solvent changes or pump priming without sending solvent to the columns. Priming is typically done by opening the auxiliary outlet and running the pump at a flow rate of 10 mL/min. This is usually accomplished with a "purge" or "prime" setting on the pump. This priming cycle is continued until there is a continuous stream of solvent coming from the outlet. Then the priming cycle is stopped, and the auxiliary outlet is closed. Failure to achieve continuous flow can indicate excessive air bubbles, a restriction in the solvent supply tubing, or worn pump seals. Most pumps can be programmed to stop in the event that the pressure of the system exceeds some predefined limit; this value should be set to one low enough to protect the columns from damage (see manufacturer's recommendations). Similarly, most pumps can and should be programmed to stop when the pressure is too low, which may be caused by either a leak or air being entrained into the solvent delivery system when solvent levels in the sol-

vent container become too low. For convenience of operation, many pumps can be programmed to start at a given time to allow for the system to be equilibrated before it is put into operation. Finally, a number of pump manufacturers have incorporated vapor sensors into the pumps to shut them down if there is a solvent leak near the pump. This can be the cause of much lost time if the operator is careless with spilled solvents from samples near the pump.

Sample Injector

The operating pressure in GPC systems can be as high as 2200 psi. Attempting to introduce a polymer solution into the flow stream with a conventional syringe is not feasible. Therefore, high-pressure injectors are used to introduce the sample. The typical manual or semiautomated injector consists of two independent flow paths, which can be re-routed via a high-pressure sliding seal. In Figure 5.7, the left schematic represents the position of a six-port valve during sample loading. The light gray path is the low-pressure path for the sample, from the syringe or auto-injector, through the sampling loop and out to waste. The dark gray path represents the high-pressure flow path, from the pump to the column. The sample is introduced into the sampling loop in a sufficient quantity to flush out any retained material from a previous injection. The valve is then rotated to the inject

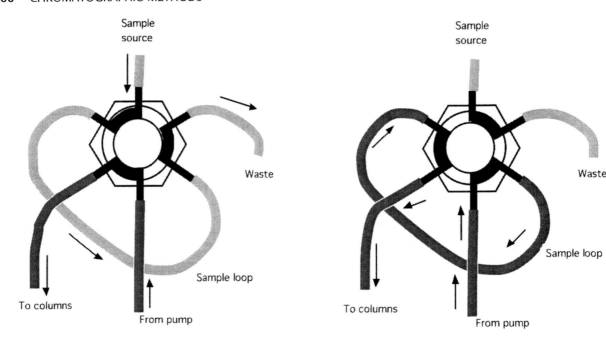

Sample Loading Position   Sample Inject Position

Figure 5.7 Six-port sampling valve configured with sample loop. Key: light gray = low-pressure path; dark gray = high-pressure path.

position. Now the sample loop is in the high-pressure flow path, and the sample introduction path is bypassed. All material in the sample loop is introduced into the solvent stream. Other configurations exist, but the basic principle remains the same. If the polymers being analyzed are soluble in solvents only at high temperatures, then the entire injection valve assembly also must be heated. Typically, the injection and detection system may be installed inside the same oven that houses the columns. Samples are loaded into the injector via syringe. In most cases, a blunt-tip needle must be used to avoid damaging the sliding high-pressure seal. *Do not use syringes designed for gas chromatography!!* Samples are typically made up as 0.1% to 0.5% (w/w) solutions of the polymer in the mobile phase. One must avoid making up samples in a different solvent, for then the polymer may precipitate in the injector or, worse, in the columns. All samples, regardless of purity, should be filtered through a 0.22- or 0.45-μm filter before they are injected, to minimize chances of plugging the column.

For laboratories with a high sample throughput, an auto-injector system is often used in conjunction with a data acquisition system. The only difference is that the auto-injector withdraws the samples from vials. Be sure to observe all the other concerns mentioned in this chapter.

Columns

To provide for the size exclusion process to occur efficiently, the stationary phase must provide several key properties:

- Minimal specific interactions
- Well-defined pore sizes
- Minimal void volume
- Tolerance to a broad range of samples, solvents and temperatures

Modern GPC columns meet most of these requirements. The typical GPC column used for polymers that are soluble in organic solvents is fabricated from polystyrene cross-linked with divinyl benzene (PS/DVB). The use of a nonpolar material like polystyrene minimizes specific interactions. In its early stages, GPC used stationary phases based on porous glass beads, and their highly polar nature led to significant interference from adsorption processes. These were minimized by chemical treatment and the addition of modifiers to the solvent to passivate the surfaces of the glass. Such treatments are not necessary for the PS/DVB-based columns. However, prolonged use, especially at high temperatures or in prolonged contact with aggressive solvents, can lead to the development of polar sites on the surface of the column. Therefore, these materials have a finite useful working time and must be replaced periodically. For aqueous GPC, polyhydroxylated or sulfonated cross-linked phases are used.

The porosity of the polymer is controlled during manufacture by the addition of a nonsolvent (referred to as a *porogen*) that serves to control the porosity during the cross-linking processes. After polymerization, the gel beads are cleaned, and the porogen is extracted to leave insoluble polymer beads with pore diameters ranging from 5 to 10^6 Å (0.5 to 10^5 nm). The physical dimensions of the beads are also controlled during the polymerization process and can range from 3 to 10 μm in diameter as the application requires. The control of the bead's diameter is the key to minimizing void volume. It can be shown that packing of particles is most efficient when they are of nearly equivalent size and of the smallest possible diameter. The lower limit for bead size is imposed by the increase in the pressure required to create a reasonable solvent flow rate through a packed column of these beads. Above a critical point, the hydrostatic pressure of the solvent permanently deforms the gel beads. Most commercial columns will withstand up to 2100 psi (150 bar) of pressure and tolerate flow rates up to 3.0 mL/min. *However, the pressure limit is clearly stated on commercially prepared columns and should be adhered to if the columns are to provide reproducible results.* Also note that most columns have a designated flow direction. Switching the direction of flow through the column can disturb the integrity of the packing material and degrade column performance.

While the PS/DVB beads are not soluble in the mobile phase, they do swell to an equilibrium diameter in a solvent. The extent of swelling represents a balance between the bead/solvent interactions and the elastic properties of the bead. Typical commercial GPC columns come packed in solvents such as toluene or tetrahydrofuran (THF). To minimize changes in the void volume and damage to the column, the column must always be maintained in a solvent-swollen condition. When not in use, the columns should be well sealed with column caps. Failure to do this can lead to column shrinkage or cracking. Re-swelling does not always restore the column to its original condition. While columns can be "re-packed", their performance is rarely equivalent to that of the original column.

One can change solvents for a given application, either to allow for dissolution of a polymer not soluble in the "as-shipped" solvent or to allow for better sensitivity by the detector (see below). There are two extremes. For more "robust" packing materials, one simply pumps through the new solvent as a relatively slow flow rate (0.1–0.5 mL/min) for a sufficient time to exchange two column volumes (time for the lowest molecular material to completely elute) of solvent. For example, if the column volume of the column is 18 mL, then at 0.5 mL/min flow the new solvent is introduced for 72 min. If the solvents are very different in their polarity or in interactions with the stationary phase, an intermediate solvent, typically acetone, is used. In some cases, the viscosity of the solvent at room temperature does not allow for establishing reasonable flow rates at the column's pressure limit. In this case, the columns can be heated, usually in a constant temperature oven designed to hold both multiple columns and the sample injector. Solvent heating is also used when polymers are soluble in the solvent only

at high temperatures. This is common for the GPC of polyolefins. In these analyses, dichlorobenzene is typically used as the solvent, and temperatures as high as 150 °C are used. Since most polymer/solvent interactions are temperature dependent, controlling column temperature is crucial for reproducible results. *As stated, carefully follow the manufacturers' suggestions for limits of temperature and range of applicable solvents.*

The pore size of a GPC column determines over what range of molecular weights the elution volume will have a predictable relationship to hydrodymic volume and therefore to molecular weight. For example, Waters, Inc., μ-Styragel columns have the ratings shown in Table 5.1.

Standard analytical GPC columns come in 300- or 600-mm lengths and typically have a 7.5-mm diameter. The greater the length of the columns, the greater the resolution of the columns for a given molecular weight range. When the range of molecular weight in the sample exceeds the linear range of a single column, a common approach is to use multiple columns in series. Adding columns of different pore sizes extends the linear range for molecular weight discrimination, but there is a practical limit. Typical operating pressures for a single 300-mm length GPC column operating at a flow rate of 1.0 mL/min is 750 psi (~50 bar). Each new column added in series increases the overall pressure and the experimental time. With three such columns, the system pressure becomes 2250 psi. This is at the typical limiting pressure for commercial GPC columns. Two or three columns in series is often a practical limit. Other manufacturers of single-range GC columns include Shodex and Polymer Laboratories.

Another more recent development is the availability of high-efficiency "mixed-bed" columns. In these, gel beads with a range of pore sizes are packed into the same column. This has become possible with tighter control on bead diameter. To increase resolution, one simply uses multiple mixed-bed columns. For example, Polymer Laboratories PLgel MIXED-C analytical columns claim a linear molecular weight from 200 to 2,000,000 g/mol. Their MIXED-D columns are optimized for a linear range from 200 to 400,000 g/mol, more optimal for lower molecular weight materials. Finally, MIXED-E columns are used for oligomers, with a

Table 5.1 Relationship of pore size to linear calibration ranges for Waters μ-Styragel columns

Pore size (in Å)	Approximate MW fractionation range
10^6	1×10^5 to 1×10^7
10^5	1×10^4 to 1×10^7
10^4	7×10^3 to 2×10^6
10^3	4×10^2 to 4×10^5
500	1×10^2 to 8×10^4
100	$<1 \times 10^2$ to 3×10^3

linear molecular weight range up to 30,000 g/mol. Typically two to three 300-mm columns are used in series for common analytical situations. Mixed-bed columns tend to be somewhat more expensive than single-porosity columns, but their more general usefulness may make them more economical overall. Other manufacturers of mixed-bed columns include Waters and Shodex.

Detectors

The determination of molecular weight averages using GPC will require knowing both the molecular weight (M_i) and the mass (w_i) of a given fraction of polymer eluting from the GPC columns. What is usually detected is the mass concentration, $c_i = w_i/V$ (in units of g/mL). If one looks at the equations defining the molecular weight averages, the volume can be factored out and the same relationships hold.

$$\overline{M} = \frac{\sum\limits_{i=1}^{N} M_i w_i}{\sum\limits_{i=1}^{N} w_i} = \frac{\sum\limits_{i=1}^{N} M_i w_i \frac{1}{V}}{\sum\limits_{i=1}^{N} w_i \frac{1}{V}} = \frac{\sum\limits_{i=1}^{N} M_i \frac{w_i}{V}}{\sum\limits_{i=1}^{N} \frac{w_i}{V}} = \frac{\sum\limits_{i=1}^{N} M_i c_i}{\sum\limits_{i=1}^{N} c_i} \quad (10)$$

The same is true for the other weighted averages.

The most general detector used in GPC is a differential refractive index (DRI) detector. In one common configuration, the DRI detector consists of a two-compartment optical cell (see Figure 5.8). The first compartment holds pure solvent, and the second compartment is in the flow path of the GPC system. The two compartments are separated by a thin glass membrane set at an angle. Light is directed through the cell, and it impinges on the split diode detector. This detector produces a signal proportional to the difference in light levels falling on both sides. As shown in the

drawing, when the refractive indices of the fluids in both compartments are matched, and the optical null plate has been properly adjusted (either manually or automatically), equal amounts of light fall on both halves of the detector and a zero signal is recorded (light gray line). When the refractive index of the liquid in the sample side differs from that of the reference side, then the beam is deflected and we get a measurable signal (dark gray line). The typical modern DRI has internal valving that allows filling the reference side with pure solvent. This is done when switching solvents or when there is concern that the chemical composition of the reference solvent has changed. The latter is common for ethereal solvents, where peroxide formation leads to changes in the refractive index. Finally, the thin glass membrane that separates the two chambers is not robust enough to withstand high pressures. This means that either the DRI detector is the only detector in the system, or it is the last in a series. Any backpressure on the outlet of detector can damage the cell. Since one side of the flow cell is at atmospheric pressure, there is a rapid pressure drop as a sample enters the cell. This can cause dissolved gasses to form bubbles (similar to the bends deep-sea divers experience after rapid decompression) that then cause large detection artifacts. Even with all these stated problems, DRI is most often the detector of choice for compounds without specific chromophores.

The refractive index of the solution of polymer is given by the equation

$$n = n_0 + \frac{dn}{dc} c$$

or

$$\Delta n = n - n_0 = \frac{dn}{dc} c \quad (11)$$

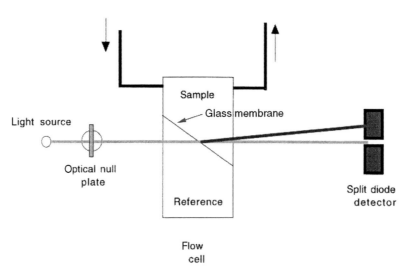

Figure 5.8 Diagram of a typical differential refractive index (DRI) detector. Key: light gray = zero signal recorded; dark gray = measurable signal recorded.

where n_0 is the refractive index of the solvent, and dn/dc is the specific refractive index increment. The latter may be positive or negative, depending on the solvent/polymer combination in question. Thus DRI signals, which are proportional to Δn, can be positive or negative, but the absolute value is proportional to concentration. For example, polystyrene has a positive dn/dc value in toluene and poly-(dimethylsiloxane) (PDMS) has a negative value. The DRI detector is nonspecific, and current instruments can detect 100 ng of materials. Due to a slight dependency of dn/dc on molecular weight, these detectors suffer from decreased sensitivity at lower molecular weight. The high sensitivity also causes the detector to be sensitive to changes in the density of the sample that are induced by pressure or temperature variations. Another major drawback is the need to carefully select the solvent/polymer combination for detectability. Using the pair of polymers in tetrahydrofuran, a common GPC solvent, polystyrene is easily detected. However, PDMS has nearly the same refractive index as THF, which causes dn/dc to be vanishingly small for that pair. The analysis of copolymers can also be complicated by molecular weight, especially in the case of block copolymers, and by a dependence of dn/dc on composition.

A more recently developed nonspecific detector is the evaporative light-scattering (ELS) detector. In this system, the eluent from the column is passed through a heated nozzle and then atomized in a steam of inert gas. The process rapidly evaporates the solvent and leaves the solute as a finely divided aerosol. This aerosol passes in front of a light source and causes scattering of the light. The amount of scattering is proportional to the mass of solute. The sensitivity of the ELS detector is independent of the solvent used, as long as it is volatile. Volatile additives are not detectable using this system. Nonvolatile solvent additives will give an overall background signal that may interfere with the quantitative response of the detector for analyte peaks. There is also anecdotal information that ELS detection suffers from a lack of response to very high molecular weight materials. Due to their considerably higher cost, evaporative light-scattering detectors are not yet as common as DRI detectors.

Adsorption spectroscopic detectors work on the principle that the polymer will have a chromophore on each monomer unit whose molar absorptivity is independent of the overall molecular weight of the polymer. Following Beers law

$$A = \varepsilon c \ell \qquad (12)$$

where ε is the molar absorptivity of the chromophore, c is the molarity of the chromophore, and ℓ is the path length, and assuming that $\varepsilon \ell$ is a constant, the absorbance is proportional to the molarity of the chromophore, which is proportional to the mass of material. The major drawback to absorption spectroscopy as a detection system is finding a solvent system that does not absorb in the region where the chromophore absorbs. Absorption detectors can be based on filter monochromators, where a wavelength is selected by dichroic filters; on grating monochromators, where a wavelength can be selected by moving the grating; or on a diode array detector, where the full spectrum can be recorded at each time interval. The latter detector can be used to examine changes in polymer composition with molecular weight, as well as to identify additives by their spectroscopic features. For molecular weight analysis, a wavelength is chosen which is not interfered with by solvent absorption and which is due to a chromophore that will not be changing with molecular weight.

All of the detectors described here provide information about concentration only. As stated earlier, to calculate molecular weight averages, we also need information about the molar mass of the eluting material. This is most commonly done by establishing the relationship between elution volume and molar mass through calibration and will be described here in detail. In recent years, two additional types of detectors have been marketed; they provide information proportional to both concentration and molar mass. The first of these is shown in Figure 5.9. The detector provides a measure of the intrinsic viscosity of a given polymer fraction via the relationship

$$[\eta]_i = \frac{1}{c_i}\log\left(\frac{\eta_i}{\eta_0}\right) \cong \frac{1}{c_i}\log\left(\frac{\Delta p_i}{\Delta p_0}\right) = \frac{1}{c_i}\left(\frac{\Delta p_i - \Delta p_0}{\Delta p_0}\right) \qquad (13)$$

where Δp_i and Δp_0 are the pressure drops measured for flow of eluting materials across a narrow capillary for the sample and pure solvent, respectively. The data from a second detector of the type described provide the concentration. The Mark-Houwink equation

$$[\eta] = KM^a \qquad (14)$$

provides for the relationship between the intrinsic viscosity and the molar mass of the eluting material, provided that values for the Mark-Houwink coefficients, K and a, are known. In their absence, a universal calibration approach can be used (see later in this chapter or reference 2). For the practical use of on-line viscometric detectors, there must be

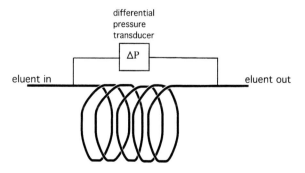

Figure 5.9 Schematic for on-line viscometric detector.

minimal variation is flow rate and solvent pressure arising from the solvent delivery system.

The other type of molar mass–sensitive detector is based on Rayleigh light scattering of polymer solutions. The standard intensity unit for scattering experiments in the Rayleigh ratio, R_θ, given by the expression

$$R_\theta = r^2\left(\frac{I_\theta}{I_0 V}\right) \tag{15}$$

where r is the distance from sample to detector, I_θ is the intensity of light scattered at some angle θ, I_0 is the incident light intensity, and V is the scattering volume sampled. For a dilute solution of polymer, the excess Rayleigh ratio

$$\overline{R}_\theta = R_\theta(\text{solution}) - R_\theta^0(\text{solvent}) \tag{16}$$

due to scattering of the solute molecules is related to their concentration by the expression

$$\frac{K_\theta c}{\overline{R}_\theta} = \frac{1}{M} + 2A_2 c \tag{17}$$

where A_2 is the second virial coefficient (a measure of solvent/polymer interactions), and K_θ is an optical constant for a given polymer/solvent combination given by

$$K_\theta = \frac{2\pi^2 n_0^2}{\lambda_0^4 N_A}\left(\frac{dn}{dc}\right)^2 \tag{18}$$

where n_0 is the refractive index of the solvent, λ_0 is the *in vacuo* wavelength of light utilized, N_A is Avagadro's number, and dn/dc is the specific refractive index increment for the solvent/polymer system. At the low concentration used in GPC, the second term of eq 17 is assumed to be zero, and, therefore, the measured value of $\overline{R}_\theta$ is proportional to cM. Once the optical constant K_θ is determined, and with the concentration provided via a mass detector, the output of a light-scattering detector provides a direct measure of M. Equation 17 assumes values of θ very close to zero, or molecules whose root mean square radius of gyration, $\langle s^2 \rangle$, is less than $1/20$ of the wavelength of light used. When the first condition is met, the technique is referred to as *low-angle light scattering* (LALS). The more common acronym LALLS (low-angle laser light scattering) derives from the use of a low-power laser to provide the incident light. The full light scattering expression takes the form

$$\frac{K_\theta c}{\overline{R}_\theta} = \frac{1}{M}\left(1 + \frac{16\pi^2}{3\lambda^2}\langle s^2 \rangle_z \sin^2\left(\frac{\theta}{2}\right)\right) + 2A_2 c \tag{19}$$

where $\langle s^2 \rangle_z$ is the z-average mean square radius of gyration for the polymer. This parameter derives from a statistical mechanical treatment of the most-probable conformation a polymer chain will adopt and is closely related to the its hydrodynamic volume. Again, at low concentration, the last term drops out, and if we collect scattering intensity at multiple angles, the angular dependency can be used to simultaneously determine M_i, as well as $\langle s^2 \rangle_{zi}$. This latter

value can give indications of changes in polymer topology (degree of chain branching, cyclics, etc.) as a function of molecular weight. A number of manufacturers have developed light-scattering detectors. Most of the current commercially available models are multiangle and require specialized software to take full advantage of their potential. The major problem with light-scattering detectors is caused by intense scattering due to particulates. This can introduce large noise spikes into the detector data that must be digitally filtered to make the data analysis manageable. Careful purification of samples and solvents minimizes these problems.

Frequently, GPC systems are outfitted with multiple detectors in series. To minimize artifacts, the connections between the column and between each detector should be kept as short as possible and should use the same diameter tubing to minimize band broadening. Again, it must be emphasized that the DRI should be the last detector in a series to avoid backpressure. The same is true for an ELS detector, but for the reason that the solvent is removed and the sample reduced to an aerosol. There will be an offset in the peak positions for the same peak as the solvent successively moves through each detector. For this reason, a separate calibration equation (see following section) must be created for each detector.

Analyzing the Sample

First, the sample must be prepared by dissolving the polymer in the eluent at a concentration of 0.1% to 0.5% w/v. This is most conveniently done by mechanically breaking the sample up into as small a size as possible, placing it in a vial, adding the solvent, sealing it, and allowing it to stand overnight. Dissolution of a polymer, especially one of high molecular weight or high crystallinity, or both, can be quite slow. For samples that will not dissolve because of their high crystallinity, one approach is to melt the polymer and then quench it in ice water. This is more effective than using liquid nitrogen, as the heat capacity of water is higher. The polymer will now have lower crystallinity, and dissolution may be more rapid. High shear stirring can cause chain scission and reduce molecular weight and should therefore be avoided. Heating may be used, but the user must cool the sample back to room temperature to ensure that precipitation does not occur after injection into the GPC system. Once the polymer is dissolved, it should be filtered to remove particulates and fillers. This is most easily accomplished with a syringe-mounted, in-line disposable filter of pore size 0.22 to 0.45 μm. If you are using an "exotic" solvent, be sure that the filter materials (membrane and housing) are compatible with the solvent.

Before injecting the sample, the chromatographic system needs to be stabilized. Typically, the pumping system is started at the beginning of the working day, and at least one to two column volumes of solvent are passed through the system to remove any degraded materials that may have accumulated. In some laboratories, the pumping system is on continuously and slowed down to 100 µL/min overnight. In this instance, the flow rate need only be increased to 1.0 mL/min and the detector outputs monitored for stable baselines. Some programmable pumps can be set up to do this automatically. If a DRI detector is used, its reference cell should be flushed and then the baseline monitored until it is stable. When you use a UV/vis detector, its lamp should be started and allowed to stabilize as well. The same goes for any temperature controls that require change of settings. If the baseline for the DRI detector continues to drift, this is an indication of a temperature or pressure instability and should be investigated. Pressure-induced drifting can be caused by small leaks in the system. Large, regular variations in the baseline of the DRI detector can also be caused by "pump noise" or a bubble in the sample cell. In either case, the reference cell should be flushed again and any further drift investigated.

Once the system has stabilized, calibration samples should be run. While recalibration on a daily basis is not usually necessary if the same sample solvent/column/temperature conditions are used, injecting a "standard sample" at the beginning of a day's operation will point out any system problems. Obtaining the various molecular weight averages presented in eq 1 requires carrying out the indicated summations. Therefore, for each of the fractions eluting from the column, we need to obtain their molecular weight, as well as their concentration. The latter is provided by the DRI, ELS, or absorption detectors. Looking at Figure 5.4, we see that in region 2 there is an approximately log/linear relationship between elution volume and molecular weight. Put in functional form, we obtain the expression

$$\log(M) = a_0 + a_1 V_e + a_2 V_e^2 + a_3 V_e^3 + \cdots \qquad (20)$$

For a given column or set of columns, the constants a_0, a_1, a_2, a_3, . . . are obtained by a polynomial least-squares regression. This is accomplished by obtaining chromatograms of polymer samples of known molecular weight and low polydispersity whose molecular weights span the potential range of interest. This can be done by multiple injections of individual standard samples or by the judicious choice of mixed standards whose peaks are at least partially resolved. A number of such calibration standards are available commercially. These include polymers such as polystyrene, methyl methacrylate, and polyethylene oxide. Figure 5.10 provides typical GPC chromatograms for two such injections. The upper trace represents the DRI response of an injection of a mixture of 400, 200, and 5 kg mol⁻¹ narrow polydispersity polystyrene (0.5% w/v each) standards onto a PL-gel MIXED-C 300-mm column, eluted with THF at a flow rate of 1.0 mL/min. The lower trace is from a mixture of 300, 250, and 50 kg mol⁻¹ standards.

Modern chromatographic software typically identifies a peak's position in units of elution time. If the flow rate is

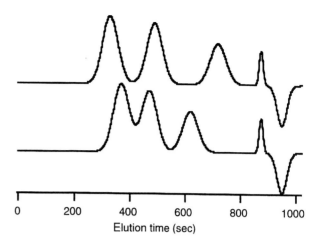

Figure 5.10 GPC chromatograms for two mixtures (see text for description).

constant, this is equivalent to elution volume. Frequently, an internal standard of low molecular weight is added to act as a marker for flow rate variation. The software will normalize the x-axis of the data to force this peak to have the correct elution volume. In these traces, the narrow peak at ~900 s is due to added toluene. The chromatographic software will typically locate each of the peak maxima in a given chromatogram and provide an opportunity for the operator to include it in the calibration procedure, or to reject it. The user then provides the molecular weight of the identified peaks. In some packages, the user must manually locate the peak maxima as well. The collection of calibration peaks obtained from one or more injections is then submitted to a calibration procedure that fits the data to eq 20. The user can usually specify the order of the fitting polynomial to provide the best fit. One requires at least one more standard than the order of the polynomial to avoid numerical problems with the fitting process. A visual inspection of the quality of calibration curve is important, for by using higher-order polynomials, a "bad" data point can be forced to fit, and erroneous results will ensue. If a point looks improper, it can be removed from the calibration set, or rerun. It should also be noted that the calibration curve is only valid within the molecular weight range of the standards used. Extrapolation above or below it, especially using higher-order fitting polynomials will lead to incorrect values of molecular weight for fractions eluting outside the calibration range.

As stated earlier, GPC separates polymer molecules by their hydrodynamic volume, not their molecular weight. Thus, calibration based on one type of polymer does not translate into a correct molecular weight for a polymer whose relationship between molecular weight and hydrodynamic volume differs from the calibrant. Thus, if we determine the molecular weight of polycarbonate using polystyrene standards, we determine a *relative* molecular weight. If they were available, we could calibrate with narrow polydispersity polycarbonate standards. There are occasions where narrow molecular weight standards are not available for the polymer at hand, but if a sample with well-characterized values for $\overline{M_n}$ and $\overline{M_w}$ is available, then we can take the complete distribution curve and fit it using nonlinear regression to a theoretical model for the distribution. One such model has the form

$$M(v) = D_1 \exp(-D_2 v)$$

where $M(v)$ is the integrated distribution as a function of elution volume, v, and D_1 and D_2 are the fitting coefficients. Software for calibration with broad molecular weight standards is less commonly available, so most workers use the narrow molecular weight standard approach in the absence of more sophisticated detectors.

It is useful to explain the large negative peak appearing at 950 sec. This is often referred to as the *system peak* and is due to dissolved gasses and other low molar mass materials that have a much lower refractive index than the solvent and elute last. When using UV/vis detection, this peak will often be large and positive. It is almost always the last peak eluting.

Once the calibration has been performed and the relationship between elution time and molecule weight established, the samples can be analyzed. Figure 5.11 shows the GPC chromatogram for a polystyrene sample injected under the same conditions as described above for the calibration mixture in Figure 5.10.

The horizontal dashed line is a "baseline" relative to which all intensities are measured. This is set either manually using a cursor or from parameters supplied to the software package. The two vertical lines define the limits over which the intensity data is summed. Taking h_i to be the baseline-corrected detector signal for the ith fraction,

$$\overline{M_n} = \frac{\displaystyle\sum_{i=start}^{i=stop} h_i}{\displaystyle\sum_{i=start}^{i=stop} \frac{h_i}{M_i}} = \frac{\displaystyle\sum_{i=start}^{i=stop} kc_i}{\displaystyle\sum_{i=start}^{i=stop} \frac{kc_i}{M_i}} = \frac{\displaystyle\sum_{i=start}^{i=stop} \frac{kw_i}{V}}{\displaystyle\sum_{i=start}^{i=stop} \frac{kw_i}{M_i V}}$$

$$= \frac{\dfrac{k}{V} \displaystyle\sum_{i=start}^{i=stop} w_i}{\dfrac{k}{V} \displaystyle\sum_{i=start}^{i=stop} \frac{w_i}{M_i}} = \frac{\displaystyle\sum_{i=start}^{i=stop} w_i}{\displaystyle\sum_{i=start}^{i=stop} \frac{w_i}{M_i}} \qquad (21)$$

where k is a proportionality constant for the relationship between detector signal and concentration, and V is unit volume. The limits "start" and "stop" correspond to the limits defined by the vertical dashed lines (integration limits). The value for M_i is obtained by using the calibration equation and the elution volume for the ith data point. For the calculation of $\overline{M_\eta}$, the Mark-Houwink "a" exponent must be provided. Depending on the software, each intensity value is used, or data points are "bunched" to speed processing time. With modern personal computers, the latter is unnecessary.

With the advent of the on-line viscometer and the light-scattering detector, calibration with polymer standards of differing chemistry from the polymer of interest is less of a problem. For the light-scattering detector, a polymer with known $\overline{M_w}$ and dn/dc is used to calibrate the optical constant for the detector. Thereafter, all one requires for a new polymer is a determination of dn/dc and the refractive index of the solvent, n_0. The light-scattering detector's output is proportional to $c_i M_i$, and in combination with the output of a concentration detector such as a differential refractometer, M_i and c_i are available for every data point.

With the viscometric detector used in conjunction with a concentration detector, we obtain at every data point $[\eta]_i$ and c_i. Using the Mark-Houwink coefficients, $[\eta]_i$ can be

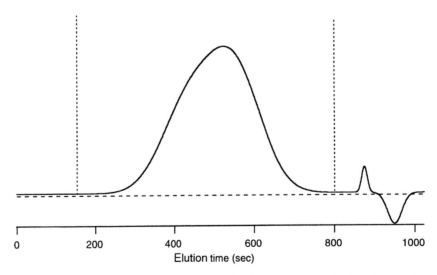

Figure 5.11 GPC chromatogram of a typical polydisperse sample. Key: horizontal dashed line = "baseline" for data analysis; vertical dashed lines = upper and lower limits for summations for the "GPC integration".

converted to M_i. For many copolymers and newly synthesized polymers where the Mark-Houwink coefficients are not available, a *universal calibration method* can be used with a viscosity detector. Two molecules of equal hydrodynamic volume should elute at the same elution volume and the product $[\eta]M$ is proportional to the hydrodynamic volume. Equation 20 can be recast in terms of the product ηM to yield the calibration equation

$$\log([\eta]M) = b_0 + b_1V_e + b_2V_e^2 + b_3V_e^3 + \ldots \quad (22)$$

Using the viscometric detector and small number of standard polymers with known intrinsic viscosity and Mark-Houwink coefficients, the user can create a universal calibration curve onto which nearly all linear polymers fall. For the sample polymer, the intrinsic viscosity value for a given elution time is provided, and, using the universal calibration curve, the molecular weight can be extracted. Deviations from the universal calibration curve are caused by changes in chain branching and can be used as a diagnostic.

Independent of the method of detection or calibration, there is often the need for judgment in the analysis of GPC data. For example, in the data set shown in Figure 5.12, the user must decide whether to use the "a" set of baseline and limits, including the small peak at ~600 sec, or to remove this from the determination (based, perhaps, on its identification as an additive or contaminant) and use set "b". A more difficult situation arises when the lowest molecular weight materials co-elute with the system peak or when there is a contaminant with a *dn/dc* value of opposite sign, as shown in Figure 5.13. The lowest molecular weight fractions are truncated, and those just before are artificially

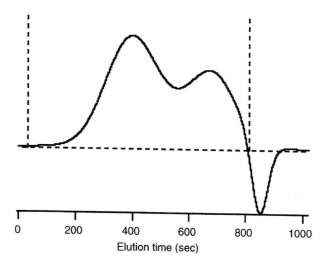

Figure 5.13 Example of GPC chromatogram in which the component eluting at 850 sec has a *dn/dc* value of opposite sign.

lowered in intensity. Overall, the effect is to increase the apparent molecular weight. Note that the molecular weight distribution in Figure 5.13 is *bimodal*—that is, it has two distinct maxima. This is due to either two chemically unique polymers, two different reaction mechanisms, or, in the case of chain-growth (addition) polymerization, two separate initiation events. One means of verifying the first circumstance is to take advantage of multiple detectors. For instance, in a copolymer of polystyrene and poly(vinyl acetate), if the polymer composition is homogeneous, then the signals from a DRI or an ELS detector should be identical to

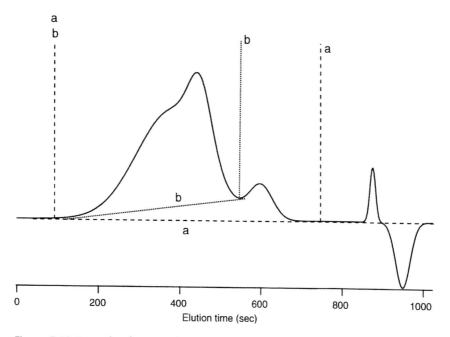

Figure 5.12 Example of a GPC chromatogram where user intervention is required in the data analysis; user must decide to use either "a" or "b" baseline and limits.

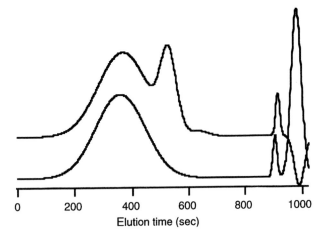

Figure 5.14 Example of a dual detector GPC analysis where a non-UV/absorbing component is detected. Key: upper trace = DRI signal; lower trace = UV/vis signal.

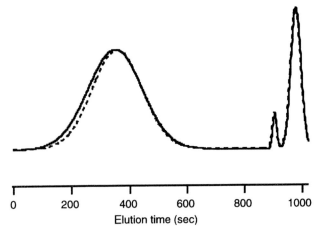

Figure 5.16 Example of small changes in MDW detected via GPC. Key: solid line = lower impact strength; dashed line = quality-control sample.

that of a UV/vis detector (accounting for the time lag between detectors in series). If the polymers are compositionally distinct, the two traces should be different.

In the chromatogram in Figure 5.14, the upper trace is the DRI detector and the lower is a UV/vis detector set at 260 nm. Note that the peak in the DRI trace at 550 s is absent in the UV/vis trace, indicating that the polymer eluting at this elution time is not UV/vis absorbing. One should be cautious, however. In the trace in Figure 5.15, the DRI and UV/vis traces were identical. In this case, the large peak at ~200 is an artifact due to using a column bank whose exclusion limit is too low. We get a "pile-up" of all molecules with molecular weights above this limit. To eliminate this as a possibility, add in a column to extend the upper range of molecular weight.

In some applications, the shape of the distribution is more critical than the average molecular weights. This is commonly the case in quality-control applications. As an example, Figure 5.16 shows the GPC traces for two different

batches of polypropylene. Components fabricated from the material represented by the solid trace have exhibited a lower impact strength. The dashed trace is for the GPC of a quality-control retain sample, run under the same conditions. In addition to the differences in the weight averages, the actual distribution indicates that the problem material has an increased proportion of high molecular weight polymer, which may contribute to decreased performance.

References

1. Young, R. J.; Lovell, P.A. *Introduction to Polymers*, 2nd ed.; Chapman and Hall: New York, 1991.
2. Yau, W. W.; Kirkland, J. J.; and Bly, D. D. *Modern Size Exclusion Liquid Chromatography: Practice of Gel Permeation and Gel Filtration Chromatography*; John Wiley & Sons: New York, 1979.

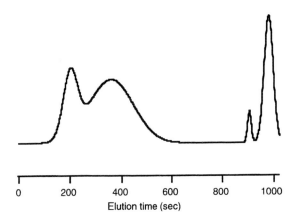

Figure 5.15 Example of a GPC chromatogram where high molecular species co-elute at a single elution time, giving rise to a "pile-up" peak.

Further Reading

Barth, H. G., Mays, J. W., Eds. *Modern Methods of Polymer Characterization*; John Wiley & Sons: New York, 1991; Chapters 1 and 2.

Crompton, T. R. *Practical Polymer Analysis*; Plenum: New York, 1993; pp 274–300.

Provder, T., Ed. *Size Exclusion Chromatography: Methodology and Characterization of Polymers and Related Materials*; ACS Symposium Series No. 245; American Chemical Society: Washington, DC, 1984.

Sandler, S. R.; Karo, W.; Bonesteel, J.-A.; Pierce, E. M. *Polymer Synthesis and Characterization: A Laboratory Manual*; Academic Press: New York, 1998; pp 140–151.

Schröder, E.; Müller, G.; Arndt, K.-F. *Polymer Characterization*; Hanser Publishers: New York, 1988; pp 166–182.

Stephens, M. P. *Polymer Chemistry: An Introduction*; Oxford University Press: New York, 1999, Chapter 2.

6

High-Performance Liquid Chromatography

LEE N. POLITE

Importance of HPLC as an Analytical Technique

High-performance liquid chromatography (HPLC) is one of the most widely employed analytical techniques in the polymer, chemical, pharmaceutical, environmental, biotechnology, and forensic industries. Any sample that can be put into solution may be analyzed. HPLC may be compared to the related and older techniques, gas chromatography (GC) or vapor phase chromatography (VPC). Because VPC requires that all analytes and matrix components be volatile, it is useable for only about 20% of known organic compounds. HPLC is much more versatile than VPC due to the wide range of mobile phase solvents available to the chromatographer, and this versatility is the primary reason for its growing popularity and importance to analytical chemistry. HPLC is particularly advantageous for polar biomolecules and high molecular weight polymers.

We have established that HPLC is such a popular technique due to its versatility. But why is HPLC so much more versatile than VPC for industrial analyses? The answer comes from the underlying operational requirements of the two techniques. Gas chromatography (GC) requires all analytes to be volatile, while HPLC requires the solubility of the analytes in the mobile phase (Figure 6.1).

Volatility is inversely related to both the molecular weight and the polarity of the analytes. Therefore, VPC is mainly chosen when the analytes and matrix are low molecular weight and nonpolar. HPLC dominates the rest of the spectrum of analytes such as polar biomolecules and high molecular weight polymers.

Classification

Low-Pressure versus High-Pressure Liquid Chromatography

Low-pressure column chromatography opened to the atmosphere was the original approach developed by Mikhail Tswett in 1906. This approach is still used today, primarily in the biological laboratory for the purification of samples that require only modest resolution. The technique usually employs a large-diameter, vertical column (>1 cm ID) loosely packed at ambient pressures with relatively large particles (>37 mm). Gravity moves the mobile phase through the column (Figure 6.2). (See Tables 6.1 and 6.2 for the advantages and disadvantages of low-pressure liquid chromatography.)

High-performance liquid chromatography has the advantages of the low-pressure techniques but avoids their limitations by mechanizing the process. HPLC is designed around a column that has been tightly packed (>6000 psi) with

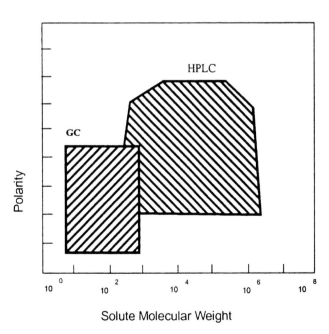

Figure 6.1 Analytes for high-performance liquid chromatography (HPLC) versus gas chromatography (GC).

109

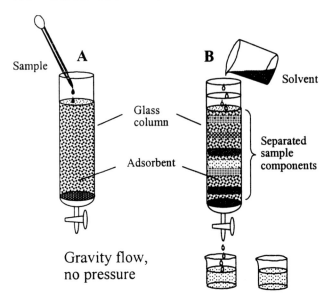

Figure 6.2 Low-pressure liquid chromatography (LC) schematic.

Table 6.2. Disadvantages of low-pressure liquid chromatography

Disadvantage	Explanation
Slow analysis	Flow through the column is limited to the force of gravity
Low resolution	Large particles and low-density packing process results in low efficiency
Difficult quantitation	Lack of on-line detector requires that each fraction must be collected and analyzed

still provides useful information in the modern laboratory. Both PC and TLC rely on the same separation mechanism as HPLC: the relative affinity of analytes partitioning between the mobile phase and the stationary phase. These techniques can be used easily in such applications as monitoring organic reactions for starting material, intermediates, and final products. They also can be used as a quick screening technique to search for known impurities and to evaluate mobile phases for use in HPLC.

small particles (<10 mm). This column design results in high-resolution separations but adds demanding instrumental requirements to the system. The tightly packed column does not allow for mobile phase *flow* through a reliance on gravity; this results in the need for a high-pressure pump and, accordingly, a specialized injection device that isolates the sample (ambient pressure) from the high-pressure flow (up to 6000 psi). After the column, an on-line detector is introduced, thus allowing for the continuous monitoring of the separation. Finally, to minimize the dispersion of the sample as it travels through the system, the components are connected with low-volume tubing and connectors (Figure 6.3).

Column versus Planar Liquid Chromatography

Most of the discussion in this text focuses on the use of column liquid chromatography (especially HPLC). However, there is another approach to LC: *planar chromatography*. Planar chromatography includes the techniques of paper chromatography (PC) and thin layer chromatography (TLC). These two techniques are generally portrayed as old-fashioned approaches. While PC is rarely used today, TLC

Isocratic versus Gradient Elution

Isocratic elution refers to the technique of using constant solvent composition throughout the chromatographic analysis. During gradient elution, the mobile phase is changed from a weak to a strong solvent. Gradients are generally chosen for samples with large numbers of components or those in a dirty or unknown matrix.

If a sample contains analytes that have widely divergent affinities for the column, a gradient is useful in shortening the analysis time and improving the shape of the peaks. Figure 6.4 is a chromatogram obtained in an isocratic analysis of a complex mixture. The first few peaks elute too closely to the void volume of the column; this suggests that the mobile phase is too strong. Also, the last peaks are short and broad with very long retention times; this suggests that

Table 6.1 Advantages of low-pressure liquid chromatography

Advantage	Explanation
Low cost	No expensive instruments to buy or maintain
Simple to use	No complicated components to assemble
High mass throughput	Large amount of stationary phase handles a high quantity of sample

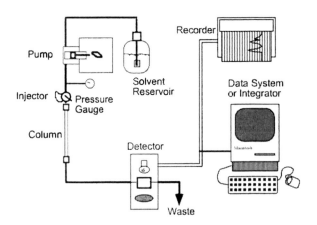

Figure 6.3 High-performance liquid chromatography (HPLC) schematic.

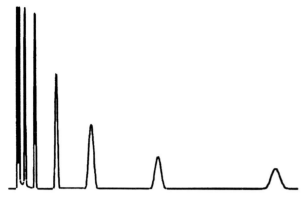

Figure 6.4 Isocaratic elution.

Table 6.3. Advantages of high-performance liquid chromatography

Advantage	Explanation
High speed	Analysis times measured in minutes or seconds (Fig. 6.6)
High resolution	Columns tightly packed with small, uniform particles (Fig. 6.7)
High sensitivity	Parts-per-million (ppm) to subparts-per-billion (ppb) detection limits (Fig. 6.8)
High accuracy	High precision sampling devices and good standards yield accurate numbers
Automated systems	Unattended operation, from sample preparation to report generation

the mobile phase is too weak. The solution to these problems is to begin with a weaker solvent and gradually strengthen the solvent throughout the course of the analysis. This is the definition of a solvent gradient. A chromatogram obtained in a gradient analysis of the same sample is shown in Figure 6.5.

With the solvent gradient, the resolution of the early eluting peaks is improved. The widths of the later peaks are decreased, while their heights are increased. The overall gradient separation yields more consistent peak widths, improved sensitivity, and shorter analysis times.

Advantages and Disadvantages of HPLC

See Tables 6.3 and 6.4 for the advantages and disadvantages of high-performance liquid chromatography; see Figures 6.6–6.8 for examples of high-speed, high-resolution, and high-sensitivity HPLC analyses, respectively.

Table 6.4. Disadvantages of high-performance liquid chromatography

Disadvantage	Explanation
Expensive instrumentation	Typical HPLC systems cost $30K to $50K
Experience required	Complex three-way chemical interactions (analyte and stationary phase; stationary phase and mobile phase; mobile phase and analyte) and complex instrumentation
No universal and sensitive detector	Maybe someday LC/MS will fill this void
Expensive supplies	Columns, fittings, and consumables are expensive
Requires spectroscopy for confirmation	Retention time characteristic of compound, not unique to that compound; extra information is needed (spectroscopic) for identity confirmation

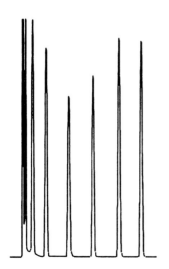

Figure 6.5 Gradient elution.

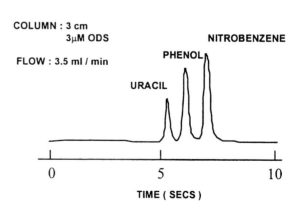

Figure 6.6 Fast HPLC: high-speed HPLC analaysis, small particles, short column, fast flow rate.

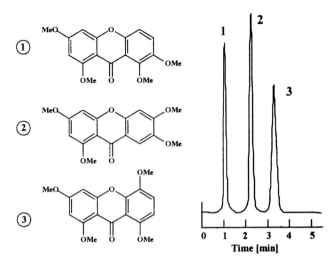

Figure 6.7 High-resolution HPLC of isomeric xanthane derivatives.

Column Methods

Normal Phase

Normal phase, defined as a polar stationary phase with a nonpolar mobile phase, is designated as such because it was the first phase to be developed. Normal phase is typically used for nonpolar to semipolar analytes. It is not generally chosen for polar analytes due to their high affinity for the surface, resulting in unacceptably long retention times.

Two types of stationary phases are used in the normal phase mode: either raw silica gel (liquid solid chromatography; LSC) or a modified silica surface (bonded phase chromatography; BPC). LSC is also referred to as *adsorption chromatography*. The mechanism of separation is the reversible adsorption of the analytes on the polar, weakly acidic sur-

face of silica gel. The mobile phase is a nonpolar solvent with a polar modifier added to control the separation. LSC is especially useful for the separation of positional isomers and "group-type" analyses (i.e., the separation of saturated, unsaturated, and aromatic hydrocarbons).

BPC is a newer approach to normal-phase HPLC. In this approach, silica is used as a support material, and polar functionalities are covalently bonded to its surface. Typical BPC normal-phase columns include cyanopropyl, aminopropyl, and diol phases. One might question why it is useful going through all of this chemistry to modify a polar surface just to end up with another polar surface. The reason is that this chemical modification yields a column that is durable and more stable than the original silica. Raw silica suffers from its surface activity being modified by adsorbed water. Even humidity changes in the laboratory can lead to retention time shifts on raw silica. Therefore, it is usually recommended that raw silica or LSC methods be transferred onto BPC packing such as cyanopropyl.

Reversed Phase

Reversed phase (RP) is designated as such because the polarities of the mobile and stationary phases are reversed, as compared to normal phase. Therefore, reversed-phase HPLC is defined as a nonpolar stationary phase (i.e., a saturated C_{18} hydrocarbon bound to the support) and a polar mobile phase (i.e., methanol or acetonitrile and water). RP can be applied to the determination of polar, semipolar, and even nonpolar analytes. RP is the most popular mode of HPLC, accounting for the majority of the column sales. The reasons for its popularity include its versatility and ability to handle polar analytes. The importance of polar compounds to the many industries was previously noted, but the fact that RP can handle such a wide variety of compounds makes it an especially useful technique.

RP packings are made by chemically bonding a hydrophobic molecule onto the silica surface. Figure 6.9 illustrates the process of bonding a trichlorooctadecylsilane onto the silica surface. Chlorine reacts with the Si—OH, thereby eliminating HCl and forming a thermally and hydrolytically stable Si—O—Si bond, thus modifying the polar silica surface into a nonpolar C_{18} surface. Unreacted chlorine on silane is hydrolyzed and forms new silanol groups.

During this reaction, a certain percentage of silanols are left unreacted due to steric hindrance of the large C_{18} chains. Residual silanols are especially undesirable when acidic or basic compounds are the analytes and are often referred to as "hot spots" or "active sites" due to their polar, weakly acidic nature. Therefore, acids and bases may adsorb due to hydrogen bonding and acid-base interactions, resulting in tailing peaks. Bases are especially vulnerable to this interaction, which is why column manufacturers offer special RP columns referred to as "base deactivated".

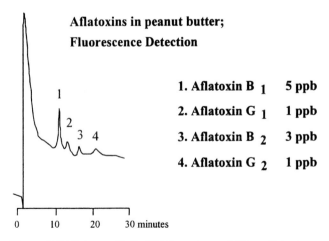

Aflatoxins in peanut butter;

Fluorescence Detection

1. Aflatoxin B_1 5 ppb
2. Aflatoxin G_1 1 ppb
3. Aflatoxin B_2 3 ppb
4. Aflatoxin G_2 1 ppb

Figure 6.8 High-sensitivity HPLC: parts-per-billion aflatoxin analysis by HPLC.

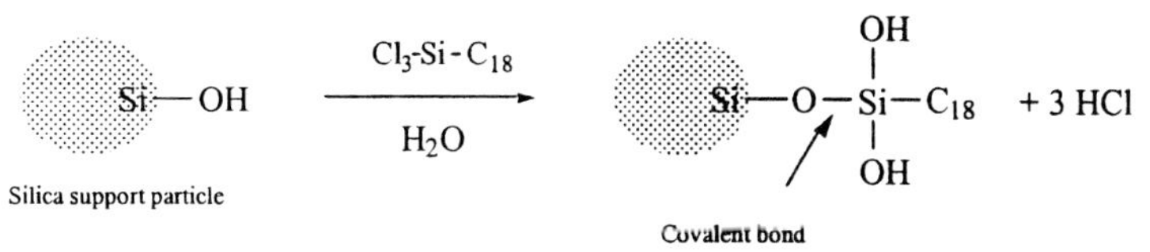

Figure 6.9 Synthesis of bonded phase chromatography (BPC) packing material: covalent attachment of the stationary phase yields a thermally and hydrolytically stable bonded phase.

The process of deactivation is complex, variable, and usually proprietary. One of the most widely used approaches is known as *end-capping* and is worth noting. This technique uses the same chemistry as described above to react a C_{18} with the silica surface. Instead of using trichlorooctadecylsilane, trimethylchlorosilane (a silicon atom with three methyl groups and one chlorine), a much smaller molecule is used to minimize steric hindrance. This allows the trimethyl group to find its way to the surface and "cap off" the remaining unreacted silanols. The resulting columns are called "end-capped", and columns made from them are more desirable for analyses of acids or bases.

Ion Exchange

The Basics

Ion-exchange chromatography (IEC) is an HPLC separation technique that takes advantage of the charge on an analyte. Typically, one thinks of ion exchange as the mode of choice when dealing with inorganic anions (i.e., fluoride, chloride, nitrate, sulfate, and phosphate) or inorganic cations (i.e., lithium, sodium, ammonium, potassium, magnesium, and calcium). But ion exchange is also useful for organic compounds that can be ionized. Essentially, any organic acid or base can be converted into its corresponding anion or cation by adjusting the pH of the mobile phase. Organic acids can be ionized into anions simply by raising the pH, thus removing a hydrogen ion and leaving a net negative charge on the molecule:[1]

$$RCOOH \Leftrightarrow RCOO^- + H^+$$

Likewise, organic bases can be ionized into cations by lowering the pH in order to protonate the molecule, leaving a net positive charge:

$$RNH_2 + H^+ \Leftrightarrow RNH_3^+$$

Once the organic molecule has been ionized, it behaves just like an inorganic anion (for acids) or inorganic cation (for bases).

For our analytes that have been converted into charged species, a stationary phase of the opposite charge is used. A positively charged stationary phase is chosen for anions, and a negatively charged stationary phase is chosen for cations. The stationary phases for ion-exchange chromatography (called *resins*) are typically based on a copolymer of polystyrene-divinylbenzene. This polymer support material is then chemically modified to impart a positive or negative charge. The separation mechanism for ion exchange is the reversible adsorption of the charged analytes on a stationary phase of opposite charge. The ions are then selectivity removed from the stationary phase by a mobile phase that contains ions of the same charge as the analytes. This "competition" for sites is the basis of the separation in ion exchange. Therefore, the technique takes advantage of the coulombic interaction (attraction of opposite charges) between the analytes and the stationary phase.

Ion Chromatography

Ion chromatography (IC) is an ion-exchange technique invented by Dow Chemical Company scientists Small, Stevens, and Bauman in 1975.[2] Technically, IC is an ion-exchange separation followed by a chemical suppression of the mobile phase in order to allow detection with a conductivity detector. A conductivity detector is used in ion exchange due to its universal response for ions. If an analyte is ionic enough to be separated by this approach, it will be detected. Conductivity detectors also are durable, very sensitive, and fairly simple in design. The only problem preventing their widespread use is the fact that the mobile phase is also ionic. The ionic character of the mobile phase is necessary for the separation as it competes for the same charged sites as the analytes, but the ionic character is undesirable when the mobile phase reaches the conductivity detector. An ionic mobile phase would produce an unacceptably high background in the detector, leading to high background noise and low sensitivity.

The elegant solution developed by those scientists was to simply follow the separator column with a "suppressor" column of opposite charge to the separator column. For example, in anion chromatography, the separator is positively charged. The anion-suppressor column is simply a high-capacity cation-exchange column (positively charged). The suppressor column removes the counter ions, thus removing the conductivity. In the anion-exchange example shown in Figure 6.10, the mobile phase is NaOH. The OH^- is the ion responsible for the separation, and the Na^+ is the counter ion. If the mobile phase were to enter the conductivity detector, the NaOH would produce an unacceptably

<u>Mobile Phase Reaction</u>

$$Na^+ OH^- + Resin^-H^+ \quad \rightarrow \quad HOH + Resin^-Na^+ \quad \rightarrow Detector$$

High Conductivity NaOH Enters $\quad \rightarrow \quad$ Near Zero Conductivity Water Elutes

<u>Analyte Reaction</u>

$$Na^+Cl^- + Resin^-H^+ \quad \rightarrow \quad HCl + Resin^-Na^+ \quad \rightarrow Detector$$

Lower Conductivity NaCl Enters $\qquad\qquad$ Higher Conductivity HCl Elutes

Figure 6.10 Ion chromatography suppression reactions: mobile phase and analyte.

high background. If, however, the mobile phase (and column effluent) is first passed through a cation-exchange column in the hydrogen form (suppressor), the Na^+ would be exchanged for H^+, and the OH^- would join the H^+ to form HOH, or pure water. Therefore, the ionic character of the mobile phase is present for the separation but is removed before the detector.

The added feature of this procedure is what happens to the analytes. In this case, the anionic analytes undergo the same exchange in the suppressor. The analyte, therefore, will travel through the detector as an acid salt instead of its Na^+ salt. The resulting signal is approximately 350% greater due to the higher mobility of the H^+ ion as compared to the Na^+ ion. The ultimate result of this technology is high efficiency (>5000 theoretical plates), excellent linearity (6 orders of magnitude), and high sensitivity detection (parts per billion by direct injection). Figure 6.11 illustrates an impressive separation of 34 anions in 15 minutes.

Ion Pair Chromatography

While ion chromatography requires a specialized column and potentially a dedicated system, ion pair chromatography (IPC) may be performed with existing reversed-phase columns and equipment. The steps involved in IPC are straightforward. The first step is to adjust the pH to ionize the analytes. This generally is done by adjusting the pH at least 2 units above the pKa for acids, or 2 pH units below the pKb for bases. The next step is to add an oppositely charged modifier to the aqueous portion of the mobile phase. This modifier is referred to as the *ion pair reagent*. Typical ion pair reagents are quartenary ammonium salts for negatively charged analytes, and alkyl sulfonic acids for positively charged analytes.

Once the ion pair reagent has been added to the mobile phase, separation is carried out under typical reversed-phase conditions (C_8 or C_{18} column, methanol/water or acetonitrile/water mobile phase). The separation mechanism is a combination of analyte complexation and stationary phase modification. The analyte complexation is due to the coulombic attraction of the oppositely charged analyte and ion pair reagent molecule. Once the analyte is introduced into the mobile phase, the analyte complexes with the oppositely charged ion pair reagent, forming a neutral complex. This complex then travels through the column, as would any neutral compound.

The stationary-phase modification occurs due to the alkyl portion of the ion pair reagent partitioning into the stationary phase. This leaves the charged head group exposed to the analytes, thus creating a pseudo ion-exchange column. It is not clear which mechanism dominates, but the technique works.

Size Exclusion Chromatography

Size exclusion chromatography (SEC) is a technique that separates analytes based solely on their size in solution (hydrodynamic volume). It is common, but not entirely correct, to refer to SEC separations as "molecular weight separations". The actual mechanism is quite straightforward. It is the exclusion of the *solvated* molecules from the pores of the stationary phase. Therefore the larger molecules elute early, and the smaller molecules elute late. The elution orders and even retention times are predictable due to the fact that there is no sample-stationary phase interaction (i.e., no adsorption, partitioning, ion exchange, hydrophobic interaction, etc.). Figure 6.12 illustrates the separation mechanism.

The large circles represent the largest molecules in the sample. These molecules are so large that they are unable to enter any of the pores of the stationary phase. Therefore, they travel the shortest distance through the column and elute first. We call this the total exclusion limit of the column. Any molecule this size or larger will elute at the same time. The very small circles at the bottom of the pores represent the smallest molecules in the sample. These molecules are so small that they easily enter the pores of the stationary phase. Therefore, they spend the longest time in the column and elute last. We call this the *total permeation*

All anion concentrations are 5 mg/L unless noted

1. Isopropyl methylphosphonate
2. Quinate
3. Fluoride   (1 mg/L)
4. Acetate
5. Propionate
6. Formate
7. Methylsulfonate
8. Pyruvate
9. Chlorite
10. Valerate
11. Monochloroacetate
12. Bromate
13. Chloride   (2 mg/L)
14. Nitrite
15. Trifluoroacetate
16. Bromide   (3 mg/L)
17. Nitrate   (3 mg/L)
18. Chlorate   (3 mg/L)
19. Selenite
20. Carbonate   (trace)
21. Malonate
22. Maleate
23. Sulfate
24. Oxalate
25. Ketomalonate
26. Tungstate   (10 mg/L)
27. Phthalate   (10 mg/L)
28. Phosphate   (10 mg/L)
29. Chromate   (10 mg/L)
30. Citrate   (10 mg/L)
31. Tricarballylate   (10 mg/L)
32. Isocitrate   (10 mg/L)
33. cis-Aconitate   } (10 mg/L)
34. trans-Aconitate

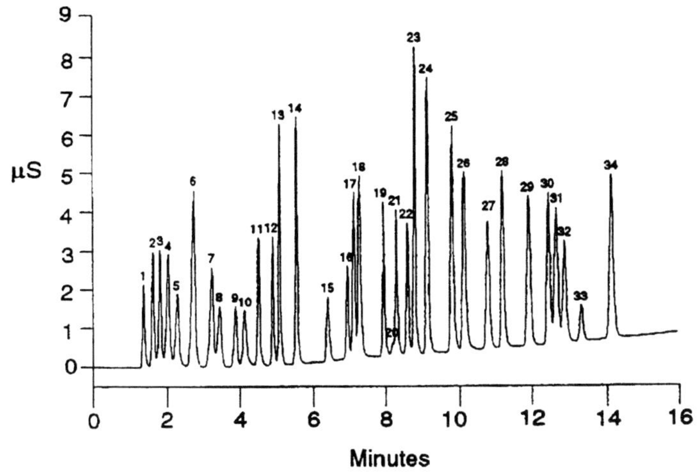

Figure 6.11 Gradient ion chromatography separation.

limit of the column. Any molecule this size or smaller will elute at the same time. The time between the total exclusion and total permeation limits of the column is referred to as the *selective permeation* area of the column. There are two different types of SEC: gel permeation chromatography (GPC) and gel filtration chromatography (GFC). The mechanisms for GPC and GFC are identical. The two techniques differ due to their targeted analytes. Table 6.5 summarizes the differences between GPC and GFC. The advantages and disadvantages of SEC are summarized in Table 6.6.

Planar Methods: TLC and PC

"Quick and Dirty" Procedures

As mentioned here earlier, TLC and PC are generally thought of as old-fashioned versions of HPLC. Although PC is rarely used in the modern analytical laboratory, TLC is still a viable approach for "quick and dirty" measurements, such as monitoring of organic reactions. TLC has the ad-

vantage of being much less expensive than HPLC and potentially faster for multiple samples.

Plates

TLC is similar to HPLC in that the separation depends on the sample analytes distributing themselves between the mobile phase and the stationary phase. In the case of TLC, the stationary-phase particles are coated as a thin layer onto a flat glass or polymer plate. The most popular mode of TLC is normal phase. Therefore, the stationary phase is generally raw silica, although other materials are available. The mobile phase is nonpolar (like hexane), and polar modifiers (like methanol) are added to adjust the elution strength.

Sample Application

Typical TLC applications require only qualitative information about the sample. Therefore, samples are generally not

• Smaller molecules penetrate smallest pores, retained longest
• Larger molecules excluded
• Separation on basis of hydrodynamic volume, not MW

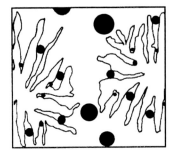

Figure 6.12 Size exclusion chromatography (SEC) separation mechanism.

measured but are simply qualitatively "spotted" onto a plate with a small glass capillary. Sensitivity can be enhanced by spotting the sample several times onto the same spot, allowing the sample solvent to dry between each application.

Visualization of Spots

There are several approaches to detecting the resultant "chromatogram" (spots). The simplest approach is simply to use the naked eye to find the individual components. This approach only works, however, when the individual components absorb light in the visible region (colored compounds).

A second approach is to shine a UV lamp on the plate. If a component absorbs UV light, it will appear as a dark spot on the plate. If a compound fluoresces, it will appear bright and colored. Another approach is to spray the plate with sulfuric acid and bake the plate in an oven. This technique will "char" the organic compounds, resulting in black spots wherever an organic analyte elutes. There are yet other approaches involving, for example, coating the plate before or after the separation in order to derivatize the silica or the analytes.

Automation and Special Equipment

The TLC process can be made more reproducible and less labor intensive by automating some of the steps. The sample application can be automated through the use of an autosampler, which deposits a known amount of the sample onto a particular area or "lane" of the plate. The added advantage to this approach is the improved quantification.

One can also automate the detection portion of the procedure. Instead of simply looking at the resultant plate, a densitometer can provide quantitative information about the position and intensity of the individual spots. A densitometer shines UV light onto the plate and then scans it, measuring the exact amount of light reflected by each small area of the plate. Therefore, if a sample component absorbs UV light, the reduction of the amount of light reflected will be proportional to the concentration of the analyte. The output of a densitometer resembles a chromatogram, with the x-axis being distance from the origin (providing the qualitative information) and the y-axis being intensity of light absorbed (providing the quantitative information).

High-Performance Thin Layer Chromatography

High-performance thin layer chromatography (HPTLC) is the ultimate form of TLC, combining the high-resolution advantages of HPLC and the multiple simultaneous sample capacity of TLC. The high resolution is accomplished by coating the HPTLC plates with HPLC packing material. These particles are much smaller, with a tighter size distribution (5 ± 0.5 μm as compared to 40 ± 4 μm) than in traditional TLC. To take advantage of these better plates, the spotting process must be automated to provide narrower initial bands and reproducible sizes. The detection process is also automated with a high-resolution densitometer.

Advantages and Disadvantages of TLC

Table 6.7 summarizes the advantages and disadvantages of TLC.

Instrumentation for HPLC

Pumps

For the theory of HPLC to become reality, stringent requirements are put on the individual components. The pump is given the difficult task of providing the force necessary to push the mobile phase through the tightly packed column. The pressure required may be as high as 6000 psi due to the small-diameter packing material and length of the column. In addition to providing the force, the pump must

Table 6.5. Differences between gel permeation chromatography (GPC) and gel filtration chromatography (GFC)

	GPC	GFC
Mobile phase	Organic solvents (THF, methylene chloride, etc.)	Aqueous buffers
Stationary phase	Rigid, cross-linked polystyrene-divinyl benzene	Soft, hydrophilic gel (Sephadex)
Samples	Synthetic polymers (polystyrene, polyethylene, polypropylene, etc.)	Water-soluble biopolymers (proteins, peptides, oligonucleotides, etc.)

Table 6.6. Advantages and disadvantages of size exclusion chromatography (SEC)

Advantage	Disadvantage
Handles high molecular weight samples with short run times	Limited peak capacity due to low resolution of SEC technique
Predictable separation times and elution order	Cannot resolve similar sized compounds (~10% difference in molecular weight is required)
No on-column sample loss or reaction	Different optimization strategy needed (pore size of column most important parameter)
Simple method development, no gradients	Must be able to dissolve the sample

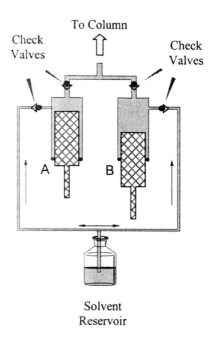

Figure 6.13 Parallel dual piston pump.

maintain an accurate and constant flow of mobile phase to provide reproducible retention times. Finally, the pump must deliver accurate mobile-phase compositions and gradients to establish the correct eluent strength for the HPLC separations. The pump accomplishes these tasks through a variety of designs.

Most HPLC pumps are based on the reciprocating piston design (Figure 6.13). Although single-headed reciprocating piston pumps are still available, the majority of pumps employ one of two versions of a dual-headed reciprocating piston design. The first style of dual-headed design places the two pump heads in parallel so that they operate 180 degrees out of phase, allowing one pump head to deliver high-pressure mobile phase while the other pump head refills with mobile phase. In this design, the sapphire piston of pump head no. 1 pushes forward, displacing its contents into the flow stream. At the end of its stroke, piston no. 1 must now reverse directions and refill with mobile phase. While this piston is refilling, piston no. 2 delivers its contents into the flow stream. Therefore, there is always one piston delivering mobile phase onto the column, resulting in a nearly pulse-free flow.

Table 6.7. Advantages and disadvantages of thin layer chromatography (TLC)

Advantage	Disadvantage
No capital expenses (no instruments to buy)	Medium to low resolution technique (cannot resolve similar compounds)
Lower operating expenses (cheaper supplies)	Only qualitative information
Faster analysis (if multiple samples are analyzed simultaneously, under identical conditions)	Slower analysis (if only one sample)
Easy to learn and use	Stationary phase exposed to elements (plate conditioning required to improve reproducibility)

Another popular pump design is the series dual-reciprocating piston pump (Figure 6.14). In this design the pump heads are assembled in series, rather than parallel. The first piston is typically twice the volume of the second piston, allowing the first piston to deliver high-pressure flow onto the column while simultaneously refilling the second piston. When the first piston is empty, the second piston assumes the function of delivering mobile phase while the first piston refills. Once again, this design results in a pump delivering nearly pulse-free flow.

Small pulsations are difficult to eliminate due to the differences in pressure between the flowing stream (high pressure) and the contents of the pump head at the beginning of its outward stroke. Even though liquids are only slightly compressible, this pressure difference is enough to cause a slight change in pressure each time the pump changes from piston no. 1 to piston no. 2. These pulsations are further minimized through the use of mechanical or electronic pulse dampers. They are noticeable only under the highest sensitivity settings of the HPLC system.

Sample Introduction Devices

HPLC presents more of a challenge than GC when it comes to introducing the sample. In GC, a small piece of rubber (the septum) is all that is required to isolate the sampling device (syringe) and the flowing stream. In HPLC, the flowing stream is typically under several thousand pounds per square inch (psi) of pressure. Therefore, the flow path must be mechanically isolated from the sample. To accomplish this, a multiport switching valve is often used.

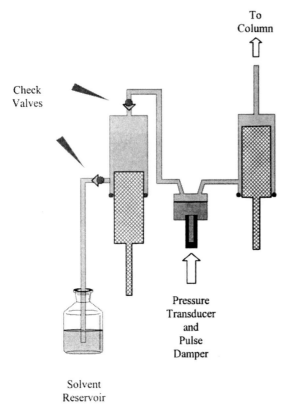

Figure 6.14 Series dual piston pump.

Whether the HPLC system has a manual or automated sampling system, the same approach is followed. The valve has two positions: load and inject. In the load position, the flow is diverted directly from the pump to the column, thus leaving the sample loop at ambient pressure (Figure 6.15). The sample loop is then loaded with sample via a manual syringe (manual injector) or through the use of an auto-

mated loading device (autosampler). In the case of manual injections, a fully loaded loop injection results in the highest reproducibility. This type of injection is accomplished by loading the sample loop with at least four times the volume of the loop. For example, a 20-μL loop should be filled with at least 80 μL. All of the excess volume goes to flush out the current contents of the loop and ensures a reproducible injection volume.

Tubing and Connectors

To minimize band broadening and maintain the efficiency of the column, the extra column volume must be minimized. The important section or "critical path" of the HPLC system includes everything from the injection loop to the detector cell. As the peaks travel along this path, they undergo longitudinal diffusion, leading to peak broadening and the loss of efficiency. If the peaks encounter any voids or unswept volumes, the peaks will also tend to tail due to the phenomenon of infinite dilution. In general, the number of fittings and the length and volume of the tubing should be minimized in this critical path. The total recommended length of tubing should not exceed that listed in Table 6.8 for each individual column dimension.

Detectors

Detectors must be chosen based on the sensitivity, selectivity, stability, and information requirements of the HPLC method. Each detector has certain strengths and weaknesses that must be understood before a detector is selected.

Ultraviolet Absorbance Detectors

Ultraviolet (UV) detectors account for a majority of the HPLC detector sales due to their versatility, sensitivity, and

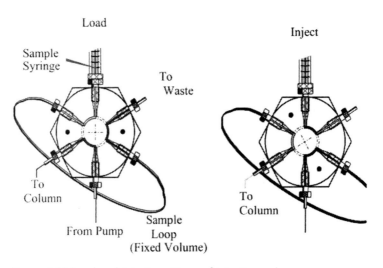

Figure 6.15 Load and inject positions of injection valve.

Table 6.8. Maximum allowable tubing length

Column characteristics				Maximum length of tubing (cm)		
Length (mm)	ID (mm)	Particle size (μm)	Efficiency (theoretical plates)	0.007″ ID	0.010″ ID	0.020″ ID
50	4.6	3	6,600	33	14	—
100	4.6	3	13,300	67	27	—
150	4.6	5	12,000	167	68	—
250	4.6	10	10,000	556	228	14
250	4.6	5	20,000	278	114	—

ID = internal diameter.

stability. Compounds absorb UV light in proportion to their concentration in solution according to Beer's Law:

$$A = a \times b \times c$$

where A = absorbance, a = molar absorptivity of the particular compound, b = the cell path length, and c = the compound's concentration.

Fixed wavelength UV detectors. The simplest version of the UV detector is the fixed wavelength detector (Figure 6.16). This detector employs a UV light source, typically a low-pressure mercury vapor lamp. The mercury lamp provides several distinct lines of UV radiation, with the 253.7 nm wavelength being the most intense. The light from the source is then passed through a filter to remove the extraneous wavelengths and allowing the 254-nm light to pass through the reference and sample cells. If no analyte is present in the sample cell, the sample and reference energies will be equal, resulting in a flat baseline. If a UV 254-nm absorbing analyte enters the sample cell, the amount of light passing through that cell will decrease, according to Beer's Law, resulting in a peak.

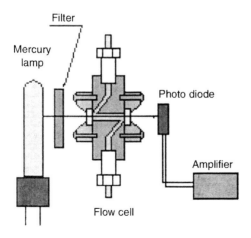

Figure 6.16 Fixed wavelength ultraviolet (UV) detector.

Variable wavelength UV detector. Due to the fact that not all compounds absorb UV light at 254 nm, variable wavelength detectors (VWDs) (Figure 6.17) allow the option of choosing the wavelength. This is accomplished by adding a monochromator to the detector design. The monochromator starts with a continuum UV source, such as a deuterium lamp, producing a broad band of radiation from 190 to more than 800 nm. The light is then reflected from a grating, which separates the light into its various wavelengths. The grating is placed on a moveable platform, allowing the user to choose any single wavelength from the spectrum.

Photodiode array detector. The photodiode array detector (PDA) (Figure 6.18) takes the UV detector one step further than the variable wavelength detector by allowing the user access to all of the wavelengths simultaneously. This is accomplished by starting with a continuum source as in the VWD and passing the entire spectrum of light through the detector cell. The light is then reflected from a grating as in the VWD. In this case, the grating does not move, and the single detector is replaced with a multitude of individual photodiode detectors. These detectors are arranged on a single chip, referred to as a *photodiode array*. The advantage of this detector is that the user is given access to the entire UV spectrum all of the time. A variety of tasks can be performed, such as peak purity (comparing UV spectra at various points along the peak), compound confirmation (adding spectral information to the retention time), and reprocessing any single wavelength as a chromatogram.

Fluorescence detector. The fluorescence detector (Figure 6.19) is one of the most selective and sensitive detectors in all of chromatography, allowing sub-ppb analyses. A small percentage of molecules have the ability to fluoresce. This phenomenon occurs when electrons are excited from their ground state to a higher energy level. The electrons will naturally return to their lower energy level, but to do so they must give off the extra energy. This energy is given off in the form of a photon of specific energy (wavelength). The fluorescence detector takes advantage of this phenomenon

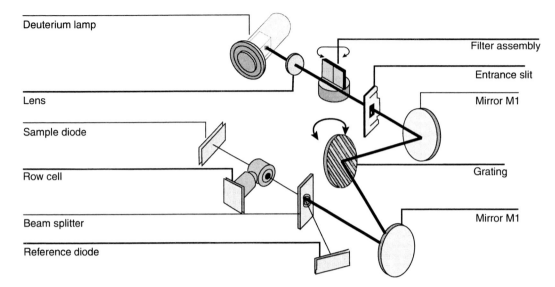

Figure 6.17 Variable wavelength ultraviolet (UV) detector.

by exciting the analytes with UV light. The detector cell is then monitored at 90 degrees from the exciting beam in order to measure the light emitted at a particular wavelength. The excitation wavelength and emission wavelength may be separately selected to optimize detection and quantification of an individual analyte.

Refractive index detector. Refractive index (RI) detectors (Figure 6.20) are the most universal and least sensitive of all the readily available HPLC detectors. They measure the minor refractive index changes that occur when the concentration of the analytes changes in the mobile HPLC effluent. These detectors are universal because the refractive index of a solution will change if any of the properties are varied. These properties include temperature, density, and concentration. Therefore, if any analyte passes through the detector cell, the change in refractive index will be recorded as a chromatographic peak. The most common RI design employs two cells separated by a glass partition. The reference side is filled with mobile phase. It is important that this mobile phase is identical to the column effluent. Therefore, fresh mobile phase is pumped through the reference cell each day before any analyses are begun. If the reference cell and sample cell contain the identical solution, the detector light passes through undeflected, resulting in a flat baseline. If the sample cell contains something other than pure mobile phase (i.e., a sample component), the incident light is refracted, resulting in a peak.

RI detectors are highly susceptible to slight variations in flow rate, detector cell pressure, and temperature. It is not unusual to wait several hours to allow an RI detector to reach temperature equilibration.

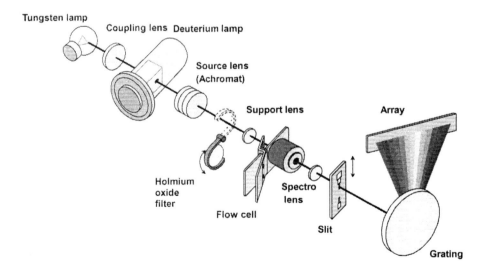

Figure 6.18 Photodiode array detector.

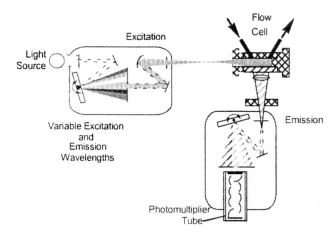

Figure 6.19 Fluorescence detector.

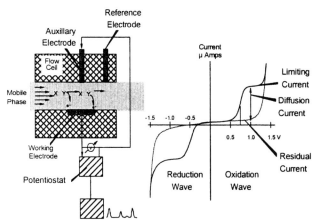

Figure 6.21 Electrochemical detector.

Electrochemical detector. Along with the fluorescence detector, the electrochemical detector (ECD) (Figure 6.21) is one of the most selective and sensitive HPLC detectors. An ECD is chosen when the analytes can be oxidized or reduced at a reasonable voltage. The ECD employs reference, counter, and working electrodes. A potential difference is set between the working and reference electrodes. When an electroactive compound passes over the working electrode, oxidation or reduction takes place, resulting in the transfer of electrons between the working electrode surface and the analyte. This flow of electrons is measured as a current and forms the basis for the chromatographic peak. ECDs are commonly used in the determination of catacholamines at the low ppb range.

Conductivity detector. Conductivity is an extremely sensitive and universal detection mode for ions. The conductivity de-

tector (Figure 6.22) takes advantage of the ions that carry a charge in solution. The conductivity is measured by placing two closely spaced electrodes in the flow path of the column effluent and measuring the resistivity of the solution. Some detectors also include a temperature-measuring device (thermister) and the associated circuitry needed to correct for the resistance variation due to slight temperature changes. Conductivity detectors combined with chemical suppression have demonstrated sub-ppb detection limits and 6 orders of magnitude of linearity for common ions.[3]

Mass spectrometer. Liquid chromatography/mass spectrometry (LC/MS) (Figure 6.23) is regarded by some to be the most powerful analytical technique in the world. This fast-growing approach combines the versatility of the HPLC separation with the structural identifying power of mass spectrometry. The difficulty in mating these two techniques stems from the inherent incompatibility of a very dilute

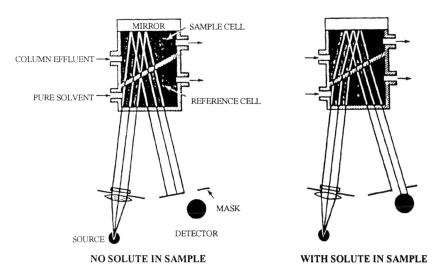

Figure 6.20 Refractive index (RI) detector schematic.

Schematics

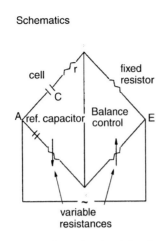

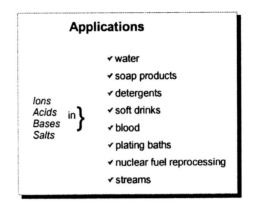

Figure 6.22 Conductivity detector.

sample in a large volume of flowing liquid (HPLC effluent) and the high vacuum requirements for the mass spectrometer. Although a large number of interface approaches have been introduced over the years, the current popular approaches include atmospheric pressure chemical ionization (APCI), and electrospray ionization (ESI). Both of these techniques ionize the analytes as they exit the HPLC column, and then transport these charged species into the mass spectrometer by employing an electrode and taking advantage of coulomic attraction. This transport approach minimizes the number and types of unwanted molecules from entering the mass spectrometer and therefore results in a more stable and robust technique.

The electrospray ionization process uses electrical fields to generate charged droplets and subsequent analyte ions by ion evaporation for mass spectrometry analysis. For higher flow rates, the nebulization may be pneumatically assisted.

APCI is a gas phase chemical ionization (CI) process in which the solvent acts as the CI reagent gas to ionize the sample.

Troubleshooting

It is impossible to create an exhaustive list of all symptoms and problems that can occur in HPLC. There are, however, a number of common causes that account for a majority of the HPLC symptoms. Here is a list of those symptoms and the associated causes.

Symptom: Periodic Pressure Fluctuation

Cause no. 1: Leaky check valve. In order for the check valve to work correctly, the internal parts (ball and seat) must be able to form a tight seal. If foreign objects (air or dirt) are introduced into the check valve, it will leak. The leaks do

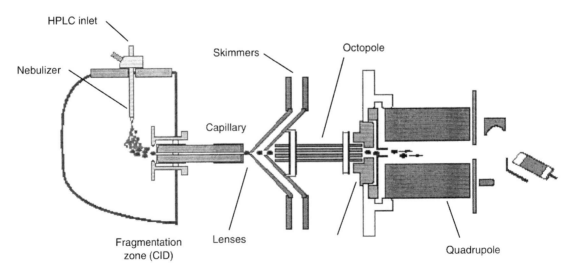

Figure 6.23 Mass spectrometry detector API-electrospray interface.

not produce any visible signs such as drops of liquid; only regular or periodic pressure perturbations.

Solution no. 1. After making sure that there is no source of air or dirt entering the pump, open the purge valve and purge the pump at a high flow rate. It may help to gently tap the check valves with a small wrench or screwdriver. This will help to dislodge any trapped air bubbles. If this does not fix the problem, remove the check valve and either clean it through sonication in dilute nitric acid or replace it with a new check valve.

Cause no. 2: Leaky pump seals. If the pump seals are leaking badly, liquid will be visible coming from the back of the pump heads.

Solution no. 2. The seals have a limited lifetime and need to be replaced as a part of the routine preventative maintenance of the HPLC system. Although seals may last over 12 months, it is a good idea to replace them at regular intervals (every 6 months). Follow the manufacturer's direction for replacing the seals, along with any associated back-up seals, o-rings, or wear retainers.

Symptom: High Pressure

There is a clog somewhere in the system. Disassemble the HPLC components, starting with the last component—usually the detector. Keep removing components until the high-pressure problem goes away, and then focus the troubleshooting efforts on that piece of equipment.

Cause no. 1: Clog in one of the following filters.

1. Any filter in the pump (some pumps have a filter built into the purge valve)
2. High-pressure in-line filter (comes after the injector)
3. Inlet frit of the guard column
4. Inlet frit of the analytical column

Solution no. 1. Clean the filter by sonication for 10 minutes in dilute nitric acid (6 molar), or replace the filter.

Cause no. 2: Clogged tubing due to mobile phase buffer precipitation. Check the entire injector and the tubing leading from the injector to the column. If you have been using buffer and have recently introduced a high concentration of organic solvent, the buffer may have precipitated in the lines.

Solution no. 2. Wash the system with pure water. Start at a very low flow rate (0.1 mL/min) until the pressure decreases.

Cause no. 3: The sample has precipitated in the lines.

Solution no. 3. Wash the system with a high concentration of the sample solvent. Then make several large volume injections of that same solvent. This should dissolve any precipitated sample. From now on, dissolve the sample in the mobile phase to avoid precipitation.

Symptom: Low Pressure

Cause no. 1: Leak.

Solution no. 1. Concentrate on the high-pressure part of the HPLC. This includes everything from the pump heads to the top of the column. Find the leak (drips or puddle) and tighten the fitting(s).

Cause no. 2: No solvent. This happens to the best of us . . . you ran out of solvent!

Solution no. 2. Refill the solvent container, then purge the lines and pump with the fresh mobile phase.

Symptom: Poor Peak Area or Height Reproducibility

Cause: Problem with the autosampler.

Solution. Concentrate your efforts on the injection system. There may be a clog in the syringe needle or a leaky line between the syringe (or metering device) and the sample vial. Replace the needle or tighten the leaky tubing.

Symptom: Noisy Baseline

Confirmation Step. Shut off the pump in order to rule it out as a cause. If the noise remains after the pump has been shut off, it is probably a problem with the detector.

Cause no. 1: Dirty detector cell.

Solution no. 1. Clean the cell with dilute (1 molar) nitric acid. This is most easily accomplished with a 10-mL syringe and an adapter that allows the syringe to connect directly to the cell inlet. Inject 5 mL into the cell and allow it to stand for 10 minutes. Follow that with an additional 5 mL, and rinse the cell with water. DO NOT PUMP THIS SOLUTION THROUGH YOUR COLUMN.

Cause no. 2: Old detector lamp.

Solution no. 2. Replace the lamp.

Capillary Electrophoresis

All current forms of column chromatography (HPLC, GC, SFC, IC, GPC, etc.) are pressure driven. This means that the techniques rely on a pressure differential across the column to drive the mobile phase. This pressure-driven flow causes the undesirable effect of a parabolic flow profile. The parabolic flow profile is due to the resistance to flow along the walls of the tubing. Therefore, the flow through pressure-driven columns causes peak spreading or band broadening due to this bullet-shaped flow. Even though we introduce the sample into the system as a narrow, straight band, the radial center of the band travels faster then the radial edges due to this phenomenon. Up until recently, this undesirable effect of the parabolic flow profile has been considered to be unavoidable.

Capillary electrophoresis (CE) avoids this problem by driving the mobile phase with a voltage difference instead of a pressure difference. This voltage differential results in electro-osmotic flow. The first step in creating this type of flow is to start with a fused silica column. The surface of fused silica is similar to the silica support material used for HPLC columns, which means the surface is covered with Si−OH or silanol groups. A basic buffer is passed through the column in order to remove the hydrogen ions, resulting in a net negative charge on the surface. The column is then filled with an aqueous buffer. The positive ions in the buffer associate themselves with the negatively charged surface. Then a high-voltage potential difference is applied across the column. The positive buffer ions move toward the negative electrode, dragging with them the aqueous buffer (mobile phase). This establishes the electro-osmotic flow of mobile phase through the column, resulting in a flat flow profile. Therefore, there is no band broadening due to the flow through the column. This ultimate result is a technique with the solvating advantages of HPLC and the capillary efficiency of GC. This technique in its simplest form is called *capillary zone electrophoresis* (CZE) and has already demonstrated efficiencies in excess of 1 million theoretical plates. This is compared to 25,000 plates for HPLC and 400,000 plates for capillary GC.

CE Systems

A CE instrument is similar to an HPLC. We start with a "pump". Here the pump is actually a high-voltage power supply capable of delivering 30,000 volts at a very low amperage (micro amps). The positive electrode is placed at the beginning of the system in a buffer reservoir. A piece of fused-silica tubing (<75 μm ID) is placed in this reservoir. This end of the capillary also serves as the injection device when it is placed into the sample vial.

There are several different modes of capillary electrophoresis that can be applied to a variety of separation problems. Here are a few of the most popular techniques and their applications.

Capillary Zone Electrophoresis

The electro-osmotic flow is adjusted so all analytes move toward the negative electrode. Even though negatively charged species will have the tendency to move toward the positive electrode, the electro-osmotic flow is adjusted to be greater than the negative flow, resulting in a net flow toward the negative electrode. CZE, also called *free-zone electrophoresis*, is the simplest form of CE and separates analytes based on their charge and size in solution. The higher the positive charge, the faster the analytes move toward the negative electrode. Also, the larger the molecule, the slower it travels due to the drag-induced resistance to flow. Therefore, positively charged analytes elute first (further separated by size), and those negatively charged elute last (again further separated by size), but neutral molecules all elute as a single peak.

Micellar Electro-Kinetic Chromatography

Obviously a technique that cannot separate neutral molecules would have only limited application as an analytical technique. Introducing a micelle to the buffer system via micellar electro-kinetic chromatography (MEKC) extends this technique to neutral molecules. Micelles are formed when surfactants are added above their critical micelle concentration (CMC). The surfactants reach a high enough concentration that they start to interact with each other. The long alkyl chains group together in the nonpolar interior of the micelle. The polar head-groups form the hydrophilic exterior. As the analytes travel through the column, they partition in and out of the micelle interiors, thus forming the basis of neutral separation. The micelles can be thought of as a pseudo-reversed-phase stationary phase. The major difference is that the "stationary" phase is not stationary. The ultimate result is a separation of positive and negative analytes, as described above for CZE, with the addition of neutral separation through micellular interaction.

References

1. McNair, H. M.; Polite, L. N. *Am. Lab.* **1988,** *December,* 116–121.
2. Small, H.; Stevens, T. S.; Bauman, W. S. *Anal. Chem.* **1975,** *47,* 1801.
3. Polite, L. N.; McNair, H. M.; Rocklin, R. D. *J. Liquid Chrom.* **1987,** *10*(5), 829–838.

Selected Bibliography

Analytical Chemistry in a GMP Environment: A Practical Guide; Miller, J. M., Crowther, J. B., Eds.; Wiley: New York, 2000.

Dolan, J. W.; Snyder, L. R. *Troubleshooting LC Systems: A Systematic Approach to Troubleshooting LC Equipment and Separations;* Humana Press: Clifton, NH, 1989.

Gooding, K., Regnier, F., Eds. *High Performance Liquid Chromatography of Biological Macromolecules: Methods and Applica-*

tions; Chromatography Science Series, 51; Marcel Dekker: New York, 1990.

Horvath, C., Ed. *High Performance Liquid Chromatography: Advances and Perspectives;* Academic Press: San Diego, 1980; Vol. 2.

Jandera, P.; Churacek, J. *Gradient Elution of Column Liquid Chromatography;* Journal of Chromatography Library Series, 31; Elsevier: New York, 1985.

Lindsay, S. *High Performance Liquid Chromatography,* 2nd ed.; Wiley: New York, 1992.

McMaster, M. C. *HPLC Practical Users' Guide;* VCH: New York, 1994.

Meyer, V. *Practical High-Performance Liquid Chromatography;* Wiley: New York, 1993.

Riley, C. M.; Lough, W. J.; Wainer, I. W. Eds. *Pharmaceutical and Biomedical Applications of Liquid Chromatography;* Elsevier: New York, 1994.

Sadek, P. C. *HPLC Solvent Guide;* Wiley: New York, 1996.

Scott, R. P. W. *Liquid Chromatography Column Theory;* Wiley: New York, 1991.

Scott, R. P. W. *LC for the Analyst;* Marcel Dekker: New York, 1994.

Scott, R. P. W. *Chromatographic Detectors: Design Function and Operation;* Marcel Dekker: New York, 1995.

Snyder, L. R.; Kirkland, J. J. *Introduction to Modern Liquid Chromatography,* 2nd ed.; Wiley: New York, 1979.

Snyder, L. R; Kirkland, J. J. *Practical HPLC Method Development;* Wiley: New York, 1988.

Swadesh, J. K., Ed. *HPLC Practical and Industrial Applications;* CRC: New York, 1996.

Szepesi, G. *How to Use Reversed Phase HPLC;* VCH: New York, 1992.

Walker, J. Q., Ed. *Chromatography Fundamentals, Applications, and Troubleshooting;* Preston Publications: Niles, IL, 1996.

7

Gas Chromatography

THOMAS H. RICHARDSON
BRIAN K. KOCHANOWSKI
CHRISTOPHER R. MUBARAK
STEPHEN L. MORGAN

A wide variety of distinct chemical substances are collectively called *polymers*. The common aspects of these macromolecular substances are their high molecular weights and their backbone structure that consists of repeating characteristic molecular substructures. To a chemist, any substance can be a target for analysis—polymers no less so than simpler, smaller molecules—by virtually any technique available. However, the method selected is often the "easiest" with respect to the physical state of the sample; it is not intuitive that solids would be amenable to methods intended for gases or highly volatile liquids. Because gas chromatography (GC) and GC/mass spectrometry (MS) require the sample to be both volatile and thermally stable during analysis, such analyses must be preceded by steps to generate species with these properties.

The traditional approach to polymer analysis typically involves hydrolysis (or other chemical treatment) to break up the polymer. The monomeric or oligomeric components of the polymer are then analyzed directly by liquid chromatography (LC), or they are derivatized to volatile components that are analyzed by GC. In recent years, however, the state of the art of GC has progressed to a point where nontraditional samples can be handled, including solid, nonvolatile polymers. This chapter focuses on analytical pyrolysis, or thermal degradation in the absence of oxygen, as a technique for extending the range of GC or GC/MS to enable analysis of these otherwise intractable materials. Pyrolysis transforms the original polymer into a mixture of products that retain information about the parent structures from which they are derived. The chromatogram of pyrolysis products is called a *pyrogram*; pyrolysis products that correlate well with polymer structures are often referred to as *chemical markers*. In some cases, pyrolysis may circumvent the need for chemical derivatization and increase the speed of analysis.

Coupling pyrolysis to high-resolution capillary GC separation (Py-GC), followed by detection using either flame ionization or mass spectrometry (Py-GC/MS), has proven to be a powerful tool for characterization and identification of a wide variety of polymeric materials. Several advantages recommend the use of pyrolysis GC for polymer analysis: minimal sample preparation is usually required, and only an extremely small amount of sample is needed. GC combined with analytical pyrolysis and MS is a valuable tool for compositional analysis, structural characterization, and stability profiling of polymers. Information about polymers that is acquired using pyrolysis GC and GC/MS complements information obtainable from other analytical methods such as size exclusion chromatography, thermogravimetry, differential thermal analysis, and spectroscopy, as well as from mass spectrometric methods such as matrix-assisted laser-desorption ionization (MALDI). This chapter summarizes practical considerations in the use of analytical pyrolysis GC for polymer characterization and reviews selected applications.

Practical Aspects

Decomposition Methods for Polymer Analysis

Polymers in and of themselves are not amenable to GC; in general, a controlled decomposition to smaller molecules is performed and it is these decomposition products that are actually analyzed and interpreted to give knowledge of the precursor polymer. On occasion, headspace[1,2] and thermal desorption[2-6] analyses of polymers at moderate (up to 300 °C) temperatures have been conducted.[7,8] In general, more extensive polymer decomposition is necessary to obtain structural information. Stepwise pyrolysis experiments have been performed[9-12] in which the initial temperature steps are sufficiently low to permit headspace or desorption analysis, with the final steps at pyrolysis temperatures of 400 to 1000 °C. In some instances, thermogravimetric data have been simultaneously obtained.[13]

A variety of decomposition procedures have been employed for GC analysis of polymers. Although "simple" fusion and chemical separation (before the chromatographic step) have been reported,[14] perhaps the most commonly used method is analytical pyrolysis: the rapid heating of a sample in an inert atmosphere followed by collection and analysis of the thermal decomposition products.[15,16] Pyrolysis in oxygen-containing atmosphere, followed by analysis of the partially oxidized products, also has been conducted,[17–19] and "chemolysis" procedures, involving a chemical reaction between the polymer sample and a suitable reagent, have been developed. These methods are often performed simultaneously with pyrolysis but, in some instances, the modification chemistry is done before the pyrolysis.[20–22] Figure 7.1 shows a systems diagram for a pyrolysis-gas chromatographic instrument consisting of a source of carrier gas to transport sample components through the system, appropriate filters and flow and pressure controllers, a pyrolysis subsystem interfaced to the GC inlet, a chromatographic oven containing the chromatographic separation column, and a detector (e.g., a flame ionization detector [FID] or a mass spectrometer).

Analytical Pyrolysis Equipment and Methods

Many different pyrolysis and combustion methods have been reported.[23,24] Filaments heated by electrical resistance[25] and flash (Curie point) induction heating[26,27] are most common; microfurnace pyrolyzers are also used, and laser pyrolysis[5,28–30] has been reported. The basic design of filament and Curie point pyrolyzers are illustrated in Figure 7.2. Table 7.1 lists selected companies that supply pyrolysis equipment designed for interfacing to GC and GC/MS.

Resistively heated filament pyrolyzers (Figure 7.2A) consist of a platinum ribbon or coil, across which a capacitor is discharged to heat to the final pyrolysis temperature (e.g., at 20 °C/ms up to 1000 °C). For control of filament pyrolysis systems, the ribbon or coil may be an element in a

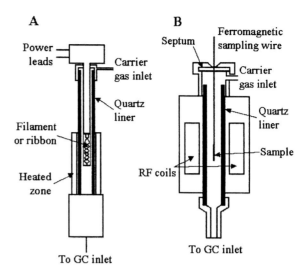

Figure 7.2 Pyrolysis interfaces: (A) resistively heated pyrolyzer, and (B) Curie point pyrolyzer.

Wheatstone bridge circuit; other designs use a thermocouple and feedback circuit to enable fine control over the filament temperature. The sample is placed directly on a platinum filament or in a quartz tube or boat inside a platinum coil. For liquid samples or suspensions of solids, a solution or suspension is placed on a ribbon or in a coil, and the solvent is evaporated before pyrolysis. Solid polymer samples can be sandwiched between quartz wool plugs inside the quartz tube to reduce extraneous nonvolatile material from leaving the tube during pyrolysis. For best reproducibility, placement of sample with respect to the quartz tube or the ribbon should be the same for all samples to ensure identical heating profiles. With quartz tubes, the sample never comes into direct contact with the pyrolyzer filament, as it does with ribbon filaments. Ribbon filaments often exhibit memory effects with polar components, are harder to clean, and typically have a shorter lifetime. Quartz tubes may be reused after cleaning. An autosampler for resistive filament pyrolysis is currently available from one manufacturer.

Curie point pyrolyzers (Figure 7.2B) employ a ferromagnetic metal wire (ca. 0.5 mm in diameter) that is inductively heated by a radio-frequency field. When the wire reaches the Curie point temperature, it becomes paramagnetic and stops inducting power. The temperature at which the wire stabilizes (the Curie point) is a function of the type of metal. For example, the Curie temperatures of cobalt, iron, and nickel are 1128, 770, and 358 °C, respectively. Wires made from alloys of these metals produce intermediate temperatures. For example, the commonly used nickel-iron wire has a Curie point of 510 °C. A polymer sample may be ground to a powder, suspended in a solvent by sonication, and then coated on the wire by depositing drops of suspension. The solvent is evaporated while rotating the wire in a stream of dry nitrogen. Carbon disulfide is a good

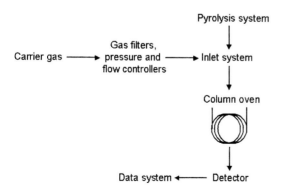

Figure 7.1 A systems diagram for a pyrolysis-gas chromatographic instrument.

Table 7.1 Suppliers of pyrolysis equipment for interfacing to gas chromatographic or mass spectrometric instrumentation

Supplier	Model	Pyrolyzer type
CDS Analytical, Inc. 465 Limestone Road Oxford, PA 19363 Phone: 610-932-3636; Fax: 610-932-4158 URL: http://www.cdsanalytical.com/ E-mail: sales@cdsanalytical.com	Pyroprobe 1000, Pyroprobe 2000, AS-2500 Pyrolysis Autosampler	Filament
Frontier Laboratories, Ltd. 1-8-14, Saikon Koriyama, Fukushima, Japan 963-8862 Phone: 81(024)935-5100; Fax: 81(024)935-5102 URL: http://www.frontier-lab.com	Thermal Analysis System 2020	Furnace
GSG Analytical Instruments, Ltd. St. James Court Wilderspool Causeway Warrington WA4 6PS, United Kingdom Phone: 44-01925 418 044; Fax: 44-01925 413 172 URL: http://www.gsg-analytical.com/ E-mail: sales@gsg-analytical.com	CPP 1040 Pyrolyser, Pyromat	Curie point
Humble Instruments & Services, Inc. 218 Higgins Street Humble, Texas 77338 Phone: 281-540-6050; Fax: 281-540-2864 URL: http://www.humble-inc.com/ E-mail: sales@humble-inc.com	Thermal Extraction-Pyrolysis Inlets (TEPI)	Furnace
Japan Analytical Instrument Co., Ltd. 208 Musashi, Mizuho, Nishitama Tokyo 190-1213, Japan Phone: 81-42-557-2331; Fax: 81-42-557-1892 URL: http://www.jai.co.jp/ E-mail: n-oguri@jai.co.jp	JHP-3S	Curie Point
LECO Corporation 3000 Lakeview Avenue St. Joseph, MI 49085-2396 Phone: 616-985-5496, USA toll-free: 800-292-6141; Fax: 616-982-8977 URL: http://www.leco.com/ E-mail: info@leco.com	VTF-900	Furnace
Pyrol AB P.O. Box 766 S-220 07 Lund, Sweden Phone: 46-46-13-97-97; Fax: 46-46-13-96-98 URL: http://www.pyrolab.com E-mail: info@pyrolab.com	Pyrola 2000	Filament

Table 7.1 Continued

Supplier	Model	Pyrolyzer type
SGE, Inc. 2007 Kramer Lane Austin, Texas 78758 USA Phone: 512-837-7190, USA toll-free: 800-945-6154; Fax: 512-836-9159 Web site: http://www.sge.com/ E-mail: usa@sge.com	Pyrojector II	Microfurnace
Shimadzu Scientific Instruments, Inc. 7102 Riverwood Drive Columbia, Maryland 21046 Phone: 410-381-1227, USA toll free: 800-477-1227; Fax: 410-381-1222 URL: http://www.ssi.shimadzu.com/index.html E-mail: info@shimadzu.com	PYR-4A Pyrolizer	Furnace

solvent to use for sampling onto Curie point wires because it evaporates easily and leaves no residue. As mentioned already, placement of the sample should be done reproducibly to reduce variability in heating profiles.

Furnace pyrolyzers allow the sample to be pushed or dropped into a heated chamber where pyrolysis takes place. Careful control and monitoring of temperatures are required to achieve rapid rise times and well-defined pyrolysis times.

Long-term reproducibility is affected by eventual deterioration of resistive filaments or sample wires. All components exposed to a sample during pyrolysis (e.g., GC injection port liners and quartz sample tubes) often require acid cleaning, solvent washing, and oven drying. Active pyrolyzer elements (coils and ribbons in filament pyrolyzers) can be heated without the sample to remove contamination (1000 °C for several seconds is usually adequate). Curie point wires are inexpensive enough to be discarded after each use.

Following loading of the sample, the filament probe or Curie point wire is inserted into a heated interface that is connected to the GC injection port or Py-MS vacuum. Concern has been expressed[31] for potential variation in decomposition behavior, depending on the size of the sample and the nature of the heating. Depending on the experimental apparatus, samples may range in size from a few tenths of micrograms to several grams. The best reproducibility is often achieved with sample sizes in the lower micrograms range. Larger sample sizes make controlled heat transfer and uniform heating difficult to achieve, may deposit unpyrolyzed or nonvolatile residue onto the walls of the pyrolysis interface, and tend to promote secondary reactions, resulting in more complicated and less reproducible pyrograms. Most systems are designed to minimize secondary reactions by using a nominally inert gas flow to sweep pyrolysis products quickly into the gas chromatograph (or other instrument) for immediate analysis[32] or into a collection system

for off-line analysis.[33] If analytical profiling is the goal, sample amounts applied to the tube are typically in the range of 10 to 100 µg. Acquisition of quality mass spectra for all peaks (including minor pyrolysis products) may require samples of several hundred micrograms. Alternate sample introduction methods include aerosols, bulk batches, and melts (continuous steady-state heating).

Manufacturers make performance claims of rise times varying from several thousand to several hundred thousand °C/s to final temperatures as high as 1500 °C. Actual rise times and final temperatures experience by the analytical sample are often less, depending on the nature of the heating mechanism, amount of the sample, positioning of the sample, contact between sample and heating elements, volume and insulation of the pyrolysis chamber, and other variables. Some instruments implement monitoring of pyrolysis temperatures.

While the literature may guide choice of conditions, optimal pyrolysis temperatures and ramp rates for a particular polymer sample must be determined by experiment. Lower temperatures induce less fragmentation and decrease the amount of pyrolysates. Higher temperatures ensure more complete fragmentation but may reduce structural information. Pyrolysis at a variety of temperatures provides selective information about degradation and may reveal different aspects of polymer structure. The time profile of pyrolysates at different temperatures also can provide kinetic information on the operative pyrolysis mechanisms.

Off-line trapping procedures may permit a gross prefractioning of the decomposition products[34-38] into "volatiles" and "condensables". Some investigations have employed hydrogen as a carrier gas with the intent that reactions between the pyrolysis products and the hydrogen will occur[39,40] and, as noted, an oxidizing carrier gas has been used.

The physical relationships of pyrolysis system and chromatographic column, transfer lines, and dead volumes have been of some concern with respect to assuring complete recovery of the pyrolysis products and residual sample components.[32,41,42] Differences between results from different pyrolyzer systems depend on the pyrolysis products examined and may be confounded with other instrumental differences, including the design of the transmission system to the detector.

Although lengthy sample pretreatment, derivatization, or clean-up steps, are not required for analytical pyrolysis investigations of polymers, the assumption that any sample pretreatment should be avoided is naive. If the characteristic discriminating information is known to reside in specific polymer components, a carefully designed pretreatment step may be able to isolate that feature of the macromolecule. This approach will generally give better reproducibility because the background of extraneous chemical components not relevant to decision-making is eliminated. Chemolysis methods have been developed in response to a need for such simpler decomposition mixtures.[43–45] The typical polymer pyrolysis produces a wide range of product compounds, in part due to secondary reactions of the initial fragments. Chemolysis techniques provide a reagent to react quickly with the primary pyrolysis products and, in effect, quench uncontrolled secondary reactions. The most common reagent for this purpose has been tetramethylammonium hydroxide,[46,47] which yields methylated pyrolysis products. However, proton[43,48] and Lewis acids,[49–51] proton-donor solvents,[37,52] and alkali solutions[53] have also been employed, as have alkali fusions.[54–57] Bromination of pyrolysis products upon injection into the gas chromatograph has also been reported.[58] Another interesting development has been the use of near-critical water vapor for so-called hydrous pyrolysis experiments.[59–63]

As an alternative, or perhaps even as a supplement, to chemolysis, derivatization methods have been applied to preserve knowledge of reactive sites on the polymer before the pyrolysis process. Wang et al.[20] have achieved success by reacting the carboxylic acid moiety of fumaric and itaconic acid–containing polymers with primary amines, while Mao et al.[40] have reduced vinyl chloride polymers to foster preservation of short-chain branching characteristics. In other work, to promote the production of specific decomposition products,[43] nylon fibers were pretreated with phosphoric acid before pyrolysis.

The nature of the polymer under investigation is also of concern. Investigations into reaction kinetics and pathways and product distributions have used purified authentic samples or freshly synthesized samples. A considerable amount of work has been addressed to mixtures, either of two or more polymers or of a polymer with a filler or other additives.[64,65] This latter work is valuable for several reasons, not the least of which is that the majority of polymer usage in society is not as pure materials but as compounded mixtures.

The analysis of mixtures of polymers has received some attention. Cortes et al.[66] have subjected poly(styrene-co-acrylonitrile) to size exclusion chromatography; the elutant fractions were transferred to a pyrolyzer, and the monomeric decomposition products were analyzed by GC. In this experiment, the focus was to ascertain product uniformity over the range of molecular sizes; however, the concept could be generalized to more complicated mixtures of polymers. The interaction of different polymers in a pyrolysis mixture has also been studied, with attention not only to variations in decomposition products but also toward altered decomposition kinetics, both of which are of interest to the waste processing industry.[67] The effect of catalysts (or potential catalysts) on the pyrolysis process has also been studied,[13,28,68] especially when a focus of the work is toward waste polymer processing.

GC and MS

In the half-century since its first description, GC has become a versatile tool for analysis of volatile mixtures. Modern instruments with specialized injectors and detectors, together with high-efficiency capillary columns, permit the analysis of materials that previously would not be considered. However, the method is not all-encompassing. Molecules of too great a molecular weight, and thus unable to become gaseous at the operating temperatures, cannot be analyzed. Despite this limitation, GC remains one of many valuable tools used in the characterization of chemical systems.[37,38]

Once the sample has been pyrolyzed, volatile fragments are swept from the heated pyrolysis/injection port by carrier gas onto a gas chromatograph (or other analytical tool; most commonly, a stand-alone mass spectrometer[70,71]) and separated into its components. In Py-GC, if the column is kept relatively cold at the start of the run (e.g., as in a programmed temperature run), pyrolysates will be focused on the head of the column until they are eluted by the increasing temperature. The amount of pyrolysis products transferred to the chromatographic column depends on whether a split injection or a direct injection without splitting is used. For highest sensitivity, direct flow without splitting is preferred; good chromatography may dictate using a reasonable split ratio (e.g., 10–20 parts split flow, 1 part column flow).

Popular choices for Py-GC include columns packed with porous polymers, conventional packed columns, and open tubular capillary columns. Porous polymer columns are suitable for analysis of low molecular weight volatile pyrolysates, but they provide only limited higher molecular weight information. Although some conventional packed columns have been used in polymer pyrolysis studies, GC applications have converted almost completely to the use of fused-silica, open-tubular capillary columns. Because polymer pyrolysis products can be labile polar species, the column material and chromatographic column should be as inert as possible. Fused-silica capillary columns coated with "bonded" phases satisfy these needs. Fused-silica columns offer improved resolution, increased inertness, and better analytical precision than packed columns.

The chromatographic column is the heart of the GC system: interactions between mixture of pyrolysis products and the stationary phase determine the elution order of components. The choice of a stationary phase to achieve a particular separation follows the rule of "like dissolves like". Nonpolar or slightly polar stationary phases tend to elute sample components in boiling point order; chemical interactions determine elution order on polar stationary phases. Choice of chromatographic stationary phases for fused silica capillary columns can range from nonpolar (e.g., 100% methyl silicone) to polar [e.g., poly(ethylene glycol)] phases. While matching the polarity of column and analytes to be separated is a major factor for a successful separation, the high resolving power of capillary columns means that the choice of stationary phase is not as critical as with packed columns. Selective stationary phases may still be required to deal with demanding separations involving enantiomers, polar compounds, and certain complex mixtures.

The superior resolution per unit time available with capillary columns means that adequate resolution with faster analysis times can often be achieved using shorter columns or with thinner films of stationary phase. Despite the complexity of the pyrolysis product distribution, fast GC techniques have the potential for significant effects on polymer analysis in time-critical situations such as process monitoring.

Figure 7.3 compares conventional and narrow bore short column pyrograms for poly(ethylene-*co*-methyl acrylate). Pyrolysis of the copolymer produces a chromatogram that has the typical repeating pattern of triplets from polyethylene with a distribution of varying chain lengths of dienes, alkenes, and alkanes. Figure 7.3 (top) shows the result of a conventional Py-GC analysis using a mixture coated column with a mixture of 5% phenyl and 95% methyl polysiloxane. This fused-silica column was 30 m in length, the internal diameter (ID) was 0.25 mm, and the stationary phase film thickness (df) was 0.25 μm. Figure 7.3 (bottom) shows the same pyrogram produced with the same stationary phase on a column that is shorter (10 m), with smaller internal diameter (0.1 mm) and thinner film thickness (0.10 μm). Both pyrograms show the expected triplet groups from polyethylene, well separated in the first half of each pyrogram and gradually merging into one peak, thus coeluting all three components later in the pyrogram. Despite a loss of resolution, the fast GC separation obtains similar discriminating information in 12 min, compared to 55 min by conventional GC. The more practical concern is whether, in achieving fast analysis times, the analyst is forced to compromise resolution or sample capacity to the

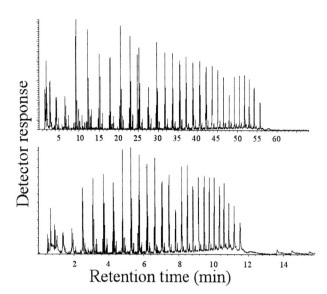

Figure 7.3 Pyrograms of poly(ethylene-*co*-methyl acrylate) polymer using (*top*) a conventional column (30 m × 0.25 mm ID × 0.25 μm df, 5% phenyl–95% methyl polysiloxane) and (*bottom*) a short, narrow column (10 m × 0.1 mm ID × 0.10 μm df, same phase). Other conditions identical for both runs include: filament pyrolysis a 900 °C for 10 s, ramp rate 20 °C/ms, interface 280 °C; 40 °C oven temperature held for 30 s, then ramped at 25 °C/min to 300 °C and held at that temperature.

extent that the information content is degraded. The lower sample capacity of small ID columns may also be a limiting factor if characteristic peaks required for analysis are only present in small amounts.

Another consideration is that correlation of peaks produced by different stationary phases can be difficult without mass spectra. Different stationary phases process analytes differently; even with similar columns, even similar instruments may filter the pyrolysis product mixture differently. For example, polar and reactive pyrolysis products may react with active metal or other sorption sites in one chromatographic system, yet be unaffected by another system.

Contamination in analytical pyrolysis systems is a common problem. Extraneous peaks that mysteriously appear in some pyrograms have determinate causes that, if understood, can be eliminated. Contamination in pyrolysis is often caused by carry-over of higher molecular weight, polar, or less volatile material that remains behind in the interface, connecting tubing, or other parts of the chromatographic system. Nonvolatile residue is often unavoidable, depending on the nature of the sample. For example, after pyrolysis of poly(vinyl chloride), char has been extracted and separately analyzed by GC/MS to identify numerous toxic, chlorinated polycyclic aromatics.[69] The buildup of such tarry products inside the system after a period of use may confound long-term reproducibility, especially when peaks from these components are flushed from the system during normal temperature programming. Trouble-shooting includes investigating different pyrolysis conditions to pyrolyze the sample more effectively and evaluating the pyrolysis system for contamination. Wiping the inner surfaces of the pyrolysis interface with a cotton swab may reveal if residue accumulation is present. The pyrolysis interface and the GC interface can both be cleaned by using various solvents (e.g., dilute nitric acid, hexane, chloroform, and acetone, in that order). Overnight heating (600–700 °C) of solvent-washed pyrolysis elements, heating in a water-saturated stream of hydrogen at 550 °C, or chemical polishing followed by washing with acetone and heating have been recommended, although manufacturers' guidelines should be followed. Direct flame heating should be avoided because it can oxidize metal and contaminate surfaces. Quartz sample tubes should be cleaned in dilute nitric acid at 100 °C overnight, rinsed with methanol and acetone, then dried in an oven.

Contamination in Py-GC or Py-MS systems may be caused by a cold spot in the transmission system that traps less-volatile residue. To see if any temperature zones are set too low, temperature settings should be increased by small increments (without exceeding temperature limits of the column or instrument), and the results should be checked. Wrapping connecting tubing with heating tape or insulation may eliminate cold spots. Condensation losses due to cold sports can severely limit the ability to observe higher molecular weight and less-volatile components. Heated fused-silica transfer lines can be inserted near the pyrolyzer element at one end and threaded straight through to the flame ionization detector flame tip or the MS ion source at the other end. In this manner, contact with active surfaces is minimized, and more complete transfer of pyrolysis products from pyrolysis interface to detector is achieved. Py-MS systems have incorporated glass reaction tubes, expansion chambers, heated walls, and positioning of the pyrolysis reactor adjacent to the ion source.

A valuable habit to acquire is the analysis of a standard sample at regular intervals between other samples. When "ghost peaks" appear in a pyrogram, analysis of the standard sample as a quality check, coupled with one or more of the ideas described above, may help to resolve the contamination problem.

The most common detector for GC, the FID, measures variations in the ionization current as components pass through a hydrogen-oxygen flame. A schematic of an FID detector is shown in Figure 7.4. The carrier gas and eluting sample components are mixed with hydrogen, and the mixture burns at the tip of the jet with air (2100 °C). Positive ions, negative ions, and electrons produced in the flame give an ion current (ca. 10^{-11} A baseline) that is measured by establishing an electrical field (400 V) between the negatively charged jet tip and the positively charged collector electrode. The resulting current is then amplified and digitized for output. The FID is destructive. The FID linear dy-

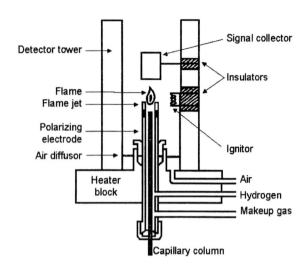

Figure 7.4 Schematic of a flame ionization detector.

namic range is 10^6 to 10^7, and its sensitivity depends on ionization efficiency and the number of oxidizable carbon atoms in an organic compound. The minimum detectable quantity (MDQ) for methane using an FID is in the range of 10^{-11} to 10^{-12} g/s. Whereas organic compounds are universally detected by the FID, compounds such as H_2O, N_2, and O_2 have little response. The FID is popular because of its low cost, ease of use, relatively high sensitivity, and universal detection of organic compounds.

Although advantages and limitations[5,72] have been noted, many contemporary gas chromatographic systems use mass spectrometers as detectors and, as such, permit reasonably unequivocal qualitative identification of the decomposition products by comparison with on-line mass spectral libraries. Py-MS produces a pyrogram that is the sum of all ion currents from all compounds in the pyrolysis product mixture superimposed in a single mass spectrum. For that reason, interpretation of Py-MS results from a complex sample can be more difficult than interpretation of a Py-GC/MS pyrogram, in which pyrolysis products are separated by chromatography before MS detection.

Pyrolysis products are swept from the GC into the MS ion source (at 150–250 °C) as gases. The ion source must be kept at a low pressure (e.g., 10^{-6} Torr) so that the mean free path for electrons and ions is large enough to avoid collisions between ions and other molecules. Such collisions cause ions to lose charge or might cause ion-molecule reactions that complicate the resulting spectrum.

The electron impact (EI) source is perhaps the most common ion source, although a variety of ion sources are available. In EI, electrons produced from a heated tungsten or rhenium filament are accelerated at 70 eV in vacuum toward an anode. Sample molecules are introduced perpendicular to this beam, and ionized fragments are produced. Positively charged fragments are pushed toward the mass analyzer by a repeller voltage. EI tends to fragment sensitive compounds severely; often, the molecular ion is not observed in the mass spectrum. Although an accelerating voltage of 70 eV is standard, lower voltages may produce more molecular ions and fewer fragment ions and thus provide better information on molecular weight.

Mass analyzers are of several general types: magnetic sectors, quadrupoles, time-of-flight (TOF), ion traps, and tandem. The electron multiplier is a common transducer for the detection of positive ions in mass spectrometers.

In magnetic-sector instruments, ions are accelerated from the ion source by potentials of 3 to 5 kV. All ions are accelerated with the same kinetic energy, but ions of different mass will have different velocities. Charged particles follow a curved path through a magnetic field, with the degree of curvature related to the momentum of the ion. By changing the field strength of the magnet, ions of different m/z values can be focused at the detector. A magnetic-sector instrument has a maximum m/z value that can be detected, depending on the radius of curvature for the magnet, field strength, and acceleration potential. Large pumping systems, long ion paths requiring precision alignment, and frequent maintenance of high-voltage sources characterize these instruments. Magnetic sectors are often combined with electric sectors to form double-focusing mass spectrometers with resolving power better than 1 part in 100,000; several different geometries are available for various applications.

Perhaps more quadrupole mass spectrometers are in use than all other types of mass spectrometers combined. Ions are accelerated from the ion source at 10 to 15 V. Opposing rods (stainless steel, molybdenum, or glass coated with gold) are connected together, and a combination of direct current and radio frequency potentials create a "mass filter" that selectively passes only the ions of desired m/z to the detector. Scan rate is a function of the time required for an ion to reach the detector; a single-charge ion of m/z 800 with energy 10 eV typically takes about 100 μs. Although this limits the ability to scan rapidly, quadrupoles can be scanned faster than magnetic sector instruments, making them popular for capillary GC applications.

The TOF MS separates ions by their travel time from ion source to detector. Ions are accelerated out of the ion source with a potential of 3 to 5 kV into a drift tube. Ions with different mass-to-charge ratios travel at different velocities and reach the detector at different times. Ions of lower mass have higher velocities and reach the detector first. An intriguing characteristic of TOF MS is that very high masses can be detected if one is willing to wait long enough. TOF MS is particularly promising as a detector for fast GC because it can be operated in an extremely fast, pulsed mode (e.g., up to 500 full scan spectra per second in some commercial instruments).

The ion trap stores ions formed during ionization in separate orbital paths, depending on their m/z values. Ions are destabilized 1 m/z value at a time and moved toward the top and bottom end-caps. Ions that move to the bottom end-cap reach the detector. Because a higher proportion of ions that are formed reach the detector, ion traps can have higher sensitivity than beam-type instruments have.

Finally, various forms of tandem MS have been developed; these involve coupling two or more mass analyzers sequentially. Using such an MS-MS instrument, ions separated on the first analyzer can be trapped and fragmented further, and then have their structure determined on the second analyzer. Of the current trends in GC likely to affect the analysis of polymers, those worthy of note include the increased availability of fast GC instrumentation, the growing spread of less-expensive desktop mass spectrometers as detectors for GC, and an increasing sophistication in data analysis enabled by faster computers.

Qualitative and Quantitative Analysis

As with any analytical problem, the first rule is to understand the sample: its origin, physical properties, and chemistry. The analyst must interpret what is known about the sample in relation to the questions posed, must understand thoroughly the structural chemistry involved and its relationship to polymer properties of interest, and must be prepared to take advantage of those aspects of macromolecular structure that lend themselves to analysis by chemical or thermal decomposition. That a polymer produces a particular pyrolysis product or other chemical marker as a result of a decomposition is, by itself, not proof of the value of that component for answering any particular question about the structure of the parent polymer. Validation of the relevance of that analytical information to the polymer structure is required. Standard reference materials of known composition and structure must repeatedly be analyzed to ensure that repeatable results can be obtained.

On occasion, combined Fourier transform infrared (FTIR) spectroscopy and MS have been used to identify the separated products, as has FTIR[3,73] by itself. Selective atomic emission and flame photometric detectors have been used[20,74-78] with good results when the analyte contains appropriate atoms, either by natural occurrence or from a chemolysis process. In similar fashion, some studies have reported good results with use of an auxiliary FID.

Traditional quantitative analysis can be performed using external calibration, but many analyses have used a variation on standard addition: a known amount of a reference polymer is added to the sample before the decomposition process is begun; this polymer is later identified and measured, either as itself or through its characteristic decomposition products.[11,20,42,79-81]

A problem that the analyst may encounter is the differentiation of samples of a polymeric species distinguished only by synthesis method or manufacturer or other nonstructural characteristic. Price et al.[5] noted that authentic but different polypropylene samples were distinguishable only by low-temperature, thermal-desorption GC/MS. Subtle differences due to antioxidants and polymerization initiators were lost in a high-temperature Py-MS evaluation.

In a similar fashion, two polymers may be sufficiently alike that their decomposition product mixtures cannot be distinguished by Py-MS, whereas Py-GC/MS does permit adequate identification. Mundy[72] has identified several such systems, including polyethylene versus ethylene-propylene-diene terpolymer (EPDM) and polypropylene versus poly(ethylene-*co*propylene).

A copolymer synthesis can be engineered to favor either random or block or even block-random arrangement of the two monomers; such a situation is exemplified by poly(ethylene-*co*-vinylcyclohexane). However, unambiguous identification of the product from the polymerization step can be difficult by traditional chemistry. Use of Py-GC/MS has enabled identification and quantification of mixed oligomers from the pyrolysis step.[82]

In addition to the chemical identification of the product mixture from polymer decomposition, the kinetics and mechanisms of these processes is also of interest. Traditional bulk thermogravimetric methods provide only a gross understanding of decomposition rates. Because chromatographic analysis of the product mixtures results in a knowledge of individual products rather than bulk loss, the kinetics and mechanism for many polymer decompositions including polystyrene,[31,33,83–85] poly(methyl methacrylate) and other alkyl methacrylate polymers,[31,83–86] polyisobutylene,[33] and natural rubbers[33,85] have been characterized. More detail on determination of degradation rate constants is given in the applications section that follows.

A major challenge for the future is how to take advantage of the mass of data generated in a typical GC/MS experiment. New uses will be developed for the wealth of information generated by the pattern and amounts of products generated when a polymer is chemically or thermally degraded. Techniques to extract efficiently from this data the maximum information relevant to solving a particular polymer problem must be further developed.

Selected Applications of Polymer Analysis

The use of polymers in contemporary society has become so pervasive that virtually any analytical situation may require the evaluation of a polymeric system. In this section, applications of GC for a variety of polymer situations are reviewed. Although the polymer classifications and distinctions used here may seem arbitrary or confusing, typically these follow from the precursor chemicals, the synthesis chemistry, or the distinct chemical nature of the product's chemical moieties. In this review, the classifications for the various polymers as used by the original authors have generally been preserved. Reference should be made to standard texts[87–89] for more specific details and elaboration.

Characterization of Waste Materials

The disposal of polymeric waste materials is an environmentally sensitive issue. Rather than disposing of these materials in a landfill, pyrolysis of waste polymer can provide hydrocarbon feedstock for reuse in synthesis or end use as a fuel. The composition of a pyrolysis product mixture depends on the original polymer mix, the pyrolysis conditions, and the nature of any added substances. In a detailed analysis over a range of pyrolysis temperatures, Hawley-Fedder et al.[90] identified over 100 products from mixtures of polyethylene, polystyrene, and poly(vinyl chloride). Levendis and co-workers[91] have reported comparable results in aerosol pyrolysis and examined the postpyrolysis chemistry of a variety of polymers.[92] Blazso[93] pyrolyzed mixtures of polyethylene and phenolformaldehyde and observed significant amounts of polycyclic aromatic hydrocarbons (PAHs) at higher temperatures. PAH production was reduced, however, by incorporating iron or copper and their oxides into the reaction mixtures. In low-temperature pyrolysis of bulk mixtures of polyethylene and polystyrene, Simard et al.[35] observed changes in the amounts of various aromatic products with variations in the polymer composition. Koo et al.[94] used HPLC followed by capillary GC/MS to evaluate both aliphatic and aromatic distributions from mixed polyethylene-polystyrene waste with similar results.

Pyrolysis of chlorine-containing polymers, most commonly poly(vinyl chloride) and poly(vinylidene chloride), produces significant amounts of hydrogen chloride and chlorinated aromatic compounds.[95–97] Similar products have been obtained when these chloropolymers are mixed with polyethylene.[98] However, production of both hydrogen chloride and aromatic hydrocarbons is suppressed by adding iron, iron oxide, or other materials to the reaction mixture.[99–102] Similarly, decreased aromatic production in derivatized poly(vinyl chloride)[103] was attributed, in part, to reduced chlorine levels. More interestingly, enhanced levels of aromatic compounds have been observed by pyrolyzing polyethylene in the presence of iron and copper chlorides.[104] Lovett et al.[105] have investigated catalytic upgrading of product mixtures from mixed polymer pyrolysis. Several studies have examined possibilities for catalytic degradation of polyethylene[106–108] and polypropylene.[108–110] Attention has been given to the highly toxic, but nonvolatile, residue from PVC pyrolysis.[69,111]

Analytical pyrolysis has been applied to polymer recycling in the laboratory, pilot plant, and production facility. Two areas have been targeted: (1) elastomers in automobile tires and other rubber products, and (2) thermoplastics, including polyethylenes, poly(vinyl chloride)s, and polystyrenes.[112–115] For example, Williams et al.[114] pyrolyzed polystyrene at 500 °C followed by secondary cracking at 700 °C, producing styrene, benzene, xylene, toluene, alkylated styrene derivatives, and styrene oligomers, along with abundant levels of PAHs (Figure 7.5). While pyrolysis of epoxy resins and polyurethane and other thermosetting polymers has been studied in the laboratory,[116,117] the nature of these compounds precludes easy batch processing.[118,119]

Analysis of Environmental Pollutants

The role polymers play as environmental pollutants is closely related to waste disposal problems. Investigations have focused on detection and quantification of polymer contaminants at relatively low concentrations in environmental samples including sediments,[120,121] soils,[122] and water.[120] Because of biodegradation and weathering, polymers in the environment may not be present in their original form, but might be detected as decomposition products. An example is the identification and tracking of the products of starch-modified polymers[8,123-127] that have been engineered to photodegrade or biodegrade when they are abandoned in the environment.

Analysis of phenolic thermosetting polymers is addressed further in the next section of this chapter. However, there is an associated environmental aspect: because of their use as binders in automobile asbestos-containing brake pads, these polymers have been identified in road dust and quantified to estimate the vehicular contribution of asbestos to the environment.[128-130] Polymeric materials are often used as mold binders in the sand-casting[131] and ceramics industries.[132,133] The decomposition product distributions of these binders have also been studied. Improper disposal of superabsorbent diapers leads to sodium polyacrylate contamination in the environment. Identification of this possible pollutant in samples can be achieved through analysis for its major pyrolysis product of 1,3-cyclopentadiene.[134]

Authentication of Identity for Artifacts and Trace Materials

In contemporary society, authentication of materials is a continuing concern. Automobile accident investigations depend to a great extent on the identification of missing or at-fault vehicles through knowledge and analysis of paints and their constituent polymers. Identification of paint components is also of importance in quality control of manufacturing processes, in evaluation of performance under normal usage conditions, and in substitution of formulation

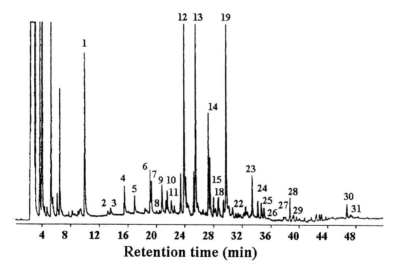

Figure 7.5 Gas chromatogram of polystyrene-derived pyrolysis oil. Tentative peak identification: (1) naphthalene, (2) 2-methyl naphthalene, (3) 1-methyl naphthalene, (4) biphenyl, (5) methylbiphenyl, (6) alkylated styrene, (7) alkylated styrene, (8) alkylated styrene, (9) fluorene, (10) methylfluorene, (11) methylfluorene, (12) styrene oligomer, (13) phenanthrene, (14) alkylated styrene, (15) 3-methylphenanthrene, (18) 1-methylphenanthrene, (19) phenylnaphthalene, (22) pyrene, (23) terphenyl, (24) terphenyl, (25) methylfluoranthene, (26) benzo(b)fluorene, (27) benzophenanthrene, (28) benzanthracene, (29) chrysene, (30) benzo(e)pyrene, (31) benzo(a)pyrene. Conditions: batch pyrolysis at 500 °C, followed by secondary cracking at 700 °C; DB-5 column, programmed at 70 °C for 2 min, ramped at 5 °C/min to 280 °C, and held for 25 min. (Reprinted from reference 114, with permission from Elsevier Science. Copyright 1993.)

components.[135-141] In addition to GC/MS identification of pyrolysis products and parent polymers, much of this work has been accompanied by statistical analysis of the data.

Other situations of legal interest involving polymeric systems have included copy toners[142] and adhesives.[143] A potential difficulty with much of this investigative work is that the sample under investigation may be considered to be a proprietary material by its manufacturer.[138,143] As a result, identification is strictly by comparison and degree of similarity to library compounds or authentic reference samples.[144] However, lack of standard reference materials for proprietary polymers and formulations often confounds unequivocal identification.

Nylon fibers,[43] as well as wool, silk, and similar proteins,[53] have been identified by Py-GC/MS at levels less than optimum for infrared (IR) spectroscopy. Of particular interest to wool analysis[53] is identification of aged wool samples; normal denaturing, which obscures traditional analyses, does not affect Py-GC results.

Other forensic studies have examined chars and soots in fire debris[145-147] as evidence for arson investigations. Decomposition products from some synthetic polymeric carpet materials have been found to mimic common accelerant mixtures.[145] Because these materials are retained on the char, false positives by arson canines have been noted. Like activated charcoal, unoxidized condensed carbon soot from combustion will absorb substances from the environment. Analysis of these absorbed materials has provided a trail to identify the nature of the precursor polymers[145] and trace progression of the fire by distinguishing among possible polymer precursors.[146]

In a similar fashion, historic or cultural artifacts that either are polymeric or incorporate a polymeric component can be authenticated by Py-GC. Shredrinsky et al.[148] reported on the analysis of natural and imitation ambers. Although many twentieth-century substitutes such as polystyrene and polyesters can be easily identified, rare red Russian amber and contemporary Bakelite can yield remarkably deceptive pyrograms (Figure 7.6); however, other materials used in forgeries of amber objects could be easily differentiated. Varnish and similar protective coatings for paintings and sculpture[149,150] have also been analyzed by Py-GC. Although a competent analysis can be performed with a relatively small amount of material and hence minimal invasive damage to the artifact, any damage can be a cause for concern. IR spectroscopy offers a nondestructive alternative that avoids the need to take and prepare a sample.[151] IR spectroscopy has also been used as a parallel technique in some investigations.[149]

Analysis of Explosives and Propellants

The polymeric nature of many engineered combustibles is not immediately obvious. Formulations of polybutadiene and its terpolymer with acrylonitrile and acrylic acid are

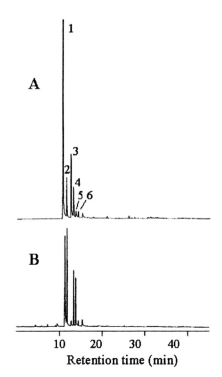

Figure 7.6 Py-GC results from (A) Bakelite (Union Carbide, 1964); (B) "rare red Russian amber" (ca. 1910). Peak identification: (1) phenol, (2) o-cresol, (3) p-cresol, (4) 2,6-dimethylphenol, (5) 2,4-dimethylphenol, (6) 2,4,6-trimethylphenol. Conditions: filament pyrolysis at 650 °C; 50 m × 0.25 mm ID × 0.5 μ df SE-54 column; oven temperature held at 50 °C for 1 min and programmed at 8 °C/min to 325 °C and held for 10 min. (Reprinted from reference 148, with permission from Elsevier Science. Copyright 1993.)

commonly used as binders in solid rocket motors. Rao and co-workers[152,153] have addressed various aspects of this chemistry and presented pyrograms of these polymers. In a parallel study,[154] both FTIR and Py-GC were used to measure the butadiene/acrylonitrile ratio of the butadiene-acrylonitrile-methacrylic acid terpolymer; these measurements correlated well with separately measured mechanical properties of the polymer.

Rietjens et al.[155-157] analyzed a series of "plastic-bonded-explosives" (PBX) by Py-GC. Due to the proprietary nature of these formulations of polybutadiene, antioxidant, and explosive, pyrograms were not published. Twenty-seven pyrolysis products were identified, and the result of extensive multivariate data analysis was reported. Features of this work included correlations of the pyrolysis product distribution with sample age, effects of added antioxidant, degree of cross-linking in the polymeric binder, and inhomogeneity of the PBX sample.

Analysis of Textiles, Papers, and Other Consumer Goods

Polyurethanes, formed from the reaction of a diisocyanate and a diol, have been used in a variety of consumer products such as foams, elastomers, fibers, and paints. Thermal degradation of these materials is of interest, as are efforts to reduce their combustibility. Several studies have examined model polyurethanes[116,117,158] to link the product mixtures with the starting polymer. Regenerated isocyanates and glycols are typically observed in the product mixture, but they often undergo secondary reactions,[159,160] especially at higher pyrolysis temperatures.[153,161] Extensive work has been reported in which the formulation contains specific additives to either quench or redirect the combustion chemistry.[162–165]

Studies of cloth and textiles have included the traditional wool and cotton, but more recently have included polyesters and other synthetic fibers. The identification of the fabric is generally a forensic concern, but the primary consumer interest is the various finishing treatments that have been applied. Examples include a shrinkage control for wool[166] and fire retardants for polyesters[167] and cottons.[168–170] Acrylic copolymers have been used for fabric stabilization during production processing and consumer cleaning.[171]

Aging and degradation of polymers in fiber-optic data-communication systems requires the analytical capability to identify or characterize polymer components. Krüsemann and Jansen[3] used thermal desorption-GC, along with other techniques, to examine a variety of poly(methyl methacrylate) fibers. Aged fibers were identified by the additional variety of decomposition compounds that are not present in new fibers.

As noted with respect to polyurethanes, fire retardancy is a concern for many polymer systems. Identification of decomposition products for treated and untreated polymers can aid the characterization of the formulation of polymer and additives. Systems studied include polystyrene[172,173] and its copolymers,[174,175] poly(ethylene-co-vinylacetate),[176] epoxy resins,[177] and polyesters.[178]

In addition to the cellulose[179] content itself, additional polymeric materials are found in papers, as naturally occurring lignins and rosins and as added sizing agents and inadvertent contaminants. Van Loon et al.[180–183] have also identified and quantified chlorolignosulfonic acids in pulp mill wastewater. Lignins are the "glue" that holds wood together and is separated from the cellulose in the pulping process. Interest in the chemical structure of lignins has been explored,[184,185] as this traditional waste material is becoming a valuable chemical feedstock.

Sizing agents are added to paper pulp to minimize ink blotting through the capillary action of the cellulose fibers. Although contemporary acid-free papers avoid their use, traditional sizing agents have been natural rosins. These formulations are acidic because of the use of aluminum sulfate (alum) to fix the rosin in the cellulose matrix. For the analysis of the fraction of added rosin retained in the product, methanolysis-GC/MS methods have been developed;[186–188] they are potentially superior to traditional wet chemical methods in terms of specificity and speed of analysis. Py-GC has been employed to quantify neutral sizing agents,[189] including polystyrene[190] and polyamide-epichlorohydrin copolymer.[191]

In recent decades, in a remarkably successful effort to minimize pollution, the paper pulping industry has developed closed-loop water usage. Unfortunately, such systems can concentrate contaminating materials, and these materials may subsequently appear as impurities in the paper product. The focus of much of this work has been toward identifying polymeric contaminants and their origins. Contaminants have been traced to equipment failure,[192,193] incomplete lignin removal,[193,194] and inadequacies of the water-purification systems.[195] Figure 7.7 shows Py-GC/MS results from several synthetic polymers collected from a pulp-mill cleaner unit.[194] The distinctive chromatograms obtained from the different rubber polymers could be used as fingerprints to determine the sources of the contaminants.

Numerous other consumer products involve the use or incorporation of polymeric materials at some point in their manufacture. Further, these polymers may or may not remain in the final commercial product. For example, lubricating oils consist of a hydrocarbon mixture plus selected additives that provide desired properties such as viscosity. Viscosity is usually achieved by adding poly(alkyl methacrylates), although poly(styrene-hydrogenated butadiene) has also been used. After separation of the polymeric component from the hydrocarbon, these polymeric additives have been identified with a confidence that permits both forensic identification and the correlation of lubrication performance with composition.[196]

In similar fashion, additives, polymeric and otherwise, have been incorporated into Portland cement to enhance desired properties in the finished concrete. Py-GC/MS analysis of pulverized samples permits identification of many of these substances, including the polymers. A point of concern was possible matrix effects on the quantification of these materials; synthetic mixtures using a variety of cement precursors produced no variation in detection and measurement.[197]

A variety of polymeric materials have been incorporated into laundry detergent formulations. Polycarboxylates have been used to replace traditional phosphates as builders and have been successfully detected in detergent solution using in situ methylation Py-GC/MS.[198] In contrast, dye transfer inhibitors, specific polymers that minimize dye transfer between articles in a wash load, have been detected in synthetic samples by direct pyrolysis of the detergent powder.[199]

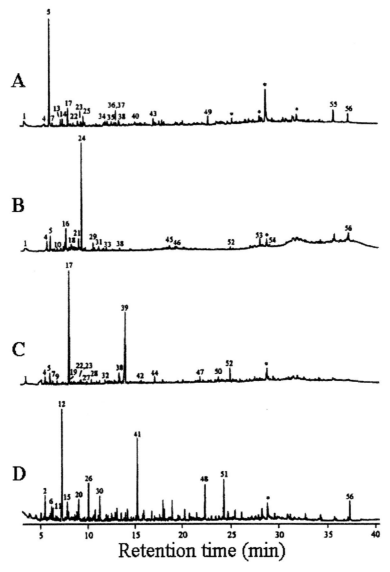

Figure 7.7 Py-GC/MS chromatograms from four synthetic rubber samples (A–D) collected from a pulp-mill cleaning unit. Major peaks: (5) styrene, (24) dipentene, (17) α-methylstyrene, (39) dimer of 2-chloro-1,3-butadiene, (12, 41, 48, 51) unidentified. Conditions: Curie point pyrolysis at 610 °C; 30 m × 0.25 mm ID × 0.25 μ df DB-5 column; oven temperature programmed from 40 °C (1 min) to 300 °C (20 min) at 6 °C/min. (Reprinted from reference 194, with permission from Elsevier Science. Copyright 1999.)

Two aspects of polymer usage in the pharmaceutical industry have been noted. Polystyrene-absorbent residue, from its use as a substitute for activated charcoal, has been quantified in formulations of vitamins and antibacterials.[200] Various acrylates are used as controlled-release coatings for many drugs; quantification of the amount of polymer correlates well with the rate of drug release.[201]

Other polymeric substances have been used as absorbents and clarifying agents in the manufacturing process. The avowed intent is that these will be fully removed before the product is shipped and that the analytical process expects to find no residual material. One such polymer is poly(vinyl pyrrolidone), which has been used in the brewing industry. A pretreatment Py-GC/MS method has been developed[202] to identify the monomer, dimer, and trimer decomposition products of the polymer; the proteinaceous component of the sample beer does not produce these molecules. This new procedure is both more specific and manipulatively simpler than the traditional colorimetric method. Other studies of polypyrrolidones[203–204] have addressed more traditional questions of structure and properties. Methacrylic absorbents have also been studied.[205]

The ubiquitous plastic films used to display and protect foodstuffs have been studied. Separate experiments have determined that the preservation aspects of poly(ethylene-*co*-vinylacetate) are enhanced when electron beam irradiation is performed.[206] Py-GC/MS was among the analytical methods used to determine the nature of this change in the altered film. The yield of acetic acid was greatly reduced relative to unirradiated films, indicating that the former carbonyl moiety and available reactive sites on adjacent polymer strands were cross-linked.

A parallel problem is the analysis of preservatives. Butylated hydroxytoluene has been a common foodstuff additive, but it has also been used in polymer formulations to minimize premature oxidation. This small molecule can survive the pyrolysis event and is observed as a contaminant in the chromatography.[208]

Phenolic and epoxy resins have become popular engineering materials. Py-GC analysis of these materials typically finds phenolic and aromatic hydrocarbons and, at higher temperatures, considerable light aliphatic hydrocarbons.[209-211] Subtle differences in the aromatic product mixtures can permit differentiation of the two classes.[210] Since these polymers are often prepared as needed in the field rather than in bulk in a production facility, variation in pyrolysis decomposition properties due to the identity and amount of curing agent and the extent of curing time have been investigated.[30,209,212-217]

At one time, the substitution of polymers for traditional materials in consumer products was cause for derision. In the intervening years, a considerable body of work has been devoted to accurately replicating the traditional substances with synthetic polymers. One such endeavor has been the development of an economical substitute for the natural urushiol lacquers. The similarity of the natural and synthetic lacquers has been well characterized by Py-GC/MS.[218-220]

The polyesters encompass a large group of compounds, the common feature being the ester-linkage between the monomers. The reactivity and stability of these bonds have been the attention of several investigations[221-224] with implications for the longevity of various fabricated articles. A commonly used product is the composite of polyester with glass fiber. Both natural and artificially aged samples have been analyzed with attention to the compositional changes accompanying the physical degradation.[225] An interesting possibility is the use of selected polyesters as solvents for solid electrolytes; however, further investigation suggests that the presence of the electrolyte greatly reduces the stability of the polymer.[226]

Characterization of Natural Products

Numerous naturally occurring materials are polymeric in nature and have been studied. In addition to the cotton[168] and wool[53,166] already noted, cellulose,[61,179,227] carbohydrates,[228]

humics,[21,229,230] lignins,[62,231] and other biologically relevant materials such as saccharides and lipids[7,232-238] are polymeric and have been investigated.

So that knowledge of functional moieties is preserved, much of this work has been done under chemolysis conditions with tetramethylammonium hydroxide. Some work has been done with tetrabutylammonium hydroxide to avoid potential confusion with the methoxyl groups present in the original molecules.[229]

The abundant polymer found in crustacean shells and insect cuticles is chitin. This polymer, as well as its derivative chitosan and compounded copolymers, has received recent attention.[26,79] Sato et al.[239] found that reactive pyrolysis in the presence of oxalic acid permits quantitative determination of the degree of acetylation of chitin. The pattern of pyrolysis products (containing acetonitrile, acetic acid, acetamide, and levoglucosenone), shown in Figure 7.8, is much simpler than when chitin is pyrolyzed alone. Acetamide originates from the *N*-acetyl groups and levoglucosenone is derived from the glycan backbone.

The naturally occurring polymeric hydrocarbons include coals,[22,81,240,241] kerogens,[12,22,81,241-246] and asphaltanes.[81,243,247,248] Interests in these substances have been varied, especially because, as with all natural products, the properties of these materials can vary with the source. Py-GC/MS studies have been conducted with particular attention given to molecular structures and carbon isotope distributions. Since these substances hold promise as precursors to synthetic liquid and gaseous hydrocarbons, the thermal decomposition products of these substances are important.

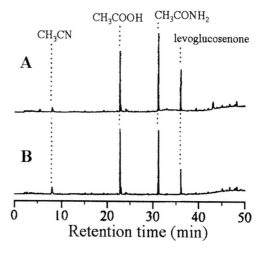

Figure 7.8 Py-GC/MS results for (A) chitin, and (B) *N*-acetyl-D-glucosamine in the presence of oxalic acid. Conditions: microfurnace pyrolysis at 450 °C; metal capillary column coated with poly(ethylene glycol) programmed from 35 °C (5 min) to 220 °C at 5 °C/min. (Reprinted from reference 239, with permission from American Chemical Society. Copyright 1998.)

Chemcial Studies of Polymer Structure and Degradation

In addition to the "obvious practical applications" mentioned, numerous investigators have applied GC to the basic nature and behavior of fragmented polymeric materials. In most instances, the analyte polymer is a model compound for the system of interest and, as such, has been deliberately synthesized or "cleaned-up" before pyrolysis and follow-up analytical procedures are undertaken. A common thread among these investigations is that, although complementary techniques may also be used, GC provides an efficient tool for the separation of these complex pyrolysis product mixtures.

These studies have focused on a variety of topics. A frequent endeavor is the determination of "traditional" kinetics and mechanisms of the decomposition process. This work has been approached from two directions: (1) to investigate what happens in pyrolysis of a selected polymer and (2) to develop models describing some measured process. In the former instance, many investigations have explored the behavior of mixtures of polymers, or of some polymer with one or more additives, in an attempt to quantify synergistic effects.

Audisio and Severini[249] prepared poly(4-vinylpyridine) from the monomer, both by radical polymerization and by anionic polymerization. Although the pyrograms of the two products were virtually identical, detailed analysis of the oligomer yields permitted a differentiation between head-to-head and head-to-tail monomer sequencing in the parent polymer and hence a knowledge of differences of the two polymerization reactions.

The study of two polymers that differ only in the isomeric nature of the monomer can provide interesting comparisons. Schulten et al.[250] compared Kevlar, poly(1,4-phenylene-1,4-phthalamide), with Nomex, poly(1,3-phenylene-1,3-phthalamide), under comparable experimental conditions. Flash Py-GC/MS of the two polymers gave nearly identical results, consisting primarily of molecular fragments smaller than the monomer. A ramped pyrolysis-field ionization MS analysis of the two polymers produced distinctive distributions of products that included significant amounts of the monomers and fragments of the dimers.

Using Py-GC as one of several techniques, thermosetting resins based on furfuraldehyde have been synthesized and structurally characterized.[251–253]

Determination of Polymer End Groups

Analysis for the chemical identity of the end groups of a polymer is an area of intense interest.[254–262] Polymer synthesis typically begins when an initiator molecule is stimulated to release a single-electron, or free radical. The radical attacks bonds of monomer molecules, sometimes with the aid of chain transfer agents, and formation of the polymer chain commences via a free-radical mechanism. A polymer molecule may have a variety of end groups, depending on which agents terminate polymerization. At times, on termination, portions of initiator or chain-transfer molecules will become incorporated into the end group moiety.

The determination of the chemical nature of the end-group entities (which are, presumably, the initiator and quencher molecules) is important in both the confirmation that the polymer synthesis has proceeded as planned and in the characterization of the polymer for further applications. Characterization of end groups can lead to clarification of the mechanisms that govern polymer chemistry. Knowledge of end groups also facilitates design of new polymers with improved properties for specific applications. A macromolecule's degree of polymerization can be calculated from the relative amounts of end groups found in the pyrogram.[254] Chain-transfer constants (C_s), a measure of how fast monomer molecules are being added into the polymer chain, may also be obtained and are comparable with those attained by titration and size exclusion chromatography (SEC) methods.[258]

The relative signal strengths of the end groups and the amounts of bulk polymer (monomers, dimers, trimers, etc.) lead to a surprisingly accurate estimate of the molecular weight of the polymer that is independent of external standards,[254] but which can be verified against SEC.[74,255–260] The only significant limitation is that the pyrolysis behavior of the polymer in question must be well characterized. An example of this approach is seen in the work of Ito et al.,[255] who used peak intensities of end groups to determine molecular weights of polystyrene up to 1 million daltons. As the polymer chains get longer, fewer end groups are present than in an equal-sized sample of lighter polymers, and the intensity of these peaks decreases. Estimation of molecular weights greater than 1 million was less reliable because the scarcity of end groups makes their corresponding GC peak areas small.

As another example, in poly(methyl methacrylate) (PMMA), the initiator molecule azobis(isobutyronitrile) (AIBN) and the chain transfer reagent thioglycolic acid (TGA), contain nitrogen and sulfur species, respectively, that were not easily detected by FID.[74] Use of FID with a nitrogen-phosphorus detector (NPD) or a sulfur-selective flame photometric detector (FPD) allowed for simultaneous analysis of these species. Benzothiophene, which generated a single peak in the chromatographs of both detectors, was used as an internal standard (0.01 mg added to 0.5 g of polymer sample). Chromatograms illustrating simultaneous FPD and FID detection are illustrated in Figure 7.9. Methyl methacrylate (MMA) is the primary pyrolysate; peaks due to CH_3CH_2SH and other sulfur-containing products were attributed to the end groups. Degree of polymerization was estimated by correlating the intensity of these sulfur-containing pyrolysis products, ratioed to the internal standard, against the MMA monomer yield.

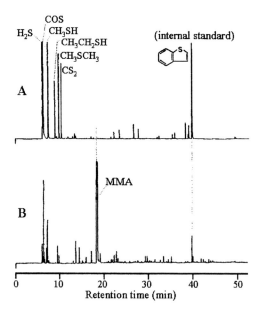

Figure 7.9 Simultaneous pyrograms of poly(methyl methacrylate) prepolymer by (A) flame photometric detector (FPD), and (B) flame ionization detector (FID). Conditions: microfurnace pyrolysis at 700 °C; 50 m × 0.2 mm ID × 0.33 μ df dimethylsilicone fused silica column; oven temperature programmed from 0 °C to 250 °C at 4 °C/min. (Reprinted from reference 74, with permission from American Chemical Society. Copyright 1994.)

Other studies address the needs of the synthetic chemist when the polymer in question is to be used as a reagent and it is desired to know if the nature of the end groups are such that the subsequent chemistry would be favorable.[261-264] The related situation in which the end groups are different has also been investigated.[152]

Structure Correlations and Quality Control for Styrene-Butadiene Rubber

Poly(styrene-co-butadiene) (SBR), or tire tread rubber, is a commonplace commercial polymer with specific properties that follow from the amount of styrene incorporated into the polymer chain. A variety of standard analytical procedures such as refractive index, infrared spectroscopy, and ultraviolet and NMR spectroscopy have been accepted for its analysis. However, these methods typically require removal of the carbon black and other additives used in compounding the product. Py-GC has the potential for direct analysis of SBR,[265] but Py-GC methods need to be correlated with established methods such as refractive index (ASTM method D1416). At least a dozen compounds, in addition to the liberated monomers, have been identified by GC/MS in the pyrolysis product mixture.[27,266]

Using both laboratory preparations and commercial samples of SBR, Ghebremeskel and co-workers[266] made correla-tion plots of results from Py-GC/MS compared to refractive index and infrared spectroscopy measurements and obtained excellent agreement. In similar fashion, comparison pyrolyses of the raw copolymer with and without the compounding additive (Figure 7.10) show virtually no differences, confirming measurements previously made by Matheson et al.[267]

Analytical pyrolysis studies of SBR copolymer are of continuing interest beyond possible use as a quality-control tool. Commercial disposal of used tires is an important problem; laboratory[78,268,269] and pilot plant [270,271] pyrolyses have been performed with the goal of obtaining a hydrocarbon mixture suitable for either synthetic reuse or use as a fuel oil. Although pyrolysis compositions analyzed by GC showed variation as a result of differing pyrolysis conditions, yields of the various fractions were similar to those found for petroleum-derived fuel oils.

Characterization of Other Commercial Rubbers

Many commercial polymers consist of a formulation of the identified polymer with proprietary compounding materials. Wantanabe et al.[272] examined two different poly(acrylonitrile-co-butadiene)s, or NBR rubbers, under thermal desorption conditions to identify differing amounts of the plasticizers dioctyl adipate, dioctyl phthalate, and dioctyl sebacate. NBR copolymers exhibit good oil-resistance but poor heat stability, but if these formulations are hydrogenated, the heat stability improves. Together with infrared and proton NMR spectroscopy, Py-GC has been used to follow the chemical modification.[273]

Lehrle and co-workers[274-277] have extensively studied polyisobutylene (butyl rubber) under a variety of pyrolysis conditions to obtain product distribution and mechanisms for the decomposition process. Results are indicative of random scission and show decomposition multiplets of oligomers similar to those observed with other ethylene derivatives (Figure 7.11).[277] Comparable results were obtained by Dubey et al.[73] using both FTIR and MS.

A concern for any analytical procedure is its reproducibility at a later date. Roussis and Fedora[278] presented replicate pyrograms of cis-polybutadiene and trans-polyisoprene along with identification of products. Zaikin et al.[279] examined butadiene-isoprene copolymers and noted that, although quantitative measurements can be performed, calibration is necessary because pyrolysis leads to mixed dimers and trimers, as well as monomers. Not all copolymer systems give complicated pyrolysis chemistry. In analyzing nitrile and chloroprene rubbers and their blends, Fuh and Wang[280] noted that these vulcanized and filled rubbers are not amenable to FTIR or NMR spectroscopy and presented pyrograms of these materials. Neat samples were sufficiently different and pyrograms were simple enough that blends could be quantified.

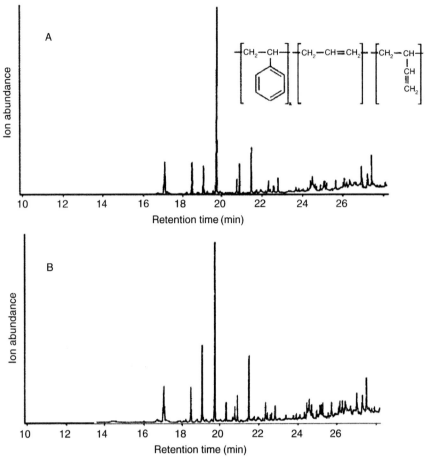

Figure 7.10 Pyrograms of styrene-butadiene rubber (SBR) (A) without and (B) with carbon black additive. Inset shows structure of SBR. Conditions: microfurnace pyrolysis at 600 °C; 30 m × 0.25 mm ID × 0.25 μ df DB-5MS column; oven temperature –30 °C for 15 min, ramped to 150 °C at 10 °C/min. (Reprinted from reference 266, with permission from American Chemical Society. Copyright 1996.)

Ethylene-propylene-diene rubbers exhibit distinct characteristic chemical and weathering resistance. The desired properties of these rubbers can be tailored through the proper selection of the butadiene derivative and monomer ratios; however, traditional chemical verification of the composition is difficult. Py-GC has shown promise for quantitative analysis[281,282] of the butadiene moiety and polymer composition.

Analytical Pyrolysis of Silicon-Containing Polymers

The Py-GC analysis of various silicon-containing polymers has been reported in recent years. In the majority of these reports, the polymer of interest was specifically synthesized for the analysis.

Zaikin et al.[283] polymerized, and then pyrolyzed, various trialkylvinylsilanes. Although a focus of the work was the further decomposition of the pyrolysis fragments within the ionization chamber, the pyrolysis products themselves (reconstituted monomers, dimers, and trimers, as well as eliminated silyl substituents) are consistent with normal addition polymerization of the monomers, as opposed to isomerization of the monomer and incorporation of the silicon atom into the polymer backbone. This characteristic is further evidenced in the work of Rao et al.,[284] who prepared and pyrolyzed poly(tetramethyldisilylene-co-styrene).

Haken et al.[14,48,54,55,285] have examined the silicone-polyester resins formed from the reaction of polysiloxanes with various polyols, dicarboxylic acids, and vegetable oils. The exceptional stability of these product molecules is evidenced by the need to use either alkali or acid fusion at high temperatures and pressures, followed by solvent extraction and derivatization GC.

Polysiloxanes are amenable to normal Py-GC analysis. Dai et al.[286] prepared and pyrolyzed a variety of benzyl ether and biphenyl mesogen derivatives of polysiloxane. These liquid crystalline polysiloxanes were stable below 250 °C

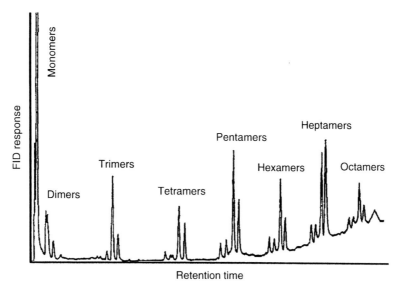

Figure 7.11 Pyrogram of polyisobutylene (9000 MW). Conditions: resistive filament pyrolysis at 610 °C for 10 s; 20 m × 0.53 mm AT-1 fused silica column, programmed from 40 °C (2 min) to 250 °C (5 min) at 10 °C/min. (Reprinted from reference 277, with permission from Elsevier Science. Copyright 1995.)

but produced a range of decomposition products at high temperatures. Nitrogen analogues of silicone, polysilazanes, have been prepared and pyrolyzed by Breuning;[287] a variety of products were observed, and tentative assignments were reported. Polysilanes have also been studied: beginning with various dialkyldichlorosilanes, Menescal et al.[288] prepared and pyrolyzed a variety of polysilane copolymers. GC analysis of the pyrolysis product mixtures found primarily tetramer products, including several cyclic tetrasilanes.

The propensity of silicon to form four bonds and to incorporate itself into hydrocarbon chains has led to its use in constructing model compounds for specific and controlled architecture. The degradation mechanisms of linear and star-shaped polyisoprenes incorporating silyl end groups and star cores have been investigated[38] using GC as one of several techniques for characterizing the pyrolysis products.

Sulfur-Containing Polymers as Models for Coal and Analysis of Sulfur Additives

Aromatic sulfur-containing polymers have been used to model the organosulfur components of coals and hydrocarbons. In a Py-GC/MS kinetic analysis of selected polythiophenes, Selsbo and Ericsson[289] found that the predominant sulfur product is hydrogen sulfide, while polysulfones[75] yield sulfur dioxide. Carbon disulfide is a minor product in each system. In contrast, the Py-GC/MS analysis of poly(phenylene sulfide)[290] and poly(xylylene sulfide)[291] produced mixtures of monomers and isomerized oligomers in addition to hydrogen sulfide. A variety of sulfur-free aromatic hydrocarbons were observed in the product mixtures, but polycyclic aromatic compounds were rarely encountered.

Sulfur, together with peroxides, has also been among the vulcanizers employed for elastomers. An assortment of these elastomers have been analyzed by both thermal desorption-headspace-GC and Py-GC, as well as by other methods, to determine incorporated cross-linking reagents and the nature of the successful cross-linking.[292] Polysulfides[293] have also been studied. The pyrolysis products are typically cyclicized monomers, and the isomeric distribution tends to vary with the selection of the polymerization reagent.

Identification of Fluorine-Containing Polymers

The great majority of fluorine-containing polymers are insoluble. This aspect, plus the wide use of fillers in commercial samples, can lead to significant analytical limitations on the use of traditional spectroscopic analysis. However, studies with a variety of fluoro-copolymers have demonstrated that Py-GC can yield credible results.[294]

Isemura et al.[295] correlated amounts of oligomers in the pyrolysis products of various poly(ethylene-co-tetrafluoroethane)s with the authentic composition, thus permitting facile sample identification. Zigel et al.[296] obtained comparable results for ternary copolymers of tetrafluoroethane-ethylene-hexafluoropropylene. The decomposition chemistry of these systems was also investigated. In similar fashion, Rizvi et al.[297] studied the decomposition of several chlorotrifluoroethylene copolymers.

Analysis of Nitrogen-Containing Polymers

A variety of nitrogen-containing polymers have been studied by Py-GC. Polyacrylonitriles[18,298,299] and polyacrylamides[300,301] typically yield the nitrile monomer, as well as some dimer and higher oligomers; polyaniline[302] also produces monomers and dimers. Thermosetting polymers such as polyimides[303-306] produce a variety of aromatic hydrocarbons and amines, while polypyrolones[307] undergo massive fragmentation. Aromatic polyamides[57,134,250] produce significant amounts of aminobenzoic acids, among other products.

A selection of polyaramides[308] was pyrolyzed with and without added tetramethylammonium hydroxide. Under methanolysis conditions, pyrolysis occurred at a lower temperature and led to a variety of benzoic acid derivatives. The higher temperature neat pyrolysis produced benzenes and anilines.

Structural Characterization of Polystyrenes

Possibly the most commonly recognized polymer, polystyrene, and its copolymers and derivatives have received considerable attention. In general, the predominant polystyrene decomposition is "unzipping" to yield the styrene monomer and oligomers,[309] but numerous small molecules have also been identified.[58,310] Investigations on this polymer system have included characterization of the stereoregularity through structural analysis of the product oligomers[311] and the decomposition mechanism.[37,52,312-314] Studies have included the deuterated polymer,[315] decomposition in solvent solution,[316,317] and decomposition of brominated[318] and chlorinated[319] polystyrene.

Various styrene derivatives have been investigated, including poly(methyl styrene),[313,316] polynitrostyrene,[320] polychlorostyrene,[321] and polybromostyrene,[322] as have numerous copolymers and polymeric blends. The copolymer with butadiene has been noted above. Other copolymer systems include styrene with maleic anhydride,[323,324] methyl methacrylate,[266,325,326] methacrylonitrile,[299,327] carbon monoxide,[328,329] and the cross-linking divinylbenzene.[162,330] An interesting study compared polystyrene and poly(styrene-*co*-vinycyclohexane) with poly(vinylcyclohexane).[331] The expected stability of the saturated six-membered ring is mirrored in the decomposition products of this "saturated polystyrene".[332]

Several studies have focused on optimizing product yields under waste recovery conditions. Among the investigations have been moderate-temperature hydrous pyrolysis[35,59,333] and acid-catalyzed pyrolysis.[49,50,51,108,334] The former experiments greatly favor regeneration of the monomer, while benzene is the predominant product from the latter procedures. It is worth noting that while laboratory investigations can work rapidly with milligram samples, the amounts that are characteristic of a waste recovery situation can require pyrolysis times of several hours.

Analytical Studies of Polymethacrylates

A variety of polymethacrylate polymers and copolymers have been studied. Since the properties of the commercial polymer can follow from the degree of stereoregularity, the decomposition as a function of stereoregularity of polymethyacrylate[335] and poly(methacrylic acid)[336] has been studied. In a similar sense, the effects of cross-linking[337-339] and polymerization reagents[340] have received attention, as has the mechanism of the decomposition itself.[341-344] McNeil and co-workers have studied several copolymer methacrylate systems, including poly(*tert*-butyl aziridine-*co*-methyl methacrylate)[345,346] and poly(vinyl acetate-*co*-methacrylic acid).[347,348] The terpolymer methyl-methacrylate-butadiene-styrene also has been investigated.[349]

Other acrylates have been studied, including copolymers consisting of varying amounts of vinylidene cyanide and methyl acetoxyacrylate.[350] This investigation indicated that while the decomposition products of two different copolymers may be identical, distinct degradation processes are possible due to the difference of strictly alternating versus numerous homosequences in the polymer. Similar results were obtained in the poly(ethyl acrylate-*co*-butyl methacrylate) and poly(styrene-*co*-ethyl acrylate-*co*-ethyl methacrylate) systems[351] in which the relative amounts of monomer were reflected in the pyrograms. The degradation mechanism of poly(allyl methacrylate-*co*-methyl methacrylate) has been investigated.[352]

Analysis of Copolymer Components in Assorted Vinyl Polymers

In addition to the systems such as poly(vinyl chloride) and polystyrene that have been described, studies have been made of a variety of polymers that have the vinyl moiety as a monomeric entity, as exemplified by poly(vinyl butyrate)[15] and poly(ethylene-*co*-vinyl acetate).[353-357] Upon pyrolysis, these polymers lose the pendant ester and produce a variety of aliphatic and aromatic hydrocarbons. The ester group appears as either the free acid or aldehyde.

Other vinyl systems studied include poly(vinylidene cyanide-*co*-cyanovinylacetate),[358] poly(vinylacetate-*co*-crotonic acid),[359] poly(isopropenyl acetate-*co*-maleic anhydride),[360] and poly(allyl acetate-*co*-maleic anhydride).[361] The predominant pyrolysis product is acetic acid, accompanied by assorted hydrocarbons.

Kinetic Studies on Polymer Degradation

The performance of polymers for industrial use often depends on their thermal stability, and accurate thermal degradation rate constants are often needed to classify their behavior. Thermal degradation mechanisms and corresponding data from controlled pyrolysis studies are an important source of models for damage to polymers. Because certain end groups are more stable toward degradation than others are, pyrolysis of organic molecules ranging from small to polymeric sizes can provide information on

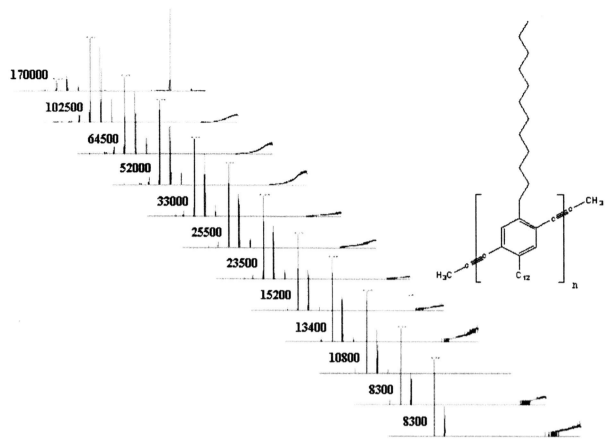

Figure 7.12 Sequential pyrolysis of dodecyl poly(p-phenyleneethylene) (PPE) (structure shown at right). The numbers are the maximum peak ion abundances of dodecane in each pyrogram. Conditions: filament pyrolysis at 600 °C for sequential 10 s intervals; 30 m × 0.25 mm ID × 0.25 μm df 5% phenyl–95% methyl polysiloxane coated capillary column; held at 50 °C for 1 min and ramped at 13 °C/min to 300 °C with a final hold time of 1 min.

Table 7.2 Pyrolysis data for the dodecane degradation peak at a retention time 5.3 min from dodecyl PPE following pyrolysis at 600 °C for sequential 10-s intervals. See Figure 7.12 for pyrograms

Cumulative time(s)	Dodecane peak area	Cumulative area	Fraction conversion (m/m_∞)	$-\ln(1 - (m/m_\infty))$
10	262441	262441	0.332203797	0.403772238
20	153502	415943	0.526510127	0.747624753
30	95375	511318	0.647237975	1.041961599
40	74599	585917	0.741667089	1.353506172
50	52631	638548	0.808288608	1.651764202
60	37530	676078	0.855794937	1.936518942
70	34902	710980	0.899974684	2.30233196
80	22082	733062	0.927926582	2.630069989
90	18212	751274	0.950979747	3.015521736
100	14509	765783	0.96934557	3.48497808
110	10636	776419	0.982808861	4.063361188
120	9603	786022	0.994964557	5.291253765

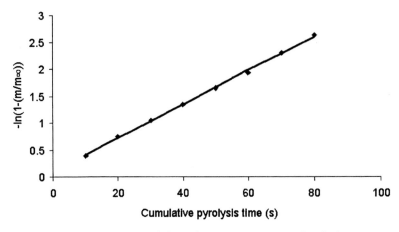

Figure 7.13 Determination of degradation rate constant for dodecane from dodecyl PPE using data in Table 7.2. The observed rate constant, determined from the fitted slope is 0.03136 ± 0.00036 s^{-1} (estimate $\pm$ 1 SD) with a coefficient of determination (R^2) of .9996.

the correlations between molecular structure and stability.[362,363]

Several methods have been developed for the estimation of degradation rate constants.[83] Bate and Lehrle have studied the effects of polymer blending on degradation rates,[364] and Lehrle et al. recently studied the degradation of Nylon-6.[362] Ericsson has done activation energy determination on cis-1,4-polybutadiene.[363] Perng et al. have studied the thermal decomposition kinetics of poly(phenylene sulfide).[365] Other recent kinetic studies have involved epoxy[366] and polyurethane copolymers.[367] One approach for estimation of rate constants from pyrolysis data, popularized by Ericsson,[363,368] involves sequential pyrolysis of the same polymer sample at the same temperature. The product yield is measured for each run, until no further product is produced. The pyrolysis time of each pyrolysis step is small enough that the pyrolysis reactions do not go to completion. First-order behavior implies that the rate of product formation, dm/dt, is proportional to the mass remaining at any time, $(m_\infty - m)$:

$$\frac{dm}{dt} = k_{obs}(m_\infty - m)$$

where k_{obs} is the degradation rate constant, m is the mass of pyrolysis product produced at any time t, and m_∞ is the total mass of pyrolysis product produced after 100% conversion. Integration of this equation leads to:

$$-\ln\left(1 - \frac{m}{m_\infty}\right) = k_{obs}t$$

in which the expression m/m_∞ is termed the fractional conversion. The rate constant, k_{obs}, is obtained by plotting $\ln[1 - (m/m_\infty)]$ against the cumulative pyrolysis time and taking the slope of the trend line. The goodness of fit of the fitted

straight line confirms first-order behavior, which typically applies only when there is no temperature gradient across the sample (reduced by using smaller sample sizes), the sample residue does not change dramatically upon repeated pyrolysis, and pyrolysis products are not formed from secondary reactions. Such sequential methods suffer some difficulties.[83] The more temperature cycles experienced by the sample, the greater the possibility of modification of the sample between pyrolysis runs. The resulting high-conversion behavior of the sample could be quite different from that of a re-pyrolysis of original samples, as done in other methods. For example, although a pyrolysis system may rapidly heat the sample, between pyrolysis runs, the filament cools down at a slower rate and the sample receives heat for an extended time in which degradation may occur and distort subsequent conversion behavior.

As an example of the determination of degradation rate constants for polymers using the Ericsson method, Figure 7.12 shows Py-GC/MS runs for a sample of dodecyl poly-(p-phenyleneethylene) (PPE) (structure also shown in Figure 7.12). PPEs have applications in the optical and electronic industries as plastic lasers, organic semiconductors, photonic devices, and light emitting diodes. PPEs have been recently synthesized by alkyne metathesis of 1,4-dipropynylated benzenes.[369] A dodecyl PPE sample was pyrolyzed sequentially 12 times at 600 °C for 10 s each time, and the relative yield of dodecane (peak at 5.3 min) was measured for each pyrogram with the results shown in Table 7.2. A plot of $\ln[1 - (m/m_\infty)]$ against cumulative pyrolysis time, with the asymptote (m_∞) estimated from the final cumulative area, is shown in Figure 7.13. Because the plot for all 12 data points showed curvature at the higher levels of cumulative time, the rate constant was estimated by the slope determined from only the first 8 data points.

The activation energy of a degradation reaction can be

estimated by fitting experimental measurements of the rate constant at different temperatures to the Arrhenius equation.[363]

Further Reading

A recent review by Smith et al.[370] describes pyrolysis as "one of the most important techniques in the study of synthetic polymers", and summarizes recent improvements in pyrolysis GC instrumentation and methods for polymer analysis. Sufficient resources are available in the literature of gas chromatography and mass spectrometry, as well as in existing texts,[371–377] for the analyst to pursue further instrumental and practical details. Resources are also available in existing analytical pyrolysis texts[378–382] for the analyst to pursue further learning about this useful technique. This chapter has reviewed the recent Py-GC and Py-GC/MS literature by presenting specific examples of polymer characterization methods for a variety of polymers. The diversity of the examples shown provide the beginner an entry point, as well as the experienced analyst a guide, for exploiting gas chromatography in polymer applications.

Acknowledgments This work was supported in part by Office of Naval Research Grant N00014-97-1-0806 and by the Office of Justice, award 97-LB-VX-0006, National Institute of Justice. Points of view are those of the authors and do not necessarily represent the official position of the U.S. Department of Justice. Financial Support for Tom Richardson during a sabbatical leave from The Citadel (Charleston, SC) is also acknowledged.

References

1. Kaljurand, A. M.; Elomaa, M.; Plit, A. *Anal. Chim. Acta* **1991**, *248*, 271–276.
2. Boden, B. F. *Anal. Proc.* **1991**, *28*, 344.
3. Krüsemann, P. V. E.; Jansen, J. A. J. *J. Chromatogr., A* **1998**, *819*, 243–248.
4. Tayler, P. J.; Price, D.; Milnes, G. J.; Scrivens, J. H.; Blease, T. G. *Int. J. Mass Spectrom.* **1989**, *89*, 157–169.
5. Price, D.; Milnes, G. J.; Tayler, P. J.; Scrivens, J. H.; Blease, T. G. *Polym. Degrad Stab.* **1989**, *25*, 307–323.
6. Lattimer, R, P. *J. Anal. Appl. Pyrol.* **1993**, *26*, 65–92.
7. Karlsson, S.; Hakkarainen, M.; Albertsson, A.-C. *J. Chromatogr., A* **1994**, *688*, 251–259.
8. Karlsson, S.; Albertsson, A.-C. *J. Macromol. Sci., Pure Appl. Chem.* **1995**, *A32*, 599–605.
9. Tsai., C. J.; Perng, L. H.; Ling, Y. C. *Rapid Commun. Mass Spectrom.* **1997**, *11*, 1987–1995.
10. Hedrick, S. A.; Chuang, D. S. C. *Thermochim. Acta* **1998**, *315*, 159–168.
11. Washall, J. W.; Wampler, T. P. *J. Chromatogr. Sci.* **1989**, *27*, 144–148.
12. Ishiwatari, M.; Sakashita, H.; Tatsumi, T.; Tominaga, H. *J. Anal. Appl. Pyrol.* **1993**, *24*, 273–290.
13. Sikabwe, E. C.; Negelein, D. L.; Lin, R.; White, R. L. *Anal. Chem.* **1997**, *69*, 2606–2609.
14. Haken, J. K.; Harahap, N.; Burford, R. P. *Prog. Org. Coat* **1989**, *17*, 277–285.
15. White, R. L. *J. Anal. Appl. Pyrol.* **1991**, *18*, 269–276.
16. Wampler, T. P.; Levy, E. J. *J. Anal. Appl. Pyrol.* **1987**, *12*, 75–82.
17. Hauk, A.; Sklorz, M.; Bergmann, G.; Hutzinger, O. *J. Anal. Appl. Pyrol.* **1994**, *28*, 1–12.
18. Usami, T.; Itoh, T.; Ohtani, H.; Tsuge, S. *Macromolecules* **1990**, *23*, 2460–2465.
19. White, R. L.; Sikabwe, E. C. *J. Polym. Sci., Part A.: Polym. Chem.* **1992**, *30*, 2781–2790.
20. Wang, F. C. Y.; Green, J. G.; Gerhart, B. B. *Anal. Chem.* **1996**, *68*, 2477–2481.
21. del Rio, J. C.; McKinney, D. E.; Knicker, H.; Nanny, M. A.; Minard, R. D.; Hatcher, P. G. *J. Chromatogr., A* **1998**, *823*, 433–449.
22. Hold, I. M.; Schouten, S.; Van Kaam-Peters, H. M. E.; Damste, J. S. S. *Org. Geochem.* **1998**, *28*, 179–194.
23. Westerhout, R. W. J.; Kuipers, J. A. M.; van Swaaij, W. P. M. *Ind. Eng. Chem. Res.* **1998**, *37*, 841–847.
24. Maeno, S.; Eddy, C. L.; Smith, S. D.; Rodriguez, P. A. *J. Chromatogr., A* **1997**, *791*, 151–161.
25. Hammond, T.; Lehrle, R. S.; Shortland, A. *Polym. Degrad. Stab.* **1995**, *50*, 169–173.
26. Stankiewicz, B. A.; van Bergen; P. F.; Smith, M. B.; Carter, J. F.; Briggs, D. E. G.; Evershed, R. P. *J. Anal. Appl. Pyrol.* **1998**, *45*, 133–151.
27. Kusch, P. *Chem. Anal. (Warsaw)* **1996**, *41*, 241–252.
28. Stout, S. A.; Hall, K. *J. Anal. Appl. Pyrol.* **1991**, *21*, 195–205.
29. Price, D.; Milnes, G. J.; Gao, F. *Polym. Degrad Stab.* **1996**, *54*, 235–240.
30. Creasy, W. R. *Polymer* **1992**, *33*, 4486–4492.
31. Hancox, R. N.; Lamb, G. D.; Lehrle, R. S. *J. Anal. Appl. Pyrol.* **1991**, *19*, 333–347.
32. Whiton, R. S.; Morgan, S. L. *Anal. Chem.* **1985**, *57*, 778–780.
33. Lehrle, R.; Atkinson, D.; Cook, S.; Gardner, P.; Groves, S.; Hancox, R.; Lamb, G. *Polym. Degrad Stab.* **1993**, *42*, 281–291.
34. Pacakova, V.; Leclercq, P. A. *J. Chromatogr.* **1991**, *555*, 229–237.
35. Simard, Y. D. M.; Kamal, M. R.; Cooper, D. G. *J. Appl. Polym. Sci.* **1995**, *58*, 843–845.
36. McNeill, I. C.; Mahmood, T. *Polym. Degrad. Stab.* **1998**, *60*, 449–458.
37. Swistek, M.; Nguyen, G.; Nicole, D. *J. Anal. Appl. Pyrol.* **1996**, *37*, 15–26.
38. Chiantore, O.; Guaita, M.; Lazzari, M.; Hadjichristidis, N.; Pitsikalis, M. *Polym. Degrad. Stab.* **1995**, *49*, 385–392.
39. Usami, T.; Gotoh, Y.; Takayama, S.; Ohtani, H.; Tsuge, S. *Macromolecules* **1987**, *20*, 1557–1561.
40. Mao, S. C.; Ohtani, H.; Tsuge, S.; Niwa, H.; Nagata, M. *Polym. J.* **1999**, *31*, 59–83.
41. Onishi, A.; Endo, M.; Uchino, S.; Harashima, N.; Oguri, N. *J. High Resolut. Chromatogr.* **1993**, *16*, 353–357.

42. Wang, F. C. Y. *J. Chromatogr., A* **1997**, *786*, 107–115.

43. Takekoshi, Y.; Kanno, S.; Kawase, S.; Kiho, T.; Ukai, S. *Jpn. J. Toxicol. Environ. Health* **1996**, *42*, 28–31.

44. Ishida, Y.; Kawaguchi, S.; Ito, Y.; Tsuge, S.; Ohtani, H. *J. Anal. Appl. Pyrol.* **1997**, *40-1*, 321–329.

45. Challinor, J. M. *J. Anal. Appl. Pyrol,* **1994**, *29*, 223–224.

46. Challinor, J. M. *J. Anal. Appl. Pyrol.* **1991**, *20*, 15–24.

47. Challinor, J. M. *J. Anal. Appl. Pyrol.* **1989**, *16*, 323–333.

48. Haken, J. K.; Harahap, N.; Burford, R. P. *J. Chromatogr.* **1988**, *452*, 37–42.

49. Lin, R.; White, R. L. *J. Appl. Polym. Sci.* **1997**, *63*, 1287–1298.

50. Carniti, P.; Gervasini, A.; Beltrame, P. L.; Audisio, G.; Bertini, F. *Appl. Catal., A* **1995**, *127*, 139–155.

51. Zhu, X. S. *Polym. Degrad. Stab.* **1997**, *57*, 163–173.

52. Swistek, M.; Nguyen, G,; Nicole, D. *J. Appl. Polym. Sci.* **1996**, *60*, 1637–1644.

53. Takekoshi, Y.; Sato, K.; Kanno, S.; Kawase, S.; Kiho, T.; Ukai, S. *Forensic Sci. Int.* **1997**, *87*, 85–97.

54. Haken, J. K.; Harahap, N.; Burford, R. P. *J. Chromatogr.* **1988**, *441*, 207–212.

55. Haken, J. K.; Harahap, N.; Burford, R. P. *J. Chromatogr.* **1987**, *287*, 223–232.

56. Haken, J. K.; Harahap, N.; Burford, R. P. *J. Coat. Technol.* **1990**, *62*, 109–113.

57. Harahap, N.; Burford, R. P.; Haken, J. K. *J. Chromatogr.* **1989**, *477*, 53–57,

58. Hathcock, S. L.; Bertsch, W. *J. High Resolut. Chromatogr.* **1993**, *16*, 651–659.

59. Beltrame, P. L.; Bergamasco, L.; Carniti, P.; Castelli, A.; Bertini, F.; Audisio, G. *J. Anal. Appl. Pyrol.* **1997**, *40-1*, 451–461.

60. Cansell, F.; Beslin, P.; Berdeu, B. *Environ. Prog.* **1998**, *17*, 240–245.

61. Bobleter, O. *Prog. Polym. Sci.* **1994**, *19*, 797–841.

62. Kubikova, J.; Zeman, A.; Krkoska, P.; Bobleter, O. *Tappi J.* **1996**, *79(7)*, 163–169.

63. Wu. C. H.; Chang, C. Y.; Lin, J. P. *J. Environ. Eng.* (New York) **1998**, *124*, 892–896.

64. Ivanova, T. L.; Ryabikova, V. M.; Popova, G. S.; Zigel, A. N.; Goldenberg, A. L. *J. Anal. Chem. USSR* (Engl. Transl.) **1998**, *43*, 571–577; *Zh. Anal. Khim.* **1988**, *43*, 721–727.

65. Ryabikova, V. M.; Zigel, A. N.; Ivanova, T. L.; Popova, G. S. *Polym. Sci. USSR* (Engl. Transl.) **1989**, *31*, 237–242; *Vysokomol. Soedin A* **1989**, *31*, 212–216.

66. Cortes, H, J.; Jewett, G. L.; Pfeiffer, C. D.; Martin, S.; Smith, C. *Anal. Chem.* **1989**, *61*, 961–965.

67. McCaffrey, W. C.; Brues, M. J.; Cooper, D. G. *J. Appl. Polym. Sci.* **1996**, *60*, 2133–2140.

68. Maksimova, N. I.; Krivoruchko, O. P.; Sidel'nikov, V. N. *Russ. J. Appl. Chem.* (Engl. Transl.) **1988**, *71*, 1386–1392; *Zh. Prikladnoi Khim.* **1988**, *71*, 1315–1320.

69. McNeill, I. C.; Memetea, L.; Mohammed, M. H.; Fernandes, A.R.; Ambidge, P. *Polym. Degrad. Stab.* **1998**, *62*, 145–155.

70. Hacaloglu, J.; Ersen, T. *Rapid Commun. Mass Spectrom.* **1998**, *12*, 1793–1795.

71. Hacaloglu, J.; Ersen, T.; Ertugrul, N.; Fares, M. M.; Suzer, S. *Eur. Polym. J.* **1997**, *33*, 199–203.

72. Mundy, S. A. J. *J. Anal. Appl. Pyrol.* **1993**, *25*, 317–324.

73. Dubey, V.; Shrivastava, R. K.; Tripathi, D. N.; Semwal, R. P.; Gandhe, B. R.; Vaidyanathaswamy, R. *J. Anal. Appl. Pyrol.* **1993**, *27*, 207–219.

74. Ohtani, H.; Luo, Y. F.; Nakashima, Y.; Tsukahara, Y.; Tsuge, S. *Anal. Chem.* **1994**, *66*, 1438–1443.

75. Almen, P.; Ericsson, I. *Polym. Degrad. Stab.* **1995**, *50*, 223–228.

76. Saitoh, E.; Hoshi, M.; Kondoh, Y.; Tokuda, H.; Matsumoto, K. *Bunseki Kagaku* **1993**, *42*, 119–125.

77. Wang, F. C. Y. *J. Chromatogr., A* **1996**, *753*, 101–108.

78. Williams, P. T.; Bottrill, R. P. *Fuel* **1995**, *74*, 736–742.

79. Sato, H.; Tsuge, S.; Ohtani, H.; Aoi, K.; Takasu, A.; Okada, M. *Macromolecules* **1997**, *30*, 4030–4037.

80. Venema, A.; Jelink, T. *J. Anal. Appl. Pyrol.* **1992**, *24*, 191–196.

81. Eglinton, T. I.; Larter, S. R.; Boon, J. J. *J. Anal. Appl. Pyrol.* **1991**, *20*, 25–45.

82. Zaikin, V, G.; Mardanov, R. G.; Kleiner, V. I.; Krentsel, B. A.; Bobrov, B. N. *J. Anal. Appl. Pyrol.* **1993**, *26*, 185–190.

83. Bate, D. M.; Lehrle, R. S.; Pattenden, C. S.; Place, E. J. *Polym. Degrad. Stab.* **1998**, *62*, 73–83.

84. Bate, D. M.; Lehrle, R. S. *Polym. Degrad. Stab.* **1996**, *53*, 39–44.

85. Lehrle, R. S.; Atkinson, D. J.; Bate, D. M.; Gardner, P. A.; Grimbley, M. R.; Groves, S. A.; Place, E. J.; Williams, R. J. *Polym. Degrad. Stab.* **1996**, *52*, 183–196.

86. Ouillon, I.; Raihane, M.; Zerroukhi, A.; Boinon, B. *Macromol. Chem. Phys.* **1997**, *198*, 3425–3439.

87. Stevens, M. P. *Polymer Chemistry: An Introduction,* 3rd ed.; Oxford University Press: Oxford, England, 1999.

88. Seymour, R. B.; Carraher, C. E. Jr. *Polymer Chemistry. An Introduction;* Marcel Dekker: New York, 1992.

89. Ebewele, R.O. *Polymer Science and Technology;* CRC Press: Boca Raton, FL, 2000.

90. Hawley-Fedder, R. A.; Parsons, M. L.; Karasek, F. W. *J. Chromatogr.* **1987**, *387*, 207–221.

91. Panagiotou, T.; Levendis, Y. A.; Carlson, J.; Dunayevskiy, Y. M.; Vouros, P. *Combust. Sci. Technol.* **1996**, *116*, 91–128.

92. Wheatley, L.; Levendis, Y. A.; Vouros, P. *Environ. Sci. Technol.* **1993**, *27*, 2885–2895.

93. Blazso, M. *J. Anal. Appl. Pyrol.* **1993**, *25*, 25–35.

94. Koo, J. K.; Seo, Y. H.; Kim, S. W. *Chemosphere* **1991**, *22*, 887–893.

95. Pasek, R. J.; Chang, D. P. Y.; Jones, A. D. *Hazard. Waste Hazard Mater.* **1996**, *13*, 23–38.

96. McNeill, I. C.; Memetea, L.; Cole, W. W. *Polym. Degrad. Stab.* **1995**, *49*, 181–191.

97. Blazso, M. *Rapid Commun. Mass Spectrom.* **1998**, *12*, 1–4.

98. Blazso, M.; Zelei, B.; Jakab, E. *J. Anal. Appl. Pyrol.* **1995**, *35*, 221–235.

99. Blazso, M. *J. Anal. Appl. Pyrol.* **1995**, *32*, 7–18.

100. Pielichowski, K.; Hamerton, I. *Polymer* **1998**, *39*, 241–244.

101. Blazso, M.; Jakab, E.; *J. Anal. Appl. Pyrol.* **1999**, *49*, 125–143.

102. Carty, P.; White, S.; Price, D.; Lu, L. *Polym. Degrad. Stab.* **1999**, *63*, 465–468.

103. Cascaval, C. N.; Robila, G.; Stoleriu, A. *J. Appl. Polym. Sci.* **1995**, *56*, 889–894.

104. Blazso, M.; Zelei, B. *J. Anal. Appl. Pyrol.* **1996**, *36*, 149–158.

105. Lovett, S.; Berruti, F.; Behie, L. A. *Ind. Eng. Chem. Res.* **1997**, *36*, 4436–4444.

106. Beltrame, P. L.; Camiti, P.; Audisio, G.; Bertini, F. *Polym. Degrad. Stab.* **1989**, *26*, 209–220.

107. Uemichi, Y.; Hattori, M.; Itoh, T.; Nakamura, J.; Sugioka, M. *Ind. Eng. Chem. Res.* **1998**, *37*, 867–872.

108. Audisio, G.; Bertini, F.; Beltrame, P. L.; Carniti, P. *Makromol. Chem., Macromol. Symp.* **1992**, *57*, 191–209.

109. Negelein, D. L.; Lin, R.; White, R. L. *J. Appl. Polym. Sci.* **1998**, *67*, 341–348.

110. Zhao, W. W.; Hsegawa, S.; Fujita, J.; Yoshii, F.; Sasaki, T.; Makuuchi, K.; Sun, J. Z.; Nishimoto, S. *Polym. Degrad. Stab.* **1996**, *53*, 129–135.

111. McNeill, I. C.; Mohammed, M. H.; Cole, W. J. *Polym. Degrad. Stab.* **1998**, *61*, 95–108.

112. Williams, E. A.; Williams, P. T. *J. Chem. Technol. Biotechnol.* **1997**, *70*, 9–20.

113. Williams, E. A.; Williams, P. T. *J. Anal. Appl. Pyrol.* **1997**, *40-1*, 247–363.

114. Williams, P. T.; Horne, P. A.; Taylor, D. T. *J. Anal. Appl. Pyrol.* **1993**, *25*, 325–334.

115. Williams, P. T.; Williams, E. A. *Energy Fuels* **1999**, *13*, 188–196.

116. Ohtani, H.; Kimura, T.; Okamota, K.; Tsuge, S. *J. Anal. Appl. Pyrol.* **1987**, *12*, 115–133.

117. Yoshitake, N.; Furukawa, M. *J. Anal. Appl. Pyrol.* **1995**, *33*, 269–281.

118. Grittner, N.; Kaminsky, W.; Obst, G. *J. Anal. Appl. Pyrol.* **1993**, *25*, 293–299.

119. Takamoto, D. Y.; Petrich, M. A. *Ind. Eng. Chem. Res.* **1994**, *33*, 3004–3009.

120. Porschmann, J.; Kopinke, F. D.; Remmler, M.; Mackenzie, K.; Geyer, W.; Mothes, G. *J. Chromatogr., A* **1996**, *750*, 287–301.

121. Fabbri, D.; Trombini, C.; Vassura, I.; *J. Chromatogr. Sci.* **1998**, *36*, 600–604.

122. White, D.; Beyer, L. *J. Anal. Appl. Pyrol.* **1999**, *50*, 63–76.

123. Albertsson, A.-C.; Barenstedt, C.; Karlsson, S. *J. Chromatogr., A* **1995**, *690*, 207–217.

124. Albertsson, A.-C.; Barenstedt, C.; Karlsson, S. *Acta Polym.* **1994**, *45*, 97–103.

125. Albertsson, A.-C.; Karlsson, S. Polyethylene Degradation and Degradation Products. In *Agricultural and Synthetic Polymers: Biodegradability and Utilization*; Glass, J. E.; Swift, G., Eds.; ACS Symposium Series 433; American Chemical Society: Washington, DC, 1990; pp 60–64.

126. Hakkarainen, M.; Albertsson, A.-C.; Karlsson, S. *J. Chromatogr., A* **1996**, *741*, 251–263.

127. Simkovic, I.; Francis, B. A.; Reeves, J. B. *J. Anal. Appl. Pyrol.* **1997**, *43*, 145–155.

128. Saito, T. *Anal. Chim. Acta* **1993**, *276*, 295–302.

129. Saito, T. *J. Anal. Appl. Pyrol.* **1995**, *32*, 171–178.

130. Nishino, R.; Saito, T. *Bunseki Kagaku* **1996**, *45*, 915–920.

131. Lytle, C. A.; Bertsch, W.; McKinley, M. *J. Anal. Appl. Pyrol.* **1998**, *45*, 121–131.

132. Falk, G.; Frisch, B.; Thiele, W. R. *Ceram. Forum Int.* **1997**, *74*, 648–652.

133. Falk, G.; Frisch, B.; Thiele, W. R. *Ceram. Forum Int.* **1997**, *74*, 726–730.

134. Buzanowski, W. C.; Cutie, S. S.; Howell, R.; Papenfuss, R.; Smith, C. G. *J. Chromatogr., A* **1994**, *677*, 355–364.

135. Wilcken, H.; Shulten, H.-R. *Fresenius' J. Anal. Chem.* **1996**, *355*, 157–163.

136. Wilcken, H.; Shulten, H.-R. *Anal. Chim. Acta* **1996**, *336*, 201–208.

137. Levy, E. J. The Analysis of Automative Paints by Pyrolysis Gas Chromatography. In *Analytical Pyrolysis*; R. Jones, C. E.; Cramers, C. A., Eds.; Elsevier: New York, 1976; pp 319–335.

138. Rodgers, W. R.; Ellis, T. S.; Cheever, G. D.; Louis-Ferdinand, R; Thorton, D. P.; Somers, N. *J. Coat. Technol.* **1994**, *66*, 27–33.

139. di Pasquale, G.; la Rosa, A. D.; Recea, A.; di Carlo, S; Bassani, M. R.; Facchetti, S. *J. Mater. Sci.* **1997**, *32*, 3021–3024.

140. Wampler, T. P.; Bishea, G. A.; Simonsick, W. J. *J. Anal. Appl. Pyrol.* **1997**, *40-1*, 79–89.

141. Challinor, J. M. *J. Anal. Appl. Pyrol.* **1991**, *18*, 233–244.

142. Lennard, C. J.; Mazzella, W. D. *J. Forensic Sci. Soc.* **1991**, *31*, 365–371.

143. Lorinci, G.; Matuschek, G.; Fekete, J.; Gebefugi, I.; Kettrup, A. *Thermochim. Acta* **1995**, *263*, 73–86.

144. Challinor, J. M. *J. Anal. Appl. Pyrol.* **1993**, *25*, 349–360.

145. Tranthimfryer, G. J.; Dehaan, J. D. *Sci. Justice* **1997**, *37*, 39–46.

146. Voorhees, K. J. Analysis of Soot Produced from the Combustion of Polymeric Materials. In *Fire and Polymers II: Materials and Tests for Hazard Prevention*; Nelson, G. L., Ed.; ACS Symposium Series 599; American Chemical Society: Washington, DC, 1995; pp 393–406.

147. Pinorini, M. T.; Lennard, C. J.; Margot, P.; Dustin, I.; Furrer, P. *J. Forensic Sci.* **1994**, *39*, 933–973.

148. Shedrinsky, A. M.; Grimaldi, D. D.; Boon, J. J.; Baer, N. S. *J. Anal. Appl. Pyrol.* **1993**, *25*, 77–95.

149. van den Berg, J. D. J.; Boon, J. J.; van den Berg, K. J.; Fiedler, I.; Miller, M. A. *Anal. Chem.* **1998**, *70*, 1823–1830.

150. Chiavari, G.; Fabbri, D.; Mazzeo, R.; Bocchini, P.; Galletti, G. C. *Chromatographia* **1995**, *41*, 273–281.

151. Golloch, A.; Heidbreder, S.; Lühr, C. *Fresenius' J. Anal. Chem.* **1998**, *361*, 545–546.

152. Rao, M. R.; Radhakrishnan, T. S. *J. Appl. Polym. Sci.* **1990**, *41*, 2251–2263.

153. Scariah, K. J.; Sekhar, V.; Rao, M. R. *Eur. Polym. J.* **1994**, *30*, 925–931.

154. Rao, M. R.; Sebastian, T. V.; Radhakrishnan, T. S.; Ravindran, P. V. *J. Appl. Polym. Sci.* **1991**, *42*, 753–766.

155. Rietjens, M. *Anal. Chim. Acta* **1995**, *316*, 205–215.

156. Rietjens, M.; Wils, R. R. *J. Propellants, Explos., Pyrotech.* **1995**, *20*, 232–237.

157. Rietjens, M.; Wils, E. R. *J. Propellants, Explos., Pyrotech.* **1995**, *20*, 182–186.

158. Furukawa, M.; Yoshitake, N.; Yokoyama, T. *J. Chromatogr.* **1988**, *435*, 219–224.

159. Ravey, M.; Pearce, E. M. *J. Appl. Polym. Sci.* **1997**, *63*, 47–74.

160. Matuschek, G. *Thermochim. Acta* **1995**, *263*, 59–71.
161. Lattimer, R. P.; Polce, M. J.; Wesdemiotis, C. *J. Anal. Appl. Pyrol.* **1998**, *48*, 1–15.
162. Mao, S.; Tsuge, S.; Ohtani, H.; Uchijima, H.; Kiyokawa, A. *Polymer* **1998**, *39*, 143–149.
163. Ravey, M.; Keidar, I.; Weil, E. D.; Pearce, E. M. *J. Appl. Polym. Sci.* **1998**, *68*, 217–229.
164. Ravey, M.; Weil, E. D.; Keidar, I.; Pearce, E. M. *J. Appl. Polym. Sci.* **1998**, *68*, 231–254.
165. Gao, F.; Price, D.; Milnes, G. J.; Eling, B.; Lindsay, C. I.; McGrail, P. T. *J. Anal. Appl. Pyrol.* **1997**, *40-1*, 217–231.
166. Cutler, E. T.; Magidman, P.; Marmer, W. N. *Text. Chem. Color.* **1993**, *25*, 27–29.
167. Kubokawa, H.; Tslunesada, T.; Hatakeyama, T. *Text. Res. J.* **1999**, *69*, 121–128.
168. Chung, H. L.; Aldridge, J. C. *Anal. Instrum.* **1992**, *20*, 123–135.
169. Price, D.; Horrocks, A. R.; Akalin, M.; Faroq, A. A. *J. Anal. Appl. Pyrol.* **1997**, *40-1*, 511–524.
170. Faroq, A. A.; Price, D.; Milnes, G, J.; Horrocks, A. R. *Polym. Degrad. Stab.* **1991**, *33*, 155–170.
171. Casanovas, A. M.; Rovira, X. *J. Anal. Appl. Pyrol.* **1987**, *11*, 227–232.
172. Zhu, X. S. *J. Fire Sci.* **1996**, *14*, 443–465.
173. Zhu, X. S.; Elomaa, M.; Sundholm, F.; Lochmuller, C. H. *Macromol. Chem. Phys.* **1997**, *198*, 3137–3148.
174. Boscoletto, A. B.; Checchin, M.; Milan, L.; Pannocchia, P.; Tavan, M.; Camino, G.; Luda, M. P. *J. Appl. Polym. Sci.* **1998**, *67*, 2231–2244.
175. Luijk, R.; Gover, A. J.; Eijkel, G. B.; Boon, J. J. *J. Anal. Appl. Pyrol.* **1991**, *20*, 303–319.
176. Marchal, A.; Delobel, R.; Le Bras, M.; Leroy, J. M. *Polym. Degrad. Stab.* **1994**, *44*, 263–273.
177. Cyrys, J.; Lenoir, D.; Matuschek, G.; Kettrup, A *J. Anal. Appl. Pyrol.* **1995**, *34*, 157–172.
178. Sato, H.; Kondo, K.; Tsuge, S.; Ohtani, H.; Satao, N. *Polym. Degrad. Stabil.* **1998**, *62*, 41–48.
179. Pastorova, I.; Botto, B. E.; Arisz, P. W.; Boon, J. J. *Carbohydr. Res.* **1994**, *262*, 27–47.
180. van der Hage, E. R. E.; van Loon, W. M. G. M.; Boon, J. J.; Lingeman, H.; Brinkman, U. A. T. *J. Chromatogr.* **1993**, *634*, 263–271.
181. van Loon, W. M. G. M.; Boon, J. J. *Trends Anal. Chem.* **1994**, *13*, 169–176.
182. van Loon, W. M. G. M.; Boon, J. J.; de Groot, B. *Anal. Chem.* **1993**, *65*, 1728–1735.
183. Bulterman, A. J.; van Loon, W. M. G. M.; Ghijsen, R. T.; Brinkman, U. A. T.; Huitema, I. M.; DeGroot, B. *Environ. Sci. Technol.* **1997**, *31*, 1946–1952.
184. van der Hage, E. R. E.; Mulder, M. M.; Boon, J. J. *J. Anal. Appl. Pyrol.* **1993**, *25*, 149–183.
185. van der Hage, E. R. E.; Boon, J. J. *J. Chromatogr., A* **1996**, *736*, 61–75.
186. Ishida, Y.; Ohtani, H.; Kato, T.; Tsuge, S.; Yano, T. *Tappi J.* **1994**, *77(3)*, 177–183.
187. Ishida, Y.; Ohtani, H.; Kato, T.; Tsuge, S.; Yano, T. *Tappi J.* **1994**, *77(7)*, 177–211.
188. Wang, L.; Ishida, Y.; Ohtani, H.; Tsuge, S. *Anal. Sci.* **1998**, *14*, 431–434.
189. Yano, T.; Ohtani, H.; Tsuge, S.; Obokata, T. *Analyst* (Cambridge, U.K.) **1992**, *117*, 849–853.
190. Ishida, Y.; Ohtani, H.; Tsuge, S.; Yano, T. *Anal. Chem.* **1994**, *66*, 1444–1447.
191. Yano, T.; Ohtani, H.; Tsuge, S.; *Tappi J.* **1991**, *74(2)*, 197–201.
192. McGuire, J. M.; Lynch, C. C. *Anal. Chem.* **1996**, *68*, 2459–2463.
193. del Rio, J. C.; Gutierrez, A.; Gonzalez-Vila, F. J. *J. Chromatogr., A* **1999**, *830*, 227–232.
194. del Rio, J. C.; Gutierrez, A.; Gonzalez-Vila, F. J.; Martin, F. *J. Anal. Appl. Pyrol.* **1999**, *49*, 165–177.
195. Sitholé, B. B.; Allen, L. H. *J. Pulp Pap. Sci.* **1994**, *20*, J168–J172.
196. Cook, S.; Lehrle, R. *Eur. Polym. J.* **1993**, *29*, 1–8.
197. Jeknavorian, A. A.; Mabud, M. A.; Barry, E. F.; Litzau, J. J. *J. Anal. Appl. Pyrol.* **1998**, *46*, 85–100.
198. Kawauchi, A.; Uchiyama, T. *J. Anal. Appl. Pyrol.* **1998**, *48*, 35–43.
199. Uchiyama, T.; Kawauchi, A.; DuVal, D. L. *J. Anal. Appl. Pyrol.* **1998**, *45*, 111–119.
200. Muguruma, S.; Uchino, S.; Oguri, N.; Kiji, J. *Chromatographia* **1998**, *47*, 203–208.
201. Asperger, A.; Engewald, W.; Wagner, T. *J. Anal. Appl. Pyrol.* **1999**, *49*, 155–164.
202. Cheng, T. M. H.; Malawer, E. G. *J. Am. Soc. Brew. Chem.* **1996**, *54*, 85–90.
203. Iskander, G. M.; Ovenell, T. R.; Davis, T. P. *Macromol. Chem. Phys.* **1996**, *197*, 3123–3133.
204. Cheng, T. M. G.; Malawer, E. G. *Anal. Chem.* **1999**, *71*, 468–475.
205. Cascaval, C. N.; Poinescu, I. *Polym. Degrad. Stab.* **1995**, *48*, 55–60.
206. Matsui, T.; Shimoda, M.; Osajima, Y. *Polym. Int.* **1992**, *29*, 85–90.
207. Matsui, T.; Shimoda, M.; Osajima, Y. *Polym. Int.* **1992**, *29*, 91–95.
208. Franich, R. A.; Kroese, H. W.; Lane, G. *Analyst* (Cambridge, U.K.) **1995**, *120*, 1927–1931.
209. Chang, C.; Tackett, J. R. *Thermochim. Acta* **1991**, *192*, 181–190.
210. Cascaval, C. N.; Rosu, D. *Rev. Roum. Chim.* **1991**, *36*, 1331–1336.
211. Cascaval, C. N.; Rosu, D.; Agherghinei, I. *Polym. Degrad. Stab.* **1996**, *52*, 253–257.
212. Rao, M, R.; Alwan, S.; Scariah, K. J.; Sastri, K. S. *J. Therm. Anal.* **1997**, *49*, 261–268.
213. Nakagawa, H.; Wakatsuka, S.; Tsuge, S.; Koyama, T. *Polym. J.* **1988**, *20*, 9–16.
214. Nakagawa, H.; Wakatsuka, S.; Ohtani, H.; Tsuge, S.; Koyama, T. *Polymer* **1992**, *33*, 4556–4562.
215. Nakagawa, H.; Tsuge, S.; Koyama, T. *J. Anal. Appl. Pyrol.* **1987**, *12*, 97–113.
216. Galipo, R. C.; Egan, W. J.; Aust, J. F.; Myrick, M. L.; Morgan, S. L. *J. Anal. Appl. Pyrol.* **1998**, *45*, 23–40.
217. Yamada, T.; Okumoto, T.; Ohtani, H.; Tsuge, S. *J. Anal. Appl. Pyrol.* **1995**, *33*, 157–166.
218. Niimura, N.; Miyakoshi, T.; Onodera, J.; Higuchi, T. *J. Anal. Appl. Pyrol.* **1996**, *37*, 199–209.

219. Niimura, N.; Miyakoshi, T.; Onodera, J.; Higuchi, T. *Int. J. Polym. Anal. Chem.* **1998**, *4*, 309–322.

220. Niimura, N.; Miyakoshi, T.; Onodera, J.; Higuchi, T. *Rapid Commun. Mass Spectrom.* **1996**, *10*, 1719–1724.

221. Sueoka, K.; Nagata, M.; Ohtani, H.; Nagai, N.; Tsuge, S. *J. Polym. Sci., Part A: Polym. Chem.* **1991**, *29*, 1903–1908.

222. Sivasamy, P.; Palaniandavar, M.; Vijayakumar, C. T.; Lederer, K. *Polym. Degrad. Stab.* **1992**, *38*, 15–21.

223. Ohtani, H.; Fujii, R.; Tsuge, S. *J. High Resolut. Chromatogr.* **1991**, *14*, 388–391.

224. Haken, J. K.; Iddamalgoda, P. I. *Prog. Org. Coat.* **1995**, *26*, 101–111.

225. Hakkarainen, M.; Gallet, G.; Karlsson, S. *Polym. Degrad. Stab.* **1999**, *64*, 91–99.

226. Huang, F.; Wang, X. Q.; Li, S. J. *J. Anal. Appl. Pyrol.* **1991**, *22*, 139–151.

227. Chang, C. Y.; Wu, C. H.; Hwang, J. Y. *J. Environ. Eng. (New York)* **1996**, *122*, 299–305.

228. Fabbri, D.; Helleur, R.; *J. Anal. Appl. Pyrol.* **1999**, *49*, 277–293.

229. Gonzalez-Vila, F. J.; del Rio, J. C.; Martin, F.; Verdejo, T. *J. Chromatogr., A* **1996**, *750*, 155–160.

230. Tanczos, I.; Rendl, K.; Schmidt, H. *J. Anal. Appl. Pyrol.* **1999**, *49*, 319–327.

231. Rodriguez, J.; Hernandez-Coronado, M. J.; Hernandez, M.; Bocchini, P.; Galletti, G. C.; Arias, M. E. *Anal. Chim. Acta* **1997**, *345*, 121–129.

232. Lehrle, R.; Williams, R.; French, C.; Hammond, T. *Macromolecules* **1995**, *28*, 4408–4414.

233. Kopinke, F. D.; Remmler, M.; Mackenzie, K.; Moder, M.; Wachsen, O. *Polym. Degrad. Stab.* **1996**, *53*, 329–342.

234. Odermatt, J.; Meier, D.; Mauler, D.; Leicht, K. *Papier* **1998**, *52*, 598–602.

235. Kopinke, F. D.; Mackenzie, K. *J. Anal. Appl. Pyrol.* **1997**, *40-1*, 43–53.

236. Lehrle, R. S.; Williams, R. J. *Macromolecules* **1994**, *27*, 3782–3789.

237. Gelin, F.; de Leeuw, J. W.; Damste, J. S. S.; Derenne, S.; Largeau, C.; Metzger, P. *J. Anal. Appl. Pyrol.* **1994**, *28*, 183–204.

238. de Leeuw, J. W.; Baas, M. *J. Anal. Appl. Pyrol.* **1993**, *26*, 175–184.

239. Sato, H.; Mizutani, S.; Tsuge, S.; Ohtani, H.; Aoi, K.; Takasu, A.; Okada, M.; Kobayashi, S.; Kiyosada, T.; Shoda, S. *Anal. Chem.* **1998**, *70*, 7–12.

240. Damste, J. S. S.; de las Heras, F. X. C.; de Leeuw, J. W. *J. Chromatogr.* **1992**, *607*, 361–376.

241. Greenwood, P. F.; George, S. C.; Hall, K. *Org. Geochem.* **1998**, *29*, 1075–1089.

242. Eglinton, T. I. *Org. Geochem.* **1994**, *21*, 721–735.

243. del Rio, J. C.; Martin, G.; Gonzalez-Vila, F. J.; Verdejo, T. *Org. Geochem.* **1995**, *23*, 1009–1022.

244. Kralert, P. G.; Alexander, R.; Kagi, R. I. *Org. Geochem.* **1995**, *23*, 627–639.

245. Goni, M. A.; Eglinton, T. I. *J. High Resolut. Chromatogr.* **1994**, *17*, 476–488.

246. Tegelaar, E. W.; Noble, R. A. *Org. Geochem.* **1994**, *22*, 543–574.

247. Calemma, V.; Rausa, R. *J. Anal. Appl. Pyrol.* **1995**, *40-1*, 569–584.

248. Artok, L.; Su, Y.; Hirose, Y.; Hosokawa, M.; Murata, S.; Nomura, M. *Energy Fuels* **1999**, *13*, 287–296.

249. Audisio, G.; Severini, F. *J. Anal. Appl. Pyrol.* **1987**, *12*, 135–147.

250. Schulten, H.-R.; Plage, B.; Ohtani, H.; Tsuge, S. *Angew. Makromol. Chem.* **1987**, *155*, 1–20.

251. Sanchez, R.; Hernandez, C.; Rieumont, J. *Polym. Degrad. Stab.* **1998**, *61*, 513–517.

252. Sanchez, R.; Hernandez, C.; Jalovszky, G.; Czira, G. *Eur. Polym. J.* **1994**, *30*, 37–42.

253. Sanchez, R.; Hernandez, C.; Keresztury, G. *Eur. Polym. J.* **1994**, *30*, 43–50.

254. Ohtani, H.; Takehana, Y.; Tsuge, S. *Macromolecules* **1997**, *30*, 2542–2545.

255. Ito, Y.; Ohtani, H.; Ueda, S.; Nakashima, Y.; Tsuge, S. *J. Polym. Sci., Part A: Polym. Chem.* **1994**, *32*, 383–388.

256. Ito, Y.; Tsuge, S.; Ohtani, H.; Wakabayashi, S.; Atarashi, J. I.; Kawamura, T. *Macromolecules* **1996**, *29*, 4516–4519.

257. Ito, Y.; Ogasawara, H.; Ishida, Y.; Ohtani, H.; Tsuge, S. *Polym. J.* **1996**, *28*, 1090–1095.

258. Tsukahara, Y.; Nakanishi, Y.; Yamashita, Y.; Ohtani, H.; Nakashima, Y.; Luo, Y. F.; Ando, T.; Tsuge, S. *Macromolecules* **1991**, *24*, 2493–2497.

259. Ohtani, H.; Tanaka, M.; Tsuge, S. *Bull. Chem. Soc. Jpn.* **1990**, *63*, 1196–1200.

260. Ohtani, H.; Tanaka, M.; Tsuge, S. *J. Anal. Appl. Pyrol.* **1989**, *15*, 167–174.

261. Ohtani, H.; Ueda, S.; Tsukahara, Y.; Wantanabe, C.; Tsuge, S. *J. Anal. Appl. Pyrol.* **1993**, *25*, 1–10.

262. Baudry, A.; Dufay, J.; Regnier, N.; Mortaigne, B. *Polym. Degrad. Stab.* **1998**, *61*, 441–452.

263. Roland, A. I.; Stenzel, M.; Schmidt-Naake, G. *Angew. Makromol. Chem.* **1998**, *259*, 69–72.

264. Sawaguchi, T.; Seno, M. *J. Polym. Sci., Part A: Polym. Chem.* **1998**, *36*, 209–213.

265. Dean, L.; Groves, S.; Hancox, R.; Lamb, G.; Lehrle, R. S. *Polym. Degrad. Stab.* **1989**, *25*, 143–160.

266. Ghebremeskel, G. N.; Sekinger, J. K.; Hoffpauir, J. L.; Hendrix, C. *Rubber Chem. Technol.* **1996**, *69*, 874–884.

267. Matheson, M. J.; Wampler, T. P.; Simonsick, W. J. *J. Anal. Appl. Pyrol.* **1994**, *29*, 129–136.

268. Lin, J. P.; Chang, C. Y.; Wu, C. H. *J. Chem. Technol. Biotechnol.* **1996**, *66*, 7–14.

269. Williams, P. T.; Taylor, D. T. *Fuel* **1993**, *72*, 1469–1474.

270. Cunliffe, A. M.; Williams, P. T. *J. Anal. Appl. Pyrol.* **1998**, *44*, 131–152.

271. Williams, P. T.; Bottrill, R. P.; Cunliffe, A. M. *Process Saf. Environ. Prot.* **1998**, *76*, 291–301.

272. Watanabe, C.; Teraishi, K.; Tsuge, S.; Ohtani, H.; Hashimoto, K. *J. High Resolut. Chromatogr.* **1991**, *14*, 269–271.

273. Kondo, A.; Ohtani, H.; Kosugi, Y.; Tsuge, S.; Kubo, Y.; Asada, N.; Inaki, H.; Yoshioka, A. *Macromolecules* **1988**, *21*, 2918–2924.

274. Grimbley, M. R.; Lehrle, R. S. *Polym. Degrad. Stab.* **1995**, *49*, 223–229.

275. Grimbley, M. R.; Lehrle, R. S. *Polym. Degrad. Stab.* **1995,** *48,* 441–455.

276. Lehrle, R. S.; Pattenden, C. S. *Polym. Degrad. Stab.* **1999,** *63,* 321–340.

277. Grimbley, M. R.; Lehrle, R. S.; Williams, R. J.; Bate, D. M. *Polym. Degrad. Stab.* **1995,** *48,* 143–149.

278. Roussis, S. G.; Fedora, J. W. *Rapid Commun. Mass Spectrom.* **1996,** *10,* 82–90.

279. Zaikin, V. G.; Mardanov, R. G.; Yakovlev, V. A.; Plate, N. A. *J. Anal. Appl. Pyrol.* **1992,** *23,* 33–42.

280. Fuh, M. R. S.; Wang, G. Y. *Anal. Chim. Acta* **1998,** *371,* 89–96.

281. Yamada, T.; Okumoto, T.; Ohtani, H.; Tsuge, S. *Rubber Chem. Technol.* **1990,** *63,* 191–201.

282. Yamada, T.; Okumoto, T.; Ohtani, H.; Tsuge, S. *Rubber Chem. Technol.* **1991,** *64,* 708–713,

283. Zaikin, V. G.; Filippova, V. G.; Semenov, O. B. *Org. Mass Spectrom.* **1991,** *26,* 751–756.

284. Rao, M. R.; Packirisamy, S.; Ravindran, P. V.; Narendranath, P. K. *Macromolecules* **1992,** *25,* 5165–5170.

285. Haken, J. K.; Harahap, N.; Burford, R. P. *J. Coat. Technol.* **1987,** *59,* 73–78.

286. Dai, R.; Ling, Y.; Luo, A.; Fu, R.; Zhang, S.; Xie, G.; Jin, S. *J. Anal. Appl. Pyrol.* **1997,** *42,* 103–111.

287. Breuning, T. *J. Anal. Appl. Pyrol.* **1999,** *49,* 43–51.

288. Menescal, R.; Eveland, J.; West, R.; Blazso, M. *Macromolecules* **1994,** *27,* 5893–5899.

289. Selsbo, P.; Ericsson, I. *Polym. Degrad. Stab.* **1996,** *51,* 83–92.

290. Cohen, Y.; Aizenshtat, Z. *J. Anal. Appl. Pyrol.* **1993,** *27,* 131–143.

291. Montaudo, G.; Puglisi, C.; de Leeuw, J. W.; Hartgers, W.; Kishore, K.; Ganesh, K. *Macromolecules* **1996,** *29,* 6466–6474.

292. Affolter, S. *Kautsch. Gummi Kunstst.* **1997,** *50,* 216–225.

293. Radhakrishnan, T. S.; Rao, M. R. *J. Appl. Polym. Sci.* **1987,** *34,* 1985–1996.

294. Ryabikova, V. M.; Ivanova, T. L.; Zigel, A. N.; Pirozhaya, L. N.; Popova, G. S. *J. Anal. Chem. U.S.S.R. (Engl. Transl.)* **1988,** *43,* 871–877; *Zh. Anal. Khim.* **1988,** *43,* 1093–1099.

295. Isemura, T.; Jitsugiri, Y.; Yonemori, S. *J. Anal. Appl. Pyrol.* **1995,** *33,* 103–109.

296. Zigel, A. N.; Ryabikova, V. M.; Pirozhnaya, L. N.; Popova, G. S.; Madorskaya, L. Y. *Polym. Sci. U.S.S.R. (Engl. Transl.)* **1991,** *33,* 1224–1229; *Vysokomol. Soedin A* **1991,** *33,* 1321–1327.

297. Rizvi, M.; Munir, A.; Zulfiqar, S.; Zulfiqar, M. *J. Therm. Anal.* **1995,** *45,* 1597–1604.

298. Ryabikova, V. M.; Zigel, A. N.; Popova, G. S. *Polym. Sci. U.S.S.R. (Engl. Transl.)* **1990,** *32,* 822–828; *Vysokomol. Soedin A* **1990,** *32,* 882–887.

299. Yang, M.; Shibasaki, Y. *J. Polym. Sci., Part A: Polym. Chem.* **1998,** *36,* 2315–2330.

300. Tutas, M.; Saglam, M.; Yuksel, M. *J. Anal. Appl. Pyrol.* **1991,** *22,* 129–137.

301. Ishida, Y.; Tsuge, S.; Ohtani, H.; Inokuchl, F.; Fujli, Y.; Suetomo, S. *Anal. Sci.* **1996,** *12,* 835–838.

302. Borros, S.; Munoz, E.; Folch, J. *J. Chromatogr., A* **1999,** *837,* 273–279.

303. Torrecillas, R.; Regnier, N.; Mortaigne, B. *Polym. Degrad. Stab.* **1996,** *51,* 307–318.

304. Torrecillas, R.; Baudry, A.; Dufay, J.; Mortaigne, B. *Polym. Degrad. Stab.* **1992,** *54,* 267–274.

305. Johnson, E. T. *J. Appl. Polym. Sci.* **1992,** *44,* 1905–1911.

306. Li, F. M.; Huang, L. Y.; Shi, Y.; Jin, X. G.; Wu, Z. Q.; Shen, Z. H.; Chuang, K.; Lyon, R. E.; Harris, F. W.; Cheng, S. Z. D. *J. Macromol. Sci., Phys.* **1999,** *B38,* 107–122.

307. Jiang, Z.; Jin, X. G.; Gao, X. S.; Zhou, W.; Lu, F.; Luo, Y. *J. Anal. Appl. Pyrol.* **1995,** *33,* 231–242.

308. Venema, A.; Geest, C. A. B. *J. Microcolumn Sep.* **1995,** *7,* 337–343.

309. Audisio, G.; Bertini, F. *J. Anal. Appl. Pyrol.* **1992,** *24,* 61–74.

310. Hathcock, S. L.; Bertsch, W. *J. High Resolut. Chromatogr.* **1993,** *16,* 609–614.

311. Nonobe, T.; Ohtani, H.; Usami, T.; Mori, T.; Fukumori, H.; Hirata, Y.; Tsuge, S. *J. Anal. Appl. Pyrol.* **1995,** *33,* 121–138.

312. Gardner, P.; Lehrle, R. *Eur. Polym. J.* **1993,** *29,* 425–435.

313. Fares, M. M.; Yalcin, T.; Hacaloglu, J.; Gungor, A.; Suzer, S. *Analyst* (Cambridge, U.K.) **1994,** *119,* 693–696.

314. Carniti, P.; Beltrame, P. L.; Armada, M.; Gervasini, A.; Audisio, G. *Ind. Eng. Chem. Res.* **1991,** *30,* 1624–1629.

315. Ohtani, H.; Yuyama, T.; Tsuge, S.; Plage, B.; Schulten, H.-R. *Eur. Polym. J.* **1990,** *26,* 893–899.

316. Murakata, T.; Saito, Y.; Yosikawa, T.; Suzuki, T.; Sato, S. *Polymer* **1993,** *34,* 1436–1439.

317. Sato, S.; Murakata, T.; Baba, S.; Saito, Y.; Wantanabe, S. *J. Appl. Polym. Sci.* **1990,** *40,* 2065–2071.

318. Pielichowski, K.; Stoch, L. *J. Therm. Anal.* **1995,** *45,* 1239–1243.

319. Nakagawa, H.; Tsuge, S.; Mohanraj, S.; Ford, W. T. *Macromolecules* **1988,** *21,* 930–933.

320. Zuev, V. V.; Zgonnik, P. V.; Turkova, L. D.; Shibaev, L. V. *Polym. Degrad. Stab.* **1999,** *63,* 15–17.

321. Bertini, F.; Audisio, G.; Kiji, J. *J. Anal. Appl. Pyrol.* **1994,** *28,* 207–217.

322. Bertini, F.; Audisio, G.; Kiji, J. *J. Anal. Appl. Pyrol.* **1995,** *33,* 213–230.

323. Cascaval, C. N.; Chitanu, G.; Carpov, A. *Thermochim. Acta* **1996,** *275,* 225–233.

324. Wang, F. C. Y. *J. Chromatogr., A* **1997,** *765,* 279–285.

325. Gardner, P.; Lehrle, R.; Turner, D. *J. Anal. Appl. Pyrol.* **1993,** *25,* 11–24.

326. Bate, D. M.; Lehrle, R. S. *Polym. Degrad. Stab.* **1997,** *55,* 295–299.

327. Shibasaki, Y.; Yang, M. *J. Therm. Anal.* **1997,** *49,* 71–77.

328. Bertini, F.; Audisio, G.; Kiji, J. *J. Anal. Appl. Pyrol.* **1999,** *49,* 32–42.

329. Kiji, J.; Okano, T.; Chiyoda, T.; Bertini, F.; Audisio, G. *J. Anal. Appl. Pyrol.* **1997,** *40-41,* 331–345.

330. Nakagawa, H.; Matsushita, Y.; Tsuge, S. *Polymer* **1987,** *28,* 1512–1516.

331. Zaikin, V. G.; Mardanov, R. G.; Kleiner, V. I.; Krentsel, B. A.; Plate, N. A. *J. Anal. Appl. Pyrol.* **1990,** *17,* 291–304.

332. Zaikin, V. G.; Mardanov, R. G.; Kleiner, V. I.; Krentsel, B. A. *Polym. Sci. U.S.S.R.* (Engl. Transl.) **1990**, *32*, 948–954; *Vysokomol. Soedin A* **1990**, *32*, 1014–1020.

333. Bertini, F.; Audisio, G.; Beltrame, P. L.; Bergamasco, L.; Castelli, A. *J. Appl. Polym. Sci.* **1998**, *70*, 2291–2298.

334. Audisio, G.; Bertini, F.; Beltrame, P. L.; Carniti, P. *Polym. Degrad. Stab.* **1990**, *29*, 191–200.

335. Nonobe, T.; Tsuge, S.; Ohtani, H.; Kitayama, T.; Hatada, K. *Macromolecules* **1997**, *30*, 4891–4896.

336. Lazzari, M.; Kitayama, T.; He, S. G.; Hatada, K.; Chiantore, O. *Polym. Bull.* **1997**, *39*, 85–91.

337. Bate, D. M.; Lehrle, R. S. *Polym. Degrad. Stab.* **1998**, *62*, 67–71.

338. Andrzejewska, E.; Kusch, P.; Andrezejewski, M. *Polym. Degrad. Stab.* **1993**, *40*, 27–30.

339. Lomakin, S. M.; Pavlova, O. V.; Oskina, O. Y.; Sivergin, Y. M.; Zaikov, G. E. *Polym. Degrad. Stab.* **1992**, *37*, 217–221.

340. Ohtani, H.; Ishiguro, S.; Tanaka, M.; Tsuge, S. *Polym. J.* **1989**, *21*, 41–48.

341. Lehrle, R. S.; Place, R. J. *Polym. Degrad. Stab.* **1997**, *57*, 41–48.

342. Lehrle, R. S.; Place, E. J. *Polym. Degrad. Stab.* **1997**, *56*, 221–226.

343. Lehrle, R. S.; Place, E. J. *Polym. Degrad. Stab.* **1997**, *56*, 215–219.

344. Zdonnik, P. V.; Zuev, V. V.; Stempanov, N. G.; Turkova, L. D.; Sazanov, Y. N.; Shibaev, L. A. *Polym. Sci. U.S.S.R.* (Engl. Transl.) **1993**, *35*, 697–700; *Vysokomol. Soedin A&B* **1993**, *35*, B260–B263.

345. Zulfiqar, S.; Zafar-uz-Zaman, M.; Munir, A.; McNeill, I. C. *Polym. Degrad. Stab.* **1997**, *55*, 275–279.

346. Zulfiqar, S.; Zarar-ur-Zaman, M.; Munir, A.; McNeill, I. C. *Polym. Degrad. Stab.* **1995**, *50*, 33–37.

347. McNeill, I. C.; Ahmed, S.; Memeta, L. *Polym. Degrad. Stab.* **1995**, *48*, 89–97.

348. McNeill, I. C.; Memeta, L.; Ahmed, S.; Cole, W. W. *Polym. Degrad. Stab.* **1995**, *48*, 395–410.

349. Oguchi, R.; Shimizu, A.; Yamashita, S.; Yamaguchi, K.; Wylie, P. *J. High Resolut. Chromatogr.* **1991**, *14*, 269–272.

350. Ouillon, I.; Zerroukhi, A.; Raihane, M.; Boinon, B. *Polym. Degrad. Stab.* **1998**, *61*, 519–525.

351. Mao, S.; Ohtani, H.; Tsuge, S. *J. Anal. Appl. Pyrol.* **1995**, *33*, 181–194.

352. Zulfigar, S.; Masud, K.; Piracha, A.; McNeill, I. C. *Polym. Degrad. Stab.* **1997**, *55*, 257–263.

353. Haussler, L.; Pompe, G.; Albrecht, V.; Voigt, D. *J. Therm. Anal. Calorim.* **1998**, *52*, 131–143.

354. McGrattan, B. J. Decomposition of Ethylene Vinyl Acetate Copolymers Examined by Combined Thermogravimetry, GC, and Infrared Spectroscopy. In *Hyphenated Techniques in Polymer Characterization*; Provder, T.; Urban, M. W.; Barth, H. G., Eds.; ACS Symposium Series 581; American Chemical Society: Washington, DC, 1994; pp 105–115.

355. McGrattan, B. J. *Appl. Spectrosc.* **1994**, *48*, 1472–1476.

356. Sultan, B. A. S.; Sorvik, E. *J. Appl. Polym. Sci.* **1991**, *43*, 1737–1745.

357. Hrdina, K. E.; Halloran, J. W.; Oliveira, A.; Kaviany, M. *J. Mater. Sci.* **1998**, *33*, 2795–2803.

358. Raihane, M.; Laleque, N.; Boinon, B. *Thermochim. Acta* **1995**, *265*, 1–13.

359. NcNeill, I. C.; Ahmed, S.; Memetea, L. *Polym. Degrad. Stab.* **1994**, *46*, 303–314.

360. McNeill, I. C.; Ahmed, S.; Rendall, S. *Polym. Degrad. Stab.* **1998**, *62*, 85–95.

361. McNeill, I. C.; Ahmed, S.; Gorman, J. G. *Polym. Degrad. Stab.* **1999**, *63*, 265–271.

362. Lehrle, R. S.; Parsons, I. W.; Rollinson, M. *Polym. Degrad. Stab.* **2000**, *67*, 21–33.

363. Ericsson, I. *J. Chromatogr. Sci.* **1978**, *16*, 340–344.

364. Bate, D. M.; Lehrle, R. S. *Polym. Degrad. Stab.* **1998**, *62*, 57–66.

365. Perng, L. H. *Polymer Degrad. Stab.* **2000**, *69*, 323–332.

366. Lin, J.-F.; Ho, C.-F.; Huang, S. K. *J. Appl. Polym. Sci.* **2000**, *77(4)*, 719–732.

367. Sekkar, V.; Ninan, K. N.; Krishnamurthy, V. N.; Jain, S. R. *Eur. Polym. J.* **2000**, *36*, 2437–2448.

368. Ericsson, I. *J. Anal. Appl. Pyrol.* **1985**, *8*, 73–86.

369. Kloppenburg, L.; Jones, D.; Bunz, U. H. F. *Macromolecules*, **1999**, *32*, 4194–4203.

370. Smith, P. B.; Pasztor, A. J., Jr.; McKelvey, M. L.; Meunier, D. M.; Froelicher, S. W.; Wang, F. C.-Y. *Anal. Chem.* **1999**, *71*, 61R–80R.

371. Grob, R. L., Ed. *Modern Practice of Gas Chromatography*, 3rd ed.; Wiley Interscience: New York, 1995.

372. Lee, M. L.; Yang, F. J.; Bartle, K. D. *Open Tubular Column Gas Chromatography: Theory and Practice*; John Wiley & Sons: New York, 1984.

373. McNair, H. M.; Miller, J. M. *Basic Gas Chromatography*; John Wiley & Sons: New York, 1997.

374. Poole, C. F.; Poole, S. K. *Chromatography Today*; Elsevier: Amsterdam, The Netherlands, 1991.

375. Robards, K.; Haddad, P. R.; Jackson, P. E. *Principles and Practice of Modern Chromatographic Methods*; Academic Press, London, 1994.

376. Scott, R. P. W. *Chromatographic Detectors: Design, Function, and Operation*; Marcel Dekker: New York, 1996.

377. Smith, R. M. *Gas and Liquid Chromatography in Analytical Chemistry*; John Wiley & Sons: New York, 1988.

378. Brown, R. F. C. *Pyrolytic Methods in Organic Chemistry: Application of Flow and Flash Vacuum Pyrolytic Techniques*; Academic Press: New York, 1980.

379. Irwin, W. J. *Analytical Pyrolysis*, Marcel Dekker: New York, 1982.

380. Meuzelaar, H. L. C. Haverkamp, J.; Hileman, F. D. *Pyrolysis Mass Spectrometry of Recent and Fossil Biomaterials*; Elsevier: Amsterdam, The Netherlands, 1982.

381. *Analytical Pyrolysis: Techniques and Applications*; Voorhees, K. J., Ed.; Butterworths & Co.: London, 1984.

382. *Applied Pyrolysis Handbook*; Wampler, T. J., Ed.; Marcel Dekker: New York, 1995.

8

Field-Flow Fractionation Characterization

MARCUS N. MYERS

Field-flow fractionation (FFF) is a versatile and powerful separation and characterization method for polymers and colloidal particles that can be applied to a wide size range of molecules from 1000 Da up to particles with diameters of 100 μm.[1,2] A wide range of materials has been studied, including biopolymers, other water-soluble polymers, lipophilic polymers, emulsions, and many of types of colloid particles. Analyses have been carried out in both organic and aqueous carrier solutions. Separation behavior follows molecular weight, particle size, density, charge, and chemical composition. The method is capable of selectivity that is several times that of other separation procedures.

The basic difference between FFF and other chromatographic methods is that the usual packed column is replaced by a simple channel. The associated equipment such as pumps and detectors are essentially the same as illustrated in Figure 8.1. The separation takes place in a thin ribbonlike open channel, with a field perpendicular to the channel. The simple geometry of the channel allows exact calculation of the flow hydrodynamics in the channel, unlike the very complex flow patterns in packed columns. Any field that will interact with the polymer or particle can be used, provided a method of fabricating an appropriate channel can be devised. The field can be changed rapidly, allowing a single channel to be used for a variety of sample types and sizes—including particles and polymers—and operating methods. Thermal, centrifugal, flow, and electrical fields are those currently used, although concentration, magnetic, shear, and dielectric fields have also been studied. The subtechniques of the greatest use for polymer analyses are thermal and flow FFF. The fields required for the use of sedimentation (centrifugal) FFF for polymers are very high, and only those materials with molecular weights over about 10^9 are practical.[3] However, the separation potential for colloidal particles by this method is very high as the differentiation depends on the third power of the diameter, whereas the other FFF methods depend on the first power.

Electrical FFF has been demonstrated for separating proteins, but the most common use is analysis of colloidal particles.[4,5] Thermal FFF has generally been used for lipophilic polymers, while flow FFF has been used for lipophobic or water-soluble polymers. Flow FFF has also been used successfully for some lipophilic polymers,[6–8] but the major restriction is the availability of membranes with appropriate size cutoff properties and compatibility with organic solvents. A very wide range of polymer types have been studied with both of these techniques; a partial list is given in Table 8.1. Both thermal FFF and flow FFF are used for characterizing colloidal particles, and particles 1 to 100 mm in diameter have been routinely analyzed by flow FFF.

FFF is an elution method that allows subsequent analysis by other methods, including light scattering, viscometry, and mass spectrometry. Parameters such as diffusion coefficients (D), thermal diffusion coefficients (D_T),[9–11] and reaction kinetics[12] are also determined with FFF. Since the field is controlled externally to the channel, the possibility of field programming in both thermal and flow FFF allows the characterization of samples with a very broad distribution. A high field can be applied to obtain retention of small components, followed by a decreased field to obtain larger materials in a reasonable time with good detection. This capability provides separation with high resolution over a very wide size range, all in a single channel, unlike other methods that require the use of several columns to cover broad sample ranges. The FFF methods are particularly effective at higher molecular weight ranges.

The open channel allows injection of samples with a minimum of preparation other than dissolving the polymer or suspending the particles. This property allows the characterization of polymers with microgels present in the sample.[12–15] Both the polymer and microgel distributions can be obtained in the same sample under the proper operating conditions. Composite materials have also been analyzed.[16] The open channel has additional advantages of very low

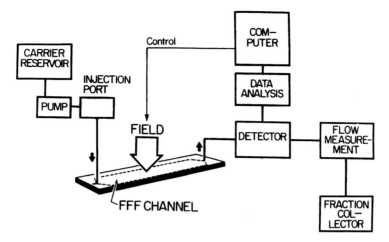

Figure 8.1 The field-flow fractionation (FFF) system is primarily different from typical high-performance liquid chromatography (HPLC) only in that the HPLC column is replaced by an FFF channel, with external control of the field applied to the channel.

shear of the sample (often 3 orders of magnitude less than gel permeation chromatography [GPC]) as it traverses the channel. Much lower interaction of the sample with the surfaces in the channel is observed, compared with the necessary contact for separation with the high surface area packing material in other methods. Thermal FFF can also supply chemical information about copolymers.[17-19]

The object of this chapter is to give an introduction to FFF as a detailed overview. Operational details and many applications are given in several references, since space limitations will not allow them here. Much of this information is found in the *FFF Handbook*.[20] Only a sampling of the studies that have been done is given. Commemorative issues in honor of J. Calvin Giddings containing papers on or related to FFF are in the *Journal of Microcolumn Separation*[21,22] and the *Journal of Liquid Chromatography and Related Technologies*.[23] Descriptive publications on FFF are also found in several journal articles.[1,2,5,24-28] A quite complete compilation of literature references to FFF can be found at www.rohmhaas.com/fff/, which contains very current information on publications from most of the active work being done at this time, as well as the earlier work in this field.

Basic Theory

The separation occurs in a very thin (127–254 µm thick) ribbonlike channel, with the field imposed at 90 degrees to the channel face, as shown in Figure 8.2. Flow lines in the channel are laminar, with a parabolic profile across the

Table 8.1 Partial list of applications in field-flow fractionation (FFF)

Material type	Substances studied
Flow FFF	
Water-soluble polymers	Polyacrylates, including polyacrylic acid and polyacrylamide; sulfonated polystyrenes; poly(vinyl pyrolidone); humic acids and fulvic acids from natural waters, soil, peat bogs
Biological material	Plasma components, including lipoproteins; a wide variety of proteins; protein aggregates; a variety of viruses; t-RNA; DNA, polysaccharides such as xanthan and guar gums, and starch; plasmids; yeast microsomes and cells; HeLa cells; red blood cells; unicellular algae; bacteria
Particulates	Pollens, nano particles; latex beads; chromatographic silica supports; ground coal and limestone; colloidal silica; paint components and pigments
Thermal FFF	
Polymers	Microgels in polymers; polystyrenes; polyacrylates; polycarbonates; poly(methyl methacrylates); polyisoprene; poly(tetrahydrofuran); polyurethanes; polyvinyl polymers; nitrocellulose and other derivatives of cellulose; polyethylene; copolymers such as styrene-acrylonitrile, butadiene-acrylonitrile-methylacrylic acid, styrene-butadiene; polysaccharides such as starch, pullalans, dextrans, Ficoll; composite materials with particles and polymers; asphaltenes; asphalts; crude oil
Particulates	Wear particles in oil; latex beads

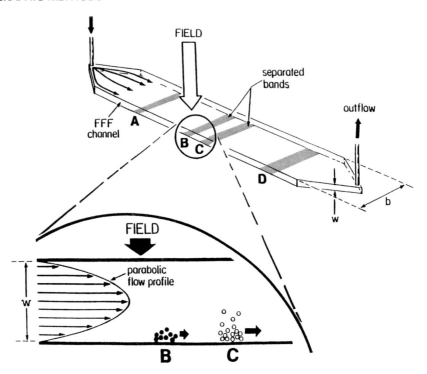

Figure 8.2 The field-flow fractionation (FFF) separation occurs in a thin ribbonlike channel with laminar flow down the channel and a field imposed 90° to the channel face. Molecules or particles are carried toward the accumulation wall and are then opposed by diffusion or Brownian motion. The net result is a steady-state zone with an average height that depends on the characteristics of the sample material. (Reproduced with permission from L. F. Kesner and J. C. Giddings, in *High Performance Liquid Chromatography*, edited by P. R. Brown and R. A. Hartwick, Eds., Wiley: New York, 1989, chapter 15, pp 601–641. Copyright 1989 Wiley-Interscience.)

channel thickness and with the maximum flow in the center of the channel and zero flow at the channel walls. The injection of a sample into the channel distributes the sample across the channel thickness. Sample components are subject to forces and displacements that can be calculated. The field induces the movement of the sample toward one wall, the velocity depending on the interaction of the field with the sample molecules or particles. Opposing this migration is the diffusion of molecules. The result is zones whose thickness depends on the sum of these two movements. The height of the zone determines the velocity of the carrier flow lines in the channel. The velocity of the zone down the channel will be higher for molecules with high diffusion rates since their thicker zone will intercept streamlines that move at higher velocities than those intercepted by molecules that form a zone nearer the wall. This differential in zone velocity results in the smaller molecules eluting first and separation is obtained. This is the opposite order observed in size exclusion chromatography (SEC). Because of the well-defined channel and the physicochemical properties, the method can be treated theoretically. The concentration of the molecules across the channel thickness will be exponential

$$c_x = c_o e^{-x/\ell} = c_o e^{-x/\lambda w} \tag{1}$$

where c_x is the concentration at position x across the channel, c_o is the concentration at the accumulation wall or the wall to which the sample is driven by the field, ℓ is the center of gravity or the mean thickness of the zone, w is the channel thickness, and λ is the dimensionless retention parameter described by ℓ/w. Coupling this equation with the flow velocity equation

$$v(x) = 6<v>(x/w - x^2/w^2) \tag{2}$$

where $v(x)$ is the flow velocity at x, and $<v>$ is the average velocity of the carrier, we eventually obtain the general equation for FFF

$$V^0/V_r = t^0/t_r = R = 6\lambda \left[\coth(1/2\lambda) - 2\lambda\right] \tag{3}$$

where V^0 is the elution volume for a nonretained solute, V_r is the elution volume for the retained solute, and t^0 and t_r are the corresponding retention times. At high retention, V^0/V_r approaches 6λ. The values of λ are related to the physicochemical properties of the material being analyzed, which includes the diffusion coefficient, D. The effect of these properties on separation will be determined by the type of field (or FFF subtechnique) being used. Thermal FFF requires a temperature differential across the channel that supplies the field needed. The upper wall of the channel is heated, and the lower wall is cooled. In flow FFF, the field is supplied by a cross flow through the lower wall of the channel, which is permeable to solvent flow, but retains the sample components.

The conversion of the observed retention parameter, λ, to molecular size requires a relation between the two parameters. Einstein showed that the diffusion coefficient could be expressed by $D = kT/f$, and Stokes found that f, the friction coefficient, was described by $f = 6\pi\eta r$, with r being the particle radius or the effective radius of a polymer molecule or particle in solution, assuming a spherical particle and η is the viscosity of the solvent. The resulting equation

$$D = kT/6\pi\eta r \qquad (4)$$

with k being Boltzman's constant and T the temperature, gives an avenue to relate D to molecule or particle size. The expression for λ in flow FFF then becomes

$$\lambda = DV^0/V_c w^2 \qquad (5)$$

with V_c as the cross flow rate. The corresponding relation for thermal FFF is

$$\lambda = D/D_T\Delta T \qquad (6)$$

where D_T is the thermal diffusion coefficient, and ΔT is the temperature difference between the hot wall and the cold walls of the channel.

For polymers, the equation

$$D = A\ M^{-b} \qquad (7)$$

can be obtained from the equivalent sphere model, where M is the molecular weight, A depends on the polymer type, and b depends on the molecular conformation in solution. Combining eqs 5 and 7 for flow FFF,

$$\lambda = A\ M^{-b}V^0/V_c w^2 \qquad (8)$$

the values of A and b can be obtained from the intercept and slope of the measured log λ versus log M.

Thermal FFF requires the inclusion of D_T and ΔT. From eqs 6 and 7

$$\lambda = A\ M^{-b}/D_T\Delta T = \phi\ M^{-n}/\Delta T \qquad (9)$$

where ϕ includes A and D_T, and n includes b and any molecular weight dependence for D_T. A more complete discussion of D_T and n is given in the section, Thermal FFF.

A number of factors cause peak broadening. The most important of these is the nonequilibrium contribution, which comes from the distribution of molecules across the channel in the solute zone. Expressing this effect in terms of the nonequilibrium plate height, H_n, we obtain

$$H_n = \chi w^2 <v>/D \sim 24\lambda^3 w^2 <v>/D \qquad (10)$$

with $<v>$ being the average channel velocity, χ is a complicated nonequilibrium factor (29) that reaches a limit at values of $\lambda < 0.1$, such that the right-hand expression holds with an error of less than 1%. Other contributions include longitudinal diffusion (the spreading due to the dispersion during the relaxation process) and instrument contributions such as the volumes external to the actual channel, often called "dead" volumes. Spreading in the zone due to actual differences in molecular size, designated by H_p, are real separations but appear as apparent band broadening in a sample, which has a range in size. This apparent contribution can be given by

$$H_p \sim LS^2(\mu - 1)$$

where $\mu = M_w/M_n$, or the ratio of the weight-average molecular weight to the number-average molecular weight, and L is the effective channel length. The selectivity, S, can be expressed as

$$S = d\ln V_r/d\ln M \sim d\ln \lambda/d\ln M \qquad (11)$$

The theoretical selectivity in FFF is considerably higher (0.5–0.7) than that achieved in SEC (0.05–0.15). This gives rise to good separation in FFF with a much lower number of plates (N) than that necessary to achieve the same separation in SEC. Resolution, R_s, is defined as $R_s = \Delta V_r/4\sigma_v$, where ΔV is the volume difference between two peaks, and σ_v is the standard deviation of the peak. In terms of selectivity,

$$R_s = S\ (N^{1/2}/4)(\Delta M/M) \qquad (12)$$

This higher resolution is demonstrated in Figure 8.3 with the number of plates larger for SEC while the resolution is greater in the FFF fractogram. The reversed order of elution is also apparent.

Longitudinal diffusion is usually very small unless the analysis time becomes excessively long. Relaxation contributions can be minimized as discussed in the section, Relaxation. Special modifications of flow FFF, such as "asymmetric flow FFF"[30] and hollow fiber flow FFF,[31,32] require some modification of the above equations, which are for the "symmetric FFF" or the normal FFF systems.

Programming

Smaller molecules and particles require high fields to achieve separation since their diffusion rates are high. In contrast, larger molecules and particles have low diffusion rates and require much lower fields to gain the desired separation. If the field is optimized for the smaller material, the larger size material will be retained excessively long, and the zone may be so dilute as to be difficult to recognize or detect. If the analysis is designed for the larger material, the smaller material will be poorly separated, if at all.

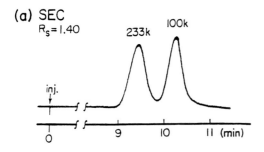

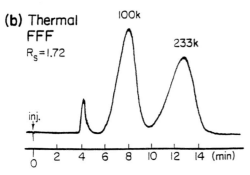

Figure 8.3 FFF has much higher selectivity than SEC, thus providing better resolution with a much smaller number of plates. The separation in (a) by SEC was done in a relatively high efficient column yielding a large number of plates. The FFF separation shown in (b) with a much smaller number of plates still can give a higher resolution. (Reproduced with permission from reference 68. Copyright 1986 Elsevier.)

The field in FFF is applied externally to the channel, allowing the field strength to be varied quickly. This capability enables the programming of the field to achieve separations that are rapid and with high selectivity over a wide distribution in molecular weight or particle size during a single analysis. Normally, the analysis begins with a high field, which may be held for a period of time to obtain separation of the smaller materials. The field is then reduced with time, according to the desired program. A wide variety of programs have been used. Most of these will give some of the desired results. Step programs, where the field is kept at one level, then abruptly changed to another level, have proven useful for a sample when two fairly narrow components that are quite different in size are present. Linear and parabolic programs have also been demonstrated. A program in which the field decays exponentially has some advantages since molecular weight can be readily obtained, as shown by Kirkland et al.[33-35] This program does suffer the disadvantage that the selectivity varies over the course of the analysis, with the highest selectivity being obtained for the smaller molecular weight material.

The power program developed by Williams has proven most useful.[36] Williams used fractionating power (F) as a more specific means to describe the ability to separate two components. This can be defined as the resolution (R_s) of two molecules whose molecular weights differ by $\delta M/M$.

$$F_M = \frac{R_S}{\delta M/M}$$

Using this program provides a constant fractionating power throughout the analysis when the proper power is selected.

$$G(t) = G(0)\{(t_1 - t_a)/(t - t_a)\}^p \qquad (13)$$

$G(t)$ is the field at time t, $G(0)$ is the beginning field, t_1 is the time the beginning field is held before decay begins, t_a is the asymptotic time, and p is the power. These parameters must satisfy the conditions $t \geq t_1 \geq t_a$, $t_1 \geq 0$, and $p > 0$. For flow and thermal FFF, the optimum condition for uniform resolution will be obtained is $t_a = -pt_1$ and $p = 2$. A fractogram from a power-programmed thermal FFF run is given in Figure 8.4.

Channel flow can also be programmed to decrease the time required for an analysis by increasing the flow rate according to a desired program.[37] The number of theoretical plates, N, depends on the flow rate

$$N = LD/(\chi w^2 <v>) = LD/(24\lambda^3 w^2 <v>) \qquad (14)$$

As the flow velocity increases during the program, the number of plates and resolution will decrease. From eq 14, we also see that reducing the channel thickness by one-half enables increasing the flow velocity by a factor of 4, while maintaining the same number of plates. Thinner channels require less time for obtaining equilibrium. A thermal FFF channel with a thickness of 51 mm has been successfully used with excellent separations in very short times, as shown in Figure 8.5. Limitations on the ability to fabricate perfectly flat surfaces may restrict the practical application of thinner channels. Nonflat surfaces in very thin channels will result in differential flow rates across the breadth of the channel and can cause a severe loss in resolution. Since the very thin channels have very low void volumes, the volumes in the system external to the channel become very important, as significant zone spreading can negate the enhanced separation obtained in the channel. These problems are less severe in thicker channels. The channel thickness used in the current FFF apparatus is somewhat of a compromise. In flow FFF, the membranes are often compressible and give reduced channel thickness that may be difficult to predict. Noncompressible membranes can be installed with a flatter surface and a more defined thickness.

The composition of the solvent may also be programmed, similar to that done in high-pressure liquid chromatography (HPLC). The solvent influences the retention of lipophilic polymers because of the chemical effect inherent in D_T in thermal FFF. Variation in ionic strength also affects the retention of water-soluble polymers and charged particles in flow FFF.[38-40] Thus programming enhances the versatility of FFF and provides a wide size range for analysis in the same channel.

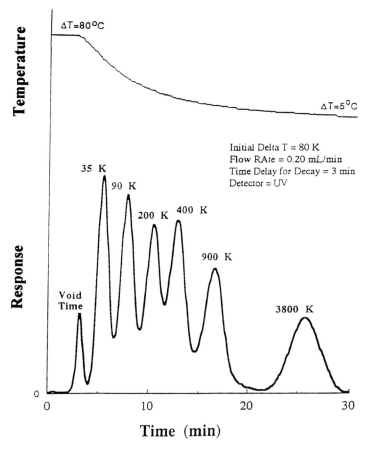

Figure 8.4 The use of programming allows the analysis of broad molecular weight samples in a short time. The ΔT was programmed in this example of the separation of narrow polystyrenes in THF using the power program. Key: initial $\Delta T =$ 80 °C, $t_1 = 3$ min., $t_a = -6$ min, flow rate = 0.20 mL/min.

Relaxation

All FFF techniques require sufficient time immediately after sample introduction for the sample components to reach their equilibrium position before separation can proceed. This time, called *relaxation time*, can range from seconds to many minutes, depending on the FFF technique and the characteristics of the sample. At very low channel flow rates and high fields, this position may be reached a short distance down the channel. However, at higher channel flow rates, the sample may not reach the equilibrium position before eluting from the channel, with a resulting loss in separation, unless means are taken to ensure that the relaxation requirements are met. The relaxation times required in both thermal and flow FFF are usually reasonably short, of the order of 1 to 5 minutes for most samples. At high cross-flow rates in flow FFF, the required time may be only a few seconds. Longer relaxation times may be used, but some samples may tend to adsorb on the channel surface if too long an exposure is allowed.

Several techniques have been used to obtain the proper relaxation times, including "stop flow", "frit inlet", "pinched inlet", and "focusing".

Stop flow allows the sample to flow just onto the channel; then the channel flow is stopped for a sufficient time to allow the field to drive the sample molecules to their quasi equilibrium near the wall. For thermal FFF, this time in seconds may be estimated from

$$t_{\text{relth}} = 2w^2/D_{\text{T}}\Delta T \tag{15}$$

For flow FFF, the corresponding time is

$$t_{\text{relf}} = 2V^0/V_{\text{c}} \tag{16}$$

where V_{c} is the cross channel flow rate. Flow is then resumed. One disadvantage of this method is that the interruption of flow can cause a disturbance in the baseline, which can last long enough to distort or even mask early-eluting sample molecules, particularly in flow FFF. The disturbance generally is not a problem with thermal FFF.

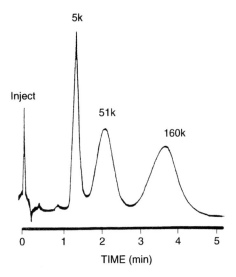

Figure 8.5 The separation of polystyrene in a 51-μm thick channel with $\Delta T = 60\ °C$ and a flow rate of 0.35 mL/min. The speed of analysis is increased by using thinner channels. (Reproduced with permission from M. Martin and J. C. Giddings, *J. Chromatogr.* **1978**, *158*, 419–435. Copyright 1978 Elsevier.)

flow, is eliminated. An additional advantage is gained for those samples that have a tendency to adsorb on the surface since they do not spend time forced near or to the channel wall. A drawback is some zone broadening.

Pinched inlet, a third method for approaching the equilibrium position, occurs when the upper wall of the first few centimeters of the channel protrudes into the channel, causing the flow to be close to the accumulation wall and near the desired position for the sample zone.[46,47] This method has the advantage of noninterrupted flow during the run, with no perturbation of the detector baseline. This method has been demonstrated but has not been developed as fully as the other methods because of the difficulty of fabricating channels with this configuration.

The focusing approach, applicable only to flow FFF, is a fourth method of establishing the equilibrium zone.[48] As samples enter the channel, there is some unavoidable spreading of the sample over the surface of the channel. Ideally, a delta function sample (zero volume sample) is desired, with the sample zone being as narrow as possible at the start of the analysis. This narrow sample zone can be approached in flow FFF by using focusing techniques. In this method, the sample is allowed to enter the channel with an opposing channel flow. Because the lower wall is permeable and allows the entering carrier solutions to exit through this wall, the two flow rates are selected so that the sample zone will concentrate at the position where the net velocity down the channel is zero. The result is a very narrow zone at that position, since all of the sample upstream as well as any downstream from the zero velocity position will be swept to that position. The time required to obtain such a zone could be estimated from

$$t_{\text{focr}} = (w^2/6D)\ \ln(LV_{\text{in}}/w_b V_c) \qquad (17)$$

Frit inlet, a modification of the channel, accomplishes essentially the same result as stop flow by introducing a second flow into the channel near the sample inlet, as shown in Figure 8.6.[41–45] In this case, a secondary flow is introduced near the injection point, which forces the sample into flow lines very near the position gained by relaxation or stop flow. The advantage obtained by this approach is that the time lost, as well as the detector instability caused by stop

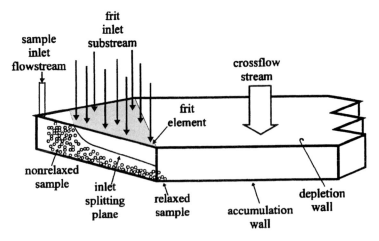

Figure 8.6 The frit inlet method of relaxation relies on a high second flow near the sample introduction point to force the sample near the final relaxed position close to the accumulation wall.

For a conventional (symmetrical) flow FFF system. V_{in} is the flow rate at the channel inlet. L is the effective channel length, and w_b is the zone bandwidth in mm. For an asymmetric FFF system with a trapezoidal channel,

$$t_{foca} = (w^2/6D) \ln(A_c V_{in}/a_z V_c) \qquad (18)$$

where A_c is the channel area, and a_z is the zone area.

There are various ways to manipulate the flow streams to achieve focusing. One troublesome problem occurs when the flows are changed from focusing to analysis mode. In the conventional mode, a large pressure pulse is produced in the detector as flow is resumed, resulting in a large signal, which may require up to several minutes to return to baseline. Analytes, which elute early, may not be detected because of this overwhelming signal from the pressure change. A patent pending method developed by FFFraction-ation, LLC., largely eliminates this problem. (See Figure 8.7.) Flow is directed through the detector at channel pressure throughout the focusing operation and the change to analysis mode. Care must be taken during focusing that the sample is not adsorbed on the membrane surface by overly long focusing times. For some samples, a broader sample band might be necessary to ensure that little adsorption

takes place. The ability to concentrate the sample by this method allows the injection of large volumes of very dilute samples for analysis.[49] Injection volumes of 1 liter and larger have been successful.

Frit Outlet

As samples elute from the FFF channel, each sample component becomes diluted. The original sample mass is already small to avoid overloading, particularly with high molecular weight polymers, so the diluted component is present in concentrations that are often below the sensitivity limits of a desired detector. Since the sample zone is only a few micrometers thick and confined close to one wall, most of the channel thickness is occupied by the carrier only. The frit outlet takes advantage of this condition by withdrawing solvent from the side opposite to that of the zone, as illustrated in Figure 8.8. By careful fabrication and control, the sample zone may be concentrated as it elutes from the channel.[45,51] Concentration factors of 10/1 are quite readily obtained in flow FFF, while factors of 50/1 and higher have been achieved. The frit outlet method is demonstrated in Figure 8.9 and has been found particularly useful for coupling FFF with multiangle light-scattering (MALS) and refractive index (RI) detectors.

Sample Preparation

Sample preparation is essentially the same as that used for HPLC or GPC. The sample is placed in a vial, the solvent is introduced, and the vial is tightly closed. Typical concentrations are in the range of a few milligrams per milliliter. For most samples, dissolution occurs readily if the proper solvent is chosen. However, higher molecular weight polymers

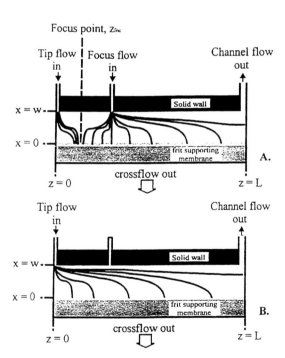

Figure 8.7 Focusing relaxation uses a flow opposing the sample flow as shown in (A). The method illustrated was developed at FFFractionation, LLC., and provides much less detector disturbance than the usual method, which has the focus flow come from the channel outlet. After the sample has reached the focusing point, the focusing flow is stopped and the tip flow rate is increased to the total flow rate of the tip and focus flow, as shown in (B).

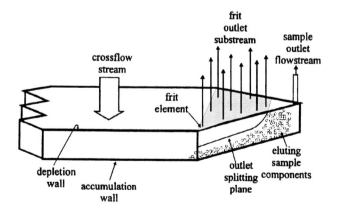

Figure 8.8 The frit outlet mode of operation uses the fact that the sample zone lies very close to the accumulation wall. The rest of the channel is carrier solution, so the withdrawal of the upper portion of the channel contents allows a concentration of the material in the sample zone.

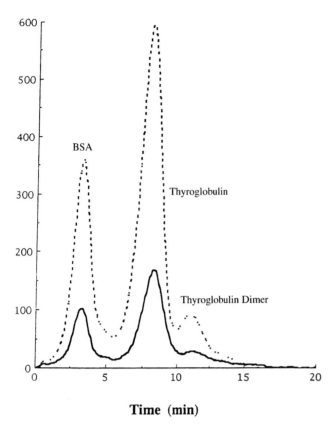

Time (min)

Figure 8.9 The concentration of the eluting sample by the frit outlet method is illustrated with protein samples. Separation conditions: channel flow rate = 2.4 mL/min, cross flow rate = 2.8 mL/min, detector flow rate for the standard channel = 2.4 mL/min, and detector flow rate with frit outlet = 0.68 mL/min.

often dissolve slowly, requiring setting overnight or longer. Gentle agitation with mild heating may be required. Samples such as polyolefins often require dissolution at temperatures above 140 °C, and the sample must remain at that minimum temperature to prevent precipitation during subsequent testing.[52] Antioxidants such as Irganox 1010 (penta erythrityl tetrakis[3-(3,5-di-*tert*-butyl-4-hydroxyphenyl) propionate]) (Ciba-Geigy) and an inert gas such as nitrogen over the sample are used to prevent sample degradation from oxidation for polyolefin samples. Unless particles are visible, filtration, with the resulting high shear in the filter, is not usually required, since the restricting limit is the cross-sectional area of the tubing used for the inlet and outlet to the channel. Some rather milky-looking samples have been successfully analyzed. If microgels are being studied, filtration is not desired since some or all of the microgels may be removed in the process. When a light-scattering detector is being used, extreme care should be used to remove any microparticulate material from the solvent used in dissolution of the sample and from the carrier solvent used to introduce the sample into the channel. Normally these solvents are filtered through a 20-nm filter.

Water-soluble polymers such as the polyacrylates can be difficult to satisfactorily disperse. These types of polymers tend to form gel-like solutions, which are difficult to analyze for molecular weights since FFF requires the molecule to behave independently from neighboring molecules. As small a sample mass as can still be well detected should be used in the analysis to maximize the sample dispersion.

Ancillary Equipment

The basic difference between FFF and chromatography methods is the separation channel. The ancillary equipment is very similar. The metering pumps for FFF are usually the same as those used for HPLC. The pumps should be as nearly pulseless as possible to prevent or reduce baseline noise. The pulse damping obtained with a packed column in HPLC is not present in FFF, and the addition of a pulse damper is often very helpful. Sufficient pressure is necessary for the pump check valves to function properly. Small backpressure regulators are available, and they provide very reliable operation over a range of flow rates. The placement of backpressure regulators may be governed by the type of detector; some detectors may be damaged if they are used above a given pressure.

The flow rates for thermal FFF are between 0.02 and 1 mL/min in normal operation. Flow FFF usually requires flow rates between 0.2 and 10 mL/min. The pressure drop across a FFF channel is very low.

Very low volume inlets and outlets to the channel are provided for carrier flow. These may be in either the top or the bottom walls in thermal FFF; to avoid rupturing the membrane, however, they should be in only the top wall in flow FFF. The volume must be kept low to prevent peak broadening in these "dead" volumes. Typical total volumes for these access tubings are 5 to 50 µL for channels with a volume of 1 to 5 mL. The size of these inlet and outlet tubings can be limited when high flows are being used, as the pressure drop increases with the inverse of the fourth power of the radius from Poiseuille's equation $P \sim v/r^4$. Where v is the velocity and r is the radius of the tube. A basic rule is to use the smallest diameter with the shortest length of tubing that is suitable for the flow rates to be used and the physical size of the instruments in the system.

Most HPLC detectors have been used for FFF. Sensitivity is quite important in FFF, as the most efficient separations are obtained with small sample mass. An additional problem for detectors is the dilution produced during the analysis. Techniques such as frit outlet can increase the concentration of the sample as it exits the channel to partially compensate for the analysis dilution.

The most frequently used detectors in FFF have been ultraviolet (UV) monitors, which have quite high sensitivity for those samples that contain a UV chromophore. Variable wavelength detectors (VWD) are useful for a wide variety of materials. Limitations include the need to use a solvent that is transparent to the wavelength of light used. Particulate samples also can be measured with UV monitors because the particles scatter light, thereby decreasing the signal. The latter type of sample may require corrections for light-scattering differences with the variation of the particle size in particle characterization.

Refractive index (RI) detectors are used as mass detectors and are somewhat universal for polymeric systems. Solvents with a refractive index that is sufficiently different from that of the polymer are required. The requirement for high sensitivity may be difficult for many of the current refractive index detectors to meet. This type of detector is often coupled with MALS when absolute molecular weight measurements are required. Particle measurement by this type of detector is not practical.

The coupling of FFF with MALS provides an excellent method for obtaining absolute molecular size.[14, 53–56] The use of a FFF system provides optimal conditions for light-scattering measurements as the sample eluting from the channel is virtually monodisperse, which allows use of the full theoretical advantage of MALS. The requirement for small sample size for optimal separation can complicate the coupling of FFF apparatus to light-scattering detectors such as MALS. A mass-sensitive detector such as an RI detector is required to obtain absolute molecular weight distributions. The small optimal mass sample may challenge the sensitivity of these detectors. Other detectors, such as evaporative light-scattering (sometimes called evaporative mass) detectors or UV detectors, may be used because they may give higher sensitivity if an accurate value for the injected amount is obtained; these detectors may not be linear, and they may not respond according to absorptive properties of the sample. The use of a frit outlet to concentrate the sample eluting from the channel will usually provide concentrations suitable for any of these detectors. A larger sample can sometimes be used even when the separation is less than optimal because the MALS detector can determine the size directly, and the zone broadening will not affect the measurement appreciably.

When a MALS detector or similar instrument is used, careful filtration and degassing of the solvent is usually required. When a light-scattering detector is used with programming in flow FFF, sometimes contaminants that were on the membrane may be observed as the field decreases. This usually can be prevented by the use of small-pore-diameter pore filters (~20 nm) in the system before the channel to prevent these contaminants from entering. Some of these very small pore filters are quite fragile and may fracture at the high flows used at the start of the programmed run. A pressure gauge just before the filter may give an indication of breakage, and a large reduction in pressure will indicate failure. The use of larger pore-diameter filters might be required in some circumstances. When larger pore filters are used, such breakage is rare, but the very small contaminants are not removed. Thermal FFF has not been observed to have this problem, and since the flow rates in this method are usually quite low, the use of small-pore-diameter filters poses little problem.

Evaporative light-scattering detectors have proven very useful in FFF analysis by providing reasonable sensitivity. Limitations are that the carrier must be completely volatile under the operating conditions. However, analyses have been done using trichlorobenzene (TCB) and dimethylsulfoxide (DMSO) as solvents. On the occasions where nonvolatile salts are used in the carrier, such a detector is not appropriate. Some questions of linearity are present with this type of detector, but over fairly narrow ranges, the response is satisfactory.

Detectors based on viscosity changes due to the presence of a solute have been useful in combination with thermal FFF to measure both the average molecular weight and the average composition of statistical copolymers.[57–59] This type of detector is especially sensitive to higher molecular weight materials, but it has a lower ability to measure smaller molecules.[60,61]

FFF equipment has been coupled with other types of equipment, such as hydrodynamic chromatography,[62] inductively coupled plasma (ICP), and mass spectrometers, to obtain additional information on sample composition with size and other parameters.[63]

Thermal FFF

Some of the advantages of thermal FFF for characterizing polymers are the following: (a) a single channel system is usable for most polymers and solvents; (b) there is no packing to degrade; (c) the open channel minimizes shear degradation; (d) the method is adaptable to various speed and resolution requirements; (e) the field strength can be tuned and programmed for different molecular weight ranges; (f) compositional information may be available; (g) ultrahigh molecular weight materials are analyzed; (h) the determination of microgels and particles can be obtained; (i) the method has been applied to polymer-particle composites and other complex mixtures.

Thermal FFF provides both size distribution information and some chemical characteristics of a given sample. Since the method has very low shear, more accurate information may be provided for those samples where shear degradation is a problem with more conventional methods, as the shear in thermal FFF channels is 2 to 3 orders of magnitude less than that found in high-efficiency GPC columns. Higher molecular weight molecules are particularly vulnerable to shear.[64] Fortunately, the larger molecules are also more easily separated by thermal FFF.[65,66] Thermal FFF can provide weight-average molecular weight, M_w, and number-average molecular weight, M_n, values, as well as size-distribution curves for lipophilic polymers. The most convenient method for measuring D_T for polymers in solution uses thermal FFF coupled with another method, such as SEC or dynamic light scattering, to obtain D. The method gives superior selectivity, up to five times that obtained with SEC for polymers with a molecular weight over 100,000,[67,68] while decreasing in resolution for molecules of lesser size with a lower practical limit of about 5000, although lower molecular weight materials have been analyzed under very high fields.[69] This increased selectivity for the higher molecular weight ranges enables excellent separation with much lower number of plates than that obtained in SEC. (See Figure 8.3.) The accuracy of these values depends on the quality of the calibration curves used. Since calibration depends on the physicochemical properties of the polymer and the solvent, accurate calibration curves obtained on one channel can be used in other channels. Studies to date show the range of errors is usually plus or minus 5% or less. Since there is a chemical component to the separation, valuable information on copolymers and polymer blends can often be obtained.[9,10,70] Microgel content has been measured quickly and routinely.[13,14,71] Information about the size and distribution of microgels is also available since the method can separate and characterize colloidal particles.[72] This capability has been applied to composites consisting of polymers and particles.[16] Coupling thermal FFF with MALS has been used to provide absolute size distributions of polymers. Thermal FFF has not been successful in the analysis of polymers in an aqueous carrier, although some rather poor separations have been obtained.[73] However, some good colloidal particle separations have been carried out in aqueous systems. The measurement of size distribution for colloidal particles has been done in both aqueous and organic solvents.

The migration of molecules and particles in a thermal field is determined by the thermal diffusion (described by D_T, the thermal diffusion coefficient) of the molecule or particle and the field provided by the ΔT between the hot and cold wall. Thermal diffusion is very poorly understood for processes in liquids. The current theories fail to adequately describe or predict the values observed for this phenomenon.[9] Thermal diffusion depends on the chemical type of polymer and the solvent being used.[9,70,74,75] Values can vary widely, depending on the polymer-solvent pair, as shown in Table 8.2. This chemical effect can be useful in separating mixed polymers that have similar molecular weights, which cannot be separated by SEC, as shown in Figure 8.10. The addition of salt can cause a large change in retention for some polymers, as seen in Figure 8.11. A similar chemical dependence is observed when working with particles.[76,77] The retention of a particle of a given size can be radically affected by changing the surface characteristics of the particle surface, as is observed for silica and polystyrene in Figure 8.12. Thermal diffusion has also been shown to be sensitive to temperature.[78,79] D_T increases at the rate of approximately 0.3% per degree K for polystyrene in tetrahydrofuran (THF). Enhanced retention can sometimes be obtained by using a mixture of solvents that give an effectively higher D_T than either of the pure solvents in the mixture and could be useful in programming the solvent composition, as mentioned previously in this chapter.[75,80] Molecular weight has been found to have little or no affect on D_T for the polymers studied to date.[9]

Unlike other FFF techniques where retention information can often be translated to size characterization from first principles, thermal FFF requires calibration to obtain precise molecular weights because of this lack of understanding of D_T. Calibration curves in thermal FFF are different from those obtained in GPC. In the latter method, each column is calibrated with narrow standards, and, in theory, the curve applies to all the polymers that behave similarly to the standards used; the standard is usually a polystyrene, as these are easily obtained. Problems with this method, which lead to the need for new calibration or column replacement, include nonideal behavior such as adsorption on the column packing material, shear caused by passage through the column (particularly for higher molecular weight material), changes in the column characteristics caused by settling, and packing deterioration with age or changing solvents. The chromatographic calibration is not transferable from one system to another. However, thermal FFF calibration curves do apply to any thermal FFF channel for a given polymer-solvent pair, since the curve depends only on the physicochemical properties of the given pair, and not on the channel characteristics.[81]

For all polymers studied to date, the molecules move toward the cold wall of the channel, and the velocity of the movement depends on the thermal gradient that is applied. While the thermal gradient is slightly nonlinear, with temperatures that range across the channel due to small changes in thermal conductivity, a good approximation for λ is given by eq 6. Both D and D_T vary with the solute/solvent (or carrier) couple being applied. However, as indicated, D_T has been shown to be essentially independent of the size and shape of the polymer molecule, although it does depend on the chemical composition of the molecule and the solvent.[9] Thus, for a given type of macromolecule, separation depends on D. Equation 9 expresses the relationship between the observed l and molecular weight. Rearranging and taking the log, we obtain

$$\log \lambda \Delta T = \log \phi - n \log M \qquad (19)$$

which gives a straight line when plotting $\log \lambda \Delta T$ versus $\log M$, the slope of the line is n, and the intercept is $\log \phi$. A modification of eq 19, which does not require extrapolation over a large range, is preferred.

$$\log \lambda \Delta T = \log (\phi_6) - n \log (M/10^6) \qquad (20)$$

where $\phi_6 = \phi/10^6$. Equation 20 provides the needed relationship between observed retention and molecular weight for a calibration curve. Ideally, narrow molecular weight

Table 8.2 Thermal diffusion coefficients, D_T, for poly(methyl methacrylate), PMMA; poly-α-methyl styrene, PαMS; polystyrene, PS; polybutadiene, PB; poly(tetrahydrofuran), PTHF; in several solvents ($D_T \times 10^7 [cm^2/sK]$)

Solvent	PMMA[a]	PαMS[a]	PS[a]	PI[a]	PB[b]	PTHF[b]
Tetrahydrofuran (THF)	1.33	1.27	1.00	0.57	0.25	0.52
Benzene	1.37	1.02	0.89	0.44		
Toluene	1.63	1.19	1.03	0.69	0.33	0.70
Ethylbenzene			0.95			
Cyclohexane			0.66		0.20	0.28
Methylethyl ketone (MEK)	1.48		1.39			
Ethylacetate			1.16		0.95	
Dioxane					0.18	

a. Data from reference 9.
b. Data from reference 74.

distribution standards are used. Unfortunately, the number of polymers for which this type of standard is available is limited. However, using four or more broad molecular weight distribution standards that are well characterized can also provide excellent calibration curves.[82, 83] A method for determining molecular weight distributions without standards based on known physicochemical values for polymers appears promising.[84] The proper use of a light-scattering detector, such as MALS, can be used to obtain molecular sizes for samples for which narrow standards are not available, and these results may be applied to develop calibration curves for that polymer.

The temperature differential results in a range of viscosities for the solvent across the channel due to the variation of viscosity with temperature. As a result, there is a change in flow velocity (and the velocity profile) across the channel

thickness. These effects can be approached by theory, and exact predictions of the velocity at any position in the channel can be calculated.[85,86] These corrections are quite complicated but are incorporated in the software for analysis that has been produced commercially. Since the curve depends on D and D_T, both of which depend on temperature, the temperature of the sample zone becomes important. This effect can be very large, as shown in Figure 8.13. A practical place to obtain the temperature of the zone is the cold wall. The average or center of the zone is only a few micrometers from this surface, but the actual temperature of the center of gravity of the zone varies slightly for each molecular weight (because the height of the zone will be

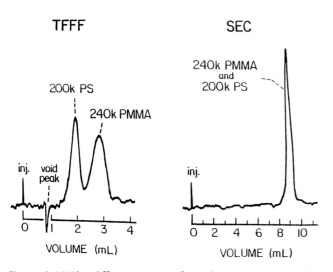

Figure 8.10 The differences in D_T for polymers can be used to obtain separations in thermal FFF that are not achieved in SEC. The same mixture of two narrow polystyrene and poly(methyl methacrylate) was not separated with SEC, but was easily separated by FFF. (Reproduced with permission from reference 70. Copyright 1986 American Chemical Society.)

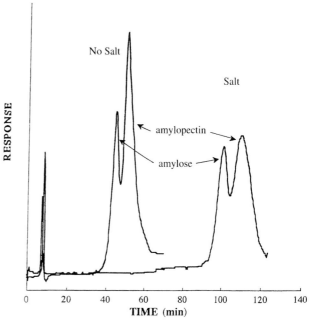

Figure 8.11 Retention of starch is greatly affected by the addition of 0.0001 M LiNO$_3$. The starch was dissolved in dimethylsulfoxide (DMSO). Key: flow rate = 0.1 mL/min, $\Delta T = 70$ °C. (Reproduced with permission from reference 129. Copyright 1994 Marcel Dekker, Inc.)

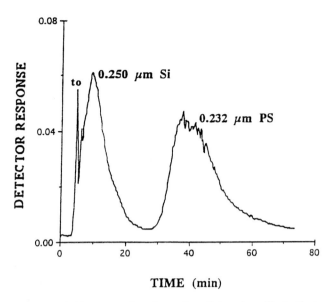

Figure 8.12 The type of surface of particles also affects their retention. The silica (Si) particles are considerably less retained than the polystyrene (PS) particles, even though they are approximately the same size. Key: $\Delta T = 45$ °C, flow rate = 0.2 mL/min. (Reproduced with permission from reference 76. Copyright 1995 Elsevier.)

different for different molecular weights). Theoretical work by Martin et al.[87] indicates this equilibration temperature, T_{eq}, will be found at about 2λ instead of λ. Calculation of T_{eq} is somewhat involved, in contrast to T_c, which can be directly measured. No gain in accuracy of molecular weight determinations has been found using T_{eq} over that obtained using T_c. A calibration curve obtained from T_c instead of T_{eq} will take care of such effects. Studies have been carried out on a few solvent-polymer pairs, and corrections to the

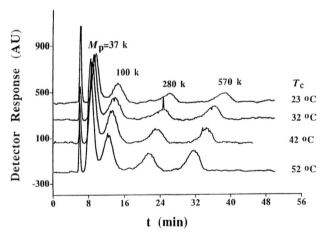

Figure 8.13 The cold wall temperature influences retention, as shown here for a mixture of polystyrene standards run with different cold wall temperatures. (Reproduced with permission from reference 89. Copyright 1999 American Chemical Society.)

curves have been obtained, which allow the use of different channels within a range of cold wall temperatures.[88, 89] Including a small correction for thermal conductivity, eq 20[24] becomes

$$\log \lambda\Theta\Delta T = \log \phi_6 + m'\log (T_c/298) - n\log (M/10^6)$$
$$\Theta = 1 + (1/k_c)(d\kappa/dT)(\Delta T/2) \qquad (21)$$

where m' is an empirical constant that depends on the polymer and solvent, κ_c is the thermal conductivity of the solvent at the cold wall, and $d\kappa/dT$ is the change in conductivity with temperature. The division of T_c by 298 relates the temperature to approximately the normal cold wall temperature without extrapolation over a large range. Values of m', n, and ϕ for three polymers in THF are given in Table 8.3.

The addition of a polymer to a solvent usually increases the viscosity of the solution dramatically. As the sample is introduced into the channel, the sample zone will experience a change in viscosity that will also affect the velocity of the zone. Large sample masses can result in very severe changes in retention of the zone, as well as changes in peak shape. Generally, retention becomes longer, and the peak may actually become double. The effect increases with increasing molecular weight. The solution to this problem is to use the smallest mass sample that can be detected readily by the detector being used. Studies also have been done which give correction equations related to sample mass for the calibration curve for three polymers in THF.[90] Corrections for sample mass are of the form

$$n = A'm^2 + B'm + C' \qquad (22)$$
$$\phi_6 = \alpha m^2 + \beta m + \varepsilon \qquad (23)$$

where m is the sample mass in micrograms for analysis at a given temperature, and A', B', C', α, β, and ε are empirical constants. Values for these constants for three polymers in THF are given in Table 8.4 for sample masses up to 20 μg.

The relationships discussed hold for constant temperature operation. The use of programming in thermal FFF is often very helpful to obtain rapid analysis with good separation.[90,91] Samples containing materials ranging from a few

Table 8.3 Thermal field-flow fractionation (FFF) universal calibration constants and temperature correction parameters for the PMMA-THF, PS-THF, and PI-THF[a] systems (89)

Polymer	Solvent	$\phi_{6,298}$	m'	n	n_2
PMMA	THF	1.08	1.49	0.625	
PS	THF	1.35	3.04	0.587	
PI	THF[a]	2.32	3.68	0.706	0.299

Key: Poly(methyl methacrylate) (PMMA), Polystyrene (PS), Polyisoprene (PI), Tetrahydrofuran (THF)

a. PI-THF data were better fit by the equation

$$\log (\lambda\theta_c\Delta T) = \log \phi_{6,298} + m\log(T_c/298) - n\log(M/10^6)[1 + n_2\log(M/10^6)]$$

Table 8.4 Sample mass corrections for n and ϕ in the universal calibration equation for poly(methyl methacrylate) (PMMA), polystyrene (PS), and polyisoprene (PI) in tetrahydrofuran (THF)

Polymer	Solvent	A′	B′	C′	α	β	ε
PMMA	THF	−0.00004	0.00313	0.621	0.00013	−0.0115	1.15
PS	THF	−0.00012	0.00495	0.584	0.00079	−0.0276	1.47
PI	THF	−0.00018	0.0086	0.614	0.0053	−0.0344	2.57

Data from reference 90.

thousand to several million in molecular weight may be separated in less than 30 min. Measuring calibration curves for programmed runs is necessary at present because the temperatures of the system are continually changing during the runs. Insufficient data are currently available to obtain corrections to the universal calibration curves for such changing conditions.

Thermal FFF Apparatus

The basic channel for thermal FFF consists of two highly polished flat metal bars clamped on a thin spacer that defines the separation channel, with the surfaces of the metal bars forming the upper and lower walls of the channel. The top bar is heated, and the bottom bar is cooled. A schematic of a typical channel is shown in Figure 8.14. The spacer is usually polyamide or Mylar, from 125 to 250 μm thick. The surfaces of the bars must be very flat to avoid differential flow rates of the carrier across the channel. The surface also must be very smooth, as texture in the surface has been shown to have significant effects on retention and band broadening. The bars are typically made from very pure copper to ensure the highest thermal conductivity and the most uniform temperature distribution along and across the bar. The bars have a thin coating of nickel or chromium to provide chemical resistance to solvents and to protect the surfaces. Electrical resistance heaters capable of delivering up to 6000 watts usually heat the hot bar (upper wall of the channel). Other methods of heating have been used, but, to date, the electrical method has been the easiest and most practical for rapid control, particularly during programming. The cold bar (lower wall of the channel) is cooled by passing coolant through passages milled into the bar. Tap water has frequently been used as the coolant. Chillers have been valuable in areas where water is limited. They can also often provide cooler wall temperatures, as well as give better control of the cold wall temperature during programming.

Temperature of the hot and cold walls is measured by platinum resistance thermometers (RTD) inserted into the copper bars to approximately 1 mm from the channel surface. The computer controls ΔT by varying the electrical heater input. The temperature, time, and response data are collected by the software during the analysis. The flow rate data may also be collected when pumps that can be interfaced with the computer are used; otherwise, the flow rate data also must be manually supplied.

Because the boiling points of several desirable solvents are about 60 °C, the channels are usually operated at a

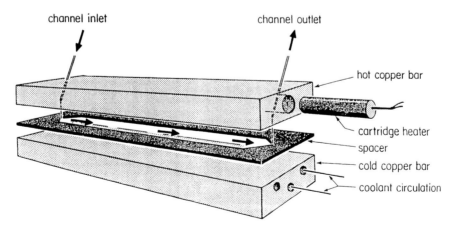

Figure 8.14 Schematic of a thermal FFF channel. (Reproduced with permission from J. C. Giddings, K. C. Caldwell, and L. F. Kesner, in *Molecular Weight Determination*, edited by A. R. Cooper, Wiley-Interscience: New York, 1989. Copyright 1989 John Wiley & Sons.)

pressure of 7 atm or higher to allow channel usage at temperatures much higher than the normal boiling point of the solvent. A rough estimation of the approximate boiling point increase with pressure can be obtained using the Clausius-Claperon equation.

Sample Screening

The first consideration with a new sample is determining an appropriate solvent. In thermal FFF, the solvent used must dissolve the polymer, have boiling points sufficiently high to allow the application of a useful field, and have other properties compatible with the type of detector desired. In addition, the chemical composition of the sample may indicate the solvent that will give the largest value of D_T in order to obtain the maximum resolution during the analysis. All components of carriers used with an evaporative light-scattering detector must be volatile at the operating capabilities of the detector. If the sample supplier can give some indication of the sample's characteristics, time is reduced for determining the operational parameters. A usual approach for an unknown sample is first to make a run at a low ΔT to determine if the sample will elute at a reasonable time. This will give some information on the broadness of the material. Subsequent conditions can then be estimated. When little is known about a sample, more rapid method development sometimes can be obtained by using field programming to scan the sample and obtain some measure of the size ranges in the sample. Low molecular weight materials will require high field strength, while higher molecular weight may need minimal fields, as discussed under programming. For example, molecular weights of 10^6 may be separated with fields as low as 5 °C. Very broad samples often require programming for their characterization in a reasonable time. Better separation is often obtained with narrow samples by using a constant field.

The detector of choice will depend on the sample. The detector with the highest sensitivity is usually preferred, as discussed earlier. However, detectors with special properties may be desirable even if the sensitivity will be somewhat lower. The type of detector will also depend on the chemical properties of the polymer. For example, it is obvious that a UV detector is not appropriate for a polymer without a UV chromophore.

Analysis of the data can usually be done readily by using commercially available software. The software requires the calibration curve constants for the particular solute-solvent pair. With samples for which the calibration constants have not been determined, the use of a MALS in connection with the FFF system can provide the necessary values. An approximate calibration for some materials can also be obtained if the Mark-Houwink values are available by using the expression

$$M = (kT/6\pi\eta D_T \Delta T)^{3/a+1} (10\pi N_A/3K)^{1/a+1} (1/\lambda)^{3/a+1}$$

where k is Boltzman's constant, N_A is Avogadro's number, and K and a are Mark-Houwink values.[60] This expression has the largest errors in the lower molecular weight ranges (where less retention is found). For polystyrene in THF, the error is less than 5% for λ values less than 0.15 with a cold wall temperature of 298 K and a sample mass of 1 μg. The equation also has been shown to give rough estimates of M values for poly(methyl methacrylate) (PMMA) and polyisoprene (PI) in tetrahydrofuran (THF). Limitations are that no correction for temperature and sample mass effects are included, and temperature variation of the Mark-Houwink values may occur. The determination of molecular weight using a viscosity detector and Mark-Houwink values has been studied.[93]

While molecular weight distributions can be obtained from FFF measurements alone, the addition of a MALS detector in connection with a mass detector, such as an RI detector, yields absolute values of molecular weight.[14]

Particle sizes can also be obtained by making a calibration curve in much the same manner as for polymers.[76, 77] The values of D_T will vary with the particle surface composition and the solvent. It is difficult to estimate these values since so few measurements of D_T for particles have been made.

Flow FFF

Flow FFF is the most universal of all the FFF techniques because all materials in the channel will be carried toward one wall by the fluid transport across the channel. Flow FFF has many of the same advantages as thermal FFF provides, except that the only chemical property which influences separation is the polymer-solvent effect on D. Diffusion coefficients can also be obtained from flow FFF measurements. Flow FFF is primarily used in aqueous systems. However, some successful separations have been accomplished in organic solvents,[94,95] even at fairly high temperatures (~420 K).[8] The limitation for work in organic solvents has been the availability of membranes that are compatible with the organic solvents and have suitable cutoff diameters in the pores of the membrane.

The diffusion of the molecules opposing the transport of the particles will result in the zone formation whose height depends only on the diffusion rate. In flow FFF, only the solute-solvent effect on diffusion is important. The Stokes-Einstein equation for diffusion, eq 4, is used for particles and is useful for examining polymers that can be considered to assume a globular form in solution.

Equation 8 gives the relationship between λ and molecular weight. As was done with thermal FFF, rearranging and plotting the log $(\lambda\ V_c\ w^2)/(V^0)$ versus log M

$$\log (\lambda\ V_c\ w^2/V^0) = \log A - b \log M \qquad (24)$$

gives the slope b and the intercept A.

Flow FFF has proven very useful in studying environmental samples for humic and fulvic acids and colloidal particles present in river, lake, and estuary waters, shown in Figure 8.15.[96-99] Biological polymers and particles such as proteins[100-102] (Figures 8.16, 8.17), lipoproteins,[103,104] DNA[105,106] (Figure 8.18), polysaccharides,[100,107] starch,[108,109] cellulosic materials, and virus have also been characterized.[100,109] Considerable work has been done with synthetic water-soluble polymers.[39, 51, 53,54,110-115] Some examples are given in Figures 8.19, 8.20, and 8.21. Proteins in cereals, particularly wheat, are being analyzed.[116-119] Modifications of proteins have been followed.[119] Reactions between proteins and polymers have been verified, as in Figure 8.22 where polyglutamic acid and immunogloblin IgG have formed a conjugate.

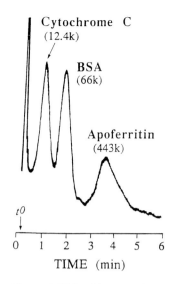

Figure 8.16 Rapid separation of proteins carried out with flow FFF.

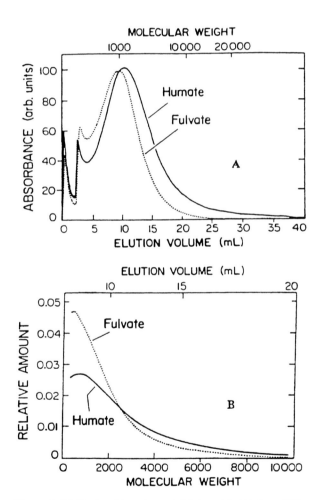

Figure 8.15 Study of humic and fulvic acid: (A) flow FFF fractogram; (B) relative molar mass distributions obtained from (A). (Reproduced from R. Beckett, J. C. Bigelow, Z. Jue, and J. C. Giddings, in *Influence of Aquatic Humic Substances on Fate and Treatment of Pollulants*, edited by P. Mc Carthy and I. H. Suffet, ACS Advances in Chemistry Series 219, American Chemical Society: Washington DC, 1989; chapter 5. Copyright 1989 American Chemical Society.)

The same requirement exists for small sample mass in flow FFF as in thermal FFF. However, the peak retention in flow FFF often differs from thermal FFF in that the peak maximum elutes earlier for too large a sample mass of water-soluble polymers than the peak for the optimal mass sample. This behavior is opposite that obtained with thermal FFF. The difference has been attributed to exclude volume effects. Charged polymers show a considerable reduction in the sample mass effect in solvents with high ionic strength.[84] The sample solutions should also be quite dilute (0.05–0.5 g/L), with injected volumes of 1 to 10 μL.

As indicated, the ionic strength and pH of the solvent can have an appreciable effect on retention of many water-soluble polymers, including proteins, DNA, and other charged polymers.[39,110-112,114,115] These parameters affect the conformational condition of the polymers in solution and the shielding of charge on the molecule, and they can even lead to aggregation of the solute.

The asymmetric flow FFF systems often can give faster analysis with better resolution for some polymers. There is some loss in versatility from that available in the symmetric system. However, either system will give excellent and similar results for most samples. Method development will be necessary in either case to obtain optimal results. Both types are often connected in series with a MALS detector[53-56] to achieve a very powerful system for obtaining information about a polymer, which may be difficult or impossible to obtain by other means, particularly in the time scale available.

Flow Apparatus

A flow FFF system is shown schematically in Figure 8.23. A normal or symmetrical flow FFF channel is shown sche-

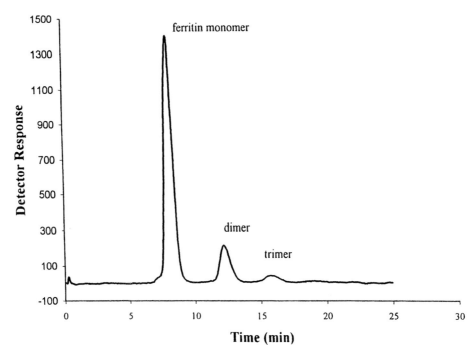

Figure 8.17 Separation of ferritin and its aggregates with asymmetric flow FFF using focusing. Key: cross flow = 4.4 mL/min, channel flow = 1.3 mL/min, PLGC membrane.

matically in Figure 8.24. In the ordinary or symmetric system, the channel is formed by clamping a thin (127–254 μm) spacer and the appropriate membrane between two blocks that are machined to hold porous ceramic frits, which form the upper and lower walls of the channel. The channel form is cut in the spacer before clamping. The frits have a pore size of 1 to 5 μm and are sealed into the blocks

to prevent leaks. The clamping blocks have reservoirs for distribution of the cross-flow liquid evenly across the channel surface. The blocks are usually made from PMMA, but stainless steel and other materials have been used. The frits are typically made from alumina, but polypropylene and other frit materials also have been used successfully.

The asymmetric channel is an important variation in flow FFF channel construction. In this case, the channel

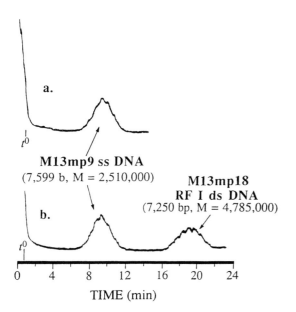

Figure 8.18 Examination of DNA with flow FFF. (Reproduced with permission from reference 112. Copyright 1992 Marcel Dekker, Inc.)

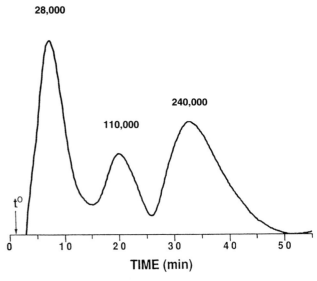

Figure 8.19 Fractogram of a cationic polymer (PVP) obtained with flow FFF. (Reproduced with permission from reference 111. Copyright 1992 American Chemical Society.)

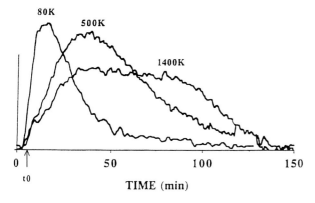

Figure 8.20 Fractograms of neutral polyacrylamides separated by flow FFF. The sample size was 1.4–2.4 µg to avoid overloading. Detection was at a wavelength of 200 nm. (Reproduced with permission from reference 114. Copyright 1997 John Wiley & Sons.)

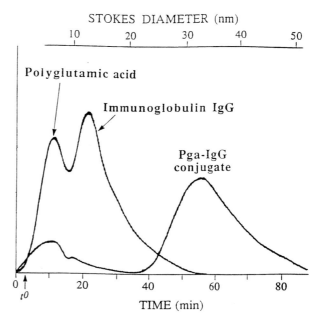

Figure 8.22 Separation of protein, polymer and their conjugate by flow FFF. (Reproduced with permission from reference 112. Copyright 1992 Marcel Dekker, Inc.)

has only one porous wall; the other wall is usually made of a transparent material. Because the channel flow also provides the cross flow, the channel flow rate decreases with distance down the channel. This flow rate decrease is often somewhat compensated by using a tapered, or "trapezoidal", channel. The asymmetric system is otherwise very similar to the symmetric system, as illustrated in Figure 8.25. The same procedures can be used with both systems, including focusing, frit inlet, and frit outlet.[51,120-122]

Hollow fiber flow FFF uses a microporous hollow tube, which is sealed coaxially inside a second tube.[32, 113,123] The outside tube is fitted to allow the cross flow from the inside tube an exit from the system, while the sample is injected into the center tube where the separation takes place. The channel flow rate decreases with distance down the tube, as occurs in asymmetric flow FFF, because the cross flow also comes from the channel flow.

Accurate flow control is critical for reproducible results with all of the flow FFF methods. The use of backpressure

regulators and switching valves can be helpful. Needle valves for flow control are and have been used, but often a considerable time is required for the flow to stabilize after changes have been made. Traps to remove any gas bubbles combined with a flow restrictor to dampen any pressure pulses have proven beneficial in many instances.

The critical component of the system is the membrane used. A variety of membranes have been used for particular analyses. A partial list of membranes and their suppliers is given in Table 8.5. Membranes must be flat with a smooth surface, have a high flux, and be compatible with the carrier solvent. Minimum interaction of the membrane surface with the solute is necessary. Low compressibility is desirable, as channel thickness may be affected when the membrane is clamped. The manufacture's given molecular weight cutoff is only an estimate of the smallest polymer molecule that will be retained by the membrane. Several examples of retention of molecules much smaller than the reported cutoff have been observed. In some instances, removing small components by allowing them to pass through the membrane has been advantageous, such as the removal of small plasma proteins in the analysis of lipoproteins in blood.[103]

Sample Screening for Flow FFF

Flow FFF usually is used with aqueous carriers. Frequently, the addition of some salt, such as $NaNO_3$ or $LiNO_3$ to 1 to 100 mmol concentration, is necessary to enhance the solubility of the polymer or to lessen interactive charge effects.

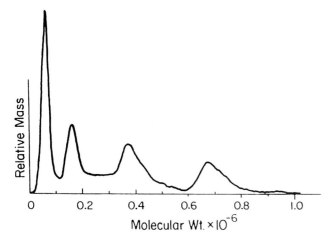

Figure 8.21 Molecular weight distributions of the anionic polymer polystyrene sulfonate determined with flow FFF.

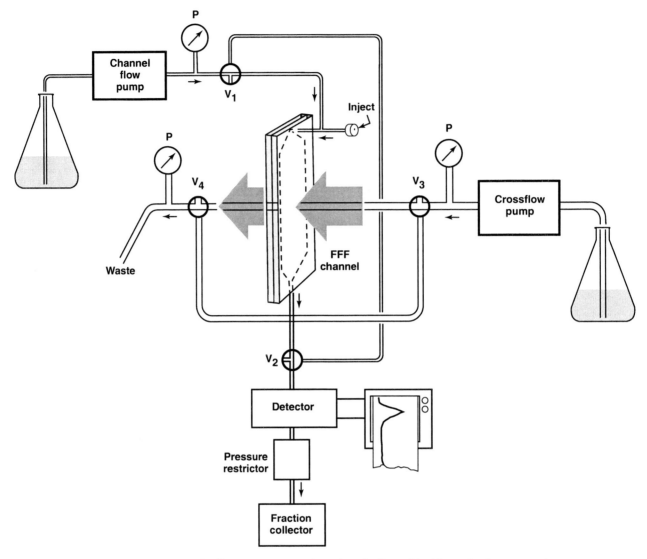

Figure 8.23 Overview of a flow FFF system. The valves (indicated by V) are for flow control and for switching flow paths during operation. Otherwise, the only basic difference between FFF and HPLC is the channel instead of a column.

As mentioned earlier, some samples are difficult to dissolve without obtaining a gel-like solution, which will not separate well in FFF. The smallest sample mass that can be reasonably detected should be used. Initial test runs on the sample are made by using a very low cross flow and channel flow to obtain information on the breadth of the distribution of the sample components and conditions needed for sample separation. Using the low cross flow lessens the chance of compressing the sample to the point that aggregation or gel formation may occur. In these initial tests, stop flow is also avoided, if possible, for the same reason that low cross flow is used. Proper operating conditions can then be estimated. For many samples, high cross flow may not be appropriate and can lead to misleading results. Likewise, large sample mass can give rise to inaccurate results,

and some of the sample may even be lost in the channel or may not undergo separation.

Microgels

FFF is one of the few, if not the only, rapid and easy method of determining microgels in polymer mixes. FFF channels are able to handle rather large particles without plugging, thus allowing the injection of unfiltered samples. The most likely site for plugging is the inlet or outlet lines since, to minimize extra channel volumes, these typically have diameters of about $127 \, \mu m$. In practice, particles as large as 100 mm have been analyzed (with larger tubing diameters required for the larger particle sizes). With this capability,

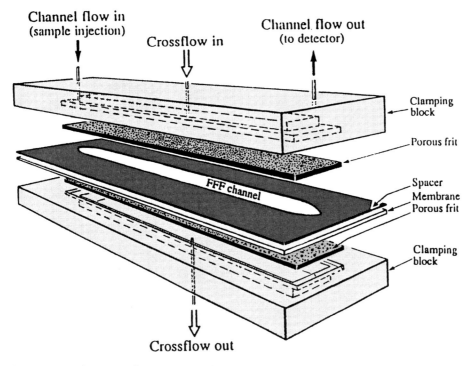

Figure 8.24 Schematic of a symmetric flow FFF channel. (Reproduced with permission from reference 103. Copyright 1993 Pergamon Journals Ltd.)

composite materials and polymers containing microgels can be studied. The microgel content of polymers has been determined by at least two methods, both using thermal FFF. In one, the fractogram of the unfiltered polymer solution is compared with that obtained after filtering the solution. This method has proven useful for PMMA, as well as for synthetic and natural rubbers.[12,13] It has also been used to study the effect of exposure of a polymer to an electron beam.[12] The other method relies on the separation of the polymer-microgel sample into the polymer and the microgel fraction during the course of the analysis, as shown in Fig-

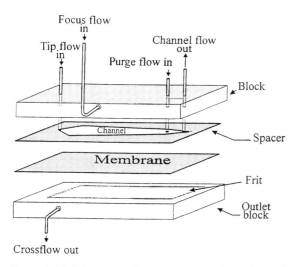

Figure 8.25 Schematic of an asymmetric flow channel. In this configuration, the top wall is solid. The channel in this illustration is tapered to partially compensate for the change in flow rate as the zone passes down the channel.

Table 8.5 Ultrafiltration membranes used in flow field-flow fractionation channels

Membrane	Supplier
Regenerated cellulose-YM	Amicon
PLGC	Millipore
RC70PP	Dow Danmark
Nadir UF-c-10	Hoechst AG
Polysulfone-PM	Amicon
PTGC	Millipore
IRIS 3038	Rhone-Poulenc
Polycarbonate-YC	Amicon
Cellulose acetate-CA 5000	Hoechst
CA 1000	Osmonics
Polyamide	Hoechst
Acrylic copolymer	Amicon
Polyethersulfone Nadir	Hoechst Celanese
Fluoropolymer-DDS	De Danske
Polyaramide-UF-PA-50H	Hoechst
Polyphenylene oxide-MPF-u20-s	Kirgat Weizman
Poly(vinylidene flouride)-AMT	Advanced Membrane Technology
PTFE/TE 30	Schleicher and Schuell
Cellulose nitrate	Schleicher and Schuell
Poly(ethylene terephthalate)-PETP	Cyclopore

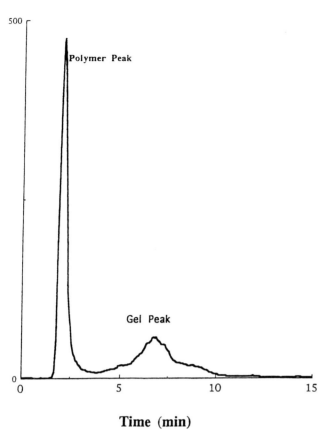

Time (min)

Figure 8.26 FFF can be used to determine microgel content in a polymer, as illustrated for a polyacrylonitrile/butadiene polymer. The solvent was methyl ethyl ketone (MEK). Key: ΔT = 20 °C, flow rate = 0.36 mL/min.

ure 8.26. The area under the microgel portion can be used to determine the quantity of gel present by using a calibration curve that was determined from samples containing known amounts of gel; these samples were obtained by the more traditional methods such as solvent extraction. Additional information about the microgel size can be obtained since both polymer and particles can be analyzed. Often the polymer and microgel tend to blend from very large polymers into the smaller microgels.[124] One must be cautious that high flow rates do not cause hyperlayer behavior of the larger particles, thus resulting in elution at the same volume as smaller size material. The coupling of a light-scattering detector, such as a MALS detector, with FFF can be very useful for analysis of microgels since the FFF system delivers monodisperse material to the MALS, which can then give absolute size for the molecule or particle.[124]

Conventionally, the analysis of a composite material that contains both particles and polymer components usually requires the use of several analysis methods. However, due to the wide analysis range of FFF, thermal FFF has given information on both the polymer and particle-size distributions, as illustrated in Figure 8.27 for an acrylonitrile-butadiene-styrene (ABS) composite.[16]

While most of the work on microgels has been done with thermal FFF, the use of flow FFF should be as effective for the study of microgels in hydrophilic polymers.

Copolymers

Copolymers have very interesting behavior in thermal FFF. Block copolymers are retained, according to which type of polymer is the most soluble in the carrier solvent being used. This implies that the more soluble polymer is on the outer surface of the molecule, and the other polymer is contained inside. Retention of random copolymers is between that obtained with the individual polymers, according to the proportion of each type of polymer present in the molecule.[10,17,18,19] Coupling thermal FFF with other techniques, such as SEC or hydrodynamic chromatography, mass detectors, and viscosity or MALS detectors, has been used to obtain both compositional and molecular weight information.[62,125–127]

Branched Polymers

As shown by the Stokes-Einstein equation for D (eq 4), the effective radius of the polymer or particle, r, will determine the separation. Linear polymer molecules will have a larger effective diameter than that of a branched molecule of the same molecular weight, since the latter is a more compact

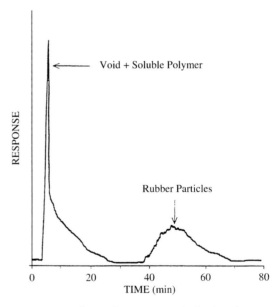

Figure 8.27 Since thermal FFF can retain both polymers and particles, it becomes possible to accomplish the separation of a composite of polybutadiene particles from polystyrene/acrylonitrile in an ABS rubber as shown here. (Reproduced with permission from reference 16. Copyright 1996 John Wiley & Sons.)

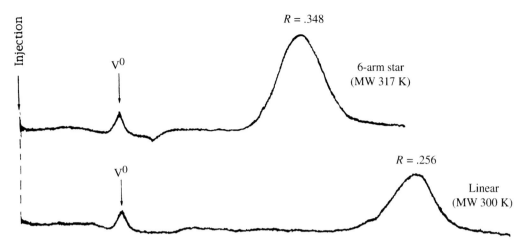

Figure 8.28 The effect of branching in FFF is shown with the retention of the six-armed star being less than that of the linear polystyrene polymer.

molecule. Schmipf et al.[128] carried out an extensive study on the effect of branching on polystyrene retention, showing that the most highly branched molecules were less retained than corresponding linear molecules, and that this trend was the inverse of the diffusion rates as determined by other methods, as expected. (See Figure 8.28.) A more recent study by Lou et al[129] with the polysaccharides pullulans, dextrans, and Ficolls, observed the same trend, with the linear pullulans having the greatest retention, followed by the partially branched dextrans; the highly branched Ficolls gave the lowest retention for comparable molecular weight materials, as shown in Figure 8.29.

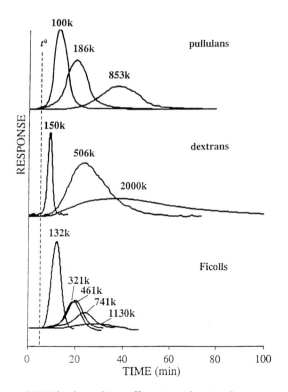

Figure 8.29 The branching effect is evident in these polysaccharides separated in DMSO by thermal FFF. The pullulans are linear, the dextrans are partially branched, and the Ficols are highly branched. (Reproduced with permission from reference 129. Copyright 1994 Marcel Dekker, Inc.)

References

1. Giddings, J. C. *Chem. Eng. News* **1988**, *66*, 34–45.
2. Myers, M. N. *J. Micro. Sep.* **1997**, *9*, 151–161.
3. Kirkland, J. J.; Yau, W. W.; Doener, W. A.; Grant, J. W. *Anal. Chem.* **1980**, *52*, 1944–1954.
4. Caldwell, K. D.; Kesner, L. F.; Myers, M. N.; Giddings, J. C. *Science* **1972**, *176*, 296–298.
5. Caldwell, K. D.; Gao, Y.-S. *Anal. Chem.* **1993**, *65*, 1764–1772
6. Schimpf, M. E. *Trends Polym. Sci.* **1993**, *1*, 74–78.
7. Kirkland, J. J.; Dilks, C. H. *Anal. Chem.* **1992**, *64*, 2836–2840.
8. Miller, M. E., Giddings, J. C. *J. Microcol. Sep.* **1998**, *10*(1), 75–78.
9. Schimpf, M. E.; Giddings, J. C. *J. Polym. Sci.: Part B: Polym. Phys. Ed.* **1989**, *27*, 1317–1332.
10. Schimpf, M. E.; Giddings, J. C. *J. Polym. Sci.: Part B: Polym. Phys.* **1990**, *28*, 2673–2680.
11. Jeon, S. J.; Lee, D. W. *J. Polym. Sci.: Part B: Polym. Phys.* **1995**, *33*, 411–416.
12. Lee, S. In *Chromatography of Polymers: Characterization by SEC and FFF*; ACS Symposium Series 521; Provder, T., Ed.; American Chemical Society: Washington, DC, 1993; pp 77–88.
13. Lee, S., Molnar, A. *Macromolecules* **1995**, *28*, 6354–6356.
14. Lee, S.; Kwon, O.-S. *Polym. Mater. Sci. Eng.* **1993**, *69*, 403–409.
15. Lee, S. *Polym. Mat. Sci. Eng.* **1991**, *65*, 19–20.

16. Shiundu, P. M.; Remsen, E. E.; Giddings, J. C. *J. Appl. Polym. Sci.* **1996**, *60*, 1695–1707.

17. Van Batten, C.; Hoyos, M.; Martin, M. *Chromatographia* **1997**, *45*, 121–126.

18. Schimpf, M. E.; Wheeler, L. M.; Romeo, P. F. In *Chromatography of Polymers: Characterization by SEC and FFF*; Provder, T., Ed.; ACS Symposium Series 521; American Chemical Society: Washington, DC, 1993; pp 63–76.

19. Cho, K.-H.; Park, Y. H.; Jeon, S.; Kim, W.-S.; Lee, D. W. *J. Liq. Chrom. & Rel. Technol.* **1997**, *20*, 2741–2756.

20. *FFF Handbook*; Schimpf, M. E., Caldwell, K. D., Giddings, J. C., Eds. John Wiley and Sons: New York, 2000.

21. *J. Micro.Sep.* **1997**, *9*(3), 117–248.

22. *J. Micro.Sep.* **1998**, *10*(1), 1–166.

23. *J. Liq. Chromatogr & Rel. Tech.* **1997**, *20*, 2509–2940.

24. Giddings, J. C. *Science* **1993**, *260*, 1456–1465.

25. Giddings, J. C. *Anal. Chem.* **1981**, *53*, 1170A–1175A.

26. Giddings, J. C. *J. Chromatogr.* **1989**, *470*, 327–335.

27. Williams, K. S.; Benincasa, M. A. In *Encyclopedia of Analytical Chemistry: Instrumentation and Applications*; Meyers, R. A., Ed.; John Wiley and Sons: New York, 2000; pp 7582–7608.

28. Schimpf, M. E. *Trends Polym. Sci.* **1993**, *1*, 7478.

29. Giddings, J. C.; Yoon, Y. H.; Caldwell, K. D.; Myers, M. N.; Hovingh, M. E. *Sep. Sci.* **1975**, *10*, 447.

30. Litzen, A., *Anal. Chem.* **1993**, *65*, 461–470.

31. Jonsson, J. A.; Carlshaf, A. *Anal. Chem.* **1989**, *61*, 11–18.

32. Lee, W. J.; Min, B-R.; Moon, M-H. *Anal. Chem.* **1999**, *71*, 3446–3452.

33. Yau, W. W.; Kirkland, J. *J. Sep. Sci. Technol.* **1981**, *16*, 577–605.

34. Kirkland, J. J.; Yau, W. W. *Macromolecules* **1985**, *18*, 2305–2311.

35. Kirkland, J. J.; Rementer, S. W.; Yau, W. W. *Anal. Chem.* **1988**, *60*, 610–616.

36. Williams, P. S.; Giddings, J. C. *Anal. Chem.* **1987**, *59*, 2038–2044.

37. Botana, A. M.; Ratatanathanawongs, S. K.; Giddings, J. C. *J. Microcol. Sep.* **1995**, *7*, 395–402.

38. Benincasa, M.-A.; Giddings, J. C. *J. Microl. Sep.* **1997**, *9*, 479–495.

39. Wittgren, B.; Wahlund, K.-G., Dérand, H.; Wesslén, B. *Langmuir* **1996**, *12*, 5999–6005.

40. Wijnhoven, J. E. G. J.; Koorn, J.-P.; Poppe, H.; Kok, W. Th. *J. Chromatogr. A* **1996**, *732*, 307–315.).

41. Giddings, J. C. *Anal. Chem.* **1990**, *62*, 2306–2312.

42. Liu, M.-K.; Williams, P. S.; Myers, M. N.; Giddings, J. C. *Anal. Chem.* **1991**, *63*, 2115–2122.

43. Liu, M.-K.; Li, P.; Giddings, J. C. *Protein Sci.* **1993**, *2*, 1520–1531.

44. Moon. M. H.; Kwon, H.; Park, H., *Anal. Chem.* **1997**, *69*, 1436–1440.

45. Li, P.; Hansen, M.; Giddings, J. C., *J. Micro. Sep.* **1998**, *10*, 7–18.

46. Giddings, J. C. *Anal Chem.* **1990**, *62*, 2306–2312.

47. Giddings, J. C. *Sep. Sci. Technol.* **1989**, *24*, 755–768.

48. Moon, M. H.; Myers, M. N.; Giddings, J. C. *J. Chromatogr.* **1990**, *517*, 423–433.

49. Wahlund, K.-G.; Giddings, J. C. *Anal. Chem.* **1987**, *59*, 1332–1339.

50. Lee, H.; Williams, S. K. R.; Giddings, J. C., *Anal Chem.* **1998**, *70*, 2495–2503.

51. Wahlund, K.-G.; Winegarner, H. S.; Caldwell, K. D.; Giddings, J. C. *Anal. Chem.* **1986**, *58*, 573–578.

52. Myers, M. N.; Chen, P.; Giddings, J. C. In *Chromatography of Polymers: Characterization by SEC and FFF*; Provder, T., Ed.; ACS Symposium Series 521; American Chemical Society: Washington, DC, 1993; pp 47–62.

53. Roessner, D.; Kulicke, W.-M. *J. Chromatogr. A* **1994**, *687*, 249–258.

54. Thielking, H.; Kulicke, W.-M. *Anal. Chem.* **1996**, *68*, 1169–1173.

55. Wittgren, B.; Wahlund, K.-G. *J. Chromatogr. A* **1997**, *791*, 135–149.

56. Jiang, Y.; Miller, M. E.; Li, P.; Hansen, M. E. *Am. Lab.* **2000**, February, 98–108.

57. *FFF Handbook*; Schimpf, M. E., Caldwell, K. D., Giddings, J. C., Ed.; John Wiley and Sons: New York, 2000, p 253.

58. Yau, W. W.; Kirkland, J. J.; *Polym. Mater. Sci. Eng.* **1988**, *59*, 4–6.

59. Kirkland, J. J.; Rementer, S. W.; Yau, W. W. *J. Appl. Polym. Sci.* **1989**, *38*, 1383–1395.

60. Kirkland, J. J.; Rementer, S. W. *Anal. Chem.* **1992**, *64*, 904–913.

61. Schimpf, M. E. *Polym. Mater. Sci. Eng.* **1993**, *69*, 406–407.

62. Venema, E.; de Leeuw, P.; Kraak, J. C.; Poppe, H.; Tijssen, R. *J. Chromatgr. A* **1997**, *765*, 135–144.

63. Hassellöv, M.; Hulthe, G.; Lyvén, B.; Stenhagen, G. *J. Liq. Chromatogr. & Rel. Technol.* **1997**, *20*, 2843–2856.

64. McIntyre, D.,; Shis, A. L.; Saroca, J.; Seeger, R.; MacArthur, A. In *Size Exclusion Chromatography*; Provder, T., Ed.; ACS Symposium Series 245; American Chemical Society: Washington, DC, 1984; pp 227–250.

65. Gao, Y. S.; Caldwell, K. D.; Myers, M. N.; Giddings, J. C. *Sep. Sci. Technol.* **1984**, *19*, 667–683.

66. Giddings, J. C.; Li, S.; Williams, P. S.; Schimpf, M. E. *Makromol. Chem., Rapid Commun.* **1988**, *9*, 817–823.

67. Giddings, J. C.; Yoon, Y. H.; Myers, M. N. *Anal. Chem.* **1975**, *47*, 126–131.

68. Gunderson, J. J.; Giddings, J. C. *Anal. Chim. Acta* **1986**, *189*, 1–15.

69. Giddings, J. C.; Smith, L. K.; Myers, M. N. *Anal. Chem.* **1975**, *47*, 2389–2394.

70. Gunderson, J. J.; Giddings, J. C. *Macromolecules* **1986**, *19*, 2618–2621.

71. Nguyen, M. T.; Beckett, R.; Pille, L.; Solomon, D. H. *Macromolecules* **1998**, *31*, 7003–7009.

72. Shiundu, P. M.; Liu, G.; Giddings, J. C. *Anal. Chem.* **1995**, *67*, 2705–2713.

73. Kirkland, J. J.; Yau, W. W. *J. Chromatogr.* **1986**, *353*, 95–107.

74. Van Asten, A. C.; Kok, W. Th.; Tijssen, R.; Poppe, H. *J. Polym.Sci.: Part B* **1996**, *34*, 297–308.

75. Sisson, R.; Giddings, J. C. *Anal Chem.* **1994**, *66*, 4043–4053.

76. Shiundu, P.M.; Giddings, J. C. *J. Chromatogr.* **1995**, *715*, 117–126.

77. Jeon, S. J.; Schimpf, M. E.; Nyborg, A. *Anal. Chem.* **1997,** *69,* 3442–3450.
78. Brimhall, S. L.; Myers, M. N.; Caldwell, K. D.; Giddings, J. C. *J. Polym. Sci.: Polym. Phys. Ed.* **1985,** *23,* 2443–2456.
79. van Asten, A. C.; Boelens, H. F. M.; Kok, W. Th.; Poppe, H.; Williams, P. S.; Giddings, J. C. *Sep. Sci. Technol.* **1994,** *29,* 513–533.
80. Rue, C. A.; Schimpf, M. E. *Anal. Chem.* **1994,** *66,* 4054–4062.
81. Giddings, J. C. *Anal Chem.* **1994,** *66,* 2783–2787.
82. Nguyen, M. Y.; Beckett, R. *Polym. Int.* **1993,** *30,* 337.
83. Nguyen, M. Y.; Beckett, R. *Sep. Science and Technol.* **1996,** *31,* 453–470.
84. Reschiglian, P.; Martin, M.; Contado, C.; Dondi, F. *J. Liq. Chrom. & Rel. Technol.* **1997,** *20,* 2723–2739.
85. Gunderson, J. J.; Caldwell, K. D.; Giddings, J. C. *Sep. Sci. Technol.* **1984,** *19,* 667–683.
86. Belgaied, J. E.; Hoyos, M.; Martin, M. *J. Chromatogr. A* **1994,** *678,* 85–96.
87. Martin, M.; Van Batten, C.; Hoyos, M. *Anal. Chem.* **1997,** *69,* 1339–1346.
88. Myers, M. N.; Cao, W.-J.; Chen, C.; Kumar, V.; Giddings, J. C. *J. Liq. Chrom. & Rel. Technol.* **1997,** *20* (16&17), 2757–2775.
89. Cao, W.J.; Williams, P. S. Myers, M. N.; Giddings, J. C. *Anal. Chem.* **1999,** *71,* 1597–1609.
90. Cao, W.-J.; Myers, M. N.; Williams, P. S.; Giddings, J. C. *J. Polym. Anal. Charact.* **1998,** *4,* 407–433.
91. Giddings, J. C.; Smith, L. K.; Myers, M. N. *Anal. Chem.* **1976,** *48,* 1587–1592.
92. Giddings, J. C.; Kumar, V.; Williams, P. S.; Myers, M. N. In *Polymer Characterization: Physical Property, Spectroscopic, and Chromatographic Methods,* Craver, C. D., Provder, T., Eds.; ACS Advances in Chemistry Series 227; American Chemical Society: Washington, DC, 1990; chapter 1.
93. Kirkland, J. J.; Rementer, S. W.; Yau, W. W. *J. Appl. Polym. Sci.* **1989,** *38,* 1383–1395.
94. Kirkland, J. J.; Dilks, C. H. *Anal Chem.* **1992,** *64,* 2836–2840.
95. Wijnhoven, J. E. G. J.; van Bommel, M. R.; Poppe, H.; Kok, W. Th. *Chromatographia* **1996,** *42* , 409–515.
96. Beckett, R.; Jue, Z.; Giddings, J. C. *Environ. Sci. Technol.* **1987,** *21,* 289–295.
97. Dycus, P. J. M.; Healy, K. D.; Stearman, G. K.; Wells, M. J. M. *Sep Sci. Technol.* **1995,** *30,* 1435–1453.
98. Schimpf, M. E.; Pettys, M. P. *Colloids Surf. A: Physicochem. Eng. Aspects* **1997,** *120,* 87–100.
99. Schimpf, M. E.; Wahlund, K.-G. *J. Microcolumn Separ.* **1997,** *9*(7), 535–543.
100. Wahlund, K.-G.; Litzén, A. *J. Chromatogr.* **1989,** *461,* 73–78.
101. Litzén, A.; Wahlund, K.-G. *J. Chromatogr.* **1989,** *476,* 413–421.
102. Kassalainen, G.; Williams, S. K. R. In *Encyclopedia of Chromatography;* Cazes, J., Ed.; Marcel Dekker: New York, 2001.
103. Li, P.; Giddings, J. C. *J. Pharm. Sci.* **1996,** *85,* 895–898
104. Li, P.; Hansen, M.; Giddings, J. C. *J. Liq. Chrom. & Rel. Technol.* **1997,** *20,* 2777–2802.
105. Liu, M.-K.; Giddings, J. C. *Macromolecules* **1993,** *26,* 3576–3588.
106. Lee, H.-K.; Williams, S. K. R.; Allison, S. D.; Anchorloquy, T. J. *Anal. Chem.* **2001,** *73,* 837–843.
107. Wittgren, B.; Wahlund, K.-G. *J. Chromatog.* **1997,** *760,* 205–218.
108. Hanselmann, R.; Burchard, W.; Ehrat, M.; Widmer, H. M. *Macromolecules* **1996,** *26,* 3277–3282.
109. Kulicke, W.-M.; Heins, D.; Schittenhelm, N. *Polym. Mater. Sci. Eng.* **1998,** *79,* 427–428.
110. Giddings, J. C.; Lin, G. C.; Myers, M. N. *J. Liq. Chromatogr.* **1978,** *1,* 1–20.
111. Benincasa, M.-A.; Giddings, J. C. *Anal. Chem.* **1992,** *64,* 790–798.
112. Giddings, J. C.; Benincasa, M.-A.; Liu, M.-K.; Li, P. *J. Liq. Chromatogr.* **1992,** *15,*1729–1747.
113. Wijnhoven, J. E. G. J.; Koorn, J. P.; Poppe, H.; Kok, W. Th. *J. Chromatogr.* **1995,** *699,* 119–129.
114. Benincasa, M.-A.; Giddings, J. C. *J. Micro. Sep.* **1997,** *9,* 479–495.
115. Wittgren, B.; Wahlund, K.-G.; Dérand, H.; Wesslén, B. *Macromolecules* **1996,** *29,* 268–276.
116. Stevenson, S. G.; Preston, K. R. *J. Cereal Sci.* **1996,** *23,* 121–131.
117. Stevenson, S. G.; Preston, K. R. *J. Liq. Chrom. & Rel. Technol.* **1997,** *20,* 2835–2842.
118. Stevenson, S. G.; Uno, T.; Preston, K. R. *Anal. Chem.* **1999,** *71,* 8–14.
119. Cauchon, G. *7th Symposium on FFF,* **1998,** Feb. 8–11, Salt Lake City.
120. Litzén, A.; Wahlund, K.-G. *Anal. Chem.* **1991,** *63,* 1001–1007.
121. Wahlund, K.-G.; Giddings, J. C. *Anal. Chem.* **1987,** *59,* 1332–1339.
122. Litzén, A. *Anal. Chem.* **1993,** *65,* 461–470.
123. Jönsson, J. A.; Carlshaf, A. *Anal. Chem.* **1989,** *61,* 11–18.
124. Lewandowski, L.; Sibbald, M.; Johnson, E.; Mallamaci, M. Paper 11, Spring Technical Meeting, Rubber Division of American Chemical Society, Chicago, April 13–16, 1999.
125. Cho,K.-H.; Young, H.; Jeon, J. F.; Kim, W.-S.; Lee, D. W. *J. Liq. Chromatogr. Relat. Technol.* **1997,** *20,* 2741–2756.
126. Jeon, S. J.; Schimpf, M. E. In *Chromatography of Polymers: Hyphenated and Multidimensional Techniques;* Provder, T., Ed.; ACS Symposium Series 731; American Chemical Society: Washington, DC, 1999; pp 141–151.
127. Ratanathanawongs, S. K.; Shiundu, P. M.; Giddings, J. C. *Coll. Surf. A* **1995,** *105,* 243–250.
128. Schimpf, M. E.; Giddings, J. C. *Macromolecules* **1987,** *20,* 1561–1563.
129. Lou, J.; Myers, M. N.; Giddings, J. C. *J. Liq. Chromatogr.* **1994,** *17,* 3239–3260.

PART III

METHODS TO DETERMINE THE MOLECULAR STRUCTURE OF POLYMERS

9

Nuclear Magnetic Resonance Spectroscopy

PETER A. MIRAU

Nuclear magnetic resonance (NMR) spectroscopy has emerged as one of the premier methods for the characterization of polymers, both in solutions and in the solid state. The popularity of NMR is due to the fact that many molecular level features can be measured from the NMR spectra, including polymer microstructure, chain conformation, and dynamics. The NMR spectra of solid polymers are sensitive to these same features and also to the length scale of mixing in blends and phase-separated materials, as well as the domain sizes in semicrystalline polymers. Since the spectral features and relaxation times are affected by local interactions, they provide information about the structure of polymers on a length scale (2–200 Å) that is difficult to measure by other methods.

The first NMR studies of polymers were reported[1] only about a year after nuclear magnetic resonance was discovered in bulk matter.[2,3] It was reported that the proton linewidth for natural rubber at room temperature is more like that of a mobile liquid than of a solid, but that the resonance broadens near the glass-transition temperature (T_g). This was recognized as being related to a change in chain dynamics above and below T_g. NMR methods developed rapidly after these initial observations, first for polymers in solution and then, more recently, for polymers in the solid state.

Solution NMR is frequently used for polymer characterization because of its high resolution and sensitivity. The chemical shifts are sensitive to polymer microstructure, including polymer stereochemistry, regioisomerism, and the presence of branches and defects. These observations led to an improved understanding of both polymer microstructure and polymerization mechanisms. With the advent of higher magnetic fields and improved NMR methods and spectrometers, it has become possible to characterize even very low levels of defects in high molecular weight polymers. The peak assignments in the early studies were established by the comparison with model compounds. The developments of spectral editing and multidimensional NMR have made

it possible to assign the spectra without recourse to model compounds. The assignments can be established not only for carbons and protons but also for any silicon, nitrogen, phosphorus, or fluorine atoms that may be present. This detailed microstructural characterization has led to a deeper understanding of polymer structure-property relationships.

The solid-state NMR analysis of polymers emerged after the solution studies because solid-state NMR requires more specialized equipment and the spectra are more difficult to acquire. Solid-state NMR is now such an important method that most modern spectrometers are capable of performing these studies. The interest in the NMR of solid polymers is due in part to the fact that most polymers are used in the solid state, and in some cases the NMR properties can be directly related to the macroscopic properties. Solid-state NMR provides information about the structure and dynamics of polymers over a range of length scales and time scales. Polymers have a restricted mobility in solids, and the chemical shifts can be directly related to the chain conformation.

Solid-state NMR is also an efficient way to monitor the reactivity of polymers, since chemical changes often give rise to large spectral changes. The relaxation times in solids depend not only on the chain dynamics but also on the morphology over a length scale of 20 to 200 Å. NMR has been extensively used to measure the length scale of mixing in blends and multiphase polymers, as well as the domain sizes in semicrystalline polymers. Solid-state NMR methods have been greatly expanded with the introduction of multidimensional NMR.[4] These studies have led to a molecular level understanding of the dynamics traditionally observed by dielectric and dynamic-mechanical spectroscopy, along with a better understanding of the relationship between polymer morphology and macroscopic properties.

The most fundamental aspect of nuclear magnetic resonance is that it is possible to observe transitions between spin states of NMR-active nuclei in a magnetic field; moreover, the frequency, intensity, and width of the signals provide information about polymer structure and dynamics. The equipment required to observe NMR signals is quite complex but is routinely available in both industrial and academic laboratories. Of primary importance is a high field magnet. As will be shown later in this chapter, the frequency separation between peaks depends on the magnetic field strength, so the highest resolution is often observed with the strongest magnet. On modern spectrometers this is a superconducting magnet, which may have field strengths as large as 18 T (1 tesla = 10,000 gauss) and a resonance frequency for protons of 800 MHz. The material used to make the magnet coil is only superconducting at low temperatures, so it is housed in a dewar of liquid helium that is surrounded by a dewar filled with liquid nitrogen. The resolution also depends on the magnet field homogeneity, which is adjusted for maximum homogeneity using secondary coils called *shim coils.*

NMR-active nuclei possess angular momentum, and their behavior in a magnetic field is similar in many ways to a gyroscope. When one of these nuclei is placed in a magnetic field, it feels a torque caused by the interaction between its magnetic moment and the external magnetic field, and it wobbles or *precesses* in the same way as a gyroscope in a gravitational field. The precession frequency depends on how strong the torque is and thus depends on the nuclear properties and the field strength. As for a bar magnet, the two lowest energy states are those in which magnetic moment is aligned with or against the magnetic field. The energy difference between these states is small in nuclei, and there is a nearly equal number of nuclei parallel and antiparallel to the magnetic field. The detected NMR signal is the magnetization, which is associated with the net difference between the two states.

We can affect the net magnetization by applying a second magnetic field perpendicular to the main magnetic field, causing a change in the number of spins aligned with and against the magnetic field. Typically, this second field is applied by irradiating the sample for a short period of time with a radio frequency (rf) field at the frequency of the nuclei of interest. Disturbing the equilibrium in this way gives the sample a net magnetization that can be detected as a voltage induced in a coil placed around the sample.

While the actual NMR experiments are detected at high frequencies (hundreds of MHz), it is often easier to understand NMR experiments using the concept of a rotating reference frame. The laboratory frame is defined by the magnetic field, with the z direction oriented along the field and the x and y axes perpendicular. In that reference frame, the nuclei are precessing at high frequency. If we consider the same view from the perspective of the nuclei, it appears stationary. The concept is similar to a person standing on the earth. This person appears stationary, but to a distant observer he would appear to be moving very quickly as the

earth rotates. In the rotating reference frame, the net magnetization is aligned along the stationary z axis, and rf pulses can tip the magnetization into the xy plane.

In the rotating reference frame, the effect of pulses is to tilt the magnetization into the xy plane. Exactly how this occurs depends on the strength of the applied magnetic field, the length of the pulse, and the phase (or direction) of the applied magnetic field. One particularly useful pulse is the 90° pulse, which rotates magnetization from the z axis into the xy plane. If the 90° pulse is applied along the x axis (90_x^0 pulse), the magnetization is rotated into the yz plane; if the pulse is applied along y direction (90_y^0 pulse), the magnetization is rotated into the xz plane. The 90° pulse is often used because the signal is detected in the xy plane, so the maximum signal is observed following this pulse. The 180° pulse is also extremely useful. A 180° pulse applied along the x axis (180_x^0 pulse) to a spin system at equilibrium rotates the magnetization from the +z to the −z position. In a similar way, the same 180_x^0 pulse rotates magnetization along the y axis from the +y to the −y position.

The signal is detected as a voltage in a coil, which is caused by the precession of the magnetization in the magnetic field. The signal is detected after the pulse and is freely precessing and for this reason is known as a *free induction decay* (FID). For a pure sample with a single type of spin, such as the protons in water, a damped cosine signal would be observed. The oscillation frequency is determined by the difference in frequency between the rf pulse and the signal of interest, and the damping behavior is related to the magnetic field homogeneity and the relaxation rates.

If only a single peak were to be studied, then it would be sufficient to measure the frequency from the oscillations in the FID. However, most samples of interest contain many different frequencies, and it would be difficult to extract all of the frequency information from the FID. For this reason, the FID is converted into a frequency spectrum using the Fourier transformation. The Fourier transformation is given by

$$F(\omega) = \int_{-\infty}^{\infty} f(t)e^{-i\psi t} dt \qquad (1)$$

where $F(\omega)$ is the frequency spectrum and $f(t)$ is the time domain signal. There are many advantages to using pulsed NMR in combination with Fourier transformation. One important advantage is that all the signals can be observed in a single spectrum. Another is that it is possible to acquire many scans and add them together to increase the signal-to-noise ratio.

The basic pulsed NMR method is summarized in the diagram shown in Figure 9.1. Figure 9.1a shows the application of a 90° pulse and the resulting FID. Fourier transformation gives the frequency spectrum shown in Figure 9.1b. Among the important parameters that can be determined are the frequency, linewidth, and integrated intensity. There is no natural reference for comparing the frequencies from the samples of interest, so the frequencies are

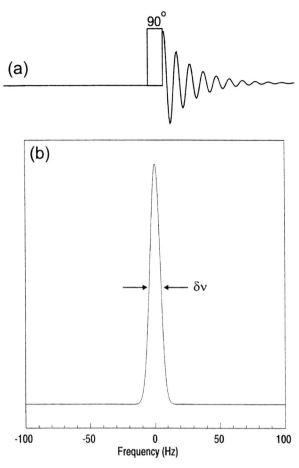

Figure 9.1 A schematic diagram illustrating pulsed NMR experiments. In (a) a short burst of radio frequency radiation at the Larmour frequency rotates the magentization from its equilibrium position into the *xy* plane, where it is recorded as a free induction decay. Fourier transformation of the free induction decay (b) gives a plot of frequency versus intensity. The important parameters are the frequency, intensity, and line width.

The simple one-pulse (or Bloch-decay) experiment shown in Figure 9.1 provides critical information for materials characterization since the chemical shifts, linewidths, and intensities are related to the polymer structure and dynamics. Additional information can be obtained using more complex sequences of pulses and delays. In relaxation experiments, for example, we can apply pulses to perturb the spins, and some time later apply another pulse to monitor the return of the spin system to equilibrium. Multidimensional (nD) NMR experiments have been used more recently to provide more detailed information about the structure, dynamics, and organization of polymers. The nD NMR experiments consist of pulses and delays like other experiments, but they contain more than one period during which the spins can freely evolve, as in the FID. The resulting data can be Fourier transformed with respect to each free evolution period. Thus an experiment with two evolution periods gives a two-dimensional (2D) spectrum. Such a spectrum contains information about the interactions between spins that is difficult to extract from the Bloch-decay spectrum.

It can be noted from the discussion here that an NMR spectrometer performs several functions in the analysis of NMR samples and requires many interconnected parts. In addition to the magnet, the spectrometer requires a probe to hold the samples in the magnetic field. The probe contains one or more rf coils that are used to apply the pulses to the samples. For some experiments, such as solid-state NMR, the probe must also be able to rapidly spin the samples. The spectrometer also must contain an rf transmitter and amplifier and a receiver to record the signals. This is typically controlled by a computer, which also stores the data and has the software needed to transform the FIDs and to plot and analyze the spectra.

Fundamental Principles

The NMR phenomenon is possible because NMR-active isotopes possess spin, or angular momentum. Since a spinning charge generates a magnetic field, there is a magnetic moment associated with this angular momentum. According to a basic principle of quantum mechanics, the maximum experimentally observable component of the angular momentum of a nucleus that possesses a spin (or of any particle or system that has angular momentum) is a half-integral or integral multiple of $h/2\pi$, where h is Planck's constant. This maximum component is I, which is called the spin quantum number or simply "the spin". In general, there are $2I + 1$ possible orientations or states of the nucleus. For spin ½ nuclei, the possible magnetic quantum numbers are $+\frac{1}{2}$ and $-\frac{1}{2}$.

When they are placed in a magnetic field, nuclei with spin undergo precession about the field direction. The frequency of this so-called Larmor precession is designated as

reported relative to some reference compound. For carbons and protons, the peak positions are reported relative to the carbon and proton frequencies of tetramethylsilane (TMS) in parts per million (ppm), which is given by

$$\delta = \frac{v - v_{ref}}{v_0} \times 10^6 \qquad (2)$$

where v and v_{ref} are the measured frequencies of the unknown peak and the reference compound and v_0 is the resonance frequency (in Hz) for the nuclei of interest (500 MHz for protons or 125 MHz for carbons in a 11.7 T magnetic field). The frequencies of the carbon and proton peaks of TMS are taken as zero ppm. The advantage of the ppm reference scale is that the chemical shifts are independent of magnetic field. The linewidth depends on the rate of decay of the FID and is related to the field homogeneity and the molecular dynamics. The integrated signal intensity depends on the number of nuclei in the sample.

ω_0 in radians per second or v_0 in Hertz (Hz), cycles per second ($\omega_0 = 2\pi v_0$). The nuclei can be made to flip over, or reverse the direction of the spin axis, by applying a second magnetic field oscillating at a frequency of ω_0 at right angles to the magnetic field. The Larmor precession frequency or resonance frequency is given by

$$\omega_0 = \gamma B_0 \qquad (3)$$

where γ is the magnetogyric ratio, and B_0 is the magnetic field strength. The two quantities that determine the observation frequency for NMR signals are the magnetogyric ratio and the magnetic field strength. Table 9.1 lists some of the nuclear properties of spins that are of interest to polymer chemists. The sensitivity depends on both the magnetogyric ratio and the natural abundance of the NMR-active nuclei. Protons have the highest sensitivity (that is, give the strongest signal, and thus are most easily detected) because they have the highest magnetogyric ratio and natural abundance. At a field strength of 11.7 T, the proton NMR signals are observed at 500 MHz. Fluorine is the second most sensitive nuclei, but it is not a common element in polymers. Most polymers contain carbon, and Table 9.1 shows that the sensitivity is very low compared to that of protons. However, the sensitivity of modern NMR spectrometers is such that carbon spectra can be routinely observed. Nitrogen is also a common element in polymers, but it is difficult to study because the most common isotope (^{14}N) has a quadrupole moment, and the lines are very broad. Sharp lines can be observed for ^{15}N, but it has a low sensitivity because of its a low magnetogyric ratio and natural abundance. ^{15}N NMR studies are possible, but usually only after isotopic labeling. The sensitivities of silicon and phosphorus are intermediate between those of protons and carbons.

NMR provides important information about the structure and dynamics of polymers because the NMR properties depend on the local atomic structure and dynamics. These properties depend on the local fields that are produced by electrons in the molecule of interest and nearby nuclei. The peak frequencies can depend on the local electron density and provide important chemical information. The peak frequency also depends on the local field from other nearby nuclei through the dipolar coupling. The strength of the coupling depends on the inverse third power of the internuclear distance and the orientation of the internuclear vector relative to the magnetic field. In solids, this effect can be as large as 50 kHz (much larger than the range of chemical shifts). Since the magnitude of the coupling depends on the distance and orientation, molecular motions can effectively average the dipolar couplings, leading to high-resolution spectra. Chain motion for polymers in solution, for example, very effectively averages these couplings such that linewidths on the order of 1 Hz can be observed.

The splitting of the energy levels in the magnetic field leads to a population difference between the upper and lower spin states that is determined by the Boltzmann distribution. When the spin system is placed in a nonequilibrium position by the application of rf pulses that interchange the populations of the upper and lower states, the spin system relaxes back toward equilibrium. This return to equilibrium is called *spin-lattice*, or longitudinal, relaxation and is characterized by the relaxation time T_1. The relaxation times provide information about the molecular dynamics of polymers since they depend on the fluctuations in the local magnetic fields caused by molecular motion. The time scale of motion is defined by the rotational correlation time τ_c. The polymer chain dynamics are often restricted in solids and the relaxation times can sometimes be very long.

A related property is the *spin-spin*, or transverse, relaxation time T_2 that is proportional to the inverse of the linewidth. As the chain motion is restricted in solid polymers, the linewidths become broader because the dipolar interactions persist, thus limiting the resolution that can be obtained. Since chain motion is a thermally activated process, it is sometimes possible to acquire the spectrum at higher temperature to reduce the linewidth.

NMR is often used to study the microstructure of polymers because the frequencies (or chemical shifts) are sensitive to the local structure and microstructure. Not only is there a difference in chemical shifts between carbons and protons (which resonate at 125 and 500 MHz in a 11.7 T magnetic field), but there is also a difference in chemical shifts in a polymer chain between neighboring carbons and

Table 9.1 Nuclear properties of spins that are of interest in polymer science

Isotope	Abundance (%)	Spin	Sensitivity[a]	Frequency (MHz)[b]
^{1}H	99.98	½	1.0	500.
^{19}F	100.0	½	0.83	470.2
^{29}Si	4.7	½	0.078	99.3
^{31}P	100.0	½	0.066	202.3
^{13}C	1.1	½	0.0159	125.6
^{2}H	0.015	1	0.00964	76.7
^{15}N	0.365	½	0.001	50.6

a. The sensitivity relative to protons.
b. The resonant frequency in a 11.7 T magnetic field.

protons that experience slightly different magnetic environments. This difference is small, measured in parts per million, relative to the frequency difference between carbons and protons, but it is usually larger than the linewidth, so many structural features can be resolved.

The origin of this variation in the chemical shift is the cloud of electrons about each of the nuclei. When a molecule is placed in a magnetic field, orbital currents are induced in the electron clouds that give rise to small, local magnetic fields that are always proportional to B_0, but opposite in direction. Each nucleus is partially shielded from B_0 by the electrons, and each resonates at a slightly higher frequency. This can be expressed as

$$\omega = \omega_0 (1 - \sigma) \tag{4}$$

where σ is the *screening constant*. It should be noted that σ is independent of B_0 but depends on the chemical structure.

The range in chemical shifts varies with the nuclei under study, and the ranges for protons and carbons are 10 ppm and 200 ppm, respectively. Any chemical group that changes the electron density will cause a change in the chemical shift; thus, the chemical shift is extremely sensitive to the polymer microstructure.

The carbon chemical shifts are particularly important for determining the polymer microstructure, since carbon is common to most polymers of interest and the lines are often well resolved. It is well known from studies on small molecules that the carbon chemical shifts depend on the chemical structure of nearby atoms.[5] The so-called α-effect arises from substituents on the carbon atom of interest and can lead to chemical shift changes on the order of 5 to 10 ppm. The β-effect arises from substituents at the next nearest-neighbor and is often in the range of 4 to 8 ppm. The combination of these two effects tends to make the carbon spectrum of polymers very sensitive to polymer stereochemistry, regioisomerism, branches, and defects. The carbon chemical shifts have been studied in a large number of small molecules, and empirical correlations between the chemical shifts and the structure have been published.[5]

The γ-effect on the carbon chemical shifts is of particular interest because the magnitude of the shift depends on the chain conformation. This effect arises from the through-space interaction of a carbon atom with its γ neighbor in the *gauche* position and is called the γ-*gauche* effect. The exact cause of this induced shift is not precisely known, but it is believed to arise from van der Waals interactions. A carbon and its γ neighbor are closest in a *gauche* arrangement (3 Å) and more distant in a *trans* conformation (4 Å), as shown in Figure 9.2. This leads to a larger induced chemical shift for the *gauche* conformation. In most polymers in solution there is free rotation about the main chain atoms, so the γ-*gauche* effect depends on the probability of finding the polymer in a *gauche* conformation. The magnitude of the γ-*gauche* effect can be estimated by calculating the fraction of bonds in the *gauche* conformation from rotational

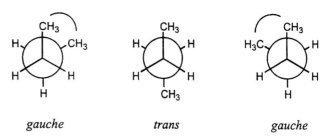

Figure 9.2 Newman diagrams illustrating the proximity of a methyl group and its γ neighbor in the *gauche* and *trans* conformations.

isomeric state models.[6] The γ-*gauche* effect also depends on the group in the γ position, and the induced shifts for methyl, hydroxyl, and chlorine groups are -5.2, -7.2, and -6.8 ppm.

The NMR signals are also affected by through-space dipolar couplings between nearby atoms. The energy separation between two states from dipolar couplings between two protons is given by

$$\Delta E = 2\mu[\boldsymbol{B}_0 \pm \boldsymbol{B}_{loc}] = \left[\boldsymbol{B}_0 \pm \frac{3}{2} \mu r^{-3} (3 \cos^2 \theta - 1) \right] \tag{5}$$

where μ is the nuclear moment, r is the distance separating the nuclei, and θ is the orientation of the vector connecting the spin pair relative to the magnetic field. This local field B_{loc} can be as large as 50 kHz, which is much greater than the proton chemical shift range. For an isolated pair of protons, the spectrum would appear as a doublet. It is more common, however, for chemically similar types of protons to experience a number of interactions at different distances and orientations, resulting in a broad featureless line. It is necessary to remove this coupling in order to observe a high-resolution spectrum. This can be accomplished in solution by rapid chain motion and in solids by high-power proton decoupling, which effectively averages the dipolar interactions to zero.

If the resolution is high enough, as for polymers in solution, the NMR spectrum is also affected by the spin state of nearby nuclei. For a carbon with a directly bonded proton, the local field it experiences will depend on whether the proton is aligned with or against the external magnetic field. Since there is an almost equal population of the protons in the $+\frac{1}{2}$ and $-\frac{1}{2}$ spin states, the carbon resonance will be split into two signals. This phenomenon is known as *scalar*, or *J*, coupling. Scalar couplings provide important information about the number of attached protons and the identity of nearby groups. This information can be used to distinguish between a methine carbon with one attached proton that appears as a doublet and a methylene carbon with two attached protons that appears as a triplet. The number of lines is given by $2nI + 1$, where n is the number of attached protons.

The scalar couplings provide information about the number of attached protons, but they also make the spectrum more complex. It is often desirable to suppress the coupling information by irradiating the protons during acquisition to obtain a spectrum with the highest resolution and a higher signal-to-noise ratio.

In a similar way, the three bond couplings between pairs of protons can also give rise to splittings. However, since the protons are separated by three bonds, the proton-proton scalar couplings are smaller (2–15 Hz) than the one bond carbon-proton couplings (120–160 Hz). In addition, the proton-proton couplings contain conformational information since they depend on the torsional angle that separates the pairs of protons. These couplings are often valuable for establishing the chemical shift assignments using multidimensional NMR.[7]

Even in the solid state, there is sometimes enough molecular motion to partially average the local dipolar interactions. Experimental studies have shown that solid polymers can experience atomic fluctuations over a wide range of time scales, including methyl group rotations, *gauche-trans* isomerizations, and other segmental motions. These molecular motions have a fundamental effect on the appearance of the spectra and the relaxation times and lineshapes. The easiest conceptual way to understand the effects of molecular motion on the NMR spectra is to consider the effects of chemical exchange.

Consider, for example, a proton that is exchanging between two sites with different chemical shifts. In the slow exchange limit, when the lifetime is much longer than the inverse of the difference in the chemical shifts, two peaks will be observed. In the fast exchange limit, the lifetime is shorter than the inverse of the difference in chemical shifts, and a single peak is observed, midway between the two chemical shifts. In the case of intermediate exchange, the lines will be broadened. Figure 9.3 shows the effect of exchange on the appearance of the spectra for a two-spin system. Therefore, the way to observe a high-resolution spectrum is to force the spin system to be in the fast exchange limit.

The time scale for exchange depends on the type of interaction and the chemical shift separation. For simple chemical exchange between conformations with a chemical shift difference of a few Hz, the fast exchange lifetime may be on the order of fractions of a second. In contrast, the dipolar couplings are on the order of 50 kHz, so much faster motion is required for these spins to be in the fast exchange limit that gives rise to a sharp, averaged lineshape.

To understand the effect of motion on linewidth, it is instructive to consider the effect of molecular motion on the dipolar couplings as given by eq 5. As the molecule moves, the orientation with respect to the field and the distances between nearby atoms will be modulated, changing the dipolar couplings and the observed line positions. If the motion is faster than the dipolar couplings, then we will ob-

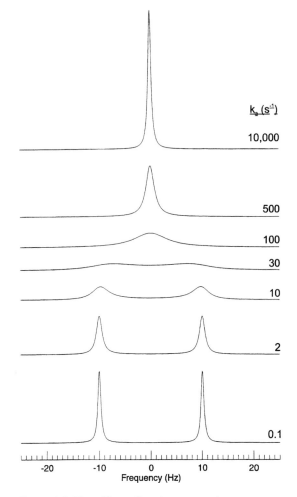

$k_a \, (s^{-1})$

10,000

500

100

30

10

2

0.1

-20   -10   0   10   20
Frequency (Hz)

Figure 9.3 The effect of exchange on the appearance of a two-line spectra as a function of the chemical exchange rate. The frequency difference between the signals is 20 Hz, and the lines broaden and finally coalesce as the exchange rate increases from 0.1 to 10,000 s^{-1}.

serve a single sharp line at the average frequency. The time scale of reorientational molecular motion for most polymers in solution is on the order of MHz, which is in the fast exchange regime for the dipolar couplings and chemical shift anisotropy. The practical result is that the lines are sharp, and high-resolution spectra can be observed for polymers in solution. Since the orientation-dependent couplings are not averaged for polymers in the solid state, some other means (magic-angle spinning or decoupling) must be used to observe the high-resolution spectra.

The additional effect of molecular motion is to change the spin-lattice and spin-spin relaxation times T_1 and T_2. Short values for the T_1 (seconds) make it possible to record spectra rapidly. T_1-type relaxation is a consequence of fluctuating local fields near the frequency of the nuclei of interest (ω_0). To the extent that these fluctuations have a component at the nuclear frequency, they will cause relaxation. The relaxation times therefore depend on the polymer motions that are described by a correlation time τ_c. The relaxation times directly influence the resolution through the spin-spin relaxation time, and for liquids the linewidth $\delta\nu$ is related to the T_2 by

$$\delta\nu = \frac{1}{\pi T_2{}^*} \qquad (6)$$

where $T_2{}^*$ is the apparent spin-spin relaxation time that contains contributions from both spin-spin relaxation and magnetic field inhomogeneity. The relationship between T_2 and $T_2{}^*$ is given by

$$\frac{1}{T_2{}^*} = \frac{1}{T_2} + \frac{\gamma\Delta\boldsymbol{B}_0}{2} \qquad (7)$$

where $\gamma\Delta\boldsymbol{B}_0/2\pi$ is the magnetic field inhomogenity in Hz. From these relationships it can be seen that long relaxation times give rise to sharper lines, and more features can be resolved in the NMR spectrum. Figure 9.4 shows a plot of the carbon T_1 and T_2 relaxation times for a CH pair as a function of the correlation time at a field strength of 11.7 T. Note that while the T_1 goes through a minimum, the T_2 continues to decrease as the correlation time becomes longer.

Another concept that is important for understanding the NMR of solids is spin exchange. This is the process by which magnetization is exchanged between pairs of nuclei that are in close proximity and experience dipolar couplings. This "flip-flop" is very efficient and occurs between nuclei of the same frequency, most commonly protons. In the dense sea of protons that is common for most organic solids, this spin exchange can lead to the flow of polarization over long distances. Spin exchange also tends to average the proton relaxation times since all the spins are coupled together and the measured relaxation time will typically be that of the fastest relaxing protons that act as a relaxation sink. Spin exchange between nuclei that differ in frequency (such as carbons and protons) can be made to occur by temporarily making them "like" spins by placing them in a strong rf field.

Solution NMR

Solution NMR is useful for polymer characterization because the chemical shifts are extremely sensitive to polymer microstructure, including stereochemistry, regioisomerism, and the presence of branches and defects. High-resolution spectra were initially not expected since it is well known that the linewidths depend on the third power of the diameter for a rigid sphere reorienting in solution.[8] This means that sharp lines are expected for small molecules, but not for high molecular weight polymers. Fortunately, the relaxation in most polymers is not due to center-of-mass reorientation but, rather, to more localized repetitive motions and to segmental motions such as *guache-trans* isomerization and librational motions. This fast molecular motion results in long T_2s and very high resolution solution spectra.

Carbon NMR has been extensively used for polymer characterization because carbon is strategically placed in most polymers to provide information about the microstructure. Carbon has a low natural abundance (1% ^{13}C), but most polymers are sufficiently soluble that it is possible to record their NMR spectrum at high concentrations ([C] > 0.05 M). The carbon spectra are often more highly resolved than the proton spectra because the carbon chemical shifts are spread over 200 ppm, rather than the 10 ppm commonly observed for protons. Nitrogen, silicon, and fluorine also have large chemical shift ranges.

Solid-State NMR

The NMR spectra of solids are fundamentally different from those of solutions because atomic motions are limited by the proximity of nearby chains that are either static or near-static. This means that the local interactions, such as the chemical shifts and dipolar couplings, are not averaged by molecular motion, and some artificial means must be used to obtain a high-resolution spectrum. It is also difficult to record the NMR spectra of solids by using the simple Bloch-decay experiment shown in Figure 9.1 since lines can be very broad and the relaxation times for solids can be very long.

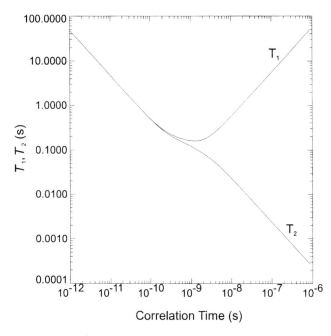

Figure 9.4 A plot of the spin-lattice (T_1) and spin-spin (T_2) relaxation times for an isolated CH pair as a function of the rotational correlation time. The relaxation times are calculated for a magnetic field strength of 11.7 T.

We introduced the chemical shift as a scalar quantity, but this description is somewhat simplified. The chemical shift is actually *anisotropic*, or directional, as it depends on the orientation of the molecule with respect to magnetic field direction. It is expressed as a tensor, a mathematical quantity that has both direction and magnitude, and is composed of three principal components:

$$\sigma = \lambda_{11}^2\sigma_{11} + \lambda_{22}^2\sigma_{22} + \lambda_{33}^2\sigma_{33} \qquad (8)$$

where λ_{ii} are the direction cosines of the principal axes of the screening constant with respect to the magnetic field. The principal axis systems may lie along the bond direction, but this is not necessarily so. The orientation of the axis system cannot be predicted a priori and must be experimentally determined.[9] By convention, $\sigma_{33} > \sigma_{22} > \sigma_{11}$. In solution, the chemical shift anisotropy is averaged by rapid molecular motion and the chemical shift is given by

$$\sigma = \frac{1}{3}(\sigma_{11} + \sigma_{22} + \sigma_{33}) \qquad (9)$$

In molecules of any degree of complexity, there will be a chemical shift anisotropy pattern for each carbon atom in the polymer. These chemical shift anisotropy lineshapes for carbons vary in width from 10 to 50 ppm (1–7 kHz at a carbon frequency of 125 MHz), so the spectrum without line narrowing will be strongly overlapped. Under these circumstances it is desirable to sacrifice the anisotropy information in order to observe a high-resolution spectrum. If we rapidly rotate the sample, the orientations and chemical shifts become dependent on time. The time average under rapid rotation is given by

$$\sigma = \frac{1}{2}\sin^2\beta(\sigma_{11} + \sigma_{22} + \sigma_{33}) + \frac{1}{2}(3\cos^2\beta - 1) \qquad (10)$$

where β is the angle between the rotation axis and the magnetic field direction. When β is equal to the so-called magic angle (54.7°, or the body diagonal of a cube), $\sin^2\beta$ is ⅔; in this case, the first term becomes equal to one-third of the trace of the tensor (i.e., the isotropic chemical shift), and the $(3\cos^2\beta - 1)$ term is equal to zero. Thus, under magic angle rotation the chemical shift pattern collapses to the isotropic average, giving the high-resolution spectrum.

The dipolar couplings are a major source of line broadening in the spectra of solids, and high-resolution spectra can only be observed if we remove this broadening. In the polymer spectra of nuclei that are present at a low natural abundance, like carbon, silicon, phosphorus, and nitrogen, the broadening is primarily due to couplings with nearby protons, and we can remove this line broadening by high-power proton irradiation. For a pair of spins (^{13}C and ^1H) with a fixed orientation relative to the magnetic field, a doublet is observed in the ^{13}C spectrum with the separation given by eq 5. The proton irradiation causes transitions between the proton spin states. If the transitions are faster than the doublet separation, then we are forcing the spin

system to be in the fast exchange limit (Figure 9.3), and we will observe a high-resolution spectrum. Homonuclear couplings can also cause line broadening, but since carbon has a low natural abundance (1.1%), there is a very low probability that two ^{13}C atoms will be close enough to appreciably broaden the lines.

In certain cases it is desirable to observe the proton spectra of solids. This is a difficult experiment since we must remove the homonuclear (proton-proton) dipolar couplings while trying to observe the proton spectrum. This can be accomplished using the combined rotation and multiple-pulse sequence (CRAMPS) shown in Figure 9.5.[10] The idea behind this approach is to apply a dense series of pulses that are designed to eliminate the proton-proton dipolar couplings while preserving scaled-down chemical shift interactions. A key feature of this pulse sequence is that it is necessary to sample the signal in the intervals between the pulses. The proton linewidth using CRAMPS is typically on the order of 1 to 3 ppm. Thus, the resolution in the CRAMPS spectrum is much lower than in the solution state spectrum.

Another consequence of the restricted atomic motion in solids is that the spin-lattice relaxation times for dilute spins (^{13}C, ^{15}N, ^{29}Si, and ^{31}P) can be very long (Figure 9.4), making it difficult to acquire a spectrum in a reasonable amount of time. This is particularly troublesome for carbons in crystalline polymers where the relaxation times can be on the order of hundreds of seconds, especially for those carbons without directly bonded protons. We can overcome this limitation using a method called *cross polarization*.

Cross polarization works by forcing the proton and carbon signals to precess at the same frequency in the rotating frame even though they do not have the same frequency in the laboratory frame. The means of doing this was demonstrated by Hartmann and Hahn in 1962,[11] when it was shown that energy transfer between dipolar-coupled nuclei with widely differing Larmor frequencies can be made to occur when

$$\gamma_C \mathbf{B}_{1C} = \gamma_H \mathbf{B}_{1H} \qquad (11)$$

and the so-called *Hartmann-Hahn condition* is satisfied. When this condition is satisfied, then it can be shown that the projection of both spins along $\mathbf{B}_0$ is periodic with the same frequency. Thus, in a reference frame that includes the $\mathbf{B}_0$ direction, both kinds of spins appear to have the same frequency. In a sense, they become "like" spins, and they can undergo carbon-proton spin exchange, which enables them to share (or transfer) spin polarization. Since γ_H is four times γ_C, the Hartmann-Hahn match occurs when the strength of the applied carbon field $\mathbf{B}_{1C}$ is four times the strength of the applied proton field $\mathbf{B}_{1H}$. When the proton and carbon rotating frame energy levels match, polarization is transferred from the abundant protons to the rare carbon-13 nuclei.

Because polarization is being transferred from the protons to the carbons, it is the shorter T_1 of the protons that deter-

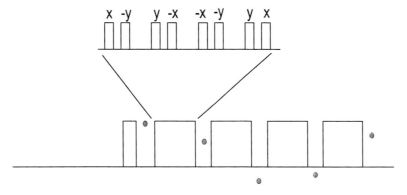

Figure 9.5 The pulse sequence diagram for combined rotation and multipulse NMR spectrsocopy. After the initial 90° pulse, the data points (represented by dots) are sampled stroboscopically after the application of the 8-pulse sequence. The 8-pulse sequence averages the dipolar interactions and scales the chemical shift; the magic-angle spinning averages the chemical shift anisotropy.

mines the repetition rate for signal averaging. The short T_1 for protons in solid polymers is due in part to rapid spin exchange. Since the magnetization is efficiently transferred among the protons, the average proton relaxation time is determined by the fastest relaxing protons. In most polymers, there are rapidly rotating groups, such as methyl groups, that have short relaxation times and act as relaxation sinks and cause the average proton relaxation time to be short. It should also be noted that cross polarization works best when there are strong heteronuclear dipolar couplings. Crystalline polymers and glasses below T_g are efficiently cross polarized, while elastomers are not. Cross polarization can also be used to enhance the sensitivity of silicon, nitrogen, and phosphorus spectra.

The cross-polarization pulse sequence (Figure 9.6) begins with a $90°_x$ pulse to the protons to tip them along the y axis. The phase of the proton $\boldsymbol{B}_{1H}$ field is then shifted by 90° and the protons are *spin-locked* along the y axis, where they precess with a frequency $\omega_H = \gamma_H \boldsymbol{B}_{1H}$. At the same time, the

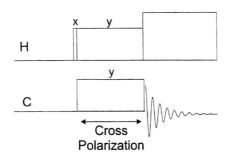

Figure 9.6 The pulse sequence diagram for cross polarization. Magnetization is exchanged between the protons and carbons during the cross-polarization time, which is on the order of 1 ms for crystalline polymers and 5 ms for rubbery polymers.

carbons are put into contact with the protons by turning on the carbon field $\boldsymbol{B}_{1C}$. This causes the carbon magnetization to grow up via spin exchange with the protons in the direction of the spin-lock field. The carbon spin-locking field is then turned off, and the spectrum is recorded with proton decoupling.

The signal intensity during cross polarization represents a compromise between several processes, including the buildup of intensity from carbon-proton cross polarization and the decay of proton and carbon magnetization in the spin locking fields. The intensity at cross polarization contact time t is given by

$$M(t) = \frac{M_o}{T_{1C}} \cdot \frac{e^{-t/T_{1\rho}(H)} - e^{-t(1/T_{CH}+1/T_{1\rho}(C))}}{1/T_{CH} + 1/T_{1\rho}(C) - 1/T_{1\rho}(H)} \quad (12)$$

where T_{CH} is the carbon-proton cross polarization time constant, and $T_{1\rho}(C)$ and $T_{1\rho}(H)$ are the carbon and proton rotating-frame spin-lattice relaxation time constants. It should be noted that the $T_{1\rho}(C)$ in eq 12 is the relaxation time measured in the presence of proton decoupling and differs from the relaxation measurement of $T_{1\rho}(C)$ that is measured in the absence of proton decoupling. In most cases $T_{1\rho}(C)$ is much greater than T_{CH}, so eq 12 simplifies to

$$M = \frac{M_o}{T_{1C}} \cdot \frac{e^{-t/T_{1\rho}(H)} - e^{-t/T_{CH}}}{1/T_{CH} - 1/T_{1\rho}(H)} \quad (13)$$

The signal intensity during cross polarization builds up rapidly from cross polarization and then slowly decays due to $T_{1\rho}(H)$ relaxation. The signal intensity depends on the relaxation rate constants, which, in turn, depend on the chain dynamics. It is often observed that the buildup and decay of magnetization is different for crystalline, amorphous, and rubbery polymers, and cross polarization is one way to separate the signals from polymers with different chain dynamics.

Multidimensional NMR

One of the fundamental limitations in polymer NMR studies is the signal overlap that results from the repeating sequence nature of polymer chains. This makes it difficult to observe the features of interest, such as defect sites and the like, that may not give well-resolved signals. One way to overcome these limitations is to expand the chemical shifts into two or more frequency dimensions using multidimensional (nD) NMR. There are many kinds of nD NMR experiments that can be categorized as correlated or resolved experiments. In *correlated* experiments, the resonance frequency of one signal is related to those of its neighbors, and molecular connectivities (or distances between atoms) can be determined. In *resolved* experiments, the frequency axes show different interactions. In one kind of resolved experiment, for example, the carbon chemical shift may appear along one axis and the proton-carbon scalar couplings along the other. Extension of these same principles leads to three-dimensional (3D) NMR experiments in which there are three independent frequency axes. This has the potential of providing still greater resolution for those cases in which the 2D spectrum exhibits extensive overlap.

nD NMR experiments are related to the more familiar experiments in that they consist of a series of pulses and delays. They differ from the more common experiments in that we allow the spin system to *evolve* instead of immediately transforming the FID following a pulse. The pulse sequence for nD NMR experiments can be divided into four periods: preparation, evolution, mixing, and detection. During the preparation period, the spins are allowed to come to equilibrium; that is, the populations of the energy levels are allowed to equilibrate with their surroundings. This interval allows the establishment of reproducible starting conditions for the rest of the experiment. During the evolution, or t_1, period, which is systematically incremented over the experiment, the spins evolve under all the forces acting on the nuclei, including the chemical shifts, dipolar couplings, and scalar couplings. One of the things that commonly happens during the t_1 period is *frequency labeling,* where the spins are labeled (i.e., prepared in a nonequilibrium state) by their frequency in the NMR spectrum. The mixing period may consist of either pulses or delays and results in the transfer of magnetization between spins that have been frequency labeled in the t_1 period. The final period, t_2, is the signal acquisition that is common to all pulsed NMR experiments. The second frequency is introduced by systematically incrementing the t_1 period. Data collection in a 2D experiment consists of gathering many FIDs, each obtained with a different value of t_1. The free induction decays are transformed with respect to t_2 to obtain a set of spectra in which the peak intensities or phases are modulated as a function of the t_1 delay. Fourier transformation with respect to t_1 converts these frequency modulations into peaks in the 2D spectrum, which is actually a surface in 3D space that

can be represented either as a stacked plot or as a contour plot. nD NMR experiments are performed by incorporating multiple evolution and mixing periods before acquisition. In a 3D experiment, for example, the sequence might be preparation-evolution (t_1), then mixing-evolution (t_2), and finally mixing-detection (t_3).

Many 2D NMR experiments have been used to study the structure and dynamics of polymers. Among the most commonly used methods for polymers in solution are COSY (correlated spectroscopy)[12] and TOCSY (total correlation spectroscopy)[13] that are used to correlate protons using two- and three-bond proton scalar couplings. NOESY (nuclear Overhauser effect spectrsocopy),[14] or 2D exchange spectroscopy, provides similar information about proton connectivities, but uses through-space dipolar interactions to correlate pairs of protons that are close in space. Heteronuclear multiple-quantum correlation (HMQC)[15] spectroscopy provides a correlation between the carbon and proton chemical shifts.

Figure 9.7 shows the pulse sequence diagram for COSY spectroscopy and the 2D spectrum of polycaprolactone (**1**).

Polycaprolactone

1

This particular variant of COSY uses a double-quantum filter[16] to suppress the intense peaks along the diagonal so that the correlations between protons that are close in frequency can be more easily observed. The correlations in the COSY spectra are between pairs of protons that have scalar couplings between 2 and 15 Hz. Among the couplings that can be observed in polymers are the two- and three-bond scalar couplings, such as those between the methylene protons in polycaprolactone. It is possible to follow the magnetization transfer along the chain to establish all of the proton assignments for polycaprolactone.

TOCSY also uses the through-bond scalar couplings to correlate nearby protons, but the cross peaks are not restricted to directly coupled pairs of protons.[13] The pulse sequence diagram and the TOCSY spectrum for polycaprolactone (**1**) are shown in Figure 9.8. A key part of the sequence is the so-called mixing period, during which a multiple-pulse decoupling sequence known as MLEV-17[17] promotes magnetization exchange between coupled protons. We observe all of the cross peaks that were visible in the COSY spectrum, but we also see peaks between pairs of protons (such as *a* and *c* or *b* and *d*) that are not directly coupled. The TOCSY spectrum is extremely useful for identifying groups of protons that are not directly coupled to each other, such as those in polymer side chains.

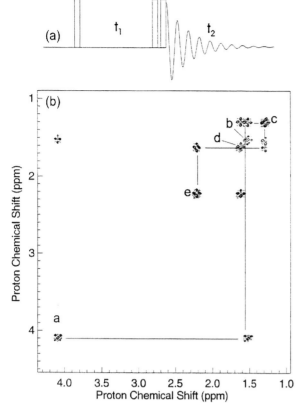

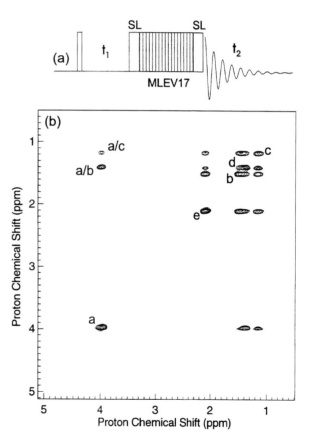

between nearby protons. The cross peaks can be from both direct and relayed correlations, and it is difficult to observe correlations between pairs of protons that are separated by more than 5 Å since the cross peak intensity depends on the inverse sixth power of the internuclear separation. The 2D exchange experiments are useful for making the assignments in polymers where the scalar couplings along the chain are interrupted, as in methyl methacrylates where the main chain methylene protons are not scalar coupled to any other protons. Figure 9.9 shows the correlations between the aromatic protons of styrene and the methoxyl protons of the methyl methacrylate that are a consequence of the through-space dipolar interactions. These experiments can also be used to monitor intermolecular interactions between polymer chains.[18]

The HMQC experiments are valuable for correlating each carbon chemical shift with the chemical shift of the nearest protons, and the pulse sequence diagram and the HMQC spectrum for polycaprolactone are shown in Figure 9.10.

Figure 9.7 (a) The pulse sequence diagram for double-quantum-filtered correlated spectroscopy (COSY); (b) the 500-MHz two-dimensional spectrum for polycaprolactone. The peaks along the diagonal correspond to the peaks in the one-pulse spectrum, and the off-diagonal cross peaks arise from correlations between coupled protons. The data set contained 1024 points in the t_2 dimension and 512 points in the t_1 dimension and required 6 hours to acquire. The letters correspond to the assignments illustrated in the text.

2D NOESY, or exchange spectroscopy, provides information about the through-space dipolar couplings that can be used to identify protons in close proximity.[14] The pulse sequence diagram and the 2D exchange spectrum for poly (styrene-co-methyl methacrylate) (**2**)

Poly(styrene-co-methyl methacrylate)

2

are shown in Figure 9.9. A key part of the sequence is the mixing time during which the magnetization is exchanged

Figure 9.8 (a) The pulse sequence diagram for two-dimensional total correlated spectroscopy (2D TOCSY); (b) the 500-MHz TOCSY spectrum for polycaprolactone. Note the appearance of cross peaks between both directly coupled peaks (a/b) and peaks that are not directly coupled (a/c). The TOCSY mixing time was 35 ms, and the spin-lock (SL) time was 2 ms. The spectrum was acquired in 6 hours using the same spectral parameters as in Figure 9.7.

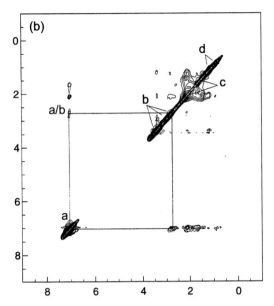

Figure 9.9 (a) The pulse sequence diagram for two-dimensional nuclear Overhauser effect spectrsocopy (NOESY) NMR; (b) the two-dimensional spectrum for the poly(styrene-*co*-methyl methacrylate) copolymer. The cross peaks arise from magnetization exchange during the mixing time τ_m. The peaks labeled a/b are assigned to styrene aromatic groups in close proximity to the methyl methacrylate methoxy group. The other assignments are shown in the text. The mixing time was set to 0.1 s.

Multiple-quantum coherences (a property of coupled spin systems) are created by the first proton pulse, a delay of $1/(2J_{CH})$ and the following carbon pulse, where J_{CH} is the one-bond carbon proton scalar coupling constant. These coherences evolve during the t_1 period and the 1800 proton pulse refocuses the proton couplings in the t_1 dimension. The carbon pulse and delay transfers the magnetization back to the protons, where the spectrum is recorded with carbon decoupling by using some multiple-pulse decoupling sequence such as Globally Optimized Alternating-Phase Rectangular Pulses (GARP).[19] The end result is the correlation of the carbon chemical shift and proton chemical shifts of the directly bonded protons. If some of the peaks in the carbon spectrum have been assigned, then we can establish the proton assignments from the correlation. This experiment is now more commonly used than the similar experiment that detects the carbon spectrum (HETCOR, or heteronuclear correlation) because the proton signals are detected with a higher sensitivity.

The NMR of Polymers in Solution

One of the main uses of NMR in polymer science is materials characterization, which provides the link between the synthesis of new materials and the structure-property relationships. In many polymers the thermal, mechanical, optical, and electronic properties depend on the chain microstructure, which can be characterized in great detail by ^{1}H, ^{13}C, ^{29}Si, ^{19}F, ^{31}P, and ^{15}N NMR in solution.

The level to which the microstructure can be elucidated depends on the resolution and the methods available to establish the resonance assignments. The resolution depends on several factors, including the nuclei under observation, the chain dynamics, the concentration, the temperature,

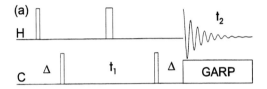

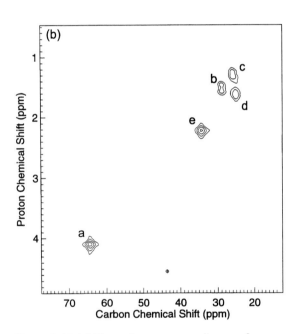

Figure 9.10 (a) The pulse sequence diagram for heteronuclear multiple-quantum correlation (HMQC); (b) the two-dimensional carbon-proton correlation spectrum for polycaprolactone. The delay time Δ is set to $\frac{1}{2}$ J_{CH}, and the 180° pulse in the middle of the t_1 is used to remove carbon-proton coupling in the t_1 dimension. The spectrum was recorded at a field of 11.7 T, corresponding to a carbon frequency of 125 MHz and a proton frequency of 500 MHz. The carbon and proton sweep widths were set to 25 kHz and 6 kHz, and the data set contained 512 points in the t_1 dimension and 1024 points in the t_2 dimension. The spectrum was recorded in 8 hours, and the assignments are illustrated in the text.

and the solvent. Well-resolved peaks are most often observed for nuclei with a wide range in chemical shifts, such as ^{13}C, ^{19}F, ^{29}Si, ^{31}P, and ^{15}N. However, the sensitivity for these nuclei is not as high as for protons (Table 9.1), so high-solution concentrations, often on the order of 10 to 30 wt%, are required. Fortunately, many polymers are soluble to this degree, so this does not present a serious experimental limitation.

An extremely detailed microstructural characterization of polymers is possible when the lines are well resolved. As noted earlier, the linewidth depends inversely on the spin-spin relaxation time T_2, which depends on the correlation time τ_c. For rigid molecules, the T_2 scales with the third power of the molar volume so that large molecules are expected to have extremely broad lines. Fortunately, most polymers are not rigid but experience segmental flexibility on length scales that range from one bond to several monomer units. This decreases the effective correlation time so that the observed linewidths are often in the range of 1 to 10 Hz, and high-resolution spectra can be observed. The side chains are less constrained than the main chain, and these resonances are often sharper.

The first step in the NMR characterization of polymers is establishing the resonance assignments. A variety of methods have been used to establish resonance assignments, including the use of model compounds, the γ-*gauche* effect, the calculation of expected chemical shifts from empirical rules, and nD NMR. The comparison of the polymer spectra with model compounds or polymers was one of the first methods used to establish resonance assignments. The assignments for polymer sterosequences, for example, can be made by comparison to sterochemically pure polymers, if these materials are available.[20] Stereosequence assignments have also been made by comparing the polymer spectra with those of small molecules in which the various diastereomers have been separated and compared.[21] The synthesis of new materials is an extremely difficult and time-consuming way to establish the resonance assignments. More commonly, the resonance assignments can be established with 2D and 3D NMR methods.

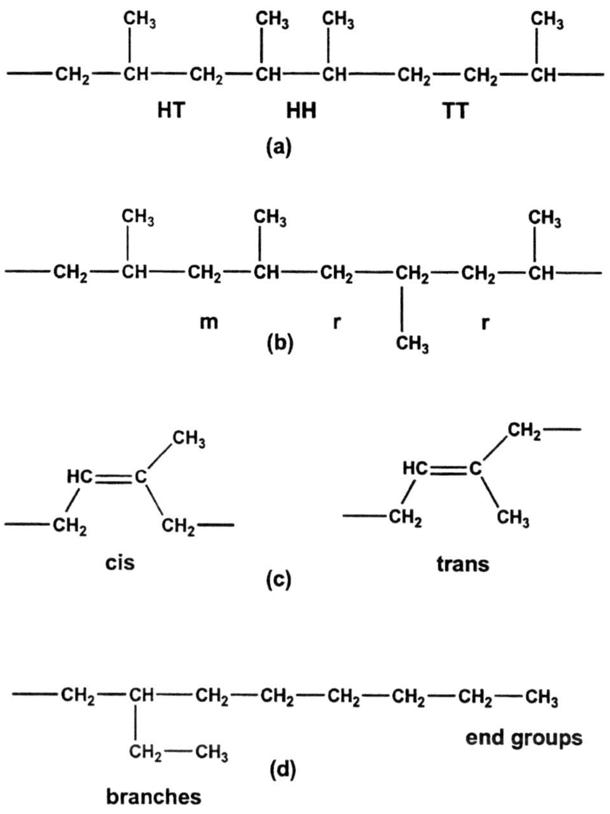

Polymer Microstructure

NMR is a powerful tool for the characterization of polymer microstructure because the chemical shifts are sensitive to the local magnetic environment that changes with the polymer microstructure. Figure 9.11 shows some possible polymer microstructures, including directional or regioisomerism, stereochemical isomerism, geometric isomerism, branches, and end groups. These microstructures can be distinguished by a number of features. In carbon NMR, for example, the carbon chemical shifts are sensitive to the identity of neighboring atoms. In the head-to-tail monomer shown in Figure 9.11a, the methine carbon that has two methylene neighbors will have a very different chemical

Figure 9.11 A chemical diagram showing defects and isomers commonly observed in polymers. Polymer microstructures often include (a) regioisomerism where directional defects lead to head-to-tail (HT), head-to-head (HH), and tail-to-tail (TT) monomer units in the chain. The placement of stereochemical groups on the same (meso, or m) or opposite (racemic, or r) side of the chain give rise to (b) stereochemical isomerism. Geometric isomerism (c) and (d) branches and defects can also be observed in polymer chains.

shift from the methine carbon in a head-to-head defect that has one methine and one methylene neighbor. In a similar way, the methylene chemical shifts will be very different for the head-to-tail, head-to-head, and tail-to-tail units.

Regioisomerism also changes distances and scalar couplings between nearby protons on the polymer chains. In the head-to-head defect, for example, there are two nearby methine protons, and the three-bond proton-proton *J* couplings between them can be measured by 2D COSY or TOCSY NMR. Since methine-methine couplings would be observed only for the head-to-head defects, they have a unique signature in the 2D spectrum. In a similar way, the couplings between neighboring methylene protons are indicative of tail-to-tail defects. In some cases these defects bring pairs of protons together that are not scalar coupled to each other, such as the two methyl groups in the head-to-head defects. These can be identified from cross peaks in the 2D NOESY spectra, since cross peaks can only be observed between pairs of protons that are separated by less than 5 Å.[22]

Stereochemical isomerism has a smaller effect on the NMR spectrum than regioisomerism has because the sequence of carbon chemical types is preserved, and only the substituent positions on asymmetric carbons are altered. Since the changes are more subtle than the change in carbon type, they are often better observed in the carbon spectrum that has a larger chemical shift range. The assignments are often more difficult to establish than they are for the regioisomers, but the assignments can be established using model compounds and polymers,[21] chemical shift calculations,[5] and 2D NMR methods.[7]

The relative orientation of substituent groups in neighboring monomer units is referred to as a *dyad*. If the substituents are on the same side of the chain, they are referred to as *meso* or *m*-centered dyads. If the substituents are on opposite sides of the chain, they are *racemic* or *r*-centered dyads. The relative orientations of larger stereochemical sequences are referred to as triads, tetrad, petads, and so on. In favorable cases, the peaks from long stereosequences, such as *mmrmmr*, can be resolved and assigned. One important consequence of the stereochemistry is the magnetic equivalence of the methylene protons in vinyl polymers. In *m*-centered sequences, the methylene protons are nonequivalent because the side groups are closer to one of the methylene protons than the other. These *m*-centered methylene protons can be identified from the COSY spectrum since the two-bond coupling constants are much larger than for the three-bond couplings (15 Hz vs. 2–8 Hz). They can also be identified from the heteronuclear correlation spectrum, since the methylene carbon resonances in *m*-centered stereosequences will be correlated with two proton chemical shifts. Once the *m*-centered and *r*-centered sequences have been identified, 2D NMR can be used to correlate longer sequence units. The *mm* sequences, for example, can be correlated with the *mmm* and *mmr* tetrad sequences.

Geometric isomerism can also be easily identified from the NMR spectrum, since there are large differences between the chemical shifts for the isomers. The carbon chemical shifts for some of the carbons in *cis*-polyisoprene and *trans*-polyisoprene, for example, can differ by more than 10 ppm.

Solution NMR spectroscopy has been extensively used to identify branches and end groups in polymers. In polyethylene, for example, methyl resonances can arise only from branches and end groups, and these signals are well resolved from the main-chain methylene signals. The assignments for the branches and end groups can be established using chemical shift calculations and 2D NMR. However, because often these defects are only present at very low levels, it is more common to establish the chemical shift assignments using model compounds and chemical shift calculations.

Table 9.2. The carbon chemical shifts for head-to-tail (HT), head-to-head (HH), and tail-to-tail (TT) defects in polypropylene

	Carbon chemical shift (ppm)		
Carbon	HT	HH	TT
Methine CH	28.5	37.0	—
Methylene CH$_2$	46.0	—	31.3
Methy CH$_3$	20.5	15.0	—

Regioisomerism

Regioisomerism has been effectively studied by NMR because these defects give rise to large chemical shift changes and because the defects sometimes have a unique signature in the 2D NMR spectrum. ^{13}C NMR has been used to investigate regioisomerism in polypropylene, since the inverted monomers have very different chemical shifts relative to the head-to-tail monomers. This is illustrated in Table 9.2, which lists the carbon chemical shifts for the methine, methylene, and methyl carbons in the head-to-tail, head-to-head, and tail-to-tail monomers. The chemical shifts for particular carbons can differ by more than 10 ppm in the carbon spectrum.

The properties of poly(vinyl fluoride) are well known to depend on the number of regiodefects in the chain, and the defect level has been investigated by ^{19}F NMR. Figure 9.12 compares the fluorine NMR spectrum of commercial poly-

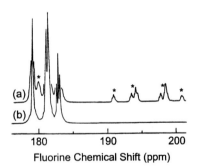

Figure 9.12 The 188 MHz ^{19}F spectrum for (a) commercial and (b) isoregic poly(vinyl fluoride). The peaks between 178 and 183 ppm arise from stereochemical isomerim, while the peaks between 190 and 202 ppm arise from regioisomerism. The spectrum in (a) is plotted at a higher gain level to show the peaks from regioisomerism. The commercial polymer contains about 5% inverted monomer units.

(vinylidine fluoride) with a regioregular material.[23] The regio-defects are well resolved from the main-chain signals in the NMR spectrum. Additional lines are resolved in the fluorine spectra from stereochemical isomerism.

Stereochemical Isomerism

The NMR spectrum is very sensitive to stereochemical isomerism in a variety of vinyl and other types of polymers. The sensitivity of the proton and carbon NMR spectrum to this isomerism is illustrated in Figure 9.13, which compares the proton and carbon spectrum for poly(vinyl acetate). Three groups of signals in the proton spectrum are assigned to the methine, methylene, and methyl protons. The stereochemistry of poly(vinyl acetate) causes the methine protons in *m*-centered and *r*-centered sequences to be slightly shifted from each other, but the shifts are not large enough for separate peaks to be resolved. There is some fine structure in the methylene signal, but, again, the stereosequences are not well resolved. By comparison, the carbon spectrum has much better resolution. It can be seen in Figure 9.13b that the methine peak is split into five signals.

While the carbon spectrum often shows the best resolution, proton NMR can also be used to investigate stereochemical isomerism in polymers. This is illustrated in Figure 9.14, which compares the 500-MHz proton spectrum for isotactic and predominantly syndiotactic poly(methyl methacrylate). As noted here earlier, the methylene protons in *meso* and *racemic* stereosequences can be distinguished because they are nonequivalent in the *meso* stereosequence. In isotactic poly(methyl methacrylate), the geminal methylene signals are separated by 0.65 ppm, while they are overlapped in the syndiotactic polymer.

The stereosequence distributions can be characterized by the relative intensities of the assigned peaks. We calculate from the peak intensities in Figure 9.14 that the predominantly syndiotactic poly(methyl methacrylate) has 5% *mm*, 30% *mr*, and 65% *rr* stereosequences. If the peaks are more highly resolved, as for the carbon spectrum, then higher-order stereosequence can be measured. If higher-order stereosequences can be quantified, then it is possible to study the polymerization mechanism. In this way it is often possible to distinguish between a polymerization following Bernoulian statistics from other statistical models.

Geometric Isomerism

The carbon chemical shifts are very sensitive to geometric isomerism. This is illustrated in Figure 9.15, which compares the 100-MHz carbon spectrum for *cis*-polyisoprene and *trans*-polyisoprene. Although there are small chemical shift changes for the carbons attached by double bonds, there are large changes in the methyl and methylene carbon signals. The assignments for geometric isomers can be

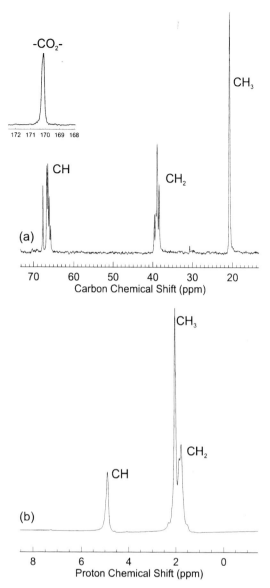

Figure 9.13 Comparison of the (a) the 125-MHz carbon and (b) the 500-MHz proton spectra of poly(vinyl acetate). The carbon spectrum appears to be more highly resolved because of the larger chemical shift range for carbon and the greater sensitivity of the carbon chemical shifts to the polymer microstructure. Note that some of the peaks (such as the CH) are much more sensitive to stereochemical isomerism than are others (such as the CH₃).

established by comparing the spectra with model compounds and chemical shift calculations. The spectra for polymers with geometric isomerism can be very complex. In polybutadiene, for example, many resonances will be observed from the *cis*, *trans*, and 1,2- vs. 1,4-addition of monomers.

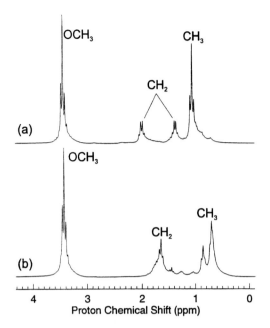

Figure 9.14 The 500-MHz proton NMR spectra of (a) isotactic poly(methyl methacrylate) and (b) predominantly syndiotactic poly(methyl methacrylate). Note the nonequivalent methylene protons in the isotactic polymer.

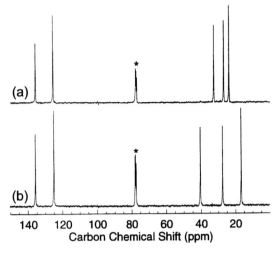

Figure 9.15 The 100-MHz ^{13}C NMR spectra for (a) cis-polyisoprene and (b) trans-polyisoprene. Note that geometric isomerism leads to large chemical shift changes in the aliphatic (20–40 ppm) region. The peak marked * arises from the chloroform solvent peak.

Branching and End Groups

Branches and end groups have been extensively studied in polymers because their presence can affect the macroscopic properties. These defects often give rise to large chemical shift changes because the chemical structure is quite different from the main chain polymer. However, these peaks are sometimes difficult to detect if they are present at very low concentrations. The effect of branches on the carbon spectrum of polyethylene (**3**) is shown in Figure 9.16.[24] The main-chain carbon resonance for polyethylene is observed near 30 ppm, and if we plot the spectrum with a much higher gain, we can begin to see many of the peaks arising from branches and end groups. These peaks can be as

signed by chemical shift calculations and by 2D NMR methods. A nomenclature has been adopted to describe the branches and end groups, and this is shown in structure (**3**). The type (length of the branches, etc.) and concentration of defects has a large effect on the properties of polyethylene.

NMR can also provide information about the polymer molecular weights (M_n) and the chain ends because, like branches, the end groups are often chemically distinct from the main-chain atoms. This is illustrated in Figure 9.17, which compares the 100-MHz carbon spectra of poly(propylene oxide) with molecular weights of 400 and 4000. Many peaks appear in the spectrum of the low molecular weight polymer, and by a simple inspection of the spectra it is difficult to determine if they arise from end groups, branches, or stereochemical isomerism. The main-chain

3

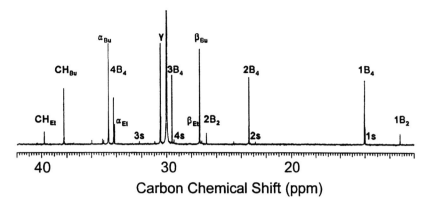

Figure 9.16 The 188-MHz carbon NMR spectrum of an ethylene/1-hexene/
1-butadiene copolymer used to study branching in polyethylene at 120 °C.
The assignments are illustrated in the text.

resonances can easily be identified by comparison with the higher molecular weight material.

Copolymers

Solution NMR methods are also used to investigate the structure of copolymers. The NMR spectra of copolymers are often complex because they can have all of the previously mentioned types of isomerism in addition to monomer sequence distributions. Moreover, there are a number of different copolymer architectures, including random, alternating, block, and graft copolymers. The spectra for the random copolymers are often the most complex because of the sequence distributions and other types of isomerism. The reactivity ratios may not be the same for the two monomers, so the polymers may have some blocky character. Higher-resolution spectra are observed for alternating copolymers since each monomer has a similar local environment, but the spectra can still be complex from stereochemical or other types of isomerism. The spectra for block copolymers are often similar to those of the homopolymers, with the exception of the peaks from monomers that are near the junctions between the blocks. If the blocks have a high molecular weight, it is often difficult to observe the signals from the junction region.

Figure 9.18 shows the 500-MHz proton NMR spectra for random and alternating copolymers of styrene and methyl

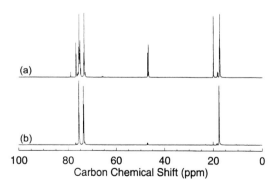

Figure 9.17 The 100-MHz carbon NMR spectrum for poly(propylene oxide) of molecular weight (a) 400 and (b) 4000. The peaks from the end groups decrease in intensity relative to the main-chain peaks as the molecular weight increases. The molecular weight can be calculated from the relative intensity of end group to main-chain signals.

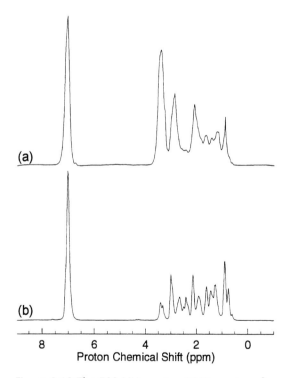

Figure 9.18 The 500-MHz proton NMR spectra of (a) random and (b) alternating stryene-methyl methacrylate copolymers. The lines in the random copolymer are broadened by monomer sequence distributions, as well as by stereochemistry.

methacrylate. While it is possible to observe some features in the random copolymer, such as the aromatic signals from the polystyrene and the methoxyl peaks from the poly(methyl methacrylate), there is extensive overlap from sequence and stereochemical isomerism. The spectrum for the alternating copolymer shows much higher resolution, and well-resolved peaks are observed for many of the signals. These peaks can be assigned with 2D exchange spectra, as shown in Figure 9.19, where correlations can be observed between pairs of protons that are close in space. Correlations can be observed between aromatic protons and methoxyl protons when they are on the same side of the chain, as in the co-isotactic sequences. The methylene protons are also well resolved, and it is possible to measure the chain conformation by measuring the relative distances between the pairs of protons.[25]

The Solid-State NMR of Polymers

Solid-state NMR has emerged as an important method for polymer characterization because most polymers are used

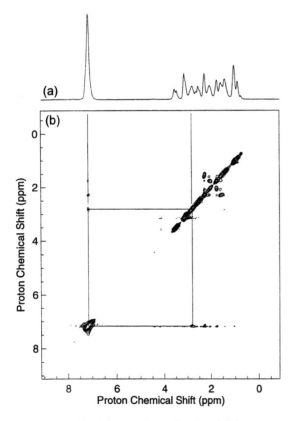

Figure 9.19 The (a) one-dimensional and (b) two-dimensional exchange NMR spectra acquired with a 0.5 s mixing time for poly(styrene-alt-methyl methacrylate). The closest contact between the styrene aromatic protons at 7 ppm and the methyl methacrylate methoxy protons at 3 to 4 ppm are for the co-isotactic sequences. This is illustrated by the box in (b).

in the solid state, and these NMR methods provide the link between the chemical structure and microstructure of polymers and their mechanical, electrical, and optical properties. In addition, the chains are not dynamically averaged as they are in solution, so the chemical shifts can provide important conformational information, and the proton relaxation rates can provide information about the morphology. Solid-state NMR also can be used to monitor the reactivity of polymers.

Solid-state NMR developed more slowly than solution NMR, in part because the lines are broader in solids and more difficult to observe. In the carbon spectrum, for example, the lines are broadened by the combination of chemical shift anisotropy and dipolar broadening. The width of the chemical shift anisotropy depends on the carbon type and can be as large as 70 ppm. Thus, for a polymer with several carbon types, these resonances would all overlap, resulting in a broad, featureless spectrum. The dipolar broadening can be as large as 50 kHz, which is much larger than the chemical shift range. The broad line spectra can be used to study the chain dynamics, but they cannot provide high-resolution structural information.

In order to obtain molecular-level information about the structure of polymers in the solid state, it is necessary to obtain a high-resolution spectrum. This can be accomplished with the combination of magic-angle sample spinning to average the chemical shift anisotropy and high-power decoupling to remove the dipolar broadening. High-resolution spectra can be observed when (1) the spinning rate is fast compared to the breadth of the chemical shift anisotropy lineshape and (2) the decoupler power is greater than the dipolar linewidth.

The effect of magic-angle sample spinning on the carbon spectra of poly(bisphenol A carbonate) (**4**) is shown in Figure 9.20. The anisotropies for the aromatic carbons are on the order of several kHz, so the spinning must be faster than this to obtain the spectrum without spinning sidebands. At spinning speeds that are on the same order as the anisotropy, the chemical shift anisotropy line shape is broken up into a series of sidebands that are separated from the isotropic chemical shift by the spinning frequency. This is illustrated in Figure 9.20a. A spinning speed of 3.5 kHz is fast enough to average the anisotropies for the methyl and quaternary carbons, but many sidebands are observed for the aromatic and carbonyl carbons. In fact, at this spinning

Poly(bisphenol A carbonate)
4

nuclear vector to the magnetic field as does the chemical shift anisotropy. This suggests that if we can perform magic-angle averaging, either by applying a series of pulses or by spinning fast enough, then we would be able to obtain a high-resolution proton spectrum. The proton chemical shift range is smaller than for carbons, but protons are common to almost all polymers and they can be detected with a high sensitivity.

High-resolution proton spectra can be acquired using either multiple-pulse NMR in conjunction with relatively slow magic-angle sample spinning or very rapid magic-angle sample spinning to average the dipolar couplings. Multiple-pulse NMR works by applying a series of pulses so that, on average, the proton-proton dipolar couplings are all at the magic angle (Figure 9.5). By itself, however, the multiple-pulse sequence does not eliminate broadening from the anisotropic chemical shift, and it must be combined with sample spinning using the CRAMPS method (discussed here earlier). Alternatively, we can use very fast magic-angle sample spinning. Probes are currently available that spin as fast as 35 kHz, which is fast enough to provide substantial narrowing for many polymers.[26] If there is some fast molecular motion, as for polymers above T_g, then slower spinning speeds can also give rise to high-resolution spectra.[27] This is illustrated in Figure 9.21, which shows the solid-state proton spectrum for a poly(styrene-*co*-butadiene) copolymer at ambient temperature with 15 kHz magic-angle sample spinning. Once a high-resolution spectrum is obtained with fast magic-angle spinning, it is possi-

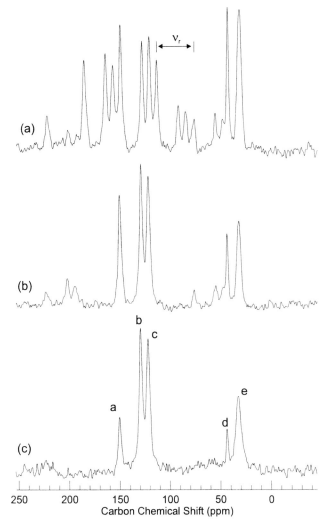

Figure 9.20 The effect of spinning speed on the 100-MHz solid-state carbon NMR spectra of polycarbonate acquired with a 1-ms cross-polarization time. The data are shown for spinning speeds of (a) 3500, (b) 7300, and (c) 9500 Hz, and the spinning sidebands are separated from the main peak by the spinning speed (v_r) Note that in the cross-polarized spectrum that the intensities are not related to the number of carbon atoms.

speed it is difficult to separate the peaks and the sidebands. By acquiring spectra at different spinning speeds, it is possible to distinguish the isotropic peaks from the spinning sidebands. As the spinning speed is increased to 7.3 kHz, the sidebands are shifted from the isotropic peaks and are reduced in intensity, and they are almost completely absent from the spectrum acquired with a spinning speed of 9.5 kHz.

For the observation of carbon-13 and many other nuclei, high-powered proton irradiation is used to remove the dipolar broadening and get high-resolution spectra. However, as shown by eq 5, the dipolar broadening has the same $(3\cos^2\theta - 1)$ dependence on the orientation of the inter-

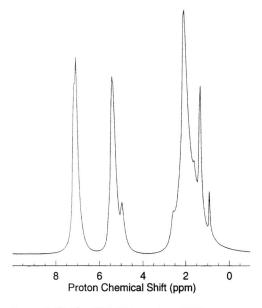

Figure 9.21 The 500-MHz proton NMR spectrum of poly(styrene-*co*-butadiene) acquired with magic-angle spinning at 15 kHz. The well-resolved peaks include the styrene aromatic protons at 7 ppm and the double-bonded protons at 5 ppm.

ble to use the common solution NMR methods, such as 2D exchange NMR, to measure both the local structure and intermolecular distances.[27]

Cross polarization is often used to measure the carbon, silicon, phosphorus, or nitrogen spectra of solid polymers because the spectra can be detected with a higher sensitivity. One important feature of cross polarization is that the rate of magnetization buildup and decay depends on the chain dynamics. This is illustrated in Figure 9.22 for a hypothetical polymer that contains a rigid and a mobile phase. The rigid (or crystalline) phase cross polarizes most quickly and decays most rapidly. Therefore, the crystalline component can be enhanced in a spectrum that is acquired with a short cross-polarization time. With a very long cross-polarization time, the signals arise primarily from the mobile phase. By observing the spectra as a function of cross-polarization time, it is possible to distinguish between the phases in complex materials. However, since the signal intensity depends on the strength of the dipolar couplings and the relaxation times, the spectra are not quantitative.

Chain Conformation

The NMR spectra of solids provide important information about the chain conformation of polymers. As for polymers in solution, the chemical shifts are determined by the local magnetic environment. Unlike the solution spectra, the

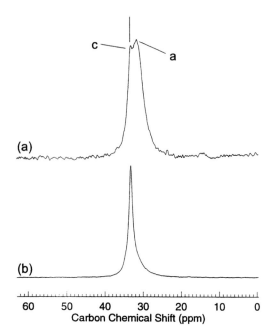

Figure 9.23 The 100-MHz carbon NMR spectra of polyethylene acquired (a) without and (b) with cross polarization. The notations c and a refer to the crystalline and amorphous phases. These spectra are not quantitative, and the relative amounts of the crystalline and amorphous phases cannot be determined from these spectra.

chains are restricted and the chemical shifts are not averaged. This means that the carbon chemical shifts can be used to determine the solid-state structure.

The effect of chain conformation on the chemical shifts in the solid state is illustrated in Figure 9.23, which shows the carbon spectrum of polyethylene acquired with and without cross polarization. The largest peak in the cross-polarized spectrum is a sharp resonance near 33 ppm that is assigned to the crystalline polyethylene in the all-*trans* conformation, and the broader resonance located 2.5 ppm to higher field is assigned to the amorphous material. Note that in the Bloch-decay spectrum acquired without cross polarization the largest peak is from the amorphous material. The ^{13}C relaxation times for crystalline polyethylene are very long, and the intensity of the crystalline peak is attenuated in the spectra acquired with a short delay time between scans. These data illustrate how the spectra of a particular phase in a complex material can be visualized with the proper choice of NMR parameters and experiments. This is very useful, but extreme care must be taken in the quantitative analysis of these materials. For these reasons, it should also be noted that the crystalline and noncrystalline peak intensity ratios are also not quantitative.

The chemical shifts for the crystalline and amorphous fractions of polyethylene are in general agreement with

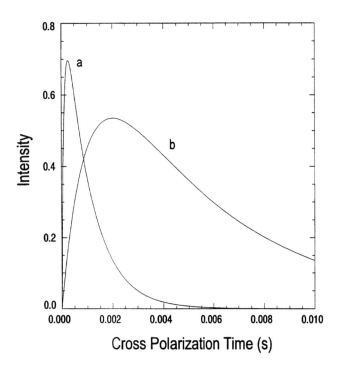

Figure 9.22 A schematic plot of the buildup and decay of magnetization during cross polarization for (a) rigid and (b) mobile material. The choice of a short cross-polarization time emphasizes the rigid phase, while a longer time gives a spectrum that is predominantly from the mobile phase.

those calculated from the γ-*gauche* effect.[6] In polyethylene melts and solutions, about 40% of the bonds are expected to be in the *gauche* conformation. This corresponds to fractional populations of 0.36, 0.48, and 0.16 for the *tt*, *gt* + *tg*, and *gg* conformations. If the value for the γ-*gauche* effect for a carbon is 5.2 ppm and the *tt*, *gt* + *tg*, and *gg* conformations have zero, one, and two γ-*gauche* interactions, respectively, this corresponds to an upfield shift of 4.1 ppm. While this prediction is in the right direction, the magnitude of the shift is considerably less than expected. The discrepancy between the calculated and observed chemical shifts may be due to a smaller fraction of chains in the *gauche* conformation or to other factors, such as intermolecular interactions, that may also affect the chemical shifts in the solid state. In the solid state it has been reported that the chemical shifts for the interior methylene carbons in *n*-alkanes can vary as much as 1.3 ppm, depending on the unit cell type, even though the isolated chain conformation is the same, namely all-*trans*.[28]

Another example of conformational effects on the solid-state NMR spectra is shown in Figure 9.24, which compares the carbon spectra for isotactic and syndiotactic polypropylene. Both polymers are crystalline, but the isotactic polymer adopts a ...*gtgtgt*... 3₁ helical conformation while the syndiotactic polymer forms a 2₁ helix with a ...*ggttggtt*... conformation.

The NMR spectra for the methylene carbons are very different in syndiotactic and isotactic polypropylene. In syndi-

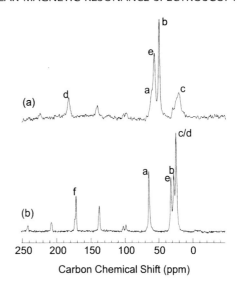

Figure 9.25 The 100-MHz solid-state carbon spectra of (a) amorphous poly(methyl methacrylate) and (b) semicrystalline polycaprolactone. Note the differences in linewidth for the crystalline and amorphous polymers. The assignments are illustrated in the text.

otactic polypropylene, half of the methylene groups lie along the interior of the helix and are in a *gauche* arrangement with both of their carbon neighbors in the γ positions, while half are on the exterior of the helix and are *trans* to their γ neighbors.[6] In isotactic polypropylene, the methylene groups have an equivalent environment: they are *trans* to one γ neighbor and *gauche* to another. For this reason, a single resonance is observed for the methylene groups in the isotactic polymer, while two resonances separated by 8.7 ppm are observed for the syndiotactic material. This difference in chemical shift for the methylene carbons in the syndiotactic polymer is approximately as large as two γ-*gauche* effects.[6] The methylene resonance for isotactic polypropylene appears midway between the two peaks in syndiotactic polypropylene, as expected for a methylene group that has one γ-*gauche* interaction.

As a general rule, the spectra for the crystalline phases are more highly resolved than the amorphous phases in semicrystalline polymers. These differences can be attributed to differences in the chain conformation and microstructure of the crystalline and noncrystalline phases. The chains in the crystalline phase all have the same conformation and are uniformly affected by the γ-*gauche* effect, while the lines are broadened in the amorphous phase from the distribution of chemical shifts that results from a distribution in conformations. This effect is illustrated in Figure 9.25, which compares the solid-state carbon spectra for semicrystalline polycaprolactone (**1**) and amorphous poly(methyl methacrylate) (**5**).

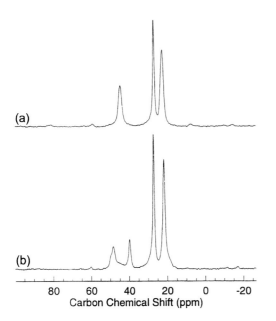

Figure 9.24 The 100-MHz solid-state carbon NMR spectra of (a) isotactic polypropylene and (b) syndiotactic polypropylene acquired with a 1-ms cross-polalization contact time and magic-angle spinning at 3.5 kHz. Note the differences in the methylene peaks (35–45 ppm) that result from the γ-*gauche* effect.

Poly(methyl methacrylate)

5

Note that the main-chain signals are broader for poly-(methyl methacrylate).

The carbon chemical shifts are very sensitive to chain conformation. This makes solid-state NMR a good method to probe the structure of crystalline polymers that can adopt multiple conformations. Figure 9.26 shows the 50-MHz carbon solid-state carbon spectra of poly(diethyl oxetane) (**6**)

Poly(diethyl oxetane)

6

for samples crystallized at 60 °C, 35 °C, and 0 °C.[29] Form I is obtained at high temperature and is characterized by an

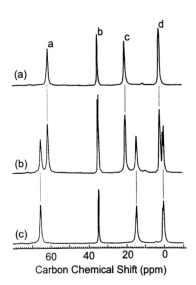

Figure 9.26 The 50-MHz solid-state carbon NMR spectra of poly(diethyl oxetane) crystallized at (a) 60 °C, (b) 35 °C, and (c) 0 °C. Note that peaks for both the low- and high-temperature phases are observed in the spectrum of the polymer crystallized at 35 °C.

all-*trans* chain conformation and a melting temperature of 73 °C. Form II is obtained from the low temperature crystallization and has a ...*ttggttgg*... conformation and a melting temperature of 57 °C. The spectra of samples crystallized at high or low temperature are characterized by a single component, while the samples crystallized at intermediate temperatures contain a mixture of the two forms.

The chemical shifts observed for forms I and II of poly (diethyloxetane) are not all in good agreement with those calculated from the γ-*gauche* effect.[29] The chemical shifts for the quaternary and methyl carbons are close to the calculated values, but the shifts for the methylene carbons are not. The main-chain methylene is predicted to be more upfield in form II, but resonantes 3.9 ppm downfield, and the side chain methylene is shifted upfield by 5.9 ppm instead of downfield. This shows that factors other than the γ-*gauche* effect can affect the chemical shifts in the solid state. Among the possible factors that can affect the chemical shift are interchain packing and changes in the valence angles in the different conformations. It is unlikely that the chemical shift differences can be attributed solely to chain-packing effects since much smaller chain-packing effects are observed in model hydrocarbon polymers.[28]

The [29]Si chemical shifts are extremely sensitive to chain conformation and have been used to study both silane and siloxane polymers. Silane polymers are of interest as potential materials for microlithography because of their unusually long wavelength UV absorption due to "sigma conjugation".[30] Two crystalline forms of polysilanes (**7**)

Poly(alkyl silane)

7

have been reported, and the chain conformation is sensitive to the side chain groups. Figure 9.27 compares the 39-MHz ^{29}Si NMR spectra of polysilanes with hexyl, pentyl, and butyl side chains.[31] In all cases two helical forms are observed: a well-ordered form I and a disordered form II. In form I for the poly(di-*n*-hexylsilane) the chain adopts an all-*trans* conformation and shows resonances around −22 ppm, while the form I for both the poly(di-*n*-pentylsilane) and poly(di-*n*-butylsilane) adopts a 7_3 helical conformation and shows resonances around −25 ppm. The chemical shifts for the disordered form II helix appear in the range of −23 to −24 ppm.

In addition to the semicrystalline and amorphous polymers, several other polymer morphologies have been studied by solid-state NMR. These include phase separated polymers,[32] elastomers,[33] swollen polymers,[34] polymer blends,[35] and inclusion compounds.[36] Figure 9.28 shows the carbon

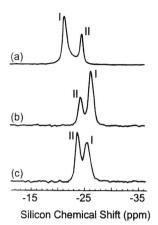

Figure 9.27 The 39-MHz ^{29}Si solid-state NMR spectra of (a) poly(di-*n*-hexyl silane), (b) poly(di-*n*-pentyl silane), and (c) poly(di-*n*-butyl silane) acquired with cross polarization and magic-angle sample spinning. The peaks are listed for the ordered form I and the disordered phase II. The relative intensity of the form I and form II peaks depends both on temperature and on the cross-polarization dynamics.

Silicon Chemical Shift (ppm)

spectra of the inclusion complex formed by α-cyclodextrin (α-CD) (**8**)

6
CH$_2$OH

α-CD

8

and polycaprolactone (**1**). α-CD is a pore-forming molecule that consists of six glucose units arranged around a 5.7 Å pore.[37-39] The pores can be filled with polymers such as polycaprolactone[40] and nylon-6.[41] Information about the chain conformation is obtained by comparing the polymer chemical shifts in the inclusion complex with those from the crystalline phase, which exists in an all-*trans* conformation. The chemical shifts in the inclusion complex are very similar to the semicrystalline polymers, suggesting that the

chains in the channels are also in the all-*trans* conformation.

Elastomers were among the earliest polymers studied by solid-state NMR.[1] These materials are above T_g at ambient temperature, and the broadening from both chemical shift anisotropy and dipolar interactions is partially averaged by chain motion. In some cases, such as *cis*-1,4-polybutadiene, it has been reported that high-resolution spectra can be obtained with magic-angle spinning alone.[42] These materials can be considered as solution-like in that the broadening is mostly averaged by chain motion. This makes it possible to use the solution NMR methods for spectral editing for the analysis of elastomers.[43] It is possible to use fast magic-angle spinning NMR to obtain high-resolution proton spectrum, as shown in Figure 9.21 for the poly(styrene-*co*-butadiene) copolymer. Magic-angle spinning has also been used to observe the proton and carbon spectra of swollen gels.[34]

Polymer Morphology

Solid-state NMR is extensively used to probe the domain sizes in semicrystalline polymers and the length scale of mixing in block copolymers, blends, and phase-separated mixtures. These studies use proton spin diffusion to estimate the length scale of mixing. The general strategy is to create a polarization gradient between spins in different regions and to monitor the return of the spin system to equilibrium via spin diffusion. The protons can relax either by spin-lattice relaxation or by mutual spin flips with nearby protons. The elementary process of mutual spin flips that underlies spin diffusion is very efficient and can lead to magnetization flow over long length scales. The transfer of magnetization can be described by the diffusion equation, and is given by

$$\dot{m}(r,t) = D\nabla^2 m(r,t) \qquad (14)$$

where D is the diffusion coefficient, and $\dot{m}(r,t)$ is the magnetization density as a function of position and time.

The production of a polarization gradient between different regions of a sample is most often based on differences in the relaxation times. If one part of the sample relaxes more rapidly than another, then it is possible to prepare the spin system in a nonequilibrium state. One way to achieve this selective excitation of one phase over another is the Goldman-Shen experiment[44] shown in Figure 9.29. This experiment relies on a difference in the T_2 relaxation times to selectively saturate the spins in a rigid phase while leaving the mobile phase only slightly perturbed. The experiment begins with a 90° pulse that tips the proton magnetization into the *xy* plane. If the spins from the rigid phase decay quickly, then a short time later only the signal from the mobile phase will remain. These spins are then returned to their equilibrium position with a 90° pulse of the opposite phase, and the magnetization is detected after a period for spin diffusion.

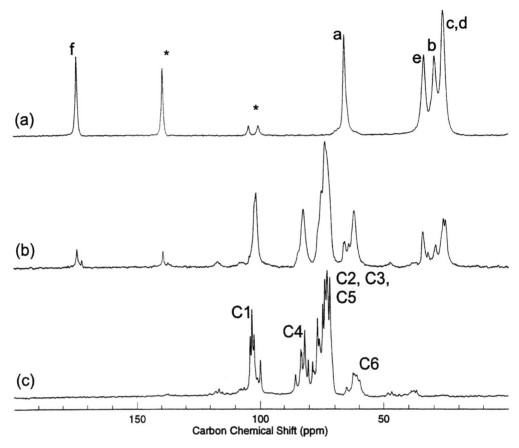

Figure 9.28 The 100-MHz solid-state carbon NMR spectra of (a) polycaprolactone, (b) the polycaprolactone/α-cyclodextrin inclusion complex, and (c) α-cylclodextrin acquired with cross polarization and magic-angle sample spinning. The many peaks observed for each carbon in the the pure cyclodextrin in (c) arise from different molecules in the unit cell. Note that the polymer peaks can be easily identified in the complex. The assignments are shown in the text.

The fundamental process in spin diffusion is the mutual spin flips between nearby protons that leads to magnetization flow from the more mobile phase with a longer T_2 to the more rigid phase with a shorter T_2. The rate-limiting step is the magnetization transfer from the mobile to the rigid phase. The spin diffusion coefficient depends on the spin density, the average distances between spins, and the chain dynamics. In one early study, the diffusion coefficient was estimated as[45]

$$D = \frac{13a^2}{T_2} \quad (15)$$

where a is the average distance between protons, and T_2 is the spin-spin relaxation time for the mobile phase. More recently, these diffusion coefficients have been measured in block copolymers with known morphologies. For glassy polymers, such as polystyrene and poly(methyl methacrylate), the diffusion coefficient is on the order of 0.8×10^{-15} m^2/s.[46] For more mobile polymers the diffusion coefficient is given by $D = 4.4 \times 10^{-4}/T_2 + 0.26$ m^2/s for polymers with T_2s between 1000 and 3500 Hz and by $D = 8.2 \times 10^{-6}/T_2^{1.5} + 0.007$ m^2/s for polymers with T_2s in the range of 0 to 1000 Hz.[47]

The return to equilibrium following a perturbation is given by the recovery factor $R(t)$ that is defined as

$$R(t) = 1 - \frac{M(t) - M(\infty)}{M(0) - M(\infty)} \quad (16)$$

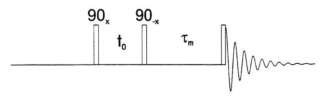

Figure 9.29 The pulse sequence diagram for the Goldman-Shen experiment. The magnetization from the rigid phase dephases during the time t_0 from T_2 relaxation. The magnetization from the mobile phase does not, and is returned to the z axis with the 90° pulse. The spectrum is sampled after the spin diffusion delay time τ_m.

where M is the measured magnetization of the rigid component. For the simplest case of one-dimensional diffusion, the recovery factor is given by[45]

$$R(t) = 1 - \varphi(t) \qquad (17)$$

where

$$\varphi(t) = e^{Dt/\bar{b}^2} erfc(Dt/\bar{b}^2) \qquad (18)$$

where $\bar{b}$ is the average domain size and $erfc$ is the complement of the error function. The recovery factor for diffusion in two and three dimensions is given by

$$R(t) = 1 - \varphi(t)^2 \qquad (19)$$

and

$$R(t) = 1 - \varphi(t)^3 \qquad (20)$$

respectively.

It is often observed that the data can be fit using the 2D or 3D models, but this description is not always physically reasonable, and more complex models must be used. These models can include an interface with a diffusion coefficient that is intermediate between the rigid and the mobile phases. These results show that the spin diffusion data alone are not sufficient to completely determine the dimensionality and domain sizes. However, if some other information is known, such as the dimensionality, then the spin diffusion data can be used to obtain accurate measurements of the domain sizes.

Given the uncertainties in the models used to fit the data, it is often best to estimate the domain sizes from the initial rate of spin diffusion. This procedure is illustrated in Figure 9.30, which shows an extrapolation of the initial rate of spin diffusion to the plateau value to obtain t_m^*. The domain size d is given by

$$d = \left(\frac{\rho_{HA}\varphi_A + \rho_{HB}\varphi_B}{\varphi_A\varphi_B} \right) \frac{4\varepsilon\varphi_{dis}}{\sqrt{\pi}} \frac{\sqrt{D_A D_B}}{\rho_{HA}\sqrt{D_A} + \rho_{HB}\sqrt{D_B}} \sqrt{t_m^*} \quad (21)$$

where ρ is the proton density, φ is the volume fraction, and ε is the dimensionality. For the simple case of $\rho_{HA} = \rho_{HB}$, $\varphi_A = \varphi_B$, and $D_A = D_B$, this equation reduces to

$$d = \sqrt{\frac{16Dt_m^*}{\pi}} \qquad (22)$$

A variety of experiments have been used to measure domain sizes in polymers. These pulse sequences use different methods to create a population difference between the parts of the sample. The Goldman-Shen pulse sequence (Figure 9.29) creates a difference in spin populations based on differences in the T_2 relaxation rates for the rigid and mobile phases. This is an effective way to create a population difference, but the resolution is limited since the proton spectrum is observed, and the lines can be very broad in solids. It is possible to monitor spin diffusion with a higher resolution

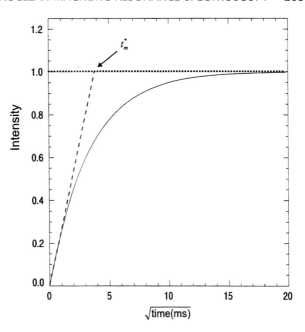

Figure 9.30 A schematic plot of magnetization recovery in a spin diffusion experiment. The initial buildup is extrapolated to the plateau value to obtain t_m^*, which is used to calculate the length scale of spin diffusion and the domain sizes.

by combining this pulse sequence with cross polarization, as shown in Figure 9.31a. This variant of the Goldman-Shen sequence has the advantage that the carbon spectrum is observed with cross polarization and magic angle spinning. Another approach is to precede the cross polarization with a dipolar filter, as shown in Figure 9.31b.[48] The dipolar filter is a train of pulses that saturates the protons in the rigid phase while only slightly perturbing the spins from the mobile phase. The dipolar filter part of the experiment is a 12-pulse homonuclear decoupling sequence. To discriminate between the mobile and rigid phases, the delay between pulses is made longer, so the decoupling of the rigid phase is very inefficient and leads to saturation.

Both low-density and high-density polyethylene have been studied by the dipolar filter, and Figure 9.32 shows a plot of the spin diffusion data for low-density polyethylene. The data could not be adequately fitted to a simple two-phase model, demonstrating that there is an interfacial component that was not detected in the earlier experiments. Direct evidence for the presence of an interfacial layer was obtained by monitoring the intensity of the amorphous component as a function of the spin diffusion time. After the rigid protons are saturated by using the dipolar filter, the intensity of the amorphous phase reaches a maximum after 5 ms, indicating that there is spin diffusion between the mobile amorphous material to a less mobile amorphous material, the interface. At a longer delay time (100 ms) there is extensive spin diffusion to the crystalline phase. The diffusion behavior was fitted to a three-phase model con-

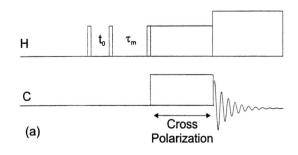

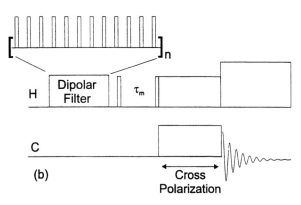

Figure 9.31 The pulse sequence diagrams for experiments to measure proton spin diffusion using (a) the Goldman-Shen experiment with detection by cross polarization and (b) the dipolar filter.

taining crystalline, amorphous, and interfacial phases. Despite the large differences in the crystallite thicknesses for the high-density (40 ± 10 nm) and low-density (9 ± 2.5 nm) polyethylene, the thickness of the interfacial layer was shown to be 2.2 ± 0.5 nm in both samples.

Block copolymers are another class of polymers with ordered solid-state morphologies. For some block copolymers, such as styrene-*b*-butadiene or styrene-*b*-isoprene, there are large differences in the mobilities of the blocks, and the domain sizes can effectively be studied using the dipolar filter. This is illustrated in Figure 9.33, which shows the dipolar filter experiment for a poly(styrene-*b*-isoprene-*b*-styrene) triblock copolymer.[50] In this experiment the proton spectrum was recorded using proton multipulse NMR (CRAMPS) for line narrowing after the dipolar filter and a spin diffusion delay. At the shortest mixing times, the styrene signals have been saturated by the dipolar filter and they build up after a spin diffusion delay time. Figure 9.34 shows a plot of the recovery factor as a function of the square root of the mixing time. We obtain a value of 0.019 s for τ_m^* by extrapolation to the plateau value, and from eq 21 we esti-

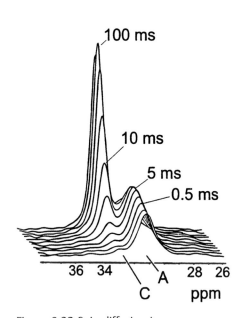

Figure 9.32 Spin diffusion in semicrystalline polyethylene measured using the dipolar filter. The crystalline peak has been saturated with the dipolar filter and grows back from proton spin diffusion.

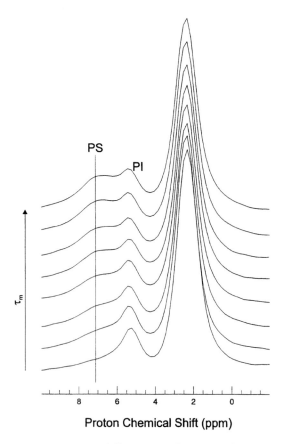

Figure 9.33 Spin diffusion in poly(styrene-*b*-isoprene-*b*-styrene) block copolymers measured by solid-state proton NMR. Spin diffusion is monitored from the recovery of the polystyrene aromatic signals at 7 ppm. The proton spectrum was recorded at 400 MHz using CRAMPS, after the styrene signals were saturated with the dipolar filter.

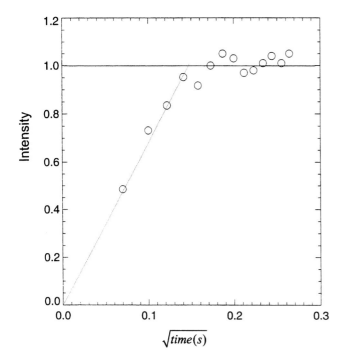

Figure 9.34 A plot of magnetization recovery for the poly(styrene-*b*-isoprene-*b*-styrene) block copolymer. The characteristic spin diffusion time was obtained by extrapolating the intial rate to the plateau.

mate the size of the dispersed domain to be 5 nm. Films with very different surface morphologies give similar domain sizes, showing that the morphological differences are not related to differences in the phase separation on the molecular length scale.

Spin diffusion in block copolymers has been studied by a number of methods that differ in the way that selective saturation was achieved. In a study of poly(styrene-*b*-methyl methacrylate) polymers, this selection was based on proton chemical shift differences.[46] This method makes it possible to study phase-separated polymers that have similar chain dynamics and relaxation times. The NMR measurements were performed on samples that had been characterized by X-ray scattering, so the results could be directly compared.

Polymer blends are another class of solid polymers that have been extensively studied by solid-state NMR. In favorable cases it is possible to identify the interacting groups that cause the polymers to mix on the molecular level[51–53] and to measure the length scale of mixing in blends that are partially mixed.[54,55] These results can be compared with other measurements, such as scattering methods, dielectric and dynamic-mechanical spectroscopy, and differential-scanning calorimetry (DSC) measurements.

Blend miscibility is frequently a consequence of strong intermolecular interactions, such as ionic or hydrogen-bonding interactions. In certain cases these interactions are strong enough to cause shifts in the NMR spectrum. This is

the case for the carbonyl peak in the NMR spectrum of the poly(ethylene oxide)/poly(acrylic acid) blend.[56] A new peak is observed at 185 ppm that can be assigned to intermolecular hydrogen bonding. In cases where the induced shifts are not as large as for the poly(methylacrylic acid)/poly(ethylene oxide) blend, complex formation is often accompanied by line broadening from a distribution of chemical shifts.[57]

One feature that distinguishes miscible blends from other solid polymer mixtures is that the chains are mixed on a very short length scale. The length scale of mixing can be measured by the methods discussed in this chapter for semicrystalline polymers and block copolymers, as well as several other methods. Perhaps the best method to demonstrate intimate mixing is intermolecular cross polarization using a mixture of protonated and deuterated polymers.[58] Since the deuterated polymer contains no protons, its carbon signals can only appear in the cross-polarized spectrum via intermolecular cross polarization if the length scale of mixing is small enough. Such intermolecular cross polarization has been reported for the blends of polystyrene and poly(vinyl methyl ether).[59]

The length scale of polymer mixing in blends and other phase-separated mixtures can also be measured by the proton T_1 and T_1 relaxation rate rates,[60] where $T_{1\rho}$ is the proton relaxation rate in the presence of a spin-locking rf field. If the chains are mixed on the spin diffusion length scale, then magnetization transfer from the slower to the more rapidly relaxing chains is a very effective relaxation mechanism. In such cases, the relaxation times of both chains are equal to a weighted average of the values for the individual chains a and b and are given by

$$k = k_a \frac{N_a \varphi_a}{N_a + N_b} + k_b \frac{N_b \varphi_b}{N_a + N_b} \qquad (23)$$

where k, k_a, and k_b are the relaxation rates (either $1/T_1$ or $1/T_{1\rho}$) for the blend and the pure polymers, N_a, and N_b are the number of protons, and φ_a and φ_b are the mole fractions. The length scale of mixing that can be probed in such experiments depends on the relaxation times. The proton T_1 relaxation is often slow in polymers (0.5–5 s) and can be used to probe mixing on a length scale up to 500 Å. The $T_{1\rho}$ relaxation is faster (0.5–5 ms) and probes mixing on a length scale of less than 20 Å. In those cases where the length scale of mixing is much longer than the spin diffusion length scale, then the measured relaxation rates will be equal to the values of the pure polymers. In cases where there is partial mixing, the relaxation behavior may be more complex, and multiexponental relaxation may be observed.[58] One limitation of this approach to measuring polymer mixing is that the relaxation rates for the pure polymers must differ by more than a factor of 2 to accurately measure the averaging of the relaxation rates.

Table 9.3 shows the relaxation times for the poly(ethylene oxide)/poly(methacrylic acid) blend.[56] In the bulk states

Table 9.3. The solid-state proton spin-lattice and rotating-frame spin-lattice relaxation times for poly(ethylene oxide) and poly(methacrylic acid) in bulk and in the miscible blend at 237 K

Sample	T_1 (s)	$T_{1\rho}$ (ms)
Bulk		
Poly(ethylene oxide)	3.6	20.7
Poly(methacrylic acid)	0.4	2.2
Blend		
Poly(ethylene oxide)	0.50	2.1
Poly(methacrylic acid)	0.52	2.2

there are large differences between the T_1 and $T_{1\rho}$ values for the two polymers, while identical values are observed for the blend. This shows that the chains are mixed on a length scale of less than 20 Å.

Polymer Reactivity

Solid-state NMR is also a valuable technique for measuring the reactivity in polymers. The reactions in polymers lead to large changes in the chemical structure that can be detected by chemical shift changes in the high-resolution spectrum. This is illustrated in Figure 9.35 for an acrylate formulation. The polymer was prepared by photochemical polymerization of a mixture of acrylates, so monomers with double bonds are consumed to form the polymer. The protons attached to carbons with double bonds are well resolved from the other protons and can be used to monitor the extent of reaction. This formulation has a low T_g, so it is possible to obtain high-resolution spectrum at ambient temperature with fast magic-angle spinning alone. The spectrum in Figure 9.35 was obtained with 15-kHz magic-angle sample spinning, and the signals from the olefinic protons are well resolved at 5 to 6 ppm. We can estimate that about 15% of the monomer is unreacted after photopolymerization.

NMR studies of the reactivity of polymers are particularly important for insoluble materials, such as polyimides and other high-temperature polymers. Figure 9.36 shows how solid-state silicon NMR can be used to monitor the reactivity in poly[(phenylsilylene)ethynylene-1,3-phenyleneethynylene] (**9**).[61]

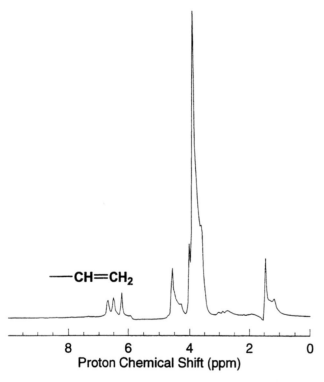

Figure 9.35 The solid-state proton NMR spectrum for an acrylate formulation obtained with 15 kHz magic-angle sample spinning. Note the well-resolved peaks at 6 to 7 ppm that are assigned to unreacted monomer.

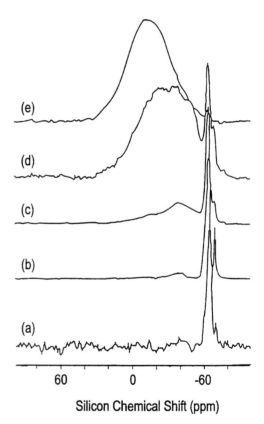

Figure 9.36 The solid-state ^{29}Si NMR spectra for poly-[(phenylsilylene)ethynylene-1,3-phenyleneethynlylene] (a) before curing and following curing at (b) 150 °C, (c) 200 °C, (d) 300 °C, and (e) 400 °C.

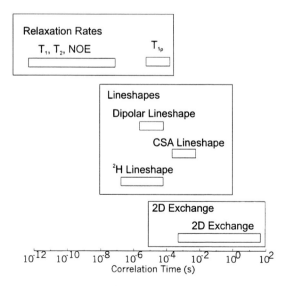

Poly[phenylsilylene)ethynylene-1,3-phenyleneethylnylene]
9

The uncured material shows a relatively sharp peak at −60 ppm that is consumed during the cure. The broad lines in the final material result from a variety of chemical products.

Polymer Dynamics

NMR has long been used to study the dynamics of polymers, since the relaxation times and line shapes are extremely sensitive to the chain motion that occurs over many orders of magnitude. It is well known from NMR and other studies that the chain dynamics of polymers vary from picoseconds to seconds. The motions include very rapid librational motions and *gauche-trans* isomerizations of polymers in solution, the slower motions in solid polymers, and the very slow diffusion of entire chains in the crystalline phase. The NMR method of choice for investigating the dynamics depends on the time scale of motion. Figure 9.37 shows a plot of some of the relaxation and lineshape measurements that can be used to probe the dynamics of polymers over a time scale from picoseconds to seconds.

There are three categories of experiments that have been used to probe the dynamics of polymers: relaxation rate

Figure 9.37 A plot showing the range in correlation time that can be measured by NMR relaxation rates, lineshapes, and two-dimensional (2D) exchange NMR.

measurements, lineshape measurements, and multidimensional exchange experiments. Some of the relaxation measurements, such as the T_1, T_2, and nuclear Overhauser effect (NOE) are sensitive to high-frequency motion. These relaxation rates have been extensively used to monitor the solution dynamics of polymers. The lineshapes and some relaxation time measurements, such as $T_{1\rho}$, are sensitive to chain dynamics on the kHz time scale and are used to probe the slow dynamics of solid polymers. The multidimensional exchange experiments measure the change in orientation of a polymer chain over a much longer time scale, on the order of milliseconds to seconds, and probe the slowest dynamics that can be measured by NMR.

Nuclear relaxation is most often caused by fluctuations in local magnetic fields that result from molecular motion. These fluctuations in the local magnetic fields can occur over a broad range of frequencies, and to the extent that these fluctuations have components at the resonance frequency for the nuclei of interest, they will cause relaxation. The distribution of motional frequencies is called the *spectral density function* and is given by

$$J(\omega) = \frac{1}{2} \int_{-\infty}^{+\infty} G(\tau)e^{-i\omega\tau}d\tau \qquad (24)$$

where $G(\tau)$ is the autocorrelation function for an internuclear vector—such as a ^{13}C-^1H vector—that is given by

$$G(\tau) = \langle F(t)F^*(t+\tau) \rangle \qquad (25)$$

The brackets denote the ensemble average over a collection of nuclei, and F represents a function, related to spherical harmonics, describing the position and motion of the molecule. $J(\omega)$ may be thought of as expressing the power available at frequency ω to relax the spins in question. The spectral densities and autocorrelation functions are Fourier inverses of each other in the time and frequency domains, respectively. If $G(\tau)$ decays to zero in a short time, this corresponds to a short correlation time τ_c, which means that molecular motion is rapid and that the molecules have only a short memory of their previous state of motion.

In order to model the relaxation of polymers and other molecules, we must adopt a dynamical model for $G(\tau)$. The simplest model is that of a rigid sphere immersed in a viscous continuum that is reoriented in small diffusive steps. The correlation time can be thought of as the interval between these alterations in the state of motion of the molecule. For such a molecule, the loss of memory of the previous motional state is exponential with the time constant τ_c

$$G(\tau) = e^{-(\tau/\tau_c)} \qquad (26)$$

and the spectral density function becomes

$$J(\omega) = \frac{\tau_c}{1 + \omega^2\tau_c^2} \qquad (27)$$

Figure 9.38 shows a plot of the spectral density function as

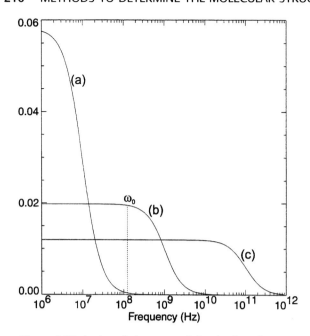

Figure 9.38 A plot of the spectral density functions as a function of frequency for (a) short, (b) intermediate, and (c) long values of the correlation time. The relaxation time is determined by the density at the Larmour frequency ω_0.

a function of frequency for three values of τ_c. A long value of τ_c corresponds to the molecular motion of a large molecule, a stiff chain, or a small molecule in a viscous medium; a short value of τ_c corresponds to the rapid motion of a small molecule or a very flexible polymer chain. Also shown in Figure 9.38 is the plot for an intermediate case, when $\tau_c\omega_0 \cong 1$, where ω_0 is the resonant frequency of the observed nuclei. The areas under the curves are the same for all three correlation times, which means that the power available to cause relaxation is the same, and only the distribution varies with τ_c. For short correlation times, the component at ω_0 is weak, and for the long τ_c, the frequency spectrum is so broad that no one component, particularly that at ω_0, is very intense. At some intermediate value of τ_c, the intensity at ω_0 will be at a maximum. Since the relaxation rate depends on the component at ω_0, it will be at a maximum for the intermediate value of τ_c, and the relaxation rates will be much slower for very fast or very slow motion. The exact dependence of the relaxation rate on the correlation time will depend on which experiment is performed and which relaxation rates are measured. A critical part of measuring the molecular dynamics of polymers is choosing the proper model for the correlation functions and spectral densities. It is well established that a single exponential correlation function is not a good model to describe the molecular motion of an object as complex as a polymer chain.[62]

Understanding the relaxation mechanisms for the nuclei of interest is an important first step in studying the molecular dynamics of polymers. Depending on the nuclei and its chemical environment, there can be several mechanisms that contribution to the observed relaxation rate, including dipolar interactions, chemical shift anisotropy or quadrupolar relaxation.[7] The observed relaxation rate is the sum of all of these contributions and is given by

$$\frac{1}{T_1} = \frac{1}{T_1^{DD}} + \frac{1}{T_1^{CSA}} + \frac{1}{T_1^{Quad}} \tag{28}$$

The dipolar contribution to the spin-lattice and spin-spin relaxation for a carbon that is relaxed by n directly bonded protons is given by

$$\frac{1}{T_1^{DD}} = \frac{n}{10}\left(\frac{\mu_0}{4\pi}\right)^2 \frac{\gamma_C^2\gamma_H^2\hbar}{r^6} \tag{29}$$

$$\times \{J(\omega_H - \omega_C) + 3J(\omega_C) + 6J(\omega_H + \omega_C)\}$$

and

$$\frac{1}{T_2^{DD}} = \frac{n}{20}\left(\frac{\mu_0}{4\pi}\right)^2 \frac{\gamma_C^2\gamma_H^2\hbar}{r^6} \{4J(0) \tag{30}$$

$$+ J(\omega_H - \omega_C) + 3J(\omega_C) + 6J(\omega_H) + 6J(\omega_H + \omega_C)\}$$

where μ_0 is the vacuum magnetic permeability, γ_H and γ_C are the magnetogyric ratios for protons and carbons, ω_H and ω_C are the resonant frequencies in rads for protons and carbons, and r is the internuclear distance. The CH distance is taken as 1.08 Å for methine and methylene carbons and as 1.09 Å for aromatic carbons. The nuclear Overhauser enhancement, which is the change in carbon intensity when the protons are irradiated, is also due to dipolar interactions and is given by

$$\text{NOE} = 1 + \frac{\gamma_H}{\gamma_C} \frac{\{6J(\omega_H + \omega_C) - J(\omega_H - \omega_C)\}}{\{J(\omega_H - \omega_C) + 3J(\omega_C) + 6J(\omega_H + \omega_C)\}} \tag{31}$$

The first step in the relaxation studies for polymers in solution is to choose a nucleus for which the relaxation mechanism is well understood and well characterized. ^{13}C is often the nucleus of choice because it is a common element in polymers, and the relaxation is generally dominated by dipolar interactions with the directly bonded protons. The relaxation in aromatic, carbonyl, and double-bonded carbons is sometimes more complex because they may have contributions from relaxation by chemical shift anisotropy. Quadrupolar nuclei, such as deuterium, can also be used since the relaxation is dominated by the quadrupolar interactions.[7]

As a general rule, it is difficult to prove that any particular motional model for $G(\tau)$ is an accurate representation of the polymer chain dynamics. The most common strategy is to make as many different types of relaxation measurements as possible over a range of magnetic field strengths, temperatures, and viscosities to demonstrate that the data are consistent with a particular model. This approach is necessary because the relaxation measurements only sample the spectral densities at a few points. The T_1s, T_2s, and NOEs all have a different dependence on the spectral densities, and the motional model must be consistent with all of the data.

NMR studies of the molecular dynamics of solid polymers use these same types of measurements in addition to several others. Solids differ from solutions in that molecular motion is usually neither fast enough nor isotropic enough to average the dipolar interactions, the chemical shift anisotropy, or the quadrupolar interactions; hence, information about the molecular dynamics can be extracted from the lineshapes. One dramatic manifestation of this effect is the comparison of the lineshapes for polymers above and below the glass-transition temperature, since the lines are narrowed above T_g by chain motion. Other types of motions that are anisotropic and fast on the time scale of the chemical shift anisotropy or the quadrupolar couplings, such as methyl rotation, lead to characteristic changes in the lineshapes. In deuterium NMR, for example, the quadrupolar lineshapes resulting from aromatic ring flips or methyl group rotations can easily be recognized.[63]

The observation of NMR lineshapes, particularly those due to quadrupolar couplings in deuterated polymers or the chemical shift anisotropy in the carbon, silicon, phosphorus, or nitrogen spectra, are a rich source of information about the molecular dynamics of polymers. The lineshapes have a characteristic appearance in the absence of molecular motion, and the changes due to the molecular dynamics depend on the geometry, amplitude, and frequency of molecular motion. Solution-like spectra are obtained above T_g, since fast and large amplitude motions with a nearly isotropic distribution effectively average the lineshape.

Deuterium NMR is often used to study polymer dynamics because the relaxation is dominated by the quadrupolar coupling and the electric field gradient tensor orientations are generally well known.[63] Deuterium has a low sensitivity, so polymers with site-specific labels are required. This is a disadvantage in that new polymers must be synthesized but an advantage in that the labels are incorporated into the polymer at well-defined sites, such as the main-chain or the side-chain atoms. In the absence of molecular motion, the frequency of a given deuteron is given by

$$\omega = \omega_0 + \delta \, (3\cos^2\theta - 1 - \eta\sin^2\theta \, \cos 2\phi) \qquad (32)$$

where ω_0 is the resonance frequency, $\delta = 3\hbar^2 qQ/8\hbar$, $e^2qQ/\hbar$ is the quadrupolar coupling constant, and η is the asymmetry parameter; the orientation of the magnetic field in the principal axis system of the electric field gradient tensor is specified by the angles θ and ϕ. For C–D bonds in rigid solids, $\delta/2\pi = 62.5$ kHz and $\eta \approx 0$. Thus, two lines are observed corresponding to the transitions for each deuteron, as shown in Figure 9.39a, and the separation between the lines depends on the orientation of the CD bond vector rela-

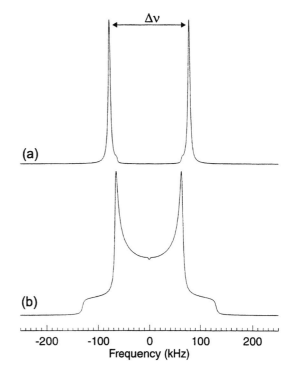

Figure 9.39 A schematic diagram showing the ^{2}H NMR spectra for (a) a single deuteron and (b) an isotropic sample. Each deuteron is split into a doublet from the quadrupolar coupling, and the powder lineshape in (b) results from adding the spectra for an isotropic distribution of sample orientations.

tive to the magnetic field. In isotropic samples, averaging over all possible orientations gives rise to the well-known "Pake" spectrum, as shown in Figure 9.39b.

Molecular motion in the polymer will cause the lineshapes to change in a way that depends on the geometry and time scale of the molecular motion. If the motion is rapid on the time scale defined by the coupling constant, $1/\delta$, it is said to be in the fast motion limit ($\tau_c < 10^{-7}$ s). This leads to a characteristic change in line shape given by

$$\omega = \omega_0 \pm \bar{\delta} \, (3 \, \cos^2 \theta - 1 - \bar{\eta} \, \sin^2 \theta \, \cos 2\phi) \qquad (33)$$

where $\bar{\delta}$ and $\bar{\eta}$ are the coupling constant and asymmetry parameter for the *averaged* electric field gradient tensor. It is important to note that $\bar{\eta}$ may be different from zero even though $\eta = 0$. This is illustrated in Figure 9.40, which shows the deuterium lineshapes for kink 3-bond rotation, crankshaft 5-bond motion, and 180° aromatic ring flips.[63] These motions give rise to $\bar{\eta}$ values of 1, 0, and 0.6 and illustrate how different types of molecular motion result in characteristic deuterium lineshapes.

^{2}H NMR is a powerful method for the study of the molecular dynamics of polymers, but caution must be used since it is possible for different motions to give rise to the same

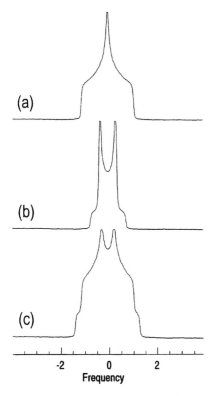

Figure 9.40 The deuterium lineshapes for deuterons undergoing (a) three-bond kink motion, (b) crankshaft 5-bond motion, and (c) 180° flips. The spectrum (c) is often observed for deuterons attached to aromatic rings.

served in the 2D exchange experiments, so off-diagonal ridge patterns are observed, rather than resolved cross peaks as in solution 2D exchange NMR. The shape of the ridge pattern depends on the jump angle, and the intensity depends on the relative ratio of the correlation time and the mixing time.

The Dynamics of Polymers in Solution

It was recognized in the early NMR studies of polymers that the chains must be extremely flexible to give such high-resolution spectra. This observation was contrary to the expected behavior, since the rotational correlation time for a rigid body is given by

$$\tau_r = \frac{2M[\eta]\eta_0}{3RT} \qquad (34)$$

where M is the molecular weight, $[\eta]$ is the intrinsic viscosity of the polymer, and η_0 is the solvent viscosity. In the absence of segmental dynamics that are rapid on the time scale of τ_r, broad lines would be expected for even moderate molecular weight polymers. Furthermore, it is frequently observed that the linewidths for the side-chain resonances are sharper than those for the main-chain resonances, demonstrating that the side chains experience additional motion relative to the main chain, presumably from rotation about the carbon-carbon bonds. Finally, in some polymers, such as the polybutadienes, different relaxation times are observed for neighboring main-chain atoms.[68] Any successful model for the chain dynamics must simultaneously account for all of these observations.

The contribution of segmental motion to the relaxation of polymers can be experimentally demonstrated by measuring the relaxation times as a function of molecular weight, as shown in Figure 9.41 for polystyrene.[69] At low molecular weights the relaxation times are sensitive to the degree of polymerization, but only small changes in the T_1s are observed for polymers with molecular weights greater than 10,000 g/mol. At molecular weights above 10,000 g/mol, the relaxation must therefore be due exclusively to segmental motion.

The chain dynamics of polymers depend on the chemical structure and microstructure. To illustrate this relationship, it is instructive to compare the carbon relaxation times for a series of polymers to evaluate how changes in the chemical structure affect the dynamics before we consider in detail the models used to interpret the relaxation results. We can grossly estimate the effect of polymer structure on the dynamics by comparing the carbon relaxation times for the methylene groups in a series of related polymers, as shown in Table 9.4. The chain dynamics are in the fast motion limit for the polymers listed in Table 9.4, so a longer relaxation time corresponds to a more mobile polymer chain. By comparing polyethylene, polypropylene, and poly(vinyl chloride), for example, we can see that side chains impede

values for $\bar{\delta}$ and $\bar{\eta}$, as well as the same lineshape. For example, the same $\bar{\eta} = 1$ pattern is obtained for a two-site jump model with a jump angle of 90° when the population of site 2 is twice that of site 1, a two-site jump with equal populations and an angle of 109°, or with restricted diffusion about a tetrahedral axis with a distribution width of 70°.[65] The accuracy of motional models can often be distinguished by measuring T_{1Q}, the relaxation of the spin-aligned state.[66] The dynamics measured by averaging the chemical shift anisotropy lineshapes are similar in concept but differ in detail.

The lineshapes are insensitive to very slow motions, and other methods must be used to measure the chain dynamics in the slow motion limit. The dynamics can be measured using the solid-state analogue of the 2D NOESY or exchange NMR.[67] The strategy is to correlate the chemical shifts before and after a mixing time. In the deuterium and the chemical shift anisotropy lineshapes, the frequency is determined by the orientation of the bond vector or the anisotropy tensor relative to the magnetic field. If the chain conformation changes during the mixing time, then a different orientation (and frequency) will be detected in the second dimension. The correlation for the entire lineshape is ob-

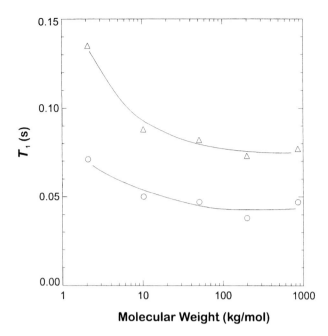

Figure 9.41 A plot of the dependence of the carbon relaxation times in polystyrene on the molecular weight for the (△) methine and (○) methylene carbons. The relaxation times were measured at 15 MHz for 20% (w/v) solutions at 44 °C.

the main-chain mobility. The comparison of polystyrene and polypropylene shows that larger side chains are more restrictive than smaller ones. The comparisons of poly(ethylene) with poly(ethylene oxide) and polypropylene with poly(propylene oxide) show that the introduction of heteroatoms leads to an increase in chain mobility. The comparison of polyethylene and polybutadiene shows that double bonds increase the chain mobility. Although motion is retarded for the double-bonded carbons, the lowered barrier for rotation at the allyic bonds more than makes up for this. The relaxation times also depend on the tacticity.

A number of motional models have been developed to relate the measured relaxation times to the chain motion.

The models range from isotropic reorientation to complex models that incorporate features of the chain structure and microstructure. One early approach was to model the conformational transition in polymers as jumps on a diamond lattice.[70] This model has some chemical realism since the jumps are restricted to the tetrahedral angles. A variety of statistical models also have been developed that use the Cole-Cole,[71] the Fuoss-Kirkwood,[72] or the log-χ^2 distribution functions[73] to calculate the spectral densities and relaxation times. These models can give a good fit to the data, but they provide little physical insight into the chain dynamics.

The relaxation times also can be modeled using the so-called model-free approach[74] where it is assumed that the correlation function can be separated into a slower [$G_0(t)$] and faster [$G_i(t)$] correlation time given by

$$G(t) = G_0(t)G_i(t) \qquad (35)$$

These models also provide a good fit to the data but provide little insight into the molecular-level dynamics.

Another approach is to model conformational transitions in polymers as a damped diffusional process along the chain. Several types of transitions are classified in the Hall-Helfand theory,[75] depending on the displacement of the polymer tails. Some transitions (crankshaft motions) lead to changes in the polymer tails, while others do not. The autocorrelation function is given by

$$G(t) = e^{t/\tau_2}e^{t/\tau_1}I_0(t/\tau_1) \qquad (36)$$

where I_0 is the modified zero-order Bessel function, τ_1 is the correlation time for correlated jumps responsible for orientation diffusion along the chain, and τ_2 is the correlation time corresponding to damping from either nonpropagative specific motions or distortions of the chain with respect to its most stable local conformation. Modifications to this model have been introduced that account for correlated motions for non-neighboring bonds, but these models have a large number of parameters. This model has been used with the first-order modified Bessel functions to account for the higher-order correlations while reducing the number of

Table 9.4. The carbon relaxation times for a variety of polymers in solution

Polymer	Solvent	T (°C)	Concentration (wt%)	T_1 (s)
Polyethylene	ODCB[a]	30	25	1.24
Polypropylene	ODCB	30	25	0.39
Poly(ethylene oxide)	C_6D_6	30	5	2.80
Poly(propylene oxide)	$CDCl_3$	30	5	1.00
cis-Polybutadiene	$CDCl_3$	54	20	3.00
trans-Polybutadiene	$CDCl_3$	54	20	2.38
Poly(vinyl chloride)	DMSO-d_6	50	20	0.13
Polystyrene	Toluene-d_8	30	15	0.10
Poly(methyl methacrylate) (syndiotactic)	$CDCl_3$	38	10	0.80
Poly(methyl methacrylate) (isotactic)	$CDCl_3$	38	10	0.12

a. ortho-dichlorobenzene.

parameters.[76] The fits to the observed data can be improved by introducing an additional term for fast anisotropic reorientation of the CH vector in a cone of half-angle θ about the rest position using[77,78]

$$G(t) = (1 - a)e^{-t/\tau_1}e^{-t/\tau_2}I_0(t/\tau_1) + ae^{-t/\tau_0}e^{t/\tau_1}e^{t/\tau_2}I_0(t/\tau_1) \quad (37)$$

where τ_0 is correlation time associated with the local anisotropic reorientation and

$$1 - a = \left[\frac{\cos\theta - \cos^3\theta}{2(1 - \cos\theta)} \right]^2 \quad (38)$$

Good fits were obtained to the relaxation data for several polymers in the bulk and in solution, and the authors suggest that such local anisotropic fluctuations are a general feature of polymer relaxation.[78,79]

The NMR studies of bulk and solution poly(methyl vinyl ether) illustrate how these models can be applied to the dynamics of polymers. The data sets for these studies consist of relaxation measurements as a function of spectrometer frequency and temperature. Figure 9.42 shows a plot of the relaxation times for bulk poly(methyl vinyl ether) as a function of inverse temperature at field strengths of 25 and 62 MHz.[79] A T_1 minimum is observed at both frequencies. The position and depth of the T_1 minimum is extremely sensitive to the dynamics model, and the data are well described by the Hall-Helfand correlation function modified to allow for faster anisotropic motion as shown by eqs 37 and 38. The lines through the data points are for a model in which $a = 0.4$, $\tau_1/\tau_0 = 200$, and $\tau_2/\tau_1 = 2$. The relaxation times for

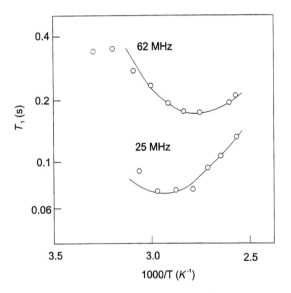

Figure 9.42 The dependence of the carbon spin-lattice relaxation time for the methylene carbons of poly(vinyl methyl ether) on the inverse temperature measured at 62 and 25 MHz. The solid line is the calculated relaxation rates using the model described in the text.

poly(methyl vinyl ether) in solution are fit by these same parameters except $\tau_1/\tau_0 = 400$.

The NMR studies have shown that the relaxation in polymers is primarily due to segmental motions and librations, and they have shown that the relaxation times do not depend on the molecular weight or apparent viscosity. However, the dynamics are expected to depend on the solvent viscosity. The dynamics of poly(ethylene oxide) and polyisoprene have been studied as a function of viscosity, and the relationship between the correlation time and the solvent viscosity is given by[80]

$$\tau_c = A\eta^\alpha e^{E_a/RT} \quad (39)$$

where the coefficient α is determined empirically. A value of 0.41 was reported in a study of polyisoprene in 10 solvents covering a factor of 70 in viscosity.

Polymer Dynamics in the Solid State

Solid-state NMR has been used to study the dynamics of a wide variety of polymers. These studies are frequently driven by the desire to understand the mechanical properties of polymers that can sometimes be directly related to the dynamics measured by NMR.[81] Solid polymers experience chain motion over many orders of magnitude, from the fastest libration motions that can occur on the picosecond time scale, to the chain diffusion between crystallites that can take many seconds to occur.

The fastest molecular motions in solid polymers can be studied by the T_1 and T_2 relaxation rates that are used to measure the solution dynamics. Often there are large differences in the relaxation rates for the crystalline and amorphous phases. In polyethylene, for example, the chain motion is restricted in the crystalline phase, and the carbon T_1 relaxation times may be as long as hundreds of seconds, depending on how the sample was prepared.[82] The amorphous regions are much more mobile and may have carbon relaxation times on the order of hundreds of milliseconds. One consequence of this difference is that the NMR spectra of the crystalline and amorphous regions can be selected based on the T_1 values. If we obtain a carbon single-pulse spectrum of polyethylene with a short delay time between pulses, for example, we will observe a spectrum that is enriched in the amorphous fraction, because the signals from the crystalline phase will not have a chance to relax between acquisitions.

In most solid-state experiments we obtain high-resolution spectra by removing the line broadening from the dipolar couplings and the chemical shift anisotropy with proton decoupling and magic-angle spinning. Under certain conditions, the chain motion can interfere with the spin-state averaging associated with the proton decoupling, thereby leading to broadened signals. The interference between chain motion and decoupling is known as Rothwell-Waugh broadening[83] and is given by

$$\frac{1}{T_2} = \frac{\gamma_C^2 \gamma_H^2 \hbar^2}{5r^6} \left(\frac{\tau_c}{1 + \omega_2^2 \tau_c^2} \right) \qquad (40)$$

where ω_2 is the decoupler field strength. When $\omega_2^2 \tau_c^2 \simeq 1$, the line width will be at a maximum and the signal will be broadened beyond detection. This effect is illustrated in Figure 9.43 for semicrystalline polypropylene. Three lines are observed in the carbon spectra at 300 K and at 17 K from the methine, methyl, and methylene carbons. However, at 105 K the methyl signal broadens and cannot be detected. The point of maximum broadening occurs when the correlation time is approximately the same as ω_2 ($2\pi \times 50$ kHz). In a similar way, chain motion on the same time scale as magic-angle spinning can also lead to line broadening. In this case, the time scale of chain motion must be on the order of several kHz to interfere with the magic-angle spinning. Interference between chain motion and magic-angle spinning has been reported for poly(oxymethylene).[84]

The conformational transitions in polymers occur over a range of timescales in amorphous polymers. The most rapid transitions, such as ring flips in polycarbonate,[85] have been observed to interfere with the line narrowing methods, and broadening similar to that shown above for semicrystalline polypropylene has been observed. In other cases, such as glassy, atactic polypropylene, the conformational transitions are so slow that they can be probed using 2D exchange NMR. It has been reported that at 250 K the methylene signal for atactic polypropylene is split into three signals that are assigned to the *trans-trans*, *trans-gauche*, and *gauche-gauche* conformations.[86] As the temperature is increased to 262 K, the methylene signal is broadened because the conformational transitions are on the intermediate timescale. The dynamics of the conformational interchange can be probed using 2D exchange NMR, and Figure

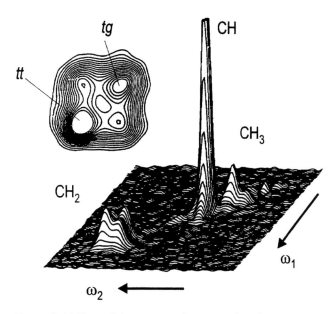

Figure 9.44 The solid-state two-dimensional exchange spectra for atactic polypropylene at 250 K acquired with a 0.5 s mixing time. The single pulse spectrum appears along the diagonal, and the off-diagonal peaks arise from exchanging carbons. The inset contour plot shows the methylene region and the chemical shift assignments for the *tt* and *tg* conformations.

9.44 shows the exchange spectrum acquired with a 0.5 s mixing time for atactic polypropylene at 250 K.[86] In addition to the peaks along the diagonal, cross peaks connecting the *trans-trans* and *trans-gauche* conformations can be observed. The rate of exchange can be measured by acquiring the 2D exchange spectra as a function of mixing time.

Deuterium NMR is often used to measure the chain dynamics of polymers because the deuterium lineshape is very sensitive to both the time scale and the amplitude of atomic motion. This is illustrated in Figure 9.45, which shows the deuterium lineshape for site-specifically labeled nylon 66 samples containing deuterium at different sites along the chain.[87] As shown in Figure 9.38(b), a "Pake" pattern is expected for deuterium in the absence of motion. All the nylon sites shown in Figure 9.45 have partially averaged lineshapes, but the lineshapes are different from each other. The lineshapes can be modeled to extract both the timescale and the amplitude of atomic motion.

Certain types of motions give rise to distinctive lineshapes in the deuterium spectra that can be used to identify the type of molecular motion. This is illustrated in Figure 9.46, which shows the deuterium NMR spectrum for ring-deuterated polystyrene.[63] These rings are not rigid but undergo flips about the C1-C4 axis, and this motion give rise to the characteristic lineshape shown in Figure 9.39a. Since only a fraction of the polystyrene rings undergo these flipping motions, the spectrum is a composite of the signals from rigid and flipping rings. The fraction of rings undergo-

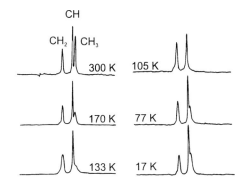

Figure 9.43 The effect of temperature on the solid-state carbon spectra of polypropylene. Note the disappearance of the methyl peak at 105 K.

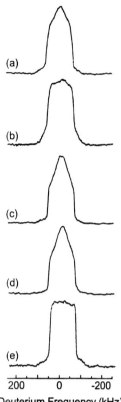

Figure 9.45 The ^{2}H NMR spectra of nylon 66 deuterated on (a) C_3 and C_4 and (b) C_2 and C_5 carbons of the adipoyl moity, and on the (c) C_3 and C_4, (d) C_2 and C_5, and (e) C_1 and C_6 carbons of the diamine moity. Note how the lineshape (and dynamics) vary with position along the chain.

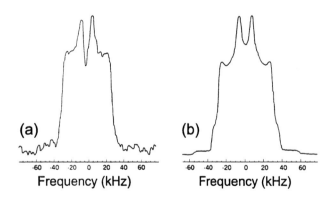

Figure 9.46 Comparison of the (a) experimental and (b) simulated deuterium NMR spectra for ring-deuterated polystyrene. The sharper features near the center of the spectrum are characteristic of deuterons undergoing 180° ring flips. Inputs for the simulation include the fraction of rings undergoing flips and the flip rate.

ing flips can be determined by fitting the lineshape for a rigid and mobile component. Since the flipping rings have a shorter T_1, the spectrum of the component with the flipping rings can be enhanced by acquiring spectra with a short delay time between scans.

The dynamics of polystyrene over a broad range of timescales has been studied by deuterium NMR, using both ring-deuterated and main-chain-deuterated polymers. Information about the dynamics on the MHz-to-seconds timescale was acquired from spin-lattice relaxation, broadline NMR, solid-echo NMR, and 2D exchange NMR. Figure 9.47 shows a plot of the correlation times for polystyrene measured by the various methods.[88] It can be seen that some of the data lie on the line for the -transition in polystyrene measured by other methods. These data are fit to the Williams-Landel-Ferry equation[89] describing the temperature dependence of polymers above T_g. The correlation times determined from the spin-lattice relaxation do not lie on this line but are correlated with the β-transition.

2D ^{2}H exchange NMR has also been used to identify molecular motions in the crystalline phases of polymers. It is known from dielectric and dynamic-mechanical spectroscopy that some crystalline polymers undergo motions in the crystalline phase that are believed to arise from translocations. It is possible to study this phenomenon by 2D exchange NMR, and Figure 9.48 shows the 2D exchange spectrum for poly(vinylidine fluoride) acquired at 370 K with a mixing time of 0.2 s.[90] The 2D exchange spectrum

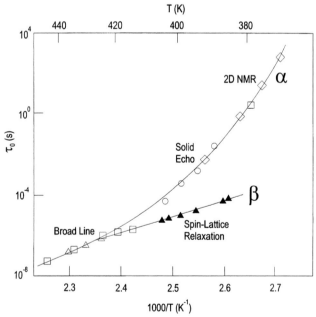

Figure 9.47 A plot of the dependence of the polystyrene correlation times measured by deuterium NMR on inverse temperature. The broadline spectra, the solid echo, and the two-dimensional (2D) NMR data lie along the line for the α transition. The temperature dependence of the spin-lattice relaxation lies along the line for the β transition.

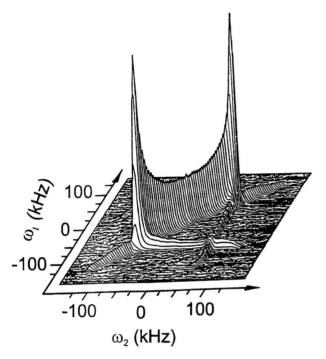

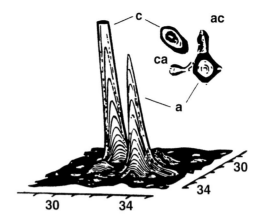

Carbon Chemical Shift (ppm)

Figure 9.49 The solid-state carbon two-dimensional exchange spectrum for polyethylene at 363 K acquired with a mixing time of 1 s. The inset shows a contour plot of the same data. Key: c = crystalline phase; a = amorphous phase.

Figure 9.48 The deuterium two-dimensional exchange spectrum for deutereated poly(vinylidine fluoride) acquired at 370 K with a mixing time of 0.2 s. The off-diagonal ridges are very sensitive to the geometry, amplitude, and rate of reorientation.

shows a ridge pattern that depends on the jump angle between sites in the crystalline lattice. The best fit data are for a two-site jump angle of 67 ± 30. These results show that poly(vinylidine fluoride) undergoes a transition between the *tgtg⁻* and *g⁻tgt* conformations.

Solid-state NMR with cross polarization can also be used to probe the lower-frequency chain dynamics using 2D exchange types of experiments that correlate the frequency before and after a mixing time. This is illustrated in Figure 9.49, which shows the 2D exchange spectrum for semicrystalline polyethylene at 363 K with a mixing time of 1 s. Two intense peaks are observed along the diagonal; these peaks are assigned to the crystalline and amorphous phases. Two cross peaks are shown connecting the signals that arise from chain diffusion between the crystalline and amorphous domains. Thus, these 2D exchange experiments can probe the ultraslow dynamics of polymers.

In summary, the NMR relaxation times, lineshapes, and 2D exchange spectra are sensitive to the chain dynamics over many orders of magnitude. An understanding of the relationship between the chemical structure and microstructure can be obtained by modeling the relaxation of polymers in solution. The dynamics are slower for solid polymers, and the geometry, rate, and amplitude of molecular motion can be determined from the lineshape and 2D exchange spectra. In many cases this makes it possible to gain a molecular level assignment for the dynamics tradi-

tionally observed by dielectric and dynamic-mechanical spectroscopy.

References

1. Alpert, N. L. *Phys. Rev.* **1947**, *72*, 637–638.
2. Purcell, E. M.; Torrey, H. C.; Pound, R. V. *Phys. Rev.* **1946**, *69*, 37–38.
3. Bloch, F.; Hansen, W. W.; Packard, M. E. *Phys. Rev.* **1946**, *69*, 127.
4. Schmidt-Rohr, K.; Speiss, H. W. *Multidimensional Solid-State NMR and Polymers;* Academic Press: New York, 1994.
5. Breitmaier, E.; Voelter, W. *Carbon-13 NMR Spectroscopy;* 3rd ed.; VCH: Weinheim, 1987.
6. Tonelli, A. *NMR Spectroscopy and Polymer Microstructure: The Conformational Connection;* VCH Publishers: New York, 1989.
7. Bovey, F. A.; Mirau, P. A. *NMR of Polymers;* Academic Press: New York, 1996.
8. Bovey, F. *Nuclear Magnetic Resonance Spectroscopy;* 2nd ed.; Academic Press: New York, 1988.
9. Mehring, M. *High Resolution NMR in Solids;* Springer-Verlag: Berlin, 1983.
10. Taylor, R.; Pembleton, R.; Ryan, L.; Gerstein, B. *J. Chem. Phys.* **1979**, *71*, 4541.
11. Hartmann, S. R.; Hahn, E. L. *Phys. Rev.* **1962**, *128*, 2042–2053.
12. Aue, W.; Bartnoldi, E.; Ernst, R. *J. Chem. Phys.* **1976**, *64*, 2229.
13. Bax, A.; Davis, D. *J. Magn. Reson.* **1985**, *63*, 207–213.
14. Jeneer, J.; Meier, B.; Bachmann, P.; Ernst, R. *J. Chem. Phys.* **1979**, *71*, 4546.
15. Muller, L. *J. Amer. Chem. Soc.* **1979**, *101*, 4481–4484.

16. Rance, M.; Sorensen, O. W.; Bodenhausen, G.; Wagner, G.; Ernst, R. R.; Wuthrich, K. *Biochem. Biophys. Res. Comm.* **1983**, *117*, 479–485.

17. Bax, A.; Davis, D. *J. Magn. Reson.* **1985**, *65*, 355–360.

18. Mirau, P.; Tanaka, H.; Bovey, F. *Macromolecules* **1988**, *21*, 2929–2933.

19. Shaka, A. J.; Barker, P. B.; Freeman, R. *J. Magn. Reson.* **1985**, *64*, 547–552.

20. Schilling, F. C.; Tonelli, A. E. *Macromolecules* **1980**, *13*, 270.

21. Zambelli, A.; Locatelli, P.; Bajo, G.; Bovey, F. A. *Macromolecules* **1975**, *8*, 687–689.

22. Kharas, G. B.; Mirau, P. A.; Watson, K.; Harwood, H. J. *Polym. Int.* **1992**, *28*, 67–74.

23. Cais, R. E.; Kometani, J. M. In *The Syntheisis of Novel Regioregular Poly(vinyl fluorides) and Their Characterization by High-Resolution NMR*; Randal, J. C., Ed.; American Chemical Society: Washington, DC, 1983; Vol. 247, pp 153–165.

24. Liu, W.; Ray, D. G.; Rinaldi, P. L.; Zens, T. *J. Magn. Reson.* **1999**, *140*, 482–486.

25. Mirau, P.; Bovey, F.; Tonelli, A.; Heffner, S. *Macromolecules* **1987**, *20*, 1701–1707.

26. Brown, S. P.; Schnell, I.; Spiess, H. W. *J. Am. Chem. Soc.* **1999**, *121*, 6712.

27. Mirau, P. A.; Heffner, S. A. *Macromolecules* **1999**, *32*, 4912–4916.

28. Vander Hart, D. L. *J. Magn. Reson.* **1981**, *44*, 117–125.

29. Tonelli, A. E.; Gomes, M. A.; Tanaka, H.; Cozine, M. H. In *Solid State 13C NMR Studies of the Structures, Conformations, and Dynamics of Semicrystalline Polymers*; Mathias, L. J., Ed.; Plenum Press: New York, 1991; pp 81–105.

30. Schilling, F.; Bovey, F.; Lovinger, A.; Zeigler, J. *Macromolecules* **1986**, *19*, 2660–2663.

31. Bovey, F. A.; Schilling, F. C. In *The Solid State ^{29}Si and ^{13}C NMR of poly(di-n-alkyl silanes)*; Mathias, L. J., Ed.; Plenum Press: New York, 1991; pp 295–304.

32. Dumais, J. J.; Jelinski, L. W.; Leung, L. M.; Gancarz, I.; Galambos, A.; Koberstein, J. T. *Macromolecules* **1985**, *18*, 116–119.

33. English, A. D. *Macromolecules* **1985**, *18*, 178–181.

34. Stover, H. D. H.; Frechet, J. M. J. *Macromolecules* **1989**, *22*, 1574–1576.

35. Mirau, P. A.; White, J. L.; Heffner, S. A. *Makromol. Symp.* **1994**, *86*, 181–191.

36. Tonelli, A. E. *Polym. Int.* **1997**, *43*, 295–309.

37. Szejtli, J. *Chem. Rev.* **1998**, *98*, 1743–1753.

38. Saenger, W. *Angew. Chem. Int. Ed.* **1980**, *19*, 344–362.

39. Schneider, H. J. *Angew. Chem. Int. Ed.* **1991**, *30*, 1417–1436.

40. Huang, L.; Allen, E.; Tonelli, A. E. *Polymer* **1998**, *39*, 4857–4865.

41. Huang, L.; Allen, E.; Tonelli, A. E. *Polymer* **1999**, *40*, 3211–3221.

42. English, A. D.; Debowski, C. *Macromolecules* **1984**, *17*, 446–449.

43. Nielsen, N. C.; Bildsoe, H.; Jakobsen, H. J. *Macromolecules* **1992**, *25*, 2847–2853.

44. Goldman, M.; Shen, L. *Physical Review* **1966**, *144*, 321–331.

45. Cheung, T.; Gerstein, B. *J. Appl. Phys.* **1981**, *52*, 5517–5528.

46. Clauss, J.; Schmidt-Rohr, K.; Spiess, H. W. *Acta Polymer.* **1993**, *44*, 1–17.

47. Mellinger, F.; Wilhelm, M.; Spiess, H. *Macromolecules* **1999**, *32*, 4686–4691.

48. Egger, N.; Schmidt-Rohr, K.; Blumich, B.; Domke, W. D.; Stapp, B. *J. Appl. Polym. Sci.* **1992**, *44*, 289.

49. Blumich, B.; Hagemeyer, A.; Schaefer, D.; Schmidt-Rohr, K.; Spiess, H. W. *Adv. Mater.* **1990**, *2*, 72–81.

50. Marjanski, M.; Srinivasarao, M.; Mirau, P. A. *Solid State NMR* **1998**, *12*, 113–118.

51. Simmons, A.; Natansohn, A. *Macromolecules* **1991**, *24*, 3651–3661.

52. White, J. L.; Mirau, P. A. *Macromolecules* **1993**, *26*, 3049–3054.

53. White, J. L.; Mirau, P. A. *Macromolecules* **1994**, *27*, 1648–1650.

54. Caravatti, P.; Neuenschwander, P.; Ernst, R. *Macromolecules* **1985**, *18*, 119–122.

55. Campbell, G. C.; VanderHart, D. L. *J. Magn. Reson.* **1992**, *96*, 69–93.

56. Miyoshi, T.; Takegoshi, K.; Hikichi, K. *Polymer* **1995**, *37*, 11–18.

57. Grobelny, J.; Rice, D.; Karasz, F.; MacKnight, W. *Macromolecules* **1990**, *23*, 2139–2144.

58. Stejskal, E.; Schaefer, J.; Sefcik, M.; McKay, R. *Macromolecules* **1981**, *14*, 275–279.

59. Gobbi, G.; Silvestri, R.; Thomas, R.; Lyerla, J.; Flemming, W.; Nishi, T. *J. Polym. Sci. Part C. Polym. Lett.* **1987**, *25*, 61–65.

60. McBrierty, V.; Douglass, D.; Kwei, T. *Macromolecules* **1978**, *11*, 1265–1267.

61. Kuroki, S.; Okita, K.; Kakigano, T.; Ishikawa, J.; Itoh, M. *Macromolecules* **1998**, *31*, 2804–2808.

62. Heatley, F. *Prog. NMR Spectroscopy* **1979**, *13*, 47–85.

63. Spiess, H. *Coll. Polym. Sci.* **1983**, *261*, 193–209.

64. Pake, G. E. *J. Chem. Phys.* **1948**, *16*, 327.

65. Hirschinger, J.; English, A. D. *J. Magn. Reson.* **1989**, *85*, 542–553.

66. Speiss, H. *J. Chem. Phys.* **1980**, *72*, 6755–6762.

67. Spiess, H. W. *Chem. Rev.* **1991**, *91*, 1321–1338.

68. Gronski, W. *Makromol. Chem.* **1976**, *177*, 3017.

69. Allerhand, A.; Hailstone, R. K. *J. Chem. Phys.* **1972**, *56*, 3718–3720.

70. Jones, A. A.; Stockmayer, W. H. *J. Polym. Sci. Polym. Phys. Ed.* **1977**, *15*, 847.

71. Cole, K. S.; Cole, R. H. *J. Chem. Phys.* **1941**, *9*, 341.

72. Fuoss, R. M.; Kirkwood, J. G. *J. Am. Chem. Soc.* **1941**, *63*, 385.

73. Schaefer, J. *Macromolecules* **1973**, *6*, 882–888.

74. Lipari, G.; Szabo, A. *J. Am. Chem. Soc.* **1982**, *104*, 4546–4559.

75. Hall, C. K.; Helfand, E. *J. Chem. Phys.* **1982**, *77*, 3275–3282.

76. Viovy, J. L.; Monnerie, L.; Brochon, J. C. *Macromolecules* **1983**, *16*, 1845.

77. Howarth, W. O. *Faraday Trans. 2* **1979**, *75*, 863.

78. Dejean de la Batie, R.; Laupretre, F.; Monnerie, L. *Macromolecules* **1988**, *21*, 2052.

79. Dejean de la Batie, R.; Laupretre, F.; Monnerie, L. *Macromolecules* **1988**, *21*, 2045–2052.

80. Glowinkowski, S.; Gisser, D. J.; Ediger, M. D. *Macromolecules* **1990**, *23*, 3520–3530.

81. Schaefer, J.; Stejskal, E. O.; Buchdahl, R. *Macromolecules* **1977**, *10*, 384–405.

82. Axelson, D. E.; Mandelkern, L.; Popli, R.; Mathieu, P. *J. Polym. Sci. Polym. Phys. Ed.* **1983**, *21*, 2319–2335.

83. Rothwell, W. P.; Waugh, J. S. *J. Chem. Phys.* **1981**, *74*, 2721–2732.

84. Veeman, W. S.; Menger, E. M.; Ritchey, W.; de Boer, E. *Macromolecules* **1979**, *12*, 924–927.

85. Lyerla, J. R. In *High Resolution NMR of Glassy Amorphous Polymers;* Komoroski, R. A., Ed.; VCH Publishers: Deerfield Beach, FL, 1986; pp 63–120.

86. Zemke, K.; Chmelka, B. F.; Schmidt-Rohr, K.; Spiess, H. *Macromolecules* **1991**, *24*, 6874–6876.

87. Miura, H.; English, A. *Macromolecules* **1988**, *21*, 1543–1544.

88. Pschorn, U.; Rossler, E.; Sillescu, H.; Kaufmann, S.; Schaefer, D.; Spiess, H. *Macromolecules* **1991**, *24*, 398–402.

89. Williams, M. L.; Landel, R. F.; Ferry, J. D. *J. Am. Chem. Soc.* **1955**, *77*, 3701–3707.

90. Hirschinger, J.; Schaefer, D.; Spiess, H.; Lovinger, A. *Macromolecules* **1991**, *24*, 2428–2433.

10

Vibrational Spectroscopy

DAVID W. BALL

In the modern scientific vernacular, it is understood that *vibrational spectroscopy* is usually synonymous with *infrared spectroscopy*. This is not always the case. Electronic transitions of f-block atoms and ions appear in the infrared portion of the spectrum, and transitions due to rotations of gas-phase molecules are often detected in an infrared spectrum. Raman spectroscopy is another spectroscopic technique that, in some circumstances, gives us information about the vibrations of molecules; however, I will not discuss this technique here. For our purposes, in this chapter I focus on vibrations of atoms in molecules and how those vibrations interact with infrared light.

Infrared (IR) spectroscopy is a very powerful and useful technique that is commonly used in laboratories. Like any technique, it is not meant to be used independently of other methods of analysis. Measurement of the vibrational spectrum of a sample provides useful qualitative and quantitative information. This information is best used in conjunction with the results of other techniques. Given its ease of use and its ubiquity, IR spectroscopy is a standard technique that can be very helpful in any analysis.

This chapter is divided into three major sections. First, I review the theoretical basis of vibrational spectroscopy. Next, I consider experimental aspects of the technique. After that, I discuss some interpretative considerations of a vibrational spectrum.

Theoretical

The Electromagnetic Spectrum

The modern concept of an electromagnetic spectrum is somewhat recent in the development of modern science. In classical and preclassical times, the term *light* referred exclusively to visible light. In the mid 1660s, Isaac Newton had shown that white light was actually the combination of all visible colors, and his interpretation of his experimental results supported the particle hypothesis of light. In 1803, English physician and scientist Thomas Young conclusively demonstrated that light had wave properties, and the identification of light as a wave was unquestioned until the development of modern science, which began in 1900 with Max Planck's theory of light.

Expansion of the definition of light started in 1800 when Sir William Herschel, the astronomer who discovered Uranus, used a thermometer to measure the relative amounts of heat that accompanied the different colors of (visible) sunlight. Curiously, in the invisible area *beyond* the color red, Herschel found the greatest temperature effect. He concluded that there was some component of sunlight that was not visible but still had some radiative effect. By 1850, Italian physicist Macedonio Melloni had shown that this invisible light had the same properties as visible light, like interference and polarization and reflection and refraction. These discoveries set the stage for James Clerk Maxwell's theory of electromagnetic radiation, espoused between 1864 and 1873, which established visible light and invisible infrared light as aspects of the same general phenomenon, just having different wavelengths or frequencies. The generality of electromagnetic radiation was firmly established when Heinrich Hertz discovered radio waves, a longer-wavelength form of light, in the 1880s.

The infrared portion of the spectrum lies just below red visible light (*infra*, L., under or beneath), as shown in Figure 10.1. While the boundaries of the different regions of the electromagnetic spectrum are vague, the general boundaries of the infrared part are about 5×10^{11} to 4×10^{14} s^{-1}. In wavelength units, this range is 600 μm to 0.75 μm. Visible light lies at higher frequencies, and microwaves are at lower frequencies. As with any form of electromagnetic

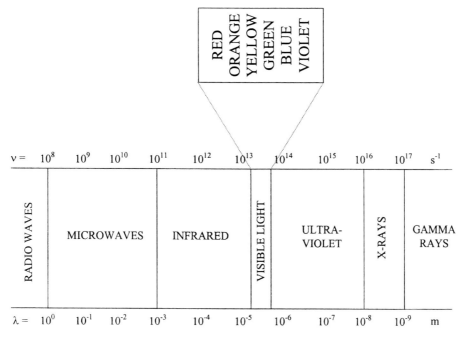

Figure 10.1 The electromagnetic spectrum. Not all regions were discovered at the same time, nor were they used with the same efficacy.

radiation, the frequency ν and wavelength λ of infrared light are related to the speed of light, c, by the equation

$$c = \lambda \cdot \nu \qquad (1)$$

The speed of light is a universal constant, equal to 3.00×10^8 m/s. Equation 1 allows one to convert back and forth between light's wavelength and frequency.

In 1900, German physicist Max Planck deduced that the energy of a light wave was not proportional to the amplitude of the light, as it was with other types of waves (sound, water, etc.). Rather, the energy of light waves was proportional to its *frequency*. Planck proposed a simple expression for the energy of a light wave:

$$E = h \cdot \nu \qquad (2)$$

where h is called *Planck's constant* and equals 6.626×10^{-34} J $\cdot$ s. One interpretation of eq 2 is that a light wave is acting as a package of energy that has a certain, specific value, depending on its frequency. This package of energy interpretation reawakened the description of light as a particle, and science uses the term *photon* to refer to a single light wave-particle. This wave-particle duality of light is not considered a problem for modern science. Light has some properties that are best described as wavelike and some properties that are best described as particle-like.

According to eq 2, infrared photons have energies of about 3.3×10^{-22} J to 2.7×10^{-19} J per photon. This is not a lot of energy, but as we will see, it is enough energy to excite the vibrations of molecules.

Many solid materials are efficient absorbers of infrared light. In a majority of cases, the energy of the absorbed light is ultimately converted into heat energy. Because of these two facts, infrared light is sometimes incorrectly thought of as heat. Infrared light is a form of electromagnetic radiation that can be efficiently perceived as heat.

Classical and Quantum-Mechanical Treatment of Vibrations

Although the behavior of molecules is best treated with quantum mechanics, many ideas about molecular vibrations are based in classical mechanics. In 1678, English physicist Robert Hooke proposed a mathematical model of simple vibrational motion. According to Hooke, the force F on a mass experiencing an oscillating motion is directly proportional to its displacement from an equilibrium position Δx, and the force acts in the direction opposite the motion. Mathematically, *Hooke's law* is written as

$$F = -k \cdot \Delta x \qquad (3)$$

where k is called the *force constant* of the oscillator. Hooke also defined the characteristic potential energy V for an oscillator. It is

$$V = \tfrac{1}{2} k \cdot (\Delta x)^2 \qquad (4)$$

Figure 10.2 shows a plot of this potential energy function. Oscillators that have this potential energy function, and whose behavior follows eq 3, are called ideal or *harmonic* oscillators.

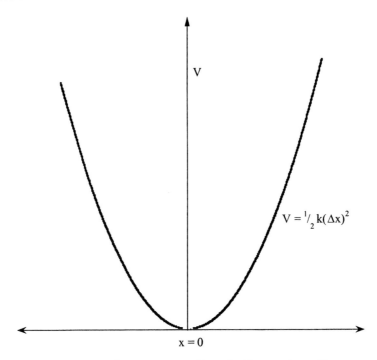

Figure 10.2 Potential energy curve of an ideal harmonic oscillator.

Quantum mechanics treats the behavior of atom-sized matter differently than bulk matter. This is primarily due to the fact that matter has wavelike behavior (much as light has matter-like behavior). The crux here is that the wavelike behavior of matter is inversely proportional to its mass, as illustrated by French physicist Louis de Broglie in 1923 with his expression

$$\lambda = h/mv \tag{5}$$

where λ is the wavelength of the matter, h is Planck's constant, m is the mass of the matter, and v is its velocity. For bulk, macroscopic objects, the de Broglie wavelength is inconspicuously small. But for microscopic objects like electrons and atoms, the de Broglie wavelength is sufficiently large that the matter's behavior must be expressed in terms of waves.

In 1925–1926, Erwin Schrödinger and Werner Heisenberg independently developed wave mechanics, or *quantum mechanics*, to describe the behavior of matter at the atomic and subatomic levels. It remains the most successful theory of the behavior of matter to date. Central to quantum mechanics is the requirement that electronic or atomic behavior must be described by a wave equation called a *wavefunction*. (Wavefunctions can be as simple as sine or cosine functions, exponential functions, or polynomial functions. We will not consider the explicit forms of these wavefunctions here.) The exact form of a wavefunction must satisfy a differential equation called the *Schrödinger equation*, and this equation ultimately yields the total energy that accompanies the matter's behavior.

One part of the Schrödinger equation is the potential energy imposed on the matter's behavior; for an atomic-scale harmonic oscillator, the potential energy from Hooke's law is used. Thus, quantum mechanics explicitly borrows ideas from classical mechanics (although it does recast these ideas into a different perspective). When the mathematics of the Schrödinger equation are applied, quantum-mechanical harmonic oscillators have certain specific wavefunctions and certain specific energies. Because harmonic oscillators have an energy that is specified by the wavefunctions, we say that the energy of a harmonic oscillator is *quantized*.

Modern atomic theory describes matter as being composed of atoms and molecules. Atoms combine to make molecules using chemical bonds, and these bonds act as "springs", while the atoms are the masses of an approximate harmonic oscillator. Therefore, the vibrations of atoms in molecules can be approximated as quantum-mechanical harmonic oscillators, and their behavior is described by the wavefunctions of harmonic oscillators. In addition, the energies of vibrating atoms are quantized, and the amount of vibrational energy is given by the Schrödinger equation. Quantum mechanics provides a particular expression for the amount of energy E in a vibration:

$$E = h\nu(v + \tfrac{1}{2}) \tag{6}$$

where h is Planck's constant, ν is the classical frequency (in units of s^{-1} or Hz), and v is a non-negative integer called the *vibrational quantum number*.* The possible values of v—

*Do not confuse the two variables ν (the Greek letter *nu*) and v (the English letter *vee*) in equation 6!

0, 1, 2, 3, . . . —are dictated by the mathematics of the Schrödinger equation.

Note how the classical description of harmonic oscillators imposes itself on the quantized values for the energy in quantum mechanics: eq 6 includes the value of the *classical* frequency of the oscillator. Classically, the frequency ν of a harmonic oscillator is given by the expression

$$\nu = \frac{1}{2\pi}\sqrt{\frac{k}{m}} \tag{7}$$

where k is the classical force constant, and m is the mass of the oscillator. Equation 7 assumes that only a single mass is moving back and forth in an oscillating motion. In molecules, atoms are moving in an oscillating motion with respect to each other. In this case, eq 7 doesn't use the mass of the atoms, but the *reduced mass* of all the atoms involved in the vibration. Reduced masses can be calculated for any number of atoms, but they are determined most easily for two atoms vibrating about a common center. In this case, the reduced mass μ is given by

$$\frac{1}{\mu} = \frac{1}{m_1} + \frac{1}{m_2} \tag{8}$$

where m_1 and m_2 are the two masses of the atoms involved in the diatomic vibration. Individual atomic masses must be used in eq 8, not molar masses. For multiple atoms involved in a vibration, eq 7 becomes

$$\nu = \frac{1}{2\pi}\sqrt{\frac{k}{\mu}} \tag{9}$$

Frequencies of vibrations in molecules are on the order of 10^{14} or 10^{15} s^{-1}, or femtoseconds per vibration.

According to eq 6, the lowest possible value of the vibrational quantum number is zero. However, when $v = 0$, the vibrational energy is *not* zero, but rather equal to $\frac{1}{2}h\nu$. The presence of this so-called *zero point energy* is a purely quantum-mechanical phenomenon. It implies that the minimum-possible energy is $\frac{1}{2}h\nu$, not zero (the minimum-possible energy for a classical harmonic oscillator), and further implies that a quantum-mechanical oscillator is always in motion, even at the lowest possible energies.

Selection Rules and Units

Almost every kind of spectroscopy uses light to go from one energy state to another. Since atomic or molecular energy states are dictated by wavefunctions, another way to consider spectroscopy is that it uses light to change the state of a species from one wavefunction to another.

Not all wavefunctions can be accessed using light and starting from any random initial wavefunction, however. There are certain rules in quantum mechanics that limit the possible initial and final wavefunctions involved in a spectral transition. These rules are called *selection rules*.

Selection rules differ, depending on the type of atomic or molecular motion being considered. Furthermore, they presume that we can consider the particular behavior of interest separately from the other types of behavior of the species; that is, we can consider electronic behavior separate from rotational behavior separate from vibrational behavior, and so on. This is somewhat of an approximation, but a very good one in most cases. In this case, this allows us to consider the selection rule for possible, or *allowed*, vibrational transitions. This will give us an idea of what to expect for the vibrational spectrum of any given molecule.

There are two types of selection rules: *specific selection rules* that can be expressed in terms of quantum numbers, and *gross selection rules* that are described in terms of the properties of the molecule. In the case of vibrational spectra, both are easily defined. The specific selection rule is determined using the tools of calculus; we will not consider that here. However, the basic rule is simple: allowed vibrational transitions are those in which the vibrational quantum number changes by +1 or −1. That is,

$$\Delta v = \pm 1 \qquad \text{for an allowed transition} \tag{10}$$

The positive sign is imposed if the vibration is absorbing a photon and going to a higher quantum-numbered wavefunction (i.e., an absorption spectrum), and the negative sign is used if the molecule is going to a lower quantum-numbered wavefunction (i.e., an emission spectrum).

Because we have an expression for the energy of a vibration from quantum mechanics (eq 6), we can also determine an expression for the change in energy involved in an allowed vibrational transition. The derivation is straightforward and left as an exercise; the simple result is

$$\Delta E = h\nu \tag{11}$$

This simple expression is amazing: the energy difference for an allowed vibrational transition is equal to the classical frequency of the molecular vibration, given by eq 7, times Planck's constant.

There are several other conclusions from this simple expression. First, note that it is independent of the quantum number v. This means that for an ideal harmonic oscillator, the energy of transition between any two adjacent vibrational wavefunctions is exactly the same. Second, because $E = h\nu$ for a photon, eq 11 implies that the photon of light absorbed or emitted has the same frequency as the vibration itself.

This last point brings us to units used in expressing in energy changes in infrared or vibrational spectroscopy. It was already mentioned here that there are 10^{14} or 10^{15} vibrations per second between any two (or more) atoms. The conclusion in the previous paragraph means that the frequency of light used in infrared spectroscopy has that same frequency: 10^{14} or 10^{15} s^{-1}. This is about a thousand billion billion waves per second.

Rather then expressing the number of light waves in a given unit of *time*, we can also express the number of light waves that, in head-to-tail fashion, give a given unit of *length*. The number of light waves per given unit of length is called the *wavenumber* of the light. The SI unit for wavenumber is (meter)$^{-1}$, understood to be the number of waves per meter. The wavenumber, $\tilde{\nu}$ (note the tilde above the Greek letter, to differentiate this variable from the frequency ν having units of s^{-1}), is the reciprocal of the wavelength of the light:

$$\tilde{\nu} = \frac{1}{\lambda} \qquad (12)$$

Mentally comfortable numbers* are obtained for vibrational spectra if we express the number of vibrations occurring per centimeter of length, rather than per meter of length. Thus, in vibrational spectroscopy, it is common to see spectra or lists of spectral transitions in units of waves per centimeter, or cm^{-1}. In terms of this unit, infrared spectroscopy ranges from about 100 cm^{-1} (i.e., one hundred waves per centimeter) to about 6000 cm^{-1} (i.e., six thousand waves per centimeter). The energy of transition is directly proportional to wavenumber, so that large-wavenumber transitions are large-energy vibrational transitions.

Wavelength is also commonly used to express the light needed to induce vibrational changes, and the unit that yields the most useable numbers for the infrared region (see footnote) is the micrometer, or *micron*. The micron is sometimes abbreviated as μ, rather than its correct μm, and persons new to infrared spectroscopy should be aware of this shorthand. In these units, infrared spectra range from about 1.6 to 1000 μm in wavelength. In this case, however, energy is *inversely* proportional to the unit, so that high wavelength light is low-energy, and low wavelength light is high-energy. Again, persons new to infrared spectroscopy need to be aware of these relationships.

We will conclude this section by considering the gross selection rule for vibrations. The specific selection rule is expressed in terms of the vibrational quantum number, while the gross selection rule is expressed in terms of the behavior of the molecule as it vibrates. In this case, we need to consider a property of the molecule itself: its dipole moment.

A dipole moment is a partial separation of charge in a molecule. For example, in poly(perfluoroethylene) $(-CF_2CF_2-)_x$, the fluorine atom is more electronegative than the carbon atom it is bonded to and thus attracts the electrons in the covalent bond more. The fluorine atom is,

overall, slightly negatively charged while the carbon atom, being relatively deprived of the electrons in the covalent bond, is slightly positively charged. This separation of charge causes a dipole moment, and we say that the C—F bond is a *polar* bond.

During the course of any vibration, a molecule changes shape—slightly. As it changes shape, the dipole moment of the bond or the complete molecule, either zero or nonzero, may change or may stay the same. It is this idea that gives us a gross selection rule for an allowed vibrational transition: for a vibration to absorb infrared light, the vibration must be accompanied by a changing dipole moment.

Figure 10.3 illustrates this idea for the triatomic molecules H_2O. All three illustrated vibrations of the molecule change dipole moment, so all three vibrations will absorb infrared light. Generally, however, this gross selection rule is only useful for small molecules. Nevertheless, for vibrations of larger molecules like polymers, the concept of changing dipole moment can be applied if we knew what the vibration looked like. Figure 10.3 also shows a portion of a poly(perfluoroethylene) molecule and a vibration that involves a stretching of the carbon-fluorine bonds. Because F is so much more electronegative than carbon, this motion undoubtedly causes a *change* in a dipole moment, and therefore the vibration should absorb infrared radiation.

Ideal versus Real Vibrations

The preceding section implicitly assumed that the vibrations of molecules could be treated as ideal harmonic oscillators. In assuming so, we were able to apply the conclusions of quantum mechanics to vibrations and make predictions about the behavior of vibrating molecules when exposed to infrared light.

While these conclusions are useful to a very good approximation, they are just that: an approximation. The vibrations of molecules are not ideal harmonic oscillators; rather,

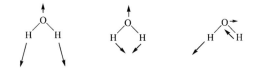

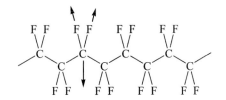

Figure 10.3 Vibrations of (*top*) H_2O and (*bottom*) poly(perfluoroethylene).

---

*This phrase might sound trite, but understand that expressing values of energies of vibrational transitions in units of s^{-1} of J or J/mol gives the same information. There is obviously some need to express values in terms of numbers that are more easily comprehendable to the human mind—but that discussion is more the purview of psychology, not spectroscopy!

they are real systems that behave in a real fashion. Use of the harmonic oscillator system to describe real vibrations has its utilities, but it also has its limitations.

The first limitation we must consider is the form of the vibrational potential energy itself. The ideal harmonic oscillator's potential energy function is a simple square function that increases monotonically as the oscillator gets farther and farther away from its equilibrium position. Vibrations in molecules cannot act like that in the extremes of position. For example, as two atomic nuclei get very close to each other, internuclear repulsion between the two positively charged nuclei act to increase the potential energy toward infinity. Physically, there comes a distance at which any two nuclei must begin moving away from each other. At the other extreme, as two nuclei get so far away from each other, the bond between them begins to break, so that at a sufficiently large distance the two atoms are no longer bonded, and there is no opposing force to get them closer together. A better approximation of a potential energy function for a real vibration looks like the curve shown in Figure 10.4.

Equations that mimic the form of Figure 10.4 are called *Morse potentials*. The mathematical form of a Morse potential is

$$V = D_e(1 - e^{-a(r-r_e)})^2 \qquad (13)$$

where D_e and a are constants that are characteristic of the molecule. D_e is called the *dissociation energy* and is the energy difference between the potential energy minimum and the asymptotic limit as r, the distance between the atoms, goes to infinity (see Figure 10.4). The constant a is determined by fitting to experimentally measured values of the molecular vibrations.

When a Morse potential is used quantum-mechanically, we can derive a slightly different mathematical expression for the vibrational energy of an oscillator. We get

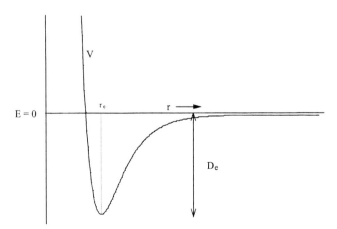

Figure 10.4 A better potential energy curve for the vibrations of a real molecule. This type of curve is more closely approximated by a Morse potential, given by eq 13.

$$E = h\nu_e\left(v + \tfrac{1}{2}\right) - h\nu_e x_e\left(v + \tfrac{1}{2}\right)^2 \qquad (14)$$

where there is now a second term in the equation for the energy. The new variable, x_e, is called the *anharmonicity constant* for the vibration. Vibrations that are described with a Morse potential energy (or other functions) are called *anharmonic oscillators*.

The other major consideration for real vibrations can be illustrated with eq 14. Changes in energy, which is what spectroscopy measures, are no longer independent of the vibrational quantum number v. That is, the value of ΔE, and the corresponding frequency of light that is absorbed, depends on the quantum numbers of the states involved in the transition.

Usually this is not a major concern, since most absorptions occur when a vibration goes from the $v = 0$ quantum state to the $v = 1$ state. However, in situations where molecules are already in some vibrational excited state, the frequency of light absorbed will be different from what is expected by $\Delta E = h\nu$.

Another point for real vibrations is that the selection rule $\Delta v = \pm 1$ is not rigidly followed. Changes in v of 2, 3, or higher can happen, although they are usually less likely and thus weaker absorptions in a spectrum. Such transitions are called *overtones* and are typically less than 2 times, 3 times, or other exact multiples of the $\Delta v = \pm 1$ transitions (the *fundamental* vibrations) because of anharmonicity effects.

Normal Modes of Vibration

When we speak of the vibrations of a molecule, what are we speaking of exactly? Any molecule has translational motion in three dimensions, as well as three independent rotational motions. (Linear molecules have only two rotational motions.) A molecule with N atoms has $3N - 6$ (or $3N - 5$ for linear molecules) independent vibrational motions during which the center of mass of the molecule does not move.

The particular motions of an individual vibration do not have to be random, however. The $3N - 6$ vibrations of any molecule can always be defined as specific motions of the individual atoms, all moving at the same frequency. These motions are called the *normal modes* of vibration. (The mathematical ability to do this for vibrations is covered in advanced treatises on molecular vibrations.)

The useful thing about the normal modes is that these are the motions whose classical frequencies are required for the quantum-mechanical description of molecular vibrations; hence, the frequencies of the normal modes dictate the frequency of light absorbed by vibrations in a molecule, as indicated by eq 11. The importance of normal modes can be expressed by the idea that all possible vibrational motions of any molecule can be described as a certain combination of the $3N - 6$ normal modes of the molecule (much like any point in three-dimensional space can be expressed as the proper combination of x, y, and z coordinates).

The other useful thing about normal modes is that *an infrared spectrum only shows absorptions or emissions whose frequency positions correspond to the frequencies of the normal modes.* This fact alone makes the normal modes of vibration a central idea in understanding the vibrations of a molecule.

For small molecules, the normal modes can be easily illustrated and described, and examples can be found in textbooks of vibrational spectroscopy. For larger molecules, the concept of normal modes of vibration remains, but our ability to describe them gets muddied. This is because each normal mode is, technically, a vibration of all atoms in a molecule. How easy is it to describe the normal modes of, say, benzene, which has 12 atoms? Or how do we describe the normal modes of poly(perfluoroethylene), whose molecules can include thousands and thousands of atoms?

For large molecules, we concede that a completely accurate description of an individual vibration is difficult, so we don't even try. Instead, we are satisfied with using an *approximate* description of the vibration. For example, in poly-(perfluoroethylene), a vibration may involve many atoms but mostly consists of a C and an F atom moving back and forth. We use the approximate description "C–F stretch" to describe this vibration, even though other atoms in the molecule move slightly. The illustration of the C–F stretch in Figure 10.3 demonstrates this very idea: several atoms of the molecule are moving, but the C–F stretching motion dominates the motion. Thus, the use of approximate descriptions for vibrations of large molecules is a recognized and useful practice in vibrational spectroscopy. Descriptions like "bending," "wagging," "in-plane," "out-of-plane," "libration," and other phrases are used to verbally illustrate the vibrational motion of interest.

Although the utility of normal modes may be restricted to smaller molecules, the ability to separate the possible vibrations of a molecule into normal modes allows us to categorize molecular vibrations into useful groupings. These categories are actually very useful in understanding the spectrum of a compound, as discussed in the next section.

Characteristic and Fingerprint Regions

Although each molecule, no matter how large, has its own normal modes with their own characteristic frequencies, molecules with similar arrangements of atoms have normal modes that have similar frequencies. This allows us to group various types of vibrations into characteristic ranges of frequencies. For example, molecules that have an O–H bond show an absorption around about 3500 to 3300 cm^{-1}, which is characteristic of the O–H stretching vibration. Molecules having C–H bonds have vibrational absorptions between 3300 and 2900 cm^{-1}, characteristic of the C–H stretching vibration. Molecules with a carbonyl group (C=O) have a characteristically intense absorption at about 1700 cm^{-1}. Because absorptions in characteristic regions of the infrared spectrum usually correspond to particular bonding arrangements within molecules, the pattern of absorptions can give us clues about the structure of an unknown sample.

The concept of the fingerprint region takes uses this concept. The part of the infrared spectrum corresponding to about 1500 to 700 cm^{-1} (or about 6.6 to 14.3 μm) is called the fingerprint region because only slight differences in the bonding arrangement in a molecule can lead to noticeable differences in an infrared spectrum. Because of these differences, this region serves as a "fingerprint" of any particular compound and can be used for identification purposes. The combined analysis of absorptions in characteristic regions and the fingerprint region goes a long way—with data from other types of analyses—in identifying a particular compound.

Table 10.1 lists some values for characteristic regions of various vibrations (mostly stretching vibrations). Ranges are given because there is going to be some variation in the exact frequency of light absorbed by a particular arrangement of atoms, depending on the circumstances (i.e., phase and temperature of sample, bonding of atoms in the molecule, sample preparation). The ranges listed in the table should be used with caution because the presence of differing bonding arrangements within any particular molecule will affect the exact value of the vibrational frequency. *Correlation charts* are also widely available that give ranges of absorption frequencies for various classes of molecules. Such charts can be particularly helpful in identifying an unknown compound.

Experimental

Most vibrational spectra are measured in a manner similar to that of other spectra: a sample is placed in a chamber of a spectrometer where it interrupts the path of an infrared

Table 10.1 Examples of characteristic frequencies

Type of vibration	Frequency range (cm^{-1})
C–H stretch (alkane)	2800–3000
C–H stretch (alkene)	2900–3100
C–H stretch (alkyne)	3200–3300
C–H stretch (aromatic)	3000–3100
O–H stretch	3100–3700
N–H stretch	3050–3500
C≡C, C≡N stretch	2200–2300
S–H stretch	2550–2600
P–H stretch	2350–2440
Si–H stretch	2090–2260
C=N, C=O stretch	1600–1700
C=C stretch (aliphatic)	1600–1700
C=C stretch (aromatic)	1460–1600
C–H bend (aliphatic)	1350–1450
C–H bend (aromatic)	1000–1200
C–C stretch	1300–1500
C–O–C stretch	~1200

light beam, and the wavelengths of infrared light are scanned through the sample. A detector, attached to some output device like a printer, measured the amount of light absorbed or not absorbed. The result is a spectrum.

Actually, as in all spectroscopic techniques, there are some unique aspects to measuring vibrational spectra. More than any spectroscopic technique, modern vibrational spectroscopy has embraced the Fourier transform as part of its repertoire. There are preferred sources and detectors for infrared spectra; and there are recommended sample holders, solvents, and sample preparations. Finally, there are several common ways of displaying vibrational spectra that workers in the field seem to find more conventional.

Dispersion Spectrometers

From a classical perspective, all spectrometers are the same. A general diagram of any classic absorption spectrometer is represented in Figure 10.5a. It consists of a source that gives off an acceptable intensity of the desired wavelength(s) of electromagnetic radiation; an (optional) optical system that separates, focuses, and/or delivers the light to the sample; a sample compartment or chamber in which the sample is positioned; additional (and also optional) optics for delivering transmitted or emitted light from the sample to a detector; and a detector that is sensitive to the wavelength(s) of light used.

Figure 10.5a notwithstanding, there are extreme variations in virtually all of the components, depending on the wavelength region of interest. Hence, the reader should be aware that what is discussed in this chapter applies to spectrometers for studying vibrational spectra and will probably not apply to any other form of spectroscopy. Readers interested in details of other spectroscopic methods should consult the appropriate chapter in this or other references.

The optical system in classic spectrometers deserves a more detailed discussion, because many of its components are represented in many types of spectrometers. The first item of optics in a spectrometer system is usually a slit, a device that blocks the light from the source except for a small vertical width (labeled S1 in Figure 10.5b). The slit opening may or may not be variable. The light coming through the slit is directed using lenses or, preferably, mirrors. (Mirrors are usually preferred because they don't absorb as much light and they don't introduce as much aberration as lenses do.) Ultimately, the transmittive or reflective optics direct the light through or onto a special optical component called a *monochromator*, labeled M in Figure 10.5b.

The purpose of the monochromator is to separate the light from the source, which is usually composed of a range of wavelengths, into its constituent wavelengths. That is, the monochromator *disperses* the light spatially, giving rise to the name *dispersive spectrometer* to this type of instrument. For example, a prism can separate different wavelengths of light because the different wavelengths have slightly different indices of refraction; they are refracted by different amounts, leading to different wavelengths of light going in slightly different directions. A diffraction grating does the same thing, but takes advantage of slightly different angles of constructive interference for different wavelengths of light. In either case, as Figure 10.5c shows, the monochromator takes the range of wavelengths of light from the source and separates them.

The monochromator is usually mounted on a pivot so it can turn slightly, allowing it to direct light of any wavelength toward the sample. This process is aided physically by a second slit (S2), which serves to block all but a certain interval from proceeding toward the sample. Strictly speaking, the only way to send a single wavelength of light through the sample would be to have an infinitely narrow slit S2. This would mean that the light energy through the sample would be infinitely small, giving no opportunity to measure a spectrum. Therefore, the finite slit width S2 always allows a *range* of wavelengths through toward the sample.

Older UV-visible and infrared spectrometers use prisms as their monochromators. For UV-visible spectrometers, high-quality quartz is used, while for IR spectrometers an ionic salt like NaCl or KBr can be used. Newer spectrometers of this type use gratings as the dispersive element. Gratings have major advantages over salt prisms for the IR region, including durability and relative ease of construction. Details on grating theory can be found in many analytical chemistry or optics textbooks.

The Fourier Transform Infrared Spectrometer

Modern vibrational spectrometers are very different from the first spectrometers that were available in the 1930s and 1940s. This difference is not due to the revolution in electronics, which has touched all forms of spectroscopy. In the case of IR spectrometers, almost all no longer have a classic mirror-and-dispersive-element system. Instead, modern IR spectrometers have an *interferometer*, they measure directly a signal called an *interferogram*, and they digitally calculate a spectrum using a mathematical function called a *Fourier transform*.

Figure 10.6 shows a diagram of an interferometer. It was invented by Albert A. Michelson in 1891 and figured prominently in his and Edward W. Morley's experiments on the ether drift. Collimated light from the source approaches a beam splitter B and is directed to two different mirrors. One of the mirrors, M1, is fixed in position, while the other mirror, M2, moves back and forth in the line of the light path. Light traveling in both paths is reflected back by the mirrors, reaches the beam splitter B again, where it recombines and is split to the original source and to the sample. At this point, only the light reflected toward the sample is of interest.

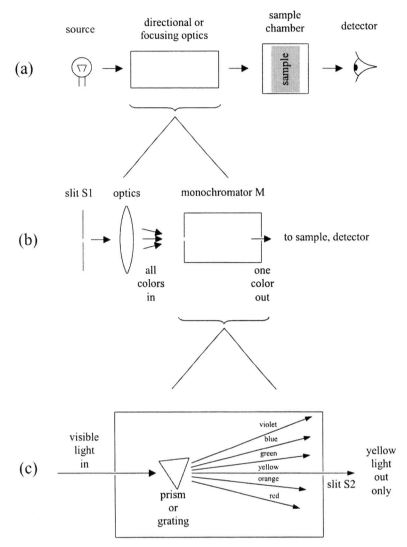

Figure 10.5 (a) A simple block diagram for an absorption spectrometer; later figures expand on key parts of this general diagram. (b) The initial optics of an absorption spectrometer are intended to select a wavelength (or range of wavelengths) of light to send through the sample. (c) The monochromator is the key component of an absorption spectrometer; it disperses light on the basis of wavelength and is designed to allow only a small wavelength range to pass through a sample at any one moment. The diagram shows a prism as the critical dispersive element, but, in most modern monochromators, a ruled grating is actually used. The optical path is usually more complicated (which is why a prism is shown in the diagram), but the concept of wavelength dispersion and selection is the same.

The difference in the two path lengths to either mirror is called the *retardation* and is symbolized by δ. Because light travels back and forth between the mirror and beam splitter, the retardation is two times the mirror movement. When δ is exactly zero, the two split beams have the same path lengths; when the beams are recombined, they are in perfect phase with each other and so interfere constructively. *This is the case for every wavelength of light in the beam.* The net result is that at $\delta = 0$, the output beam through the sample is relatively bright.

However, as the mirror moves away from equidistant path lengths and $\delta \neq 0$, the signal traces a complex path and quickly degrades to some nearly constant level. This is *because all wavelengths of light are passing through the interferometer at the same time* and, if the retardation is not exactly zero, *there is quick and massive destructive interference* of the wavelengths of light. Thus the recombined beam contributes little to the intensity of the resulting light coming out of the interferometer. A plot of the light intensity versus the retardation is called an *interferogram*, as shown in Figure

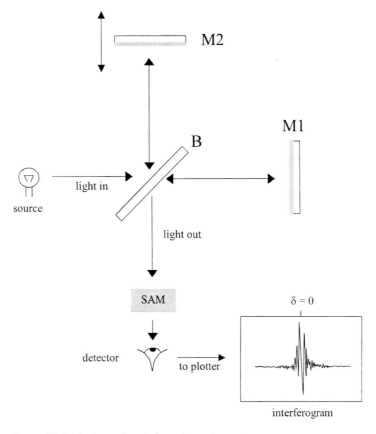

Figure 10.6 Most modern infrared spectrometers use an interferometer instead of a monochromator. The resulting signal, the interferogram, is shown at the lower right and is mathematically transformed into a vibrational spectrum.

10.6. The burst of intensity at zero retardation is called the *centerburst*. It is important to keep in mind what an interferogram is: it is a graph of intensity of light versus retardation, or a plot of intensity I versus position in cm (as measured from $\delta = 0$).

In 1892, Lord Rayleigh recognized that an interferogram was mathematically related to a spectrum through a mathematical function called a Fourier transform. For any one-dimensional function $f(a)$ that is defined over $-\infty \leq a \leq +\infty$, the Fourier transform of $f(a)$, denoted F, is defined as

$$F = \frac{1}{\sqrt{2\pi}} \int_{-\infty}^{+\infty} f(a) \cdot e^{iax} dx \qquad (15)$$

The *Fourier cosine transform* F of $f(a)$, for $0 \leq a \leq +\infty$, is defined as

$$F = \frac{1}{\sqrt{2\pi}} \int_{0}^{+\infty} f(a) \cdot \cos(ax) dx \qquad (16)$$

The Fourier sine transform is defined similarly, except using a sine function. It can be shown that $f(a)$ is also the Fourier transform of F; that is, determining the Fourier transform of a function twice regenerates the original function.

In the case of an interferogram, we have a function of intensity versus the retardation (which has units of distance), so the interferogram can be denoted $I(\delta)$. It is related to the intensity function of the source/optics of the spectrometer in terms of wavenumber, which we will denote $B(\tilde{\nu})$. For any particular wavenumber $\tilde{\nu}$, $I(\delta)$ also varies sinusoidally depending on the retardation δ, because of destructive and constructive interference. Thus, for any particular wavenumber, the intensity at any retardation is given by

$$I(\delta)_{at}\tilde{\nu} = B(\tilde{\nu}) \cdot \cos(2\pi\tilde{\nu}\delta) \qquad (17)$$

Thus, $I(\delta)$ *is the Fourier cosine transform of* $B(\tilde{\nu})$. The total intensity is determined by summing the individual intensities at all wavelengths. The sum over all infinitesimal wavelengths is an integral; eq 17 can be generalized as

$$I(\delta) = \int_{-\infty}^{+\infty} B(\tilde{\nu}) \cdot \cos(2\pi\tilde{\nu}\delta) d\delta$$

Because the functions $I(\delta)$ and $B(\tilde{\nu})$ are Fourier transforms of each other, the light intensity versus wavenumber—the *spectrum*—is the Fourier transform of $I(\delta)$:

$$B(\tilde{v}) = \int_{-\infty}^{+\infty} I(\delta) \cdot \cos(2\pi\tilde{v}\delta)d\delta \qquad (18)$$

Therefore, to determine a spectrum from an interferogram, we need to take the Fourier transform of the interferogram. In eq 18, $I(\delta)$ is a function of intensity versus units of distance, while $B(\tilde{v})$ is a function of intensity in terms of inverse distance (i.e., wavenumbers, or cm^{-1}).

Determining the Fourier transform (or FT) of a simple function is a straightforward exercise in calculus. Determining the FT of a complex function like an interferogram is conceptually very difficult. Thus, while the theory of FT infrared, or FTIR, spectroscopy was well known, in practice it was quite challenging, even with the introduction of computers and digitized intensity-versus-retardation. In order to calculate the FT of an interferogram consisting of N digitized points, at least N^2 calculations were necessary. According to the *Nyquist criterion*, a sinusoidal signal can be unambiguously sampled if the sampling frequency, h, were equal to two times the interval to be sampled. If the resolution* of the spectrum is given by $\Delta\tilde{v}$, the number of points N needed to sample the interval is

$$N = \frac{\text{interval}}{\Delta\tilde{v}} \qquad (19)$$

To sample a 1000-cm^{-1} interval at a resolution of 1 cm^{-1} thus requires 1000 points, requiring a minimum of *1 million* calculations—and this doesn't come close to the intervals and resolutions needed for many spectroscopic needs.

In 1965, Cooley and Tukey described a new algorithm called a fast Fourier transform (or FFT). If the number of points to be sampled were some power of 2, or

$$N = 2a \qquad a = \text{some integer} \qquad (20)$$

then Cooley and Tukey showed that the Fourier transform could be expressed in terms of a matrix that could be factorized into smaller matrices; these smaller matrices could then be used to calculate the FT with a smaller number of calculations. It can be shown that that while a "normal" Fourier transform requires N^2 calculations, a fast Fourier transform only requires $Na/2$ calculations. Thus, the FFT is much more computationally efficient. Since the development of the fast Fourier transform, FTIR has become the preferred form of infrared spectroscopy.

FTIR has other advantages over dispersive infrared spectrometers. *Fellgett's advantage* is the recognition that all wavelengths of light are sampled simultaneously, because all wavelengths of light contribute to the shape of the interferogram. (This is also known as the *multiplex advantage*.) An overall brighter signal is measured, as opposed to the signal of a dispersed, slit-limited light beam. Although it is easier to detect a bright signal, it is possible to use dimmer signals to measure a similar-quality spectrum. Furthermore, a brighter signal tends to overwhelm signal fluctuations that constitute *noise*.† FTIR spectra are therefore easier to detect and are generally not as noisy as dispersive spectra.

In a related issue, FTIR can produce less noisy spectra because spectra are originally collected in digital form (i.e., as pairs of numbers [intensity, retardation]) rather than in analog form (i.e., as a tracing on a piece of paper). Measuring multiple interferograms and averaging them acts to reduce random noise in the interferogram and, hence, the calculated FTIR spectrum. To do this, the spectrometer must know, *as exactly as possible*, the retardation of the moving mirror over multiple cycles, or *scans*. To determine this, most interferometers have a small helium-neon (He-Ne) laser that points into the interferometer, parallel to the source beam. The monochromatic light of the laser, having a wavelength of 632.8 nm, produces a regular sinusoidal interference pattern that passes through the zero every 316.4 nm. By using this interference pattern as an AC signal, a small detector can determine the position of the moving mirror to 0.000 032 cm using the IR centerburst or an auxiliary white light to generate a centerburst as a benchmark position. This ability to accurately determine the position of the moving mirror allows an FTIR spectrometer to average multiple interferograms accurately. Random noise is therefore reduced by a factor proportional to the square root of the number of scans.

Finally, mathematical optics can be used to calculate a factor called *throughput* for an interferometer and compare it to the throughput for a grating spectrometer. Throughput is a measure of the power delivered by an optical system to the detector, and it can depend on many components of the overall optical system, as well as on the frequency of the light itself. However, in the infrared region of the electromagnetic spectrum, the maximum throughput of an interferometer-based spectrometer can be shown to be 1 to 2 orders of magnitude higher than the maximum throughput of a grating-based spectrometer. This is known as *Jacquinot's advantage*. Overall, FTIR spectrometers are generally much more sensitive than dispersive spectrometers and have the capability of measuring IR spectra that are much less affected by noise. The chief disadvantage to FTIR spectrometers is cost. Construction of high-performance interferometers is relatively costly, compared to making a grating. Interfacing an optics bench with a computer is necessary for an FTIR spectrometer, while a dispersive instrument needs only a chart recorder. However, considering the advantages that the FT method has over a dispersive instrument, the increased cost has not been an

---

*Resolution is the consideration of how close two peaks in a spectrum can be and still be distinguished as two separate peaks. A detailed discussion of resolution will not be given in this chapter.

†Noise is another topic that, while important in spectroscopy, will not be discussed here.

impediment to allowing FTIR spectrometers attain a huge monopoly in the field.

Sources and Detectors

All spectrometers need a light source. For vibrational spectroscopy, a source is needed that gives off a useable quantity of infrared radiation. Generally, any solid that can be heated electrically to 1000 to 1500 °C and is air-stable at that temperature can be used. Small incandescent lamps, with filaments of nichrome or other high-melting material, can be used. Ceramic sources are more common. The two common ones are the Nernst glower and (more common) the Globar. A Nernst glower is a rare-earth oxide rod with metal wire electrodes. Current is passed through the rod, generating heat. Because the rare-earth oxide mixture has a negative temperature coefficient of resistance (i.e., it has a lower resistance at higher temperatures), it must be preheated in order to pass enough current to maintain its temperature.

The Globar is similar in construction but composed of silicon carbide. It has a positive temperature coefficient of resistance and a better output of lower-energy infrared light. Both the Nernst glower and the Globar suffer from fragility due to brittleness. Also, because they operate in air, they have a shorter lifetime than lamps. However, their superior output makes them the preferred sources.

There is more choice in detectors due not just to technology but to spectral region considerations. While spectral output can be adjusted relatively easily by changing the temperature of the source, spectral detection is more challenging because of the lower intensities of infrared sources and the lower energy of the infrared photon. Further, no single detector is equally sensitive to all wavelengths of light, even if the light is confined to the infrared region. Early detectors, called *thermal detectors*, took advantage of IR's relatively easy conversion to heat. These detectors include thermocouples whose voltage is temperature-dependent; bolometers, whose electrical resistances vary with temperature; and the Golay cell, a gas-filled chamber whose internal pressure also depended on the temperature. These detectors depended on their ability to convert the incident infrared radiation into heat, which was then measured in terms of some other physical observable (i.e., voltage, resistance, or pressure changes). Such detectors are slow and relatively insensitive, but they do have the advantage of applicability to a wide range of infrared light frequencies.

A related type of detector takes advantage of the so-called pyroelectric effect. Certain crystalline materials, liked doped triglyceryl sulfate (dTGS), will experience changes in polarization if the temperature changes quickly. Thus, a thin crystal of such a material, placed between IR-transparent electrodes, will act as a variable capacitor under the influence of incident infrared radiation. These detectors are fast and can be operated at room temperature, since their detection depends on a modulation of the temperature, not the temperature itself. Pyroelectric detectors are common in modern spectrometers.

Except for the pyroelectric sort, thermal detectors have largely been replaced by devices that can detect the low energy of the individual infrared photon. *Photonic detectors* are based on some semiconductive material that has a low enough energy gap for the IR photon to induce an electron into the semiconductor's conduction band. Conductivity of the active material is measured electronically, so these detector are also referred to as *photoconductive*. Because some minimum amount of energy is necessary to excite an electron, these detectors have some threshold wavelengths below and above which they are insensitive. Depending on the semiconductor material used, thermal energy may be sufficient to excite a large number of electrons into the conduction band. While for some detector materials this simply contributed to the noise of the measurement, for others it can mask the effect of the incident infrared light. Therefore, some of these detectors must be cooled (typically with liquid nitrogen) before they can be used for IR detection. Mercury-cadmium-telluride (HgCdTe, or MCT) detectors are common liquid-N_2-cooled detectors; indium antimonide (InSb) is another, and it can operate at room temperature.

Most infrared spectrometer manufacturers give buyers several options for detectors when a spectrometer is purchased, depending on the buyer's needs. However, the vast majority of spectrometers have a single detector. Users of any infrared spectrometer should be aware of the detector type the spectrometer has and, if necessary, be prepared to support it (i.e., with a liquid nitrogen supply).

Other Instrumental Concerns

It is highly probable that an experimenter will be using a computer-operated FT-based infrared spectrometer. Given that this is highly probable, we should mention several other instrumental concerns and settings that will affect the measurement and the quality of the spectrum.

Single-Beam versus Double-Beam Measurements

The Fourier transform infrared spectrometer is almost without exception a *single-beam instrument*. This means that there is a single light path through the spectrometer optics and that a single spectrum is measured at a time. This contrasts with *double-beam spectrometers*, which split the light beam and send one beam through a sample area and one beam through a reference area. The double-beam setup allows an experimenter to measure the spectra of a sample and a blank (typically an equivalent sample holder plus the solvent or support matrix). These spectrometers also have two detectors, and the spectrum is generated automatically as the electronics of the detector system compare two light intensities. Any differences in the light intensities are presumed to be caused by the sample of interest, and this differential light intensity at the two detectors becomes the spectrum.

FTIR spectrometers have a single optical path, requiring that a single-beam *background spectrum* be measured and saved in the computer's memory, either on a computer disk or in a reserved area of the computer's operational memory. After the background spectrum has been measured and saved, a single-beam *sample spectrum* can be measured. After measuring the sample spectrum, the computer can be instructed (usually automatically) to numerically compare, or ratio, the sample spectrum to the background spectrum. After this numerical comparison, the computed "double-beam" spectrum can be displayed. (Single-beam FTIR spectra can be measured and displayed, but ratioing sample to background spectra makes the resulting spectrum easier to view and interpret.) There are different ways of displaying an infrared spectrum; they will considered shortly.

Because of the need to measure background spectra separately from sample spectra, there are several precautions an experimenter should consider:

1. Background and sample spectra should be measured under as similar conditions as possible. The same sample holder, path length, solvent, solid-state matrix, and instrument settings should be used for both spectra (although not all of these may be applicable or even feasible for all spectroscopic measurements).

2. Background spectra and sample spectra should be measured as close together in time as possible. Drifts in the electronics' behavior, environmental conditions, variances in source and detector performance, and other variable factors can influence a vibrational spectrum whose components were measured a long time apart. Minimizing the time between background and sample single-beam spectra will minimize those effects. (In an extreme—and maybe apocryphal—instance, a colleague once mentioned that spectrum quality could be compromised by a person using the elevator, when the elevator shaft was next to the room in which spectra were being measured.)

3. Regions of the spectrum in which very little light reaches the detector (due either to low source intensity, limited transmission range of optical materials, or strong absorption by the sample or solvent) may exhibit nonreproducible spectral features: very large positive or negative absorptions, or even spurious peaks. This is because tiny *absolute* changes in the light reaching the detector (which may be caused by random fluctuations, or noise) can represent large *relative* differences between background and sample spectra, leading to large spectral features that may or may not be due to your sample. Care should be exercised when evaluating spectra in regions where little light reaches the detector.

The best rule of thumb is that the conditions for sample spectra should be exactly the same as possible as conditions for background spectra—except, of course, for the presence of the sample.

Purging

A related issue concerns the purging of infrared-absorbing gases from the sample chamber, both in measuring background and sample spectra. Of the common gases in air, water vapor and carbon dioxide are strong IR absorbers. As vapors, they show up in the gas phase as series of lines occurring in a wide grouping around 1600 cm^{-1} and 3400 cm^{-1} for water, and around 2330 cm^{-1} for CO_2. (CO_2 also has vibrations around 667 and 1330 cm^{-1}, but they don't appear strongly in a vibrational spectrum.) Appearance of these absorptions in a spectrum is a nuisance, especially if they tend to obscure absorptions that arise from the sample of interest.

There are two major ways to minimize or eliminate these absorptions from the ratioed spectrum. The accepted way is to *purge* the sample chamber with a non-IR-active gas—typically argon or dry, CO_2-free nitrogen—until the water vapor and carbon dioxide are flushed from the optical path. Most IR spectrometers are required to have some purging gas flowing through them anyway, to promote smooth interferometer motion or to project the optics from water vapor, or both. The same gas flow can be used to flush out the sample chamber to remove IR-active atmospheric gases.

Purging the sample chamber takes time: 10, 15, 30 min or more, depending on the gas flow rate. However, unless your sample chamber can be evacuated, this is the only way to remove the atmospheric contaminants.

The second method takes advantage of the warning in the last section to measure your background and sample spectra under as exactly the same experimental conditions as possible. Thus, if you measure your background spectrum with the sample chamber exposed to atmosphere, then measure your sample spectrum with the chamber exposed to atmosphere. If the two spectra are measured close together in time, the ambient amounts of water vapor and carbon dioxide in the air won't have changed, and the absorptions due to H_2O and CO_2 in the atmosphere will ratio exactly and will not show up as signals in your spectrum. Or, you may close the sample chamber's cover and purge for 1 min exactly; if you do that for both background and sample, the absorptions for the atmospheric contaminants will have "ratioed out" in the final computed spectrum. Caution must be taken if doing this, because humidity changes throughout the day will show up if background and sample single-beam spectra are measured too far apart in time.

While, frankly, to some people, this second method might not be considered an acceptable technique, there may be no avoiding it. There are some optical setups designed to fit into a sample chamber that prohibit closing and sealing the sample chamber, so measurement of the IR spectrum while exposed to the ambient atmosphere is inescapable. Even construction of a custom-made purgable shroud may not be possible. When measuring IR spectra on samples whose compositions change over time (like pressure-induced decomposition of samples inside diamond anvil cells), an experimenter may not be able to wait for a complete purge. While the details of the spectral measurement might be dictated by the experiment, the experimenter should take every effort to ensure that an IR spectrum is as clean and as spectrally uncontaminated as possible.

Finally, be aware that *purge gases should never be completely shut off* when you are finished measuring spectra, because the fragile optics of the interferometer may require a constant, if low-level, purge. (In some spectrometers, the purge gas also provides "lubricant" to air bearings attached to the moving mirror of the interferometer. Shutting off the purge gas can ruin the optical mounting!) Always determine the necessary purge status of the instrument before shutting off any purge gases.

Number of Scans and Resolution

For an FT spectrometer, the two instrumental settings that affect the spectrum quality most are the number of scans to measure and the required resolution of the spectrum. The number of scans is easier to address. Because the random noise inherent in a series of N measurements decreases with $(N)^{1/2}$, the larger number of interferograms that can be measured and averaged, the less noisy a spectrum will be. Thus, a spectrum of 100 averaged scans will be 10 times less noisy than a spectrum from a single scan. It will take about 100 times as long to measure the spectrum, however. A trade-off ensues, between the amount of time it takes to measure a spectrum and the minimum acceptable noise of the spectrum. In most cases, this is not too much of an issue, as Fellgett's advantage virtually assures a better-quality spectrum from an FTIR than from a dispersive instrument. Only for samples where the light throughput is unusually low does a large number of averaged scans become necessary. Many spectrometers have operating software that is pre-set to a default number of scans, like 16.

Resolution is a different issue. Resolution is mathematically defined as the ratio of the light wavelength λ and the spectral bandwidth, $\Delta\lambda$:

$$\text{resolution} = \frac{\lambda}{\Delta\lambda} \tag{21}$$

Practically speaking, resolution is a measure of how close two spectral signatures can be and still be differentiated as separate signatures. It also affects the apparent width of an individual absorption. Paradoxically, a higher resolution is expressed as a smaller number.

There are two different criteria for applying resolution to a spectrum. The *baseline criterion* requires that the spectrum signal reach the baseline of the spectrum in order to consider the two separate signals truly resolved. This criterion is illustrated in Figure 10.7. The *Rayleigh criterion* requires that if two signals are overlapping, each peak maximum must be at a wavelength where the other signal has already reached baseline. This criterion, illustrated in Figure 10.8, is less stringent and probably is used more by practicing scientists. As mentioned, resolution also affects signal width: a signal whose width is smaller than a spectrum resolution will have a larger than necessary width, reflecting the resolution of the spectrum and not the true width of the signal.

For dispersive spectrometers, resolution is usually controlled by adjusting a manual slit that allows for passage of a selected wavelength range, $\Delta\lambda$, through the spectrometer. Because spectrometer optics may not treat light of all wavelengths exactly the same, the resolution of a given spectrum may vary from one end of the spectral range to the other. Also, smaller slits may make for higher-resolution spectra, but at the cost of a decreased amount of light passing through the spectrometer optics at any one time, thereby leading to noisier spectra. Thus, for dispersive spectrometers, resolution and noise must be balanced against each other.

The resolution issue is slightly different for Fourier transform spectrometers. An FT spectrometer measures an interferogram and digitizes the signal into individual, discrete points. Ultimately, a signal is simply a series of points, (x,y), in a data file. The more data points that are collected, the better that the transformed interferogram will be able to accurately and uniquely represent the true spectrum. Thus, the more data points that are collected, the better the resolution of the spectrum can be.

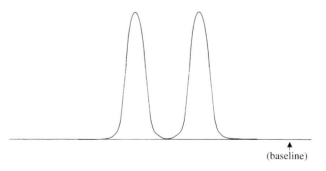

(baseline)

Figure 10.7 The baseline criterion for resolution requires that the signal return to baseline before rising to indicate another absorption. By this criterion, these two absorptions are resolved.

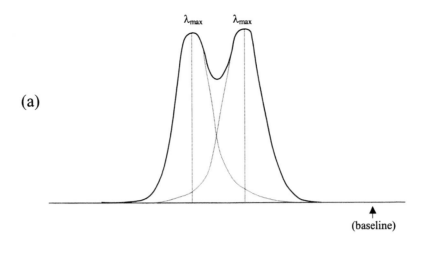

(a)

λ_{max} λ_{max}

(baseline)

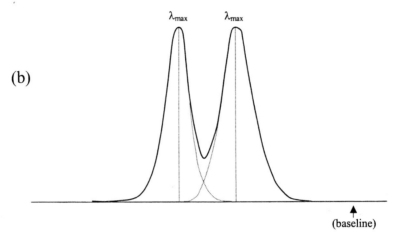

(b)

λ_{max} λ_{max}

(baseline)

Figure 10.8 The Rayleigh criterion for resolution requires that the signal from an overlapping absorption not overlap the wavelength of the other absorption's maximum. Therefore, (a) shows two signals that are not resolved, while (b) shows two signals that are resolved.

In addition, because the interferogram is a composite waveform that has information from all wavelengths simultaneously, the more the interferogram can be measured, the better chance that the transformed spectrum can represent individual (i.e., closer-spaced) signals. The width of the interferogram is dictated by the amount of movement of the moving mirror in the interferometer (M2 in Figure 10.6). Thus, the greater the travel of the moving mirror, the greater the resolution of the transformed spectrum.

It can be shown for a Fourier transform spectrometer that the optical path difference δ (which will be twice the mirror motion, because light travels back *and* forth between the beam splitter and the moving mirror) and the wavenumber resolution Δṽ are related by a simple inverse:

$$\Delta \tilde{v} = \frac{1}{\delta} \qquad (22)$$

Therefore, for a spectrum to have a resolution of 1 cm^{-1}, the moving mirror must provide an optical path difference of 1 cm. A higher resolution, capable of separating signals closer together, requires a larger optical path difference. Many modern spectrometers are capable of providing <1 cm^{-1} resolution (0.25 cm^{-1} is common), but the relationship in eq 22 points out another trade-off: increasing resolution requires larger mirror movements. Mirror movement, the heart of the interferometer function, must be very finely controlled if one is to measure reproducible interferograms, making large mirror motions mechanically challenging at best. This leads to the concept "increasing resolution is increasingly expensive", a slight revision of an old microscopist adage.

In any event, for a FTIR spectrum, resolution is an instrumental parameter set by the user, using the computer program that controls the optical instrument. Settings of the number of data points and the mirror travel distance are automatically adjusted for any given resolution. Common resolutions are 1, 2, 4, 8, 16, and 32 cm^{-1} and, as mentioned, most instruments have subunit wavenumber resolution available. A higher resolution means detection of different signals closer together, but at a cost of more time for the interferometer mirror to move back and forth. Even for an FTIR, there are trade-offs.

Spectrum Output Type

The other major choice an experimenter has it how the final spectrum will look. A spectrum is a graph of signal versus some characteristic of light (wavelength, frequency, energy, wavenumber). Experimenters can choose to display or plot a spectrum in various ways, depending on the units for the x axis (typically the light characteristic axis) and the way the signal is presented.

The issue of units was discussed in an earlier section. For an infrared spectrum, typical x-axis units are microns (a unit of wavelength) or wavenumbers (a unit of inverse wavelength). Because wavenumbers are directly proportional to energy, this x-axis unit is preferred by some. Because wavelength is *inversely* proportional to energy, a spectrum plotted by linear wavelength is *not* linear in terms of energy or wavenumber. Thus, comparison of spectra displayed or plotted in terms of wavelength and wavenumber will be complicated by the fact that the two abscissa units are not directly proportional to each other. In addition, many spectrometers also allow the option to expand the lower-energy part of a spectrum by a factor of 2 (or more). This also creates confusion for the unaware: spectra might be displayed on one scale between, say, 4000 and 2000 cm^{-1}, and on an expanded scale between 2000 and 500 cm^{-1}. Caution in interpreting spectra is advised. The typical ranges for an infrared spectrum are 4000 to 400 cm^{-1} (the so-called mid-IR, and these are typically left and right limits of a default spectrum) and 2.5 to 25 μ (again, typically the left and right limits, respectively). They can be changed using the controlling program of any FTIR.

There is also choice on how the y axis is defined. The two most common ordinates for absorption spectroscopy are *transmittance* and *absorbance*.

Figure 10.9 shows the passage of incident light through a sample. If the sample absorbs some of the power of the incident beam, then the transmittance T of the sample is defined as

$$T = \frac{P}{P_0} \qquad (23)$$

where P_0 is the power of the incident beam, and P is the power that passes through the sample. By definition, the transmittance can have a maximum of 1 (corresponding to no light absorbed by the sample) and a minimum of zero (corresponding to all light absorbed by the sample). In practice, rather than using a decimal value for transmittance, the fraction in eq 23 is multiplied by 100% to get a percentage of transmittance, whose value can range from 0% to 100%. Typically, transmittance spectra are displayed or

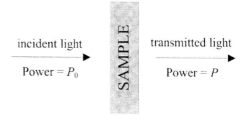

$$T = \frac{P}{P_0}$$

Figure 10.9 The definition of transmittance.

plotted with 0% on the bottom and 100% on the top; the *baseline* of a nonabsorbing sample is at the top, at 100%. Absorption of light of different wavelengths is manifested by having "dips" in the signal. Examples of transmittance spectra are shown in Figure 10.10a,b. Figure 10.10a shows a transmittance spectrum versus wave*number* units, while Figure 10.10b shows the same spectrum, but plotted versus wave*length* units. Note the difference in proportion: while one abscissa unit is proportional to energy (expressed in wavenumber units), the other is inversely proportional to energy (expressed in wavelength units). Nonetheless, they are the same spectrum. (Spectra displayed in this mode are occasionally—and erroneously—referred to as "transmission spectra".)

Absorbance, A, is a logarithmic scale that is defined in terms of the transmittance:

$$A = -\log T = -\log\left(\frac{P}{P_0}\right) \qquad (24)$$

As its name implies, a sample that absorbs no light will have an absorbance of 0. A sample that absorbs 90% of the light will have an absorbance of 1 [i.e., $-\log(0.1)$, which equals 1], a sample that absorbs 99% of the light will have an absorbance of 2 [$-\log(0.01)$], and so on. Samples that absorb all light will have a numerical absorbance of infinity, but many spectrometers have a maximum absorbance value that they will display by default (like 8). Since absorbance spectra are displayed with 0 as the bottom minimum and some positive absorbance as a maximum, absorbance spectra have a baseline at the bottom of the display, at an absorbance value of 0. Absorption of light is manifested as upward peaks, as shown in Figure 10.11. (The sample in Figure 10.11 is the same as that in Figure 10.10a,b; note the similarities and differences among these three spectra.)

Absorbance has an advantage over transmittance in that it can be shown to be proportional to two major experimental parameters: the concentration of the sample, c, and the distance that the light travels through the sample, b. This

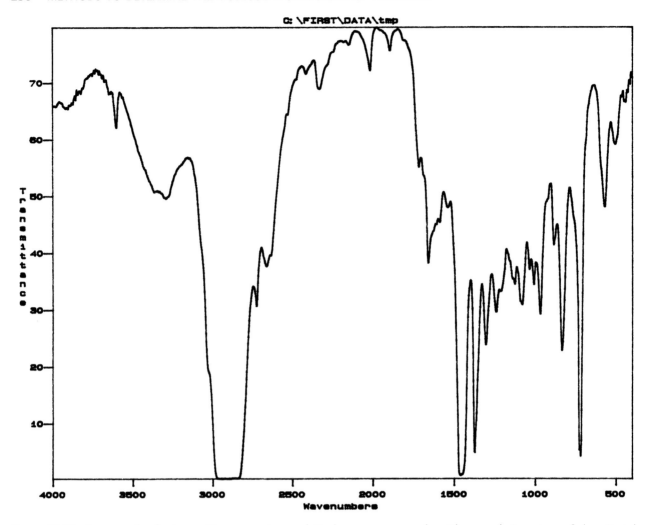

Figure 10.10a An example of a transmittance spectrum, plotted versus wavenumber. The sample is a piece of clear Scotch tape. Downward peaks denote the transmission of less light of that wavenumber passing through a sample.

distance is also referred to as the *pathlength*. The proportionality constant that relates absorbance to b and c is called the *absorptivity*, a, and the mathematical expression between these quantities is

$$A = abc \tag{25}$$

If the pathlength b were expressed in units of centimeters and the concentration c were given in units of mol/L (or molarity), this expression is modified to

$$A = \varepsilon bc \tag{26}$$

where ε is the *molar absorptivity* or the *molar extinction coefficient*. Equations 25 and 26 don't hold for every circumstance (for example, they don't apply very well to extremely dilute or concentrated samples, or samples in which the pathlength is unusually large), but under the right circumstances they provide the basis for *quantitative* spectroscopic measurements. Equation 26 is one form of the *Beer-Lambert law*, or more simply, *Beer's law*.

In both cases of transmittance and absorbance, the computer program that operates the spectrometer must compare a digitized background spectrum to a digitized sample spectrum. A parameter specifies the spectrum output type, and the computer can easily calculate a transmittance or absorbance spectrum. Indeed, doing this calculation is probably the fastest part of the computer's functions; collecting the interferogram and calculating the Fourier transform of the interferogram are more timely and intensive processes. Once a spectrum is calculated, it is usually a trivial function to convert the spectrum to another format.

One other common calculation that can be performed on a spectrum is called a *baseline correction*. Matter can also scatter light, even if it doesn't absorb light. Short-wavelength light, having a higher wavenumber, is scattered more efficiently than long-wavelength, low-wavenumber light. (This is why the sky is blue.) If the background spectrum is measured for no material and the sample spectrum is measured for some material object (as might happen

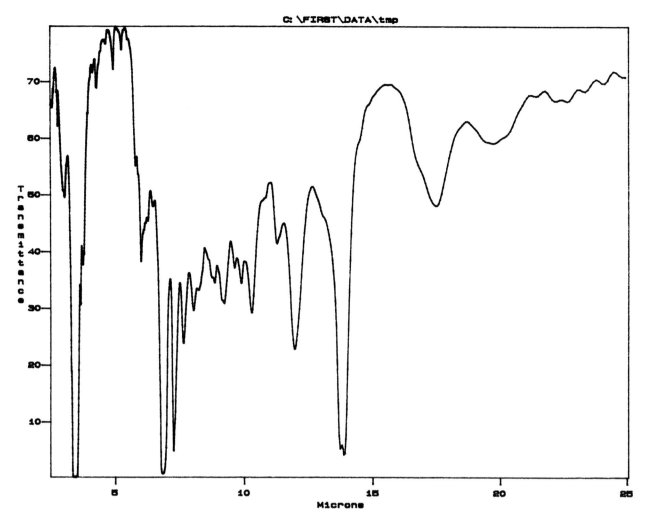

Figure 10.10b The same spectrum, but plotted versus wavelength units. The spectrum is distorted from that in Figure 10.10a because wavelength is *inversely* proportional to energy. Transmittance spectra are commonly plotted versus wavelength.

when the sample is, say, a thin polymer film), the sample will scatter more high-wavenumber light than low-wavenumber light. The result is a spectrum imposed on a sloping baseline, as shown in Figure 10.12a. For better comparison of many spectra, it is desirable to have a more horizontal baseline, and most FTIRs have the capability to mathematically correct the baseline to a more horizontal slope. Figure 10.12b shows the same spectrum, but after baseline correction.

In most cases, spectrometers either perform a baseline correction based on pre-set parameters, or they require the user to designate two points in a spectrum between which to correct the baseline. Extreme care must be taken in performing baseline corrections, because an improper correction can actually introduce artifacts into the spectrum that might be misinterpreted as originating from the sample under investigation. Most important, poor baseline corrections can improperly change the intensities of absorptions, leading to incorrect conclusions based on absolute or relative

intensity. Experimenters are urged to review the manuals that accompany the spectrometer for details about how the FTIR software handles baseline corrections.

Sample Preparation

Sample preparation is an important part of any spectroscopic measurement. Vibrational spectra may have the advantage of having the largest number of ways to prepare a sample. We will discuss only the major sample preparation techniques here, especially those that might be useful for polymeric samples.

Neat Samples

If a sample has the right physical form, it is possible to mount it directly onto a holder and measure its vibrational spectrum directly. Typically this works only if the sample is a thin film. For example, a thin film of polystyrene can be

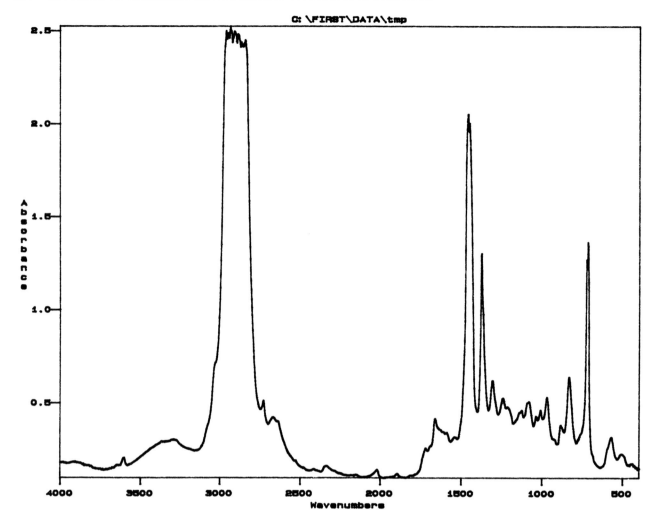

Figure 10.11 An example of an absorbance spectrum of the same material used for Figures 10.10a and 10.10b. Upward peaks denote the absorption of light of that wavenumber.

used to calibrate infrared spectrometers. The polystyrene film is mounted over a hole in a small piece of cardboard, which is stood inside the spectrometer's sample chamber.

KBr Pellets

One standard technique for preparing IR spectrum samples is to mix a powdered sample with IR-grade KBr (potassium bromide) in about a $1:100$ ratio and grind in a mill or small mortar and pestle (agate ones are best, as they are nonporous; ceramic mortars and pestles are easily contaminated and difficult to clean). Then, the intimate mixture is placed in a specialty press and subjected to pressure. Under pressure, the KBr flows to produce a transparent matrix, or *pellet*, in which the sample is embedded. The infrared spectrum of the sample can be measured while the pellet is still transfixed in the press. The press can be cleaned out by rinsing with water and then drying thoroughly.

Polymer samples can be prepared for spectrum measurement if the polymer exists as a powder, or if flat, thin sheets

of polymer can be laid so that they are embedded parallel to the pellet surfaces. Large pieces (millimeter size or larger) of polymers cause problems because they might make a very inhomogeneous sample.

Finally, experimenters should be advised that there is a special grade of KBr marketed especially for the production of pellets for vibrational spectroscopy. While "normal"-grade KBr can be used, it is typically more granular and not as pure, and also it may contain significant amounts of adsorbed water. (KBr is hygroscopic, so IR-grade KBr should be stored in an oven or a dessicator, and should be exposed to atmosphere as little as possible. Some pellet presses allow one to press under vacuum, which helps draw out as much CO_2 and H_2O from the pellet as possible.)

Mull and Salt Windows

While the technique of mull and salt windows is probably used less than KBr pellets for polymers, it bears a quick mention. If the sample is finely powdered, a small amount

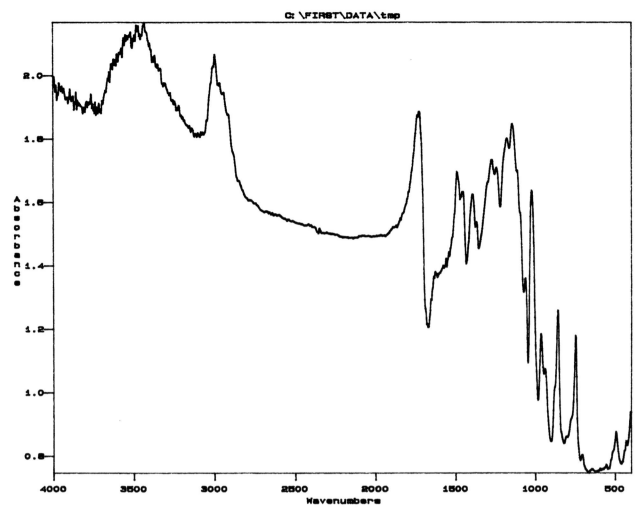

Figure 10.12 (a) A spectrum of poly(ethyl methacrylate) in a potassium bromide pellet shows a sloping baseline caused by unequal scattering of shorter-wavelength light.

can be mixed with several drops of a heavy hydrocarbon or fluorocarbon oil to make a *mull*. A small amount of this mull can be sandwiched between two salt plates, typically NaCl (useful down to ~600 cm^{-1}) or KBr (useful down to ~250 cm^{-1}). (Use gloves or finger cots when handling salt plates!) The mull is usually sticky enough to keep the two plates together, and the assembly can be placed on a regular IR sample holder so that the vibrational spectrum can be measured. Background spectra of the mulling oil should be measured (and should be evaluated ahead of time to see if the oil spectrum may interfere with the polymer spectrum). Salt plates can be round, square, or rectangular and have varying thicknesses. A typical salt window is one inch (25.4 mm) in diameter and has a thickness of 3 to 5 mm. Plates should be cleaned with a nonaqueous solvent, and if scratched or clouded they easily can be repolished using a paper towel wetted with water or alcohol. When they are no longer useful as windows, KBr plates can be ground up for use in pellets.

Other IR window materials include CsI (which is very soft), AgCl (which is UV-sensitive), ZnSe, or KRS-5 (crystalline thallium halide salts that should be handled *very* carefully, as thallium compounds are very toxic). Each material has its own advantages and disadvantages (such as solvent compatibility, acid or base resistance, and characteristic spectral ranges). Selection of the appropriate window material is best done by considering these properties, which are tabled in a variety of handbooks and catalogues. As such, they will not be listed here.

Dissolution and Liquid Cells

If a sample can be dissolved in an appropriate solvent, the resulting solution can be used as a sample for a vibrational spectrum. If this is the preferred technique, two considerations are the *solvent* and the *liquid cell*.

The first consideration is the solvent. First of all, is there an available solvent that will dissolve the polymer of inter-

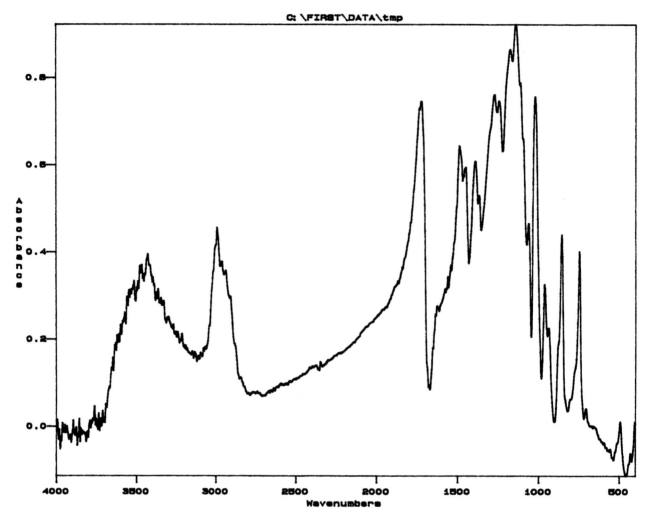

Figure 10.12(b) The same spectrum, after a mathematical baseline correction has been applied. Note the shape of the overall baseline, which has changed slightly as part of the correction. Also note that the y axis has changed, as have the absolute intensities of the individual absorptions.

est? If so, the next concern is the infrared spectrum of the solvent itself. Unlike salt windows or powders that do not absorb in the mid-infrared region of the spectrum (because they are composed of individual ions, rather than covalently-bonded molecules), liquid solvents by their very nature are molecular, and the atoms in the molecules will vibrate and absorb infrared light. The important question is, will the vibrational absorptions of the solvent interfere with the vibrations of interest of the polymer?

All solvents have windows of transparency where they themselves don't absorb, so this solvent can be used (if it dissolves the polymer) if the absorptions of interest are within those windows. Table 10.2 lists some common solvents and their windows of transparency. Water absorbs very strongly around 3500 and 1600 cm^{-1} but might be used as a solvent if the windows of the cell are not water-soluble. (CaF$_2$ is relatively water-insoluble and can be used as a window for aqueous samples.) Chlorinated hydrocar-

bons have large regions of relative transparency, while aliphatic and aromatic hydrocarbons have some useful ranges. Carbon disulfide, CS$_2$, has a large window at low wavenumber (i.e., high wavelength), but its low flash point and high flammability require extra caution.

On the assumption that a suitable solvent is found and a solution is made, a specially constructed *liquid sample cell* is required. Figure 10.13 shows a schematic of such a cell. It has two salt windows (which may have to be changed if an aqueous solution is used), gaskets to make a good nonleaking seal, and a spacer. The depth of the spacer can be highly variable, depending on whether the sample or solvent absorbs strongly or only weakly, or depending on the concentration of the solution. (See the preceding discussion of Beer's law.) Some cells come with interchangeable spacers; sometimes different cells are available for different sample thicknesses. In any event, you should try to use the shortest pathlength cell that gives you an acceptable spec-

Table 10.2. Various solvents and their infrared (IR) windows of transparency

Solvent	Window range(s) (cm^{-1})
Water, H_2O	1700–3000, 400–1400
Carbon tetrachloride, CCl_4	850–5000, exc. ~1400
Carbon disulfide, CS_2	600–1300
Chloroform, $CHCl_3$	800–5000, exc. 3000, 1250, 850–950
Cyclohexane, C_6H_{12}	1500–2500, 1100–1200, 700–840
Tetrachloroethylene, C_2Cl_4	900–5000, exc. ~1250, 1075
Diethyl ether, $(C_2H_5)_2O$	1500–2600, 300–800
Dimethyl sulfoxide, $(CH_3)_2SO$	3100–4000, 1500–2800
Fluorolube	1300–4000
Nujol	3000–4000, 1500–2800, 300–1300,
Methylene chloride, CH_2Cl_2	800–5000, exc. ~3000, 1250

trum. This is because, typically, the solvent absorbs infrared light, and so the less solvent in the pathlength, the better.

Liquid cells for infrared spectroscopy typically have a small pathlength and must be filled using a small syringe. The syringe fits into a port at the bottom of the cell. A similar port sits at the top of the cell. The cell should be filled at the bottom port to allow any air to escape through the top. After the cell is filled, the ports can be closed with stoppers and the infrared spectrum measured. After the spectrum is measured, either the cell should be rinsed thoroughly with pure solvent or the cell should be disassembled and cleaned. When the cell is reassembled, care should be taken to properly align the gaskets and spacers.

Films

If a polymer is soluble in a high vapor pressure solvent, a film can be made by dripping the solution onto a salt (i.e., KBr) pellet and evaporating the solvent. Gentle heat can be applied to encourage evaporation. After the solvent has completely evaporated, the salt window can be placed in a sample holder, and the IR spectrum of the polymer film can be measured.

This method has the advantage of not requiring the solvent to have a window of transparency in order to observe the spectrum of the polymer. It has two disadvantages: the need to find a solvent that will dissolve the polymer, and the need to *completely* evaporate the solvent before a good spectrum can be measured. A common error is to measure a vibrational spectrum before the solvent has evaporated completely, resulting in absorption bands that are mistakenly attributed to the polymer sample. To avoid this, try reheating the film/plate combination a second time and measuring the spectrum again. If the spectra are identical, one can conclude that the solvent has probably completely evaporated. Another suggestion is to re-form the sample using a different solvent and remeasure the spectrum. If the two spectra are the same, it is highly probable that neither solvent was left behind in detectable quantities.

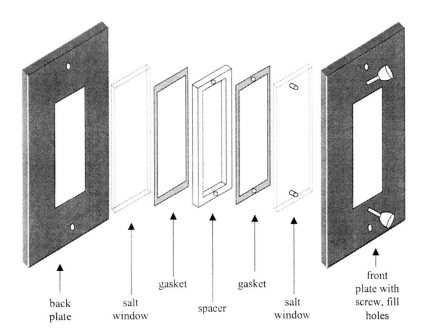

back plate   salt window   spacer   gasket   gasket   salt window   front plate with screw, fill holes

Figure 10.13 A schematic of a liquid sample cell. The gaskets seal the individual parts, while the spacer between the gaskets can have varying depths to match the absorption strength of any given sample. Holes in some (but not all) of the pieces allow the liquid sample to be introduced between the salt windows. These holes must be aligned precisely to allow the liquid sample to be properly introduced.

A similar experimental procedure can be used with an *attenuated total reflection* (ATR) apparatus, if one is available. The ATR technique takes advantage of the fact that, at or above certain angles of incidence, light is completely reflected at an interface (rather than being partially reflected and partially transmitted; it is this effect that allows one to see the sky reflected on water at a distance, but to see through water that is close by). An infrared-transparent plate, called the ATR element, is cut so that the incoming IR light reflects totally at any subsequent element/air interface, as shown in Figure 10.14. The light continues to reflect off the parallel surfaces of the ATR element, displaying what is called *total internal reflection*. Eventually, the light escapes the ATR element at the other end and, with the appropriate optics, is directed toward a detector.

Even though the light is totally reflected at the element/air interface, the light wave itself actually penetrates *beyond the physical boundary of the ATR element*. If there is another material on the other side of the ATR element, it might absorb some wavelengths of light so that an absorption spectrum of that material can be measured. Thus, if a solution of polymer were dripped onto an ATR element and the solvent were evaporated, a spectrum of the polymer could be measured.

While the optics of ATR are more complex than dripping a solution onto a salt window, ATR has the advantage of not *needing* a solution to provide the sample. A solid sample, pressed against an ATR element, can provide an infrared spectrum. This technique has the advantage of being potentially nondestructive, if a flat surface of polymer is available. The disadvantages of this "film" techniques are that (1) a flat sample is best (so looking at polymer coatings of curved parts won't work); (2) the sample must be pressed against a brittle salt, bringing with it the danger of breaking an expensive ATR element; and (3) the ATR optical setup itself is expensive and may have to be custom-designed for a particular infrared spectrometer.

There are other film options. If a useful solvent cannot be found and the polymer will flow at temperatures less than ca. 300 °C, one can make stand-alone films by applying heat and pressure simultaneously (by, perhaps, warming a pellet press while exerting moderate pressure). This method can make a thin film that can be mounted over a hole so that an IR spectrum can be measured. This is a good method of preparing samples of irregular polymer samples (like chunks or pellets), but it is limited to low-melting or low-softening materials.

Interpretational

NOTE: In this section, all spectra will be absorbance spectra, plotted in units of wavenumbers. If you are considering a spectrum that is not displayed in these terms, be aware that some of the quantitative aspects of the following discussion may not be applicable, even though some of the qualitative aspects may still be valid. *Caveat spectre.*

The infrared spectrum of any particular compound—its characteristic wavelengths, and the amounts of light absorption—is essentially unique. However, it is extremely difficult to identify a particular compound solely on the basis of its infrared spectrum. One need only look at the vibrational spectra of the long-chain hydrocarbons—the spectra are all very similar—to understand that specificity works only one way. The infrared spectrum of some unknown compound, then, might *at most* uniquely identify that compound. But it will *at least* provide some information about that compound that can be used to support an identification. Do not fall into the trap of relying too heavily on a vibrational spectrum to identify a sample absolutely. (Of course, the same caveat holds true for any other single analytical technique.)

Any spectrum yields two types of information: the energies (or wavelengths, or wavenumbers or frequencies) of the spectral transitions, and the relative intensities of each absorption. Both are important data. How the data are used depends on the reason for measuring an infrared spectrum. Are you attempting to quantify a known material, or are you attempting to identify an unknown material? We shall consider each circumstance separately.

Quantifying a Known Material

If an infrared spectrum is measured to quantify a known material, then it is assumed that the experimenter has a

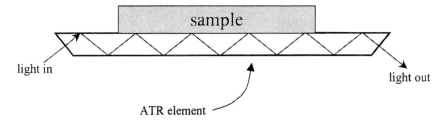

Figure 10.14 In an ATR (attenuated total reflection) element, infrared light experiences total reflection at the element's boundaries but extends slightly into a sample, and so it can record a vibrational spectrum of the material. The multiple reflections improve the spectrum quality.

reference spectrum available; that is, the experimenter has a spectrum of a known material against which to compare the sample of interest. The spectrum of a sample may be measured to confirm its identity (but see caveat above!), look for the presence of impurities, or confirm a desired concentration. In any case, comparison of a spectrum with a known composition to a spectrum of a sample is the key idea.

Confirming Identity

If the reason for measuring the spectrum is for identification purposes, a spectrum of a known material is a must. Many spectrometers now come with software libraries of known materials that can be used to compare to a sample. (This is also useful when trying to identify an unknown material, but the experimenter should recognize the limitations. This will be discussed later in this section.) There is also a huge volume, the *Aldrich Library of Infrared Spectra*, that can be perused to find spectra of known compounds. However, such libraries might be limited in use due to differences in spectrometer, spectrum conditions, sample preparation, and the like. The best way to confirm the identity of a sample is to measure the spectrum of a known material in the same form as your sample and have it available for comparison. In this way, variations in the spectrum due to experimental factors are minimized. Agreement in the positions of the absorptions and their relative intensities is usually considered good evidence that the samples are the same material, especially if there is already reason to believe that they are.

Checking for Impurities

Different materials have different spectra; that's a fundamental truism about spectroscopy. A pure material will have a single characteristic spectrum, while an impure material—with impurities present—may have different spectral signatures. Therefore, comparison of spectra of a pure material and some sample might indicate the presence of impurities by the sample spectrum having additional absorptions or having absorptions of obviously different intensity or shape. If the absorptions of the sample and the impurities coincide, impurities may still be detected by differences in relative intensities or peak shapes.

The key phrase, however, is "may have different spectral signatures"—note the presence of the word "may". Impurities might indeed show up in a vibrational spectrum. Then again, they might not. Extreme care must be exercised before basing conclusions about impurities solely on the basis of an infrared spectrum. First, an impurity might not be present in sufficient quantities to appear in a vibrational spectrum. While some impurities may absorb infrared light very strongly and be detectable at very low concentrations, other impurities may absorb relatively weakly and be virtu-

ally invisible, even in substantial quantities. If the impurities are absorbed in the same regions as the sample, they may be completely obscured by sample absorptions and be virtually undetectable.

Second (but related to the first point), an impurity might not absorb infrared light at all, or perhaps not in the wavelength range being measured. For example, if you are looking for atmospheric impurities, you don't want to use IR spectroscopy to detect the presence of gaseous N_2 or O_2 in a sample; these homonuclear diatomic molecules don't absorb infrared light at all. If you're looking for the presence of simple ionic salts, vibrational spectroscopy is not a useful tool since most simple ionic compounds don't absorb infrared light. (In fact, simple ionic compounds, like NaCl, KBr, etc., are used as infrared-transparent optics like windows and lenses *because* of this fact.) These compounds *do* absorb very low energy infrared light due to vibrations of the ions within the solid-state lattice, but these lattice modes absorb outside the normal range of vibrational spectra. (Indeed, "far-IR" spectroscopy—typically frequencies less than 400 cm^{-1}, and usually lower—is necessary to detect these low-energy vibrations, and this usually requires a spectrometer specially designed to probe such low-energy transitions.)

"Absence of evidence" does *not* mean "evidence of absence". The presence of additional absorptions in a vibrational spectrum, when compared to a known pure spectrum, may imply the presence of an impurity in the sample. The absence of additional absorptions does not confirm the lack of such impurities.

Concentration Determinations

Beer's law, eqs 25 and 26, is applicable to almost every form of spectroscopy and defines the basis for using light-matter interactions in quantitative measurements. Again, however, there are some cautions.

Beer's law is not always followed in regimes of extremely low or extremely high concentration. If the sample has a low concentration of absorbing units, the light that is absorbed may be so little that other factors like noise or instrument drift may obscure the amount of light absorbed. Also, there is usually a lower limit on absorption below which any spectrometer may not be able to detect with certainty. Very tiny amounts of light absorption are suspect, simply because they would be of the same magnitude as other factors and thus be indistinguishable as true absorption of light.

At very high concentrations of absorbing materials, there are several problems. What if the sample absorbs *all* of the light of a particular wavelength? The immediate answer is that the transmittance should be zero, the absorbance infinity. But if all of the light is absorbed before it leaves the sample, how can you do accurate quantitative measurements of that sample, using that wavelength of light? All of the light may have been absorbed in the first micron of sample pathlength; it may have been absorbed in the first millimeter of pathlength, or it may have been absorbed in the last millimeter of pathlength. There would be no way of knowing when it was absorbed, and hence no way of judging how concentrated the sample really is.

There is another factor. Absorption and emission are opposite processes. If light of a certain wavelength is absorbed, light of that wavelength can also be emitted. Usually, the major difference is that light being absorbed is coming from a particular direction (i.e., as directed by the optics of the spectrometer), whereas emission can be in any direction. If there is more light in the incoming beam than can be absorbed, then usually the remaining light will overwhelm any light that might be re-emitted by the atoms and molecules of the sample. However, if virtually all of the light is absorbed, then the light re-emitted by the sample can interfere with the accurate measurement of the original light beam's remaining intensity. This is another mechanism by which measurements of very strongly absorbing samples can provide potentially deceiving spectra.

There is also the issue that absorbance, one of the common ways to display a spectrum, is a logarithmic scale. At 99.99% transmittance, the value of the absorbance is 4, while at 99.999% transmittance, the absorbance is 5. A change of less than 0.01% in the amount of light transmitted—and the transmission of light is the fundamental measurement in spectroscopy—yields a 25% change in the value of the absorbance. Thus, for concentrated samples there can be relatively large changes in the absorbance in a spectrum, even though there are relatively small changes in the *absolute* amount of light that is absorbed.

For these reasons, one should exercise care when evaluating a spectrum in which the absorptions exceed 2 absorbance units. Tiny changes in measurements of a concentrated sample can manifest themselves as huge changes in an absorption spectrum, which may not be physically relevant (because of the presence of other random factors like noise, instrumental factors, random error, etc.).

The ultimate lesson is that quantitative measurements are best done for not-too-dilute, not-too-concentrated samples. Only experience will tell you what the appropriate concentration regime is for any particular sample. Experimenters should be able to explore the options of increasing or decreasing the sample concentration, amount, or thickness to obtain spectra that yield useful absorption information.

The next issue that must be considered for quantitative infrared measurements is the need for accurate, quantified standards for comparison. A series of known amounts, concentrations, or thicknesses should also be measured. Because infrared spectra of complex materials are usually a collection of absorptions, the experimenter might want to evaluate several absorptions to use for comparison. The best absorption to use as a standard for comparison shows a linear relationship between absorption area* and amount and is relatively isolated from other absorptions to minimize interference from other vibrations.

Finally, good quantitative measurements require consistency of procedure. There is no easier way to invalidate any quantitative measurement than to be inconsistent in the measurements.

Qualifying an Unknown Material

If vibrational spectroscopy is one technique being used to determine the identity of an unknown sample, there will be some differences in approaching the spectrum. First, however, we will assume that the experimenter can measure the infrared spectrum of the sample, by whatever means, and obtain an image of the spectrum either on a display screen or on paper. The ability to manipulate the spectrum image, in terms of expansion and contraction of each axis, and record different regions on paper, will be a useful skill.

Finding a Match

To identify an unknown material from its vibrational spectrum, first check the software library of spectra that is available with most FTIRs. There may also be extra software libraries available for purchase, some of which focus on certain groups of materials like polymers or illegal drugs. Each program has its own instructions or commands for performing such a computer-based comparison, so it is useful to be familiar with your own software. Using the software is usually fairly straightforward and should produce a series of suggested "matches". While it may be unlikely to find an exact match for any random sample (there are over 20 million known compounds!), such matches may help in identifying what type of compound your material is, or what functional groups might exist. Even if a conclusive match is available, the experimenter must still apply some judgment in comparing any two spectra for identification purposes.

Some FTIR software libraries allow experimenters to add their own spectra to a database. If this is so, an experimenter might want to consider measuring a range of known samples on the chance that an unknown might be matched to a previously measured known material.

---

*Notice we say *area*, not *height*. In an ideal situation, absorptions are infinitely thin lines, and the amount of light absorbed by a particular vibration is directly related to the height of the absorption. However, in a real situation, absorptions have definite shapes and finite widths. In this case, the better indication of the amount of sample is the area of the absorption, not the height. (In some cases, however, height is a good enough indicator of amount, but an experimenter should validate this before basing conclusions on it.)

Samples that are mixtures may be difficult to identify because their spectra are combinations of two or more "pure" spectra. Varying concentrations of different components will yield varying relative intensities, further confusing the identification process. Much judgment is required in these circumstances, too.

If a match isn't made using a software library, there are hard-copy libraries available, like the *Aldrich Library* and the *Sadtler Handbook of Spectra*. While looking through these references can be time-consuming, it may be the best opportunity to identify an unknown material.

Narrowing the Field

Even if an exact match cannot be found, a vibrational spectrum can provide a lot of evidence for or against an identification. This is because characteristic bonding arrangements of atoms in molecules have characteristic vibrations, and the presence or absence of these vibrations in a spec-

trum are clues to the molecular structure of a sample. We can also take advantage of the characteristic vibrational absorption regions, as listed in Table 10.1.

An example is the best way to illustrate the use of a vibrational spectrum for identification purposes. Figure 10.15 shows an IR absorption spectrum of a polymer film. First, note the presence of two groups of absorptions in the C–H stretching region. The absorptions in the range of 2800 to 3100 cm^{-1} suggest alkanic C–H stretches, while the absorptions between 3000 and 3100 cm^{-1} suggest aromatic C–H stretches. Thus, the appearance of these features suggests an aromatic/aliphatic polymer. But note the absence of a strong absorption around 1700 cm^{-1}, which is characteristic of a carbonyl group. One can immediately conclude that this polymer probably isn't a polyester. The lack of any absorptions at energies higher than the C–H stretches suggests that this polymer does not contain O–H or N–H groups, either.

Below 1500 cm^{-1}, the vibrational spectrum is a complex group of absorptions at varying intensities. These absorp-

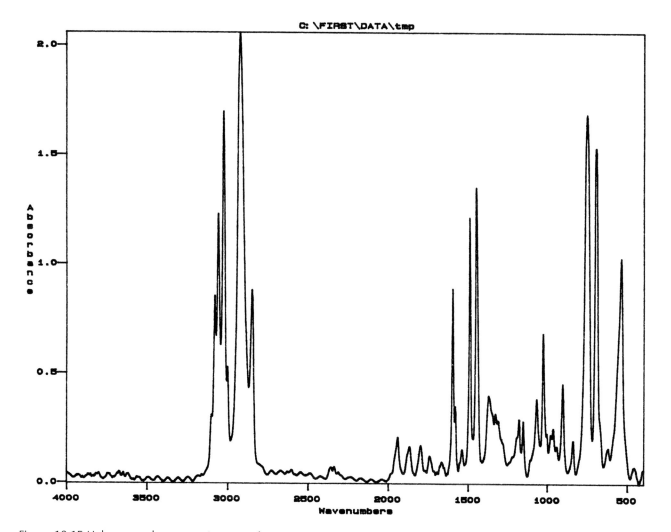

Figure 10.15 Unknown polymer spectrum no. 1.

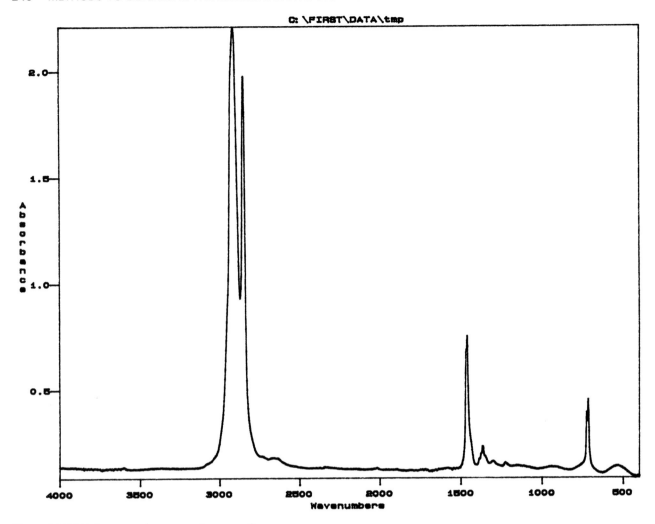

Figure 10.16 Unknown polymer spectrum no. 2.

tions represent molecular vibrations that would be described as carbon-carbon stretches, carbon-hydrogen bends, various vibrations of an aromatic ring, and other motions. Because of the complexity, it would be difficult to uniquely identify this compound unless one had other known materials' spectra for comparison. In addition, one might hope to get lucky and find a match in a computer or printed database. In the case of this particular material, a diligent search should be fruitful, for this material is a fairly common polymer. It is polystyrene, which is commonly used as a calibration material for vibrational spectrometers.

As a second case study, a sample of a plastic wrap found in a laboratory stockroom was used to measure an infrared spectrum. The sampling technique was straightforward: a small piece of the plastic was cut and mounted in a disposable cardboard holder that has a strategically placed hole in it. When placed in a standard IR sample holder, the hole in the cardboard holder was in the IR light path.

Figure 10.16 shows the spectrum of the plastic, and it looks very simple: an intense absorption appearing at 3000 cm^{-1} is the C—H stretch, a medium absorption at 1500 cm^{-1} is probably a C—C stretch, while the medium absorption at 750 cm^{-1} is probably a CH_2 bending motion. This last absorption could also be a C—Cl stretch, if there were any reason to believe that this sample derived from a chlorinated monomer [i.e., poly(vinyl chloride), or PVC]. There are other, weaker absorptions near the C—C stretch, probably representing similar molecular motions (like C—C—C bends).

Given the lack of other indicative absorptions, the spectrum suggests a relatively simple material. Perhaps the simplest polymer is polyethylene, composed of repeating CH_2 units. If the absorption at 750 cm^{-1} is a C—Cl stretch, then PVC may be a viable identification. The best thing to do is to combine the vibrational spectrum with other tests or perform a comparison with known spectra, or both. Comparisons using the *Aldrich Library of IR Spectra* suggest an identical match to polyethylene—although, again, this should be confirmed with additional evidence.

Vibrational spectra are quick and simple measurements to make but may be difficult to interpret. Without a confirming spectrum of polystyrene, the material that yielded the spectrum in Figure 10.15 would be unidentifiable, except for a few general statements about the presence or absence of certain functional groupings. No doubt these general statements are important, but they do not absolutely and conclusively identify the specific material. Caution and confirmation are necessary.

Summary

Vibrational spectroscopy is a powerful tool. As one of the most widespread methods of instrumental analysis, the vibrational spectrum is used to identify unknown samples, quantify amounts, ensure quality control, and verify identity. The specificity of the vibrational spectrum of a known material is a valuable tool, and the ability of a vibrational spectrum to help identify an unknown material is also useful. Properly performed and interpreted, a vibrational spectrum is an important datum in understanding matter.

Acknowledgments Thanks to Shannon Gomez for plotting examples of spectra that were used in this chapter.

Additional Reading

The Aldrich Library of Infrared Spectra, 3rd ed.; Aldrich Chemical Company: Milwaukee, 1981.

Griffiths, P. R.; de Haseth, J. A. *Fourier Transform Infrared Spectroscopy;* Wiley Interscience: New York, 1986.

Nakamoto, K. *Infrared and Raman Spectra of Inorganic and Complex Compounds*, 5th ed.; John Wiley & Sons: New York, 1997.

Simons, W. W., Ed. *Sadtler Handbook of Infrared Spectra;* Sadtler Research Laboratories: Philadelphia, 1978.

Smith, B. C. *Fundamentals of Fourier Transform Infrared Spectroscopy;* CRC Press: Boca Raton, FL, 1996.

11

Mass Spectrometry

THOMAS P. WHITE
ANTHONY R. DOLAN
E. PETER MAZIARZ III
TROY D. WOOD

Synthetic polymers play pivotal roles as materials for a wide variety of applications in modern society. Unlike biological polymers such as enzymes and nucleic acids, which are usually monodisperse, synthetic polymers are largely polydisperse (i.e., they consist of an ensemble of polymer chains with a distribution of molecular weights). Molecular weight and molecular weight distribution are key, fundamental properties of a synthetic polymer, as the combined influence of the constituents relates to such properties as tensile strength, brittleness, osmotic pressure, melt viscosity, solubility, adhesion, and abrasive and chemical resistance.[1] Because of these correlations, evaluation of the oligomeric distribution—including number-average molecular weight (M_n), weight-average molecular weight (M_w), and polydispersity (D)—for a particular polymer is crucial for a complete understanding of its physico-chemical and mechanical properties. As such, mass spectrometry (MS) is becoming an indispensable tool to the polymer chemist.[2]

Unlike the classical techniques such as ebulliometry, cryometry, and osmometry (for determination of M_n) and low-angle light scattering (for determination of M_w), which determine the molecular weights of synthetic polymers indirectly, the technique of MS is used for direct molecular weight determination. In addition, MS has the advantage that it can be used to determine aspects of *structural* information on polymer molecules. For example, repeat unit sequence and composition, end group chemistry, and the presence of polymer impurities (or additives) can all be determined using MS.[3–8] In addition, MS analysis requires minimal sample, especially in contrast to nuclear magnetic resonance (NMR), and data collection can take as little as a few minutes.

Early MS analyses of polymers were limited to species that were relatively volatile, using an ionization method called pyrolysis.[9] Pyrolysis generates thermal decomposition products that can be quite useful in elucidating structural features of a synthetic polymer. However, as the polymer chemist is aware, few synthetic polymers have sufficient vapor pressure or thermal stability to be analyzed as *intact* oligomers by pyrolysis. Indeed, most of the commercially important synthetic polymers and polymer blends consist of relatively nonvolatile synthetic polymers; therefore, historically the analysis of these synthetic polymers as ions in the gas-phase by MS has been impeded.

However, the last two decades have seen major advances in the development of ionization techniques that are capable of volatilizing and transferring intact polymer molecules into the gas-phase as ions. These techniques are often referred to as being "soft" (relative to pyrolysis and electron impact) because they do not deposit a great deal of energy into the internal modes of the polymer molecule, which might cause it to dissociate. These soft ionization techniques include desorption chemical ionization (DCI),[10] field desorption (FD),[11,12] ^{252}Cf plasma desorption (PD),[13,14] fast atom bombardment (FAB),[15,16] secondary ion mass spectrometry (SIMS),[17–19] laser desorption (LD),[20] matrix-assisted laser-desorption ionization (MALDI),[21,22] and electrospray ionization (ESI).[23–25]

While advances in ionization methods have been made, additional advances in mass analyzers used to detect ions of such massive species as synthetic polymers have been developed concurrently. Notable among these with regard to synthetic polymer analysis are time-of-flight (TOF) mass spectrometry and Fourier transform ion cyclotron resonance mass spectrometry (FT-ICRMS), most often described in the literature by its shorter moniker, Fourier transform mass spectrometry (FTMS).

The development of ionization methods capable of generating ions from intact macromolecules and the advances in mass analyzer technology have significantly changed the field of synthetic polymer characterization. As evidence of the growing application of MS in synthetic polymer analysis, recent issues of the *Journal of the American Society for Mass Spectrometry* (Vol. 9, No. 4, 1998) and *European Mass*

Spectrometry (Vol. 4, No. 6, 1998) were entirely devoted to the subject of synthetic polymer analysis by MS.

In this chapter we discuss the different ionization methods, mass analyzers, and hyphenated techniques (like gel permeation chromatography) coupled to MS at the disposal of the polymer chemist and under what types of situations each might prove useful. Selected applications of MS in elucidating elements of polymer structure are then discussed, followed by conclusions.

Experimental Methods

A method for volatilization of the sample into the gas phase and its ionization, coupled with a means for mass analysis, are central to mass spectrometry. In this section, we cover the ionization methods and mass analyzers that are available to the modern polymer chemist for use in mass spectrometry.

Ionization Sources

Electron Impact

In electron impact (EI) ionization, the ion source is a relatively small, enclosed box, which is heated and contains an electron beam. The electron beam is produced by heating a filament that is made from tungsten or rhenium wire. With electron impact, two magnets are positioned to align the electron beam. The electron beam is accelerated through a potential of approximately −70 V (in positive ion mode), which leads the beam into the source.[26-29] When a gaseous sample is introduced into the source, it can interact with the electrons in the beam. If electrons pass within close proximity of an uncharged neutral molecule, an electron can be ejected. This phenomenon can occur only if the wavelength of the electron is similar to bond lengths within the analyte. If this does indeed occur, the wave becomes complex, and an energy transfer takes place; in this case, the frequency's energy corresponds to a transition in the molecule.[26] In addition, this interaction then causes the molecule to lose an electron and form a positively charged radical ion. This occurrence is considered the molecular ion reaction, and the event is referred to as electron impact, even though no impact actually occurs.[30] This is shown below, where A−B is the analyte molecule:

$$(A{-}B) + \text{electron impact} \rightarrow (A{-}B)^{+\bullet} + 2e^{-}$$

In addition, if enough excess energy is present within the molecule, a second reaction may occur. The molecular ion may fragment from excess energy, and if there is sufficient energy, all of the ions will be converted into fragments:[27]

$$(A{-}B)^{+\bullet} \rightarrow A^{+} + B^{\bullet}$$

The fragmentation is unique for different substances. The fragmentation pattern is what leads to a mass spectrum.

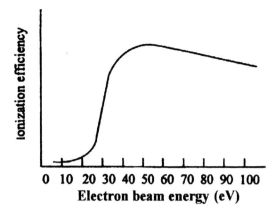

Figure 11.1 Characteristic ionization efficiency curve for positive ion formation by electron impact. (Adapted with permission from reference 29. Copyright 1987 John Wiley and Sons.)

The best electron impact sources may ionize one in 1000 molecules.[26,27] Although the probability of ionization differs among analytes, it may be the highest for electron energies between 50 and 100 eV[27,30] (see Figure 11.1).[29] Once the ions are formed, they are directed toward the exit slit. This is accomplished by using an accelerating electric field and a repelling plate to guide the ions out of the source. A schematic diagram of this source is shown in Figure 11.2. Nier has described this type of source in detail.[31,32]

For some molecules, negative radical ions (M⁻) can be formed. This event occurs when a molecule (M) gains an electron. However, most mass spectrometry is conducted in the positive ion mode because positive ions are easier to produce than negative ions. This is because the probability of a molecule losing an electron is about 100 times greater than an electron being gained by a molecule.[29]

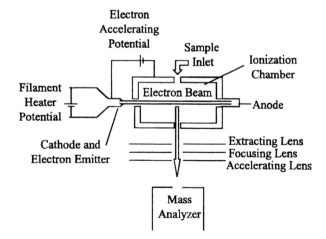

Figure 11.2 An electron impact ionization source. (Adapted with permission from reference 26. Copyright 1996 John Wiley and Sons.)

Electron impact ionization is one of the most widely used methods of ionization because it easily creates both molecular and fragment ions. There are some problems associated with this technique, however. In some cases, it is difficult to measure the relative molecular mass of a molecular ion; this is because the molecular ion may fragment before leaving the source. Another problem is that it is difficult to differentiate between isomers because, more often than not, the isomers have very similar fragmentation patterns. Another problem is thermal decomposition of the analyte before or after the molecule enters the source from the high temperatures needed for vaporization.[29] Furthermore, if a sample is not volatile, then acquisition of a mass spectrum becomes a difficult task with EI. If the polymer can be dissolved, then the sample can be loaded with a syringe and injected into the source chamber so that analysis can take place. In some instances, thermal energy can be added to increase the volatility of a sample in order to obtain a mass spectrum. Thus, the number of polymers that can be analyzed with this technique is limited. For the most part, electron impact ionization is useful for end group and fragmentation pattern analysis for volatile polymers and for analyzing volatile additives.

Chemical Ionization

Chemical ionization (CI) is similar to EI, except ion-molecule reactions are used to ionize the sample rather than the electron beam.[33] CI uses a reagent gas, typically methane or ammonia, which will eventually form a set of ions that are nonreactive with the reagent gas. These ions are initially formed by using an electron beam, as in EI. After forming an initial ion, they will react with more reagent gas, forming another ion that will not undergo any further reactions with itself. This ion will then react with the sample and form the gas phase ions, which will enter the mass analyzer. This sequence of events is shown below, using methane as an example of the carrier gas:

$$CH_4 + e^- \rightarrow CH_4^{+\bullet} + 2e^{-\bullet}$$
$$CH_4^{+\bullet} + CH_4 \rightarrow CH_5^+ + CH_3^\bullet$$
$$CH_5^+ + M \rightarrow CH_4 + MH^+$$

where M is the analyte molecule of interest. These equations represent the primary reaction sequence that occurs; however, there are many minor reactions that compete with these.[34]

The abundance of the sample ions present in the mass analyzer depends on the rate of the ion/molecule reactions, the concentration (or partial pressures) of the reacting species, and the time available for the reaction to occur. The reagent gas is kept in a large excess relative to the analyte to ensure that the initial electron beam will preferentially ionize the reagent gas and not the analyte. This is accomplished by using partial pressures of reagent gas around 0.5 to 1 Torr and a pressure of 10^{-3} Torr for the analyte vapor. The only requirement of the analyte is that it must be able to be vaporized into the gas phase. The analyte can be introduced into the ion source through a heated inlet with the aid of a solvent, as a solid on a direct insertion probe, or after a gas chromatograph.[34] These possibilities make it possible to analyze solid samples, as well as liquid and gaseous ones, with CI. The mass analyzer region is kept at a pressure of $\leq 10^{-6}$ Torr to prevent excess ion scattering, and the ion source region is held below 10^{-4} Torr to prevent damage to the electron filament. In order for this to be achieved, the ion source must be as gas-tight as possible. This explains why there is a greater amount of pressure in a CI source than in an EI source for the same amount of sample material. CI must use electron energies up to 400 eV to allow the electrons to penetrate further into the ion source. Recall this is much larger than the 70 eV, which is typical for EI experiments. The whole CI technique, including its principles and the theory of operation, is summarized succinctly in a review article.[35]

The formation of the sample ion is accomplished through proton-transfer, acid-base, or hydride-transfer reactions. These processes are based on the relative proton affinity of the analyte versus the reagent gas.[36] The ion formation process is usually accomplished through the transfer of a hydrogen ion to the sample molecule; hence, the molecular weight of the analyte is simply one dalton (Da) less than the mass of the highest mass peak in the spectrum. However, care must be taken when interpreting a spectrum because some reagent gases will react differently and lead to adduction. Therefore, the mass of the reagent gas must also be subtracted from the highest mass peak. Since these reactions are not highly exothermic, the product ions will not always possess enough energy to decompose; therefore, the fragmentation of the ions will be limited.[37] Since the reagent gas can be any one of a variety of different gases (e.g., methane, isobutane, ammonia), this incorporates a type of tuning that must be included in an experiment. By differing the reagent gas, more or less fragmentation can be obtained in the spectrum.

Using oxidizing reagent gases presents a problem in that these gases lead to a limited lifetime of the electron filament. For these harsh conditions, two different electron sources have been used: either a Townsend discharge ion source[38] or a glow discharge ion source.[39] These also allow for the source to be held at a lower temperature. There is a problem coupling the CI technique to a sector instrument of which one should be aware. There is a possibility of electrical discharge through the gas in the ion source and the ground, which is usually located in one of the inlet lines.[34] This problem is solved by placing a glass tube packed with glass wool to reduce the velocity of the ions in the ionization source.[40] The use of quadrupole mass spectrometers alleviates this problem since lower potentials are applied than is done when using sector instruments.

A variation of CI is atmospheric pressure chemical ionization (APCI), where the ions are generated in a flowing carrier gas that is held at atmospheric pressure. The ions are generated either by β particles from a ^{63}Ni source[41,42] or by corona discharge.[43] APCI with a low volume chamber and usually a corona discharge source is ideally suited for coupling to GC[44] and LC.[45,46] Figure 11.3 is a typical low-volume APCI source used for coupling to LC.

Another alternative type of CI is called direct CI, or desorption CI (DCI); this approach uses a dispersion of the sample on a suitable probe that can be inserted into the reagent ion plasma in the ion source.[47] This technique is good for involatile and thermally labile molecules, which are difficult to vaporize into the gaseous phase. It is also an excellent ionization source for polymeric samples that are not easily dissolved in polar solvents and for polymers that pyrolyze before vaporization can occur, as is the case in other ionization techniques.[48] The direct exposure probe can take on many different designs,[49–52] but basically it uses a loop of electrically resistant wire that has approximately 50 to 100 ng of sample loaded onto it, and this wire is inserted into the reagent plasma or electron beam. By slowly heating the wire, the sample will be desorbed into the beam, where it will be ionized and passed on into the mass analyzer.

CI is better than EI for polymer analysis because it can be used to ionize higher molecular weight compounds and polyfunctional compounds. CI can be used for both qualitative and quantitative studies. It provides a built-in type of tuning with the use of different reagent gases. CI usually produces less fragmentation of the sample than in EI experiments. CI spectra contain quasi-molecular ions of the sample, and the spectra are easier to interpret than are EI spectra. However, the two techniques are quite complementary to one another, and the combination of EI and CI is a useful tool for polymer characterization.[53]

Field Ionization/Desorption

Field ionization (FI) and field desorption (FD) have both been coined "soft", or mild, ionization techniques. This term "soft" refers to a minimal amount of energy being transferred and left with the molecule when ionization occurs. In the field ionization process, the energy that is transferred is only a percentage of an electron volt.[27] In addition, the amount of fragmentation that is normally observed (with EI and CI) is decreased.[27,54,55] FI and FD have been traditionally treated in a similar manner,[54] although the mechanisms for ionization are different. An in-depth look at the mechanisms of ionization for FD and FI has been published.[54,56,57] The mechanism for FI has been developed on the basis of the wave function of matter by Kirchener (1954), Muller and Bahadur (1956), and Ingham and Gomer (1955).[54] Furthermore, FD has been explained from field ionization theory and based on the mathematical derivation by Muller (1956)[58] and Gomer (1959).[59]

Field ionization and field desorption both operate by the same basic process. When an electric potential is applied across a needle and a flat plate (counterelectrode) many equipotential lines create an electric field and crowd in on the needle (see Figure 11.4a). A potential difference between the needle and the electrode can create a field up to approximately 10^8 V/cm for FD[27,54,60,61] and potentials above this value for FI.[54] From Figure 11.4a it can be seen that the potential near the needle tip is more intense (the lines are closer together) than the potential farther away from the tip (lines farther apart). When this occurs, a high electric field is created at the tip of the needle. In addition, when a sample is placed on the tip of the needle it will feel this high potential, and it can be ionized. For example, when a neutral molecule is placed on the tip of the needle (which is an electrode with a positive charge), the electrons in the molecule are attracted to the positive charge. Thus, the positively charged nucleus is repelled and causes distortion of the molecule's electric field (see Figure 11.4b). Furthermore, this distortion decreases the energy that the molecule needs to hold onto an electron, and ionization can occur.[57] The next event to take place is that the molecule's electron is found next to the positive electrode, and the electron is removed (Figure 11.4c). This is defined as *electron tunneling*.[55,57] The molecule is now a positive ion radical, M$^{+\bullet}$, and is guided into the ion source by a negatively charged counterelectrode. The counterelectrode has a slit within it, which allows the ions to pass through to the mass analyzer of the mass spectrometer.

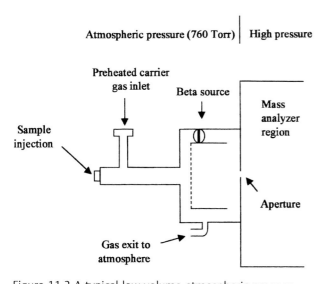

Atmospheric pressure (760 Torr) | **High pressure**

Preheated carrier gas inlet

Beta source

Mass analyzer region

Sample injection

Aperture

Gas exit to atmosphere

Figure 11.3 A typical low-volume atmospheric pressure chemical ionization (APCI) source, showing how the ionization source is at ambient pressure while the mass analyzer is under vacuum. The beta source, which generates the ions in this device, is ideal for coupling MS to other analytical techniques. (Adapted with permission from reference 41. Copyright 1973 American Chemical Society.)

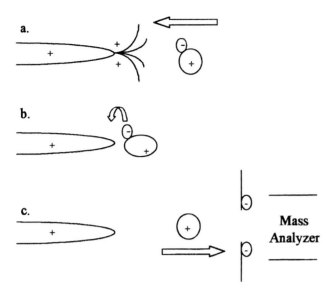

Figure 11.4 Field ionization process: (a) electrons of a molecule are attracted to the electrode, and an electric field crowding in on the emitter creates an intense electric field at the tip of the needle; (b) electron tunneling and distortion of a molecule's electric field due to repulsion between the positively charged nucleus and the positively charged emitter occurs; (c) positive ion is repelled by the electrode.

To create a potential that is intense enough to ionize the sample, one of two approaches may be used. The first approach is to generate and use an extremely high voltage, but this method can have its drawbacks: for example, high voltages can be dangerous and can also be difficult to generate and maintain. The second approach is to make the curvature of the needle tip as small as possible in order to create a high electric field while using a lower voltage. The field strength (F), is equal to the voltage (V) divided by the radius (r) and the geometric factor (k).[62] For example, if k is equal to 1, V is 1000 volts, and r is 10^{-6} m, then the field strength is 10^9 V/m. However, if the radius is 10^{-4} m, to create the same field strength, a potential (V) of 100,000 V is needed. It is easy to see from this example why the radius of curvature is an important issue.

As discussed, the ionization needle needs to be small (~0.001 cm in length, with a radius of ~0.1 nm).[27,60] However, using only one small needle would result in a low ion yield per unit time. To combat this problem of small ion yields, many small needles could be used. Indeed, this was eventually accomplished by growing small tips or "whiskers" on a 10-µm tungsten wire.[54,63] The whiskers are sometimes constructed of silicon,[64] are metallic, or are made from pyrolytic carbon.[63]

For our purposes, the only difference with these two ionization methods is in the way that the samples are examined. For FI, the sample is heated and volatilized in a vacuum so that it can be positioned toward the ionization surface. However, in FD, the sample is not heated and vola-

tilized but directly placed on the ionizing surface.[54] In FD, this process may be accomplished by dissolving the sample in a suitable solvent, and then placing the sample onto the ionizing surface. For example, Schulten and Plage dissolved aliphatic polyamide samples in hexafluoro propanol and applied them to a FD emitter.[65] The solvent is then allowed to evaporate. In addition, the emitter can be loaded by dipping the emitter into the sample mixture.[66] Furthermore, in both cases, once the analyte is positioned toward the ionizing surface, it is subjected to a high field potential so ionization can take place.

FI is a technique that works quite well for volatile and thermally stable samples. Unfortunately, the sample must be volatile, and this does indeed limit its use to analyze polymers. However, FD can be used for samples that are either nonvolatile or thermally labile, or both.[27,60] Hence, this is an advantage over other ionization techniques such as EI, CI, and FI.[54] In addition, FD offers the ability to ionize a large range of compounds, including solids such as peptides, sugars, inorganic and organic salts, oganometallics, polymers, and organic polymer additives.[54,63]

FD is a good ionization technique for low molecular weight polymers, both polar and nonpolar;[63] in addition, FD is good for high molecular mass nonpolar compounds.[61] There are some drawbacks with this technique, however. For some samples, a persistent spectrum cannot be produced. In addition, when an emitter is turned on, ions are formed, but the ion currents seem to decrease quickly, which does not allow much time to change the experimental conditions and optimize the signal.[60,63] Another problem associated with this technique is that the quality of the data obtained is highly dependent on the heating current of the emitters. Thus, the preparation time of the emitters is time consuming.[27]

Californium-252 Plasma Desorption

Californium-252 (^{252}Cf) plasma desorption is a unique ionization/desorption technique that is adequate for analyzing nonvolatile molecules.[67] One method of vaporizing a sample is to bombard the sample surface with high-energy (approximately 100 MeV) heavy ions. The radioactive nuclide ^{252}Cf is a source of such ions as a result of spontaneous fission.[68] The fission of ^{252}Cf results in an asymmetric split into two ions, which will travel in almost exactly opposite directions. There are over 40 different pairs of fragment ions with different energies. Two examples are ^{142}Ba^{+18} (79 MeV) and ^{106}Tc^{+22} (104 MeV), and ^{100}Sr^{+20} (110 MeV) and ^{150}Ba^{+20} (75 MeV). These ions contain enough energy to ionize and vaporize the sample without decomposing it.[69]

Figure 11.5 is a diagram of a ^{252}Cf-plasma desorption ion source that is connected to a mass analyzer. The inherent features of the nuclear fission process of ^{252}Cf make time-of-flight an ideal mass analyzer to be coupled with this ion source. The ^{252}Cf fission fragment source is the heart of the

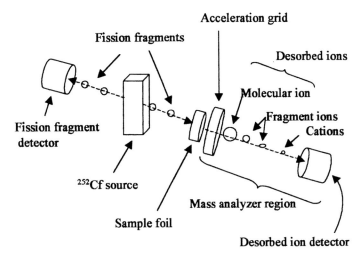

Figure 11.5 A ^{252}Cf ionization source coupled to a mass analyzer. This diagram shows that different size ions are formed from the bombardment of the sample with fission fragments. In a TOF mass analyzer, the lighter mass ions will reach the detector faster than the heavier ones. (Adapted with permission from reference 67. Copyright 1983 American Chemical Society.)

instrument. After fission, the two ions travel in opposite directions. One ion heads toward a fission fragment detector that is used to mark the time when the ionization of the sample begins. The other ion travels toward a metal foil (e.g., Ni) or substrate (e.g., nitrocellulose) upon which a thin film of the sample to be analyzed has been coated. This ion will deposit its energy, which is available as translational motion, into the sample on the order of picoseconds[70] leading to desorption of many charged species and up to a few thousand neutral molecules and many photons.[71] On this time scale, there is not enough time for the molecule to be excited into an excited state, so fragmentation and decomposition of the sample is minimized. Similar fragmentation patterns are seen as in CI using *t*-butyl ions.[72] These ions will then enter into the mass analyzer. The major problem with ^{252}Cf plasma desorption is that there is a low ion yield of quasi-molecular ions. Therefore, it is very important to have a low ion background with high transmission and to perform single ion counting in order to obtain a usable mass spectrum.[73]

The method of preparing the sample is crucial for this type of experiment, since ^{252}Cf plasma desorption is a surface phenomenon and the yield of ions is proportional to the area of the surface actually covered by the sample molecules.[70] Dissolving the sample in an appropriate solvent and depositing it onto a Ni foil leads to nonuniform sampling. The electrospray technique[74,75] provides a uniform molecular film that gives comparable intensities as the evaporation method, but using smaller amounts of sample.[70] Sometimes sodium and lithium ions as salts are added so that their adducts can be seen in the resulting spectrum.[76] Depending on the method chosen, the foil containing the sample must

be precisely aligned with the fission fragment source to ensure that the ionization event will take place. Since the excited surface area is so small, approximately 5000 Å^2 per fission fragment,[77] after the experiment has been run, most of the sample has not been affected and can be recovered.

This technique, like SIMS, can be used to analyze polymer surfaces.[78,79] The main problem with SIMS profiling is that there is a buildup of charge on the surface of the polymer that needs to be neutralized.[80] However, spectra can be obtained without neutralizing the sample surface with ^{252}Cf plasma desorption with a sampling depth of a few monolayers.[76,81]

Fast Atom Bombardment

Fast atom bombardment (FAB) is considered a soft ionization technique, just like CI and FD, because it does not induce a large amount of fragmentation into the process of ionizing the sample. This ionization source overcomes the problem with the previously described ionization techniques in that the sample must be in the gas phase before it can be ionized.[82] This was difficult to do with thermally unstable compounds and compounds with a general polar nature using other ionization techniques.[83] FAB uses a liquid matrix solution to alleviate this problem. FAB uses a high-energy beam of atoms to bombard a surface containing the matrix and the sample; the matrix absorbs most of the energy that generates a collision cascade, which results in the formation and desorption of many different ions and neutral species from the surface.[84] These ion beams usually persist for reasonable lengths of times, several tens of minutes to hours, so many types of analysis can be carried out.[85]

The basic design of a FAB ion source is shown in Figure 11.6. The whole process starts with an ion gun that accelerates inert gas ions to approximately 2–10 keV.[86] The inert gas chosen can be either argon or xenon; however, because xenon has larger mass and momentum, it gives a larger sample ion intensity signal.[87] This ion beam then enters into a high-pressure collision chamber that contains more of the inert gas. As a result of resonant charge exchange and a very minimal loss of forward momentum, the beam of ions is converted into a beam of "fast" (energetic) atoms.[88] Any excess ions that are present in the beam that did not undergo charge exchange are removed with the aid of a deflector, which attracts the ions out of the beam. Now the beam of atoms, with the same amount of kinetic energy as the initial ion beam, is directed onto the tip of an insertion probe that contains the sample plus the matrix. An angle of incidence of approximately 70° appears to be optimal.[87] This results in the formation of a secondary ion beam that contains ions of both the matrix and the sample of interest, which are directed on into the mass analyzer region. One disadvantage of this technique is that the beam of atoms cannot be focused and once formed, its translational energy cannot be changed.[89]

A more recent approach for generating the atom beam is to use a saddle field discharge source.[90] In this type of

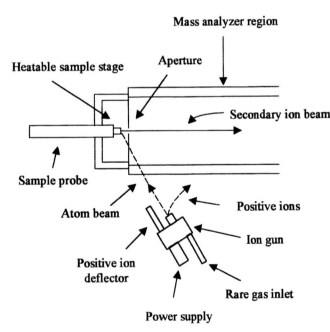

Figure 11.6 A fast atom bombardment (FAB) ionization source: schematic diagram. The beam of atoms generated by the ion gun bombards the sample, which is affixed onto the end of the same probe, and the desorbed ions enter into the mass analyzer region. (Adapted with permission from reference 87. Copyright 1982 American Association for the Advancement of Science.)

configuration, electrons oscillate between two cathodes as a result of a DC field. As molecules of gas enter the source at a low pressure, there is a high probability of them being ionized because of the long path length traveled by the electrons. The ions from the noble gas are accelerated and participate in a charge exchange process with the electrons, similar to the process previously discussed that uses the ion gun to produce the beam of atoms.

FAB has a relatively easy sample preparation procedure in comparison to some of the other ionization techniques. The sample can be introduced into the ion source on a sample insertion probe through a vacuum lock. It is deposited onto the end of the probe from solution. This solution usually consists of an appropriate matrix, which is able to either dissolve or disperse the sample. Samples are usually prepared as a 1% to 10% by weight solution in the chosen matrix.[63] Some important properties of the matrix include its boiling point, vapor pressure, viscosity, surface tension, and proton affinity. Because the ability to handle and dispose of the sample is essential, it is important to investigate the persistence of the sample solution in the source of the mass spectrometer. The only sure way to establish a suitable matrix is to try it because the range of sample-solvent interactions is very broad.[89] Some typically used matrices for synthetic polymer samples include glycerol, thioglycerol, and 3-nitrobenzyl alcohol.[91] As in ^{252}Cf plasma desorption, ions such as Na^+ are added to induce cation attachment by means of dissolving them as salts in a volatile liquid, such as methanol, and applying this solution to the tip of the probe, as well as to the sample.[16]

Liquid samples have an advantage over samples prepared through evaporation. If a bombarding atom strikes a spot on the surface that has already been ionized, no ions will be generated in a solid sample because it was previously removed; however, liquid is in a constant state of flux, so there will never be any "holes" on the surface. This can also be accounted for because of the viscosity of the matrix. The use of viscous matrices allows for a random "movement" of the sample on the end of the probe tip during an experiment. By slightly altering the probe, continuous-flow FAB or dynamic FAB can be performed.[92,93] This is shown in Figure 11.7. The silica capillary is connected to a source of the sample and transports it to the tip of the capillary. With low flow rates, a continuous supply of sample is always available at the tip, and there is no longer any need to remove the tip to replace the sample and then perform a vacuum pump-down cycle. This source is an ideal interface for both liquid chromatography (LC) and capillary electrophoresis (CE) to MS. Spectral interpretation is made slightly easier with this technique since pseudo-molecular ions along with some fragment ions are observed. The major fragmentation pathway is at a carbon-heteroatom bond, if one is present.[94] Adduct ions can form between the analyte and the FAB matrix, or reactions between these two species can occur; therefore, it is necessary to perform the experi-

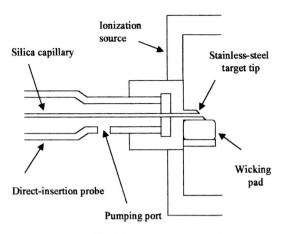

Figure 11.7 Modified fast atom bombardment (FAB) source, allowing for a continuously flowing sample to be ionized. The sample flows down the silica capillary to the tip, which is positioned in the path of the ion beam. The wicking pad absorbs any excess sample so that the thickness of the sample remains constant while the sample is being ionized. (Adapted with permission from reference 89. Copyright 1985 Academic Press.)

ment in the presence of a deuterated matrix.[95] Even though the presence of known matrix peaks allows for a built-in mode of calibration, it is not a very good quantitative technique. FAB routinely gives M_n and M_w values that are lower than those determined by other ionization methods.[16,95]

FAB seems to be an adequate ionization source for the analysis of polymers. Since little modification or expense is needed to alter an existing EI or CI ion source, modification of instrumentation to use these sources with FAB is facile. Also, because ionization can occur directly from the solid[96] and volatilization is unnecessary, this technique can be used in situations where other ionization methods cannot. Considering that the preparation of the field emitters in FD is a time-consuming process, FAB seems to have an inherent advantage over FD. However, care must be taken in analyzing the results obtained with a FAB ionization source. FAB has supplanted FD for routine analysis of polar materials in almost all areas except for the analysis of synthetic polymers.[27] FAB and FD mass spectrometry are an excellent complementary pair of techniques to use in analyzing industrial polymer chemicals.

Secondary Ion Mass Spectrometry

Secondary ion mass spectrometry (SIMS) is very similar to FAB. The major difference is that the beam that bombards the surface containing the sample consists of ions in SIMS, whereas in FAB the beam is composed of atoms. Another difference is that a liquid matrix is not typically used to prepare the polymer sample for a SIMS experiment; however, in liquid SIMS (LSIMS), a matrix is frequently used in the analysis of biopolymers. Even though these do not seem to be major differences in the two techniques, they result in quite different spectra obtained from the same sample.

The basic instrumentation for this ion source is identical to FAB, except that after the high-energy ions are generated from an ion gun, there is no reason for them to undergo charge exchange with more gas. Therefore, the deflector plate (which is used in FAB to remove any excess ions before striking the sample) is not needed in SIMS. A thermionic source[97] has recently become popular for generating a stable current of alkali ions.[98] The source is held at the primary accelerating voltage while it is heated by using a resistive heater. By simply changing the emitter source, one is able to change the type of ion beam that is generated. Cesium sources[99] are popular because the cesium ion beam can be accelerated to a very high kinetic energy.[100]

SIMS has three distinct modes of operation—static, dynamic, and imaging—each requiring its own slightly different type of ion source.[101] Dynamic SIMS uses high-current densities to erode the sample by sputtering and emitting secondary ions. Typically, cesium ion sources are used. Due to this large current, the sensitivity of this technique is in the picomole to femtomole range.[102] Dynamic SIMS is normally used for elemental analysis.[103] This technique has developed along with the aid of liquid metal ion sources that are based on a field evaporation or ionization process. Since the beam can be focused into a small spot, by scanning the ion beam across a surface, both chemical information and physical or topographical information can be obtained, provided the angle of incidence of the ion beam and the angular distribution of the sputtered ions is known. Static SIMS, sometimes referred to as molecular SIMS, uses low-current densities to directly desorb molecules from the surface.[104] Normally, noble gas ion sources are used to achieve these low-current requirements; as a result, the surface is not considerably altered. The secondary ions come from approximately the uppermost 2 to 5 nm of the surface.[105] This surface analytical technique is an attractive procedure for the analysis of surface layers and the surfaces of polymers.[106] Static SIMS is able to detect all elements, including hydrogen, unlike X-ray photoelectron spectroscopy (XPS) and Auger electron spectroscopy (AES).[107,108]

Sample preparation involves evaporating the polymer sample from a suitable solvent such as methanol, toluene, or tetrahydrofuran, which has a concentration of approximately 10^{-2} to 10^{-3} M with respect to the repeat unit.[109] Roughly 1 to 7 μL of sample are evaporated onto a target that can be placed on the end of an insertion probe. This target is made of a metal, usually silver, which was etched with nitric acid.[110] However, samples that cannot be dissolved in a suitable solvent can still be analyzed, simply by affixing the sample to the probe with the aid of double-sided sticky tape.[111] The generation of the secondary ions in static SIMS is usually the result of energy and momentum transfers, as well as fragment formation.[112] For larger molecular weight polymers, fragmentation (rather than desorption of intact oligomers) is the main mechanism of ion formation.[102] This fragmentation usually occurs through the scission of a carbon-heteroatom bond instead of a carbon-carbon bond. The fragmentation pattern can be extremely useful in determination of repeat unit composition and sequence. In SIMS spectra of polymers, doubly charged ions are rarely ever observed.[113] This is because the formation of cyclic fragments is a more favorable process for high molecular weight samples. To help ease the task of spectral interpretation, several compilations of SIMS spectra of many commonly encountered polymer surfaces are available and can be used for comparison.[114,115] In the static SIMS mode, the use of either Ar^+ or Xe^+ provides good molecular fragment secondary ion intensities,[116] which is helpful for end-group analysis, copolymer composition, and investigating the principles of polymer adhesion.

SIMS has some inherent built-in disadvantages when compared with its related technique, FAB. Since a beam of ions is used rather than atoms, there is a buildup of charge on the surface that can affect the sputtering of the secondary ions. This also leads to an uncertainty of the surface potential.[80,117] Also, it is much easier to direct a neutral beam into a source rather than a charged beam. For FAB, there is no need to switch the operating character of the ion source when switching between positive and negative ion modes. SIMS also results in a high damage rate of the sample because of the ion beam.[118] With all these problems involved in SIMS analysis, it still must be used in order to achieve molecular specificity at high spatial resolution for polymer analysis.[119] Both SIMS and FAB can be combined in the analysis of a polymer sample since the two techniques are so similar, and a wealth of information can be obtained from each of them. These two ionization techniques are extremely complementary to one another.

Laser Desorption

Laser desorption (LD) was developed because there was no suitable ionization technique to study thermally labile and high-mass compounds by mass spectrometry. Techniques such as EI and CI can be used for polar substituents of nonvolatile samples, but as the size of a sample increases and the number of end groups increases, the use of these ionization sources becomes limited. Another technique that could be used is FD, but FD results can be reproduced with only moderate success.[120]

The basic operation of an LD system allows one to identify mass and to measure the intensity of ions that are produced by a high-power-density laser. The typical lasers that are used for the measurements and analysis of polymers are visible lasers, Nd:YAG, nitrogen, and CO_2. The main advantage of this technology comes from the ability to focus the laser beam to a small spot on the sample. When the laser hits the sample, a burst of desorbed neutrals and some ions are generated. A small packet of ions begins to travel from

the surface of the sample toward the mass spectrometer. Some of these ions may have been formed by gas-phase proton-transfer reactions or by cationization due to the presence of alkali metal ions. As the ions accelerate, they have similar velocity distributions, but because they have different masses, they will have different kinetic energies. Thus, the ions will then reach the detector at different times. In addition, as the ions move toward the detector, they will reach the detector in packets that contain the same mass. This allows a means of separating and detecting ions of different mass, which finally creates a mass spectrum.

In past years, many experiments have been accomplished to study and understand the laser desorption process.[121–124] However, it is difficult to completely understand this process because of the numerous parameters that can be manipulated. These parameters include laser power density, photon wavelength, electronic absorption, variety and number of molecules contained within the sample, thickness of the sample, orientation of the incoming laser beam to the sample surface, and use of multilaser systems to achieve ionization.[103,120] These parameters can all affect the mass spectrum, and changing one (or all) of them may change the desorption mechanism. It has been found that if heating is rapid, the production of intact molecules is favored over thermal degradation. However, it has also been found that the use of high-power-density lasers tends to produce more fragmentation.[120]

There has been much interest in the formation of ions in the gas phase from the sample using LD, and the process is not yet exactly known. Through various experiments, the ion-formation mechanism is being studied and is continuously being modified. As was mentioned earlier, the difficulty in understanding this process comes from the numerous parameters that can be manipulated. One reason for the interest in this process is that it may ultimately lead to more control in acquiring the desired information from a mass spectrum. The following mechanism was developed by Hercules et al.[125] to obtain a better understanding of the ion-formation process.

As the laser beam strikes the sample surface, high kinetic energy particles are emitted. The approximate energy the particles have as they leave the surface is in the range of tens of electron volts.[120] The first ions that are emitted are usually small molecular or atomic fragments. In Figure 11.8, one can see four distinct regions that can define the LD process. The first region (region 1) is the direct ionization region. Here the extremely high temperatures create atomic and molecular fragments. Adjacent to region 1 is region 2, where secondary ionization takes place. Secondary ionization is where the ions are desorbed from the surface of the sample and then ionized. Also in this region, which is spatially adjacent to where the laser strikes the surface, there tends to be a thermal gradient that is created by two factors: the first is from a shock wave that is created

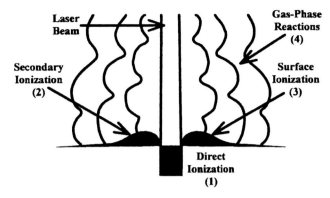

Figure 11.8 Important regions created during the laser desorption process: (1) direct ionization, where laser beam directly strikes sample surface for ionization; (2) secondary ionization, a region adjacent to direct ionization region, where ions are desorbed from sample surface; (3) surface ionization, another region adjacent to direct ionization region, where ion may be formed from solid phase; and (4) gas-phase reactions regions above the sample surface. (Adapted with permission from reference 125. Copyright 1982 American Chemical Society.)

from the impact of the laser beam, and the second is from collapsing plasma. It is believed that higher molecular weight ions and neutrals are emitted from this region. This region tends to be cooler and increases the possibility of formation of intact molecules. In addition, some reactions may occur between the sample and gas phase ions in this region. In region 3, also immediately adjacent to the laser pulse, some surface ionization may take place. This is where some debate comes into play in laser desorption mass spectrometry. The controversy lies in where the ion is being formed. Is the ion being formed in the gas phase, or is it being directly ionized from the solid phase? It has been determined that both surface ionization and secondary ionization do, in fact, take place, and the extent of each depends on the type of sample. Region 4 is the region where gas phase reactions take place. This occurs in an area above region 1, where the laser has vaporized the sample.

The first LD mass spectrometers created a high ion production because long-duration, large-diameter laser pulses were used. This led to a broad kinetic energy distribution of ions and space charge broadening, which created limited mass resolution.[126] This main disadvantage of LD is what led to the development of the first LAMMA (laser-assisted microprobe mass analyzer) system. As research and development proceeded, the aim was to decrease the laser pulse length and spot diameter. This was done by attempting to focus the laser beam and by the use of ion transmission configurations for ion production and excitation.[126] The first reported true LAMMA instrument made use of a Q-switched ruby laser that created laser pulses in the nanosecond range and beam diameters in the 20 μm range.[127] This instrument used a transmission configuration and a time-of-flight mass analyzer. The second advancement in the

production of LAMMA instruments came by modifying the laser focusing system. By using microscope optics to guide and focus the laser beam, Hillenkamp et al.[128] created beam diameters of 0.5 μm. In addition, this advancement led to the ability to manipulate the position of the laser beam to scan the surface of a sample. Thus, like SIMS, LAMMA can be used for imaging on surfaces.

Sample preparation can be crucial in obtaining useful information in LD. More often than not, the best mass spectra are obtained by directly depositing a solid sample from solution onto a metal disk (i.e., Al). In addition, sometimes a salt is added. For example, Liang et al.[129] prepared ~3.0 μL of aqueous Triton polymer to 0.5 mL of reagent-grade methanol saturated with KBr to obtain an LD/FTICR mass spectrum. However, if the polymer is insoluble, a different technique must be used. Brown et al.[130–132] compressed a powered polymer between a stainless steel probe tip and a piece of weighing paper, then removed the paper so that the sample was exposed. For samples that did not adhere to the tip, a piece of double-sided tape was used.

LAMMA-TOF systems are only good for low molecular weight polymers because pyrolytic disintegration into carbon clusters is often observed; however, FT-LAMMA instruments work for polymers up to several thousand daltons.[133] Both SIMS-TOF and MALDI are more effective techniques for studying polymer systems because of the ability to effectively minimize fragmentation and for the availability of commercial instruments.

Matrix-Assisted Laser-
Desorption Ionization

Although LD is a useful technique in ionization and volatilization of synthetic polymers, generally molecules above approximately 10 kDa become difficult to ionize and detect with LD. A fundamental breakthrough that has revolutionized synthetic polymer analysis by MS was the introduction of "matrix-assisted" laser-desorption ionization, or MALDI. MALDI is a novel soft ionization technique that generates high mass, predominately singly-charged ions by irradiating a solid analyte or matrix mixture with a pulsed laser beam. Although ultraviolet lasers are generally used, recently infrared wavelength lasers have found increasing use for desorption purposes. In MALDI, the matrix, by its large excess, minimizes analyte fragmentation that results from laser radiation by absorbing the majority of the incident energy. Therefore, as a prerequisite, the matrix must contain a chromophore with high molar absorptivity at the operating laser wavelength. The complex mechanisms involved in the production of gas phase ions during the MALDI process are not well understood, but it is widely accepted that analyte ionization occurs as a result of collisions of neutral analyte with a charging agent (e.g., Na^+) in the initial plume of ablated material. Several review articles have appeared in the recent literature regarding MALDI fundamentals, as well as areas of application.[134–136]

MALDI is most often coupled to pulsed time-of-flight (TOF) mass analyzers, and several groups have demonstrated the usefulness of MALDI-TOF for measuring molecular weight distributions of large synthetic polymers. In the first communication of MALDI-TOF, Karas and Hillenkamp reported the extraordinary signal-to-noise (S/N) ratio achieved for the analysis of the biopolymer bovine serum albumin (~67.5 kDa).[21] Later Danis et al.[137] reported on the MALDI-TOF analysis of synthetic polymers, poly(styrene sulfonic acid) and poly(methyl methacrylate) (PMMA) with approximate nominal masses of 400 and 256 kDa, respectively. More recently, Schriemer et al.[138] communicated results for MALDI analysis of poly(styrene) (PS) with masses as large as 1.5 MDa!

However, despite the ability to couple TOF to intrinsically pulsed ionization schemes readily, TOF mass analyzers suffer from either modest resolving power (RP) or a limited mass accuracy (even for reflectron TOF analyzers, which use electrostatic lenses for temporal focusing of ions; see mass analyzer section), especially for polymers with molecular masses under 10 kDa. These limitations frequently prohibit detailed polymeric analysis such as the determination of monomeric units (e.g., in copolymers and block copolymers) and end-group chemistries. To overcome these shortcomings, some researchers have begun to employ MALDI with the high-resolution FTMS mass analyzer. In one of the first reports of polymer analysis by MALDI-FTMS, Castoro et al.[139] demonstrated an average RP of 3200, sufficient to resolve oligomers (44 Da repeat unit) of 10 kDa nominal mass poly(ethylene glycol) (PEG).

One of the most crucial aspects of MALDI analysis involves careful sample preparation to achieve a homogeneous mixture of analyte within a target matrix.[140] Some of these considerations include proper choice of matrix, solvent, deposition method, and matrix/analyte molar ratio, as well as their absolute concentrations, pH, and addition of alkali metal salts as charging agents. In addition, the resulting matrix should favor cocrystallization as it minimizes unwanted chemically or photochemically induced reactions between analyte, matrix, and charging agent. As a result, numerous small model organic compounds with high molar absorptivities at key laser lines have been tested in a combinatorial approach to determine the methods for producing matrices that have the desired performance characteristics. For example, Latourte and coworkers investigated 11 different MALDI matrices with varying matrix/analyte molar ratios for the analysis of a fluorinated polymer and found that a 2,5-dihydroxybenzoic acid (DHB) matrix with a matrix/analyte molar ratio of 1000 : 1 provided the optimum signal intensity.[141] In another study, Dey and coworkers report using DHB as a matrix for the analysis of PEGs with a matrix/analyte molar ratio of 5000 : 1 for single-component polymer analysis and a 10000 : 1 : 1 : 1 : 1

molar ratio for quaternary-component polymer mixtures.[142] Schriemer and coworkers used pure trans-retinoic acid as a matrix, THF as solvent, and AgNO₃ as a charge agent for the analysis of PS polymers with nominal molecular weights ranging from 96 to 1500 kDa.[138] Schriemer et al. emphasized the need to expel solvent impurities (e.g., water, peroxides, acids, and antioxidants) that previously led to signal suppression. In addition, they found that with increasing molecular weight PS polymers, higher matrix/analyte ratios were required to maintain good signal. These ratios varied from 3000 : 1 for low molecular weight polymers to $10^6 : 1$ for higher molecular weight samples.[138] The large excess of matrix used for higher molecular weight samples is thought to prevent polymer entanglement or the formation of regions of microcrystallinity, which are common causes of target inhomogeneity. As a final note, the matrix/analyte solution can be placed on the metal probe tip by either a dried droplet approach or an electrospray deposition method. In more recent MALDI applications, the electrospray deposition method is preferred as it produces a more uniform layer of matrix/analyte mixture that improves spot-to-spot reproducibility and more reliability in quantitative analyses.

Despite the many matrix-associated experimental pitfalls that must be navigated for successful analysis, MALDI has many distinct features that make it a premier ionization technique for polymer diagnostics. Key among these features is the fact that specific analyte polarity is not a critical factor, allowing a wide range of water-soluble, polar organic–soluble, and nonpolar organic–soluble polymers to be analyzed by MALDI-MS. Further, assuming no adverse interactions with the matrix to result in inhomogeneity, MALDI is relatively insensitive to the concentration of the charging agent. This lends the versatility to examine polymers that are difficult to ionize by charge adduction. Another feature is that MALDI predominately generates singly charged species, thereby greatly simplifying spectral interpretation.

Electrospray Ionization

Like MALDI, electrospray ionization (ESI) is also a revolutionary soft ionization technique with several key features that account for its widespread application toward macromolecular systems, both synthetic and natural. These include minimal or zero fragmentation (unless so promoted by ancillary means), high ionization efficiency, and ion charge multiplicity, which yields a high mass range even for limited mass-to-charge range analyzers.[143] In fact, ESI-MS has become the method of choice for analysis of large, nonvolatile, and thermally labile compounds, as well as the touchstone for comparison with alternative soft ionization techniques. Fenn and coworkers used ESI with a quadrupole mass analyzer that has an upper scan limit of m/z

1500 for the analysis of PEG polymers with nominal molecular weights up to 5 MDa.[143]

The conventional ESI mode-of-operation involves assisted desolvation and delivery of ions preformed in solution into the gas phase. In practice, a syringe pump is used to continuously infuse a solution at typical flow rates of 60 to 500 μL h^{-1} through a small-diameter (50 μm i.d.) needle. Application of a high electric field (2–5 kV) results in the formation of an aerosol of charged droplets that are desolvated by a combination of heat, vacuum, and inert gas flow. The precise nature of the mechanics that result in the formation of highly charged single molecules still remains a topic of debate.[144,145] Thus, the interested reader is referred to reviews by Kebarle and Tang[146] and Nohmi and Fenn,[143] where arguments for the opposing theories of "charged residue" (as proposed by Dole)[23] and "ion evaporation" (proposed by Iribarne and Thomson)[147] are developed and assessed.

Both positive and negative ion ESI modes are possible; this is determined from the polarity of the applied electric field. ESI is distinct from other ionization sources in its unique and powerful ability to generate multiply charged analyte ions. This feature is particularly advantageous when analyzing species (certain proteins and enzymes) that possess perfectly conserved sequences and therefore the same molecule-to-molecule mass. The redundant molecular weight information obtained from the multiple charge state distributions allows for highly accurate molecular weight computations for the analyte of interest. Conversely, this feature can be problematic for analysis of synthetic polymers that contain an ensemble of polymer chains and subunits possessing a distribution of molecular weights. In this case, the multiplicity of charges placed on each ion during the ESI process results in complex mass spectra that consist of overlapping polymeric distributions as a result of multiple oligomeric masses, which, in turn, possess multiple charge states. A detailed analysis (e.g., accurate mass measurements, end-group determination) of such complex spectra requires sufficient RP to determine the charge state(s) for each of the ions of interest or for all the ions if weight-average and number-average molecular weights (M_w and M_n) are calculated. Spectral complexity becomes exacerbated with increasing molecular weight, because distributions of multiple charge states increasingly overlap in the mass-to-charge (m/z) scale. Increasing spectral complexity with finite and instrument-dependent RP eventually obviates full deconvolution of the data and, hence, the complete extraction of useful information about the polymer. This limit is rapidly approached for low-resolution mass analyzers, in which case, *detailed* analyses are limited to low molecular weight polymers. However, ESI is particularly well suited for use with FTMS, which has been demonstrated to provide superior RP and allows, for example, the deconvolution of all charge states (+19 to +28) that result from the analysis of 23 kDa PEG.[148]

In an extreme case, a polymer may be entirely insoluble in *all* possible solvents, precluding ESI analysis. Even considering proper polymer dissolution, analysis may still be hampered if the chemistry of the polymer is not conducive to charge adduction (positive ion detection) or atom depletion (negative ion detection). Such demands keep researchers in ESI polymer MS studies in perpetual search of experimental conditions that strike an acceptable compromise between polymer solubility and ionization source compatibility.

The most fundamental mechanism of ESI involves disruption of a liquid surface into a spray of charged droplets under the influence of an external electric field. Therefore, solvent properties such as electrical conductivity, surface tension, viscosity, density, and dielectric constant heavily influence spray characteristics. In general, polar or semipolar solvents are used because their solution properties are conducive to both stable sprays and hosting charge formation between neutral analyte and charging agents. Such solvent systems are ideal for biological studies in that biomolecules are readily soluble in hydrophilic (polar) solvents and they contain residues (e.g., lysine or arginine amino acid residues) that may be readily charged by in situ acid-base chemistry. For similar reasons, many of the early applications of ESI-MS toward synthetic polymer analysis involved the study of hydrophilic polymers such as PEG. Other synthetic polymers, however, present much more challenging scenarios because their solubilities often limit their application to "ESI-unfriendly" solvents of a more nonpolar nature. For these cases, solvent conditions (e.g., polar cosolvent modifiers) and experimental parameters (e.g., flow rate, applied potential) must be judiciously selected so that acceptable polymer and charging-agent solubility and ESI-compatibility are both achieved. For example, Jackson and coworkers reported on a binary solvent system of methanol (MeOH) and tetrahydrofuran (THF) (1 : 1, v/v) for polymer additive dissolution that entailed addition of just enough water cosolvent to dissolve the ammonium chloride charging reagent.[3] Hunt and coworkers determined optimum experimental conditions for analysis of 16 different polyester resins by investigating various solvent systems, including those composed of acetone, methanol, isopropanol, THF, and methyl ethyl ketone, as well as the effects of alkali metal cation addition.[149] Numerous polymer systems have been studied by ESI, including polyester resins,[149,150] PMMA,[151,152] poly(tetrahydrofuran),[151] PS,[151,153] poly(sulfide),[6,154] poly(dimethylsiloxane) (PDMS),[155–157] poly(amidoamine) starburst dendrimers,[158] octylphenoxy poly(ethoxy) ethanol,[159] and PEG.[148,149,160]

Mass Analyzers

After the analyte of interest has been transformed into gas phase ions, the next event is to detect the ions. A variety of mass analyzers have been developed to produce a mass spectrum to do this. In this section, sector, time-of-flight, quadrupole, quadrupole ion trap, and Fourier transform ion cyclotron resonance mass analyzers are discussed.

Sector

Once the ions leave the source, they move toward the accelerating plates. These plates create a region of electric potential of approximately 4000 to 8000 V, which accelerates the ions.[161,162] As the ions move out of this region, they have now acquired kinetic energy (KE) and make their way into the mass analyzer region of the mass spectrometer. As the ions enter the mass analyzer, they have a KE (E_k) that can be described as ½ mv^2. The amount of E_k that is gained is proportional to the potential difference of the accelerating plates. This can be described by the following equation:

$$qV = \tfrac{1}{2}\,mv^2 = E_k$$

In this equation, q denotes the total charge of the ion, V denotes the accelerating voltage (or the potential difference between the accelerating plates), m is the mass of the ion, and v is the velocity of the ion.[161,163,164] In addition, the total charge of the ion is described by:

$$q = ze$$

where z denotes the number of charges on the ion, and e is 1.6×10^{-19} coulombs.

Magnetic

Once the ions have gained E_k, they move into a magnetic field. As the ions travel through the field, a magnetic force acts on them, and this may change the direction of their flight path. If the force acts on them at a right angle, the ions will begin to follow a circular orbit, and to maintain this orbit, the centrifugal force has to be equal to magnetic force that is exerted on the ion. Thus, the radius of this arc can be described by:

$$qvB = mv^2/r \qquad \text{or} \qquad r = mv/qB$$

where r is the radius of curvature, and B is the magnetic field strength.[161,163,164] Furthermore, by combining the previous two equations, the velocity term can be eliminated, and the following relationship is obtained:

$$m/q = B^2r^2/2V$$

In addition, if B and V are held constant, then the radius of curvature depends solely on m/q. Hence, this allows a means of separating ions of different mass and charge (see Figure 11.9). These instruments are considered disperse.[163] However, it seems more reasonable to guide the ions to one area so that they converge on one point to be collected and detected. Indeed, by holding r, the radius, constant and changing B or V, one can bring all ions of different m/q into focus in sequential order.[164] This type of instrument is

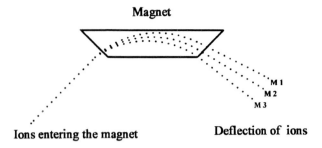

Figure 11.9 Ions being deflected as they move through a magnetic field, and the separation of ions due to different values of m/q (M1 > M2 > M3).

considered scanning. In addition, if the magnetic field is held constant, and the accelerating voltage is varied, the instrument is considered voltage scanning. In contrast, if the accelerating voltage is held constant and the magnetic field is varied, the instrument is considered magnetic scanning. In early instruments, the magnetic field was held constant and the voltage was varied to bring the ions into focus. The main reason for this was that it was more cost efficient to scan the voltage. However, by scanning the magnetic field, one can achieve superior sensitivity.[161]

The first magnetic sector instruments were developed by Dempster (1918),[165] Bleakney (1932),[166] Tate, Smith, and Vaughan (1935),[167] and Nier (1936).[168] These first instruments were of the 180° sector configuration. However, as years passed, other configurations were developed. A 60° magnetic sector instrument has been described by Nier.[31,32] In addition, 90° sector instruments were also developed and have been documented.[169]

Electrostatic

Another type of sector mass analyzer is the electrostatic, or electric, sector. This type of analyzer works on the same basic process as the magnetic sector; however, it uses an electric field instead of a magnetic field. As mentioned earlier, as the ions leave the accelerating plates, there is an E_k that is associated with them. As the ions move between a pair of plates that are maintained at a high potential difference, the ions are deflected toward one of the plates (the charge of the ion depends on which plate). The force (F) that is exerted on the ions can be given as:

$$F = qE$$

where q is still the total charge of the ion, and E is the potential difference between the pair of plates.[161,163,164] As mentioned earlier, as a force acts on the moving ions at a 90° angle, the ions tend to take on a centrifugal orbit, which can be explained by the centrifugal force. Hence, to maintain this centrifugal orbit, the electric force that acts on the ions equilibrates the centrifugal force. This is given as:

$$qE = mv^2/r$$

In addition, by combining this equation with the E_k, one arrives at:

$$r = 2V/E$$

Up to this point, in both magnetic and electric sector mass analyzers, it has been assumed that all the ions with the same m/q have the same kinetic energy after acceleration. However, this is not the case.[161,163,164] It turns out that for a given m/q, there tends to be a spread of E_k as a result of the Boltzmann distribution.[161] In addition, this spread is a consequence of exactly where in the source the ions of the same m/q are formed. Ions of the same m/q that are formed in different positions within the source experience slightly different potentials, which then leads to an E_k spread. In addition, the ions of the same m/q may enter the field of the sector with different trajectories.[163] This may then create a slightly different radius for the ions with the same m/q. These factors ultimately limit the resolving power of the mass spectrometer.

To improve the resolving power of the mass spectrometer, two sectors have been placed in sequence to create a double-focusing sector mass spectrometer. Some of the first reports of these instruments were described by Mattauch (1936)[170] and Brainbridge and Jordan (1936).[171] More often than not, the electric sector is placed in front of the magnetic sector and acts as the energy-focusing sector (see Figure 11.10). Furthermore, once the ions are energy focused, they then move through the magnetic sector and are mass analyzed.[161,162] These types of instruments have resolving powers up to 1 million.[161]

Time-of-Flight

Mass spectrometers that have been equipped with time-of-flight (TOF) mass analyzers are relatively inexpensive, simple devices. These instruments can obtain good resolution

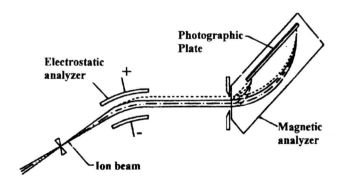

Figure 11.10 Double-focusing mass spectrometer. Photographic plate detectors are currently no longer in use; they may be replaced with a photocathode or a faraday cup. (Adapted with permission from reference 161. Copyright 1987 John Wiley and Sons.)

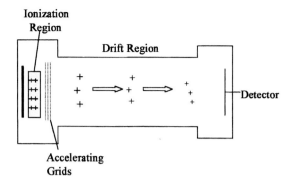

Figure 11.11 A linear time-of-flight (TOF) mass spectrometer. Notice that the lighter ions reach the detector before the heavier ions do. (Adapted with permission from reference 100. Copyright 1995 Academic Press.)

($>10^3$ in linear mode and $>10^4$ with a reflectron)[172,173] and can analyze practically every range of mass simultaneously.[174] The most basic TOF mass spectrometers are those with a linear arrangement (see Figure 11.11). However, by using an electric field called a reflectron to change the direction of the ions, one can increase the resolution of these instruments.[175] TOF mass analyzers have been around for many years. The first proposal of a time-of-flight analyzer was by Stephens in 1946,[176] and the first experimental documentation occurred in 1948.[177] In addition, in 1955 Wiley and McLaren described an improved resolution TOF-MS.[178]

In recent years, with the development of matrix-assisted laser-desorption ionization (MALDI), there has been considerable interest in TOF as the analyzer. More often than not, this type of analyzer is used with a source that produces bundles of positive ions by pulses of electrons or secondary ions or by using a laser pulse.[179,180] The bundle of ions that are formed are then accelerated by a potential of approximately 1000 to 10,000 volts. Once the ions are accelerated, they then enter a region that does not contain any external field; this area is termed the *drift tube*.[180] The ions then move on toward the detector, where they can be mass analyzed. With this technique, the sensitivity can be quite high. This arises from ions being produced in relatively small time spans and in theory all being detected. As stated earlier in this chapter, if we assume that all the ions exiting the source have the same kinetic energy, then the lighter ions will reach the detector faster than the heavier ions.[180] Recall from the magnetic sector section that the kinetic energy of an ion accelerated by an electric field can be given by:

$$E_k = \tfrac{1}{2}\,mv^2 = qV$$

In addition, the velocity (v) of an ion can be described as:

$$v = (2qV/m)^{1/2}$$

where q denotes the total charge of the ion, V denotes the accelerating potential, and m is the mass of the ion. In addition, since the velocity is equal to the length of the flight tube (d) over the time it takes the ion to travel through the flight tube (t):

$$v = d/t$$

One can arrive at the following:

$$t = d(m/2qV)^{1/2}$$

In addition, to calculate m/z (remembering that $q = ze$), is related to time and can be given by:[175,179]

$$t^2 = m/z(d^2/2Ve) \quad \text{or} \quad m/z = 2Ve(t/d)^2$$

This equation is what allows one to create a mass spectrum by measuring the flight time of an ion, which is a function of its m/z.

The time-of-flight mass analyzer can be improved with an electrostatic reflectron. The first reflectron was introduced by Mamyrin et al. in 1973,[181] and a parabolic electrostatic reflectron was documented by Steadman and Syage in 1993.[182] By using a reflectron, a compensation for the distribution in kinetic energies can be obtained.[175,179] As seen in Figure 11.12, the location of the reflectron is at the end of the flight tube. The reflectron is constructed of a series of grids, which creates a sequential increase in potential until the voltage is greater than that of the source. As the ions penetrate the reflectron, they eventually reach a distance x, where they stop and E_k goes to zero. The ions then reverse direction and make their way back out of the reflectron.[175] As this occurs, the ions then regain the E_k that they have lost upon entering the reflectron and acquire the same E_k as before, but in the opposite direction.[175,179]

The penetration (x) into the reflectron can be described as follows:

$$x = E_k/qE$$

where E_k is kinetic energy, q is the total charge of the ion, and E is the electric field in the reflectron.[179] The ions that contain a smaller E_k will not travel into the reflectron as far as those that have a larger E_k. In addition, the ions with a larger E_k will take a longer path to the detector. For example, if the flights of two ions, A and A', have the same m/q and leave the source at the same time, with A having a greater kinetic energy than A', A will have a greater velocity and a longer flight path than A'. Thus, A and A' will arrive at the detector at the approximate same time, and the kinetic energy spread will be corrected (see Figure 11.12). Thus, the mass resolution of the TOF-MS is improved. However, it has been shown that by using a reflectron, a drop in the sensitivity may result.[181]

As seen in Figure 11.12, there are two detectors. One is the neutral detector, and the other is the reflected ion detector. The neutral detector is in position to detect all neutral ions that will not be affected by the reflectron. Thus, this increases the resolution of the mass spectrometer. The neutrals usually result from metastable ion dissociation as the ions travel through the flight tube.[175]

As mentioned earlier, time-of-flight mass analyzers have the ability to analyze a total mass distribution simultaneously. However, they do not possess the ability for selective ion storage before mass analysis. Thus, the ability to eliminate or eject unwanted ions from a mass spectrum cannot be accomplished. Nevertheless, this technique does offer some advantages. For example, it is a relatively simple mass analyzer: there are no moving parts or slits and no scanning electric fields. In addition, it is relatively inexpensive and easy to maintain. An added benefit is the rapid spectral acquisition that can be obtained with TOF, allowing one to rapidly collect many spectra for improved S/N.

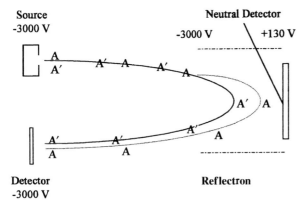

Figure 11.12 A time-of-flight (TOF) mass spectrometer containing a reflectron. A and A' are two ions of identical m/q but with A at slightly greater kinetic energy than A'. (Adapted with permission from reference 179. Copyright 1996 John Wiley and Sons.)

Quadrupole

Quadrupole mass analyzers are quite different from sector-based and time-of-flight mass analyzers. Quadrupoles rely on only electric fields to help guide and sort the ions into the detector. Therefore, quadrupoles resolve ions based on their mass-to-charge ratio rather than on the basis of momentum or E_k.[183] A typical quadrupole mass analyzer is shown schematically in Figure 11.13. As its name implies, it is composed of four identical rods. Theoretically, these rods should have a hyperbolic shape; however, it is much easier and cheaper to use circular rods, which do not decrease the resolution of the analyzer very much.[184,185] The resolution with hyperbolic rods is about two times greater than the resolution obtained with round rods.[186] Electrode shape, electrode spacing, charging effects, and nonsinusoidal waveforms all lead to distortions in the peak shapes and peak splitting.[187,188] The rods are arranged parallel to one another, as seen in Figure 11.13. Static, direct current (DC) and alternating radio frequency (RF) electric potentials are applied to opposite pairs of rods. This results in a fluctuating

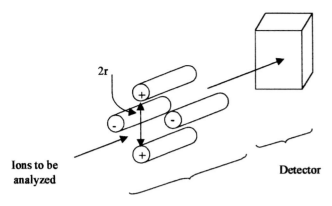

Quadrupole mass analyzer

Figure 11.13 The quadrupole mass analyzer consists of four parallel rods spaced 2r apart. Previously generated ions must be able to pass through the quadrupole mass analyzer in order to be detected. If an ion comes into contact with a rod, it will be neutralized and subsequently not be detected.

electric field. If the ions are able to traverse the length of the analyzer without striking any of the rods, they will reach the detector. However, if they strike a rod, they will be neutralized and pumped away.

Before entering the mass analyzer, the ions that are generated in the ion source must be focused. This is accomplished by placing ion lenses between these two devices, which focuses the ions into a concise ion beam and improves the mass resolution of the spectrum.[186,189] Another purpose of this device is to apply a small voltage to the beam, which then "pushes" the ions down the analyzer. The voltage induces a small amount of forward momentum to start the ions on their path toward detection. The instrument radius, ion mass, ion energy, and frequency of the applied field are all used to optimize the lens system to couple the ion sources to quadrupole mass analyzers.[190]

The actual trajectory of an ion in a quadrupolar field is very complex. It can be described through the use of differential equations, more specifically the Mathieu equation. A detailed discussion of these equations will not be examined here;[191,192] we will look at a much more simplified version.[183,193] The potential, Φ, that is applied to the rods can be given by:

$$\Phi = U + V \cos \omega t$$

where U is the DC voltage, V is the RF voltage, and w is the frequency. This potential is applied to two opposite rods, while the negative of this potential is supplied to the other two rods. Some typical values for these parameters are as follows: $\omega = 1-2$ MHz, $U = 500-1000$ V, and V up to 3000 V.[194] Two other parameters, a and q, need to be explained before the ion trajectory can be discussed:

$$a = 4eU/\omega^2 r_o^2 m$$
$$q = 2eV/\omega^2 r_o^2 m$$

where e is the charge of an electron, r_o is the radius of the circle at the closest distance of approach of the rods, and m is the mass of the ion. The rods are approximately 6 to 15 mm in diameter, and their length is about 50 times the radius of the rod.[195] Figure 11.14 is an enlarged section of a graph derived from the Mathieu equations, which shows the region of stable ion trajectory. The area under the curve shows the masses of ions, which will have a stable trajectory. Anything outside this area will strike a rod and be neutralized. The mass scan line is what determines the total resolution that will be obtained. Ideally, one would like this line to intersect right at the apex of the graph; however, this is not practically used because of fluctuations in the voltages, which could move the scan line so that nothing at all will be passed through the quadrupole and detected. After choosing the resolution desired, while maintaining the same U/V ratio (which is the slope of the line), both the DC voltage and the RF field are varied, while the frequency is held constant, resulting in the whole mass scale being scanned and producing a mass spectrum.[191] So in actuality, a quadrupole mass analyzer is more properly called a quadrupole mass filter. Based on the electric fields applied, the x to z direction serves as a high-pass mass filter and the y to z direction is a low-pass mass filter. Anything that is able to pass through both filters is the area under the curve in Figure 11.14 and will be detected.

Several other types of quadrupole analyzers and different modes of operation enable various pieces of information to be obtained. For example, the graph from Figure 11.14 is an enlarged section of the Matthieu equation. In the full-scale version of this graph, there are other regions that contain stable ion trajectories.[191] These values could be used in an analysis, as well as the one shown in Figure 11.14.

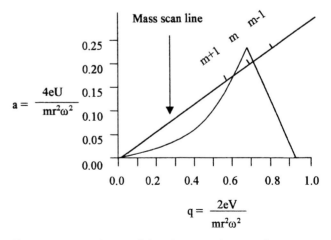

Figure 11.14 Mathieu stability diagram, showing the region most used for quadrupole mass analysis. The slope of the scan line determines the amount of resolution that will be obtained. (Adapted with permission from reference 183. Copyright 1986 American Chemical Society.)

Although these regions contain higher mass resolution, they can only be used to measure low molecular weight samples;[195–197] therefore, they may provide an alternative means of performing end-group analysis on polymer samples. The quadrupole can be operated in an RF-only mode.[198,199] This would essentially take the scan line in Figure 11.14 and convert it to a horizontal line on the x-axis. This would allow for the passage of almost all ions through the analyzer region. Only some very small mass ions would be filtered out in this mode of operation. This is frequently used when coupling a GC to this type of analyzer.[200] Another example includes a monopole mass analyzer,[201] which uses one rod and a 90° bent channel to analyze the ions instead of the four rods used in a quadrupole. This analyzer has the disadvantages that the peaks take on an asymmetric shape (with tailing involved) and that there is a low percentage of ion transmission through it.[202] These problems can be minimized by using fringe field effects.[190,203]

By coupling quadrupole analyzers together, a new type of experiment can be performed, tandem mass spectrometry. When three of these analyzers are combined, the instrument has been termed a "triple quadrupole" mass analyzer.[204] Tandem mass spectrometry, or MS/MS, can be performed on an unknown sample to help discover its identity. In the first quadrupole a particular ion, the ion of interest, can be selected and sent into a second RF-only quadrupole region where gas phase collisions between ions and neutrals can occur to produce product ions. These collisions occur between the ion of interest and a collision gas, such as argon or helium. The resulting product ions can be sent to a third quadrupole analyzer, where they are mass analyzed. The fragmentation information that is obtained can be useful in identifying the initially selected ion. There are several other methods or variations in the above-mentioned experiment, all of which result in valuable information about a sample.[194]

The quadrupole mass analyzer has many advantages over sector-based analyzers. These include a smaller size since no magnet is necessary, a linear mass scale, fast scanning abilities, lower relative cost, and robustness; furthermore, since they are not operated at high voltages, arcing is not a major problem.[192,205] These types of mass analyzers are used in a large number of industrial facilities, and they are ideal for interfacing with ESI. The major disadvantages of quadrupole analyzers include an ultimately low attainable RP and a low sensitivity for high mass ions. However, some currently available commercial instruments have included a combination of the quadrupole and TOF mass analyzers, called a QTOF. This has resulted in an increase in resolution and the ability to perform tandem MS.[206]

Quadrupole Ion Trap Mass Spectrometry

In addition to the beam-type mass analyzers described here—sector, TOF, and quadrupole—mass analyzers based on the trapping of ions also exist and should be considered for use by the practicing polymer chemist. The first of the ion trapping mass analyzers to be developed was the quadrupole ion trap. The quadrupole ion trap is similar to the quadrupole in that it uses a combination of superimposed DC and RF electric fields.[207,208] However, in the case of the quadrupole, these fields confine the ions in only two dimensions. In the case of the quadrupole ion trap, a combination of two "cap" electrodes and one "ring" electrode effectively confine ions in three dimensions and can be envisioned as a three-dimensional analog to the quadrupole. A background bath gas (usually helium) is used to cool ions and keep them within its confines until they are to be detected.

The quadrupole ion trap shares many analytical attributes with the quadrupole, including the fact that ion traps are readily suited for coupling to ESI sources. In addition, because it is an ion-trapping device, a quadrupole ion trap can be used for multiple stages of ion isolation or dissociation and for tandem MS experiments. Fragmentation patterns generated by tandem MS can be particularly useful for determination of end group and repeat unit sequence. Tandem MS capability is the chief advantage of quadrupole ion traps over single quadrupoles. Another significant advantage of quadrupole ion traps is their small size, making them much more easily introduced into laboratory settings than other larger MS systems. While quadrupole ion traps are more expensive than single quadrupoles, they are significantly less expensive than the triple quadrupoles used in tandem MS.

Fourier Transform Mass Spectrometry

Fourier transform mass spectrometry (FTMS) is an innovative ion-trapping technique that confines ions through the action of a DC electric field coupled with a magnetic field. Confinement and subsequent detection of ions is based on the principle of ion cyclotron resonance (ICR), thus the technique is sometimes also referred to as FT-ICRMS in the literature. Any discussion of FTMS should begin with the principle of ion cyclotron motion. An ion moving in the presence of spatially uniform electric ($\mathbf{E}$, in $V\ m^{-1}$) and magnetic ($\mathbf{B}$, in tesla) fields in an ICR ion trap is subjected to a Lorentz force (Figure 11.15):

$$\text{Force} = m \times a = q\mathbf{E} + q\mathbf{v} \times \mathbf{B}$$

in which m is ionic mass (kg), a is acceleration (m s^{-2}), q is ionic electrostatic charge (C), $\mathbf{v}$ is ion velocity (m s^{-1}), and the magnetic field is directed toward the positive z-direction: $\mathbf{B} = B_0\mathbf{k}$. If r is the ion cyclotron orbital radius, then angular acceleration about the z-axis

$$\frac{d\mathbf{v}}{dt}$$

is equal to

and

$$m = \frac{d\mathbf{v}}{dt} = m\frac{v_{xy}^2}{r} = q\mathbf{E} + q\mathbf{v} \times \mathbf{B}$$

In the absence of an electric field, **E**, this can be simplified to the following.

$$\frac{m}{q} = \frac{B_0 r}{v_{xy}}$$

Cyclotron motion is illustrated in Figure 11.15. Because the angular frequency, ω, can be expressed as

$$\omega = \frac{v_{xy}}{r}$$

we arrive at the classical ion cyclotron equation:

$$\omega_c = \frac{qB_0}{m}$$

In terms of Hz, the ion cyclotron equation becomes:

$$\nu_c = \frac{qB_0}{2\pi m}$$

Ion cyclotron motion is measured on the detection electrodes of the ICR trap (Figure 11.16). Ion cyclotron motion

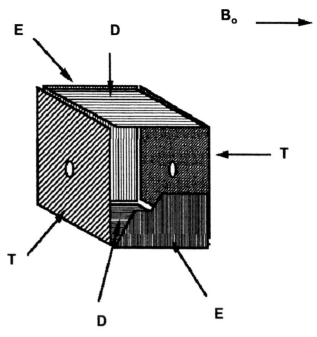

Figure 11.16 The cubic ion cyclotron resonance (ICR) trap consists of three pairs of parallel electrodes: excitation (E), detection (D), and trap (T). A direct current potential is applied to the trapping electrodes, which intersect the magnetic field (or z) axis and confine ions in the axial direction. Although other trap geometries exist, all require trapping fields and electrodes for excitation and detection. (Reproduced with permission from reference 363. Copyright 1999 Society for Applied Spectroscopy.)

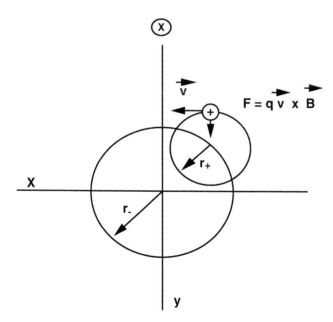

Figure 11.15 Ion cyclotron and magnetron orbits about the magnetic field axis (which points into the plane of the paper). The trajectory of an ion moving in the xy plane is "bent" into a circle by the Lorentz magnetic force. The resulting cyclotron orbit (with radius r_+) precesses about the magnetic field axis, which is magnetron motion (with radius r_-). (Reproduced with permission from reference 363. Copyright 1999 Society for Applied Spectroscopy.)

is the constrained motion of an ion about a magnetic field line within the xy plane and is described by its characteristic frequency. However, the unperturbed ion cyclotron frequency described in the cyclotron equation is not actually measured in an experiment because an ion's "natural" cyclotron frequency is subject to frequency shifts that are induced by trapping potential, space charge (number of ions which affect overall charge observed by individual ions), and image charge effects. Several different mass calibration formulas have been developed to take these frequency shifts into account. As long as the magnetic field in which ions are confined is relatively homogeneous (within a few ppm), frequency can be measured very accurately; consequently, so can mass-to-charge ratio, as the two are related by the cyclotron equation. Thus, accurate mass measurement is one of the major advantages of FTMS.

Since the initial demonstration of FTMS by Comisarow and Marshall in 1974,[209] countless researchers have made contributions to this technology. Such innovation and improvement has frequently involved developments in instrumental hardware or algorithm software with the aim of tailoring the technique for macromolecular analysis. Paralleling application refinements, FTMS has become famous for its ultrahigh mass resolving power[210,211] as well as its facility for multistage tandem mass spectrometry experiments.[210] Compared with the 10^2 to 10^3 RP typical of quadrupole and time-of-flight mass spectrometers, FTMS routinely offers an RP of about 10^5, allowing isotopic peak resolution and accurate mass measurements and making molecular composition determination a reality. For polymer analysis,[212,213] accurate mass capabilities are especially useful for end-group analysis, determination of the degree of polymerization, identification of repeat unit sequence in homopolymers and block copolymers,[214–216] and providing evidence for the existence of additives,[217] reactants, or impurities.[148,157] The information-rich capacity of FTMS gives it unparalleled potential as a quality-control tool for verifying synthetic pathways and assaying batch-to-batch compositional variations within polymer blends.

Further developments in FTMS have been shaped by the need for soft ionization such that intact molecular ions could be generated from nonvolatile, thermally labile, and fragile species. Three principle soft ionization techniques have been interfaced to FTMS for polymer analysis to date including LD,[129–132,216,218,219] MALDI,[139,142,212–214,220–224] and ESI.[148,156,157,215,216,225–227] In the successful design of the interface between such an ionization source and the FTMS, there is a significant probability that ions may be deflected from their trajectories leading into the ion trap. The two instrumental designs that meet these demands are referred to as "external" or "internal" sources with respect to the high magnetic field. Complete descriptions of various formulations of external and internal sources in FTMS instrumentation will not be discussed in detail here.

MALDI-generated polymer mass spectrum. For ESI, the propensity to generate multiply charged analyte ions results in overlapping polymeric charge state distributions. Calculation of M_n and M_w requires the assembly of a composite M_r distribution by deconvolution of all charge state distributions. This process requires that the overlapping charge state manifold peaks be well resolved from one another. O'Connor et al.[148] reported an ESI-FTMS mass spectrum of 14 kDa nominal mass PEG (Figure 11.17a) containing 12 charge state distributions ranging from +7 to +18. Specifically, they showed that an RP of 10^5 was sufficient to determine the charge state (and, hence, mass) for every peak used to assemble the M_r distribution that contained oligomers with 256 to 319 repeat units (n ranges from 256 to 319), as illustrated in Figure 11.17b. The calculated values of M_n, M_w, and D were 12,661, 12,663, and 1.002, respectively. In the same study, an RP of 5×10^4 was sufficient to identify 47 oligomers in 10 charge states for 23 kDa PEG.[148] Manual interpretation of such data is both labor intensive and time consuming. In order to complement the fast analysis potential of FTMS, it is imperative to incorporate automated deconvolution software into the analysis procedure. For MALDI, generation of +1 ions occurs almost exclusively and, therefore, the peaks in the mass spectrum are a reflection of the experimental mass distribution profile. Hence, deconvolution of multiple charge states is not required, thus simplifying the analysis of MALDI spectra.

Applications of Mass Spectrometry in the Analysis of Synthetic Polymers

Molecular Weight Distribution Measurements

Traditionally, gel permeation chromatography (GPC) analysis is used to determine polymer molecular weight distributions by comparing the retention volume to those of known, preferably monodisperse, well-defined standards. Unfortunately, GPC analysis is too often plagued by a lack of appropriate standards that can correlate well with the hydrodynamic volume of rigid polymers. The structurally independent nature of MS overcomes this limitation, permitting analysis of polymers even in the low kDa regime where GPC has limitations. Furthermore, the possible sensitivity, precision, assay time, and dynamic range for GPC are far inferior to those readily achieved by MS. Recently, ESI-MS and MALDI-MS have shown promise as candidates to determine polymer mass distributions due to their capacity to generate further compositional information (e.g., end group and repeat unit identification) from a single measurement. Various data-reduction algorithms may be used to obtain a relative mass (M_r) distribution from an ESI- or

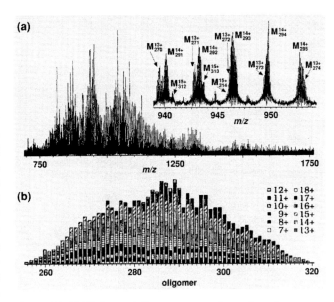

Figure 11.17 High-resolution mass spectrometry of a synthetic polymer: (a) ESI-FTMS spectrum of PEG 14,000 (*inset:* several isotopically resolved oligomers and their charge states); (b) deconvolution of data in (a) show relative oligomer distribution as a function of charge state. (Reproduced with permission from reference 148. Copyright 1995 American Chemical Society.)

Mass Discrimination

To construct the true polymer molecular weight distribution, it is important that every ion be detected with an equal probability. Unfortunately, mass discrimination effects that occur throughout the course of an ESI or MALDI experiment may result in misrepresented molecular weight distributions for polydisperse polymers. For MALDI and LD, mass-dependent desorption/ionization processes tend to underrepresent higher mass components, for a more polydisperse sample, relative to the lower mass components.[228] In ESI, modulation of lens potentials can differentially focus charge state distributions.[149] As a general rule of thumb, with increasing mass, oligomers have a higher propensity to accommodate additional charges. Therefore, depending on the focused charge state distributions, the high mass components can be either over- or underrepresented. Mass discrimination during ion transmission (time-of-flight effect) from the source to the detector has also been observed in ion-trapping instruments like FTMS.[220]

Hyphenated Separation and MS Techniques

Because of the complexity of synthetic polymer samples, mass spectrometry is often coupled to separation processes, either on-line or off-line. Table 11.1 lists some of the separation and MS techniques that have been coupled and employed in synthetic polymer analysis. Hyphens are used in the abbreviations of these techniques, and so as a class they are referred to as "hyphenated techniques". Several of these are described in detail in this section.

Hyphenation of GPC with ESI and MALDI

As discussed by Montaudo et al., MS generally fails to transfer and detect a range of oligomers with a polydispersity greater than 1.20.[228] For ESI, as polymer size increases, so does the number of possible charge states for individual oligomers. Since each charge state of each oligomer has an associated isotopic distribution, potentially thousands of different mass-to-charge values can be convolved to yield a complex (and possibly intractable) mass spectrum.[152] With increasing spectral overlap, deconvolution software becomes unreliable, and manual interpretation is much too tedious and time consuming. Furthermore, the average S/N ratios for peaks in a polydisperse polymer will often be relatively small, limiting analytical sensitivity. A common way to overcome this limitation is to employ a technique like GPC before undertaking mass analysis. The principle on which GPC separation occurs is based on differences in the hydrodynamic volume of different analytes in solution (i.e., size exclusion chromatography; SEC). The combination of GPC with both ESI-MS[159,216,229–231] and MALDI-MS[228,232–236] has been recently reported by several groups.

The solution infusion characteristics common to GPC and ESI suggest the possibility of on-line hyphenation of these techniques. In 1993 Prokai and Simonsick were the first to demonstrate on-line coupling of ESI to GPC.[159] This combination offers a great reduction in spectral complexity by allowing direct mass measurement of each individual oligomer as it elutes from the GPC column. In addition, this dramatically increases sensitivity in FTMS because the ion signal can be "concentrated" into a fewer number of oligomers rather than the complete polydisperse sample. In 1998 Aaserud and Simonsick were the first to demonstrate the power of an on-line coupled GPC-ESI-FTMS system.[216] One of the great strengths of this hyphenated technique over other GPC detection methods (e.g., refractive index, quadrupole ESI-MS) is the high resolution and chemical information FTMS can provide about the polymer of interest. In this report, it was determined that peaks shifted 56 Da higher than the oligomer peaks were due to branched structures that resulted from the reaction of epichlorohydrin (present in the synthesis of the prepolymer) with secondary hydroxyl groups followed by expulsion of HCl (36 Da). Because of the high mass RP of FTMS, it is trivial to determine the difference between isotope peaks and, hence, the charge state for the ion products. A further advantage of GPC-ESI-FTMS stated by the authors is an improved GPC calibration method for obtaining M_n and M_w molecular weights over the conventional PS calibration method.[216] Calibration curves are generated by plotting elution time (volume) versus molecular weight as determined by the FTMS. For quantification purposes, the refractive index detector is still used, although for simple systems the FTMS offers an additional means of quantifying the amount of a given oligomer. Recently, these authors have developed a micro-SEC ESI-MS system with a quadrupole ion trap and an FTMS.[230]

The advantages of using GPC off-line with ESI-FTMS were investigated by Shi et al.[215] to separate high and low molecular weight copolymers of glycidyl methacrylate (GMA) and butyl methacrylate (BMA). In this study, GPC fractions of isobaric GMA/BMA copolymers were analyzed by ESI-FTMS to distinguish between isotopic distributions of isobars having the same nominal mass. Figure 11.18 illustrates ESI-FTMS mass spectra for selected GPC fractions. An RP of 500,000 was achieved for the data in Figure 11.18 and was sufficient to determine the extent of copolymerization for the detected oligomers up to 7 kDa.

Studies that involve GPC coupled to MALDI-MS techniques have traditionally been performed by collecting GPC fractions off-line followed by MALDI analysis.[228,235] For example, Montaudo et al. collected GPC fractions of polydisperse polymers and copolymers and analyzed each by MALDI-TOF.[228,235] The value of M_n as measured by MALDI-TOF were recorded for each fraction and then used to calibrate GPC curves. The calibrated GPC curves were sub-

Table 11.1 Techniques coupled to MS for synthetic polymer analysis

Technique to be coupled	Ionization method[a]	Mass analyzer[b]	Polymer(s) analyzed[c]	References
Pyrolysis	EI	Quad., Sector	Library of 150 polymers	(290)
	EI, CI, FAB	Sector	Lignin polymers	(291)
	EI	Quad.	PVC blends	(292)
	EI, CI	Sector	PS, PVC, PMMA	(53)
	EI	N/A	PS	(293)
	CI	Quad.	PMMA, PET	(294)
	N/A	Quad.	Polythiophene	(295–297)
Gel permeation chromatography (GPC)	MALDI	TOF	PPG	(298)
	MALDI	TOF	PDMS	(233)
	MALDI	TOF	Poly(vinyl acetate)	(299)
	MALDI	TOF	PMMA	(234, 299)
	MALDI	TOF	Copolyester resin	(237)
	ESI	TOF	PMMA, polyester resin	(229)
	ESI	Quad., FTMS	Octylphenoxypoly(ethoxy)ethanol	(230)
	N/A	Sector	Polystyrenesulfonate	(300)
Supercritical fluid chromatography (SFC)	EI	Sector	Polymer additives	(301)
	EI, CI	Sector	Polymer additives	(302)
	EI, CI	Quad.	Polymer additives	(303)
	EI, CI	N/A	Polymer additives (alkyl ketones)	(304)
	APCI	Quad.	Polymer additives	(305)
	APCI	Sector	PEG, PS	(306)
	PD	N/A	propoxylated dimethylpyrazoles	(307)
Capillary electrophoresis (CE)	MALDI	Review article		(308)
	MALDI	TOF	PEG, ethoxylated surfactants	(309)
	ESI	Sector	Polynitriles	(310)

a. EI = electron impact; CI = chemical ionization; FAB = fast atom bombardment; MALDI = matrix-assisted laser-desorption ionization; ESI = electrospray ionization; APCI = atmospheric pressure chemical ionization; PD = plasma desorption

b. Quad = quadrupole; TOF = time-of-flight

c. PVC = poly(vinyl chloride); PS = polystyrene; PMMA = poly(methacrylate); PET = poly(ethylene terephthalate); PPG = poly(propylene glycol); PDMS = poly(dimethylsiloxane); PEG = poly(ethylene glycol)

sequently used for measuring M_n and molecular weight distribution for a number of unfractionated samples.[228,235]

The incompatibilities associated with hyphenation of GPC on-line with MALDI are due to both complex sample preparation schemes (matrix formulation) and the intrinsically pulsed nature of the laser source. However, more recently Fei and Murray demonstrated an on-line GPC-MALDI-TOF system in which effluent from the GPC column was mixed with matrix solution and sprayed into a TOF-MS.[232] Here, ion formation was induced by irradiation of the spray with a Nd:YAG laser. Nielen has developed a novel micro-SEC MALDI-MS in which the MALDI matrix is coaxially added to the SEC column effluent and directly spotted onto the MALDI targets using a robotic interface.[237]

The combination of GPC with ESI or MALDI, when achieved, will accelerate the analysis of large, polydisperse polymers. GPC fractions generally have narrow molecular weight distributions ($D < 1.05$), which makes them more amenable to further MS analysis. In this way, mass spectral results from several GPC fractions can be assembled, resulting in more accurate polymer molecular weight distributions.

Supercritical Fluid Chromatography/ Mass Spectrometry

Supercritical fluid chromatography (SFC) is a hybrid technique that combines the best features from both GC and LC. A supercritical fluid is defined by a phase diagram, which is shown in Figure 11.19, and is used as the mobile phase. Carbon dioxide is the most commonly used mobile phase in SFC. In the supercritical fluid region, there is a continuous transition from gas-like to liquid-like properties—there is no phase change.[238] Due to the liquid-like interactions with the analyte, species with volatilities too low for GC can be eluted in SFC.[239] Separations of relatively nonpolar species can also be performed more quickly in SFC than in LC because of the higher diffusion coefficients in supercritical fluids.[240] Due to the lower viscosities of supercritical fluids, the pressure drop required to produce a mobile-phase flow rate

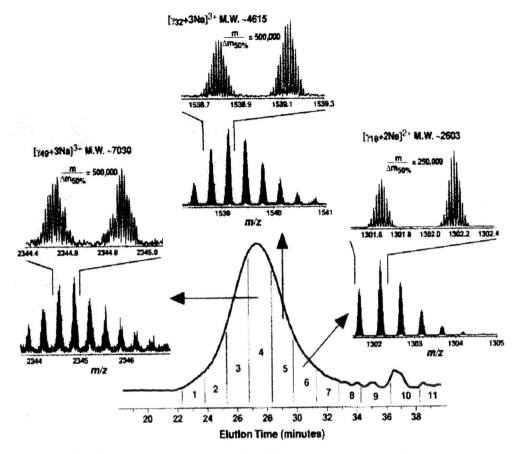

Figure 11.18 Gel permeation chromatographic (GPC) separation of GMA/BMA copolymers. *Insets:* Ultrahigh mass resolving power of GPC fractions 4, 5, and 6. The lower spectrum in each set corresponds to the isotopic peak distribution, while the upper insets represent isobaric distributions of the selected isotopic clusters. (Reproduced with permission from reference 215. Copyright 1998 American Chemical Society.)

is lower, so longer columns can be used in SFC, thus resulting in higher total plate counts than in LC.[241] Other advantages of SFC include an inexpensive and environmentally friendly mobile phase and easy recovery of the sample after analysis.[242]

The properties of supercritical fluids that result in important chromatographic advantages are also important for the transport and gas phase introduction of analyte molecules into a mass spectrometer.[243] The actual interfacing of SFC with a mass spectrometer is not simple. The optimum SFC-MS configuration in terms of interface type, ionization technique, and mass analyzer is desgned around the column type.[244] The choice of whether to use a packed column or a capillary column is very important.[245] Both types of columns have their own inherent advantages and disadvantages.[246] It is easier to couple a capillary column to an ionization source directly.[247] However, packed columns allow for the addition of a cosolvent to the mobile phase to help "tune" the mobile phase and the separation.

Another key component for coupling these two techniques is a restrictor. The purpose of a restrictor is to maintain a high mobile phase density along the entire analytical column. This ensures that the whole column is filled with supercritical mobile phase and that no solute precipitates from the supercritical fluid during the separation. The three most commonly used restrictors are the multipath frit,[248] the robot-pulled tapered restrictor,[249] and the integral restrictor.[250] The choice of restrictor flow rate is another difficult decision to be made: fast-flowing restrictors do not plug as often, but with the higher velocity comes less chromatographic efficiency and a higher gas load introduced into the mass spectrometer.[240,251]

Other techniques that can be used to eliminate the mobile phase before introduction to the ionization source include the moving belt[252] and particle beam[253] interfaces. These techniques are considered indirect coupling techniques and have many inherent difficulties.[243,254] A recent review article that describes the developments in SFC-MS coupling ex-

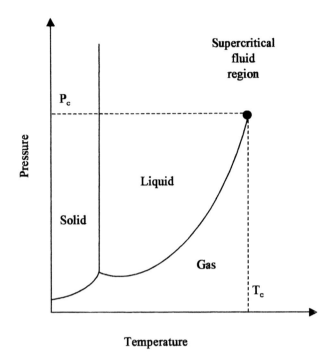

Figure 11.19 A typical phase diagram in supercritical fluid chromatography (SFC). The solid lines indicate phase boundaries. The critical pressure, P_c, and the critical temperature, T_c, define the supercritical fluid region. For carbon dioxide, the P_c is 72.8 atm and the T_c is 31.1 °C.

plains the use of various ionization sources and the means of their coupling.[255] Atmospheric pressure chemical ionization (APCI) has an advantage over other ionization sources in that when it is coupled with capillary SFC, the conditions of the separation have little or no effect on the ionization process.[256] Therefore, column changes and restrictor maintenance is easy since there is no disruption of the vacuum system.

SFC-MS is a very useful technique for the analysis of polymer mixtures. SFC has been coupled to most of the ionization sources and all the mass analyzers described previously. Care must be taken when physically coupling the two techniques to ensure that there is a minimal amount of dead volume. One must also be careful to avoid the pitfalls of the mass spectrometer, the ionization source, the SFC separation, and, most crucially, the problems associated with coupling everything. After these situations are addressed, the inherent advantages of the coupling between SFC and MS can be exploited in the analysis of synthetic polymers.

Capillary Electrophoresis/
Mass Spectrometry

Since the first report in 1987 by Smith et al.,[257] capillary electrophoresis (CE) coupled to mass spectrometry (CE/MS) has gained considerable interest. CE is a technique that in-

volves applying a high potential across a buffer-filled small diameter capillary (typically ≤ 100 μm i.d. and 50–100 cm in length). The separation is based on different migration rates of charged species as they move through the capillary when an electric field is applied. Unlike LC, CE has no longitudinal molecular diffusion or mass transport, which results in sharper peaks and greater sensitivity.[258]

Traditionally, CE detection is preformed by on-line UV absorbance or fluorescence.[257] However, these detection schemes are limited to analytes that absorb in the UV range, fluoresce, or can be labeled with a fluorescing or absorbing species.[257,259,260] However, by usng a mass spectrometer as the detector, a larger range of analytes can be analyzed. In addition, MS provides both structural and molecular weight information, while providing selectivity and specificity. In the last decade, several review articles have been published on CE/MS.[258,261-263]

When coupling CE to MS, there are two main compatible ionization techniques, ESI and FAB.[264] APCI[265] and MALDI[266] are also compatible with MS. When developing a CE/MS instrument, the main problem is the interface between the CE capillary and the ionization source of the MS. Most CE devices involve submersing both ends of the capillary into a conducting buffer to encourage an electroosmotic flow, and this creates a problem when one wants to use an on-line system. However, Smith and coworkers addressed this problem by creating the first CE/ESI/MS interface.[257] This interface used a metal-coated capillary, which made contact to a metal sheath capillary. The sheath capillary acts as both the electrospray source and as the CE cathode so that a second buffer reservoir was not needed. In addition, other types of CE/ESI/MS interfaces such as sheath flow (coaxial), liquid junction, and sheathless have been described in detail.[258,263] Aside from the CE/ESI interface, Jorgenson and coworkers developed the first on-line CE/FAB interface, and this has been discussed in the literature.[267] In addition to the variety of ionization sources that have been used with CE/MS, CE/MS has been performed with a number of mass analyzers. These include quadrupole,[257,268] triple quadrupole,[269,270] FTMS,[271] ion trap,[272] TOF,[273-275] and sector.[276]

Although most CE/MS work has been done on biopolymers, some work has been on synthetic polymers. For example, Barry et al. have shown that PEG can be analyzed with a CE/MALDI/TOF instrument.[277] Furthermore, the future looks promising for the analysis of synthetic polymers with this technique. This can be attributed to the separation efficiency, speed, and potential sensitivity of CE in sequence with the quality of information that can be gained with an MS detector. However, one might wonder why more work has not been done with CE/MS. One factor that makes CE/MS difficult is that CE separations tend to have temporally narrow peaks, and MS usually scans a mass range over a period of time. Thus, if the analyte of interest is not being passed to the detector at the appropriate time, the peak may be missed. Additional problems that may be encountered with CE/MS are limited dynamic range, limited number of buffers, and the low concentrations that can be used with this system. The buffers are limited to ones that are volatile, and the concentrations are typically in the low millimolar range.[278]

Determination of Polymer Composition

While direct determination of molecular weight is an important advantage of MS in synthetic polymer analysis, elucidation of molecular structure is a particularly enchanting feature of MS. From fragmentation patterns and accurate mass measurements, it is possible to determine end-group identities, repeat unit length and sequence, degree of copolymerization, and the presence of possible additives or impurities that might be present. Table 11.2 lists mass spectrometry ionization methods and mass analyzers that have been used to determine the structural features of synthetic polymers. In the last several years, much of the focus has been on using MALDI and ESI in structural analysis of synthetic polymers, so the remainder of this section focuses on applications using these two techniques.

In the last several years, MALDI-TOF has been at the forefront of analyzing synthetic polymers for composition information. The sheer volume of published research applications is too large to cover completely here, so only a few selected applications are discussed. Weidner et al. used MALDI-TOF to examine the hydrolytic degradation products of poly(ethylene terephthalate) and showed products with different end groups based on the observed oligomer masses; this interesting result suggests that MALDI-TOF may play an important role in examining polymer degradation mechanisms.[279] Using post-source decay (PSD) fragmentation analysis on a MALDI-TOF, seven different end groups were inferred for a polycarbonate sample by Przybilla et al.[280] Gooden et al. showed that the cyclization in hyperbranched pentafluorophenyl-terminated poly(benzyl ether)s could be determined by MALDI-TOF;[281] diagnostic of the cyclization was the loss of 20 Da hydrogen fluoride (HF) from the acyclic species, owing to expulsion of HF from the polymer chain ends during the cyclization process. Chaudhury et al. examined biocatalytic polyester synthesis of activated diesters and diols;[282] they noted that competition between transesterification and hydrolysis reactions produced polyesters with both acid and hydroxyl end groups and that the introduction of an acid end group would terminate the elongation of the polyesters. Nagasaki et al. used MALDI-TOF mass measurements to characterize a PEG with methacryloyl and formyl end groups.[283] Weidner and Kuhn also used MALDI-TOF to examine end-group derivatives of PEG.[284] Montaudo et al. also used MALDI-TOF to distinguish between cationic and anionic end groups of PEG[285] and in the characterization of various end groups from nylon 6.[286]

A major strength of using MALDI in concert with FTMS is the high mass accuracy that can be achieved. A graphical method for determining end-group composition and repeat units was developed by deKoster and collaborators for MALDI-FTMS.[212] A series of PEG 1000 and 4000 polymers with different end-group chemistries were used in this study to illustrate their technique. Linear regression analysis of a plot of experimental masses for a homologous series as a function of degree of polymerization (n) reveals the mass of the repeat unit (slope), as well as the total mass for the end groups and adduct cations (y-intercept). In a follow-up investigation, they were able to measure oligomer masses with a mass accuracy of 1 to 5 ppm for synthetic polymer distributions up to m/z 5000, which were generated by external MALDI-FTMS.[214] By averaging over all oligomers present in the ensemble and then extrapolating linearly, the end-group masses were determined to within 10 to 100 ppm. This degree of mass accuracy is frequently more than sufficient for unambiguous end-group determination. External source MALDI-FTMS was also recently employed in the structural characterization of poly(oxyalkylene)amines in the range m/z 500 to 3500.[287] Here, the authors report oligomeric and end-group mass determination to within 20 and 50 mDa, respectively. Again, a linear regression method was used to deduce monomer repeat unit mass and end-group masses. Easterling and coworkers tested external calibration for an internal MALDI-FTMS source using poly(methyl methacrylate) (PMMA) 6000, obtaining a 8.1 mDa mean square error for repeat unit mass and 23.6 mDa error for end-group mass.[223] These same authors also noted that by accounting for space charge effects that reduce overall mass accuracy, they were able to achieve a mean squared error of 1.4 ppm for oligomers in PMMA-6000.[223]

One may exploit the high mass accuracy limit of MALDI-FTMS in the characterization of copolymers, including identification of the composition of repeat units within individual polymers. For example, van Rooij et al.[214] analyzed block length copolymers of poly(oxypropylene) and poly(oxyethylene) and were able to identify the number of each monomer type present in each oligomer. Aaserud and Simonsick also recently described infrared LD-FTMS analysis of a number of copolymers used in automotive coatings with molecular weights below 10 kDa;[216] again the mass accuracy of FTMS helped to identify the end-group composition and number of each type of monomer unit. Other copolymer systems studied by FTMS to date include those of glycidyl methacrylate/butyl methacrylate (GMA/BMA), methyl methacrylate (MMA)/BMA, and hydroxyethyl methacrylate (HEMA)/BMA.[216] Simonsick and Ross showed that the high mass accuracy capability of LD-FTMS with potassium cationization is a valuable technique in characterizing other types of industrial polymers, such as poly(ethylene oxide) methyl ester styrene oligomers and tetrafluoroethylene-ethylene oxide copolymers.[288]

An interesting application of LD-FTMS was shown in the characterization of three forms of poly(2-vinylthiophene) prepared by chemical and electrochemical means.[289] The

Table 11.2 Synthetic polymer classes analyzed by mass spectrometry methods

Method of ionization	Type of mass analyzer[a]	Polymer(s) analyzed[b]	References[c]
Electron impact (EI)	TOF and Quad.	PMMA	(311)
	Sector	PDMS	(312)
Chemical ionization (CI)	Sector	PS, PEG, polysiloxane, polynorbornene	(48)
	Quad.	Polysiloxane	(313)
	Sector	Perfluorinated polyethers	(314)
	N/A	Polyurethanes	(315)
Field desorption (FD)	Quad.	Polyquinones	(316)
	Sector	Polyamides	(65)
	Sector	PET	(317)
	Sector and TOF	Polypropylene additives	(318)
	N/A	Phenylacetylene	(319)
	Sector	Mesitylacetylene	(320)
	N/A	Polyethylene	(321)
Fast atom bombardment (FAB)	Sector	Trifluoro-phenylpropyne	(91)
	Sector	PEG	(16, 322, 323)
	Sector	Polyesters	(324, 325)
	Sector	Aniline and (epoxypropoxy)phenyl propane	(326)
	Sector	Polyamides	(324)
	Sector	PPG	(16, 323)
Secondary ion mass spectrometry (SIMS)	Review article		(327)
	TOF	PEG, PPG	(102)
	TOF	Poly(alkyl methacrylates)	(111)
	TOF	PDMS	(113, 328)
	TOF	PDMS blends	(329)
	TOF	PS	(328–331)
	TOF	Polyesters, polyurethanes	(332)
	TOF	PVC	(333)
	TOF	Perfluorinated polyethers	(328)
	TOF	Polypropylenes	(334)
	TOF	Polybutadienes	(335)
	TOF	PMMA	(328, 333)
^{252}Cf plasma desorption (PD)	TOF	PS	(78, 336)
	TOF	Poly(ether etherketone)	(337)
	TOF	PS/poly(vinyl methyl ether) blends	(336, 338)
	TOF	PET	(78, 81)
	TOF	Polyethers	(76)
	TOF	Polypropylene	(81)
	TOF	PMMA, poly(butyl methacrylate)	(78)
Laser desorption (LD)	Review article		(339)
	FTMS	PS, polyisoprene, polybutadiene, polyethylene	(340)
	TOF	PS	(341)
	TOF	PEG	(20, 341)
	TOF	Poly(ethylene imine), PPG	(20)
	FTMS	Poly(ethylene oxide) methyl ether stryene, tetrafluoroethylene copolymers	(288)
	FTMS	PET, polypropylene, acrylonitrile-butadiene-styrene	(342)
	FTMS	Poly(phenylene sulfide), polyaniline, poly(vinylphenol), polypyrene	(132)
	FTMS	Poly(phenylene)	(130)
	FTMS	Aromatic-conducting polymers	(131)
	TOF	Polythiophene	(343)
	LAMMA	PDMS, PS	(344)
Matrix-assisted laser-desorption ionization (MALDI)	Review article		(345, 346)
	FTMS	Polyisoprene, Polybutadiene	(347)
	TOF	PS	(347–351)
	TOF	PMMA	(351–353)
	TOF	Poly(butyl methacrylate)	(354)
	TOF	Polyglycols, PDMS	(355)
	TOF	Polyesters	(356)
	TOF	Polythiophene	(357)
	N/A	Polythiophene	(358)

Table 11.2 Continued

Method of ionization	Type of mass analyzer[a]	Polymer(s) analyzed[b]	References[c]
Electrospray ionization (ESI)	Review article		(359)
	Quad.	PEG	(149)
	FTMS	PEG	(148)
	TOF	Fluorinated PEG	(141)
	N/A	PMMA	(4)
	Sector	PMMA	(152)
	FTMS	Glycidyl methacrylate/butyl methacrylate copolymers	(215)
	FTMS	Polyamidoamine	(360)
	Quad.	Polyester resins	(150)
	TOF	Polyesters	(361)
	N/A	PS	(153)
	Quad.	Polysulfides	(154)
	FTMS	Dimethylaniline	(362)
	FTMS	PDMS	(157)

a. TOF = time-of-flight; Quad. = quadrupole; FTMS = Fourier transform mass spectrometry; LAMMA = laser-assisted microprobe mass analyzer
b. PMMA = poly(methyl methacrylate); PDMS = poly(dimethylsiloxane); PS = polystyrene; PEG = poly(ethylene glycol); PET = poly(ethylene terphthalate); PPG = poly(propylene glycol); PVC = poly(vinyl chloride)

high resolution of FTMS was critical in drawing conclusions of the reaction mechanisms of oligomer formation for each of the three forms. Since conductive polymers may be prepared by these methods (and such polymers are usually insoluble in their conductive forms), MALDI-FTMS has bright promise for the characterization of conductive polymers.

Tandem mass spectrometry combined with MALDI-FTMS is also useful for structural analysis of synthetic polymers. Pastor and Wilkins coupled MALDI-FTMS with sustained off-resonance irradiation-collisionally activated dissociation (SORI-CAD) to structurally analyze several polar and nonpolar polymers.[224] Single oligomers were isolated within the FTMS trap and subjected to dissociation. One series of product ions generated included the loss of the end groups by chain cleavage, suggesting that this method could be used for end-group identification.[224]

An emerging technique for determining polymer end groups, as well as repeat unit composition analysis, is ESI-FTMS. An advantage of ESI-FTMS for synthetic polymer characterization lies in its ability to detect small "impurities" within the polymer sample. For example, O'Connor and McLafferty showed that PEG containing less than 0.02% propylene oxide monomer units could be detected readily and distinguished from PEG containing only ethylene oxide monomer units due to the high mass accuracy capabilities of ESI-FTMS.[148] In a compelling demonstration, Shi et al. showed that with the high mass accuracy and resolving power advantages of FTMS, copolymers formed by two isobaric monomers could be distinguished.[215] A copolymer consisting of the isobaric monomers GMA and BMA, which differ in mass by 0.036 Da (CH_4 vs. O), was investigated by ESI-FTMS at 9.4 tesla. Lower-resolution mass analyzers would not be able to distinguish between

these monomer units when analyzing a GMA/BMA copolymer for composition. However, ESI-FTMS unequivocally resolved isobaric components of these copolymers up to 7 kDa with a resolving power of approximately 500,000.[215] This enabled identification of all five possible products predicted in the GMA/BMA free-radical polymerization process and confirmed that the GMA monomer was less reactive than its BMA counterpart.

A currently active research area in polymer chemistry centers on synthesis of dendrimers. In contrast to linear polymers, dendrimer construction begins with a polyfunctional "core" group, around which successive layers (generations) of monomer units are joined. Recently, Pasa-Tolic et al. used ESI-FTMS to characterize dendrimers of polyamidoamine (PAMAM).[158] Generations 1 to 5 of the PAMAM dendrimers were analyzed by ESI-FTMS, allowing for characterization of the polydispersities of these dendrimers and of the structure and composition of surface-modified dendrimers. The utility of high resolution mass spectrometry in the characterization of dendrimers should grow substantially as new dendrimer formulations are achieved.

Case Study: Fragment Ions or Silicone Contaminants? An Examination by ESI-FTMS

An advantage of ESI stems from the ability to modulate potentials within the source to induce nozzle-skimmer dissociation (NSD). Such an event can provide useful structural information about the analyte of interest. Maziarz and coworkers demonstrated the utility of NSD to determine the end-group chemistry of a 2.5 kDa nominal mass aminopropyl-terminated poly(dimethyl siloxane) (AP-PDMS) sample.[157] Figure 11.20a,b represents ESI-FTMS mass spectra of a 30 μM AP-PDMS sample recorded before and after NSD

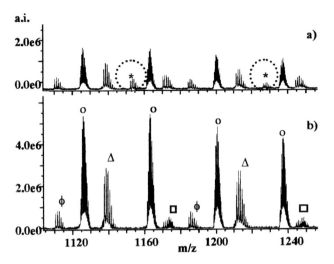

Figure 11.20 Electrospray ionization and Fourier transform mass spectrometry (ESI-FTMS) spectrum of 2500 aminopropyl-terminated poly(dimethylsiloxane) (PDMS): sum of 25 scans (a) under nozzle-skimmer dissociation and (b) in the absence of nozzle-skimmer dissociation conditions. Asterisk (*) indicates new oligomeric distribution resulting from the loss of a propylamine terminal group from intact AP-PDMS oligomers (Δ). (Reproduced with permission from reference 157. Copyright 1999 American Chemical Society.)

conditions, respectively. A new oligomeric distribution, evident in Figure 11.20a, is indicated by an asterisk (*). It was determined from exact mass calculations (7 ppm) that the asterisked distribution resulted from the loss of a propylamine terminal group from intact AP-PDMS oligomers (Δ). Accurate mass measurement, enabled by the high RP of FTMS analysis, yielded less than 10 ppm error from theoretical masses for all oligomeric distributions. In addition, NSD was used, in combination with simple bond energy calculations, to determine the origin of two other oligomeric distributions known not to represent intact AP-PDMS polymer. Of these, the (□) distribution (3.2 ppm) was determined to be a fragment of intact AP-PDMS polymer due to its increase in signal intensity with induced NSD conditions. Correspondingly, the (φ) distribution (0.4 ppm) was determined to be an impurity due to its decrease in signal intensity with induced NSD conditions. It should be noted that the (□) and (φ) distributions are independent of one another as no reasonable fragmentation route can relate them.

Conclusions

By merit of its high sensitivity, broad dynamic range, specificity, selectivity, and information-rich nature, mass spectrometry has become an indispensable analytical tool for the analysis of a multitude of polymeric materials. The development of soft ionization sources that are capable of transferring intact polymers from the solution phase or solid state into the gas phase as ions has dramatically increased the use of MS for synthetic polymer analysis. In addition, advances in mass analyzers have also influenced the use of MS in synthetic polymer characterization. The unlimited mass range of TOF, coupled with its high spectral acquisition rate and its potential for high-resolution analysis, has made TOF a key element in synthetic polymer characterization, both for molecular weight distributions and in obtaining structural information. The ultrahigh resolution capabilities of FTMS make possible accurate mass measurements with less than 10 ppm deviation from theoretical masses on a routine basis. This attribute makes molecular structural determination of polymers a reality and has been used to identify end-group and repeat unit chemistries and copolymer sequence orders. With the ability to obtain such detailed information, the use of FTMS for polymer characterization should grow in the near future.

Acknowledgments We gratefully acknowledge financial support from the following: the Center for Biotechnology (State University of New York at Stony Brook) for an Innovative Technology Grant; the American Society for Mass Spectrometry (through the Exxon Education Foundation) for a Research Award (to T.D.W.); the donors of the Petroleum Research Fund, administered by the American Chemical Society; Bausch & Lomb, Inc.; and the State University of New York at Buffalo. A.R.D. also acknowledges financial support from a graduate fellowship through the Environment and Society Institute at SUNY-Buffalo.

References

1. Carraher, C. E., Jr. In *Polymer Chemistry: An Introduction*, 4th ed.; Marcel Dekker: New York, 1996; pp 81–103.
2. Montaudo, G. *Trends Polym. Sci.* **1996**, *4*, 81–86.
3. Jackson, A. T.; Buzy, A.; Jennings, K. R.; Scrivens, J. H. *Eur. Mass Spectrom.* **1996**, *2*, 115–127.
4. Jackson, C. A.; Simonsick, W. J., Jr. *Curr. Op. Sol. St. Mat. Sci.* **1997**, *2*, 661–667.
5. Saf, R.; Mirtl, C.; Hummel, K. *Acta Polymer* **1997**, *48*, 513–526.
6. Mahon, A.; Kemp, T. J.; Buzy, A.; Jennings, K. R. *Polymer* **1997**, *38*, 2337–2349.
7. Lattimer, R. P.; Harris, R. E. *Mass Spectrom. Rev.* **1985**, *4*, 369–390.
8. Schulten, H. R.; Lattimer, R. P. *Mass Spectrom. Rev.* **1984**, *3*, 231–315.
9. Radhakrishnan, T. S.; Rama Rao, M. *J. Polym. Sci.* **1981**, *19*, 3197–3208.
10. Vincenti, M.; Pelizzetti, E.; Guarini, A.; Costanzi, S. *Anal. Chem.* **1993**, *64*, 1879–1884.
11. Lattimer, R. P.; Harmon, D. J.; Hanson, G. E. *Anal. Chem.* **1980**, *52*, 1808–1811.
12. Matsuo, T.; Matsuda, H.; Katakuse, I. *Anal. Chem.* **1979**, *51*, 1329–1331.

13. Macfarlane, R. D. *Trends Anal. Chem.* **1988**, *7*, 179–183.
14. Loo, J. A.; Wang, B. H.; Wang, F. C.-Y.; McLafferty, F. W.; Klymko, P. *Macromolecules* **1987**, *20*, 698–700.
15. Montaudo, G.; Scamporrino, E.; Puglisi, C.; Vitalini, D. *Macromolecules* **1988**, *21*, 1594–1598.
16. Lattimer, R. P. *Int. J. Mass Spectrom. Ion Processes* **1983/ 1984**, *55*, 221–232.
17. Benninghoven, A.; Sichtermann, W. K. *Anal. Chem.* **1978**, *50*, 1180–1184.
18. Bletsos, I. V.; Hercules, D. M.; Magill, J. H.; vanLeyen, D.; Niehuis, E.; Benninghoven, A. *Anal. Chem.* **1988**, *60*, 938–944.
19. Benninghoven, A. *Angew. Chem. Int. Ed. Engl.* **1994**, *33*, 1023–1043.
20. Cotter, R. J.; Honovich, J. P.; Olthoff, J. K.; Lattimer, R. P. *Macromolecules* **1986**, *19*, 2996–3001.
21. Karas, M.; Hillenkamp, F. *Anal. Chem.* **1988**, *60*, 2299–2301.
22. Beavis, R. C.; Chait, B. T. *Anal. Chem.* **1990**, *62*, 1836–1840.
23. Dole, M.; Mack, L. L.; Hines, R. L.; Mobley, R. C.; Ferguson, L. D.; Alice, M. B. *J. Chem. Phys.* **1968**, *49*, 2240–2249.
24. Yamashita, M.; Fenn, J. M. *J. Phys. Chem.* **1984**, *88*, 4451–4459.
25. Mack, L. L.; Kralik, P.; Rheude, A.; Dole, M. *J. Chem. Phys.* **1970**, *52*, 4977–4986.
26. de Hoffmann, E.; Charette, J. J.; Stroobant, V. In *Mass Spectrometry: Principles and Applications*; Wiley: New York, 1996; pp 9–11.
27. Chapman, J. R. In *Encyclopedia of Anlytical Science*; Townshend, A., Ed.; Academic Press: New York, 1995; Vol. 5, pp 2803–2811.
28. Elliott, R. M. In *Mass Spectrometry*; McDowell, C. A., Ed.; McGraw-Hill Series in Advanced Chemistry; McGraw-Hill: New York, 1963; pp 69–103.
29. Davis, R.; Frearson, M. In *Mass Spectrometry*; Prichard, F. E., Ed.; Analytical Chemistry by Open Learning; Wiley: New York, 1987; pp 43–53.
30. Barker, J. In *Mass Spectrometry*, 2nd ed.; Ando, D. J., Ed.; Analytical Chemistry by Open Learning; Wiley: New York, 1999; pp 19–26.
31. Nier, A. O. *Rev. Sci. Instrumen.* **1940**, *11*, 212–216.
32. Nier, A. O. *Rev. Sci. Instrumen.* **1947**, *18*, 398–411.
33. Munson, M. S. B.; Field, F. H. *J. Am. Chem. Soc.* **1966**, *88*, 2621–2630.
34. Harrison, A. G. In *Chemical Ionization Mass Spectrometry*, 2nd ed.; CRC Press: Boca Raton, FL, 1992; pp 49–82.
35. Munson, B. *Anal. Chem.* **1971**, *43*, 28A–40A.
36. Davis, R.; Frearson, M. In *Analytical Chemistry by Open Learning*; Prichard, F. E., Ed.; Wiley: New York, 1987; pp 54–63.
37. March, R. E.; Hughes, R. J. In *Quadrupole Storage Mass Spectrometry*; Winefordner, J. D., Ed.; Wiley: New York, 1989; pp 214–216.
38. Hunt, D. F.; McEwen, C. N.; Harvey, T. M. *Anal. Chem.* **1975**, *47*, 1730–1734.
39. Schneider, B.; Breuer, M.; Hartmann, H.; Budzikiewicz, H. *Org. Mass Spectrom.* **1989**, *24*, 216–218.
40. Illes, A. J.; Bowers, M. T.; Meisels, G. G. *Anal. Chem.* **1981**, *53*, 1551–1552.
41. Horning, E. C.; Horning, M. G.; Carroll, D. I.; Dzidic, I.; Stillwell, R. N. *Anal. Chem.* **1973**, *45*, 936–943.
42. Carroll, D. I.; Dzidic, I.; Stillwell, R. N.; Horning, M. G.; Horning, E. C. *Anal. Chem.* **1974**, *46*, 706–710.
43. Dzidic, I.; Carroll, D. I.; Stillwell, R. N.; Horning, E. C. *Anal. Chem.* **1976**, *48*, 1763–1768.
44. Horning, E. C.; Carroll, D. I.; Dzidic, I.; Haegele, K. D.; Horning, M. G.; Stillwell, R. N. *J. Chromatogr. Sci.* **1974**, *12*, 725–729.
45. Carroll, D. I.; Dzidic, I.; Stillwell, R. N.; Haegele, K. D.; Horning, E. C. *Anal. Chem.* **1975**, *47*, 2369–2373.
46. Kambara, H. *Anal. Chem.* **1982**, *54*, 143–146.
47. Cotter, R. J. *Anal. Chem.* **1980**, *52*, 1589A–1606A.
48. Vincenti, M.; Pelizzetti, E.; Guarini, A.; Costanzi, S. *Anal. Chem.* **1992**, *64*, 1879–1884.
49. Hansen, G.; Munson, B. *Anal. Chem.* **1978**, *50*, 1130–1134.
50. Cotter, R. J. *Anal. Chem.* **1979**, *51*, 317–320.
51. Hansen, G.; Munson, B. *Anal. Chem.* **1980**, *52*, 245–248.
52. Bruins, A. P. *Anal. Chem.* **1980**, *52*, 605–607.
53. Shimizu, Y.; Munson, B. *J. Polym. Sci. Polym. Chem. Ed.* **1979**, *17*, 1991–2001.
54. Prokai, L. In *Field Desorption Mass Spectrometry*; Brame, E. G., Ed.; Practical Spectroscopy Series; Marcel Dekker: New York, 1990; Vol. 9, pp 1–37.
55. Busch, K. L.; Glish, G. L.; McLuckey, S. In *Mass Spectrometry/Mass Spectrometry: Techniques and Applications of Tandem Mass Spectrometry*; VCH: Deerfield, FL, 1988; p 158.
56. Schulten, H.-R. *Int. J. Mass Spectrom. Ion Phys.* **1979**, *32*, 97–283.
57. Beckey, H.-D. In *Field Ionization Mass Spectrometry*; Pergamon Press: New York, 1971; pp 127–156.
58. Muller, E. W. *Phy. Rev.* **1956**, *102*, 618–624.
59. Gomer, R. J. *J. Chem. Phys.* **1959**, *31*, 341–345.
60. Davis, R.; Frearson, M. In *Mass Spectrometry*; Prichard, F. E., Ed.; Analytical Chemistry by Open Learning; Wiley: New York, 1987; pp 63–73.
61. de Hoffmann, E.; Charette, J. J.; Stroobant, V. In *Mass Spectrometry: Principles and Applications*; Wiley: New York, 1996; p 24.
62. Halliday, D.; Resnick, R.; Krane, K. S. In *Physics*, 4th ed.; Wiley: New York, 1992; pp 657–658.
63. Lattimer, R. P.; Harris, R. E. *Rubber Chem. Technol.* **1985**, *58*, 577–621.
64. Matsuo, T.; Matsuda, H.; Katakuse, I. *Anal. Chem.* **1979**, *51*, 69–72.
65. Schulten, H.-R.; Plage, B. *J. Polym. Sci. Part A Polym. Chem.* **1988**, *26*, 2381–2394.
66. Evans, W. J.; Decoster, D. M.; Greaves, J. *Macromolecules* **1995**, *28*, 7929–7936.
67. Macfarlane, R. D. *Anal. Chem.* **1983**, *55*, 1247A–1264A.
68. Metta, D.; Diamond, H.; Barnes, R. F.; Milsted, J.; Gray, J. J.; Henderson, P. J.; Stevens, C. M. *J. Inorg. Nucl. Chem.* **1965**, *27*, 33–39.
69. Torgerson, D. F.; Skowronski, R. P.; Macfarlane, R. D. *Biochem. Biophys. Res. Commun.* **1974**, *60*, 616–621.

70. Macfarlane, R. D.; Torgerson, D. F. *Science* **1976**, *191*, 920–925.

71. Biersack, J. P.; Fink, D.; Mertens, P. *J. Nucl. Mater.* **1974**, *53*, 194–200.

72. Meot-Ner, M.; Field, F. H. *J. Am. Chem. Soc.* **1973**, *95*, 7207–7211.

73. Macfarlane, R. D.; Torgerson, D. F. *Int. J. Mass Spectrom. Ion Phys.* **1976**, *21*, 81–92.

74. Bruninx, E.; Rudstam, G. *Nucl. Instrum. Methods* **1961**, *13*, 131–140.

75. McNeal, C. J.; Macfarlane, R. D. *Anal. Chem.* **1979**, *51*, 2036–2039.

76. Chait, B. T.; Shpungin, J.; Field, F. H. *Int. J. Mass Spectrom. Ion Processes* **1984**, *58*, 121–137.

77. Chatterjee, A.; Maccabee, H. D.; Tobias, C. A. *Radiat. Res.* **1973**, *54*, 479–494.

78. Quinones, L.; Schweikert, E. A. *J. Vac. Sci. Technol. A* **1988**, *6*, 946–949.

79. Jordan, E. A.; Macfarlane, R. D.; Martin, C. R.; McNeal, C. J. *Int. J. Mass Spectrom. Ion Processes* **1984**, *53*, 345–348.

80. Brown, A.; Vickerman, J. C. *Surf. Interface Anal.* **1986**, *8*, 75–81.

81. Macfarlane, R. D.; McNeal, C. J.; Martin, C. R. *Anal. Chem.* **1986**, *58*, 1091–1097.

82. Morris, H. R.; Dell, A.; Etienne, A. T.; Judkins, M.; McDowell, R. A.; Panico, M.; Taylor, G. W. *Pure Appl. Chem.* **1982**, *54*, 267–279.

83. Taylor, L. C. E. *Ind. Res. Devel.* **1981**, *23*, 124–128.

84. Williams, D. H.; Bradley, C.; Bojesen, G.; Santikarn, S.; Taylor, L. C. E. *J. Am. Chem. Soc.* **1981**, *103*, 5700–5704.

85. de Hoffmann, E.; Charette, J. J.; Stroobant, V. In *Mass Spectrometry: Principles and Applications*; Wiley: New York, 1996; pp 21–24.

86. Barber, M.; Bordoli, R. S.; Sedgwick, R. D.; Tyler, A. N. *J. Chem. Soc. Chem. Commun.* **1981**, *7*, 325–327.

87. Rinehart, K. L. *J. Science* **1982**, *218*, 254–260.

88. Barber, M.; Bordoli, R. S.; Sedgwick, R. D.; Tyler, A. N. *Nature* (London) **1981**, *293*, 270–275.

89. Busch, K. L. In *Encyclopedia of Analytical Science*; Townshend, A., Ed.; Academic Press: New York, 1995; Vol. 5, pp 2845–2850.

90. Godechot, X.; Bernardet, H.; Lejeune, C. *Rev. Sci. Instrum.* **1990**, *61*, 2608–2613.

91. van Breemen, R. B.; Huang, C. H.; Bumgardner, C. L. *Anal. Chem.* **1991**, *63*, 2577–2580.

92. Ito, Y.; Takeuchi, T.; Ishii, D.; Goto, M. *J. Chromatogr.* **1985**, *346*, 161–166.

93. Caprioli, R. M.; Fan, T.; Cottrell, J. S. *Anal. Chem.* **1986**, *58*, 2949–2954.

94. Morris, H. R.; Panico, M.; Barber, M.; Bordoli, R. S.; Sedgwick, R. D.; Tyler, A. *Biochem. Biophys. Res. Commun.* **1981**, *101*, 623–631.

95. Lattimer, R. P.; Schulten, H. R. *Int. J. Mass Spectrom. Ion Processes* **1985**, *67*, 277–284.

96. Surman, D. J.; Vickerman, J. C. *J. Chem. Soc. Chem. Commun.* **1981**, *7*, 324–325.

97. Heinz, O.; Reaves, R. T. *Rev. Sci. Instrum.* **1968**, *39*, 1229–1230.

98. Hues, S. M.; Colton, R. J.; Wyatt, J. R.; Schultz, J. A. *Rev. Sci. Instrum.* **1989**, *60*, 1239–1244.

99. Teodoro, O. M. N. D.; Catarino, M. I. S.; Moutinho, A. M. C. *Appl. Surf. Sci.* **1993**, *70–71*, 291–294.

100. Thompson, P.M. In *Encyclopedia of Analytical Science*; Townshend, A., Ed.; Academic Press: New York, 1995; Vol. 5, pp 2924–2934.

101. Wolstenholme, J.; Walls, J. M. *Res. Devel.* **1986**, *28*, 58–61.

102. Hittle, L. R.; Altland, D. E.; Proctor, A.; Hercules, D. M. *Anal. Chem.* **1994**, *66*, 2302–2312.

103. Hanley, L.; Kornienko, O.; Ada, E. T.; Fuoco, E.; Trevor, J. L. *J. Mass Spectrom.* **1999**, *34*, 705–723.

104. Day, R. J.; Unger, S. E.; Cooks, R. G. *Anal. Chem.* **1980**, *52*, 557A–572A.

105. Wilson, R. G.; Stevie, F. A.; Magee, C. W. In *Secondary Ion Mass Spectrometry: A Practical Handbook for Depth Profiling and Bulk Impurity Analysis*; Wiley: New York, 1989; pp I1–I10.

106. Campana, J. E.; Ross, M. M.; Rose, S. L.; Wyatt, J. R.; Colton, R. J. In *Ion Formation from Organic Solids*; Benninghoven, A., Ed.; Springer Series in Chemical Physics No. 25; Springer-Verlag: New York, 1983; pp 144–155.

107. Benninghoven, A.; Mueller, K. H.; Schemmer, M. *Surf. Sci.* **1978**, *78*, 565–576.

108. Gardella, J. A. Jr.; Hercules, D. M. *Anal. Chem.* **1981**, *53*, 1879–1884.

109. Bletsos, I. V.; Hercules, D. M.; vanLeyen, D.; Benninghoven, A. *Macromolecules* **1987**, *20*, 407–413.

110. Colton, R. J.; Murday, J. S.; Wyatt, J. R.; DeCorpo, J. J. *Surf. Sci.* **1979**, *84*, 235–248.

111. Gardella, J. A. Jr.; Hercules, D. M. *Anal. Chem.* **1980**, *52*, 226–232.

112. Benninghoven, A. *Surf. Sci.* **1975**, *53*, 596–625.

113. Dong, X.; Proctor, A.; Hercules, D. M. *Macromolecules* **1997**, *30*, 63–70.

114. Davis, R.; Frearson, M. In *Handbook of Static Secondary Ion Mass Spectrometry*; Briggs, D., Brown, A., and Vickerman, J. C., Eds.; Wiley: New York, 1989; p 156.

115. Henderson, A. SurfaceSpectra: Home of the Static SIMS Library. http://www.surfacespectra.com.

116. Van Ooij, W. J.; Brinkhuis, R. H. G. In *Secondary Ion Mass Spectrometry, SIMS VI*; Benninghoven, A., Huber, A. M., and Werner, H. W., Eds.; Wiley: New York, 1988; pp 635–638.

117. Briggs, D.; Brown, A.; Van den Berg, J. A.; Vickerman, J. C. In *Ion Formation from Organic Solids*; Benninghoven, A., Ed.; Springer Series in Chemical Physics No. 25; Springer-Verlag: New York, 1983; pp 162–166.

118. Briggs, D.; Hearn, M. J. *Vacuum* **1986**, *36*, 1005–1010.

119. Briggs, D. In *Ion Formation from Organic Solids*; Benninghoven, A., Ed.; Springer Series in Chemical Physics No. 25; Springer-Verlag: New York, 1983; pp 156–161.

120. Nuwaysir, L. M.; Wilkins, C. L. In *Lasers and Mass Spectrometry*; Lubman, D. M., Ed.; Oxford University Press: New York, 1990; pp 291–315.

121. Posthumus, M. A.; Kistemaker, P. G.; Meuzelaar, H. L. C. *Anal. Chem.* **1978**, *50*, 985–991.

122. van der Peyl, G. J. Q.; Isa, K.; Haverkamp, J.; Kistemaker, P. G. *Int. J. Mass Spectrom. Ion Phys.* **1983**, *47*, 11–14.

123. Lindner, B.; Seydel, U. *Anal. Chem.* **1985**, *57*, 895–899.

124. Karas, M.; Bachmann, D.; Hillenkamp, F. *Anal. Chem.* **1985**, *57*, 2935–2939.

125. Hercules, D. M.; Day, R. J.; Balasanmugam, K.; Dang, T. A.; Li, L. P. *Anal. Chem.* **1982**, *54*, 280A–305A.

126. Denoyer, E.; Van Grieken, R.; Adams, F.; Natusch, D. F. S. *Anal. Chem.* **1982**, *54*, 26A–41A.

127. Fenner, N. C.; Daly, N. R. *Rev. Sci. Instrum.* **1966**, *37*, 1068–1070.

128. Hillenkamp, F.; Kauffmann, R.; Nitsche, R.; Unsold, E. *Appl. Phys.* **1975**, *8*, 341–348.

129. Liang, Z.; Marshall, A. G.; Westmoreland, D. G. *Anal. Chem.* **1991**, *63*, 815–818.

130. Brown, C. E.; Kovacic, P.; Wilkie, C. A.; Cody, R. B. J.; Kinsinger, J. A. *J. Polym. Sci.: Polym. Lett. Ed.* **1985**, *23*, 453–463.

131. Brown, C. E.; Kovacic, P.; Wilkie, C. A.; Cody, R. B. J.; Hein, R. E.; Kinsinger, J. A. *Syn. Met.* **1986**, *15*, 265–297.

132. Brown, C. E.; Kovacic, P.; Welch, K. J.; Cody, R. B.; Hein, R. E.; Kinsinger, J. A. *J. Polym. Sci.: Part A: Polym. Chem.* **1988**, *26*, 131–148.

133. Scrivens, J. H. In *Encyclopedia of Analytical Science*; Townshend, A., Ed.; Academic Press: New York, 1995; Vol. 5, pp 3022–3027.

134. Hillenkamp, F.; Karas, M.; Beavis, R. C.; Chait, B. T. *Anal. Chem.* **1991**, *63*, 1193A–1203A.

135. Wu, K. J.; Odom, R. W. *Anal. Chem.* **1998**, *70*, 456A–461A.

136. Karas, M.; Gluckmann, M.; Schafer, J. *J. Mass Spectrom.* **2000**, *35*, 1–12.

137. Danis, P. O.; Karr, D. E. *Macromolecules* **1995**, *28*, 8548–8551.

138. Schriemer, D. C.; Li, L. *Anal. Chem.* **1996**, *68*, 2721–2725.

139. Castoro, J. A.; Köster, C.; Wilkins, C. *Rapid Commun. Mass Spectrom.* **1992**, *6*, 239–241.

140. Busch, K. L. *Spectroscopy* **1999**, *14*, 14–16.

141. Latourte, L.; Blais, J. C.; Tabet, J. C.; Cole, R. B. *Anal. Chem.* **1997**, *69*, 2742–2750.

142. Dey, M.; Castoro, J. A.; Wilkins, C. L. *Anal. Chem.* **1995**, *67*, 1575–1579.

143. Nohmi, T.; Fenn, J. B. *J. Am. Chem. Soc.* **1992**, *114*, 3241–3246.

144. Fernandez de la Mora, J.; Van Berkel, G. J.; Enke, C. G.; Cole, R. B.; Martinez-Sanchez, M.; Fenn, J. B. *J. Mass Spectrom.* **2000**, *35*, 939–952.

145. Cole, R. B. *J. Mass Spectrom.* **2000**, *35*, 763–772.

146. Kebarle, P.; Tang, L. *Anal. Chem.* **1993**, *65*, 972A–986A.

147. Iribarne, J. V.; Thomson, B. A. *J. Chem. Phys.* **1976**, *64*, 2287–2294.

148. O'Connor, P. B.; McLafferty, F. W. *J. Am. Chem. Soc.* **1995**, *117*, 12826–12831.

149. Hunt, S. M.; Sheil, M. M.; Belov, M.; Derrick, P. J. *Anal. Chem.* **1998**, *70*, 1812–1822.

150. Hunt, S. M.; Binns, M. R.; Sheil, M. M. *J. Appl. Polym. Sci.* **1995**, *56*, 1589–1597.

151. Nielen, M. W. F. *Rapid Commun. Mass Spectrom.* **1996**, *10*, 1652–1660.

152. McEwen, C. N.; Simonsick, W. J., Jr.; Larsen, B. S.; Ute, K.; Hatada, K. *J. Am. Soc. Mass Spectrom.* **1995**, *6*, 906–911.

153. Festag, R.; Alexandratos, S. D.; Joy, D. C.; Wunderlich, B.; Annis, B.; Cook, K. D. *J. Am. Soc. Mass Spectrom.* **1998**, *9*, 299–304.

154. Mahon, A.; Kemp, T. J.; Buzy, A.; Jennings, K. R. *Polymer* **1996**, *37*, 531–535.

155. Baker, T. R.; Pinkston, J. D. *J. Am. Soc. Mass Spectrom.* **1998**, *9*, 498–509.

156. Yan, W.; Ammon, D. M., Jr.; Gardella, J. A., Jr.; Maziarz, E. P., III; Hawkridge, A. M.; Grobe, G. L., III; Wood, T. D. *Eur. Mass Spectrom.* **1998**, *4*, 467–474.

157. Maziarz, E. P., III; Baker, G. A.; Wood, T. D. *Macromolecules* **1999**, *32*, 4411–4418.

158. Pasa-Tolic, L.; Anderson, G. A.; Smith, R. D.; Brothers, H. M.; Spindler, R.; Tomalia, D. A. *Int. J. Mass Spectrom. Ion Processes* **1997**, *165*, 405–418.

159. Prokai, L.; Simonsick, W. J., Jr. *Rapid Commun. Mass Spectrom.* **1993**, *7*, 853–856.

160. Fenn, J. B.; Mann, M.; Meng, C. K.; Wong, S. F.; Whitehouse, C. M. *Science* **1989**, *246*, 64–71.

161. Davis, R.; Frearson, M. In *Mass Spectrometry*; Prichard, F. E., Ed.; Analytical Chemistry by Open Learning; Wiley: New York, 1987; pp 24–37, 91–97.

162. Rose, M. E.; Johnstone, R. A. W. In *Mass Spectrometry for Chemists and Biochemists*; Cambridge University Press: Cambridge, 1982; pp 28–30.

163. de Hoffmann, E.; Charette, J. J.; Stroobant, V. In *Mass Spectrometry: Principles and Applications*; Wiley: New York, 1996; pp 66–73.

164. Farmer, J. B. In *Mass Spectrometry*; McDowell, C. A., Ed.; McGraw-Hill Series in Advanced Chemistry; McGraw-Hill: New York, 1963; pp 7–44.

165. Dempster, A. J. *Phys. Rev.* **1918**, *11*, 316–325.

166. Bleakney, W. *Phys. Rev.* **1932**, *40*, 496–501.

167. Tate, J. T.; Smith, P. T.; Vaughan, A. L. *Phys. Rev.* **1935**, *48*, 525–531.

168. Nier, A. O. *Phys. Rev.* **1936**, *50*, 1041–1045.

169. Graham, R. L.; Harkness, A. L.; Thode, H. G. *J. Sci. Instrum.* **1947**, *24*, 119–128.

170. Mattauch, J. *Phys. Rev.* **1936**, *50*, 617–623.

171. Brainbridge, K. T.; Jordan, E. B. *Phys. Rev.* **1936**, *50*, 282–296.

172. Cotter, R. J. *Anal. Chem.* **1999**, *71*, 445A–451A.

173. Chernushevich, I. V.; Ens, W.; Standing, K. G. *Anal. Chem.* **1999**, *71*, 452A–461A.

174. Qian, M. G.; Lubman, D. M. *Anal. Chem.* **1995**, *67*, 234A–242A.

175. Cotter, R. J. *Anal. Chem.* **1992**, *64*, 1027A–1039A.

176. Stephens, W. E. *Phys. Rev.* **1946**, *69*, 691.

177. Cameron, A. D.; Eggers, D. F. *J. Rev. Sci. Instrum.* **1948**, *19*, 605–607.

178. Wiley, W. E.; McLaren, I. H. *Rev. Sci. Instrum.* **1955**, *26*, 1150–1157.

179. de Hoffmann, E.; Charette, J. J.; Stroobant, V. In *Mass Spectrometry: Principles and Applications*; Wiley: New York, 1996; pp 59–66.

180. Barker, J. In *Mass Spectrometry*; 2nd ed.; Ando, D. J., Ed.; Analytical Chemistry by Open Learning; Wiley: New York, 1999; pp 82–84.

181. Mamyrin, B. A.; Karataev, V. I.; Smikk, D. V.; Zagulin, V. A. *Soviet Phys. JETP* **1973**, *37*, 45–48.

182. Steadman, J.; Syage, J. A. *Rev. Sci. Instrum.* **1993**, *64*, 3094–3103.

183. Miller, P. E.; Denton, M. B. *J. Chem. Ed.* **1986**, *63*, 617–622.

184. Farmer, J. B. In *Mass Spectrometry*; McDowell, C. A., Ed.; McGraw-Hill Series in Advanced Chemistry; McGraw-Hill: New York, 1963; pp 34–38.

185. Marmet, P. *J. Vac. Sci. Technol.* **1971**, *8*, 262.

186. Dawson, P. H. *Int. J. Mass Spectrom. Ion Phys.* **1974**, *14*, 317–337.

187. Dawson, P. H.; Whetten, N. R. *J. Vac. Sci. Technol.* **1969**, *6*, 97–99.

188. Dawson, P. H.; Whetten, N. R. *Int. J. Mass Spectrom. Ion Phys.* **1963**, *3*, 1–12.

189. Steel, C.; Henchman, M. *J. Chem. Ed.* **1998**, *75*, 1049–1054.

190. Dawson, P. H. *Int. J. Mass Spectrom. Ion Phys.* **1975**, *17*, 423–445.

191. March, R. E.; Hughes, R. J. In *Quadrupole Storage Mass Spectrometry*; Winefordner, J. D., Ed.; Wiley: New York, 1989; pp 31–52.

192. Ellefson, R. E. In *Foundations of Vacuum Sciences and Technology*; Lafferty, J. M., Ed.; Wiley: New York, 1998; pp 456–460.

193. Campana, J. E. *Int. J. Mass Spectrom. Ion Phys.* **1980**, *33*, 101–117.

194. de Hoffmann, E.; Charette, J. J.; Stroobant, V. In *Mass Spectrometry: Principles and Applications*; Wiley: New York, 1996; pp 41–51.

195. Dawson, P. H.; Bingqi, Y. *Int. J. Mass Spectrom. Ion Processes* **1984**, *56*, 25–39.

196. Dawson, P. H.; Bingqi, Y. *Int. J. Mass Spectrom. Ion Processes* **1984**, *56*, 41–50.

197. Grimm, C. C.; Clawson, R.; Short, R. T. *J. Am. Soc. Mass Spectrom.* **1997**, *8*, 539–544.

198. Yost, R. A.; Enke, C. G.; McGilvery, D. C.; Smith, D.; Morrison, J. D. *Int. J. Mass Spectrom. Ion Phys.* **1979**, *30*, 127–136.

199. Thomson, B. A. *Can. J. Chem.* **1998**, *76*, 499–505.

200. Rose, M. E.; Johnstone, R. A. W. In *Mass Spectrometry for Chemists and Biochemists*; Cambridge University Press: Cambridge, 1982; pp 32–34.

201. von Zahn, U. *Rev. Sci. Instrum.* **1963**, *34*, 1–4.

202. Dawson, P. H.; Whetten, N. R. *Rev. Sci. Instrum.* **1968**, *39*, 1417–1422.

203. Dawson, P. H. *J. Vac. Sci. Technol.* **1971**, *8*, 263–265.

204. Reinsfelder, R. E.; Denton, M. B. *Int. J. Mass Spctrom. Ion Phys.* **1981**, *37*, 241–250.

205. Davis, R.; Frearson, M. In *Analytical Chemistry by Open Learning*; Prichard, F. E., Ed.; Wiley: New York, 1987; pp 98–102.

206. Morris, H. R.; Paxton, T.; Dell, A.; Langhorne, J.; Berg, M.; Bordoli, R. S.; Hoyes, J.; Bateman, R. H. *Rapid Commun. Mass Spectrom.* **1996**, *10*, 889–896.

207. Cooks, R. G.; Kaiser, R. E., Jr. *Acc. Chem. Res.* **1990**, *23*, 213–219.

208. March, R. E. *Int. J. Mass Spectrom. Ion Processes* **1992**, *118/119*, 71–135.

209. Comisarow, M. B.; Marshall, A. G. *Chem. Phys. Lett.* **1974**, *25*, 282–283.

210. McLafferty, F. W. *Acc. Chem. Res.* **1994**, *27*, 379–386.

211. Winger, B. E.; Hofstadler, S. A.; Bruce, J. E.; Udseth, H. R.; Smith, R. D. *J. Am. Soc. Mass Spectrom.* **1993**, *4*, 566–577.

212. de Koster, C. G.; Duursma, M. C.; van Rooij, G. J.; Heeren, R. M. A.; Boon, J. J. *Rapid Commun. Mass Spectrom.* **1995**, *9*, 957–962.

213. van Rooij, G. J.; Duursma, M. C.; Heeren, R. M. A.; Boon, J. J.; de Koster, C. G. *J. Am. Soc. Mass Spectrom.* **1996**, *7*, 449–457.

214. van Rooij, G. J.; Duursma, M. C.; de Koster, C. G.; Heeren, R. M. A.; Boon, J. J.; Schuyl, P. J. W.; van der Hage, E. R. E. *Anal. Chem.* **1998**, *70*, 843–850.

215. Shi, S. D.-H.; Hendrickson, C. L.; Marshall, A. G.; Simonsick, W. J., Jr.; Aaserud, D. J. *Anal. Chem.* **1998**, *70*, 3220–3226.

216. Aaserud, D. J.; Simonsick, W. J., Jr. *Prog. Org. Coatings* **1998**, *34*, 206–213.

217. Asamoto, B.; Young, J. R.; Citerin, R. J. *Anal. Chem.* **1990**, *62*, 61–70.

218. Nuwaysir, L. M.; Wilkins, C. L. *Anal. Chem.* **1988**, *60*, 279–282.

219. Hogan, J. D.; Laude, D. A., Jr. *Anal. Chem.* **1992**, *64*, 763–769.

220. O'Connor, P. B.; Duursma, M. C.; van Rooij, G. J.; Heeren, R. M. A.; Boon, J. J. *Anal. Chem.* **1997**, *69*, 2751–2755.

221. Pastor, S. J.; Wood, S. H.; Wilkins, C. L. *J. Mass Spectrom.* **1998**, *33*, 473–479.

222. Pastor, S. J.; Castoro, J. A.; Wilkins, C. L. *Anal. Chem.* **1995**, *67*, 379–384.

223. Easterling, M. L.; Mize, T. L.; Amster, I. J. *Int. J. Mass Spectrom. Ion Processes* **1997**, *169/170*, 387–400.

224. Pastor, S. J.; Wilkins, C. L. *Int. J. Mass Spectrom.* **1998**, *175*, 81–92.

225. Gard, E. E.; Green, M. K.; Warren, H.; Camara, E. J. O.; He, F.; Penn, S. G.; Lebrilla, C. B. *Int. J. Mass Spectrom. Ion Processes* **1996**, *158*, 115–127.

226. Simonsick, W. J., Jr.; Aaserud, D. J.; Grady, M. C.; Prokai, L. *Polym. Prepr. (Am. Chem. Soc., Div. Polym. Chem.)* **1997**, *38*, 483–484.

227. Maziarz, E. P., III; Baker, G. A.; Lorenz, S. A.; Wood, T. D. *J. Am. Soc. Mass Spectrom.* **1999**, *10*, 1298–1304.

228. Montaudo, G.; Garozzo, D.; Montaudo, M. S.; Puglisi, C.; Samperi, F. *Macromolecules* **1995**, *28*, 7983–7989.

229. Nielen, M. W. F.; Buijtenhuijs, F. A. *Anal. Chem.* **1999**, *71*, 1809–1814.

230. Prokai, L.; Aaserud, D. J.; Simonsick, W. J., Jr. *J. Chromatogr. A* **1999**, *835*, 121–126.

231. Prokai, L.; Kim, H. S.; Zharikova, A.; Roboz, J.; Ma, L.; Deng, L.; Simonsick, W. J., Jr. *J. Chromatogr. A* **1998**, *800*, 59–68.

232. Fei, X.; Murray, K. K. *Anal. Chem.* **1996**, *68*, 3555–3560.

233. Montaudo, G.; Montaudo, M. S.; Puglisi, C.; Samperi, F. *Rapid Comm. Mass Spectrom.* **1995**, *9*, 1158–1163.

234. Montaudo, G.; Montaudo, M. S.; Puglisi, C.; Samperi, F. *Int. J. Polym. Anal. Charact.* **1997**, *3*, 177–192.

235. Montaudo, M. S.; Puglisi, C.; Samperi, F.; Montaudo, G. *Rapid Commun. Mass Spectrom.* **1998**, *12*, 519–528.

236. Kassis, C. E.; DeSimone, J. M.; Linton, R. W.; Remsen, E. E.; Lange, G. W.; Friedman, R. M. *Rapid Commun. Mass Spectrom.* **1997**, *11*, 1134–1138.

237. Nielen, M. W. F. *Anal. Chem.* **1998**, *70*, 1563–1568.

238. Bartle, K. D. In *Encyclopedia of Analytical Science*; Townshend, A., Ed.; Academic Press: New York, 1995; Vol. 5, pp 4849–4856.

239. Chester, T. L.; Pinkston, J. D.; Owens, G. D. *Carbohydr. Res.* **1989**, *194*, 273–279.

240. Pinkston, J. D.; Chester, T. L. *Anal. Chem.* **1995**, *67*, 650A–656A.

241. Berger, T. A.; Wilson, W. H. *Anal. Chem.* **1993**, *65*, 1451–1455.

242. Sheeley, D. M.; Reinhold, V. N. *J. Chromatogr.* **1989**, *474*, 83–96.

243. Wright, B. W.; Kalinoski, H. T.; Udseth, H. R.; Smith, R. D. *J. High Resolut. Chromatogr. Chromatogr. Commun.* **1986**, *9*, 145–153.

244. Anton, K. In *Encyclopedia of Analytical Science*; Townshend, A., Ed.; Academic Press: New York, 1995; Vol. 8, pp 4856–4862.

245. Arpino, P. *Fres. J. Anal. Chem.* **1990**, *337*, 667–685.

246. Berry, A. J.; Games, D. E.; Mylchreest, I. C.; Perkins, J. R.; Pleasance, S. *J. High Resolut. Chromatogr. Chromatogr. Commun.* **1988**, *11*, 61–64.

247. Smith, R. D.; Udseth, H. R. *Anal. Chem.* **1987**, *59*, 13–22.

248. Markides, K. E.; Fields, S. M.; Lee, M. L. *J. Chromatogr. Sci.* **1986**, *24*, 254–257.

249. Chester, T. L.; Innis, D. P.; Owens, G. D. *Anal. Chem.* **1985**, *57*, 2243–2247.

250. Guthrie, E. J.; Schwartz, H. E. *J. Chromatogr. Sci.* **1986**, *24*, 236–240.

251. Pinkston, J. D.; Delaney, T. E.; Morand, K. L.; Cooks, R. G. *Anal. Chem.* **1992**, *64*, 1571–1577.

252. Berry, A. J.; Games, D. E.; Perkins, J. R. *J. Chromatogr.* **1986**, *363*, 147–158.

253. Edlund, P. O.; Henion, J. D. *J. Chromatogr. Sci.* **1989**, *27*, 274–282.

254. Ventura, M. C.; Farrell, W. P.; Aurigemma, C. M.; Greig, M. J. *Anal. Chem.* **1999**, *71*, 2410–2416.

255. Arpino, P. J.; Haas, P. *J. Chromatogr. A* **1995**, *703*, 479–488.

256. Tyrefors, L. N.; Moulder, R. X.; Markides, K. E. *Anal. Chem.* **1993**, *65*, 2835–2840.

257. Olivares, J. A.; Nguyen, N. T.; Yonker, C. R.; Smith, R. D. *Anal. Chem.* **1987**, *59*, 1230–1232.

258. Ding, J.; Vouros, P. *Anal. Chem.* **1999**, *71*, 378A–385A.

259. Mikkers, F. E. P.; Everaertes, F. M.; Verheggen, T. P. E. M. *J. Chromatogr.* **1979**, *169*, 11–20.

260. Jorgenson, J. W.; Lukacs, K. D. *Science* **1983**, *222*, 266–272.

261. Cai, J.; Henion, J. *J. Chromatogr. A* **1995**, *703*, 667–692.

262. Severs, J. C.; Smith, R. D. In *Electrospray Ionization Mass Spectrometry: Fundamentals, Instrumentation, and Applications*; Cole, R. B., Ed.; Wiley: New York, 1997; pp 343–382.

263. Smith, R. D.; Wahl, J. H.; Goodlett, D. R.; Hofstadler, S. A. *Anal. Chem.* **1993**, *65*, 574A–584A.

264. Wienberger, R. In *Practical Capillary Electrophoresis*; Academic Press: New York, 1993; pp 246–256.

265. Takada, Y.; Sakairi, M.; Koizumi, H. *Anal. Chem.* **1995**, *67*, 1474–1476.

266. Preisler, J.; Foret, F.; Karger, B. L. *Anal. Chem.* **1998**, *70*, 5278–5287.

267. Moseley, M. A.; Deterding, L. J.; Tomer, K. B.; Jorgenson, J. W. *J. Chromatogr.* **1989**, *480*, 197–209.

268. Smith, R. D.; Olivares, J. A.; Nguyen, N. T.; Udseth, H. R. *Anal. Chem.* **1988**, *60*, 436–441.

269. Figeys, D.; van Oostveen, I.; Ducret, A.; Aebersold, R. *Anal. Chem.* **1996**, *68*, 1822–1828.

270. Kelly, J. F.; Ramaley, L.; Thibault, P. *Anal. Chem.* **1997**, *69*, 51–60.

271. Hofstadler, S. A.; Wahl, J. H.; Bruce, J. E.; Smith, R. D. *J. Am. Chem. Soc.* **1993**, *115*, 6983–6984.

272. Sheppard, R. L.; Henion, J. *Anal. Chem.* **1997**, *69*, 2901–2907.

273. Wu, J. T.; Qian, M. K.; Li, M. X.; Zheng, K.; Haung, P.; Lubman, D. M. *J. Chromatogr. A* **1998**, *794*, 337–389.

274. Deforce, D. L.; Raymackers, J.; Meheus, L.; Wijnendaele, F. V.; De Leenheer, A.; Van den Eeckhout, E. G. *Anal. Chem.* **1998**, *70*, 3060–3068.

275. McComb, M. E.; Krutchinsky, A. N.; Ens, W.; Standing, K. G. *J. Chromatogr.* **1988**, *800*, 1–11.

276. Perkins, J. R.; Tomer, K. B. *Anal. Chem.* **1994**, *66*, 2835–2840.

277. Barry, J. P.; Radtke, D. R.; Carton, W. J.; Anselmo, R. T.; Evans, J. V. *J. Chromatogr. A* **1988**, *800*, 13–19.

278. Tomer, K. B.; Deterding, L. J.; Parker, C. E. In *Advances in Chromatography*; Brown, P. R., Grushka, E., Eds.; Marcel Dekker: New York, 1995; Vol. 35, pp 53–99.

279. Weidner, S.; Kuhn, G.; Friedrich, J.; Schroder, H. *Rapid Commun. Mass Spectrom.* **1996**, *10*, 40–46.

280. Przybilla, L.; Rader, H. J.; Mullen, K. *Eur. Mass Spectrom.* **1999**, *5*, 133–143.

281. Gooden, J. K.; Gross, M. L.; Mueller, A.; Stefanescu, A. D.; Wooley, K. L. *J. Am. Chem. Soc.* **1998**, *120*, 10180–10186.

282. Chaudhury, A. K.; Beckman, E. J.; Russell, A. J. *Biotechnol. Bioeng.* **1997**, *55*, 227–239.

283. Nagasaki, Y.; Ogawa, R.; Yamamoto, S.; Kato, M.; Kataoka, K. *Macromolecules* **1997**, *30*, 6489–6493.

284. Weidner, S.; Kuhn, G. *Rapid Commun. Mass Spectrom.* **1996**, *10*, 942–946.

285. Montaudo, G.; Montaudo, M. S.; Puglisi, C.; Samperi, F. *Macromolecules* **1995**, *28*, 4562–4569.

286. Montaudo, G.; Montaudo, M. S.; Puglisi, C.; Samperi, F. *J. Polym. Sci. A-Polym. Chem.* **1996**, *34*, 439–447.

287. van der Hage, E. R. E.; Durrsma, M. C.; Heeren, R. M. A.; Boon, J. J.; Nielen, M. W. F.; Weber, A. J. M.; de Koster, C. G.; de Vries, N. K. *Macromolecules* **1997**, *30*, 4302–4309.

288. Simonsick, W. J., Jr.; Ross, C. W., 3rd *Int. J. Mass Spectrom. Ion Processes* **1996**, *158*, 379–390.

289. O'Malley, R. M.; Randazzo, M. E.; Winzierl, J. E.; Fernandez, J. E.; Nuwaysir, L. M.; Castoro, J. A.; Wilkins, C. L. *Macromolecules* **1994**, *27*, 5107–5113.

290. Qian, K. N.; Killinger, W. E.; Casey, M.; Nicol, G. R. *Anal. Chem.* **1996**, *68*, 1019–1027.

291. van der Hage, E. R. E.; Boon, J. J. *J. Chromatogr. A* **1996**, *736*, 61–75.

292. Sato, H.; Tsuge, S.; Ohtani, H.; Aoi, K.; Takasu, A.; Okada, M. *Macromolecules* **1997**, *30*, 4030–4037.

293. Fabbri, D.; Trombini, C.; Vassura, I. *J. Chromatogr. Sci.* **1998**, *36*, 600–604.

294. Adams, R. E. *Anal. Chem.* **1983**, *55*, 414–416.

295. Yigit, S.; Hacaloglu, J.; Akbulut, U.; Toppare, L. *Syn. Met.* **1997**, *84*, 205–206.

296. Hacaloglu, J.; Yigit, S.; Akbulut, U.; Toppare, L. *Polymer* **1997**, *38*, 5119–5124.

297. Vatansever, F.; Akbulut, U.; Toppare, L.; Hacaloglu, J. *Polymer* **1996**, *37*, 1103–1107.

298. Yun, H.; Olesik, S. V.; Marti, E. H. *J. Microcol. Sep.* **1999**, *11*, 53–61.

299. Danis, P. O.; Saucy, D. A.; Huby, F. J. *Polym. Prepr.* **1996**, *37*, 311–312.

300. van der Hage, E. R. E.; van Loon, W. M. G. M.; Boon, J. J.; Lingeman, H.; Brinkman, U. A. T. *J. Chromatogr.* **1993**, *634*, 263–271.

301. Bucherl, T.; Gruner, A.; Palibroda, N. *Packaging Technol. Sci.* **1994**, *7*, 139–148.

302. Bucherl, T.; Eschler, M.; Gruner, A.; Palibroda, N.; Wolff, E. *J. High Resolut. Chromatogr. HRC* **1994**, *17*, 765–769.

303. van Leuken, R.; Mertens, M.; Janssen, H. G.; Sandra, P.; Kwakkenbos, G.; Deelder, R. *J. High Resolut. Chromatogr. HRC* **1994**, *17*, 573–576.

304. Calvey, E. M.; Begley, T. H.; Roach, J. A. G. *J. Chromatogr. Sci.* **1995**, *33*, 61–65.

305. Carrott, M. J.; Jones, D. C.; Davidson, G. *Analyst* **1998**, *123*, 1827–1833.

306. Matsumoto, K.; Nagata, S.; Hattori, H.; Tsuge, S. *J. Chromatogr.* **1992**, *605*, 87–94.

307. Pinkston, J. D.; Hentschel, R. T.; Lacey, M. P.; Keough, T. *Fres. J. Anal. Chem.* **1992**, *344*, 447–452.

308. Murray, K. K. *Mass Spectrom. Rev.* **1997**, *16*, 283–299.

309. Barry, J. P.; Radtke, D. R.; Carton, W. J.; Anselmo, R. T.; Evans, J. V. *J. Chromatogr. A* **1998**, *800*, 13–19.

310. Stockigt, D.; Lohmer, G.; Belder, D. *Rapid Commun. Mass Spectrom.* **1996**, *10*, 521–526.

311. Krajnovich, D. J. *J. Phys. Chem. A* **1997**, *101*, 2033–2039.

312. Dagger, A. C.; Semlyen, J. A. *Polymer* **1998**, *39*, 2621–2627.

313. Ranasinghe, A.; Lu, L.; Majumdar, T. K.; Cooks, R. G.; Fife, W. K.; Rubinsztajn, S.; Zeldin, M. *Talanta* **1993**, *40*, 1233–1243.

314. Guarini, A.; Guglielmetti, G.; Vincenti, M.; Guarda, P.; Marchionni, G. *Anal. Chem.* **1993**, *65*, 970–975.

315. Mumford, N. A.; Chatfield, D. A.; Einhorn, I. N. *J. Appl. Polym. Sci.* **1979**, *23*, 2099–2115.

316. Blazso, M.; Jakab, E.; Szekely, T.; Plage, B.; Schulten, H.-R. *J. Polym. Sci. Part A Polym. Chem.* **1989**, *27*, 1027–1043.

317. Harrison, G. G.; Taylor, M. J.; Scrivens, J. H.; Yates, H. *Polymer* **1997**, *38*, 2549–2555.

318. Jackson, A. T.; Jennings, K. R.; Scrivens, J. H. *Rapid Commun. Mass Spectrom.* **1996**, *10*, 1449–1458.

319. Kojima, Y.; Matsuoka, T.; Sato, N.; Takahashi, H. *J. Polym. Sci. Part A Polym. Chem.* **1995**, *33*, 2935–2940.

320. Kojima, Y.; Matsuoki, T.; Takahashi, H. *J. Appl. Polym. Sci.* **1999**, *72*, 1539–1542.

321. Evans, W. J.; DeCoster, D. M.; Greaves, J. *J. Am. Soc. Mass Spectrom.* **1996**, *7*, 1070–1074.

322. Lattimer, R. P. *Int. J. Mass Spectrom. Ion Processes* **1992**, *116*, 23–36.

323. Lattimer, R. P. *J. Am. Soc. Mass Spectrom.* **1992**, *3*, 225–234.

324. Ballistreri, A.; Garozzo, D.; Giuffrida, M.; Montaudo, G. *Anal. Chem.* **1987**, *59*, 2024–2027.

325. Ballistreri, A.; Montaudo, G.; Impallomeni, G.; Lenz, R. W.; Ulmer, H. W.; Fuller, R. C. *Macromolecules* **1995**, *28*, 3664–3671.

326. Klee, J. E.; Hagele, K.; Przybylski, M. *Macromolec. Chem. Phys.* **1995**, *196*, 937–946.

327. Zhuang, H.; Gardella, J. A., Jr. *MRS Bull.* **1996**, *21*, 43–48.

328. Deimel, M.; Rulle, H.; Liebing, V.; Benninghoven, A. *Appl. Surf. Sci.* **1998**, *134*, 271–274.

329. Chen, J. X.; Gardella, J. A., Jr. *Macromolecules* **1998**, *31*, 9328–9336.

330. van den Eynde, X.; Bertrand, P. *Surf. Inter. Anal.* **1999**, *27*, 157–164.

331. van den Eynde, X.; Reihs, K.; Bertand, P. *Macromolecules* **1999**, *32*, 2925–2934.

332. Cohen, L. R. H.; Hercules, D. M.; Karakatsanis, C. G.; Rieck, J. N. *Macromolecules* **1995**, *28*, 5601–5608.

333. Briggs, D.; Fletcher, I. W.; Reichlmaier, S.; Agulo Sanchez, J. L.; Short, R. D. *Surf. Inter. Anal.* **1996**, *24*, 419–421.

334. Xu, K. Y.; Gusev, A. I.; Hercules, D. M. *Surf. Inter. Anal.* **1999**, *27*, 659–669.

335. Hittle, L. R.; Hercules, D. M. *Surf. Inter. Anal.* **1994**, *21*, 217–225.

336. Park, M. A.; Cox, B. D.; Schweikert, E. A. *J. Vac. Sci. Technol. A* **1991**, *9*, 1300–1306.

337. Nsouli, B.; Rumeau, P.; Allali, H.; Chabert, B.; Debre, O.; Oladipo, A. A.; Soulier, J. P.; Thomas, J. P. *Rapid Commun. Mass Spectrom.* **1995**, *9*, 1566–1571.

338. Cox, B. D.; Park, M. A.; Kaercher, R. G.; Schweikert, E. A. *Anal. Chem.* **1992**, *64*, 843–847.

339. Zenobi, R. *Int. J. Mass Spectrom. Ion Processes* **1995**, *145*, 51–77.

340. Kahr, M. S.; Wilkins, C. L. *J. Am. Soc. Mass Spectrom.* **1993**, *4*, 453–460.

341. Schriemer, D. C.; Li, L. *Anal. Chem.* **1996**, *68*, 250–256.

342. Xiang, X.; Dahlgren, J.; Enlow, W. P.; Marshall, A. G. *Anal. Chem.* **1992**, *64*, 2862–2865.

343. Komolov, A. S.; Sinichenko, V. V.; Monakhov, V. V. *Phys. Low-Dimen. Struc.* **1997**, *7*, 49–53.

344. Holm, R.; Karas, M.; Vogt, H. *Anal. Chem.* **1987**, *59*, 371–373.
345. Wilkins, C. L.; Pastor, S. *Polym. Prepr. (Am. Chem. Soc., Div. Polym. Chem.)* **1996**, *37*, 284–285.
346. Hillenkamp, F. *Adv. Mass Spectrom.* **1995**, *13*, 95–114.
347. Pastor, S. J.; Wilkins, C. L. *J. Am. Soc. Mass Spectrom.* **1997**, *8*, 225–233.
348. Arakawa, R.; Watanabe, S.; Fukuo, T. *Rapid Commun. Mass Spectrom.* **1999**, *13*, 1059–1062.
349. Rashidezadeh, H.; Guo, B. C. *J. Am. Soc. Mass Spectrom.* **1998**, *9*, 724–730.
350. Dourges, M. A.; Charleux, B.; Vairon, J. P.; Blais, J. C.; Bolbach, G.; Tabet, J. C. *Macromolecules* **1999**, *32*, 2495–2502.
351. Jackson, A. T.; Jennings, K. R.; Scrivens, J. H. *J. Am. Soc. Mass Spectrom.* **1997**, *8*, 76–85.
352. Sakurada, N.; Fukuo, T.; Arakawa, R.; Ute, K.; Hatada, K. *Rapid Commun. Mass Spectrom.* **1998**, *12*, 1895–1898.
353. Larsen, B. S.; Simonsick, W. J. J.; McEwen, C. N. *J. Am. Soc. Mass Spectrom.* **1996**, *7*, 287–292.
354. Danis, P. O.; Karr, D. E.; Simonsick, W. J. J.; Wu, D. T. *Macromolecules* **1995**, *28*, 1229–1232.
355. Williams, J. B.; Gusev, A. I.; Hercules, D. M. *Macromolecules* **1996**, *29*, 8144–8150.
356. Williams, J. B.; Gusev, A. I.; Hercules, D. M. *Macromolecules* **1997**, *30*, 3781–3787.
357. Fall, M.; Aaron, J. J.; Dieng, M. M.; Jouini, M.; Aeiyach, S.; Lacroix, J. C.; Lacaze, P. C. *J. Chimie Physique Physico-Chimie Biologique* **1998**, *95*, 1559–1562.
358. Visy, C.; Lukkari, J.; Kankare, J. *Macromolecules* **1994**, *27*, 3322–3329.
359. Hop, C. E. C. A.; Bakhtiar, R. J. *J. Chem. Ed.* **1996**, *73*, A162–A169.
360. Tolic, L. P.; Anderson, G. A.; Smith, R. D.; Brother, H. M.; Spindler, R.; Tomalia, D. A. *Int. J. Mass Spectrom.* **1997**, *165*, 405–418.
361. Guittard, J.; Tessier, M.; Blais, J. C.; Bolbach, G.; Rozes, L.; Marechal, E.; Tabet, J. C. *J. Mass Spectrom.* **1996**, *31*, 1409–1421.
362. Maziarz, E. P., III; Wood, T. D. *J. Mass Spectrom.* **1998**, *33*, 45–54.
363. Lorenz, S. A.; Maziarz, E. P., III; Wood, T. D. *Appl. Spectrosc.* **1999**, *53*, 18A–36A.

12

Steady-State Fluorescence

CARIN A. HELFER
FRANCISCO MENDICUTI
WAYNE L. MATTICE

Fluorescence spectroscopy has proven to be a very powerful technique for studying polymers because a wide variety of information about polymer systems can be obtained.[1-6] Modern instrumentation can detect fluorescence emission easily. Therefore fluorescence is an appropriate technique for detection of trace amounts of emitters in polymers. These emitters may be present as independent small molecules, such as plasticizers or stabilizers, or as intrinsic minor constituents of the polymer, as with a chain that is labeled with special groups at its ends. Most of the emitters in polymeric systems are aromatic rings or other unsaturated and conjugated species. If the experiment is to provide accurate information about the concentration of the emitter, the efficiency of its emission (or quantum yield for fluorescence) must be known under the conditions of the experiment. Because quantum yields for fluorescence can be quite sensitive to the environment of the emitter, quantitative studies of emitter concentration must be performed under carefully controlled conditions.

Often fluorescence studies are performed to learn about the physical organization of a polymeric system. For example, excimers and exciplexes, which are excited state complexes, have been used to study properties of polymers, including chain relaxation in polymer blends and conformations of aromatic vinyl polymers such as polystyrene. Excimers are sometimes called a "4 Å ruler" because they are sensitive to this short distance scale. Energy transfer between "donor" and "acceptor" chromophores can be used to estimate either distances within a polymer chain or the separations of selected points on two different polymer chains. This process is sensitive to distances an order of magnitude larger than those probed by excimers. In addition, conformational transitions in a polymer have been studied using fluorescence depolarization measurements, and polymer compatibility and the interpenetration of polymer chains have been studied with fluorescence quenching experiments.

When a molecule interacts with a photon of the appropriate frequency, absorption occurs. After absorption of the photon, many possible events can return the system to the ground state. These various processes are discussed in detail in the section on theory in this chapter. At this point, one of these possibilities, which is luminescence, will be mentioned. In this case, a photon may be emitted from an electronically excited state. One of the two types of luminescence is known as fluorescence. In fluorescence, the emission of the photon occurs when the transition is between states of the same multiplicity. In this type of luminescence, the electrons are paired, which means the electron in the higher energy orbital has the opposite spin orientation from the electron in the ground state. The transition from the excited state to the ground state is an allowed transition in fluorescence, where the spins are paired. Therefore, the transition occurs with lifetimes on the order of nanoseconds.

The second type of luminescence is phosphorescence, which is the emission of a photon that results from a transition between states of different multiplicity. The electrons have the same spin orientation, and a change in the electron spin is required for a return to the ground state. Therefore, unlike fluorescence, which is a very fast process, phosphorescence is comparatively slow, with lifetimes on the order of microseconds to seconds.

Fluorescence spectroscopy has several advantages that make the technique ideal for studying polymers.[1-10] The bond rotations of polymers in low viscosity solvents occur on a timescale of nanoseconds. Therefore, the timescale of fluorescence spectroscopy, which is nanoseconds, makes the technique well suited for studying conformational changes in polymers. In addition, fluorescence spectroscopy is sensitive to all of the processes that occur during the excited state lifetime of a molecule, which can include molecules up to 100 Å away. Finally, another advantage of fluorescence is the high sensitivity of the technique so that extremely small polymer samples can be used in very dilute solutions.

In studies of polymers using photophysical techniques, synthetic polymers are often classified into two types, according to the density and the distribution of chromophores in the chain. A chromophore, as defined by Berlman in the *Handbook of Fluorescence Spectra of Aromatic Molecules*, is a "molecule or group of atoms which give rise to a particular absorption band system".[11] The two classifications of polymers are as follows:

1. Polymers with a low chromophore density. The photophysical properties for these polymers are often similar to the properties of the free chromophore. These groups are attached covalently to the polymer chain as a *label* or may be added as a small isolated molecule as a *probe* to the polymer system. An example of a labeled polymer is a poly(oxyethylene) chain with a naphthalene unit covalently attached at one end. If a small amount of free naphthalene is dissolved in poly(oxyethylene), the naphthalene is considered to be a probe.
2. Polymers with chromophores that are regularly distributed along the polymer chain, intrinsically forming a part of it. These chromophores occur generally at a high local concentration. In these polymers, the spectral emission properties of the chromophore are usually strongly affected by intrachromophore interactions. An example is poly(vinyl naphthalene).

In this chapter, we discuss the theory, the instrumentation necessary for measurements, sample preparation, precautions, and results with their interpretations for using fluorescence spectroscopy to study polymer conformations. In addition, a real-life example is included to aid in the further understanding of this technique. Phosphorescence will not be discussed further, because it has been used much less frequently than fluorescence in the applications to polymers. However, for information on phosphorescence, one can review *Polymer Photophysics and Photochemistry* by James Guillet.[5] In addition, this chapter includes only steady-state fluorescence measurements and excludes time-resolved measurements.

There are three justifications here for the focus on steady-state measurements. First, steady-state fluorometers are much more widely available than are instruments capable of time resolution. Second, the differences in the complexity of the two types of instrumentation imply that a novice, working alone, has greater probability of success with the steady-state measurements. Finally, most time-resolved studies are performed on polymers that have previously been characterized fully by steady-state methods. Two useful references that the reader may review for an introduction to time-resolved fluorescence spectroscopy are *Principles of Fluorescence Spectroscopy* by Joseph Lakowicz[3] and *Polymer Photophysics and Photochemistry* by James Guillet.[5]

Theory

Absorption

In order for absorption of light by a molecule to occur, the following two conditions are necessary:[5]

1. $h\nu = E_n - E_m$, where h is Planck's constant, ν is the frequency, m denotes the initial state and n denotes a higher energy state, and E is energy
2. The transition moment must be nonzero.

In describing the absorption of light, the Beer–Lambert equation is used[3,5]

$$\log I_0/I = \varepsilon cd = \text{optical density} \qquad (1)$$

where I_0 is the incident light intensity, I is the exiting light intensity, ε is the decadic molar extinction coefficient (liter $mol^{-1}cm^{-1}$), c is the concentration of the chromophore (mol $liter^{-1}$), and d is the sample thickness (in cm). The reorganization of electron density in a chromophore upon absorption of light occurs on a timescale of approximately 10^{-15} sec. This process is fast enough that negligible change in the nuclear coordinates can occur (the Franck-Condon principle).[2,3,5]

Jablonski Diagram

Following the absorption of a photon, a number of processes can occur that reduce the energy and return the molecule to the ground state. These processes are often described with the aid of a Jablonski diagram, an example of which appears in Figure 12.1.

Absorption usually initiates from V_0 of the singlet ground state because $E_{V1} - E_{V0}$ is much greater than the product of Boltzmann's constant and temperature, k_BT, for most chromophores. After absorption, the molecule may be in an excited vibrational state of an excited electronic state. The excited molecule can undergo collisions with surrounding molecules in a dense medium (and polymers will certainly be studied in a dense medium, not in vacuo). These collisions cause a loss of excess vibrational energy. This process, which is known as vibrational relaxation, leaves the molecule in the ground vibrational state within the electronic state. If the energy difference between S_2 and S_1 is not too large, internal conversion, which is another radiationless transition, can occur. Generally, the energy difference between S_1 and S_0 is large enough so that internal conversion does not return the molecule to the ground electronic state S_0 from the ground vibrational state of S_1. Another process that may occur is intersystem crossing to the triplet excited state, T_1. The triplet excited state is a metastable state, since the molecule can remain in this state for a relatively long time.[11] From T_1, the molecule may return to the ground electronic state, S_0, with the emission of a photon. This radiative transition is phosphorescence. However, as men-

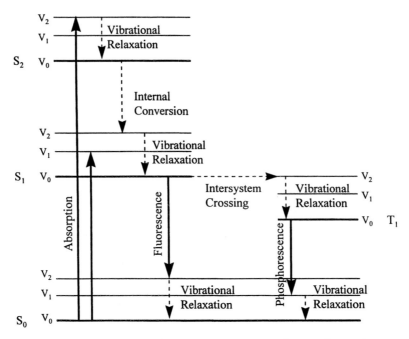

Figure 12.1 Example of a Jablonski diagram. Key: S_n denote the singlet electronic state, with S_0 the ground electronic state for the molecule. Within an electronic state, the states labeled V_0, V_1, and V_2 refer to the vibrational states, with 0 being the ground vibrational state. For our purposes we can ignore the ground state vibrational energy. T_1 is the first excited triplet state. The arrows show some of the processes that can occur. Nonradiative processes (without the involvement of a photon) are shown with dashed arrows. The solid arrows represent radiative processes (with the absorption or emission of a photon).

tioned previously, this transition is a slow process since the electron must undergo a change in spin.

More processes can be included in a Jablonski diagram. The electronically excited molecule may undergo decomposition (as with photoactivated initiators of polymerization), leading to photochemistry. The excited chromophore may donate its excitation to other components of the system. This donation may be nonradiative insofar as the initially excited molecule is concerned, but it may induce a radiative process in the acceptor of the excitation. For present purposes, it is important to recognize that Figure 12.1 does not include all competitive processes that might produce nonradiative relaxation of the singlet electronic excited state of the chromophore responsible for the initial absorption of light.

Fluorescence

Finally, the molecule may fluoresce from the ground vibrational state of S_1 to the ground electronic state S_0 with a photon being emitted. The timescale for the transition is on the order of nanoseconds. Fluorescence will usually only occur from the ground vibrational state of S_1 because of the more rapid occurrence of vibrational relaxation and internal conversion (Kasha's rule[5,12]). However, fluorescence

emission will return the system to any of the vibrational levels of the ground electronic state. Therefore, a distribution of wavelengths for the fluorescence transition occurs, which yields the fluorescence spectrum.[4] The fluorescence emission spectrum is red-shifted from the spectrum that describes the initial absorption of light.

The parameters fluorescence quantum yield, Q, and fluorescence lifetime, τ, are often encountered in discussions of fluorescence. The fluorescence quantum yield of a substance is defined as the number of photons emitted as fluorescence divided by the number of photons absorbed. In terms of unimolecular rate constants, if one defines Γ to be the rate constant for the radiative processes and k the rate constant for the unimolecular nonradiative processes, the quantum yield is[3]

$$Q = \frac{\Gamma}{\Gamma + k} \qquad (2)$$

The fluorescence lifetime is the average time a molecule spends in the excited state before returning to the ground state. Using the rate constants defined above, the fluorescence lifetime is given as[3]

$$\tau = \frac{1}{\Gamma + k} \qquad (3)$$

Various bimolecular processes can occur that cause a decrease in the expected emission intensity. Collectively, these processes are known as quenching. Some quenching processes that are useful in studying polymer conformations include excimer formation, which occurs at a distance between the chromophores of about 3 to 4 Å, and energy transfer, which occurs at a distance of 10 to 100 Å.[2]

Excimers

In an excimer, a complex forms from the interaction between a chromophore in the lowest excited electronic singlet state (A*) and another chromophore in the singlet ground state (A).[3-6] The excited molecule and the ground state molecule are the same or similar species. The excited state complex that is formed is lower in energy than the separate species. Therefore, the excimer is a more stable structure than the separate chromophores in the same electronic state. Excimers are only stable in the excited state, because they bind the two chromophores so closely that the configuration would be repulsive if they were both in the electronic ground state. Excimers can only be detected by fluorescence emission and are not observed in the absorption spectra. In the fluorescence emission spectrum, a new, broad, structureless emission band is observed that is centered at a lower energy (red-shifted), but higher wavelength than the the monomer band. In discussions of excimers, the term *monomer* is used to describe the single chromophore, and the term *dimer* is used to describe the complex of two chromophores. No fine structure is observed for the excimer emission because, as mentioned previously, the ground state is unstable.[11]

A critical requirement for the formation of excimers is the formation of a coplanar sandwich-like structure of the two chromophore groups that are separated by approximately 3 to 4 Å (Figure 12.2). If a conformational transition is necessary to form this complex, the transition that will yield the correct geometry needs to occur within the lifetime of the excited state. Only a small variation in the position of the two chromophores in this sandwich-like structure can occur. Therefore, since the relative position of the chromophores to each other is quite well-defined, excimer formation is useful in studying the conformations of polymers. For dilute solutions of polymers containing chromophores that are regularly distributed along the chain, the excimers that form are due mainly to intramolecular interactions between the nearest neighbor chromophores, if the necessary geometry is sterically allowed. At very high dilution, only intramolecular excimer formation is observed.[5]

The geometric criteria for the formation of an excimer are apparent from classic studies of small molecules such as naphthalene[13] and paracyclophanes.[14] The influence on excimer emission of the incorporation of two phenyl rings into a small hydrocarbon in 13 different ways was clearly demonstrated in classic work by Hirayama.[15] This work led to Hirayama's rule that 1,3-diaryl substitution is particu-

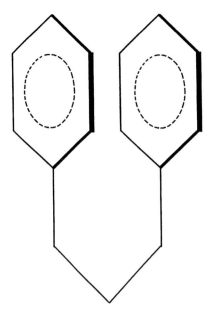

Figure 12.2 Ideal excimer structure for 1,3-diphenylpropane, obtained with *gauche* states of opposite sign at the two CH_2—CH_2 bonds.

larly favorable for intramolecular excimer formation. The importance of vinyl polymers creates great interest in the excimer formation by 1,3-diarylpropanes,[16] with aryl being a common chromophore such as phenyl,[17] pyrene,[18-20] or carbazole.[21,22] Extension to the 2,4-diarylpentanes,[16] where aryl might be phenyl[23-25] or pyrenyl,[26-29] reveals important differences in the meso and racemo stereoisomers, which are shown in Figure 12.3. Typically, excimer emission is

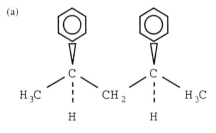

(a)

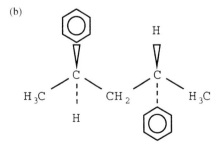

(b)

Figure 12.3 Stereoisomers of 2,4-diphenylpentane: (a) meso; (b) racemo.

more intense in the meso stereoisomer. For this reason, it is often the meso stereoisomer that is selected when a 2,4-diarylpentane is to be used as a probe of the mobility of a polymer matrix in which it is dispersed.[30-32] Often another group, X, has been substituted for the central methylene in the 1,3-diarylpropane,[24,33] with X = oxygen or a substituted nitrogen, and aryl including phenyl or naphthyl. Other pertinent model compounds include small molecules that are models for higher oligomers of vinyl polymers[23,34,35] and bis (aryl) alkanes with larger numbers of methylene groups separating the two chromophores.[18]

Energy Transfer Process

Singlet-singlet energy transfer between a donor (D) and an acceptor (A) by the Förster[37] or resonance mechanism occurs when an excited donor, D*, nonradiatively transfers its excitation energy to a ground state acceptor, A.

$$D^* + A \rightarrow D + A^* \qquad (4)$$

The energy transfer mechanism takes place via long-range dipole-dipole or Coulombic interaction between a D* and A pair. As mentioned, this phenomenon is sensitive to larger distances than those probed by excimers and has a wider time scale.[3-6,37] Two types of energy transfer can occur in a polymer system. The first type is intermolecular, which involves the transfer of energy from a small donor molecule to a polymer acceptor. This type of energy transfer can also occur from the polymer chain donor to a small molecule acceptor, which results in the quenching of the fluorescence emission from the polymer. Or it may occur from one polymer molecule to another.

The second type of energy transfer is intramolecular. This type can involve both a donor chromophore and an acceptor chromophore that are located on the polymer chain, as shown in Figure 12.4. In type 2 polymers, which contain identical chromophores that are regularly distributed along the chain, the excitation can be localized for some finite period of time on a particular chromophore before it is transferred to another chromophore. This transfer can then proceed to yet another chromophore, and so on, as a multistep energy transfer mechanism. This process is known as *energy migration*.[3,5,6] For dilute solutions of polymers in a good solvent, energy migration is an intramolecular process, often modeled as a random walk of the excitation along the chain.[38]

For a D-A pair separated by a distance R, the rate of the energy transfer, k_{ET}, is given by the Förster equation[2-6,37]

$$k_{ET} = \frac{9000(\ln 10)\kappa^2 Q_D}{128\pi^5 n^4 N R^6 \tau_D} J \qquad (5)$$

where Q_D and τ_D denote the quantum yield and lifetime for fluorescence of D in the absence of A; n is the refractive index of the medium; κ^2 is the orientation factor, which is a parameter that takes into account the relative orientation of the absorption and emission transition moments of A and D, respectively; and N is Avogadro's number. J is the overlap integral between the normalized fluorescence intensity of the donor $I_D(\lambda)$ and the extinction coefficient of the acceptor $\varepsilon_A(\lambda)$, defined as

$$J = \int \lambda^4 I_D(\lambda) \, \varepsilon_A(\lambda) \, d\lambda \qquad (6)$$

where, by normalization,

$$\int I_D(\lambda) \, d\lambda = 1 \qquad (7)$$

The value of κ^2 can range from 0 to 4 for specific orientations of the donor and acceptor with respect to one another. In almost all studies of polymeric systems, the orientation is unknown and is likely to be random. The value of κ^2 is ⅔ when there is a random orientation of D and A. This value of κ^2 is usually assumed unless there is specific information that demonstrates a nonrandom orientation of these chromophores in the system under study.

An important parameter used as a measure of the distance at which D can transfer its excitation energy to A is the Förster radius, R_0, for a D-A pair. At this distance, the transfer rate D → A is equal to the decay rate of D in the absence of A. R_0 is defined as[2-6,37]

$$R_0^6 = \frac{9000(\ln 10)\kappa^2 Q_D}{128\pi^5 n^4 N} J \qquad (8)$$

The efficiency of energy transfer (ϕ_{ET}) is given by

$$\phi_{ET} = \frac{(R_0/R)^6}{1 + (R_0/R)^6} \qquad (9)$$

which represents the fraction of radiation absorbed by D that is transferred to A. Equations 5, 8, and 9 are useful for a fixed D-A distance, R. Otherwise, these equations may be modified to take into account the D-A distribution of distances.

A corollary of eq 9 is the operational definition of R_0 as being the separation of the chromophores at which the efficiency of energy transfer is ½. The sixth power dependence in this equation causes the efficiency to rise strongly as R falls below R_0, and to decrease strongly as R rises

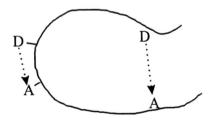

Figure 12.4 Intramolecular energy transfer in a polymer chain. Key: D = donor chromophore; A = acceptor chromophore.

above R_0. The efficiency decreases from 90% to 10% when R/R_0 changes from 0.693 to 1.44.

Experimental verification of these predictions from theory has been achieved with study of bichromophoric model compounds in which the donor and acceptor are held at known separation by their attachment to specific sites on a rigid polycyclic framework[39] or at opposite ends of a helical polymer, poly-L-proline, of carefully controlled degree of polymerization.[40,41]

Numerical values of Förster radii, R_0, have been tabulated for thousands of donor-acceptor pairs.[42] Values of R_0 for the transfer between like molecules that are models for the chromophores in common aromatic vinyl polymers (7.2 Å for toluene, 8.25 and 11.75 Å for 1- and 2-methyl naphthalene, 21.34 Å for N-methyl carbazole) support the notion that energy migration should be active in polystyrene, poly(vinyl naphthalene), and poly(vinyl carbazole). Larger Förster radii can be obtained with nonidentical pairs (R_0 for toluene → 1-methylnaphthalene is 19.83 Å, but R_0 for toluene → toluene is only 7.2 Å) because the overlap integral, J, is often small for self-transfer, due to weak overlap of absorption and emission spectra for the same chromophore.

Fluorescence Anisotropy

Upon excitation of an immobile isolated chromophore with polarized light, the emission will be polarized, also.[3,43] If the emission takes place quickly, before the fluorophore relaxes its geometry or transfers its energy, the degree of polarization of the emitted radiation will depend only on the angle between the absorption and transition moments. However, if during the lifetime of the excited state any extrinsic phenomena occur that modify the orientation of transition moments, such as rotational diffusion, conformational changes, and the like, there will be a loss of the degree of polarization of the emission. Depolarization may also occur if the excitation hops from one chromophore to another, if the two chromophores are not identically oriented.

The parameters *polarization* and *anisotropy* of fluorescence give the degree of polarization of the emitted radiation. Both terms express the same phenomenon, however, anisotropy has become more commonly used because many theoretical equations tend to be more simply written when using anisotropy instead of polarization.[3] Upon excitation with vertically polarized light, anisotropy, r, and polarization, P, are defined as

$$r = \frac{I_\parallel - I_\perp}{I_\parallel + 2I_\perp} \tag{10}$$

$$P = \frac{I_\parallel - I_\perp}{I_\parallel + I_\perp} \tag{11}$$

where $I_\parallel$ is the fluorescence intensity of the vertically ($\parallel$) polarized emission and $I_\perp$ is the fluorescence intensity of the

horizontally ($\perp$) polarized emission.[3] The two measures of fluorescence depolarization are interconverted as

$$r = \frac{2P}{3 - P} \tag{12}$$

Methods

Instrumentation

In Figure 12.5, a schematic diagram of an SLM 8000 spectrofluorometer is shown.[3,44] This instrument is similar to the one used in the real-life example that will be presented later in this chapter. A high-pressure xenon arc lamp is a commonly used light source because it yields high-intensity light at wavelengths greater than 250 nm. In addition, this light source has a fairly constant output in the wavelength range from 270 to 700 nm.[3] A few precautions are necessary when handling a xenon arc lamp. First, since the gas in the lamp is under high pressure, a risk of the lamp exploding exists. Therefore, safety glasses and protective clothes should be worn when handling the lamp. The manual accompanying the fluorometer should be studied before attempting to change the lamp. In addition, when the lamp is lit, it should never be viewed directly because the intense brightness could damage the eyes. Finally, the quartz envelope should not be touched because fingerprints on the quartz envelope could char, which will cause the lamp to fail.[3]

The monochromators are used to select the excitation and emission wavelengths. Motorized monochromators are useful when scanning the excitation and emission wavelengths. Usually, the slit width can be varied. If a smaller slit size is used, higher resolution results,[3] but less light intensity will pass through the monochromator. By using a larger slit size, the intensity of the light that passes through is higher, and the signal-to-noise ratio is also higher.[3] Stray light, which is any light at wavelengths other than the selected wavelength that passes through the monochromator, is one source of interference in fluorescence measurements. Therefore, the quality of the monochromators is a very important factor when making fluorescence measurements.[2]

The sample chamber is contained in the optical module. Inside this chamber is a holder for the cuvette, which contains the sample to be analyzed. Following the light path through the optical module, shutters are found before and after the sample chamber. The shutters close off the excitation or emission channels. A beam splitter is used to reflect some of the excitation light to a reference cell. By dividing the resulting fluorescence intensity of the sample by the fluorescence intensity of the reference, one can correct for variations in the lamp intensity. The polarizers are also found in the optical module in both the excitation and emission light paths. When obtaining anisotropy measure-

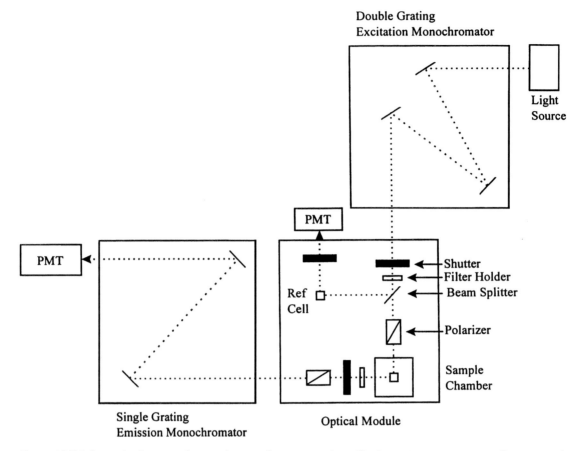

Figure 12.5 Schematic diagram of a steady-state fluorometer (specifically, an SLM 8000 spectrofluorometer), similar to the one used in the real-life example. PMT = photomultiplier tube.

ments, it is important to have accurate positioning of these polarizers.[3] Finally, in order to minimize interference from stray light, optical band pass filters can be used in addition to the monochromators.

The photomultiplier tubes are the components that actually detect the fluorescence. The photomultiplier tubes may run in either photon-counting mode or analog mode.[3] In photon-counting mode, the individual anode pulses are counted. This mode works best when the signal level is low. In analog mode, the individual pulses are averaged. Therefore, analog mode yields better precision.[3]

When evaluating the instrumentation for fluorescence measurements, one needs to remember the following two points:[3] fluorescence is a highly sensitive technique, and there is no ideal spectrofluorometer. Since fluorescence is such a sensitive technique, one needs to avoid interference in the measured fluorescence that arises from light leaks in the instrument itself. In addition, interferences can arise from the sample, such as solvent impurities or turbid solutions, which will be addressed here later.

Spectrofluorometers do not give true excitation or emission spectra.[3] In an ideal spectrofluorometer, the light source would have a constant output at all wavelengths,[3] but in reality, the light intensity from the light source is a function of wavelength. Another distortion in the measured fluorescence of a sample occurs because the monochromators do not pass photons at the same efficiency for all wavelengths.[3] Also, the efficiency of the monochromators depends on the polarization of the light. Finally, the efficiency at which the photomultiplier tube detects photons is also a function of wavelength. By dividing the sample fluorescence intensity by the fluorescence intensity of a reference, the distortions due to the fluctuation in the intensity of the lamp can be corrected. Rhodamine B in ethylene glycol (3 g/L) is a commonly used reference.[3] A standardized light source is of assistance in correcting emission spectra for wavelength dependencies in other parts of the instrument, such as the photomultiplier. In addition, to best compare fluorescence spectra, the spectra should be recorded under identical conditions. The effect of polarization can be eliminated by operating at "magic angle" conditions, in which the excitation light is vertically polarized and the emission polarizer is set at 54.70 from vertical.[3] Under these conditions, the signal is actually proportional to the total fluorescence intensity ($I_{\parallel} + 2I_{\perp}$). One should note, however, that the presence of polarizers in both the excitation and emission light paths will cause the signal to decrease at least fourfold.[3] For additional details on the instrumentation, one should review the book by Lakowicz.[3]

Sample Geometry

Several different geometries may be used during fluorescence measurements. The most commonly used geometry, which is right angle observation of the center of a cuvette that is centrally illuminated, is the geometry shown in the diagram of an SLM 8000 spectrofluorometer in Figure 12.5. When using this geometry, it is best to work with dilute solutions because intermolecular interactions, inner filter effects, and self-absorption can be avoided at very low concentrations.[3] Inner filter effects cause a decrease in the observed fluorescence emission, which can be a problem when the sample has a very low quantum yield. When using high optical densities, the fluorescence emission spectra can also be distorted. If one must work with more concentrated solutions, off-center illumination or front-face illumination is useful.[3] The path length is decreased in off-center illumination. Front-face illumination works best when the illuminated surface is 30° (or 60°) from the incident beam because a small amount of the reflected light passes through the emission monochromator.[3] Angles close to 45° are avoided to minimize the amount of light that reflects directly into the emission channel, which increases the chance of intrinsic depolarization of fluorescence! This geometry can be used for polymer films or very concentrated solutions. As mentioned, if possible, dilute solutions should be used.

Sample Preparation

Since fluorescence is such a highly sensitive technique, all of the materials used in preparing the samples, including any glassware and the cuvettes, should be thoroughly cleaned to avoid fluorescence emission from a source other than the sample of interest. Disposable pipets are useful for transferring solutions and the solvents. High-grade quartz fluorescence cuvettes should be used to hold the samples during the experiments. The solvents should be spectroscopic grade, if possible. Before performing the actual fluorescence measurements, the solvents should be monitored for any fluorescent impurities. Freshly prepared samples should be used to avoid any photochemical reactions that may occur. If the solutions need to be stored for any length of time, the solutions should be stored in a dark place. In order to obtain a large enough signal with weak emitters, molecular oxygen may need to be removed from the system because it is a known quencher of fluorescence.

As mentioned, dilute solutions should be used if the fluorescence quantum yield is sufficient. At the excitation wavelength of interest, the measured optical density of the samples should be around 0.1. At much lower concentra-

tions, the signal is weaker than desired. At higher concentrations, artifacts may be generated by the "inner filter" effect, when the incident beam is attenuated significantly as it traverses the sample. If the sample dissolves well in the solvent being used, the sample and solvent can be placed together directly in the cuvette and then mixed to measure the UV absorbance. The concentrations can then be adjusted to obtain an optical density in the appropriate range.

Raman scattering from the solvents can be a problem if the instrument gain needs to be increased because of a low quantum yield of the sample.[3] Fortunately, this source of scattering can easily be identified because it occurs at a constant wavenumber difference from the excitation light. Therefore, by changing the excitation wavelength, one can easily determine if a peak is due to the fluorescence of the sample or due to Raman scattering.

Experimental Results

When performing fluorescence experiments, it is useful to initially obtain the absorption spectrum for the sample of interest. A typical absorption spectrum is shown in Figure 12.6. By performing absorption measurements, one can be certain that the solutions are sufficiently dilute (optical densities of approximately 0.1 or less) to avoid the previously mentioned problems that can arise from more concentrated solutions. In addition, the wavelength of maximum absorption can be determined. However, the wavelength of maximum absorption can also be obtained from a fluorescence excitation spectrum if a single electronic transition occurs in the spectral region of interest.

Fluorescence Spectrum

Fluorescence results are generally presented as a plot of the normalized fluorescence intensity as a function of wave-

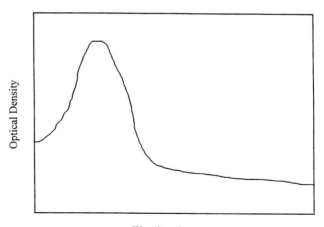

Figure 12.6 Typical absorption spectrum. The example is fisetinidol, a monomer unit of condensed tannins.

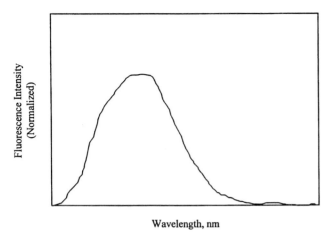

Figure 12.7 Typical fluorescence emission spectrum. The example is fisetinidol.

length, λ, in nm or wavenumber, $\bar{v}$, in cm^{-1}. Most commercial instrumentation provides the results in wavelength.[3] However, one can easily convert the units of wavelength to wavenumber. A typical fluorescence emission spectrum is shown in Figure 12.7. Usually, the computer that controls the operation of the fluorescence equipment also handles the data analysis and printing.

One may be interested in obtaining both the excitation and emission spectra. Before measuring the sample's fluorescence, a solvent blank should be scanned for the wavelength range of interest. In order to record the excitation spectrum, the excitation wavelengths are scanned with the emission monochromator set somewhere within the emission spectrum. The fluorescence emission spectrum is the wavelength distribution of the emission with a constant excitation wavelength.[3] In order to obtain the emission spectrum, one selects the excitation wavelength and scans the emission wavelengths. The excitation wavelength should be within the absorption spectrum of the transition responsible for the emission.

Generally speaking, the absorption spectrum and the fluorescence emission spectrum should be mirror images of each other, as is shown in Figure 12.8. This occurs because the same transitions are involved in both the absorption and emission processes. Deviations in the mirror image rule can be useful to determine excited state reactions that may be occurring, such as excimer formation.[3] The two spectra in Figure 12.8 are drawn with the same height at the maximum in order to facilitate the comparison of their shapes. In fact, only about 1 in 10 absorbed photons leads to fluorescence in this system: that is, Q is on the order of 0.1.

Quantum Yield

Menduciti and Mattice[45] reported a method for improving the reliability of quantum yield measurements when using dilute solutions of weak emitters. The quantum yield for fluorescence can be written as

$$Q \sim A^{-1}\int I \, d\lambda \qquad (13)$$

where A denotes the absorbance of the sample at the excitation wavelength, I is the fluorescence intensity at wavelength λ, and the integral is performed over the emission band. The software for the fluorescence instrumentation should contain some type of data manipulation feature that allows for the integration of the spectrum. If $\int I \, d\lambda$ is plotted as a function of A for samples of different concentrations, the quantum yield, Q, is proportional to $d(\int I \, d\lambda)/dA$, which is the slope of the straight line obtained from the plot. Quantum yields for fluorescence are usually calculated by comparing the emission of the sample with that of a standard. A commonly used standard is quinine sulfate in 1.0N sulfuric acid, which is known to have a quantum yield of 0.546 at 25°C when the excitation wavelength is in the range of 220 to 340 nm.[46,47] Therefore,

$$Q = Q_s d(\int I \, d\lambda)/dA/[d(\int I \, d\lambda)/dA]_s \qquad (14)$$

where the subscript s denotes the standard.

Figure 12.9 shows $\int I \, d\lambda$ as a function of absorbance for both fisetinidol in 1,4-dioxane and quinine sulfate in 1.0N sulfuric acid. The failure of the lines to extrapolate through the origin indicates the presence of a constant interference. This interference is canceled out from the calculations when using the above method for the quantum yield calculations.

For a series of condensed tannin oligomers, specifically epicatechin oligomers with $(4\beta \rightarrow 8)$ interflavan bonds, Cho and coworkers[48] observed that the quantum yields depended on the molecular weight of the oligomers as

$$Q = k/x \qquad (15)$$

where x is the number-average degree of polymerization and k is a constant that depends on the solvent. A migration of the electronic excitation appeared to occur from one chromophore to another in the polymer chain. The migration increases the chance of encountering a quencher, therefore, causing the quantum yield to decrease.[48] In another context, this quenching via energy migration is exploited in photon-harvesting polymers.[38,49] The absorbed energy is not emitted by the polymer responsible for the initial absorption but is instead transferred to a trap.

Excimer Formation

Since the timescale for excimer formation is on the same order as rotations about bonds in common flexible polymers in media of low viscosity, studies of excimer formation can be used to determine if transitions due to rotations about C–C bonds can easily occur.[5] As mentioned previously, the formation of an excimer can be determined by analyzing the fluorescence emission spectrum. A broad, structureless band that is red-shifted from the monomer's emission band will be observed. Again, the term "monomer" is used to describe the single chromophore.

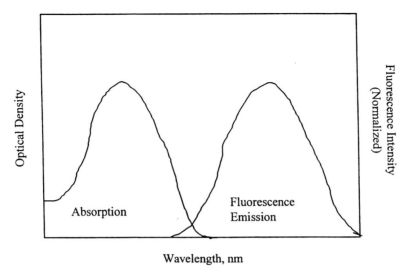

Figure 12.8 Absorption and fluorescence emission spectra. The example is fisetinidol.

Often there is not a clean separation of the emission bands from the monomer and the excimer. In such systems, a simple way of evaluating the amount of intramolecular excimers is by the corrected ratio I_D/I_M according to

$$\frac{I_D}{I_M} = \frac{I_P - I_S}{I_{norm}} \qquad (16)$$

where I_P is the normalized fluorescence intensity for the polymer at the selected wavelength for the dimer band, I_S is the normalized intensity obtained with the single chromophore at the same wavelength, and I_{norm} is the intensity at the wavelength used for normalization of the emission spectra. The real-life example at the end of this chapter will include additional information on the analysis of excimer formation.

Energy Transfer

From eq 8, the ratios of the Förster radii for nonradiative singlet energy transfer by two molecules, $D^* \to A$ and $D'^* \to A'$, in the same environment can be written as

$$(R_{0\,D\text{-}A}/R_{0\,D'\text{-}A'})^6 = \kappa^2_{D\text{-}A}\, Q_D\, J_{D\text{-}A}/\kappa^2_{D'\text{-}A'}\, Q_{D'}\, J_{D'\text{-}A'} \qquad (17)$$

where all magnitudes were already defined. A random distribution of isolated chromophores has the ratio $\kappa_A^2/\kappa_B^2 = 1$. Using a reference $D'\text{-}A'$ pair for which $R_{0\,D'\text{-}A'}$ is known, it is possible to estimate $R_{0\,D\text{-}A}$ from the experimental quantum yields for fluorescence of D and D' and the J values of both systems. The naphthalene to naphthalene $(N \to N)$ self-transfer (with Förster radii in the range of 7.35 Å[50] to 6.69 Å[11]) can be used as a reference. Experimental quantum yields can be obtained by the procedure described previously by Mendicuti and Mattice[45] and J by the graphic integration of eq 6. In some applications of energy transfer measurements, it is simply the ratio of the emission intensities of the donor and acceptor that is characterized as a function of the perturbation of the system that is under investigation.[51]

Fluorescence Anisotropy

Since most fluorometers have only a single emission channel, the most common method for obtaining both parameters is one named by Lakowicz[3] as the "single-channel or L-format method". Recalling that the anisotropy, r, that is defined in eq 10 can be written as

$$r = \frac{(I_\parallel/I_\perp) - 1}{(I_\parallel/I_\perp) + 2} \qquad (18)$$

and using

$$\frac{I_\parallel}{I_\perp} = \frac{1}{G}\frac{I_{VV}}{I_{VH}} \qquad (19)$$

r can be defined operationally as

$$r = (I_{VV} - GI_{VH})/(I_{VV} + 2GI_{VH}) \qquad (20)$$

where I_{xy} is the intensity of the emission that is measured when the excitation polarizer is in position x and the emission polarizer is in position y (V for vertical, H for horizontal), and G is given as I_{HV}/I_{HH}, which is a parameter that corrects for any depolarization produced by the optical system. Therefore, the anisotropy is obtained by measuring the four quantities I_{VV}, I_{VH}, I_{HV}, and I_{HH}. The expected range of r is 0 to 0.4. Any value outside of these limits indicates an artifact, like light scattering from the sample, reabsorption of the emitted radiation, misalignment of the polarizers, stray light, and so on. Many of the modern spectrofluorometers that have optical components controlled by a computer and the appropriate software allow for G and r to be obtained simultaneously, which yields a single-point anisotropy value for a pair of excitation and emission wave-

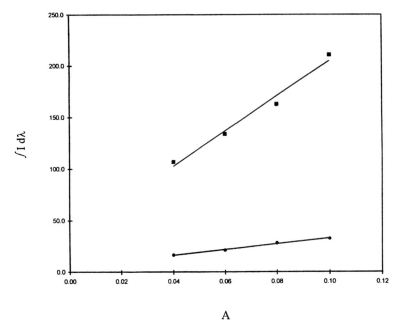

Figure 12.9 The integrated area of the fluorescence emission as a function of absorbance for fisetinidol in 1,4-dioxane (●) and in quinine sulfate in 1.0N sulfuric acid (■).

Real-Life Example

Interpretation of Intramolecular Excimer Formation and Nonradiative Energy Transfer in Naphthalene-Containing Polyesters and Their Bichromophoric Model Compounds

lengths as an average of a certain number of measurements. By recording the excitation and emission anisotropy (or polarization) spectra, additional information of the excited states can also be obtained.[3]

In the introduction, two types of polymers as classified by Phillips,[4] Guillet,[5] and Rabeck[6] were defined. We focus our interest on the type 2 polymers. We can also distinguish the situation of an idealized isolated polymer chain, such as a dilute solution of a polymer in a good solvent, from a concentrated solution of a polymer where intermolecular chain interaction and then interchromophore interactions take place. Many fundamental photophysical studies are more easily interpreted if the measurements avoid the perturbation produced by intermolecular interactions. Such is the case with the investigation of polymer conformation and dynamics of isolated chains. In order to interpret the results of such studies easily, it is often useful to relate the fluorescence properties of polymers to those of the small-molecule-analog that closely resemble the polymer repeat unit, such as a bichromophoric compound, an oligomer, and the like.

Intramolecular excimer formation and singlet-singlet nonradiative energy transfer (and/or energy migration) have been widely used in the characterization of these polymers.[3-6] The dependence of both mechanisms on the distance and orientation between chromophores makes the efficiency of both processes strongly dependent on the conformation and dynamics of the chain.

Flexible aromatic polyesters with repeating sequences that can be represented by $-[A-B_n]$, where B_n is a flexible spacer between rigid aromatic chromophore groups A, are examples of such a type of polymer. Many macroscopic properties of these polymers depend on the nature of A and B and on the small integer n.

For the dilute solution of these polymers, both the intramolecular excimer formation and intramolecular energy transfer processes depend on the relationship between the fluorescence lifetime of A, τ_f, and the time required for the conversion of B_n from one rotational isomer to another, τ_r. If τ_f is equal to or greater than τ_r the flexible spacer B_n can change during τ_f from a conformation in which excimer formation or energy transfer is not possible to one in which it is, because either the distance or the orientation, or both, between chromophores is now adequate. As a consequence, chain dynamics increase the efficiencies of both processes.[52-54] Another situation is presented when τ_f is much less than τ_r, where the chain dynamics is suppressed. The values of τ_f and τ_r may satisfy the last condition when τ_f is short, in a rigid medium whose viscosity is very large, or when experimental results are extrapolated to infinite viscosity. The dependence of both properties on n could be susceptible to rationalization by using an equilibrium model, such as the rotational isomeric state (RIS) model,[55] in the absence of chain dynamics and molecular dynamics simulations in the presence of it.

However, even when τ_f is much less than τ_r, the intramolecular excimer formation mechanism is a little more complicated because excimers can also be populated via energy migration. By a multistep intramolecular energy transfer process, the excitation energy can reach one of a pair of chromophores in the chain that has a conformation appropriate for excimer formation. According to this, the interpretation of the intramolecular excimer formation by an equilibrium model (RIS)[55] may be possible in polymers where τ_f is short and with a small Förster radius R_0 for energy transfer to a ground state A, so that energy migration does not occur. These requirements are met in polyesters derived from tere-, iso-, and phthalic acids.[56–61] In systems with a long τ_f and a large R_0, the influence of intramolecular energy transfer and dynamics on excimer formation can be suppressed by studying the bichromophoric compounds A-B_n-A instead of the polymers in a rigid medium or extrapolation to a medium of infinite viscosity.[62–72]

A prototype of -[A-B_n], polymers with a large τ_f and a large R_0, is poly(ethylene naphthalene-2,6-dicarboxylate), known as PEN, whose structure is one related to poly(ethylene terephthalate). In general, information about the microscopic structure of these polymers can be obtained from studies of intramolecular energy transfer and intramolecular excimer formation by the fluorescence technique in PEN and other members of the series with a different number of methylene groups n. The properties can be rationalized by using molecular dynamics or by an equilibrium RIS model.

Here, we review a study of the intramolecular energy transfer and intramolecular excimer formation processes by fluorescence on polyesters and bichromophoric compounds, where A contains a naphthalene chromophore, B is a methylene group, and n ranges from 2 to 6. The study is mainly focused on high-viscosity media where the dynamics are suppressed. A theoretical rationalization of the experimental trend of both properties with n has also been performed by using an RIS model.[55]

Methods

Sample Preparation

The compounds, abbreviated as NMnN and Pn, are depicted in Figure 12.10. Model compounds MN and DMN, containing a single chromophore, are also depicted. They are used as references in order to locate unambiguously the monomer band. Synthesis and purification were performed starting from 2,6-naphthalene dicarboxylic acid chloride or 2-naphthoyl chloride and HO-$(CH_2)_n$-OH glycols by the method described elsewhere.[62,65]

Dilute samples of the naphthyl-containing molecules, NMnN, Pn and their model compounds, dispersed in glassy poly(methyl methacrylate) (PMMA) were prepared by the thermal polymerization of very dilute solutions (optical density in the range 0.2–0.3 at the wavelength used for excitation) of the compounds in distilled methyl methacrylate in the absence of any initiator. Approximately 3 mL of each fresh solution is introduced into a 2.5- to 3-cm diameter test tube used as a mold. To minimize the formation of voids and quenching of the free radicals at the end of the chain, molecular oxygen is removed by using an ultrasonic bath for approximately 45 min and bubbling dry nitrogen through the solutions for about 10 min. It is important to avoid noxious vapors. Therefore, this operation and further polymerization should be performed in an extractor hood. Cover the tubes with adequate stoppers to avoid evaporation.

The polymerization is carried out in two steps simultaneously for all samples. To obtain PMMA samples with similar optical characteristics, it is important to perform the preparation of all samples whose anisotropies are going to be compared simultaneously and at the same experimental conditions. First, the temperature is increased slowly up to about 70 °C for approximately 3 h. Second, the temperature is increased further up to 90 °C and maintained for approximately 12 h. The boiling point of methyl methacrylate is around 100 °C, so care should be taken when adjusting this temperature. After polymerization, the vitrified transparent samples are retrieved by breaking the test tube.

Excimer Formation

As mentioned, a simple way of evaluating the amount of intramolecular excimers is by the corrected ratio I_D/I_M due to the emission of the monomer band at the wavelength selected for the dimer emission band, according to:

$$\frac{I_D}{I_M} = \frac{I_{NMnN,400} - I_{MN,400}}{I_{norm}} \quad (21)$$

where $I_{NMnN,400}$ is the normalized fluorescence intensity for any NMnN at 400 nm (dimer band), $I_{MN,400}$ is the normalized intensity obtained with MN at the same wavelength, and I_{norm} is the intensity at the wavelength used for normalization of the emission spectra. Fluorescence intensity is measured in arbitrary units. Normalization permits comparison of the shapes of two or more such spectra by scaling them so that they all have the same intensity at the wavelength chosen for normalization. A similar equation can be used for evaluation of I_D/I_M for the DMN and the Pn.

Fluorescence Anisotropy

Energy transfer and energy migration between identical chromophores can be inferred from studies of fluorescence depolarization. When an isolated polymer chain or bichro-

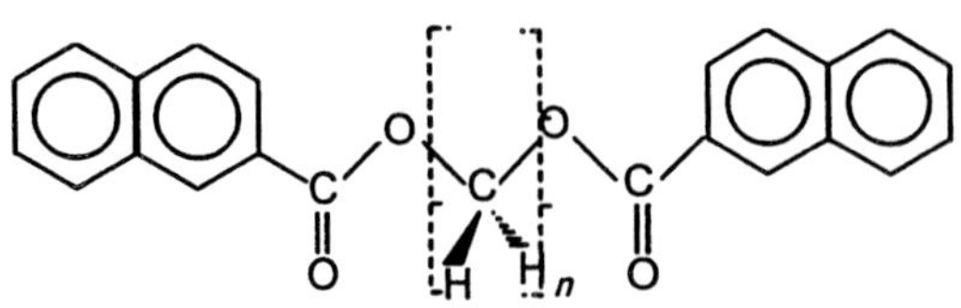

NM*n*N

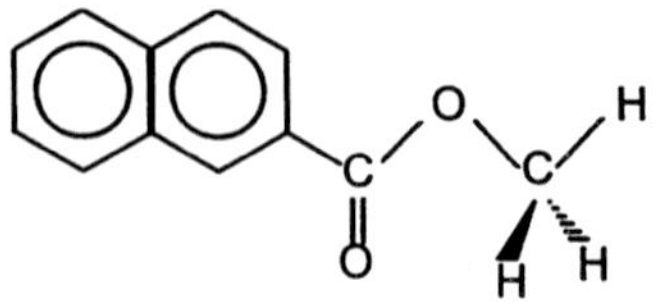

MN

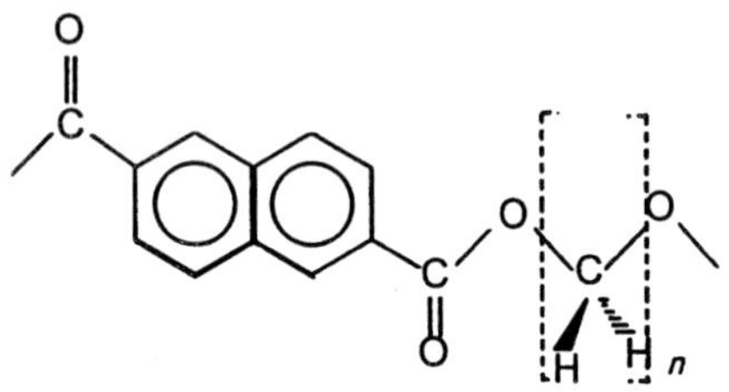

Pn

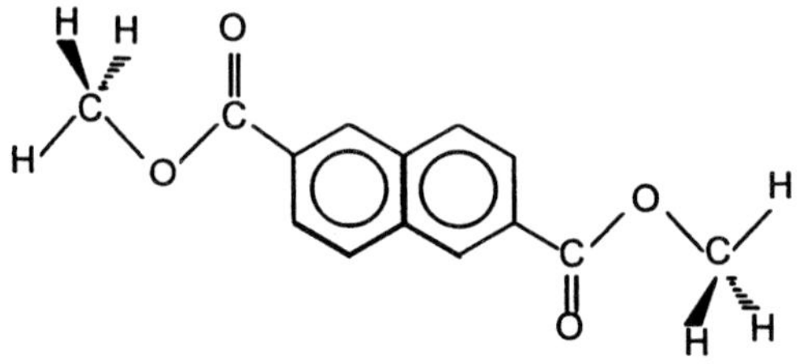

DMN

Figure 12.10 Structures of NM*n*N, MN, P*n*, and DMN.

mophoric molecule is placed in a rigid medium, such as a solid PMMA matrix, the only extrinsic cause of depolarization may be due to an intramolecular energy transfer process, and the polarization will decrease as the efficiency of such a process increases.

Instrumentation

The fluorescence measurements are performed at room temperature using an SLM 8100 AMINCO fluorometer, with a similar scheme to the SLM 8000 depicted in Figure 12.5, equipped with a cooled photomultiplier and a double monochromator in the excitation path. The excitation and emission slit widths of 4 nm give a good signal-to-noise ratio. Here the slit width is defined in terms of the range of wavelengths that pass through the slit, and not in terms of the physical gap between the two edges of the slit. Expression of the slit width in terms of wavelength is useful because it provides information about the finest spectral features that can be resolved in the measurement. The excitation wavelengths were 292 and 294 nm for NM*n*N and P*n*, respectively. Right-angle geometry and rectangular 10-mm path quartz cells were used for measurements of dilute liquid solution samples. Front-face illumination with the incident beam of the samples forming a 60° angle with the plain surface of the solid sample was used.

Results

Emission Spectra

Bichromophoric Compounds Figure 12.11 depicts emission spectra for MN and NM*n*N compounds in methanol ($\eta \approx$ 0.55 cp) and hexanol ($\eta \approx 4.6$ cp) at 25 °C and in PMMA at room temperature. All spectra were normalized to the maximum of the monomer band obtained for MN. The band displaced to the red, relative to the monomer band observed for MN for each compound, is due to intramolecular excimer emission. Inspection of Figure 12.11 shows that excimer emission for a particular sample is always higher in the less viscous solvent, and there is a strong dependence on n. In methanol, the I_D/I_M ratio changes with n as $3 > 6 > 4 > 5 > 2$. In hexanol, there is an odd-even effect, with the largest values for I_D/I_M observed at $n = 3$ and 5. In PMMA an odd-effect also appears, but the intensity of the excimer band is weaker than in the measurements performed in the alcohols.

To study the dependence of I_D/I_M on η, we normally use several systems to manipulate η: (a) methanol-ethylene glycol mixtures at 25 °C, (b) ethylene glycol at 15 to 50 °C, and (c) the series of alcohols $H(CH_2)_mOH$, $m = 1 - 6$ at 25 °C. Linear extrapolation of the behavior observed at high η, when depicting I_D/I_M versus $1/\eta$, yields $(I_D/I_M)_{\eta \to \infty}$, the dimensionless ratio extrapolated to a solvent of infinite η, where the dynamic contribution to intramolecular excimer formation is suppressed. An odd-even effect is always obtained in the present systems, with the maximum values at $n = 3$ and 5. In general, although I_D/I_M quantitatively depends on the nature of the solvent, two different behaviors of I_D/I_M with n can be observed when using solvents of low and high η. The two types are illustrated in Figure 12.11 by the results in methanol on the one hand and results for hexanol, PMMA, or $(I_D/I_M)_{η→∞}$ on the other hand.

Polymers Figure 12.12 depicts the normalized emission spectra for DMN and Pn in methanol, 1,2-dichloroethane, and PMMA. Emission from intramolecular excimers can sometimes be resolved into one or two bands, depending on the nature of the solvent. The influence of the nature of the solvent on the I_D/I_M trend with n is larger for Pn than for NMnN. For a particular solvent and for spacers of a specific size, the value of I_D/I_M is larger for Pn than for NMnN. In the case of the smaller spacers P2 and P3, the values of I_D/I_M in PMMA are very similar. These results suggest that there is a mechanism for populating the excimer in Pn that is not accessible in NMnN, and this mechanism is more effective for the smaller size spacers. Energy migration is an example of this mechanism. Figure 12.13 summarizes these results.

Fluorescence Anisotropy

Figure 12.14 depicts the emission anisotropy spectra for five bichromophoric compounds, NMnN, in a solid PMMA matrix at room temperature. Also depicted in the figure are the normalized emission spectra for one member of the series with n equal to 3 and MN in PMMA at room temperature.

Looking at the emission spectra first, there is a spectral region, at approximately 350 to 400 nm, where the results of the anisotropy are more reliable because the intensity of the emission is sufficient to provide a reasonable signal-to-noise ratio. Outside this range, the anisotropy emission spectra have a high noise, and some of the anisotropy values are even larger than the theoretical limit of 0.4 for a random orientation of independent chromophores.[3] These unreliable nonphysical values are ignored in the following analysis. In the spectral region where the data are reliable, we focus on the relative values of anisotropy r for MN and NMnN. The anisotropy is always larger for MN containing a single chromophore. Values of r for NMnN, from the average in the range 350 to 400 nm, show that r varies as $2 > 3 > 4 > 5 > 6$ as n increases from 2 to 6. However, due to the closeness of the anisotropy spectra, single-point anisotropy values could give a different behavior. Similar conclusions can be reached from the analysis of r for Pn. Nevertheless, r values for Pn at each n are smaller than the ones obtained for NMnN. Table 12.1 provides the average values of the anisotropies for NMnN and Pn.

Assuming that the depolarization observed for a model compound with a single chromophore like MN (DMN) is due to intrinsic reasons, the experimental efficiency of intramolecular energy transfer, ϕ_{exp}, for NMnN (Pn) could be estimated as

$$\phi_{exp} = (r_{MN} - r_{NMnN})/r_{MN} \qquad (22)$$

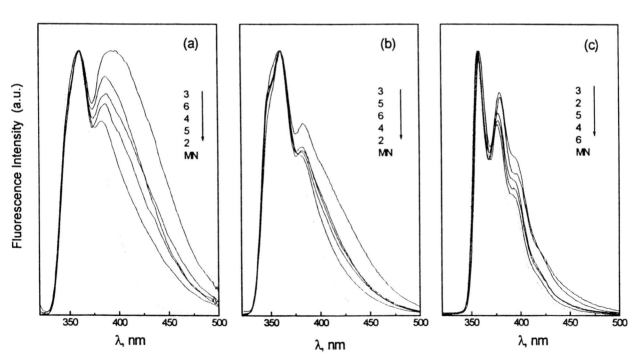

Figure 12.11 Emission spectra normalized at the maximum of the monomer band for MN and NMnN with $n = 2$ to 6 in (a) methanol at 25 °C, (b) hexanol at 25 °C, and (c) poly(methyl methacrylate) (PMMA) at room temperature. The excitation wavelength is 292 nm.

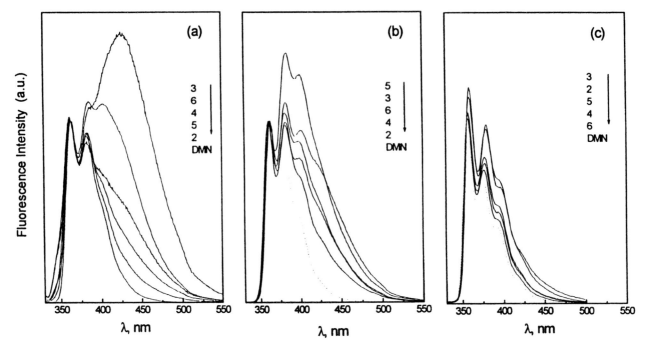

Figure 12.12 Emission spectra normalized at the maximum of the monomer band for DMN and P*n* with *n* = 2 to 6 in (a) methanol at 25 °C, (b) 1,2-dichloroethane at 25 °C, and (c) poly(methyl methacrylate) (PMMA) at room temperature. The excitation wavelength is 294 nm.

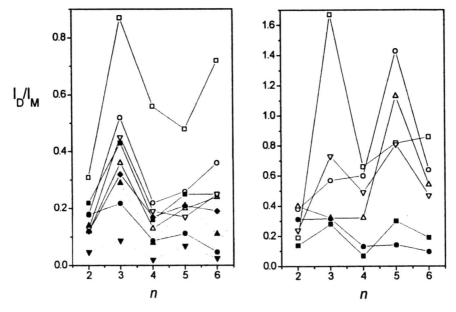

Figure 12.13 Ratios of I_D/I_M for NM*n*N (*left panel*) and P*n* (*right panel*) in different systems. Key: methanol (□), 1,2-dichloroethane (○), 1,1,2,2-tetrachloroethane (△), methyl methacrylate (▽), 1-hexanol (◆), ethylene glycol (■), $(I_D/I_M)_{\eta-\infty}$ from ethylene glycol-methanol mixtures (▲), $(I_D/I_M)_{\eta-\infty}$ from *n*-alcohols (▼) at 25 °C and rigid PMMA matrix (●) at room temperature.

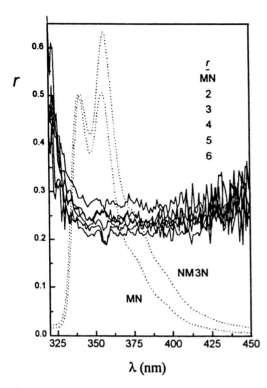

Figure 12.14 Fluorescence emission anisotropy spectra in a solid poly(methyl methacrylate) (PMMA) matrix at room temperature for MN and NMnN with n = 2 to 6, and normalized emission spectra (dashed line) for NM3N and MN in a solid PMMA matrix at the same experimental conditions ($\lambda_{\text{excitation}}$ = 292 nm).

where r_{MN} and $r_{\text{NM}n\text{N}}$ are the anisotropies observed at each wavelength for MN (DMN) and NMnN (Pn), respectively. The results of ϕ_{exp}, are summarized in Table 12.1. Efficiencies for Pn are larger than for NMnN compounds. However, the relative values of ϕ_{IET} varies with n as 2 > 3 > 4 > 5 > 6 for both NMnN and Pn.

Experimental Förster Radius

The Förster radius R_0 for the transfer between MN (DMN) chromophores in NMnN (Pn) is obtained by using eq 17. It is necessary to obtain J and Q_D. Figure 12.15 depicts the overlap between absorption and emission spectra from both MN and DMN as model compounds for NMnN and Pn containing a single chromophore. Graphical integration by a representation of $\lambda^4 I(\lambda)\varepsilon(\lambda)$ versus λ will provide J. The units of J are $M^{-1}cm^3$ when ε and λ are $M^{-1}cm^{-1}$ and cm, respectively. Experimental Q_D are obtained by the procedure described previously by Mendicuti and Mattice.[45] Measurements of Q and J for dilute solutions of naphthalene, MN and DMN, using methanol and propanol as solvents, and naphthalene as the reference compound give the ratios collected in Table 12.2. The averaged values of R_0 for the MN → MN and DMN → DMN transfers are ~12 and ~13

Å, respectively. We take these values to be R_0 for the intramolecular energy transfer between the end-to-end 2-naphthoyl groups for NMnN and between the 2,6-dinaphthoyl groups for Pn.

Theoretical Rationalization: Equilibrium RIS Model

The fragments studied were NMnN compounds for n ranging from 2 to 6. As was described more extensively,[56,57,62-70] the amount of intramolecular excimer formation and the efficiency of intramolecular energy transfer were obtained by a discrete enumeration of each conformation allowed to the flexible spacer Bn. The bond lengths, bond angles, and first- and second-order interaction energies were already used previously.[67,69,70] The rotational isomers for the O–CH$_2$ and CH$_2$–CH$_2$ bonds are the classic rotational isomers *trans*, *gauche*$^+$, and *gauche*$^-$. Each of these rotational isomers is split into three isoenergetic states at 0 and 0 ± 20°, where 0 is the torsion angle used in the customary rotational isomeric state model, as a crude approximation to the nonactivated torsional oscillations that can occur within each rotational isomeric state at the temperature of interest. There are also two equally weighted rotational isomers, with torsion angles of 0° and 180°, at each C^{ar}–C* bond, where C^{ar} denotes a carbon atom in the ring and C* is the carbonyl carbon atom. A total of $2^2 9^{n+1}$ conformations was generated for each bichromophoric compound.

For the evaluation of intramolecular excimer formation, each conformation was examined for fulfillment of the conditions for the formation of a sandwich geometry between six-membered rings of the pair of end-naphthalene rings. The criteria used for the formation of this sandwich geometry between chromophores are $3.35 < d_z < 4.0$ Å, $0 < d_{xy} < 1.4$ Å, and $0 < \Psi < 40°$, where d_z is the shortest of the distances between the center of mass of a six-membered ring in one naphthalene unit and the mean plane of a six-

Table 12.1 Averaged values of r and ϕ_{exp} obtained in the range of 350 to 420 nm for emission for NMnN and Pn

n	r		ϕ_{exp}	
	NMnN	Pn	NMnN	Pn
2	0.212 ± 0.015	0.050 ± 0.009	0.22	0.61
3	0.226 ± 0.009	0.057 ± 0.009	0.17	0.56
4	0.233 ± 0.007	0.069 ± 0.015	0.15	0.46
5	0.243 ± 0.012	0.071 ± 0.015	0.11	0.45
6	0.248 ± 0.008	0.077 ± 0.014	0.10	0.40
RC[a]	0.273 ± 0.008	0.129 ± 0.033	—	—

a. RC denotes the reference compounds MN and DMN for NMnN and Pn, respectively.

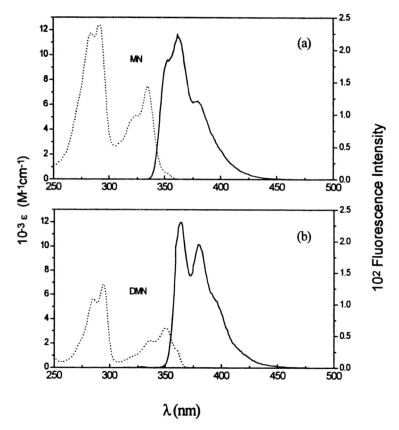

Figure 12.15 Absorption (*light dashed line*) and emission (*heavy solid line*) spectra, the first one as ε vs. λ, and the second one normalized to 1, for (a) MN and (b) DMN in propanol at room temperature, showing the overlap between absorption and emission spectra for the MN → MN and DMN → DMN transfer processes. (The emission curve in part a is adapted from reference 70. Copyright 1996 by John Wiley and Sons.)

membered ring in another naphthalene unit, and vice versa; d_{xy} is the lateral offset of the two six-membered rings that define d_z; and Ψ is the angle between the normals to the planes of the rings. For the calculation of the efficiency of intramolecular energy transfer two parameters, denoted p_R and ϕ_{ET}, are obtained. p_R denotes the probability for finding the centers of mass of the two naphthalene rings separated by a distance no larger than R, and ϕ_{ET} is the effi-

ciency of energy transfer, eq 9. In the frozen system, this equation can be rewritten as

$$\phi_{ET} = \sum_{1}^{N} p\left[1 + \kappa^2 R^6/(2/3)R_0^6\right]^{-1} \qquad (23)$$

where p denotes the probability of each of the N-generated conformations, and R is the distance between the centers of two naphthalene rings.

The left panel of Figure 12.16 depicts the probability p_{exc} of conformations whose geometry fulfill the excimer criteria, as a function of n, at 400 and 500K using an equilibrium RIS model.[44] The trend of this probability with n reproduces the odd-even effect in the experimental results for the variation of I_D/I_M with n at infinite (and high) η.

The right panel of Figure 12.16 depicts the dependence of ϕ_{ET} on n for several different postulates about the value of R_0 ranging from 3 to 19 Å. An odd-even effect is predicted for ϕ_{ET} for R_0 smaller than 8 Å. The monotonic decrease in ϕ_{ET} as n increases from 2 to 6 does not set in until the assumed R_0 is as large as 11 Å. These theoretical results, together with the experimentally determined R_0 of 12

Table 12.2 Quantum yields, Q, and overlap integrals, J, for MN → MN and DMN → DMN energy transfer processes relative to the one for the self-transfer naphthalene → naphthalene, measured in methanol and 1-propanol at 25 °C

Solvent	D	A	Q_D/Q_{naph}	$J_{D \to A}/J_{naph \to naph}$
Methanol	DMN	DMN	10.02	3.60
Methanol	MN	MN	6.40	4.19
1-Propanol	DMN	DMN	9.32	3.59
1-Propanol	MN	MN	6.74	3.84

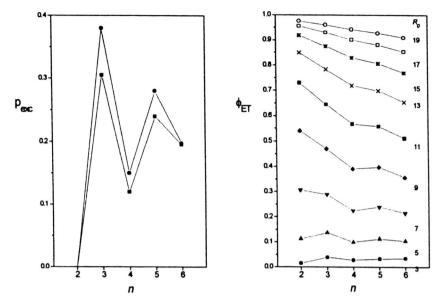

Figure 12.16 Probabilities of potential excimer-forming conformations p_{exc} versus n at 400K (■) and 500K (●) (*left panel*), and efficiency of Förster transfer ϕ_{ET} calculated using different postulates of the values of R_0 ranging from 5 to 19 Å (*right panel*).

and 13 Å for MN → MN and DMN → DMN transfers for NMnN and Pn, respectively, agree with the observed trend of ϕ_{exp} with n for NMnN and Pn in a PMMA solid matrix, which shows a monotonic decreasing as n increases. Similar results are obtained by evaluating p_R.

Conclusions

1. All NMnN and Pn compounds show, to a greater or lesser degree, a broadening of the monomeric band, which, at the experimental conditions, is attributed to the intramolecular formation of excimers.
2. The extent of intramolecular excimer formation, measured by the I_D/I_M ratio, depends strongly on the chromophore structure and on the conformation and dynamics of the flexible spacer Bn.
3. Because of the large τ_f of the MN and DMN chromophores, the I_D/I_M ratio is strongly dependent on η. As the viscosity of the medium decreases, the amount of intramolecular excimer emission increases, due to an enhanced dynamic contribution to the excimer formation.
4. Although I_D/I_M quantitatively depends on the nature of the solvent, I_D/I_M shows two different behaviors with n when using solvents of low and high viscosities for NMnN. However, the I_D/I_M trend with n for Pn also seems to depend on the nature of the solvent.
5. Pn show a larger amount of intramolecular excimer emission than NmnN with the same n. Intramolecular excimers in Pn could also be populated via energy migration, which is more effective for the smaller values of n.
6. The RIS equilibrium model allows the calculation of the probability of conformations that fulfill the severe geometric requirements expected for the formation of an excimer. Theoretical results obtained thereby reproduce the experimental trend of I_D/I_M with n in media of high viscosity.
7. R_0 for the MN and DMN chromophores implied in the intramolecular energy tranfer for NMnN and Pn compounds have been estimated. Results are approximately 12 and 13 Å for MN and DMN, respectively.
8. The experimental fluorescence anisotropies r obtained in a rigid PMMA matrix show that intramolecular energy transfer is more effective as n decreases. For a particular n, intramolecular energy transfer is more effective in Pn than the NMnN, since a multistep intramolecular energy transfer process takes place in Pn.
9. The RIS equilibrium model also permits calculating several parameters related to the efficiency of intramolecular energy transfer. Theoretical results agree with the experimental decreases of the efficiency of the intramolecular energy transfer process as n increases for NMnN and Pn compounds and the experimental R_0 values.

Acknowledgments This research was supported by the NSF/ EPIC Center for the Molecular and Microstructure of Composites (CMMC), National Science Foundation Division of Material Research 9523278, and Comisión Interministerial de Ciencia y Tecnología grant PB-97-0778 (to F.M.). F. Mendicuti thanks M. L. Heijnen for assistance with the preparation of the manuscript.

References

1. Morawetz, H. *Science* **1979**, *203*, 405–410.
2. Winnik, M. A., Ed.; *Photophysical and Photochemical Tools in Polymer Science: Conformation, Dynamics, and Morphology*; D. Reidel: Dordrecht, 1982.
3. Lakowicz, J. R. *Principles of Fluorescence Spectroscopy*; Plenum Press: New York, 1983; pp 1–496.
4. Phillips, D. Ed.; *Polymer Photophysics: Luminescence, Energy Migration and Molecular Motion in Synthetic Polymers*; Chapman & Hall: London, 1985; pp 2–3.
5. Guillet, J. *Polymer Photophysics and Photochemistry*, Cambridge University Press: Cambridge, 1985; pp 1–391.
6. Rabeck, J. F. *Mechanisms of Photophysical Processes and Photochemical Reactions in Polymers: Theory and Applications*; Wiley: Chichester, 1987.
7. Frank, C.; Harrah, L. *J. Chem. Phys.* **1974**, *61*, 1526–1541.
8. Ishii, T.; Handa, T.; Matsunaga, S. *Macromolecules* **1978**, *11*, 40–46.
9. Frank, C. W.; Gashgari, M. A. *Macromolecules* **1979**, *12*, 163–165.
10. Morawetz, H. In *Photophysical and Photochemical Tools in Polymer Science: Conformation, Dynamics, and Morphology*; Winnik, M. A., Ed.; D. Reidel: Dordrecht, 1982; pp 1–13.
11. Berlman, I. B. *Handbook of Fluorescence Spectra of Aromatic Molecules*; Academic Press: New York, 1971; pp 12–29, 421.
12. Birks, J. B. *Photophysics of Aromatic Molecules*; Wiley-Interscience: London, 1970; p. 144.
13. Braun, H.; Förster, Th. *Ber. Bunsenges. Phys. Chem.* **1966**, *70*, 1091.
14. Vala, M. T. Jr.; Haebig, J.; Rice, S. A. *J. Chem. Phys.* **1965**, *43*, 886–897.
15. Hirayama, F. *J. Chem. Phys.* **1965**, *42*, 3163–3171.
16. De Schryver, F. C.; Collart, P.; Vandendriessche, J.; Goedeweeck, R.; Swinnen, A.; van der Auweraer, M. *Acct. Chem. Res.* **1987**, *20*, 159–166.
17. Chakraborty, T.; Lim, E. C. *J. Chem. Phys.* **1995**, *99*, 17505–17508.
18. Zachariasse, K.; Kühnle, W. *Z. Phys. Chem. Neu Folge* **1976**, *101*, 267–276.
19. Zachariasse, K. A.; Duveneck, G.; Busse, R. *J. Am. Chem. Soc.* **1984**, *106*, 1045–1051.
20. Zachariasse, K. A.; Striker, G. *Chem. Phys. Lett.* **1988**, *145*, 251–254.
21. Johnson, G. E. *J. Chem. Phys.* **1975**, *63*, 4047–4053.
22. Buchberger, E. M.; Mollay, B.; Weixelbaumer, W. D.; Kauffmann, H. F. *J. Chem. Phys.* **1988**, *89*, 635–652.
23. Bokobza, L.; Jasse, B.; Monnerie, L. *Eur. Polym. J.* **1977**, *13*, 921–924.
24. Bokobza, L.; Jasse, B.; Monnerie, L. *Eur. Polym. J.* **1980**, *16*, 715–720.
25. De Schryver, F. C.; Moens, L.; van der Auweraer, M.; Boens, N.; Monnerie, L; Bokobza, L. *Macromolecules* **1982**, *15*, 64–71.
26. Zachariasse, K. A.; Duveneck, G.; Kühnle, W.; Reynders, P.; Striker, G. *Chem. Phys. Lett.* **1987**, *133*, 390–398.
27. Reynders, P.; Dresskamp, H.; Kühnle, W.; Zachariasse, K. A. *J. Chem. Phys.* **1987**, *91*, 3982–3992.
28. Collart, P.; Toppet, S.; De Schryver, F. C. *Macromolecules* **1987**, *20*, 1266–1271.
29. Reynders, P.; Kühnle, W.; Zachariasse, K. A. *J. Chem. Phys.* **1990**, *94*, 4073–4082.
30. Pham-Van-Cang, C.; Bokobza, L.; Monnerie, L.; Vandendriessche, J.; De Schryver, F. C. *Polym. Comm.* **1980**, *27*, 89–92.
31. Jing, D. P.; Bokobza, L.; Monnerie, L.; Collart, P.; De Schryver, F. C. *Polymer* **1990**, *31*, 110–114.
32. Freeman, B. D.; Bokobza, L.; Sergot, P.; Monnerie, L.; De Schryver, F. C. *Macromolecules* **1990**, *23*, 2566–2573.
33. Goldenberg, M.; Emert, J.; Morawetz, H. *J. Am. Chem. Soc.* **1978**, *100*, 7171–7177.
34. Itagaki, H.; Horie, K.; Mita, I.; Washio, M.; Tagawa, S.; Tabata, Y.; Sato, H.; Tanaka, Y. *Macromolecules* **1989**, *22*, 2520–2525.
35. Itagaki, H.; Horie, K.; Mita, I.; Washio, M.; Tagawa, S.; Tabata, Y.; Sato, H.; Tanaka, Y. *Macromolecules* **1990**, *23*, 1686–1690.
36. Reynders, P.; Kühnle, W.; Zachariasse, K. A. *J. Am. Chem. Soc.* **1990**, *12*, 3929–3939.
37. Förster, Th. *Ann. Phys.* **1948**, *2*, 55.
38. Webber, S. E. *Chem. Rev.* **1990**, *90*, 1469–1482.
39. Latt, S. A.; Cheung, H. T.; Blout, E. R. *J. Am. Chem. Soc.* **1965**, *87*, 995–1003.
40. Stryer, L.; Haugland, R. P. *Proc. Natl. Acad. Sci. USA* **1967**, *58*, 719–726.
41. Stryer, L. *Science* **1968**, *162*, 527–533.
42. Berlman, I. B. *Energy Transfer Parameters of Aromatic Compounds*; Academic Press: New York, 1973.
43. Weber, G. *Biochem J.* **1952**, *51*, 145–167.
44. *SLM 8000C Photon Counting Spectrofluorometer Operator's Manual*; SLM Instruments: Urbana, IL, 1988; p 18.
45. Mendicuti, F.; Mattice, W. L. *Polym. Bull.* **1989**, *22*, 557–563.
46. Melhuish, W. H. *J. Phys. Chem.* **1961**, *65*, 229–235.
47. Melhuish, W. H. *J. Opt. Soc. Am.* **1962**, *52*, 1256–1258.
48. Cho, D.; Mattice, W. L.; Porter, L. J.; Hemingway, R. W. *Polymer* **1989**, *30*, 1955–1958.
49. Guillet, J. *Trends in Polym. Sci.* **1996**, *4*, 41–46.
50. Berlman, I. B. *Energy Transfer Parameters of Aromatic Compounds*; Academic Press: New York, 1973; p 308.
51. Morawetz, H. *Science* **1988**, *240*, 172–176.
52. Haas, E.; Katchalski-Katzir, E.; Steinberg, I. Z. *Biopolymers* **1978**, *17*, 11–31.
53. Stryer, L. *Ann. Rev. Biochem.* **1978**, *47*, 819–846.
54. Martinho, J. M. G.; Castanheira, E. M. S.; Reis e Sousa, A. T.; Saghbini, S.; André, J. C.; Winnik, M. A. *Macromolecules* **1995**, *28*, 1167–1171.
55. Mattice, W. L.; Suter, U. W. *Conformational Theory of Large Molecules: The Rotational Isomeric State Model in Macromolecular Systems*; Wiley: New York, 1994.

56. Mendicuti, F.; Viswanadhan, V. N.; Mattice, W. L. *Polymer* **1988**, *29*, 875–879.
57. Mendicuti, F.; Patel, B.; Viswanadhan, V. N.; Mattice, W. L. *Polymer* **1988**, *29*, 1669–1674.
58. Mendicuti, F.; Patel, B.; Waldeck, D. H.; Mattice, W. L. *Polymer* **1989**, *30*, 1680–1684.
59. Patel, B.; Mendicuti, F.; Mattice, W. L. *Polymer* **1992**, *33*, 239–242.
60. Mendicuti, F.; Mattice, W. L. *Polymer* **1992**, *33*, 4180–4183.
61. Mendicuti, F.; Mattice, W. L. *Makromol. Chem.* **1993**, *194*, 2851–2860.
62. Mendicuti, F.; Patel, B.; Mattice, W. L. *Polymer* **1990**, *31*, 453–457.
63. Mendicuti, F.; Patel, B.; Mattice, W. L. *Polymer* **1990**, *31*, 1877–1882.
64. Mendicuti, F.; Saiz, E.; Zúñiga , I.; Mattice, W. L. *Polymer* **1992**, *33*, 2031–2035.
65. Mendicuti, F.; Saiz, E.; Mattice, W. L. *Polymer* **1992**, *33*, 4908–4912.
66. Gallego, J.; Mendicuti, F.; Saiz, E.; Mattice, W. L. *Polymer* **1993**, *34*, 2475–2480.
67. Mendicuti, F.; Saiz, E.; Mattice, W. L. *J. Polym. Sci., Part B: Polym. Phys.* **1993**, *31*, 213–220.
68. Bravo, J.; Mendicuti, F.; Mattice, W. L. *J. Polym. Sci., Part B: Polym. Phys.* **1994**, *32*, 1511–1519.
69. Martin, O.; Mendicuti, F.; Saiz, E.; Mattice, W. L. *J. Polym. Sci., Part B: Polym. Phys.* **1995**, *33*, 1107–1116.
70. Martin, O.; Mendicuti, F.; Saiz, E.; Mattice, W. L. *J. Polym. Sci., Part B: Polym. Phys.* **1996**, *34*, 2623–2633.
71. Gallego, J.; Mendicuti, F.; Saiz, E.; Mattice, W. L. *Comput. Polym. Sci.* **1994**, *4*, 7–11.
72. Martin, O.; Mendicuti, F.; Saiz, E.; Mattice, W. L. *J. Polym. Sci., Part B:* **1999**, *37*, 253–266.

PART IV

METHODS TO CHARACTERIZE
THE MOLECULAR ORGANIZATION
OF POLYMERS

13

Thermal Analysis

ALAN T. RIGA
WEI-PING PAN
JOHN CAHOON

Thermal analysis measures the physical properties, thermal transitions, and chemical reactions of a polymer as a function of temperature, heating rate, and atmosphere. The major thermal analysis techniques used to characterize polymers and the principal property each measures are differential scanning calorimetry (DSC) for heat flow; thermogravimetric analysis (TGA) for mass loss; differential thermal analysis (DTA) for temperature difference; thermomechanical analysis (TMA) for dimensional stability; dynamic mechanical analysis (DMA) for stiffness and damping; and dielectric thermal analysis (DETA) for permittivity and conductivity. The field of thermal analysis also embraces evolved gas analysis (EGA), which is made up of several techniques that are frequently coupled to thermal analytical instruments for the purpose of analyzing gaseous decomposition products. The most important EGA techniques are mass spectrometry (MS) and Fourier transform infrared spectroscopy (FTIR). Microthermal analysis (MTA) is the name given to atomic force microscopy coupled to DSC, TMA, or DTA measurements.

Together, the composition, structure, and morphology of a polymer control its properties and performance, as shown in Figure 13.1. Their interrelationships affect its processing, ultimate properties, and end-use performance. Compounding with fillers and additives introduces additional variables; but often it is essential to allow the development of specific properties that are not intrinsic in the natural polymer.

The main properties that characterize all polymer molecules are their size, molecular weight, molecular weight distribution, chain stiffness or rigidity, physical state (amorphous, semicrystalline, or crystalline), and the nature of any cross-linked network structure. These molecular components can be used to classify all polymers; they are pictured in Figure 13.2 and described in Table 13.1.[1]

The practical information that may be obtained from polymer thermal analysis is summarized in Table 13.2. The examples described illustrate the type of information that can be gained when using a particular thermal analytical technique.[1] Oxidative thermal techniques[2-4] monitor the effect of antioxidants and other stabilizers. Thermal decomposition spectra are used to predict longevity and durability of polymers. Reaction kinetics at high temperatures can yield valuable information about, for example, face shields on a reentry vehicle.[5] Thermal analysis is used routinely in quality-control protocols for both incoming and outgoing polymer parts.

Thermal analytical methods are invaluable research, quality-control, and performance measuring tools for polymers. Polymers are particularly suited to these types of analyses, for they may be amorphous with a glass transition temperature or semicrystalline with a melting temperature, or they may contain both amorphous and crystalline phases. Thermal analysis can easily determine the proportions of each phase and the physical properties of the sample.

Thermal analysis methods measure a property of a sample as a function of temperature or time while the sample is subjected to a constant or variable programmed temperature. Thermal methods, individually and in combination, accurately and reproducibly measure many physical and chemical properties of polymers and explain polymer processes at temperatures ranging from -140 to $1200\ °C$.[1-10]

Examples of the physical properties measured by these thermal analytical techniques are heat of fusion, heat of crystallization, specific heat capacity, volatilization, oxidation onset temperature (OOT), oxidation induction time (OIT), thermal conductivity, coefficient of thermal expansion, dynamic modulus, and thermal conductivity.[11-24] Physical transitions may also be exothermic or endothermic. The principal exothermic transition is crystallization, and crystalline transitions also may be endothermic or exothermic. Endothermic transitions include adsorption, vaporization, glass transition, melting, sublimation, and absorption. As the sample changes heat capacity and passes through its glass-

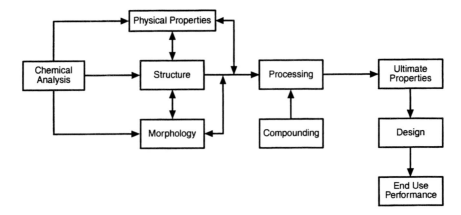

Figure 13.1 Structure, property, and performance relationships in polymers.

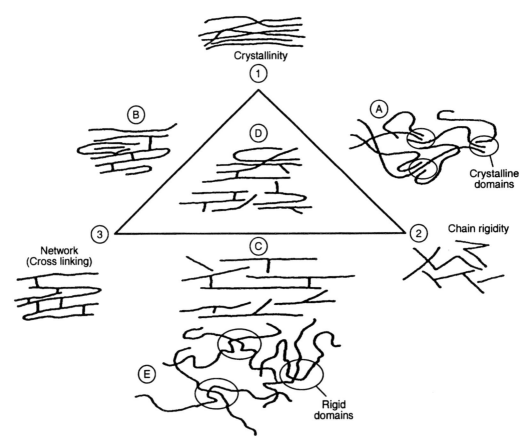

Figure 13.2 Basic elements of engineering polymers. Key: A = crystalline domains in a viscous network—see Table 13.1 for details; B = moderate cross linking, with some crystallinity; C = rigid chains, partly cross linked; D = crystalline domains with rigid chains between them and cross-linking chains—see Table 13.1 for details; E = rigid-chain domains in a flexible-chain matrix—see Table 13.1 for details; 1 = flexible and crystyallizable chains; 2 = cross-linked amorphorous networks of flexible chains—see Table 13.1 for details; 3 = rigid chains.

Table 13.1 Basic elements of engineering polymers

Location	Characteristics	Examples[a]
1	Flexible and crystallizable chains	PE PP PVC PA
2	Cross-linked amorphous networks of flexible chains	Phenol-formaldehyde cured rubber Styrenated polyester
3	Rigid chains	PIs (ladder molecules)
A	Crystalline domains in a viscous network	PET Terylene (dacron) Cellulose acetate
B	Moderate cross linking, with some crystallinity	Chloroprene rubber Polyisoprene
C	Rigid chains, partly cross linked	Heat-resistant materials
D	Crystalline domains with rigid chains between them and cross-linking chains	High-strength and temperature-resistant materials
E	Rigid-chain domains in a flexible-chain matrix	Styrene-butadiene-styrene, triblock polymer Thermoplastic elastomer

a. PE = polyethylene; PP = polypropylene; PVC = poly(vinyl chloride); PA = polyamide; PI = polyimide; PET = poly(ethylene terephthalate)

transition temperature, there is an endothermic baseline shift.[22-23]

Chemical transitions detected in a DSC analysis may be exothermic or endothermic. Exothermic transitions include chemisorption, oxidative degradation, combustion, polymerization or cure, and catalytic reactions. Decomposition, oxidation-reduction reactions, dehydration, and solid state reactions may be exothermic or endothermic. Desolvation and dehydration are endothermic.[15-19,22-23]

Differential Scanning Calorimetry

Differential scanning calorimetry (DSC) is used to determine a wide range of physical properties of materials, including

Table 13.2 Practical information derived from polymer analysis methods

Test method	Property	Practical information provided
Differential thermal analysis (DTA)	Transition temperatures, glass transition temperature, T_g; melt/crystallization temperature, T_m/T_c	Phase changes, T_g, T_m, T_c
Differential scanning calorimetry (DSC)	Heat of polymerization, heat of fusion	Phase changes, reaction kinetics, oxidative stability
Modulated temperature DSC (MTDSC)	T_g, T_m, T_c, heat capacity, oxidation onset time and temperature	Degree of cross linking, degradation, inhibitor content, and effectiveness
Thermogravimetric analysis (TGA) and Derivative TGA (DTGA)	Composition, weight loss with time or temperature, mass loss spectra	Thermal and oxidative stability, volatilization kinetics
Dynamic mechanical analysis (DMA)	Elastic modulus, loss modulus, tan delta	Mechanical properties, phase transitions, damping, softening, cross-linking effects
Thermal mechanical analysis (TMA)	Penetration temperature, expansion coefficient, T_g, T_m	Dimensional stability, T_g, T_m, phase changes, modulus, compliance, deflection temperature under load
Mechanical spectroscopy	Viscosity, normal stress, shear elastic, and loss modulus	Rheological properties, flow behavior melt or solution elasticity, yield stress
Evolved gas analysis (EGA)		
Infrared spectroscopy (IR)	Chemical functional groups	Molecular information
Fourier transform IR (FTIR)	Chemical functional groups	Functional group identification
Mass spectroscopy (MS)	Mass/charge of ions produced	Molecular structure
Microthermal analysis (micro-TA)	Micron level topography, DSC and TMA on selected areas at high magnification	Phase changes, T_g, T_m, T_c; curing and polymerization, degree cross-linking, degradation

the glass-transition temperature T_g, the melting temperature T_m, and solid-solid transitions. In this technique, a sample and a reference material are subjected to a controlled temperature program. When a phase transition such as melting occurs in the sample, an input of energy is required to keep sample and reference at the same temperature. This difference in energy is recorded as a function of temperature to produce the DSC trace.[21] For example, the DSC curve of poly(ethylene terephthalate) is shown in Figure 13.3.

Differential scanning calorimeters operate in one of two modes—the differential temperature mode or the temperature servo measurement mode. Because the temperature difference across the sample seldom exceeds a few tenths of a degree, the DSC instrument operates well within classical calorimetry conditions and produces valid thermodynamic data.[24]

An example of a differential temperature instrument is the TA Instruments DSC 910 system, and Figure 13.4 shows a cross-sectional view of its DSC cell. A constantan disc transfers heat to the sample and reference positions and is also one element of the two temperature-measuring thermoelectric junctions. The polymer sample and a reference material are placed in individual containers that rest on two constantan discs on raised platforms. Heat is transferred through the discs and into the sample and reference

via the containers. The heat flow to the sample and reference is monitored by chromel-constantan thermocouples, which are formed by the junction of the constantan disc and a chromel disc that covers the underside of each platform. As the sample is heated at a constant programmed rate, its temperature increases until it reaches its melting point. Without additional energy, the programmed input of heat would be used to melt the sample, and its temperature would lag behind that of the reference. Therefore, additional heat is supplied through a circuit separate from the temperature program to maintain sample and reference at the same temperature, and it is the energy in this separate circuit that is plotted against temperature. Thus this portion of energy is used only to effect a phase change and is not used to heat the sample. When the amplified and scaled signal in milliwatts is integrated over time, a value for the heat in millijoules associated with the transition is obtained.[24]

The second type of instrument, the temperature servo system, is shown in Figures 13.5 and 13.6.[26] In this design, the sample and reference pans are placed in close contact with platinum resistance thermocouples, which are used to measure their temperature continuously. In addition, the unit has two individual heaters that control the heat flow

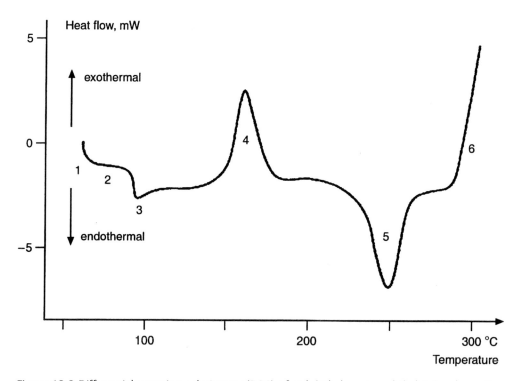

Figure 13.3 Differential-scanning calorimetry (DSC) of poly(ethylene terephthalate): schematic diagram. Key: (1) a start-up deflection that is proportional to the heat capacity of the sample; (2) a heat flow signal that depends on the heat capacity and heating rate; (3) the glass-transition temperature (T_g) of the amorphous portion of the polymer; (4) an exotherm due to cold crystallization or devitrification; (5) the melting point (T_m) of the crystalline portion of the sample; and (6) oxidative decomposition of the sample.

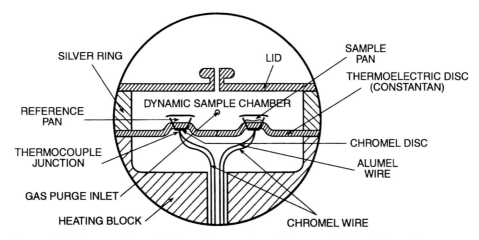

Figure 13.4 Cross section of differential-scanning calorimetry (DSC) cell (heat flux type).

to these samples. When the sample and reference are temperature programmed through the melt, heat to the sample melts the sample, but heat to the reference raises its temperature. The differential temperature signal ΔT developed between sample and reference is amplified, scaled, and displayed.[26,27]

Calibration of DSC

It is necessary to calibrate the DSC instrument with known standard materials. A compilation of known melting temperatures and heats of fusion for many standards is given in Table 13.3,[33] where 15 organics, metals, and standard materials are listed. The values cited in the table can be used to establish the DSC cell constants that are used to convert measured values such as temperature and heats of fusion and crystallization to accurate physical properties.

Calibration from −100 to 600 °C is performed by heating the calibration material at a controlled rate in a controlled atmosphere through a known thermal transition. ASTM Method E-697[28] is used to calibrate temperature (x-axis), and ASTM Methods E-968,[29]) and D-3417 and D-3418[30]

are used to calibrate heat flow (y-axis). ASTM Method E-1860[32] is used to establish elapsed time.

For low-temperature calibration, most prefer to use the melting temperature and heat of fusion of spectroscopic grade organic chemicals, for example, octane at −58 °C, decane at −30 °C, or dodecane at −10 °C. However, the most common melting temperature standards are indium at 156.6 °C, tin at 231.9 °C, and zinc at 419.5 °C. For a two-point temperature calibration, pure materials in the temperature range of interest are used. One can calibrate a DSC with indium and zinc and, with some degree of confidence, extrapolate the temperature correction factors about 100 °C in each direction, giving a calibrated operating range from about 50 to 500 °C.

Sample Preparation

Sample pans should have high thermal conductivity, low heat capacity, and low cost. Pans may be made from aluminum, copper, gold, platinum, glass, carbon, stainless steel, or mild steel, but typically solvent-cleaned aluminum pans are used. The selection of the sample pan depends on specimen size and shape, the maximum temperature of the experiment, the possibility of chemical reaction between pan and sample, and whether or not the pan needs to retain vapors. The pan can be open, covered with a lid and crimped, or hermetically sealed to preserve the sample. Specialized pans may contain a laser-drilled pinhole that functions as a molecular leak, may be designed for an autosampler, or may have a raised center to fit over a thermal sensor. Purge gases, usually nitrogen or helium, are used in the sample and reference compartments at a flow rate of 50 mL/min.

Pans as received must be cleaned to remove oils or lubricants used in manufacturing. After organic oils and lubricants are removed with xylene or acetone, the pan is dried

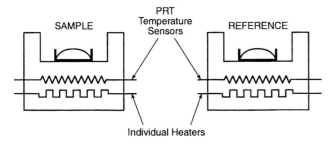

Figure 13.5 Temperature servo cell: schematic diagram. Key: PRT = platinum resistance thermocouples; SAMPLE = sample sensor; REFERENCE = reference sensor.

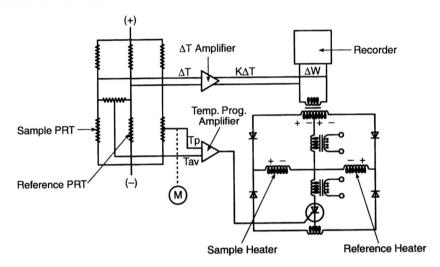

Figure 13.6 Temperature servo system: overall schematic diagram. Key: PRT = platinum resistance thermocouple; ΔT = differential temperature signal between sample and reference; $K\Delta T$ = temperature difference uncalibrated; ΔW = energy difference in watts; T_p = programmed temperature; T_{av} = averaged temperature; M = meter.

with nitrogen and stored according to ASTM Method E-1858.[34]

Samples should be kept thin to minimize any heat transfer problems that may arise from irregular sample size and shape. Therefore, the lightest, flattest pan possible is used, and the sample shape is selected to ensure good contact with the pan. Typical sample mass is 10 to 15 mg for polymers and is measured to the nearest 0.1 mg. Replicates should be the same size, within 10%, and of uniform size and shape. If the sample contains volatiles, put one or more

pinholes in the lid of the pan before crimping in order to facilitate evaporation.

Operation

Typical DSC survey conditions for thermoplastic and thermoset polymers are a sample size of 10 to15 mg, a heating rate of 10 °C/min, a nitrogen (thermal properties) or oxygen (oxidative properties) atmosphere with a flow rate of 50 mL/min, and temperature range from ambient to 400 °C.

The optimization of DSC is critical in acquiring accurate and precise data. Inspect the DSC cell and ensure that it is in good condition. For example, clean the thermal sensors as needed by heating to 500 °C in air to burn off any decomposed materials. Use a brush to further wipe off any debris that will interfere with the transfer of heat flow from the sample to the thermal sensor. Always use a purge gas. Typical inert gases are nitrogen and helium; the latter enhances heat transfer. Oxidative gases are oxygen or air containing 21% oxygen.

Some workers use an internal melting point standard (selected from Table 13.3) to verify transition temperature accuracy. To do this, place the sample in a DSC pan. Invert the lid, and either crimp it in place or force it into the pan so that good contact is achieved on the top and bottom of the sample. Weigh the internal standard—for example, indium metal—into the inverted lid. Record the DSC thermal curve. Superimposed on the curve will be the melting profile of indium, as shown in Figure 13.7.

The effects of heating and oxidation during sample preparation must be minimized, for any localized heating or stress in the sample produced by mechanical treatment

Table 13.3 Enthalpy and temperature calibration of various materials

Material	Melt temperature (°C)	Heat of fusion (J/g)	Standard deviation (+/−)
n-Pentane	−132.66	36.51	0.02
n-Heptane	−90.56	138.62	
Cyclohexane	−87.06	78.70	
n-Octane	−56.76	180.00	
n-Decane	−26.66	199.87	
n-Dodecane	−9.65	214.35	
Water	0.01	335	0.6
Cyclohexane	6.54	30.91	
Diphenyl ether	26.87	101.15	0.10
n-Octadecane	28.24	238.76	
Benzoic acid	122.37	147.4	0.1
Indium	156.5985	28.57	0.17
Tin	231.928	60.6	0.2
Lead	327.502	23.1	0.3
Zinc	419.527	108	0.6

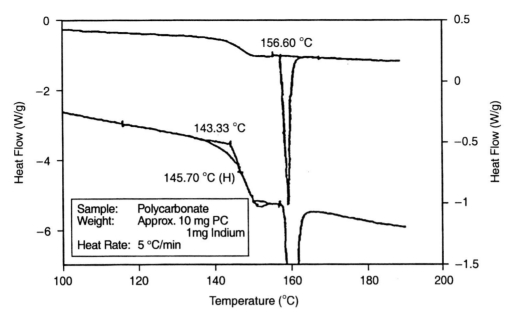

Figure 13.7 Use of internal melting temperature standard in differential-scanning calorimetry (DSC).

may produce artifacts in the DSC curve. It is better to cut out a representative sample rather than crush it, which may impart stresses to the sample. Cutting a sample from the bulk with a razor blade or clipper imposes a minimum stress to the sample. Cryogenic grinding in a freezer mill creates easily handled powders and minimizes treatment effects. Alternatively, films can be heat-pressed to the desired thickness and then punched out. Heating through transitions erases the thermal history imparted to a sample at lower temperatures.

A DSC baseline plot should be flat. The latter can be achieved through calibration with empty pans and application of the electronic correction. All samples have a baseline with an endothermic slope due to the increasing specific heat capacity of the sample with increasing temperature (see Figure 13.8). DSC baseline curvature seen in Figure 13.9 cannot normally be corrected with calibration. However, if necessary, it can be eliminated, with baseline subtraction. Baseline slope and curvature can cause an error if peaks are integrated over a broad temperature range and

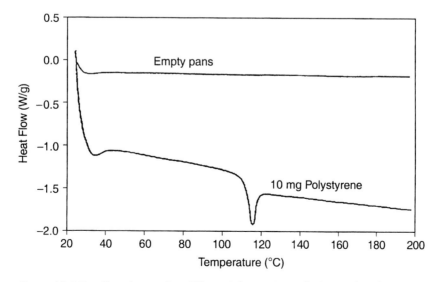

Figure 13.8 Baseline slope using differential-scanning calorimetry (DSC).

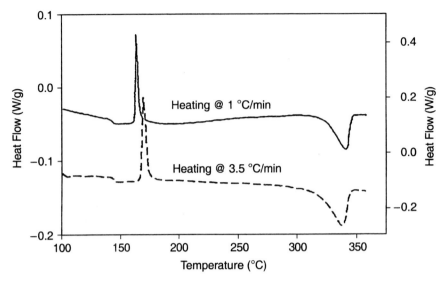

Figure 13.9 Baseline curvature using differential-scanning calorimetry (DSC).

can make detection of a weak T_g difficult. Sample size is chosen to achieve a heat flow of 0.1 to 10 mW in going through a transition, as shown in Figure 13.10. Differences in specific heat between the polymer sample and the reference may be minimized, if necessary, by adding aluminum to the reference pan. This improves the overall baseline, as shown in Figure 13.11.

Begin a DSC analysis about 40 °C below the first event of interest. A heating rate of 5 to 10 °C/min gives four to eight minutes for the instrument to equilibrate before reaching the temperature of interest. Adjust sample weight, heating

rate, and purge gas, if necessary, to improve sensitivity or resolution, as indicated in Table 13.4.

Knowing the upper stability limit of a sample extends the life of a DSC cell. Select a final temperature that does not cause decomposition of the sample in the calorimeter. Degradation products can condense in the cell and cause corrosion of the cell and baseline problems. The use of hermetically sealed aluminum pans can minimize corrosion and destruction of the cell liner. If necessary, use sealed glass ampules or stainless steel pans that can take pressures greater than 1000 psi in order to study decomposition. A

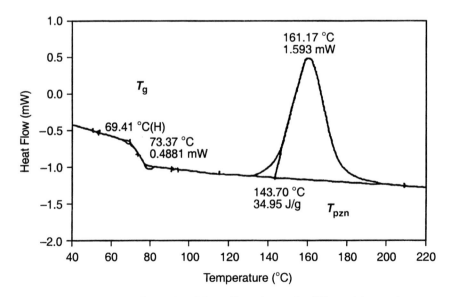

Figure 13.10 Transition (T_g and T_{pzn}) heat flow change in differential scanning calorimetry (DSC). Straight line is the baseline fitted by the instrument; curved line is the sample data.

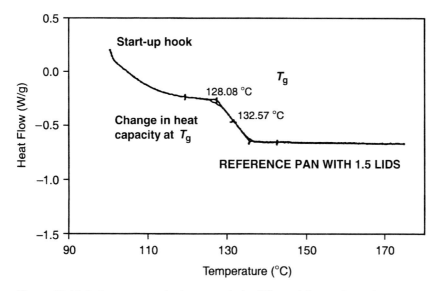

Figure 13.11 Reference matched to sample in differential scanning calorimetry (DSC): T_g, change in heat capacity at T_g, and start-up hook. Curved and straight lines as in Figure 13.10.

special pressure cell is needed for both high-pressure and low-pressure studies in nitrogen, air, or oxygen[35–37]

Parameters measured to quantify the stability of a polymer include oxidation induction time (OIT) and oxidation onset temperature (OOT). A 1- to 3-mg sample is weighed into an open pan, as specified in ASTM Method E-1858 on OIT.[38] The polymer sample is heated and held at constant temperature, either 175 °C at 35 MPa oxygen or 195 °C at 14 kPa of air or oxygen. The extrapolated onset time is defined as the OIT. A second method, ASTM Method E-2009 on OOT is based on a scanning DSC or a programmed differential scanning calorimeter (PDSC).[39] The OOT is defined as the extrapolated onset temperature, in a plot of heat flow versus temperature. A thermogravimetric analyzer coupled with a differential thermal analyzer, TGA-DTA, has also been used to determine the OOT of hydrocarbons in air. In this study, the sample size was 10 mg, the flow rate of air was 250 mL/min, the heating rate was 10 °C/min, and the sample pans were aluminum or platinum.[40–41]

Table 13.4 Optimization: DSC resolution and sensitivity

Parameters	To increase sensitivity	To increase resolution
Sample weight	Increase	Decrease
Heating rate	Increase	Decrease
Purge gas	Nitrogen	Helium

Basic heat flow equation:
$$dQ/dT = C_p \times dT/dt + f(T,t)$$
heat flow = heat capacity × heating rate + kinetic component

Analyzing Thermoplastics and Thermosets

When analyzing thermoplastics by DSC, the experimental program should include three cycles at 10 °C/min: heating, cooling, and heating.[42] The results of the first heating depend on the unknown thermal history of the material and are unreliable for analysis. The cooling cycle gives the sample a known thermal history, and the scan can be used to compare polymer crystallization properties. Finally, the results of the second heating are a function of the material and the known thermal history imparted during the cooling cycle, and are useful for comparison of polymers, as shown in Figures 13.12 to 13.14.

When analyzing thermoset polymers by DSC, the experimental protocol should begin with an annealing period, should progress to the maximum analysis temperature, and then progress to a very rapid cooling (quenching). Finally, the analytical curve is obtained by heating at a rate of 10 to 20 °C/min. Annealing at approximately 25 °C above the onset of the T_g eliminates effects of enthalpic relaxation. The heating portion of the program measures the T_g and any residual cure (exothermic polymerization). Do not heat through the decomposition onset temperature. Quench the sample, usually with liquid nitrogen or a dry ice – acetone slurry, thereby giving the sample a known thermal history. During the final heating the T_g of the fully cured sample is measured, as shown in Figures 13.15 to 13.17.

Modulated Temperature DSC

Modulated temperature DSC (MTDSC) is a recent innovation in DSC. It provides the same qualitative and quantitative in-

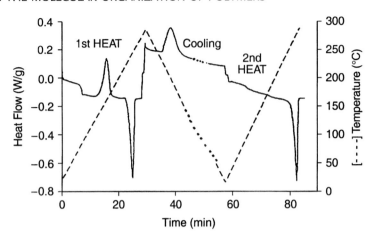

Figure 13.12 Thermoplastic differential-scanning calorimetry (DSC): 1st and 2nd runs with heat flows and cooling. Key: solid line = heat flow; dashed line = temperature.

formation about physical and chemical changes as conventional DSC, and it also provides unique thermochemical data that are unavailable from conventional DSC.[25,44-53] The effects of baseline slope and curvature are reduced, increasing the sensitivity of the system. Overlapping events such as molecular relaxation and glass transitions can be separated. Heat capacity can be measured directly with MTDSC in a minimum number of experiments. Both DSC and MTDSC differentiate first-order thermodynamic changes such as polymer fusion temperature and second-order thermodynamic changes such as glass-transition temperature T_g.

Both MTDSC and DSC measure the difference in heat flow to a sample and to an inert reference. The sample and reference cells are identical. However, MTDSC uses a different heating profile. Whereas DSC measures heat flow as a function of a constant rate of change in temperature, MTDSC superimposes a sinusoidal temperature modulation on this rate. The sinusoidal change in temperature permits the measurement of heat-capacity effects simultaneously with the kinetic effects.

Typical experimental procedures for an initial MTDSC experiment include a heating rate from isothermal to 5 °C/

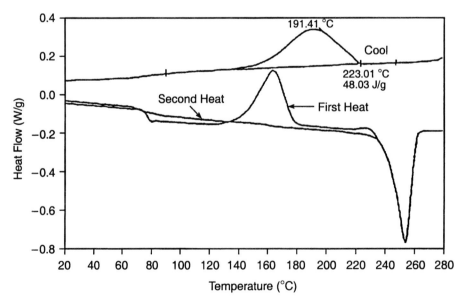

Figure 13.13 Thermoplastic differential-scanning calorimetry (DSC): temperature and heat flows.

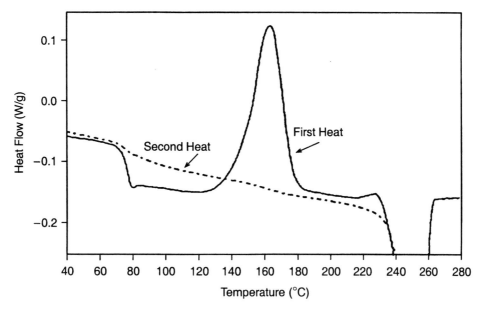

Figure 13.14 Thermoplastic differential-scanning calorimetry (DSC): 1st and 2nd runs with heat flows. Key: solid line = first heat flow; dashed line = second heat flow.

min and a modulation amplitude from 0.01 to 10 °C. The modulation period can vary from 10 to 100 seconds or, expressed as a frequency, from 10 to 100 mHz.

A useful comparison of MTDSC and DSC is the examination of poly(ethylene terephthalate) (PET) by the two methods. The MTDSC procedure is as follows: (1) equilibrate at 0 °C; (2) modulate 1 °C every 40 s; (3) hold at 0 °C for 5 min; and (4) ramp at 5 °C/min to 280 °C. The correspond-

ing DSC procedure is as follows (1) equilibrate at 0 °C; (2) hold at 0 °C for 5 minutes; and (3) ramp at 20 °C/min to 280 °C. The DSC curve shown in Figure 13.18 records only the total heat flow to the sample. The MTDSC curve shown in Figure 13.19 reveals all thermal events occurring in the sample. The total heat flow is the same in both techniques at the same average heating rate. In MTDSC, the reversing heat flow is the heat capacity component of the total heat

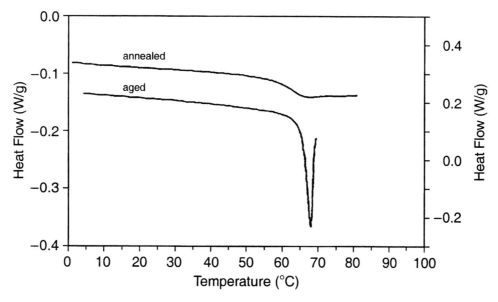

Figure 13.15 Annealing effect on the glass-transition temperature using differential-scanning calorimetry (DSC).

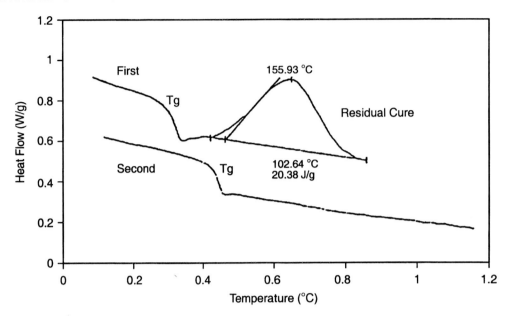

Figure 13.16 Thermoset: comparison of 1st and 2nd heatings using differential-scanning calorimetry (DSC).

flow, and the nonreversing heat flow is the kinetic component of the heat flow.

MTDSC and Polymer Blends

The MTDSC of an as-received molded blend of PET and acrylonitrile-butadiene-styrene (ABS) is represented in Figure 13.18. The total heat flow curve (middle) indicates the onset of the T_g of PET at about 60 °C and an exothermic event beginning at about 97 °C. In the nonreversing curve (top), the exothermic event is the cold crystallization of PET, and this masks the T_g of the ABS. However, the reversing heat flow thermal curve (bottom) clearly differentiates the T_gs of ABS and PET. The extrapolated onset of the T_g of ABS, denoted T_g(e), is 104.5 °C, and the T_g(e) of PET is 67.0 °C.

In Figure 13.19 the T_gs and melt temperature of another polymer blend are clearly delineated in their first heat by MTDSC. This blend contains PET, polycarbonate (PC), and high-density polyethylene (HDPE). The exothermic PET cold crystallization overlaps the melting peak of HDPE and confounds the total heat flow curve (middle). The reversing heat flow thermal curve (bottom) shows a T_g for PET at approximately 72 °C, a peak melt temperature for HDPE at 123 °C, and a T_g for PC at approximately 138 °C. The nonreversing heat flow, representative of a standard DSC experiment, shows only an enthalpic relaxation endotherm at approximately 70 °C and the PET crystallization peak at near 115 °C.

Specific Heat Capacity by DSC and MTDSC

ASTM Method E-1269 covers the determination of specific heat capacity C_P of polymer solids and liquids by conventional DSC.[54] The normal operating range of the test is from −100 to 600 °C. First, the total DSC heat flow of an empty sample and reference pan is measured. A second run is made with a sample in the sample cell. Then the C_P as a function of temperature is determined by subtracting the first curve from the second. A sapphire standard is used to calibrate the instrument.

Measurements of C_P by MTDSC are direct; since the specific heat varies linearly over the desired temperature range.[55-56] To determine C_P by MTDSC, first calibrate the cell with sapphire every 10 °C and correct the reversing heat

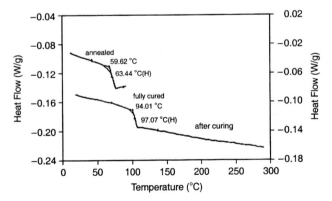

Figure 13.17 T_g shift as a result of curing in differential-scanning calorimetry (DSC).

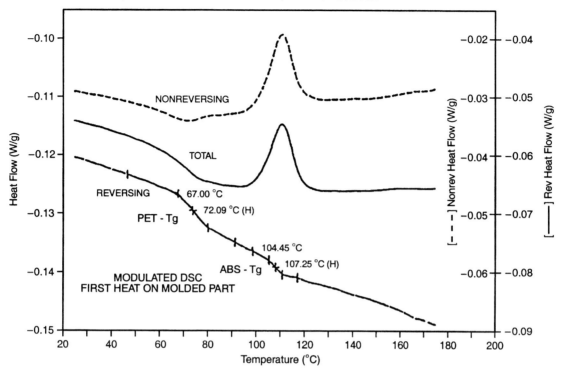

Sample: PET-ABS BLEND (MOLDED PART)
Size: 0.4500 mg
Method: MOSC 1/60 @ 2 °C/MIN
Comment: 25 N2 PUR; MOSC 1/60 @ 2 °C/MIN TO 300 °C; PART AS RECEIVED

DSC

File: C: BLEND.03
Operator: THOMAS
Run Date: 9-Feb-94   15:09

Figure 13.18 Modulated temperature in differential-scanning calorimetry (MTDSC): poly(ethylene terephthalate)/acrylonitrile-butadiene-styrene (PET/ABS) blend "as received".

flow output (the calorimetric portion of the total heat flow). A high molecular weight NIST polystyrene standard may be used to verify the sapphire calibration. Calibrated C_P measurements are repeatable within 5% at a specific temperature.

Heat Transfer of Engineering Plastics: Reference Polymers

The determination of thermal conductivity and diffusivity of a polymer by MTDSC are described in ASTM Method E-1952.[57] This test is most useful for rigid engineering polymers.

Another approach to obtain fundamental heat transfer properties is the measurement of effusivity, for effusivity squared is equal to the product of thermal conductivity, density and diffusivity.[58] Effusivity measured with a Thermal Probe[59] was used by the Society of Plastics Engineers to rank several standard reference polymers, and results are given in Table 13.5. The polymers are identified by their resin kit number.[60] The heat-transfer properties of these polymers indicate that high-density polyethylene (HDPE), ethylene–vinyl acetate (EVA), and the acetal polymers are

relatively more heat conducting (Eff = 870–950) than are polystyrene or an ABS–poly(vinyl chloride) (PVC) blend (Eff = 600–750). Polystyrene in the form of Styrofoam™ insulation has an effusivity value of 440. Heat transfer in oriented or layered structures is rapidly and precisely determined by the Thermal Probe technique.

Dynamic Differential Scanning Calorimetry and Pressure DSC

Dynamic differential scanning calorimetry (DDSC) is a variant of modulated temperature DSC. Whereas MTDSC superimposes a sinusoidal modulation on the rate of increase in temperature, DDSC imposes a saw-tooth modulation. Pressure DSC (PDSC) is a system that studies heat flow as a function of temperature or time under high pressure or partial vacuum, and this approach is particularly useful for evaluating the oxidative behavior of polymers. The optimum conditions for DDSC, PDSC, DSC, and DTA using a Perkin Elmer DSC are a sample size of 1 to 75 mg, depending on the sensitivity required;[43] a standard heating rate of 20 °C/min (but rates from 0.1 to 100 °C/min may be used

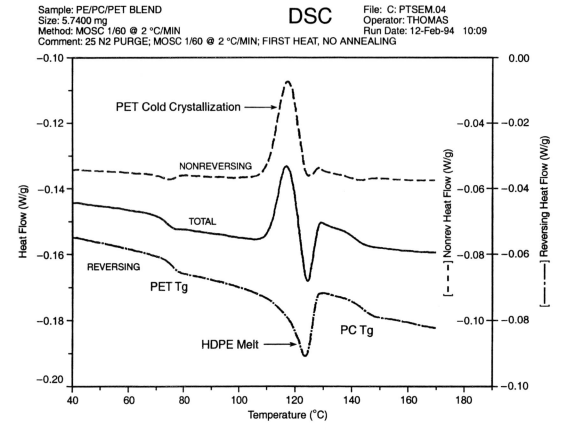

Figure 13.19 Characterization of a polymer blend using modulated temperature differential-scanning calorimetry (MTDSC).

depending on the application); a preheat before data collection for 30 s at 2 °C per s; and cooling after data collection at 1 °C. A wide range of other values works, too, but this is a good starting point and one that leads to specific heats within a few percentage points without any special calibration. A power-compensated DSC is described in Figures 13.5 and 13.6.

Differential Thermal Analysis

Differential thermal analysis measures the temperature difference between a sample and an inert reference as a function of time and temperature.[20] DTA measurements provide information about transition temperatures, time-based phenomena, and thermodynamic properties. Data from DTA are much less quantitative than are data from DSC, and for this reason DTA is used infrequently.

Thermogravimetric Analysis

Thermogravimetric analysis (TGA), known also as thermogravimetry (TG), measures the change in weight of a poly-

mer as a function of temperature or time. The sample, in nitrogen or in an oxidizing atmosphere, is brought from ambient to a constant high temperature by a controlled temperature program and held at that temperature. [24,61-71]

A thermogravimetric analyzer consists of a balance, a furnace, an environmental control, a sample container, a temperature probe, and optional coupled devices for evolved gas analysis (EGA). The usual operating temperature range is ambient to 1000 °C. TGA methods can determine loss of water, loss of solvent, loss of plasticizer, pyrolysis products, oxidative and thermal stability, weight percent filler, and weight percent ash.[72] TGA can determine the composition of multicomponent systems, estimate the lifetime of a polymer product, measure decomposition kinetics, and evaluate the effect of reactive or corrosive atmospheres on polymers. The heating rate can be slowed to near isothermal conditions when the sample is losing mass. Interfacing TGA with mass spectrometry or Fourier transform infrared spectroscopy facilitates the identification of volatile constituents and decomposition products.[70-71] Careful isothermal experiments can produce reliable kinetic data.

ASTM Method E-1582 is used to calibrate the temperature scale of a TGA analyzer over the range from 25 to 1000 °C.[73] This calibration may be accomplished by the use

Table 13.5 Effusivity (thermal conductivity) of reference polymers from the Society of Plastics Engineers Resin kit at 24 °C

Rank	Polymer	Resin kit no.	Effusivity $(W\sqrt{s}/m^2{*}K)$
1	High-density polyethylene	Control	954 ± 1.6
2	Ethylene vinly acetate	34	925
3	Acetal copolymer	32	896
4	Acetal homopolymer	34	894
5	p-phenylene sulfide	33	834
6	Polypropylene engineering plastic	26	758
7	Poly(vinyl chloride)	29	704
8	Acrylonitrile-butadiene-styrene high impact	7	696
9	Acrylonitrile-butadiene-styrene medium impact	6	688
10	Polystyrene-medium impact	2	661
11	Polystyrene-general purpose	1	651
12	Acrylonitrile-butadiene-styrene/ poly(vinyl chloride)	40	577

of either melting point standards, in the form or wires, or magnetic transition standards. The TGA mass scale is calibrated as described in ASTM Method E-2040.[74]

Knowledge of the approximate transition and decomposition temperatures of the sample is needed to select the temperature range, scanning rates, and peak temperature for an initial TGA analysis. The heating rate may vary from 1 to 20 °C/min. If filler or ash content is to be measured, then heating at 50 to 100 °C/min to the end temperature is appropriate.

Values measured in TGA may include the onset temperature and the initial temperatures. The onset temperature T_o (see Figure 13.20) is defined as the temperature at which the extrapolated initial sample weight line and the line of steady weight loss with temperature intersect. It is possible

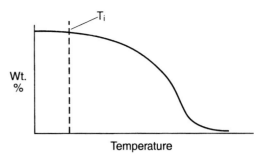

Figure 13.21 Initial temperature (T_i), or the temperature at which weight loss is first noticeable, using thermogravimetric analysis (TGA).

to have several onset temperatures, each representing a different decomposition process. The initial temperature T_i, as shown in Figure 13.21, which is considered to be the temperature at which weight loss is first noticeable, is not reproducible and should not be used. The percentage of change in weight (ΔY), as shown in Figure 13.22, is accurate and reproducible.

The first derivative of the weight loss curve, known as the derivative TGA or DTG, is repeatable, reproducible, and very useful in revealing polymer properties. For example, DTG can show the evolution of absorbed moisture, as well as the number of steps in the decomposition of a polymer. An example of a DTG curve is shown in Figure 13.23.

A procedure for compositional analysis by TGA is described in ASTM Method E-1131. Typically, a polymer sample (3–20 mg) is weighed into a platinum boat or pan, the scale is set at 100%, a nitrogen flow rate of 50 mL/min is established, and the TGA is programmed to heat from ambient to 500 °C, or to the point where the polymer decomposes.

The relative thermal stability of poly(vinyl chloride) (PVC), poly(methyl methacrylate) (PMMA), high-pressure polyethylene (HPPE), poly(tetrafluoroethylene) (PTFE), and polyimide (PI) was ascertained by TGA, as shown in Figure 13.24. Clearly, PVC is the least stable and PI is the most stable.

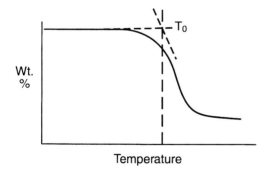

Figure 13.20 Onset temperature (T_0), or the temperature at which the extrapolated initial sample weight line and the line of steady weight loss with temperature intersect, using thermogravimetric analysis (TGA).

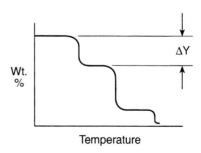

Figure 13.22 Weight percent change (ΔY) using thermogravimetric analysis (TGA).

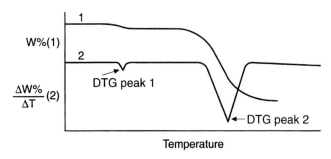

Figure 13.23 Thermogravimetry (curve 1) and derivative thermogravimetry (DTG); peaks may represent separate events or may be the net weight change from two or more events, and the relative size of a peak indicates the relative rate of polymer weight loss. TGA and DTG imply a two-component system. Key: W% = weight %; 1 = TGA (curve 1); 2 = DTG (first derivative of curve 1); $\Delta W\%/\Delta T$ = change in weight %/change in temperature.

Table 13.6 Temperature (°C) in heat-sensitive poly(vinyl chloride) using two approaches

	Differential scanning calorimetry (DSC)		Thermogravimetric analysis (TGA)
	1st run	2nd run	1st run
$T_g(e)$	69	72	
$T_g(h)$	72	77	
T deco	245	249	281

$T_g(e)$ = glass-transition temperature, extrapolated-onset temperature; $T_g(h)$ = glass-transition temperature at half height; T(deco) = decomposition temperature DSC and TGA, extrapolated onset.

TGA and DSC were used to evaluate a heat-sensitive PVC, and results are given in Table 13.6. The measurement shown in Figure 13.24 correlates well with the decomposition temperature cited in Table 13.6. The temperature at the half-height of the change in specific heat capacity at T_g is denoted $T_g(h)$; for PVC, this increased by 5 °C upon heat cycling.

Dynamic Mechanical Analysis

DMA is a thermal technique in which the storage modulus ε' (elastic response) and loss modulus ε'' (viscous response) of a polymer are measured as a function of temperature or time as the polymer is deformed under an oscillatory load (stress) at a controlled (isothermal or programmed) temperature in a specified atmosphere.[22,26] The storage modulus is related to stiffness, and the loss modulus to damping and energy dissipation. Glassy, viscoelastic, elastic, and liquid polymers can be differentiated by DMA, and some details of

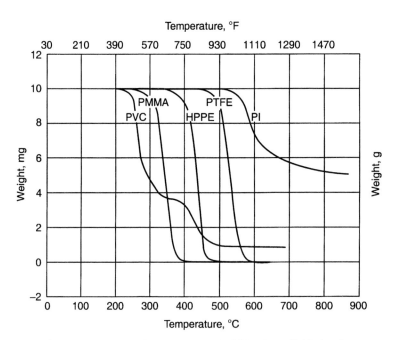

Relative thermal stability of polymers by TGA; 10 mg (0.15 g) at 5 °C/min (9 °F/min) in nitrogen; HPPE, high-pressure polyethylene

Figure 13.24 Relative thermal stability of polymers by thermogravimetric analysis (TGA). Key: PVC = poly(vinyl chloride), PMMA = poly(methyl methacrylate), HPPE = high-pressure polyethylene, PTFE = poly(tetrafluoroethylene), and PI = polyimide.

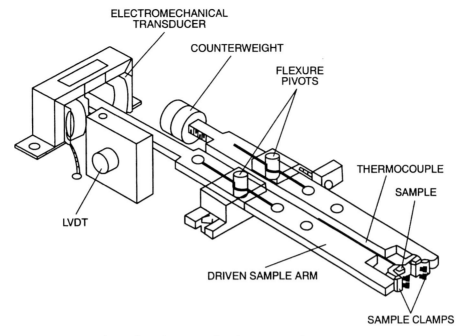

Figure 13.25 Mechanical configuration for dynamic mechanical analysis (DMA). Key: LVDT = linear variable differential transformer.

polymer structure can be inferred from the results. DMA is particularly useful for evaluating viscoelastic polymers that have mechanical properties which exhibit time, frequency, and/or temperature effects.[75-86] DMA is the most sensitive of all thermal analytical techniques.

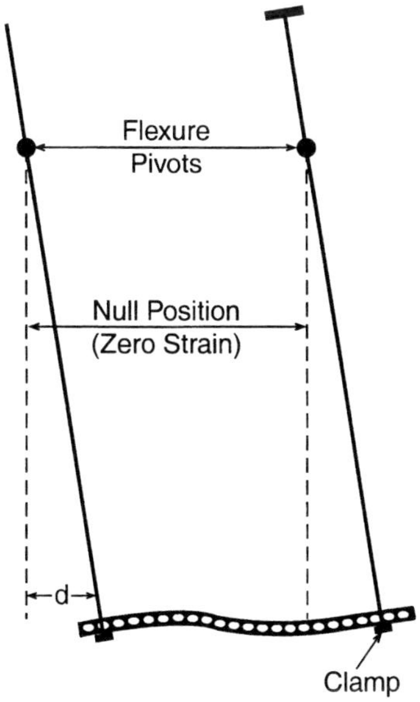

In DMA the sample is clamped between the ends of two parallel arms, as shown in Figure 13.25. The arms are mounted on low-force flexure pivots, which allow motion in only the horizontal plane. The distance between the arms is adjustable by means of a precision mechanical slide to accommodate a wide range of sample lengths from less than 1 mm up to 65 mm. An electromechanical motor attached to one arm drives the arm/sample system to a pre-selected strain or amplitude. As the arm/sample system is displaced, the sample undergoes a flexural deformation, as shown in Figure 13.26. A linear variable differential trans-

Figure 13.26 Flexural deformation in dynamic mechanical analysis (DMA). Key: d = displacement strain.

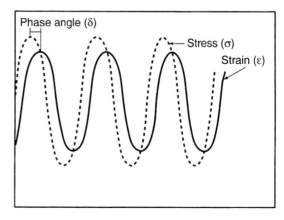

Figure 13.27 Fixed frequency using dynamic mechanical analysis (DMA).

former mounted on the driven arm measures the sample response, strain, and frequency, as a function of the applied stress, and it provides feedback control to the motor. The sample is positioned in a temperature-controlled chamber that contains a radiant heater and a coolant distribution system. This heating system is precise and gives accurate control of the sample temperature. An adjustable thermocouple mounted close to the sample provides precise feedback information to the temperature controller, as well as readout of sample temperature. The system uses cold nitrogen gas for subambient operation.

The four modes of DMA operation are fixed frequency, resonant frequency, creep relaxation, and stress relaxation. In the fixed-frequency mode, illustrated in Figure 13.27, applied stress forces the polymer sample to undergo sinusoidal oscillation at a frequency and amplitude of strain selected by the analyst. In viscoelastic polymers, the maximum strain and maximum stress do not occur at the same time. A phase lag or shift, defined as the phase angle δ, is measured and used with known sample geometry and driver energy to calculate the viscoelastic properties of the polymer sample.

In the resonant frequency mode illustrated in Figure 13.28, the sample is allowed to oscillate at its natural resonance frequency under set conditions. This mode provides higher damping sensitivity than the fixed-frequency mode.

The creep mode, illustrated in Figure 13.29, is used to measure sample creep and strain as a function of time and temperature at a selected stress. Using an isothermal step method, the sample is allowed to equilibrate to the relaxed state at each selected temperature. After equilibration, the polymer is subjected to a constant stress. The resulting deformation strain is recorded as a function of time for a given period. The sample is then allowed to recover in an un-

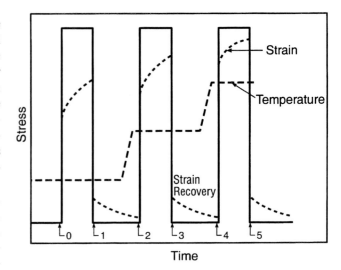

Figure 13.29 Three cycles of creep and strain (dotted line) measurements as a function of time and temperature at a selected stress (solid line), using dynamic mechanical analysis (DMA). Key: numbers 0, 2, 4 = application of stress; numbers 1, 3, 5 = removal of stress.

stressed state. Sample recovery (strain) is recorded as a function of time. When the measurement at one temperature is complete, the temperature is changed and the experiment is repeated. Creep studies are valuable for determining the long-term flow properties of a polymer and its ability to withstand loading and deformation influences. Lastly, the stress relaxation mode, illustrated in Figure

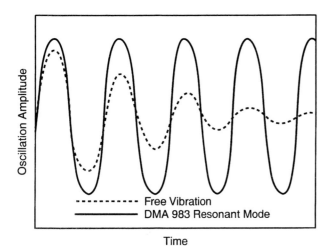

Figure 13.28 Resonant frequency using dynamic mechanical analysis (DMA).

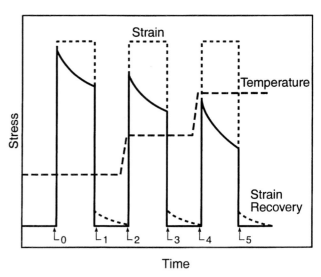

Figure 13.30 Three cycles of stress decay (solid line) measured as a function of time and temperature at a set strain (dotted line), using dynamic mechanical analysis (DMA). Key: numbers 0, 2, 4 = application of strain; numbers 1, 3, 5 = removal of strain.

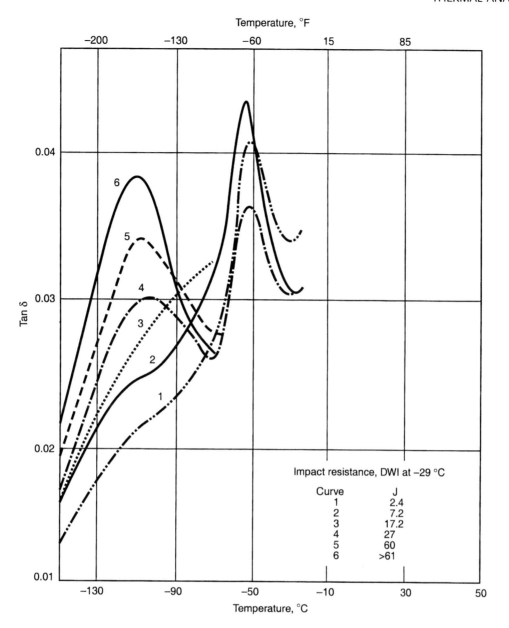

Comparative damping of impact-modified polypropylene by DMA. Size, 3.18 mm (0.125 in.) thick, 12.1 mm (0.48 in.) wide, 19.1 mm (0.75 in.) long; programmed at 5 °C/min (9 °F/min), amplitude at 0.4 mm (0.016 in.).

Figure 13.31 Comparative damping of impact-modified polypropylene using dynamic mechanical analysis (DMA).

13.30, provides definitive information for prediction of long-term performance of polymers by measuring the stress decay as a function of time and temperature at a set strain.[87]

The standard test method for the temperature calibration of a DMA is ASTM Method E-1867.[88] The temperature range for this test is −150 to 500 °C. A relationship is de-

veloped between the DMA temperature sensor and the melting of standards. This is accomplished by loading a melting temperature standard into a polymer or aluminum foil and submitting it to a mechanical oscillation at either fixed or resonant frequency. The extrapolated onset of melting is identified by a rapid decrease in the ordinate signal, the apparent storage modulus. This onset temperature is

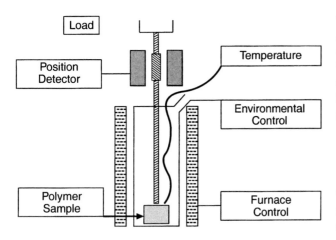

Figure 13.32 Instrumentation for thermal mechanical analysis (TMA).

used for temperature calibration with two melting temperature standards, for example, *n*-octane at −56.8 and benzoic acid at 122.4 °C.

A relationship was developed among the amount of polybutadiene in polypropylene blends, the DMA tan delta height (comparative damping), and the impact resistance according to ASTM D4247-96. The correlation was developed between the tan delta curve at 158 to 165 K (−115 to −108 °C) and the impact resistance determined by the ASTM Dart Weight Impact (DWI) test, at 244 K (−29°C), as shown in Figure 13.31. The maximum tan delta value was observed with a DWI energy-to-rupture value of greater than 61 J and the minimum had a DWI value of 2.4 J. The impact properties were directly related to the amount of rubber phase (as polybutadiene) present in the blends.

Thermomechanical Analysis

Thermomechanical analysis (TMA) is an analytical technique in which the dimensional change or deformation of a polymer under a static nonoscillatory load is measured as a function of temperature or time while subjected to a controlled temperature program in a given atmosphere, typically nitrogen or air.[22,24,26]

TMA measures the linear coefficient of expansion, the glass-transition temperature in the expansion or force condition, dimensional stability, softening or yield temperatures, and the heat-distortion temperature.[89-93] Creep and tensile modulus can also be calculated as a function of time or temperature. Applied properties derived from TMA studies are viscosity, gel time, and temperature, as well as creep/stress relaxation and creep recovery. Linear or volumetric changes in polymers can be calculated from TMA data.

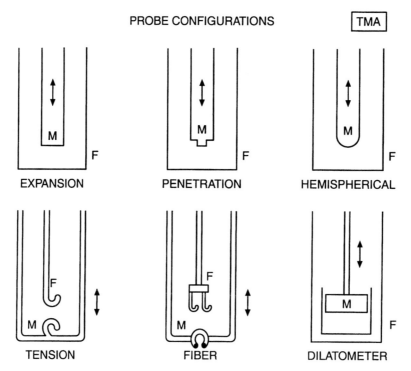

Figure 13.33 Probe configurations for thermal mechanical analysis (TMA).
Key: F = fixed member; M = movable member; arrows indicate vertical probe motion.

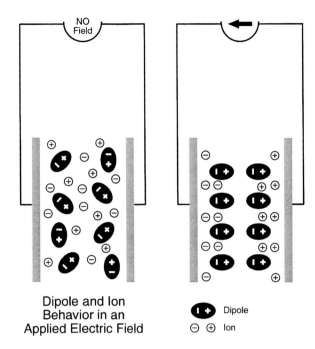

Figure 13.34 Dipole and ion behavior with and without an applied electric field using dielectric thermal analysis (DETA): *left*, in zero field, particles are randomly distributed; *right*, in an applied field, particles are oriented.

A typical TMA is based on a movable-core linear variable differential transformer whose output is proportional to displacement of the core, as shown in Figure 13.32.[89] The latter moves in response to dimensional changes in the polymer. An electromechanical coil applies force, and a precise low-mass furnace controls the heat. The instrument can be cooled and heated in a controlled manner. TMA probes operate in either compression or tension, and Figure 13.33 illustrates the variety of probes in common use. Expansion

probes are used to measure the linear coefficient of expansion, T_g, and the compressive modulus of polymers. The penetration probe has an extension on its tip that concentrates the force on a small area of the sample's surface. Precise melting and softening temperatures can be determined with this probe. Coatings and films can be evaluated without removal from the substrate.[90-93]

The temperature scale of a TMA apparatus is calibrated from −50 °C to greater than 800 °C by the procedure in ASTM Method E-1363.[94] Melting standards are used to delineate a mechanical change at a specific temperature, for example, the melting of indium at 156.6 °C.

ASTM Method E-831[95] is used for both the determination of the linear thermal expansion of a polymer by TMA at constant heating rate and also the calibration of the TMA measurement scale from −120 to 600 °C. The dimensional change can be calibrated using a rigid flat piece of either aluminum or platinum that has a known coefficient of linear expansion over the temperature range studied.

Polymer analysis by TMA is routine and easy. A flat piece or a film of the polymer is placed on a quartz stage. The quartz or steel probe is adjusted for zero dimensional change at the surface of the test specimen. The furnace is brought into position around the quartz stage and specimen, and a nitrogen purge is initiated. The TMA is usually heated at 5 °C/min through the polymer transitions, T_g, and melt.

Dielectric Thermal Analysis

Dielectric thermal analysis (DETA), known also as dielectric analysis (DEA), measures the permittivity, capacitance, and dielectric loss of a polymer under an oscillating electric field as a function of temperature or time. The polymer sample

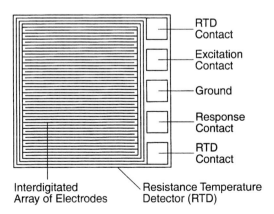

Figure 13.35 Ceramic single-surface sensor for dielectric analysis (DEA) or dielectric thermal analysis (DETA).

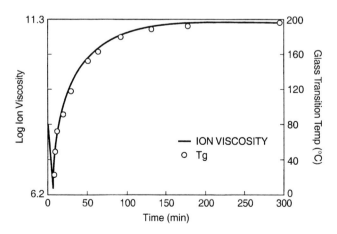

Figure 13.36 Comparison of ion viscosity and T_g for isothermal cure: dielectric thermal analysis (DETA) enables curing to a specific T_g.

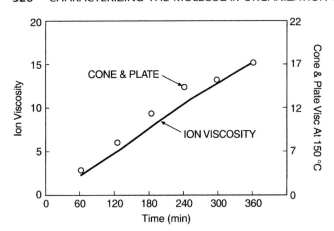

Figure 13.37 Correlation of cone-and-plate viscosity and ion viscosity for batch reaction: dielectric thermal analysis (DETA) tracks the polymerization process in a batch reactor.

time or temperature. ("Dielectric constant" is an obsolescent term for the permittivity at frequencies greater than 5000 Hz.) Permittivity is a measure of the alignment of dipolar groups in the polymer (see Figure 13.34). The loss factor, e'', another fundamental property measured by DETA, is a measure of the energy that is expended by aligning dipoles and moving ions in the polymer. The conductivity of the sample is the product of the loss factor times the frequency times a geometric cell constant. In the DETA single-surface cell shown in Figure 13.35, an alternating current voltage is applied across interdigitated electrodes at varying frequencies, and the ratio of e''/e' is the tan delta value.

The conductivity of a material decreases as the monomer cures or polymerizes. When the polymerization is followed by DETA, the viscosity change, the rate and extent of reaction, and the diffusion and electrical properties of the cured polymer are obtained.[97-102]

is subjected to a controlled temperature program in an inert atmosphere.[22,96]

One of the basic advantages of DETA is ability to measure the permittivity e' at a variety of frequencies as function of

A typical DETA survey experiment uses either a single-surface gold ceramic sensor or a parallel plate gold sensor, as shown in Figure 13.35. A film covering the entire sensor is cast from an appropriate solvent and all solvent is evaporated, or a film is pressed into close contact with the sensor

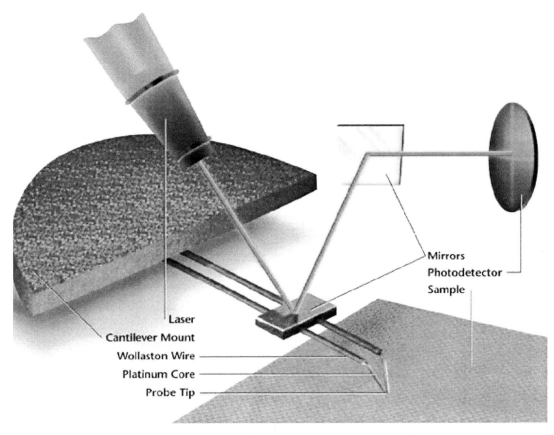

Figure 13.38 Instrumentation of microthermal analysis (micro-TA): the Wollaston wire thermal probe acts as both a heater and a thermometer.

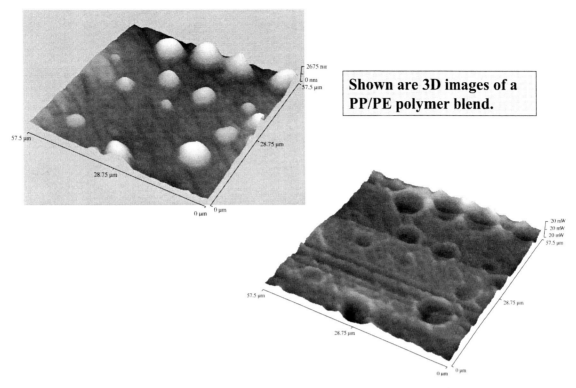

Shown are 3D images of a
PP/PE polymer blend.

Figure 13.39 Three-dimensional microthermal analysis (micro-TA) of a polypropylene/polyethylene blend; surface topography (*top*); thermal conductivity at 50 °C (*bottom*). (From TA Instruments with permission.)

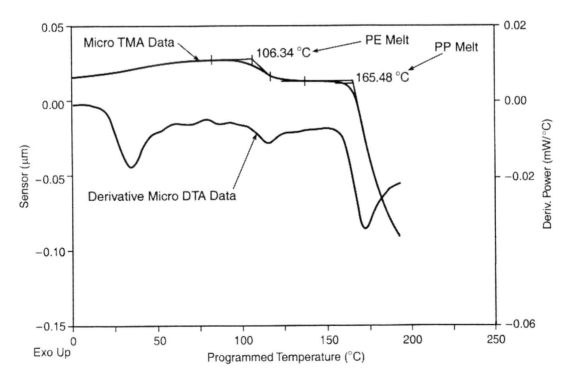

Figure 13.40 Microthermal analysis (micro-TMA) and differential thermal analysis (micro-DTA) of a polymer blend. (From TA Instruments with permission.)

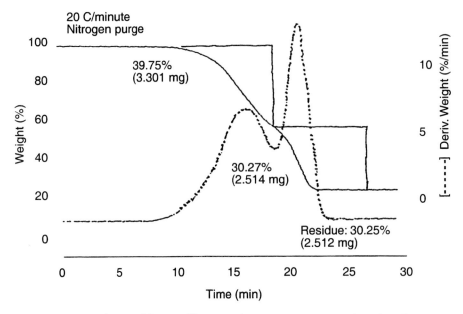

Figure 13.41 Urethane rubber profile using thermogravimetric analysis (TGA).

array. A temperature ramp at 2 to 5 °C/min from ambient to 300 °C and a nitrogen atmosphere at a flow rate of 50 mL/min are used.[99-100]

The method for temperature calibration of dielectric analyzers is ASTM Method E-2038.[103] Pure benzoic acid powder is placed either on a single-surface sensor or between two parallel plates. The instrument is purged with nitrogen, and the melting is observed as a dramatic and rapid increase in the permittivity, or electrical conductivity. ASTM Method

E-2039 is a protocol for determining and reporting dynamic dielectric properties[104] such as permittivity, loss factor, conductivity, and tan delta values as a function of frequency and temperature over the range from −100 to 300 °C.

DETA calibration and properties have been addressed by two ASTM methods.

Isothermal polymer cure can be monitored by DETA.[105] The logarithm of the ion viscosity, which is the inverse of the measured ion conductivity, is measured and is seen to

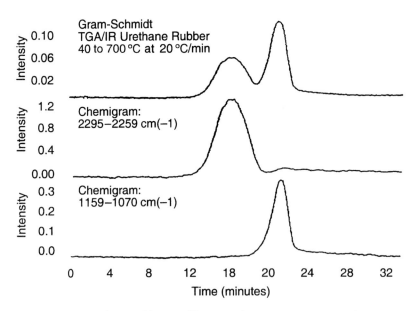

Figure 13.42 Urethane rubber profile using thermogravimetric analysis (TGA).

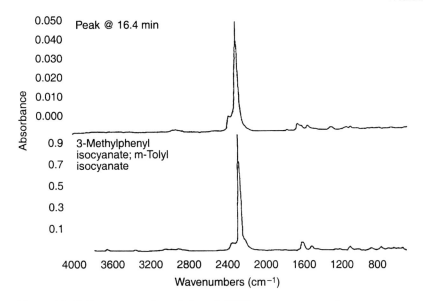

Figure 13.43 Fourier transform infrared (FTIR) spectrometry of gasses evolved from a urethane rubber at 16.4 min after the onset of the analysis in Figure 13.42.

track the T_g as measured by DSC. Therefore, one may use DETA to cure a polymer to a specific T_g, as shown in Figure 13.36. There is an excellent correlation between the cone-and-plate viscosity and the ion viscosity, as shown in Figure 13.37. DETA can easily be used to follow the polymerization process in a batch reactor at lab scale or production scale.

Microthermal Analysis

Microthermal analysis (micro-TA or μ-TA) maps polymer surfaces at high magnifications and rapidly determines surface transition temperatures by DSC, TMA, or DTA. This technique measures the thermal properties of a discrete localized area of a sample and can differentiate between bulk

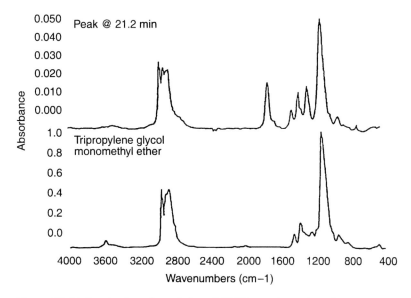

Figure 13.44 Fourier transform infrared (FTIR) spectrometry of gasses evolved from a urethane rubber at 21.2 min after the onset of the analysis in Figure 13.42.

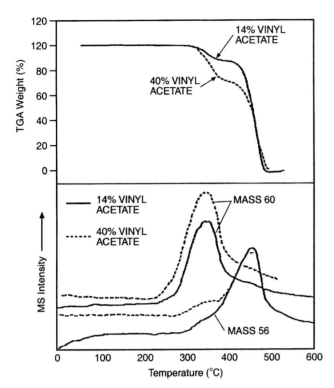

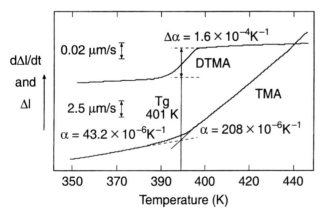

Figure 13.45 Thermogravimetric analysis (*top*) and mass spectrometry (*bottom*) (TGA/MS) of poly(ethylene-*co*-vinyl acetate).

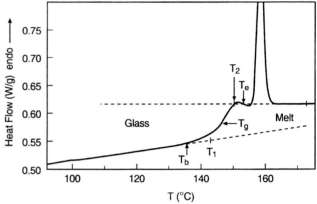

Figure 13.47 Melting of indium on top of a 0.6-mm polycarbonate disk: internal standard using differential-scanning calorimetry (DSC). Key: T_1 = extrapolated onset temperature; T_2 = end temperature for T_g event; T_b = temperature at first deviation of the baseline; T_e = end temperature after endothermic processing peak; T_g = glass transition temperature = half height of change in heat flow. (After H. E. Bair, reference 110.)

and surface properties. The surface properties of a polymer have a critical influence on its stability, reactivity, and performance. Whereas conventional thermal analysis averages the bulk response, micro-TA provides a unique means to study the surface transition temperatures and to map their changes over the surface of a polymer.[106]

In the microthermal analysis instrument, a Wollaston wire thermal probe acts as both heater and thermometer,[107] as shown in Figure 13.38. The thermal analysis measurements, TMA or DSC, in the microthermal analysis apparatus can be calibrated as described above, using melting temperature standards or polymers with known thermal properties. Quite often, poly(ethylene terephthalate) is used since it has some interesting transitions, among them T_g, cold crystallization, and fusion.

A polypropylene/polyethylene blend was evaluated by micro-TA, and data are given in Figures 13.39 and 13.40. The surface topography and thermal conductivity micro-

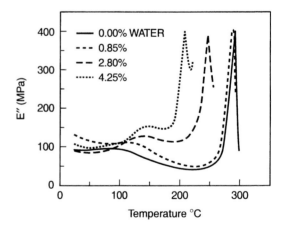

Figure 13.46 Determination of T_g by thermal analysis: dynamic thermomechanical analysis (DTMA) and thermomechanical analysis (TMA). (After B. Wunderlich, reference 109.)

Figure 13.48 E'' versus temperature for samples of Torlon poly(amideimide) with various moisture contents using dynamic mechanical analysis (DMA). (After R. Chartoff, reference 111.)

graphs give three-dimensional images of this polymer blend. Thermal analysis by micro-TMA and micro-DTA detected a local artifact that was identified as polyethylene by its melting point of 106 °C. The polypropylene melting point of 165 °C was measured after the probe penetrated through the polyethylene to the polypropylene base.

Evolved Gas Analysis

TGA and Fourier Transform Infrared Spectroscopy

A sample of polyurethane rubber was examined by TGA and Fourier transform infrared spectroscopy (FTIR) analysis of the effluent gases, and the results are shown in Figures 13.41 through 13.44. The TGA-DTG profile indicated two major mass losses near 16 min (39.8 weight percent) and near 21 min (30.3 weight percent). A Gram-Schmidt reconstruction of the TGA-DTG curve was accomplished by monitoring the effluent gases in the infrared bands at 2295 to 2259 cm^{-1} and at 1159 to 1070 cm^{-1}, and in this manner two chemigrams associated with the TGA peaks were generated. Complete infrared spectra were then measured at the top of each peak in the chemigrams. The urethane rubber volatiles emitted at 16.4 min were identified as 3-methyl phenyl isocyanate, and volatiles emitted at 21.2 minutes were shown to be tripropylene glycol monomethyl ether. The IR-structure and TGA-property techniques give the polymer analyst the opportunity to fully understand the decomposition mechanism.

TGA and Mass Spectrometry

Another evolved gas analysis technique is the use of mass spectrometry to monitor the effluent from the decomposition of poly(ethylene-*co*-vinyl acetate) (EVA), as shown in Figure 13.45. TGA differentiated between two EVA polymers that contained either 14% or 40% of vinyl acetate monomer. Mass spectrometry (MS) of the volatiles detected acetic acid (mass number 60) and butene (mass number 56). Thus, with the appropriate standards, TGA-MS can rapidly determine the acetate content in EVA.

Measurement of the Glass-Transition Temperature

The T_g of thermoplastics such as polystyrene and polycarbonate are distinguishable from the T_g of elastomers such as polybutadiene and polystyrene-*co*-butadiene. Two T_g values are observed in a blend of immiscible polymers with a T_g difference of greater than 10 °C.

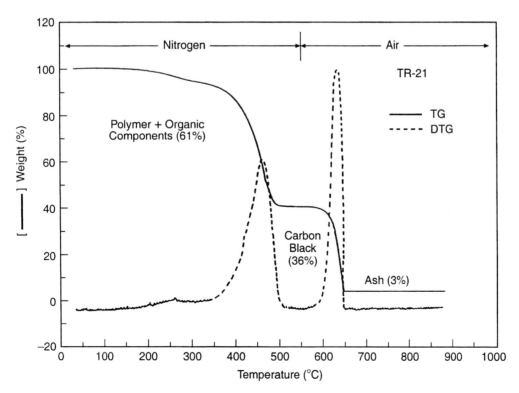

Figure 13.49 Compositional analysis of carbon black filled polymer using thermogravimetry (TG) and dynamic thermogravimetry (DTG).

Table 13.7 ASTM thermal analysis techniques used in polymer characterization, ASTM Committee D20 on Plastics

ASTM index no.	Volume	Title of standard test method or practice
D-3417	08.02	Standard Test Method for Enthalpies of Fusion and Crystallization of Polymers by DSC
D-3418	08.02	Standard Test Method for Transition Temperatures of Polymers by Thermal Analysis
D-3895	08.02	Test Method for Oxidative Induction Time of Polyolefins by DSC
D-4065	08.02	Practice for Determining and Reporting DMA Properties
D-4092	08.02	Terminology Relating to DMA Measurements of Plastics
D-4591	08.03	Test for Temperatures and Heats of Transitions by DSC (Fluoropolymers)
D-5023	08.03	Test Method for Measuring the Dynamic Mechanical Properties of Plastics using Three-Point Bending
D-5024	08.03	Test Method for Measuring the Dynamic Mechanical Properties of Plastics in Compression
D-5026	08.03	Test Method for Measuring the Dynamic Mechanical Properties of Plastics in Tension
D-5028	08.03	Test for Curing Properties of Pultrusion Resins by DSC
D-5805	09.01	Test for Rubber from Synthetic Sources—Carbon Black in Master Batches by TGA

Table 13.8 ASTM thermal analysis techniques used in polymer characterization, ASTM Committee E37 on Thermal Measurements

ASTM index no.	Volume	Title of standard test method or practice
E-0473	14.02	Standard Definitions of Terms Relating to Thermal Analysis
E-0537	14.02	Thermal Stability by DTA and DSC
E-0698	14.02	Test for Arrhenius Kinetic Constants of Thermally Unstable Materials using DTA and DSC
E-0793-95	14.02	Standard Test Method for Heats of Fusion and Crystallization by DSC
E-0794-95	14.02	Standard Test Method for Melting and Crystallization Temperature by Thermal Analysis (DSC and DTA)
E-0831-93	14.02	Standard Test Method for Linear Thermal Expansion of Solid Materials by Thermomechanical Analysis
E-0928-96	14.02	Standard Test Method for Mol Percent Impurity by DSC (Crystalline Monomer Purity)
E-0967-97	14.02	Standard Practice for Temperature Calibration of DSC and DTA Analyzers
E-0968-99	14.02	Standard Practice for Heat Flow Calibration of DSC
E-1131-98	14.02	Test Method for Compositional Analysis by Thermogravimetry
E-1142-97	14.02	Terminology Relating to Thermophysical Properties
E-1269-99	14.02	Test Method for Determining Specific Heat Capacity by Differential Scanning Calorimetry
E-1356-98	14.02	Test Method for Glass Transition Temperatures by DSC or DTA
E-1363-97	14.02	Test Method for Temperature Calibration of Thermomechanical Analyzers
E-1545-95a	14.02	Test Method Assignment of the Glass Transition Temperature by Thermomechanical Analysis
E-1582-93	14.02	Practice for Calibration of Temperature Scale for Thermogravimetry
E-1640-99	14.02	Test Method for Assignment of the Glass Transition by Dynamic Mechanical Analysis
E-1641-99	14.02	Test Method for Decomposition Kinetics by Thermogravimetry
E-1824-96	14.02	Test Method for Assignment of the Glass Transition Temperature using Thermomechanical Analysis under Tension
E-1858-97	14.02	Test Method for Determining Oxidation Induction Time of Hydrocarbons by DSC (PDSC)
E-1860-97a	14.02	Test Method for Elapsed Time Calibration of Thermal Analyzers
E-1867-97	14.02	Test Method for Calibration of Dynamic Mechanical Analyzers
E-1877-00	14.02	Practice for Calculating Thermal Endurance of Materials from Thermogravimetric Decomposition Data
E-1952-98	14.02	Test Method for Thermal Conductivity and Thermal Diffusivity by Modulated Temperature DSC
E-2009-99	14.02	Test Method for the Determination of the Oxidation-Onset Temperature of Hydrocarbons (Including Polyolefins) by DSC
E-2038-99	14.02	Test Method for Temperature Calibration of Dielectric Analyzers
E-2039-99	14.02	Practice for Determining and Reporting Dynamic Dielectric Properties
E-2040-99	14.02	Test Method for Mass Scale Calibration of Thermogravimetric Analyzers
E-2041-99	14.02	Test Method for Estimating Kinetic Parameters by Differential Scanning Calorimeter
E-2046-99	14.02	Test Method for Reaction Induction Time by Thermal Analysis

Table 13.9 Abbreviations and acronyms

Abbreviation	Definition
DSC	Differential scanning calorimetry
DTA	Differential thermal analysis
T_g	Glass-transition temperature
MTDSC	Modulated temperature DSC
DDSC	Dynamic DSC
TMA	Thermal mechanical analysis
DMA	Dynamic mechanical analysis
DETA	Dielectric thermal analysis
DEA	Dielectric analysis
C_p	Specific heat capacity
ΔH	Heat reaction or fusion
T_m	Melt temperature
$T_{o/e}$	Extrapolated-onset temperature
T_m	For pure materials
T_p	Peak temperature, T_m for mixtures
T_c	Crystallization temperature
$T_g(e)$	Glass-transition temperature, $T_{o/e}$
$T_g(I)$	Glass-transition temperature at the inflection temperature
ΔC_p at T_g	Change in C_p at the T_g
ΔT at T_g	Change in temperature at T_g
TGA	Thermogravimetric analysis
TG	Thermogravimetry
DTGA	Differential TGA
OIT	Oxidation induction time
OOT	Oxidation-onset temperature
e''	DETA loss factor
e'	DETA permittivity
G'	DMA shear modulus
G''	DMA
E''	DMA loss modulus
E'	DMA Young's modulus
Loss tangent	Ratio of E''/E' or G''/G' or e''/e'
Tan Δ	Loss tangent function
Eff	Effusivity
k	Thermal conductivity
rho	Density
D	Thermal diffusivity
Micro-TA	Microthermal analysis
M-DTA	Micro-DTA
M-TMA	Micro-TMA

T_g by DSC

Glass transition temperatures measured by DSC are dependent on the temperature program and on local physical and chemical variations in samples of the same polymer. Diluents and fillers in a polymer always affect the T_g, and moisture is particularly troublesome in T_g measurements.[108] Hysteresis phenomena and physical aging cause T_g to vary with time.[109] Complications resulting in symmetric and asymmetric broadening of the T_g are linked to nanometer and micrometer scale phase separations. Last and most significant, the thermal history of the sample influences the extrapolated T_g, as shown in Figure 13.46.

T_g by DSC Using an Indium Standard

The melting of a 0.06-mm disk of indium placed on top of a 0.56-mm thick disk of polycarbonate is shown in Figure 13.47.[110] The materials were heated at 15 °C/min, and melting of indium was recorded at 156.6, 157.2, and 157.0 °C, depending on the indium being placed below, in the middle, or on top of the polycarbonate sample. The T_g of polycarbonate was 146.6 °C, as compared to 142 °C for a measurement with an adiabatic calorimeter. The ΔC_P at the T_g was 0.25 J/g per °C for both determinations.

T_g by DMA

Many factors are involved in measuring accurate T_g values with DMA.[111] They may be classified as instrumental factors, the frequency of the applied stress, material characteristics, and choice of T_g criterion. Material characteristics include the degree of crystallinity in crystalline polymers, the degree of cross-linking in thermosets, the thermal and mechanical history of the material, and, possibly, moisture effects. For example, loss modulus-versus-temperature measurements for poly(amideimide) with various moisture contents is given in Figure 13.48, where it is seen that the T_g decreases from 300 to 200 °C as the moisture content increases to 4.2%. The T_g criterion relates to the choice of the viscoelastic function used to establish the T_g. For example, the peak value of the modulus-versus-temperature curve gives a different value from the peak tan delta value.

T_g by DETA

Permittivity and dielectric loss increase sharply at T_g. When a poly(vinyl chloride) and an epoxy resin were evaluated by DETA, the dielectric T_g approached the DSC T_g as the frequency of the dielectric measurement decreased. For a variety of epoxies, the T_g by DSC was 5 °C lower than the DETA value at 1 Hz.[112]

Analysis of an Unknown Polymer by Thermal Techniques

Weigh a 10- to 20-mg sample into a DSC pan, and examine it at 100× magnification with a reflecting polarizing microscope. Note the color and morphology of the polymer. Next, evaluate the sample by DSC or TMA from −100 to 120 °C in nitrogen. Determine the T_g and any other transitions that may be associated with the thermal history of the sample, such as moisture evolution or stress relaxation exotherms. Reexamine the sample under the microscope, and note any changes in polymer form, crystallinity, or color. Reweigh the sample; the loss of weight represents the loss of water and, possibly, of solvent, plasticizer, or other volatiles. Examine the sample in the DSC from 80 to 350 °C in nitrogen. Determine its melting range, processing temperatures (indicated by exothermic stress-relief events), and exothermic heats of reaction and temperatures of cross-linking. Calculate the degree of cure by comparison to heats associated with complete cure. Program cool the DSC from 300 to −100 °C in nitrogen. For semicrystalline or crystalline polymers, observe the temperature and heat of recrystallization. Note the T_g on cooling. Heat the DSC sample once again from −100 to 300 °C in nitrogen. Redetermine the melting temperature and the T_g now that the sample is free from artifacts introduced by processing. Cool the DSC sample to room temperature and transfer it to the TGA pan. Examine the sample from 300 to 900 °C in nitrogen and determine its thermal degradation characteristics. Capture effluents from the TGA in a liquid nitrogen cold finger and analyze them by MS or by FTIR spectroscopy.[113]

Following is an example using TGA for the compositional analysis of a filled polymer. A number of elastomers and carbon-filled rubber composites used in track pads for Army tanks were analyzed by TGA.[114] The test specimens contained styrene-butadiene rubber (SBR), butyl rubber (BR), and natural rubber (NR). TGA was used to determine the percentage of highly volatile organics, elastomers, carbon black, and inorganic residue in each sample. The data illustrate that the percentage of organics and carbon black were consistent with known formulations, but the percentage of inorganic-ash residue showed some variation. DTGA curves were used to determine the type of elastomer contained in each track pad, and DSC was used to detect uncured polymers. A typical TG-DTG is given in Figure 13.49.

Resources, Abbreviations, and Acronyms

The North American Thermal Analysis Society and the International Congress on Thermal Analysis and Calorimetry are resources for scientists who seek a high level of understanding of the many thermal analysis methods. ASTM Committee E-37 on Thermal Measurements and Committee D-20 on Plastics are sources of practical standardized thermal analysis procedures, as well as new method development. These two committees have developed a number of protocols used in polymer thermal analysis, and these are summarized in Tables 13.7 and 13.8. Abbreviations and acronyms used in this chapter are summarized in Table 13.9.

References

1. Riga, A. T.; Collins, E. In *Engineering Plastics*; Dostal, C., Ed.; ASM International: Novelty, OH, 1988; pp 824–837.
2. Riga, A.; Collins, R.; Mlachak, G. *Thermochim. Acta* **1998**, *226*, 201–210.
3. Riga, A.; Patterson, G., *Thermochim. Acta* **1993**, *324*, 135–149.
4. Patterson, G.; Riga, A; Bomback, B., *Thermochim. Acta* **1994**, *243*, 277–288.
5. Xie, W.; Xie, Y.; Pan, W-P.; Riga, A., *Thermochim. Acta* **2000**, *357–358*, 239–250.
6. Wunderlich, B. The Basis of Thermal Analysis. In *Thermal Characterization of Polymeric Materials*; Turi, E., Ed.; Academic Press: New York, 1997; pp 206–472.
7. Prime, B. Thermosets. In *Thermal Characterization of Polymeric Materials*; Turi, E., Ed.; Academic Press: New York, 1997; pp 1380–1766.
8. Chartoff, R. Thermoplastic Polymers. In *Thermal Characterization of Polymeric Materials*; Turi, E., Ed.; Academic Press: New York, 1997; pp 484–744.
9. Gallagher, P. Thermoanalytical Instrumentation. In *Thermal Characterization of Polymeric Materials*; Turi, E., Ed.; Academic Press: New York, 1997; pp 2–205.
10. Bair, B. Thermal Analysis of Additives in Polymers. In *Thermal Characterization of Polymeric Materials*; Turi, E., Ed.; Academic Press: New York, 1997; pp 2264–2420.
11. Turi, E. *Thermal Characterization of Polymers*; Academic Press: San Diego, CA, 1981.
12. Slade, P.; Jenkins, L. (Eds.) *Thermal Characterization Techniques*; Marcel Dekker: New York, 1970.
13. Wendlandt, W. *Thermal Analysis*, 3rd ed.; Wiley: New York, 1986.
14. Wateson, E.; O'Neill, M.; Justin, J.; Brenner, N. *Anal. Chem.* **1964**, *36*, 1233–1240.
15. Bair, H. *Polym. Eng. Sci.* **1970**, *10*, 247–253.
16. Casseino, M.; Blevins, D.; Sanders, R. *Thermochim. Acta* **1996**, *284*, 145–153.
17. Nielsen, L. E. *Mechanical Properties of Polymers*; Reinhold Publishing: New York, 1962.
18. Tomason, J. L. *Polymer Composites* **1990**, *11*, 105–112.
19. Boyd, R. H. *Polymer* **1985**, *26*, 323–329.
20. ASTM Method E-0473; *(DTA) Definitions of Terms Relating to Thermal Analysis*; American Society for Testing and Materials: West Conshohocken, PA, 1998; 14.02, 226.
21. ASTM Method E-0473; *(DSC) Definitions of Terms Relating to Thermal Analysis*; American Society for Testing and Materials: West Conshohocken, PA, 1998; 14.02, 226.
22. Sestak, J.; Stepanek, B. *J. Thermal Anal.* **1994**, *43*(I–IV), 371–374;
23. Billmeyer, F. W. *Polymer Science*; Wiley-Interscience: New York, 1971.
24. Speyer, R. *Thermal Analysis of Materials*; Marcel Dekker: New York, 1994.
25. Riga, A. T.; Collins, R. Differential Scanning Calorimetry and Differential Thermal Analysis. In *Encyclopedia of Analytical Chemistry*; Meyers, R., Ed.; John Wiley & Sons: Chichester, U.K., 2000; pp 13147–13179.
26. Perkin Elmer; *Thermal Analysis Newsletter* No. 9, 1970. Perkin Elmer; *DSC-2 Manual*; Perkin Elmer Corporation, Norwalk, CT, 1972. Cassel, B. *Materials Analysis in the QC and Analytical Laboratories*; Perkin Elmer Seminar CD; Perkin Elmer Corporation, Norwalk, CT, 1999. Cassel, B. Private communication; Perkin Elmer Corporation: Norwalk, CT, 2000.

27. Riga, A. T.; Collins, R. Differential Scanning Calorimetry and Differential Thermal Analysis. In *Encyclopedia of Analytical Chemistry*; Meyers, R., Ed.; John Wiley & Sons: Chichester, U.K., 2000; pp 13155–13158.

28. ASTM Method E-967. *Calibration of the DSC Temperature*; American Society for Testing and Materials: West Conshohocken, PA, 2000; Vol. 14.02, pp 630–633.

29. ASTM Method E-968. *Calibration of Heat Flow in a DSC and PDSC*; American Society for Testing and Materials: West Conshohocken, PA, 2000; Vol. 14.02, pp 634–638.

30. ASTM Methods D3417 *Heat of Fusion and Crystallization of Polymers by Thermal Analysis* and D3418 *Transition Termperatures of Polymers by Thermal Analysis*; American Society for Testing and Materials: West Conshohocken, PA, 2000; Vol. 14.02.

31. Hourston, D.; Song, M.; Hammiche, A.; Pollock; H.; Reading, M. *Polymer* **1997**, *38*(1), 1–7.

32. ASTM Method E-1860. *Calibration of Lapsed Time in a Thermal Analysis Instrument*; American Society for Testing and Materials: West Conshohocken, PA, 2000; 14.02.

33. TA Instruments Co. *Enthalpy of Melting for Standards*, Version 2.5; Thermal Analysis Application Library; TA Instruments: New Castle, DE, 1998.

34. Riga, A.; Patterson, G. Factors Affecting Oxidation Properties in DSC Studies. *Thermochim. Acta* **1993**, *226*, 201–210. ASTM Method E-1858. *Oxidation Induction Time by DSC and PDSC of Hydrocarbons and Polyolefins*; American Society for Testing and Materials: West Conshohocken, PA, 2000; Vol. 14.02.

35. Riga, A. T.; Collins, R. M.; Patterson, G. Thermal Oxidative Behavior of Readily Available Reference Polymers. In *Oxidative Behavior of Materials by Thermal Analytical Techniques* (STP 1326); Riga A. T., Patterson, G. H., Eds.; American Society for Testing and Materials: West Conshohocken, PA, 1997; pp 91–101.

36. Blaine, R. L.; Lundgren, C. J.; Harris, M. B. Oxidation Induction Time-Review of DSC Experimental Effects. In *Oxidative Behavior of Materials by Thermal Analytical Techniques* (STP 1326); Riga, A. T., Patterson, G. H., Eds.; American Society for Testing and Materials: West Conshohocken, PA, 1997; pp 3–15.

37. Cassel, R. B.; Salamon, A. W.; Curran, G.; Riga, A. Comparative Techniques for OIT Testing. In *Oxidative Behavior of Materials by Thermal Analytical Techniques* (STP 1326); Riga, A. T., Patterson, G. H., Eds.; American Society for Testing and Materials: West Conshohocken, PA, 1997; pp 164–171.

38. ASTM Method E-1858. *Oxidation Induction Time by DSC and PDSC of Hydrocarbons and Polyolefins*; American Society for Testing and Materials: West Conshohocken, PA, 2000; Vol. 14.02, pp 1097–2001.

39. ASTM Method E-2009. *Oxidation Onset Temperature by DSC and PDSC of Hydrocarbons and Polyolefins*; American Society for Testing and Materials: West Conshohocken, PA, 2000; Vol. 14.02, pp 728–737.

40. Riga, A. T.; Patterson, G. H. Development of a Standard Test Method for Determining OIT by DSC and PDSC. In *Oxidative Behavior of Materials by Thermal Analytical Techniques* (STP 1326); Riga, A. T., Patterson, G. H., Eds.; American Society for Testing and Materials: West Conshohocken, PA, 1997, pp 151–163.

41. Riga, A.; Kovach, P. *J. Thermal Anal. Calorimetry* **1997**, *49*, 425–435.

42. TA Instruments Co. *Thermoplastics by DSC*, Version 2.5; Thermal Analysis Application Library; TA Instruments Co.: New Castle, DE, 1998.

43. Cassel, B. Private communication; Perkin Elmer Corporation: Norwalk, CT, 1999. Schawe, J.; Hohne, G. *J. Therm. Anal.* **1996**, *46*, 893–904.

44. Wunderlich, B.; Boller, A.; Okazaki. I.; Kreitmeier, S. *J. Thermal Anal.* **1996**. *47*, 1013–1026.

45. Song, M.; Hourston, D.; Pollock, H.; Schafer, F.; Himmiche, A. *Polymer* **1997**, *38* (3), 503–507.

46. Jin, Y.; Bonilla, J.; Lin, G.; Morgan, J.; McCracken L.; Carnahan, J. *J. Thermal Anal.* **1997**, *38*(12), 3025–3034.

47. van Ekenstein, G.; ten Brinke, G.; Ellis, T.; *Poly. Mater. Sci. Eng.* **1997**, *76*, 219–220.

48. Hourston, D.; Song, M.; Hammiche, A.; Pollock, H.; Reading, M. *Polymer* **1997**, *38*(1), 1–7.

49. Hu, X.; Breach, C.; Young, R. *Polymer* **1997**, *38*(4), 981–983.

50. Sauerbrunn, S.; Thomas, L. *Amer. Lab.* **1995**, *27*, 19–22.

51. Luget, A.; Wilson, R. *Thermochim. Acta* **1994**, *238*, 295–307.

52. Reading, M. *Trends Polym. Sci.* **1993**, *1*, 248–253.

53. Androsch, R.; Moon, I.; Pyda, M.; Wunderlich, B. *Thermochim. Acta* **2000**, *357–358*, 267–278.

54. ASTM Method E-1269. *Specific Heat Capacity by DSC*; American Society for Testing and Materials: West Conshohocken, PA, 2000; Vol. 14.02, pp 787–791.

55. Wunderlich, B. *Thermal Analysis*; Academic Press: Boston, 1990.

56. Wunderlich, B.; Chen, W. *Macromol. Chem. Phys.* **1999**, *200*, 283–311.

57. ASTM Method E-1952. *Thermal Conductivity and Diffusivity by DSC*; American Society for Testing and Materials: West Conshohocken, PA, 2000.

58. Riga, A.; Mathis, N.; Chandler, K. Design and Selection of Engineering Plastics and Automotive Fluids by Heat Transfer Methods. Presented at IUPAC Conference on Chemical Thermodynamics, Halifax, NS, Canada, 2000.

59. Mathis, Nancy. Private communication; Mathis Instruments: New Brunswick, NS, Canada, 2000. Riga, A.; Mathis, N.; Chandler, K. Design and Selection of Engineering Plastics and Automotive Fluids by Heat Transfer Methods. Presented at IUPAC Conference on Chemical Thermodynamics, Halifax, NS, Canada, 2000.

60. Society of Plastics Engineers. Educational Resin Kit (50 Polymers), 14 Fairfield Drive, Brookfield, CT 06804–0403, 1994.

61. Earnest, C. M. The Modern TG Approach to the Compositional Analysis of Materials. In (ASTM STP 997) *Compositional Analysis by Thermogravimetry*; Earnest, C. M., Ed.; American Society for Testing and Materials: West Conshohocken, PA, 1988; 1–18.

62. Charsley, E.; Warrington, S. B. Industrial Applications of Compositional Analysis by TGA. In (ASTM STP 997) *Compositional Analysis by Thermogravimetry*; Earnest, C. M., Ed.; American Society for Testing and Materials: West Conshohocken, PA, 1988; 19–27.

63. Gillmor, J.; Seyler, R. J. Using Chemistry in Compositional Analysis by TGA. In (ASTM STP 997) *Compositional Analysis by Thermogravimetry*; Earnest, C. M., Ed.; American Society for Testing and Materials: West Conshohocken, PA, 1988; 38–47.

64. Lorigan Allen, P. TGA of Sheet Molding Compound Materials to Determine Distribution of Compound Components in Molded Parts. In (ASTM STP 997) *Compositional Analysis by Thermogravimetry*; Earnest, C. M., Ed.; American Society for Testing and Materials: West Conshohocken, PA, 1988; 70–84.

65. Pamphilis, N. A. Compositonal Analysis of Light Duty Motor Bearings in Fuel Pumps. In (ASTM STP 997) *Compositional Analysis by Thermogravimetry*; Earnest, C. M., Ed.; American Society for Testing and Materials: West Conshohocken, PA, 1988; 117–131.

66. Kau, H. T. Determination of the Degree of Mixing of Sheet Molding Compound Paste Using TGA. In (ASTM STP 997) *Compositional Analysis by Thermogravimetry*; Earnest, C. M., Ed.; American Society for Testing and Materials: West Conshohocken, PA, 1988; 98–116.

67. Lann, P.; Derejanik, T. S.; Snyder, J.; Ward, W. *Thermochim. Acta* **2000**, *357–358*, 225–230.

68. Peebles, L. H. Degradation of Acrylonitrile Polymers. In *Encyclopedia of Polymer Science and Technology*; Mark, H., Ed.; Interscience Publishing: New York, 1976; Vol. 1 (Suppl.), pp 1–52.

69. Reich, L.; Levi, D. Thermogravimetry. In *Encyclopedia of Polymer Science and Technology*; Mark, H., Ed.; Interscience Publishing: New York, 1971; Vol. 14, pp 1–41.

70. Ozawa, T. *Bull. Chem. Soc. Japan* **1965**, *58*, 1881–1892.

71. Flynn, J. H.; Wall, L. A. *J. Polym. Sci.* **1966**, *4*, 323–341.

72. Uptmor, M.; Blaine, R. *Thermal Analysis Operating Manual*, TA-081. TA Instruments: New Castle, DE, 2000.

73. ASTM Method E-1582. *Temperature Calibration in a TGA*; American Society for Testing and Materials: West Conshohocken, PA, 2000.

74. ASTM Method E-2040. *Mass Scale Calibration in a TGA*; American Society for Testing and Materials: West Conshohocken, PA, 2000. ASTM Method E-1131. *Compositional Analysis by TGA*; American Society for Testing and Materials: West Conshohocken, PA, 2000.

75. Kargin, V. A.; Slonimsky, G. C. Mechanical Properties of Polymers. In *Encyclopedia of Polymer Science and Technology*; Mark, H., Ed.; Interscience Publishing: New York, 1968; Vol. 8, pp 445–468.

76. Ferry, J. D. *Viscoelastic Properties of Polymers*; Wiley: New York, 1961; pp 203–209.

77. Tobolsky, A. V. *Properties and Structure of Polymers*; Wiley: New York, 1960; pp 72–83.

78. Hazer, B. Synthesis and Characterization of Block Copolymers. In *Handbook of Polymer Science and Technology*; Cheremisinoff, N. P., Ed.; Marcel Dekker: New York, 1989; Vol. 1, pp 133–176.

79. Cheremisinoff, N. P. Techniques for Property Characterization. In *Handbook of Polymer Science and Technology*; Cheremisinoff, N. P., Ed.; Marcel Dekker: New York, 1989; Vol. 1, pp 471–504.

80. Kovarskii, A. Techniques for Studying Polymer Relaxation. In *Handbook of Polymer Science and Technology*; Cheremisinoff, N. P., Ed.; Marcel Dekker: New York, 1989; Vol. 1, pp 643–676.

81. Briedis, I. P. Viscoelastic and Molecular Characterization of Commercial Polymers. In *Handbook of Polymer Science and Technology*; Cheremisinoff, N. P., Ed.; Marcel Dekker: New York, 1989; Vol. 1, pp 605–642.

82. Garn, P. D. *Thermal Analytical Methods of Investigation*; Academic Press: New York, 1965.

83. Ramachaandran, S.; Gao, H. W.; Christiansen, E. B. Effect of Polymer Structure in Polymer Flow. In *Handbook of Polymer Science and Technology*; Cheremisinoff, N. P., Ed.; Marcel Dekker: New York, 1989; Vol. 3, pp 199–270.

84. Failla, M. D.; Carella, J. M. Properties of Solid State Extrudates of Polyethylene. In *Handbook of Polymer Science and Technology*; Cheremisinoff, N. P., Ed.; Marcel Dekker: New York, 1989; Vol. 3, pp 503–540.

85. Cheremisinoff, N. P.; Cheremisinoff, P. Properties and Applications of Fiber Re-enforced Plastics. In *Handbook of Polymer Science and Technology*; Cheremisinoff, N. P., Ed.; Marcel Dekker: New York, 1989; Vol. 4, pp 421–498.

86. Foreman, J. *DMA of Polymers*; TA 236; TA Instruments: New Castle, DE, 2000.

87. Blaine, R, *DMA How it Works*; TA 229; TA Instrument: New Castle, DE, 2000.

88. ASTM Method E-1867. *Temperature Calibration of a DMA*; American Society for Testing and Materials: West Conshohocken, PA, 2000.

89. Neag, C. M. TMA in Material Science. In *Material Characterization by TMA (STP 1136)*; Riga, A., Neag, C. M., Eds.; American Society of Testing and Materials: West Conshohocken, PA, 1991; pp 3–21.

90. Riga, A. T.; Collins, E. A. Material Characterization by TMA: Industrial Applications. In *Material Characterization by TMA (STP 1136)*; Riga A., Neag, C. M., Eds.; American Society of Testing and Materials: West Conshohocken, PA, 1991; pp 71–83.

91. Schoff, C. K.; Kamarchik, P. Applications of TMA in Organic Coatings. In *Material Characterization by TMA (STP 1136)*; Riga A., Neag, C. M., Eds.; American Society of Testing and Materials: West Conshohocken, PA, 1991; pp 150–160.

92. Weidemann, H.; Riesen, R.; Roller, A. Elasticity Characterization of Material during Thermal Treatment by TMA. In *Material Characterization by TMA (STP 1136)*; Riga, A., Neag, C. M., Eds.; American Society of Testing and Materials: West Conshohocken, PA, 1991; pp 84–99.

93. Riga, A. T. *Polymer Eng. Sci.* **1974**, *14*, 764–771.

94. ASTM Method E-1363. Temperature Calibration of a TMA; American Society for Testing and Materials: West Conshohocken, PA, 2000.

95. ASTM Method E-831. Linear Thermal Expansion by

TMA; American Society for Testing and Materials: West Conshohocken, PA, 2000.

96. Hedvig, P. *Dielectric Spectroscopy of Polymers*; McGraw-Hill: New York, 1977.

97. Runt, J. P.; Dielectric Studies of Polymer Blends. In *Dielectric Spectroscopy of Polymeric Materials*; Fitzgerald, J. J., Runt, J. P., Eds.; American Chemical Society: Washington, DC, 1997; pp 283–302.

98. Kranbuehl, D. E. Dielectric Monitoring of Polymerization and Cure. In *Dielectric Spectroscopy of Polymeric Materials*; Runt, J. P., Fitzgerald, J. J., Eds.; American Chemical Society: Washington, DC, 1997; pp 303–328.

99. Riga, A.; Cahoon, J.; Lvovich, V. Characterization of Surfactants and Dispersants by DETA. In *Characterization of Materials by Dynamic and Modulated Thermal Analytical Techniques* (STP 1409); Riga, A., Judovits, L., Eds.; American Society for Testing and Materials: West Conshohocken, PA, 2000.

100. Riga, A.; Cahoon, J.; Pialet, J. Characterization of Electrorheological Components by DETA. In *Characterization of Materials by Dynamic and Modulated Thermal Analytical Techniques* (STP 1409); Riga, A., Judovits, L., Eds.; American Society for Testing and Materials: West Conshohocken, PA, 2000.

101. Riga, A. T.; Collins, R. M.; Patterson, G. Thermal Oxidative Behavior of Readily Available Reference Polymers. In *Oxidative Behavior of Materials by Thermal Analytical Techniques* (STP 1326); Riga, A., Patterson, G. H., Eds.; American Society for Testing and Materials: West Conshohocken, Pa, 1997; pp 91–101.

102. Blaine, R. L.; Lundgren, C. J.; Harris, M. B. Oxidation Induction Time-Review of DSC Experimental Effects. In *Oxidative Behavior of Materials by Thermal Analytical Techniques* (STP 1326); Riga, A. T., Patterson, G. H., Eds.; American Society for Testing and Materials: West Conshohocken, Pa, 1997; pp 3–15.

103. ASTM Method E-2038. Temperature Calibration of a DETA; American Society for Testing and Materials: West Conshohocken, PA, 2000.

104. ASTM Method E-2039. Reporting Dynamic Dielectric Properties; American Society for Testing and Materials: West Conshohocken, PA, 2000.

105. Shepard, David. Private communication; Micromet/Netzsch Instrument Inc.: Burlington, MA, 2000.

106. Reading, M.; Price, D.; Pollock, H.; Hammiche, A.; Murray, A. Recent Progress in Microthermal Analysis. *Am. Lab.* **1999**, 11–13.

107. TA Instruments. *Characterization of Surface Morphology Changes by Micro-Thermal Analysis*; TA 258; TA Instruments: New Castle, DE, 2000.

108. Seyler, R. Opening Discussion. In *Assignment of the Glass Transition* (STP 1249); Seyler, R. J., Ed.; American Society for Testing and Materials: West Conshohocken, PA, 1994; pp 13–15.

109. Wunderlich, B. The Nature of the Glass Transition and its Determination by Thermal Analysis. In *Assignment of the Glass Transition* (STP 1249); Seyler, R. J., Ed.; American Society for Testing and Materials: West Conshohocken, PA, 1994; pp 17–31.

110. Bair, H. Glass Transition Measurements by DSC. In *Assignment of the Glass Transition* (STP 1249); Seyler, R. J., Ed.; American Society for Testing and Materials: West Conshohocken, PA, 1994; pp 50–74.

111. Chartoff, R. P.; Weissman, P. T.; Sircar, A. The Application of DMA Methods to T_g Determination in Polymers: An Overview. In *Assignment of the Glass Transition* (STP 1249); Seyler, R. J., Ed.; American Society for Testing and Materials: West Conshohocken, PA, 1994; pp 88–107.

112. Bidstrup, S. A.; Day, D. R. Assignment of the Glass Transition Temperature using Dielectric Analysis: A Review. In *Assignment of the Glass Transition* (STP 1249); Seyler, R. J., Ed.; American Society for Testing and Materials: West Conshohocken, PA, 1994; pp 108–119.

113. Riga, A. T.; Brentgardner, D., *Instrumental Analysis of Polymers*; ASTM Technical Professional Training; American Society for Testing and Materials: West Conshohocken, PA, May 1999.

114. Macaione, D. P.; Sacher, R. E.; Singler, R. E. TGA Characterization of Elastomers and Carbon Filled Rubber Composites for Military Applications. In *Compositional Analysis by TGA* (STP 997); Earnest, C. M., Ed.; American Society for Testing and Materials: West Conshohocken, PA, 1988; pp 59–69.

14

X-ray Diffraction

RYONG-JOON ROE

X-ray diffraction is a technique by which the structure of materials at the angstrom or nanometer length scale is determined. Soon after the discovery of X-rays in 1895 by Röntgen, Laue in Germany and Bragg in England realized that X-ray diffraction can be a valuable tool for exploring the arrangement of atoms in materials. Much of our knowledge on organization of atoms in materials, both crystalline and noncrystalline, owes to X-ray diffraction. The technique has been widely used for the study of the structure of solid polymers from the very inception of polymer science.

X-rays are electromagnetic waves which, when incident on a material, are scattered as a result of their interaction with electrons in the material. The X-ray waves scattered from different electrons interfere with each other and give rise to a diffraction pattern in which the intensities vary with scattering angle. The diffraction pattern thus reflects the spatial distribution of electrons in the material, and an analysis of the diffraction pattern therefore leads us to information on the arrangement of electrons and atoms. The particular usefulness of X-rays, in comparison to other electromagnetic waves in the spectrum, arises from the fact that the wavelength of X-rays used for diffraction studies is typically of the order of 1 Å ($=10^{-10}$ m), which is of the same order of magnitude as the interatomic distances in condensed matter.

Scattering of light is also widely used for the study of structure of polymers. Light is also an electromagnetic wave, but of wavelengths that are about 3 orders of magnitude larger than X-ray wavelengths. As a result, light scattering provides information on structures much larger in scale than can be obtained by X-ray diffraction.

Another technique of scattering for structural studies, gaining importance in recent years, is that of neutron scattering. Neutrons, although more usually thought of as particles, exhibit wavelike characteristics as well. The wavelength of neutrons depends on their velocity. For use in structural studies, high-speed neutrons generated in a nu-

clear reactor are first "moderated" to much lower velocities, which correspond to wavelengths on the order of 1 Å. There are many similarities between the X-ray and neutron-scattering techniques, both in the principles involved and in the methods of analysis of the data, and most of the materials discussed in this chapter are equally applicable to neutron scattering. However, there are also sufficient differences between the two to make them complementary to each other. The most important difference arises from the fact that hydrogen and deuterium atoms differ greatly from each other in their ability to scatter neutrons, whereas X-rays are scattered only very weakly from either of these isotopes. This makes possible the trick of *deuterium labeling*— that is, either labeling some part of hydrogen-containing molecules or labeling only some of the molecules in a sample by selectively substituting hydrogens with deuteriums. Such selective labeling allows the deuterated atoms to stand out in scattering neutrons without significantly affecting other physical properties of the molecules in most cases of interest.

Fundamentally, the X-ray diffraction pattern we observe reflects the distribution of electron densities in the material. Since electrons are grouped into atoms, the study of an X-ray diffraction pattern leads us to information about the arrangement of atoms in the sample. In the case of crystalline materials, the diffraction phenomenon can be described conveniently in terms of the Bragg law:

$$\frac{2 \sin \theta}{\lambda} = \frac{1}{d_{hkl}} \qquad (1)$$

where θ is one half of the angle between the incident and diffracted beams, λ is the wavelength of the X-rays, and d_{hkl} is the interplanar spacing between the crystallographic planes that have Miller indices (hkl). Figure 14.1 illustrates the physical factors that underlie eq 1. The diffraction pattern from a large crystal with a perfect crystalline order consists of a series of sharp Bragg peaks, each correspond-

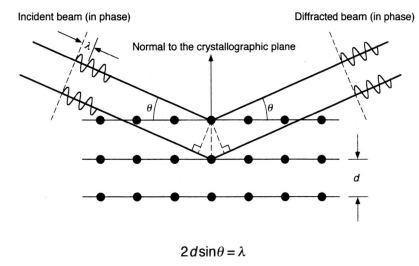

Figure 14.1 Geometry of diffraction illustrating the Bragg law and the relationship among the incident and diffracted beams and the normal to the set of crystallographic planes (*hkl*) that give rise to the diffraction.

ing to a different Miller plane and located at the diffraction angle 2θ, thus satisfying eq 1. In contrast, the diffraction pattern from a noncrystalline material is smeared out, without any readily recognizable features except an extremely broad maximum, commonly referred to as "amorphous halo".

Strictly speaking, the word *scattering* refers to the primary process of the X-rays being scattered by an electron and the word *diffraction* to the process in which the waves scattered by different electrons interfere with each other and produce the interference pattern. As long as the sample contains more than one electron, as is always the case, what we measure is the diffraction pattern. In usual practice, however, the distinction between the two words has shifted, and the word *diffraction* now tends to be used more narrowly and only when the pattern obtained contains discrete diffraction peaks, whereas the word *scattering* is instead used when the pattern is diffuse. In the absence of a clear-cut logic to differentiate them, the two words are used more or less interchangeably in this chapter.

X-ray diffraction (or scattering) techniques are usually classified into wide-angle and small-angle methods. In the former, the interest is in measuring the intensity of scattering at angles larger than a few degrees; in the latter, the measurement is performed at much smaller scattering angles. It is easily seen from the Bragg law (eq 1) that with X-rays of λ approximately 1 Å, the wide-angle scattering pattern arises from structures (or interplanar spacing d) of around 1 nm or smaller in size, whereas the small-angle scattering pattern contains information about the structure that is larger than about 1 nm. Although the basic principles governing the two cases are the same, the instrumental requirement can be very different because of the need to have an extremely well collimated incident beam for the small-angle scattering. The method of analysis of the data

is also often different, reflecting the different size scales and the different approximations and assumptions involved.

There are many excellent sources from which more detailed introductions to the X-ray diffraction techniques can be obtained. References 1 through 5 discuss X-ray diffraction by materials in general, while references 6 through 11 are concerned with the technique applied to polymers. References 12 and 13 are addressed specifically to the problems of small-angle scattering. Although reference 14 deals with neutron scattering only, it discusses many aspects of analysis that apply equally well to X-ray scattering. For those new to the X-ray diffraction techniques, reference 5 offers a particularly readable introduction to the principles and basic experimental techniques. The areas of applications it discusses are geared more toward metallurgical studies, however. Applications to polymers and discussions of more recent development, as well as more advanced topics, can be found in reference 11.

Experimental Procedure

X-ray Source

X-rays are generated in the laboratory in an X-ray tube. In the tube electrons, emitted from a hot tungsten filament, is accelerated toward a metal target via a potential difference (usually in the range of 35 to about 50 kV) between the target and the filament. On striking the target, part of the kinetic energy of the electrons is converted to X-rays and the rest to heat, which must be removed by means of cooling water to prevent the target metal from melting. In a sealed X-ray tube, the filament and the target metal are stationary, whereas in a rotating anode tube, a band of target metal is affixed on a cylindrically shaped support that is rotated rapidly. Since in the latter a fresh target metal surface is constantly brought to the position exposed to the accelerated electrons, the need for cooling is less acute, and a higher electron current can be tolerated, leading to a much larger output of X-rays.

A vastly higher intensity of emitted X-rays is attained in a synchrotron radiation source. In a synchrotron, a beam of charged particles, usually electrons, is accelerated to a speed approaching that of light and is made to circulate on a closed orbit in a storage ring. The closed orbit is achieved by bending the path of the high-velocity electrons with a series of magnets. A charged particle that is accelerated emits electromagnetic radiation, and bending is a form of acceleration. Thus the action of the bending magnets to produce the closed orbit is itself sufficient to produce intense radiation. The emitted radiation, however, is usually further enhanced by means of additional magnetic devices (termed wigglers and undulators), inserted in the orbit, that disturb the electron paths locally for the sole purpose of inducing radiation.

The wavelength spectrum of the radiation emitted by a synchrotron source covers an extremely wide range of electromagnetic waves that extends from the infrared ($\lambda \sim 10^{-4}$ m) to the hard X-rays ($\lambda \sim 0.1$ Å $= 10^{-11}$ m). The spectrum of the X-rays generated by a tube depends on the nature of the target metal in the tube, and for polymer studies a copper target tube serves the purpose well, although a molybdenum target tube is occasionally called for. The spectrum consists of two intense, narrow *characteristic lines*, called $K\alpha$ and $K\beta$, which are superimposed on the background of a broad *continuous* (or *white*) *radiation*, as illustrated in Figure 14.2. The range and spectral shape of the continuous radiation depend somewhat on the tube high voltage employed, whereas the wavelengths of the characteristic lines are determined by the electronic structure of the target metal atoms and are therefore specific to the target metal used in the tube. With a copper tube, the $K\alpha$ line is at 1.542 Å and the $K\beta$ line is at 1.392 Å.

Monochromater and Filter

Most X-ray diffraction studies of polymers require the use of monochromatic radiation. Monochromatization of the broad-spectrum radiation emitted by a tube or a synchrotron radiation source is best achieved by use of a crystal monochromator. When a single crystal, with its surface cut parallel to a set of (hkl) crystallographic planes of interplanar spacing d_{hkl}, is placed at an angle θ to the broad-spectrum incident beam, strong diffraction occurs only for a beam containing a narrow range of wavelengths centered around the λ that satisfies the Bragg law (eq 1). For precision work, the use of such a crystal monochromator is imperative, but the trade-off is that a large fraction of the incident beam intensity is thereby lost. For less demanding

work, the use of a filter instead of a crystal monochromater is permissible, if an X-ray tube is used as the source in the laboratory. The purpose of the filter, called a β filter, is to eliminate the $K\beta$ characteristic line as much as possible without attenuating the $K\alpha$ line appreciably at the same time. In the case of the copper target radiation, insertion of a Ni foil filter about 0.02 mm thick reduces the intensity ratio of the $K\beta$ to $K\alpha$ lines from the initial 0.13 to 0.002, while the intensity of the $K\alpha$ line itself is reduced by a factor of 2.4. However, some of the white radiation is still transmitted, and the effect of its presence contributes to the diffuse background that is spread out over a range of diffraction angles. The effect of the white radiation may be further reduced by means of a pulse-height discrimination when a proportional counter is used as the detector, as is explained shortly.

Detectors

There are many different types of detectors currently in use, but we may classify them into *point detectors* and *position-sensitive detectors*. A point detector measures the intensity of the scattered beam arriving at a given point or area where the detector window is placed, and to obtain a diffraction curve the detector must be moved successively to different positions. A *scintillation counter* and a *proportional counter* are the two point detectors used today in most X-ray work. A position-sensitive detector measures the intensities of the beam falling simultaneously at many different positions on the detector window and gives the intensity data as a function of the position. A position-sensitive detector can be a *linear detector* or an *area detector*, depending on whether the position information is one-dimensional or two-dimensional. A photographic film is an example of an area detector. There are several area detectors of recent inventions that are vastly superior to photographic film.

To understand the working of most detectors, it is more convenient to think of an X-ray beam as a stream of X-ray photons. The energy E of an X-ray photon is related to the frequency ν and the wavelength λ of the X-ray wave by

$$E = h\nu = hc/\lambda \qquad (2)$$

where c is the speed of light, and h is Planck's constant. The intensity of a beam has nothing to do with the energy of the individual photons it contains. A beam of higher intensity means that there are more photons (of an identical energy) streaming in the beam per second.

In the proportional counter, the element sensing the arrival of an X-ray photon is an inert gas (argon or xenon in most cases) that becomes ionized, whereas in the scintillation counter the sensing element is a solid scintillater that produces a flash of light on receiving an X-ray photon. In both cases, detection of an X-ray photon results in the generation of an electric pulse in the accompanying electronic circuitry. The number of such pulses *counted* per fixed

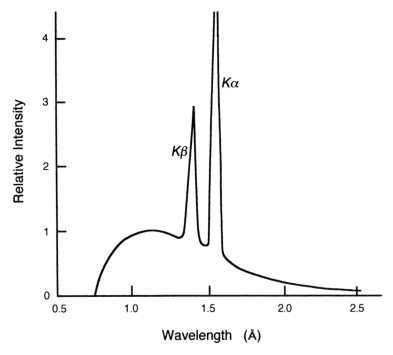

Figure 14.2 X-ray spectrum from a copper target tube: the characteristic lines $K\alpha$ and $K\beta$ are superimposed on the continuous radiation (relative intensities not to scale).

length of time is then proportional (depending on the detection efficiency of the detector) to the number of photons that impinge on the detector window during the time period and therefore is proportional to the intensity of the beam.

An important point here is that the size of the pulses generated depends on the energy of the photons (hence the wavelength of the X-rays) detected. By counting only those pulses with height falling within a certain range, in effect one can detect only those X-rays with wavelengths that fall within a selected range. Such a *pulse-height discrimination* cannot be a perfect means of monochromatization, however, since the proportionality between the incident photon energy and the height of the pulse generated is only approximate. Nevertheless, the proportionality is maintained much better with the proportional than with the scintillation counter.

A position-sensitive detector allows measurement of diffracted beam intensities over a range of diffraction angles at the same time, thus speeding up the measurement greatly in comparison to the use of a point detector. Some linear detectors and area detectors are based essentially on the same principles as the proportional counter is, but in these, the counts of pulses are sorted and recorded on different channels in the memory, depending on the positions in the window where the photons were detected. A *photographic film* is a kind of area detector of long-standing tradition that is still in wide use. Its most important virtue is low cost. The *image plate* has a thin coating of phosphor

crystals that produces a latent image of the diffraction pattern when exposed to X-rays, much as in an undeveloped sheet of photographic film. The diffraction pattern is subsequently retrieved by illuminating the plate, pixel by pixel, with a laser beam and reading the latent image. Compared to photographic film, the image plate does not require a dark room for development, the data are retrieved in digital form, and the plate can be reactivated and reused. In another design of an area detector, the X-rays that fall on the detector window are first converted into a pattern of visible light by means of a phosphor plate, in much the same way as in the scintillation counter. The pattern of visible light is then detected by means of a vidicon TV camera or a CCD (charge-coupled device) chip, as in a digital camera.

Diffractometer and Diffraction Geometry

Polymer samples are most often available in the form of a thin film or sheet. Even with a fiber sample, it is usual to have a bundle of fibers spread in the form of a film on a support. In a camera or a diffractometer, one usually has the choice of mounting such a sheet sample either in reflection or in transmission geometry, as illustrated in Figure 14.3.

In the diffractometer, the sample remains at the center while the detector is rotated around it on the *diffractometer circle*. In symmetrical reflection geometry, the sheet sample mounted at the center is irradiated by an incident beam that diverges from a single, small source. The rays diffract

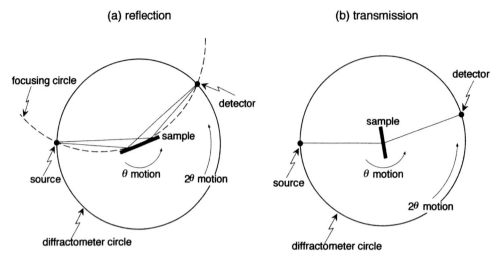

Figure 14.3 Reflection and transmission geometry of sample placement in a diffractometer.

at the same 2θ angle, but from different points on the sample surface, then all converge to a single point again at the detector window. Such a focusing effect is not available in the transmission geometry, and therefore a good angular resolution requires the incident beam to be fairly well collimated. The rotation of the detector arm is usually coupled to the rotation of the sample holder in a 2 : 1 angular ratio, so that the sample, once mounted in a symmetrical reflection or transmission geometry, can maintain the symmetry even while the detector arm is rotated.

The reflection geometry allows the use of a divergent incident beam, without loss of angular resolution, and therefore gives a much stronger diffracted beam intensity than does the transmission geometry used under otherwise the same condition. The reflection geometry, however, requires a much larger sample with a good flat surface, especially at relatively low 2θ angles. When the sample is anisotropic—for example, as a result of a stretching or other type of deformation—the structure in the sample is different in different directions. In such cases, symmetric reflection geometry examines the direction normal to the sheet surface, whereas symmetric transmission geometry examines the direction parallel to the sheet surface. This fact must be kept clearly in mind in cutting and mounting the sample in the diffractometer.

Absorption

When a material is irradiated with an X-ray beam, a part of the beam is scattered in many different directions, a part is transmitted through the material unmodified, and the rest is absorbed by the material and converted to different forms of energy, mostly heat. If I_0 is the intensity of the incident beam and I is the combined intensity of the transmitted and diffracted beams, then the two are related to the thickness x of the material by

$$I = I_0 e^{-\mu x} \tag{3}$$

where μ is called the *linear absorption coefficient* and has dimension [length^{-1}]. The linear absorption coefficient depends on the atomic constituents of the material, as well as its physical state (liquid, solid, density, etc.).

A more fundamental quantity expressing the loss of X-ray energy intercepted by a material is the *mass absorption coefficient*, which is represented by the somewhat awkward symbol (μ/ρ). The mass absorption coefficient is a number unique to each element (for a given X-ray wavelength), usually expressed in units of cm^2/g, and can be found tabulated in reference books such as the *International Tables of Crystallography*.[15] For a material with a known atomic composition the linear absorption coefficient can be calculated by

$$\mu = \rho[(\mu/\rho)_1 w_1 + (\mu/\rho)_2 w_2 + \cdots] \tag{4}$$

where ρ in the beginning of the right-hand side is the mass density of the material, and $(\mu/\rho)_i$ and w_i are the mass absorption coefficient and mass fraction of the ith atomic constituent in the material. The mass absorption coefficient increases rapidly with increasing atomic number, and therefore for any material containing a heavy atom such as a halogen or a metal, the overall linear absorption coefficient is large and is governed mostly by the mass absorption coefficient of the heavy atom.

The absorption coefficient depends also on the wavelength of the X-rays. In general, the coefficient decreases rapidly as the wavelength is decreased. In other words, as the X-ray photons are made more energetic, they become more penetrating through matter. An exception to this general rule occurs at wavelengths specific to individual elements. The wavelength is called the "absorption edge", but it will not be explained in this short article. With metals or other materials containing high proportions of heavy atoms, it becomes necessary to use X-ray tubes that emit shorter wavelength characteristic radiations, such as either a molybdenum- or a silver-targeted tube.

In choosing the thickness of the sample to be studied, the effect of absorption must be taken into consideration. In transmission geometry, the thicker the sample, the more material is available to scatter X-rays, but the beam is more attenuated as a result of absorption. The maximum scattered intensity is obtained (when the scattering angle 2θ is relatively small) with the sample thickness equal to $1/\mu$. The mass absorption coefficients, for $CuK\alpha$ X-rays, of elements H, C, N, and O are 0.391, 4.51, 7.44, and 11.5 cm^2/g, respectively, and therefore for polymers containing only these light atoms and having the density not far from 1 g/cm^3, the linear absorption coefficient μ is around 4 cm^{-1} or a little larger. The optimum thickness therefore turns out to be around 2.5 mm or a little less. If the material contains heavy atoms, the value of μ increases appreciably, and the sample thickness must be reduced correspondingly.

The preceding several paragraphs give the most essential points that need be understood in doing X-ray measurements. More detailed discussions of the experimental techniques are found in references 4, 5, 7, and 11.

Safety

Before closing this section, a word about safety is in order. In the operation of any X-ray equipment, strict precautions must be taken to avoid exposure to an X-ray beam. It is obvious that the operator of the instrument should not place any part of his or her body in the path of the incident or scattered X-ray beams. This is easier said than done, because X-rays are invisible and easily scattered in both unwanted and unsuspected directions. All X-ray instruments must be shielded with protective sheeting such as lead sheets or lead glass panels as much as possible, and warning lights should be wired to light up automatically whenever the X-ray generator is energized. All personnel working in the room where X-ray equipment is located are required to wear a radiation badge to monitor exposure to stray X-ray radiation. The biologically harmful effect of exposure to X-rays is known to be cumulative, and therefore any person working with repeated X-ray measurements over some length of time should pay special attention to the question of safety.

Basic Principles of X-ray Scattering and Diffraction

In an X-ray diffraction measurement, the sample is irradiated with a beam of X-rays, and the intensity of scattered X-rays is measured as a function of scattering direction. The scattering angle—that is, the change in the direction of the scattered beam from that of the incident beam—is customarily denoted as 2θ (see Figure 14.4). If $\mathbf{S_0}$ is the unit vector

in the incident beam direction and $\mathbf{S}$ is the unit vector in the scattered beam direction, the relationship between the incident and scattered beams can be conveniently expressed by means of the *scattering vector* $\mathbf{s}$, which is defined as

$$\mathbf{s} = \mathbf{S}/\lambda - \mathbf{S_0}/\lambda \tag{5}$$

The magnitude of $\mathbf{s}$ is related to the scattering angle 2θ by

$$s = |\mathbf{s}| = 2(\sin\theta)/\lambda \tag{6}$$

Some workers prefer to use the scattering vector $\mathbf{q}$, which is defined by

$$\mathbf{q} = 2\pi\mathbf{s} \tag{7}$$

in place of $\mathbf{s}$ to represent the scattering geometry. Symbols $\mathbf{h}$, $\mathbf{Q}$, $\mathbf{k}$, and $\mathbf{\kappa}$ are also used in place of $\mathbf{q}$ by some workers.

If an (unpolarized) X-ray beam of intensity I_0 irradiates an electron, the intensity I scattered at angle 2θ and measured at distance r from the electron is given, according to the classical theory, by

$$I = I_0 \frac{K}{r^2}\left(\frac{1 + \cos^2 2\theta}{2}\right) \tag{8}$$

which is called the *Thomson formula*. Here the constant K is given by

$$K = e^4/m^2c^4 \tag{9}$$

where e and m are the charge and the mass of an electron, and c is the velocity of light. The square root of K is often referred to as the *classical electron radius* and has the numerical value 2.82×10^{-15} m. The factor $(1 + \cos^2 2\theta)/2$ arises from a polarization effect and will be somewhat different if a partially polarized incident beam such as that reflected from a monochromator crystal is used.

When the intensity of X-rays scattered from a sample is determined, it is often expressed in arbitrary units, such as counts per second. A more rational measure of the intensity is to compare it against the intensity that would have been scattered by a single electron, positioned in the place of the sample, under otherwise exactly the same conditions. The ratio of these two intensities is called the *absolute intensity* or the *intensity in electron units*.

When the sample contains more than one electron, as is always the case, the X-ray waves scattered coherently from individual electrons interfere with each other. The amplitude $A(\mathbf{s})$ of the X-ray beam that is realized in the direction $\mathbf{s}$ as a result of such interference effect depends on the distribution $\rho(\mathbf{r})$ of the electron density in the sample. Theory shows that the amplitude of the diffracted beam $A(\mathbf{s})$ in electron units is related to $\rho(\mathbf{r})$ by

$$A(s) = \int\rho(\mathbf{r})\exp(-2\pi i\mathbf{r}\mathbf{s})d\mathbf{r} \tag{10}$$

where i is the imaginary number equal to $\sqrt{-1}$, $d\mathbf{r}$ denotes the volume element ($=dxdydz$) in three-dimensional space, and the integration sign actually represents a triple integral

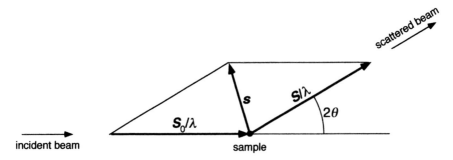

Figure 14.4 Definition of scattering vector **s**; **S**$_0$ and **S** are unit vectors in the direction of the incident and scattered rays.

in x, y, and z directions. The amplitude $A(\mathbf{s})$ in eq 10 is therefore given as a complex number, which is specified by its real and imaginary parts or, alternatively, by its modulus (or absolute magnitude) and the phase angle. What is measured experimentally is the intensity $I(\mathbf{s})$, which is given by the square of $A(\mathbf{s})$:

$$I(\mathbf{s}) = |A(\mathbf{s})|^2 = A(\mathbf{s})A^*(\mathbf{s}) \qquad (11)$$

where $A^*(\mathbf{s})$ is the complex conjugate of $A(\mathbf{s})$.

The amplitude in electron units of scattering from an atom, due to the presence of all the electrons belonging to it, is called the *atomic scattering factor f* of the atom. Its magnitude depends on the scattering direction, or **s**. In the forward direction ($\mathbf{s} = 0$), the waves scattered from all the electrons have the same phases and simply add to each other, and the atomic scattering factor f is therefore equal to the atomic number Z. As the scattering angle 2θ deviates from zero, the phases of the waves emanating from different electrons begin to deviate from each other, and, as a result, the magnitude of f is a steadily decreasing function of the scattering angle 2θ. The values of the atomic scattering factor f can be found tabulated in reference 15. In Figure 14.5

the atomic scattering factors for hydrogen, carbon, and oxygen are plotted as a function of $s = 2(\sin\theta)/\lambda$.

By recognizing that $\rho(\mathbf{r})$ in eq 10 can be factored into groups of electrons, each belonging to an atom, eq 10 can be rewritten as

$$A(\mathbf{s}) = \sum_{n=1}^{N} f_n \exp(-2\pi i \mathbf{s} \mathbf{r}_n) \qquad (12)$$

where f_n and $\mathbf{r}_n$ are the atomic scattering factor and the position of the nth atom in a sample containing the total of N atoms. Equation 12 is the fundamental equation that shows the relationship between the amplitude of the scattered X-rays, on the one hand, and the structure of the material expressed in terms of the positions of all the atoms in it, on the other hand.

Wide-Angle Studies

In the beginning of this chapter, I mentioned that X-ray scattering studies are usually classified into wide-angle and small-angle studies. In the wide-angle studies, in which scattered intensity is measured at scattering angles 2θ larger than a few degrees, the aim is to obtain information about the structure of the material on the size scale of a few angstroms or less. Thus, if the material is crystalline, the study will lead to information about the crystalline structure: that is, the size and shape of the crystalline lattice and the arrangement of atoms in the unit cell. If the material is noncrystalline, the wide-angle study leads to information about the statistical arrangement of atoms in the neighborhood of another atom. The scattering pattern from an amorphous material is devoid of details at casual glance, however, except for a broad peak commonly called the *amorphous peak* or *amorphous halo*. Although much useful information about the arrangement of atoms and molecules in an amorphous material can be derived from the analysis of such broad diffraction curves, the task is involved and is therefore not usually undertaken for the routine characterization purposes with which this chapter is concerned. The rest of this section is therefore devoted to describing the studies of crystalline polymers by wide-angle scattering.

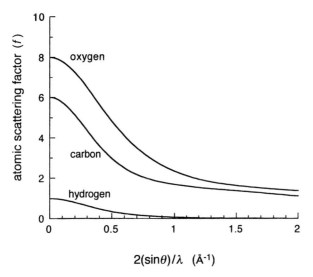

Figure 14.5. Atomic scattering factors f for hydrogen, carbon, and oxygen atoms, plotted against $2(\sin\theta)/\lambda$.

For a perfectly crystalline material, the scattering of X-rays is observable only in the sharp diffraction peaks at angles that satisfy the Bragg relation. This is true irrespective of whether the crystal consists of atoms, ions, small molecules, or large molecules such as polymers and biological macromolecules. With polymers, however, the crystal is almost never perfect, and the various kinds of imperfections present in the crystalline lattice and the additional fact that crystals are very small in size lead to appreciable broadening of the diffraction peaks. Moreover, even a highly crystalline polymer contains at least some noncrystalline fraction that gives rise to the background of a broad, amorphous scattering. This means that the diffraction pattern from a semicrystalline polymer consists of a series of crystalline diffraction peaks of some width superimposed on top of a broad amorphous scattering pattern. Figure 14.6 gives an example of a diffractometer scan of a fairly highly crystalline linear polyethylene. The broadening of the diffraction peaks with a consequent tendency to overlap with neighbors and the superposition of the amorphous background make interpretation of the diffraction pattern from polymers more difficult. Often the result may not be as precise as in the analysis of the diffraction data from other crystalline materials such as minerals or small molecule crystals, but the amount of information that can be derived from X-ray diffraction studies of polymers is still gratifyingly large.

Identification of Polymers

The details of the diffraction pattern from a crystalline material are unique to that material and can be used as a sort of fingerprint for its identification. By "diffraction pattern" we mean the positions of diffraction peaks and their relative heights. The diffraction peak positions are directly related to the shape and size of the unit cell of the crystalline lattice, and the relative heights among these diffraction peaks reflect the positions of various atoms within the unit cell. No two crystalline substances have exactly the same lattice constants or the same atomic content in the unit cell. Therefore, the diffraction pattern is a unique characteristic of the given crystalline material. If the crystalline structure of a material changes as a result of a polymorphic transition, the diffraction pattern changes. In other fields of study dealing with highly crystalline materials such as metals and ceramics, identification of materials by means of their diffraction patterns has been one of the most important applications of X-ray diffraction techniques. An extensive compilation of diffraction characteristics of a large number of inorganic and small molecule organic crystals is available, for example in a CD-ROM format, to serve as the databases with which individual research workers can compare their own observed diffraction pattern for identification purposes. The practice is not as widespread in the study of polymers, and often the diffraction pattern by itself may not be sufficient to identify the polymer unambiguously. However, when augmented by some other pieces of information, such as the overall atomic composition and the transition temperature, identification of unknown polymers can usually be accomplished.

At present there exists no extensive database that lists the diffraction patterns of a large number of crystalline polymers. However, the lattice constants, defining the size and shape of the unit cell, of most of the known crystalline polymers have been compiled in reference books.[16-18] Once the lattice constants—that is, the lengths of the unit cell edges, a, b, and c and the angles α, β, and γ between the unit cell axes—are known, the interplanar spacing d_{hkl} of the (hkl) crystallographic plane can easily be calculated. From the knowledge of d_{hkl} the diffraction angle 2θ is then obtained by means of the Bragg equation. The following

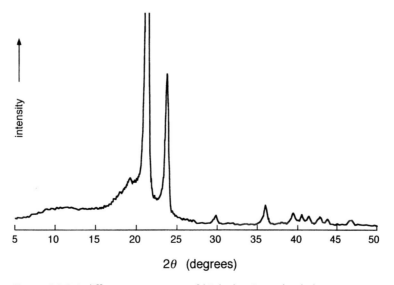

Figure 14.6 A diffractometer scan of high-density polyethylene.

formula for d_{hkl} is the one applicable to a triclinic crystal system:[3]

$$\frac{1}{d_{hkl}^2} = \left[\frac{h^2\sin^2\alpha}{a^2} + \frac{k^2\sin^2\beta}{b^2} + \frac{l^2\sin^2\gamma}{c^2} + \frac{2hk}{ab}(\cos\alpha\,\cos\beta - \cos\gamma)\right.$$

$$\left. + \frac{2kl}{bc}(\cos\beta\,\cos\gamma - \cos\alpha) + \frac{2lh}{ca}(\cos\gamma\,\cos\alpha - \cos\beta)\right] \Big/$$

$$(1 + 2\cos\alpha\,\cos\beta\,\cos\gamma - \cos^2\alpha - \cos^2\beta - \cos^2\gamma) \quad (13)$$

This is the most general expression for calculating d_{hkl}, but it is reduced to a much simpler form for other crystal systems of higher symmetry, when the consequence of the symmetry is fully taken into account. For example, for an orthogonal crystal system for which $\alpha = \beta = \gamma = 90°$, eq 13 becomes

$$\frac{1}{d_{hkl}^2} = \frac{h^2}{a^2} + \frac{k^2}{b^2} + \frac{l^2}{c^2} \quad (14)$$

Similar expressions applicable to other (cubic, tetragonal, hexagonal, and monoclinic) crystal systems can be found in most books on X-ray diffraction.[3,4,5,7,8]

Lattice Constants

As described in the preceding section, given the knowledge of the lattice constants a, b, c, α, β, and γ, one can readily calculate the interplanar spacing d_{hkl} and the diffraction angle $2\theta_{hkl}$ for any Miller planes (hkl). Conversely, when the diffraction angles $2\theta_{hkl}$ for some number of Bragg peaks are available from experiment, it should then be possible to proceed in reverse and determine the precise values of the lattice constants. To accomplish this, the number of Bragg peaks for which the precise 2θ value is measured must be not less than the number of the unknown lattice constants that are to be determined.

The first task in the process of determining the lattice constants is to "index" the Bragg peaks: that is, to identify the Miller indices h, k, and l for each of the observed diffraction lines. This is by no means a simple task, especially if at the outset nothing is known about the lattice structure of the material. Fortunately for polymer scientists these days, the need to undertake the indexing from scratch seldom arises, since for most polymers we encounter, the lattice parameters have already been determined at least approximately and are available either in the original literature or in reference books.[16–18] With the approximate lattice constants thus available, the indexing is accomplished simply by finding the values of h, k, and l that give the best match between the observed 2θ and the one calculated by means of eq 13 or its equivalent such as eq 14. A set of equations, each for one of the observed (hkl) peaks, is then set up with the lattice constants as the unknowns, and the set is solved for the precise values of the unknowns that satisfy all these equations simultaneously.

Obviously, the Bragg angles must be measured accurately. There are X-ray cameras, such as the Guinier-deWolf camera, that are designed specifically for the purpose of obtaining high accuracy in the Bragg angle measurement. When the angles are measured by a diffractometer, certain precautions must be taken to minimize systematic errors.[9,11] If at all possible, any clearly recognizable Bragg peaks in the high-angle region should be included in the measurement. The reason for this is that differentiation of the Bragg eq 1 gives

$$\frac{\Delta d}{d} = -\cot\theta\,\Delta\theta \quad (15)$$

showing that the relative error $\Delta d/d$ in the planar spacing d due to the error $\Delta(2\theta)$ in the diffraction angle decreases as 2θ increases toward 180°. With polymers, however, the paucity of well-defined Bragg peaks at high angles makes this strategy difficult to apply in many cases.

The precision measurement of the lattice parameters gives us clues about the factors that lead to the change in the lattice structure or the nature and magnitude of the forces that induce it. Volume expansion by changes in temperature and densification by application of pressure are reflected in the lattice parameter changes. Polymer crystals are inherently anisotropic, and the lattice parameter change accompanying volume change is the least in the chain direction, which is usually the c axis direction. Changes in the lattice parameters have also been observed when the polymer is deformed—for example, by drawing. If there is a way to estimate the true stress applied to individual crystals in such a deformation process, this will lead to estimation of the true modulus of the crystalline polymer.

Incorporation of defects into the crystalline lattice reveals itself by variations in the lattice parameters. The defects may occur as a result of many different causes. Differences in the lattice parameters of the same polymer crystal obtained under different crystallization conditions may be thought of as arising from the different extent of defects embedded in the resulting crystals. Crystals obtained by rapid cooling show clear difference in the lattice constants from crystals obtained by slow cooling or by annealing. Defects can also be introduced into the crystals by adding occasional branches in the polymer molecule or by copolymerizing the polymer with a small amount of noncrystallizable comonomer.

A good review of the literature concerning the variations in the lattice parameters under various conditions is given in reference 9.

Degree of Crystallinity

X-ray diffraction provides the most direct means of revealing the presence of a crystalline order in the material. The diffraction pattern from a crystalline material is fundamentally different from the pattern obtained with a noncrystalline material. When the sample contains both crystalline and noncrystalline fractions, the determination of the degree of crystallinity requires first separating these two types of diffraction patterns and then evaluating the relative contribution of each to the overall. Thus in principle the determination of the degree of crystallinity by means of X-ray diffraction is a very straightforward task, although in practice there are some additional complications that have to be resolved.

The first task in the determination of the degree of crystallinity is to separate the component of diffraction due to the crystalline fraction from the amorphous background. Although the diffraction pattern from a large, perfect crystal is expected to consist of a series of sharp Bragg peaks, with polymers the diffraction peaks are invariably broadened; as a result, many of the peaks overlap with neighboring ones. The overlap is more pronounced at high diffraction angles where peaks due to different crystallographic planes tend to crowd together. Figure 14.6 shows the diffraction pattern obtained with a high-density polyethylene, and despite its relatively high degree of crystallinity, an appreciable degree of overlap among some of the peaks is evident.

The separation of the crystalline from amorphous scattering can be aided greatly when the shape of the purely amorphous intensity curve is known independently. The amorphous curve can be obtained, for example, if the polymer is prepared in a totally amorphous state by rapid quenching or if the scattering curve of the polymer is measured at several temperatures above its melting point and the result is extrapolated to room temperature. The amorphous intensity curve is then scaled and fitted to the observed diffraction pattern under study, so that these two will coincide at regions of 2θ angles away from the influence of Bragg peaks. Figure 14.7 illustrates such a fit, where curve C is the diffraction pattern obtained with a semicrystalline isotactic polystyrene sample, curve A is the intensity curve obtained with the same polymer prepared in a totally amorphous state by quenching, and curve B is the latter scaled to coincide with curve C at several low points between broadened Bragg peaks.

Once the separation into the crystalline and amorphous scattering has been achieved, the degree of crystallinity can be evaluated by comparing the contribution to the total scattering intensity from the crystalline and amorphous components, respectively. To carry out such an evaluation rigorously, some additional requirements must be met, including the following:

1. The intensity must be measured in as wide a range of diffraction angles 2θ as possible.
2. What is to be compared is not simply the sum of intensity $I(s)$ but, rather, the sum of $s^2I(s)$. In other words, the degree of crystallinity, w_c, is represented by

$$w_c = \frac{\int_0^\infty s^2 I_c(s)ds}{\int_0^\infty s^2 I_{total}(s)ds} \qquad (16)$$

where $I_c(s)$ and $I_{total}(s)$ are the crystalline and total scattered intensity as a function of s.

3. Allowance must be made for the fact that the effect of thermal vibrations of atoms around their mean positions is to let some fraction of the crystalline scattering intensity $I_c(s)$ to be diverted from the Bragg peaks to the diffuse background. A rigorous procedure to take account of all these fine points has been suggested by Ruland.[19]

For many practical purposes, one may be interested in a measure of the degree of crystallinity that is less rigorous but is simpler to determine than is implied in the preceding paragraph. When a series of samples of the same polymer that differ only in the degree of crystallinity are being examined, one may simply compare the height of the broad amorphous peak or the area under one or more of the prominent Bragg peaks. In such a comparison, the intensity data must of course be obtained under closely replicating experimental conditions with samples of closely matching thickness. The measure of crystallinity thus obtained may be called a *crystallinity index* to emphasize the point that the numbers are only relative indicators and not absolute measures of the degree of crystallinity.

Preferred Orientation

In an isotropic polymer, that has not been subject to any deformation either in the melt state or after crystallization, crystallites are oriented in all directions with equal probabilities. The X-ray diffraction picture from such a sample, taken with a flat film, will consist of a series of concentric circles of different diameters, each corresponding to a Bragg peak. Such a picture is called a *powder diagram*, alluding to the pattern that is obtained with a powder of mineral or small molecule crystals. However, if the polymer is highly deformed uniaxially by stretching, the diffraction pattern obtained will resemble that in Figure 14.8, in which diffraction spots are arranged on *layer lines*. Such a pattern is called a *fiber diagram*. In the uniaxially oriented polymer, individual crystallites tend to orient themselves often in such a way that the polymer chain direction aligns parallel with the *fiber axis* (the stretch direction), but they maintain random orientation around the fiber axis.

If the extent of deformation of the polymer is less severe, the individual circular ring in the powder diagram is broken up, not to diffraction spots, but to partial arcs, the length of which indicates the extent of the crystallite orientation that is imparted. Measurement of the intensity of scattering along the arc then provides the orientation distribution of the crystallographic planes relative to the fiber axis. Of course, instead of using a stationary sample and a stationary film, one can obtain a more precise measurement by using a diffractometer in which the orientation of the sample mounted is systematically varied. Obviously,

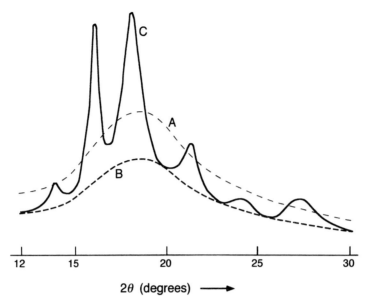

2θ (degrees) ⟶

Figure 14.7 Separation of a diffraction pattern from semicrystalline polymer into its crystalline and amorphous contributions: curve C is the pattern obtained from an isotactic polystyrene; curve A is from a totally amorphous isotactic polystyrene; curve B is the latter scaled to fit the former at intensity minima between crystalline peaks.

such measurement of the orientation distribution of crystallographic planes can also be made with samples deformed biaxially as in molded sheets.

Figure 14.9 gives an example of the orientation distribution of crystallographic planes, in this case (110) planes, measured with a uniaxially stretched polyethylene. For such a measurement, the detector in the diffractometer is

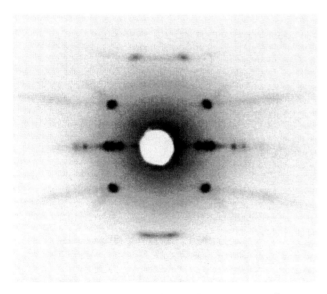

Figure 14.8 A fiber diagram obtained with a uniaxially oriented isotactic polypropylene. (Courtesy of E. S. Clark.)

kept stationery at a Bragg angle 2θ appropriate to the (*hkl*) crystallographic plane concerned, and the diffracted beam intensity is measured as the polymer sample is successively placed in different orientations with respect to the diffractometer geometry. The angle Θ in Figure 14.9 refers to the angle between the fiber axis of the sample and the active direction in the diffractometer—that is, the direction bisecting the angle subtended, at the sample position, by the X-ray source and the detector (see Figure 14.3). The diffracted beam intensity is proportional to the number of (*hkl*) crystallographic planes in the sample having their normal directions aligned parallel to the active direction. In an isotropic sample crystallographic planes are oriented in all directions with equal probabilities, and therefore a plot such as in Figure 14.9 would exhibit a constant value independent of Θ.

In a uniaxially deformed polymer, the orientation distribution of crystallographic planes is cylindrically symmetric around the fiber axis, and the measurement needs to be made only as a function of a single angular variable Θ. In a sample deformed biaxially, in contrast, the orientation distribution of crystallographic planes no longer possesses such a fiber symmetry, and specification of the distribution then requires determining the intensity $I(\Theta, \Phi)$ as a function of two angles, Θ and Φ. The result of such a measurement is usually presented as a stereographic projection, which is called a *pole figure* by those working with metals and minerals. Collection of data to construct such a pole figure is tedious and involves determining the diffraction intensities

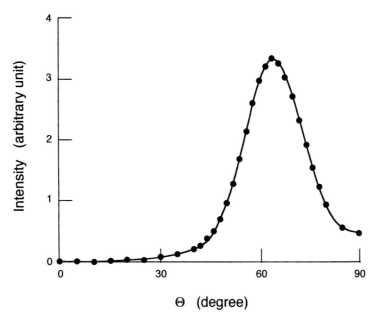

Figure 14.9 Orientation distribution of (110) plane normals in a polyethylene sample stretched uniaxially. (Adapted from reference 20.)

with the sample oriented as a function of two angular variables, Θ and Φ. There are attachments to diffractometer sample holders, called pole figure devices, which have been devised to automate much of such measurements.

Instead of presenting the complete distribution curve, one might occasionally be interested in specifying a single number characterizing the "average" of such a distribution. Such an average is often given by $\langle\cos^2\Theta\rangle$ calculated from the observed intensity $I(\Theta)$ by

$$\langle\cos^2\Theta\rangle = \frac{\int_0^{\pi/2}\cos^2\Theta\, I(\Theta)\,\sin\Theta\, d\Theta}{\int_0^{\pi/2} I(\Theta)\,\sin\Theta\, d\Theta} \tag{17}$$

or more commonly by the *Hermans orientation factor f*, which is defined as

$$f = \frac{1}{2}(3\langle\cos^2\Theta\rangle - 1) \tag{18}$$

Two or more such orientation parameters pertaining to different crystallographic planes, the number depending on the symmetry of the crystal structure, are required to specify uniquely the state of preferred orientation of crystallites in an anisotropic polymer sample.

Certain other methods yield information about the orientation distribution of polymer segments in anisotropic polymer samples, including those that rely on infrared dichroism, fluorescence polarization, and nuclear magnetic resonance (NMR). Nevertheless, the X-ray method is unique in that it measures the orientation distribution of only the crystalline components. When combined with these other methods that measure the overall segment orientation distribution, the X-ray method therefore opens the possibility of allowing determination of the segment orientations in the crystalline and noncrystalline regions separately.

More detailed discussion of the measurement of preferred orientation can be found in references 7, 8, 9, and 11.

Crystal Imperfections and Small Crystal Size

The diffraction pattern from a large, perfect crystal is expected to consist of a series of sharp Bragg peaks. The diffraction pattern actually observed with polymers, even when the sample is highly crystalline, consists of peaks much broader than this ideal suggests, however. The line broadening arises as a result of three separate factors: instrumental broadening, crystalline lattice imperfections, and small crystal size. For example, instrumental broadening of diffraction lines may result from the failure to achieve a perfect collimation of the primary beam, from the contamination of the monochromatized incident beam by broad spectrum radiations, or from the detector slit or window that is allowed too wide an opening. The need to make corrections to instrumental broadening is not unique to X-ray measurement, of course, and methods that are applicable to correcting the instrumental broadening of, for example, spectral lines can be equally adapted to the correction of X-ray diffraction line shapes.

One type of lattice imperfection present in all materials is the one that is due to thermal vibration of atoms. As a result of the vibration, atoms are displaced from their ideal lattice positions. The effect of such thermal vibrations is to reduce the intensity of Bragg peaks by the Debye-Waller factor D:

$$D = \exp\left(-\frac{4}{3}\pi^2\langle u^2\rangle s^2\right) \tag{19}$$

where $\langle u^2 \rangle$ is the mean square displacement of the atoms from their ideal lattice positions. In view of the factor s^2 in the exponent, it is seen that the higher-order Bragg peaks are progressively more attenuated. However, the Debye-Waller factor does not in fact predict a broadening of the Bragg peaks but, instead, predicts a progressive reduction in the height (or the integrated intensity) of the peaks.

Among the many different types of lattice imperfections that can occur in crystals, the one called the *imperfection of the second kind*, or *paracrystalline disorder*, is encountered frequently and has been studied extensively. The term *paracrystal* was coined by Hosemann[21] and refers to the type of crystals in which the neighboring unit cells are not strictly identical but vary randomly according to some statistics. The effect of this type of imperfection, unlike the one due to thermal vibrations, is to make the diffraction lines broader and, in particular, to make the line width (the full width Δs at half height) increase in proportion to s^2.

The diffraction line is also broadened when the crystals are very small. The line width Δs is inversely proportional to the crystal size, and the broadening becomes noticeable when the crystal is on the order of 1000 Å or smaller. The small crystal size effect is to broaden all Bragg peaks to the same extent (in terms of Δs), irrespective of the scattering angle. The two effects, due to imperfections and small crystal size, can therefore be separated by plotting the width of the diffraction lines against s^2. The extent of the paracrystalline disorder can be evaluated from the slope of such a plot, whereas the crystal size can be obtained from the line width extrapolated to s^2 equal to zero.

Crystal Structure Analysis

The term *crystal structure analysis* denotes the process of determining the positions of atoms contained in a crystal by a long sequence of calculations, starting from the observed intensities of the Bragg diffraction peaks. This is the technique developed and refined to a high degree of sophistication, leading to the large body of knowledge we now have on the crystal structures of a large number of materials including biological macromolecules. It most eloquently demonstrates the power of X-ray diffraction as a means of uncovering the structure of materials.

The basic data that go into the crystal structure analysis are the intensities $I(hkl)$ of all observable Bragg peaks, as many of them measured as possible if the structure is to be determined to good resolution. The inverse Fourier transform of the basic equation (eq 10) can be written, when dealing with a crystalline material, as

$$\rho(x,y,z) = \sum_{h=-\infty}^{\infty} \sum_{k=-\infty}^{\infty} \sum_{l=-\infty}^{\infty} A(hkl) \exp[2\pi i(hx + ky + lz)] \quad (20)$$

where $\rho(x,y,z)$ is the electron density at coordinates x, y, z within the unit cell, and $A(hkl)$ is the amplitude of the (hkl)

Bragg reflection. As stated earlier, the amplitude $A(hkl)$ is a complex number, and its magnitude is given by the square root of the observed intensity $I(hkl)$ in view of eq 11. The phase angles, however, are not available directly from measurement of individual Bragg reflections and must be deduced by some other independent, often ingenious, method. Once the phase angles are deduced, the summation indicated by eq 20 leads to the electron density map $\rho(x,y,z)$ from which high-density peaks can be identified as the atomic center positions.

The crystal structure determination requires preparation of a good, single crystal that gives a large number of good, clearly distinguished Bragg reflections. The major task in the crystal structure analysis, however, lies in solving the phase problem—that is, finding ways to figure out the phase angles of the observed Bragg reflections. Over the years, several ingenious methods have been devised, some, for example, relying on preparation of a crystal similar to the one under study but containing heavy metal atoms, and others relying on statistical analysis of the distribution of intensities among a range of Bragg reflections. Solving the structure of a large, complicated crystal represents an intellectual accomplishment of the first class, and the results obtained with large, complicated biological molecules have been the most spectacular and in the past have been the basis for several Nobel prizes.

Of course, the same technique can be applied to the study of polymer crystals but often with some difficulties because of the characteristics specific to polymers. The main difficulty arises from the fact that polymer crystals are seldom prepared to the high degree of crystalline perfection that is possible with crystals of other types of materials. As a result, the number of Bragg peaks that can be observed with precision is severely limited. This leads to a reduced resolution of the deduced atomic positions. Yet, an amazing amount of detail emerges frequently, even from the analysis based on such limited numbers of observable Bragg reflections.

The task of crystal structure analysis requires a fairly dedicated effort and is beyond what can be considered a routine characterization technique. The need for crystal structure analysis of polymers, fortunately, seldom arises these days, since the structure of most of the crystalline polymers that have been synthesized and known to the polymer science community has already been determined and can be found in the literature. In this introductory article we therefore refrain from discussing this topic any further. Readers interested in the technique as applied to polymers should consult references 6, 7, 8, 11, and 22.

Small-Angle Scattering

X-ray diffraction occurs basically as a result of the inhomogeneities in electron density in the sample. This basic physical fact is most concisely expressed in eq 10, which shows that the amplitude of scattered X-rays, $A(\mathbf{s})$, is the Fourier transform of the electron density distribution $\rho(\mathbf{r})$ as a function of position vector $\mathbf{r}$. Inhomogeneities in electron density existing in materials arise from several different physical causes. On the smallest scale, the inhomogeneity is due to the fact that electrons are grouped into atoms, and atoms are not uniformly distributed in space. This type of electron density fluctuation has a typical size scale of the order of 1 Å, and the corresponding diffraction phenomenon then occurs in the wide-angle region. Many materials, however, contain larger-scale electron density fluctuations in addition to such atomic scale fluctuations. An example could be the electron density difference between the two polymers in an immiscible blend, between polymer molecules and the surrounding solvent in a solution, or between neighboring microdomains in a microphase-separated block copolymer. The electron density fluctuations of this kind are usually in the size range of 10 to 1000 Å, and their presence manifests itself as the scattering pattern observable in the small-angle region. Measurement of such a small-angle scattering requires the use of small-angle equipment especially designed for this purpose and available in many modern laboratories.

There is no fundamental difference between small-angle and wide-angle scattering as far as the principle of X-ray scattering and diffraction is concerned. Some of the experimental requirements are, however, different between the two, as will be briefly mentioned in the following discussion. The methods of data interpretation and analysis can also differ somewhat between wide-angle and small-angle studies because of differences in emphasis or differences in the approximations and assumptions employed. For example, in small-angle studies, $\sin\theta$ can always be approximated by θ, and the polarization factor $(1 + \cos^2 2\theta)/2$ in eq 8 can be equated to 1. This section gives a brief outline of the types of information that can be obtained by small-angle studies and the types of analysis that are necessary to extract such information from the observed data. More detailed information on the methods of analysis of small-angle data can be found in references 7, 9, 11, 12, and 13.

Experimental Requirement for Small-Angle Scattering Measurement

To be able to measure the intensity of scattering accurately at a small angle 2θ, the most important requirement is that the incident beam itself is well collimated and does not contain rays that diverge by more than the same small angle 2θ. Otherwise, the divergent rays of the primary beam will contaminate the scattered beam. Since the scattered beam is always many orders of magnitude weaker than the primary beam, even a small contamination will be overwhelming. The primary beam divergence

is reduced by making the diameter of the pinhole collimator smaller, but very fine collimation inevitably blocks all but a small fraction of the energy in the incident beam from being transmitted toward the sample and makes the scattered intensity very weak. This difficulty can be mitigated by the use of a more intense X-ray source, such as the rotating anode X-ray tube or the synchrotron radiation source.

To get around the dilemma of finer collimation versus weaker beam intensity, slit rather than pinhole collimation is often used. A slit essentially corresponds to a large number of pinholes placed closely along a line, and the scattering from each pinhole is superimposed together to produce an overall scattering pattern of much higher intensity that, however, is somewhat smeared. The slit-smeared scattering pattern must then be de-smeared by using one of the number of mathematical procedures proposed for the purpose. In recent years, the increasing availability of higher-intensity X-ray sources has made the need for slit collimation less compelling.

Some of the quantities evaluated in small-angle studies require that the scattered intensity be available in absolute (electron) units. This is accomplished if the instrument in use is calibrated to allow conversion of the observed intensities to absolute intensities. Such a conversion factor can be determined if a calibration sample of known scattering power is available. The reference sample can be a "primary standard" for which the scattering power is known on theoretical grounds (some highly scattering gases or liquids of well-defined composition and density). Alternatively, a "secondary standard" (often a piece of polymer), which itself has been calibrated by means of a primary standard, can be more convenient to use.

Traditionally, small-angle studies are made with the sample mounted in transmission geometry. But it is possible to mount the sample in reflection geometry, in which case the incident beam merely glances the surface region before being reflected off. Such a measurement is therefore called a *reflectivity* measurement, and recently it has been found to be a valuable tool in studying the structure of the very surface layer of the sample. More information about the reflectivity measurement technique can be found in references 11 and 14.

Equipment for small-angle studies is perhaps less widely available than are wide-angle instruments. Nevertheless, the method is gaining acceptance in scientific communities for studying polymers, as well as biological materials and metals and ceramics. A number of commercial instruments of newer designs are also available in the market. Some of the small-angle instruments set up in government laboratories, such as Oak Ridge National Laboratory and National Institute for Standard and Technology, and in some synchrotron facilities are often available to outside users on a collaboration basis.

Types of Structures Amenable to Small-Angle Studies

As stated, a sample that contains electron density inhomogeneities on the size scale of 10 to about 1000 Å can potentially be studied by using the small-angle technique. The lower limit, 10 Å, is not a limit of any significance but is simply a point at which the distinction between the small-angle and wide-angle techniques becomes irrelevant. The upper limit, 1000 Å, is set by the finest collimation that can be attained in practice without an unduly elaborate endeavor. A sample that possesses any kind of electron density inhomogeneity in this size range can be studied by small-angle scattering, but several specific types of electron density inhomogeneities are particularly amenable to theoretical analysis and interpretation. Four such structures are the (1) dilute particulate system, (2) irregular two-phase system, (3) soluble blend system, and (4) periodic system. The characteristics of these different types of structures and the method of analysis specific to them are described individually in the following sections. The first step in the small-angle study of a sample is therefore taken by choosing one of these models on the basis of other independent information available about the nature of the sample.

Dilute Particulate System

In the *dilute particulate system*, particles (polymer molecules, colloidal particles, fine filler particles, etc.) of one material are dispersed into the continuous matrix of a second material. The size of the particles is in the range of 10 to about 1000 Å, as mentioned, and the concentration of the particles is sufficiently dilute so that it can be assumed that the positions of individual particles are random and uncorrelated with each other. When these assumptions are satisfied, the scattered intensity pattern can be analyzed on the basis of the *Guinier law* and yields information about the average size of the particles. The Guinier law states that the intensity $I(s)$ of scattering at small angles can be approximated by

$$I(s) = N(\Delta\rho)^2 v^2 \exp\left(-\frac{4}{3}\pi^2 s^2 R_g^2\right) \qquad (21)$$

Here N is the number of particles in the scattering volume (which is the volume of the sample contributing to the observed scattered intensity), $\Delta\rho$ is the difference in electron density between the particle and the continuous matrix, v is the volume of the particle, and R_g is the radius of gyration of the particle, which is a parameter characterizing the size of the particle as explained shortly below. Equation 21 suggests that if we plot the logarithm of the intensity $I(s)$ against s^2, we should obtain a linear plot with the slope equal to $-(4/3)\pi^2 R_g^2$, from which the radius of gyration R_g is evaluated.

The radius of gyration is a parameter that is most appropriate to characterizing the size of a particle when the shape of the particle is irregular. It is the root-mean-square distance of all mass points from the center of mass of the particle, the term *mass* meaning the electron mass in the context of X-ray scattering. The radius of gyration can be defined by several, slightly different expressions, depending on whether the electron density within the particle is considered to be spread out continuously or concentrated in discrete subparticles (such as the segments in a polymer molecule).

If a polymer molecule is considered to consist of Z segments, with the electron mass of a segment concentrated at the segment center located at $\mathbf{r}_i$ ($i = 1 \ldots Z$), the radius of gyration is given by

$$R_g^2 = \frac{1}{Z}\sum_{i=1}^{Z}(\mathbf{r}_i - \mathbf{r}_{com})^2 \qquad (22)$$

where $\mathbf{r}_{com}$ is the position of the center of electron mass of the molecule. For a Gaussian chain, the radius of gyration is known to be equal to $1/\sqrt{6}$ times the end-to-end distance on the average.

If the particle is a solid body of uniform, constant electron density within it, the radius of gyration can be calculated by

$$R_g^2 = \frac{1}{v}\iiint_v [(x - x_{com})^2 + (y - y_{com})^2 + (z - z_{com})^2]dx\,dy\,dz \qquad (23)$$

where x_{com}, y_{com}, and z_{com} are the Cartesian coordinates of the center of electron mass of the particle, v is the volume of the particle, and the triple integration is throughout within the boundary of the particle. Applying eq 23 to a solid sphere of radius R, it can be easily seen that

$$R_g = \sqrt{3/5}R \qquad (24)$$

whereas for a thin rod (or needle) of length L and negligible cross-section, we obtain

$$R_g = \frac{1}{\sqrt{3}}\frac{L}{2} \qquad (25)$$

Figure 14.10a shows the intensity of scattering $I(s)$ calculated for a Gaussian chain, a solid sphere, and a thin rod, plotted against R_g times s. In Figure 14.10b the same results are plotted in terms of the logarithm of $I(s)$ against $(sR_g)^2$. It is clearly seen that for small s all the plots are straight lines of the same slope, in agreement with the Guinier law.

In eq 21, intensity $I(s)$ is given in electron units. Of the proportionality constant in front of the exponential factor, the electron density contrast $\Delta\rho$ between the suspended particles and the matrix can be calculated if the chemical compositions of these materials are known. For a fixed concentration (by mass) of the suspended particles, the product of N and v remains constant, irrespective of the size of the particles. It then follows that from the measurement of $I(s)$

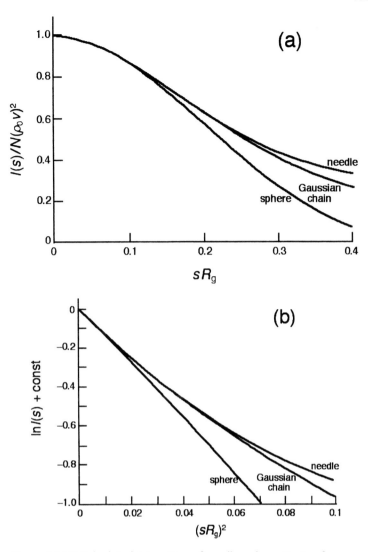

Figure 14.10 Calculated intensities of small-angle scattering from independent particles having the shape indicated. The data in plot (a) are replotted in plot (b) with the axes given in ln $I(s)$ versus $(sR_g)^2$, as suggested by the Guinier law (eq 21).

one can also determine the particle volume v, which of course is also a measure of the size of the particles. Although this is true, the particle volume v is more difficult to determine than the radius of gyration R_g. To determine the latter, it is sufficient to have the intensity measured in any arbitrary units, whereas the determination of the former requires that the intensity be determined in absolute units, a procedure that requires a little more care in practice.

For the Guinier law (eq 21) to be strictly valid, certain conditions must be met:

1. As stated earlier, the system must be dilute with regard to the concentration of the particles.
2. The law is valid only for $s \ll 1/R_g$.

3. The system must be isotropic. This means that, if the particles are not spherically symmetric, they must be tumbling to take many different orientations during the time of X-ray measurement, or, alternatively, if they are stationary, different particles are orientated in different, random directions so that the sample as a whole is isotropic.

For the Guinier law (eq 21) to be applicable, it is not necessary that all the particles are identical. For particles of different sizes and shapes, the radius of gyration that is evaluated from the slope of the plot of ln $I(s)$ against s^2 is the z-average of the radius of gyration. By contrast, the particle volume that is obtained from the measurement of the absolute intensity $I(s)$ is a weight-average of v. The astute reader may have noticed that the Guinier law in fact expresses the same physical basis as that which underlies the determination of molecular weight and radius of gyration of polymer molecules in solution by light scattering.

Irregular Two-Phase System

When the concentration of the dispersed phase cannot be considered dilute, the Guinier law is no longer applicable. Two different possibilities then arise. If the two substances are molecularly dispersed into each other—as, for example, in a solution of a polymer in solvent or in a blend of two miscible polymers—continuous variations in the electron density inhomogeneity pervades throughout the material as a result of local concentration fluctuations. Alternatively, the system may be considered an irregular mixture of two substances or two phases, with the electron density within each phase remaining fairly constant. Such a system may be found, for example, in a phase-separated polymer blend, a dispersion of filler particles in a polymer matrix to a moderate to high concentration, or a semicrystalline polymer in which the crystalline and amorphous phases are irregularly intermixed. The second type of material belongs to the irregular two-phase system considered in this section, and the first type of material is discussed shortly under the heading of miscible blend system.

In discussing the scattering curve from the irregular two-phase system, it is convenient first to examine the scattering from an idealized system called the *ideal two-phase model*. The model assumes that the material consists of two phases, each of which has a strictly constant electron density within its boundaries, and that the phase boundary is

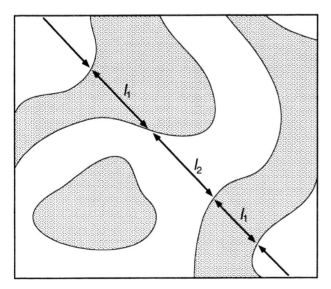

Figure 14.11 Model of an irregular two-phase system: a straight line passing through the material is cut into cords of lengths l_1 and l_2, respectively, in the two phases.

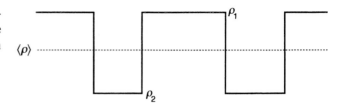

Figure 14.12 Example of an electron density profile in the ideal two-phase model. In each phase, the electron density remains constant at either ρ_1 and ρ_2, and the transition between the two phases is abrupt.

sharp, with no transition zones. Figure 14.11 depicts such a system; here, the two phases are irregularly dispersed in each other. The electron density profile along any straight-line passing through the material may then look like the one shown in Figure 14.12. In real two-phase material, the transition between the phases may be less abrupt, and within each phase there will always be some electron density fluctuations, as in any pure amorphous or crystalline materials. How these deviations in real materials from the ideal two-phase model affect the scattering patterns will be discussed shortly below.

The total scattering power of any sample can be represented by a quantity called the *invariant*, Q, which can be determined by summing the intensity (in electron units) observed in all scattering directions, or more exactly by

$$Q = 4\pi \int_0^\infty s^2 I(s)\,ds \qquad (26)$$

The invariant Q represents the mean square deviation, $\langle(\rho - \langle\rho\rangle)^2\rangle$, of the electron density ρ from the mean $\langle\rho\rangle$ throughout the sample. For an ideal two-phase system the mean square electron density fluctuation turns out to be

$$\langle(\rho - \langle\rho\rangle)^2\rangle = (\Delta\rho)^2 \phi_1 \phi_2 \qquad (27)$$

so that if the invariant Q is measurement experimentally according to eq 26, it can be related to the electron density difference $\Delta\rho = \rho_1 - \rho_2$ between the two phases and the volume fractions ϕ_1 and ϕ_2 of the phases. Thus, for example, for a semicrystalline polymer, for which the electron densities of the crystalline and amorphous phases are known, the measurement of the invariant Q leads to the determination of the degree of crystallinity. Or, in another example, the possibility of a partial miscibility in an incompatible polymer blend can be confirmed by comparing the observed invariant with the value of the quantity in eq 27 calculated on the basis of the electron density difference and the relative amounts of the two pure polymers that were used to prepare the blend.

For a sample obeying the ideal two-phase model, the scattering curve at the tail end (that is, at relatively large values) of s in the small-angle region is known to obey the *Porod law*

$$I(s) = K/s^4 \qquad (28)$$

where K is a constant. One of the factors that are included in the constant K is the total area S of the interfacial boundary between the two phases. The interfacial area S is a useful parameter to characterize the material, since a large S means the two phases are more finely divided and intimately intermixed into each other. Another way of expressing the same fact is that the average of the cord lengths l_1 and l_2 (see Figure 14.11), into which any straight line passing through the material is cut, is related to S by

$$\langle l_1 \rangle = 4\frac{V}{S}\phi_1 \qquad \text{and} \qquad \langle l_2 \rangle = 4\frac{V}{S}\phi_2 \qquad (29)$$

We now have to examine how the scattering curve obtained with a real material can be interpreted on the basis of the theoretical result derived from the ideal two-phase model discussed above. In any amorphous material there always exist some electron density inhomogeneities, which arise from the grouping of electrons into atoms and also from the fluctuation of atomic densities due to their thermal motions. As a result of these inhomogeneities, even a pure amorphous material gives rise to a broad scattering pattern that is spread out over all scattering angles and has a broad maximum known as the *amorphous halo*. In a real two-phase material, each of the phases inevitably contains such density inhomogeneities within itself, and its consequence is that a similar broad "background" scattering is superimposed on the scattering curve due to the ideal two-phase structure. Therefore, before the observed scattering curve can be interpreted in terms of the ideal two-phase model, the broad background due to the internal density fluctuations must be removed. This is usually accomplished by measuring the background scattering in the range of s beyond the small-angle region, and then extrapolating it back to the small-angle region and subtracting the extrapolated background from the observed curve. Although some degree of uncertainty is unavoidable in such a background correction, the procedure seems to work well in most practical situations.

The diffuseness of the interfacial boundaries between the two phases is to modify the scattering curve primarily in the s region where the Porod law applies. The theoretical treatment for two-phase structures with diffuse interfaces is fairly straightforward and leads to a modified Porod law in which the diffuseness of the interfaces enters as a parameter. Examining the deviation of the observed scattering curve from the prediction of the unmodified Porod law therefore provides a means of determining the thickness of the interface boundaries. Carrying out such an analysis and arriving at reliable value of the interface thickness, however, is not an easy task in practice and requires a scattering intensity data that are obtained with a great deal of care and precision. In the interest of keeping this chapter manageably short, we will not discuss this aspect further, and interested readers are advised to seek more detailed accounts in references 9 and 11.

Soluble Blend System

A polymer solution or a truly compatible polymer blend is a single-phase, homogeneous material in the thermodynamic sense. Yet its concentration on a local scale is not truly constant but varies continuously from place to place. The electron density fluctuation resulting from such a concentration fluctuation thus gives rise to a diffuse scattering of X-rays, and this scattering is much stronger than the scattering from a pure liquid or amorphous polymer that is due solely to density fluctuations caused by thermal motions of atoms. The extent of the concentration fluctuation depends on the degree of compatibility between the solvent and the solute or between the two polymer components. Consequently, the intensity of scattering from such a compatible binary system is intimately related to the thermodynamics of mutual solubility or compatibility of the two components.

Because the thermodynamic properties of a binary system is completely described by the free energy of mixing the two components, the intensity of scattering can also be expressed in terms of the free energy of mixing. Indeed, the connection between the scattering phenomenon and the thermodynamics can be summarized by

$$\frac{I(0)}{V(\Delta\rho)^2} = kT / \frac{\partial^2 g_m}{\partial \phi_1^2} \qquad (30)$$

where $\Delta\rho$ is the difference in the electron density between the two pure components, V is the volume of the sample contributing to the scattering, g_m is the free energy of mixing per unit volume, and ϕ_1 is the volume fraction of one of the components. $I(0)$ in eq 30 is the intensity that is obtained by extrapolating the observed intensities to zero scattering angle. The reason the extrapolated zero angle intensity shows up in eq 30 is that thermodynamics relates properties of materials of macroscopic size, and the corresponding information is contained in the scattering curve in the limit of $s \rightarrow 0$.

If we substitute the Flory-Huggins free energy expression for g_m, eq 30 is recast into an equation relating $I(0)$ to the Flory interaction parameter χ. Thus, by measuring the variation in $I(0)$ with temperature, for example, one can determine the temperature dependence of χ. In the thermodynamics of binary systems, it is known that the spinodal point, at which the mixture as a single phase becomes no longer stable, is reached when the second derivative of the free energy of mixing becomes zero. From eq 30, it is therefore seen that the extrapolated zero-angle intensity $I(0)$ diverges. This is in agreement with what one would expect on a physical ground because as the mixture approaches the spinodal point and is rendered less stable, the extent of concentration fluctuation becomes pronounced not only on a local scale but also on a macroscopic, thermodynamic scale.

Periodic System

Many polymeric material systems contain ordered, periodic structures with a periodicity in the range of 10 to

1000 Å. Ordered block copolymers of various morphologies, in which spherical, cylindrical, or lamellar microdomains are arranged on a macrolattice, are examples. Semicrystalline polymers often consist of parallel stacks of folded-chain lamellar crystals, with the lamellar thickness typically on the order of 100 Å. In nature, many biological and mineral systems have a structure that repeats itself into a periodic organization. The scattering of X-rays from such periodic systems can be treated in essentially the same way as in the study of the scattering from a crystal. The difference in the size scale between the crystalline lattice and the macrolattice considered here of course leads the Bragg peaks to show up in different regions of the scattering vector s.

All the methods of analysis discussed in the section on wide-angle scattering are also applicable in principle to systems of the size scale considered here. The degree of order found in these macrolattices is almost always much inferior to those found in crystalline lattices, however. As a result, the diffraction peaks from macrolattices are afflicted much more by the effects of lattice imperfections. In fact, it is rare to see a periodic polymer system that clearly exhibits more than a few orders of diffraction peaks in the small-angle region. The amount of information that can be extracted about the structure of the macrolattice is correspondingly more limited in comparison to what is obtained from analysis of a crystalline diffraction pattern. Despite such limitations, traditionally, the X-ray diffraction technique has served as an indispensable tool in the study of such periodic polymer systems.

Concluding Remarks

In this chapter, the types of structural information that can be derived from X-ray diffraction study of polymers have been briefly described, and the kinds of analysis and interpretation necessary to extract such information have been discussed. In presenting these results, I have made no attempt to explain or justify the cited statements and equations in terms of underlying physical principles. Readers who are interested in these principles or in the derivation of many of the equations quoted should consult my book, published recently.[11]

You have undoubtedly noticed that in X-ray studies the structural information sought is not available immediately in the observed data but, instead, can be obtained only after the raw data have been subjected to several steps of analysis and interpretation. This need for analysis is in great contrast to some other techniques such as electron microscopy where the images obtained often give the desired structural information directly. Fundamentally, this is because the data obtained by diffraction studies contain information in reciprocal space (the space spanned by the scattering vector

s), whereas we usually observe and understand the structure as a phenomenon in direct space (the space spanned by the position vector r).

Another important difference between the study of structure by microscopy and by a diffraction method is that the diffraction data include information about the average of all the structures contained in a scattering volume, whereas from a microscopic image the usual tendency is for the observer to select just a few features that are more noticeable and to draw inferences on the assumption that the observed features are indeed representative of all. Some element of selection is involved in electron microscopy also because of the fact that the observation must be from either a very thin sample or from the surface region of the sample. In fact, microscopy and diffraction studies are complementary to each other, and by combining the two techniques, the chance of misinterpretation is substantially reduced.

References

1. James, R. W. *The Optical Principles of the Diffraction of X-rays*; Ox Bow Press: Woodbridge, CT, 1982.
2. Guinier, A. *X-ray Diffraction in Crystals, Imperfect Crystals, and Amorphous Bodies*; Freeman: San Francisco, 1963.
3. Warren, B. E. *X-ray Diffraction*; Addison-Wesley: Reading, MA, 1969.
4. Klug, H. P.; Alexander, L. E. *X-ray Diffraction Procedures for Polycrystalline and Amorphous Materials*, 2nd ed.; Wiley-Interscience: New York, 1974.
5. Cullity, B. D. *Elements of X-ray Diffraction*, 2nd ed.; Addison-Wesley: Reading, MA, 1978.
6. Vainshtein, B. K. *Diffraction of X-Rays by Chain Molecule*; Elsevier: Amsterdam, 1966.
7. Alexander, L. E. *X-Ray Diffraction Methods in Polymer Science*; Wiley-Interscience: New York, 1969.
8. Kakudo, M.; Kasai, N. *X-ray Diffraction by Polymers*; Elsevier: Amsterdam, The Netherlands, 1972.
9. Baltá-Calleja, F. J.; Vonk, C. G. *X-ray Scattering of Synthetic Polymers*; Elsevier: Amsterdam, 1989.
10. Roe, R. J. In *Encyclopedia of Polymer Science and Engineering*, 2nd ed.; Kroschwitz, J. I., Ed.; Wiley: New York, 1989; Vol. 17, p 943.
11. Roe, R. J. *Methods of X-ray and Neutron Scattering in Polymer Science*; Oxford University Press: New York, 2000.
12. Guinier, A.; Fournet, G.; Walker, C. B.; Yudowitch, K. L. *Small-Angle Scattering of X-rays*; Wiley: New York, 1955.
13. Glatter, O.; Kratky, O. *Small Angle X-ray Scattering*; Academic Press: Orlando, FL, 1982.
14. Higgins, J. S.; Benoît, H. C. *Polymers and Neutron Scattering*; Clarendon Press: Oxford, 1994.
15. *International Tables for Crystallography*, Vol. C; Kluwer Academic: Dordrecht, 1992.
16. Brandrup, J.; Immergut, E. H. *Polymer Handbook*, 3rd ed.; Wiley: New York, 1989.

17. Wunderlich, B. *Macromolecular Physics*, Vol. 1; Academic Press: New York, 1973.

18. Clark, E. S. In *Physical Properties of Polymers Handbook*; Mark, J. E., Ed.; American Institute of Physics: Woodbury, NY, 1996; pp 409–416.

19. Ruland, W. *Acta Cryst.* **1961**, *14*, 1180–1185.

20. Krigbaum, W. R.; Roe, R. J. *J. Chem. Phys.* **1964**, *41*, 737–748.

21. Hosemann, R.; Bagchi, S. N. *Direct Analysis of Diffraction by Matter*; North-Holland: Amsterdam, 1962.

22. Tadokoro, H. *Structure of Crystalline Polymers*; Wiley: New York, 1979.

PART V

ANALYSIS OF POLYMER SURFACES

15

Surface Energy

MICHAEL J. OWEN

Nowadays, one is most likely to assess the surface energy of a polymer by using computer imaging technology to measure the contact angle of liquid drops placed on a solid sample of that polymer. A few clicks of the mouse define the drop shape on the video monitor, and a couple of key strokes later, a surface energy value is available to several decimal places. Evidently, this is one of the simplest operations in polymer surface characterization, but in reality it is one of the most difficult to interpret satisfactorily. Because contact angles are so surface sensitive, the result obtained is critically dependent on the quality of the three phases (gas, liquid, solid) involved, particularly so on the purity of the liquid and the smoothness and homogeneity of the solid surface. Immediate questions come to mind: What liquids should be chosen? How large a drop should be used, and how should it be placed? How long must one wait before processing the measurements? And so on. Less obvious are questions concerning the rigor of the analysis: What equation was used to obtain the surface energies? Are there more appropriate alternatives to use in certain cases?

The aim of this chapter is to provide some answers to these and other related questions. The subject is important because there is no more fundamental surface property than surface energy. Wettability critically matters when polymeric materials are brought together. Phenomena of major technological importance in the polymer industry, such as adhesion and release, and degree of dispersion of one polymer in another, are directly affected by this property. Much of the discussion will involve surface energies of solid polymers derived from contact angle data because this measurement represents the bulk of the available data. However, liquid surface energies and non-contact-angle approaches will also be considered.

Definition of Surface Energy

Surface energy is the result of the imbalance of attractive forces between molecules at the surface. In the interior of the material, each unit is acted on equally in all directions, but this is not so at the surface with air, other gases, or vacuum. The same is true at the interface between condensed (liquid or solid) phases where an interfacial energy arises from the force imbalance. By surface energy, we more correctly mean the surface free energy (γ_{SV}) of a material that is defined[1] as the change in the total surface free energy (G) per unit change in surface area (A) at constant temperature (T), pressure (P), and moles (n):

$$\gamma_{SV} = (\partial G/\partial A)_{T,P,n} \qquad (1)$$

For a pure liquid, the surface area can be changed under the above conditions. The free energy per unit area is then numerically and dimensionally identical to the surface tension, expressed as a force per unit length. For solids, however, surface area cannot generally be changed at constant chemical potential—that is, in a way that allows an equal number of the same molecules to be present in the surface after an increase in surface area as there was before expansion. Elastic forces complicate the issue, and the surface state after extension may be far from equilibrium. Consequently, the surface energies of liquid and solid polymers are necessarily measured in very different ways. Liquid surface tensions are directly measured, and values are usually independent of the specific technique chosen, provided equilibrium is established, whereas solid surface energies are commonly derived from contact angle measurements using semiempirical equations, thus producing values that

Table 15.1 Methods for measuring liquid polymer surface and interfacial tension[5]

Method	Applicability to surface tension	Applicability to interfacial tension	Other requirements
Drop profile methods:			
Pendent drop	Yes	Yes	Known density
Sessile drop	Yes	Yes	Known density
Rotating bubble	Yes	Viscosity limited	Known density
Wilhelmy plate	Yes	No	Zero contact angle[a]
Capillary height	Viscosity limited	No	Known density and zero contact angle
Detachment methods:			
du Nouy ring	Viscosity limited	No	Zero contact angle
Drop weight	Yes	No	Known density
Maximum bubble pressure	Viscosity limited	No	Known density

Adapted from reference 5.

a. Zero contact angle means the liquid under investigation must completely wet the plate or ring of the measuring apparatus.

depend on the choice of contact angle test liquids and interpretative equations.

Given that surface tension normally applies to liquids and surface-free energies to solids, it is best to quote the former values (SI units) in mN/m and the latter in mJ/m^2. (A plausible exception for the critical surface tension of wetting [see section on Solid Polymer Surface Energy] can be made, considering how this property is named and measured.) Surface and interfacial energies are related to the contact angle θ by the Young equation[2]

$$\gamma_{SV} - \gamma_{SL} = \gamma_{LV} \cos \theta \qquad (2)$$

where the subscripts SV, LV, and SL refer to the solid/vapor and liquid/vapor surfaces and the solid/liquid interface, respectively. Liquids that do not fully wet out the solid must be selected so that there is a definite point of contact at the edge of the drop. The angle between the liquid at the three-phase (liquid, solid, vapor) point of contact with the solid, measured through the liquid phase (not the vapor phase) is the contact angle θ.

First described in 1805 using neither mathematical symbols nor diagrams, this fundamental relationship could not be experimentally verified for a liquid in contact with a solid until 1971 because before then no technique for the direct measurement of their interfacial tension was available. In that year Johnson, Kendall, and Roberts[3] applied their contact mechanics approach (see section on Non-Contact-Angle Approaches to Solid Surface Energy) to the contact of natural rubber spheres, both dry and under water. They predicted a water contact angle of 64° in excellent agreement with the experimentally measured value of 66°. Note that we have chosen a classical reference for the Young equation and for other fundamental equations that follow. For a modern treatment of these fundamental equations of surface science, the reader is directed to the well-known textbook of Adamson and Gast,[4] where, for example, a simple derivation of eq 2 is given based on surface free-energy changes.

Liquid Polymer Surface Tension

A wide range of methods is available for the determination of surface tension of liquids. In principle, any of this multitude of techniques could be used with polymers. In practice, those techniques are preferred that are particularly suited to viscous materials and do not perturb the surface during measurement, thus accommodating lengthy equilibration times. Table 15.1, adapted from one given by Wu[5] in his comprehensive 1974 review of interfacial and surface tension of polymers, summarizes the suitability of the more common methods. In this table the requirement of zero contact angle means the liquid under investigation must completely wet the plate or ring of the measuring apparatus. This is why high-surface-energy metals such as platinum are used for the plate or ring.

It is evident that the techniques based on drop profiles where the surface tension is calculated from the shape of pendent (or pendant)* or sessile drops are best. Such drops are shown in Figure 15.1 In particular, the pendent drop method is the most simple, versatile, and reliable method for both the surface and interfacial tension of polymer liquids and melts. This method has been used extensively by Wu[6] and Roe,[7] who have provided much of the data available for polymers. Included among the outstanding advantages of the pendent drop method listed by Ambwani and Fort[8] are the following:

1. Accuracy
2. Complete mathematical analysis available (no empirical correction factors)
3. Rapid measurement possible (shortest drop age ca. 5 sec)

Pendent (adj.) seems preferable to *pendant* (n.). However, pedantic insistence on adjectives would oblige us to use the unfamiliar term, *surficial tension*, rather than *surface tension*, so we do not insist. (Strangely, the more correct *interfacial tension* is used!)

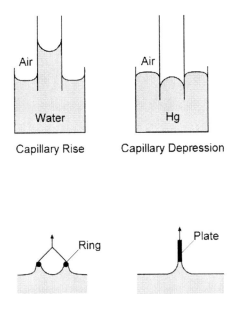

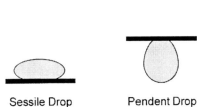

Figure 15.1 Some solid-liquid-vapor contact configurations.

4. Measurement does not perturb surface (excellent for dynamic aging studies)
5. Small samples required
6. Independent of contact angle with supporting capillary
7. Video image (or photograph) can provide permanent record
8. Suitable for both surface and interfacial tension determination

The main disadvantages apparent to Ambwani and Fort were

1. Surface ages less than 5 sec not accessible
2. Some setup effort and expense required
3. No commercial equipment available
4. No practical, current guide to use available

Time has not much changed this third drawback for polymers. Companies such as Material Interface Associates, Inc. have provided custom-built equipment, but to my knowledge, there is no readily available pendent drop apparatus that is well suited for polymers.

Forming drops that remain stable on aging can also be a problem (vibration insulation is essential), and it is also wise to check by imaging suitably calibrated graticules that there are no excessive distortion errors in the optical system (a comment that also applies to contact angle imaging techniques). As contact angle apparatus need to have drop formation devices, it is possible in principle to use them to obtain pendent drops for surface or interfacial tension measurement. Koberstein and coworkers[9] have done this for polymer surface and interfacial tensions using a Rame-Hart Model C-2033 environmental chamber mounted on a NRL 100–00 contact-angle goniometer.

The equation for the shape of a free surface of liquid under the influence of surface tension and gravity originated with Young and Laplace.[2] Later, Bashforth and Adams[10] provided a dimensionless form. The expression is an intractable second-order, second-degree differential equation that cannot be solved analytically in finite terms but can be numerically integrated. This is no problem for a modern computer, but it was a lifetime's challenge in 1883. Historically, the mathematical difficulties were circumvented by providing a specific solution for particular cases, the most common of which is the selected plane approach of Andreas, Hauser, and Tucker.[11] They showed that the surface tension can be obtained from the following equation:

$$\gamma = \rho g (d_e)^2 / H \tag{3}$$

where d_e is the maximum (equatorial) diameter of the drop. $1/H$, a drop shape factor, comes from tables of $1/H$ and S, another shape factor equal to d_e/d_s where d_s is the diameter of the pendent drop in a selected plane, a distance d_e above the apex of the drop, as shown in Figure 15.2. Various versions of these shape factor tables are available.[12] Note also that the density of the polymer must be known or measured to obtain the surface tension. With the advent of digital image processing techniques, it is now possible to directly compare drop shapes with those stored or generated in computer memory. The use of many experimental points for shape analysis reduces the errors associated with the selected plane technique. Koberstein and his coworkers[9] have described how this can be done for polymers.

Bulk polymer properties generally depend linearly on the reciprocal of molecular weight. This relationship applies only rarely to surface tensions, the most notable example is poly(tetrafluoroethylene) (PTFE).[13] Two more generally applicable equations have been proposed for the surface tension dependence on molecular weight:

$$\gamma^{-1/4} = (\gamma_\infty)^{-1/4} + k_1/M_n \tag{4}$$

$$\gamma = \gamma_\infty - k_2/(M_n)^{2/3} \tag{5}$$

Equation 4, the Wu equation, can be derived from MacLeod's equation for simple liquids, $\gamma = \gamma_0 \rho^4$. γ_0 is a constant, independent of temperature. Experimentally, the MacLeod exponent varies between 3.0 and 4.2 for polymers. Although eq 5, the LeGrand and Gaines equation,[14] is wholly empirical, the two equations produce values for surface ten-

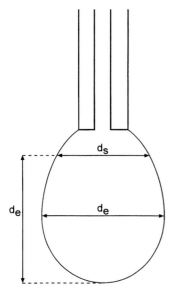

Figure 15.2 Pendent drop surface tension measurement. Key: d_e is the maximum (equatorial) diameter of the drop; d_s is the diameter of the pendent drop in a selected plane, a distance d_e above the apex of the drop.

density effect. The temperature coefficients, $-d\gamma/dT$ (which is also the surface entropy) are in the range of 0.05 to 0.09 mN/(m K). This is a lower range than that for small liquids and has been attributed to conformational restrictions in polymer chains.[5]

The question arises as to how reasonable an approximation to solid surface energy is an extrapolation of directly measured surface tension data, either from elevated temperatures or to infinite molecular weight from low molecular weight liquid surface tension data. Leaving aside any shortcomings of the equations used in the extrapolation, there is the complication that property extrapolations change slope or may even jump as they pass through phase transitions, so clearly this is not a rigorous process.

Wu[5] has shown that surface tension jumps at the crystal-melt transition but not at the glass-transition temperature (surface entropy, or the surface tension/temperature slope, jumps at both transitions). Sometimes these changes are small, and for short extrapolations the approximations involved are reasonable, certainly no more suspect than the shortcomings of contact angle analysis (see next section). However, the researcher should always consider the magnitude of the phase transition effects involved. Thermal expansion coefficients of the various phases are often available, and a quick calculation can put these procedures in perspective. One significant effect to consider concerns crystalline and amorphous phases. The surface of a partially crystalline polymer is likely to be amorphous, as the lower density of the latter phase should result in a lower surface energy. However, if the solid has been treated, for example, by nucleation against a high-energy substrate, to produce high surface crystallinity, extrapolation of amorphous melt data could be significantly in error.

There is also the obvious, practical factor that often the investigator does not have access to lower molecular weight fractions of the material of interest. It may also be in a cross-linked thermoset form. In such cases, which naturally are the majority, the solid must be examined as provided, usually by a contact angle approach.

sion at infinite molecular weight within 1 to 2 mNm of each other, as shown in Table 15.2. McLeod's equation implies that the variation with molecular weight is solely a density effect. As density usually increases with molecular weight so, too, does surface tension. There are examples of molecular weight independent surface tensions, for example, poly(ethylene glycols) and poly(propylene glycols).[15,16] This is an end-group effect; when the terminal hydroxyl groups are replaced by non-hydrogen-bonding groups such as methyl ether, the normal temperature dependence is observed.

Surface tensions of polymers generally show a linear decrease with increasing temperature. This is also primarily a

Table 15.2 Liquid polymer surface tension at infinite molecular weight[5]

Polymer	Temperature (°C)	γ_∞ from eq 4[a] (mN/m)	γ_∞ from eq 5[b] (mN/m)
Poly(dimethylsiloxane)	20	21.3	20.4
Perfluoroalkanes	20	25.9	25.8
Polystyrene	176	30.0	29.5
Polyisobutylene	24	35.6	34.6
n-alkanes	20	37.8	36.9
Poly(oxyethylene)	24	44.4	42.3

a. Wu equation.
b. LeGrand and Gaines equation.

Solid Polymer Surface Energy

The first major contribution to the quantification of surface energies of solid polymer surfaces was made by Zisman and coworkers.[17] They found that when the cosine of the contact angles of liquids placed on a polymer surface are plotted against their surface tensions, a nearly linear plot is obtained. The intercept of this line with the surface tension axis at the value of $\cos \theta = 1$ is known as the critical surface tension of wetting of the polymer, γ_c, (note that "of wetting" has subsequently been frequently omitted). It is a measure of the surface energy of the polymer in that it demarcates those liquids that spread on its surface from those that do not. However, it is not equal to the surface-free energy because it neglects the interfacial tension that need not be zero even when θ is zero. An example for the low-surface-energy fluoropolymer, poly[(1H,1H,2H,2H-heptadecafluorodecyl)methylsiloxane] (PHDFDMS) obtained using n-alkane liquids,[18] is given in Figure 15.3.

Shafrin and Zisman[17] found that γ_c varies systematically with the chemical constitution of the surface and was thus able to develop a guide to the relative impact on surface energy of constitutive groups and even atoms in a polymer surface. This idea that we can conceive of separate surface energies for each of the different parts of complex molecules, and that the surface energy of such a molecule is determined by the composition and orientation of the outermost groups independent of the underlying components, is known as Langmuir's principle of the independence of surface action.[19] It is only a first approximation, but it enables reasonable predictions of polymer wettability to be made from a knowledge of the chemical composition of its outermost layer.

The order of increasing surface energy for atoms—or, in other words, the effectiveness of substitution of individual atomic species in increasing the wettability of organic surfaces—is[17]

$$F < H < Cl < Br < I < O < N$$

The effect of substituent groups on polymer surface energy is shown in Table 15.3. Besides critical surface tension of wetting, this table also shows data from another contact angle approach introduced by Owens and Wendt that is discussed later in this section (see eq 13). It is compiled from data found in reference 20. It is not meant to be comprehensive, but provides a sampling of technologically important polymers to illustrate the range of surface energies available. In common with all other contact angle approaches to solid polymer surface energy that have since been developed, the precise values of critical surface tension of wetting depend on the choice of liquids made.

Where possible, the values in Table 15.3 have been chosen so that each column consists of comparable data, but the concerned reader is advised to check this compilation from the original literature. These variables should always be accounted for when making comparisons. The strength of the Zisman approach is the wide range of polymers for which data are available, and it is broader by far than for any subsequent approach. More extensive compilations than the selection in Table 15.3 are available.[21,22]

Young's equation (eq 2) can be combined with the Dupre equation for the thermodynamic work of adhesion[23]

$$W_{SL} = \gamma_{SV} + \gamma_{LV} - \gamma_{SL} \qquad (6)$$

to give

$$\gamma_{LV}(1 + \cos \theta) = W_{SL} \qquad (7)$$

If we can find a way of expressing W_{SL} in terms of the surface energies of the two phases S and L, then we have a way of estimating the solid surface energy. Girifalco and Good[24] did this by expressing W_{SL} in geometric mean form as

$$W_{SL} = 2\Phi(\gamma_{SV}\gamma_{SL})^{0.5} \qquad (8)$$

where Φ is a correction factor for intermolecular interactions. It is equal to unity if the intermolecular forces acting across the interface are alike as, for example, for an n-alkane on polyethylene. In other cases it can range down to 0.5. Although Φ can be calculated, this necessity has limited the usefulness of this approach. Combination of eqs 7 and 8 yields

$$\gamma_{LV}(1 + \cos \theta) = 2\Phi(\gamma_{SV}\gamma_{LV})^{0.5} \qquad (9)$$

and for the case where Φ is unity reduces to

$$\gamma_{SV} = \gamma_{LV}(1 + \cos \theta)^2/4 \qquad (10)$$

This is a useful way of estimating surface energies of low surface energy polymers, or at least the London dispersion force component of surface energy (see eq 12), from one liquid, usually n-hexadecane. According to eq 9, $\cos \theta$ will

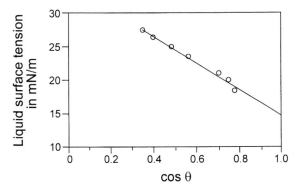

Figure 15.3 Critical surface tension of wetting plot for poly[(1H,1H,2H,2H-heptadecafluorodecyl)-methylsiloxane] (PHDFDMS). (Reproduced with permission from reference 18. Copyright 1993 Wiley-VCH.)

Table 15.3 Solid surface energies of polymers

Polymer	Critical surface tension of wetting, γ_c (mN/m)	γ_{SV}, by eq 13 (mJ/m^2)
Poly(hexafluoropropylene)	16.2	12.4
Poly(tetrafluoroethylene)	18.3	19.1
Paraffin wax	23	25.4
Poly(dimethylsiloxane)	24	22.8
Poly(vinyl fluoride)	28	36.7
Polypropylene	29	29.0
Polystyrene	32.8	42.0
Polyethylene	33	33.1
Poly(vinyl chloride)	39	41.5
Poly(methyl methacrylate)	39	40.2
Poly(hexamethylene adipamide)	42.5	47.0
Poly(ethylene terephthalate)	43	47.3

Data obtained from references 20 and 21.

be unity when $\Phi^2\gamma_{SV} = \gamma_{LV}$. This is also the condition for the critical surface tension of wetting, so we obtain the relationship

$$\gamma_c = \Phi^2\gamma_{SV} \qquad (11)$$

showing that only when the interaction parameter is equal to unity can critical surface tension of wetting be equated with solid free-surface energy.

A more useful way ahead was provided by Fowkes,[25] who suggested that the surface energy of a solid or liquid is made up additively of components that correspond to intermolecular interactions

$$\gamma = \gamma^d + \gamma^p + \ldots \qquad (12)$$

As many as seven terms have been suggested, but a common simplification is to consider only two: the component resultant from electrodynamic London dispersion forces common to all matter, and the so-called polar component that incorporates all the other interactions.

Two equations of this type have been widely used to interpret contact angle data of polymers. The Owens and Wendt[26] equation (eq 13), independently derived by Kaelble,[27] is based on the geometric mean approach, and the Wu[4] equation (eq 14) uses a harmonic mean combining rule:

$$\gamma_{LV}(1 + \cos\theta) = 2(\gamma_{LV}{}^d\gamma_{SV}{}^d)^{0.5} + 2(\gamma_{LV}{}^p\gamma_{SV}{}^p)^{0.5} \qquad (13)$$

$$\gamma_{LV}(1 + \cos\theta) = 4\gamma_{LV}{}^d\gamma_{SV}{}^d/(\gamma_{LV}{}^d + \gamma_{SV}{}^d)$$
$$+ 4\gamma_{LV}{}^p\gamma_{SV}{}^p/(\gamma_{LV}{}^p + \gamma_{SV}{}^p) \qquad (14)$$

The two unknowns $\gamma_{SV}{}^d$ and $\gamma_{SV}{}^p$ require a minimum of two contact angle determinations. Water and methylene iodide are the most popular choice, water having a large polar component of surface energy whereas methylene iodide is primarily a dispersion force liquid. Both liquids have high surface tension, thus increasing the chances that they will

form a finite contact angle on a given solid and not wet it out completely and thereby making an analysis by these equations impossible. Examples of solid surface energy data obtained using eq 13 are included in Table 15.3. A comparison of eqs 13 and 14, taken from Wu,[28] is shown in Table 15.4. Note that Tables 15.3 and 15.4 have been deliberately taken from different compilations to illustrate some of the variations that are apparent in the literature. One notable difference in the application of these equations is that the geometric mean approach (eq 13) consistently yields a smaller polar component of surface energy than that given by the harmonic mean approach (eq 14).

Despite its popularity, methylene iodide is not an ideal liquid to use. Its light sensitivity makes it difficult to keep pure, and it must be stored in a light-tight container and stabilized, usually by copper wire. Moreover, it is known to be unsuitable for use on fluorinated surfaces,[29,30] and the latest work[30] shows that it has a considerable polar component when measured on such surfaces. A truly apolar liquid such as n-hexadecane would be a better choice, but its low surface tension severely restricts the number of polymers on which it forms a finite, measurable contact angle. As contact angle liquid choices depend on the material used, it is important that these choices be unambiguously reported, along with the actual contact angle data, the derived surface energies, and the methodology used in those derivations. It is surprising how often one or more of these essential pieces of information are not provided in the literature. Without all of them, recalculations for meaningful comparisons with new data are impossible.

As neither of these frequently encountered geometric-mean or harmonic-mean combining rules of surface interactions is exact, researchers continue to refine the theory. No totally satisfactory approach has yet emerged, but the most promising progress is being made through the application of Lifshitz's theory of van der Waals interactions. This

Table 15.4 Comparison of eqs 13 and 14 (all values in mJ/m^2; measurements made at 140 °C)

Polymer	γ^d, eq 13	γ^p, eq 13	γ^d, eq 14	γ^p, eq 14
Poly(hexafluoropropylene)	12.0	0.8	14.0	3.0
Poly(tetrafluoroethylene)	18.6	0.5	20.5	2.0
Paraffin wax	25.4	0	31.0	0
Poly(vinyl fluoride)	31.3	5.4	27.2	11.2
Polyethylene	33.2	0	35.3	0.8
Poly(methyl methacrylate)	35.9	4.3	31.1	10.1
Poly(vinyl chloride)	40.0	1.5	35.8	6.1
Poly(ethylene terephthalate)	37.8	3.5	32.8	9.3
Polystyrene	41.4	0.6	38.4	4.2
Poly(hexamethylene adipamide)	34.1	9.1	29.3	15.4

Data from reference 29.

work has resulted in one equation that has been applied to a considerable range of polymers, and it is also an equation that will be encountered as an option in the software of some commercially available contact angle instruments. This is the Good, Chaudhury, and van Oss equation.[31] This theory treats interfacial forces as composed of two major components that arise from a modification of earlier approaches. One component, known as the LW (Lifshitz–van der Waals) component of the interaction, is the dispersion forces with a small contribution from Debye (induction) and Keesom (dipolar) forces. These are the terms that can correctly be treated by the familiar geometric mean combining rule as in eqs 8 and 13. The nondispersive component of surface energy, termed AB, arises from Lewis acid-base or electron donor-acceptor polar interactions. In the Good, Chaudhury, and van Oss case, approximate acid-base surface parameters are defined: δ^+ is the electron acceptor term, and δ^- is the electron donor term, such that

$$W_{\text{LS}}^{\text{AB}} = \gamma_{\text{L}}(1 + \cos\theta) - 2(\gamma_{\text{L}}^{\text{LW}}\gamma_{\text{S}}^{\text{LW}})^{0.5} = \delta_1^+\delta_2^- + \delta_1^-\delta_2^+ \quad (15)$$

Contact angle approaches to surface energy determination remain a controversial subject.

The reader may want to consult recent reevaluations of contact angle data, such as those of Kwok and Neumann.[32,33]

Contact Angle Measurement

Liquids and solids can come together in a wide variety of configurations, some of which are shown in Figure 15.1, so it is no surprise that there is a similar diversity of contact angle measurement techniques including:

1. Sessile drop method
2. Captive bubble method
3. Wilhelmy plate method
4. Capillary rise method
5. Level-surface techniques

The sessile drop and captive bubble techniques are essentially the same technique with the liquid and vapor phases interchanged. It is also possible to use these techniques with two immiscible liquid phases. Some sessile drop instruments such as the Rame-Hart NRL A100 Contact Angle Goniometer are equipped with a goniometer eyepiece for direct observation of the contact angle. This requires the operator to place the crosshair tangentially to the drop at the point of contact. This requires some practice (Neumann and Good[34] recommend a self-training procedure in which goniometer readings are compared to those calculated from drop dimensions). Other instruments use a variety of ways of obtaining the contact angle from the dimensions of the drop. These might appear to be less subjective ways of obtaining the required data, but the user should be aware of the assumptions behind the chosen equations. For example, if the simple spherical cap equation is used:

$$\sin\theta = 2hr/(h^2 + r^2) \quad (16)$$

the assumption is that the distorting effect of gravity is negligible, and the drops used must be very small, of the order of 10^{-4} mL, corresponding to a drop radius of the order of 0.03 mm.

Modern instruments are based on curve fitting software that is not always readily disclosed by the manufacturer. The saving grace is that the tangent is displayed on the monitor, thus enabling the operator to directly judge the quality of the result. There is the seemingly less subjective technique of Fort and Patterson,[35] in which the light source and viewer are mounted on a lever that pivots on the axis of the contact angle to be measured. The angle from the vertical at which light reflected from the drop is extinguished is equal to the contact angle. Probably, one defect that has limited its more widespread use is that angles greater than 90° cannot be measured.

Equations such as eq 16 do not require knowledge of the liquid drop surface tension. With the advent of appropriate computer software, it is possible to design instruments that measure surface tension and contact angle simultaneously. Neumann et al.[36] have described a technique called axisymetric drop shape analysis–profile (ADSA-P) that does this and also computes drop volume, drop radius, and drop surface area. Their method uses a number of randomly selected points (e.g., 20) along the digitized profile of a sessile drop, which the computer fits to the Young and Laplace[2] equation of capillarity with surface tension and drop apex curvature as the adjustable parameters (density also needs to be known).

When the Wilhelmy plate method[4] is used to measure the surface tension of liquids, it is usually assumed that the contact angle is zero. This is why high-surface-energy materials such as flame-cleaned platinum are used. However, the technique can also be used with liquids of known surface tension to derive the contact angle. Use of the Wilhelmy technique has increased in recent years, being particularly popular with investigators interested in dynamic contact angle hysteresis effects (see next section on Contact Angle Hysteresis). The solid under investigation must be available in a suitable plate form for this technique, but successive cycles of immersion (advancing angle) and retraction (receding angle) can readily be performed with an instrument such as the Cahn DCA 322 dynamic contact angle balance. Capillary rise is another contact-angle-dependent method of measuring surface tension that has been used to measure contact angles. Various geometries can be used: cylinders (vertical cylindrical axis), slits (vertical parallel flat plates), and single, vertical plates.

The level-surface methods using a tilted plate (seesaw configuration) or a cylinder (horizontal cylindrical axis, or floating log configuration) are not as widely used, probably because larger volumes of liquid are required than sessile drops. With plates, the tilt angle is adjusted by rotating the plate about its horizontal axis to obtain meniscus-free contact with the liquid surface. In the techniques using horizontal cylinders or spheres, these components are partially immersed in the liquid and raised or lowered until contact with the liquid surface exhibits no curvature. Dynamic contact angle hysteresis can be investigated by rotating the cylinder or sphere.

The definitive article on the practical measurement of contact angles and associated pitfalls has already been written by Neumann and Good;[34] my comments here rely heavily on their wisdom. I focus here on the sessile drop method. In many ways, this method is fraught with more difficulties than many of the other techniques, but it is the most widely used approach, mainly because of the ease with which a wide diversity of samples can be accommodated. It is not always possible to fashion the material under investigation into a Wilhelmy plate or a capillary tube, but it is usually possible to place a liquid drop on it. Contact angles on fibers and powders pose special problems that are not discussed in detail here. Fibers are best investigated by a tensiometric technique. Powders are packed into a column and the height to which liquids rise in the column or the pressure required to prevent such displacement is measured. Uniform preparation of such columns is difficult and time-consuming. It is a mistake to compress the powder into a pellet and attempt to measure sessile drops thereon; this approach leads to uncontrolled roughness variations and renders such data useless. This phenomenon of capillary rise in a porous medium is known as wicking.

What Size Drop Should Be Used?

The size of the drop depends on the nature of the sample and how the contact angle is measured, but it is not often a crucial issue. Drop size effects have been reported, both with contact angle decreasing with decreasing drop size and vice versa. The former dependence appears to be related to contact angle hysteresis. If hysteresis is small, the effect is not noticeable for drops that are a convenient size for measurement. However, when hysteresis is appreciable, the effect is noticeable below about 2 mm diameter. Alternatively, if the contact angle is being derived from equations such as the spherical cap equation, it is preferable to have small drops so that the gravitational distortion is negligible. Some recent careful studies[36] on self-assembled monolayers of alkyl thiols on gold, using the ADSA-P technique, showed contact angles decreasing as the radius of the three-phase line of contact for a sessile drop increased from 1 to 5 mm. This variation is attributed to a positive line tension with values of the order of 1 μJ/m trending toward larger values for higher liquid surface tensions. In practice, it is usual to choose a constant drop size, say 2 μL. One can always do an initial drop size study on a new system to ascertain if it is particularly sensitive to this variable and proceed accordingly.

How Should the Drop Be Placed on the Solid?

The drop must be placed very carefully. It must be brought slowly and steadily to the surface. Neumann and Good[34] describe a technique employed in Zisman's laboratory using a flame-cleaned platinum wire, but today syringes are generally used. Most instruments have micromanipulators that permit steady raising of the sample stage or lowering of the syringe support. Naturally, scrupulous cleanliness of the syringes and checking of the purity of the liquid, particularly if dedicated syringes are used to store each liquid, is essential. Some practitioners employ tapping or other vibrating, supposing this will facilitate attainment of equilibrium. However, this can lead to an angle intermediate between advancing and receding and should be resisted. It is wise to avoid vibration problems by using an air-supported table or other vibration insulating system.

How Long before Making the Measurement?

Obviously, the time required is another factor that depends on the system used. As equilibrium values of liquid surface tension are used in the thermodynamic analysis of data, it is important to leave sufficient time for a stable equilibrium contact angle to be attained. However, if swelling of the polymer sample by the probe liquid occurs, clearly, the shortest time possible is the most suitable. In practice, when measurements are made on both sides of a drop several times, swelling problems are usually easy to spot if the operator is alert to the possibility. For volatile liquids, there is also the possibility that evaporation can change the drop environment from advancing to receding or intermediate (see next concern).

Should the System Be Enclosed?

Some laboratories conduct all contact angle studies in a clean room, whereas others install their instruments in lamellar flow hoods and others obtain adequate data with open stages exposed to the laboratory air. An enclosure not only protects the sample from contamination from dust and aerosols in the atmosphere but also enables the drop to be in equilibrium saturation with its vapor. In practice, this is not essential for water and higher boiling point liquids, but it is needed for n-alkanes such as hexane and octane. Enclosed environmental chambers usually allow temperature-dependence studies to be performed, although such investigations are evidently relatively rare. In summary,[34] if high-precision results are required, enclosure will be needed, but open stages may often be used, except with volatile liquids, particularly if other causes of contact angle variation such as roughness or other heterogeneity are known or suspected to be present.

Contact Angle Hysteresis

Contact angle hysteresis is simply defined, but it is a complex problem that is not fully understood. It is the difference between the advancing contact angle, θ_a, and the receding contact angle, θ_r. The former refers to the contact angle on previously unwetted surface, while the latter refers to previously wetted surfaces. For example, with the sessile drop technique for the advancing angle, the drop is increased in size so that the area of contact with the solid is increased to ensure the three-phase line of contact is with unwetted surface. Conversely, the receding angle is obtained by extracting sufficient liquid from the drop so that the line of contact recedes to a position on the previously wetted surface. The effect can be simply demonstrated by tilting the stage on which the drop and sample rest until the drop just starts to roll down the tilted sample or by moving the stage and sample laterally with respect to the drop (see Figure 15.4). For tangenting by eye, it is preferable to leave the syringe needle in the drop for these operations as its presence is too remote to affect the contact angle. The terms "advancing" and "receding" are confusing as they imply dynamic values, whereas quasi-equilibrium values are usually desired. Semantically, "advanced" and "receded" would be better usage, but the practice is too ingrained to be changed.

Contact angle hysteresis is so common that it should be expected, and it is good practice to measure advancing and receding values for at least one liquid, usually water, on each new sample. Among the various causes of contact angle hysteresis are the following:

1. Surface roughness
2. Chemical heterogeneity
3. Surface reorganization
4. Surface deformation
5. Drop size
6. Extraction of leachable species
7. Deposition of contaminants
8. Penetration by contact liquid
9. Evaporation of contact liquid
10. Chemical attack

The effect of submicroscopic roughness is described by Wenzel's relationship:[34]

$$r = A/A_0 = \cos \theta / \cos \theta_0 \qquad (17)$$

where the roughness ratio r is the ratio of the actual area A to the apparent area A_0 (i.e., the projected area on a plane parallel to the apparent surface). θ is the observed macroscopic contact angle and θ_0 is the intrinsic contact angle that would be obtained if the surface were flat. Surfaces with r equal to unity are rare. The effect of this relationship is that if the smooth material gives an angle greater than 90° roughness will increase it further whereas if the intrinsic angle is less than 90° roughness will cause the observed angle to decrease. For example, smooth silica plates made hydrophobic with agents such as hexamethyldisilazane have contact angles of ca. 105° but similarly treated silica powders compressed into pellets have contact angles as high as 150°. Johnson and Dettre[37] have provided a useful analysis of this effect. It is frequently found that filled polymers exhibit considerably more contact angle hysteresis than the unfilled polymer even though techniques such as XPS do not detect any filler in the outermost surface region. Microroughness induced by the filler is a probable explanation. An extreme example is the vapor deposition of plasma-polymerized poly(tetrafluoroethylene)[38] where the microroughness caused by the deposition process results in a water contact angle of 170°.

The effect of chemical heterogeneity is described by Cassie's relationship:[34]

$$\cos \theta = a_1 \cos \theta_{01} + a_2 \cos \theta_{02} \qquad (18)$$

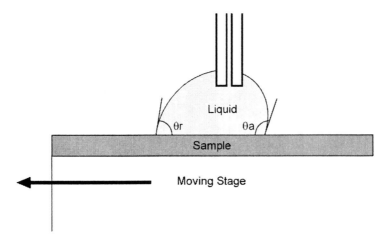

Figure 15.4 Contact angle hysteresis. Key: θr is the receding contact angle; θa is the advancing contact angle.

where a_1 is the fraction of surface area with intrinsic angle θ_{01}, and a_2 is the fraction with intrinsic angle θ_{02} ($a_1 + a_2 = 1$). It is generally held that the advancing angle is sensitive to the lower surface energy component of a heterogeneous surface, whereas the receding angle responds particularly to high surface energy contaminants. Both Cassie's and Wenzel's laws have been systematically derived from a generalized Young's equation and an additional law proposed for drops on surfaces that are both rough and chemically heterogeneous.[39]

The dynamic nature of polymer surfaces and their ability to reorganize in response to changing environments is well established through the work of Andrade et al.[40] and Holly and Refojo.[41] There have been many demonstrations of the possible differences in surface composition between polymer surfaces in air and their interfaces with higher surface energy substrates. Water, in particular, enables low levels of hydrophilic groups to rotate into the interface and markedly lower interfacial tension. For soft polymers, a local surface deformation can occur at the three-phase line of contact[42] that can influence the dynamic contact angle hysteresis. For example, an application of the Wilhelmy plate technique to study water contact angle hysteresis of this kind is given by Tomasetti et al.[43] They showed that the wetting hysteresis of blends of rigid polypropylene and a soft ethylene-propylene copolymer can be understood by considering viscoelastic energy dissipation and the static contact angles. The energy dissipation depends on the elastic modulus. The static contact angle is determined by the more hydrophobic phase in the case of advancing angles, and the less hydrophobic phase in the case of receding angles.

The effect of drop size has already been discussed in the preceding section. It should also be obvious that rigorous efforts to exclude contaminants from the air, polymer samples, contact angle liquids, and their various containers is essential. Frequent checks of liquid purity by surface tension measurement should be routine. Penetration by the contacting liquid often presents difficulties with polymers. Water can hydrate the surface layers of polar polymers or be more extensively absorbed. Swelling by n-alkanes of less polar hydrocarbon-based polymers, many of which have glass-transition temperatures below room temperature and ample free volume, is a particular problem. Not only do the lower homologs readily penetrate the polymer, but also they can be rapidly lost by evaporation. This can be avoided by using an enclosed environmental chamber. The drop shrinkage from evaporation causes the contact angle to move from the advancing value to the receding value, and the situation is further complicated if any drop size effects are involved. In addition to the variety of physical causes of hysteresis listed, chemical attack of the polymer by the contact angle liquid is also a possibility that should be considered. Happy and rare is the investigator whose surface does not exhibit contact angle hysteresis and who need not consider the relevance of these diverse possible explanations. For further information on contact angle measurement and interpretation, the reader is referred to three special issues of the *Journal of Adhesion Science and Technology* on apparent and microscopic contact angles.[44-46]

Non-Contact-Angle Approaches to Solid Surface Energy

As surface energies play a role in a variety of polymer behaviors, it is not surprising that researchers have used such phenomena to estimate the magnitude of the surface energy. Phenomena that have been exploited in this way include onset of crazing and healing of scratches, engulfment of particles by an advancing solidification front, heat of fusion of powdered polymers, and stress/strain behavior near the melting point. The most significant progress in recent years in estimating the surface thermodynamic quantities of solids that is not based on contact angles has come from the mechanics of contact deformations. In particular, the Johnson, Kendall, and Roberts[3] (JKR) theory has been established as a meaningful technique for polymer surfaces. Essentially, the JKR theory is a modification of the energy-balancing method of predicting the contact deformation of curved solids that was developed by Hertz and modified for surface forces in a manner originating in the Griffith theory of fracture of brittle solids.

Figure 15.5 shows the contact between a deformable semispherical solid with radius of curvature R and a flat plate,[29] thus resulting in the formation of a circular region of radius a whose size depends on the surface forces and the external applied load P. For this sphere on plate geometry, the thermodynamic work of adhesion W (see eq 6) can be obtained from

$$a^{3/2}/R = (1/K)(P/a^{3/2}) + (6\pi W/K)^{1/2} \qquad (19)$$

where K is the composite elastic modulus. Chaudhury[1,29] has reviewed the use of this technique. He has shown that poly(dimethylsiloxane) (PDMS) is a very suitable substrate for such studies and how the PDMS surface can be modified by plasma oxidation followed by adsorption of silane self-assembled monolayers to vary the nature of the contacting surfaces. When the two contacting surfaces are the same, the work of adhesion is twice the surface energy of the material under investigation.

The surface forces apparatus, which was designed to measure forces between surfaces as a function of distance, also has a suitable geometry and construction for determining the work of adhesion by the JKR approach when the surfaces are in contact.[47,48] JKR determinations of the surface energy of several polymers are now available, as

Table 15.5 Polymer surface energies using the Johnson-Kendall-Roberts (JKR) approach

Polymer	γ_{JKR} (mJ/m^2)	Reference
Poly(dimethylsiloxane)	23	29
Poly(4-methyl 1-pentene)	26	47
Poly(vinyl cyclohexane)	28	47
Polyethylene	33	48
Poly(n-butyl acrylate)	34	49
Natural rubber (cis 1,4 polyisoprene)	35	3
Polystyrene	44	47
Poly(methyl methacrylate)	53	47
Poly(ethylene terephthalate)	61	48
Poly(2-vinyl pyridine)	63	47

shown in Table 15.5. Comparison with Table 15.3 shows reasonable agreement between the JKR and contact angle methods for the lower surface energy polymers where the dispersive force components of surface energy dominate, but much poorer agreement as the polar nature of the polymers increases. This conclusion that the present use of contact angle measurements to accurately assess the surface energy of polymers is limited to those with apolar surfaces is echoed in other recent evaluations.[30] The poly(ethylene terephthalate) studies of Mangipudi et al.[48] are on samples that were biaxially stretched during sample preparation, so this surface should be semicrystalline in contrast to the amorphous surfaces used for most contact angle studies. Despite this, Mangipudi et al.'s contact angle studies on the same samples used in the JKR studies gave values similar to other contact-angle studies such as those listed in Table

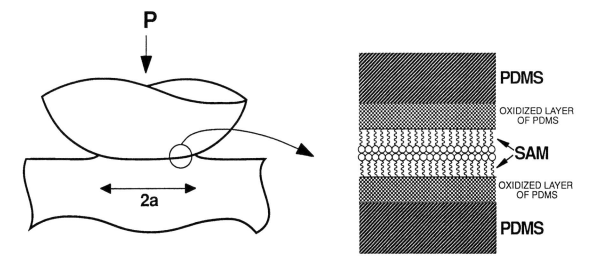

Figure 15.5 Contact of poly(dimethylsiloxane) (PDMS) semispherical cap and flat plate. The size of the radius a depends on the surface forces and the external applied load P. An enlargement of the interfacial contact region is shown on the right to demonstrate how modified surfaces using self-assembled monolayers (SAM) can be achieved. (Reproduced with permission from reference 29. Copyright 1992 American Association for the Advancement of Science.)

Table 15.6 Solid surface energies (all values in mJ/m^2)

Polymer	Surface energy (Owens-Wendt approach)	Critical surface tension of wetting (Zisman)
PMTFPS	13.6	21.4
PDMS	23.5	24

PMTFPS = poly[methyl(3,3,3-trifluoropropyl)siloxane]
PDMS = poly(dimethylsiloxane)

15.3. This implies little difference between amorphous and semicrystalline surfaces in this case and does not offer an explanation of the high JKR surface energy of this polymer.

Illustrative Practical Example

An object lesson on the need for careful interpretation of contact angle characterizations of polymers is provided by the fluorosilicone polymer poly[methyl(3,3,3-trifluoropropyl)siloxane] (PMTFPS). Consultation of a polymer data handbook[20] provides the data shown in Table 15.6. Application of the Owens-Wendt geometric mean analysis to water and methylene iodide contact angle data suggests a surface energy two thirds that of the most common silicone polymer, poly(dimethylsiloxane) (PDMS), whereas the Zisman critical surface tension of wetting values obtained with *n*-alkanes indicate the two polymers are much closer in surface energy. Which of these interpretations is the more correct?

Let us look more closely at the actual contact angle values behind the surface energy calculations. Some typical values obtained in our laboratories are given in Table 15.7 using liquids whose measured surface tensions are in close agreement with accepted literature values. In each case, we have chosen high molecular weight, cross-linked polymer gums. These gums can be smoothly cleaved, and they avoid the complications of added fillers and filler treatments and possible associated contaminants. The freshly cleaved surfaces are rigid enough to provide excellent substrates for contact angle measurement and can be shown to be adequately smooth by optical microscopy and scanning electron microscopy (SEM). X-ray photoelectron spectroscopy (XPS) analysis showed both materials to be free from significant contamination, having actual surface compositions

Table 15.7 Contact angle data (all values in degrees)

Polymer	Water	Methylene iodide	*n*-hexadecane
PMTFPS	104	90	51
PDMS	102	69	36

PMTFPS = poly[methyl(3,3,3-trifluoropropyl)siloxane]
PDMS = poly(dimethylsiloxane)

close to theoretical expectation. Table 15.7 shows it is only the contact angle of water that is similar for these two polymers; PMTFPS is significantly less wetted by the other two contact angle test -liquids.

There are indications in the literature[29,30] that methylene iodide is a much poorer choice than *n*-hexadecane for fluorinated surfaces. On this basis, we would choose the Zisman critical surface tension data as more meaningful and conclude that these two polymers are somewhat similar in surface energy. Indeed, if we apply the Owens and Wendt methodology to the water and *n*-hexadecane data, we get better agreement: 23.0 mJ/m^2 for PDMS and 19.1 mJ/m^2 for PMTFPS. But matters are not as simple as they might appear. Methylene iodide may be a "bad actor" on PMTFPS, but are *n*-hexadecane and the other *n*-alkanes really good choices for PDMS? It turns out that the *n*-alkanes rapidly swell PDMS. In fact, one of the primary reasons for developing PMTFPS was to produce a silicone that was much less prone to swelling by hydrocarbon fuels, oils, and solvents. It is only the instantaneous contact angle of many of the *n*-alkanes that is similar for the two polymers. Their response thereafter to hydrocarbons is quite different. Only in a very narrow initial sense can they be said to have similar oleophobicity, and practical hydrocarbon solvent resistance is a very different matter. By contrast, both polymers exhibit stable water contact angles greater than 90° and can safely be said to have similar degrees of hydrophobicity.

Turning to liquid surface tensions produces yet another surprise. Both polymers retain their liquid nature to high molecular weight as a consequence of the extreme flexibility of the siloxane backbone and the low intermolecular forces between groups that are inherent in both hydrocarbon and fluorocarbon substituents. Thus the LeGrand-Gaines extrapolation to infinite molecular weight is not an

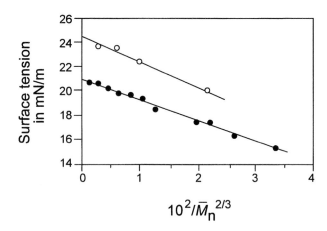

Figure 15.6 Surface tension of liquid poly(dimethylsiloxane) (PDMS) (solid circles) and poly[methyl(3,3,3-trifluoropropyl)siloxane] (PMTFPS) (open circles). (Reproduced with permission from reference 50. Copyright 1993 Wiley-VCH.)

extreme one, and, as shown in Figure 15.6,[50] despite the fewer data points for PMTFPS, there is no doubt that it has a higher liquid surface tension than PDMS has (recall that for solid surface energies, we were trying to resolve whether PMTFPS has a similar or lower value than PDMS has).

Considering Zisman's order of the effect of substituent groups, when we go from PDMS to PMTFPS, we remove one methyl and replace it by one trifluoromethyl of lower surface energy and two methylene groups of higher surface energy. The net effect in the liquid state is a modest, yet significant, rise in surface energy. Note that an uncompensated dipole between the fluorocarbon and hydrocarbon entities has also been introduced. Is some difference implied in orientation or packing of the trifluoropropyl pendent groups in the liquid and solid states? Maybe, but the most conservative position is to assert that no explanation is necessary. We have learned that solid surface energy values depend on the choices of contact angle test liquids and interpretative equations. They may only be compared with identical treatments of other polymers, and certainly numerical equivalence with directly measured liquid values is not to be expected. The fact that there is much closer numerical equivalence in the case of PDMS must be considered coincidental. In this regard, note that poly(tetrafluoroethylene) (PTFE) also shows a considerably higher liquid surface tension value at infinite molecular weight from extrapolation of n-perfluoroalkane data than do any of the contact angle derived solid surface energies. Further speculation is left to the reader. Some, unfamiliar with organosiloxane chemistry, might wonder why not replace the CH_3- group directly with CF_3- to obtain improved hydrocarbon resistance. However, the ethylene bridge is essential to obtain a polymer with adequate thermal and hydrolytic stability.

Summary

The aim of this chapter is to enable readers to make practical determinations of the surface energy of polymers they are investigating. If their material is in liquid or melt form, then, in principle, there is no problem in obtaining a meaningful quantitative value of the surface energy, although there may be practical difficulties in gaining access to a pendent drop or sessile drop apparatus that can cope with the viscosity and temperature range necessary. Should the material be solid, absolute values of surface energy are hard to come by. Presently, the most rigorous way of obtaining meaningful data seems to be the possiblity of fabricating the material into forms suitable for JKR contact mechanical evaluation.

However, the reality is that most practitioners will have solid samples that are not amenable to the JKR technique and must resort to contact angle techniques. All of the equations involved in such analyses are semiempirical. The values obtained for surface energy depend on the choice of equations and liquids employed. If the polymer is a nonpolar hydrocarbon-based material, most of the approaches described here will give comparable results, but higher-surface-energy and more polar polymer data will demonstrate numerous discrepancies. No one approach is "the best". There is a current trend to favor that introduced by van Oss and coworkers[31] based on Lifshitz's theory, but this, too, is not an exact theory. Absolute values of the electron acceptor and donor terms used in eq 15 are not known. One practical argument for using the geometric mean approach of Owens and Wendt[26] is that there is a great deal of reported data available for comparative purposes, provided that the inherent limitations in its use are recognized. An acceptable pragmatic compromise is to ensure that the actual contact data are also reported, and this will enable researchers to apply future improved equations, despite the literal validity of purist positions such as Erbil's,[51] who asserts that "the geometric mean approach has been in constant use for about 25 years even though many articles have been published proving it is incorrect. Most of the leading scientists in the surface science field abandoned the use of the Owens and Wendt approach for this reason".

The unfortunate reality is that no wholly rigorous contact angle approach has yet arisen to replace the Owens and Wendt approach. Della Volpe et al.[52] state in their discussion of contact angles that "all the calculated values of the surface energy of solids published in the literature are 'apparent'. This conclusion can be difficult to accept, although it is a simple truth". But this is true of other polymer properties and is no reason to not make the measurements. Kwok and Neumann[33] offer hope that "literature contact angles do not have to be discarded completely; they can be used to determine solid surface tensions, with caution". Even if the absolute value of the derived surface energy may be in doubt, contact angle studies of polymer surfaces can be very informative. Accept that neither the theory nor your sample is perfect, and welcome contact angle hysteresis as an insight into that imperfection.

References

1. Chaudhury, M. K. *Mat. Sci. Eng.* **1996**, *R16*, 97–159.
2. Young, T. In *Miscellaneous Works*, Peacock, G., Ed.; J. Murray: London, 1855; Vol. 1.
3. Johnson, K. L.; Kendall, K.; Roberts, A. D. *Proc. R. Soc.* (London) **1971**, *A324*, 301–313.
4. Adamson, A. W.; Gast, A. P. *Physical Chemistry of Surfaces*, 6th ed.; John Wiley and Sons: New York, 1997.
5. Wu, S. *J. Macromol. Sci.-Revs. Macromol. Chem.* **1974**, *C10*(1), 1–73.
6. Wu, S. *J. Colloid Interface Sci.* **1969**, *31*, 153–161.
7. Roe, R-J. *J. Phys. Chem.* **1968**, *72*, 2013–2017.
8. Ambwani, D. S.; Fort, Jr. T. In *Surface and Colloid Sci.*; Good, R. J., Stromberg, R. R., Eds.; Plenum Press: New York, 1979; Vol 11, pp 93–119.

9. Anastasiadis, S. H.; Chen, J. K.; Koberstein, J. T.; Sohn, J. E.; Emerson, J. A. *Polym. Eng. Sci.* **1986**, *26*, 1410–1418.

10. Bashforth, F.; Adams, J. C. *An Attempt to Test the Theories of Capillary Action;* Cambridge University Press: Cambridge, UK, 1883.

11. Andreas, J. M.; Hauser, E. A.; Tucker, W. B. *J. Phys. Chem.* **1938**, *42*, 1001.

12. Padday, J. F. *J. Electroanal. Chem. Interfacial Electrochem.* **1972**, *37*, 313–316.

13. Dettre, R. H.; Johnson, Jr. R. E. *J. Colloid Interface Sci.* **1969**, *31*, 568–569.

14. LeGrand, D. G.; Gaines, Jr. G. L. *J. Colloid Interface Sci.* **1969**, *31*, 162–167.

15. Bender, G. W.; LeGrand, D. G.; Gaines, Jr. G. L. *Macromolecules*, **1969**, *2*, 681–682.

16. Rastogi, A. K.; St. Pierre, L. E. *J. Colloid Interface Sci.* **1971**, *35*, 16–22.

17. Shafrin, E. G.; Zisman, W. A. *J. Phys. Chem.* **1960**, *64*, 523.

18. Kobayashi, H.; Owen, M. J. *Makromol. Chem.* **1993**, *194*, 259–267.

19. Langmuir, I. *J. Amer. Chem. Soc.* **1916**, *38*, 2286.

20. Owen, M. J. In *Physical Properties of Polymers Handbook;* Mark, J. E., Ed.; AIP Press: Woodbury, NY, 1996; pp 669–676.

21. Shafrin, E. G. In *Polymer Handbook*, 2nd ed.; Brandrup, J., Immergut, E. H., Eds.; Wiley: New York, 1975; Vol 3, pp 221–228. [Later editions of this handbook do not list critical surface tensions of wetting but focus on the harmonic mean and geometric mean approaches described next in this chapter.]

22. Shafrin, E. G. In *Handbook of Adhesives*, 2nd ed.; Skeist, I., Ed.; Van Nostrand Reinhold: New York, 1977; pp 67–71.

23. Dupre, A. *Theorie mechanique de la chaleur;* Gauthier-Villars: Paris, 1869; p 369.

24. Girifalco, L. A.; Good, R. J. *J. Phys. Chem.* **1957**, *61*, 904.

25. Fowkes, F. M. *Ind. Eng. Chem.* **1964**, *56*, 40.

26. Owens, D. K.; Wendt, R. C. *J. Appl. Polym. Sci.* **1969**, *13*, 1741–1747.

27. Kaelble, D. H. *J. Adhesion* **1970**, *2*, 66–81.

28. Wu, S. *Polymer Interface and Adhesion;* Marcel Dekker: New York, 1982.

29. Chaudhury, M. K.; Whitesides, G. M. *Science* **1992**, *255*, 1230–1232.

30. Balkenende, A. R.; van de Boogaard, H. J. A. P.; Scholten, M.; Willard, N. P. *Langmuir* **1998**, *14*, 5907–5912.

31. Good, R. J.; Chaudhury, M. K.; van Oss, C. J. *Adv. Colloid Interface Sci.* **1987**, *28*, 35–64.

32. Kwok, D. Y.; Neumann, A. W. *Coll. Surf.* **2000**, *161*, 31–48.

33. Kwok, D. Y.; Neumann, A. W. *Coll. Surf.* **2000**, *161*, 49–62.

34. Neumann, A. W.; Good, R. J. In *Surface and Colloid Science;* Good, R. J., Stromberg, R. R., Eds.; Plenum Press: New York, 1979; Vol. 11, pp 31–91.

35. Fort, Jr. T.; Patterson, H. T. *J. Colloid Sci.* **1963**, *18*, 217.

36. Amirfazli, A.; Kwok, D. Y.; Gaydos, J.; Neumann, A. W. *J. Colloid Interface Sci.* 1998, *205*, 1–11.

37. Johnson, Jr. R. E.; Dettre, R. H. In *Surface and Colloid Science;* Matijevic, E., Ed.; Wiley-Interscience: New York, 1969; Vol. 2, pp 85–153.

38. Washo, B. D. *Org. Coatings Appl. Polym. Sci. Proc.* **1982**, *47*, 69–72.

39. Swain, P. S.; Lipowsky, R. *Langmuir* **1988**, *14*, 6772–6780.

40. Andrade, J. D.; Smith, L. M.; Gregonis, D. E. In *Surface and Interfacial Aspects of Biomedical Polymers;* Andrade, J. D., Ed.; Plenum Press: New York, 1985; pp 249–292.

41. Holly, F. J.; Refojo, M. F. *J. Biomed. Mater. Res.* **1975**, *9*, 315–326.

42. Shanahan, M. E. R.; Carre, A. *Langmuir* **1994**, *10*, 1647–1649.

43. Tomasetti, E.; Rouxhet, P. G.; Legras, R. *Langmuir* **1998**, *14*, 3435–3439.

44. Mittal, K. L.; van Ooij, W. J. *J. Adhesion Sci. Technol.* **1999**, *13*(10), 1067–1240.

45. Mittal, K. L.; van Ooij, W. J. *J. Adhesion Sci. Technol.* **1999**, *13*(12), 1363–1533.

46. Mittal, K. L.; van Ooij, W. J. *J. Adhesion Sci. Technol.* **2000**, *14*(2), 145–314.

47. Mangipudi, V. S.; Tirrell, M.; Pocius, A. V. *Proc. Annu. Meet. Adhes. Soc.* **1996**, *19*, 7–9.

48. Mangipudi, V.; Tirrell, M.; Pocius, A. V. *J. Adhes. Sci. Technol.* **1994**, *8*, 1251–1270.

49. Ahn, D.; Shull, K. R. *Langmuir* **1998**, *14*, 3646–3654.

50. Kobayashi, H.; Owen, M. J. *Makromol. Chem.* **1993**, *194*, 1785–1792.

51. Erbil, H. Y. In *Handbook of Surface and Colloid Chemistry*, Birdi, K. S., Ed.; CRC Press: Boca Raton, FL, 1997; pp 265–312.

52. Della Volpe, C.; Deimichei, A.; Ricco, T. *J. Adhesion Sci. Technol.* **1988**, *12*, 1141–1180.

16

Electron, Ion, and Mass Spectrometry

ILARIO LOSITO
LUISA SABBATINI
JOSEPH A. GARDELLA JR.

In the past three decades the need for chemical information related to polymer surfaces has been driven by the development of complex polymeric materials in different fields of application, such as biomaterials, coatings, adhesives, and composites.[1] Besides chemical composition, more sophisticated information about surface structure is required in these cases: depth profiles, functional group orientation, domain size, molecular weight distributions at or near the surface, and the like.

A number of surface analytical techniques, whose first developments date back to the end of the 1950s, are now available to meet this demand. Among them, X-ray photoelectron spectroscopy (XPS; also known as ESCA, electron spectroscopy for chemical analysis), whose first applications to polymer surfaces were described in the early years of development,[2] can be considered the leading technique, and it is used almost as a routine analytical tool in this field.

Different reasons contribute to this primary role, including the ability to provide elemental and, in many cases, molecular information at a relatively low level of complexity, along with the low damage induced on polymer samples, at least if compared with other surface spectroscopies. Of course, the low degree of sample pretreatment required before analysis makes XPS, and most surface spectroscopies as well, particularly suitable for the characterization of polymers, which, in many cases, are hardly soluble in common solvents. Moreover, recent instrumental improvements, in terms of both energy and spatial resolution and increased signal-to-noise ratio, have strengthened XPS analytical capabilities, facilitated shorter acquisition times, and reduced sample degradation.[3]

Other surface analytical techniques, like secondary ion mass spectrometry (SIMS) and ion scattering spectrometry (ISS), based on ion sputtering and scattering from solid surfaces, respectively, have also been applied to polymer analysis, although less extensively than XPS. For the SIMS technique, the introduction of low primary ion dosages (the

static mode[4]), time-of flight mass analyzers, having good mass resolution at high mass ranges, and imaging capabilities have greatly enhanced its analytical potential for complex polymers and its importance among surface techniques. In contrast, ISS has a limited but peculiar field of application, being able to provide special information on polymer structures, like functional group orientation or segregation at the outermost layers in multicomponent materials.[5]

The main characteristics of XPS, SIMS, and ISS are summarized in Table 16.1. In this chapter, these techniques are presented as the main analytical tools applied to investigate the chemical composition and structure of polymer surfaces.

As a general scheme, a description of the theoretical aspects and instrumentation commonly used is given in the first part of each section. Practical aspects related to sample analysis (preparation, mounting, control of degradation) are then presented, with special attention to polymer samples. The last part of each section is focused on applications to specific, "real" polymers, as examples of how even complex analytical problems related to polymer surfaces can be solved by XPS, SIMS, and ISS. As a conclusion, the possible future perspectives of the three techniques in the field of polymer analysis are presented.

X-ray Photoelectron Spectroscopy

Fundamental Aspects

Starting from the pioneering work by Siegbahn and his co-workers at Uppsala University,[6] XPS has grown rapidly in the last three decades, becoming the most important technique for surface analysis. Its theoretical basis can be found in Einstein's work, early in this century, on the photoelectric effect, but only in the late 1960s did progress in ultra-

Table 16.1 Basic characteristics of X-ray photoelectron spectroscopy (XPS), secondary ion mass spectrometry (SIMS), and ion scattering spectrometry (ISS)

	XPS	SIMS	ISS
Conditions of analysis	High vacuum	Ultra high vacuum	Ultra high vacuum
Resolution	0.6 eV	0.1-1 amu	Variable
Element range	No detection for H, He	all elements plus isotopes	No detection for H, He,
Molecular information	Chemical shifts	Molecular weight, fragmentation	—
Detection limit	% of monolayer	ppm/ppb (elemental), 0.01 monolayer (molecular)	% of monolayer
Lateral resolution	5 μm	10 nm (atomic)	None
Depth sensitivity	50-100 Å	10 Å	3-5 Å
Sample damage	Low (higher with un-monochromatized X-ray sources)	High (can be lowered under static conditions)	High
Major uses	Elemental and chemical analysis, electronic structure	Low detection limits, molecular ions fragmentation	Atomic orientation

high vacuum technology make this phenomenon applicable also to the chemical analysis of solid surfaces. Indeed, an XPS analysis is performed by irradiation of a sample with X-rays, under ultra-high vacuum conditions. Electrons are emitted from the sample's atoms as a result of their interaction with X photons (photoemission), bearing qualitative and quantitative information on the surface chemical composition.

The surface sensitivity of the technique arises from the relatively small inelastic mean free path (IMFP) of electrons inside matter in a condensed phase. The IMFP (also indicated by λ), corresponds to the average distance (in nanometers) covered by an electron between successive inelastic collisions and depends both on the material and on the electron energy,[7,8] which generally falls in the range of 10 to 40 Å. As a consequence, although X-ray photons usually penetrate into a solid sample to a depth of several microns, only electrons emitted in the topmost atomic layers are able to overcome the sample's work function and to emerge, without significant loss of their initial kinetic energy (KE). Other electrons, subjected to inelastic scattering after photoemission, can also escape from the sample, but they contribute to the usually structureless background of an XPS spectrum.

The XPS sampling depth is then defined[7] as the sample thickness from which 95% of the detected photoelectrons originate. It corresponds to three times the IMFP, thus a surface layer about 100 Å thick can be usually analyzed by XPS, although information from deeper layers can be obtained if higher-energy X-rays are adopted (as shown later in this chapter).

In a typical XPS experiment, electrons are collected and their intensity is measured as a function of the relevant KE. As a first approximation, the following equation can be used to correlate the KE to the binding energy (BE) of the photoemitted electrons:

$$E_B = h\nu - E_K \qquad (1)$$

where $h\nu$ = X-ray photon energy.

Typically, as noted by Siegbahn,[6] soft X-ray sources, such as AlKα (1486.6 eV) and MgKα (1253.6 eV), have the necessary intensity (flux) and narrow linewidth to be useful; however, other X-ray sources are also used for particular problems, such as to probe to deeper depths. The characteristics of the most common X-ray sources for XPS are reported in Table 16.2. For each element in the sample, with the exception of hydrogen and helium, electrons can be photoemitted from one or more atomic levels, each having a specific binding energy. An XPS spectrum, then, is essentially a plot of the intensity of photoemitted electrons versus their kinetic or binding energy.

In a typical experiment, a low-resolution "survey" spectrum is usually obtained in the first stage of analysis, collecting electrons over the entire range of binding energies accessible with the X-ray source (usually from 0 to a few thousands of eV). An example is given in Figure 16.1.

Table 16.2 Characteristics of X-ray sources for photoelectron spectroscopy (XPS)

Anode material	Energy/ eV	Linewidth/ eV	Intensity relative to MgKα
MgKα	1253.6	0.7	1.0
AlKα*	1486.6	0.85	0.5
SiKα	1739.5	1.0	0.19
ZrLα	2042.4	1.7	0.05
AuMα	2122.9	2.4	0.03
AgLα*	2984.3	Not available	0.02
TiKα	4510.9	2.0	0.05
CrKα	5417.0	2.1	Not available

*Often used with a monochromator.

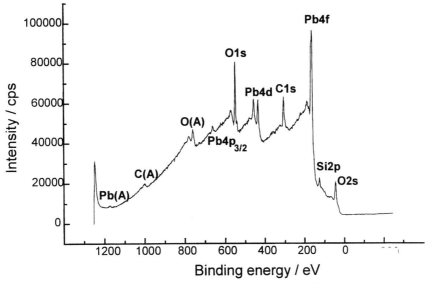

Figure 16.1 Example of X-ray photoelectron spectroscopy (XPS) survey scan. Key: (A) = Auger peaks.

Peaks with highest intensities usually correspond to primary photoemission transitions (those having the highest cross-sections for photoionization) and can be used for a first identification of the elements on the sample surface. This is performed by comparing the relevant BE with tabulated reference values,[9] provided the former has been corrected for an instrument-based work function and the energy scale has been calibrated using an internal or external standard (see later in this chapter).

At least two other types of lines are also present in a typical XPS survey spectrum and can be of help in confirming the preliminary identification of the elements in the sample: auger lines and secondary photoemission lines—that is, lines related to transitions with lower probability.

A good correlation between the BE values of more than one peak and the corresponding values reported in the literature usually rules out errors in spectral interpretation, thus providing a more accurate assignment of the signal to a particular element.

A typical example of this approach involves elements for which the most intense line corresponds to the 2p core level, but a significant signal is also observed for the 2s line, including Si, P, S, and Cl. A special case is that of Al, for which the peak intensity of the 2s signal can be greater than the 2p line. In this case, a higher degree of information can be obtained if the relative intensities of the two lines are also considered. Provided their binding energies are not very different (see below), their intensity ratio is related to their photoemission cross-sections, thus having elemental specificity. The binding energies and the sensitivity factors[10] (the latter are related to the photoemission cross-section and are used to normalize XPS signal intensities, as explained later in this chapter) for the 2p and 2s

photoelectron levels of the elements cited are reported in Table 16.3.

Auger lines are another useful feature of an XPS survey spectrum. They arise from a complex relaxation phenomenon following photoionization that involves three electronic levels. After photoemission of a core electron from a level A in an atom, a vacancy is generated, and an electron on a higher level (B) undergoes a transition to fill the hole. A certain amount of energy is then released and can be used in the emission of another electron, from the same or a higher level (C). This is called "Auger electron", after the name of the physicist who discovered this effect,[11] and its

Table 16.3 Binding energies (eV) and sensitivity factors relevant to 2p and 2s photoelectron levels of selected elements

Photoelectron line	Binding energy/eV[a]	Sensitivity factor[b]
Si2p	103	0.27
Si2s	153	0.26
Al2p	74	0.185
Al2s	119	0.23
P2p	134	0.39
P2s	191	0.29
S2p	166	0.54
S2s	229	0.33
Cl2p	201	0.73
Cl2s	270	0.37

a. Values taken from reference 9 and referred to C1s = 287 eV.
b. Factors taken from reference 10 and obtained empirically on two different XPS spectrometers operated at constant pass energy (see text for details).

kinetic energy can be expressed, in a condensed way, by the relation $KE_{ABC} = E_A - E_B - E_C^*$, where E_A, E_B, and E_C^* are the binding energies of the three levels involved in the process (E_C is starred to indicate that its value is referred to an ionized atom, due to the presence of a hole in the B level).

Since only the differences between three binding energies levels are involved in determining the final KE, Auger lines recorded with two different X-ray sources (which is an easy procedure when a "twin anode" source, like Mg/Al, is available) have the same kinetic energy and can easily be recognized if the spectra are superimposed. Like the core lines, Auger signals have elemental specificity, thus helping in the determination of the atomic composition of the sample.

Once obtained, the information originating from the XPS survey spectrum can be enriched by looking at the specific lines in more detail, under higher resolution conditions.

Multiple, signal averaged spectra for a single transition can lead to a precise determination of the peak binding energy (which can provide chemical shift information) and peak area, the latter being the first stage for a quantitative (elemental) surface analysis of the sample. Alternatively, additional structures accompanying the photoelectron signals are evidenced in high-resolution spectra. These can provide additional characteristic chemical information. The most common ones arise from coupling between the orbital and the spin angular momenta of electrons (spin-orbit coupling), when the angular quantum number of the emitted electron is equal or greater than 1 (i.e., orbitals p, d, and f are involved). In these cases, two different final states are possible (since the unpaired electron resulting from the photoemission process can have two different spin numbers), and two signals are observed. Usually they are distinguished by adding a subscript to the orbital name, corresponding to the total angular quantum number ($J = 1 + s$, 1 and s being the orbital and spin quantum numbers, respectively). As an example, if a p core electron is considered, the subscripts 1/2 (1 = 1, $s = -\frac{1}{2}$) and 3/2 (1 = 1, $s = \frac{1}{2}$) are used. The intensities of the two peaks are proportional to $2J + 1$, and for a p electron their ratio is 2, for a d electron 3/2, and so on.

The extent of spin-orbit splitting strongly depends on the element: sometimes it is so small that two almost overlapped peaks are observed, or, equivalently, a single, asymmetric peak (as for the Mg2p, Si2p, Al2p signals) is obtained, unless very high resolutions are available (e.g., when X-ray synchrotron radiation is used). However, many times, the energy difference between peaks is large enough to be characteristic of chemical state of the element.

Other common additional structures in the XPS signal are due to the "shake-up" and "shake-off" processes. A shake-up process occurs when the emission of a photoelectron from a core level is accompanied by an electronic transition between two higher energy levels of the ionized atom. The final result is then a decrease in the kinetic energy of

the photoelectron (the difference being used for the transition). In a shake-off process, a portion of the kinetic energy of the photoelectron is used to remove another electron from the ionized atom. Shake-up/off phenomena then result in additional peaks with apparently higher binding energies. They are very common for transition and f-series elements, their intensity being sometimes almost comparable to the primary photoelectron peak in these cases. As an example, shake-up peaks relevant to Cu2p are shown in Figure 16.2.

Other, less common structures can be found in XPS high-resolution spectra due to phenomena such as multiplet splitting and plasmon loss; interested readers can find a detailed description of these and other XPS additional spectral features in fundamental books like that in reference 12.

A special case is represented by the so-called ghost lines, which are due to contamination of the X-ray source. If a material other than the primary emitter is present on the source anode, X-radiation with a different energy is also generated, and photoelectrons are emitted by it from the sample. Other lines due to the same elements are then found at a different energy in the spectrum, thus complicating its interpretation. X-rays emitted by the copper substrate of aged Mg or Al anodes are often responsible for ghost lines, which can even be of help for the identification of elements in a sample, although this is not a conventional procedure.

Intensity of XPS Signals and Quantification

As already cited, a surface quantitative analysis is also possible by XPS, due to the correlation existing between the intensity of a photoelectron signal, I, and the concentration of the relevant element, reported as atomic density, N. The general relation between I and N for a signal A can be expressed by means of the following equation:[12]

$$I_A = \sigma_A D(E_A) \int_{\Gamma=0}^{\pi} \int_{\phi=0}^{2\pi} L_A(\Gamma) \int_{y=-\infty}^{+\infty} \int_{x=-\infty}^{+\infty} J_0(x,y) T(x,y,\Gamma,\phi,E_A)$$

$$\times \int_{z=0}^{+\infty} N_A(x,y,z) \exp[(-z/\lambda(E_A)\cos\theta]dx\ dy\ dz\ d\phi\ d\Gamma \quad (2)$$

where σ_A is the photoemission cross section for the core level considered, D is the electron detector efficiency (dependent on the pass energy of electrons, E_A), J_0 is the X-ray flux on the analyzed surface, L_A is a function related to the dependence of photoemission on the angle Γ between the X-ray and the electron direction, T is the transmission function of the electron analyzer (also depending, among other factors, on E_A), λ is the inelastic mean free path, θ is the take-off angle (that is, the angle between the sample surface and the direction of the analyzed electrons), and ϕ is the dihedral angle between two planes: the first is defined by the X-rays' and the photoelectron's directions; the second is defined by the X-rays' direction and the normal to the sample surface passing through the point of X-rays incidence.

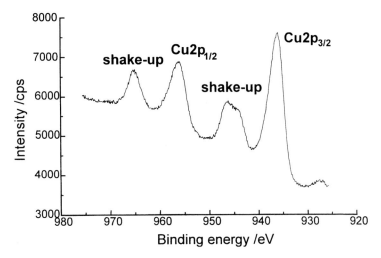

Figure 16.2 Detailed spectrum for the Cu2p signal of copper(II) oxide; the shake-up features are also indicated.

If the elemental composition of the sample is homogeneous throughout the sampling depth, the integral in eq 2 can be calculated; and the following equation is obtained, in which the L_A and J_0 functions are considered to be constants:

$$I_A = \sigma_A D(E_A) L_A J_0 N_A \lambda(E_A) \cos\theta \int_{y=-\infty}^{+\infty} \int_{x=-\infty}^{+\infty} T(x,y,E_A) dx\ dy \quad (3)$$

The presence of several instrumental factors in both equations makes the calculation of absolute values for N almost impossible. Rather, either atomic ratios or percentages are calculated from the signal intensities, using the "sensitivity factor" approach. Indeed, some of the factors in eq 2 or eq 3—for example, the detector efficiency (provided the final energy of electrons reaching the detector is the same, which is a frequent condition) and the X-ray flux—can be canceled if two intensities are ratioed. Alternatively, the parameters that depend on the specific photoelectron line are taken into account by the *sensitivity factor*, which includes contributions of σ, λ, T, and, eventually, L (the latter depends on the core orbital excited and on the usually negligible crystallinity of the sample).

As a result, if two signals A and B are considered, the atomic ratio between the relevant elements in the sample is provided by the ratio of their intensities I divided by the relevant sensitivity factors S:

$$\frac{N_A}{N_B} = \frac{I_A/S_A}{I_B/S_B} \quad (4)$$

One of the first sets of XPS sensitivity factors was obtained by Wagner and coworkers on a specific instrument by using a semiempirical approach.[10]

It has been shown[13] that, due to similarities in the KE-dependence of the analyzer transmission functions, the same set can be used on different spectrometers with reasonable accuracy. Indeed, many commercial XPS instru-

ments have their own set of sensitivity factors, usually implemented into the data analysis software, so that the atomic composition of a sample can be directly obtained from the signal intensities. Usually, the lowest concentrations that can be determined are of the order of a few tenths of atomic percent.

Depth Profiling

The basic assumption of eq 2 is that the sample composition is homogeneous throughout the XPS sampling depth (usually 100 Å), but this is not the general case. It is very common, especially in polymer surface analysis, that a concentration profile exists near the surface, and different methods have been developed in order to obtain this kind of information by XPS.

A typical, nondestructive approach is based on angle-resolved measurements. The general principle is illustrated in Figure 16.3. In this case, XP spectra are recorded at different take-off angles: since the inelastic mean free path (and then the sampling depth) does not depend on this angle, the rotation of the sample results in different sampling depths (estimated along the normal to the sample surface). Lower take-off angles correspond to smaller sampling depths, as shown in the figure.

Due to the presence of an attenuation effect (expressed by the negative exponential function in eq 1), quantitative information obtained at a certain angle is actually a convolution of those signals that arises from all the sample layers between the corresponding sampling depth and the surface. Nevertheless, special deconvolution programs have been developed to recover the $N(x)$ function, starting from angle-resolved measurements (see reference 14 as an example).

The dependence of the inelastic mean free path of electrons on their energy is exploited in other nondestructive methods to obtain depth information. For example, the in-

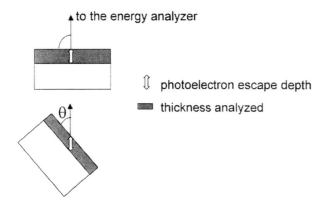

to the energy analyzer

⇕ photoelectron escape depth

▨ thickness analyzed

θ

Figure 16.3 Principle of angle-resolved X-ray photoelectron spectroscopy (XPS) measurements. Key: the thin arrow shows the direction of photoelectron collection, θ is the take-off angle; lower θ values correspond to smaller sample thicknesses analyzed.

tensities of two photoelectron lines of the same element but that have very different kinetic energies can be influenced by the vertical distribution of the element in the sample. The photoelectron line with lower kinetic energy will be more "surface sensitive", due to the lower escape depth of the corresponding electrons. If a fluorinated polymer is considered, the ratio between the F1s and F2s line intensities can provide useful information if the fluorine depth distribution is not homogeneous. If a decrease in fluorine has occurred, starting from the external surface (i.e., as a consequence of a plasma treatment on a fluoropolymer), the F1s/F2s ratio will show a decrease with plasma exposure time, since the F1s line (lower kinetic energy) is more surface sensitive than F2s.

The same principle is also involved in the energy-dependent depth profiling, which is performed by using different X-ray sources on the same sample. For example, TiKα ($h\nu = 4510.9$ eV) sources are used to obtain information on deeper layers of polymer samples,[15] for they provide sampling depths almost twice those of the more common Mg/Al Kα sources. Using higher energy sources also provides access to Auger lines that are not observed with MgKα or AlKα, which can be helpful in assigning chemical states through the Auger parameter (see later in this chapter).

Signal-to-noise (S/N) problems due to the low X-ray flux of Ti sources have been recently overcome with the advent of multichannel detectors, which reduce the acquisition time required to obtain a good S/N ratio. Acquisition time can be a critical parameter when XPS is applied to polymers, due to X-ray damage.

Depth profiling based on sputtering is quite different from the approaches presented so far and can provide information on the composition of deep sample layers.[16] By definition, sputtering is a destructive technique, since layers are progressively removed from the sample "in situ"—that is, under the ultra-high vacuum conditions of the XPS spec-

trometer. Ar or Ne ion beams, with energies between 1 and 10 keV, are typically used for this purpose, and acquisition stages are alternated with sputtering ones during the experiment, thus obtaining the composition of the sample as a function of sputtering times. Provided the relationship between eroded depth and sputtering time is known, a concentration profile can be obtained for the sample at depths much higher than those accessible with nondestructive approaches.

The assumption of this approach is that, ideally, the original sample composition is not altered while it is bombarded with the ion gun. Yet, depending on the sputtering yields, which are different for each element (and often influenced by the chemical environment, too), a modification of the initial composition may occur. The surface will become enriched with the element most difficult to remove, and thus a correction of the initial data has to be made. This drawback (and other collateral phenomena) has limited the application of sputtering as a routine technique for depth profiling in XPS. This is particularly evident in surface analysis of polymer samples, for which surface damage and alterations are a major concern.

XPS Instrumentation

As shown in the schematic drawing of Figure 16.4, five main components can be recognized in a typical XPS spectrometer:

- An ultra-high vacuum apparatus (e.g., the Rowland Circle and electron analyzer)
- Sample introduction and preparation chambers, which may contain sample holders and manipulators

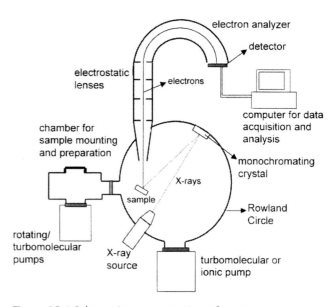

Figure 16.4 Schematic representation of an X-ray photoelectron spectroscopy (XPS) instrument equipped with a monochromatized X-ray source.

- An X-ray source
- A kinetic energy analyzer and a detector for photo-electrons
- A computer-based system for hardware control and data acquisition

A brief description of these components follows; details can be found in fundamental books (see reference 12).

Vacuum System

Ultra-high vacuum conditions are required to minimize the collisions between photoemitted electrons and residual gas molecules inside the spectrometer, which would cause losses in the photoelectron initial kinetic energies. For similar reasons, the X-ray source is also kept under ultra-high vacuum.

Turbomolecular, diffusion, cryogenic, or ion pumps provide ultimate pressures around $10^{-8}/10^{-9}$ torr, under which an XPS analysis is usually performed; rotating pumps (providing high vacuum conditions) are also used, connected either to the turbomolecular ones or directly to the sample introduction chamber. The sample chamber is opened to the atmosphere when a new sample has to be introduced into the spectrometer. After high vacuum conditions are restored (usually by means of a rotating pump), a sample holder transfers the sample into the analysis chamber, where a manipulator enables translations and rotations of the sample stage according to alignment and analysis (i.e., angle-resolved measurements) requirements. In modern instruments, these operations, including remote control of the valves placed between the different chambers, can be accomplished by a computer.

Sometimes a sample carousel is also available, enabling the introduction of more than one sample at the same time. In this case, the samples can be sequentially analyzed by rotating (either manually or automatically) the carousel, thus exposing a different sample to the X-ray source and to the electron analyzer. The main advantage consists in reducing the time required to reach vacuum conditions suitable for XPS analysis, which is, in many cases, the most time-consuming step.

X-ray Sources

Two kinds of X-ray sources are commonly used in XPS spectrometers: conventional and monochromatised. In a conventional source, electrons emitted by a heated filament are accelerated by a high voltage (>10 kV) against a target anode. The collision produces a sudden decrease of the electron energy, which is turned either into heat (the anode is water-cooled to avoid melting) or into a broadband radiation, called Bremsstrahlung. At the same time, electronic transitions are induced in the target atoms and X-ray fluorescence lines characteristic of the anode material are emitted.

An aluminum foil (<10 μm in thickness) is used as the source "window", filtering high-energy electrons and preventing the Bremsstrahlung radiation from striking the sample. The radiation of conventional sources is then represented by the anode X-ray fluorescence radiation, usually a main line accompanied by "satellites", or lines with higher energy but lower intensity. Both stimulate the emission of photoelectrons, thus satellite peaks are to be expected in an XP spectrum obtained with a unmonochromatized source.

Another important aspect is the X-ray radiation linewidth. In the case of Mg and Al, which are the most common materials used in sources for XPS, the most intense radiation emitted is actually an overlap of two major lines, $K\alpha_1$ and $K\alpha_2$. This results in inherent linewidths of 0.7 and 0.85 eV, respectively, which represent a limit for spectral resolution.

Monochromatized sources, in which X-radiation is back diffracted by a quartz crystal, have been introduced[17] to improve spectral resolution. Aluminum is commonly adopted as the target material: due to the AlKα wavelength and the lattice spacing of the 1010 planes in quartz, the Bragg condition is satisfied for an angle of 78°. In this case, the inherent linewidth can be reduced to less than 0.4 eV.

The introduction of the Rowland Circle monochromator mounting for the quartz crystals, along with focused electron guns for the source itself, has brought remarkable improvements also in X-ray focusing, leading to spot sizes as small as 10 μm for monochromatized X-rays. A major drawback is represented by the lower flux intensities, if compared with conventional sources, but they are partially compensated by improving photoelectron collection and detection (see later in this chapter).

Warning: As already cited, all kinds of X-ray sources involve the use of very high voltages for their operation. Although safety devices are always implemented in commercial XPS instrumentation to prevent the accidental exposure of users to electrical hazard, care must be taken when operating the radiation source.

Electron Energy Analyzer

The electron analyzer is one of the main components of an XPS spectrometer. It consists of a system of electrostatic lenses and a kinetic energy filter. The lenses collect and focus photoelectrons from the sample, driving them into the energy filter, where electrons are separated according to their kinetic energy before entering the detector.

The most common KE filter, in modern XPS instrumentation, is the hemispherical sector analyzer, which consists of two concentric hemispheres with entrance and exit slits. A potential difference is applied between the two hemispheres, so that only electrons with a particular kinetic energy can escape the analyzer through the exit slit, without impinging on one of the two hemispheres. A potential scan is then required, in order to enable electrons with different energies to enter into the detection system, thus obtaining a spectrum. Two approaches are usually adopted.

In the fixed retarding ratio (FRR) mode, photoelectrons are retarded in the same way (a common retarding ratio is 3) by electric fields applied through the electrostatic lens system, then a potential scan is applied to the sector analyzer. In the fixed analyzer transmission (FAT) mode, a potential scan is applied to the lenses, so that all the photoelectrons, although they have different KE values when they leave the sample, enter the analyzer with the same energy, called "pass energy" (E_0).

The resolution for a hemispherical sector analyzer, indicated as $\Delta E/E_0$, where ΔE is the spectral resolution, is a function of the geometrical characteristics, being directly proportional to the width of the entrance slit and inversely proportional to the radius of the hemispheres. Higher E_0 leads to poor spectral resolution, as ΔE is increased. For this reason, the FAT mode using relatively low pass energies (from a few to 50 eV) is adopted to obtain high-resolution spectra. It should be pointed out that the spectral resolution is the same for all the regions in this case, E_0 being a constant.

Large electron analyzers have been introduced in the last few years in XPS instrumentation, thus providing either higher spectral resolutions without using low pass energies (with their low S/N ratios), or higher throughputs at the same resolution.

Detectors

The typical detector for photoelectrons is an electron multiplier, usually a channeltron, placed at the exit slit of the analyzer. The use of multichannel detectors (MCDs) has recently become very common. The basic principle of MCDs is that photoelectrons with different energies will exit the energy analyzer in different positions on the same plane. If detecting plates are placed in those positions, the photoelectron spectra will be obtained at the same time, instead of scanning the potential difference applied to the analyzer (or to the electrostatic lenses). This results in lower acquisition times or higher S/N ratios, if the same time is adopted.

Software for Data Acquisition and Analysis

Most commercial XPS spectrometers now available can be almost completely computer-operated. The acquisition software is usually able to control the X-ray source, the energy analyzer, and the detector, thus acquiring spectra according to the sequence decided by the operator. When a rotating carousel is available, the acquisition software is usually able to rotate the carousel, and more samples can be analyzed sequentially in a fully automated procedure.

The data analysis software for XPS spectrometers includes a basic procedure for elemental surface analysis: the areas of all the (primary) photoelectron lines are used for the quantitation, usually based on sensitivity factors already implemented in the software. Subtraction of baseline, X-ray satellites and background, and calculation of first and second derivatives usually can be performed on high-resolution spectra. At a higher degree of sophistication, curve synthesis is used to reconstruct the real spectrum by adding several peaks, corresponding to different chemical states (e.g., oxidation state and functional groups). The principle underlying this procedure is the XPS "chemical shift", which will be discussed later in more detail.

Experimental Issues Related to XPS Analysis

Sample Preparation and Mounting

The absence of significant sample pretreatment procedures, with the relevant problems (e.g., finding a suitable solvent for the material to be analyzed), is one of the characteristics XPS shares with many other surface techniques.

Samples as different as powders, free-standing films, or coatings on different substrates obtained by solvent, dip, or spin casting or by electropolymerization, plasma synthesis, and the like, can easily be mounted on the sample stage of an XPS spectrometer. This is true also for particular samples like fibers, rods, and pottery fragments.

Conducting or double-adhesive tapes are commonly used to attach the samples to the sample stub; yet, in case of very thin, free-standing films, this approach can be quite complicated, because undulations may be formed. As an alternative, metal (steel) masks can be used to fix the sample to the stub: they are blocked on the latter by means of screws and have one or more slots, enabling the XPS analysis of different portions of the same sample, if necessary.

Powders can be pressed as pellets before being analyzed, using the same kind of press adopted for IR spectroscopy. In case of thermally stable polymers, the powders can be hot pressed so that a film is produced. In other cases, especially when a small amount of sample is available, a powder can be mounted by a small piece of adhesive tape or by means of special sample holders, which prevent sample grains from falling into the analysis chamber. In both cases, however, the high roughness of the sample surface may affect the quality of XPS data.

All the samples must always be carefully dried before the introduction into the XP spectrometer, because the presence of solvent traces can make the recovering of ultra-high vacuum conditions inside the instrument very difficult. Sometimes even dried samples may degas when brought under vacuum conditions. In this case, cooling the sample holder (e.g., with liquid nitrogen) can help in reaching the pressure conditions suitable for XPS analysis. This precaution can be useful also when degassing occurs as a consequence of X-ray exposure, due to the release of degradation products from the sample surface. In any case, since X-ray sources can be damaged if used under poor vacuum conditions, most commercial instruments are equipped with safety devices that turn off the source automatically when a sudden increase of the pressure, due to degassing, occurs in the analysis chamber.

Besides sample mounting, two aspects have to be carefully considered for polymer samples during an XPS experiment: sample charging and degradation arising from the analysis conditions (X-ray exposure, in particular).

Sample Charging

As mentioned, the binding energy values obtained from eq 1 can be used as an approximation. In fact, being referred to the vacuum level of the spectrometer (E_{vac}^{sp}), they depend on the work function of the spectrometer (Φ_{sp}), as shown in the scheme of energetic levels reported in Figure 16.5. In order to compare data obtained from different instruments, the BE values reported in the literature are then usually referred to the Fermi level (E_F).

If a conducting sample is mounted in the spectrometer, and a good electrical contact exists between them (usually both the sample holder and the spectrometer are grounded), their Fermi levels are the same, and the following equation can be used to calculate the binding energy of a core photoelectron (indicated as BE_F; that is, referred to the Fermi level):

$$BE_F = h\nu - KE - \Phi_{sp} \qquad (5)$$

Actually the kinetic energy scale of an XPS spectrometer is

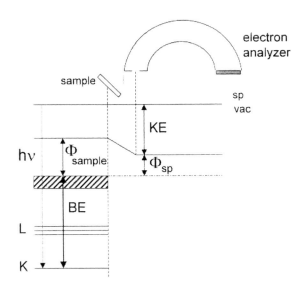

Figure 16.5 Diagram of energy levels for a conducting sample mounted in an X-ray photoelectron spectroscopy (XPS) spectrometer (sample and spectrometer grounded). Key: BE is binding energy, KE is kinetic energy; sp is spectrometer; vac is vacuum; Φ_{sp} is the work function of the spectrometer; hν is photon energy; L and K are typical core energy levels.

calibrated so that Φ_{sp} is already enclosed in the KE values (or, equivalently, the BE obtained are already referred to the Fermi level). This is accomplished by analyzing standards for which accurate values of the binding energies are known, like silver (Ag$3d_{5/2}$ level) and gold (Au$4f_{7/2}$) (ASTM E42 standard[18]).

As a part of the calibration procedure, the linearity of the BE scale is also checked, using two signals with very different BEs, usually Cu$2p_{3/2}$ from metallic copper, together with Au$4f_{7/2}$ or Cu$3s$. The binding energy for C1s, a signal arising from hydrocarbon contamination on these conducting samples, is also measured, as it can be very useful for calibration purposes, when insulating samples are taken into account. In this case, the positive charge developed on the surface as a result of photoelectron emission is not compensated through electrical contact with the spectrometer. Polymers are a typical example of this kind of materials.

If ultra-thin (e.g., with thickness comparable to the XPS sampling depth) polymer films are analyzed using an un-monochromated X-ray source, the positive charge can be neutralized by secondary electrons. These are generated by the interaction of the Bremsstrahlung component of X-radiation with the aluminum window of the source, or emitted by the film substrate itself, when hit by the primary X-radiation.

When this *internal* neutralization is impossible (e.g., when monochromatized sources are used or thick samples are analyzed), high BE values are observed, due to the positive charge on the sample. In this case, a low-energy (usually lower than 20 eV) electron gun (flood gun) can be used as a charge-compensating device. Care must be taken to avoid overcompensation, which would produce unreasonably low values for the binding energies, and to keep the linewidths as narrow as possible.

In recent XPS instrumentation, self-adjusting systems for charge compensation based on an electron gun and magnetic lenses have been introduced.[19] However, degradation of polymers due to the application of a flood gun can be a concern. An alternative approach to overcome the problem consists in using an internal BE reference; actually, this method has to be used even when a neutralization occurs, since a potential of a few electron-volts (positive or negative, according to the different conditions) can be present on an irradiated insulating sample. Emission of C1s photoelectrons from contamination carbon is usually adopted for this purpose, and its binding energy usually is fixed at 285 to 284.8 eV. Deliberate contamination of different polymer samples with hexatriacontane (a saturated linear hydrocarbon, $C_{36}H_{74}$) for BE referencing purposes has been reported by Beamson and Briggs.[20]

The method of internal charge correction is especially useful when aliphatic carbons (similar to the contamination carbon of XPS instruments) are present in the sample. But when aromatic carbons are included in the polymer, great care must be taken since a slight BE difference (about 0.2 eV) from aliphatic carbon has been reported.[21]

Another case is represented by polymers that have only carbon species characterized by high binding energy C1s photoelectrons, like cellulose or fluorinated polymers (e.g., Teflon). In these samples, contamination carbon, if present (which is a common condition), is clearly distinguished as a separate peak, then the BE scale correction is relatively simple. However, if no preliminary information is available on the chemical nature of carbon in the sample, other internal BE references should be adopted.

An important issue related to the electrical properties of insulating samples is *differential charging*, which occurs when a sample is not completely in electrical contact with the spectrometer. Photoelectrons arising from regions with different charge have slightly different apparent BEs, thus generating distorted peak shapes or much larger linewidths. In special cases, *vertical* differential charging is observed, because the outermost layer of the sample can be partially charge-compensated by secondary electrons present in the analysis chamber, whereas deeper layers show higher charging.

Sample Damage

Although XPS is generally considered the least destructive among the common surface analytical techniques, degradation of polymer samples must always be considered. Evident effects (changes in color or morphology) or variations in the chemical composition (like loss of chlorine and fluorine from chlorinated and fluorinated polymers, respectively) are often observed after exposure of polymers to X-rays during an XPS experiment.[22-25]

In their handbook on XPS of polymers, Beamson and Briggs[20] reported a "degradation index" for the most common polymers, based on the percentage of decrease of a degradation-related parameter (i.e., the halogen/C atomic ratio for chlorinated and fluorinated polymers) after 500 min of exposure to an X-ray source operated at 1.4 kW.

It should be pointed out that X-rays are just one of the sources of sample damage, and probably not the most important, during an XPS experiment. Other phenomena, like heating or interaction with secondary electrons or Bremsstrahlung, can be even more dangerous for a sensitive sample. Indeed, studies on sample degradation have shown that lower damage is induced by small spot sources, in which Bremsstrahlung and secondary electrons are not present and the separation of the X-ray source from the sample eliminates heat transfer.

Chemical Shift and Molecular Information by XPS

Chemical Shift

Since the very first applications of XPS,[26] it was evident that the chemical environment of an atom has a strong influence on the binding energies of its photoelectrons. Differences in the oxidation state, the molecular environment (e.g., the functional groups in organic samples), or even the crystallographic site may result in changes in the BE values for the same element. This phenomenon is usually known as *chemical shift* (as in NMR spectroscopy). XPS provides this powerful information, which extends its capabilities well beyond surface elemental analysis. BE values reported[9] for core electrons of some elements in different compounds are shown in Table 16.4.

A simple explanation of the chemical shift can be given by considering Koopmans's theorem,[27] which states that the binding energy of a photoelectron corresponds to the difference between the energies for the $(n-1)$-electron final state and the n-electron initial state of the atom. Several phenomena can influence the final state after photoemission, in particular the process known as *relaxation*—that is, the rearrangement of the other electrons in the ionized atom (atomic relaxation) or in the neighboring neutral atoms (extra-atomic relaxation), which tends to minimize the final state energy. Other final state effects involve correlation and relativistic energies, although these effects are typically negligible.

Relaxation phenomena are responsible for the discrepancies observed when experimental BE data are compared with orbital energies (with negative sign) obtained for a

Table 16.4 Compilation of binding energy values for primary photoelectron lines of different elements in inorganic compounds

Element/ photoelectron line	Compound	Binding energy (eV)
Ca2p	Ca	346.3
	CaF_2	347.8
	$CaSO_4$	348.0
Cu2p	Cu	932.6–932.7
	Cu_2O	932.5
	CuO	933.7
Fe2p	Fe	706.7–707
	FeO	709.4
	Fe_2O_3	710.8–710.9
N1s	Si_3N_4	397.4
	$(NH_4)_2SO_4$	401.3
	$PbNO_3$	407.2
Na1s	Na	1071.4–1071.8
	NaF	1071.0–1071.2
	Na_2SO_4	1071.2
O1s	CuO	529.6
	FeO	529.8
	K_3PO_4	530.4
	$CaCO_3$	531.4
	KNO_3	532.7
	$LiClO_4$	533.4
S2p	S	164.0–164.1
	FeS	161.6
	Na_2SO_4	168.8

Data taken from reference 9.

core electron by means of the Hartree-Fock method. The BE differences between photoelectrons of the same element in different chemical states (i.e., the chemical shift) usually can be predicted using the respective Hartree-Fock energies. This means that final state effects have almost the same influence, even if different chemical states are considered. The trend of BE values for different species of the same element can be predicted by using a more classical approach, known as the charge potential model.[28] Given that BE_0 is the binding energy for the neutral atom, the model can be summarized in the following equation:

$$BE = BE_0 + kq_i + \sum_{j \neq i} q_j/r_{ij} \qquad (6)$$

where q_i is the valence charge of the atom, q_j are those on the surrounding atoms, r_{ij} represents the distance between the i and j atoms, and k is a constant. The last term in eq 6 is usually known as the Madelung potential (by analogy with the Madelung energy of a solid) and abbreviated to V_i.

The BE difference for a photoelectron of atom i in two different chemical environments can be then expressed as:

$$BE(1) - BE(2) = k[q_i(1) - q_i(2)] + V_i(1) - V_i(2) \qquad (7)$$

Provided that the Madelung potentials of the two chemical states are comparable, it can be predicted that if $q_i(1) > q_i(2)$, then $BE(1) > BE(2)$. This means that the higher the oxidation state (or, better, the charge) of an atom, the higher is the BE; this is found experimentally in many cases, as shown in Table 16.4.

There are remarkable exceptions to this rule, the most famous being the Cu(0)-Cu(1) couple: the 2p photoelectrons for these chemical states show the same BE.[29] In other cases, higher oxidation states correspond to lower BEs (for example Co(II) and Co(III))[30] because of the Madelung potentials and the final state effects, which are neglected in the charge potential model.

Auger Parameter

When BE values do not allow a distinction between different chemical states, additional information drawn from XPS spectra can be of help. In particular, Auger peaks are usually influenced more than photoelectron core level signals. The Auger parameter α (see reference 31 for a review) has been introduced to quantify this effect and is expressed by the following equation:

$$\alpha = KE(Auger) - KE(photoelectron) \qquad (8)$$

or, in the modified form:

$$\alpha' = \alpha + h\nu = KE(Auger) + BE(photoelectron) \qquad (9)$$

In most cases, even when the BEs for two chemical states of the same element are the same, differences in the KE(Auger) allow a distinction between them.

Compilations of α and α' values (often in the form of two-dimensional plots) are available in the literature[9] and can

be used for chemical assignments. It is worth noting that, by definition, the Auger parameter is not influenced by sample charging, as any charging effect is cancelled by subtracting two KE values.

Molecular Information for Polymers

In polymers knowing the BE shift due to different chemical environments is the key to getting structural information from XPS data. Table 16.5 shows a compilation of BE values for the most common elements found in polymeric materials, taken from reference 20.

When several chemical species are present in the same polymer—for example, different functional groups of carbon, like C—OH, C=O, COOH—the differences in the respective BEs can be exploited to draw qualitative and quantitative information. Curve fitting is a common way to obtain this information. The spectrum is fitted by using a group of peaks, which are related to the different chemical species in the sample. As a general procedure, the first stage of curve fitting is performed by "manually" adjusting peak parameters like centroid position and height, whereas others, like width and shape (pure Gaussian or mixtures of Gaussian and Lorentzian functions are commonly used in XPS curve fitting) are fixed. Values reported in the literature or obtained on standard samples (e.g., monofunctional polymers) can be adopted.

In the second stage, the first trial fitting is optimized by means of special programs that modify a number of parameters at the same time in order to find the best fit. In many cases, these programs (usually implemented in the software of modern XPS instruments) also provide a detailed statistical report related to the best fit.[32] For example, the results of curve fitting for the C1s spectrum of a monofunctional polymer, poly(vinylmethylketone), are reported in Figure

Table 16.5 Binding energies for C1s in different functional groups on polymers

Functional group	Binding energy (eV)
CH_2	285.0 (BE reference)
C=C	284.73 ($n = 4$)
C—OH	286.55 ($n = 5$)
Epoxide	287.02 ($n = 1$)
C=O	287.70 ($n = 3$)
C—O—*C=O	288.99 ($n = 21$)
C—N	285.94 ($n = 9$)
N—C=O	288.11 ($n = 6$)
N—(C=O)—N	285.84 ($n = 1$)
C—S	285.37 ($n = 2$)
$C—SO_2$	285.38 ($n = 2$)
C—F	287.91 ($n = 1$)
CF_2	290.90 ($n = 1$)
CF_3	292.69 ($n = 2$)

For data determined from several polymers, average values are reported (taken from reference 20).

16.6. Three peaks can be detected by curve fitting and assigned to three chemical states of carbon, according to their BEs. The area ratios allow a quantitative analysis of the polymer structure.

Chemical Derivatization-XPS

Although generally useful, curve fitting cannot be adopted when species with very close chemical shifts are present in a sample. This can be a major concern for many complex polymers. Some BE data for C1s, N1s, and O1s, reported by Beamson and Briggs[20] on monofunctional polymers, are reported in Table 16.6. It is evident that using even the highest resolutions achievable with monochromatic sources, as well as curve-fitting procedures, some functional groups cannot be distinguished and quantified separately when they are present in the same sample.

Complex polymeric materials, like those obtained after surface treatment (e.g., corona discharge, exposure to plasma, chemical etching) of simpler polymers, or electrosynthesized polymers are a typical case. The need for reliable analytical information on these materials has led to the introduction of chemical derivatization in XPS.[33,34] The general principle of this method is modifying a functional group on the polymer surface with a "chemical label" containing an element that can easily be detected by XPS (i.e., having a high cross-section for photoionization). A chemical label that is not present in the sample is desirable, though not mandatory. Detection of the XPS signal of the labeling element can be used as a qualitative evidence for the presence of the group of interest, and, provided the derivatization reaction is specific and complete, it also enables a quantification, based on the reaction stoichiometry.

Some of the most common reactions adopted in CD-XPS of polymers are shown in Figure 16.7.

Specificity and completeness of the reaction are just two of the many criteria for application of chemical derivatization to XPS of polymers. Undesired modifications induced by the reactions—for example, those due to solvent effects—have to be carefully considered in each case. These aspects have been extensively studied in the past two decades, and much information is available on application of CD-XPS, even for very complex materials (for a review on the topic see reference 34).

Applications of XPS to Complex Polymeric Materials

Electrosynthesized Polymers

Polymers obtained by electropolymerization represent a huge field of investigation in current materials chemistry

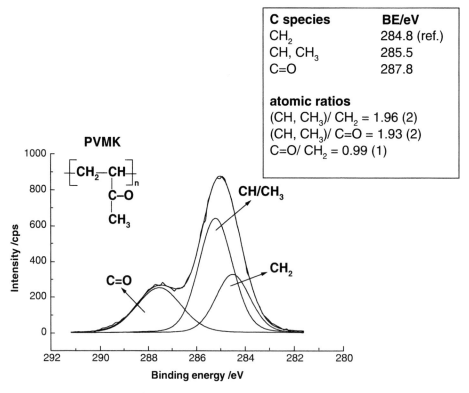

Figure 16.6 C1s spectrum for poly(vinyl-methyl-ketone), PVMK. Peaks resulting from curve fitting and their resultant are also plotted. The positions (binding energies) and atomic ratios (the theoretical values are shown in the brackets) of carbon peaks are reported in the inset.

Table 16.6 Examples of very close binding energies values in polymers

Photoelectron line	Functional group	Binding energy/eV
C1s	C−O−C	286.45
	C−OH	286.55
	*C−O−C=O	286.64
	C=O	287.90
	O−C−O	287.93
	HO−C=O	289.26
	O=C−O−C=O	289.41
N1s	C−N−C	399.87
	O=C−N	399
	N−(C=O)−N	399.89
O1s	C=O	532.33
	C−O−(C=*O)−R[a]	532.41
	C−OH	533.09
	O−C−O	533.05
	C−*O−(C=O)−Ar[a]	533.04

a. R = aliphatic group, Ar = aromatic group

Data taken from reference 20, BE reference: 285.0 eV for aliphatic carbon.

and physics, from both a fundamental and a technological point of view. Aspects like electrical conductivity and selective permeability are closely related to the polymer structure and are of crucial interest for applications in electronics (data storage, electrochromic devices, light-emitting diodes), energy sources (lightweight batteries), electrochemistry (protection from corrosion), and chemical sensors.

Electropolymerization can be considered a special kind of electrolysis, in which the monomer at the working electrode (e.g., platinum, gold, graphite) is transformed into a reactive species, usually a radical cation, which then reacts with other monomers to produce a polymer. In most cases, very stable thin or ultra-thin films can be obtained on the substrate, and one of the peculiar advantages of electrosynthesis is using the amount of charge to control film thickness. At the same time, multilayer structures can conceptually be obtained by using several electropolymerization stages with different monomers, provided the films previously obtained are electrical conductors, which is a very common property for these materials.

A number of monomers have been considered for electropolymerization: pyrroles (since the end of the 1960s[35]), anilines, phenols, phenilenediamines, and thiophenes, each with many different substituents. The literature about these materials has been growing accordingly in the last three decades, and references can be consulted for a review about synthesis, properties, and applications.[36-38]

Surface analytical techniques, in particular XPS, have played a key role in the characterization of these materials (for reviews, see references 39 and 40), one of the main reasons being their ability to analyze samples without dissolving them, which can be a very difficult, if not impossi-

ble, task for many electrosynthesized polymers. Among electrosynthesized polymers, polypyrrole and poly(1,2-diamino-benzene), electrosynthesized on platinum substrates from aqueous solutions (electrolyte KCl, or phosphate and acetate buffers), have been studied for their application in chemical sensors.[41-45] In the following, the surface characterization of polypyrrole by means of conventional and chemical derivatization-XPS will be presented as a case study, in order to show the XPS analytical potential for complex polymers.

Two different types of polypyrrole have been investigated: pristine polypyrrole (PPY) and overoxidized polypyrrole (PPYox). PPY is the material first obtained in electrosynthesis, and shows good electrical conductivity; PPYox is obtained by applying to PPY the electrosynthetic potential adopted for (+0.7 V vs. an Ag/AgCl/KCl$_{sat}$ reference electrode) for a prolonged time (5 h). This process results in a loss of electrical conductivity and in a dramatic change in permselectivity, which is very useful for its application in chemical sensors.[42,44]

XPS elemental analysis[46] shows the presence of oxygen and chlorine in PPY, along with carbon and nitrogen, the constituents of the monomer. The amount of oxygen is remarkably increased after overoxidation; at the same time, chlorine, corresponding to chloride anions acting as "dopants" in the pristine material (and involved in its electroactivity), almost disappears. In order to get structural information on PPY and PPYox, curve fitting has been applied to high-resolution spectra of carbon, nitrogen, and oxygen.

Figure 16.8 shows typical C1s, N1s, and O1s spectra for PPY and PPYox and the results of curve fitting (peaks and resultants). The relevant BE values have been used to make a first assignment of each peak to functional groups or additional spectral features (shake-ups due to the presence of aromatic systems) on the two polypyrroles. The results obtained from two sets of PPY and PPYox samples[47] have been reported in Table 16.7, with chemical assignments and quantitative determinations (atomic percent of each peak).

The correction for the binding energy scale was accomplished using the main N1s component, related to nitrogen in pyrrolic rings, as a reference, because neither of the carbons in the pyrrole ring can be classified as aliphatic, whereas the BE for pyrrole nitrogen is well known in the literature.[48]

Some of the structures that can be hypothesized in polypyrrole, starting from XPS data and comparison with information available in the literature, are shown in Figure 16.9. Indeed, much structural information can be drawn from conventional XPS.[47] Oxygenated functionalities (C−OH and C=O) are already found in the pristine polymer, but each increases remarkably after overoxidation, thus also explaining the increase in total oxygen.

Charged nitrogen decreases at the same time, and this finding is closely related to the loss of electrical conductivity after overoxidation, as charged species of nitrogen are in-

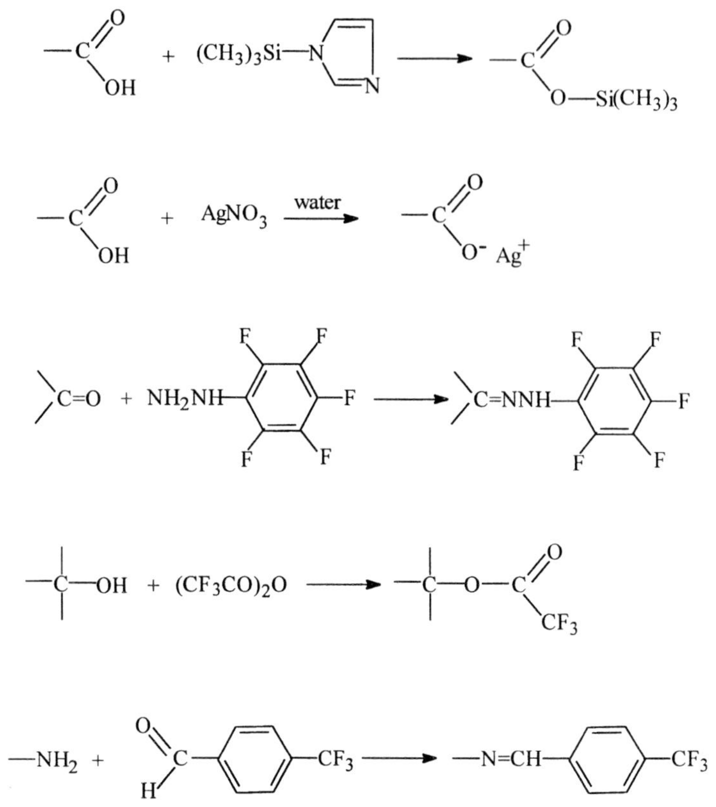

Figure 16.7 Examples of reactions commonly adopted for chemical derivatization–X-ray photoelectron spectroscopy (CD-XPS).

volved in the mechanisms of intrachain and interchain conductivity of polypyrrole.[36] Their disappearance (balanced by the increase of imine functionalities) also explains the decrease in the amount of chloride ions, which act as their counterions, being embedded in the PPY matrix from the electrolyte during polymerization.

Another important consequence of overoxidation is the decrease of the number of α carbons—those directly linked to nitrogen in the pyrrole ring, which are also involved in C–C bonds between rings along the polymer chain.[36]

As a first hypothesis, the increase during overoxidation of groups containing oxygen, especially C=O, could involve more α than β positions, as usually reported in the literature. As shown in Figure 16.9, this process would involve breaking of interring C–C bonds (or chain interruptions), with negative consequences on electrical conductivity. It could also involve the formation of COOH/COO⁻ groups, whose presence can be invoked to explain some C1s data on PPYox and the charge balance of the polymer.

Conventional XPS provides no further evidence about these hypotheses, as no significant difference exists in BE values for C=O groups in α and β position, or for N linked to a C=O (amide) compared to regular pyrrolic N. Moreover, a cross-check of C/O functionalities, made by comparing their quantitative determinations from the relevant components in C1s and O1s fits, showed an excellent agreement

ment[47] for C=O in both PPY and PPYox, but only in PPYox for C–OH, whose amount in PPY seemed to be overestimated when using the C1s component.

Chemical derivatization-XPS has been used to investigate the issues unexplainable by curve fitting on PPY and PPYox.[47,52] Reactions for C–OH, C=O, and COOH groups were performed using fluorine- or titanium-containing reactants as "labels", and are shown in Figure 16.10; the main results are summarized in Table 16.8.

The amount of each group was calculated from the XPS signal of the labeling element on the derivatized sample, starting from the stoichiometric relations (e.g., 5 F atoms are introduced for each reacted C=O group, using O-pentafluorobenzylhydroxylamine, PFBHA, as a reactant). Besides evaluating the influence of reaction time on the yield, which is mandatory whenever a new polymer is derivatized, any collateral modification caused by the reactions (i.e., increase of the amount of other elements, like C and O, contained in the derivatizing agents, possible side effects due to solvent exposure, possible side-reactions) was considered. This check led to exclude a common reaction for C=O groups—that is, derivatization with pentafluorophenylhydrazine (PFPH), from the possible reactions, due to a side reactivity of PFPH with imine groups on PPYox.[52]

The derivatization of C=O groups on PPY with PFBHA is complete, whereas only some of them (approximately corre-

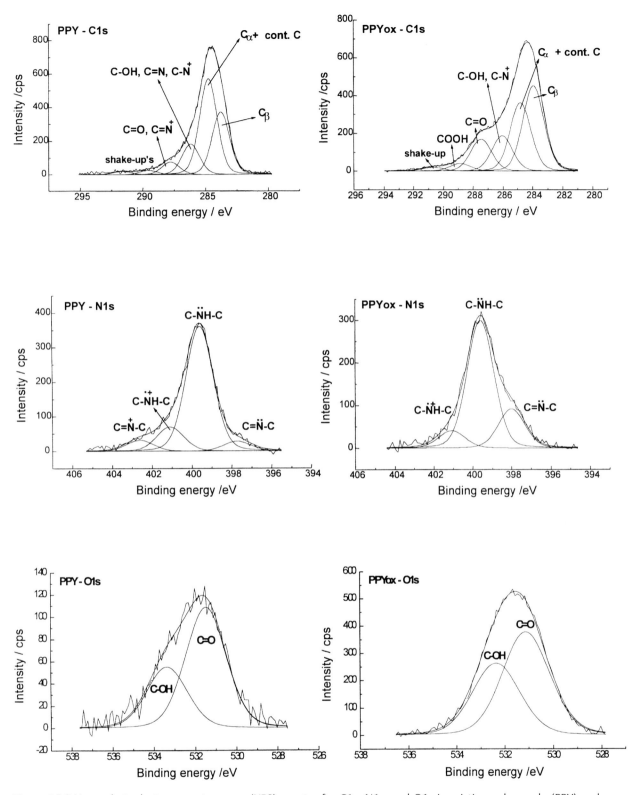

Figure 16.8 X-ray photoelectron spectroscopy (XPS) spectra for C1s, N1s, and O1s in pristine polypyrrole (PPY) and overoxidized polypyrrole (PPYox). Peaks obtained from curve fitting and their resultants are also plotted.

Table 16.7 Results obtained from curve fitting of C1s, N1s, and O1s spectra of polypyrroles

	PPY (8 samples)			PPYox (10 samples)		
	BE (eV)	Assignment	%	BE (eV)	Assignment	%
C1	283.83 ± 0.05	Cβ	31 ± 2	284.00 ± 0.06	Cβ	37 ± 5
C2	284.81 ± 0.03	Cα, cont. C	48 ± 3	284.87 ± 0.13	Cα, cont. C	31 ± 3
C3[a]	286.29 ± 0.06	C$-$OH	5.7 ± 0.7	286.14 ± 0.19	C$-$OH	8 ± 2
		C$=$N	1.1 ± 0.5		C$=$N	3.7 ± 0.4
		C$-$N$^+$	7.0 ± 1.1		C$-$N$^+$	3.0 ± 0.6
C4[a]	287.83 ± 0.11	C$=$O	4.2 ± 0.6	287.54 ± 0.15	C$=$O	13.1 ± 0.8
		C$=$N$^+$	0.9 ± 0.5			
C5	289.7 ± 0.2	shake-up 1	1.8 ± 0.6	289.0 ± 0.4	COOH, COO$^-$	3.3 ± 0.8
C6	291.5 ± 0.4	shake-up 2	1.7 ± 0.5	291.0 ± 0.3	shake-up	1.4 ± 0.4
N1	397.68 ± 0.12	C$=$N	5 ± 2	397.97 ± 0.03	C$=$N	19 ± 2
N2	399.6	$-$NH$-$ (pyrrole)	76 ± 3	399.6	$-$NH$-$ (pyrrole)	74 ± 3
N3	401.11 ± 0.09	C$-$N$^+$	14.5 ± 1.8	401.0 ± 0.3	C$-$N$^+$	7.7 ± 1.6
N4	402.71 ± 0.17	C$=$N$^+$	4.5 ± 1.2			
O1	531.53 ± 0.10	C$=$O	67 ± 4	531.23 ± 0.13	C$=$O, COO$^-$	63 ± 5
O2	533.18 ± 0.11	C$-$OH	33 ± 4	532.63 ± 0.12	C$-$OH	37 ± 5

Pristine polypyrrole (PPY); overoxidized polypyrrole (PPYox).
a. The amount of carbon-nitrogen functionalities was established from the corresponding components of N1s curve fitting.

sponding to the groups usually present before overoxidation) are derivatized in PPYox, even when much longer reaction times are adopted [10 h vs. 1 h, which is the time for complete reaction of a standard polymer for C=O, poly-(vinyl methylketone)]. This can be explained with the different reactivity of C=O groups, according to the position of carbon on the pyrrole ring: carbonyls on the α position are actually amide C=O, which cannot react with hydroxylamines to generate oximes. The result confirms the hypothesis that new C=O groups produced from overoxidation are formed in the α position of pyrrolic rings.

Derivatization of COOH with trifluoroethanol confirms that this functionality is present in PPYox, but shows also that some of these groups are ionized and not suitable for esterification with CF$_3$CH$_2$OH. Indeed, a contribution from carboxylates could be hypothesized already from the global charge balance of PPYox, as chloride ions are removed during overoxidation, but positively charged nitrogen is still present in the sample, although at a decreased level. The presence of COO$^-$ groups can be of primary importance for the permselectivity characteristics of PPYox. Its rejection of many anionic species has been exploited for use as a membrane in biosensors.[42,44]

Data relevant to derivatization of C$-$OH groups with titanium di-isopropoxide-bis-2,4-pentandionate (TAA) are also shown in Table 16.8. As for PFPH, the application of the most utilized reagent for CD-XPS on hydroxyl groups, trifluoroacetic anhydride (TFAA), was prevented by a side reaction between CF$_3$COOH, released as a by-product, and the basic imine groups.[52] Nevertheless, use of TAA supports the hypothesis, based on conventional XPS, that the estimate of C$-$OH groups in PPY from the C1s fitting is probably

influenced by a spectral feature related to the band structure of pristine PPY, a shake-up with unusually low binding energy, whose presence had been already suggested in the literature.[53]

A curious aspect, although not completely unexpected, arises from application of the TAA reaction to PPYox: C$-$OH groups are completely derivatized in this case, but it seems that all the coordination sites of titanium in the complex are occupied by C$-$OH groups of the polymer. This finding arises from a check of the reaction stoichiometry and suggests that conformational changes occur in the polymer matrix, probably enabling C$-$OH groups of different chains to be cross-linked by titanium.

Plasma Modified Polymers Used as Biomaterials

Surface modifications by means of plasma of fluoropolymers (e.g., fluorinated ethylene propylene, FEP) used in biomaterial applications, have been the object of extensive work in the last few years (see, for example, references 54–60). In particular, these materials have been proposed as the basis for potential neural prosthetics. In these devices, *minimal peptide sequences*, or oligopeptides, which are known to promote neuronal cells attachment, are linked covalently to the modified fluoropolymer surface.

To create linking sites for the peptide sequences, two main modifications have been accomplished on the fluoropolymer surface: introduction of hydroxyl functional groups in a radio frequency glow discharge (RFGD) H$_2$-methanol (vapor) plasma, and subsequent reaction of these new OH groups with siloxanes containing amine groups. Litho-

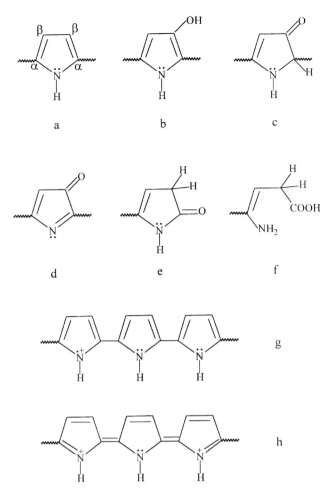

Figure 16.9 Possible structures on polypyrrole chains hypothesized from the literature and X-ray photoelectron spectroscopy (XPS) data. Structures e and f imply chain breaking. Structures g and h are called *polaron* and *bipolaron*, respectively (the positive charge(s) resonate along three connected pyrrolic rings), and are involved in the electrical conductivity and electroactivity of pristine polypyrrole (PPY).

graphic masking techniques have been applied to develop patterns of these functional groups in microdomains at the polymer surface.[56] XPS was chosen to characterize the surface of these polymers, together with other techniques like SIMS, IR spectroscopy, and scanning electron microscopy.[54–60,60–62]

Relative atomic surface concentrations of FEP as a function of RFGD exposure time have been obtained by XPS. A decrease of the fluorine content is observed with increasing time of exposure to plasma. In contrast, the amount of oxygen increases in the first 3 min of plasma treatment, but beyond this time no significant modification of the oxygen concentration occurs. Further information can be obtained by comparing the detailed spectra of C1s for unmodified and plasma treated FEP (FEP-OH).[62] Indeed, the RFGD pro-

cess leads to a shape change in the C1s spectrum, which can be attributed, using curve fitting, to the presence of C—C and C—OH groups. The XPS data were exploited to find the optimal conditions to produce high OH coverages on the fluoropolymer surface. The OH groups are then reacted with aminopropyltrietoxy silane (APTES),[55] whose molecules condensate to form an interconnecting siloxane network, with pendant aminopropyl groups. Several NH2 functionalities are then available on the fluoropolymer surface for subsequent reaction (i.e., formation of an amide) with the C-terminus (COOH) of oligopeptide sequences. Quantitative information on the amount of silanization, as a function of reaction time, is obtained by XPS, using silicon and nitrogen as markers for the APTES molecule.

As an alternative strategy for attaching the N-terminus of oligopeptides to the fluoropolymer surface, carbonyldiimidazole (CDI) has been used as a coupling reagent.[62] In this case, a preliminary reaction between FEP-OH and CDI occurs, namely an esterification of the OH groups on FEP by means of carboxy-imidazolyl groups. The "activated" FEP surface is then reacted with the oligopeptide, leading to a linkage between its N-terminus and modified OH groups. Moreover, XPS survey spectra of FEP films show incorporation of nitrogen (which is absent both on unmodified FEP and on FEP-OH) after reaction with CDI, as expected. Furthermore, the presence of C=O, C—N, and C—C groups, which are contained in the CDI structure, can be determined by curve fitting of the C1s spectrum.

After reaction with the oligopeptide, a significant increase is observed in the amounts of the C—C and C=O components of the C1s spectrum, which is consistent with immobilization of the oligopeptide molecules. This is also confirmed by the increase of the N-C component (due to amides) in the N1s spectrum, when it is compared with the relevant spectrum of FEP-CDI. These modifications in the peak shapes are also accompanied by an increase in the nitrogen content.

The simplest route to oligopeptides immobilization on RFGD-modified FEP surfaces is direct reaction between OH groups and the C-terminus of oligopeptides, which has been accomplished for the tyrosine-isoleucine-glycine-serine-arginine (YIGSR) and the isoleucine-lysine-valine-alanine-valine (IKVAV) sequences,[60–61] both isolated from laminin, the protein responsible for the interface in neural/muscle cell attachment. In this case, a base-catalyzed esterification was used to covalently bind the C-terminus of a peptide to the surface hydroxyl groups of FEP-OH. The percentages of oxygen and nitrogen at the surface were monitored by XPS. An RFGD treatment time between 0.5 and 1 min produced the most reactive surface for the YIGSR sequence, in accordance with the OH groups increase.

However, it is worth noting that FEP-OH samples obtained after longer times of exposure to plasma show a remarkable decrease in the reaction efficiency. This could be

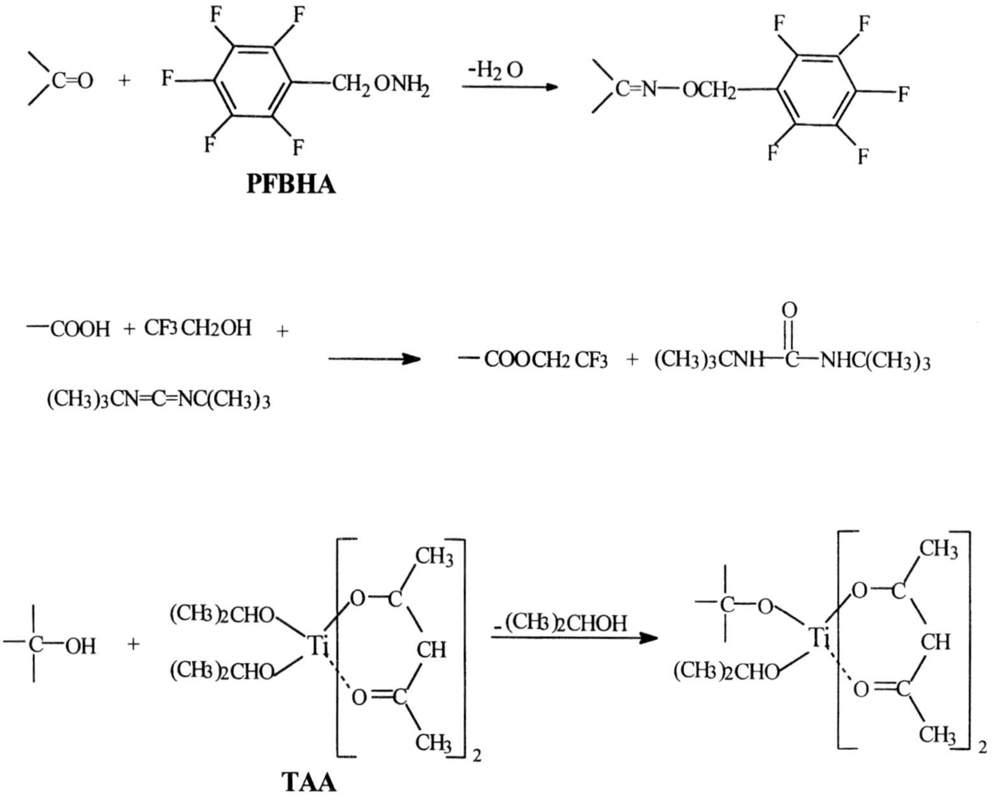

Figure 16.10 Derivatization reactions adopted successfully on polypyrroles: pentafluorobenzylhydroxylamine (PFBHA), titanium di-isopropoxide-bis-2,4-pentandionate (TAA).

due to more extensive defluorination occurring after long RFGD treatments times, as shown by XPS. The consequent decrease in electron density localized near the FEP-OH surface probably lowers the nucleophilic character of the OH groups, thus reducing their reactivity.

Future Perspectives

The further development of XPS techniques for surface analysis of polymers will build on advances in instrumentation,[17] which bring higher signal to noise, speed, spatial and energy resolution, and simplicity to the analysis. For example, for the past several years, new instruments have been developed by all major manufacturers to speed and simplify XPS analysis for quality-control applications. These instruments can analyze either multiple samples or spots on a sample with computer control and visual/microscope identification, with rapid sample introduction of hundreds of samples at a time. This means that off-line quality control is possible with XPS analysis. Applications to semiconductor wafer processes, organic contamination, patterning, polymer coating processing, and aging, weathering, and defect analysis are all possible with this instrumentation.

Expanded use of the speed and computerized control for multiple, rapid sample analysis is one of the major future

perspectives that has been foreseen, but not fully realized, for over 20 years. It will also mean that more manufacturers of polymers, paints, coatings, and other polymer materials will use XPS as a primary analytical tool outside of the research environment. Another area of future development is the ability to analyze more complex polymer samples. Water-borne polymers, latex samples, hydrogel polymers (e.g., soft contact lenses) and other materials in contact with wet environments require special sample handling methods to handle water or freeze polymer configurations into place. Placing a sample below its glass transition temperature T_g is a common means to hold a particular surface configuration and this approach is in broad use.[63,64] Still another area for future expansion of the XPS technique for polymer analysis is the use of rapid angle-dependent measurements for depth profiling polymers without ion sputtering. Methods to interpret the configurations and depth profiles[14,64] will become more common as faster instruments and simpler algorithms are developed.

A related but less realized development is the use of higher energy X-ray sources to probe deeper into samples.[17] Again, this area has been open for over 15 years, but more applications need to be developed to see this in more general use. For example, TiKα sources[15] have been used successfully to probe 200 Å in depth, complementing angle-dependent measurements that probe over the first 100 Å.[14,64] Many multicomponent polymers, or materials with fillers and additives, require a surface analysis that is effective at these depths.

At the same time, polymeric materials applications like coatings, microscopic inclusions of polymers in semiconductors or latexes, and studies of particles and fillers embedded in polymers have been facilitated by the improved spatial resolution of XPS. The ability to map surfaces with resolution in the submicron range is growing with new instrumentation, which focuses X-rays or image electrons on the analysis plane, leading XPS to complement microscopies such as atomic force microscopy (AFM), transmission electron microscopy (TEM), and scanning electron microscopy (SEM). Again, correlation with simpler imaging technologies is a welcome approach, but spatial resolution for both mapping and small samples/defects is a major area of growth of the instrumentation.

While all fields of growth mentioned in the past few paragraphs are built on common technological application, there are still prospects for more complex photoemission experiments to be applied to polymer materials. For example, synchrotron radiation still presents a major advantage for energy selectivity and focus, but as yet it is mainly unrealized in fundamental studies of polymer surface structure.

Semiconducting and conducting polymers could also benefit by a clearer understanding of surface electronic structure readily available from synchrotron experiments.

Secondary Ion Mass Spectrometry

Fundamental Principles

Secondary ion mass spectrometry (SIMS) is a surface analytical technique based on the generation of (secondary) ions by bombardment of a solid surface by incident (primary) ions that have energies in the kiloelectronvolt range.[65-68] Momentum is transferred from the primary ion to the solid surface, which then perturbs the lattice structure, generating a "collision cascade", whose penetration depth and intensity will depend on the mass, energy, and angle of incidence of the primary ion. A detailed theory about this process has been developed by Sigmund.[69]

The cascade is due to the recoiling of the sample's atoms, which are displaced from their original positions in the lattice when they are hit by the incident ion; other atoms are then displaced by them, and a sequence of collisions is generated. While propagating through the sample, the cascade can finally reach the sample's surface, breaking the bonds

Table 16.8 Results obtained from chemical derivatization–X-ray photoelectron spectroscopy (CD-XPS) and from curve fitting on polypyrroles

Derivatization	Sample[a]	Data from CD-XPS	Data from curve fitting	
C=O groups		% C=O[b]	% C=O[b]	
with PFBHA	PPY	3.8	3.6	
	PPYox	2.2	13	
	PPYox	3.1	13	
	PPYox (reaction time: 10 hrs)	3.0	12.7	
	PPYox (solvent water)	3.1	12.8	
COOH groups		COOH/N[c]	COOH/N[c]	COO⁻/N[c]
with CF₃CH₂OH	PPYox	0.11	0.16	0.02
	PPYox	0.05	0.19	0.12
	PPYox (reaction time: 24 hrs)[d]	0.10		0.06
	PPYox	0.05	0.16	0.08
C–OH groups		C–OH/N	C–OH/N	C–OH/N
with TAA	PPY	0.07	0.31	0.08
	PPYox	0.32[e]	0.33	0.29

Pentafluorobenzylhydroxylamine (PFBHA); titanium di-isopropoxide-bis-2,4-pentandionate (TAA); pristine polypyrrole (PPY); overoxidized polypyrrole (PPYox).

a. If not specified, results from different polymer films are shown in the table.

b. Percent of carbonylic carbon over total C.

c. Atomic ratios; the amount of ionised carboxylic groups was estimated from the total charge balance of the polymers (neutralization of positively charged nitrogens).

d. Reaction performed on the previous sample by extending the reaction time.

e. This value was obtained considering four OH groups of PPYox coordinated to each titanium atom (see text for details).

between atoms, clusters of atoms, or molecules. Atomic, cluster, molecular, or fragment ions are then emitted from the surface. A number of other phenomena occur with the collision cascade, leading also to the emission of atomic and molecular ions (positive and negative) and electrons; back-scattered primary ions are also observed. A schematic representation of the whole process is showed in Figure 16.11. The secondary ions, bearing chemical information related to the sample surface, can then be collected and focused into a quadrupole, time-of-flight, or magnetic sector mass spectrometer, and a mass spectrum is finally obtained.

Herzog and Honig were first credited with coupling ion sputtering with mass spectrometry in the 1940s and 1950s.[70,71] In their experimental conditions, the sample was eroded, allowing the determination of its elemental composition as a function of depth. This SIMS mode is usually known as *dynamic*, characterized by primary ion current densities as high as 1 mA/cm^2.

In the late 1960s, a different SIMS experiment, able to increase the surface sensitivity of the technique by minimizing surface erosion, was described by Benninghoven.[66-68] This particular mode was called *static*, and its development has played a key role in the application of SIMS for surface analysis. The fundamental definition for static SIMS arises from the following equation:

$$P = \Sigma \sigma_i / A \qquad (10)$$

where P is the probability of prebombardment of the primary ion beam, σ_i is the damage cross-section, and A is the target area.

Secondary ions originate from an area not previously bombarded by the incident ion beam only when P is much less than 1 (i.e., $\Sigma \sigma_i \ll A$). As a general indication, static conditions can be achieved when the ion beam current density is lower than 1 nA/cm^2 and the total ion dose is less than 10^{13} ions/cm^2. It can be predicted that the lifetime of

a single monolayer of the sample, under these conditions, is 1 h, although other factors, like mass and energy of the incident primary ion beam, should be taken into account.

The formation of ions is governed by two stages: desorption of the neutral atom, and formation of a simple positive or negative ion, which are related to the surface-binding energy and to the ionization potential (electronegativity), respectively. As a consequence, ions that originate from metal atoms are usually positive, whereas elements with high electronegativities, such as oxygen and halogens, form negative ions. Polyatomic cluster ions originate from the combination of oxygen and halogen with metals and may have a positive or negative overall charge; it is commonly assumed (and experimental evidence supports this thesis) that they are generated by recombination above the surface.[65]

In most cases, the formation of atomic and polyatomic ions is deeply influenced by matrix effects, and several theories have been proposed for specific classes of materials. As an example, the yields of atomic ions from metals and alloys are generally low if the surface is not oxidized, and the probabilities of formation for cluster ions M_x^+ are even smaller (especially for high values of x). Surface oxidation leads to remarkable increases of the yield (2–3 orders of magnitude).

The sputtering of intact molecular ions from the surface is possible under static conditions, since complex fragmentations due to primary ion beam damage become less likely; according to Figure 16.11, this phenomenon occurs at some point within the collision cascade regime, but away from the point of primary impact, where fragments are formed.

In general, secondary molecular and quasi-molecular ions originate from low-energy ion-molecule collisions;[72] the parent ions also can undergo a unimolecular dissociation, producing a daughter ion and a neutral loss species, which can be a small stable molecule, like H_2O, CO_2, or NH_3. Different mechanisms have been hypothesized in the literature to explain the generation of molecular secondary ions, which is particularly important when polymers are analyzed by SIMS.

The first model is based on the protonation or deprotonation (or both) of neutral molecules (M) to produce $(M + H)^+$ and $(M - H)^-$ ions.[73] The second involves cationization or anionization of organic molecules, such as the interaction between the latter and a metal cation, to form a solvated metal (M_e) complex $(M + M_e)^+$. Electron transfer processes may also produce ionic species, namely radicals, M^-* and M^+*; finally quasi-molecular ions can be directly sputtered from preexisting organic cations and anions, as in cases of direct sputtering from the surface of a salt.[66,72,74] The latter is the most efficient mechanism, followed by cationization/ anionization and electron transfer.

When polymers are involved, different types of ions can be considered, roughly divided either by mass range (small, medium, large) or by mechanism of formation (fragmenta-

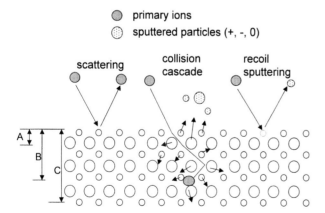

primary ions
sputtered particles (+, -, 0)

scattering   collision cascade   recoil sputtering

Figure 16.11 Schematic representation of interactions between primary ions and a solid surface. Key: A, escape depth; B, implantation depth; C, depth of collision cascade.

tion and rearrangement, desorption and cationization of oligomers, simple bondbreaking/charge stabilization). Aspects related to secondary ion formation in polymer analysis by SIMS will be discussed in the paragraph dedicated to SIMS applications to specific polymers.

Instrumentation

Four components can be considered the basic constituents of a SIMS spectrometer: a primary ion source, a system for secondary ions collection (extractor), a mass analyzer, in which ions are separated according to the m/z ratio as in ordinary mass spectrometry, and a detector.

Like XPS instruments, SIMS spectrometers must be operated under ultra-high vacuum conditions, which are achieved using similar pumping systems (mainly ionic and turbomolecular pumps). Due to the number of possible combinations between ion sources and mass analyzers, and to their inherent complexity, a detailed description of current SIMS instrumentation goes beyond the aims of this chapter. The following discussion presents the most common instrumental configurations adopted in both dynamic and static SIMS, emphasizing advantages and drawbacks, according to the materials analyzed. More specific information about SIMS instruments can be found in fundamental books, like references 66 and 75, and the others cited here.

Dynamic SIMS

Two main instrumental configurations are generally adopted for current dynamic SIMS measurements: the ion microscope and the ion microprobe, shown schematically in Figure 16.12. In both cases, depth and spatial resolution have to be optimized to produce three-dimensional concentration profiles.

For the ion microscope, the essential factors are the collection and focusing of the secondary ions after emission from the sample surface, which is typically accomplished using a combination of immersion lens and transfer optics.[66] Two detection systems are adopted: the first is an ion multiplier, providing the collective signal for optimization and depth profiling; the second system generates an ion image, using a scintillator plate or a resistive anode detector. Charge coupled device (CCD) cameras have been also introduced for collection of ion images, allowing digital integration, stacking, and further processing to produce three-dimensional depth profiles.

The ion source is usually a mass-filtered, high-current-density source, chosen from different designs, including the Cesium source and the duoplasmatron gun,[66] which produces a high current output of oxygen or noble gases ions. In the mass analyzer, the spatial distribution of the secondary ions must be preserved; a double sector (electrostatic/magnetic) analyzer is usually adopted for this purpose, in spite of the fact that the mass range is usually limited to 1

to 200 m/z. Time-of-flight analyzers have been developed for applications involving higher mass ranges.

With the ion microprobe, images are generated sequentially by scanning a focused ion beam; the focal characteristics of the primary ion source and the coordination with the signal collection are critical instrumental features. Ion focusing optics include condenser and objective lenses; voltage waveforms for the ion beam scanning are usually coupled to the detection electronics. The crater created by the ion beam is a major concern. It is desirable to eliminate signals that emerge from the crater edge, for a more consistent and definitive depth can be probed only in the middle. Some kind of electronic gating is therefore used on the ion signal as it emerges from the mass analyzer, in order to prevent signals that arise from the crater edge to be collected. Because less stringent requirements are put on spatial resolution for mass analyzers in the ion microprobe configuration, quadrupole mass analyzers are usually chosen, with the advantage of a superior throughput over the sector mass analyzer.

Static SIMS

Two sets of instrumental configurations are commonly used also for static SIMS, as shown in Figure 16.13; they involve a combination of two types of primary ion beam sources and two types of mass analyzers.

As a source, either noble gas ion beams or focused ion beams produced from liquid metals, like Ga,[66,75] or desorption from a solid state source (Cs) are adopted. Focused beams are operated at very low ion currents, although current densities are quite high. Quadrupole and time of flight (TOF) systems are used as mass analyzers. The former offered a good compromise between transmission and resolution at the outset of the development of dynamic SIMS techniques. As workers tried to generalize a common configuration for both static and dynamic SIMS, the extension of quadrupoles to higher mass ranges proved to be suitable for many static SIMS experiments. However, the practical limit of 1000 to 2000 m/z and the decreasing transmission at higher mass was too limiting for many molecular applications.

The TOF analyzer was then introduced: in this case, the mass limit can be as high as 10,000 to 20,000 m/z, with high transmission and good resolution; at lower resolutions, even higher mass ranges can be analyzed. Of course, for static SIMS of polymers, the need to provide detection to these high mass ranges with good transmission is crucial.

Analytical Information Available with SIMS Experiments

Dynamic SIMS

One of the main aspects of dynamic SIMS application is trace analysis, which is needed to detect impurities, un-

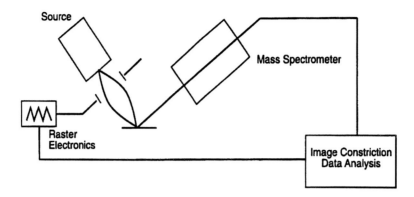

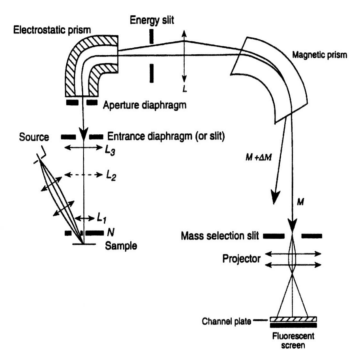

Figure 16.12 Typical instruments configurations for the ion microprobe (*top*) and the ion microscope (*bottom*).

wanted or unknown additives, and species added in very small amounts to modify the properties of materials (like dopants for semiconductors). The ability of SIMS to detect even ppb amounts of an element at the sample's surface meets this demand. In many cases, also the location of traces is a main concern; thus an in-depth and spatially resolved trace analysis is required. Areas of application like layered materials, chemical vapor deposition, molecular beam epitaxial growth, and oxide layers on semiconductors are examples of this kind of analytical problem.

The combination of in-depth and spatially resolved analytical information can be obtained with both the instrumental configurations adopted for dynamic SIMS. Ion microscopy is potentially able to provide at the same time three-dimensional profiles for elements both at majority and

trace concentrations. Although the large amount of data required to construct this kind of information can be a limit, many applications have been reported for semiconducting materials and biological samples.[76]

The ion microprobe has better spatial resolution (submicron features can be determined),[66] yet three-dimensional concentration profiles are complicated by the need to collect signals sequentially. Depth resolution of lateral data can also be degraded by the change in depth during the collection of a good signal for the element of interest.

It should be pointed out that the ability to detect species at majority, minority, and trace concentrations does not necessarily translate into an easy ability to quantify them. While signal levels in SIMS range over 6 orders of magnitude, the yields of elemental ions may vary over 4 orders of magnitude and, furthermore, matrix effects complicate the conversion of signal intensities to concentrations through yields.

SIMS practitioners have focused on quantification of a few commonly analyzed systems. In these areas, the use of standard addition techniques, external and internal standards, and other approaches allow concentrations of dopant species to be analyzed routinely. When new experimental systems are considered, a large amount of certification, qualification, and reference development work needs to be performed for quantification. Further details on quantification with SIMS is presented in the next section of this chapter.

Static SIMS

Static SIMS studies of small molecules on metal or semiconductor surfaces were the basis of the early development of the method.[66] The technique is extremely valuable for the determination of the degree of coverage, especially at ex-

tremely low concentrations (submonolayer); a detection limit of 0.01 monolayer for molecular species has been estimated. Several studies have followed the generation of secondary molecular ions as a function of coverage, noting that signal levels respond linearly up to one monolayer.

On well-defined surfaces, the structure, orientation, and adsorption sites have been shown to influence the dynamics of secondary ion generation.[65] In particular, the angle and direction of ion generation have been used to understand the orientation of molecules on surfaces, although only a few SIMS instruments are designed to differentiate the angle of secondary ion emission. As larger molecules are probed from surfaces, the problems of predicting molecular ion formation mechanisms complicate spectral interpretation. Understanding the generation of molecular ions from surfaces[77] in SIMS and in biomolecular mass spectrometry[78,79] is based on the close relationship of the SIMS experiment to mass spectrometries coupled to fast atom bombardment (FAB-MS),[80] plasma desorption (PDMS),[81] and laser desorption (LDMS).[82] These methods have been particularly useful

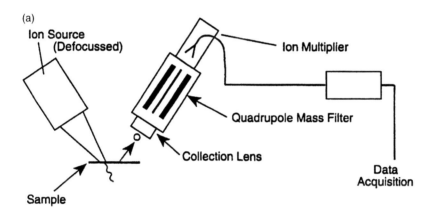

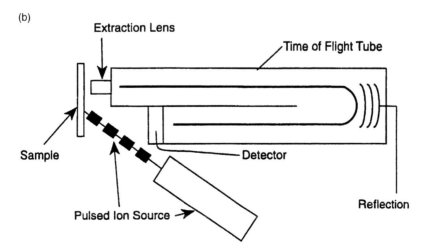

Figure 16.13 Typical configurations for quadrupole (a) and time-of-flight (b) static secondary ion mass spectrometry (SIMS) instrumentation.

for the mass spectrometry of high molecular weight biological molecules, allowing molecular weight determination, identification, and attempts to sequence nucleotides and proteins within the mass spectrometer.

The basic principles of ion formation from atomic species have been recognized as the same for molecular secondary ions. This allows the derivation of molecular structure information based on molecular ion and fragmentation information similar to that inherent in traditional mass spectrometry.

Along with such surface specific molecular structure analysis, the excellent detection limits and ability for simultaneous analysis has allowed mixture analysis, which, as shown next, can be very important for polymeric materials.

SIMS Analysis of Polymers

By comparison with metals, alloys, semiconductors, and even glasses or catalysts, polymeric materials present inherent analytical difficulties, especially when practical or "technical" polymers are considered; in this case, additives and stabilizers are commonly found in the sample. Moreover, a polymeric material usually involves a distribution of molecular weights, or the presence of amorphous and disordered phases in contact with microcrystalline structure. A further, fundamental aspect is the insulating character of polymers. When low-energy ion beams (500 eV to a few keV) impinge on their surface, an electrostatic charge is developed, and this charge rapidly prevents the acquisition of reliable spectral information. Use of an electron gun for neutralization, as in XPS analysis, can be then adopted as a possible solution. Sputtering is always a concern when polymeric materials are analyzed by SIMS: even under mild ion beam conditions, damage still occurs, in terms of molecular bonds breaking and reforming.

The first application of the static SIMS to the analysis of polymer surfaces was the characterization of low mass (0–150 amu) fragment ions, in order to distinguish structure and isomerism of butyl methacrylates that have different side chain structures.[83,84] Subsequent improvements in SIMS experimental conditions (lower ion dosages) have led to a reduction in surface damage, due to primary ion bombardment, allowing the analysis of higher mass fragments that are related to the original structure of the polymer.[85]

Generally speaking, each new technical advance in minimizing damage due to the ion beam has lead to drastic improvements in the information available with SIMS on polymeric surfaces. Besides data related to the surface structure, like the presence and concentration of functional groups, which can also be obtained by other techniques like XPS (especially in concert with infrared or Raman spectroscopy), many other aspects have been considered in SIMS literature about polymeric materials. For example, TOF-SIMS has been applied to study high mass ions from nylons, polystyrenes, poly(dimethylsiloxanes), and surface-modified polymers,[86–90] showing an accurate molecular weight distribution for high mass oligomeric ions, or suggesting the sensitivity of fragmentation patterns to intrachain bonding in nylons and to composition in polyurethanes.

Other examples of the analytical capabilities of SIMS in polymer surface analysis are given in more detail in the paragraph dedicated to specific applications.

Quantitative Analysis by SIMS

As anticipated, the presence of matrix effects and the difficulties in measuring absolute ion yields make quantitation in SIMS very complicated. By analogy with XPS, a complete equation[91] can be written for the ion current $I_{A(B)}{}^{\pm}$ due to a mass number of element A in a matrix B:

$$I_{A(B)}{}^{\pm} = I_P S_{tot} A c_A T_A D_A \alpha_{A(B)}{}^{\pm} \int_E^{E+\Delta E} \int_\Omega n_A dE \ d\Omega \qquad (11)$$

All the factors influencing the intensity of SIMS signals are enclosed in this equation: I_P is the primary ion current; S_{tot} is the total sputtering yield; A is the area on the sample from which secondary ions can be collected; c_A is the concentration of element A in matrix B; T_A and D_A are the spectrometer transmission and the detector efficiency for the particular mass from element A; $\alpha_{A(B)}{}^{\pm}$ the probability of positive or negative ionization of element A in matrix B; E and $E + \Delta E$ represent the energy range selected for secondary ions; Ω is the solid angle of acceptance of ions; and n_A is the energy and angular distribution of sputtered material.

Some terms in eq 11 can be considered constant, leading to the following simplified form:

$$I_{A(B)}{}^{\pm} = F_A S_{tot} \alpha_{A(B)}{}^{\pm} c_A \qquad (12)$$

where F_A contains the instrumental factors related to element A (namely to a particular ion—i.e., to a specific mass). The main problem in quantification with SIMS is due to the term $\alpha_{A(B)}{}^{\pm}$, in which the matrix effect is expressed.

Several successful approaches have been developed for quantitative analysis, based on two general strategies: the application of physical theories, describing the sputtering/ionization process, and the use of standards, to provide empirical calibrations and determination of sensitivity factors.

Among the theories describing the yields of secondary ions from various matrices, one of the most widely used is the so-called local thermal equilibrium (LTE) model, first proposed by Anderson and Hinthorne[92] and then simplified by Morgan and Werner.[93] The LTE model assumes that ions are formed in the selvedge, at equilibrium, by a process involving electron loss (for positive ions) or gain (for negative ions):

$$A \rightleftarrows A^+ + e^- \qquad \text{or} \qquad A + e^- \rightleftarrows A^- \qquad (13)$$

The equilibrium constants for these "reactions" can be expressed as:

$$K^+ = [A^+][e^-]/[A] \text{ and } K^- = [A^-]/[A][e^-] \quad (14)$$

or, using the Saha-Eggert ionization equation, as:

$$K^{+ \text{ or } -} = (2\pi m_e kT/h^2)^{3/2} [2Z^{+ \text{ or } -}/Z_0] \exp[-(E_i - \Delta E_i)/kT] \quad (15)$$

where k is Boltzmann's constant, h is Planck's constant, m_e is the mass of the electron, the Zs represent the partition functions of ions ($Z^{+ \text{ or } -}$) or atoms (Z_0), E_i is the ionization constant, and ΔE_i is the reduction in ionization energy due to the plasma and is related to $[e^-]$ and to the plasma temperature T.

Considering that $[A^+]$ and $[A^-]$ are proportional to I_A^+ and I_A^-, respectively, and $[A]$, the concentration of atoms in the plasma, may be taken as proportional to c_A, the concentration of the element A at the surface (thereby neglecting the number of ionized atoms), the quantity $\ln(I_A^{+ \text{ or } -} Z_0/c_A Z^{+ \text{ or } -})$ should be related linearly to E_i.

In practice, multiple measurements (corresponding to different E_i values) are performed, and a fitting procedure is based on a linear equation is then adopted to obtain estimates for $[e^-]$ and T, which are used (although their physical significance is not clear) as calibration constants related to the instrumental conditions and sample type.

The LTE model has been shown to yield concentrations within a factor of 2 of the true concentration in a variety of systems, in particular glasses and semiconductors. Nevertheless, complex empirical approaches have to be used, as a correction, if matrices with significantly different composition, in respect to the calibration samples, are considered; this explains the rapid growth in the application of fully empirical methods. One empirical method is based on the "absolute" sensitivity factor, which relates the ion current to the concentration of the element, and, referring to eq 3, is given by $F_A S_{tot} \alpha_{A(B)}^{\pm}$. Provided the concentration is not high enough to alter the matrix, a linear plot should be obtained when ion currents for standard samples are reported versus the element concentration. An alternative approach is based on ratios of the ion currents for two elements, one of which is usually the main matrix element: relative sensitivity factors are then obtained, and the effects on ion currents due to variations in the matrix itself are compensable.

Of course, the main demand on quantification techniques with standards is the need for a set of surface homogeneous standards relevant to the sample system. Glasses are particularly useful from this point of view, as they contain high oxygen concentrations, lending themselves to advantageously high ion yields.

A further development of quantification methods, for depth profiles, involves the knowledge of the total concentration of an implanted dopant. In this experiment, the signal (total ion current) is integrated until it disappears at some depth. Calibrating the total signal to the amount of implanted species, the individual amounts are then calculated at various depths. Although this approach relies on an independent determination of the total amount of species implanted, it is used routinely for many processes that involve semiconductor dopants.

When molecular ions are considered, both probability of ion formation and the depth from which an ion can originate influence the quantitation. Most molecular ion quantitation has focused on measurement of relative intensity at a suitably low dosage, where the signal is considered stable; in particular, the ion current is ratioed to an internal standard, also considered stable. This approach has been used successfully for the relative concentration of species in simple systems, like submonolayer to monolayer coverages of small molecules on metal substrates.[65,66,94,95] A major complication for the broad application of this approach is the lack of knowledge about a suitable primary ion dose for quantitative signal stability. In order to overcome this problem, a modification of the integration approach adopted for atomic ions was proposed by Clark and Gardella[96] and Cornelio and Gardella.[97]

The basic idea is measuring the total molecular ion signal, until the signal disappears, thus eliminating the concern about finding a specific primary ion dosage where the secondary molecular ion signal is stable.

Once a suitable concentration of molecular species has been determined, extending the knowledge to different depths is not as straightforward as depth profiling atomic species in dynamic SIMS. In general, the sampling depth of static SIMS is thought[67] to be on the order of a monolayer to greater than 20 Å, but studies on LB film models, using alternating layer model systems,[98,99] showed that sampling depth depends on the ion stability. In particular, ions formed after complex processes, such as cationization, originate from the topmost monolayer; chemically stable ions, like $(M-H)^-$ from acids can originate from much deeper depths, even two to five monolayers.

Applications of SIMS to Specific Polymers

A general investigation on the relationship existing between surface structure and bulk composition of complex polymeric materials has been performed at the State University of New York–Buffalo, in collaboration with the Naval Research Laboratory, in the last years. Under this project TOF-SIMS has been applied to the analysis of multicomponent polymers designed for minimal biological adhesion.

TOF-SIMS data relevant to segmented copolymers synthesized from poly(dimethylsiloxane), used as a difunctional macromonomer, and urea-urethane monomers (PU-PDMS), will be presented in the following as an example of static SIMS capabilities for polymer surface analysis.

Due to the lower surface energy of PDMS, siloxane segments are expected to segregate at the surface, giving the material its minimal adhesion properties. Indeed, angle-dependent XPS measurements show that a gradient in concentration exists for PDMS at the polymer surface, providing information relevant to the composition and thickness of both the surface PDMS-rich layer and the subsurface urea-urethane rich region.[100] However, no information can be obtained from XPS on the molecular weight distribution at the surface, which could explain surface properties that are not predicted by compositional differences. TOF-SIMS was then employed as a new approach to obtain this kind of information for PU-PDMS films.[64] Submonolayer films of PDMS were prepared on a silver substrate (to enhance cationization of polymer fragments) and analyzed as a reference.

The TOF-SIMS spectrum, in the high mass range (600–2000 m/z), for PDMS with molecular weight of 1000 is shown in Figure 16.14a. Regular groups of four peaks (i.e., having the same mass separations between the component peaks and similar intensity ratios) can be seen in the spectrum. Moreover, the separation between a particular peak in a group and the corresponding ones in the preceding and successive groups is always the same. This finding can be interpreted by considering that each group corresponds to

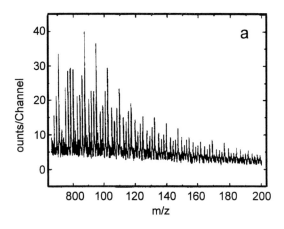

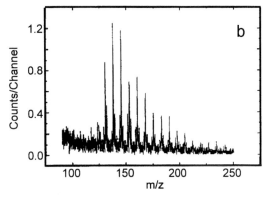

Figure 16.14 High-mass time-of-flight–secondary ion mass spectrometry (TOF-SIMS) spectra for (a) a submonolayer film of poly(dimethylsiloxane) (PDMS) (MW = 1000) on Ag in the range of 600 to 2000 mass per charge (m/z); (b) a thick film of polyurethane-PDMS (PDMS MW = 2400) on Al in the range of 900 to 2500 m/z).

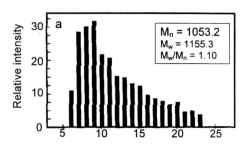

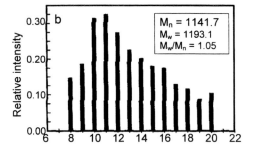

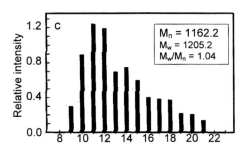

Figure 16.15 Comparison of molecular weight distributions obtained from time-of-flight–secondary ion mass spectrometry (TOF-SIMS) (m represents the DMS repeat units in fragments detected): (a) PDMS (MW = 1000); (b) PDMS segments segregated at the surface of the thick film of PU-PDMS (PDMS MW = 1000); (c) PDMS segments segregated at the surface of the thick film of PU-PDMS (PDMS MW = 2400).

a PDMS molecule of a particular length; one of the four peaks is the polymer molecule cationized by Ag^+, and the remaining three are related fragments. The difference between two contiguous groups in the spectrum is then a DMS unit. If the peak intensities are analyzed as a function of the number of dimethylsiloxane units (m), calculated from the m/z ratio of the corresponding peak, a molecular weight distribution is obtained, as shown in Figure 16.15a. The MW value thus found is consistent with the expected one.

The main assumption made with this approach is that intact polymer chains are turned into secondary ions, so that the SIMS spectrum mirrors the actual molecular weight distribution of the polymer. Generally speaking, this is not expected to be the primary process when thick, continuous polymer films are analyzed by SIMS, because chain entanglement and intrachain interactions usually result in high fragmentation probabilities.[88] Nevertheless, the volatility of the siloxane segment in PU-PDMS results in a higher probability for the generation of high mass ions from the surface, despite solid-state effects.

The same approach for the MW measurement was then adopted for the PU-PDMS copolymers, considering two average molecular weights for the PDMS prepolymer: 1000 and 2400 Da. In this case, the materials were analyzed as solvent-cast thick (about 50 µm) films. The high-mass TOF-SIMS spectrum for the PU-PDMS film with PDMS MW of 2400 is shown in Figure 16.14b, and the molecular weight distributions for both the copolymers are reported in Figure 16.15b and c.

Differently from the PDMS prepolymer, the mass range in which a distribution is observed for the copolymers is much lower than the effective molecular weights of the copolymers, then the ion series observed in the TOF-SIMS spectra of PU-PDMS samples arise probably from fragmentation processes. In particular, they correspond to the distribution of siloxane segments at the polymer surface, and it is interesting to note that the change in molecular weight for the PDMS prepolymer does not significantly affect the polysiloxane distribution at the surface of the copolymer. This results suggests that the surface of the copolymer with PDMS MW of 2400 should not have the different properties expected for longer PDMS segments.

Future Perspectives

The recent growth of TOF-SIMS measurements on polymers has been explosive. Its ability to provide high mass resolution and high spatial resolution for high mass polymers has led to wide acceptance of SIMS for surface studies of polymers.

The development of new ion sources continues. Molecular ion beams promise to distribute damage more broadly, improve ion yield (compared to common atomic ion sources), and improve detection of higher mass signals, although a specific application to polymers has yet to be fully developed. One area that will need to be further examined is the application of TOF-SIMS to study surface reaction chemistry.[101] Biodegradable polymeric materials are finding wide use in tissue engineering, drug delivery, and many other applications for environmental handling of polymers. Evaluating the kinetics of polymer surface reactions should be made possible by mass spectrometric techniques, because of the high degree of information within a mass spectrum.

All such applications will await the defeat of a major limitation, the desorption and detection of higher mass oligomeric ions and fragment ions. Presently, ions up to 10,000 Da can be detected by SIMS analysis, but rarely from thick polymer films, likely due to chain entanglement, which prevents easy desorption. If improved sample handling methods can be developed, more easily desorbed polymers (because of less chain entanglement and more volatility) will be studied. Such materials (i.e., siloxanes, fluorocarbons) are used as plasticizers, additives for lubrication, biocompatibility, aging and weathering stability, and so on. Analysis of the distribution of these materials is a major application that will drive the need to extend the mass range to higher masses. The possibility of determining the molecular weight distribution at a surface of a polymer will always be limited by problems with desorption.

Quantitative analysis with SIMS also remains a growth area, with great promise for the future. Modern instrumentation provides more reproducible signals with less sample damage, allowing more dependable, precise signal averaging and thus more dependable (relative) quantitative analysis. Much work still needs to be done on ion formation probabilities and understanding the relationship between signal intensity and concentration, or signal intensity and oligomeric distribution (for molecular weight analysis).

High spatial resolution with low sample damage is still a tricky possibility in practical SIMS analysis of polymers, but as with XPS, all the applications mentioned above will benefit from mass spectrometry. For example, compounds with equivalent carbon-hydrogen-nitrogen-oxygen-sulfur (CHNOS) content often are difficult to differentiate with XPS, but are easy with SIMS, since different molecular weights, fragmentation patterns, and the ability to measure smaller concentrations in mixtures are all advantages of SIMS over XPS. Combining high spatial resolution with these advantages suggests increasing application of SIMS in the analysis of weathering, defects, impurities, and materials processing applications.

Ion Scattering Spectrometry

Fundamental Principles

When an ion beam impinges on a solid surface, several phenomena may occur. One such phenomenon is scattering of primary ions. This process is exploited by ion scattering spectrometry (ISS) to draw information on the outmost layers of solid samples.

In 1957 Brunee[102] reported that alkali metal ions, with energies between 400 and 4000 eV, were scattered by a molybdenum plate and that the energies of the scattered ions showed a maximum. In the following years, experiments performed using H^+, He^+, N^+, O^+, and Ar^+ as primary ions with different energies showed that the energy maximum was related, in a particular experimental configuration, to the masses of the primary and of the target atoms.[103] At the end of the 1960s, the analytical potential of ion scattering was finally demonstrated by Smith.[104] Resolved peaks were observed for scattered noble gas ions from polycrystalline Mo and Ni with adsorbed CO. It was possible to assign them to the different atoms on the surface, and even obtain quantitative information on the adsorbed CO, starting from the relative intensities of signals due to carbon and oxygen. In other work,[105] the different faces of a CdS single crystal were distinguished by using the intensities of peaks due to cadmium and sulphur; the ability to provide crystallographic information on surfaces has since become one of the unique attributes of ISS.

From a theoretical point of view, the processes involved in ion scattering on a solid surface can be described accurately by means of quantum mechanics; nevertheless, in the conditions adopted for an ISS experiment, classical mechanics provide a good approximation. The laws of kinetic energy and momentum conservation allow the prediction of the energies for both incident ion and target atom after a binary elastic collision in which the kinetic energy of the incident ion is transferred to the target atom without any change in internal energy. Inelastic collisions are also possible, and, in this case, kinetic energy is partly spent for processes like ionization or electronic excitation. A schematic representation of the elastic scattering process, in which the target atom is at rest before the collision, is provided in Figure 16.16.

From the equations of conservation of kinetic energy and momentum, the following relation can be obtained:

$$\frac{E_1}{E_0} = \frac{[\cos\theta \pm (A^2 - \sin^2\theta)^{1/2}]}{(1+A)^2} \quad (16)$$

where E_1 and E_0 are the final and initial energies of the incident particle, respectively; A is the ratio of M_1, the incident ion mass, to M_2 the mass of the target atom; θ and is the angle of scattering. The model assumes that both particles can be considered to be rigid spheres. The minus sign has to be used when A is greater than 1, but both signs are valid for 1 greater than or equal to A greater than or equal to sin θ. According to eq 16, given a value for θ, the energy spectrum of the scattered ion depends only on the surface elemental composition of the sample.

The approximation that atoms in the lattice are free is assumed in the model of binary collisions and has been used successfully in interpreting ISS data obtained with light noble gases ions. This finding can be explained by at least three reasons: (1) the energy (hundreds of eV) involved in the collision is much higher than the lattice binding energy of the target atom (5–20 eV); (2) the short time of the collision ($10^{-15} - 10^{-16}$ s for ion beams with energies between 500 and 5000 eV), compared to the vibration period of the target atom (10^{-13} s); and the dependence of the interaction potential on the distance.

A fundamental aspect for analytical applications of ISS is the energetic resolution of peaks corresponding to elements with similar masses. The following relationship between mass (of the target atom) and energy (of the scattered ion) resolution has been found:[106]

$$\frac{\Delta E_1}{E_1} = g(A,\theta)\frac{\Delta M_2}{M_2} \quad (17)$$

where $g(A,\theta)$ is a function increasing with the angle of scattering and decreasing with A. This means that energy resolution can be improved, for a certain mass difference and angle of scattering, by using higher mass incident ions—for example, lowering A. A limit for this procedure is the increase of sputtering yields when A is less than or equal to 1.

For quantitative analysis, the following general equation can be written to express the intensity of an ISS signal:[107]

$$I_s = I_0 SN\sigma(1-P)R\Delta\Omega TD \quad (18)$$

in which I_0 is the incident beam intensity, S is the shadowing coefficient (see below), N is the density of atoms on the sample surface, σ is the differential scattering cross section, P is the probability of primary ion neutralization, R is a parameter related to the sample roughness, $\Delta\Omega$ is the solid angle of acceptance, and T and D are the transmission function and the detection efficiency of the ion analyzer, respectively.

A detailed analysis of the different contributions to eq 18 is beyond the aims of this chapter (a complete treatment, both on quantitative and on general aspects of ISS, can be found in reference 107). However, a brief description of the "shadowing" effect[108] can be useful to understand some of the peculiar applications of ISS.

As shown in Figure 16.17a, a "shadow cone" can be determined if a center of scattering is considered on the sample's surface; this feature is a consequence of the interaction between incoming ion and target atom in ISS, which can be described by means of a screened Coulomb potential.[109] Any atom of the sample lying inside the shadow cone of another one cannot interact with the incident ion and will be "invisible" in the ISS spectrum. However, if the angle of incidence is changed, the position of the shadow cone is also modified, as shown in Figure 16.17b, and atoms previously hidden can now interact with the incident ions.

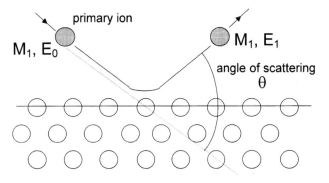

Figure 16.16 Schematic representation of ion scattering on a solid surface. Key: M_1 = mass of scattering ion; E_0 = initial energy of scattering ion; E_1 = energy after scattering.

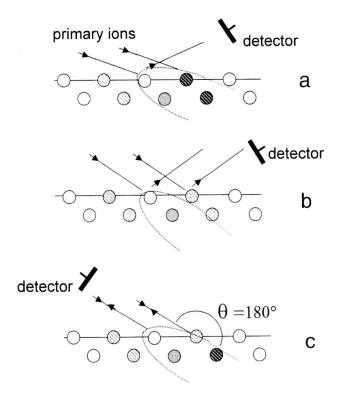

Figure 16.17 Shadowing effect in ion scattering spectrometry (ISS): (a) and (b) show scattering angles where darker atoms cannot scatter primary ions. (a) Shows a shallow angle where adjacent atoms are "in the shadow"; (b) shows a higher angle of incidence where the same two atoms are available for scattering; in (c) the collection of ions scattered at an angle of 180° is exploited in determining interatomic distances.

A particular value of the incidence angle can be found, for which an atom on the surface lies on the shadow cone due to the adjacent atom. Provided the shape of the shadow cone is known and the scattering angle adopted is almost 180° (Figure 16.17c), ISS enables a measurement of the interatomic distance. This particular approach, known as impact collision ISS,[110,111] has proved to be very useful for crystallographic applications.

Along with the shadowing coefficient, neutralization probabilities and surface roughness make "absolute" ISS quantitation, based on eq 18, very difficult; the influence of these factors has been discussed thoroughly in the literature.[112-114] The common approach for quantitation is then based on sensitivity factors, which have been applied especially in the surface analysis of binary alloys.[115]

ISS Instrumentation

Two general configurations have been adopted for ISS spectrometers, as represented schematically in Figure 16.18.

The most versatile arrangement, allowing variations of the scattering angle, is shown in Figure 16.18a; here, a mass filter is put between the ion source and the target in order to select a particular mass from a mixtures of isotopes. This feature can also be exploited to change the pri-

mary ion, when more gases (e.g., different noble gases) are used within the ion source simultaneously.

In ISS, as for static SIMS, the possible damage induced on the sample's surface is a major concern. Thus, the ion source is usually operated at very low currents ($10^{-9} - 10^{-8}$ amperes), and the primary ion energy is typically not higher than a few keV. Details on sources (and on the other components) can be found in fundamental texts like reference 107.

A hemispherical-sector analyzer is generally used to select the energies of the scattered ions;[116,117] this is due to its good energy resolution and also to the fact that, in most cases, ISS analysis is performed in XPS spectrometers by mounting the primary ion gun on the analysis chamber. An example of this multitechnique approach is the Leybold LHS10 spectrometer. In other cases a cylindrical mirror analyzer (CMA) has been adopted for energy selection: Figure 16.18b shows a typical configuration, in which the primary ion source is coaxial with the CMA.[118,119] Ions scattered by the sample are collected through a circular slit, at an angle of about 40° with the ion gun axis, and, after energy selection, detected on a ring detector. Compared to the hemispherical analyzer, the CMA provides a higher throughput, since the solid acceptance angle is 360°; however, energy resolution is lower.

In the case of polymers, a flood gun is also mounted on an ISS spectrometer to neutralize positive charge eventually generated on insulating samples.

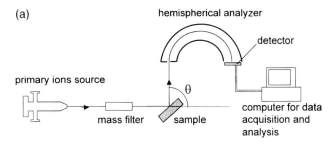

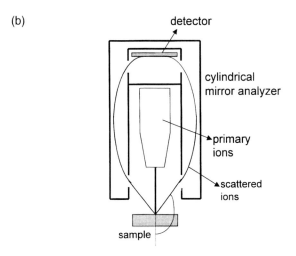

Figure 16.18 General configurations adopted for ISS spectrometers: with hemispherical ion analyzer (a) and with cylindrical mirror analyzer (b).

Applications of ISS to Polymer Surfaces

Due to its extreme surface sensitivity, ISS has been extensively applied in fundamental investigations of adsorption and desorption of surface layers,[120,121] segregation in alloys,[122,123] and surface reconstruction[124,125] and in the characterization of applied materials, especially catalysts.[107]

Application of ISS to polymers has been developed following two main directions: investigation on the functional group orientation, using the shadowing effect, and detection of surface chemical composition and reactivity due to molecular orientation. Early work was accomplished on methacrylates, in order to find low damage conditions that were able to preserve slight changes in surface conformation.[126,127] Actually, no significant differences between surface and bulk composition were observed with glassy methacrylates, but further work[128] proved that ISS could distinguish between conformations of stereoregular polymers (atactic, syndiotactic, and isotactic). In particular, changes in the ratio between the intensities of ions scattered from carbon and oxygen on poly(methyl methacrylates), and copolymers between them and poly(methacrylic acid) could be explained by shielding and shadowing of functional groups in different conformations.[127] By analogy, the Si/O ratio is influenced by the conformation of methyl and phenyl substituents on polysiloxanes.[128]

As a further application of ISS to the determination of side group orientation in polymers, data relevant to atactic isomers of poly(2-vinyl pyridine), P2VP, and poly(4-vinyl pyridine), P4VP, can be considered.[129] Molecular mechanics

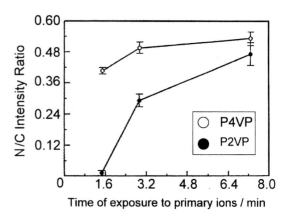

Figure 16.20 Plot of nitrogen/carbon scattered ion intensity ratios (see text) versus time of exposure to the primary ions for poly(2-vinyl-pyridine) (P2VP) and poly(4-vinyl-pyridine) (P4VP).

calculations predict that nitrogen atoms in the pyridine ring should be shielded by carbon atoms along the polymer backbone in P2VP, but not in P4VP. Indeed, ISS data, reported in Figure 16.19, show that no scattering of primary ions ($^3He^+$) from N atoms occurs in P2VP, whereas nitrogen is detected in P4VP. It is also interesting to consider the evolution of ISS spectra with time. The change in the ratio of ions scattered from nitrogen to ions scattered from carbon (the N/C intensity ratios) at different times of exposure to the primary ions are reported in Figure 16.20. An increase of the signal due to primary ions scattered by nitrogen in P2VP is observed and can be explained with sputtering of the polymer, leading to a similar surface composition for both isomers after long times.

A modification of pyridine rings orientation in P2VP can be obtained by increasing the steric hindrance of nitrogen, thus leading to a rotation of the rings away from the backbone. This effect was accomplished by quaternization of nitrogen with different alkyl groups; as shown in Figure 16.21, higher N/C ratios are observed with ISS on modified P2VP after short times. Of course, the shielding effect due to the alkyl group has to be considered, and it can explain the lower ratio observed for butyl groups than for ethyl ones. As for normal P2VP, the occurrence of sputtering leads to similar surface compositions for all the polymers after long times of exposure to the primary ions beam.

ISS also has been used, in conjunction with other surface spectroscopies, to obtain depth profiles of multicomponent polymeric materials. For example, the surface analysis of random block copolymers of bisphenol-A-polycarbonate (BPAC) and dimethyl siloxane (DMS), by means of XPS and ISS, can be cited.[130] Due to the lower surface energy of DMS, a surface enrichment of this component is expected; indeed, both XPS and ISS show a remarkable increase of the DMS percentage, as compared to the bulk composition. However, the degree of enrichment observed with ISS, in the top 3 to

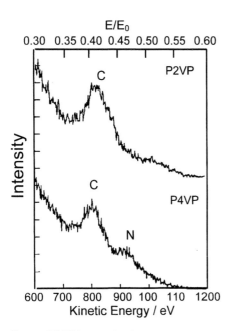

Figure 16.19 Ion scattering spectrometry (ISS) spectra for poly(2-vinyl-pyridine) (P2VP) and poly(4-vinyl-pyridine) (P4VP).

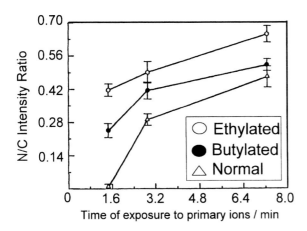

Figure 16.21 Plot of nitrogen/carbon scattered ion intensity ratios (see text) versus time of exposure to the primary ions for poly(2-vinyl-pyridine) (P2VP) modified by different alkyl groups.

5 Å region, is higher, thus suggesting two possible models for surface morphology, both characterized by domains of BPAC and DMS arranged perpendicularly to the surface. According to the first model, the increase in DMS concentration at the topmost layers can be explained by raising of DMS regions above the adjacent BPAC regions, which would be hidden inside the DMS "shadow". The alternative model predicts an extreme enhancement of DMS surface segregation, leading to a uniform siloxane layer, a few Ångstrom thick, on top of the vertical domains of the two polymers.

Polymer blends obtained from homopolymers of BPAC and DMS have also been studied using ISS and XPS.[131] Both techniques show that different blends, with DMS percentages lower than 11% (by weight), always have the same surface concentration of DMS, which is 85% by weight. The formation of a compatible mixture of the two polymers in that range of bulk compositions has been postulated. It is worth noting that at bulk concentrations of DMS higher than 11%, a complete phase separation occurs, and the surface is 100% DMS.

Future Perspectives

Applications of low-energy ISS to polymers have been few and far between in the past 10 years. However, adapting TOF detectors to ISS is a major unrealized area of application. Currently, one TOF-SIMS manufacturer allows the reconversion of the instrumentation to also accomplish low-energy ISS. With the higher signal throughput inherent in the TOF analyzer, many experiments that are difficult to do with current (adapted XPS) instrumentation would be easier. However, the fundamental limitation of ISS of polymers is the need for basic knowledge of the structure and composition of the surface, as well as the need to examine a series

of treatments or similar surfaces to determine relative differences. Therefore, ISS will remain a technique that can reveal interesting information about the fundamental configurations at the topmost surface of polymer materials, but not a technique where applications in practice can drive further need for the instrumentation and use of the technique, as is the case for XPS and SIMS.

The study of inelastic collision ISS (ICISS) for functional group orientation at polymer surfaces remains a tantalizing application, as does the use of ISS to examine liquid crystalline polymers of different structures and phases.

References

1. Garbassi, F.; Morra, M.; Occhiello E. *Polymer Surfaces: From Physics to Technology*; John Wiley and Sons: Chichester, 1994.
2. Clark, D. T. In *Handbook of X-Ray and Ultraviolet Photoelectron Spectroscopy*, Briggs, D., Ed.; Heyden: London, 1977, p 211.
3. Gardella, J. A. Jr. *Anal. Chem.* **1989**, *61*, 589.
4. Benninghoven, A. *Z. Phys.* **1970**, *230*, 403.
5. Vargo, T. A.; Gardella, J. A. Jr.; Schmitt, R. L.; Hook, T. J.; Salvati, L. Jr. In *Surface Characterization of Advanced Polymers*; Sabbatini, L., Zambonin, P. G., Eds.; VCH: Weinheim, 1993; p 163.
6. Siegbahn, K.; Nordling C. N.; Fahlman, A.; Nordberg, R.; Hamrin, K.; Hedman, J.; Johansson, G.; Bermark, T.; Karlsson, S. E.; Lindgren, I.; Lindberg, B. *ESCA: Atomic, Molecular and Solid State Structure Studied by Means of Electron Spectroscopy*; Almqvist and Wiksells: Uppsala, 1967.
7. Powell, C. J. *J. Electron. Spectrosc. Rel. Phenom.* **1988**, *47*, 197.
8. Seah, M. P.; Dench, W. A. *Surf. Interface Anal.* **1979**, *1*, 2.
9. Moulder, J. F.; Stickle, W. F.; Sobol, P. E.; Bomben, K. D. *Handbook of X-ray Photoelectron Spectroscopy*; Perkin Elmer Corporation: Eden Prairie, MN, 1992.
10. Wagner, C. D.; Davis, L. E.; Zeller, M. V.; Taylor, J. A.; Raymond, R. H.; Gale, L. H. *Surf. Interface Anal.* **1981**, *3*, 211.
11. Auger, P. *J. Phys. Radium*, **1925**, *6*, 205.
12. Briggs, D.; Seah, M. P., Eds. *Practical Surface Analysis*; Wiley: New York, 1992; Vol. 1.
13. Seah, M. P.; Jones, M. E.; Anthony, M. T. *Surf. Interface Anal.* **1984**, *6*, 242.
14. Zhuang, H.; Gribbin Marra K.; Ho, T.; Chapman, T. M.; Gardella, J. A. Jr. *Macromolecules* **1996**, *29*, 1660.
15. Vargo, T. G.; Gardella, J. A. Jr. *J. Vac. Sci. Technol. A*, **1989**, *7*(3), 1733.
16. Hofmann S. In *Practical Surface Analysis*; Briggs, D., Seah, M. P., Eds.; Wiley: New York, 1990; Vol. 1, p 143.
17. Gardella, J. A. Jr. *Anal. Chem.* **1989**, *61*, 589.
18. ASTM. E 902-94, Standard Practice for Checking the Operating Characteristics of X-Ray Photoelectron Spectrometers and E 996-94, Standard Practice for Reporting Data in Auger Electron Spectroscopy and X-ray Photo-

electron Spectroscopy. *Annual Book of ASTM Standards.* ASTM: West Conshohocken, PA, 1999.

19. Fulghum, J. E. *J. Electron. Spectrosc. Rel. Phenom.* **1999,** *100,* 331.
20. Beamson, G.; Briggs, D. *High Resolution XPS of Organic Polymers*; Wiley: New York, 1992.
21. Beamson, G.; Clark, D. T.; Kendrick, J.; Briggs, D. *J. Electron Spectrosc. Rel. Phenom.* **1991,** *57,* 79.
22. Pepper, S. V.; Wheeler, D. R. *J. Vac. Sci. Technol.* **1982,** *20,* 226.
23. Chaney, R.; Barth, G. *Fres. Z. Anal. Chem.* **1987,** *329,* 143.
24. Chang, H. P.; Thomas, J. H. III *J. Electron Spectrosc. Rel. Phenom.* **1982,** *26,* 203.
25. Le Moel, A.; Duraud, J. P.; Balanzat, E. *Nucl. Instr. Methods Phys. Res.* **1986,** *B18,* 59.
26. Hendrickson, D. N.; Hollander, J. M.; Jolly, W. L. *Inorg. Chem.* **1969,** *8,* 2642.
27. Koopmans, T. S. *Physica,* **1934,** *1,* 104.
28. Siegbahn, K.; Nordling, C.; Johansson, G.; Hedman, J.; Heden, P. F.; Hamrin, K.; Gelius, U. I.; Bergmark, T.; Werme, L. O.; Manne, R.; Baer, Y. *ESCA Applied to Free Molecules*; North Holland: Amsterdam, 1969.
29. McIntyre, N. S.; Cook, M. G. *Anal. Chem.* **1975,** *47,* 2208.
30. Brundle, C. R.; Chuang, T. J.; Rice, D. W. *Surf. Sci.* **1976,** *60,* 286.
31. Wagner, C. D.; Joshi, A. *J. Electron Spectrosc. Rel. Phenom.* **1988,** *47,* 283.
32. Ansell, R. O.; Dickinson, T.; Povey, A. F.; Sherwood, P. M. A. *J. Electroanal. Chem.* **1979,** *98,* 79.
33. Briggs, D. In *Practical Surface Analysis*; Briggs, D., Seah, M. P., Eds.; Wiley: New York, 1990; Vol. 1, p 448.
34. Chilkoti, A.; Ratner, B. D. In *Surface Characterization of Advanced Polymers*; Sabbatini, L., Zambonin, P. G., Eds.; VCH: Weinheim, FRG, 1993; p 221.
35. Dall'Olio, D.; Dascola, Y.; Varacca, V.; Bocchi, V. *Comptes Rendus Acad. Res. Ser.* **1968,** *C267,* 433.
36. Skotheim, T. A., Elsenbaumer, R. L., Reynolds, J. R., Eds. *Handbook of Conducting Polymers*, 2nd ed.; Marcel Dekker: New York, 1998.
37. Scrosati, B., Ed. *Applications of Electroactive Polymers*; Chapman and Hall: London, 1993.
38. Salaneck, W. R.; Clark, D. T.; Samuelsen, E. J., Eds. *Science and Applications of Conducting Polymers*; Adam Hilger: Bristol, 1991.
39. Malitesta, C.; Morea, G.; Sabbatini, L.; Zambonin P. G. In *Surface Characterization of Advanced Polymers*; Sabbatini, L., Zambonin, P. G., Eds.; VCH: Weinheim, FRG, 1993; p 181.
40. Sabbatini, L.; Malitesta, C.; De Giglio, E.; Losito, I.; Torsi, L.; Zambonin, P. G. *J. Electron Spectr. Rel. Phen.* **1999,** *35,* 100.
41. Malitesta, C.; Palmisano, F.; Torsi, L.; Zambonin, P. G. *Anal. Chem.* **1990,** *62,* 2735.
42. Centonze, D.; Guerrieri, A.; Malitesta, C.; Palmisano, F.; Zambonin, P. G. *Fres. J. Anal. Chem.* **1992,** *342,* 729.
43. Centonze, D.; Malitesta, C.; Palmisano, F.; Zambonin, P. G. *Electroanalysis* **1994,** *6,* 423.
44. Palmisano, F.; Guerrieri, A.; Quinto, M.; Zambonin, P. G. *Anal. Chem.* **1995,** *67,* 1005.
45. Losito, I. Ph.D. thesis, University of Bari, Bari, Italy, 1998.
46. Palmisano, F.; Malitesta, C.; Centonze, D.; Zambonin, P. G. *Anal. Chem.* **1995,** *67,* 2207.
47. Malitesta, C.; Losito, I.; Sabbatini, L.; Zambonin, P. G. *J. Electron Spectr. Rel. Phen.* **1995,** *76,* 629.
48. Pfluger, P.; Street, G. B. *J. Chem. Phys.* **1984,** *80,* 544.
49. Niwa, Y.; Kobayashi, H.; Tsuchiya, T. *J. Chem. Phys.* **1974,** *60,* 799.
50. Kumar, S. N.; Bouyssoux, G.; Gaillard, F. *Surf. Interface Anal.* **1990,** *15,* 531.
51. Wagner, C. D. *NIST X-ray Photoelectron Spectroscopy Database,* Vers. 1.0; National Institute of Standard and Technology: Washington, DC, 1989.
52. Malitesta, C.; Losito, I.; Sabbatini, L.; Zambonin, P. G. *J. Electron Spectr. Rel. Phen.* **1998,** *97,* 199.
53. Tourillon, G.; Jugnet, Y. *J. Chem. Phys.* **1988,** *89,* 1905.
54. Vargo, T. G.; Gardella, J. A. Jr.; Meyer, A. F.; Baier, R. E. *J. Polym. Sci., Polym. Chem.* **1991,** *29,* 555.
55. Hook, D. J.; Vargo, T. G.; Gardella, J. A. Jr.; Litwiler, K. S., Bright, F. V. *Langmuir,* **1991,** *7,* 142.
56. Vargo, T. G.; Thompson, P. M.; Gerenser, L. J.; Valentini, R. F.; Aebischer, P.; Hook, D. J.; Gardella, J. A. Jr. *Langmuir,* **1992,** *8,* 130.
57. Bright, F. V.; Litwiler, K. S.; Vargo, T. G.; Gardella, J. A. Jr. *Anal. Chim. Acta,* **1992,** *262,* 323.
58. Valentini, R. F.; Vargo, T. G.; Gardella, J. A. Jr.; Aebischer, P. *Biomaterials,* **1992,** *13,* 183.
59. Ranieri, J. P.; Bellamkonda, R.; Jacob, J.; Vargo, T. A.; Gardella, J. A. Jr.; Aebischer, P. *J. Biomed. Mater. Res.* **1993,** *27,* 917.
60. Vargo, T. G.; Bekos, E. J.; Kim, Y. S.; Ranieri, J. P.; Bellamkonda, R.; Aebischer, P.; Margevich, D. E.; Thompson, P. M.; Bright, F. V.; Gardella, J. A. Jr. *J. Biomed. Mater. Res.* **1995,** *29,* 767.
61. Ranieri, J. P.; Bellamkonda, R.; Bekos, E. J.; Vargo, T. G.; Gardella, J. A. Jr.; Aebischer, P. *J. Biomed. Mater. Res.* **1995,** *29,* 779.
62. Ranieri, J. P.; Bellamkonda, R.; Bekos, E. J.; Gardella, J. A. Jr.; Mathieu, H. J.; Ruiz, L.; Aebischer, P. *Int. J. Devl. Neuroscience,* **1994,** *12,* 725.
63. Hawkridge, A. M.; Sagerman, G. and Gardella, J. A. Jr., *J. Vac. Sci. Technol., Part A* **2000,** *18*(2), 567.
64. Zhuang, H. Z.; Gardella, J. A. Jr. *MRS Bull.* **1996,** *21,* 43.
65. Garrison, B. J.; Winograd, N. *Science* **1982,** *226,* 805.
66. Benninghoven, A.; Rudenauer, F. G.; Werner, H. W., Eds. *Secondary Ion Mass Spectrometry: Basic Concepts, Instrumentation Aspects, Applications and Trends*; Wiley: New York, 1987.
67. Benninghoven, A. Z. *Physik* **1970,** *230,* 403.
68. Benninghoven, A. *J. Vac. Sci. Technol.* **1985,** *A3,* 451.
69. Sigmund, P. In *Inelastic Ion-Surface Collisions*; Tolk, N. H., Tully, J. C., Heiland, W., White, C. W., Eds.; Academic Press: New York, 1987, p 121.
70. Herzog, R. F. K.; Viehbock, F. *Phys. Rev.* **1949,** *76,* 855.
71. Honig, R. E. *J. Appl. Phys.* **1958,** *29,* 549.
72. Day, R. J.; Unger, J. A.; Cooks, R. G. *Anal. Chem.* **1980,** *52,* 557A.
73. Benninghoven, A.; Lange, W.; Jirikowsky, M.; Holtkamp, D. *Surf. Sci.* **1982,** *123,* L721.
74. Pachuta, S. J.; Cooks, R. G. *Chem. Rev.* **1987,** *87,* 647.

75. Briggs, D.; Seah, M. P., Eds. *Practical Surface Analysis*; Wiley: New York, 1992; Vol. 2.

76. Mantus, D. S.; Morrison, G. H. *J. Vac. Sci. Technol.* **1989,** *A8*, 2209.

77. Colton, R. J.; Campana, J. E.; Kidwell, D. A.; Ross, M. M.; Wyatt, J. R. *Appl. Surf. Sci.* **1985,** *21*, 168.

78. Chait, B. T.; Standing, K. G. *Int. J. Mass Spectrom. Ion Phys.* **1981,** *40*, 185.

79. Demirev, P.; Olthoff, J. K.; Fenselau, C.; Cotter, R. J. *Anal. Chem.* **1987,** *59*, 1951.

80. Barber, M.; Bordoli, R. J.; Sedgwick, R. D.; Tyler, A. N. *Nature* **1981,** *293*, 270.

81. Macfarlane, R. D.; Torgerson, D. F. *Science* **1976,** *191*, 920.

82. Hillenkamp, F.; Karas, M.; Holtkamp, D.; Klusener, P. *Int. J. Mass Spectrom. Ion Proc.* **1986,** *69*, 265.

83. Gardella, J. A. Jr.; Hercules, D. M. *Anal. Chem.* **1980,** *52*, 226.

84. Gardella, J. A. Jr.; Hercules, D. M. *Anal. Chem.* **1981,** *53*, 1879.

85. Briggs, D.; Hearn, M. J. *Vacuum* **1986,** *36*, 1005.

86. Bletsos, I. V.; Hercules, D. M.; Greifendorf, D.; Benninghoven, A. *Anal. Chem.* **1985,** *57*, 2384.

87. Bletsos, I. V.; Hercules, D. M.; van Leyen, D.; Benninghoven, A. *Macromolecules* **1987,** *20*, 407.

88. Bletsos, I. V.; Hercules, D. M.; Magill, J. H.; van Leyen, D.; Niehuis, E.; Benninghoven, A. *Anal. Chem.* **1988,** *60*, 938.

89. Bletsos, I. V.; Fowler, D.; Hercules, D. M.; van Leyen, D.; Benninghoven, A. *Anal. Chem.* **1990,** *62*, 1275.

90. Bletsos, I. V.; Hercules, D. M.; van Leyen, D.; Hagenoff, B.; Niehuis, E.; Benninghoven, A. *Anal. Chem.* **1991,** *63*, 1953.

91. Riviere, J. C. *Surface Analytical Techniques*; Clarendon Press: Oxford, 1990; p 504.

92. Anderson, C. A.; Hinthorne, J. *Anal. Chem.* **1973,** *45*, 1421.

93. Morgan, A. E.; Werner, H. W. *Anal. Chem.* **1976,** *48*, 699.

94. Lange, W.; Jirikowski, M.; Benninghoven, A. *Surf. Sci.* **1975,** *79*, 549.

95. Tamaki, Sichtermann, W.; Benninghoven, A. *Jap. J. Appl. Phys.* **1984,** *23*(5), 544.

96. Clark, M. B. Jr.; Gardella, J. A. Jr. *Anal. Chem.* **1990,** *62*, 870.

97. Cornelio, P. A.; Gardella, J. A. Jr. *J. Vac. Sci. Technol.* **1990,** *A8*(3), 2283.

98. Cornelio-Clark, P. A.; Gardella, J. A. Jr. *Langmuir,* **1991,** *7*, 2136.

99. Johnson, R. W. Jr.; Cornelio-Clark P. A.; Gardella, J. A. Jr. In *Macromolecular Assemblies*; Stroeve, P., Balazs, A., Eds.; ACS Symposium Series; American Chemical Society: Washington, DC, 1992.

100. Chen, X.; Gardella, J. A. Jr.; Ho, T.; Wynne, K. J. *Macromolecules* **1995,** *28*, 1635.

101. Chen, J. X.; Gardella, J. A. Jr., *Macromolecules* **1999,** *32*, 7380.

102. Brunee, C. *Z. Phys.* **1957,** *147*, 161.

103. Panin, B. V. *Sov. Phys. JEPT* **1962,** *42*, 313; idem **1962,** *15*, 215.

104. Smith, D. P. *J. Appl. Phys.* **1967,** *38*, 340.

105. Smith, D. P. *Surf. Sci.* **1971,** *25*, 171.

106. Taglauer, E.; Heiland, W. *Appl. Phys.* **1976,** *9*, 261.

107. Niehus, H. In *Practical Surface Analysis*; Briggs, D., Seah, M. P., Eds.; Wiley: New York, 1992; Vol. 2, p 507.

108. Aono, M.; Zaima, C.; Otani, S.; Ishizawa, Y. *Jap. J. Appl. Phys.* **1981,** *20*, L829.

109. Buck, T. M. In *Methods of Surface Analysis*; Czanderna, A. W., Ed.; Elsevier: Amsterdam, 1975; chap. 3.

110. Aono, M.; Hou, H.; Oshima, C.; Ishizawa, Y. *Phys. Rev. Lett.* **1982,** *49*, 567.

111. Niehus, H.; Comsa, G. *Surf. Sci.* **1984,** *140*, 18.

112. Baun, W. L. *Appl. Surf. Sci.* **1977,** *1*, 81.

113. Baun, W. L. In *Quantitative Surface Analysis of Materials*; ASTM STP 643; McIntyre, N. S., Ed.; American Society for Testing and Materials: Philadelphia, 1978; p 150.

114. Powell, C. J., *Appl. Surf. Sci.* **1980,** *4*, 492.

115. Swartzfager, D. G. *Anal. Chem.* **1984,** *56*, 55.

116. Martin, P. J.; Netterfield, R. P. *Surf. Interface Anal.* **1987,** *10*, 13.

117. Taglauer, E. *Appl. Phys.* **1985,** *A38*, 161.

118. Brongersma, H. H.; Meijer, F.; Werner, H. W. *Philips Tech. Rev.* **1974,** *34*, 357.

119. McCune, R. C. *J. Vac. Sci. Technol.* **1981,** *18*(3), 700.

120. Oshima, C.; Aono, M.; Tanaka, T.; Kawai, S. *Surf. Sci.* **1981,** *102*, 312.

121. Taglauer, E.; Heiland, W.; Orsgaard, J. *Nucl. Instrum. Meth.* **1980,** *168*, 571.

122. Kang, H. J.; Shimizu, R.; Okutani, T. *Surf. Sci.* **1982,** *116*, L173.

123. Katayama, I.; Oura, K.; Shoji, F.; Hanawa, T. *Phys. Rev. B* **1988,** *38*, 2188.

124. Buck, T. M., Wheatley, G. H.; Jackson, D. P. *Nucl. Instrum. Meth.* **1983,** *218*, 257.

125. Niehus, H.; Comsa, G. *Nucl. Instrum. Meth. B* **1986,** *15*, 122.

126. Gardella, J. A. Jr.; Hercules, D. M. *Anal. Chem.* **1980,** *52*, 226.

127. Gardella, J. A. Jr.; Hercules, D. M. *Anal. Chem.* **1981,** *53*, 1879.

128. Hook, T. J.; Schmitt, R. L.; Gardella, J. A. Jr.; Salvati, L. Jr.; Chin, R. L. *Anal. Chem.* **1986,** *58*, 1285.

129. Hook, T. J.; Gardella, J. A. Jr.; Salvati, L. Jr. *Macromolecules* **1987,** *20*, 2112.

130. Schmitt, R. L.; Gardella, J. A. Jr.; Magill, J. H.; Salvati, L. Jr.; Chin, R. L. *Macromolecules* **1985,** *18*, 2675.

131. Schmitt, R. L.; Gardella, J. A. Jr.; Salvati, L. Jr. *Macromolecules* **1986,** *19*, 648.

17

Optical Methods for Three-Dimensional Visualization

Photon Tunneling Microscopy and Laser Scanning Confocal Microscopy

MOHAN SRINIVASARAO

JUNG OK PARK

Current research is interested in assessing the manifestations of a "rough surface" in a variety of fields, ranging from optics and tribology to the study of wetting phenomena.[1-4] Certainly, "rough surfaces" are ubiquitous in nature and otherwise. But before we answer questions such as how rough a given surface is, it is worthwhile to understand how a solid surface is formed. Most solid surfaces are formed as a result of one or more of the following: (a) solidification of a liquid; (b) machining or other material removal processes, followed perhaps by polishing; (c) deposition of a material on an existing solid surface; and (d) by fracture of solids.[5] Inevitably, such processes will not yield smooth surfaces or surfaces considered atomically smooth or flat. Flat surfaces that are atomically smooth can be obtained only under very carefully controlled conditions and are quite rare in nature. It then becomes very important to understand a variety of issues related to rough surfaces, not the least of which is to have a measure of surface roughness. Then one is led to ask the most obvious questions: How rough is rough? And when does one call a seemingly smooth surface smooth?

It is quite clear "smooth" and "rough", when applied to surfaces, are quite subjective and qualitative terms. For example, a polished glass surface (or, for that matter, any solid surface) may appear very smooth when the instrument probing the surface is a finger. However, an optical microscope may reveal the hills and valleys on the apparently smooth surface. The problem here is that the finger as a sensor can measure lateral roughness on the scale of about 1/2 cm and about 100 μm in the thickness direction (typical resolution of a finger!).[3] In contrast, a good optical microscope can measure lateral roughness on the scale of the resolution of a microscope, typically about $\lambda/2$, where λ is the wavelength of light, coupled with the ability to measure or distinguish 0.1 μm variation in the thickness direction. So two different people using the two different tools (a finger and a microscope, for example) will inevita-

bly come to different conclusions about the "roughness" of a given surface. As tools with higher and higher sensitivity are used to probe the surface, they reveal that scale-dependent roughness is ubiquitous in nature.[3,4]

In fact, for most solid surfaces observed under ever increasing magnification or sensitivity, one should not be surprised to find roughness all the way down to atomic dimensions, where the roughness appears in the form of atomic steps.[6] The basic nature of a solid surface can be schematically represented as shown in Figure 17.1. The roughness of a surface is often random (and disordered) and does not appear to follow any particular pattern. It is this nature of randomness and roughness on multiple scales that contribute to the complexity of surfaces.[7-9]

In all of the discussion so far, we have tacitly assumed that the complexity of the surface arises only from the topographical nature of the surface. But another form of surface complexity plays a critical role in a number of applications. It has to do with the surface composition of materials, which is usually quite different from that of the bulk of the material. In this chapter, however, we do not deal with the issue of surface composition, simply because the optical methods will not allow us to explore the compositional differences.

A variety of methods is available for the study of surface roughness. Perhaps the most widely used and practically accepted tool for such measurements is the mechanical profiler,[3,10,11] which consists of a narrow diamond stylus that is lightly traced across the surface. The motion of the stylus produces a time-varying voltage, which is proportional to the height of the surface profile. This type of an instrument runs into problems similar to using the finger as a tool to measure surface roughness, in that accurate information about the surface topography is obtained as long as the width of the stylus tip is small compared with the lateral size of the surface irregularities or surface topography. Even though the profiler provides useful information, it suffers

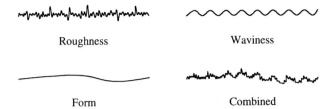

Figure 17.1 Classification of continuous surface errors.

from two drawbacks that are significant: (1) it can damage the object under investigation, and this is particularly true for objects made from the so-called soft materials; (2) less sophisticated stylus instruments use velocity (or other types of motion) transducers that do not directly yield the surface profile. Hence, their output is often limited to the average of the vertical deviation of the surface irregularities. The correlation length can be obtained by taking the autocorrelation of the time varying voltage;[10] however, this is done primarily for research instruments rather than in manufacturing environment. Furthermore, most stylus profilers provide information in one dimension only, because its speed is limited for obtaining information in two dimensions. Recent developments have improved things significantly, to the extent that the stylus instrument may be a practical alternative for measurement in two dimensions.[11]

In contrast, a variety of scanning probe microscopes have been developed that provide information in two dimensions. The atomic force microscope[12,13] is an example of such an instrument; it provides information in two dimensions with very good vertical sensitivity down to the Å level. In this article, we primarily focus on optical methods used to probe the surface. The reader is referred to many excellent review articles and books on using the atomic force microscope for studies of the surface.[12,13]

Considerable effort has gone into developing optical methods to probe the surface roughness and quantitatively measure the surface topography. Both interferometric and light-scattering methods[1,10,11] have been extensively studied from various points of view. Since the advent of lasers, the speckles produced by a coherent laser light source have received considerable attention, with special focus on the speckle patterns and their connection to surface roughness.[14] It has been recognized that one can use evanescent waves to probe the surface roughness and obtain topography of the surface. However, the use of evanescent waves limits the vertical scale that can be studied to distances on the order of a wavelength, at best.

The measurement of surface roughness using optical methods will be the theme of this chapter. The chapter is organized as follows: a brief description of statistical characterization of roughness; a discussion of imaging in general; and specific sections devoted to two imaging techniques, photon tunneling microscopy (PTM) and laser scanning confocal microscopy (LSCM), both of which provide detailed

images of roughness at different length scales and complement each other. Interference microscopy is another important technique for studying polymer surfaces. It can be used both for imaging the topography of smooth polymer surfaces in the phase shifting mode and for imaging the topography of rough polymer surfaces in the white light vertical-scanning mode.

The purpose of this chapter is to present recent advances in optical microscopy and to discuss some applications where these techniques find use in the characterization of topographical features displayed by polymeric surfaces created by a variety of physical processes. We describe here techniques that have been developed in the past decade and that have achieved the goal of enhancing both the lateral (spatial) and axial resolution.

Statistical Treatment of Surface Roughness

Roughness, as the word implies, is a measure of the topographic relief of a surface. Examples of surface relief include polishing marks on optical surfaces, machined parts, waviness on silicon surfaces, polymer film dewetting from a surface,[15] and the more exotic surface relief structures on butterfly wings, which give rise to their beautiful iridescent colors.[16] In optics, a clear distinction has been assumed to exist between surface "form" or "figure", surface polish, and surface defects or flaws. Surface form is the departure of the general shape of the surface from its intended or ideal shape (Figure 17.1), usually a plane or a sphere. Such deviations are measured using optical interferometry. Surface polish (meaning, lack of roughness!) and localized surface defects (sometimes termed "scratch and dig") are usually assessed visually in a qualitative fashion by trained inspectors without resorting to other measurement tools.

Regarding the surface roughness as the variation of surface height and assuming this variation to be random and isotropic, we can characterize the roughness in a statistical manner, using a variety of parameters and functions. A *surface parameter* is a single number that characterizes a surface, whereas a *surface function* is a set of numbers and hence contains more information. Parameters may be classified as averaging parameters (such as average roughness), extreme-value height parameters (such as the 10-point height), shape parameters (such as skewness and kurtosis), spacing parameters (such as mean spacing of profile irregularities), and hybrid parameters (such as the average slope). Likewise, there are several functions to describe the surface roughness; height and slope distribution functions, autocovariance function, autocorrelation function, and power spectral density function. We discuss only a few examples here, and interested readers are referred to other literatures for more detail.[4,10,11,17–24]

Average Parameters

The parameter most frequently used for machined surfaces[17] is the average departure of surface height from its mean value, usually called the *average roughness* (R_a) or the *centerline average roughness*. This average roughness is simply defined in terms of the surface height variations, z_i or $z(x)$, from the mean surface level:

$$R_a = \frac{1}{n}\sum_{i=1}^{n}|z_i| \qquad (1)$$

or

$$R_a = \frac{1}{L}\int_0^L |z(x)|\,dx \qquad (2)$$

where x is the coordinate along the surface, and L is the measurement length. For other parameters described in this section, we will include only the integral expressions. A typical R_a value resulting from a coarse machining is 3–10, from fine machining 1–3, and from polishing 0.02–0.4, respectively.[21]

The quantity commonly used in the optics industry is the *root-mean-square (rms) roughness*, R_q, where the squared term gives greater significance to surface variations farther away from the mean. It is also known as σ in optics and statistics, and defined as

$$R_q = \left[\frac{1}{L}\int_0^L z^2(x)\,dx\right]^{1/2} \qquad (3)$$

The relationship between R_a and R_q depends on the particular statistical distribution of surface height present in any given case. In the case of a Gaussian distribution, for example, it is noted that R_a is approximately equal to 0.8 R_q.[21] In the machining industry, it is usually assumed that R_a is approximately equal to 0.9 R_q,[18] while the multiplying factor is smaller for surfaces with abnormally high numbers of large deviations from the mean surface height.

Both R_a and R_q are in almost universal use for general quality control, because they are easy to define and measure, and they provide good general description of height variations. Unlike R_a, R_q has extra theoretical significance as the standard deviation of the height distribution. One serious drawback in using both R_a and R_q values is that they may fail to distinguish between a relatively gently undulating surface and one with a much spikier profile. Moreover, they are not intrinsic properties of the profile and depend on the measurement length, the surface area being averaged for each measurement (lateral resolution), and the distance between data points.[10,11]

Shape Parameters

Skewness, S_k, is the third central moment of the probability density, $p(z)$, of a distribution of heights[4] and indicates whether the distribution of surface heights is biased (skewed) toward positive or negative height deviations:

$$S_k = \frac{1}{R_q^3}\int_{-\infty}^{+\infty} z^3\, p(z)\,dz \qquad (4)$$

A generally smooth surface with occasional but relatively large "hills" would have a positive skewness, while a surface with pits or holes would have a negative skewness.

Kurtosis, K, is defined in terms of the fourth power of the surface height distributions:[17]

$$K = \frac{1}{R_q^4}\int_{-\infty}^{+\infty} z^4\, p(z)\,dz \qquad (5)$$

A Gaussian distribution has a kurtosis with a numerical value of 3. Profiles that have appreciable numbers of particulates and pits on a surface are said to be *leptokurtic* and have K greater than 3. Profiles that have proportionately fewer high and low extreme points than a Gaussian have a kurtosis of K less than 3 and are said to be *platykurtic*.

Skewness and kurtosis are useful parameters for evaluation of machined surfaces. The skewness, for example, offers a convenient way to illustrate load-carrying capacity and porosity.[17]

Functions

Differences in surface texture may be more systematically described by investigating the correlation between pairs of points on a profile as the separation of the pairs is varied. The *autocovariance function* was developed to describe time series, and the power spectral density function is its Fourier transform. For a profile of length L, the autocovariance function, $R(\tau)$, is defined as[4]

$$R(\tau) = \frac{1}{L-\tau}\int_0^{L-\tau} z(x)z(x+\tau)\,dx \qquad (6)$$

where $z(x)$ and $z(x+\tau)$ are pairs of heights separated by a time delay τ. The autocovariance function has dimensions of height squared. The *autocorrelation function*, $\rho(\tau)$, is a normalized form of the autocovariance,[4]

$$\rho(\tau) = \frac{R(\tau)}{R(0)} \qquad (7)$$

where $R(0)$ is the variance R_q^2 of the height distribution. The autocorrelation function is dimensionless, with an initial value of unity at zero lag distance.

Autocovariance and autocorrelation functions are a measure of how well the surface height at one point can be predicted from the height at a second point. If the two points are very close together, the heights at the two points are highly correlated. As the distance between the two points (the *lag*) is increased, the correlation decreases until, after a certain distance, there is no correlation between the two surface heights. The distance over which the correlation decays to some fraction of its initial value is called the *correlation length*, sometimes taken as a tenth, sometimes as

$1/e$. Pairs of points separated by distances greater than the correlation length are statistically independent.

The *power spectrum*, *Wiener spectrum*, or power spectral density function of the surface height variations gives the distribution of the spatial frequencies in the roughness, defined by the Fourier transform of the autocovariance:[4]

$$G\left(\frac{\omega}{2p}\right) = 4 \int_0^\infty R(\tau) \cos \omega\tau \, d\tau \qquad (8)$$

where $\omega = 2\pi/\lambda$ is an angular frequency with dimensions of reciprocal length. A single value of the power spectral density function has a dimension of height squared per unit frequency. When the power spectrum is integrated over a pass-band of surface frequencies between ω_1 and ω_2, the area under the function is the square of the root-mean-square roughness, R_q^2.

All these statistical treatments of the surface roughness give some numbers that could be useful in quality-control purposes. However, none of the parameters is an intrinsic property of a real surface, because their numerical values depend on the scale of measurement.[4,10,11,23] Besides, the various functions need to be numerically characterized in some way, if it is to be of practical use.[4] The correlation length is such an example defined from autocovariance or autocorrelation function. Nonetheless, all these numbers do not provide us with a chance to "see" any real surface roughness. Therefore, we will focus our discussion on two novel optical microscope techniques with which we can actually "see" the surface in three dimensions (3D). Although one can always attempt to connect the optical microscope images with the statistical treatments briefly mentioned above, we will not do so here because it is beyond our scope and the reader is referred to excellent references on this subject.[4,10,11,24]

Principles of Microscopic Imaging

Background

An optical microscope is perhaps one of the most widely used imaging instruments in a variety of disciplines to "see" things at high magnification. One of the limitations of traditional optical microscopy (far field techniques) is its diffraction limited resolution.[25] Despite the limited resolution, its widespread use as an imaging instrument is due to its simplicity and a unique variety of available contrast mechanisms, which include absorption, polarization, phase contrast, dark-field, and fluorescence, among others.

In view of the advantages provided by the various contrast mechanisms and its ease of use, efforts to push the resolution (both spatial and axial) of the traditional optical microscope well beyond that of the diffraction limit have produced a number of very useful techniques, such as laser scanning confocal microscopy (LSCM),[26] standing-wave fluorescence microscopy (SWFM)[27] (where one is able to visualize structure in 3D by optical sectioning of the object being viewed), solid immersion microscopy (SIM),[28] photon tunneling microscopy (PTM),[29,30] and near-field scanning optical microscopy (NSOM) at a higher spatial resolution.[31-34] All of these techniques, by virtue of using light at visible wavelengths, provide information that is not accessible by other high-resolution techniques such as electron microscopy and the family of probe microscopes (scanning tunneling microscopy and force microscopy).[35-38]

It is well recognized that scanning probe microscopes have better spatial resolution than do conventional optical far-field techniques.[38] However, it was also realized that, in principle, nonscanning devices could lead to enhanced resolution or super-resolution, especially when they use evanescent waves. This was actually demonstrated by several authors, who used evanescent waves to record grating-like objects with finer details than is obtainable by the classical resolution limit.[39-41] The existence of evanescent waves has been known since, at least, the time of Newton.[42]

Evanescent waves are formed in a variety of phenomena commonly encountered in optics. One of the most common ways of creating evanescent waves is to have the light beam totally reflected at an interface: that is, to create conditions for total internal reflection (TIR). The use of evanescent waves to observe topographical features was reported as early as 1930.[43] It appears, though, that this method did not enjoy widespread use until Harrick demonstrated in the early 1960s that evanescent waves provided a powerful means in measurement of film thickness and surface relief,[44] when he adapted the frustration of TIR for recording fingerprints and showed that the contrast of the image obtained was rather remarkable (see figure 3 in reference 44). Other early demonstrations include the use of the frustration of TIR as a tool for metrology in proximity problems[45] and the use of evanescent waves as a spectroscopic tool to probe chemisorbed molecules at a surface.[46,47]

The optical system designed by Harrick to observe surface relief or "topography" did not produce an image that is both undistorted and in focus everywhere. In an effort to circumvent this, an axially symmetric illumination was developed by McCutchen,[48,49] which provided good quality images of surface relief or the topography of the surface.

The use of evanescent waves to study biological samples was demonstrated by Ambrose,[50] who termed the frustration of TIR caused by cells adhering to surfaces as "surface contact microscopy". In recent years, this has led to a technique to excite fluorescence of species (mainly biological samples) quite close to the surface, thereby allowing one to study cell/substrate interactions. This method has also been used in polymer science to measure, among other things, the concentration profile of polymer chains close to an absorbing surface[51] and recently to demonstrate slip in flow of polymers.[52]

Evanescent waves have been used in a variety of other applications where distance measurements are needed. It was demonstrated that very small displacements could be measured,[53] and these are used in atomic force microscopes.[54] Its use in holography is the subject of an excellent review by Bryngdahl.[55,56] It has also been used to measure repulsive interactions between a colloidal particle and a surface.[57-60] Since we are primarily concerned with the 3D imaging aspects of surface roughness, we shall not discuss these applications in any detail.

The remainder of this chapter is structured in the following manner: we discuss the limits of resolution of an imaging system, primarily based on the classical Rayleigh criterion for resolution, for both spatial and axial resolution. This is followed by a section on various techniques that have appeared in the past decade to enhance both types of resolution. Then we discuss various applications of these methods to the characterization of roughness, especially for photon tunneling microscopy (PTM) and laser scanning confocal microscopy (LSCM). The general use of confocal microscopy has been in biology, and there have been many very well written books and review papers to which the reader is referred.[27,61,62] The use of confocal microscopy in the field of polymer characterization is also increasing.[63-66]

Image Formation in a Microscope and Resolution

A detailed description on image formation in a microscope can be found in many excellent references;[25,67-70] hence in this section we provide a brief account of how an image is formed and what is the limit on resolution. It should be recognized that there is a clear distinction between resolving an image and visualizing an object. Visualization of objects that are far below the resolution limit of an optical microscope, the so-called diffraction limit, is not a new event,[69,70] but resolution below the diffraction limit is.[31-33] These two issues must not be confused with each other.

The Rayleigh criterion is the distance between two just-resolvable points. It is often used to describe the resolution of optical instruments and image-forming devices. There have been several discussions on the resolution limit, claiming that the Rayleigh limit is only a convenient measure of the resolution obtainable by optical instruments and that it should be possible, in principle, to obtain resolution without limit.[67] In an elegant paper, G. Toraldo di Francia[70] has shown that even though it should be possible to obtain resolution without any limits based on mathematical grounds, the practical realization of the mathematical ideal is quite impossible. Hence, it is worthwhile to remind ourselves of the arguments that have been presented.

We begin by considering an imaging system as schematically shown in Figure 17.2. Image formation in a microscope may be understood in terms of diffraction of the incident beam on passing through the sample and recombination thereafter to give a real (and magnified) image of the object.[71]

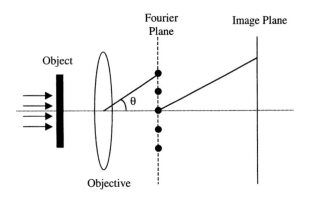

Figure 17.2 Schematic diagram showing image formation in a microscope.

Image formation in a microscope can then be considered as a double diffraction process. To illustrate this in a qualitative fashion, let's consider a diffraction grating. On passing through this grating, a parallel beam of light produces several orders of diffraction in the back focal plane (BFP) of the objective (first lens). The diffracted beams emerge at angles θ_i, given by $nd \sin \theta_i = m\lambda$, where n is the refractive index, d is the grating period of the object, m is an integer, and λ is the wavelength of light. Each diffracted order can be considered as a plane wave, and the set of plane waves produced by the diffraction grating can be collected by a lens (or an objective) having a sufficiently large numerical aperture (NA), so that they converge individually to a set of points in the focal plane (Fourier transform plane) of the objective. The diffracted light passes on to interfere, resulting in an *accurate* image of the diffraction grating, only in a plane conjugate to that of the grating (i.e., the object), which is the image plane.

A diffraction grating was chosen as the object for several reasons. Historically, a grating was used to demonstrate Abbé's theory,[69,70,72] and it also serves the purpose of illustrating that image formation in a microscope can be considered as two distinct stages: (1) production of the Fraunhofer diffraction pattern in the rear focal plane and (2) a Fourier transform of the Fraunhofer diffraction pattern to form a magnified image. In the focal plane of the lens, having produced the Fraunhofer diffraction pattern of the object, the various diffraction orders behave like a set of equally spaced point sources, and the image is their diffraction pattern. The finest structure observable is determined by the highest order of diffraction that is transmitted by the lens used for the purpose of imaging, and is a function of the numerical aperture of the objective used in the microscope. The numerical aperture (NA) of an objective is the light-gathering power and is defined as

$$NA = n \sin \theta \qquad (9)$$

where n is the refractive index of the medium in which the light is collected, and θ is the half-angle of the cone of light entering the objective.

In a light microscope, the part of the Fraunhofer diffraction most likely to be lost is the outer part, which corresponds to larger angles of diffraction, and it will be lost if the numerical aperture (the light-gathering power) of the objective is too small. This results in a loss of sharpness in the image, a deterioration of the resolution, and a loss of the finer detail, since it is the sharper boundaries and finer detail that diffract light at larger angles. It is useful to collect as many orders of diffraction as possible, since this contributes to the detailed structure of image.

To better understand the image formation, one can resort again to Fourier optics.[25,68-70] Consider the image of a grating, as described earlier in this chapter. In the back focal plane (which is the Fourier plane), we have several orders of diffraction behaving as a set of equally spaced point sources. The zero-order beam in the diffraction plane produces a uniform amplitude or uniform illumination across the image plane, which by itself carries no information about the period of the grating. Each higher-order beam produces a sinusoidal amplitude distribution. These various amplitudes are superposed in the image plane to produce the image. The amplitude distribution across the image plane can then be represented in terms of various Fourier components, or *spatial frequencies*. To observe the periodicity of the grating, it is enough to reproduce the fundamental (or the lowest spatial frequency) component in the amplitude distribution of the image. This is accomplished if the objective of the microscope can admit the zero order and *one* of the first-order beams, providing two effective sources in the Fourier plane, thus giving maxima in the intensity distribution across the image plane that correspond to the grating spaces, resulting in a resolved grating. This leads us to the Abbé principle, which states that both the zero order and one of the first-order beams must be transmitted. The limit to resolution is then given by the finest grating that can be resolved in this way. It becomes clear that finer gratings give rise to larger angles between the zero and first orders, thus requiring lenses of larger numerical aperture in order to be resolved. Hence the resolution of a microscope is a function of the numerical aperture, and the theoretical limit is given by the expression of $0.61\lambda/NA$.[71] This expression is often approximated by $0.5\lambda/NA$ and comes about when one takes account of the numerical aperture of the condenser. Further details can be found elsewhere.[73]

The preceding discussion represents an approximate theory of image formation. It is assumed that light associated with a given diffraction order is in a single direction, and all light associated with this order is either accepted or rejected by the collecting objective. The theory does not take into account the possibility that part of the light in a given diffracted order is admitted and the finite amount of light is diffracted between the principal maxima.

The preceding discussion was developed for nonluminous objects, as Abbé described his theory for the microscope. Rayleigh[74] took a different approach to the problem. Light from each point in an object that is illuminated by a condenser produces a set of circular rings (called the Airy pattern; see Figure 17.3) in the conjugate image plane, and the final image is a superposition of these Airy patterns. The effects of such superposition can be calculated if the phase relations and the degree of coherence are known, which are determined by the structure of the object and the form of illumination. The Airy pattern results from diffraction by the objective, and the structure of the object is taken into account by the phase relations between the Airy patterns. This leads again to the definition of resolution, the Rayleigh criterion, the separation of two just resolvable points. This is given when the maximum of one Airy pattern is located at the minimum of the other Airy pattern. The resultant two-point intensity will then have a minimum, between two just resolved objects, that has about 20% less intensity than the maximum.

In all of this, nothing has been said about signal-to-noise (S/N) ratio in the measurements. As with any measurement, signal-to-noise ratio plays an important role. Hence, if one can have a good signal-to-noise ratio in the measurement of the convoluted intensity distribution from two just resolved Airy patterns, a smaller separation can be estimated with adequate certainty.

We now return to the issue of resolution and its limits. It is known that an object of finite extent has a Fourier transform that is an analytic function. Given enough information, even if a part of the Fourier transform is lost (due to the presence of effective apertures) in the Fourier transform plane (back focal plane), in principle, it must be possible to recover the entire function. This has been the topic of many elegant papers,[68,69,70,75-79] presenting arguments that the Rayleigh criterion is just a practical limit and not a fundamental one. This argument is mathematically quite correct, as was pointed out by Marcuse,[80] among others. It should be recognized that these arguments are of little use in terms of going around the diffraction limit, due to the issue of noise entering into the detection scheme. It is inevitable that one will be limited by the noise in any measurement or detection scheme, and hence there is really very little hope of being able to precisely reproduce the object field from a knowledge of the image field. Nonetheless, the Rayleigh criterion is a useful limit of resolution to think about when considering any scheme for enhancement of lateral resolution.

Thus far, the discussion has been on lateral resolution. Since optical sectioning microscopy (OSM) has revolutionized biology,[61,26,27] there has been considerable effort to obtain thinner and thinner sections, and a number of methods have been developed in order to achieve high axial resolution. This form of microscopy has been primarily used in biology to study the structure of cells in three dimensions. The image formed in an optical microscope contains information from the area that is in clear focus in a thin focal plane, as well as out-of-focus information. This is espe-

(a)

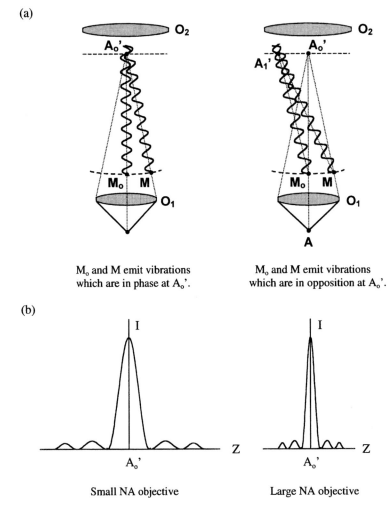

M₀ and M emit vibrations which are in phase at A_o'.

M₀ and M emit vibrations which are in opposition at A_o'.

(b)

Small NA objective

Large NA objective

Figure 17.3 (a) Schematic representation for the formation of the Airy disk pattern, and (b) light distribution in the Airy disk. The parameter Z depends not only on the geometrical distance from the center A_o' but also on the characteristics of the microscope's objective. To observe very fine details, the diameter of the central spot is to be as minute as possible. This is achieved by using a high numerical aperture objective. (Redrawn after reference 81.)

cially true if the object is much thicker than the depth of field. The depth of field can be as low as 0.15 to 0.20 μm when a high-magnification, high-NA objective is used. A vast majority of experiments use fluorescence microscopy to visualize the 3D structure of cells, and the imaging is usually done in low-light conditions. This decreases the signal-to-noise ratio and limits the axial resolution attainable.

Axial resolution is defined as the least distance between two objects separated only axially (along the optic axis of the microscope), which can be identified *not to be* in the same plane (plane normal to the optic axis of the microscope). As there is no single agreed upon formula that defines the limit to axial resolution, we use the expression given by Francon[81] for the axial setting:

$$\Delta z = \frac{\lambda}{4n \sin^2(\theta/2)} \quad (10)$$

where Δz is defined as the minimum axial separation when two objects can be distinguished as such by changing the focal plane. This expression would predict an axial setting accuracy or resolution of about 0.18 μm to 0.14 μm for high-NA lenses (NA = 1.4), which has been demonstrated by Inoue.[82]

Photon Tunneling Microscopy

In many a situation it is desirable to "see" structures selectively at surfaces, while ignoring those in the bulk. Examples of such situations abound in a variety of disciplines: cell adhesion to surfaces,[50, 83] the concentration of a polymer that is in contact with a surface,[51] the orientation of liquid crystals at a surface,[47] motion of diatoms on a surface,[83] and so on. A useful way to study these processes is selective illumination, provided by the evanescent waves created by TIR at a surface. Photon tunneling microscopy (PTM)[29,30,84,85] is a form of microscopy that uses the properties of evanescent waves to study the topographical features on a surface, or index variations at the surface of interest, which is otherwise flat. In this section, we describe the use of PTM to the study of polymeric surfaces. Its utility in the characterization of both static and dynamic structure is illustrated.

Photon tunneling microscopy, described fully elsewhere,[29,30,84,85] has all the benefits of reflected light microscopy: whole field imaging with direct visual access, and in real time, with higher than diffraction-limited lateral resolution (while less than its scanning probe cousins), thus benefiting from the supernumerical aperture afforded by the tunneling or near-field information that is normally lost in a light microscope. PTM also shares the subnanometer vertical resolution and profiling ability of its scanning probe microscopy cousins, and it has proven to be a valuable tool for imaging a variety of polymeric surfaces. Related to and descended from frustrated total internal reflection microscopy (FTIRM),[86] it is also referred to as evanescent wave or surface contact microscopy[50] when it is used in the transmission mode. In the PTM, the contiguous transducer acts like a close-packed array of probes, allowing more kinds of samples to be imaged in real time without probe errors or damage. The 3D analog display of the photometric video image facilitates real-time imaging with continuous and immediate image rotation about all axes.

Most important for polymer applications, PTM images dielectric surfaces directly without metallization, carbon shadowing, vacuum, or damaging electrons. Furthermore, surface alteration is negligible. While the transducer is in optical proximity to the sample surface, it is separated by a tunneling gap most everywhere (or else it would not function), moves with the sample, and is flexible enough to accommodate particulates, because its hardness is comparable to that of most polymers.

Brief Description of PTM

The photon tunneling microscope is based on the behavior of light at a total internal reflection (TIR) surface, or the boundary between a medium 1 with index of refraction n_1 and a medium 2 with a lower index of refraction n_2. Total reflection occurs in medium 1 if the angle of incidence is equal to or greater than the critical angle, θ_c:

$$\theta_c = \sin^{-1}(n_2/n_1) \qquad (11)$$

Probing the rarer medium near the TIR surface (hence "near-field") with a dielectric medium of refractive index n_3

greater than n_2 reveals energy coupling from the first medium into the third medium. This indicates the presence in the rarer medium of an electromagnetic wave with the same frequency as its parent wave in the denser medium, when the near field is unperturbed. This electromagnetic wave is referred to as an "evanescent wave", which does not carry any energy with it. The field strength of the evanescent wave, $E_{evanescent}$, decays exponentially as[47]

$$E_{evanescent} = E_0 \exp(-z/d_p) \qquad (12)$$

where z is the distance normal to the TIR boundary, and d_p is the penetration depth at which the field strength falls to $1/e$ of its initial value E_0:[47]

$$d_p = \frac{\lambda_1}{2\pi(\sin^2\theta - n_{21}^2)^{1/2}} \qquad (13)$$

where $n_{21} = n_2/n_1$ is the ratio of refractive indices of medium 1 and 2, and λ_1 is the illumination wavelength in medium 1. Figure 17.4 schematically illustrates the penetration depth and the exponential decay discussed here.

Evanescent waves are fully explained with classical optics,[69,70,56] and one need not invoke quantum theory, but doing so adds to the intuitive understanding of the phenomenon. The analogy to the electron barrier tunneling experiment has been elegantly made,[34,87] and the effect was termed as "optical" or "photon tunneling".[88] Equation 12 then describes the probability of finding a photon from medium 1 in the tunneling gap at some distance z normal to the TIR boundary. Perhaps one of the first "devices" to make use of "optical tunneling" is a beam splitter based on the phenomenon of frustration of total internal reflection, as schematically shown in Figure 17.5. When the air gap is uniform, depending on the separation between the two prisms, a certain fraction of light will be transmitted and the rest is reflected (Figure 17.5a). Alternatively, if the air gap is not uniform, as shown in Figure 17.5b, then the transmission and reflectance are a function of the thickness and position (y, z), giving rise to an intensity distribution.[44,49]

The use of tunneling photons or evanescent waves brings with it several remarkable properties that are of use to microscopy. First, each tunneling photon serves as a height probe of a third medium of n_3 ($> n_2$), as TIR is locally frustrated to a degree that is exponentially and inversely dependent on z. Thus when the third medium is a dielectric sample with homogeneous optical properties, the reflected light intensity is modulated, corresponding spatially to the modulation of the tunneling gap by the sample surface topography, and an image of the surface topography is formed in grayscale, wherein height information is mapped into an intensity distribution. More importantly, since the tunneling probability (potential) decays exponentially with distance z normal to the TIR boundary, the vertical resolution does not depend on the wavelength. Vertical resolution is limited only by photometric resolution, determined by sig-

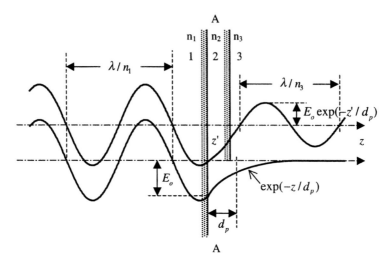

Figure 17.4 Schematic diagram showing the energetics and geometry of photon tunneling. See the text for a detailed description. (Redrawn after reference 84.)

nal-to-noise ratio of the detector. Second, photometry of this tunneling image yields quantified height measurement directly and without ambiguity, as bright area corresponds to regions farther away from the TIR surface and dark area to regions closer to the TIR surface in the intensity-mapped topography. Third, a tunneling photon has the same frequency in the air tunneling gap as in medium 1, such that high lateral resolution is achieved without oil immersion of the sample. This has also been described in terms of a "supernumerical aperture", where the collection of information in the evanescent wave, normally lost in a light microscope, results in higher resolution.

Resolution in PTM

Nassenstein[39] demonstrated that increased resolution could be obtained by using evanescent waves. He holographically recorded information with resolution greater than what was possible within the diffraction limit. The explanation for this lies in the fact that the wavelength to be considered for resolution is λ_1 (wavelength in the medium), which is λ/n, where n is the refractive index of the medium in which TIR is occurring.

PTM has been shown to have lateral resolution better than $\lambda/4$,[84,85] thus demonstrating resolution beyond the diffraction limit. A schematic diagram of a PTM is illustrated in Figure 17.6. Examination of Figure 17.6(b) reveals that the image in PTM is formed using totally internally reflected light, which leads to an annular illumination, although there is no central stop to block the central beam. Central obstruction is unnecessary since TIR is so intense that it overwhelms the central illumination. Such an illumination is known to enhance spatial resolution by a factor of 1.3 to 1.6[71] and is used in applications ranging from microlithography to astronomy. By having a centrally obstructed aperture, the diameter of the central peak of the diffraction im-

age can be made smaller than the standard Airy disk, which leads to enhanced spatial resolution.[71,82] This annular illumination can be made arbitrarily small with the hope of attaining higher resolution. However, as the diameter of the central obstruction is increased, the amount of light available for imaging is decreased, and, hence, the resolving power is limited by the amount of light available (see section on image formation). The possibility that evanescent waves generated by the sample topography or microfeatures can be converted to image-forming rays by the proximity of the transducer, leading to an additional factor for enhancement of lateral resolution, cannot be ignored. This will lead to a "supernumerical aperture" discussed above.

Configuring a reflected light microscope for photon tunneling (Figure 17.6b) is rather straightforward, requiring only TIR of the epi-illumination at a surface in the object plane. The objective is oil immersed to a transparent optical window (or transducer) and is focused at the transducer-air interface. The NA of the objective is larger than 1, in order to contain the critical angle defined by the refractive indices of the transducer and the tunneling air gap, so that TIR occurs at the transducer's distal surface. Hence, for an aqueous tunneling gap, as with in vivo biological samples, the NA should be greater than 1.33.

For photon tunneling to occur, the sample is placed in the proximity of the transducer. The window facilitates conversion of height modulation in the sample topography into intensity modulation. In addition to providing the required optical unity for TIR (now frustrated by local photon tunneling into the sample; see Figure 17.5 and Figure 17.6b), the immersion oil allows the sample and transducer to travel together, thereby eliminating damage to the sample by sliding contact and allowing complete freedom of movement about the surface. A recent innovation is the use of a proprietary (Polaroid) flexible optical membrane as the transducer, which is stiff and flat over the field of view of the microscope, but flexible enough on a larger scale to accommodate curved surfaces and even dirt on the sample. Figure 17.7 shows the TIR beam in the back focal plane of an objective with an NA of 1.4, clearly demonstrating the annular nature of the illumination.

The vertical tunneling range is about 0.75λ, or about 0.4 μm for green light. When the gap grows larger than this, the transducer serves as a soft contact interference reference surface; hence, only refocusing and suppression of TIR illumination are required for interferometry. Unlike the scanning tunneling microscopes, which operate at constant tunneling current, the photon tunneling microscope forms an image with the variations in tunneling referenced to a constant surface (the transducer). The grayscale tunneling image of the surface thus formed is viewed directly through the microscope.

The PTM gleans height information from the tunneling modulation caused by the (optically homogeneous) sample topography referenced to the transducer. This simple arrangement eliminates the need for separate vertical control instrumentation. However, in order to measure the height (not optical variations as in latent images or biological samples), the tunneling response must be calibrated to the optical property of the sample. There are a number of possibilities for doing this, but a straightforward calibration method is to render the material in a known geometry, such as a spherical surface, which provides the relation of the sample height to the grayscale tunneling image to be measured directly and also allows a check on optical homogeneity. For polymers that cannot be fashioned into a sphere, spin coating a spherical substrate with the polymer works well. Alternately, an occasional dirt particle under the flexible transducer offers the opportunity to calibrate directly on the sample by using the interference mode.

For height quantification, a CCD or vidicon detector views the tunneling image at the microscope's phototube in the normal manner, and this output is readily analyzed and displayed in any number of ways, including 3D, by any image processing program now available for PCs. Detector trade-offs must be considered, in that while the CCD has the better signal-to-noise ratio and therefore higher vertical resolution and vertical range, the vidicon has higher lateral resolution at lower cost with real-time response; with a pasicon tube, for example, the vertical resolution is about a nanometer. Fortuitously, the nonlinear vidicon response to intensity largely corrects for the tunneling's exponential response to linear height, so that in the absence of image processing to correct the grayscale, the 3D display of the topography is not distorted. Hence the vertical resolution is convolved with the detector characteristics, and detector trade-offs must be considered.

Analog has advantages in the 3D display as well. A high-resolution xy oscilloscope displays each of the video raster lines as amplitude traces, thus electronically mapping intensity back into height for a real-time 3D image. Additional analog circuits to shear or expand and collapse the multiple oscilloscope traces create an immediate control over 3D perspective with simple turns of a few knobs. The quality of reality and immediacy this display provides helps to interpret the surface and to understand it. A single trace can be isolated anywhere in the image for electronic cross-sectioning and height measurement. Figure 17.8 provides comparative images of an injection-molded optical storage structure using PTM, scanning electron microscopy (SEM), micro interferometer, and atomic force microscopy (AFM). It is very clear from the images provided that the overall correlation between the PTM and other techniques is very good. However, the use of PTM in a judicious manner can provide high-resolution images of surface topography.

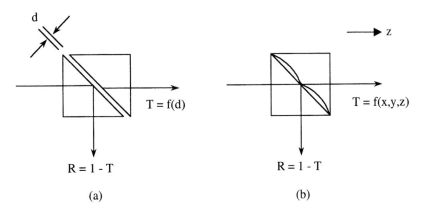

Figure 17.5 Schematic diagram of a beam splitter based on the principle of frustration of total internal reflection. (a) When the air gap is uniform, the fraction of light transmitted or reflected depends only on the separation between the two prisms. (b) If the air gap is not uniform, the transmission and reflectance are a function of the thickness and position (y, z), giving rise to an intensity distribution.

(a)

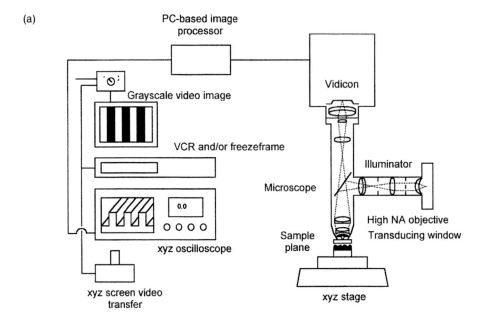

(b)

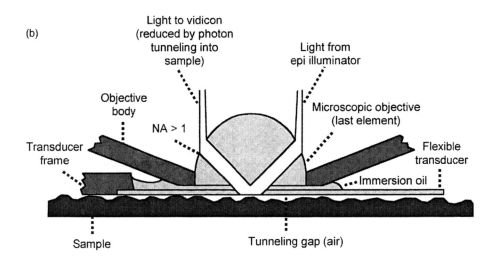

Figure 17.6 Schematic diagram of (a) the photon tunneling microscope and (b) the transducer, objective, and sample. (Redrawn after reference 84.)

Study of Film Formation

Film formation from polymer latexes and the mechanism of film formation are of considerable industrial importance and have received a lot of attention in recent years.[89-91] Polymer latexes are used in many water-based coating formulations, which contain the polymer suspension in the form of colloidal spheres. When allowed to dry, some latexes produce transparent, continuous, crack-free, and mechanically rigid films, while some others give rise to opaque powders or films. This process has been found to depend on the latex particle size and the temperature at which the film was formed, among other things. Films formed above the so-called minimum film forming temperature (MFT) usually give rise to a transparent, continuous film, while films formed below MFT usually produce rather opaque, discontinuous, mechanically brittle surfaces. It is of interest to probe the surface roughness of water-borne coating to understand the effectiveness of the coating process and also to provide information on the film formation process.

Several multistep mechanisms have been proposed to understand the film formation process and to take account of several variables that affect the final film. The process is generally thought to involve a number of successive steps, and the most widely accepted mechanism is as follows: in stage 1, the water evaporates, leading to close contact of particles; in stage 2, the particles pack and deform; and in stage 3, at temperatures higher than the glass temperature of the polymer, the particles anneal to form a mechanically rigid film. All three stages and the respective temperature ranges depend on a number of parameters, which include the latex structure, size, and composition, as well as the formulation of the dispersion.

We used 50-nm spheres of polystyrene suspended in water. A drop of the suspension was air dried on a glass plate, leading to very interesting surface topography of the film. This final film is rather opaque and is not mechanically rigid. Nomarski imaging at 100× magnification reveals a herringbone pattern of cracks (Figure 17.9a). The grayscale tunneling image from the PTM (Figure 17.9b) shows these surface height variations in high contrast, which, when viewed in 3D (Figure 17.9c–d), are revealed to be ridges, with occasional orthogonal flips across tiles defined by cracks. These features are also seen in Nomarski imaging, but with much less resolution than with the PTM. The spheres themselves are just beyond PTM's lateral resolution.

It is well known that as water evaporates, the latex particles come close together and pack to give structures (as examined by SEM and TEM), consistent with face-centered cubic (fcc) packing. The observed topography of this latex film appears to be consistent with an incomplete close-packed structure for the latex particles.

Study of Polymer Single Crystals

Many synthetic polymers can be crystallized from dilute solutions to form what are usually referred to as "polymer single crystals". The discovery of such "single crystals", made almost at the same time by Keller,[92] Till,[93] and Fischer[94] in 1957, has greatly improved the understanding of crystallization and self-assembly of macromolecules and proteins. Their electron diffraction data for polyethylene, for example, showed that the polymer chains were normal to crystal platelets formed from dilute solution, and most remarkable was the fact that the platelets were only about 100 Å thick. This led researchers to conclude that the molecules must fold sharply on themselves to give rise to lamellar periods of only 100 Å.[95,96]

The principal tools that have been used to study the morphology of single crystals are electron microscopy and diffraction of electrons or X-rays. Despite its obvious importance, electron microscopy is not without its shortcomings. These include the relatively high cost, the need for high vacuum, the inability to manipulate samples, and possible damage to samples upon irradiation with the electron beam. These limitations are overridden in many cases by the resolution achievable and by the fact that one can perform selected area diffraction to determine the unit cell structure and orientation of crystalline polymers. Nevertheless, in an effort to circumvent many of the limitations of electron microscopy, Wunderlich and coworkers[97,98] and others[99] have used several forms of interference microscopy to visualize the single crystals formed from solution.

Polyethylene (Marlex-50, $M_w = 100{,}000$) was crystallized from a dilute (0.5 wt %) solution in p-xylene at 800 °C to obtain large single crystals. Figure 17.10a shows the grayscale image of a polyethylene single crystal as viewed directly under the PTM, where the individual single crystals are visible as intensity differences. The high resolution and contrast of the image is rather remarkable, as viewed through the microscope, compared to the images shown

(a)

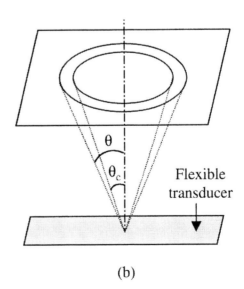

(b)

Figure 17.7 (a) Image showing the total internal reflection (TIR) light in the Fourier plane of an objective of NA = 1.4; (b) schematic showing that the TIR occurs at angles $\theta > \theta_c$.

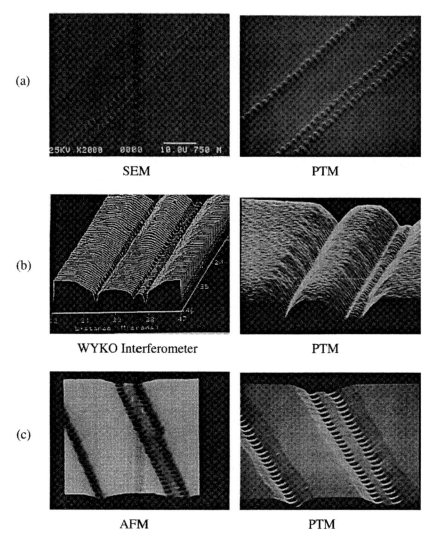

Figure 17.8 Comparison of the resolution in the images of an optical storage disk, obtained by various imaging tools: (a) scanning electron microscopy (SEM) versus photon tunneling microscopy (PTM), (b) micro interferometer versus PTM, and (c) atomic force microscopy (AFM) versus PTM. The scale bar in (a) corresponds to 10 μm.

here. The analog 3D display (Figure 17.10b) of the gray-scale image clearly shows the individual lamellae of the polyethylene crystals, with step heights of approximately 70 Å. Manipulation of the perspective is relatively simple and helps reveal the structure of the surface. It should be pointed out that the apparent height in the 3D image is exaggerated in comparison with the lateral dimensions. While the long side of the image is 25 μm across, the step heights are only on the order of 70 to 80 Å. Hence, one should be quite careful in interpreting the images shown in the figures.

Frequently, the topography of the single crystals exceeds the tunneling range. This is due primarily to the fact that many of the crystals have a center of the growth spiral that is much thicker than the rest of the crystal. This lifts the flexible transducer beyond the tunneling range. A limit to the tunneling range exists since the probability with which one frustrates TIR decays exponentially. This is typically 0.75λ, which for green light, is about 0.4 μm. In such a situation, one uses the other attributes of the microscope to image the thicker crystal—for example, by conversion of PTM to an interference microscope. This is accomplished by suppression of the TIR illumination, required for interferometry, and refocusing the microscope to image the crystal in the interference mode. Now, the flexible transducer serves as a contact interference reference surface. Figure 17.11 is an interference image of a single crystal, consisting of several lamellae, obtained using the PTM in the interference mode. The use of monochromatic light allows quantification of the height of each of the lamella.

Dewetting Studies of Polystyrene

The process of a liquid wetting or drying on a solid surface is important in many practical applications such as paints, coatings, detergents, and flow-through porous media. Thin polymer films are used in a variety of applications where their thermal stability is important and, in many instances, a smooth and stable surface is desirable. The process of a liquid film wetting or dewetting a solid surface depends on the nature of long-range interactions between the liquid and the solid. The case of a liquid wetting and spreading on a surface is relatively well understood.[100–102] Here we describe experiments designed to study the inverse process: retraction/rupture or dewetting of a thin polymer film on a glass substrate. A common example of such a process can be seen in a film of water spread on a waxed automobile, where the film spontaneously ruptures and is drawn up into small beads.

It is possible, using techniques like spin coating, to prepare smooth surfaces of controlled thickness of polymer films even on nonwettable surfaces (defined by a nonzero contact angle between the substrate and the polymer film). Films prepared on nonwettable surfaces, however, are not stable above their glass-transition temperature and will rupture spontaneously, giving rise to beads of the polymer.[103] Experimental studies of dewetting of both thick films of a nonvolatile fluid[104] and thin films (<100 nm) of polystyrene[105] have been reported. Thick films undergo the dewetting process by a nucleation and growth mechanism,[104] while thin films are expected to undergo dewetting via a spinodal type decomposition, as evidenced by the size of the droplets observed on completion of dewetting.

The experiments described next were motivated by our desire to directly observe the undulations of the film, which grows and eventually ruptures the film, leading to droplets of size h^2/a, where h is the initial film thickness and a is a molecular size.[103] We used polystyrene ($M_w = 19,000$, $M_w/M_n < 1.05$), obtained from Polymer Laboratories, UK. Thin films of polystyrene were coated onto a glass substrate from dilute solutions of polystyrene in toluene. The thickness of the films was measured to be 710 Å (within 10 Å error). The glass surface was a plano-convex lens with a focal length of 1000 mm. The uniformity of the spin-coated films was checked by differential interference contrast microscopy and was found to be within 5 nm over the entire disc (2 inches in diameter).

When the thin polystyrene film is heated to above its glass transition temperature, a variety of patterns of ripples and periodic lines arise, and the merging of these structures, giving rise to droplets, is dramatically revealed in Figure 17.12. The structures at this stage are about 0.2 µm high with remarkable periodicity.

Figure 17.13 shows a time sequence of the dewetting process. The polystyrene films in this case were heated to 165 °C, well above the glass transition temperature. As can be seen from the figure, the process of dewetting starts out

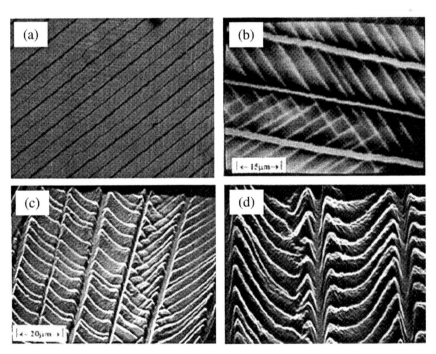

Figure 17.9 Cracks in the film formed from an aqueous suspension of 50-nm polystyrene latex spheres: (a) Nomarski image of the cracks; (b) photon tunneling microscopy (PTM) grayscale image of the same cracks; (c) and (d) two 3D images of the same film with PTM at different magnifications. (Reprinted with permission from reference 84.)

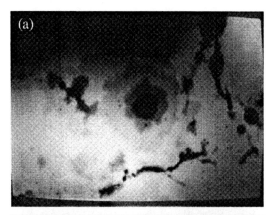

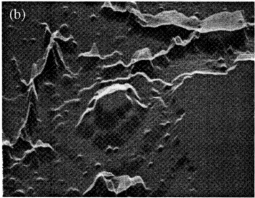

Figure 17.10 (a) Grayscale image of polyethylene single crystals as seen through photon tunneling microscopy (PTM); (b) topography (3D image) of a polyethylene single crystal. The lateral dimension of the images is 25 µm.

producing a ripple-like structure (known as a target pattern), which grows in amplitude as a function of time. The amplitude at the start of the breakup is on the order of a few nanometers, which will eventually reach a peak-to-peak height of 0.2 µm. The spacing between the ripples is about 5 µm. Several such ripples start at the same instant, leading to the observed structure. The ripples eventually grow in amplitude and spatial extent, colliding with similar structures formed a distance away. Each filament of the polymer making up the ripple undergoes a Rayleigh-type instability, giving rise to discrete droplets. In time, the entire surface of the glass substrate is filled with little beads of the polymer. The ripples are attributed to the amplifications of thermal fluctuations prior to dewetting of the film. This particular example was chosen to demonstrate the power of PTM to observe dynamic processes as they occur. The technique provides exquisite images, as well as a way of gleaning knowledge of the physics involved.

Imaging Multilayers

Many polymeric materials can be spread as a monolayer at an air/liquid interface. Such films can be transferred onto glass surfaces, layer by layer, to create unique structures.

In this case, one is interested in knowing how the transfer takes place onto the glass surface. To that end, we have imaged monolayers of a "hairy rodlike polymer" formed by Langmuir-Blodgett (LB) deposition. Multilayers of this polymer are formed on a glass plate by repeatedly dipping the glass slide through a monolayer in a Langmuir trough, as described elsewhere.[106] PTM was used to study the topographic variations that are seen in the film as the number of layers was varied spatially along the length of the slide. The boundary between 11 (~260 Å) and 21 (~462 Å) layers was chosen to study the topographic variation. Figure 17.14 shows the variation in topography at this boundary, which ideally would be sharp, had the dipping process been carried out precisely to the same depth. It is apparent that the topography at the boundary is far from sharp, and perhaps its origins can be traced to how the material gets deposited during the dipping process. Thus, the image demonstrates that one can try to understand the transfer process of monolayers simply by using evanescent waves as a probe without destroying the soft monolayer that has been deposited on a glass plate.

Diamond Turning

The process of machining an optical surface with a diamond tool can produce unique surface features that provide valuable insight into the process and the tool. The topography then is akin to an information storage system, in which, in this case, information about the process is stored as topographical features. To recover the stored information, one usually needs a profiler to obtain the topography of the object with as little damage as possible. However, many mechanical profilers do damage the surface, depending on the load used for the profiling. As an illustration of the power of the PTM to obtain such information without any damage, Figure 17.15 shows the topography of a dia-

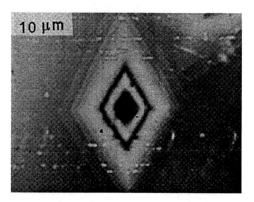

Figure 17.11 Contact interference image of polyethylene single crystals obtained by suppression of the total internal reflection (TIR) illumination in photon tunneling microscopy (PTM). (Reprinted with permission from reference 84.)

Figure 17.12 Variety of structures formed during the dewetting of polystyrene films on a glass surface, with a typical peak-to-valley distance of 0.2 μm.

mond-turned acrylic object.[107] This figure demonstrates the good centering of the diamond tool, while revealing radial tool chatter, quite similar to patterns that can be generated by a metal-working lathe. In the case shown in Figure 17.15, the chatter height is only about 0.015 μm, while the pitch of the spiral is about 10 μm with a typical height of about 0.035 μm.

Of course, one can use the more recent developments of scanning probe microscopes to obtain high-quality images of precision machined parts. We chose the case of topography measurements of precision-ground optical glass. Bennett et al.[108] used a number of methods to characterize the topography of the precision-ground glass. Figure 17.16 shows a PTM image of BaCD16 glass supplied by Hoya Corporation that was initially studied by probe microscopy. It is very clear from the image that the right-hand part (triangle on the right) of the sample was damaged by the probe microscope scan made with too large a loading. Such damage can be minimized by using a noninvasive tool like the PTM.

Laser Scanning Confocal Microscopy

As pointed out earlier in this chapter, the penetration of evanescent waves is quite shallow. If surfaces are quite rough ($> \lambda_{visible}$), then probing with evanescent waves becomes completely useless. Therefore, one has to resort to other forms of imaging devices. We now focus our attention on another form of light microscopy, known as laser scanning confocal microscopy (LSCM).

The availability of lasers and the incredible progress made in powerful, low-cost computing/data-processing systems have resulted in considerable growth of interest in scanning optical microscopy, especially in the confocal mode. One of the key advantages of this form of microscopy is its ability to provide high-resolution images of mostly transparent objects in three dimensions.[26,109]

Confocal microscopes have become an established tool in a variety of research areas, largely owing to the rapid commercial development of designs, optimized to image a rather limited range of objects characteristic of each discipline. In the field of metallurgy, electronic device manufacture, and regular industrial inspection, the prevalence of highly reflecting objects has led to the development of confocal reflection brightfield microscopes for viewing and measuring three-dimensional topographies. In biological applications, the development of a confocal microscope capable of imaging in fluorescence mode has revolutionized imaging of biological samples. Although its use in the field of polymer science is not nearly as extensive, it is gaining considerable attention from polymer scientists.[63–66] By proper choice of objects and imaging conditions, one can indeed obtain high-quality confocal images. In the following discussion, we provide some examples of structures that are made visible using two ways of forming images in a confocal microscope: reflection fluorescence and reflection brightfield (to image buried interfaces).[109]

Description of LSCM

A conventional optical instrument is a parallel processing system where an image of the entire field of view is obtained simultaneously. If one chooses not to visualize the entire field of view but requires that the system provide a good image of only one object point at a time, then the system requirement becomes considerably less severe. This can be done by illuminating the object with a tightly focused light beam, and then collecting the reflected or transmitted light from the illuminated spot. Of course, such a requirement comes with a price, which is the need to scan to build up an image of the entire object. Whether one is willing to pay the price depends on the situation and application in mind. In many cases, the answer has been a resounding yes, and therefore there has been a tremendous proliferation in the use of laser scanning confocal microscopes.

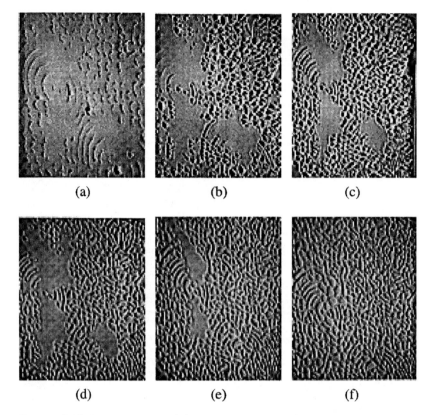

Figure 17.13 Time sequence of the various stages in the dewetting process of polystyrene films: (a) 60 s, (b) 456 s, (c) 540 s, (d) 564 s, (e) 864 s, and (f) 1200 s. Images (c) and (d) were made almost simultaneously with the gain of the vidicon adjusted (in image d) to accommodate the growing amplitude of the surface waves. Peak-to-valley height is typically 0.2 μm.

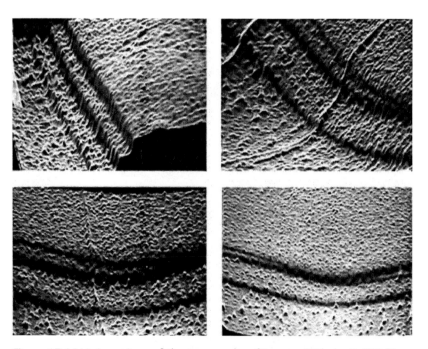

Figure 17.14 Various views of the topography of Langmuir-Blodgett (LB) films of a rod-like polymer on a glass surface. The transition is from 21 layers to 11 layers, corresponding to 7 nm in thickness. The image size is 6μm × 4 μm.

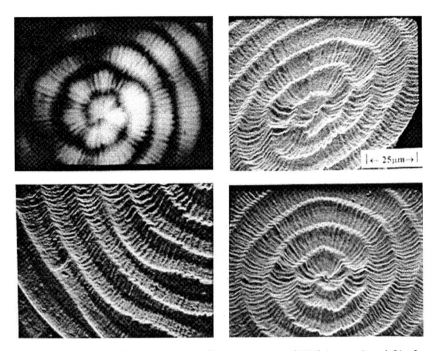

Figure 17.15 Grayscale photon tunneling microscopy (PTM) image (*top left*) of diamond-turned acrylic sheet; the rest are 3D images of the sheet, showing different perspectives. (Reprinted with permission from reference 105.)

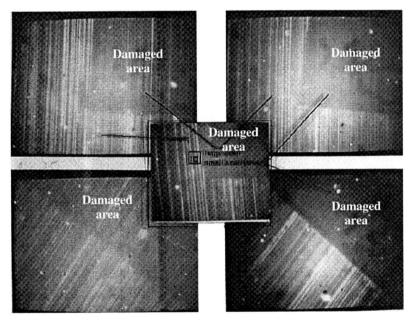

Figure 17.16 Photon tunneling microscopy (PTM) images of BaCD16 glass samples after scanning at different loads in a probe microscope. It is clear that the sample can be damaged when loading is too high.

The essential components for constructing a LSCM are some form of mechanism for scanning the light beam (usually a focused spot of a laser) relative to the specimen and appropriate detection systems to collect the light, either in the reflected or the transmitted mode. With the development of fast and affordable computers, it has become the norm to use the computer to drive both the microscope and to collect and display the image. Since the image is being acquired as discrete data points, the information about the image and the sample is obtained in digital form and is readily amenable to a variety of image enhancement and processing techniques. Quantitative measurements can also be made rather easily to a high degree of accuracy. For example, one may image the end of an optical fiber and find a variation in intensity as a function of position. This occurs because variations in refractive index lead to changes in the measured reflectivity. So measurements of variation in intensity provide a map of refractive index variation, which can be used for obtaining the refractive index profile of an optical fiber, as demonstrated by Wilson.[110]

In an optical microscope, the image contains both the in-focus and out-of-focus light, which does not allow a 3D visualization of the structure of the object under study. An important attribute of LSCM is that it can probe the 3D structure by optical sectioning of the object. Development of 3D microscopy has revolutionized biology and is becoming widely used in many other areas also, such as materials science, metrology, and medicine. Examples include the 3D

organization of single cells, cells in tissues, in vivo microscopy of the human eye,[111] and colloidal crystals.[112,113] In recent years, LSCM has been applied to study polymers, in particular for the study of morphology of polymers and polymer blends.[114–119] The difference between a conventional microscope and a confocal microscope can be understood from the schematic of a confocal microscope shown in Figure 17.17.

The essential advantage of an LSCM is its ability to discriminate against unwanted out-of-focus light that limits a normal light microscope from obtaining a 3D image of a transparent object.[61,109] This is accomplished by scanning a tightly focused light source in a single plane that is much thinner than the sample thickness. The ability to discriminate against out-of-focus light is due to the fact that the detection system is confocal to the illumination, thereby only a single location in the specimen is imaged at any given time. Since the confocal aperture admits light only from the focal plane of interest, a very thin slice (xy projection) can be obtained simply by scanning the beam in the xy plane. Since each point in a focal (image) plane is individually examined, the in-plane or spatial resolution is enhanced, with achievement of about a 200-nm resolution being typical, while the axial (z) resolution is around 500 nm. The z-resolution depends on a number of parameters, including the size of the confocal aperture and the numerical aperture of the objective used for imaging.[61,109] Since the images of the 2D (xy image) sections are obtained in digital

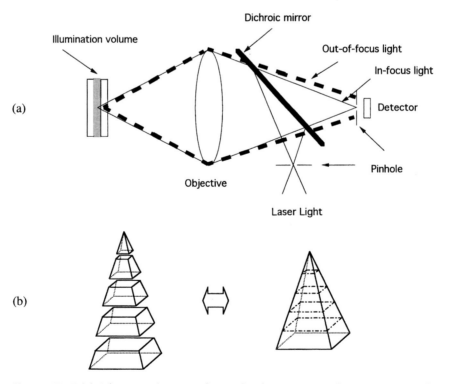

Figure 17.17 (a) Schematic diagram of a confocal microscope; (b) construction of a 3D image from a set of optically sectioned images.

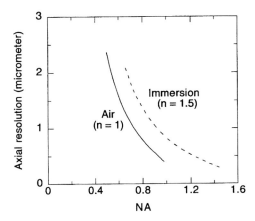

Figure 17.18 The axial optical sectioning width for imaging of plane objects as a function of numerical aperture. The axial resolution is the full width at the half-intensity points of the curves of I(u) against u when red light (0.6328 μm wavelength) is used.

form, several successive optical sections can be used to construct a 3D image of the object.

One can simply demonstrate the sectioning capability by scanning a perfect reflector axially through the focus and measuring the strength of the reflected signal, I(u). A simple paraxial theory[109] gives an expression for imaging of plane object as:

$$I(u) = \left\{ \frac{\sin(u/2)}{(u/2)} \right\}^2 \qquad (14)$$

and for imaging of point objects as:

$$I(u) = \left\{ \frac{\sin(u/4)}{(u/4)} \right\}^4 \qquad (15)$$

where u is a normalized axial coordinate that is related to the real axial distance, z, through the following relation:

$$u = \frac{8\pi}{\lambda} z n \sin^2(\theta/2) \qquad (16)$$

If one chooses the full width, Δz, at half intensity of the I(u) curves as a measure of the strength of sectioning, one can then show that the sectioning ability increases with the numerical aperture. Figure 17.18 shows a specific case of imaging plane or point objects with red light (λ = 0.6238 μm), which is calculated using the preceding arguments. It is clear that one can get much better sectioning strength by using higher NA objectives, even though one should also consider that the image intensity would decrease for high NA objectives and that optimum setting should be made by adjusting other experimental conditions.

In the case of other object features, such as points and lines, we refer the reader to a variety of references on confocal microscopy for further details.[26,27,61,109] However, we

would like to point out that the sectioning ability can also be demonstrated very clearly by plotting the total integrated intensity as a function of axial displacement denoted by u. Figure 17.19 is a plot of the total integrated intensity of a point object for a conventional and a confocal microscope as a function of the axial displacement u, calculated using eq 15. It is shown that the total integrated intensity falls off quite rapidly in the case of a confocal microscope, thus clearly demonstrating its ability to focus at specified depth and, hence, to optically section a given object, making it possible to obtain 3D images.

The axial resolution of a microscope, using the Rayleigh quarter-wave criterion,[25,27,81,82] based on the phase shift suffered by the marginal rays compared to that of the central ray passing through the lens, is given by $\Delta z = \lambda \{4n \sin^2(\theta/2)\}^{-1}$, where n is the index of refraction. As pointed out earlier, this expression would give an axial resolution of 0.17 to 0.13 μm for a high NA objective. While this has been

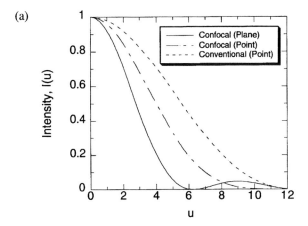

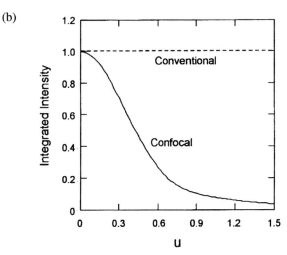

Figure 17.19 Comparison of the image obtained for a single point object in a conventional and a confocal microscope as a function of the axial displacement u: (a) the intensity profile of plane or point objects; (b) the integrated intensity (total power in the image) of a point object.

demonstrated for transmitted light imaging,[82] the limit reached in the fluorescence mode of a confocal microscope is usually on the order of 0.3 to 1.0 μm.

Imaging of Fiber in the Fluorescence Mode

Because high-performance fibers like Kevlar fibers are intrinsically fluorescent, in principle one can obtain a 3D image rather easily. Such fibers have very good mechanical properties along the fiber direction under tension while they perform rather poorly under compression or in transverse direction of the fiber. In an effort to improve mechanical properties under compression and in the transverse direction, it has been attempted to link the individual polymer chains laterally to provide better stress-bearing capability. In the example shown here, fibers made of dimethyl-substituted poly(benzobisthiazole) (DMPBZT) polymers were e-beam treated to cross-link the chains in a direction perpendicular to the fiber axis, which can enhance the compressive properties of such fibers. After the radiation treatment, we examined the fibers using an LSCM in the fluorescence mode at fairly high laser intensity. From 200 xy sections (only 4 sections are shown in Figure 17.20), we reconstructed a 3D view of the fiber surface (Figure 17.21).

Since DMPBZT fiber strongly absorbs the light, as expected, the resulting 3D images show features only near the fiber surface. Nonetheless, we were quite intrigued and surprised to find the features shown. It is seen that the surface is filled with structures, which were not anticipated, and that some part of the fiber inside is damaged. Figure 17.21 also shows the same fiber viewed from a different angle, which demonstrates that it is possible to view the fibers from the inside simply by inverting the image using the image analysis software. Although there are other techniques that could have been used to probe the surface, none would have been as nondestructive as the LSCM.

The appearance of a fiber depends primarily on how light interacts with the fiber and how much of it is reflected. The shape of the fiber cross-section can greatly affect the appearance properties of a carpet or fabric made from the yarn,[120,121] and non-round fibers are frequently used because of their special effect on the appearance of the final product. Therefore, it is important to have a 3D image of the fiber for computations of reflectivity. The reflectivity of a fiber with a given shape can then be easily computed, based on geometrical optics, primarily with ray-tracing methods. In the past, it has been quite difficult to obtain a good 3D image of the fiber for modeling purposes when fibers have a shape other than an infinite cylinder. Such 3D

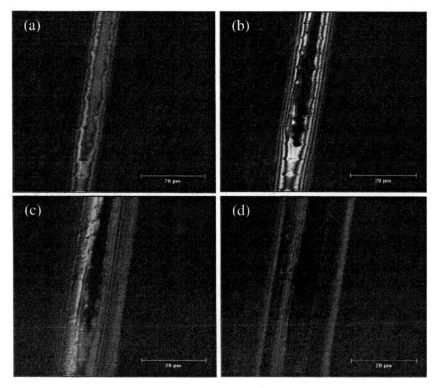

Figure 17.20 Series of optically sectioned images for e-beam treated dimethyl-substituted poly(benzobisthiazole) (DMPBZT). These images were taken by laser scanning confocal microscopy (LSCM) in fluorescence mode using 488-nm laser in high intensity. From (a) to (d), the focus loci were moved from the fiber skin toward the center of the fiber. See color insert.

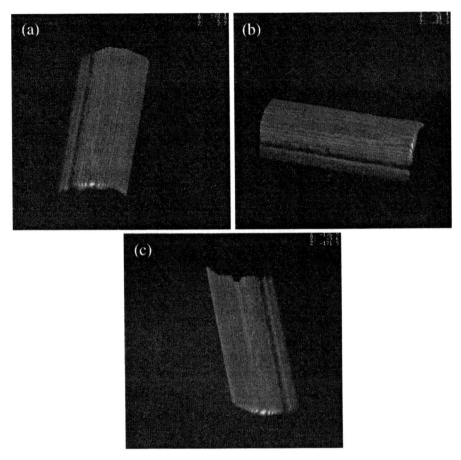

Figure 17.21 3D images near the surface of dimethyl-substituted poly(benzobisthiazole) (DMPBZT) fiber, obtained from 200 *xy* optical sections at different sample depth (four sections of which are shown in Figure 17.20). (a) and (b) are viewed from the outside of the fiber at different angles, while (c) is viewed from the inside of the fiber by reconstructing the same *xy* images with an image processing software. See color insert.

images can easily be obtained by using a laser scanning confocal microscope. A 3D image of a trilobal nylon fiber is shown in Figure 17.22a, which was reconstructed from 95 horizontal section images at different vertical positions. Since the images were acquired using the LSCM in the fluorescence mode, the fiber was dyed with fluorescein, a commonly used and widely studied fluorescent dye. Figure 17.22b is a 3D image of a nylon 66 fiber with a circular cross-section, where the radial intensity gradient is not due to the hollow interior of the fiber, but due to the dye concentration distribution in the fiber. At short dyeing times, only the skin part of the fiber is dyed and LSCM is shown to be very useful in imaging the spatial distribution of the dye in three dimensions.

Figure 17.23 shows another application in which confocal imaging proves to be quite useful to look at the buried interface. Attempts were made to form cylindrical fibers that have a hollow core, surrounded by a two-layer structure, consisting of a polyester and polyetheramide. The cross-sectional (*xz*) image shows that some fibers indeed have the cylindrical shape, but some others are flattened.

As clearly seen for both cylindrical hollow fiber and flattened fibers, all the fiber maintains two different layers, where the brighter layer corresponds to the more fluorescent polyester. For any combination of materials with fluorescence variation or refractive index variation, LSCM can provide information on buried interface.

Imaging in the Reflection Mode

Although many of the commercial instruments are designed with the ability to do imaging in the reflection mode, such imaging is largely ignored in favor of the fluorescence mode. However, because the confocal microscope can produce clear images from narrow planes of focus while eliminating the out-of-focus components, it is well suited to reflected brightfield applications, which are normally difficult or impossible to achieve with a conventional microscope. In order to be able to correctly interpret results of the images obtained, one needs to be aware of the many variables that affect the acquisition of the image. A given surface may reflect light, scatter light, or change its reflectivity due to

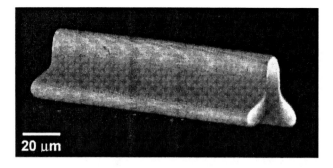

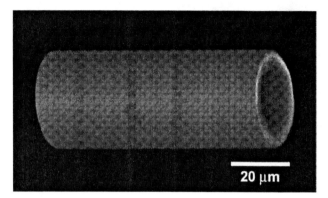

Figure 17.22 (a) 3D image of a trilobal nylon fiber, reconstructed from 95 horizontal section images at different vertical positions. Fiber was dyed with fluorescein and images were taken in fluorescence mode using 488-nm excitation. (b) 3D image of a ring-dyed nylon 66 fiber, reconstructed from 200 horizontal section images at different vertical positions. Fiber was dyed with fluorescein in infinite dyebath of 0.15 g/L fluorescein concentration (pH = 6 buffer) for 16 minutes at 95 °C. The images were taken in fluorescence mode using 488-nm excitation and collecting fluorescence emission from 500 nm to 700 nm.

variations in refractive index, or it may possess absorption and reflection characteristics that depend on the incident wavelength. In the following, we wish to illustrate the use of confocal microscopy in the reflective mode with a few examples.

It was recently reported that polymer solutions may be used to create an ordered array of air bubbles in thin films when there is airflow in a direction parallel to the solution surface under humid condition. As the solvent evaporates, the solution temperature drops quickly, due to evaporated cooling, and the moisture condenses into water droplets on the surface. The laminar flow then packs these water droplets into a hexagonal array. When the solvent density is higher than the water density, the droplets sit on the solution, and the evaporation continues until completion. When the solvent density is lower than water, however, the water droplets sink into the solution, and the process is repeated until the evaporation is over. Then the system temperature returns to room temperature, and the water droplets also evaporate, resulting in an ordered array of holes in a polymer film, either in 2D or 3D packing.[122]

Figure 17.24 is a 3D image of polystyrene film with 2D holes, which was made from 5 wt % polystyrene in carbon disulfide. The images were taken either in reflection mode (Figure 17.24a) or in fluorescence mode (Figure 17.24b), where the bright area is the polystyrene network. Some fluorescent dye molecule was added into the solution to enhance the contrast. Even though the fluorescence mode provides images with sharper image and better contrast, the reflection image itself clearly reveals the ordered array of holes that are 2 to 3 μm in size. Examples of images for the films with 3D holes, as well as the mechanism of the formation of the ordered array of holes, can be found elsewhere.[123]

To demonstrate that high-resolution images can be produced using the reflection mode of the confocal microscope, we imaged glass surfaces that were etched using either hydrofluoric acid or a hydrofluoric acid-nitric acid-acetic acid mixture. A mask was used to create the periodic structures and patterns. The high-resolution images in Figure 17.25 demonstrate that different etchants produce different etch profiles on the glass surface.

In the reflection mode, it is also possible to image buried interfaces, for example in polymer-polymer laminates, if the two materials have slightly different refractive indices. This has been demonstrated in the imaging of plant cuticles.[123] An interesting and important application of imaging buried interfaces deals with the adhesion of polymer laminates used in bottling applications, where researchers are quite interested in understanding the adhesive properties of the polymers in the laminates. Another application would be to image the interface between two miscible polymers above

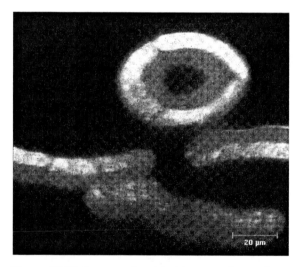

Figure 17.23 Cross-sectional (*xz*) images of complex fibers obtained by laser scanning confocal microscopy (LSCM) in fluorescein mode. The hollow fiber has void-core-sheath structure, while the flattened fibers show that the interface between two polymers is still maintained even after collapse. The brighter layer corresponds to polyester.

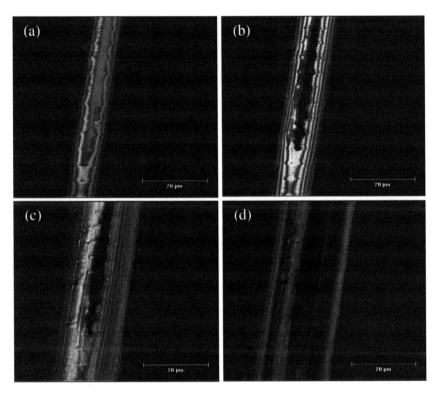

Figure 17.20 Series of optically sectioned images for e-beam treated dimethyl-substituted poly(benzobisthiazole) (DMPBZT). These images were taken by laser scanning confocal microscopy (LSCM) in fluorescence mode using 488-nm laser in high intensity. From (a) to (d), the focus loci were moved from the fiber skin toward the center of the fiber.

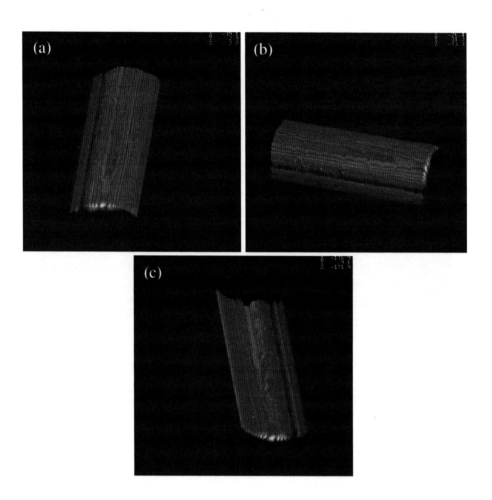

Figure 17.21 3D images near the surface of dimethyl-substituted poly(benzobisthiazole) (DMPBZT) fiber, obtained from 200 *xy* optical sections at different sample depth (four sections of which are shown in Figure 17.20). (a) and (b) are viewed from the outside of the fiber at different angles, while (c) is viewed from the inside of the fiber by reconstructing the same *xy* images with an image processing software.

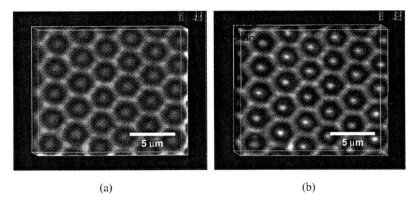

(a)   (b)

Figure 17.24 3D images of the ordered array of 2D holes in polystyrene (PS) film constructed from 100 sections of *xy* images (a) in reflection mode and (b) in fluorescence mode. The hexagonally packed holes are 2 to 3 μm in size.

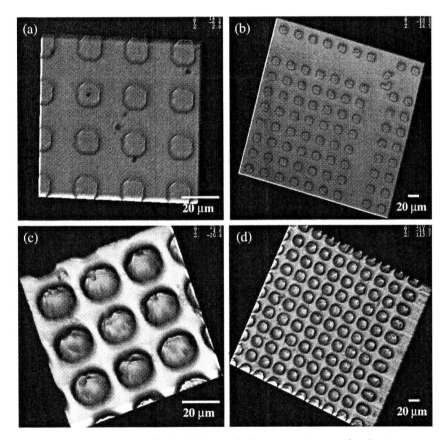

Figure 17.25 3D images of etched glasses by laser scanning confocal microscopy (LSCM). Two types of etchant were used: (a)–(b) hydrofluoric and (c)–(d) hydrofluoric-nitric acid-acetic acid mixture.

Figure 17.26 3D image of a polyester overhead transparency film scratched by a needle.

their glass transition temperature or melting temperatures where the two polymers might mix with each other. If one could image the buried interface as a function of time, it would be possible to obtain the diffusion coefficient of one polymer into the other. This might prove to be a useful tool to study diffusion of polymers in melt state.

We conclude this section with one more application of the reflection mode of imaging using the LSCM. For example, researchers wish to understand the wear behavior of polymeric surfaces since many polymer-based materials are increasingly used in small sliding parts, such as gears and bearings, largely because of the ease with which polymers can be processed into the shape that is necessary.[124–126] Polymeric materials are also being used as binders for magnetic particles in making magnetic recording media. In such applications, submicrometer- or nanometer-level wear is increasingly becoming a problem. Most of the studies of microwear processes of polymeric surfaces have been performed using an AFM. As an example to show how LSCM can be used for a wear study, the surface of a transparency polymer film was scribed using a sharp needle. The mechanical damage caused by this motion was imaged by LSCM in the reflection mode (Figure 17.26). By means of controlled scratch experiments, it would be possible to study the brittle or ductile deformation. Often, a very thin coating of abrasion-resistant material provides excellent scratch resistance, and the coating performance might easily be evaluated by imaging the scratches with LSCM. A variety of coating materials may be formulated, such that the coated film will render excellent scratch resistance, while it has different degree of flexibility for specific applications.

Concluding Remarks

In this review we have tried to provide the reader with two light microscopy techniques to measure or visualize the surface roughness in three dimensions. Both of these techniques complement each other and cover roughness values from the subnanometer range to well into the micron dimensions. We have taken the approach that this review is primarily concerned with imaging techniques; hence, we have not discussed methods dealing with scattering measurements and speckle methods. However, very good discussions of such methods can be found in a variety of places.[1,3–14]

Acknowledgments It is a pleasure to acknowledge many people who helped make this manuscript possible, especially John Guerra of Calimetrix Corporation, with whom one (M. S.) of us has collaborated for several years and continues to do so. We thank Satish Kumar of the School of Textile and Fiber Engineering at the Georgia Institute of Technology for providing radiation-treated DMPBZT fibers. We also acknowledge Peter Lillehei and Larry Bottomley (both from the School of Chemistry and Biochemistry at Georgia Institute of Technology) for providing the etched glass samples. We also thank Rob Dickson and Andrew Bartko for providing the image contained in Figure 17.7. M. S. would like to acknowledge partial financial support from the National Science Foundation under grant no. DMR-0096240 through the CAREER program.

References

1. Whitehouse, D. J., Ed. Selected Papers on *Optical Methods in Surface Metrology*; SPIE Milestone Series, Volume MS 129; SPIE: Bellingham, WA, 1996.
2. de Gennes, P. G. *Rev. Mod. Phys.* **1985**, *57*, 827.
3. Bhushan, B. In *Handbook of Micro/Nano Triboloby*; Bhushan, B., Ed.; CRC Press: Boca Raton, FL, 1999; p 189.
4. Thomas, T. R. *Rough Surfaces*, 2nd ed.; Imperial College Press: London, 1999.
5. Hull, D. *Fractography: Observing, Measuring and Interpreting Fracture Surface Topography*; Cambridge University Press: Cambridge, 1999; chap. 2.
6. Williams, E. D.; Bartlet, N. C. *Science* **1991**, *251*, 393.
7. Berry, M. V.; Hannay, J. H. *Nature* **1978**, *271*, 573.
8. Berry, M. V. *J. Phys. A* **1978**, *12*, 781.
9. Sayles, R. S.; Thomas, T. R. *Nature* **1978**, *271*, 431.
10. Bennett J. M.; Mattsson, L. *Introduction to Surface Roughness and Scattering*; Optical Society of America: Washington, DC, 1989.
11. Morrison, E. *Nanotechnology* **1996**, *7*, 37.
12. Nyffenegger, R. M.; Penner, R. M. *Chem. Rev.* **1997**, *97*, 1195.
13. Carpick, R. W.; Salmeron, M. *Chem. Rev.* **1997**, *97*, 1163
14. Asakura, T. In *Speckle Metrology*; Erf, Robert K., Ed.; Academic Press: New York, 1978; pp 11–47.
15. Reiter, G.; Sharma, C.; Casoli, A.; David, M. O.; Khanna, R.; Auroy, P. *Langmuir* **1999**, *15*, 2551.
16. Srinivasarao, M. *Chem. Rev.* **1999**, *99*, 1935.
17. Standard: Surface Texture (Surface Toughness, Waviness and Lay); ANSI/ASME B46.1–1995; American Society of Mechanical Engineers: New York, 1985.

18. Francon, M. *Laser Speckle and Applications in Optics*; Arsenault, H. H., Trans.; Academic Press: New York, 1979; pp 121–131.

19. Stout, K. J.; Obray, C.; Jungles, J. *Opt. Eng.* **1985**, *24*, 414–418.

20. Sirohi, R. S., Ed. *Speckle Metrology*; Marcel Dekker: New York, 1993.

21. Williams, J. A. *Engineering Tribology*; Oxford University Press: New York, 1994.

22. Bennett, J. M.; Mattsson, L. *Introduction to Surface Roughness and Scattering*; Optical Society of America: Washington, DC, 1989.

23. Thomas, T. R. *Precision Eng.* **1981**, *3*, 97.

24. Vorburger, T. V.; Teague, E. C. *Precision Eng.* **1981**, *3*, 61.

25. Longhurst, R. S. *Geometrical and Physcial Optics*; Wiley: New York, 1968.

26. Sheppard,C. J. R.; Wilson, T. *Theory and Practice of Optical Scanning Microscopy*; Academic Press: New York, 1985.

27. Bailey, B.; Farkas, D. L.; Taylor, D. L.; Lanni, F. *Nature* **1993**, *366*, 44.

28. Mansfield, S. M.; G. S. Kino, *App. Phys. Lett.* **1990**, *57*, 2615.

29. Guerra, J. M. *Appl. Opt.* **1990**, *29*, 3741.

30. Guerra, J. M. *SPIE Proceedings*, **1988**, *1009*, 254.

31. Betzig, E.; Lewis, A.; Harootunian, A.; Kratschmer, E. *Biophys. J.* **1986**, *49*, 269.

32. Pohl, D. W.; Denk, W.; Lanz, M. *App. Phys. Lett.* **1984**, *44*, 652.

33. Betzig, E.; Trautman, J. K.; Harris, T. D.; Weiner, J. S.; Kostelak, R. L *Science* **1991**, *251*, 1468.

34. Pohl, D. W. In *Advances in Optical and Electron Microscopy*; Sheppard, C. J. R., Mulvey, T., Eds.; Academic Press: London, 1990; pp 243–312.

35. Binnig, G.; Rohrer, H.; Gerber, C.; Weibel, E. *Phys. Rev. Lett.* **1982**, *49*, 57.

36. Binnig, G.; Rohrer, H. *Physica* **1984**, *127B*, 37.

37. Eigler, D. M.; Schweizer, E. K. *Nature* **1990**, *344*, 524.

38. Wickramasinghe, H. K., Ed. *Scanned Probe Microscopy*; American Institute of Physics: New York, 1992.

39. Nassenstein, H. *Opt. Comm.* **1970**, *2*, 231.

40. Nassenstein, H. *Optik* **1969**, *29*, 597.

41. Nassenstein, H. *Optik* **1969**, *30*, 44.

42. Newton, I. *Opticks*; Dover: New York, 1730/1979; Part I, pp 193–224.

43. Bryngdahl, O. *Prog. Opt.* **1973**, *11*, 189.

44. Harrick, N. J. *J. App. Phys.* **1962**, *33*, 2774.

45. Young, T. R.; Rothrock, B. *J. Opt. Soc. Am.* **1961**, *51*, 1038.

46. Harrick, N. J. *Phys. Rev. Lett.* **1960**, *4*, 224.

47. Harrick, N. J. *Internal Reflection Spectroscopy*; Harrick Scientific Corp.: New York, 1979.

48. McCutchen, C. W. *App. Opt.* **1962**, *1*, 253.

49. McCutchen, C. W. *Rev. Sci. Inst.* **1964**, *35*, 1340.

50. Ambrose, E. J. *Nature* **1956**, *24*, 1194.

51. Ausserre, D.; Hervet, H.; Rondelez, F. *Phys. Rev. Lett.* **1985**, *54*, 1948.

52. Migler, K. B.; Hervet, H.; Leger, L. *Phys. Rev. Lett.* **1993**, *70*, 287.

53. Allegrini, M.; Ascoli, C.; Gozzini, A. *Opt. Comm.* **1971**, *2*, 435.

54. Diaspro, A.; Aguilar, M. *Ultramicroscopy* **1992**, *42–44*, 1668.

55. Bryngdahl, O. *J. Opt. Soc. Am.* **1969**, *59*, 1645.

56. Bryngdahl, O. In *Progress in Optics*. E. Wolf, Ed.; North-Holland, Amsterdam, 1973; chap. 4.

57. Brown, M. A.; Staples, E. J. *Farad. Disc., Chem. Soc.* **1990**, *90*, 193.

58. Previe, D. C.; Luo, F.; Lanni, F. *Farad. Disc. Chem. Soc.* **1987**, *83*, 297.

59. Previe, D. C. ; Frej, N. A. *Langmuir* **1990**, *6*, 396.

60. Chew, H.; Wang, D-S.; Kerker, M. *App. Opt.* **1979**, *18*, 2679.

61. Brakenhoff, G. J.; van Spronsen, E. A.; van der Voort, H. T. M.; Nanninga, N. In *Methods in Cell Biology*; Taylor, D. L., Wang, Yu-Li, Eds.; Academic Press: New York, 1989; Vol. 30, Part B, chap. 14.

62. Lewis, A. In *New Techniques of Optical Microscopy and Microspectroscopy* ; Cherry, R. J., Ed.; CRC Press: Boca Raton, FL, 1991; chap. 2

63. Li, L.; Sosnowski, S.; Chaffey, C. E.; Balke, S. T.; Winnick, M. A. *Langmuir* **1994**, *10*, 2495.

64. Ling, X.; Pritzker, M. D.; Byerley, J. J.; Burms, D. M. *J. App. Polym. Sci.* **1997**, *67*, 149.

65. Ribbe, A. E. *Trend. Polym. Sci.* **1997**, *5*, 333–337.

66. Song, Y.; Srinivasarao, M.; Tonelli, A. E.; Balik, C. M.; McGregor, R. *Macromolecules* **2000**, *33*, 4478.

67. Goodman, J. W. *Introduction to Fourier Optics*; McGraw-Hill: New York, 1968; chap. 6.

68. Ditchburn, R. W. *Light*; Dover: New York, 1991; chap. 8.

69. Lipson, S. G.; Lipson, H. *Optical Physics*; Cambridge University Press: Cambridge, 1981; chaps. 3 and 9.

70. Toraldo di Francia, G. *J. Opt. Soc. Am.* **1969**, *59*, 799.

71. Born, M.; Wolf, E. *Principles of Optics*; Pergamon: New York, 1986; pp 418–425.

72. Porter, A. B. *Phil. Mag.* **1906**, *11*(6), 154.

73. Thomas, T. R. *Precision Eng.* **1981**, *3*, 310–315.

74. Rayleigh, Lord. *Phil. Mag.* **1896**, *42*(5), 167.

75. Lukosz, W. *J. Opt. Soc. Am.* **1966**, *56*, 1463.

76. Toraldo di Francia, G. *J. Opt. Soc. Am.* **1955**, *45*, 497.

77. Pask, C. *J. Opt. Soc. Am.* **1977**, *66*, 68.

78. McCutchen, C. W. *J. Opt. Soc. Am.* **1967**, *57*, 1190.

79. Harris, J. L. *J. Opt. Soc. Am.* **1964**, *54*, 931.

80. Marcuse, D. *Light Transmission Optics*; Van Nostrand Reinhold: New York, 1972; pp 164–173.

81. Francon, M. *Progress in Microscopy*; Row, Peterson: Evanston, IL, 1961.

82. Inoue, S. In *Methods in Cell Biology*; Taylor, D. L., Wang, Yu-Li, Eds.; Academic Press: New York, 1989; Vol. 30, Part B, Chap. 3.

83. Axelrod, D. In *Methods in Cell Biology*; Taylor, D. L., Wang, Yu-Li, Eds.; Academic Press: New York, 1989; Vol. 30, Part B, Chaps. 9 and 15.

84. Guerra, J. M.; Srinivasarao, M.; Stein, R. S. *Science* **1993**, *262*, 1395.

85. Srinivasarao, M. In *Comprehensive Polymer Science, Second Suppl*; Aggarwal, S. L., Russo, S., Eds.; Pergamon: New York, 1996; pp 163–197.

86. Temple, P. A. *App. Opt.* **1981**, *20*, 2656.

87. Zhu, S.; Yu, A. W.; Hawley, D.; Roy, R. *Am. J. Phys.* **1986**, *54*, 601.

88. Baumeister, P. W. *App. Opt.* **1967,** *6,* 897.

89. Winnik, M. A.; Wang, Y.; Haley, F. *J. Coat. Tech.* **1992,** *64,* 51.

90. Wang, Y.; Kats, A.; Juhue, D.; Winnick, M. A.; Shivers, R. R.; Dinsdale, C. J. *Langmuir* **1992,** *8,* 1435.

91. Eckersley, S. T.; Rudin, A. *J. Coat. Tech.* **1990,** *62,* 89.

92. Keller, A. *Phil. Mag.* **1957,** *2*(8), 1171.

93. Till, P. H. Jr. *J. Polym. Sci.* **1957,** *24,* 301.

94. Fischer, E. W. *Z. Naturforsch.* **1957,** *120,* 753.

95. Keller, A.; O'Connor, A. *Disc. Farad. Soc.* **1958,** *25,* 114.

96. Geil, P. H. *Polymer Single Crystals;* Wiley Interscience: New York, 1963.

97. Sullivan, P.; Wunderlich, B. *SPE Trans.* **1964,** *4,* 2.

98. Sullivan, P.; Wunderlich, B. *Polym. Lett.* **1964,** *2,* 537.

99. Reneker, D. H.; Geil, P. H. *J. Appl. Phys.* **1960,** *31,* 1916.

100. Teletzke, G. F.; Ted Davis, H.; Scriven, L. E. *Chem. Eng. Comm.* **1987,** *55,* 41.

101. de Gennes, P. G. *Rev. Mod. Phys.* **1985,** *57,* 827.

102. Daillant, J.; Bennatar, J. J.; Leger, L. *Phys. Rev.* **1990,** *A 41,* 1963.

103. Brochard-Wyart, F.; Daillant, J. *Can. J. Phys.* **1990,** *68,* 1084.

104. Redon, C.; Brochard-Wyart, F.; Rondelez, F. *Phys. Rev. Lett.* **1991,** *66,* 715.

105. Reiter, G. *Phys. Rev. Lett.* **1992,** *68,* 75.

106. Gupta, V. K.; Kornfield, J. A.; Ferencz, A.; Wegner, G. *Science* **1994,** *265,* 940.

107. Guerra, J. M. *Appl. Opt.* **1993,** *32,* 24.

108. Bennett J. M.; Namba, Y.; Guerra, J. M.; Jahanmir, J.; Balter, T. L.; Podlesny, J. C. *Appl. Opt.* **1997,** *36,* 2211.

109. Wilson, T. In *Confocal Microscopy;* Wilson, T., Ed.; Academic Press: New York, 1990; chap. 3.

110. Wilson, T.; Gannaway, J. N.; Sheppard, C. J. R. *Optical Quant. Elec.* **1980,** *12,* 341.

111. Masters, B. R. In *Confocal Microscopy;* T. Wilson, Ed.; Academic Press: London, 1990; chap. 11.

112. Dosho, S.; Ise, N.; Ito, K.; Iwai, S.; Kitano, H.; Matsuoka, H.; Nakamura, H.; Okumura, H.; Ono T.; Sogami, I. S.; Ueno, Y.; Yoshida, H.; Yoshiyama, T. *Langmuir,* **1993,** *9,* 394.

113. Ito, K.; Yoshida, H.; Ise, N. *Science* **1994,** *263,* 66.

114. Jinnai, H.; Koga, T.; Nishikawa, Y.; Hashimoto, T.; Hyde, S. T. *Phys. Rev. Lett.* **1997,** *78,* 2248–2251.

115. Ribbe, A. E.; Hashimoto, T.; Jinnai, J. H. *Mater. Sci.* **1996,** *31,* 5837–5847.

116. White, W. R.; Wiltzius, P. *Phys. Rev. Lett.* **1995,** *75,* 3012–3015.

117. Ribbe, A. E. *Trend. Polym. Sci.* **1997,** *5,* 333–337.

118. Cutts, L. S.; Roberts, P. A.; Adler, J.; Davies, M. C.; Melia, C. D. *J. Microsc.* **1995,** *180,* 131–139.

119. Blonk, J. C. G.; Don, A.; Aalst, H. V.; Birmingham, J. J. *J. Microsc.* **1993,** *169,* 363–374.

120. Rubin, B. *Adv. Mat.* **1998,** *10,* 1225–1227.

121. Rubin, B.; Kobsa, H.; Shearer, S. M. *App. Opt.* **1997,** *36,* 6388–6392.

122. Srinivasarao, Mohan; Collings, David; Philips, Alan; Patel, Sanjay. *Science,* **2001,** *292,* 79.

123. Cogswell, C. J.; Sheppard, C. J. R. In *Confocal Microscopy;* Wilson, T., Ed.; Academic Press: New York, 1990; Chatper 8.

124. Kaneko, R. *Tibology Internat.* **1995,** *28,* 33.

125. Poon, C. Y.; Bhushan, B. *Wear* **1995,** *190,* 76.

126. Kaneko R.; Hamada, E. *Wear* **1993,** *162,* 370.

18

Light Microscopy

BARBARA FOSTER

A long-time colleague, well known for his lifelong dedication to polymer science, microscopy, and microscopy education, once observed that no one under the level of Ph.D. should be allowed to use a light microscope. Needless to say, his comments caused quite a stir. When asked to defend his position, he commented, "Because then laboratory managers would take this technique seriously."

At first glance, the light microscope is deceptively easy. Just plug it in, turn it on, place a sample on the stage, and voila! an image! However, two decades of teaching in industry indicates that relatively few scientists, technicians, or engineers really understand what that image is telling them or whether the image is fraught with artifacts or misleading data. The purpose of this chapter is to equip you to meet both challenges: to really understand how to set up a light microscope so that you know what it is telling you about your sample and to avoid typical pitfalls, errors, and assumptions.

There are two very fine classics on the use of microscopy in polymer analyses: Derek Hemsley's *Applied Polymer Microscopy*[1] and Linda Sawyer's *Polymer Microscopy*.[2] This chapter is intended to complement these works. Its focus is the light microscope and its application to polymer science. By combining a strong foundation in basic principles with a wide range of practical tips, the goal here is to introduce the broad arsenal of techniques inherent in this discipline so that whether you are analyzing films, fibers, foams, block polymers, multilayer systems, or the particulate matter used as fillers and catalysts, you will be competent enough to know when to use which technique, how to prepare the sample for best viewing, and how to interpret the resulting information.

Unlike most other polymer references, this chapter contains short experiments that can be performed at your own lab bench. Among analytical techniques, microscopy is distinctive in that you, as the practitioner, are part of the system. As a result, the technique requires a unique hand-eye-brain coordination. I strongly encourage you to "play" with the techniques discussed here until they become a natural part of your problem-solving repertoire. The mental images built by these exercises will form the foundation of an irreplaceable personal library. In turn, that library will form the foundation of an ongoing collection of reference images, which, again, will speed and expand future analyses.

For polymer scientists, microscopy should be an integral part of the whole analytical scheme. No other technique provides such an interactive view into the world of polymer science. Microscopy, and especially light microcopy, provides simultaneous quantitative, image, and process data, typically with only modest sample preparation. With the movement toward integrated imaging in Fourier transform infrared (FTIR) and other spectroscopies, a working knowledge of light microscopy is an investment that will reap significant rewards.

Basic Definitions

Before we get started, it is important to have a few key definitions and terms.

Microscopy

Microscopy can be summarized by a long-standing definition from the Royal Microscopical Society as "the art and science of making fine detail visible". This definition embraces both the scientific principles behind the microscope and the artistic control of the sample preparation and imaging system. It also brings into focus the three main concerns of any microscopist:

Magnification—the size relationship between the object and the image
Resolution—the smallest separation between any two edges or features, which still permits them to be imaged as independent entities
Contrast—visibility against the background

The RMS definition of microscopy also reveals a hidden truism: the microscope's job is to translate information from an *object* to an *image*. Note that the object and its image are two independent entities. You can prove this concept quickly and effectively by looking at your finger through a magnifying glass. Start with your finger close to the back of the lens, then slowly move it further and further away from the lens. Notice that it appears to get larger and larger. In actual fact, just the image got larger; your finger (the object) stayed the same size. As you continue to increase the distance between your finger and the lens, you will notice a point at which the image disappears altogether then reappears, inverted, yet at no time did your finger leave your hand.

Our job as microscopists is to reproduce information present in the object (orientation, size, spacing, color, etc.) as accurately as possible in the image. Since this chapter deals with *light microscopy*, photons will be used as the probe to develop images for our evaluation.

Magnification

Magnification is the relationship between the image and the object from which that image was derived. As seen in eq 1, it can be expressed as either a size relationship or a distance relationship. The distance relationship will be discussed further in the section on geometric optics.

$$S_i/S_o = D_i/D_o \quad \text{(magnification)} \quad (1)$$

where S_i is the size of the image, S_o is the size of the object, D_i is the distance of the image from the lens which formed it, and D_o is the distance from the object to the lens that will image it.

The nomenclature regarding magnification is derived directly from this equation. If a lens makes an image 10 times larger than the original object, we say that the object has been magnified by a factor of 10 or "10×". There is nothing implicit in this equation, however, that says that the image cannot be smaller than the object, resulting in *minification*. As a matter of fact, the eyepiece minifies the background information so that it can fit through the small pupil in your eye. To test this observation, set up the microscope with a well-prepared, colored sample. Stretch a piece of lens tissue over the eyepiece, then move the tissue away. At some point you will see a small, white circle of light. This is the *exit pupil* of the microscope, also called the *Ramsden disk*. It carries the information that appears in the back of the objective. If you pull out the eyepiece, you can compare the size of that optical plane with the small, focused point observed on your tissue "screen": it has been minified by the eyepiece. When you are correctly aligned with the microscope, the Ramsden disk rests on the cornea of your eye.

Resolution

When asked "What is the primary function of the microscope?" nearly everyone answers "magnification". How-

ever, the microscope's resolving ability is just as important. As a matter of fact, anyone responsible for buying optics knows that high-resolution lenses come at a premium.

Mathematically, resolution is defined as a relationship between wavelength and the collecting angle of the objective (eq 2):

$$R = 1.22 \, \lambda/(\text{NA}_o + \text{NA}_c) \quad \text{(resolution)} \quad (2)$$

where 1.22 is the Bessel function related to use of round apertures in the microscope; λ is the wavelength of light, expressed in micrometers; NA_o is the numerical aperture of the objective; and NA_c is the numerical aperture of the condenser.

Note that resolution is a distance: the smallest distance by which two objects can be separated and still imaged as two, separate objects. If they are brought any closer, their information will merge and be imaged as a blob. Moreover, resolution is strongly dependent on the wavelength of light used (Figure 18.1). The shorter the wavelength, the smaller the particle that can be resolved. In polymer microscopy, particles below the resolving limit of the light microscope are typically imaged with scanning electron microscopes (SEM).

Figure 18.2 illustrates the effect of the collecting angle of the objective on resolution. Light is the messenger that carries information from the object to the image. The image is formed by interference between light diffracted by the specimen and light forming the background. Any ray diffracted outside of the collecting angle of the objective results in lost information.

The objective's collecting effectiveness is described in terms of the *numerical aperture* (NA) and relies on two parameters: the actual collecting half angle of the objective (α) and the refractive index (RI or n) of the material between the top of the specimen and the front of the objective. Both will be discussed further in the section on imaging.

Figure 18.2 also presents two other related concepts: *working distance* (WD) and the *back focal plane of the objective* (BFP_o). Working distance is the clear, working space between the front of the objective and the top of the specimen. Frequently, the larger the NA, the shorter the WD.

The back focal plane of the objective will become very familiar to you as you read this chapter. It contains the Fourier transform of the image and presents diagnostic information needed for troubleshooting imaging artifacts, as well as important polarized light data. To view it directly, simply remove the eyepiece from the microscope and peer down the tube, into the back of the objective.

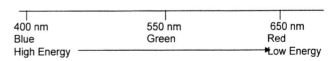

Figure 18.1 The visible spectrum.

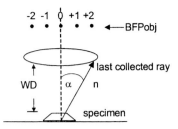

Figure 18.2 Numerical aperture depends on α, half the collecting angle of the objective, and n, the refractive index of the material between the top of the sample and the face of the objective. WD = working distance.

Contrast

Contrast is defined as visibility against the background. Mathematically, it is expressed as differences in intensity between the object and the background:

$$C = (I_b - I_o)/ I_b \qquad \text{(contrast)} \qquad (3)$$

where I_b is intensity of the background, and I_o is intensity of the object.

In actuality, since humans can detect changes in either color or intensity, we can use either mechanism for improving visibility. For polymer microscopy, contrast is a major issue since many systems contain particles, fillers, block copolymers, and multilayers. Specific features may fade into the background. Also, contrast enhancement can bring out chemical or orientation effects that may significantly affect performance. Because of the range of techniques available and its importance, contrast enhancement is discussed in a separate section.

Measurement

In this digital age of microscopy, there are very few instances where we stop with magnification, resolution, and contrast. The fourth major concern of most microscopists is quantitation: how big, how many, what volume percentage, and so on. In polymer microscopy, other factors such as orientation on both the molecular and particle level and surface roughness have a profound effect on performance. While a complete discussion of measurement is outside the scope of this chapter, details will be provided where pertinent, as well as a general overview of other techniques.

Microscopes: Structure, Function, and Alignment

Configuration

Light microscopes come in a variety of configurations to meet a variety of needs. Microscopes are divided into two sets of categories, each determined by specific design crite-

ria. Traditionally, the first is determined by the need to look at three-dimensional (3D) versus flat objects and divides microscopes into *stereo* versus *compound* microscopes.* The second group can be divided further into *inverted* versus *upright* stands, depending on whether the objectives are mounted above (upright) or below (inverted) the sample.

Stereo Microscopes

After a cursory examination by eye and with a hand lens, the knowledgeable microscopist observes the specimen under a stereo microscope like the one shown in Figure 18.3, left. This lower-magnification, 3D stereo view provides context from the bulk sample and often points to the best areas for sampling.

Internally, this microscope is actually has two microscopes, each of which presents its own image to one of your eyes. The images, separated by 10° to 12°, are combined into one, 3D stereo image by the brain. These microscopes are often mistakenly referred to as "low-power" microscopes, since they operate over limited magnifications (typically up to about 600× total magnification).

Taking photographs with a stereo microscope can be deceptive since the photo systems allow the camera to view only one of the two images. As a result, photographs taken through this type of microscope typically do not display the real-life three-dimensionality seen in direct observation. If the photo system has the capability to switch back and forth between each of the images, the camera can capture a *stereo pair* of images. Until recently, bulky viewers were needed to merge the stereo pair into one 3D image, but new technology is becoming available which can print them on an inkjet printer, producing polarized 3D prints and transparencies[3] that can be viewed with special polarizing glasses.

Compound Microscopes

The next step up from the stereo microscope is the *compound microscope* (Figure 18.3, right). Since compound microscopes are used to look at flat samples, sample preparation for this step may require either sectioning or mounting, grinding, and polishing the sample. Also, since this step involves removing a specific area from the bulk, it is important to give some thought to how representative that area is.

Compound microscopes come in a variety of configurations. For most polymer applications, *upright microscopes* (characterized by objectives looking down on the specimen, as seen in Figure 18.3, right) are the best choice. For rare instances involving bulkier, heavier samples, an inverted microscope is more appropriate.

Compound microscopes also offer two alternative viewing systems: *transmitted light* for looking though samples moun-

---

*The boundary between stereo and upright microscopes is being blurred by new technologies that are building 3D viewing systems into upright stands.

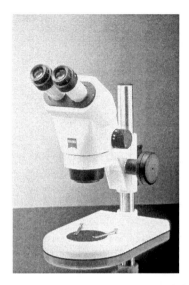 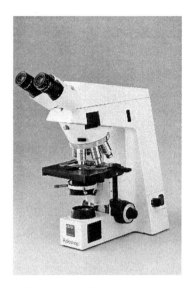

Figure 18.3 Stereo microscope (*left*); upright compound microscope (*right*). (Images courtesy of Carl Zeiss, Inc.)

ted on microscope slides (e.g., films, fibers, powder dispersions, and thin sections) or *reflected light* for examining surface and near-surface information on opaque samples. Figure 18.4 contains simple diagrams showing the difference.

For documentation, compound microscopes typically have a variety of ports to which cameras can be attached. These cameras can use conventional film (usually 35 mm), instant film in a variety of formats, or video or digital systems. As a matter of fact, in today's labs it is rare to find a microscope without some sort of electronic imaging system that feeds the image into a computer for processing, measurement, archiving, and communication over a local area network (LAN) or the Internet.

For contrast enhancement, compound microscopes can be ordered with a range of techniques from darkfield, phase contrast, and Hoffman Modulation Contrast to fluorescence, polarized light, and differential interference contrast (also known as Nomarksi or DIC). These techniques are discussed in detail later in the section on contrast.

A Quick Tour around the Microscope

Understanding the function of major components is a major advantage in building a wide arsenal of effective problem-solving techniques (Figure 18.5). Of special interest are three lenses (the objective, condenser, and eyepieces), which are responsible for conditioning the light and viewing the sample, and two irises (the field iris, FI, and aperture iris, AI), which are responsible for the optical control of the microscope. As an interesting exercise, read the next part of the chapter while you sit at your microscope. It will give you much more insight into the capabilities of your equipment.

Objective

The *objective* is the mastermind of the microscope. A quick look at its engravings (Figure 18.6) provides instant orientation as to the capabilities of a particular system, specifically magnification, resolution, engineering approach, coverslip requirements, and, often, contrast techniques.

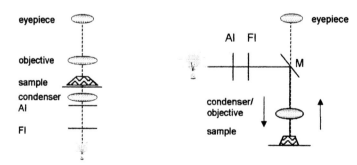

Figure 18.4 Light paths for transmitted (*left*) and reflected (*right*) light microscopes. Key: AI, aperture iris; FI, field iris; M, half-silvered mirror.

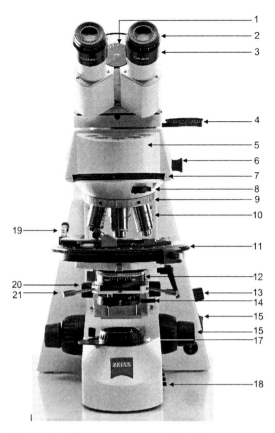

Figure 18.5 Key components of the light microscope. Key: (1) interpupillary distance setting; (2) diopter setting; (3) eyepiece (ocular); (4) analyzer (rotatable); (5) turret/fluorescence cubes; (6) polarizer/reflected light; (7) wheel to choose fluorescence cube; (8) compensator in compensator slot; (9) nosepiece; (10) objective; (11) stage (rotatable); (12) condenser/aperture iris control; (13) rheostat (illumination intensity); (14) flip out polarizer (rotatable); (15) on/off switch; (16) coaxial coarse (large) and fine (small) focus; (17) field iris (sitting on light port); (18) filters (neutral density, blue, green); (19) mechanical stage movement; (20) condenser focus; (21) condenser centration screw. (Image courtesy of Carl Zeiss, Inc.)

Table 18.1 summarizes the engravings that appear on this objective. In addition to the notations listed here, objectives engineered to look through deeper or multilayer samples will be marked *LWD* for *long working distance*. Additionally, there may be an engraving to denote a specific contrast enhancement technique. Table 18.2 lists a number of these markings.

The coarse and fine focus knobs provide the control for the stage, increasing or decreasing its distance from the objective. When focusing the objective with respect to the sample, a good rule of thumb is to *focus away*. That is, while observing the distance between the objective and sample directly (not through the eyepieces), use the coarse focus to bring the sample nearly in contact with the objective. Pay careful attention to the direction in which the knob was

turned. Then, while viewing through the eyepieces, turn the knob the other direction until the image comes into focus. Using this approach guarantees (a) location of the image, (b) efficiency of operation, and, most importantly, (c) preservation of both the objective and your sample.

Condenser and Aperture Iris

Partnering with the objective is an often-overlooked piece of glassware, the *condenser*. Since most of its responsibilities begin with that letter "C", that letter provides a convenient memory clue for the operation of this lens. Its major function is to condition the light by controlling the angle and location of light approaching the specimen. For transmitted light systems, such as the one shown in Figure 18.5, the condenser is the optical assembly mounted immediately below the stage. Note that it has its own focus knob, usually located to the left, under the stage. It may also have carriers for filters, polarizers, and auxiliary lenses.

The key control mechanism in the condenser is the *aperture iris* (also called the *condenser aperture iris*). Although adjusting this opening causes a change in intensity in the image, it should not be used for brightness control. Its true function is to control the range of angles at which light emerges from the condenser and, as a result, the coherence of the light approaching the sample. Figure 18.7 shows the condenser with the aperture iris fully opened. Notice that rays traveling through the periphery have a longer optical path than those traveling directly up the axis. As discussed in the next section, if light waves travel the same optical path (for example, up the optical axis of the microscope) and if they start out in step or *in phase* with each other, they will continue to be in phase. Conversely, waves that travel a longer optical path will get out of step with those traveling up the axis. Light in which the waves are in phase is said to be *coherent*. Lasers, for instance, produce highly coherent light.

Closing the aperture iris reduces the contribution of incoming light to just the axial rays and produces highly coherent light. This simple adjustment has two profound effects on the information in the image. First, highly coherent

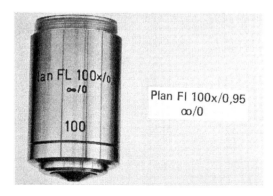

Figure 18.6 Engravings on a typical microscope objective. (Image courtesy of Carl Zeiss, Inc.)

Table 18.1 Summary of objective engravings

Engraving	Meaning	Comments
100×	Magnification	Objective establishes first step of magnification
1.4	Numerical aperture (NA)	Determines resolution of system
oel	Oil immersion	When oiled to sample with appropriate immersion oil, will resolve smaller features. High NA enables optical sectioning; lower NAs are preferable for imaging deeper sections for context
∞	Infinity-corrected optics	Engineering concept; as opposed to *fixed mechanical tubelength*[a] of 160, 170, or 210 mm
0.17	Coverslip correction	This objective requires a 0.17-mm thick (No. 1½) coverslip[b] Other notations: 0 = no coverslip; "-" = works with or without coverslip Some objectives also available with adjustable coverslip correction collars
Plan	Flat field	Entire image will be in focus, edge to edge
Apo	Apochromat; level of aberration correction	Lenses are corrected for chromatic and spherical aberrations. Apochromats are the highest corrected. See section on geometric optics

a. Mechanical tubelength is the distance from the shoulder of the objective where it screws into the nosepiece to the primary image plane, just below the seat of the eyepiece.
b. The coverslip, a small glass plate that is placed directly on the sample, is an important optical element.

light sets up the special condition for diffraction at edges in the sample. Those diffracted rays that meet in step will undergo constructive interference, creating bright fringes at that edge; those that meet one-half wave out of step will undergo destructive interference, creating dark fringes. This set of bright and dark fringes helps define edge information. When the condenser is closed too far, it is not unusual to see alternating bright and dark bands around dust particles and sharp-edged features.

As seen in eq 2, the condenser also contributes to the resolving power of the microscope. As a result, if the condenser is closed appreciably, it may reduce the ability to resolve very fine detail.

As with the objective, the condenser will have an engraving indicating its numerical aperture (NA). A reminder: the engraved NA is the maximum value and only works when the condenser is fully opened. Also, condensers marked with NAs greater than 0.9 require a drop of immersion oil between the top element of the condenser and the back of the slide in order to work at full NA.

Finally, the condenser may also contain a variety of attachments that contribute to contrast. Although some con-

densers, such as *cardiod darkfield condensers*, can be used for only one contrast method (Figure 18.8, left), others have a rotatable turret, thus allowing the user to dial in a variety of methods to match components in the objectives (Figure 18.8, right). Frequently, turret condensers will have two or three Phase or Hoffman Modulation positions and several DIC positions, each matched to components in different magnification objectives.

Eyepieces

The eyepieces are the objective's other partners. They contribute the second step of magnification and, as with the objectives, are marked with large numbers describing that value. As shown in eq 4, the calculated magnification for the microscope image, as presented to the eye, is the product of the magnification of the objective multiplied by the magnification of the eyepiece. It is always a good idea to check the front of the microscope just below the binocular body for any intermediate lenses, especially turrets with intermediate magnifiers, which also contribute to the size of the final image.

Table 18.2 Some typical contrast notations

Technique	Notation
Darkfield	German: HD ("helle und dunklefeld" = brightfield/darkfield); Japanese: D or BD (brightfield/darkfield)
Phase contrast	PH
Hoffman modulation contrast	HMC
Polarized light microscopy	Pol
Differential interference contrast/Nomarski	DIC; IK (frequently on German optics)

Note: FL refers to "Fluor" objectives, a classification of correction rather than optics used for fluorescence.

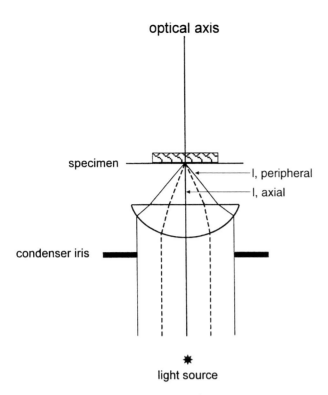

optical axis

specimen ——

——I, peripheral

——I, axial

condenser iris ——

✴
light source

Figure 18.7 The aperture iris is the key component to controlling light in the condenser. Peripheral rays travel further than rays coming directly up the optical axis.

$$M_{\text{T}} = M_{\text{objective}} \times M_{\text{eyepiece}} \times M_{\text{intermediate}}$$

(total calculated magnification)   (4)

The second function of the eyepieces is to determine the actual territory available for viewing, called the *field of view* (FOV), or the diameter of the field being observed, typically expressed in micrometers.

Eyepieces come in two different forms, Huygenian and Ramsden (Figure 18.9). To determine which type of eyepiece is on your microscope, remove the eyepiece and turn it upside down. Huygenian eyepieces have a lens mounted in the bottom, and Ramsden eyepieces have the last lens mounted about two-thirds the way up the tube. If your microscope has Ramsden eyepieces, you can locate the small ridge or baffle that determines the field of view by turning the eyepiece upside down and gently sliding your finger up the lower wall.

The baffle is located in a very specific optical plane, one that contains the actual image information. As a result, it is also a logical location for reticles and masks. These inserts can be simple mesh grids for quick particle counting, rulers for measuring length, or special patterns for determining particle orientation (e.g., glass fiber orientation and tilt), optical axis spacing, or stereology.

The *field number* (the distance, in millimeters, across the baffle opening) is typically engraved alongside the magnification. The field number is extremely useful for doing quick estimates of particle or feature size:

$$\text{FOV} = \text{field no.}/(M_{\text{objective}} \times M_{\text{intermediate}})$$

(calculating for field of view)   (5)

EXERCISE

Find the field of view for a 100× objective with a field number of 25 mm.

FOV = 25 mm/100 = 0.250 mm = 250 µm.

Equation 5 illustrates how to do this calculation. Simply divide the field number by the product of the magnifications of the objective and any intermediate lenses. To determine the size of a particle or any specific feature, estimate how many of them would fit across the diameter,

Figure 18.8 Condensers may also contribute to contrast techniques. Darkfield cardiod condenser (*left*); turret condenser with a selection of contrast options (*right*). (Images courtesy of Olympus America.)

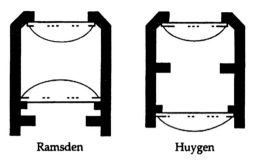

Ramsden　　　　**Huygen**

Figure 18.9 Ramsden (*left*) and Huygenian (*right*) eyepieces.

then divide that number into the FOV. For instance, if 10 particles fit across the FOV calculated above, then each feature would be 250 μm/10, or approximately 25 μm.

As with the objective and condenser, many eyepieces have their own focusing mechanism. Again, it is a good idea to evaluate your own system since either one or both eyepieces may be adjustable. Adjustable eyepieces have rotatable rings on the eyepiece or on the mount, with either a black, silver, or white ring or a fiduciary mark (dot or "0") for setting the null position. The process for focusing is described later in this chapter, in the section about setting up Koehler illumination.

The eyepieces in your microscope are optically set to view the magnified image of your specimen. However, there is a lot of information to be gained by looking at a special optical plane called the *back focal plane* of the objective (BFP$_o$). While this plane can be viewed directly by removing the eyepieces and peering deep down the tube, it is more comfortable to view that plane with an accessory called either a *centering telescope* or a *phase telescope*. To use this accessory, simply replace the eyepiece with the telescope. It has its own focusing system and should be adjusted to give a crisp, clean image of the BFP. Your system may also have *photo eyepieces*. These lenses are typically more slender than the eyepieces used for regular viewing and are placed into the phototube that leads to the camera.

Field Iris

The final component used extensively for control is the *field iris*. Usually found as an adjustable ring around the light port or adjacent to the light port, this mechanism works with the baffle in the eyepiece to determine the diameter of the FOV. When fully open, it typically sits just outside the viewable area. Its function is control of the spillover light that is responsible for glare. For highly scattering samples, the field iris can be a valuable ally: move the feature of interest to the center of the field, then close the field iris around it to remove the haze-forming contributions from the edge of the field.

The case of the dull yogurt carton illustrates the value of this technique. A polymer company that makes yogurt cartons was plagued by a glaze that did not meet specifications for shine. Using a high magnification/high numerical aperture objective and DIC (a good technique for imaging irregularities in surfaces), it was determined that the problem stemmed from small hillocks (about 1 μm, as estimated from the FOV calculation) that had formed where the clear glaze met the opaque body. These bumps were bouncing light randomly in all directions, creating both the haziness in the glaze and a serious imaging problem in the microscope. Closing the field iris down around the central 10% of the field removed the contribution from the rest of the field and permitted photography for documentation and discussion with the engineers.

Reflected Light versus Transmitted Light Systems

To transpose all this information about transmitted light to reflected light, simply "fold" the transmitted light path at the specimen (Figure 18.4). Notice that, while this action "merges" the objective and the condenser into one optical component, the functions of the two separate lenses still occur independently: light is still conditioned on the way down to the sample (the condenser), and information about the resulting interaction is collected on the way back (the objective).

One of the best examples to illustrate this behavior is the large-barreled HD (brightfield/darkfield) objective (Figure 18.10). A quick look down the back of this objective reveals an outer channel through which light travels down to the specimen (the shaded area in Figure 18.10). Small reflectors at the bottom of this channel (R) cause the light to impinge on the sample at a very high angle. The angle is so obtuse that light reflected by the surface (shown with asterisks) will not be collected by the smaller, inner channel that performs the collecting function of the objective. Only light scattered by pits, scratches, and other scattering elements (dotted lines) will be directed back into the collecting angle of the objective and be used to form the image.

The second major difference between the transmitted and reflected light paths is the relative placement of the aperture and field irises. If we followed the "folding" analogy completely, the aperture iris, which is normally mounted in the front focal plane (FFP) of the condenser, would be physically mounted in the back of the objective, a location that is difficult to access. It is much easier to insert a relay lens in the light bridge, allowing the engineers to place the aperture iris just behind the field iris. Typically, the aperture iris is noted by just an "A" or "AI", while the field iris is noted as "F" or "FI". To test which iris is which, place a well-behaved specimen on the stage and bring it into focus. Make sure that both irises are wide open. Closing the aperture iris will change intensity and bring out edge information, while closing the field iris will actually decrease the size of the field of view.

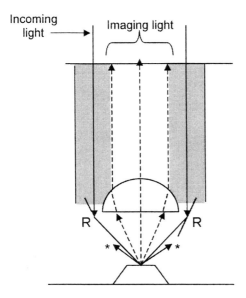

Figure 18.10 Simplified diagram of a reflected light, HD ("helle/dunkel", or brightfield/darkfield) objective. Key: shaded area is the outer channel through which light travels down to the specimen; R indicates the reflectors at the bottom of the channel; * indicates the light reflected by the surface.

A properly centered light source optimizes image quality and resolution. However, most microscopists either are not adept in this process or do not pay adequate attention to it. There are a number of ways of centering lamps. While the best reference is the instructions that came with your system, here are a few quick tricks.

To center a quartz halogen lamp in transmitted light, remove any diffusing filters from the light path, then put a piece of lens tissue over the light port. Using the controls on the lamp housing, center the image of the coiled filament. If there are two images (the real image and the mirror image), they are usually centered north to south, with the two images "knuckle-to-knuckle" in the middle of the field. On some models, there will also be a collector lens that focuses the real image. Some manufacturers recommend defocusing the images once they are centered to produce a more even field of illumination. Alternatively, remove the diffuser, objective, and eyepiece. Open the condenser iris fully. Center the image of the coiled filament that appears in the back focal plane of the objective. Reminder: centering telescopes are helpful in imaging this plane.

To center an arc lamp (usually used in reflected light), place a piece of flat white card stock (for example, the back

Other Important Terms

Now that you understand the basics, here are several important distances (Figure 18.11). *Working distance* is the free space between the front surface of the objective and the top of the specimen. Typically, the higher the magnification and/or the higher the numerical aperture, the shorter the working distance.

Mechanical tubelength is the distance from the shoulder of the objective to the seat of the eyepiece. For fixed tube-length microscopes, this distance will typically be 160, 170, or 210 mm. For infinity corrected systems, the distance is, theoretically, infinite. To determine which engineering approach was taken for your particular microscope, look for either one of the fixed tubelength numbers or a small "∞" engraved on the barrel of the objectives. Further details are provided in the section on geometric optics and cases of lenses.

A Word about Illumination

Microscope illuminators also come in a variety of forms. Most work is done with quartz halogen lamps, typically with 100 W bulbs. For fluorescence applications or for those situations requiring greater intensity, arc lamps are added. Conventional arc systems have elaborate lamp housings, often with centerable mounts for both the reflector and the arc source.

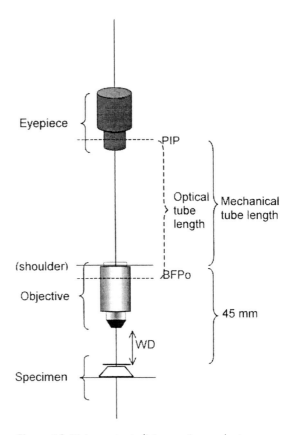

Figure 18.11 Important distances in an electron microscope. Key: PIP is the primary image plane; BFP_0 is the back focal plane of the objective; WD is the working distance.

of a business card) in place of the specimen. Remove an objective, and shine the image of the arc on the paper. (Using sunglasses or neutral density filters is a good idea to cut back the intensity.) Using the controls on the lamp housing, center the arc image. Again, (1) there may be a lamp collector lens to aid focusing of the arc image and (2) there may be controls for both the mirror and the real arc. Typically, the mirror and arc images are superimposed on each other, then put out of focus with the collector lens. See the details for your particular stand.

Also worth investigating are new precentered liquid light guide systems that provide high intensity with very even illumination, even at low magnification.

Koehler Illumination: Setting the Baseline

Proper alignment of the microscope takes less than a minute but is critical so that observation is consistent from sample to sample and day to day. There are as many "recipes" for setting Koehler illumination as there are microscopists. Since our tour around the microscope focused on the three major lenses and two irises that provide control, we will use them to describe the steps for setting Koehler.

Before beginning, it is important to determine which is your dominant eye. Unless it suffers from some sort of physical problem, this is the eye best suited for optimal alignment of your microscope and for measurement. To determine which eye is dominant:

- Make a triangular window by putting your index fingers together and thumbs together.
- Hold the "window" at arms length and, with both eyes open, focus through the window on an article across the room.
- Close your right eye and observe what happened to the object on which you were focusing. If it stayed in place, you are left eye dominant and should use that eye for setting up your microscope and for critical measurement. If the object moved, you are right eye dominant.

Before beginning Koehler, set the focus adjustments on the eyepieces to their zero points (either to the zero mark, or rotate the top until it just covers the silver, white, or black ring). Then proceed as follows:

1. Place a well-behaved specimen on the stage, in the stage clips and, while observing from the outside of the microscope, adjust the coarse focus until the specimen is nearly touching the objective. Note which way you turned the knob.
2. To set focus for the objective, while looking through the eyepieces, "focus away" (turn the knob in the reverse direction) until the image comes into sharp focus.
3. For the condenser, in *transmitted* light:

   a. Close the field iris. While observing from outside the microscope, raise the condenser, using its own focus knob, until it is just short of the specimen.
   b. While looking through the eyepieces, lower the condenser until the edges of the leaves of the field iris are sharply in focus.
   c. Use the centration screws (located about 5 o'clock and 7 o'clock on the condenser mount) to center the image of the field iris.
   d. Unless the sample is highly scattering, open the field iris to just outside the field of view.
   e. Adjust the condenser iris for best contrast (well-defined edges, clean background).

4. For the "condenser" in *reflected* light:

   a. Once the objective has been focused, the condenser is automatically in focus.
   b. Close the field iris. If it has its own centration system (two screws near the "F" or "FI", one on either side of the light bridge), center its image. Unless the specimen is highly scattering, open the field iris to just outside the field of view.
   c. Check the aperture iris. If it has its own centration system (two screws near the "A" or "AI", one on either side of the light bridge), remove the eyepiece, close the aperture iris, and center the image seen in the back focal plane of the objective. Return the eyepiece and, while looking at the image, adjust for optimum contrast (well-defined edges, clean background).

5. For the eyepieces:

   a. Adjust the interpupillary setting. Because each of us has a different pupil-to-pupil spacing, microscope binocular bodies are fitted with either sliders or wing-type binoculars. While looking at the microscope image, adjust the spacing between the two eyepieces until you see just one, round image.
   b. Check to make sure that the microscope image is crisp for your dominant eye. If not, adjust using the fine focus.
   c. Adjust the focus on the other eyepiece (the *diopter setting*) to bring the image into focus for the non-dominant eye.

In summary:

- Set the eyepieces to zero
- Focus on the specimen using the coarse focus
- For transmitted light: focus and center the condenser; for both reflected and transmitted light, set the field iris just outside of the field of view and the aperture iris for optimum contrast.
- Adjust the interpupillary and diopter settings in the eyepieces.

With practice, setting Koehler illumination should take you only 30–60 seconds. Start each day with this procedure and adjust both irises as you move from one objective to another or even one section of the sample to another. Remember, Koehler establishes a baseline that minimizes artifacts and optimizes good interpretation.

Now that you are well acquainted with the microscope, it is time to learn more about how images are formed.

Light, Matter, and Light-Matter Interactions

As indicated by its name, in light microscopy, light is the messenger and matter is the source of information that is carried to the image. A more complete understanding of each is essential for good image interpretation. This section provides a quick overview. For further information, see a standard physics review.

Light

Properties of Light

Light is typically described as electromagnetic radiation—that is, an energy wave that has both electrical and magnetic properties. For most of our work, especially in polymer microscopy, light's electrical character is more important.

As shown in Figure 18.12, light is described as a wave produced as an electrical vector builds up and then decreases—first in a positive sense, then in a negative sense. This diagram illustrates the four key parameters of light. The direction in which the wave builds up and then decreases is called the wave's *direction of vibration*, a key concept in the optical studies of polymers with polarized light. The direction in which the wave moves through space from the source to the viewer is referred to as its *direction of travel*. Notice that these two are always at right angles to each other. The distance from peak-to-peak, trough-to-trough, or any two similar points on the wave is referred to as the *wavelength* (λ) and, as shown in Figure 18.1, is related to both the color and the energy of the wave. Finally, the maximum displacement from the central refer-

ence line is referred to as the wave's *amplitude*. *Intensity* is related to amplitude squared.

Interference

If waves meet certain specific requirements as they move through space, they can interfere with one another. The requirements for *interference* are well defined. The waves must (a) originate from the same point source (b) at the same time, (c) travel in the same direction, (d) have the same direction of vibration, and (e) have the same wavelength. For optimum interference, they should also have approximately the same amplitude. Waves oscillating in synch with each other (i.e., peak-to-peak and trough-to-trough) are said to be in step or *in phase*. The resulting light is *coherent*.

Figure 18.13 illustrates the two extreme cases of interference. When waves are in step by whole wave cycles ($n\lambda$) and meet peak-to-peak, they undergo *constructive interference* (Figure 18.13, left). When they are out of step by half waves ($n\lambda/2$) and meet peak-to-trough, they undergo *destructive interference* (Figure 18.13, right). Notice that in each instance, the individual amplitudes add to form a resultant wave. In the microscope, the doubled amplitude in Figure 18.13, left, would result in a feature four times brighter. The resultant amplitude of "0" in Figure 18.13, right, would result in a dark feature (no intensity).

A logical question at this point is "What causes two waves to be out of step?" When waves interact with matter, they slow down proportionately to the refractive index of the material. If the affected waves emerge from the interaction out of step with their original partners, they are said to have suffered a *phase shift*, expressed as a fraction of the wavelength.

Interference is a powerful tool. For example, a special optical system can be constructed inside the microscope to produce sample and reference beams. Depending on the path each travels, they will produce a pattern of bright and dark fringes where they undergo, alternately, constructive and destructive interference. As the sample beam interacts with either a step height on a surface or a difference in optical density (refractive index), it will cause a shift in the fringe pattern, which can be measured. This type of quantitative microscopy, called *interferometry*, can be very useful for measuring surface roughness and wear, either of polymer films and molded parts or of molds and rollers. Since most interferometry depends on reflected light,* most polymers must be metallized, typically using a sputter coater of the type used in electron microscopy, to have sufficient reflectivity. For molds, metallized replicas of the surface can

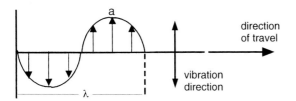

Figure 18.12 The wave nature of light and four key parameters: the direction of vibration, the direction of travel, the wavelength (λ), and amplitude (a).

*Older systems, such as the Zeiss/Jena Interphako, also do transmitted interferometry.

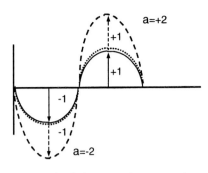 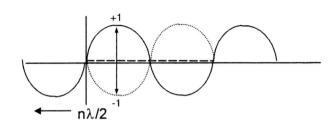

Figure 18.13 Constructive, or peak-to-peak (*left*), and destructive, or peak-to-trough (*right*), interference.

be made using quick-drying films such as gelatin or polystyrene dissolved in benzene then sputter coated.

The Color Wheel

As shown by the color wheel (Figure 18.14), regular white light is composed of the three *primary colors*: red, green, and blue. Note that the primary colors for light (red, green, blue) are different from those for pigments (magenta, cyan, yellow). Mixing neighboring primary colors for light produces the following secondary colors:

$$\text{red} + \text{blue} = \text{magenta} \qquad \text{blue} + \text{green} = \text{cyan}$$
$$\text{red} + \text{green} = \text{yellow}.$$

One of the conditions for constructive and destructive interference was that the waves must be of the same wavelength. If waves of that particular length undergo destructive interference, they fall out of the color wheel, leaving the other primary colors. For example, if green light of 550 nm undergoes destructive interference, red and blue light will be left, resulting in our seeing magenta. This color wheel will also be very useful for determining which filters can be used to suppress or enhance contrast in pigmented polymers and will form the foundation for understanding polarization colors.

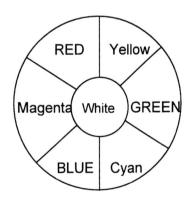

Figure 18.14 The color wheel for light. Primary colors: red, green, blue. Secondary colors: yellow, cyan, magenta.

Refractive Index

The phase shift that occurs when light passes through a material is directly related to a property of matter called *refractive index* (RI or *n*). RI is a measure of the velocity of light in a material compared to the velocity of light in a vacuum (eq 6) and is one of the most critical and fundamental concepts in all of microscopy.

$$n = \frac{V_{vacuum}}{V_{material}} \qquad (6)$$

Since there are relatively few molecules to slow light in air, the velocity of light in air is considered to be essentially the same as that in vacuum: 300,000 km/s. As a result, the RI of air, 1.0, forms the baseline of the RI continuum. Several quick examples:

EXERCISE 1:

If light moves at 300,000 km/s in air and 200,000 in glass, what is the refractive index of glass?

$$n = (300,000 \text{ km/s})/(200,000 \text{ km/s})$$
$$= 1.50 \text{ (note that } n \text{ is dimensionless)}$$

EXERCISE 2:

The refractive index of polystyrene resin is 1.67. What is the velocity of light in this resin?

Rearranging eq 6 to solve for velocity in the resin:

$$\text{Vps} = (300,000 \text{ km/s})/1.67 = 180,000 \text{ km/s}$$

From these three examples (air, glass, and polystyrene), it should be clear that lower refractive indices indicate that light is interacting less with the electrical field in matter and, therefore, moves through it faster.

Snell's Law

Refractive index is important for a number of reasons. First, a difference in RI between one material and another defines the optical boundary between the two. For example, if you immersed a glass rod (*n* = 1.5) in mineral oil (*n* = 1.521), light would not change velocity appreciably as it passed from the liquid into the glass and back again, so you would not see any boundary. The glass rod would just disappear.

When light passes from material of one RI to another RI, it can undergo a number of changes. In the simplest instance, if light just passes straight across the boundary shown (Figure 18.15, The "Normal"), it just changes velocity. In most cases, however, light will encounter the boundary at an angle. Imagine, for a moment, that the incoming beam is actually a broad wave front, sitting at right angles to the direction of travel. As it approaches the boundary, the part of the wave front that enters the material first will slow down. When one edge slows down with respect to the other, the wave front bends toward that direction, causing a change in direction. The angle of the change is governed by Snell's law (eq 7) and depends on both the angle of approach (i) and the relative RIs.

$$(\sin i)/\sin r = n_1/n_2 \qquad \text{(Snell's law)} \qquad (7)$$

where $\sin i$ is the sine of the angle of incidence; $\sin r$ is the sine of the angle of refraction; n_1 is the refractive index of material on the incoming side; and n_2 is the refractive index of material across the boundary.

Snell's law is in action all through the microscope and sample. It is responsible for the action of lenses and, in all too many cases with polymers, with the loss of light and information that are refracted out of the collecting angle of the objective.

Matter

Isotropic versus Anisotropic

Refractive index is a major criterion by which matter can be classified. Imagine that your pen represents the electric vector that was used earlier to describe the electrical character of light. If those waves were traveling from the overhead light onto this page, the vector could be vibrating north–south, east–west, or any angle in between.

As shown in Figure 18.16a, some materials present the same RI, no matter how light approaches them. As a result, their electrical environment can be depicted as a sphere. No matter how you lay your pen across the sphere, it intercepts the same amount of the electrical field. This type of material is called *isotropic*—that is, having the same ("iso") structure ("tropism") in all directions. Typical examples include glass and the amorphous regions of polymers such as polyethylene.

The second class of materials has an electrical field that differs with direction, as shown in Figure 18.16b. Since this type of material has at least two principle refractive indices, it can be depicted as an ellipse formed by connecting the ends of the vectors representing the two refractive indices. If you lay your pen along the vertical axis (north–south) there will be less interaction between the electrical field in light and the electrical field in matter. Because light will not be slowed down as much in this direction, it will have the lower RI, n_{low}. Conversely, if you turn your pen to lay along the horizontal (east–west) axis, there will be a maximum interaction between the two electrical fields, slowing light down considerably and resulting in the higher RI, n_{high}. Although you can intercept varying amounts of the ellipse at differing angles, only the maximum and minimum values are used to define the optical properties. This class of materials is called *anisotropic*, or not ("an") the same ("isotropic") with direction. Also, since there are at least two defining RIs, anisotropic materials are also said to be *birefringent* or to have the property of *birefringence*. Mathematically, birefringence can be calculated by subtracting one refractive index from the other (eq 8):

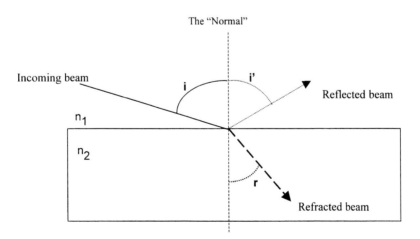

Figure 18.15 Snell's law. Key: n_1 is the refractive index (RI) of material on the incoming side; n_2 is the RI of material across the boundary; i is the angle of incidence; i' is the angle of reflection; r is the angle of refraction; and dashed and dotted arrows indicate the angles of approach.

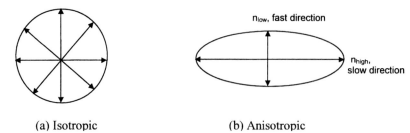

(a) Isotropic   (b) Anisotropic

Figure 18.16 Isotropic materials present a uniform electric environment to the approaching beam of light, no matter what its direction of vibration. Anisotropic materials present differing electric fields. The arrows are vectors, indicating amount of interaction for specific directions of vibration. In this figure, light traveling with a "north–south" vibration will encounter a smaller electric field than light exhibiting an "east–west" vibration. Because it slows down less, that path is named the "fast" direction. Its refractive index will be lower than the RI for the "slow" direction.

$$\Delta = |n_1 - n_2| \qquad \text{(birefringence)} \qquad (8)$$

While there can be up to three defining RIs (α being the lowest, β being the middle, and γ being the highest), these materials are still called birefringent, not trirefringent. The section on quantitative polarized light details methods for imaging these differences.

The difference in electrical environment that results in birefringence can be derived from either the atomic structure within a molecule or the orientation of molecules within the bulk, or both. When tangled polymer chains are in a random chain form, their electrical environments cancel one another. However, when polymers are stretched, drawn, or otherwise processed, the molecules align, and the resulting film, fiber, or foam will develop global differences in RI. That is, they will become anisotropic and therefore exhibit birefringence.

Materials that have two RIs may have several directions in which both RIs are visible but only one direction that exhibits a single RI. Drawn fibers are a good example (Figure 18.17, left). The fiber exhibits both RIs if it is viewed along its length, one parallel and one perpendicular to the direction of draw. However, observing the cross-section shows only one RI: $n_\perp$. This direction is referred to as the *optical axis* of the material. Since there is only one direction in which it exhibits this property, this material is said to be *uniaxial*.

Films such as mylar are typically drawn in two directions (Figure 18.17, right), resulting in three different RIs: one in each direction of draw ($n_\parallel$ and $n_\perp$) and one perpendicular to the plane of the film (n_z). Whether the film is observed as a flat cut or cross-section, there will always be two RIs evident. Materials of this type will have two unique directions, both isotropic, and are therefore called *biaxial*. The separation of the two axes, 2V, is another defining optical property. The section on polarized light in this chapter discusses experimental methods for determining the difference between uniaxial and biaxial materials, as well as 2V.

Amplitude Objects

In addition to causing changes in velocity and, therefore, refractive index, materials can also be classified by their ability to absorb energy. Since the height of the resulting wave changes on absorption, these materials are called *amplitude* or *absorption objects*.

Pigmented polymers are typical amplitude objects (Figure 18.18, left). They absorb all but one wavelength of light. For example, the red-pigmented plastics used for automotive taillight covers absorb green and blue light, allowing only red light to emerge. A magenta-colored transparent plastic absorbs only green, transmitting red + blue, or magenta. Another example, often used in microscopy, are the Wratten filters used for color-correcting photographic images.

Some polymers absorb uniformly, across the spectrum (Figure 18.18, right). The gray plastic used in sunglasses, for example, absorbs equal amounts of red, green, and blue. Neutral density filters used in the microscope to moderate light intensity are another example. An important point: neutral density filters are the preferred method for reducing

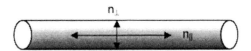

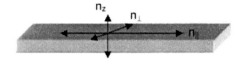

Figure 18.17 Refractive indices (n) parallel and perpendicular to direction of draw in uniaxial fiber (*left*) and biaxial film (*right*).

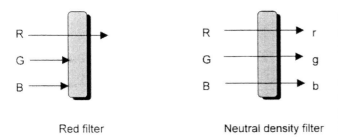

Red filter   Neutral density filter

Figure 18.18 Amplitude objects absorb a select portion of the spectrum (*left*) or equally across the spectrum (*right*). Key: R = red; G = green; B = blue. r, g, and b are used to show that equal amounts of R, G, and B have been absorbed. Arrows indicate the light path.

intensity in the microscope. Decreasing voltage at the lamp or closing down the aperture iris either lowers the color temperature of the image (making it more yellow) or increases coherence and changes edge information.

Phase Objects and Phase Gradients

As mentioned, when light crosses the boundary between one RI and other, it undergoes a change in velocity that puts it out of step or out of phase with waves that are passing through the surrounding area. Objects as simple as the polyethylene sheeting used for overhead transparencies create these phase shifts and therefore are called *phase objects* (Figure 18.19, left). If the phase object has sloped edges, it will create changing phase shifts or *phase gradients* (Figure 18.19, right). In microscopy, the difference between these two becomes the basis for decisions such as when to use specific contrast techniques: Phase contrast is used for simple phase objects whereas oblique illumination, Hoffman Modulation Contrast, and Differential Interference Contrast (DIC/Nomarski) are used to enhance gradients.

Phase shifts result when two waves travel different optical paths (OP_1 and OP_2). An *optical path* is the product of the geometric thickness, t, and the refractive index, n. Mathematically, the phase shift can be expressed as the *optical path difference*, OPD, between the two waves (eq 9).

$$OPD = OP_1 - OP_2 = t_1 n_1 - t_2 n_2$$

$$\text{(optical path difference)} \qquad (9)$$

where OP_1 is the optical path through the material; OP_2 is the optical path through the surrounding; $t_1 n_1$ is the thickness × refractive index in material; and $t_2 n_2$ is the thickness × refractive index in surrounding. This expanded form of the equation takes into account changes in either thickness or refractive index, or both. Most cases are like the overhead transparency film: the OPD is due to a change in RI and the equation can be simplified to $OPD = t(n_1 - n_2)$.

Unfortunately, most features do not behave as simple phase objects. Most have slanted or rounded edges and present changing optical paths (Figure 18.19, right). Again,

this is particularly true in polymers with fillers, glass fibers, and even droplets of one polymer phase suspended in a matrix of another polymer.

Mathematically, the equation for phase gradients is very similar to the equation for finite OPDs. To differentiate these changing path differences (OP_a, OP_b, OP_c) from the fixed path differences, the symbol φ (phi) replaces "OPD". Similarly, to account for the angle of the gradient that causes the change in thickness, the symbol α (alpha) replaces "t" (eq 10).

$$\varphi = \alpha\,(n_1 - n_2) \qquad \text{(phase gradient)} \qquad (10)$$

In this discussion of phase objects, phase shifts are derived from the differences in optical path length for light traveling within a substance compared to light traveling in the surrounding material (inside versus outside). In an earlier section, we introduced anisotropic materials: substances that have different RIs within a material. Different optical paths resulting from this internal anisotropy is called *retardation* and is covered in detail in the section on polarized light.

As mentioned earlier, as humans, we can only detect changes in color or intensity. Simple phase changes, phase gradients, and retardation are invisible to us unless they are somehow translated into one of these two properties. The section on contrast discusses a number of ways to reconfigure light microscopes to bring out this information.

Light and Matter Interactions

Basically, there are only seven mechanisms by which light can interact with matter: reflection, refraction, diffraction, absorption, fluorescence, polarization, and scatter.

Reflection

Reflection is the only interaction that is explained in terms of the particle theory of light. In reflection, light behaves as though it were a ball, bouncing off a surface. If the surface is hard and smooth with respect to the wavelength of light,

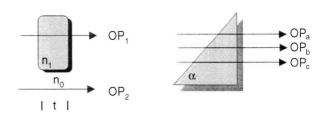

Figure 18.19 Phase object (*left*) versus phase gradient (*right*). Key: t = thickness; n_0 = refractive index of the surrounding; n_1 = refractive index of the material; OP_1 = optical path through the material; OP_2 = optical path through the surrounding; OP_a, OP_b, OP_c = varying optical paths through a phase gradient; α = gradient angle. Arrows indicate the light path.

the bounce will be well defined and the resulting image will accurately represent the object. This type of reflection is called *specular reflection* and is the ideal situation for well-polished surfaces of hard polymers.

If the surface is rough, the ball behaves as though it has shattered, sending pieces flying in all directions, causing the scatter, glare, and haze known as *diffuse reflection*. Glare and haze obscure imaging information and are the microscopist's nemesis. Both can be controlled to some extent by moving the feature of interest to the center of the field and closing the field iris to exclude the extraneous scatter. Other tricks of the trade include using reflected polarized light or putting a drop of oil or water on the surface. In the latter approach, use the fluid for which the objective is engineered. If it is not an oil- or water-immersion objective, try putting a drop of fluid directly on the surface of the polymer, then placing a coverslip on top of the preparation to keep the fluid from contaminating the objective. Also, keep in mind that some polymers may interact with oil or water.

Refraction

Refraction is the bending of light as it crosses a boundary from one refractive index to another at an angle and is governed by Snell's law.

Diffraction

Diffraction is also a change in direction, but in this case, it occurs at edges and results from the electrical interaction between the edge of the material and the edge of the wave.

Absorption

As we saw with filters of various types, matter either reduces the amplitude of a wave for selected wavelengths, resulting in a specific color, or for all wavelengths, resulting in a gray or neutral density filter.

Fluorescence

Fluorescence is an option if the molecule has mobile π electrons which are susceptible to excitation by high-energy ultraviolet (UV), blue, or green light. On excitation the electrons will jump into a higher molecular orbital, then try to find a mechanism to release the absorbed energy and return to their ground energy states. This energy is released primarily as light, with a small amount retained as heat. The small heat component reduces the energy of the emitted light in what is called a *Stokes* shift. As a result, the emitted light, or *fluorescence*, has a longer wavelength. Since the intensity of the fluorescence can be on the order of only 1/1000th to 1/10,000th the intensity of the incoming light, the microscope is fitted with special optics to separate the two signals. Many polymers exhibit natural fluorescence, or *autofluorescence*, making fluorescence an interesting technique for polymer studies.

Polarization

Anisotropic materials have two or three different RIs and, as a result, respond to polarized light. Polarized light microscopy typically starts by passing ordinary light—which contains all directions of vibration—through a material that absorbs all but one direction of vibration, creating plane-polarized light.

Scatter

Scatter is a combination of two or more of the interactions just discussed, resulting in random deflection of the light that emerges from the sample. Since scatter creates haze and glare, it is typically undesirable. The only time that scatter is useful is in darkfield microscopy, where the image is formed from selectively scattered light or in low-angle light scattering from spherulites observed in the back focal plane of the objective.

Imaging Theory

From earlier discussions, we know that the image is an entirely different entity from the object and that our job, as microscopists, is to reproduce, as faithfully as possible, the information from the object into the image. There are two parts to the story behind image formation: *geometric optics*, which has to do with how the lenses operate, and *diffraction theory*, which focuses primarily on the light-matter interaction.

Geometric Optics

Externally, the microscope seems like a "black box" with a scattering of lenses, buttons, and knobs. Internally, it is an elegantly simple array of lenses and electrical and mechanic components. Geometric optics gives us insight into the glassware.

How Lenses Work

All the optical activities in the microscope can be summarized by the four simple cases of lenses that are discussed below and shown in Figures 18.20 and 18.21. Although the lenses in the microscope are complex stacks of small lenslets, for the sake of this discussion, we will assume that the lenses in the microscope behave like simple, thin lenses that have a convex : convex shape—that is, that they gently bow outward on both faces. Because they are made of a material whose refractive index is greater than air (we will use 1.5, a reasonable approximation for many glasses), Snell's law is in operation and light waves will refract, both on entering and leaving the lens. Under the right conditions that refraction will cause the light to come to a focus and form an image.

A simple magnifying glass or even the eyepiece from your microscope can be used to demonstrate many of the principles. Use the overhead light as the object. To the lens, this object is infinitely far away. (A rule of thumb is that "infinity" is any distance greater than ten times the focal length of a lens.) In Figure 18.12, we described light as a wave. That figure shows just the cross section of the wave front. The overhead light has emitted a spherical wave front, moving out as peaks and troughs, like the ripples caused by a stone dropped into a pond. As the ripples move away from the source, the curve of the wave front flattens. In our experiment, the lens is so far from the source, it sees the waves as flat (a plane wave front). In Figure 18.20, that wave front is shown as a series of parallel lines approaching the lens. Each line represents the crest of the wave front; the space in between represents the trough.

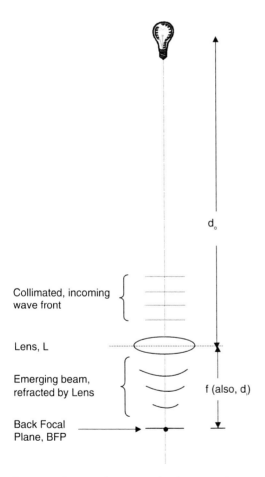

Figure 18.20 How lenses work. The incoming wave is refracted by the lens (L). The converging light forms an image at some distance, d_i, from the lens. If d_o, the distance to the source or object, is "infinite" ($>10f$), the image will form at the back focal plane (BFP), at distance f, the focal length of the lens.

As different sections of the wave front interact at different angles with the curved surface of the lens, they undergo different amounts of refraction. As a result, the emerging wave front is curved and converges to a *focal point*. To demonstrate this effect, bring the lens close to a horizontal surface and move it up and down until you focus an image of the overhead light on that surface. If the light were a single point source, you would see one clean, sharp point of light, located at the focal point. However, since your object has two dimensions, the image forms on a *focal plane*. Because that plane is behind the lens (on the side opposite of the object), that plane is called the *back focal plane*, or BFP. Because the light in this experiment came "from infinity", both the focal point and its 2D version, the focal plane, are located at the *focal length* of the lens, f.

Two other landmarks are noted on Figure 18.20: d_o is the distance from the object to the lens and d_i is the distance from the lens to the image. If you can measure d_i and d_o, you can actually calculate f from the lensmakers' formula (eq 11):

$$1/f = 1/d_i + 1/d_o \qquad (11)$$

There is also a *front focal plane* (FFP) located at exactly the same distance from the center of the lens as the BFP. You can find it on your lens by turning the lens over and repeating the same experiment. A word of caution: if you try this experiment with a piece of optics from your microscope, you may find it difficult to locate both focal planes. With stacks of lenselets, d_i and d_o are measured to the *optical center* of the lens, a location defined by the chemistry of the glass in the lenselets, their individual curvatures, and their unique spacings within the stack. The optical center may or may not be at the geometric center of the stack.

Four Cases of Lenses

Figure 18.20 illustrates the first of four simple cases of lenses: the object is at infinity. The other three cases are illustrated in Figure 18.21. Again, knowing the focal length of your lens, you can repeat these cases using your finger as the object and placing it (a) inside the FFP, (b) at the FFP, (c) at some distance beyond the FFP, and (d) at an infinite distance beyond the FFP.

Figure 18.21 uses simplified ray diagrams. The horizontal represents the *optical axis* of the system (OA). For this discussion, we will use an arrow as the object and two rays to determine if and where an image is formed. One ray will run from the object parallel to the optical axis until it interacts with the lens, at which point it will bend (refract) and move through the back focal plane. The second, called a *principal ray*, will run from the object through the center of the lens. An image will form if, and only if, the two rays cross.

In Case 1 (Figure 18.21a), the object is placed inside the FFP of the lens. Notice from the ray tracing that the rays diverge after leaving the lens. However, they will cross when projected backward (as shown by the dotted lines),

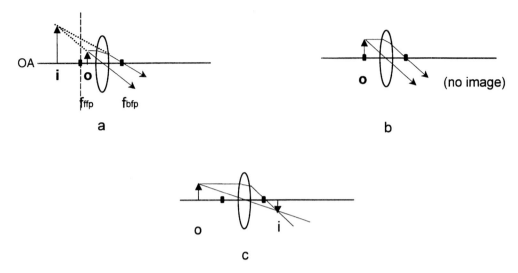

Figure 18.21 Three simple cases of lenses: (a) object inside front focal plane (FFP); (b) object at FFP; (c) object beyond FFP. Key: i = image; o = object; bfp = back focal plane. Arrows indicate light path.

indicating that the image is formed on the same side of the lens as the object. As seen in the ray tracing, the image is located in front of the lens, at a distance greater than the focal length, is magnified, and upright. Try the experiment with your finger. Is the image on the same side of the lens as the object? Is it upright and magnified?

Case 1 explains how the eyepiece works. First, the objective collects and images information from the object. This image becomes an object for the eyepiece. Its position is engineered to fall inside the $FFP_{eyepiece}$ and, as a result, it is seen as coming from a plane within the microscope. It is enlarged, upright, and *virtual*. That is, it cannot be projected onto a screen or a camera chip like a *real image*. For this reason, regular eyepieces cannot be used in photographic systems.

In Case 2 (Figure 18.21b), the object is moved to the FFP of the lens. The ray tracings become parallel (collimated) after leaving the lens, indicating that no image is formed. Optical physicists describe the light as "going to infinity". This condition is very important for the newer design of microscopes called *infinity corrected*. In this engineering approach, the specimen is placed at the $FFP_{objective}$ and the image-forming light is "sent to infinity". Infinity space offers a great advantage since optical components can be inserted into that location without shifting the size or the X, Y, or Z position of the image. Infinity-corrected optics require a partner to bring the image back into focus. That partner will be described in Case 4.

In Case 3 (Figure 18.21c), the object is moved farther away from the lens, to some finite distance beyond the FFP_{lens}. The ray tracings converge on the opposite side of the lens, indicating formation of a real but inverted image. Sliding the object back and forth along the optic axis illustrates the effect of the relative distances of object and image,

the basis for the magnification equation (eq 1). Notice that when the object is closer to the lens, d_o is smaller and d_i is larger. The result is greater magnification. Conversely, when the object is farther from the lens, d_o is larger and d_i is smaller. The result is minification. The situation that occurs when the object is at $2f$ is unique: the image forms at $2f$ on the other side of the lens and is exactly the same size as the object (1:1 magnification).

Many microscopes built before the mid-1980s (exceptions: metallographs, Reichert, and American Optical stands) were built on Case 3. The sample was placed beyond the $FFP_{objective}$ so that a real, inverted image fell at the *primary image plane* (PIP), located a few millimeters below the lip of the mounting tube for the eyepiece. Optics built to this design are called *fixed tubelength* systems, indicating that they have a specific distance from the shoulder of the objective to the primary image plane (the *mechanical tubelength*). Engravings on the objective barrels such as 160, 170, and 210 give the tubelength in mm. The difference between fixed tubelength and infinity-corrected systems raises an important purchasing issue: optics from one type of stand cannot be mixed with those from another, an issue which caused a great deal of consternation in the mid-1980s and early 1990s, when most of the major manufacturers switched from one engineering concept to the other. Many people buying new microscopes thought they could use their large collection of older, fixed tubelength objectives on their new infinity-corrected stands, only to find the two systems incompatible. Some correction collars were designed to bridge the gap, but, typically, optics designed for a specific type of stand work best only on that stand.

The last condition is actually the case used for our first lens experiment (Figure 18.20). The object has been moved so far beyond the FFP that the lens "sees" it as being at infinity and images it at the BFP. Note that Case 4 is the reverse of Case 2. Interestingly, Cases 2 and 4 always work in tandem in the microscope. Infinity-corrected systems use a *tube lens*, a "silent partner" that is often located just below the binocular body, to bring information back to a focus. The objective operates in Case 2 to send information "to infinity", and the tube lens operates in Case 4 to bring the information to a focus at the primary image plane. Some eyepieces also work this way: the eyepiece operates in Case 2 to send information to infinity, and your eye works in Case 4 to bring the information back to a focus on your retina.

Conjugate Focal Planes and the Two-Handed Exercise

Why is it important to understand how lenses work? In the simplest case, the microscope can be described as an optical bench with four lenses: a lamp collector lens, the condenser, the objective, and the eyepieces. The lenses are carefully spaced using the four conditions with two goals in mind: (1) to separate illumination information (the lamp's coiled filament or arc) from the specimen information and (2) to provide full aperture, even background illumination. Every time we align the microscope using Koehler Illumination, we are setting up that optical bench using the four conditions of lenses. The result is two independent light paths: one carrying illuminating information and one carrying specimen information. These two light paths consist of sets of *conjugate focal planes*, so called because the object in the lower plane is imaged (focused) in the next plane. Understanding where these planes fall in the microscope not only makes troubleshooting more efficient but also allows you to knowingly manipulate the microscope to bring out the best information in your sample and to insert masks and reticles for measurement, counting, and orientation evaluations.

Table 18.3 summarizes the two sets of conjugate focal planes. The individual planes are offset to show their relative positions in the microscope. By interlacing the fingers of your hands, you can model the same relationship.

Whether your microscope is a transmitted light or reflected light stand, it will have these planes. Physically locating them is informative and will speed your troubleshooting efforts in the future. You will need a well-behaved, colorful sample, a dental mirror (now available at most drugstores), and a sheet of lens tissue. If you are working with a reflected light system, use a well-polished specimen.

Place the specimen on the stage and set up Koehler illumination using a lower power objective (for example, 10×). If you are working with a transmitted light microscope, close the aperture iris in the condenser and use the dental mirror to observe the image on the closed leaves of the iris

Table 18.3 Comparison of conjugate focal planes

Illuminating set	Specimen/imaging set
	Retina/camera
Exit pupil of the microscope (*Ramsden disk*)	
	Primary image plane (a.k.a. "PIP")
BFP objective[a]	
	Specimen
FFP condenser[a]	
	Field iris (lamp collector lens)
Lamp (filament or arc)	

Note: Alternating spacing signifies that each set of planes is unique. The two sets do not coincide.

a. BFP_o and FFP_c are the same location for reflected light systems.

(the FFP condenser). If a diffuser is in place, you will only see a lighted disk.*

To find the BFP objective, remove the eyepiece and look far down into the tube. You should see the same image you saw with the dental mirror. You will also see the leaves of the aperture iris again. Test this observation by opening and closing the aperture iris. Notice that, if you have looked far enough down the tube, there is no image of the specimen, indicating that you are isolating just the illuminating information.

The final plane is the exit pupil of the microscope. You found this focal point earlier by stretching a sheet of lens tissue over the top of the eyepiece then slowly moving it away until you found a sharp point of focus. The eyepoint is typically about 18 mm above the top of the eyepiece. Notice that there will be specimen images on either side of this point, but the point itself contains either the image of the light source or the diffuser disk. If you have some way of holding the tissue in place, you can confirm that this plane is part of the illuminating set by, again, opening and closing the aperture iris. If you are working with a reflected light system, you will not be able to see the FFP_c directly, since it is located concomitantly with the BFP_o. All other observations should hold true.

The imaging set of planes actually begins with the field iris. To be completely in line with optical physics, the first plane should be located at the lamp collector lens. However, the featureless surface of the lens makes it impossible to image. The field iris is a good optical stand-in. Looking through the microscope at the image, open and close the field iris. You should be able to see the leaves of the iris closing in on the specimen image. Next, open and close the aperture iris. Notice that, while this operation changes

---

*Diffusers may be hidden in the base of the microscope and may not be easily removed.

intensity and edge information, you cannot see the image of the leaves of the iris opening and closing, indicating that it is not in the same set of planes as the specimen.

The primary image plane is located a few millimeters below the seat of the eyepiece and can be readily found by removing the eyepiece and stretching the lens tissue over the opening. You may need to either reduce the room lighting, turn up the intensity of the lamp, or both. Whether your system is infinity-corrected or fixed tubelength, the primary image will always fall in this same location. Again, you can open and close first the field iris then the aperture iris to prove which of them is and which is not in this imaging set of planes.

The final plane in this series is the retina of your eye or, alternatively, another type of detector such as a camera or photometer. The test for this last plane is obvious, since you see the image of the specimen, not the image of the light source, when you look through the microscope.

By now it should be clear that these two sets of planes are, indeed, separate from one another but interlaced through the microscope. When you establish Koehler illumination, you automatically set the planes in their proper locations. As a result, you will not see the coil of the lamp (or the Westinghouse "W" from the lamp) superimposed on the image of the specimen. It also means that you now have an effective way of locating aberrations and artifacts: anything *in focus* must be in one of the imaging planes, while anything *out of focus* must be in one of the illuminating planes. Finally, if you want to insert a ruler for measuring, a grid for counting, or a pattern for determining the orientation of glass fibers in a filled polymer, you know that it must go in one of the imaging planes. Since the field iris is hidden behind a protective glass and it would be awkward to place such a device directly on the specimen, the most logical place is the *primary image plane* (PIP). If your microscope has an eyepiece with any type of feature in it, remove it and turn it over. You will find a glass plate with that structure mounted in a location that sits neatly at the primary image plane.

Understanding how your sample behaves is another good reason for understanding conjugate focal planes. As shown in Figure 18.22, polymers can act like lenses. To create this darkfield image of the tacky adhesive on repositional notepaper, a patch stop was mounted in an illuminating plane using a three-point support. At first glance, this technique has brought out edge information from the adhesive droplets, but a closer look reveals the lensing effect. A number of the globules have small dark centers, suspended by three-point mounts. These are images of the patch stop, translated from illuminating to specimen planes!

A thick polymer layer can also shift a plane from one set to the other or, worse yet, shift a plane to an intermediate position so that the specimen information will appear constantly out of focus. Inclusions are another source of artifacts. For example, a bubble of incompletely polymerized

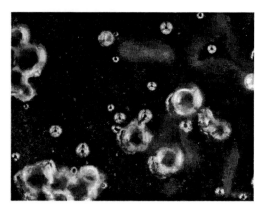

Figure 18.22 Polymers can act like lenses. (Post-it note adhesive. Reflected light darkfield image. 32× objective.)

product trapped in a matrix may produce the same lensing effect as the adhesive in Figure 18.22 and image the filament in the specimen plane. One solution for the thick layer problem may be to use long working distance objectives. There is no solution for the lensing problem, but if the image is clear and intriguing, snap a picture and submit it for the cover of the next annual report.

Lens Aberrations

Because microscope lenses are ground as sections of spheres, they suffer from several imaging problems. The two most critical are chromatic and spherical aberrations (Figure 18.23).

Chromatic aberrations are derived from refraction: long-wavelength red light resists refraction, bending less as it emerges from the lens into the air, and focusing farther downfield (Figure 18.23a). In the microscope, chromatic aberration appears as a colored red, blue, or yellow fringe at sharp edges or around fine detail. If your condenser is not marked "apl-achro" (a high highly corrected aplanatic-achromatic lens), it will demonstrate this aberration every time you set up Koehler illumination. Close the field iris and, as you focus the condenser, watch for the colored fringes at the ends of the iris. As Figure 18.23a shows, when the condenser is just above the plane of focus, you will see a blue fringe. When it is below the plane of focus (further away), you will see the red fringe. To correct for this aberration, optical designers combine lenses of different glasses with varying refractive indices, as well as different curvatures and spacings, bringing the colors of the spectrum to one point of focus.

Spherical aberrations are also derived from refraction but stem from the fact that light entering near the axis encounters less of an angle than light entering the lens at the periphery (Figure 18.23b). As a result, central rays focus farther down the optical axis than do peripheral rays. This

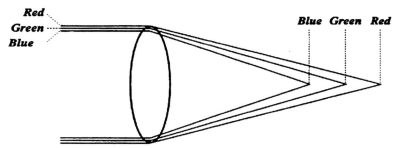

a. Chromatic aberration

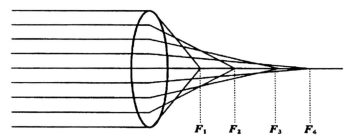

b. Spherical aberration

Figure 18.23 Microscope imaging problems: (a) chromatic aberration and (b) spherical aberration. Chromatic aberrations result from refractions' wavelength dependency. Since red refracts less than green or blue, it focuses farther away. Spherical aberrations result from the curvature of the lens. Axial rays encounter the lens straight on. Because they do not bend, they focus far downfield (F_4). Conversely, as the rays interact with the lens more toward the periphery, they encounter greater angles, suffer more refraction, and focus closer to the lens (F_3, F_2, F_1).

spread of focal points creates a *zone of confusion*. In the microscope, spherical aberration appears as a soft, out of focus image. Spherical aberration also has a chromatic component called *lateral chromatic aberration*: each of the rays, from axial to peripheral, will exhibit its own dispersion. Again, engineering a lens with a set of small lenselets of different curvature and spacing corrects the problem.

Since these corrections cost money, the manufacturers have developed classes of lenses with varying price and performance characteristics (Table 18.4). *Achromats* are good workhorses and make good choices for rapid scanning, but they typically have lower numerical apertures and are not good for high-resolution work. Although they have the best numerical apertures and correction, *apochromats* may have

lenselets that are under strain, making them unacceptable for polarized light work. *Fluorites*, especially on today's microscopes, are an excellent, middle-of-the-road choice. The bottom line: choose the lens that best fits your application.

Diffraction

Diffraction theory, the second half of the imaging story, deals with the information carried by light from the specimen.

The Diffraction Pattern

When light interacts with the specimen, it is diffracted by the fine detail. If the object is a simple one (for example, a grating or a pattern of dots), the resulting diffraction pattern can be seen in the BFP_o. Figure 18.24 shows both the sample (left) and the resulting diffraction pattern (right). The diffraction pattern carries all the information necessary to form the image. The central Zero-order spot (Figure 18.24, right) is responsible for the *background* in the image. If just this one spot is colored, then the whole background in the image will take on that color. This point seems trivial but is the basis for Rheinberg illumination (a type of color contrast invented in

Table 18.4 Classes and corrections of lenses

Lens class	Notation on objective	Chromatic corrections	Spherical corrections
Achromat	A or none	R + B	R + B
Fluorite	Fl	R + G + B	R + B
Apochromat	Apo	R + G + B + UV	R + B + G

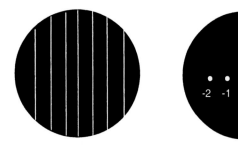

Figure 18.24 The diffraction pattern: object (ruling through chrome on glass) (*left*); resulting diffraction pattern, as seen in the back focal plane of the objective (BFP$_o$) (*right*).

the late 1800s), as well as for the black backgrounds in dark-field, fluorescence, and polarized light microscopy.

The *orientation* of the pattern is responsible for carrying orientation of the features in the object to the image. Since the diffraction pattern is actually the Fourier transform of the object, the orientation in the pattern will always be at 90° to the structure in the object.

Spacing between edges or objects is set by the spacing between any two adjacent spots in the diffraction pattern. Again, since the diffraction pattern is the Fourier transform of the object, there is a reciprocal relationship between the actual spacing in the specimen and the spacing in the pattern: very narrow spaces in the specimen will result in wide spaces in the pattern, and vice versa. This principle can be demonstrated with an Abbé diffraction test kit. The kit contains a special objective. Its collar has a slit that permits access to the BFP$_o$ and, therefore, to the diffraction pattern. For example, if a mask is inserted into the BFP$_o$ that blocks the two second-order spots (+2 and −2), the code that specifies spacing will change and the image seen in the microscope will change from a realistic coarse grating to an erroneous fine grating, exhibiting half the spacing.

Individual spots in the diffraction pattern suffer *dispersion* (the spreading of white light into a rainbow of colors). Since blue light resists diffraction, the blue edge of the spots will exhibit slightly narrower spacing than the red edge, and a smaller NA may still capture it, while losing the red. This phenomenon explains the wavelength component of the resolution equation.

Collecting three or more adjacent spots establishes the criterion for *edge fidelity*. (The only exception is the combination of −1, 0, and +1.) The more spots, the sharper and better defined the edge.

The diffraction pattern, then, carries the full code for image formation. In the microscope, the image is constructed at the PIP by the interference between the undiffracted light (information carried by the zero order) and the diffracted light from the specimen, so any information missing from the code will cause degradation or artifacts in the image.

Numerical Aperture

It should now be clear why the numerical aperture of the objective is so important: NA is a measure of the collecting angle of the objective and determines how much of the code will be available to reconstruct the object's information in the image (Figure 18.25).

Mathematically, NA has two primary components: the actual collecting half angle and the RI of the fluid between the sample and the front lens of the objective. Using eq 12, the NA can be calculated for a variety of cases:

$$NA = n \sin \alpha \quad \text{(numerical aperture)} \quad (12)$$

where n is the refractive index of the material between the specimen and the front surface of the objective, and α is half the collecting angle of the objective.

Theoretically, if the objective were touching the specimen, the half angle would be 90° (sin 90° = 1.00). If the fluid between the sample and objective was air ($n = 1.00$), the theoretical limit for this objective's NA would be 1. Replacing the air with a drop of water ($n = 1.33$) reduces refraction at the sample:air boundary and raises the theoretical limit to 1.33. Since many polymers are impervious to water, buying and correctly using a water-immersion objective would improve both the resolution and the edge information gathered from the specimen. Going one step further, the air could be replaced with immersion oil ($n = 1.5212$), raising the NA to 1.5. In actual practice, the maximum NA for a *dry objective* is about 0.95; for water immersion it is 1.25, and for oil it is 1.4. Notice that this is the maximum NA. If the objective is fitted with an internal iris, the working NA will be significantly smaller, depending on the iris setting.

Effect of NA on Edge Fidelity

Clearly, the larger the collecting ability of the objective, the more diffraction orders can fit through its aperture. If fewer than three adjacent orders are collected, edges will appear rounded and lack crispness.

Effect of NA on Resolution

To calculate resolution, substitute the NAs of the objective and condenser into eq 13. In reflected light systems, the objective and condenser are one and the same, reducing the equation to $0.61\lambda/NA_o$.

$$R = 1.22\ \lambda/(NA_o + NA_c)$$
$$\text{(numerical aperture and resolution)} \quad (13)$$

where 1.22 is the Bessel function associated with round apertures; λ is the wavelength of light used (0.550 μm is an average for mid-spectral green); NA_o is the NA of the objective; and NA_c is the NA of the condenser.

This equation leads to a couple of quick observations. First, the larger the NA, the smaller the value for R and the higher the resolution (i.e., the greater the ability to sepa-

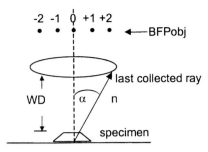

Figure 18.25 Numerical aperture, the diffraction pattern, and resolution. Key: WD = working distance; α = half the collecting angle of the objective; n = refractive index.

rately image fine detail). Mathematically, this statement supports the observation that larger NA objectives can collect the broader spaced diffraction patterns generated by finer detail. Second, the smaller the wavelength λ, the higher the resolution. Again, this supports the experimental evidence discussed earlier, since each diffraction order suffers dispersion, with blue toward the outer edge.

In many polymer applications, we are imaging either continuous structures such as films or phases, or larger structures such as fibers. Diffraction patterns from these structures are condensed and easily collected even by low numerical aperture objectives. However, if you are trying to resolve fine particles or spherulites, the need to collect more broadly spaced diffraction patterns suggests investing in larger NA objectives.

Effect of NA on Depth of Field
and Depth of Focus

NA has a significant effect on the vertical optical section from which information is collected in the specimen (*depth of field*). There are a number of approaches for expressing depth of field, but one of the more relevant, cited by Delly[4] and attributed to Shillaber, is shown in eq 14:

$$d = \lambda \left(\frac{\sqrt{n^2 - NA^2}}{NA^2} \right) \quad \text{(depth of field)} \quad (14)$$

Intuitively, it is logical that a lens that has a very large collecting cone should have a shallow cross over point at its apex. Conversely, as the collecting angle becomes more shallow, the beam should become more columnar. This situation can be demonstrated easily on a transmitted light microscope by removing the sample and the objective, placing a piece of paper vertically between the stage and the nosepiece to catch the beam emerging from the condenser. Opening and closing the condenser aperture iris illustrates this effect. When the condenser is fully open (large NA), there is a shallow crossover at the specimen plane. When

the condenser is closed (small NA), the beam forms a column or pencil of light that illuminates a large depth.

Note that *depth of field* is a term reserved for the sample only. It is often erroneously confused with *depth of focus*, the term used to describe the z-distance over which the image is in focus at the camera or photometer. Depth of focus is independent of NA. It relies solely on the square of the magnification (M^2). Using this simple relationship, it is easy to see that a lower power objective (e.g., 5× or 10×) will have a more shallow and defined depth of focus than a high power objective (50× or 100×). For this reason, low magnification objectives are always used to set the plane of focus for any viewing system (*parfocalization*). Once set for the low power objective, the camera or photometer will always be in focus for all higher magnification optics.

Sample Preparation

There are two key caveats involved in preparing polymer samples: (1) avoid strain and (2) be aware of solvent effects. Where possible, use a preparation procedure which leaves the polymer undisturbed.

Films and Fibers

If the polymer can be stretched (as, for example, films, foams, and fibers), take care not to put undue force on it. Narrow strips of double-sided tape placed at right angles to the long direction of the slide are ideal for mounting, as shown in Figure 18.26.

If you plan to do light microscopy and Fourier transform infrared (FTIR) spectrometry analyses on the same sample, there are FTIR mounts available that have holes in the middle. Again, place a band of tape on either side of the hole, then gently stretch the fiber across the opening and anchor it to the tape.

Another trick of the trade is tape available from 3M with conventional tape adhesive on one side and "tacky note" adhesive on the other. Affixing the tape to the slide with the stickier side leaves the tacky note adhesive available for convenient and repeated mounting and unmounting of fibers and films.

Particles

Particles such as filler, pigment, and catalyst present a variety of issues. First, as received from the manufacturing line,

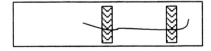

Figure 18.26 Using double-sided tape for mounting fibers or films.

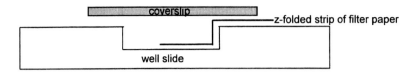

Figure 18.27 Using a well slide to grow crystals.

these samples may be too large for convenient observation. The first inclination is to crush them. However, size, shape, epitaxial growth, surface structure, and materials sticking to the surface may be important in your troubleshooting. Carefully evaluate your test procedure before beginning. Either taking a look with the stereomicroscope or recrystallizing might be a better "first step". If the particles are small enough to observe directly, sprinkle a small amount on a slide and cover with a coverslip. If they tend to clump, try one of the antistatic guns that are available from most electron microscopy sample preparation companies. If they are difficult to keep in place, try spraying the slide first with artist's fixative (used with pastels) or hair spray.

There are several tricks to recrystallizing. A first approach, if the material is water soluble and the crystals are small enough to be held to a coverslip by static electricity, is to try sprinkling a few crystals on a coverslip and gently "huffing" on them to bring the moist, warm air from deep in your lungs. When the material has recrystallized, flip the coverslip over onto a slide. Depending on their refractive index, you can place a drop of glycerin (RI = 1.47) or mineral oil (RI = 1.515) on the slide first, then place the coverslip on top. This may help image them more clearly.

A second trick is to use a well slide, coverslip, and a strip of filter paper. Fold the paper into a "Z" and place it in the well so that the lower arm rests on the bottom of the well (Figure 18.27). Sprinkle some crystals on a coverslip, place a drop of the solvent on the lower arm of the filter paper, and quickly flip the coverslip over the top of the well. The vapors from the solvent will dissolve the crystal.

A third approach is to dissolve the crystals in a small drop of solvent (typically water or aqueous solution which has been pH modified) on one end of a slide. Gently warm the solution on a hot plate until a crust forms at the edge of the droplet. Using a micro stirring rod, push the crust into the solution to seed it, place a coverslip over the top, and observe immediately under the microscope.

Multilayer Systems

Multilayer laminates, films, and fibers can be cross-sectioned but, again, the process should be carefully considered, step by step, to reduce both strain and solvent effects. Many laminates and films can be cut with a single-edged razor blade. For long fibers, the following technique is inexpensive and effective. Start with a small cork about 1″ long and ½″ wide. Use a darning needle to put a hole through the long axis. Next, thread the needle with a small bundle of the fiber under investigation. Gently pull the needle and fiber bundle through the hole. Use a single-edged razor blade to cut narrow disks from the cork, then mount the disks on a microscope slide. The cork will hold the fibers in position, and the hole provides a convenient viewing port. If your work requires a great deal of fiber observation, invest in a fiber microtome.*

Hard Polymers

Hard polymers can be treated very much like metals. If you can't observe their surfaces without preparation, they can be ground and polished using standard metallographic techniques.†

Solvent Effects

Because of their extreme susceptibility to solvent effects, it is important that you understand the chemistry of any polymer system with which you are working. If no phase in the polymer is water soluble, a drop of water on the surface of a fiber or film will often remove surface scatter and make fine structure more visible. This technique is especially effective in removing much of the lensing effect that is created by the very rounded edges of fibers. Similarly, try inert fluids such as glycerin, Nujol, mineral oil, or even immersion oil. It pays to have a selection of oils on hand, with varying RIs.

Contrast Techniques

Understanding contrast techniques—how to generate them, and how to interpret the resulting images—constitutes some of the most powerful skills available to any polymer microscopist. The right contrast technique can bring out hidden information or eliminate an obscuring artifact.

*McCrone Associates in Westmont, IL, is a good source for sample preparation equipment for polymer microscopy.

†Good sources for companies providing this type of equipment are the Microscopy Society (www.msa.microscopy.com), MicroWorld Resources and News (www.microscopy.info), and the Microscopy Vendors list (www.kaker.com/mvd).

Figure 18.28 shows the adhesive used in repositionable notes, imaged using different techniques. Notice that some techniques such as axial (b) and darkfield illumination (d) bring out edges, while others, such as oblique illumination (c), produce more 3D images by keying in on slopes. Each technique is explained in detail in this section.

As in all previous discussions, light is once again the messenger. In contrast enhancement, light is manipulated to elicit specific information about phase relationships (edges), slopes (gradients or angles), and the internal chemistry and organization (body effects). For each technique, it is important to ask three key questions:

How were the optics changed?
What effect did that change have on the light approaching the sample?
What was the resulting light and matter interaction?

One final reminder: being human, we can only detect color or changes in intensity. For those interactions outside of our perceptual capability, such as polarization and phase shifts, the microscope must be reconfigured to convert that interaction into one of the two optical effects we can detect.

Free or Inexpensive Techniques

Staining

Although polymers are often susceptible to swelling or changes in the presence of a solvent, they rarely take up stains well. One exception involves the use of black India ink or magic marker to bring out surface information, as shown in Figure 18.29. The upper part of this image has been stained with black magic marker. This technique works for two reasons. First, it reduces scattered light and therefore minimizes surface glare. Second, the ink sits in holes and boundaries, creating contrast from those features.

In another staining application, Lloyd Donaldson[5] cites adding eosin to polyurethane to stain coatings on wood. He adds the stain to the monomer before polymerization occurs. He has also used safranin to stain epoxy. In both cases, the sample was imaged via its fluorescence. Confocal microscopy was used to image individual layers and reduce scatter.

Karen Zaruba[6] uses staining to locate adhesive layers and backings on tapes and dressings and to observe their performance. For visible light microscopy, she advocates Sudan

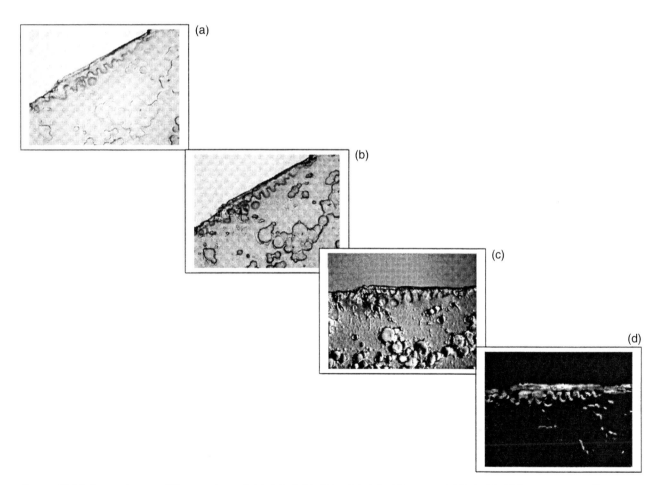

Figure 18.28 One polymer, different views: (a) brightfield, (b) axial, (c) oblique, and (d) darkfield (Post-it note adhesive. Transmitted light, 16× objective).

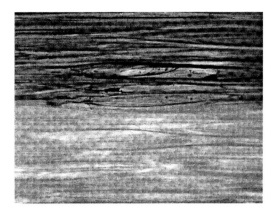

Figure 18.29 Black magic marker (*top*) improves visibility of voids on the surface of blown film. (Transmitted light brightfield, 10× objective.)

Black B in 70% ethanol, as well as some of the standards such as "epoxy tissue stain" available from typical supply houses. For fluorescence microscopy, use a 0.1% solution of aqueous Neutral Red.

Refractive Index

The ability to wisely manipulate refractive index is especially important for fibers, films, and molded parts. More closely matching the RI of the mounting medium to the polymer decreases edge contrast, while widening the RI mismatch increases contrast. This technique has special application for automated image analysis in those instances such as counting and sizing particles in a film or fiber. In Figure 18.30, a salt crystal has been mounted in air (left) and in an oil (right). The RIs of the salt, air, and oil are 1.47, 1.00, and 1.515, respectively. The large mismatch between the salt and the air produces thick edges and obscures fine detail. Conversely, the closer match refines edges

and other detail, presenting the crystal in clean, sharp relief for easy segmentation by an automated imaging system.

Axial Illumination

Every microscope is capable of axial illumination. To convert from brightfield to axial illumination, start by setting up Koehler illumination then close the aperture iris as completely as possible. A number of effects result. As illustrated in Figure 18.31, closing the aperture iris discards a large portion of the incoming light, greatly diminishing the amount of illumination to the sample. It is easy to see why inexperienced microscopists are tempted to this technique for brightness control. However, neutral density filters offer better intensity control.

This figure also demonstrates how closing this iris generates highly coherent illumination. The smaller aperture selects a narrow range of incoming light waves that are in step with each other as they travel up the axis of the microscope. At the edges of fine detail, the incoming light undergoes diffraction, which shifts the phase of that particular segment of the population. Any emerging waves that are still in step with those from the surrounding will undergo constructive interference, while those that are out of step by half a wavelength will undergo destructive interference. Used judiciously, the resulting bright and dark fringes can dramatically improve edge visibility, thus providing imaging information that would otherwise be invisible (Figure 18.32). However, closing the iris too far can produce very thick, dark fringes or repeating bright/dark bands referred to as "ringing", both of which obscure detail.

Axial illumination has additional potential. Figure 18.31a shows the brightfield beam approaching the specimen at a high angle, producing a very shallow cross-over point and illuminating only a shallow plane in the sample. This approach works well for optical sectioning: small adjustments

 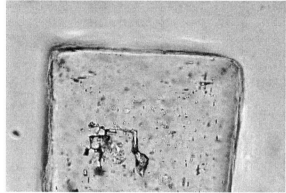

Figure 18.30 A mismatch in refractive indices produces gross edges that obscure information, while a closer match refines edges and other details: salt mounted in air (*left*, 20× objective); salt mounted in oil (*right*, 40× objective). (Brightfield, 40× objective.)

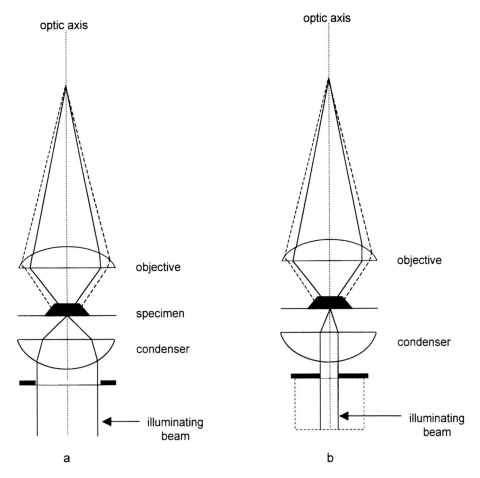

Figure 18.31 Diagram of the condenser settings for (a) brightfield and (b) axial illumination.

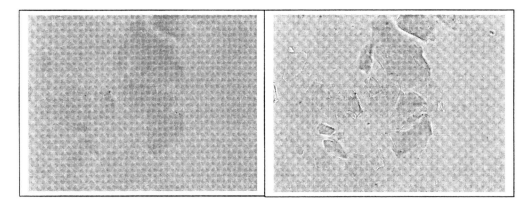

Figure 18.32 Polymer catalyst: brightfield image (*left*); axial image (*right*). (40× objective.)

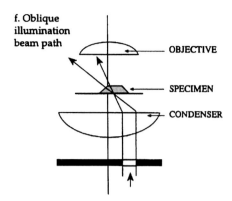

Figure 18.33 Oblique illumination. (Image courtesy of Polaroid Corporation.)

in the fine focus will bring different sample planes into focus. Conversely, the pencil-like axial beam illuminates a deep plane in the specimen, making it ideal for bringing spatially separate information into focus all at once.

As with much of microscopy, there is a trade-off: the reduced numerical aperture of the condenser reduces the resolution (see eq 2). However, given the choice between seeing the information at lower resolution and greater depth of field versus not seeing any information at all, most microscopists opt for improved contrast.

Oblique Illumination

Oblique illumination starts with the concept of axial illumination but uses a device in the front focal plane of the condenser (FFP$_c$) to offset the beam to one side (Figure 18.33). This technique provides a practical illustration of diffraction theory: *you* are controlling placement of the Zero order from the diffraction pattern.

The offset device can be something as simple as your thumb or a business card or something more elaborate such as a shim metal slide with a pinhole (an *Abbé slider*). It should be placed as close as possible to the plane of the aperture iris (FFP$_c$) and should be used with a fully opened aperture iris. If your microscope is equipped with a universal condenser with phase rings or a darkfield patch stop (or both), these annuli can be rotated into an intermediate position to create the oblique beam. For reflected light systems, there is sometimes a sliding pinhole mounted in the aperture iris plane.

Test the effectiveness of this technique by preparing a slide made by whipping a small amount of immersion or other oil into a drop of water. Cover with a coverslip and observe the sample first in brightfield, with the condenser fully open. Next, observe in axial illumination. Finally, use one of the techniques described in this section to narrow and offset the condenser aperture. One side of the oil bubble will be shaded, and the other will be bright. Your eye-brain

combination interprets this information as coming from a highly 3D sample. If you have used a movable pinhole or the phase or darkfield annulus, remove the eyepiece and look down the tube into the BFP$_o$. Especially if you used a pinhole, you will note that you have moved the zero order off to one edge of this aperture.

Oblique illumination depends on refraction (Snell's law) and detects slopes or gradients. The respective RIs of the mountant and material will cause the offset illumination to be refracted either into the collecting angle of the objective (bright edge) or out of the collecting angle (no collected light = dark edge). If the crystal has the higher index, the side facing the direction of the incoming beam will be bright. A reminder: since this technique is highly directional, only those slopes parallel to the direction of the incoming light will be affected. If your microscope is fitted with a rotatable stage, you can orient the sample to enhance selected detail and suppress other information.

Unlike most contrast techniques, oblique illumination improves both contrast and resolution. By narrowing the beam (i.e., selecting a Zero order from the combined diffraction patterns) and moving it to one edge, you are making better use of the available space along the diameter of the BFP$_o$. As shown in Figure 18.34, diffraction patterns from very fine detail, which normally escape the collecting angle of the objective, can now be captured.

Figure 18.35 shows the profound effect of oblique illumination. The sample is a polymer overlay on circuitry from an inkjet printer head. Oblique illumination was achieved by a special offset pinhole mounted in the plane of the aperture iris. The effect was enhanced by slanting the slide in the direction of the offset, using a blank microscope slide as a wedge. Figure 18.35, left, shows the image in standard brightfield. Figure 18.35, right, illustrates the three-dimensionality derived from one edge bright/one edge dark, as well as the increased depth of field that results from the narrowed aperture.

Techniques Using Special Condenser Inserts

Darkfield/Rheinberg Illumination

Darkfield illumination takes the offset concept beyond the edge of the objective's collecting angle. For lower powered

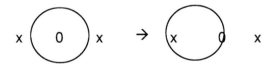

Figure 18.34 Simplified diffraction patterns seen in the BFP$_o$: brightfield (*left*) versus oblique illumination (*right*), in which the Zero order is moved to one side of the BFP$_o$, allowing the microscope to capture the two adjacent orders required to image.

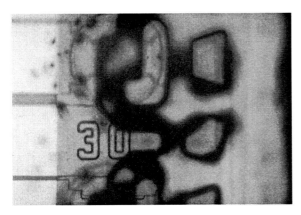

Figure 18.35 Oblique illumination improves depth, resolution, and contrast: brightfield (*left*); oblique illumination (*right*). (Polymer overlay on circuitry from an inkjet printer head, 40× objective.) (Image courtesy of Milo Overbay and Tony Fuller, Hewlett Packard, Corvallis, OR.)

objectives (10×–20×), a simple black patch stop is typically placed in the FFP_c.* If you are using a universal condenser, remove it, flip it over, and rotate through all the accessories. It is very likely that one of the positions has a very large dark center with a thin, clear ring round it: the darkfield patch stop. For critical work or higher NAs, use a special cardiod condenser. Cardiod condensers are typically oiled with immersion oil to the back of the sample to guarantee the very high angle necessary to exceed collection by high NA objectives.

Figure 18.36 illustrates how darkfield works. Any portion of the beam that does not interact with the sample reaches the objective at such a high angle that it cannot be collected. The result is a black background. Only light scattered by the specimen itself can be collected. The image is formed by interference of adjacent, non–zero order diffraction spots derived from this scattered light.

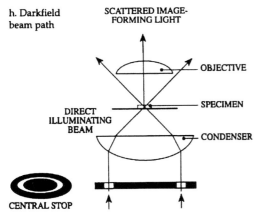

Figure 18.36 Low-power darkfield, created using a patch stop placed in the front focal plane condenser (FFP_c). (Image courtesy of Polaroid Corp.)

Since the human eye only requires two or three photons for it to sense that an object is present, any object that scatters even a few photons of light will be visible in darkfield. Notice that this technique is *detection* limited, not resolution limited. That is, light will come from objects well below the resolving power of the microscope, but there will be no information about actual size, shape, or edges.

In addition to imaging pits and scratches on polymer surfaces, darkfield microscopy is valuable for imaging inclusions like micro voids or crystals in photographic films or filler in fibers. Reflected-light darkfield is also very effective in reducing surface scatter and glare and is the technique of choice for looking through clear protective coatings at printing on packaging or the physical structures below the film such as in microelectronics. Reducing surface scatter also results in more accurate color rendition (Figure 18.37).

Rheinberg illumination is an interesting derivative of darkfield. This nineteenth-century optical staining technique works on essentially same principle but replaces the dark patch stop and bright ring with colored filters.† Several examples are shown in Figure 18.38.

As with darkfield, the background of the image carries the color of the center portion of the filter, while the scattered edge information carries the color of the outer ring (see Table 18.5). The circular filter in Figure 18.38 is nondirectional and will color all edges of any scattering feature: the background will be blue, and the edges will be yellow. The quadrant filter is directional: since scattering occurs at right angles to the pattern in the filter, the north–south quadrants will affect edges running east–west, and vice versa. This technique is useful for bringing out structures in woven fabric: fibers running north–south will be blue, while those running east–west will be red.

---

*Available from your microscope vendor.

†A set of photographic Rheinberg filters can be purchased through Microscopy/Marketing and Education. Contact the author via www.MicroscopyEducation.com.

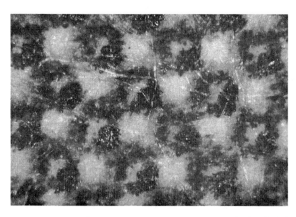 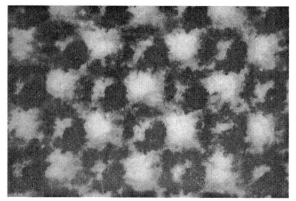

Figure 18.37 Inkjet print on paper, 40×: brightfield (*left*) and darkfield (*right*). See color insert.

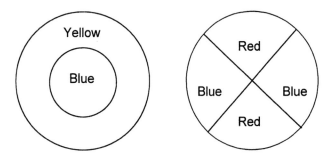

Figure 18.38 Circular and quadrant Rheinberg filters.

Combining Components in Both Condenser and Objective

Phase Contrast

Phase contrast requires four components that typically come as a single kit:

- A patch stop, called a *phase ring* or *phase annulus*, inserted into the FFP$_c$, typically in a universal condenser or on a slider containing several annuli of varying sizes
- A *phase plate*, placed in the BFP$_o$, resulting in a special phase objective that matches the phase ring (look for the PH annotation on the barrel of the objective)
- A *centering telescope*, used in place of the eyepiece, to center the image of the phase annulus with the phase plate
- A 546-nm green filter

Phase contrast summarizes the imaging theory discussed earlier. As its name implies, it detects phase shifts caused by differences in optical path. The phase annulus is designed to carefully position a ring of zero-order diffraction spots exactly into a cut on the phase plate, thus conditioning the light responsible for the background illumination (solid lines, Figure 18.39). On its way up the optical train in the microscope, a portion of this light passes through the sample and interacts with it, generating the specimen portion of the beam (dotted lines, Figure 18.39). For Phase contrast to work well, the specimen must have an optical path difference sufficient to cause the specimen light to slow down by approximately one-quarter wave (λ) compared to the background.

The phase plate is the next point of interaction. These plates are engineered to make best use of the objective NA and come in one of three sizes (PH1, PH2, PH3), marked on the barrel of the objective. The annuli have matching markings on the condenser. As shown in Figure 18.39, the specimen portion of the beam passes through the entire phase plate, while the background portion passes only through the cut. The plate is designed so that the thicker portion will slow the specimen light by another ¼ λ, making it lag behind the background illumination by a total of ½ λ, the condition for destructive interference. When these two beams interfere at the PIP to form the image, they will undergo destructive interference, creating dark regions in those portions of the image and, therefore, contrast against the background.

Table 18.5 Rheinberg filter selection and results in the image

Technique	Center top/ background color	Outer ring/ color of edges in features
Darkfield	Black/black	White/white
Rheinberg: blue center, yellow outer ring	Blue/blue	Yellow/yellow

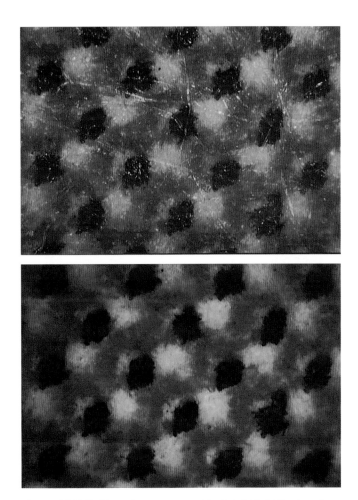

Figure 18.37 Inkjet print on paper, 40×: brightfield (*top*) and darkfield (*bottom*).

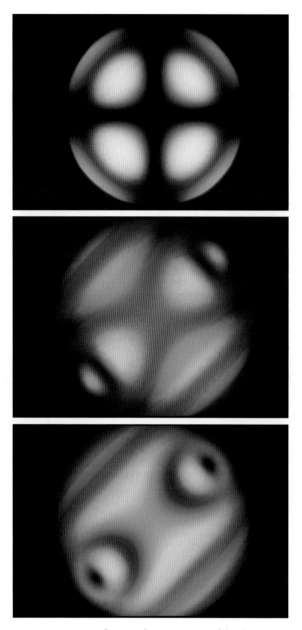

Figure 18.55 Interference figures: uniaxial (*top*);
biaxial, isogyres open (*center*); biaxial, isogyres open,
in circularly polarized light (*bottom*). (Images courtesy
of Phil Robinson, North Staffordshire Polytechnic,
retired.)

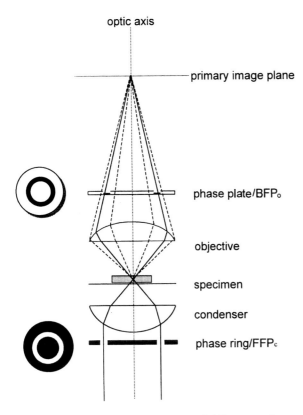

Figure 18.39 Phase contrast. Key: solid lines are the background illumination; dotted lines are the specimen portion of the beam.

The phase plate has one other feature: a neutral density filter. A quick review of interference theory indicates that the two beams which will form the image (background and specimen) should have approximately the same amplitude to cancel successfully. This filter cuts the intensity of the background to about 15% of its original intensity, a closer match to the intensity of the specimen light.

Finally, since Phase is so dependent on wavelength, the best results are achieved at the wavelength for which the system is engineered. In most cases, that wavelength is 546 nm green light, requiring insertion of a narrow band filter somewhere in the illumination path (typically over the light port).

To set up phase contrast:

1. Establish Koehler using the phase objective of your choice and the brightfield setting in the condenser.
2. Rotate the appropriate phase annulus into place and, using the special phase-centering screws on the condenser (see instructions in the kit), align the image of the phase annulus under the smoky image of the phase plate. To observe this process, replace the normal eyepiece with the centering telescope and focus

the telescope on the image in the BFP$_o$. Return the regular eyepiece for viewing.

3. Insert the green filter.

Because Phase contrast depends on the phase relation between sample and mountant, changing the refractive index of the mounting material offers an opportunity to fine tune the experiment. Phase images can be characterized by an edge-obscuring halo, and, again, a closer match in refractive index between mountant and sample reduces this artifact.

Hemsley[1] suggests several key applications for Phase contrast, including the study of multiphase polymer systems in which the polymeric phases are large enough to be resolved by the light microscope, differentiation of layers in coextruded or laminated products, and the fibrillar substructure of spherulites.

Hoffman Modulation Contrast

Coming on the scene in the early 1970s, Hoffman Modulation Contrast (HMC) is one of the newest and most useful contrast techniques for polymer microscopy. Like phase, it uses components that are inserted into the conjugate pair of focal planes composed of the FFP$_c$ and BFP$_o$ (Figure 18.40). However, like oblique illumination and DIC, HMC detects gradients and slopes, bringing out different information than phase. Like oblique illumination, it uses an offset slit to carefully position the Zero-order background light and, therefore, has a strong sense of directionality as well as the capacity for fine tuning. Unlike DIC, HMC does not rely on polarized light to set the stage for contrast. On the contrary, HMC can be used very successfully in combination with parallel or crossed polars: HMC will bring out surface structure, while polarized light will define orientation, stress effects, and amorphous versus more organized phases.

Like Phase contrast, HMC is ordered as a kit. Polymer scientists typically find the 10× and 40× Hoffman components most useful. The kit includes:

- Condenser components, usually mounted in a universal condenser, with an offset slit in the FFP$_c$, mounted in a floating ring for easy centration
- A polarizer, typically placed on the light port, to adjust the width of the slit
- Objectives fitted with *modulating* plates in the BFP$_o$
- A 546-nm green filter (usually a narrow band interference filter)
- A centering telescope

As with Phase contrast, in HMC the optical elements in the condenser and objective must be aligned. The slit in the condenser has two portions: one open, and the other covered with a piece of polarizing film.

As shown in Figure 18.41, the open portion of the slit must be positioned over the gray portion of the modulator. Follow this procedure:

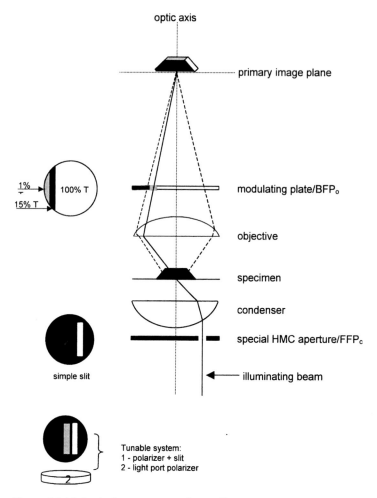

Figure 18.40 Optical components for Hoffman Modulation Contrast (HMC).

1. Start with Koehler, using the HMC objective of choice and the brightfield setting on the condenser.
2. Rotate the matching HMC slit in the condenser into position. To view the BFP$_o$ for alignment, replace the usual eyepiece with the centering telescope.
3. Using gentle finger pressure, push, pull, and rotate the floating mount on the underside of the condenser until the image of the clear portion of the slit overlays the gray sector of the modulating plate.
4. Replace the telescope with an eyepiece and observe the image.

There are several ways to fine tune an HMC image. One is to control the size of the slit. As mentioned earlier, the second half of the slit is covered with polarizing material. This portion sits over the clear sector in the modulator. A second polarizing filter is placed over the light port. Watching the BFP$_o$ as the polarizer is rotated demonstrates that this adjustment has two effects. As the filter is rotated, it variably reduces the intensity background contributed by the polarizing part of the slit, tuning the intensity of the background to the light interacting with the specimen. Moreover, at specific positions, the polarizing filter totally closes the second half of the slit, increasing coherence in the same way as closing the aperture iris would, resulting in enhanced edge information.

Because HMC enhances gradients, the relative RIs of the specimen and its mountant have an effect and can be used

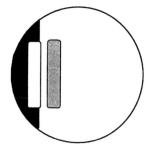

Figure 18.41 Aligning Hoffman modulation contrast (HMC).

to further improve the image. Finally, if true color is not important, optimize the image by inserting a green filter, either over the light port or in the illuminating beam. As with phase contrast, HMC works best with a 546-nm narrow band filter.

Advanced Techniques

Fluorescence

Fluorescence can be summarized as "energy in : energy out". High-energy light, typically in the UV or blue range of the spectrum, excites susceptible π electrons in the polymer, causing them to jump to a higher, less stable energy state. In an effort to return to the lower energy state, the molecule will absorb a small amount of the excitation energy as heat then release the rest as light, or *fluorescence*. Because of the small energy loss, the emitted fluorescence typically undergoes a spectral shift (Stokes shift) toward the longer (red) end of the spectrum. This wavelength difference can be put to good use, separating the incoming excitation light from the emitted fluorescence.

Fluorescence has enjoyed great success in biological research and medical diagnosis with a wide range of biopolymers, yet it is severely underused in the material science applications of polymer microscopy. Although the procedure is not inexpensive, most research-level light microscopes can be fitted with fluorescence. The basic components and process are outlined in Figure 18.42.

Converting a microscope to fluorescence typically requires the addition of a reflected light module fitted with a high-energy light source and beam splitting cubes. It also may involve purchasing additional nonfluorescing, high NA objectives.

The choice of each component is driven by the excitation and fluorescence emission characteristics of the sample. All components must be spectrally matched. For illumination, the primary choices are high-energy arc lamps filled either with mercury (HBO) or xenon (XBO). Mercury arcs have exceptionally high output in specific ranges (for example, 546 nm) and are ideal if those peaks match the excitation peaks of the polymer under study. In contrast, xenon emits lower energy but has a more evenly distributed spectrum, making it a good alternative when the fluorescence characteristics of the material are not well known.

The heart of the system is a cube containing three filters: the *exciter*, the *dichroic beam splitter*, and the *barrier filter*. As indicated by its name, the exciter selects the wavelength for the incoming radiation. As shown in the left portion of Figure 18.42, that light then proceeds to the dichroic beam splitter, which performs like an optical gate. The FT420 notation shown in this diagram indicates that this particular filter will reflect any light shorter than 420 nm blue and pass any light of longer wavelength. In step 1, then, the exciter selects a range of blue wavelengths, and the dichroic reflects them downward to interact with the sample.

In step 2 (right diagram), the sample has released its fluorescence. Because it has longer wavelength, the fluorescence passes through the dichroic, on to a barrier filter for final signal cleaning, then on to the viewing system. Because none of the initial excitation light reaches the viewer, the background is dark, a very important feature, since the fluorescence signal is typically only 1/1000th to

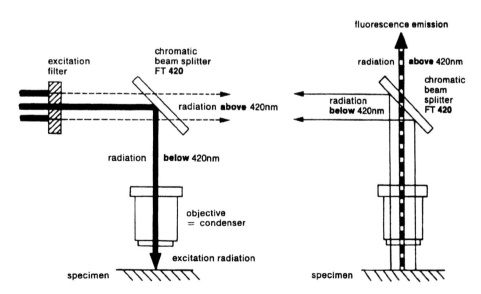

Figure 18.42 Components necessary for fluorescence microscopy. (Image courtesy of Carl Zeiss, Inc.)

1/10,000th the intensity of the excitation radiation and would be swamped by the excess light.*

Fluorescence microscopy is the most rapidly growing area in the biological sciences but has been slow to catch on in the materials arena. Because of its sensitivity and specificity, however, fluorescence microscopy can be a valuable technique for locating contamination, as well as for evaluating polymer performance. All indications point to its use in conjunction with other techniques such as FTIR for failure analysis and confocal microscopy. With new methods for 3D visualization and quantitation, the combination of confocal and fluorescence can provide interesting answers to long-standing questions such as the real structure of polyurethane foam cells and the relationship between structure and function of process parameters, such as time at temperature, humidity, and catalyst concentration. There are also interesting implications for filled polymers and films.

Polarized Light

Polarized light brings together both sides of the story: the incoming light and the internal characteristic of the material under study. Because of the complexity and extent of this information, the detailed discussion of polarized light will be deferred until the next section of this chapter.

In the simplest sense, polarized light enhances contrast based on the difference in refractive indices in at least two directions in a material. A drawn fiber, for instance, will have one RI along its length and another across its diameter. This anisotropy makes it birefringent and, therefore, responsive to polarized light.

Nomarski/DIC

DIC stands for *differential interference contrast*. It is often called *Nomarski* in honor of Dr. Georges Nomarski, the French physicist who discovered a method for economically cutting the Wollaston prisms on which the system relies. Some manufacturers still use the Nomarski designs, while others use systems designed by Françon or Smith. However, the generic term Nomarski is widely applied. Like oblique illumination, DIC enhances slopes or gradients. Since this technique is based in polarized light, the full description will be deferred to the end of the next section.

For easy reference, Table 18.6 summarizes the contrast discussion in this section.

*It is well known that grass, trees, and other chlorophyll-containing plants respond to the 546-nm green light in normal sunlight and fluoresce red. However, the red fluorescence signal is swamped by the overwhelming amount of green, reflected light. You can verify these findings if you have access to a fluorescence microscope with a green excitation cube. Tip: peel a leaf and mount a tiny section in a drop of water under a coverslip.

Polarized Light

Polarized light is one of the most valuable tools in the polymer microscopist's arsenal. It provides contrast as well as a number of optical measurements. Polarized light not only elucidates identifying parameters, it often detects subtle changes. A well-known axiom of polymer microscopy is that polymers retain their thermal and mechanical histories in their polarized light signatures. Whether you are investigating polymer films, foams, fibers, filled polymers, or catalysts, even small changes in chemistry or processing may manifest themselves as a telltale response under the scrutiny of the polarizing microscope.

Traditionally, the study of polarized light microscopy has been very challenging for two reasons. One is lack of exposure to its unique light-matter interactions. The other is its complex vocabulary. The first challenge can be met by exploring some of the basic science behind polarized light microscopy (PLM or Pol) and the second, by dissecting the vocabulary in terms of those simple principles. A major portion of this section of the chapter lays the foundation for practical use of the Pol scope by combining these two tools. The last portions put the principles and vocabulary to use in a description of an actual analysis.

Basic Principles of Polarized Light

How Polarized Light Differs from Ordinary Light

Ordinary light, such as that by which you are reading this page, contains light waves vibrating in all directions perpendicular to their direction of travel (Figure 18.43). To generate polarized light, ordinary light must pass through or be reflected by a polarizing device. That device will absorb all directions of vibration except one: the permitted direction. As a result, the light emerging from the interaction is said to be *polarized*.

Mechanisms for polarizing ordinary light include reflection from the surface of a dielectric material such as glass or a polymer film, or transmission through a material such as polarizing filters, or through beam-splitting materials such as quartz or calcite.

Typical polarizing filters are made by doping polyvinyl alcohol with potassium iodide and molecular iodine to produce an I_3-complex. Stretching the film orients the I_3-complex, creating two different electrical environments: one along the direction of stretch, and one perpendicular.

EXERCISE

You can determine the permitted direction of a polarizing filter by doing the following experiment. You will need a polarizing filter and a glass slide.

1. Place the slide flat on the table in front of you and view it at about 50° from the vertical. Catch the

Table 18.6 Summary of contrast techniques

Technique	Generated by	Operating principle	Detects	Pros/cons
1. Quick, easy, and inexpensive				
a. Axial illumination	Closed down condenser aperture	Diffraction	Edges	Greater depth of field; can be either + or –; (–) "ringing" at edges; (–) reduced resolution
b. Oblique illumination	Offset illumination in condenser	Zero order set to side of BFP_o; gradients throw light in or out of BFP_o	Gradients	Improved contrast and resolution
c. Darkfield	Patch stop in FFP_c	Background light not collected; image formed by scattered light from specimen	Any scattering element	Detects particles below limit of resolution for microscope; (–) color may come from dispersion
d. Rheinberg illumination	Colored patch and ring in FFP_c	Colors zero order background differently from scattered specimen light	Scattering elements	Very effective for unstained filamentous objects
e. Stains of colored filters	Colored filter over light port	Complementary color enhances contrast; same color minimizes	Colored objects	Inexpensive and quick
2. Condenser + objective techniques				
a. Phase contrast	Annulus or ring in FFP_c; matching phase plate in BFP_o	$\lambda/2$ condition for destructive interference between diffracted specimen light and undiffracted background; first $\lambda/4$ retardation comes from diffraction and refraction in specimen; second $\lambda/4$ created by phase plate in objective	Phase objects (body effect)	(–) halo may obscure edges and fine detail
b. Hoffman Modulation Contrast (HMC)	Offset slit in FFP_c; modulator in BFP_o	Intensity variation	Slopes (gradients)	Enhances both contrast and resolution; directional; tunable for both coherence and matching intensity of background to sample; shallow depth of field good for optical sectioning; works well with plastics
3. Advanced techniques				
a. Fluorescence	Beam splitter with filters	Energy in/energy out	Body effect (molecules with easily excited electrons)	Bright objects vs. dark background; useful for confocal
b. Polarized light	Polarizing filters above and below sample	Polarization; depending on internal electrical environment, sample responds to specific direction of vibration in polarized light	Body effect (directional variation in internal electrical environment causes variation in refractive index)	Color variation produced by relative retardation; many identifying characteristics and measurements
c. DIC	Sandwich of polars and beam splitters above and below sample; tunable retardation plate	Color or intensity variation produced by relative retardation of sheared wave fronts	Slopes	Improved contrast; shallow depth of field good for optical sectioning; (–) does not work well with plastics or birefringent materials

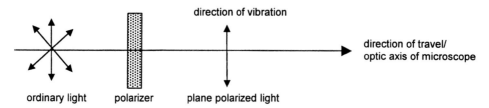

Figure 18.43 Ordinary versus polarized light.

reflected image of light coming from a window or overhead light. The light forming the image is polarized by reflection from the glass surface, with a horizontal (east–west) permitted direction of vibration (parallel to the flat surface of the slide).

2. Now view the reflection through the polarizing material. Hold it close to your eye, as though it were a sunglass lens. Slowly rotate the filter clockwise. You should observe angles at which the reflection is brighter and other angles (90° from the first) at which the reflection is darker. When the reflection is brightest, the filter is oriented so that it permits the east–west vibrations to pass. When the reflection is dimmer, the filter's permitted direction is north–south. Light from the reflection is absorbed.

Parallel versus Crossed Polars

The terms *parallel polarizers* and *crossed polarizers* are derived from the respective permitted directions of vibrations of two polarizers. You can test this difference by removing the lenses from an old pair of polarizing sunglasses. Stacking the two lenses on top of each other so that they nest orients their permitted directions of vibration parallel to each other and, therefore, permits light to pass through (*parallel polars*). Rotating one of the filters by 90 degrees orients their directions of vibration at right angles (*crossed polars*). Vibrations permitted through the first filter will be absorbed by the second, and the view through them will be dark. These effects also explain the preceding experiment.

Birefringence

As with the phase relationships, the underlying concept for everything in the world of polarized light is refractive index. Materials that respond to polarized light are those that have either two or three identifying RIs, each of which results from differences in electrical field caused by the internal arrangement of atoms, ions, or molecules. Mathematically, the difference between two internal RIs defines the material's birefringence (delta, Δ):

$$\Delta = |(n_1 - n_2)| \qquad \text{(birefringence)} \qquad (15)$$

Notations to distinguish the two RIs vary. Sometimes they are given the simple numerical subscripts shown in this equation. In polymer films and fibers, they are often given subscripts $\parallel$ and $\perp$, to show the refractive indices parallel and perpendicular to the directions of draw or spinning. Finally, they may be annotated as "e" and "o" or "ε" and "ω" (epsilon and omega) to correspond to specific optical directions.

EXERCISE

The nomenclature for e and o is actually derived from early polarized light studies of crystals. You can repeat these experiments by obtaining a small piece of calcite from a museum gift shop. Put an ink dot on a piece of paper, and place the crystal on top of it. You will notice that the image of the dot is split into two images: the e-ray and o-ray. Rotating the crystal will cause one image of the dot to "walk" or precess around the other. The image that remains stationary was originally called the *ordinary* or o-ray because it behaves as one would ordinarily expect the image to behave. The other image is called the *extraordinary* or e-ray because it is behaving in an extraordinary fashion.

Fast and Slow Directions

Electrically, birefringent materials behave very much like wood, exhibiting directions "with the grain" and "against the grain".

Remember that light and matter both have electrical character. In Figure 18.16 we saw that if an incoming beam of polarized light encountered a birefringent material so that its direction of vibration (and, therefore, its electrical field) is aligned with the lower electrical field in the material, there would be only moderate interaction. Light vibrating in this direction would travel quickly through the material. Since traveling "with the grain" is faster, this direction is called the *fast direction*, characterized by the lower RI, and given the Greek symbol alpha (α). Conversely, if the direction of vibration for the incoming beam is oriented so that it encountered a maximum interaction, the light would travel more slowly. This direction is the *slow direction* (gamma, γ) and exhibits the higher refractive index. Regardless of the internal orientation of the atomic, ionic, or molecular structure, the optical directions representing *fast* and *slow* are always at right angles to each other.

Since all materials are three dimensional, materials with only two RIs will have one cross-section in which the RIs will be the same. In materials with three RIs, the intermediate RI is called beta (β) and will, again, be oriented at right angles to the other two.

Whether "e" or "o" is the fast (α) direction defines the *optical sign* of a material. Replacing the numerical subscripts in eq 15 with these letters and removing the absolute notation provides the basis: $\Delta = (n_e - n_o)$. If e is the slow direction, n_e will be larger than n_o, producing a *positive sign of birefringence*. Similarly, if o is the slow direction, birefringence is said to be negative. "POOF" is a simple memory clue: the optical sign is "*PO*sitive if *O*-ray is *F*ast".

Positions of Brightness and Extinction

If a birefringent material such as a film, fiber, or catalyst crystal is placed between crossed polarizers and rotated (clockwise, for example), it will go through alternating bright and dark positions.

Figure 18.44 illustrates with three reference positions and shows how vector analysis can be used to predict these positions. Arrows P and A indicate the permitted directions for the polarizer and analyzer, respectively. The polarizer provides the incoming energy. Notice that P and A are at right angles to each other, indicating crossed polars. Vectors e and o show their respective directions in reference to a polymeric structure such as a sheet of film, fiber, or crystal. For the polymer to appear bright between crossed polars, some component of that energy must be accepted by e and o and have some resulting component that will pass through the analyzer. To determine if there is an emerging component, complete the parallelogram of forces then draw the diagonal (resultant, R).

EXERCISE

To find the resultant vector, complete the rectangle (parallelogram of forces) formed with the vector representing e on one leg and the vector representing o on the other. The resultant will be the diagonal of that rectangle.

In Figure 18.44a, all the energy from the polarizer goes into vector o. The material behaves as though it is isotropic. The resultant lies entirely in the plane of o, no light passes through the analyzer, and the polymer appears dark. This is called a *position of extinction*. Figure 18.44b illustrates the opposite extreme, with all the energy going into e. This is the second of four positions of extinction, and, again, the polymer appears dark. In Figure 18.44c, the polymer is rotated so that e and o are at + or −45°. In this position, the contribution of energy from the polarizer is at a maximum for both vectors. Completing the parallelogram of forces shows a maximum component, R, in the direction of the analyzer. The polymer is brightest in this position because energy from both e and o vectors can pass through the analyzer. Rotating the polymer from either a or b to c produces brightness increasing to this maximum. Rotating it from c to either position produces decreasing brightness.

Retardation

As in the discussion on phase objects, each light wave encountering a different electrical environment has its own optical path, $t \times n$, and as with phase objects, there is an optical path difference. However, with birefringent materials, the optical path difference occurs within the material and is called *retardation* (gamma, Γ). As seen in eq 16, the equations for OPD and retardation are essentially identical. The major difference is the pair of RIs under observation.

Optical path difference for phase relationships

$$OPD = t(n_1 - n_0) \qquad \text{(optical path difference) (16a)}$$

where t is the geometric thickness of material; n_1 is RI inside the material; and n_0 is RI outside the material.

Retardation for polarized light relationships

$$\Gamma = t(n_\parallel - n_\perp) = t\Delta \qquad \text{(retardation)} \qquad (16b)$$

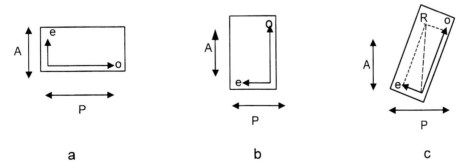

Figure 18.44 Positions of brightness and extinction: (a) and (b) extinction (polymer is dark between crossed polars); (c) maximum brightness. Key: A = analyzer; P = polarizer; R = resulting component; e and o are the vectors; arrows indicate the permitted directions of vibration.

where t is the geometric thickness of material; $n_\parallel$ is the RI parallel to the direction of draw; $n_\perp$ is the RI perpendicular to the direction of draw; and Δ is the birefringence, $|(n_\parallel - n_\perp)|$.

This equation for retardation has a hidden parameter. Because the two characteristic refractive indices are only displayed in viewing angles of the fiber, film, and the like, sample orientation is critical in the practical applications of Pol. The effect of orientation is covered in the following section under orthoscopy.

Orientation Birefringence

Many polymeric materials exhibit birefringence simply because their internal chemical structure establishes two or three different directional electrical environments. This situation is especially true of fibers, films, and crystals.

In Figure 18.45, an incoming beam of ordinary light is polarized, isolating a single electrical vector, **E**, vibrating in one plane. When that vector passes through a birefringent material (in this case, a film), it is split into two component vectors, **e** and **o**.

EXERCISE

You can prove that each beam is polarized and that they are polarized at right angles by viewing the dots produced in the crystal beam-splitting experiment described earlier through a piece of polarizing film, like the lens from polarizing sunglasses. First, do the experiment outlined earlier to determine the permitted direction of the film. Using a Sharpie marker, mark that permitted direction on one corner of the film with a two-headed arrow. To test the directions of vibration of **e** and **o**, rotate the crystal so that the image from the e-ray sits north of the o-ray. Orient your film so that its permitted direction of vibration is east–west and view the two dots. Record your observations, then rotate the crystal so that the e-ray sets to the east of the o-ray. Again, position your film east–west and observe the two dots. Repeat the experiment with the e-ray to the south and then to the

west. If your film only permits E–W vibration to pass and if **e** and **o** are, indeed, polarized at right angles to each other, only the ray with the same E–W vibration will be visible. Energy from the other will be absorbed, causing that dot to disappear in that particular orientation.

As your experiment proved, both **e** and **o** are polarized. Their permitted directions of vibration will always be at right angles to one another. As a result, they encounter different electrical environments ($n_\parallel$ and $n_\perp$), and one will slow down with respect to the other. The lag between them is the *retardation*, Γ. The longer the physical distance traveled, the greater the lag. This type of birefringence is also referred to as *orientation birefringence* and is characteristic of all anisotropic materials.

From this example, it is intuitively logical that drawing a fiber or film changes the internal orientation of the molecules, changing the relative electrical fields seen by **e** and **o**, therefore affecting both the birefringence and the resulting retardation. The greater the draw, the greater these values. For these reasons, it is critical not to introduce any stress during sample preparation. Films and fibers, especially, must be handled gently and, as they are mounted on the microscope slide for observation, not be stretched or strained.

States of Polarization

The optical path difference between **e** and **o** produces a *state of polarization* that significantly influences the direction of vibration and the shape of the path taken by the resulting electrical vector emerging from the material. All that is required is a single polarizer to establish the permitted direction of vibration for the incoming electrical vector, **E**, and a birefringent material to cause retardation between **e** and **o**. Simple polymer films like polyethylene and polypropylene are good examples.

As seen in the top two diagrams in Figure 18.46, if **e** and **o** are out of step by multiples of either whole or half wavelengths (e.g., $\lambda = 546$ nm), the resulting electrical vector seesaws through space, on the diagonal, producing *linearly polarized* light. Depending on whether the retardation is $\lambda/2$ or λ, the vector will oscillate into opposite quadrants.

If **e** and **o** are out of step by quarter wavelengths (the bottom two diagrams in Figure 18.46), the resulting electrical vector will rotate through space in a circular corkscrew pattern, generating *circularly polarized light*. This situation is highly desirable when doing automated image analysis such as particle sizing and counting with anisotropic materials. Normally, if a collection of anisotropic materials such as a random array of catalyst crystals or cut fibers is placed between crossed polars, the particles or fibers will take on varying intensities, depending on their orientation with respect to the two polarizers. Circularly polarized light removes that orientation effect. As a result, all the particles in the field will have the same intensity and will be bright against a dark background. This condition is ideal for auto-

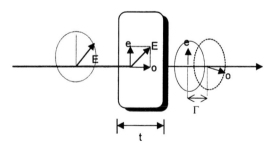

Figure 18.45 Source of retardation, Γ. Key: **E** = electrical vector; **e** and **o** = polarized components of **E**; **t** = thickness of polymer film; Γ = retardation or lag of **e** behind **o**. Arrows indicate direction of vibration.

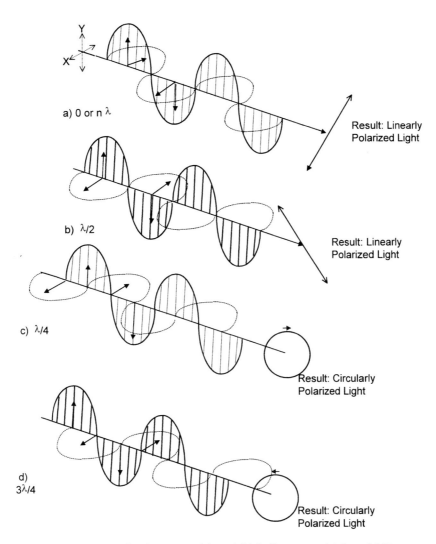

Figure 18.46 States of polarization: (a) and (b) half-wave multiples of 546 nm generate linearly polarized light, but with opposite directions of vibration; (c) and (d) quarter-wave multiples generate circularly polarized light, with opposite directions of rotation. X and Y show the respective permitted directions of vibration for the e-ray (hatched) and o-ray (white). The small arrows represent the respective positions of **e** and **o** at maximum amplitudes at various times along the wave. The resulting polarization can be found by standard vector analysis. For example, in (a) the resultant starts out at zero, reaches a maximum at +45° between **e** and **o**, then drops back to zero, reaches a maximum again at −45°, then drops back to zero. The result is the large arrow shown at the end of that diagram, which would be seesawing through space, along the line of travel, into quadrants I (+45°) and III (−45°).

mated segmentation and can be accomplished by substituting special circular polarizers for the conventional ones in your microscope.*

The transition between half and quarter multiples causes

the retardation to pass through intermediate values. The resulting electrical vector will again travel through space in corkscrew fashion but with unequal axes, producing *elliptically polarized light*. States of polarization are summarized in Table 18.7.

Source of Polarization Colors

If a second polarizer is positioned after the birefringent material it will analyze the retardation. This second polarizer

*Circular polars are available from most manufacturers or can be homemade by (1) crossing the regular polars, then (2) placing quarterwave plates (146-nm retardation material) above and below the sample, also in crossed position (+ or −45° to the polarizer). The first plate will create the circularly polarized light necessary to remove the orientation effect, while the second will reverse the process so that the analyzer can detect the normal retardation.

Table 18.7 Summary of the states of polarization

Lag between e and o[a]	Resulting state of polarization
0 or λ	Linear, vibrating NE to SW
λ/4	Circular, rotating clockwise
λ/2	Linear, vibrating NW to SE
3λ/4	Circular, rotating counterclockwise
Intermediate values	Elliptical, rotation determined by adjacent states

a. Also applies to any multiple of these values.

is constructed of exactly the same material as the first, but because its function is different, it is called the *analyzer*. The results of the analysis will be manifest as a *polarization color* (Figure 18.47).

To understand the entire process, follow the electric vector **E**. A polarizer (*P*) defines the direction of vibration for **E**—in this case, north–south. The birefringent material, *B*, splits the electric vector into an e-ray and an o-ray, vibrating at right angles to each other. Each of them encounters a different electrical environment as they pass through the material, causing one to lag behind the other (the retardation, Γ). When the light reaches the second polarizer (analyzer, *A*), those components of **e** and **o** that are parallel to the permitted direction of the analyzer (*e'* and *o'*) will pass through it. Since these components are now vibrating again in one plane (the direction permitted by the analyzer), they meet the criteria for interference.

Under the normal interference conditions, waves that are out of step by λ/2 would undergo destructive interference. As discussed in the earlier section on light, the interference color produced would be a combination of the other two primary colors. The earlier example was the destructive interference of green light (for example, at 550 nm), resulting in red + blue, or magenta.

In polarization, however, the waves emerging from the analyzer have undergone the retardation induced by the

sample coupled with a 90° rotation generated by the permitted direction of the analyzer. As a result, destructive interference occurs at *full* λ intervals rather than at the normal λ/2 condition.

The Michel-Levy Chart

The *Michel-Levy chart** encompasses both the development of Pol colors and the effects of thickness and birefringence on retardation (Figure 18.48).

Color is the first element of the chart that catches one's eye. The chart starts with stark black, representing a retardation of 0 nm, then moves through the typical Newton's series of colors. If the 0-nm black edge is placed to the left, the axis along the bottom of the chart reveals that the colors can be correlated directly to retardation values, measured in nanometers. The colors repeat, as retardation meets the full λ criterion for polarized light destructive interference. Typically, red dots are placed as reference points on this axis at points to indicate where 550-nm green light and its multiples undergo destructive interference. These dots define *polarization orders* and their corresponding shades of magenta. The range of colors from 1 to 550 nm is defined as the first order; from 550 to 1100 nm is the second order, and so on. The black at 0 nm retardation is given the special name of *Zero Order* black. Note that this is a Pol color, not to be confused with the Zero order in the diffraction pattern.

As illustrated in Figure 18.49, color spacing and intensity change from left to right. With increasing retardation, the color bands become broader due to the normal differences in wavelength for blue (short wavelength, frequent repeat), green, and red light (long wave, less frequent re-

---

*Michel-Levy charts are usually available in single quantities free of charge from your local microscope representative. As of the printing of this chapter, they were also available on the Web through Florida State University: http://micro.magnet.fsu.edu/primer/techniques/polarized/michel.html.

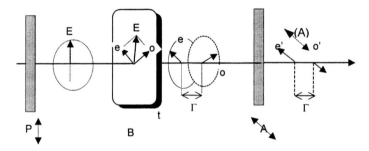

Figure 18.47 The analyzer (A) rotates **e** and **o** into the same plane for interference. Key: P = polarizer; E = electric vector in plane polarizer light; B = birefringent material; **e**, **o** = polarized components of E; Γ = lag, retardation, e', o' = components of **e** and **o**, which can pass A and interfere to form Pol colors.

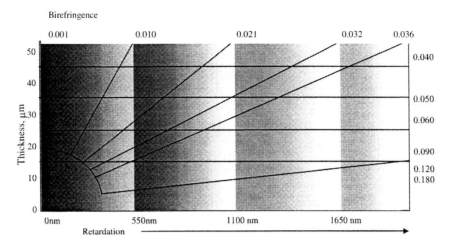

Figure 18.48 Diagrammatic version of the Michel-Levy chart: starting at 550 nm, gray boundaries identify magentas and pinks, signifying transition from one order to another.

peat). As they proceed further along the retardation axis, the bands overlap, mixing the colors and causing the shades to become paler. For example, at 550 nm, green light has undergone destructive interference, and the remaining red and blue combine to form the rich magenta known as *First Order* red. Notice that this band is narrow and well-defined, making it very useful for Pol analyses.

At the first multiple of 550 ($1100 = 2 \times 550$ nm), larger amounts of red and blue are present, but the wavelength effect is beginning to assert itself, and the color band begins to broaden. The resulting magenta is paler, forming the region called *Second Order* pink. At 1650 nm (3×550 nm), the broadening is even more prevalent, resulting in very pale *Third Order* pink.

From this figure, the source of *First Order* white is also clear: it is the location where red, green, and blue are present in equal intensity. The colors proceed to get more and more pale through about the sixth and seventh orders, where the overlap becomes so pronounced that all the colors blend together into white again. This white is referred

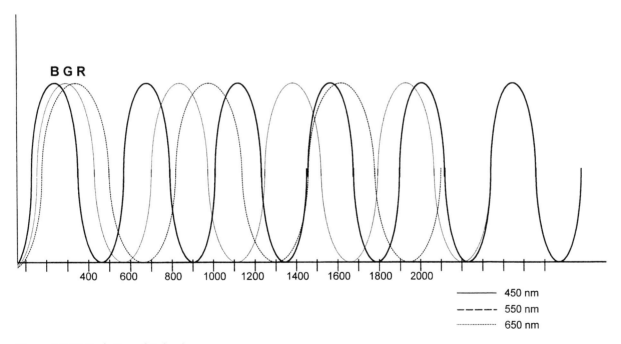

Figure 18.49 Evolution of Pol colors.

to as *High Order* white and, in the absence of defined color bands, retardation analysis must be done with special test plates called *compensators* (discussed in the next section).

In addition to an axis for retardation, the Michel-Levy chart has two other axes: one running up the left border for thickness (0 to 50 μm) and the other running around the top and down the right border for birefringence (0.001 to >0.180). Radiating lines tie this information together. If any two of the three parameters are known, the chart can be used to find the third, as shown here:

EXERCISE

A 25-μm thick polymer film, taped at its ends to a microscope slide so that it stays flat, is placed on a rotatable microscope stage. The polarizer and analyzer are inserted in appropriate slots above and below the sample, in crossed position. When the sample is rotated on the stage so that it exhibits its brightest intensity, the color produced is brilliant, deep magenta.

The analysis would proceed as follows: According the Michel-Levy chart, the observed color corresponds to a retardation of 550 nm. That color band would be followed to its intersection with the horizontal line corresponding to the thickness of 25 nm to find the nearest radiating line. Following that line to the birefringence axis reveals that this particular film has a birefringence of 0.021.

Similarly, if the two refractive indices in the film are actually measured, birefringence could be calculated. Following that radiating line to its intersection with the observed Pol color would direct you across the chart to its thickness.

Some forms of the Michel-Levy chart list typical materials around the birefringence axis. Although most of these charts emphasize minerals, they often list commonly used polymers or fibers including nylon, silk, and cellulose (cotton). Once you have become familiar with the workings of the chart, you can also expand it to include materials frequently used in your own laboratory.

Addition, Subtraction, and Compensation

Alone or coupled with special test plates, the Michel-Levy chart is a valuable diagnostic tool. The preceding discussion was based on simple observation of a single material in the Pol microscope. The concepts of *addition* and *subtraction* are necessary to extend the application to use of test plates and to more complex, multilayer systems.

Imagine two pieces of polymer film (left diagram, Figure 18.49). Each has its own thickness, its own set of fast and slow directions (shown here by short and long vectors representing the low and high refractive indices), and its own retardation. Stacking these one on top of another causes their optical properties to sum, very much like the results of a relay race. However, stacking them so that their slow directions are aligned or stacking them so that the slow direction of one overlays the fast direction of the other has a profound effect on the outcome of the race.

If you have an overhead projector (or can view the experiment against a light box or other even source of lighting), two pieces of polarizing material, and two pieces of film, you can visualize the results of the race. In this discussion, assume that the first piece of film has a retardation of 150 nm and the second has a retardation of 550 nm. First, align the films so that the slow direction from film 1 overlays the slow direction from film 2 (Figure 18.50a). Their retardations will add. To test this effect with your materials, cross the polarizers. Start by viewing each film independently in a position of brightness between the crossed polars. Each film should show a Pol color corresponding to its retardation:* the 150 nm will be soft dove gray, and the 550 nm will be *First Order* red.

Next, stack the films so that their slow directions are aligned (Figure 18.50a). The retardation from the first film will be augmented by the retardation from the second. It is

*The actual colors would be read from the Michel-Levy chart.

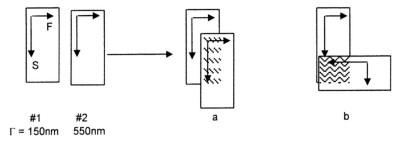

#1 #2
Γ = 150nm 550nm

a b

Figure 18.50 Examples of addition (a) and subtraction (b). Use two films. In this case, one has Γ = 150 nm and the other has Γ = 550 nm, as determined by comparing their Pol colors to the Michel-Levy chart. In (a) they are aligned with their slow directions overlapping; Γ = 150 + 550 = 700 nm (2nd 0 blue). In (b) they are aligned with slow overlapping fast: Γ = 550 − 150 = 440 nm (1st 0 yellow).

Table 18.8 Various types of compensators and test plates

Type (alternate names)	Range	Γ (nm)	Applications
λ/4 (quarter wave, mica plate)	146 nm	146	Low first-order phase differences via +/–; strained glasses, polymers with low Δ
λ (full wave, gypsum plate, first-order red plate, sensitive tint plate)	560 nm	560	First order, +/–: O, optical sign of spherulites; C, optical sign of fibers, films; can also be used to enhance contrast of inclusion droplets of gels, etc.
Quartz wedge	1–4 orders 1–7 orders	0–2200 0–3850	Variable compensation over range of wedge; O
Brace-Kohler	λ/10 λ/20 λ/30	a. 55 b. 25 c. 15	Glasses or polymers of very low strain birefringence; O
DeSernamont (λ/4 in subparallel position, combined with rotatable analyzer, monochromatic light)	1 order	Useful for suborder calculation in various orders	DeSernamont is often used (either just the first order or in conjunction with the quartz wedge) to refine a retardation reading within one of the first 7 orders O: retardation of fibers, films, foams
Ehringhaus tilting compensator	~6 orders	0–2900	Thin fibers, plastic films, O
Tilting compensators	10 orders 30 orders	0–5500 0–16,500	Thick fibers, polymer films

Note: + = addition; – = subtraction; Δ = birefringence; Γ = retardation; O = orthoscopic viewing; C = conoscopic viewing.

as though the slow runner from the first relay handed off to the slow runner from second team: the combined team would lose the race by the total of the two slow times. The retardation is the sum of the 150 nm from film 1 plus the 550 nm from film 2, or 700 nm (*Second Order* turquoise). This demonstration illustrates the effect of *addition*.

Next, stack the films in crossed position so that the fast direction of film 1 is aligned with the slow direction of film 2 (Figure 18.50b). Following the analogy of the relay race, it is as though the first runner won his heat by 150 nm but handed off to a slow runner who lost by 550 nm. The end result would be a retardation of 400 nm; the intersection where the two films cross should exhibit the *First Order* tawny yellow corresponding to that retardation. Since the total retardation moved down the Michel-Levy chart, this example illustrates the effect of *subtraction*.

Unless otherwise specified, polarized light tests are always performed by rotating the sample into a position of brightness then evaluating with a special test plate to determine if the sample is in a position of subtraction. Again, unless otherwise specified, most tests require the sample to be in a position of subtraction.

What would happen if we modified this experiment by replacing film 1 with a *First Order* red test plate? In the additive positions, the result would be *Second Order* pink, but in the subtractive position, *all* the retardation from the film would be subtracted by the test plate. The result then would be *Zero Order* black. This special case of full subtraction is called *compensation*, and, as might be expected, the test plates used in this type of analysis are called *compensators*.

Compensators come in a variety of forms. Your choice depends on the level of retardation in your sample (Table 18.8). The *First Order* red plate (also called a sensitive tint plate, quartz plate, or lambda plate) is the most common compensator. Examine yours to determine its slow direction (higher RI). It should be marked with an arrow, usually running parallel to the short end of the insert, and a "γ" or "Z" (Figure 18.51).

A Last Word on Birefringence

We can now begin to put some of these basic principles to use in an extension of the earlier discussion on birefringence. If a polymer system is stressed, both isotropic and anisotropic materials can exhibit *strain birefringence*. You can demonstrate this phenomenon by observing a piece of plastic film from a food storage bag between crossed polars. Stretching the bag causes a change in polarization color, corresponding to a change in retardation. As the film is stretched, the distance between the atoms of the material changes, affecting the electric field in the direction of the stress and, therefore, the refractive index in that direction. It is not always easy to measure the effect of strain, but reference tables can be constructed by starting with a material of known thickness and retardation, then plotting the differences in both under known loads.

Form birefringence results when block copolymers separate into domains that exhibit different RIs. The effect can be observed and measured by exposing the polymer to a solvent for one of the phases. As it swells, its birefringence will go to zero.

Figure 18.51 A *First Order* Red plate, with slow direction γ, overlaying an image of the sample as viewed between crossed polars. The vector **e**, in the sample, is shown as ε. Notice that the material has been rotated into a position of brightness.

Using Polarized Light

The Polarizing Microscope

As with fluorescence, components for the polarizing microscope have to be carefully chosen so that they do not respond to the technique itself: all optics between the polarizer and analyzer need to be strain free.

For general Pol observations, a simple polarizer can be placed over the light port, and a cap analyzer can be placed over the eyepieces. Strain introduced by the binocular beam splitter may limit use of this approach. To test, cross the polars with no sample in place and observe the background. If it does not go to black, try removing the head, placing the analyzer (in this case, a simple piece of polarizing film) underneath it, returning the head, and repeating the experiment.

For intermediate and advanced work, slots for compensators, rotating stage, and a cross-hair in the eyepiece* are a necessity (Figures 18.5 and 18.52). If your work includes *conoscopy* (tests conducted in the BFP$_o$), you will also need to invest in very high NA strain-free objectives and condensers. Fortunately, most of these observations are done at one magnification (40× is typically convenient), thus minimizing that investment. Most microscope manufacturers provide special sets of polarizing optics (marked "P" or "Pol") that meet both the high NA and strain-free requirements.

Polarizers and analyzers come in a variety of configurations. In the simplest case, they are fixed and mounted in sliders so that when they are inserted, they will automatically be crossed. If automated image analysis of particles or fibers is also part of your analytical scheme, try a set of circular polarizers to remove orientation effects.

The next step in sophistication is a simple rotatable polarizer with a fiduciary mark like a white dot or line to note the 0° position. Fully rotatable polarizers and analyzers mark the top of the line, each having 360° gradations and, in some cases, verniers for exact readings. Analyzers of this type are frequently used in conjunction with deSernamont compensators. Figure 18.52 shows both a fully rotatable polarizer (the silver ring just below the condenser) and a fully rotatable analyzer (the black bar with the wheel, inserted in the intermediate piece just below the binocular head). The black slider inserted in the 45° position is a compensator, and the double black wheel immediately below the binocular body is the Bertrand lens, a device similar to a centering telescope, used for observing the BFP$_o$. The top wheel has two settings: in and out; the lower wheel focuses the lens. This particular Bertrand system is, indeed, in the path to the camera, thus providing excellent access for documenting information in the BFP$_o$.

Because polarized light microscopy absorbs large amounts of light, it also may be worthwhile investing in a

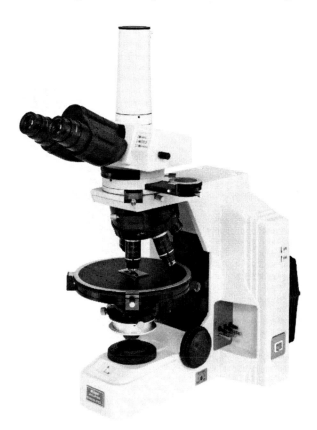

Figure 18.52 A polarizing microscope. (Image courtesy of Nikon.)

---

*Frequently, the eyepiece with the cross-hair in it will also have a nub on it so that it will lock in a specific orientation.

higher wattage lamp and lamp housing. One caveat: in most cases, the next step up from 100 w halogen is a mercury arc burner (HBO). However, the strong green component of the mercury spectrum can make accurate Pol color assessment difficult. There are two alternatives: use the mercury arc, but insert a light magenta color correction filter to bring the background back to true white, or use a xenon arc. New technologies available from companies like OptiQuip (Highland Mills, NY) make xenon an attractive solution.

Approaches to Polarized Light Observations

Organization is the key to successful polarized light microscopy. Because of the large number of tests available with this technique, it is important to have a well-developed plan. While there are many possible observations, the tests themselves flow naturally from one to another, generating a large body of information quickly. Experienced microscopists who have heavy sample loads will often tape-record their observations rather than take their hands away from the microscope to write in a notebook.

Orthoscopy versus Conoscopy

There are two types of polarized light analysis. *Orthoscopic* observations are conducted in the conventional manner, looking through the eyepieces at the sample with polarizers inserted and, for most tests, crossed. *Conoscopic* observations are conducted by viewing the back focal plane of the objective. The map shown in Figure 18.53 presents an organized way to conduct these observations.

Orthoscopic Observations

1. Establish Koehler illumination. Following the road map, start by observing the material under medium power. Note any obvious characteristics such as crystal shape and cleavage, inclusions, and color. Also note if the material is homogeneous or a mixture of different materials or phases.

2. Move the sample out of the path. Cross the polars and check to make sure that the background is rich black, indicating good-quality extinction for the background. If it is not, check rotation of the polarizer and analyzer. Also, check the markings on the objective and condenser (IK, IC, P, or Pol) to make sure that they are strain free.

3. Move the sample back into position and rotate between crossed polars. Note the positions of brightness and *extinction* (complete darkness). If the material stays dark on rotation, it is isotropic and will not respond to polarized light analyses. If it flashes bright and dark or flashes different colors, it is *anisotropic* and will be a good test object for further Pol

investigations. A common mistake for novices at this point is to think that they can rotate the polarizer or analyzer instead of the sample. Remember: these two filters set up very specific optical conditions and should remain in their specific positions unless you are otherwise directed.

Most polymer films, fibers, and crystals will go to extinction parallel to the permitted directions of the polarizer and analyzer. If they go to extinction at some other angle, align an edge with one of the crosshairs in the eyepiece. Read the position from the markings on the edge of the stage then rotate the sample into a position of extinction and take another reading from the stage. Subtracting the two readings provides the *angle of extinction*, a valuable optical property.

4. This is a good point to measure refractive indices. For isotropic materials, use the Becke Line test or other similar techniques discussed in the measurement section at the end of this chapter. For anisotropic materials, cross the polars and rotate the material into a position of extinction. As shown in Figure 18.50, this process isolates one of the two refractive indices. Remove the analyzer and measure the refractive index in this position then rotate the stage by exactly 90 degrees and repeat the process to measure the second refractive index. For materials with three refractive indices, it will be necessary to prepare another sample that captures an orientation that includes that third RI.

5. If you noticed a change in color on rotation with one polar, the material is either *dichroic* (characteristic of materials with two RIs) or *pleochroic* (characteristic of materials with three RIs). Unlike Pol colors that result from interference between crossed polars, these colors result from absorption of the plane polarized light produced by just the polar. Their palette is more like a paint palette and includes true reds, browns, etc. Again, dichroism and pleochroism are valuable optical properties.

6. Cross the polars again and rotate the material into a position of brightness. Use the Pol color and the Michel-Levy chart to estimate *retardation*. If the Pol color is white, insert the *First Order* red plate into the compensator slot and rotate the sample while viewing through the eyepieces. If the white is mid–*First Order* white (approximately 200 nm of retardation), on rotation the Pol colors will move up to *Second Order* blue-green (addition) and down to *First Order* tawny yellow. If it is *High Order* white (>*Seventh Order*), there will be no change in color.

   The round edges of fibers can present a special challenge. In general, you can follow the colored fringes from *Zero Order* black at the edge up to the maximum Pol color at the center of the fiber.

A Polarized Analysis Road Map

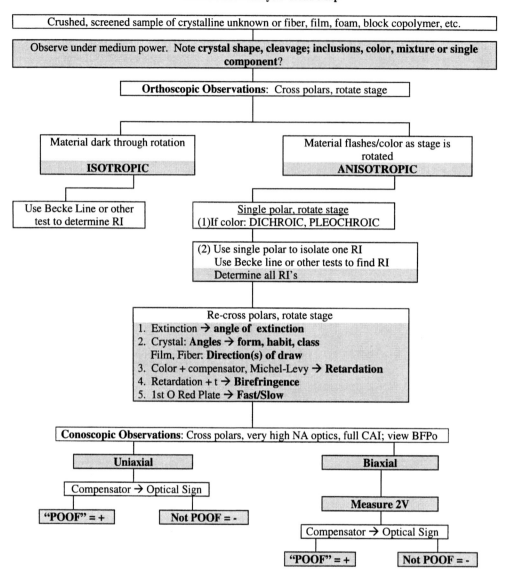

Figure 18.53 A polarized analysis road map.

Remember that thickness is also part of the equation for retardation. Look for narrowed areas (*necking*) that generate a localized change in retardation. For optical notation and calculations, use the maximum retardation values, both for the general bulk and for special cases such as necking. It is also to observe fibers in cross section, to determine if there is a *skin:core effect*.

7. If you know the thickness of the sample and have now identified the retardation, you can use the Michel-Levy chart to calculate the *birefringence*.

8. To measure the retardation more directly, use a compensator. Your initial observations of regarding the Order of the Pol color coupled with the information provided in Table 18.8 will indicate which

compensator is best to use. Each works on a slightly different premise, so read the instructions carefully. One commonality: most of the measurements are conducted in a position of maximum brightness, in a position of subtraction.

If your samples exhibit very high birefringence, the retardation induced by the compensator may conflict with the retardation from the sample, producing *dispersion of birefringence*. The most telltale sign is that *Zero Order* black (0 nm retardation) is colored. For fibers and films, cut one edge in a wedge to approximate 0 nm thickness then count up the orders to the full thickness of the sample.

9. To determine the *fast* and *slow* directions, rotate the sample between crossed polars into a position of

brightness and note its Pol color. Note the slow direction (γ or Z′) on the compensator then slowly insert the *First Order* red plate into the compensator slot and watch what happens to that Pol color. If it moves up the Michel-Levy (addition), the *slow* direction in the material is sitting parallel to the slow direction on the compensator. To confirm your observation, rotate the material by exactly 90° in either direction and observe what happens to the Pol color. You are now in a position of *subtraction*, the position most often used for compensation measurements. Both the downward movement and the upward movement should be an equal number of nanometers of retardation. This test will not work if your sample exhibits very low (<~40 nm) or very high (>~4 orders) retardation. For those ranges, combine this test with the more detailed retardation measurements described below.

In your laboratory notes, make a simple drawing of the polymer or crystal. Place vectors on the drawing to show the fast (short vector) and slow (long vector) directions.

10. Now that you have all the optical properties available through orthoscopic observation, you can draw another figure in the corner of your diagram to relate the optical properties to the physical properties. Called the *indicatrix*, this figure uses vectors to represent both the orientation and relative sizes of the refractive indices. Start by drawing vectors to represent the relative RIs in X, Y, and Z. A shorter vector indicates a faster direction. Note that these axes will always be at right angles to each other. Connect the tips of the vectors (all XY, then all YZ, then all XZ). As shown in Figure 18.53, the resulting 3D figure describes the internal electrical environment. Using your knowledge of the RIs of the material and the positions of the fast and slow rays, draw an indicatrix in one corner of the diagram of your material.

Figure 18.54 illustrates three specific indicatrices. In (a), all the refractive indices are equivalent ("α"). No matter how light approaches this indicatrix, it encounters the same electrical field. From another perspective, no matter which axis you sight down, all the refractive indices are equal and the electric field is equal. This indicatrix represents isotropic materials.

In (b), the refractive indices in Y and Z are equivalent (both α) but different from X (γ, a higher refractive index, as shown by both a later Greek letter and a longer vector). The view that includes the YZ plane is circular, indicating that this material is isotropic when viewed down the X axis. Since there is only one axis that produces this result, this type of material is called *uniaxial* and its representative figure is called the *uniaxial indicatrix*.

The last figure (c), represents a material with three refractive indices: α, β, and γ. As shown by both its placement in the Greek alphabet and the length of its vector, β represents a refractive index intermediate between α and γ. Viewing down any of the conventional axes produces only ellipses. However, there will be two directions, each at some unique angle to X, Y, and Z, which will produce circular cross-sections and, therefore, isotropic views. While these two axes cannot easily be imagined from the figure, they can be readily seen in the microscope using conoscopy (described next). This type of material is called *biaxial* and its representative figure is called the *biaxial indicatrix*. Mylar, a film that is drawn in two directions, is a good example.

Conoscopic Observations The word *conoscopic* is derived from the broad *cone* of light presented to the sample by a high NA condenser. While orthoscopic viewing provides a direct view of the sample, the conoscopic view, seen in the BFP$_o$, presents the optical transform of the sample in the form of an *interference figure*. The two basic forms of interference figures, shown in Figure 18.55, correspond to the uniaxial and biaxial indicatrices described in the preceding text.

Going back to basics makes interpreting these figures easy. The brushes in the unaxial figure represent isotropic orientations in which **e** and **o** lie parallel to the polarizer or analyzer (see Figure 18.45). Their intersection (center) coincides with the emergence of the optical axis: that one axis along which all RIs will be equivalent. Rotating the sample causes rotation around this axis. Since the electrical environment is equivalent in all directions, there will be no change on rotation for the uniaxial figure. One more important fact is that the e-rays in this material will always sit along the radii of this figure and o-rays will sit along the tangent. This fact will be important in determining optical sign.

As seen in the color version of this figure (see color insert) the concentric rings proceed in the same order as the Michel-Levy chart, starting with zero-order black in the center and marching up the chart as they move toward the periphery. This observation is consistent with light paths through high NA condensers. Light coming straight up the optical axis will have the shortest path to travel and therefore will exhibit a low Pol color. Conversely, light coming from the periphery will pass through the sample at a high

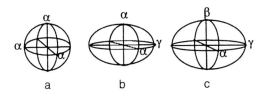

Figure 18.54 Optical properties can be summarized using the indicatrix: (a) isotropic indicatrix; (b) uniaxial indicatrix; (c) biaxial indicatrix. Key: α, β, γ = vectors representing the various refractive indices.

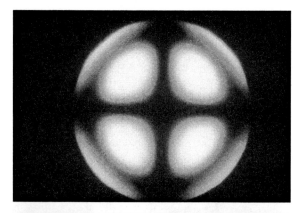

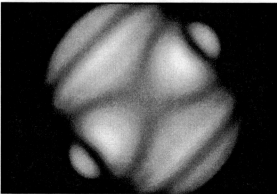

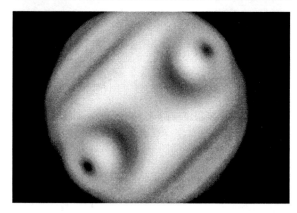

Figure 18.55 Interference figures: uniaxial (*top*); biaxial, isogyres open (*center*); biaxial, isogyres open, in circularly polarized light (*bottom*). See color insert. (Images courtesy of Phil Robinson, North Staffordshire Polytechnic, retired.)

angle, encountering an oblique cross-section of the sample, and, therefore, greater thickness, resulting in a higher Pol color.

A biaxial figure differs considerably. On rotation, its brushes (called *isogyres*) open (Figure 18.55, center) and close. As with the brushes in the uniaxial figure, the major portion of the brushes correspond to orientations in which one of the two primary RIs lies in a plane parallel to the polarizer or analyzer. At this point, the similarity ceases. Because there are three primary RIs, rotation of the sample

between crossed polars presents different relative of positions α, β, γ to the polars, thus causing the isogyres to "open" and "close". An optical axis emerges at the apex of each brush, a phenomenon easily observed in circularly polarized light (Figure 18.55, bottom). Also, instead of the single set of symmetrical concentric circles, there are two sets, each centered to one of the optic axes.

Orientation of the electrical environments with respect to the optical axis of the microscope is critical to good conoscopy. For uniaxial materials, the best view is down the material's optic axis. For biaxial materials, it is down the crystallographic Z direction. Fortunately for polymer microscopists, because of the way films, fibers, and block copolymers are manufactured, most materials are automatically oriented for the best view. If a biaxial material exhibits an off-centered figure, try sectioning to present a better figure: cut perpendicular to the Z direction.

Conoscopy requires access to the BFP$_o$, and there are a number of ways to gain that access. The simplest technique is to remove one of the eyepieces and peer down the tube. Another technique involves replacing the eyepiece with a pinhole, which can be made simply by using a hot needle to pierce the center of an eyepiece dust cap. A one-eighth-inch hole is ideal. Still another method is to use a centering telescope. If your microscope is fitted for either Phase contrast or Hoffman Modulation Contrast, chances are you received a centering telescope with that kit. Alternatively, they can be ordered from the manufacturer. Finally, a full Pol microscope may come equipped with an internal viewing system called a *Bertrand lens*. If you need to photograph and document your conoscopic observation, inquire about Bertrand lenses that sit in the pathway to both the binocular body and the camera. In simpler Pol stands, Bertrand lenses are built into the binocular body and just flip in and flip out for convenient conoscopic viewing. In more sophisticated models, the Bertrand may be focusable and may also come with auxiliary pinholes.

The following steps are used for conoscopic observations:

1. Begin as you would for orthoscopy. Start with Koehler illumination, then cross the polars with no sample in place and check for full extinction in the background. Place the sample on the stage and rotate into a position of brightness.
2. Open the condenser completely. If there is a flip-in upper element, put it in place. If you are working with an oil-immersion condenser for ultra-high NA, oil the condenser to the back of the slide with a drop of the appropriate immersion oil.
3. Use whichever technique is available for viewing the BFP of the objective.
4. To determine if a polymer is uniaxial or biaxial, rotate the sample on the stage. The uniaxial figure will stay in place, while the brushes of the biaxial figure will open and close.

5. To determine optical sign, insert a first-order red plate. My own personal preference is to test quadrant I in uniaxial materials, where the e-ray sits on the radius. Addition (change from 560 nm *First Order* red to *Second Order* blue) in this quadrant signals an optically positive material. For biaxial materials, rotate the sample until the isogyres open around 5 o'clock and 11 o'clock (NW and SE). As you insert the first-order red plate, observe the convex (top of the curve) of the lower isogyre. If addition occurs, the material is optically positive.

6. If the material is biaxial, you can also measure the angle formed by the two optic axes. Since the only angle collectable by the NA of an objective is the acute angle, you will be measuring 2V. Two methods are recommended. The simplest is to use an eyepiece reticle similar to the one shown in Figure 18.56 and estimate 2V from the curve of the isogyres. Alternatively, use an eyepiece micrometer to measure d, the distance from the apex of one isogyre to the apex of the other in a full open position, and D, the diameter of the field (Figure 18.57).[7] As seen in Figure 18.55, bottom, circular polarized light makes measuring d definitive.

As light emerges from the sample into air, refraction at that boundary will make 2V appear larger than it really is. For very critical measurements of 2V, use eqs 17a and 17b. The first calculates the true observed separation, 2E. Substituting sin E and β (the intermediate refractive index) into the second equation generates sin V. Use a trigonometry table to find V, then double to find 2V.

$$\sin E = d(\sin a)/D \qquad (17a)$$

where E is one-half the observed separation, in air; d is the distance measured from one apex to the other; a is one-half angle of the objective (for air, sin a equals NA); and D is the diameter of the back focal plane (measured with aperture iris full open).

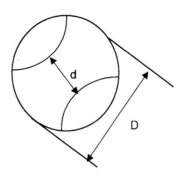

Figure 18.57 Using d and D to find 2V. Key: d = distance; D = diameter.

$$\sin V = (\sin E)/\beta \qquad (17b)$$

where V is one-half the actual separation between the optical axes, and β is the intermediate refractive index of the polymer.

Differential Interference Contrast: Taking Pol to the Next Step

Differential interference contrast, also known as *DIC* and, generically, as *Nomarski*, is a full NA technique that also generates delicate contrast, making it the most elegant of the contrast enhancement techniques. It is also an appropriate topic to summarize many of the basic principles discussed up to this point.

DIC Components

The complex stack of DIC components shown in Figure 18.58, left, can be simplified using the analogy of a sandwich. Polarizers and analyzers (crossed) are placed on either side of the sample and can be thought of as the two slices of bread. Beam-splitting prisms (often of the *Wollaston* type) are placed inside the polarizers. They are cut and mounted in crossed position and can be thought of as the "mayonnaise" and "butter". The sample is the "meat", and the compensator is the "lettuce". Note that DIC can be used in either transmitted or reflected light systems (Figure 18.58, right.

The only real caveat pertains to the sample. Because DIC is based on polarized light, many novices are fooled into thinking that, just because *they* have inserted the right components, the sample and microscope will produce a DIC image. The sample, however, may be responding only to the polarizers, producing beautiful Pol colors but none of the three-dimensionality induced by DIC.

How DIC Works

DIC detects gradients. It is derived from polarized light and is based on a two-step process, as shown in Figure 18.59.[8]

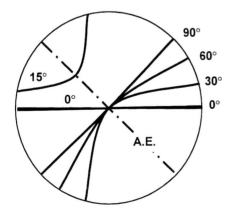

Figure 18.56 A reticle can be used to estimate 2V.

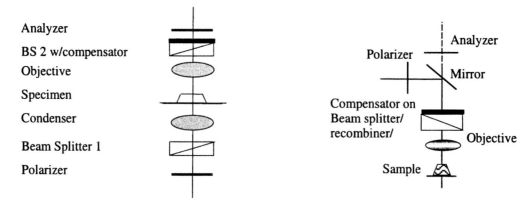

Figure 18.58 Differential interference contrast (DIC) components: transmitted light DIC (*left*) and reflected light DIC (*right*).

The initial waveform shown in the upper left diagram results from a groove or scratch in the surface of a transmitting material such as a polymer film. Because the light traveling through the scratch encounters the electric field of the film for less distance, that part of the wavefront advances further than does that portion that is traveling through the full depth of the film. Notice that the basic waveform carries the slope characteristic of the edges of a scratch.

The next diagram to the right shows the waveform sheared laterally. The source of the shear is easily determined by looking at the components below the sample. First, the incoming light is polarized. Next, it is split into e- and o-rays at the beam splitter. The lateral shear is the manifestation of these two rays. However, because the separation of the two images is smaller than the resolving power of the microscope, there is no ghost image.

Next, a compensator is inserted to create a specific retardation (the "lift" of one wave with respect to the other, as shown in the lower left diagram). The background and any parallel surface will exhibit the Pol color corresponding to the amount of retardation introduced by the compensator (Pol color 1).

Because of the slopes in the sample, the sheared/compensated image will have positive and negative gradients, each of which has a polarization color corresponding to its own retardation. As shown by the relative spacings in the diagram, gradient 2 has closer spacing, indicating a smaller amount of retardation than gradient 3. The difference in retardation will be reflected in a difference in Pol color in the image:

1. The background and any parallel surface will exhibit the Pol color that corresponds to the setting of the compensator.
2. Gradient 2 (the smaller retardation) will exhibit a color lower on the Michel-Levy chart.
3. Gradient 3 (the larger retardation) will exhibit a color higher on the Michel-Levy chart.

At this point, refer back to the Michel-Levy chart. The most sensitive tuning for DIC corresponds to a soft dove-gray background (approximately 146 nm retardation). The slope producing the smaller retardation is imaged as dark charcoal or black (~10–40 nm), while its partner is imaged as bright white (~200 nm). The bright-dark combination produces contrast by triggering a mental response that interprets this image as highly three-dimensional.

Aligning DIC for Best Results

Follow these steps to align DIC.

1. As with polarized light, start by establishing Koehler illumination and crossing the polarizers with no sample in place. If a turret condenser is used to provide a variety of contrast options, make sure that it is turned to the brightfield position. Also, confirm that the upper beam splitter is pulled out.
2. Place the sample on the stage and check for response to polarized light. If there is a response, it is very likely that DIC will not work. Tuning the compensator will create the expected addition or subtraction effects but no three-dimensionality.
3. Insert the beam splitters. Note that, because each objective has its own numerical aperture, there will be a specific beam splitter for that NA or range of NAs. Confirm (a) that the beam splitter is appropriate for the NA being used and (b) that the upper and lower beam splitter match.
4. If the compensator is a separate component, insert it. If not, adjust the upper beam splitter for optimum contrast (background is soft, dove gray).

DIC Tips

The most important caveat involved with using DIC with polymers is to beware of anisotropic materials. Figure 18.60 illustrates the problem. Imaged between crossed polars with a *First Order* red plate for contrast, Figure 18.60,

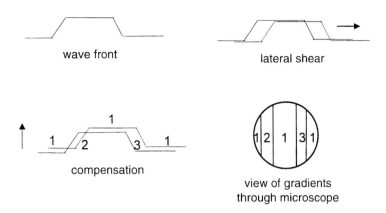

Figure 18.59 Mechanism for differential interference contrast (DIC): initial wave front (*upper left*); sheared wave front (*upper right*); sheared wave fronts displaced by compensator (*lower left*); resulting DIC image (*lower right*).

left, shows that most of the layers in this ketchup bottle cross-section contain either spherulites (upper right corner and tie layers on either side of pale gray central stripe) or anisotropic film (lower left corner) which respond to polarized light. The higher magnification DIC image (Figure 18.60, right) confirms that only the central light gray band is responding to DIC. Closer inspection reveals the gentle changes in topography (white arrow), which are delicately imaged by this technique.

DIC is highly tunable, again giving rise to the potential for misinterpretation of image detail and less-than-ideal three-dimensionality. Frequently, the compensator can be set to either positive or negative positions, as though the Michel-Levy chart had a mirror image. When tuning the compensator from one side of this chart to the other, the gradients will switch: what was bright in one setting will become dark in the other. Your eye-brain combination will become confused: what you perceive as a bump in one setting will appear as depression in the other. The key to deciphering what is real lies in your knowledge of the sample. Find some feature that you know is definitely up or down, and use its shading pattern as a guide for the rest of the features. Alternatively, first focus on the flat surface of the sample, then focus away. If the center of the feature comes into focus, it is a bump.

In addition, some DIC compensators work just in the range up to 200 nm white, others through second-order blue (about 700 nm). Gradients are best imaged when the background is set to approximately quarter-order gray (approximately 150 nm retardation). Tuning the background to *First Order* red or higher significantly reduces the sense of topography. However, it is fun and beautiful to create the color contrast that is available at higher settings, and, while it is not as scientifically informative, it can be good for your career. These color contrast images make wonderful covers for annual reports and photographs for the lobby.

Measurement

All but the most experienced light microscopists are surprised at the variety of measurements that can be obtained

 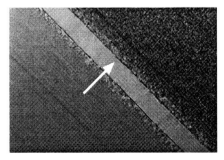

Figure 18.60 Cross-section of ketchup bottle. (Transmitted light. *Left*: crossed polars, 10× objective; *right*: DIC close-up, 40× objective.) The white box shows the area that appears in the closeup. The arrow points to the only layer responding to DIC.

from a reasonably well equipped light microscope. Many, such as length, depth, and refractive index, require very little in the way of accessories. Since most of today's microscopes are already fitted with cameras and computer connections, moving up to particle sizing, counting, morphometry, and stereology typically involves adding a software package or two. A fully equipped microscope may also have instruments capable of physical and thermal measurements, as well as microinterferometry and spectroscopy. This broad range of measuring capabilities has often prompted novice microscopists to observe "IDKYCDT!" (I didn't know you could do that!).

Linear Measurement

Linear measurement in the XY plane is one of the easiest. A small disk of glass marked with a measuring scale, called a *reticle*, can be inserted into nearly any eyepiece. Since the image of the scale must be seen at the same time as the specimen, the reticle must sit in a focal plane conjugate to the image of the specimen—that is, in the primary imaging plane. For Ramsden eyepieces, this plane is coincident with the baffle below the lower lens and is very easily reached. For Huygenian eyepieces, it sits in the middle of the eyepiece between the eye lens and field lens, and is less accessible. In either case, the necessary part can be ordered from your local microscope dealer, who is also experienced in its installation.

Since the eyepiece sees fields presented by objectives of differing magnification, any measuring scale mounted in the eyepiece must be calibrated for each magnification. Note that if the microscope has a magnification changer as well, you must calibrate for each combination of objective and magnifier. The task is easy and requires only two components: the reticle mounted in the eyepiece and a stage micrometer for reference. The most typical stage micrometers are chrome on glass, have 1 mm marked off in 100 divisions of 10 μm each, and are available from nearly any supply house.

Place the stage micrometer on the stage and align it so that its markings run parallel to those on the eyepiece reticle (Figure 18.61). Next, pick two markers on the eyepiece micrometer which match with two on the stage micrometer and set up a ratio. In the example shown here, 360 μm from the stage micrometer match with nine divisions on the eyepiece reticle:

$$360 \text{ μm}/9 \text{ eyepiece units} = 40 \text{ μm}/1 \text{ eyepiece unit}$$

$$\text{(calibration factor)} \qquad (18)$$

This ratio establishes a *calibration factor* between the known units on the stage micrometer and the unknown units on the reticle. Once you have collected the calibration factors for all objective:magnification changer combinations, write the values on a piece of paper and post them on a wall near the microscope for easy reference.

To make a real-world measurement, simply place a regu-

lar specimen on the stage. Measure in terms of eyepiece units then multiply by the calibration factor to find length in micrometers.

Depth

Depth measurements can be performed on any microscope that is fitted with a graduated fine focus mechanism. For reference and to avoid backlash, measure from the bottom to the top. (Reminder: as you focus *away*, you are moving from lower in the sample toward the top of the sample.) Record the initial reading on the fine focus, then move to the top and record the second reading. The apparent depth is the difference between these two:

$$\Delta Z' = |Z_1 - Z_2| \qquad \text{(apparent depth)} \qquad (19)$$

where $\Delta Z'$ is apparent depth, and Z_1, Z_2 are the initial and second readings.

Since Snell's law is very much in effect, you must compensate for differences in refractive index:

$$\Delta Z = \Delta Z' \ (n/n') \qquad \text{(actual depth)} \qquad (20)$$

where ΔZ is true depth, $\Delta Z'$ is apparent depth, n is the RI of the embedding or mounting medium, and n' is the RI of the immersion medium between the sample and the objective.

EXERCISE
Use a marker to put dots in the same location but on the top and bottom of a coverslip, then measure its thickness by focusing first on the bottom dot, then on the top dot, and recording the fine focus reading for each. Two reminders: (a) most no. 1½ coverslips range between 0.13 and 0.17mm in thickness and (b) the refractive index of glass is typically about 1.5212.

Refractive Index

There are a number of ways to measure RI. If the polymer with which you are working is fairly stable in a variety of solvents, you might try the basic Becke line test. This test requires a set of test liquids called Cargille oils* and a small collection of the fibers or particles to be tested.

The simplest version of this test involves closing the aperture iris. Place a few crystals or one fiber of the polymer on a slide, and top it with a drop of one of the test liquids then cover with a coverslip. (Tip: place one edge of the coverslip on the slide first, then drop the coverslip to avoid bubbles. A gentle tap with the clean eraser end of a pencil will help the liquid settle.) Focus on a well-defined edge and note the location of the bright halo just at the edge (the Becke line). Examples of the Becke line are the inner halo on the crystal in Figure 18.31, right, and outer halos on Figure 18.32, right. Focus away and note the direction in which the halo

*Cargille Laboratories, Cedar Grove, NJ. Their phone number, as of the printing of this chapter: 973-239-6633.

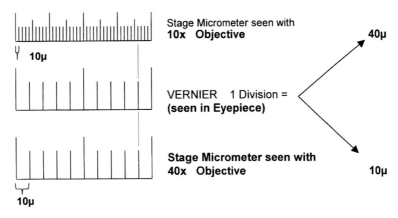

Figure 18.61 Aligning the stage micrometer with the eyepiece micrometer. (Diagram courtesy of Dr. Kenneth Piel.)

moves. When focusing away, the halo will always move toward the material of higher RI. Also, the more pronounced the edge and the halo, the greater the difference in RI between the polymer and the test oil of closer RI. Using these two pieces of information, you can choose a second oil and repeat the process. When the RI of the polymer matches that of the test oil exactly, the polymer will disappear. Since you know the RI of the oil (listed on the bottle), you now know the RI of the polymer.*

A few points to refine this test. First, RI is highly sensitive to wavelength and temperature. For critical work, use a narrow band interference filter. The RI for the sodium "D" line (589 nm) is noted on the bottle as n_D. Second, if possible, use a heating stage to keep the temperature at 25 °C. For the same reason, the sensitivity of RI to wavelength and temperature, altering either parameter provides alternative methods for measuring RI. For example, if the thermal response of a test oil is known (again, available from Cargille) and a hot stage is available that can be carefully controlled, place the polymer in a drop of the test oil, cover with a

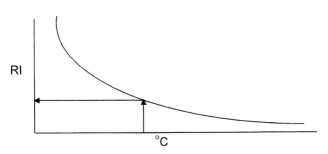

Figure 18.62 Thermal method for determining refractive index (RI).

coverslip, and watch for the temperature at which it disappears (Figure 18.62). That temperature will correspond to a specific RI. Similarly, a small monochromator can be mounted on most light ports. The procedure is similar to the thermal method, with the exception that wavelength is varied instead of temperature.

These techniques can be applied to anisotropic materials as well, using the procedures outlined in the Pol section to isolate each of the RIs before they are measured. This is another case in which a rotating stage with a goniometer around its circumference is helpful in finding the exact 90° positions. If there is a third RI, try rolling the sample to expose the third direction and repeat the test.

Image Analysis: Morphology, Sizing, Counting, and Structure-Function Relationships

A number of software packages and integrated systems are now on the market that extend normal viewing, archiving, and documentation into the realm of measurement. In the simplest case, these programs acquire an image, define the parameters necessary to differentiate the features of interest from the rest of the field (*segmentation*), then size and count the particles. To link the microscope to the computer hardware requirements include a camera and an analog-to-digital converter,† either in the camera or added to the computer. Most of these systems also have some provision for exporting data to Microsoft Excel for statistical analysis and graphing, as well as the ability to dialogue with Microsoft Word or Corel's WordPerfect for report writing. The systems open microscopy to an extensive quantitative world. If the image is well behaved, researchers can determine a

---

*For a more thorough explanation of how each of these tests works, as well as other tests, see *Practical Refractometry by Means of the Microscope* by R. M. Allen, available through Cargille Laboratories.

†If you are planning to proceed in this direction, a system integrator can provide the specifications for a number of alternatives for camera, digitizing system, software, and peripherals such as printers and storage media.

large number of parameters such as commonly measured area, perimeter, angle, and length, as well as more sophisticated measurements, including circularity (for example, in polymer beads), isotropy of alignment (for example, in graphite fibers), and percentage of the area (for example, in voids or inclusions).

A number of other options on the market extend these simple operations into full image processing and analysis. They provide mathematical and Fourier filters to solve a wide range of imaging problems, including poor edge definition, touching particles, and incomplete boundaries. Segmentation is one of the greatest challenges facing automated image analysis, and using these filters for improving the image is one of the greatest challenges in learning these upper end systems.

Finally, some additional software abstracts three-dimensional information from two-dimensional images. Called *stereology*, this analytical approach assumes that any variation in size or shape is derived from the way the physical sectioning cut the sample, not from heterogeneity in the particles or inclusions in the field of view. Strongly grounded in statistics, stereology is a powerful tool for relating two-dimensional parameters such as mean free distance across the window of a polymer foam to industrial process parameters such as humidity or time at temperature and, ultimately, to function.

Thermal and Physical Testing

There is a surprisingly large gamut of thermal and physical tests that can be performed on a light microscope to allow correlation between microscopy and more conventional chemical analyses. For polymers, one of the most logical is thermal analysis. Several companies sell hot stages with varying levels of control sophistication. The simplest stage can be homemade from a microscope slide using heat-conducting glass. For temperature variation, use alligator clips and wires to attach the slide to a rheostat. Turning up the rheostat increases current to the stage and results in increased temperature.

The most complex thermal stages operate under fully programmable control, permitting replication of glass transition, annealing, and other industrial processes. In addition to simple hot stages, Mettler (Oak Park, IL) now offers stages for conducting differential scanning calorimetry (DSC) and differential thermal analysis (DTA). Watching the sample while it goes through these processes may provide surprising information about structural changes that are not visible on the macro level. Again, since most microscopes are fitted with video systems, changes can be documented dynamically. One caveat: in order to closely control temperature in these advanced stages, most are fitted with closed chambers or large glass cover plates. As a result, they are either limited to the longer working distance, lower magnification objectives or require special long work-ing distance objectives for higher magnification. By default, most longer working distance objectives will have lower numerical apertures and, concomitantly, lower resolution.

Physical properties can also be tested under the microscope. Scientists investigating hard polymers can borrow microhardness testers from their metallographic colleagues. For tensile strength testing, Rheometrics (Piscataway, NJ) makes a special stage called a "Mini-Mat" that comes equipped with a variety of clamps for either fibers or films and which can be used either on a conventional upright microscope or under a stereo microscope. Observing the sample between crossed polarizers during tensile strength provides valuable information on stress and strain.

Microinterferometry

As the name implies, interferometry is a technique based on the interference between a sample beam and a reference beam, which creates a pattern of bright and dark fringes as a result of constructive and destructive interference, respectively. As the fringes pass over an edge or a change in RI, they are deflected. The optical path difference engendered by either of these factors can be determined by measuring the fringe spacing and fringe deflection (in μm or micrometers) using a calibrated eyepiece micrometer, then substituting into the following equation:

$$OPD = \lambda \, (\text{fringe deflection/fringe spacing})$$

(interferometry for OPD)   (21)

where λ is the wavelength of light in μm. The resulting OPD, along with other known information, is substituted into the standard equation, generating information on either thickness or refractive index:

$$OPD = t(n_1 - n_2)$$

(interferometry for thickness or RI)   (22)

where t is the thickness of material; n_1 is the RI of the polymer; and n_2 is the refractive index of the mountant.

If both RI and thickness are unknown, the double immersion technique can be used. First, measure the OPD of the sample in air ($n = 1.00$), then introduce a benign liquid of known RI (for example, water, $n = 1.33$) and remeasure OPD. This approach results in two, simultaneous equations:

$$\text{a. } OPD_{air} = t \, |(n_{sample} - n_{air})|$$
$$\text{b. } OPD_{liq} = t \, |(n_{sample} - n_{liq})|$$

Dividing (a) by (b) produces one equation with only one unknown, the refractive index of the sample:

$$n_{sample} = (OPD_{air} \times n_{liq}) - (OPD_{liq} \times n_{air})/(OPD_{air} - OPD_{liq})$$

(double immersion method for RI and thickness) (23)

Once the refractive index of the sample is calculated, substituting that value back into either (a) or (b) above will produce thickness.

For polymer films, interferometry can also be used to find simple step heights that correlate to roughness or machined texture. This measurement is conducted in reflected light. To make the surface of the film reflective, sputter coat a thin metallic film on the surface. Step heights are measured directly from OPD:

$$s = \tfrac{1}{2}\, \text{OPD} = \tfrac{1}{2}\, (\text{deflection/spacing}) \times \lambda$$

(interferometry for step height or roughness)   (24)

Several older microscopes such as the Reichert Polyvar had integrated interferometers. Today, interferometers are typically add-on accessories. Hache Instruments, for instance, provides an affordable Michelson interferometer that attaches to the $10\times$ objective. Nikon also provides a range of add ons, from single-beam Michelsons to the more sensitive multiple-beam Tolansky.

Microspectroscopy

Inclusions, coatings, fibers, and foams frequently respond to high-energy excitation and produce either Raman (vibrational) or fluorescence spectra that can be measured either as a point intensity at a specific wavelength or as a full spectrum.

UV-visible spectroscopy can also be used to differentiate between colored materials. Simple, low-cost spectrometers, such as those available through Optical Technologies, Inc. (Elmsford, NY) can be attached in lieu of the video system and can be used in quality analysis and control settings for quick analog readouts of parameters such as fiber brightness in fabrics. These spectrometers also have RS232 outputs for computer control and collection of simple UV-visible spectra.

The next step up are spectrometers integrated with software, such as those available from LightForm (Belle Mead, NJ). This system automatically reads the spectra based on differences in color or fluorescence from all components in the field (fibers, inclusions, etc.), then assigns unique pseudo-color tags to the features with the same spectral fingerprint. Finally, there are completely integrated systems such as Leica's TAS system, which can run studies in fluorescence, reflectance, or absorbance from the near UV (220 nm) well into the infrared (2200 nm) light.

Recently, an emerging trend has coupled light microscopy with either Fourier transform infrared (FTIR) or Raman spectroscopy. While established analytical firms such as Hewlett Packard and Perkin-Elmer have had spectrometers with a microscope viewing system, companies such as Thermo Nicolet, Mattson, ChemIcon, and SensIR Technologies have announced fully integrated microscope and spectrometer systems. Renishaw and Jovan Yvon have taken this technology a step further by including raman spectroscopy with confocal microscopy. All these systems run in brightfield and fluorescence, with some, such as Thermo's Continuum adding other contrast techniques such as phase contrast, polarized light, and DIC. These instruments are the harbingers of the future, combining the visualization of light microscopy with the chemical fingerprinting of spectroscopy, an unparalleled duo for polymer scientists.[9-11]

The Particle Atlas

Now that you have a large arsenal of measurement techniques at your disposal, the next step would be a reference that brings together various information into a searchable form. *The Particle Atlas* from McCrone Associates (Westmont, IL) combines light microscopy measurements such as color, fluorescence, and polarized light data with electron microscopy and microanalytical information in an easy to use, searchable CD-ROM.

References

1. Hemsley, D. *Applied Polymer Light Microscopy*; Elsevier: Amsterdam, 1987.
2. Sawyer, L.; Grubb, David T. *Polymer Microscopy*, 2nd ed.; Chapman and Hall: London, 1996.
3. Scarpetti, J. J., DuBois, P. M., Friedhoff, R. M., Walworth, V. K. Full-color 3D prints and transparencies. In *Proceedings of the IS&T 50th Annual Conference: A Celebration of All Imaging, Imaging Science & Technology*; Cambridge, MA, May, 1997; pp 639–642.
4. Delly, John Gustave. *Photography through the Microscope*. Kodak: Rochester, NY, 1988.
5. Donaldson. L. Forest Research, Rotorua, New Zealand; Private conversation, May 3, 2000.
6. Zaruba, K. 3M Company, BioMaterials Technology Center, St. Paul, MN; Private conversation, May 3, 2000.
7. McCrone, Walter C.; McCrone, Lucy B.; Delly, John Gustav. *Polarized Light Microscopy*; Ann Arbor Science Publishers: Ann Arbor, MI, 1979; pp 155–157.
8. Foster, Barbara. Notes on the use of differential interference contrast in light microscopy, *Am. Lab.* **1988**, April, 96–100.
9. Koenig, J. *Spectroscopy of Polymers*; American Chemical Society: Washington, DC, 1992.
10. Messerschmidt, R. G.; Harthcock, M. A. *Infrared Microspectroscopy: Theory and Applications*; Marcel Dekker: New York, 1988.
11. Morris, M. D., Ed. *Microscopic and Spectroscopic Imaging of the Chemical State*; Marcel Dekker: New York, 1993.

19

Atomic Force Microscopy

DONALD A. CHERNOFF
SERGEI MAGONOV

Atomic force microscopy (AFM) has developed into the leading scanning probe technique with broad applications in science and technology. Unique capabilities of AFM for material characterization at the nanometer and micrometer scale have already been recognized in the semiconductor and data storage industries. Recent AFM applications to polymers reveal that this method substantially complements other microscopic and diffraction techniques in characterization of polymer morphology, microstructure and nanostructure, and, in a number of applications, it is the only technique available.

This chapter explores the capabilities of AFM for studies of polymers. The first part includes basic ideas of scanning probe techniques and descriptions of different AFM modes. It also presents practical aspects of surface imaging, such as optimization of imaging parameters, calibration issues, tip characterization, sample preparation, and comprehensive experimental protocol. The second part contains a spectrum of AFM applications to polymer materials. It begins with AFM measurements of surface roughness. Many of these measurements are routine and simple, yet others may be rather delicate. Other applications described are molecular-scale imaging of chain macromolecules, high-resolution studies of morphology and nanostructure of various polymers, local studies of adhesive and mechanical properties, compositional imaging of heterogeneous polymer materials, probing of subsurface structures, and measurements at different temperatures. The chapter closes with a discussion of future trends in AFM of polymers.

Direct visualization of atoms and molecules became possible with the invention of scanning tunneling microscopy (STM) in 1982 and atomic force microscopy (AFM) in 1986.[1,2] During the 1990s, STM and AFM were successfully applied for studies of various surfaces and materials at the submicrometer level. In these techniques, as well as in other related scanning probe microscopy (SPM) methods, a sharp probe is positioned in the immediate vicinity of a sample, so

that interactions between the closest atoms of the probe and the sample can be detected. In STM, tunneling current between a sharp metallic probe and a conducting sample is measured. In AFM, attractive and repulsive tip-sample forces are recorded. Spatial variations of these interactions, which are detected when the probe scans above the surface (or in contact with it), are presented in STM and AFM images. In a first approximation, these images reflect sample electronic properties and surface topography on scales from hundreds of micrometers down to angstroms. Although atomic imaging sounds exotic and made STM and AFM well known, imaging on this wide range of length scales provides broad applications to materials and devices of practical interest.

STM was invented first, and atomic-resolution images have been obtained on a variety of samples (metals, semiconductors, inorganic layered materials, organic conductors, etc.).[3] This technique, however, has a serious limitation because it can be used only on conducting or semiconducting surfaces, whereas most real-world samples have insulating surfaces. Therefore, AFM, in which universal tip-sample force interactions are employed for surface imaging, became the leading scanning probe technique. AFM can be used to examine practically any material in air, in liquid, and in vacuum. In AFM images, three-dimensional surface profiles are measured with subangstrom accuracy that is rarely achieved by other microscopic methods. These topographical data are important in many applications, and they are also invaluable in other scanning probe techniques (magnetic force microscopy, electric field microscopy, scanning capacitance microscopy, scanning thermal microscopy, etc.). This makes AFM a primary SPM technique that can be easily extended to measurements of magnetic, electric, mechanical, and thermal properties on both micrometer and submicrometer scales. At present, AFM is recognized as an invaluable complementary tool to more traditional microscopic and diffraction methods, and it is used in an increasing number of research laboratories, not only in academia but also in industry. The leading role of AFM is also related to its broad industrial applications, which have been developed in recent years. Precise evaluation of surface roughness of flat silicon wafers and polysilicon surfaces, nanometer-scale measurements of step height and trench depth, and profiling photoresist structures are the main AFM applications in the semiconductor industry. Miniaturization of semiconductor components, which, step by step, approaches the 50-nm scale, limits the use of optical and electron microscopy; therefore, the importance of AFM will increase further. The data storage industry benefits from AFM studies of magnetic recording and high-resolution measurements of compact disc (CD) and digital versatile disc (DVD) patterns and of pole tip recession.

Nowadays AFM applications are widening in the polymer, rubber, paint, biomaterials, and paper industries. The main AFM applications to polymer materials can be classified into several research areas.

The first area involves the basic AFM function: high-resolution profiling of surface features. On the micrometer- and nanometer-length scales, visualization of surface topography and evaluation of surface roughness of different polymer materials help to quantify this important surface property and correlate it with other surface characteristics such as appearance, optical properties, and film formation. On the nanometer-length scale, studies extend from visualization of single macromolecules, which are adsorbed on different substrates, to imaging of nanometer-scale structures and polymer morphology.[4,5] These studies are important because they supply additional information to optical and electron microscopy studies. Histograms of contour lengths of polymer chains, which are seen in AFM images, directly indicate the molecular weight and its distribution. Therefore, AFM complements light scattering and chromatography techniques in studies of molecular weight distribution. On the subnanometer-length scale, the contact AFM mode can observe periodic lattices on some polymer samples, revealing chain packing and conformational order. These structural features are usually analyzed with X-ray diffraction techniques. However, such AFM studies of surface lattices lost some of their initial attractiveness when it was realized that AFM usually does not provide knowledge about atomic-scale structural defects. Nevertheless, AFM detection of periodic surface structures on the nanometer scale is invaluable for better understanding the results of small-angle X-ray scattering.

The second area of AFM applications to polymers includes local studies of mechanical properties, including stiffness and adhesion. Such studies can be performed in a number of ways using force curves, force volume (FV) imaging, nanoindentation technique, and lateral (frictional) measurements.[6] Most of the results obtained in this area lack a quantitative description. To approach this goal, more work is required toward a better understanding of the dynamics

of AFM probes, along with a better theoretical description of the mechanical interactions between a sharp tip and viscoelastic media.

Compositional imaging is the third area of AFM applications to polymers.[5] Although the local mechanical properties that are probed in AFM are difficult to describe quantitatively, compositional imaging successfully uses the relative differences in properties of individual components in heterogeneous polymer systems. Research and analytical laboratories in academia and industry routinely use AFM to map domains of semicrystalline polymers, block copolymers, polymer blends, and composite materials. Because chemical staining or etching is not needed for such measurements, AFM has a substantial advantage over transmission electron microscopy (TEM). Another attractive feature of compositional AFM imaging is the ability to visualize the component distribution not only in the surface plane but also at different depths near the surface. This is realized for polymer samples with a rubberlike surface material ($T_g < T$).

So far, most AFM studies of polymers have been conducted at room temperature. However, polymers exhibit thermal phase transitions, which substantially define their structure and properties. A number of thermal techniques can be applied to characterize the thermal behavior of polymers, and AFM studies performed at different temperatures will provide additional information about these transitions. The first AFM studies of polymers at elevated temperatures have been reported recently and this area will be developing further.[7,8] The same is true of studies of polymers using thermal scanning microscopy and electric field microscopy. These applications are under development, and they might be useful for mapping and identification of polymer components and locations with different thermal and dielectric properties.

AFM is an invaluable partner of optical and electron microscopy because it combines high-resolution visualization of surface structures, relatively simple or no sample preparation, and the capability to perform measurements in different media. Examination of many polymer samples with AFM does not require the elaborate sample preparation that is common for electron microscopy. Different aspects of sample preparation in AFM of polymers are summarized in the section "Practical Aspects of AFM Imaging". AFM measurements of polymer materials are usually performed in air. This is a favorable condition for examination of many polymers, which might be modified by placing them in vacuum conditions. Imaging under liquid can be performed using specially designed liquid cells. These measurements are necessary for studies of biological objects (DNA, proteins, cells, etc.) and of polymer materials used for biomedical purposes. AFM also can be applied in vacuum, and there are a number of commercial instruments for such experiments.

Despite the relative simplicity of AFM instrumentation and experiments, definite knowledge and practical skill are required to obtain comprehensive information from AFM studies. As a result, there is a need for updated coverage of AFM of polymer materials, and we hope that this chapter will be useful for this purpose.

How AFM Works

Basic Principle of SPM

The precise scanning of a sample surface with a miniature probe that senses the presence of the surface is the essence of SPM. The main components of an SPM are a piezo-ceramic scanner, a probe, a detector, and associated electronics (Figure 19.1). Scanners allow measurements on horizontal scales from hundreds of micrometers down to several nanometers and on vertical scales from 0.01 nm to 10 μm. Most of the microscope functions are controlled with a digital signal processor and computer. Microscope operations are performed with the software and the final information is presented as an image typically consisting of 512×512 data points. Vibration isolation and sometimes acoustic protection (not shown here) are required to minimize mechanical and acoustic noise during SPM measurements, especially at small scales.

In STM, a tip-sample current of several nanoamperes (nA) can be detected when a sharpened metallic probe (usually a cut or etched wire) is placed in the immediate vicinity of a conductive or semiconducting sample, and when an appropriate voltage (typically millivolts [mV] or volts [V]) is applied between these two electrodes. A current amplifier is the actual detector in STM. The necessity for sample conductivity limits STM applications. In AFM, more universal tip-sample force interactions are used for imaging. Tunneling current and repulsive force strongly depend on the tip-sample separation. This explains why the foremost atoms of the tip are mainly involved in electron transfer and tip-sample force interactions, and why extremely high resolution in vertical and lateral dimensions can be achieved in SPM. In scanning near-field optical microscopy (SNOM), optical signals are detected.[9] Most work in SNOM has involved the development of the instrumentation itself, with few applications to polymers, so discussing it is outside the scope of this chapter.

One AFM setup is shown in Figure 19.2. In this configuration, a microscopic probe, which consists of a cantilever and a tip, is attached to a cylindrical actuator (scanner); during the experiment the probe scans over the sample. In an alternative setup, the sample is fixed to the scanner and moves while the probe is immobile. Scanning electron microscope (SEM) micrographs of most commonly used AFM probes and their main characteristics are shown in Figure 19.3. An AFM probe is a complete assembly, usually consisting of a rectangular substrate (typically 3.7 mm long × 1.8 mm wide, with thickness in the range 0.3–0.6 mm), which supports a rectangular or triangular-shaped cantile-

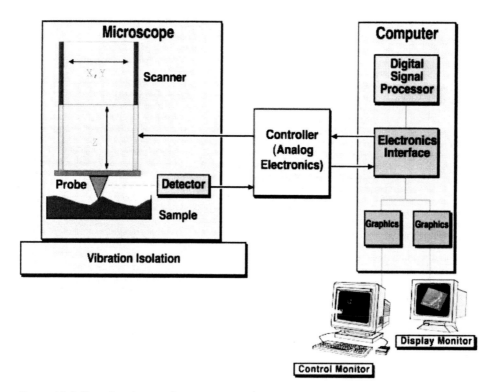

Figure 19.1 Generic scheme of a scanning probe microscope. (Courtesy of Digital Instruments/Veeco Metrology Group.)

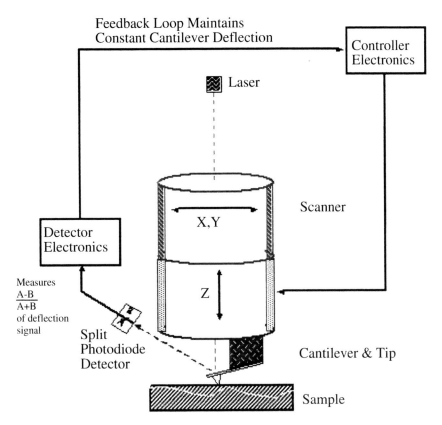

Figure 19.2 Generic scheme of an atomic force microscope. (Courtesy of Digital Instruments/Veeco Metrology Group.)

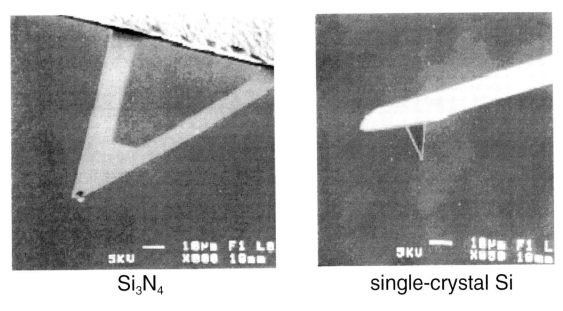

Figure 19.3 Scanning electron micrographs of atomic force microscope probes and their main characteristics. (Courtesy of Digital Instruments/Veeco Metrology Group.)

ver with a sharp, vertical pyramidal tip at one end. The substrate is fixed into a probe holder. The probe assembly (substrate, cantilever, and tip) is made by a batch microfabrication process. Monolithic silicon probes are etched from a Si wafer; in silicon nitride probes, in contrast, the cantilever and tip consist of a thin Si_3N_4 film on a glass substrate. Typical dimensions of the rectangular Si cantilevers are width 30 to 60 μm, length 100 to 400 μm, thickness 2 to 8 μm, tip height 8 to 10 μm, tip opening angle ca. 30°, and tip apex diameter 10 to 20 nm. Spring constants of the commercial probes vary in the 0.01- to 600-N/m range.

The bending of the microscopic cantilever is used to measure the forces acting between the tip and the surface. Practically, this is done using a laser beam, which is reflected from the backside of the cantilever toward a position-sensing photodetector. Because the distance from cantilever to detector is relatively long compared with the cantilever length, the optical lever magnifies a small angular deflection of the cantilever into a much larger lateral displacement. Given the intrinsic sensitivity of the photodiode to submicrometer displacements, it is easy to detect subangstrom deflection of the tip. This also accounts for the high force sensitivity (piconewton) of AFM. The high force sensitivity and the high precision motion (subangstrom) that is achieved with piezo-scanners are the key features that give AFM unique capabilities for surface imaging.

Surface imaging is initiated after the probe approaching the sample achieves an operator-defined level of the tip-sample interaction (e.g., setpoint current in STM and setpoint repulsive force in AFM). Then, the probe is moved laterally across the sample, and at every point the setpoint interaction is kept constant by adjusting the vertical position of the sample (or probe) by the scanner. The computer records the vertical displacement of the scanner at each point and displays a height image on a monitor screen with high surface regions shown brighter than low surface regions. In addition to the height image, an error signal image is often recorded. The error signal is the difference between the instantaneous value of probing interaction and the setpoint value before the vertical adjustment is made. This image resembles a map of the first derivative of the height profile and emphasizes fine details of surface structures.

AFM Operational Modes with Examples

In the development of AFM, several operational modes have been introduced. Of these imaging modes, the most commonly used are tapping mode and contact mode. In this description, we start with analysis of force curves, which are single-point measurements of the tip-sample force interactions applied for both modes. A whole spectrum of imaging modes for contact and tapping regimes are described and illustrated with practical examples. Except as noted, we discuss AFM work done in air.

Force Curves

Deflection versus Z-Travel Curves. Local tip-sample force interactions can be examined by recording deflection versus Z-travel (DvZ) curves (also known as force curves) when the probe is positioned over a single surface location. DvZ curves describe how the cantilever deflection depends on the vertical separation between the probe substrate (base) and the sample (Figure 19.4). DvZ measurements are performed by applying a periodic voltage to the piezo-scanner to activate an oscillating vertical motion with frequency in the 1 to 100 Hz range. In response to tip-sample forces, the cantilever bends (either up or down), and its deflection is recorded with a position-sensitive photodetector. In the analysis of force curves, one should remember that the cantilever deflection is a measure of the net tip-sample force, which results from contributions of attractive and repulsive force interactions.

For quantitative estimates of the tip-sample forces, one should know the cantilever stiffness (force constant). The force constant of a particular cantilever is determined by its composition (silicon or silicon nitride) and geometry. Width, length, and thickness can be measured with either an optical or scanning electron microscope. These geometric measurements can be supplemented by a measurement of the cantilever's resonance frequency. As a starting point, AFM vendors supply force constants computed either generically for a given model cantilever or estimated for a batch (wafer) based on measurements of a few probes in the batch. However, the actual spring constants might vary substantially, as much as a factor of 2, from one probe to another. For high accuracy (10%–20%), it is necessary to measure the actual cantilever used.[10]

As the base approaches the sample (piezo extension, points 1–2), the cantilever first bends toward the sample because of long-range attractive forces (point 3). On further approach (points 3–4), the tip contacts the sample, and the cantilever bends upward because of repulsion. When the motion is reversed (points 4–5), the cantilever deflection decreases and then changes to a downward bend because of the attractive forces of tip-sample adhesion and capillary

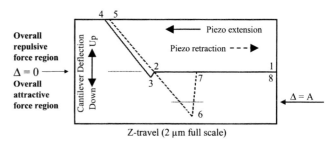

Figure 19.4 A force curve recorded in contact mode in air. $\Delta = 0$ is the zero deflection level. Imaging can be stable for setpoints above level $\Delta = A$. (Courtesy of Digital Instruments/Veeco Metrology Group.)

interactions. (Capillary forces are present because real-world surfaces in air usually are covered with a 1- to 3-nm-thick layer of water and other liquid contaminants.) As the base retracts further from the sample, the cantilever deflects further downward (point 6). The tip leaves the sample when the spring force of the cantilever overcomes the attractive forces; the cantilever deflection is essentially 0 beyond this point (point 7). Finally, the tip returns to its original position above the surface (point 8). Adhesion results in hysteresis: the triangular part of the force curve recorded during retraction has no counterpart during extension. The pull-off force, which is needed to detach the probe from the sample, can be used to estimate the local tip-sample adhesion.

Force curves can be used for various purposes. As a diagnostic of instrument performance, a force curve can indicate whether the setpoint used for imaging corresponds to a net repulsive or net attractive force; no calibration is needed to make this judgment. We note that when the setpoint is net repulsive, then imaging is stable. When the setpoint is net attractive, the imaging tends to be stable only if the negative setpoint deflection is not close to the pull-off level (point 6 in Figure 19.4).[11] This limitation of contact mode imaging in air may prevent nondestructive imaging of soft materials because the force cannot be decreased enough. In practice, imaging forces of a few nanonewtons can be achieved in such conditions with soft cantilevers (~0.01 N/m). When a sample is placed in liquid (using an AFM liquid cell), then the capillary effects that cause hysteresis of the force curves vanish, and a setpoint deflection can be chosen very close to the disengagement level. This allows lowering imaging forces 1 to 2 orders of magnitude below those in air. Additional information about AFM imaging under liquid can be found in reference 12.

Force curves can be used to investigate sample mechanical properties such as stiffness and adhesion, as well as effects of long-range electrostatic tip-sample forces. A variety of force curves, which are relevant to different tip-sample interactions, are shown in Figure 19.5a–c. Tip-sample adhesion is evaluated by the size of the hysteresis part, sample stiffness by a deflection slope, and long-range forces by curve bending before the tip-sample contact. There are also two extreme situations: when the sample is plastically deformed by applying an extreme load (nanoindentation), and when a bit of sample material becomes stuck to the tip during its contact with the sample and is extended with the retracting probe. It has been shown that single macromolecules or bundles of macromolecules can be physically or chemically attached to the tip and their stretching or fracture can be observed in the force curve.[13,14]

We now discuss tip-induced sample deformation in more detail because it is common in studies of polymer samples. If the goal is just to establish a stiffness ranking or to measure the relative stiffness of different samples, one can measure force curves on each sample, using the same probe; it is also necessary to measure a force curve on a known rigid surface, such as a silicon wafer. (One should realize that recording of force curve on a Si wafer may lead to the tip damage unless either the Z-travel or cantilever force constant is small.) The key information is the slope of the repulsive portion of each force curve. When the probe stiffness is much less than the stiffness of the sample (or its components, when the sample is heterogeneous), then the slope of the repulsive part is determined only by the cantilever stiffness. Figure 19.6a,b shows force curves that were measured on polyethylene (PE) samples with different density, using a probe with 0.1 N/m stiffness. Here, the cantilever deflection is equal to the displacement of the base. When the cantilever stiffness is higher (1–3 N/m, for example) then the slope of force curves in their repulsive parts is reduced because the tip indents the sample (Figure 19.6c,d). From these graphs, one sees that the indentation is more pronounced on the softer material, PE with lower density. The absence of hysteresis in the force curves in Figure 19.6c,d indicates that stiffer probes are less sensitive to attractive and capillary forces.

A simple analysis begins by assuming that all other elements of the mechanical system are infinitely stiff so that the overall displacement of the base can be divided into two components: cantilever deflection and specimen indentation. We first measure the slope of the DvZ curve:

$$\text{Slope} = (\text{change in deflection voltage})/(\text{change in Z-position})$$

Then we compute the apparent stiffness as tip-applied force/indentation depth, using:

$$K_{sp} = MK/(1 - M)$$

where K_{sp} is the apparent stiffness of the specimen, K is the cantilever force constant, and M is the (slope observed on specimen)/(slope observed on rigid surface). The apparent stiffness may also be called the contact stiffness. To obtain adequate sensitivity, K should be within a factor of 5 or 10 of K_{sp}. This limited dynamic range is extended by using a library of cantilevers with different force constants.

The differences between the force curves in the repulsive force regime are related to the partial deformation or penetration of the probe into the softer polymer material. In principle, force curves can be used for a quantitative evaluation of elastic and viscoelastic properties. However, a lack of theoretical description of sample viscoelastic deformation by sharp probe limits such estimates. (Note that many authors have applied the Hertz; Johnson, Kendall, and Roberts [JKR]; Derjaguin, Muller, and Toporov [DMT]; and Sneddon models of particle-surface forces for analysis of AFM results. These models were developed for conditions that are different from those in AFM of polymers, such as spherical particles with large radius of curvature interacting with smooth planar surfaces. Consequently, their use for experiments with tip indentation is not well justified.) Therefore, only

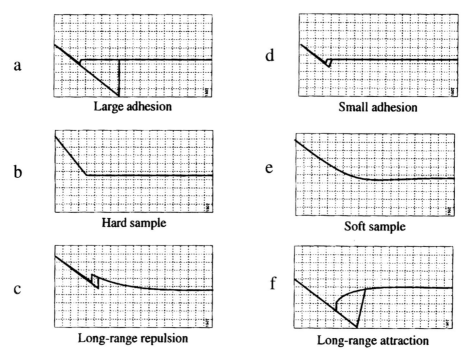

Figure 19.5 Force curves recorded on surface locations having different properties. (Courtesy of Digital Instruments/Veeco Metrology Group.)

comparative studies of different materials or different regions of a heterogeneous sample can be performed. Nevertheless, this information can be important for many practical problems. Local studies of mechanical sample properties are discussed in more detail in the Force Modulation section.

Amplitude versus Z-Travel and Phase versus Z-Travel Curves. Our discussion of force curves so far is relevant to contact mode AFM imaging, in which the tip slides along the sample surface and the (quasi-direct current [DC]) cantilever deflection is used for feedback. We now discuss two types of force curves used in tapping mode. This oscillatory mode can provide imaging in both force regimes, depending on the amount of the tip-sample interaction chosen for the feedback during scanning. In tapping mode, a small piezo-actuator incorporated into a probe holder vibrates (or drives) the probe substrate when a generator applies voltage to the piezo-actuator at various frequencies. The probe responds so that the cantilever oscillates with significant amplitude when it is driven at its resonant frequency. The same photodiode that detects quasi-DC cantilever deflection in contact mode now detects the alternating current (AC) deflection waveform of the cantilever. The signal is processed further to derive the root mean square (rms) amplitude and (optionally) the relative phase of the cantilever oscillation. Typical amplitude and phase curves as a function of frequency are shown in Figure 19.7, top. The resonant frequency of etched Si probes that are commonly used

in tapping mode experiments is in the 150- to 350-kHz range. By changing the voltage applied to the actuator, the cantilever amplitude can be chosen in a broad range from a few angstroms to hundreds of nanometers. When attractive or repulsive forces influence the probe, its resonant frequency shifts to lower or higher values, respectively. Frequency changes are accompanied by phase changes.

Amplitude, frequency, or the phase of the oscillating probe can be used to detect the tip-sample force interactions. Amplitude versus Z-travel (AvZ) curves and phase versus Z-travel (PvZ) curves are both relevant to tapping mode just as deflection curves are relevant to AFM contact mode. A set of these curves measured at different amplitudes of the freely oscillating probe (A_0) is shown in Figure 19.7, bottom. Let us assume that driving voltage is tuned at the resonant frequency of a free-oscillating probe (ω_0) and the phase level is defined as zero (Figure 19.7). When the base of the cantilever approaches the sample, the cantilever amplitude and phase are both influenced by viscous forces in the air just above the sample surface. Therefore, for more consistent results, it is desirable to retune the driving frequency when the tip is in the immediate vicinity of the sample surface. On further approach, the tip contacts the sample briefly at the bottom of its oscillation. This causes a steep decrease in amplitude with distance; this sharp indication of surface elevation enables high-resolution imaging. As the cantilever amplitude decreases, the phase first decreases (becomes negative) when the tip encounters an attractive force and then increases (becomes positive) as the force becomes repulsive. Attractive forces dominate tip-sample interactions in a large Z-travel region when A_0 is small and they are negligible for large A_0. An increased diameter of the tip's apex favors attractive tip-sample force interactions.

Knowledge of the amplitude and phase behavior allows optimization of imaging in tapping mode by the choice of appropriate magnitudes of A_0 and setpoint amplitude, A_{sp}. The latter is analogous to setpoint deflection, which is employed for feedback during scanning in contact mode. Imaging with low tip-sample force interactions (*light tapping*) is achieved by setting A_{sp} close to A_0. Such conditions are favorable for high imaging resolution, both horizontally and vertically, and for studies of soft materials. For ultimate light tapping, A_0 should be chosen as small as possible (about 2–5 nm); however, imaging sticky samples can be difficult because the probe may become trapped on the surface. For stable tapping mode imaging of sticky materials, one should use large A_0 and choose A_{sp} much smaller than A_0.

By analogy with the DvZ curves, we observe differences in AvZ and PvZ curves of different samples. Tip-induced deformation and even tip penetration through soft surface layers can be observed. AvZ and PvZ curves measured on high- and low-density samples of PE illustrate this effect in Figure 19.8a. PvZ curves vary more between low-density and high-density PE than do AvZ curves. Therefore, one should expect a high sensitivity of phase contrast to material properties of a sample. Differences between AvZ curves of the PE samples and that determined on a hard Si surface are also obvious. These effects indicate different penetration depths of the tip into these materials, which is reasonable because at ambient conditions amorphous PE is above its glass transition temperature, and tip penetration is most pronounced for ultra-low-density PE (Figure 19.8b). The way to estimate the amount of tip penetration at a given

setpoint is shown in the figure. Because tip penetration varies with setpoint, imaging of the soft sample at different A_{sp} occurs at different depths. For samples or surface layers, which are in a viscoelastic state, such imaging is nondestructive: the surface layers are restored once the probe is removed.

Imaging Procedure and Image Analysis

The high sensitivity of AFM probes to tip-sample force interactions, which is displayed in force curves, is implemented for surface imaging in a number of different modes. We previously described the general method of collecting image data in the Basic Principle of SPM section. Now we describe briefly how image information is presented and analyzed. Then we discuss imaging modes in detail.

The three-dimensional (3D) information contained in height images can be displayed in many different ways, a few of which we now illustrate. Consider a single image of data marks on an optical disc. Figure 19.9a (line plot) is a perspective view constructed as a stack of line scans (plots of Z height versus X position at a sequence of Y positions). Figure 19.9b (top view, height mode) is a plan view rendered with a false color fill, where the brightness of each pixel indicates height (white is high, black is low). Figure 19.9c (top view, slope mode) is a plan view where the false color is keyed to local slope: spots where the slope is nearly horizontal would reflect light from an imaginary light source back to the observer and are shaded bright; other spots where the slope is tilted are shaded dark. Figure 19.9d (surface plot, height mode) shows the same view as Figure 19.9a, using the same false color scheme as in Figure 19.9b. Figure 19.9e (surface plot, mixed mode) is a perspective view where the false color is computed on the basis of both height and slope. Each of these renderings has particu-

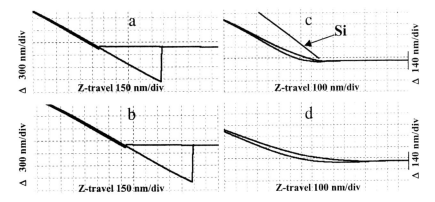

Figure 19.6 Force curves recorded on the surface of high-and low-density PE using Si probes with different stiffnesses, *K*: (a), (b) force curves obtained with the probe *K* = 0.1 N/m on high- and low-density regions, respectively; (c), (d) force curves obtained with the probe *K* = 2 N/m on high- and low-density regions, respectively. The line marked Si in (c) shows the DvZ curve for this probe on a very stiff surface, a Si wafer. (Courtesy of Digital Instruments/ Veeco Metrology Group.)

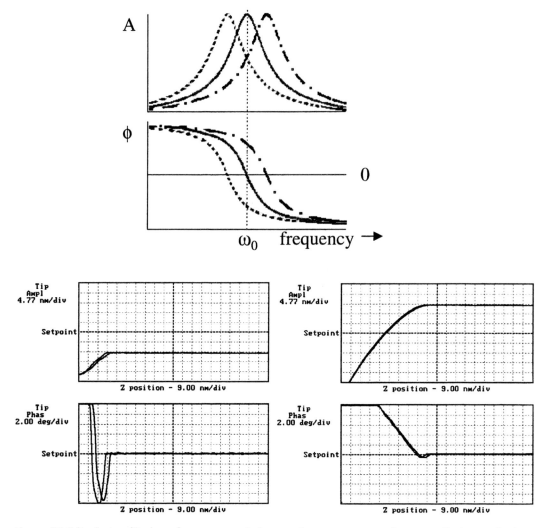

Figure 19.7 (*top*): Amplitude vs frequency and phase vs frequency curves for an oscillating probe in tapping mode. Attractive and repulsive force gradients shift the resonant frequency lower (*dotted curves*) and higher (*dotted-dashed curves*), respectively. (*bottom*): Amplitude vs Z-travel and phase vs Z-travel curves for initial amplitudes $A_0 = 10$ nm (left) and $A_0 = 100$ nm (right). (Courtesy of Digital Instruments/Veeco Metrology Group.)

lar merits. Perspective view images give a good qualitative and semiquantitative impression of the 3D character of the surface. Top view (plan view) images give the best impression of the horizontal relationships (size, shape, and position) of different features on the surface. Furthermore, top view images are used for quick distance and angle measurements: for example, one can place two cursors on the top view image and read the XY distance and height change (Z) between the two points; moreover, by drawing one or two vectors, one can read the angular orientation or angular span of a feature. Height mode coloring provides soft contrast, yet allows distinguishing height changes over long distances. Slope mode provides sharp contrast and helps in perceiving both fine and gross details. Mixed mode is a useful compromise that improves the visual appeal of perspective views; it tends to impart a glossy appearance.

It is important to note that AFM images have a special characteristic, dual magnification, that is not found in other microscopies. For image display, we select the vertical (Z) and horizontal (X,Y) ranges independently, to best present the surface structure, and these ranges can be changed at will after the image is captured. This means that a high magnification may be used on the Z axis and a moderate magnification may be used in the horizontal plane. This allows visualization of subtle height variations over a wide field of view. The ratio of vertical to horizontal magnification can be very large (1000 or more) to allow easy perception of differences between very smooth surfaces. It is not possible to visualize this using standard techniques in optical, scanning electron, and transmission electron microscopy. Finally, we remark that the use of different Z and X magnification is also customary in profilometry, which is discussed in the chapter in this volume on surface roughness (see chapter 17).

Commercial scanning probe microscopes are equipped with image analysis software. This includes image modification using low- and high-pass filtering, plane-fitting and flattening routines, and image measurement using two-dimensional (2D) Fourier transform analysis (which helps emphasize periodic surface structures), roughness and bearing evaluation, height profile (cross-section) display, and the like. AFM images can be exported in different digital formats that allow their treatment with various other image analysis packages.

Contact Mode and Lateral Force Imaging

In the contact mode, the tip comes into a permanent contact with the surface that is signaled by reaching a setpoint cantilever deflection. The latter defines a tip-sample force, which is kept constant during lateral scanning. Afterward, the tip slides on the surface, riding up and down the surface hills and valleys. When homogeneous samples with stiff-

ness higher than that of the probe are scanned, the height image reflects the true surface topography. This is not necessarily true during imaging of polymer samples, whose stiffness might be close to the stiffness of the probes (0.01–0.1 N/m) that are usually applied in contact mode studies. In other words, preloading of a sample with the probe, which is inherent to contact mode, might give erroneous conclusions about surface topography. Therefore, when applying contact mode to the study of polymer surfaces, one should compare height images recorded at different setpoint deflections. If changes are not found, then these images reflect true sample topography. Otherwise, one should reduce tip-sample force to a lower level to produce height images, which are closer to the true surface topography. Reduction of the tip-force can be achieved by minimizing setpoint deflection, by using a less stiff probe, or by performing measurements under liquid. All of these methods are commonly used by researchers involved in AFM studies of biological objects. This experimental approach should be also applied for imaging of heterogeneous polymers (polymer blends, composites, block copolymers, etc.), where stiffness differ-

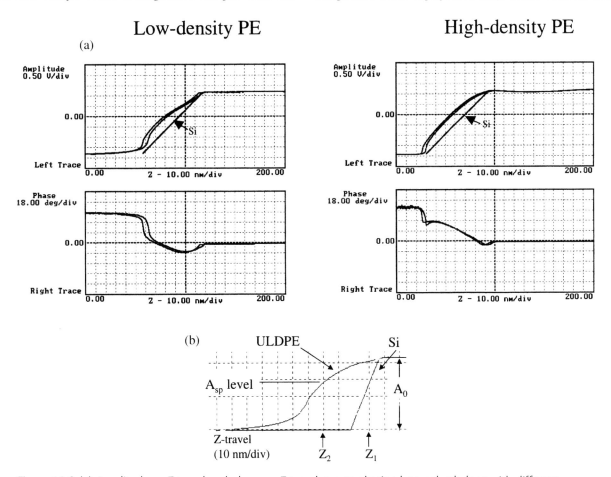

Figure 19.8 (a) Amplitude vs Z-travel and phase vs Z-travel curves obtained on polyethylene with different density, $A_0 = 75$ nm. The line marked Si shows the AvZ curve on a very stiff surface, a Si wafer. (b) Surface deformation. At setpoint level A_{sp}, the atomic force microscopy finds the surface at position Z_1 on Si and at Z_2 on ultra-low-density polyethylene. $Z_1 - Z_2$ is the amount of surface deformation or penetration by the tip. (Courtesy of Digital Instruments/Veeco Metrology Group.)

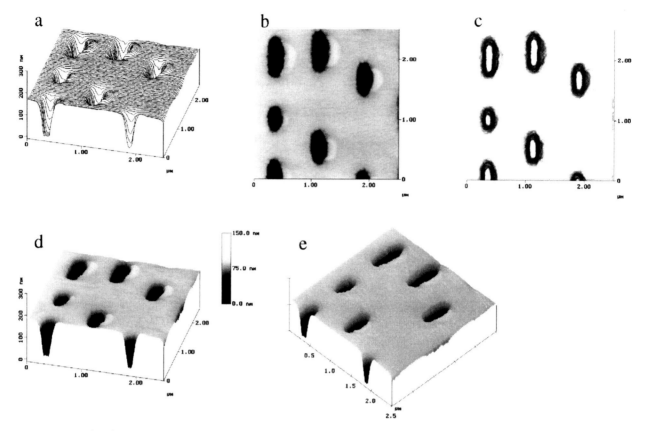

Figure 19.9 A height image (2.5-μm scan) of data pits in an optical disc is rendered in five different ways: (a) line plot; (b) top view, height mode; (c) top view, slope mode; (d) surface plot, height mode; and (e) surface plot, mixed mode. (Courtesy of Advanced Surface Microscopy.)

ences from spot to spot might lead to false judgments about surface topography.

In addition to sensing the normal component of the tip-sample force, AFM probes can sense lateral forces. When the tip scans in a direction perpendicular to the cantilever axis, the lateral tip-sample force due to friction causes the cantilever to twist. The amount of twist can be measured using a four-segment photodiode and presented as a lateral force microscope (LFM) image. This signal can be recorded simultaneously with the height image. Increasing the set-point, which means imaging with a higher preloading, can increase the twisting signal. In an LFM image, the contrast is due to variation in the relative lateral force between different surface locations, which can be related to composition. Figure 19.10a is an LFM image of a wood pulp fiber. The dark spot in the center is a low friction area, which we have assigned as a region where a thin lignin deposit coats the cellulose.

To interpret LFM images, first note that lateral forces come from several sources, including friction and topography. To identify the frictional forces, compare images or waveforms captured for opposite directions of probe motion (trace and retrace). Edges produce spikes in the lateral force signal and these have opposite signs for trace and retrace

(Figure 19.10b). Slopes project the normal force onto the horizontal plane, producing a lateral force that is the same for both directions (not shown). A true friction force will produce opposite twist directions when the probe motion is reversed, so that the apparent image contrast is reversed: in Figure 19.10c, the trace signal is lower and the retrace signal is higher in low-friction areas. Second, it is necessary to distinguish between friction force and friction coefficient. In macroscopic situations, friction force often varies linearly with normal force and is zero when the normal force (load) is zero. The proportionality constant is the friction coefficient. In AFM work, the situation is complicated by the fact that adhesive interactions between tip and sample cause significant lateral force even when the apparent normal force is zero. This means that the concept of friction coefficient needs to be modified. In a sample containing two material domains, A and B, it is possible for A to produce a higher friction force at low load than B, and for B to produce a higher friction force than A at high load. This means that the LFM image contrast can invert as the load is changed. Third, when working in air, it is difficult to control the normal force applied to the tip-sample contact because of the capillary forces mentioned (in the discussion of force curves). Therefore, the best quantitative work is done

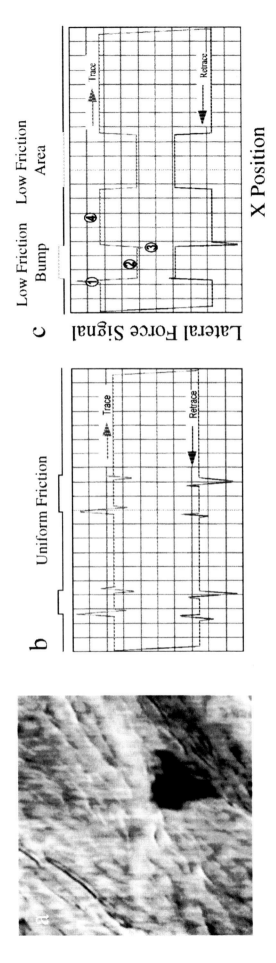

Figure 19.10 (a) Lateral force (LF) image of a wood pulp fiber. The dark spot is a low-friction area, assigned as lignin; 750 nm scan. (b) LF signal vs X position for a bumpy surface having uniform friction. (c) LF signal: The trace and retrace signals are closer together in low-friction areas. Spikes are superimposed at bump edges. [(a) Courtesy of Advanced Surface Microscopy. (b). (c) Courtesy of Digital Instruments/Veeco Metrology Group.]

in liquid, where there is no air-liquid meniscus. Fourth, to make an absolute measurement of friction forces, it is necessary to evaluate the cantilever's twisting force constant in addition to its deflection force constant. Numerical modeling is used but the results have limited accuracy.[15]

Force Modulation Imaging

Force modulation imaging is a technique for the qualitative identification of relatively hard or soft microscopic regions of composite specimens. This technique can be based on either contact mode or tapping mode.

Contact Mode. In force modulation based on contact mode (Figure 19.11a), images are captured by scanning the probe in contact with the specimen while the probe holder is vibrated vertically at a moderate frequency (e.g., 11.3 kHz). As in standard contact mode, the mean deflection signal is used to control the feedback loop and the height image is derived from the Z piezo position. Because the holder vibration does not coincide with a cantilever resonance, large-amplitude cantilever deflection is observed only when the tip contacts a solid surface. Thus, the amplitude of the AC deflection signal represents the relative stiffness of the surface. Height and amplitude images, which were recorded in force modulation on a multilayer blend of PE (narrow layers in high-density PE; wide layers in low-density PE), are shown in Figure 19.11b. The dark contrast of narrow

strips indicates a large deflection amplitude at these locations. This result is consistent with the higher stiffness of high-density PE layers. In addition to force modulation based on contact mode, there is another technique, in which force modulation is based on tapping mode. We describe this at the end of the section "Interleave Scanning, Lift Mode, and Related Techniques".

Tapping Mode and Phase Imaging

In tapping mode AFM, the change in amplitude is used for feedback to track the surface topography. Tapping mode is gentler than contact mode when applied to soft samples (which most polymers are), because it almost eliminates lateral forces. The short duration of the tip-sample contact in tapping mode (a probe spends only a part of its 150- to 300-kHz cycle interacting with the sample) restricts development of viscoelastic deformation in polymer materials. These two practical advantages of tapping mode greatly broaden the range of polymer samples accessible for AFM analysis compared with contact mode. Gentleness is illustrated in Figure 19.12. In Figure 19.12a, one sees that a liquid droplet can be imaged using tapping mode, whereas it is impossible to see this droplet in contact mode because it is swept away by the probe (image not shown). We do not know the viscosity of the oil. The relationship of droplet size, viscosity, and surface tension to the ability to image a

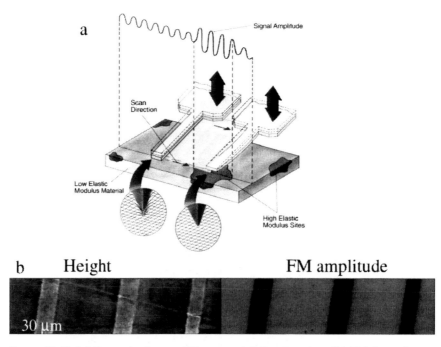

Figure 19.11 (a) General scheme of force modulation imaging. (b) Height and amplitude images of a multilayer blend of PE (30-μm scan width). In the amplitude image, the dark bands have high amplitude and correspond to higher stiffness. (Courtesy of Digital Instruments/Veeco Metrology Group.)

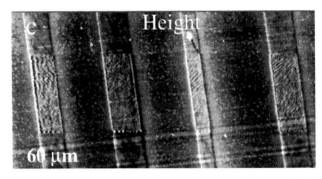

Figure 19.12 (a), (b) Tapping mode images of a nanodroplet of oil. The needle-like features are crystallites floating on the oil, 1-μm scan. (c) Tapping mode image of a multilayer polyethylene blend. A roughened strip across the middle of the image was previously scanned in contact mode, 60-μm scan width. [(a), (b) Courtesy of Advanced Surface Microscopy. (c) Courtesy of Digital Instruments/Veeco Metrology Group.]

nanodroplet using tapping mode is a topic for future research.

Another image (Figure 19.12a) was obtained on a multilayer sample, which consists of alternating layers of high- and low-density PE. This image was obtained in tapping mode on an area that includes a smaller region, which was previously scanned in contact mode. This place, which is seen as a window, has been slightly damaged in the contact mode imaging.

As tapping mode became more useful than contact mode for imaging polymers, it brought a challenge in evaluating tip-sample force interactions. Whereas cantilever deflection in contact mode is a direct measure of tip-sample force, the amplitude of an oscillating probe does not describe tip-sample forces in full. The phase is more informative in this respect. In tapping mode, the phase signal is typically recorded simultaneously with the height signal, and phase images have become extremely useful in AFM analysis of polymers. There are two main reasons for this.

First, in studies of topographic features phase images substantially complement height images by contrasting their minor structural details. Phase is very sensitive to small structural imperfections because the slope of the phase ver-

sus frequency curve is very steep near resonance (Figure 19.7, top). This is clearly seen in a height and phase image of oriented poly(vinylidene difluoride) (PVDF) film (Figure 19.13). Polymer strands with nanometer-size beads are best resolved in the phase image. Changes in relative phase are certainly produced by slopes and edges on the specimen surface, and we have found that phase images show edges more clearly (with less noise) than the slope mode rendering of corresponding topographic data. Before the introduction of phase imaging in tapping mode, amplitude images were applied for this purpose. However, phase images appear to be superior in this respect.

Second, the high sensitivity of phase contrast to material properties, which was mentioned earlier in description of PvZ curves, allows compositional mapping of multicomponent polymer systems. In other words, when the probe encounters surface regions of different composition, the phase changes, and this is reflected in phase images as profound contrast variations. The amplification of materials differences in phase images is demonstrated by images of a wood pulp fiber (Figure 19.14). The height image shows the topographic features, whereas the phase image indicates the different components present in this sample.

Figure 19.13 Height (*left*) and phase (*right*) images of poly(vinylidene fluoride) film, 5-μm scan. (Courtesy of Digital Instruments/Veeco Metrology Group.)

For several years, it has been known that in many materials, strong phase differences exist between patches of the surface having different composition. Questions remain: What causes phase contrast and how can it be applied to compositional imaging? Unfortunately, these questions are not yet fully answered. Any interaction between probe and specimen, which causes a loss of mechanical energy, can increase the damping of the oscillating probe, which in turn causes its phase to lag further behind the driver; this is registered as lower phase. Though a correlation between phase behavior and energy dissipated in the tip-sample junction in tapping mode has been established, it does not identify what material properties are mostly responsible for the energy dissipation.[16,17] Adhesion, capillary effects, and indentation forces are involved in the interplay, leading to energy dissipation; in general, it is practically impossible to quantify the individual contributions of these interactions. Furthermore, the distinction between adhesion and stiffness is unclear at a fundamental level: a low-stiffness material deforms more in response to the probe, which increases the area of contact and often increases the adhesive force. Another difficulty is that imaging can be performed in either

the attractive or the repulsive force regime, and at some imaging parameters, there are jumps from one region to another, which can cause inconsistent phase contrast. Because images are usually displayed using only relative phase changes (i.e., the difference between the phase of the cantilever free-oscillation, which can be chosen by the operator, and the phase of the cantilever interacting with the sample), the phase contrast behavior appears to be more complicated, especially when results of different users are compared. Therefore, one should be extremely careful about interpreting phase images in terms of specific materials properties (mechanical or adhesive) of components in multicomponent systems.

An empirical approach can be helpful in studies of heterogeneous samples. It might be useful to have a heterogeneous material with a known ratio of components and explore the phase behavior at different A_0 and A_{sp}. For some systems, such as PE blends and elastomer systems with fillers, such knowledge has been obtained. For example, by choosing large A_0 (80–120 nm) and $A_{sp} = 0.4A_0$, one can get a consistent correlation between brightness of phase contrast and stiffness of components in semicrystalline PE

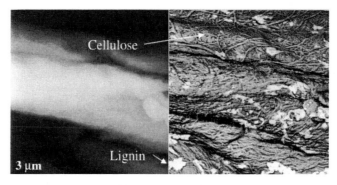

Figure 19.14 Height (*left*) and phase (*right*) image of a wood pulp fiber. Note the cellulose microfibrils and the amorphous lignin patches, 3-μm scan. (Courtesy of Advanced Surface Microscopy.)

and blends of PE with different density. In phase images recorded under such conditions, high-density regions appear brighter than low-density regions, and this is primarily related to differences of their mechanical properties, mainly stiffness. In an ultramicrotomed cross-section of a multilayer PE blend, layers of different thickness and density are present at the surface as bands of different width. Figure 19.15 shows the height and phase profiles recorded when imaging with $A_0 = 80$ nm and different values of the setpoint ratio, $r_{sp} = A_{sp}/A_0$. The height profile across several layers shows corrugation and roughness caused by imperfect ultramicrotoming. The height profile did not change noticeably when the r_{sp} was reduced from 0.97 to 0.12.[5] However, the phase profile changed dramatically, showing the largest contrast for $r_{sp} = 0.4$ to 0.5; this is the local maximum of energy dissipation. The high-phase strips are the stiffer, high-density PE layers. At r_{sp} below 0.2, AFM operation more closely resembles modulated contact than tapping, producing the largest tip-induced deformation and often damaging the surface. Thus, we classify different tapping regimes according to r_{sp}: = 0.97 (light tapping), = 0.43 (hard tapping), and = 0.12 (quasi-contact). On the basis of this example, in which stiffer layers of high-density PE are higher (brighter) in the phase profile obtained in hard tapping, one can recommend this condition as a starting point in optimization of compositional imaging of multicomponent polymer systems. In contrast, when imaging with low A_0, differences in adhesive properties of components might change the way the components appear in phase images.

We now offer some advice for the novice AFM user. Tapping mode is the first method of choice for AFM analysis of polymers because it is so gentle. Do not be afraid to use phase imaging. It first rewards you with edge sensitivity that enhances the apparent AFM resolution. Then it provides material contrast, which is extremely useful. Even if the label "brighter phase is stiffer or less adhesive" has not been established in your material system, being able to vi-sualize components A and B when you could not previously see them at all gives you valuable information about the size, shape, and distribution of the domains.

Other Imaging Techniques

FV Imaging. An FV image is a data set consisting of force curves captured at each point in a 2D grid. This data set provides replication (neighboring force curves should be similar) and correlation (specific curves can be matched to specific locations in the topographic image). Data capture proceeds as follows. The scanner moves to the first grid point and stops. The Z axis of the scanner extends the probe toward the surface until a preset trigger level for the feedback signal is reached, and the scanner then retracts. The Z position corresponding to the trigger event indicates the local elevation. During the extension and retraction of the probe, a monitored signal is recorded, just as is done for a normal force curve, as discussed above. The scanner then moves to the next grid point and the process is repeated. In FV imaging based on contact mode, deflection is usually used for both the feedback and monitored signals. In FV imaging based on tapping mode, amplitude is usually used for feedback and phase is usually the monitored signal. It is possible to use many other signal combinations.

The offline data presentation (Figure 19.16) includes the following items: (1) The height data (Figure 19.16a) are presented as a top-view image rendered in height mode. The pixel resolution is usually much less than for a normal height image because scan time here is proportional to the total number of pixels, instead of the number of scan lines, as in normal imaging. We often use 64×64 pixels to capture images in a few minutes. (2) The force curve display (Figure 19.16c) is a graph of the force curve(s) captured at one or more grid points. Recall that a force curve shows the monitored signal as a function of Z-travel. The left end of the graph is the trigger point. A cursor (vertical bar) on

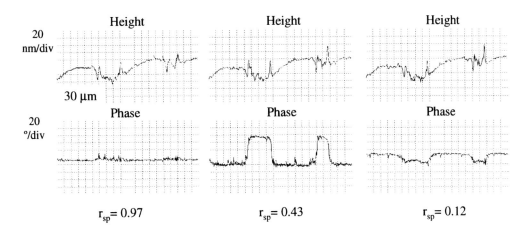

Figure 19.15 Height and phase profiles of a multilayer polyethylene blend at different setpoint ratios: $r_{sp} = A_{sp}/A_0$. (Courtesy of Digital Instruments/Veeco Metrology Group.)

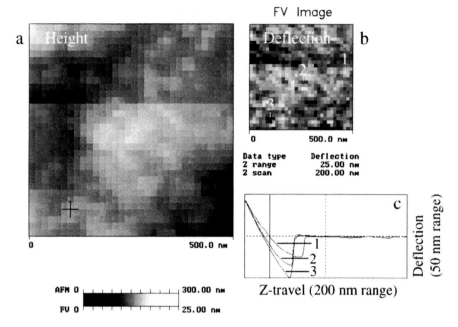

Figure 19.16 Force-volume image of a mass of latex particles, measuring the deflection signal. Locations 1, 2, and 3 in the deflection image (b) had curves with different slopes in the force curve display (c), indicating different stiffnesses. Corresponding points are marked with crosses in the height image (a), 500-nm scan. (Courtesy of Advanced Surface Microscopy.)

the graph marks the display point. (3) The FV image (Figure 19.16b) itself is a map of the signal level found at the display point on the extending or retracting force curve at each grid point. This map is rendered as a top-view image in height mode.

Figure 19.17 is a tapping mode FV image recorded using PvZ (phase) curves and a standard Si probe. The height and FV images here show the same topography and phase contrast that was seen in a standard tapping mode image of the same field of view (not shown). Because the spatial resolution was substantially better here than in the contact mode FV image, the 50-nm latex particles were distinct. The force plot display shows force curves at grid points 1 (bright phase), 2 (medium phase), and 3 (dark phase). Looking at the curves from left to right, the phase first decreased with some slope, reached a minimum, then increased, and finally became constant. The slope at the left represents a region where the tip-surface interaction is repulsive, the minimum is in an attractive region, and the tip is off the surface in the region of constant phase. Note that curve 1 had a steep repulsive slope and a shallow minimum, curve 3 had a shallow repulsive slope and a deep minimum, and curve 2 was intermediate in both properties. We interpret this to mean that particles of type 1 are stiffer (because of the steeper repulsive slope) and less adhesive (shallower minimum) than the others. This interpretation of phase in terms of stiffness is consistent with the results obtained in the contact mode FV image. Additional applications of FV imaging are discussed in reference 18.

Interleave Scanning, Lift Mode, and Related Techniques. The imaging methods we have described so far involve scanning the field of view once while capturing one or more data channels simultaneously, for example, height and phase or height and friction. To observe the effect of changing scanning conditions such as setpoint, one can simply scan the image again with the new settings and then compare the two consecutive images. Sometimes, however, it is desirable to capture images under different scanning conditions contemporaneously, that is, almost simultaneously. Interleave scanning accomplishes this. Contemporaneous data capture proceeds as follows: The probe makes two passes on each scan line. During the main pass, the scanning conditions are normal. During the interleave pass, different scanning conditions are used. For example, one might capture height and phase during the main pass at one setpoint, and capture phase during the interleave pass at a different setpoint. However, this simple example is not frequently used.

The most common use for interleave scanning is a modification of tapping mode and involves lifting the probe above the surface during the interleave scan to sense long-range forces such as electrostatic and magnetic fields. Corresponding scanning probe techniques are known as magnetic force microscopy (MFM), electric force microscopy (EFM), and surface potential (SP) imaging. Operation in EFM is accomplished as follows (Figure 19.18a). During the main pass, the probe taps the surface and we capture the topography. During the lift pass (interleave scan), the probe retraces the same path (amplitude feedback is switched off),

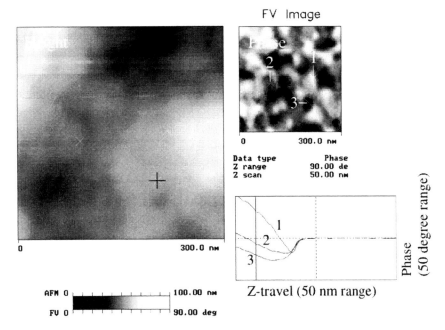

Figure 19.17 Force-volume image of a mass of latex particles, measuring the phase signal. The attractive phase region had different depths and extents at points 1, 2, and 3; 300-nm image. (Courtesy of Advanced Surface Microscopy.)

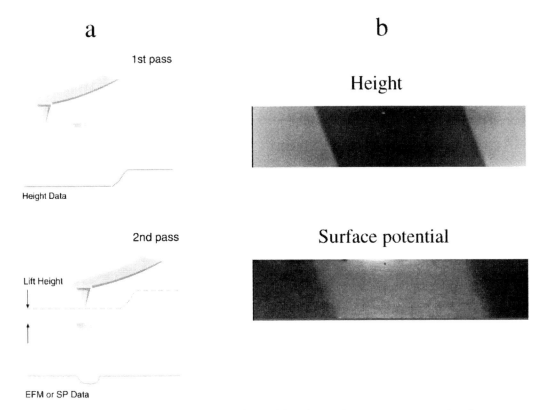

Figure 19.18 (a) Sketch illustrating lift mode used for electric force microscopy and surface potential imaging. (b) Height and surface potential images of a photoresist sample. (Courtesy of Digital Instruments/Veeco Metrology Group.)

but it is raised above the surface by a set amount (such as 50 nm) and does not tap the surface. During the lift pass, a DC voltage (usually in the range ±10 V) is applied to the probe tip and we capture the frequency data type, which registers changes in the cantilever resonance frequency. Frequency changes are proportional to the force gradient perpendicular to the surface, dF/dz, which in turn is proportional to the field gradient, dE/dZ. Higher frequency (bright) corresponds to a repulsive force gradient and lower frequency (dark) corresponds to an attractive force gradient. Note that conducting probes should be used for EFM, and this can be achieved by coating the Si probes with a conducting thin film. MFM operation is quite similar; the only difference is the use of AFM probes with a ferromagnetic coating.

EFM is used to map the vertical and near-vertical field gradient between the tip and the sample, whereas SP imaging maps the electrostatic potential on the sample surface with or without a voltage applied to the sample. As the tip travels above the sample surface in LiftMode, the tip and the cantilever experience a force if the SP is different from that of the tip. This force can be nullified by applying an appropriate voltage to the tip. Therefore, by plotting this voltage versus lateral coordinates, one can obtain an SP image. In a sample of acrylic photoresist exposed to synchrotron radiation, but not developed, it was found that the exposed area had shrunk, creating a groove (Figure 19.18b, top). Phase imaging and nano-indentation, which measure mechanical characteristics, did not show any contrast between the exposed and unexposed regions. However, EFM and SP images showed significant contrast. Synchrotron radiation creates a large number of dipoles and free radicals, which contribute to the higher potential of the groove compared to the potential of an untreated material. Therefore, the groove is brighter in the SP image (Figure 19.18b, bottom).

Applications of electric force-related scanning probe techniques to polymers are still in their infancy. EFM is applied to polymers to detect fixed charges located at or near the surface. The sign of the charge can be determined by comparing electric field images taken with different applied voltages. Another application is to detect thin spots in polymer coatings on metal substrates. When the metal is grounded, the electric field is induced by the probe (image charge), the force gradient is always attractive, and the magnitude of the effect depends approximately on the square of the applied voltage and the square of the separation of the tip from the metal surface. When the coating is thinner, the tip-metal separation is less and the frequency shift is greater. EFM has achieved the visualization of carbon nanoparticles in a carbon black filled polymer composite, so that each particle can be classified either as active (incorporated into a conducting percolation network) or passive (not connected with this network).[19]

One can also use the interleave technique to perform force modulation experiments in tapping mode. To realize this option, we operate as follows: (1) The probe cantilever is mounted in the force modulation holder, as in conventional force modulation imaging. (2) The Main scan is performed using TappingMode to capture the topography. This means that the drive frequency is at or near the resonance of the cantilever. The amplitude signal is used for feedback. Height data and, optionally, phase data are captured. (3) The Interleave scan is performed with parameters optimized for force modulation. The drive frequency is set to the resonant frequency of the piezo-actuator. At this frequency, the cantilever's oscillation amplitude will be smaller than its amplitude during the main scan. To touch the surface, we set lift < 0 (typically −60 to −5 nm). Amplitude data are captured. Force modulation based on tapping mode tends to work better on soft samples than contact mode force modulation. The interpretation of bright as soft and dark as stiff is the same for both modes (Figure 19.11b).

Practical Aspects of AFM Imaging

When one chooses AFM for material characterization, there are several questions to be considered. These questions concern sample preparation, choice of an appropriate operation mode and probe, experimental protocol for data collection, calibration, ultimate image resolution and accuracy, rational image analysis, and possible image artifacts. Several of these questions have been already addressed and others are discussed below.

Sample Preparation

In many cases, little or no sample preparation is needed for AFM examination. If the as-produced, exposed surface is of interest, polymers in the form of coatings, films, sheets, blocks, or pellets can simply be placed directly in the AFM or cut to a suitable size and mounted. A few points must be considered: sample tilt, secure mounting, and access to recessed or tilted structures. It is desirable for the scanned surface to be approximately horizontal (that is, parallel to the X, Y scanning plane). This allows large scans to be made without exceeding the Z range of the scanner, which is approximately 5 to 6 μm for most commercial microscopes. The specimen must be mounted securely, because the AFM is sensitive to small motions. If the specimen can lie perfectly flat (like a piece of paper), then it is acceptable to place it on transfer adhesive on a glass microscope slide or steel disk. However, if the bottom surface is curved (like a beverage can) or bumpy, it is better to use 5-minute epoxy or similar hard glues, because transfer adhesive creeps easily with even small stress. If the sample is mounted in a miniature vise and is compressible, do not tighten the vise jaws so much that the sample creeps during imaging. Samples with unusual geometry may require special handling. Although cylindrical fibers 10 to 200 µm in diameter (e.g., hair, optical fibers, fiber glass, individual filaments in synthetic yarns) can be secured satisfactorily on transfer adhesive, individual wood pulp fibers are twisted ribbons about 10 to 20 µm wide and make too little contact with the adhesive surface. We have had good results by pressing a fiber gently into the surface of warm Tempfix (SPI Supplies, www.2spi.com). Tempfix is a noncontaminating polymeric adhesive that is tacky at 40 to 60 °C and adheres well to many materials at room temperature. For some samples, the as-produced material has the surface of interest exposed, but it may be recessed or tilted at a large angle. In these cases, it is important to consider the geometry of the AFM probe and probe holder and how they fit in proximity with the specimen. For example, to examine the bottom of a recessed pit with a probe that is held from the right, the right wall of the pit should be cut off. To examine a textured surface where the surfaces of interest are tilted relative to the mean plane of the specimen, the specimen should be mounted on a tilted plane and oriented so that the probe scans a region near the higher edge of the specimen.

In many cases, the structures of interest are inside the bulk of the polymer. For example, the impact strength of a polymer blend is determined mainly by the bulk microstructure, not the surface structure. Usually an ultramicrotome should be used to cross-section the sample, and this is often done at low temperature. This is the same equipment that would be used to prepare a sample for TEM, but the preparation is easier here for the following reasons. First, the surface of interest is the faced-off block (not the thin section cut from the block), and it is easier to make one flat, smooth surface than two. Second, it is not necessary to stain the sample for AFM examination (and often it is preferable to not stain the sample).

A microtomed surface contains at least some debris of the fractured material. Although the debris may not interfere with imaging large structures in the range of micrometers and tens of micrometers, it can hide the true material features in the submicrometer range. In this case, one might try chemical or plasma etching, which will remove debris and also amorphous material in samples of crystalline polymers. This might change the nanoscale structures of interest. As in electron microscopy and other analytical techniques, the effects of sample preparation must be considered. A valid result is one that is consistent with data from other techniques and with a fundamental understanding of the system. The height image in Figure 19.19a shows a sample surface, prepared by cutting a pellet of commercial low-density PE with an ultramicrotome. This image indicates the presence of spherulitic structures in the bulk. However, to visualize these structures more clearly and to examine their nanometer-scale features, one should use permanganate

etching, which is often used for sample preparation in electron microscopy. This is confirmed by the height and phase images in Figure 19.19b,c. The phase image emphasizes different lamellae orientation (edge-on and flat-on) in band structures of the spherulite and grain-like lamellae substructure.[20] Sometimes it is possible to prepare the surface of interest by tearing at room temperature or by fracturing at low temperature; this depends on the mechanical properties of the material. On both microtomed and fractured surfaces, specimen preparation often leaves stray marks, such as longitudinal marks corresponding to irregularities of the knife edge or to the hard debris that may accumulate on the edge during cryo-ultramicrotomy. It is a good idea to use an optical microscope image to position the AFM probe to scan an area that has fewer stray marks. Phase images often show blend structures clearly, even when knife marks cannot be avoided. Etched surfaces of microtomed samples can be free of knife marks. Finally, note that it is possible to remove the top surface of the specimen by scraping with the AFM probe, thus exposing a surface of interest. This technique, called nanodissection, was first used to examine cell membranes.[21]

AFM studies of thin and ultrathin (thickness <100 nm) polymer films are straightforward for samples prepared by spin-casting of their solution on a flat substrate or dipping the substrate into solution. Among commonly used substrates are mica, silicon, glass, gold films with large crystal facets, or highly ordered pyrolytic graphite (HOPG). When the polymer concentration in solution approaches 10^{-5} wt %, dispersions of nanometer-scale particles and individual macromolecules can be obtained. The goals for sample preparation of such nano-objects include secure mounting, maintaining adequate dispersion and surface purity, and retaining the native structure or adopting an extended conformation. Secure mounting (good adhesion) is achieved by considering the chemical nature of the substrate and the molecule. Collagen molecules adsorb well on untreated mica but DNA does not. This is because the surface of mica is negatively charged at neutral pH, collagen is positively charged, and DNA is negatively charged. The binding of DNA is improved by pretreating the substrate with Mg^{+2} ions, or by including Mg^{+2} in the DNA buffer solution. Many additional chemical treatments exist (see the literature on self-assembled monolayers).

Good dispersion means that the individual particles are separated so that they can be examined clearly, yet dense enough that they are easy to find. For soluble molecules or uniform suspensions of particles, one approach is to incubate the substrate in the solution or suspension for a period of time and then rinse the substrate gently with pure solvent, followed by wicking off the solvent using filter paper and then air drying (if the material is to be examined dry). The surface concentration is controlled by the solution concentration and incubation time. The first experiments with

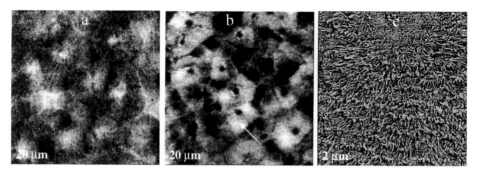

Figure 19.19 Height images of a microtomed surface of a low-density polyethylene pellet (a) before and (b) after permanganate etching. (c) Phase image of part of the spherulite indicated by an arrow in (b). (Courtesy of Digital Instruments/Veeco Metrology Group.)

a new material should include a very wide range of concentrations (factor of 1000 or more) and subsequent experiments should include a narrower range of concentration (factor of 10). When seeking an optimum concentration, it is often helpful to examine specimens prepared at concentrations differing by a factor of 10 or 3. Another approach is to spray the substrate with a fine aerosol of solution droplets and allow it to dry. However, be aware that the surface concentration will vary widely. When droplets evaporate, the highest concentration of material is deposited in a ring, with lower concentrations found inside and outside the ring. If the rings can be seen in an optical microscope or easily found in AFM scans, then it is feasible to scan several spots near a ring to find one with the desired density of particles. Oriented, dispersed specimens of rodlike molecules (such as DNA and collagen) can be prepared as follows. The substrate or molecule is modified chemically so that one end of the molecule will bind much more strongly to the substrate. The liquid solution is applied to the substrate, incubated for a short period of time, and then removed with a steady shear force (e.g., by capillary action or spinning the substrate). The shear force straightens the molecules and orients them parallel to the force vector. This greatly facilitates overall length measurements and distance measurements within each molecule.[22]

Surface purity is satisfactory when it is easy to locate and scan the molecules of interest, without interference from other material. Buffer salts and other residues can interfere with DNA examination. It can be challenging to obtain solvents with less than 1 ppm of nonvolatile contaminants, yet the concentration of the material of interest is often much less than that. This unfavorable concentration ratio is overcome by the selective binding of the molecules of interest. In addition, the contaminant deposit may be distinguished from the material of interest on the basis of morphology (e.g., dots of buffer salts vs rods of DNA) or even phase response. The following miscellaneous facts may be helpful. A droplet of volume 127 μL covers a 12.7-mm-diameter circle (a typical specimen size) to an average depth

of 1 mm. If the solvent contains 1 ppm of nonvolatile material and evaporates to dryness, then the average thickness of the residue is 1 nm. This is larger than or comparable to the diameter of most polymer chains. The octane molecule in the all-trans conformation is 1 nm long. Most organic materials have a density within 20% of 1 g/cm^3. At this density, 1 nm^3 has a mass of approximately 600 Da (atomic mass units); this value is useful for estimating the diameter of a globular polymer.

Finally, we note that many polymer samples may carry significant static charges. Electrostatic forces can adversely affect AFM operation, and can even prevent engaging the probe to the surface in contact and tapping modes. It is helpful to discharge sample surfaces using an ion source. A StaticMaster (NRD Inc., Buffalo, NY) is a convenient passive device (Po-210 radioactive source). Although it is inexpensive and considered very safe in the United States, some countries may require a special license. Alternatives include the Zerostat piezoelectric "gun" (www.2spi.com) and continuous ion blowers.

Imaging Mode and Experimental Protocol

Tapping mode with phase imaging currently is the most appropriate mode for microscopic observations of polymer samples. With a smaller level of tip-sample force interactions than contact mode, tapping mode gently images soft materials and, in combination with phase imaging, achieves compositional mapping of individual components of multicomponent polymer systems. Contact mode can also be applied for some of these purposes, yet the range of polymer samples for such applications is substantially reduced. For examination of local mechanical and adhesive properties of polymer samples, one should realize that different operating modes have different time scales of tip-sample force interactions. Tapping mode and phase imaging operates at 50 to 500 kHz, whereas contact mode, lateral force imaging, and DvZ curves operate at 1 to 100 Hz. The latter may be more suitable for many local studies of mechanical properties of polymers because the time scale of the tip-sample interaction is similar to that used in macroscopic apparatus for dynamic mechanical analysis.

When applying either tapping or contact mode, one will achieve the best resolution in visualization of top surface features by minimizing the tip-sample force interaction. Compositional imaging is best seen at elevated tip-sample forces when the difference in mechanical properties of components of heterogeneous samples is emphasized. Optimization of both goals could be facilitated by an appropriate choice of the AFM probe. Therefore, one should apply not only different imaging parameters to vary the tip-sample force but also use probes with different stiffness. These options need to be considered in defining an experimental protocol, which will lead to a number of images recorded at different parameters. Analysis of such sets of images is the recommended way to obtain full information about sample topography, surface structure, or material distribution.

An experiment in AFM starts with the engagement, which should ensure an accurate tracking of the sample surface by the probe. If we assume that the laser has been aligned on the cantilever, the reflected spot has been aligned on the photodiode, and the probe has engaged successfully to the specimen surface, then the most critical parameters are setpoint, integral gain, and scan rate. To get an idea how these parameters affect tracking accuracy, it is helpful to scan a calibration specimen that has a regular structure. Display the height signal as a function of position for both trace and retrace scan directions (this is called Scope Mode on NanoScope SPMs; other vendors may have different names for the same function). In addition, it is helpful to display the error signal (deflection for contact mode, amplitude for tapping mode). The error signal will be minimized, but not zero, and the largest error signal will be on the steepest slopes. When the scanning parameters are optimized, the shape of the trace and retrace height profiles should be very similar, although not necessarily identical. Remaining differences may be due to Z axis hysteresis of the scanner, which is observed when the sample has a significant slope with respect to the X,Y scanning plane.

In contact mode, when the setpoint (for the deflection signal) is increased, more force is applied to the surface. In tapping mode, the opposite is true: When the setpoint (for the amplitude signal) is decreased, more force is applied to the surface. In either case, when the setpoint is too close to the value seen in free air, the probe will not track the topography. In the worst case, the probe will lift off the surface and one may see a flat height profile (and a flat error signal) or the height profile may show smooth, long period waves (see discussion of optical interference fringes in the artifacts section). In a borderline case, the probe will track the surface topography on upward slopes satisfactorily, but downward slopes will appear shallower in the height profile and the error signal will be large on the downward slopes. In this situation more precise tracking is achieved by ad-

justing the setpoint to increase the force, increasing gain parameters, or decreasing the scanning rate, which all will improve the feedback.

When the integral gain is too small, the probe will not track slopes properly and downward slopes will be shallower than upward slopes. When the integral gain is too large, the height profile and the error signal will show oscillations, particularly on regions of flat topography. When the scan rate is too fast, the effects are similar to those seen with low gain with the possible addition of overshoot and oscillation. After some experience, one can optimize these parameters and achieve very gentle imaging similar to that demonstrated at $r_{sp} = 0.97$ in Figure 19.15. In that figure, note that the height profiles match nicely.

In tapping mode, one sometimes sees images that have sudden, unexpected height and phase changes of 2 to 10 nm and 10 to 100°, respectively. Because bumps may be surrounded by a sharp groove, this is called the "rings artifact" (Figure 19.20). The appearance of these features is most likely related to instabilities leading to sudden changes of the tip-sample force interaction from attractive to repulsive.[16] To avoid these effects, one can change A_{sp} (setpoint), A_0 (drive amplitude), or drive frequency. The rings artifact is more common when the drive frequency is above the resonant frequency of the probe. Static charge on the sample surface can cause an unexpected downward shift in the resonant frequency. One can either decrease the drive frequency to compensate or discharge the surface. After adjusting the drive frequency, it is usually necessary to adjust the drive amplitude or setpoint.

Calibration

Before starting meaningful experiments with AFM, one should check that the instrument is properly calibrated. Piezoelectric materials provide the exquisite positioning sensitivity that makes scanning probe techniques so useful. However, this comes with a price: The intrinsic properties of these polycrystalline materials are substantially nonlinear. In an ideal scanner, the motion would be linear with applied voltage and independent of motion on other axes. This is not the case for real scanners, which are used in SPM. Without careful design, construction, and calibration, real scanners exhibit significant nonlinearity, hysteresis, and creep on each individual axis by itself, as well as additional distortions related to coupling (cross-talk) between axes. The distortions will be manifested as incorrect magnification, change in apparent feature size with position in the image, and curvature of straight lines. These effects can be detected by scanning a test structure comprised of features of known size, shape, and spacing. Although distortions may not be apparent when scanning other samples, the measurement data may still be inaccurate.

A scanner system consists of both the scanner and the scan controller. There are two general approaches to scan-

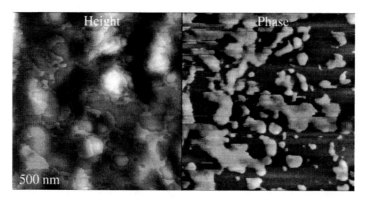

Figure 19.20 Rings artifact: abrupt changes in height and phase caused by instability of the probe-surface interaction. (Courtesy of Digital Instruments/Veeco Metrology Group.)

ner system design. In an open-loop system, the controller drives the scanner using a specific voltage profile, $V = V(time)$, to achieve a specific motion. In a closed-loop system, the scanner is fitted with position sensors. The controller reads the sensor outputs and adjusts the drive voltage to achieve the desired motion. In both cases, scan accuracy depends on accurate prior calibration using appropriate calibration standards. One typical calibration standard is a square array of pits with a pitch of 10 μm in the horizontal plane (to calibrate the X and Y axes). The pit depth is 180 nm and this is used to calibrate the Z axis. One-dimensional and 2D diffraction gratings that have smaller pitch values (150 nm or less) are available to calibrate smaller scan sizes (Advanced Surface Microscopy; www.asmicro.com) and atomic-scale scans can be calibrated using the lattice spacing of crystalline surfaces of mica and HOPG.

In one approach to calibrating the X and Y axes of open-loop systems, the first step is to select a nonlinear voltage profile that produces a linear motion as a function of time. The shape of this profile is adjusted according to scan size.[23] Subsequent steps involve a series of measurements that calibrate how the scanner sensitivity (in nm/V) depends on scan size, scan rate, and cross-coupling of the X and Y axes. The result is a set of approximately 20 calibration parameters. As the scanner ages, users usually need only adjust the four parameters (two for each axis) that describe how scanner sensitivity depends on scan size.

In closed-loop systems, scan accuracy depends not on the characteristics of the piezoelectric material but on the performance of the sensors and the accuracy and stability of the reference frame. One cannot assume that the sensor system is linear and orthogonal. Axis cross-coupling can be troubling for larger scan sizes (about 10–100 μm). Therefore, factory calibration of closed-loop systems requires the same level of care as for open-loop systems. An additional trade-off that is not immediately obvious is the fact that sensor noise can contribute to overall image noise, particularly for smaller scan sizes (about 2 μm or smaller).

In summary, users should consider what scan sizes are important for their work. A calibration reference specimen is appropriate to use if it has 1 to 50 complete pitch periods in the scan size that is being checked. Users should perform calibration checks at time intervals of 1 to 3 months or more frequently for critical work. We discuss ultra-high-accuracy calibration and measurement techniques below, in the Precision Measurements of CD and DVD Patterns section.

Image Resolution and Tip Shape Effects

Describing AFM resolution is more complicated compared to that of electron and optical microscopy because AFM provides 3D information. The key factors include the operating mode, tip geometry, sample characteristics, and the level of tip-sample force interaction. Contact AFM mode was used first and ambient-condition images that reveal surface lattices in the atomic and molecular scale were obtained on a large number of organic and inorganic crystals.[3] The fact that tip-sample forces in contact mode are relatively large (in the range from a few to tens of nanonewtons) suggests that in these measurements the tip-sample contact area involves at least tens of atoms from each side. Lattice imaging is still possible because the tip can experience periodic variations of normal and lateral forces as it moves along a periodic surface.[24] (Consider this real-life experience: driving a car on a brick road leads to periodic oscillations inside the car, which correspond to periodic corrugations of the road despite the fact that the contact area of a car tire is larger than the size of a single brick.) In this case, AFM images exhibit signatures of the periodic surface, but local defects of atomic scale are not distinguished. Figure 19.21 shows images of polydiacetylene and cyclic alkane $C_{48}H_{96}$ crystals, which reveal surface lattices of the largest planes of these crystals. The lattice parameters from the AFM images agree with the crystallographic data obtained for these compounds with X-ray diffraction. Periodic

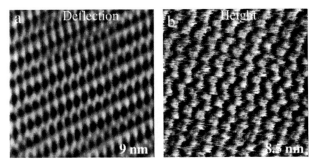

Figure 19.21 Atomic force microscopy images of polymer lattices: (a) polydiacetylene crystal; (b) cyclic alkane ($C_{48}H_{96}$) crystal. (Reproduced with permission from chapter 11 in reference 3.)

lattices of a number of highly oriented polymers have also been obtained in contact mode studies.

After the AFM capability to detect periodic lattices of organic crystals (including normal and cyclic alkanes) was demonstrated, researchers determined whether lamellar surfaces of polymer single crystals also exhibit a periodic pattern formed by molecular folds of individual chains. Although several observations of periodic patterns on a number of polymer single crystals were reported, these results might reflect only the periodic core, which is formed by chain stems. Most likely, the top surface is somewhat disordered, as is seen in tapping mode images. In contact mode AFM, where tip-sample forces are usually larger, the tip might feel the crystal lattice, which is actually under the "carpet" of the less ordered and soft top surface. On a somewhat longer length scale, this situation definitely occurs in contact mode imaging of polystyrene-b-polybutadiene-b-polystyrene (SBS) films.[25] The corresponding images reflect a microphase separation pattern (which is common for material underneath a top film surface) that consists of polybutadiene blocks. This is because the AFM probe can easily penetrate a 10- to 20-nm layer of viscoelastic polymer, which is above its glass transition temperature. This might also take place in studies of single crystals of polymers. Imaging in light tapping shows the topmost layer of SBS films and nanometer-scale grain structure of the top surface of a single crystal of PE.

Because polymer materials and their surfaces do not exhibit a perfect crystalline order, it is natural to ask about AFM resolution in imaging of these structures. It is reasonable to assume that the tip-sample contact area determines this resolution. In general, the contact area depends on the radius of the tip apex, mechanical moduli of the tip and the sample, and the tip-force chosen for imaging. Theoretical estimates for resolution on polymers yields a number approximately 10 to 20 nm, and resolution of a few nanometers can be achieved in practice. In such cases, only a part of the apex radius actually touches the sample. Because the radius of commercial probes is about 5 to 10 nm or larger,

minimizing the tip-sample force is critical for high-resolution imaging of surface features. Ambient conditions impose definite restrictions in minimizing tip-force because of capillary forces involving a liquid contamination layer found on most surfaces in air. In other conditions (under liquid, in ultra-high vacuum [UHV]), it is possible to minimize tip-sample force further to reach a contact area a few atoms in size.[26] In such cases, missing atoms have been seen in AFM images of some crystals. In other words, resolution can be as high as 0.4 to 0.5 nm.

The considerations of image resolution discussed so far are valid for relatively flat surfaces with corrugations in the subnanometer range. Larger surface corrugations might substantially limit image resolution and bring more of the tip shape into play. A qualitative description follows of several common effects that appear when the probe apex is comparable in size with surface features. Bumps protruding above a surface appear wider than they really are and pits appear narrower. Tall bumps appear wider than short bumps. Sharp angular features appear rounded. Narrow pores appear as shallow pits because the tip can penetrate the pore only a short distance. Steep sidewalls appear to have shallower slopes. In addition, the sidewall may appear to have a layered structure, which actually reflects nanometer-scale corrugations on the tip's sidewall (shank). Short objects adjacent to tall features are not detected. Any asymmetry of the tip shape near its apex might also distort the images.

A complete theoretical treatment of the geometric factors of AFM resolution exists, and it assumes only that the tip and sample surfaces are rigid. This rigorous, quantitative description of tip-shape effects uses the language of mathematical morphology. It is strictly incorrect to describe the image as a convolution of the surface with the tip. Rather, the observed image is the dilation of the surface with the tip. The true surface can be estimated by eroding the image with the tip shape, which is known either from scanning a tip characterizer or from a process of blind reconstruction.[27]

Currently, most AFM users do not use quantitative image reconstruction but instead may make an informal judgment about image quality. For one to make such a judgment, it is important to have in mind the nominal tip shape, how the probe is held in the AFM, and the corresponding measurement limitations. The most common probe used for tapping mode imaging is etched Si. The overall shape is an irregular pyramid. The nominal radius of curvature at the tip is <10 nm. Near the apex, the cross-section of the shank is triangular and it is kite-shaped further from the tip. The half angles of the shank are 17° (sides), 25° (front), and 10° (rear), where "front" is closest to the end of the cantilever. The most common probe used for contact mode imaging is oxide-sharpened silicon nitride. The overall shape is a square pyramid. The nominal radius of curvature at the tip is <20 nm. The half-angles of the pyramid faces are 35°. Specially etched probes and those made of carbon nanotubes are sharper, and they are most suitable for profiling deep trenches, grooves, and other complicated geometry. The significance of radius of curvature, R, is that a bump 1 nm high will appear to have width at least R. The significance of cross-sectional shape is that in the image of a spherical particle of height H, the particle boundary at its base has the same apparent shape as the shank cross-section a distance H from the apex. For example, a 50-nm-high particle would appear triangular when imaged with a Si tip. The shank half-angles and the tilt angle at which the probe is held in the AFM (typically 10°) combine to determine the limiting slopes that can be registered on objects taller than 50 nm or so. For example, consider a square pit with sidewalls that are nominally vertical. When imaged by an etched Si probe held at 10°, the apparent sidewall angles would be as follows: front 55° [= 90 − (25 + 10)], rear 85 to 90° [= 90 − (10 − 10)], sides 73° (= 90 − 17, unaffected by tilt angle), where labels such as "front" are chosen according to which face of the tip imaged that side of the pit. Thus, users should consider the absolute orientation of the specimen and tip when making critical measurements of steep slopes.

In an experimental approach to reliable imaging, one can begin by testing AFM probes to choose probes with the most appropriate tip characteristics. Two of our favorite test specimens are the following: gold colloid spheres (about 12 nm diameter) on a smooth substrate test the apex of the probe, and silicon spikes about 750 nm high (specimen model TGT-01; www.siliconmdt.com) test the tip and shank of the probe. This approach helps to avoid artifacts that might appear in AFM images obtained with a nonperfect probe.

Image Artifacts

All microscopes, like all scientific instruments, are subject to artifacts. In this section, we consider those aspects of AFM imaging that are important to control to produce meaningful qualitative and quantitative images. When tracking is poor, bumps are recorded with long tails. Countermeasures are discussed in the Imaging Mode and Experimental Protocol section. Magnification errors are discussed in the Calibration section. Most of the remaining common artifacts are caused by bad probes, nonideal scanner behavior, and optical effects. We prefer not to assign image variations caused by changes in tip-sample force interactions to image artifacts. However, such changes can be rather complicated and they require special considerations that fall into the areas of image analysis and rational image interpretation.

A bad tip causes many artifacts. Dull or contaminated tips produce images with broadened features and all the features in the image may look similar. Broken tips or tips with attached contamination particles might produce images with double or even triple features. In other words, each bump on the specimen surface is detected twice, as

shown in Figure 19.22. The quickest way to judge whether a given AFM image was made with a bad tip is to look at the shapes of the bumps and edges. If they are dull or doubled, then determine whether this anomaly is repeated across the image. If one suspects that the tip is not good, one may choose from three general actions: replace the probe immediately, test the probe on a favorite test specimen having a known structure (such as isolated bumps and spikes), or change the orientation of the specimen of interest and scan it again. This last case is important. If an anomaly always appears on a particular side of a given type of feature, regardless of the specimen orientation, then the tip is at fault. However, if the anomaly rotates with the specimen, then it is a true feature on the specimen surface.

The polycrystalline nature of piezoelectric materials and the fabricated shape of the scanner cause some nonideal behaviors. In the most common type of scanner, a tube, the lateral (X, Y plane) probe motion is produced by pivoting the probe. This scanning surface is not perfectly planar, but instead has a slight vertical bow, which scales approximately as (scan length)^2. This unwanted Z-motion is about 50 nm in a 100-μm lateral scan; it should be taken into consideration, especially during large scans on flat samples. Most bowing can be removed by a second- or third-order planefit of the image data. For many polymer samples, the residual bow in a 100-μm image is not noticeable. For a polished silicon wafer, however, the residual bow at a 10-μm scan size can be troublesome; this is why the most critical microroughness measurements on ultrasmooth surfaces are made at smaller scan sizes (1–5 μm).

Creep is another artifact related to the nonideal characteristics of piezoelectric scanners. In response to a DC offset in the applied voltage, the scanner quickly moves most of the distance corresponding to that voltage change and then slowly drifts (creeps) the rest of the way. Figure 19.23 shows that creep causes features to stretch in the direction of the offset for a short period of time. In this image of a multilayer structure, a DC offset was applied and then the frame was captured from top to bottom. The bending of the lines is due to the creep. Note that the creep settles out by the end of the scan. A similar effect occurs when changing from a large scan size (e.g., 25 μm) to a small scan size (e.g., 2 μm).

Optical interference fringes can affect AFM images because the AFM detects cantilever bending using light reflected from the cantilever. The detector also receives stray light scattered from the specimen being scanned. The interference of reflected and scattered rays causes a periodic signal variation as a function of scanner position. This can affect height, friction, phase, and other images. Figure 19.24a is a height image of a smooth surface that had a single large defect which scattered light. The waves in the height profile were up to 14 nm high (which is unusually large), with a horizontal period of about 2.1 μm. The period can be explained using the diagram shown in Figure 19.24b.

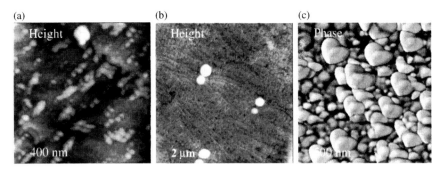

Figure 19.22 Atomic force microscopy images made with broken or contaminated tips show a repeating pattern of multiple or broadened apparent bumps for each real bump. [(a), (b) Courtesy of Digital Instruments/Veeco Metrology Group. (c) Courtesy of Advanced Surface Microscopy.]

This simple model is a particular case that uses Young's two-slit interferometer.[28] $X1$ is the location of the surface defect and $X2$ is the reflecting spot on the probe. Interference fringes occur when the pathlength difference, $L = X2 - X1$, is an integer multiple of the wavelength of light. The fringe spacing along the X axis is

$$\Delta X = \lambda/[n \times \sin(\theta)]$$

where θ (typically $24°$) is twice the tilt angle of the cantilever; λ (typically 670 nm) is the wavelength of light; and n (1 for air) is the index of refraction.

A similar analysis of Z axis motion leads to a fringe spacing of

$$\Delta Z = \lambda/\{n \times [1 + \cos(\theta)]\}$$

For the typical values shown above, the X- and Z-axis fringe spacings are expected to be 1650 and 350 nm, respectively.

In actual experience, X-axis spacings of about 1000 to 2200 nm are commonly observed, with peak-to-peak amplitude of a few nanometers; the spacing varies because the scattering geometry is more complex than the model. These effects are seen frequently on flat opaque samples, such as Si wafers, when defects are present on the surface, and on transparent samples, such as mica or clear coatings on sheet metal, when defects or rough structures are present on or below the surface. On samples that have steep bumps taller than 500 nm, Z-axis interference fringes can be seen. The Z-axis fringes are also manifested as ripples on DvZ and other force curves. Reducing the ratio of scattered to reflected light minimizes the fringe amplitude. AFM users can (1) select a cantilever that has higher reflectivity or is less transparent or (2) align the laser spot on the cantilever so that as little light as possible spills over the edge of the cantilever. Residual fringes may be further reduced by image processing: Set the AFM scan angle so that the fringes in the captured image are parallel to the image X axis and then modify the image by using the flatten filter.

For a more complete discussion of STM and AFM image artifacts, with many images, see reference 29. We note that what one person regards as an undesired artifact can be a useful tool for another person, when put under control and applied properly. For example, Z-axis interference fringes can be used as pseudo-contour lines of equal height.

Measurements of Surface Topography, Roughness, and Technological Surface Patterns

General Considerations

Examination of polymer samples with AFM reveals morphologic patterns in the 1- to 100-μm scale, which are similar to those shown in optical and electron microscopy micrographs. The main advantage of AFM studies of these structures is in the 3D nature of its images, which reveal

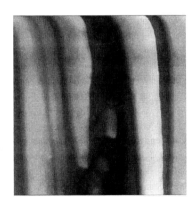

Figure 19.23 Creep artifact: Features shift in the direction of a direct current offset. Scan size: 10 μm. An offset of 8 μm to the left was applied just before the scan started at the top of the frame. (Courtesy of Digital Instruments/Veeco Metrology Group.)

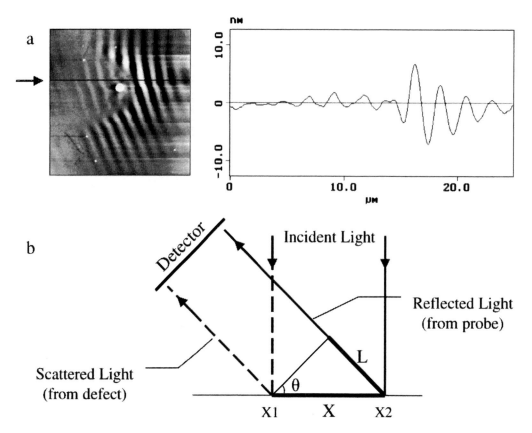

Figure 19.24 Optical interference fringes: (a) height image and profile (along line noted by arrow) of a Si chip showing wavy appearance near a large defect; (b) ray diagram for Young's two-slit interference model. (Courtesy of Advanced Surface Microscopy.)

height variations with high accuracy. We also note this limitation: rough features several micrometers high, which are common for some fracture surfaces, are less accessible for AFM, and such specimens can be better examined with scanning electron microscopy.

The examination of nanometer- to micrometer-scale structure is important for a variety of practical applications in diverse industries. AFM provides several advantages for quantitative evaluation of surface roughness as compared to stylus profilometry and optical methods. Because of the small tip-sample forces, AFM is gentler than stylus profilometry and it has better resolution than optical techniques for examining surface roughness in the nanometer scale. The latter became important in integrated circuit production, where the roughness of Si wafers is an important characteristic.[30] Roughness measurements of paint coatings with AFM indicate the stages of film formation.[31] In comparison with optical gloss measurements, AFM studies provide a more reliable correlation between roughness and appearance of commercial paints. In particular, optical gloss measurements are less accurate for transparent coatings because substrate roughness affects the results.[32]

Table 19.1 gives an overview of some practical AFM applications, including performance issues, specific examples, and the AFM approach.

Different Measurements of Surface Roughness

AFM images can give an outstanding qualitative and semiquantitative impression of surface topography. To obtain a quantitative measure of the topography, surface roughness parameters are easily calculated from the height data. These parameters describe the deviation of the surface from an ideal flat or curved form. Many different roughness parameters have been devised.[33] Here we describe several of the most common parameters. Roughness measurements are also discussed in chapter 17 in this volume. Many of the parameters shown in Table 19.2 are defined for both image areas and cross-section lines.

R_a values usually increase with the size of the measured area. This is because taller features are usually wider and they are not sampled properly in smaller scans. Therefore, when comparing roughness values found at different spots or on different specimens, the most valid comparison is made using the same size fields of view. The choice of scan sizes depends on the material, how the surface was formed or processed, and the performance properties of interest.

Evaluation of Surface Roughness with AFM

Table 19.3 shows R_a values found for a variety of surfaces. The paper industry produces a variety of products for many

Table 19.1 Some practical AFM applications in industry

Performance issue	Example	AFM approach
Defect analysis	Food can coatings	Pit or peak, particle size and shape
Appearance (haze, gloss)	Packaging	Pit or peak, height of texture features
Data storage accuracy	Optical discs	Size, shape and position of data marks
Wettability	Chemically treated textiles, implants	Coating continuity
	Medical diagnostic reagent coatings	Detect contaminant
Printability	Packaging	Detect contaminants
Adhesion failure	Coatings, joints	Determine failure mode (adhesive or cohesive) by distinguishing substrate from residue
Materials processing	Polymer blends	Phase distribution
	Wood pulp refining	Distinguish lignin from cellulose
	Molecular size distribution and state of aggregation	Image molecules directly

Key: AFM = atomic force microscopy.

Data courtesy of Advanced Surface Microscopy.

different applications at a wide range of prices. As an extreme example, ordinary white paper for photocopiers and laser printers costs $0.006/sheet, whereas photo-quality paper for inkjet printers costs about 100 times more. The difference in value is due to performance properties such as brightness, gloss, opacity, and printability. Modifying the surface of the base sheet using a suitable coating creates much of the improved performance and added value in premium paper products. Zhang et al.[34] prepared and characterized a series of coating formulations for white packaging cardboard. The coatings contained kaolin and calcium carbonate pigment particles with latex (styrene acrylic copolymer) and protein as binder. Particle size and shape is a major consideration in coating design, because this influences pore size and shape, which in turn has a strong effect on brightness and the dynamics of ink interaction with the surface (printability). Kaolin particles are mostly hexagonal plates, whereas some calcium carbonate particles have similar dimensions along all three axes and others are needle-like. For each type of particle, different grades may exist, with names such as fine, ultrafine, engineered, and precipitated. A suite of characterization techniques was applied. Standard performance tests included gloss, brightness, opacity, pick strength, and ink tack. Fundamental tests included mercury intrusion porosimetry, laser goniophotometry, macroscopic roughness by Print-Surf (Parker), macroscopic topography and roughness by Paperscape diamond stylus profilometry (IMERYS Pigments and Additives), and microscopic topography and roughness by AFM. Figure 19.25 shows a 1 mm Paperscape scan along with a 10-μm AFM scan. The Paperscape shows macroroughness caused primarily by individual wood pulp fibers (10–20 μm wide) and clumps of fibers (50 μm wide), whereas the AFM image shows the individual platelike (kaolin) and needle-like (precipitated calcium carbonate) particles making up the coating. The Paperscape image is rendered in height mode (white is high, black is low), whereas the AFM image is rendered in illumination mode, where brightness corresponds to local slope. In fact, quantitative measurements of local slope from both images provided key information. Statistical tests showed that sheet gloss was determined by Paperscape slope, AFM microroughness, and AFM slope. Print gloss was determined by sheet gloss, pore radius (from mercury porosimetry), and AFM microroughness. Thus, AFM measurements were significant in providing a fundamental understanding of this complex physical and chemical system.

Edge Roughness

Just as the AFM can sense surface roughness, so can it sense edge roughness. Whereas surface roughness is the *vertical* deviation of the real surface from an ideal form, edge roughness is the *horizontal* deviation of a feature boundary from its ideal form. In the semiconductor industry, the narrowest lines are now less than 100 nm wide. The edge roughness of the polymer photoresist used to pattern those lines must be considered, because this could limit the precision of width and overlay (placement) control. We illustrate this concept using AFM scans of a photoresist pattern having pitch less than 300 nm and line width less than 150 nm. Figure 19.26a is a 2-μm height image. The edge roughness is plainly visible, because the boundaries of the lines are slightly ragged. To obtain longer scan data, we analyzed a 10-μm image (not shown). Edge positions at 95% of the ridge height were determined on each scan line using special software.[35] We plotted edge position as a function of distance along a given line and fit the resulting curve to a polynomial. The residuals of the fit form the roughness profile, which is shown in Figure 19.26b. For photoresist patterned by noncontact optical exposure, the R_{rms} edge roughness was 4.0 nm. For comparison, we examined polycarbonate formed by high-speed injection-compression molding. The specific structure is the pre-formed groove

Table 19.2 Commonly used roughness parameters

Parameter	Definition	Comment
R_a	Average roughness, calculated as the average deviation of elevation values from the mean line or center plane, corrected for tilt	Least sensitive to noise spikes
R_{rms} (or R_q)	Root-mean-square deviation of surface elevation values from the mean elevation within a designated area or along a line; also called standard deviation; may or may not be corrected for tilt	Usually 1.1 to 1.4 × R_a
R_z	The average of the five largest peak-to-valley differences along a line or in an area, relative to the mean line or plane (corrected for tilt)	
R_{max}	Height difference between the highest and lowest points on line or in an area, relative to the mean line or plane (corrected for tilt)	Measures peaks and values; very sensitive to noise spikes; usually 5–10 × R_a
Z_{range}	Difference between maximum and minimum elevation within a designated area (not corrected for tilt)	Measures peaks and values; very sensitive to noise spikes
Surface area difference	[[(actual area of the surface)/(measurement area in xy plane)] − 1] × 100%. Larger numbers correspond to a rougher surface (more ups and downs). For most materials, this parameter ranges from 0 to 20%. Only very rough surfaces, for example, microlithographic structures or an uncured film of latex particles, give values above 20%. This parameter is influenced both by the height of surface features (as are parameters such as R_a and R_{rms}) and by the width of surface features (spatial frequencies). If two surfaces had equal R_a values, the one with more closely spaced, narrower features would have the higher value of surface area difference.	

Data courtesy of Advanced Surface Microscopy.

on a recordable CD (CD-R). In this case, we found rms edge roughness of 6.7 nm.

Precision Measurements of CD and DVD Patterns: A Case Study

For a consumer, the essential quality specification for a CD or DVD is playability: a random disc must play well in any player or disc drive. Many people who have purchased more than 20 CDs have experienced at least one playability problem, such as skipping or failure to play. The administrative costs of handling the resulting customer returns can be as high as 100 times the original manufacturing cost. To improve customer satisfaction, many details of manufacturing must be controlled.

Optical disc manufacturing depends on high-quality polymer materials and good process knowledge. One common process for making CDs can be outlined as follows. A glass disc (the master) is coated with a thin film of photoresist

Table 19.3 Roughness values for various surfaces

Material	Scan size (µm)	R_a (nm)
Cleaved mica crystal	1	0.02–0.04
Polished Si wafer	1	0.04–0.1
Polymer film (no filler particles)	5	0.5–5
Paint films (containing pigment particles)	30	10–100
Polymer tubing	100	50–150
Coated papers	10–20	25–500

Data courtesy of Advanced Surface Microscopy.

and placed in a laser beam recorder. The laser beam pulses on and off, exposing a sequence of spots in a spiral pattern; these will be the data marks. Development removes the photoresist at the exposed spots, leaving pits. The structure is metallized and the photoresist removed, creating a Ni stamper that has bumps. The stamper is placed in a mold and injection-compression molding produces polycarbonate replicas (which have pits). Each replica is coated with a thin film of Al, then thick films of lacquer and printed ink are applied, resulting in a finished disc.

Polymer properties of interest include optical characteristics, such as clarity, birefringence, bubbles and specks, and rheological characteristics, such as melt viscosity at the molding temperature. Methods for analyzing many of these bulk properties are discussed in chapter 22 of this book. The AFM is used to examine the microstructure of discs at each major stage of the process: glass master, stamper, and replica. As is true for most of the AFM applications discussed in this chapter, many types of defects or failures can be characterized by qualitative inspection of the images or by quantitative measurements that need only 5% to 10% accuracy. For example, clouds or stains covering large areas are cosmetic defects that may also cause functional problems. The cloudy appearance is due to a change in light diffraction from the data surface. The AFM image in Figure 19.9 shows that the pit and land structure was distorted, with a 12-nm-high lump appearing on what should be a flat area to the right of the pit. The root cause for this defect is called plowing, which is caused by the radial motion of stamper bumps relative to replica pits during mold release.

Figure 19.25 Height images of paper coating: (a) a 1-mm profilometer scan, showing fibers; (b) a 10-μm atomic force microscopy scan, height data rendered in slope mode, showing pigment particles. (Courtesy of Imerys.)

Many quality problems require measurements of much higher accuracy (1%) and precision (<0.3%). A publisher returned to the original CD manufacturer a batch of several thousand discs that had playability problems. In addition to replacing the discs with a new batch, the manufacturer chose to investigate the root cause of the problem. The CD engineers used the optical microscope to examine the stamper used to make the bad discs; they saw anomalies and suspected bad track pitch (see Figure 19.27b for definition). They asked one of the authors (DAC) to make AFM measurements, which confirmed the suspicion: at the anomaly, one track was misplaced by 0.2 μm, so that the track pitch was 1.4 μm on one side and 1.8 μm on the other. This information indicated that a specific piece of equipment, the laser beam recorder, needed to be repaired. Although this particular measurement did not require unusual precision or accuracy, the CD engineers requested ongoing improvements in AFM precision and accuracy to support further development of equipment to make the high-density format now known as DVD.

The DVD industry is truly the largest nanotechnology industry, measured by volume of product. The smallest features are pits (called T3 pits) about 400 nm long, 320 nm wide, and 120 nm deep, with a track pitch of 740 nm. The

size, shape, and position of these marks must be controlled on the nanometer scale. The track pitch specification requires feature placement in the radial direction with 1 standard deviation (SD) < 7 nm, and the electrical pulse jitter specification is similarly stringent in the tangential direction (leading and trailing pit edges must be placed with 1 SD < 11 nm). To make useful pass-fail judgments, the metrology tool must be at least four times more precise than these specifications. Chernoff et al.[36] discovered a problem: A standard AFM had significant differential nonlinearity, so that track pitch values measured at different spots within an image could differ by more than 5%. They were able to solve this problem by developing a procedure for correcting the measurements to the required precision.[36] Subsequently, Chernoff and Burkhead developed fully automated software for calibrating and analyzing AFM images to report track pitch, jitter, and numerous size and shape parameters.[37]

The measurement process includes the following steps:

1. Capture images of a calibration grating and one or more test specimens (10 × 10 μm images are optimal for DVD) (see Figure 19.27a,b)
2. Analyze the calibration image data to assess microscope accuracy, including the magnification error and nonlinearity
3. Develop correction functions that map apparent position to corrected position
4. Analyze the test images to locate and measure the features of interest
5. Apply the correction functions to produce corrected measurement data

Figure 19.27c shows track pitch results for two discs. Although the mean values for both were within the specification (740 ± 10 nm), the individual values for disc 1 were outside the allowed range (740 ± 30 nm). Figure 19.27d is a study of process conditions. The laser power used during mastering was varied so that different radial bands on a single stamper were recorded at distinct power levels. At each band, we captured one AFM image and measured

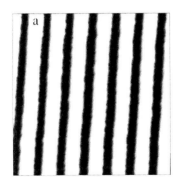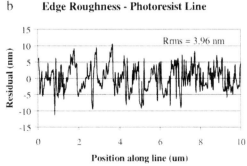

Figure 19.26 (a) Atomic force microscopy height image of photoresist lines, 2-μm scan. (b) Edge roughness profile from a 10-μm scan. (Courtesy of Advanced Surface Microscopy.)

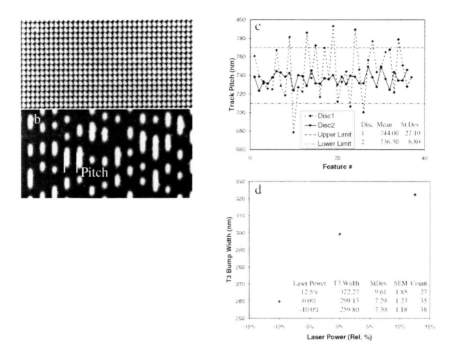

Figure 19.27 (a) Calibration grid and (b) digital versatile disc stamper, 10 × 5 μm scans. (c) Track pitch measurements for two discs. (d) Width at half height of T3 bumps on a power series stamper. (Courtesy of Advanced Surface Microscopy.)

height, width, length, and sidewall angles for all of the bumps (about 90). In the DVD format, the bumps can be classified by length into discrete groups. For the shortest group (T3), width at half height increased monotonically with laser power, from 260 nm at the lowest setting to 322 nm at the highest. Because the intrinsic width variation was significant (1 SD = 7–10 nm), it would not suffice to measure just one bump for each power level. It is only by measuring many bumps that a precise mean value can be obtained (and this precision was 1 standard error of mean (SEM) = 1.2–1.9 nm). Using the AFM as a high-precision measuring tool, optical disc engineers can fine-tune process variables and see the effect each change has on pit geometry, and they can then see how that change in pit geometry affects the electrical signals produced during playback. This capability will be even more important for future, higher density formats.

High-Resolution AFM Imaging of Polymers

Observation of Single Macromolecules

Molecular weight and conformation are two of the most fundamental characteristics of polymer molecules. Since the first STM and AFM studies of DNA macromolecules in the late 1980s, visualization of single macromolecules has developed substantially. At present, AFM is widely applied for visualization of biological macromolecules[38] and synthetic polymers.[4] Images of double-stranded DNA, which is deposited on mica, and domains consisting of single macromolecules of liquid crystalline fifth-generation carbosiloxane dendrimer on Si substrate[39] are presented in Figure 19.28 a and b, respectively. In some cases, epitaxy of adsorbed macromolecules on a substrate is in favor of the extended chain conformation. This is the case when macromolecules of jacketed polystyrene (PS), a PS molecule with minidendritic (branchlike) side groups, adopt an extended conformation because of specific registry between side-chain ethylene groups and the graphite substrate (Figure 19.28c).

The primary goals of such experiments, determination of macromolecule size and conformation, can be achieved to some extent. The main difficulties are related to possible tip-induced sample deformation (affecting height measurements) and insufficient resolution (affecting width measurements; see Image Resolution and Tip Shape Effects section). Although width and height of extended polymer macromolecules cannot be precisely estimated from AFM images, length errors are only a few percent of the overall length, which is sufficiently accurate for many purposes. Therefore, histograms of contour length are used for a quantitative evaluation of molecular weight distribution.[40]

In most cases, examination of individual macromolecules with AFM is straightforward, and the optimization of the imaging is mostly related to a minimization of the tip-force. Success in visualization of individual macromolecules with AFM is largely determined by sample preparation. Macro-

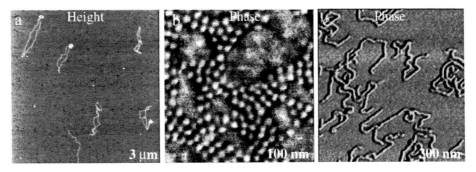

Figure 19.28 Atomic force microscopy images of single macromolecules: (a) double-stranded DNA on mica; (b) arrays of macromolecules of carbosilane liquid crystalline dendrimer on Si; (c) single chains of monodendron-jacketed polystyrene on highly ordered pyrolitic graphite. (Courtesy of Digital Instruments/Veeco Metrology Group.)

molecules deposited on a substrate from dilute solution should be properly fixed to avoid moving them by a scanning tip and should adopt the required conformation. In one case, the size of polymer coil can be of the interest, whereas an extended conformation of the macromolecule might be needed for direct measurements of its molecular weight. Rodlike macromolecules such as collagen are the easiest objects for AFM visualization. This structural protein consists of three polypeptide chains, which associate to form a triple helix (monomer) over most of their length. In type I collagen, a monomer is nominally 300 nm long × 1.5 nm diameter.[41]

The height images in Figure 19.29 show many individual collagen monomers of varying lengths, as well as small fragments and a variety of oligomeric structures. Quantitative evaluation of these components is an important tool for optimizing the digestion and purification of collagen from animal tissue for subsequent use as a medical biomaterial. On the basis of such images, histograms reflecting variations of the monomer size (contour length) can be constructed.

This task is easier if straightened collagen molecules are prepared (Figure 19.29b) using a motor-driven stretching apparatus originally developed for straightening DNA molecules.[42] In addition to facilitating the measurement of molecular length, straightening helps determine the physical locations of cleavage or binding sites, which is critical information for understanding structure-function relationships of interacting molecules. Because a tiny amount of unmodified sample can be used, the AFM method provides a useful supplement to conventional techniques such as fluorescence labeling, immunostaining, and gel electrophoresis.[22]

Visualization of Nanometer-Scale Polymer Structures

High-resolution AFM imaging opens new insights in studies of polymer chains and their self-organization. Some of the most valuable AFM applications are in studies of polymer structures on the submicrometer scale, where AFM complements not only electron microscopy but also diffraction

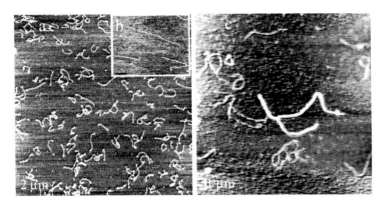

Figure 19.29 Atomic force microscopy height images of collagen molecules: (a) mostly monomers of varying length; (b) straightened monomers, bar = 100 nm; (c) oligomers. [(a), (c) Courtesy of Advanced Surface Microscopy. (b) Reproduced with permission from reference 22.]

methods in examination of nanometer-scale structures. This is demonstrated on a number of AFM images of crystalline polymers.

Images in Figure 19.30a–c show the morphology and nanoscale structures observed on a surface of high-density PE, which was crystallized from the melt on a mica substrate. Large-scale circular structures are attributes of the banded spherulite, which is a common morphology for crystalline polymers. Fine extended structures, which are preferentially oriented along tangential and radial directions, are seen in the central image. They might be assigned to edge-on oriented lamellae whose different orientation is responsible for optical birefringence patterns associated with banded spherulites. The image in Figure 19.30c shows a granular substructure of lamellar edges, which have been also observed on a number of different melt-crystallized samples with AFM.[20] Spherulitic morphology with fine linear structures, which are oriented in the tangential direction, is seen in AFM images of isotactic polypropylene (iPP) film deposited on Si substrate from solution (Figure 19.30d–f). These linear structures can be assigned either to lamellar edges or to fibrils, which both exhibit the granular substructure. Arrays of individual grains (10–20 nm in size) are seen in some places.

A number of crystalline polymers, which show spherulitic morphology, do not exhibit lamellar structures. This situation is seen in images of melt-crystallized aliphatic polyketone (Figure 19.31). Although large spherulites are clearly seen in the image in Figure 19.31, top, their interior consists completely of grains approximately 30 nm in diameter. In some locations, grains form linear sequences by partially merging with each other. AFM observations of the granular structures in a large number of crystalline polymers provided crucial evidence and helped in the development of a new approach for understanding polymer crystallinity. It was suggested that polymer crystallization is a substantially two-step process.[43] First, self-organization leads to formation of individual grains, the order of which is governed by microphase separation of crystallizable and noncrystallizable chain segments. Next, these grains build up larger one-dimensional (1D) fibrils and 2D assemblies. The size of the individual grains and the extent of their organization into larger structures completely depend on the chemical nature of the polymer chain, its ability to fold, and crystallization conditions. The corresponding nanometer-scale structures whose existence was proved by AFM are schematically shown in Figure 19.31, bottom. In an AFM study[20] of melt-crystallized low-density polyethylene (LDPE), it was shown that in light tapping the height image shows assemblies of grains, whereas extended fibrils are seen at the same location in hard tapping. These fibrils represent hard cores of 1D grain sequences, which are formed by chain folding. The exterior material is less ordered and is easily depressed by the AFM probe.

Compositional Imaging

Compositional mapping of heterogeneous polymer systems substantially impacts industrial applications of this technique. Tapping mode and phase imaging are well-suited for visualizing the size, shape and spacing of the different mate-

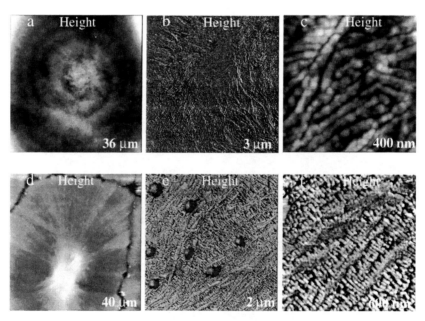

Figure 19.30 Atomic force microscopy images of melt-crystallized high-density polyethylene (a–c) and isotactic polypropylene, crystallized from solution (d–f). (Courtesy of Digital Instruments/Veeco Metrology Group.)

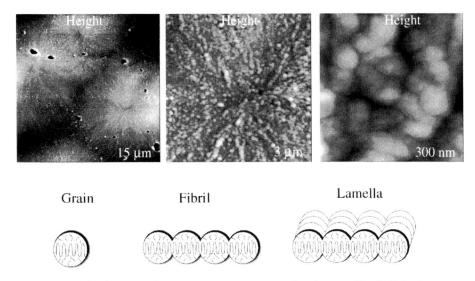

Figure 19.31 (*top*) Atomic force microscopy images of melt-crystallized aliphatic polyketone. (*bottom*) Sketch of nanoscale structures of crystalline polymers. (Courtesy of Digital Instruments/Veeco Metrology Group.)

rial domains or components in polymer blends, block copolymers, semicrystalline polymers, and composite materials. This microstructure is a major determinant of performance properties such as impact strength, viscoelasticity, wettability, printability, adhesion, and friction. Differentiation of components that is based on differences of their mechanical properties is illustrated in Figure 19.32. It presents height and phase images of a multilayer PE blend, which consist of alternating layers of polymer with higher and lower densities. Because these layers are chemically identical, TEM cannot distinguish them, whereas phase images obtained in hard tapping as well as amplitude images in force modulation are able to distinguish the layers. Bright strips in the phase images correspond to PE layers with higher density. Although direct interpretation of the phase contrast can be complicated, for many systems one can prepare reference samples and calibrate the phase response. This has been done for elastomer blends, where relative brightness in phase images decreases in the following sequence: fillers, glassy or crystalline polymers, butyl rubber, ethylene-propyl-

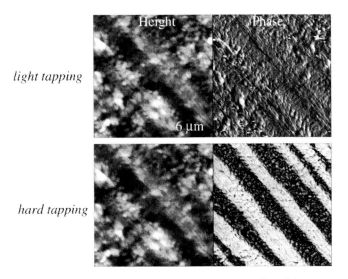

Figure 19.32 Atomic force microscopy images of multilayer blend of polyethylene components with 0.94 and 0.92 g/cm³ densities [0.94 is bright in the hard tapping phase image (*lower right panel*)]. (Reproduced with permission from reference 54.)

ene-diene monomer (EPDM), styrene-butadiene rubber (SBR), polybutadiene rubber (BR), natural rubber (NR), oil, and the like.[44]

Polymer blends are important industrially because they achieve desirable materials properties at an economical price. One example is the blending of rubber with polystyrene to combine toughness and strength. It is known that particle size, shape, and distribution must be controlled to meet performance specifications. TEM has traditionally been used to examine this microstructure. However, TEM requires preparation of thin sections by ultramicrotomy and is limited to the examination of materials that provide suitable electron contrast, either with or without staining. AFM currently is often used to image this morphology. Its advantages include easier sample preparation (staining is usually unnecessary, and if ultramicrotomy is used, a smooth-faced block is the specimen, not a thin section) and broader scope (AFM can distinguish materials, such as elastomer blends, that have no intrinsic electron contrast and that cannot be stained). In addition, regions with higher cross-linking density appear brighter in the phase images than regions with lower cross-linking density. Therefore, AFM allows direct visualization of sample regions with different cross-linking density.[45]

Impact-modified polystyrene can be produced in a variety of physical forms. For an extruded sheet of an impact-modified polystyrene blend containing filler particles, AFM images showed detailed morphology of the as-produced surface with no sample preparation. In a 10-µm image (Figure 19.33), the fine structure was elongated parallel to the Y axis. The inclusions were elliptical and were associated with trails of dark phase material. In the 3-µm phase image, the polystyrene matrix had two regions: a light gray amorphous region and a dark gray region with fine bands that appear to be crystal lamellae. The crystalline regions tended to be found immediately adjacent to the inclusions and in vertical strips aligned with the inclusions, whereas the amorphous regions tended to be found between those vertical strips. The inclusions consisted of irregular particles that protruded above the matrix in the height image (bright). Because they were bright in the phase image, we identify these particles as the relatively hard filler. Dark areas in the phase image are identified as rubber, and these were recessed below the matrix in the height image.

Spontaneous segregation of material phases often enriches one phase at the surface. Johnston et al.[46] investigated this means of surface chemical modification in a hybrid poly(dimethyl siloxane) (PDMS) system consisting of soft siloxane domains with alkyl functionality and harder siliceous domains with fluoro-alkyl functionality. They prepared thick film specimens with varying stoichiometries and characterized them with multiple techniques, including optical microscopy, AFM, X-ray photoelectron spectroscopy (XPS), and time-of-flight secondary ion mass spectroscopy (TOF-SIMS). Optical microscopy showed micrometer-scale phase separation in particular specimens. AFM height images showed well-separated surface bumps on the same length scale (Figure 19.34a). In AFM phase images, the bumps were distinctly brighter than the land, indicating that the bumps were stiffer (Figure 19.34b). Quantitative surface chemical analysis by XPS showed that the surface was enriched in the fluoro-siliceous phase. Chemical imaging of the surface using the F-mass peak in TOF-SIMS showed separated regions with high fluorine content surrounded by a matrix of low fluorine content. This confirmed the AFM assignment of the high-phase, topographic bumps as fluoro-alkyl-siliceous domains.

The pulp and paper industry is one of the largest industries involved with natural polymers. An ongoing technical challenge is to produce high-quality white paper within cost and environmental constraints. In the Kraft pulp production process, wood chips are cooked with NaOH and Na_2S to solubilize much of the brown lignin. The resulting brown pulp can be used to make brown paper bags and cardboard boxes, or bleaching the residual lignin that remains on the wood pulp fibers can further refine it. The bleached fibers can then be made into white paper. One traditional bleaching process uses chlorine-based bleaches

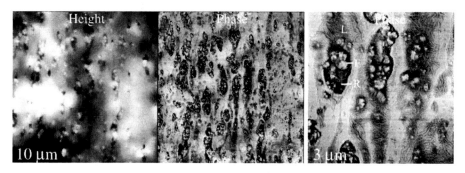

Figure 19.33 Atomic force microscopy images of high-impact polystyrene (PS). In the phase images, dark areas are rubber (R), bright irregular particles are a hard filler (F), and lamellae (L) are visible in the PS matrix. (Courtesy of Advanced Surface Microscopy.)

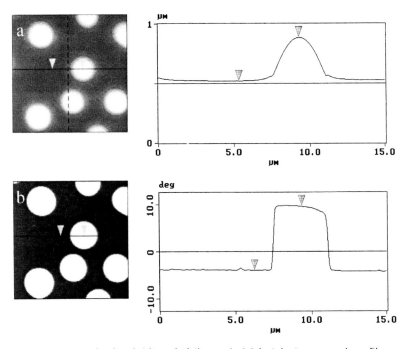

Figure 19.34 Hybrid poly(dimethylsiloxane): (a) height image and profile; (b) phase image and profile, 15-μm scan. (Courtesy of Advanced Surface Microscopy.)

to react with the lignin. However, these have been linked to formation of dioxin, a toxic pollutant. Therefore, it has been a goal in the industry since 1991 to reduce or eliminate the use of chlorine-based bleach. To optimize the bleaching process, it is necessary to characterize residual lignin in pulp. Lignin is a highly cross-linked polymer composed of phenyl propane monomers with hydroxyl, methoxyl, and a variety of other oxygen-containing functional groups. It is difficult to characterize chemically and it varies considerably depending on the tree species and strain. The reactivity of lignin in the bleaching process has been hard to predict using existing methods and, therefore, it is important to develop new techniques, such as AFM, for characterizing residual lignin.[47]

Figure 19.14 is a height and phase image of a wood pulp fiber. This fiber is eucalyptus wood, prepared in a modified countercurrent cooking (MCC) digester, and then delignified using O_2. The relatively rough topography made it difficult to see fine structure in the height image, but the phase image had rich structure. The edge- and slope-sensitivity of the phase signal defined closely spaced microfibrils, which are identified as cellulose. Bright patches in the phase image appeared amorphous and were about 10 nm thick. These are identified as lignin. It was proposed that the phase contrast in this case was based on probe-surface adhesion differences from spot to spot. The Si probe, which has available OH groups on its surface, could have relatively strong adhesion to cellulose, which has plentiful OH groups, and relatively weak adhesion to lignin, which has

fewer OH groups.[48] This image has historical interest. Captured in 1994, it is one of the first phase images made by Chernoff, who was the first AFM user outside of Digital Instruments to recognize that this technique was valuable for materials analysis in general. He pioneered the early application of phase imaging to a variety of material systems.

Pereira et al.[49] subsequently made a large, systematic AFM study of wood pulp fibers from a variety of commercial and experimental pulping and bleaching processes. Pereira found a significant correlation of phase image morphology and signal contrast with process step and with chemical analysis (methoxyl content by Fourier transform infrared spectroscopy [FTIR]) in the Aracruz commercial processes. Brown pulp fibers collected after the MCC step had approximately 50% coverage by lignin particles and patches with a wide range of sizes. The phase contrast at this stage was small (2–4°). Two bleaching processes were investigated: elemental chlorine free (ECF) and totally chlorine free (TCF). Each process had several steps. At the earliest bleaching step, the lignin particle size was reduced below 100 nm and the coverage was reduced to a few percent. By the last process steps in each case, the phase contrast (measured using bearing analysis) increased to about 12° (TCF) and 15° (ECF). Concomitantly, methoxyl phenyl propane content, which measures the amount of lignin, decreased steadily. It is important to note that the variability of individual phase measurements increased as the process moved downstream; however, one can be confident in the overall trend because of the large number of images (48) captured per sample at each process step. Pereira concluded that AFM images successfully measured decreasing amounts of lignin in samples from progressive stages of the pulping and bleaching processes, the phase contrast signal correlated with methoxyl content and bleachability performance, and AFM revealed other characteristics and peculiarities of each sample. The correlation of AFM phase signal with chemical data is considered novel, particularly in the industrial context.[49]

The above examples of AFM compositional mapping deal mostly with components located on the top surface. In some systems, where the top layer is formed of rubbery material or oil, the AFM probe can penetrate through the top layer and interact with underlying rigid structures. The penetration depth might vary from several to hundreds of nanometers based on the thickness of the soft top layer and operating conditions. This AFM capability to reveal subsurface structures is demonstrated in Figure 19.35, where height and phase images recorded on a commercial film of linear low-density polyethylene (LLDPE) are presented. In light tapping, only slight surface corrugations, without traces of lamellar structures, are visible. Hard tapping reveals the underlying lamellar aggregates, which were previously invisible because they are immersed in a viscous wax, which dominates the top surface layer. The image changes seen in this figure are completely reversible.

We close this section by considering how other materials analysis techniques can be used along with phase imaging. Phase imaging by itself does not provide direct chemical identification of the domain composition. Such identification is made by inference, using the information from the phase image along with other knowledge, such as the overall composition of the material, the process steps by which it was made, and the results of bulk and surface chemical analyses (e.g., by infrared spectroscopy or XPS). For example, if XPS shows that the average surface composition is 75% A and 25% B, then the predominant domain in the phase image is probably A. Some surface analysis techniques, such as SIMS, provide high-resolution imaging ca-

pability, and those images may be compared with phase images. When making this comparison, keep in mind that SIMS imaging resolution and signal-to-noise ratio is relatively poor on polymers and the submicrometer resolution found on metals and semiconductors is not feasible. Successful comparisons of SIMS and tapping mode phase images have been made at low magnification (50–100 μm scan size),[46] but the authors are not aware of SIMS-AFM comparisons at high magnification (such as a 3-μm scan size).

AFM Studies of Polymers at Elevated Temperatures

So far, most AFM studies of polymers have been conducted at ambient conditions. The main AFM capabilities of such studies (such as high-resolution imaging and compositional mapping) have been demonstrated and generally are understood. It is important to extend these studies to a broad temperature range because polymers exhibit various thermal phase transitions. It is both challenging and appealing to use AFM for visualization of the corresponding structural and morphologic changes accompanying these transitions. Such studies have been started and several results are presented in this section.

It has been demonstrated that microphase separation patterns of block copolymers are well distinguished in AFM height and phase images. The image in Figure 19.36(25C) shows a fingerprint-like pattern in a polystyrene-b-polybutadiene-b-polystyrene film, where bright and dark features correspond to polystyrene and polybutadiene blocks, respectively. This image was obtained in hard tapping when the probe penetrates through a top surface layer that is enriched in polybutadiene blocks. The difference in the phase contrast in Figure 19.36 is related to the fact that at room temperature, polystyrene blocks are in a glassy state and are more rigid than polybutadiene blocks, which are substantially above their glass transition temperature. As sam-

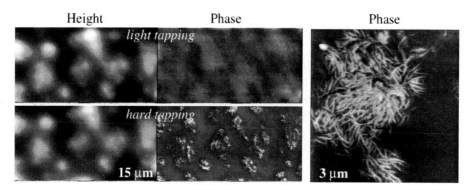

Figure 19.35 Atomic force microscopy images of a commercial linear low-density polyethylene film. The high-resolution image at the right was obtained using hard tapping. (Courtesy of Digital Instruments/Veeco Metrology Group.)

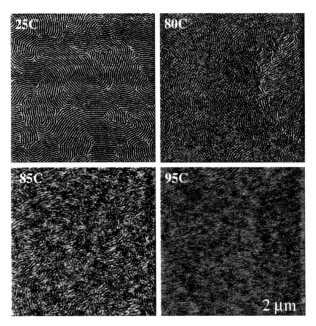

Figure 19.36 Phase images of a film of polystyrene-*b*-polybutadiene-*b*-polystyrene triblock copolymer. Phase contrast between the styrene and butadiene domains decreases as temperature increases. (Courtesy of Digital Instruments/Veeco Metrology Group.)

ple temperature increases, the contrast becomes weaker and it almost disappears at temperatures close to the glass transition point of polystyrene (which is ~100 °C). At this temperature, the microphase separation is still present in this material; however, it cannot be seen with AFM. In another example, softening of top ordered layers in fluorinated polystyrene-*b*-poly(isoprene) at the smectic phase temperature was detected by an AFM probe that could penetrate through the top layer and visualize a different ordering pattern in a subsurface layer.[25]

Pronounced structural changes during polymer melting and crystallization are shown in Figures 19.37 and 19.38. In a thin film of syndiotactic polypropylene (sPP), a number of single crystals 10 nm in height were seen (Figure 19.37a). Melting of the single crystals proceeds heterogeneously, as seen in Figure 19.37b–d, where parts of crystals disappear before their complete melting at $T = 140$ °C (Figure 19.37e). The vertical bands that appear in this image are an instrumental artifact related to optical interference and should be ignored. After cooling to 100 °C, new single crystals were found (Figure 19.37f). Although their appearance was similar to that of the initial crystals, their size and positions were different. Another example is the melting and recrystallization of an ultrathin film of LLDPE (Figure 19.38). This film was deposited on a Si substrate by dipping and it exhibits 2D spherulites (Figure 19.38a). Lamella edges, which are seen inside these spherulites, became less pronounced at $T = 80$ °C. The image recorded at $T = 125$ °C shows only a melted polymer layer without any structural features.

After the sample temperature was reduced to $T = 102$ °C, crystallization was observed (Figure 19.38d–e). Newly grown lamella occupy a substantial part of the film after 2 h of crystallization at $T = 102$ °C (Figure 19.38e). By measuring sizes of the crystalline features, one can obtain quantitative data about the crystal growth at each temperature. Complete crystallization was achieved after cooling to room temperature (Figure 19.38f).

A recent AFM study of an ultrathin film of LDPE at elevated temperatures demonstrated that monitoring of structural changes during phase transitions complements differential scanning calorimetry (DSC) studies by allowing visualization of polymer recrystallization during heating, which strongly depends on the heating rate.[50] Interplay between crystallization and dewetting in ultrathin polymer films has been demonstrated.

Local Probing of Mechanical Properties

Stiffness, hardness, toughness, and abrasion resistance are just a few of the performance properties that are important for polymers. Test instruments and methods for measuring these properties in macroscopic samples are discussed in chapter 23 of this book. If one uses the AFM probe to perform nanoindentation and nanoscratching, it is feasible to measure these properties with very high spatial resolution. In a nanoindentation experiment, the AFM horizontal scan is halted and the scanner extends, pressing the probe into the specimen surface until a set load is achieved. The scanner then retracts. From the DvZ data, one can estimate the stiffness. VanLandingham et al. discuss the detailed numerical analysis of such force curves in reference 51 and in references cited therein. By imaging the indented region and measuring the area of the indent mark, one can estimate the hardness. A nanoscratching experiment begins with the same steps as a nanoindentation experiment. However, once the probe is pressed into the surface to achieve the set load, the scanner then moves a chosen distance horizontally. The probe is lifted off the surface and the scratch is imaged. One measures the depth of the scratch and the height of material piled up adjacent to the scratch, and notes the occurrence of cracking and delamination.

The spatial resolution of the technique depends on the probe tip size and the applied load. In AFM-based nanoindentation, diamond-tipped stainless steel cantilevers and silicon probes are generally used, which have force constants in the range 100 to 300 N/m and tip widths in the range 10 to 30 nm. Loads are generally in the range 100 nN to 100 µN. At the lowest loads, the contact area can be as little as a few tens of square nanometers and the penetration depth as little as 1 to 3 nm. As a rule of thumb, the strain field during indentation extends a distance 10 times further than the actual penetration depth. The observed mechanical properties relate to the material that is included

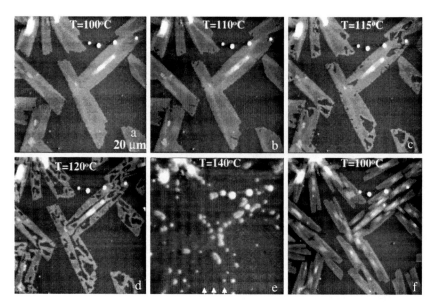

Figure 19.37 Atomic force microscopy height images showing melting and crystallization of single crystals of syndiotactic polypropylene. Arrows (in $T =$ 140 °C image) indicate optical interference fringes (artifact). (Courtesy of Digital Instruments/Veeco Metrology Group.)

within this interaction volume. This means that meaningful information can be obtained for ultrathin films that are as little as a few nanometers thick (albeit with some effect because of the substrate) and horizontal spatial resolution can be much better than 100 nm.

It is known that the strength of fiber-polymer composite materials is influenced strongly by the bonding of the fiber to the polymer matrix; a composite made of intrinsically strong fibers in an intrinsically tough matrix can perform poorly if the fiber-matrix adhesion is poor. In one common process for making a metal fiber-polymer composite, the metal fibers are first coated with a liquid solution containing epoxy prepolymers. After the material is dried, there is a solid, uncured epoxy layer on the metal fibers; this sizing layer may be 1 to 10 μm thick. The sized fibers are then placed in a bulk mixture of a second type of epoxy containing a stoichiometric amount of amine curing agent and heated to cure the bulk. During this process, the curing agent diffuses into and fully cures the sizing layer. The metal fibers are now embedded in a strong epoxy matrix. VanLandingham et al.[52] investigated stiffness variations across the coating-bulk interface in such a system. Samples consisted of a copper fiber with an 8-μm-thick epoxy coating, embedded and cured as described above. AFM specimens were fabricated by cutting a cross-section perpendicular to the fiber axis and then polishing by standard metallographic techniques, with a final grit size of 50 nm. In the nanoindentation experiment, many indents were made in the bulk epoxy far from the copper to establish the mean and SD of the bulk stiffness. The stiffness of the coating and the stiffness variation across the interface were evaluated by mak-

ing a linear sequence of indents at 0.3-μm intervals along a path extending about 12 μm from the fiber surface out into the bulk epoxy. Several parallel rows of indents were made with small longitudinal offsets so that the effective spatial resolution perpendicular to the interface was about 50 nm. A special feature of this work was the temperature study. The nanoindentation was done at several temperatures ranging from room temperature up to 120 °C. The results are shown in Figure 19.39. The fiber surface was out of range to the left in the graphs and image. The graph at each temperature shows stiffness relative to the mean stiffness of the bulk epoxy at that temperature. In addition, the solid lines indicate 95% confidence intervals for individual stiffness values for the coating and bulk epoxy. There are two regions where the stiffness values fall outside those limits: (1) the region immediately adjacent to the fiber has high stiffness caused by a proximity effect and (2) there is a transition region about 1.5 μm thick between the coating and the bulk epoxy layers. This transition region or interphase probably results from interdiffusion of the coating and bulk prepolymers, and has distinctly different characteristics. At 20 °C, the stiffness monotonically decreases as one crosses the interphase from coating to bulk. At 80 °C, the stiffness of the interphase equals that of the bulk. At 120 °C, the stiffness shows a distinct minimum in the interphase region. The AFM image at 120 °C shows that the indent marks were larger in the interphase than in the other two regions, which is consistent with the lower stiffness. By combining these experimental results with a finite element analysis of indentation response as a function of temperature,[53] they concluded that the interphase has a

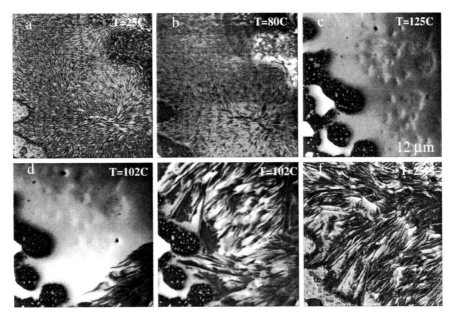

Figure 19.38 Phase images of melting and crystallization of linear low-density polyethylene, 12-μm scans. (Courtesy of Digital Instruments/Veeco Metrology Group.)

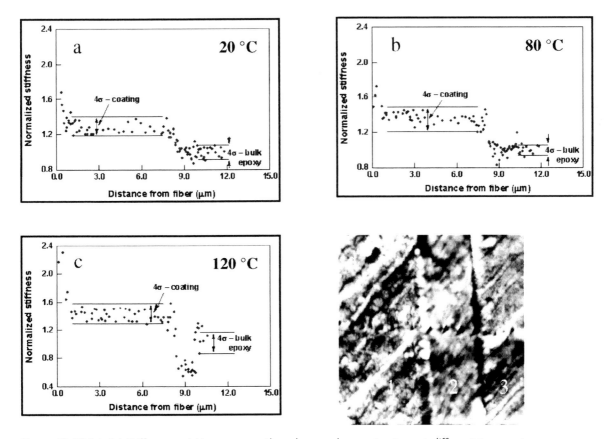

Figure 19.39 (a)–(c) Stiffness variation across a three-layer polymer structure at different temperatures. (d) Atomic force microscopy height image of indents made at 120 °C. Regions 1, 2, and 3 are the coating, interphase, and bulk epoxy, respectively, 5-μm scan. (Reproduced with permission from reference 52.)

glass transition temperature of about 100 to 120 °C, compared with 150 and 160 °C for the coating and bulk epoxies. The existence of the interphase and its behavior were unexpected. This work thus gives an important insight into the behavior of composite materials as a function of temperature.

Conclusions

AFM is the leading scanning probe technique and has become important in polymer analysis. AFM is based on simple principles and the instrumentation is versatile and reliable. Although novice users can achieve useful results, experience and attention to detail are rewarded. We discuss operating procedures and image interpretation, and point out potential pitfalls and challenges. A principle AFM application is making high-resolution 3D images of surface topography. The examination of roughness of manufactured surfaces and accurate control of microscopic patterns are two important areas of industrial AFM applications. The unbeatable force sensitivity and nanometer-scale resolution of AFM is likely to make it practically the exclusive method for evaluation of sub-100-nm structures that will be the essence of nanotechnology. Using phase contrast and other techniques that sense material properties, AFM visualizes surface structures at the micrometer and sub-micrometer scales. Often, AFM images show structures that cannot be seen easily or at all by other techniques. AFM routinely shows the organization of crystalline polymers, microphase separation of block copolymers, distribution of individual components in polymer blends and composites, and the presence of unwanted surface contaminants. These results substantially complement optical and electron microscopies, and are invaluable for optimization of polymer properties, improved understanding of fundamental structure-property relations, and for process control and troubleshooting. The recent availability of thermal accessories to scanning probe microscopes broadens AFM studies of polymers to include temperature effects. In our discussion of AFM techniques and applications, we supplement the presentation of concepts with some discussion of the scientific, engineering, and commercial motivations for the work. We also discuss how AFM results compare with other techniques. We hope that this information will be of value both to the bench scientist and to the manager of an analytical or materials characterization laboratory, whose job is to solve problems efficiently by selecting the appropriate techniques from a large number of possibilities.

The above-described applications are the main ones now, but are not the only applications that define the future of AFM. Measuring local mechanical, adhesive, electric, and thermal properties with specially designed probes is another area that is developing quickly. In the future, differences in thermal and electric properties may be employed for compositional imaging, as we now do with mechanical properties. As is true of all other small spot analysis techniques, the spatial resolution of material property maps will probably be inferior to the resolution of standard images used for visualization only. Nevertheless, such measurements will help us to understand various aspects of polymer material performance at the micrometer scale. Combined efforts of experimentalists and theoreticians are needed for better understanding and quantification of local surface properties (such as mechanical, electric, and thermal) measured with AFM on the micrometer and submicrometer scales.

References

1. Binnig, G.; Rohrer, H.; Gerber, C.; Weibel, E. *Phys. Rev. Lett.* **1982**, *49*, 57–61.
2. Binnig, G.; Quate, C.; Gerber, C. *Phys. Rev. Lett.* **1986**, *56*, 930–933.
3. Magonov, S. N.; Whangbo M.-H. *Surface Analysis with STM and AFM.* VCH: Weinheim, 1996; Chapters 6, 9, 10.
4. Sheiko S. *Adv. Polym. Sci.* **1999**, *151*, 61–174.
5. Magonov S. Atomic Force Microscopy in Analysis of Polymers. In *Encyclopedia of Analytical Chemistry*; Meyers, R. A., Ed.; Wiley & Sons: Chichester, U.K., 2000, pp 7432–7491.
6. Scanning Probe Microscopy of Polymers; Ratner B., Tsukruk V. V., Eds.; ASC Symposium Series 694; American Chemical Society: Washington, DC, 1998.
7. McMaster, T. J.; Hobbs, J. K.; Barham, P. J.; Miles, M. J. *Probe Microsc.* **1997**, *1*, 43–56.
8. Pearce, R.; Vansco, J. *Polymers* **1998**, *39*, 1237–1242.
9. Shiku, H.; Dunn, R. C. *Anal. Chem.* **1999**, *71*, 23A.
10. Cleveland, J. P.; Manne, S.; Bocek, D.; Hansma, P. K. *Rev. Sci. Instrum.* **1993**, *64*, 403–405.
11. Weisenhorn, A. L.; Maivald, P.; Butt, H.-J.; Hansma, P. K. *Phys. Rev. B* **1992**, *45*, 11226–11232.
12. Bustamante, C.; Rivetti, C.; Keller, D. *Curr. Opin. Struct. Biol.* **1997**, *7*, 709–716.
13. Lee, G. U.; Chrisey, L. A.; Colton, R. J. *Science* **1994**, *266*, 771–773.
14. Grandbois, M.; Beyer, M.; Rief, M.; Clausen-Schaumann, H.; Gaub, H. E. *Science* **1999**, *283*, 1727–1730.
15. Bliznyuk, V. N.; Hazel, J. H.; Wu, J.; Tsukruk, V. V. *Scanning Probe Microscopy of Polymers*; Ratner B., Tsukruk V. V., Eds.; ACS Symposium Series 694; American Chemical Society: Washington, DC, 1998; p 252.
16. Cleveland, J. P.; Anczykowski, B.; Schmid, A. E.; Elings, V. B. *Appl. Phys. Lett.* **1998**, *72*, 2613–2615.
17. Wang, L. *Appl. Phys. Lett.* **1998**, *73*, 3781–3783.
18. Heinz, W. F.; Hoh, J. H. *Trends Biotechnol.* **1999**, *17*, 143–150.
19. Viswanathan, R.; Heaney, M. B. *Phys. Rev. Lett.* **1995**, *75*, 4433–4436.
20. Magonov, S. N.; Godovsky Y. K. *Am. Lab.* **1999**, *31*(4), 52–58.
21. Hoh, J. H.; Lal, R.; John, S. A.; Revel, J. P.; Arnsdorf, M. F. *Science* **1991**, *253*, 1405–1408.

22. Sun, H. B.; Smith, G. N., Jr.; Hasty, K. A.; Yokota, H. *Anal. Biochem.* **2000**, *283*, 153–158.

23. (a) Elings, V. B.; Gurley, J. A. U.S. Patent 5,051,646, 1991. (b) Chernoff, D.; Lohr, J. U.S. Patent 5,644,512, 1997.

24. Landman, U.; Luedke, W. D.; Nitzan, A. *Surf. Sci.* **1989**, *210*, L177–L184.

25. Magonov S. N.; Cleveland, J.; Elings, V.; Denley, R.; Whangbo, M.-H. *Surf. Sci.* **1997, 389**, 201–211.

26. Sugawara, J.; Ohta, M.; Ueyama, H.; Morita, S. *Science* **1995**, *270*, 1646.

27. (a) Villarrubia, J. S. *J. Res. Natl. Inst. Stand. Technol.* **1997**, *102*, 425–454. (b) Villarrubia, J. S. *J. Vac. Sci. Technol. B*, **1996**, *14*, 1518–1521. (c) ASTM E1813. *Standard Practice for Measuring and Reporting Probe Tip Shape in Scanning Probe Microscopy*; American Society for Testing and Materials: West Conshohocken, PA, 2002.

28. Halliday D.; Resnick, R. *Fundamentals of Physics*; Wiley & Sons: New York, 1970; pp 705–708.

29. ASTM subcommittee E42.14. *Guide to Scanner and Tip Related Artifacts in Scanning Tunneling Microscopy and Atomic Force Microscopy*; American Society for Testing and Materials: West Conshohocken, PA, in preparation.

30. Ohmi, T.; Miyashita, M.; Itano, M.; Imaoka, T.; Kawanabe, I. *IEEE Trans. Electron. Devices* **1992**, *39*, 537–545.

31. Gilicinski, A. G.; Hegedus C. R. *Progr. Org. Coat.* **1997**, *32*, 81.

32. Magonov, S. B.; Antrim B. Unpublished work, 1999.

33. (a) Amstutz, H. *Surface Texture: The Parameters*; Technical Note MI-TP-003-0785; Sheffield Measurement Division, Warner & Swasey Co.: Dayton, OH, 1985. See also www.predev.com/smg/index.html ("Surface Metrology Guide", Precision Devices Inc., Milan, MI). (b) *Surface Texture, Surface Roughness, Waviness and Lay*; ANSI/ASME Standard B46.1–1995; ASME Press: Fairfield, NJ.

34. Zhang, Z. R.; Wygant, R. W.; Lyons, A. V.; Adamsky, F. A. How Coating Structure Relates to Performance in Coated SBS Board: A Fundamental Approach. *1999 TAPPI Coating Conference Proceedings*; TAPPI Press: Atlanta, GA, 1999; pp 275–286.

35. DiscTrack Plus Media Measurement System, Advanced Surface Microscopy, Inc. www.asmicro.com.

36. Chernoff, D. A.; Lohr, J. D.; Hansen, D.; Lines, M. High-Precision Calibration of a Scanning Probe Microscope (SPM) for Pitch and Overlay Measurements. In *Metrology, Inspection, and Process Control for Microlithography*; SPIE: Bellingham, WA, 1997; p 243.

37. Chernoff, D. A.; Burkhead, D. L. *J. Vac. Sci. Technol.* **1999**, *A17*, 1457–1462.

38. Han, W. H.; Lindsay, S. M.; Dlakic, M.; Harrington, R. E. *Nature* **1997**, *386*, 563–564.

39. Ponomarenko, S. A.; Boiko, N. I.; Shibaev V. P.; Magonov, S. N. *Langmuir* **2000**, *16*, 5487–5493.

40. Prokhorova, S. A.; Sheiko, S. S.; Ahn, C.-H.; Percec, V.; Moeller, M. *Macromolecules* **1999**, *32*, 2653–2660.

41. Piez, K. A. Collagen. In *Encyclopedia of Polymer Science and Engineering*; Wiley & Sons: New York, 1985; Volume 3, p 699.

42. Yokota, H.; Johnson, F.; Lu, H.; Robinson, R. M.; Belu, A. M.; Garrison, M. D.; Ratner, B. D.; Trask, J.; Miller, D. L. *Nucleic Acids Res.* **1997**, *25*, 1064–1070.

43. Heck, B.; Hugel, T.; Iijima, M.; Sadiku, E.; G. Strobl. *New J. Phys* **1999**, *1*, 17.1–17.29.

44. Galuska, A. A., personal communication, 1998.

45. Galuska A. A.; Poulter, R. R.; McElrath, K. O. *Surf. Interface Anal.* **1997**, *25*, 418–429.

46. Johnston E.; Bullock, S.; Uilk, J.; Gatenholm, P.; Wynne, K. J. *Macromolecules* **1999**, *32*, 8173–8182.

47. Pereira, D. Lignin Studied by Atomic Force Microscopy. Doctorate Thesis, Basel University, Basel, Switzerland, 1998, Chapter 1.

48. Chernoff, D. A. In *Proceedings Microscopy and Microanalysis*; Bailey, G. W., Ellisman, M. H.; Hennigar, R. A.; Zaluzec, N. J., Eds.; Jones & Begell Publishing: New York, 1995, pp 888–889.

49. (a) Pereira, D. E. D.; Chernoff, D.; Claudio-da-Silva, E., Jr.; Demuner, B. J. The Use of AFM to Investigate the Delignification Process: Part I—AFM Performance by Differentiating Pulping Processes; *ATIP Magazine* **2001**, *56*(2), 6–12. (b) Pereira, D. Lignin Studied by Atomic Force Microscopy. Doctorate Thesis, Basel University, Basel, Switzerland, 1998, Chapter 5, 8.

50. Godovsky, Y. K.; Magonov S. N. *Langmuir* **2000**, *16*, 3549–3552.

51. VanLandingham, M. R.; McKnight, S. H.; Palmese, G. R.; Elings, J. R.; Huang, X.; Bogetti, T. A.; Eduljee, R. F.; Gillespie, J. W., Jr. *J. Adhesion* **1997**, *64*, 31–59.

52. VanLandingham, M. R; Dagastine, R. R.; Eduljee, R. F.; McCullough, R. L.; Gillespie, J. W., Jr. *Composites A* **1999**, *30*, 75–83.

53. Bogetti, T. A.; Wang, T.; VanLandingham, M. R.; Gillespie, J. W., Jr. *Composites A* **1999**, *30*, 85–94.

54. Magonov, S. N.; Elings, V.; Whangbo, M.-H. *Surf. Sci.* **1997**, *375*, L385–L391.

PART VI

PERFORMANCE PROPERTIES

20

Behavior in Solvents

EDWARD N. PETERS

The behavior of polymers in solvents is important for both practical and theoretical reasons. Several characterization techniques of macromolecules are performed on solutions of polymers.[1,2] For example, molecular weight determinations are obtained from solution techniques such as gel permeation chromatography, dilute solution viscometry light scattering, and osmometry. Solubility is an important criterion in defining the conditions for polymerization. The solubility of polymers is an important consideration in the formulation of coatings and adhesives.[3,4] Solutions of polymers are used in the spinning of fibers and the formation of films. The effectiveness of additives such as plasticizers, antioxidants, and colorants may depend on their solubility in the polymer. On the other hand, in some applications the effect of solvents on a polymer may be undesirable. For example, crazing, environmental stress cracking, and swelling caused by the presence of solvents can have an adverse effect on a polymer.

Solubility

A polymer solution is a uniform molecular dispersion of a polymeric solute in a solvent that is usually of much lower molecular weight. The solubility of a polymer is the measure of the extent to which it and a solvent can be homogeneously mixed. Solubility can be expressed as the greatest mass of polymer that can be dissolved in a specified mass or volume of solvent at a particular temperature.

The process of dissolving a pure polymer begins with solvent molecules diffusing into the polymer. Polymer chains near the surface swell to accommodate the penetrating molecules of solvent. Individual polymer molecules may be freed from near the surface and diffuse into the solvent phase. The system eventually becomes homogeneous if the solubility limit is not exceeded. The extent to which the solution process can occur depends on the chemical nature of

the polymer (amorphous, semicrystalline, degree of crystallinity) and solvent, the molecular weight of the polymer, and the temperature.

Polymers, by virtue of their high molecular weight, are soluble only in selected solvents. Some general experimental observations on the dissolution of polymers follow:[5]

1. Like dissolves like; that is, chemical similarity of polymers and solvent increases the likelihood of solubility. For example, polar polymers are more likely to dissolve in polar liquids than in nonpolar liquids. Polymers that are predominantly aromatic are more likely to dissolve in aromatic liquids than in hydrocarbon liquids.

2. In a given solvent at a particular temperature, the solubility of a polymer will decrease with increasing molecular weight.

3. In a given solvent, the solubility of a polymer will increase with increasing temperature in most cases.

4. Solubility decreases when strong intermolecular forces are present. Thus, crystalline polymers exhibit decreased solubility and are often soluble only at elevated temperatures (sometimes above the crystalline melting point). In general, crystallinity acts like pseudo-cross-linking; in some cases, however, it is possible to find solvents strong enough to overcome the crystalline bonding forces and dissolve the polymer.

5. The rate of dissolution of branched polymer can be affected by the nature of the branching. Solubility can increase with short branches that tend to retard the polymer chains from close packing and hence allow the solvent molecules to diffuse more readily. Solubility can decrease with longer branches because the increased chain entanglement of long branches makes it harder for the polymer chains to separate.

6. Cross-linking eliminates solubility. All of the polymer chains are interconnected by chemical bonds in a

cross-linked network. Solvent molecules will diffuse into the network polymer matrix and it will swell. However, polymer chains cannot be freed into the solvent phase. At the saturation or equilibrium state, the swollen polymer cannot accept additional solvent molecules and there is a two-phase system comprising a saturated polymer gel and excess solvent.

Items 1, 2, 3, 4, and 6 are equilibrium phenomena and, therefore, in principle, can be described thermodynamically, whereas item 5 is a kinetic phenomenon and is governed by the rates of diffusion of a solvent and polymer.

Thermodynamic Basis for Polymer Solubility

The process of dissolving a polymer in a solvent is governed by the familiar free energy equation (eq 1):

$$\Delta G_m = \Delta H_m - T\Delta S_m \qquad (1)$$

where ΔG_m is the change in Gibbs free energy of mixing, ΔH_m is the enthalpy of mixing, T is the absolute temperature at which the process is carried out, and ΔS_m is the entropy of mixing.[2,6,7]

A necessary, although not sufficient condition for the solution process to be thermodynamically feasible is that ΔG_m is negative. Both ΔH_m and ΔS_m vary with temperature and concentration. The absolute temperature must be positive, and the change in entropy for a solution process is (with rare exception) positive, owing to the more random nature of solutions compared to that of the unmixed components. Although the product of T and ΔS_m is positive, the third term $(-T\Delta S_m)$ in eq 1 has a negative sign and hence it favors solubility.

Because the dissolution of a polymer is always connected with a positive entropy, the magnitude of the enthalpy is the deciding factor in determining the sign of the free energy change and hence the solubility of a polymer in a solvent. The change in enthalpy may be either positive or negative. A negative value of ΔH_m indicates that the solution of the polymer and solvent is in a lower energy state and the solubility of the polymer in the solvent is favored. A negative ΔH_m occurs where specific interactions (e.g., hydrogen bonding) are formed between the solvent and polymer molecules.

A positive ΔH_m means the solvent and polymer prefer to remain separate; that is, the pure materials are in a lower energy state. On the other hand, if ΔH_m is positive, solubility of the polymer in the solvent would be favored if ΔH_m is less than $T\Delta S_m$.

Phase Equilibria

The solubility of a binary polymer-solvent mixture can be determined by examining the free energy of the system as a function of composition.[2,4,7,8] A schematic representation

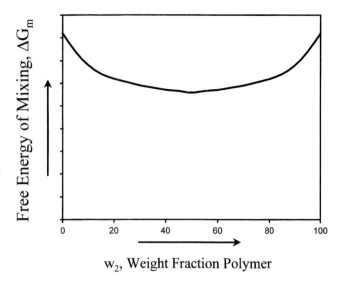

w_2, **Weight Fraction Polymer**

Figure 20.1 Free energy of mixing versus composition: single phase.

of the free energy of mixing is shown graphically in Figure 20.1, in which ΔG_m is plotted against the weight fraction w_2 of polymer. The mixing occurs at constant temperature and pressure. The curve in Figure 20.1 is characteristic of a system in which the components are miscible in all proportions. The thermodynamic requirement for the binary mixture to be stable at all compositions with no phase separation is such that a curve is everywhere concave upward that is, a negative ΔG_m exists and hence has no inflection points. Therefore, the curve represents the condition where the binary mixture is miscible in all proportions.

The case of partial miscibility is shown graphically in Figure 20.2. This curve has two upward-facing concavities

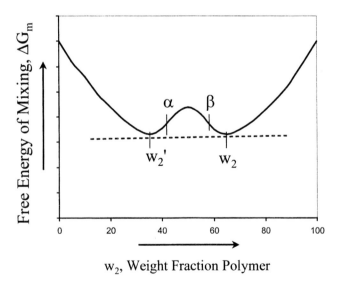

w_2, **Weight Fraction Polymer**

Figure 20.2 Free energy of mixing versus composition: partial miscibility. (Data from references 2, 8, and 9.)

separated by a convex section and two points of inflection, α and β. Between the inflection points α and β, the curve is convex or downward (positive ΔG_m) and hence a one-phase system is thermodynamically unstable. There are two regions of thermodynamic stability (soluble-one phase) at either end of the composition range, a region of total instability between α and β, and two metastable regions (between w_2' and α, and between β and w_2). It can be shown that w_2' and w_2 are located at the two points of tangency on the straight line drawn tangent to the two points on the curve, as depicted by the dashed line in Figure 20.2.[2,8,9]

The addition of more polymer to the solution beyond the concentration at inflection point α results in the formation of a second phase with concentration w_2, which is in equilibrium with the first phase at w_2', that is, the system achieves its lowest free energy by forming two phases. Thus, at concentrations between the two inflection points there can only be two phases with concentrations w_2' and w_2, where w_2' is the solubility limit of polymer in solvent and w_2 is the solubility limit of the solvent in polymer.[2,8]

The effect of temperature on solubility of a polymer-solvent system is depicted graphically in Figure 20.3. Here the ΔG_m response surface is shown for different temperatures as a function of the weight fraction of polymer. At the higher temperatures, the polymer-solvent systems are completely miscible. As the temperature is decreased, phase separation occurs at the lower temperatures.[8] Phase separation is greater at lower temperatures, that is, the two points of inflection are further apart. The temperature at which the two inflection points coincide is called the critical solution temperature, T_c. Above the T_c there is complete solubility, and below it there is partial solubility.

Another feature of Figure 20.3 is the gray area that is bordered by the inflection points. This represents the composition as a function of temperature where the region is two-phase. The curved area is sometimes referred to as the cloud-point curve because typically on cooling from the one-phase region, solutions become cloudy as droplets of a second phase form. The shapes of these solubility curves vary depending on the polydispersity of the polymer.

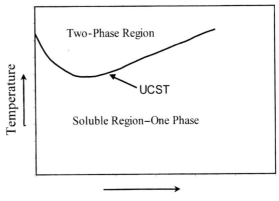

Figure 20.4 Upper critical solution temperature (UCST).

Upper and Lower Critical Solution Temperature

In the previous section, phase separation in a binary solution upon lowering the temperature is analyzed and the critical temperature defined. This can be depicted graphically in Figure 20.4 by plotting weight fraction of polymer versus the critical temperature for solubility. This curve defines the temperature for critical phase separation. The boundary between the soluble, single-phase region and the two-phase region is called the upper critical solution temperature (UCST) and lies above the two-phase region.

Several polymer-solvent systems exhibit a second critical solution temperature as the temperature of the mixture is raised.[10–12] The critical point associated with this phenomenon is called the lower critical solution temperature (LCST).[12] It is depicted graphically in Figure 20.5. Thus, the LCST lies below the two-phase system.

The existence of an LCST implies that both ΔH and ΔS are negative and hence phase separation occurs on heating. The more familiar situation observed near room temperature is the UCST, where such phase separation occurs on

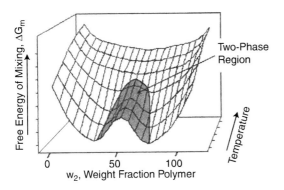

Figure 20.3 The effect of temperature on polymer solubility.

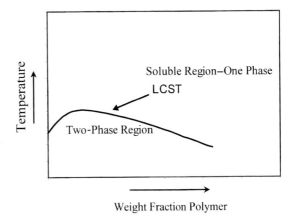

Figure 20.5 Lower critical solution temperature (LCST).

cooling. In either case, it has been reported that ΔH and ΔS must be of the same sign in order for separation to occur:[1]

If $\Delta H < 0$ and $\Delta S < 0$, then an LCST may occur; and if $\Delta H > 0$ and $\Delta S > 0$, then a UCST may occur.

Effect of Molecular Weight on Solubility

In a given solvent at a particular temperature, the solubility of a polymer will decrease with increasing molecular weight. A phase diagram of weight fraction of polymer versus temperature for a polymer with increasing molecular weights is depicted graphically in Figure 20.6. In this three-coordinate graph, the response surface is the critical solution temperature, T_C, as a function of weight fraction of polymer and molecular weight of polymer. At a given weight fraction of polymer, as the molecular weight of the polymer is increased, the temperature needed to reach the critical solution temperature increases. In other words, as the molecular weight is increased, the polymer is no longer completely soluble and the temperature must be increased to get all of the polymer in solution (i.e., to reach T_C). Hence, a higher molecular weight requires a higher the T_C. Moreover, at a given T_C where a lower molecular weight polymer is soluble, a higher molecular weight polymer is not completely miscible.

Properties of Dilute Solutions

In a fairly dilute solution there are not too many entanglements between the polymer molecules. In a very good solvent (one for which the solubility parameter very closely matches that of the polymer), the secondary forces between polymer segments and solvent molecules are strong, and the polymer molecules will assume a dispersed conformation in solution. In a poorer solvent but one that still dissolves the polymer, the attractive forces between the segments of the polymer chain will be greater than those

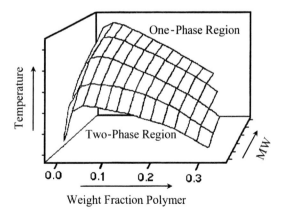

Figure 20.6 Effect of molecular weight and temperature on solubility.

between the chain segments and the solvent—that is, the chain segments prefer their own company, and the chain will form a ball. This is depicted by the simple model shown in Figure 20.7.

Prediction of Solubility

As previously discussed, the $T\Delta S_m$ term in eq 1 usually favors solubility. Thus, it is important to be able to estimate the magnitude of the ΔH_m term. For regular solutions (those in which solute and solvent do not form specific interactions), Hildebrand and Scott proposed that the change in internal energy upon solution is approximated by eq 2:[13,14]

$$\Delta H_m = V_m \left[(\Delta E_1/V_1)^{1/2} - (\Delta E_2/V_2)^{1/2}\right]^2 \varphi_1\varphi_2 \qquad (2)$$

where ΔH_m is the overall heat of mixing; V_m is the total volume of the mixture; ΔE_1 and ΔE_2 are the energy of vaporization of solvent and polymer, respectively; V_1 and V_2 are the molar volume of solvent and polymer, respectively; and φ_1 and φ_2 are the volume fraction of solvent and polymer in the mixture, respectively.

The expression $\Delta E/V$ is the energy of vaporization per unit volume. This is a measure of the strength of the intermolecular forces holding the molecules together in the liquid state. It also has been described as the cohesive energy density (CED).[15] Equation 2 can be rearranged to yield eq 3:

$$\Delta H_m/V_m\varphi_1\varphi_2 = \left[(\Delta E_1/V_1)^{1/2} - (\Delta E_2/V_2)^{1/2}\right]^2$$
$$= \left[(CED_1)^{1/2} - (CED_2)^{1/2}\right]^2 \qquad (3)$$

Thus, the heat of mixing per unit volume is equal to the square of the difference between the square roots of the CEDs of the polymer and solvent. This is a good approximation for the dissolution of polymers under most conditions, so eq 3 provides a means of estimating enthalpies of solution if the CEDs of the polymer and solvent are known.

Solubility Parameters

The square root of the CED is commonly referred to as the solubility parameter, δ. This is expressed mathematically in eq 4:

$$\delta = (\Delta E/V)^{1/2} = (CED)^{1/2} \qquad (4)$$

Information on polymer solubility can be estimated using solubility parameters.[6,7,16] Hence, eq 3 could be written as eq 5:

$$\Delta H_m/V_m\varphi_1\varphi_2 = (\delta_1 - \delta_2)^2 \qquad (5)$$

where δ_1 and δ_2 are the solubility parameters for solvent and polymer, respectively.

Inspection of eq 5 reveals that ΔH_m is minimized, and the tendency toward solubility is maximized by matching the solubility parameters as closely as possible. Thus, heat of mixing per unit volume of a polymer and solvent is dependent on the term $(\delta_1 - \delta_2)^2$. Clearly, the term $(\delta_1 - \delta_2)^2$ has to be relatively small in order for the heat of mixing not to

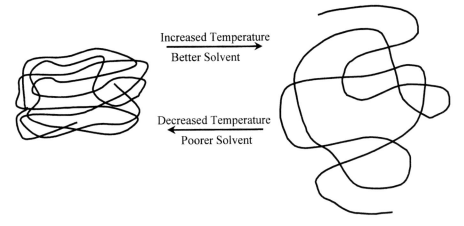

Increased Temperature
Better Solvent

Decreased Temperature
Poorer Solvent

Figure 20.7 The effect of solvent on dilute solutions.

be so large that it prevents dissolution. Hence, if the δ values of the polymer and solvent are nearly equal, then miscibility of the polymer and solvent is favored. Furthermore, if $(\delta_1 - \delta_2)^2$ equals zero, then solution is ensured by the entropy factor.

The utility of solubility parameters is that they allow the heat of mixing to be estimated in terms of a single parameter assigned to each component. This considerably simplifies the characterization and prediction of mixing.

Other applications of solubility parameters include finding compatible solvents for coating resins, predicting the compatibility of plasticizers with polymers, studying the swelling of cross-linked polymers in contact with solvents, estimating solvent pressure in devolatization and reactor equipment, and studying polymer-polymer, polymer-binary, random copolymer, and multicomponent solvent equilibria.

The solubility parameters of most solvents can be determined from their molar energies or enthalpies of vaporization, as noted above. Typical values of δ_1 for various solvents appear in Tables 20.1, 20.2, and 20.3 for poorly, moderately, and strongly hydrogen-bonding solvents, respectively.[6,7,13–21]

Mixtures of solvents are often used to dissolve polymers and are used extensively in coatings. Typically, solvents with different evaporation rates are used in coatings to facilitate a uniform film formation. Phenol, which is a strong hydrogen-bonding solvent, is a solid at room temperature; hence, a mixture of phenol with another solvent would circumvent the need to use warm phenol (temperatures above the 40 °C melting point). Moreover, a mixture of two nonsolvents could be a solvent for a given polymer if one nonsolvent δ value is lower, the other is higher than the solubility parameter of the polymer, and the solubility parameter of the mixed solvents, δ_1, is similar to that of the polymer. The solubility parameter for a mixed solvent can be approximated from the volume fraction of the components. Thus, δ_1 for a mixture of solvent A and solvent B can be obtained from eq 6:

$$\delta_1 = \delta_A \varphi_A + \delta_B \varphi_B \tag{6}$$

where φ_A and φ_B are the volume fractions of solvent A and B, respectively.

High polymers, on the other hand, cannot be vaporized without decomposition (because of their large molecular weight, polymers have enormous cohesive energies). Hence, their solubility parameters must be determined indirectly.

Several methods have been employed to determine experimentally the solubility parameter of polymers.[6,7,16–24] Most methods use a series of potential solvents with known solubility parameters. In one method, the solubility parameter δ_2 for the polymer is taken at the midpoint of the range of δ values for those liquids that completely dissolve the polymer.[25] If the polymer is not completely miscible, then δ_2 is equated to the solubility parameter δ_1 of the liquid in which it has the greatest solubility. In another method, the swelling of a lightly cross-linked polymer is measured in various liquids. The greatest swellings should be found with solvents for which $\delta_1 \sim \delta_2$.

The variation of the intrinsic viscosity of the dilute polymer solution with the solubility parameter of the solvent provides yet another method for evaluating δ_2. The viscosity is greatest when $\delta_1 \sim \delta_2$ (see Figure 20.7).[26] These and other procedures have been summarized.[6,7,16]

Solubility parameters can be calculated for polymers and solvents.[16,27,28] Molar attraction constants allow the estimation of solubility parameters merely from the structural formula of the compound and its density.[16,17,29] The molar attraction constants, F, are related to the solubility parameters by eq 7:

$$\delta = \frac{d}{M} \Sigma F \tag{7}$$

where ΣF is the sum for all the molar attraction constants for every atom-group in the molecule, d is the density, and M is the molecular weight. For polymers, the F values are added for the structural groups of the repeating unit in the

Table 20.1 Solubility parameters of nonpolar and poorly hydrogen-bonding solvents

Solvent	δ (Mpa$^{1/2}$)
n-Pentane	14.3
n-Hexane	14.9
n-Octane	15.6
Methylcyclohexane	16.0
Cycolhexane	16.8
Benzonitrile	17.2
α-Methylstyrene	17.4
Carbon tetrachloride	17.6
Ethylbenzene	18.0
Toluene	18.2
Benzene	18.8
Chloroform	19.0
Tetrahydronaphthalene	19.4
Chlorobenzene	19.4
Bromobenzene	20.3
Carbon disulfide	20.5
1,2-Dichlorobenzene	20.5
Nitrobenzene	20.5
1-Nitropropane	21.1
Bromonaphthlene	21.7
Nitroethane	22.7
Acetonitrile	24.3
Nitromethane	26.0

Data adapted from references 6, 7, and 13–21.

polymer chain and M is the molecular weight of the repeating unit.

Molar attraction constants at 25 °C are summarized in Tables 20.4 and 20.5. Small's values were determined from measurements of the heats of vaporization.[29] Hoy's values were derived from vapor pressure measurements.[17] Other molar attraction constants have been published.[16,27,28] The group contribution technique is based on the assumption that the contributions of different structural groups to the thermodynamic property are additive.

Molar attraction constants should not be used for alcohols, amines, carboxylic acids, or other strongly hydrogen bonded compounds unless such functional groups constitute only a small part of the molecule so that the proportional error is not great. In general, the accuracy is quite good to the first decimal place, which is adequate for estimating polymer solubility.

The technique of calculating the solubility parameter for a polymer using molar attraction constant is exemplified for polystyrene in Table 20.6. The solubility parameter of alternating copolymer can be calculated by using the copolymer repeat unit as that of a homopolymer. The solubility parameter of random copolymer, δ_c, may be calculated by using eq 8:

$$\delta_c = \Sigma \delta_i w_i \qquad (8)$$

where δ_i is the solubility parameter that corresponds to monomer i in the copolymer, and w_i is the weight fraction

of repeating unit i in the copolymer.[30] Calculating solubility for graft and block copolymers is more difficult because of the longer chain lengths, differences in δ for each polymer chain, or differences in polarity or hydrogen-bonding tendencies.

Solubility parameters for a large number of solvents and polymers have been complied.[6,7,16–24] A list of δ_2 for various polymers appears in Table 20.7. Sometimes, a range of δ_2 is given on the basis of the various data in the literature and the fact that solvents with a δ_1 in that range will dissolve the polymer.

In applying solubility parameters to predict solubility, miscibility would usually be favored when δ_1 and δ_2 are within 2 or 3 MPa$^{1/2}$ and there are no strong polar or hydrogen-bonding interactions in either polymer or solvent.[13,31] However, miscibility may not always be achieved for various reasons. For example, semicrystalline polymers may not always dissolve in solvents with similar solubility parameters at temperatures much below their crystal melting points. The solvent may have insufficient solvating power available to force the low energy crystalline regions to go into solution. Instead, semicrystalline polymers may only be swollen or softened by solvents. Increased temperature will facilitate miscibility.

Another reason that miscibility may not be achieved is due to differences in the specific solvent-polymer interac-

Table 20.2 Solubility parameters of polar, moderately hydrogen-bonding solvents

Solvent	δ (Mpa$^{1/2}$)
Diethylether	15.1
2,6-Dimethyl-4-heptanone	16.0
3-Methyl-1-butyl acetate	16.0
2,4-Dimethyl-3-pentanone	16.4
2-Methyl-1-propyl acetate	17.0
Ethylene glycol diethyl ether	17.0
Ethyl proprionate	17.2
Isopropyl acetate	17.2
1-Butyl acetate	17.4
Methyl proprionate	18.2
Dibutyl phthalate	19.0
Methyl acetate	19.6
Acetone	20.3
Cyclohexanone	20.3
Dimethyl carbonate	20.3
Dioxane	20.5
Diphenyl ether	20.7
Methyl formate	20.9
Cyclopentanone	21.3
Methyl salicylate	21.7
Dimethyl sulfoxide (DMSO)	24.6
N,N-Dimethyl formamide (DMF)	24.8
2,3-Butylene carbonate	24.8
Propylene carbonate	27.2
Ethylene carbonate	30.1

Data adapted from references 6, 7, and 13–21.

Table 20.3 Solubility parameters of polar strongly hydrogen-bonding solvents

Solvent	δ (Mpa$^{1/2}$)
Triethyl amine	15.1
Tributyl amine	15.8
1-Dodecanol	16.6
Piperidine	17.8
Tri(propylene glycol)	18.8
2-Ethyl-1-hexanol	19.4
3-Methyl-1-butanol	20.5
4-Methyl-1-pentanol	20.5
Acetic acid	20.7
m-Cresol	20.9
1-Octanol	21.1
2-Methyl-1-propanol	21.5
2-Ethyl-1-butanol	21.5
1-Heptanol	21.7
1-Hexanol	21.9
Pyridine	21.9
1-Pentanol	22.3
1-Butanol	23.3
2-Propanol	23.5
1-Propanol	24.3
Ethanol	26.0
Methanol	29.7
Ethylene glycol	29.9

Data adapted from references 6, 7, and 13–21.

tion. In general, "like dissolves like" and nonpolar polymers are more likely to dissolve in nonpolar solvent, polar polymers are more likely to dissolve in polar solvents, and polymers that can hydrogen bond are more likely to dissolve in solvents that can hydrogen bond. Hence, solubility may not be achieved even if $\delta_1 = \delta_2$ when the polymer and solvent have different tendencies for hydrogen bond formation. Clearly, it is important to take into consideration the influence of hydrogen-bonding interactions.

In an attempt to improve the utility of the method for more complex systems, several empirical modifications have been proposed.[6,7] In one variation, the hydrogen bonding capacity of a potential solvent is assigned to one of three categories: strong hydrogen bonding capability (e.g., acids, alcohols, and amines); moderate (e.g., esters, ethers, ketones); or poor (e.g., hydrocarbons).[6] The δ_2 value assigned to the polymeric solute is adjusted depending on the hydrogen-bonding category of the solvent. Miscibility is predicted to be most likely when δ_1 and δ_2 are similar when determined for the same category of hydrogen bonding. In another version of the solubility parameter method, δ_1 is separated into contributions from dispersion, polar, and hydrogen-bonding interactions.[32]

Some of the challenges noted here perhaps reflect the use of an oversimplified view of solubility parameters and hydrogen bonding. However, any attempt to address the inadequacy will most likely complicate the predictive method without a commensurate gain in practical utility.

Getting a Polymer into Solution

Getting a polymer into solution depends on the nature of the polymer and solvent and the molecular weight of the polymer. After a good solvent has been identified, one can encounter a few peculiarities when attempting to get a polymer into solution. Initially, the polymer in the form of a powder, flake, pellet, or some form is placed in a container with a good solvent. As the solvent molecules are absorbed into the surface, the polymer molecules at the surface can be plasticized by the solvent. Hence, the glass transition temperature, T_g, of these molecules will decrease. As the T_g drops below room temperature, the surface can become sticky and the various particles of polymer can agglomerate to produce a ball of plasticized polymer, which can stick to the sides of the container. This greatly reduces the surface area and the rate of diffusion. Moreover, this

Table 20.4 Group molar attraction constants, part I

Group	Group type	F (Hoy) (MPa$^{1/2}$ cm^{3}mol^{-1})	F (Small) (MPa$^{1/2}$ cm^{3}mol^{-1})
–CH$_3$	Aliphatic	303	437
–CH$_2$–	Aliphatic	269	272
>CH–	Aliphatic	176	57
>C<	Aliphatic	66	–190
CH$_2$=	Alkene	259	388
–CH=	Alkene	249	277
>C=	Alkene	173	39
–C≡C–	Alkyne	—	454
–CH=	Aromatic	240	—
>C=	Aromatic	201	—
(phenyl)	Phenyl	—	1500
(phenylene)	Phenylene	—	1350
(naphthyl)	Naphthyl	—	2345
–O–	Ether	235	143
–O–	Epoxide	360	—
–O–	Acetal	236	—
–COO–	Ester	668	634
>C=O	Ketone	538	562
–(CO)O(CO)–	Anhydride	1160	—
–COOH	Carboxylic acid	565	—
–OH	Primary alcohol	674	—
–OH	Secondary alcohol	592	—
–OH	Tertiary alcohol	799	—
–OH	Phenolic	350	—
–OH	Hydroxyl	—	350
–NH$_2$	Primary amine	464	—
>NH	Secondary amine	368	—
>N–	Tertiary amine	125	—
–C≡N	Nitrile	725	839
–CONH–	Amide	1135	—
–CONH$_2$	Amide, pendant	1207	—
–OCONH–	Urethane	1262	—
–S–	Thioether	428	460
–SH	Thiol	—	644

Data adapted from references 6, 7, 17, and 29.

Table 20.5 Group molar attraction constants, part 2

Group	Group type	F (Hoy) (MPa$^{1/2}$ cm^{3}mol^{-1})	F (Small) (MPa$^{1/2}$ cm^{3}mol^{-1})
Halogen groups			
–Cl	Primary chloride	420	552
–Cl	Secondary chloride	426	–
<Cl$_2$	Gem dichloride	701	532
–Cl	Aromatic	329	–
–Br	Primary bromide	528	695
–Br	Aromatic	421	–
–F	Primary fluoride	85	–
–CF$_2$–	Fluorocarbon	–	307
–CF$_3$	Fluorocarbon	–	561
Structural features			
Diene			
	Conjugation	48	51
	Cis	–15	–
	Trans	–28	–
Aromatic substitution			
	Para	82	–
	Meta	14	–
	Ortho	20	–
Cycloaliphatic			
	Six-membered ring	–48	205
	Five-membered ring	43	225
	Four-membered ring	159	–
	Bicycloheptane	46	–
	Tricyclodecane	128	–

Data adapted from references 6, 7, 17, and 29.

agglomerated mass can cause agitation problems in a flask or kettle with a mechanical stirring paddle.

When small, dilute solutions of polymer are prepared, the polymer and solvent can be placed in a bottle with an inert cap and placed on a mechanical shaker or in an ultrasonic bath. Any agglomerated material should eventually dissolve.

When larger, more concentrated polymer solutions are prepared in a flask or kettle, it is recommended that the polymer be added in small portions to a stirred solvent to minimize the effects of agglomeration. Warming the solvents will increase the rate of dissolution. After most of the polymer has dissolved, another small portion can be added and dissolved.

A procedure needs to be developed for each particular set of circumstances depending on the targeted concentration and nature of the polymer and solvent. The key is to add

the polymer at a rate without affecting agitation. It is best to be conservative the first time.

Swelling Ratio, Gelation, and Cross-Link Density

Cross-linked polymers behave differently in solvents compared to linear and branched polymers. Cross-linking reactions are those that lead to the formation of insoluble and infusible polymers in which chains are joined together to form a three-dimensional network structure. In a good solvent, cross-linked polymers form a gel but do not dissolve to form a solution. The properties of a gel depend strongly on the interaction of these two components. The liquid prevents the polymer network from collapsing into a compact mass, and the network, in turn, retains the liquid.

Cross-linked polymers are used in thermoset molding compounds, cured rubbers, paints, adhesives, and sealants.[33,34] Cross-linked polymers can also contain uncross-linked polymers. The number of uncross-linked polymers present depends on the efficiency of the cross-linking reaction and the number of cross-linkable sites on the polymer chains. Measurement of the gel content, swelling ratio, and the amount of extractable polymer provides a means for controlling the cross-linking reaction and quality of the cured resin.

Degree of Cross-Linking

Degree of cross-linking can be expressed as number of cross-links per gram or per unit volume of polymer. This is expressed mathematically in eq 9.[35]

$$C = \frac{n}{2} = \frac{d}{2M_c} \qquad (9)$$

where C is moles of cross-links per unit volume, n is the number of network chains per unit volume, d is the density

Table 20.6 Calculation of the solubility parameter for polystyrene where $d = 1.05$ g cm^{-3}

$$\left[CH_2\text{-}CH \right]$$

Structural group	Number	M (g mol^{-1})	F (Small) (MPa$^{1/2}$ cm^3 mol^{-1})	F × number
–CH= aromatic	5	65.09	249	1245
>C= aromatic	1	12.01	173	173
–CH$_2$– saturated	1	14.03	269	269
–CH– saturated	1	13.02	176	176
		M = 104.15		ΣF = 1863

$\delta = (1.05$ g cm$^{-3})(1863$ MPa$^{1/2}$ cm^3 mol$^{-1})/104.15$ g mol$^{-1} = 18.8$ MPa$^{1/2}$

Table 20.7 Solubility parameters of various polymers

Polymer	δ (MPa$^{1/2}$)
Polyethylene (PE)	16–17
Polypropylene (PP)	16–17
Polystyrene (PS)	17–22
Poly(vinyl acetate) (PVOAc)	17–19
Poly(vinyl chloride) (PVC)	17–23
Poly(bisphenol A carbonate) (PC)	19–22
Poly(methyl methacrylate) (PMMA)	22–23
Poly(vinyl alcohol) (PVOH)	25–27
Poly(hexamethylene adipate) (PA66)	28–31
Polysulfone (PSF)	20.3–21.7
Poly(ethylene terephthalate) (PET)	21.7–22.7

Data adapted from references 6, 7, and 16–24.

of cross-linked polymer, and M_c is number-average molecular weight of the polymer segments between cross-links.

Swelling in a good solvent is a method used to determine the degree of cross-linking experimentally. However, in many cases experimental values do not agree with the degree of cross-linking calculated from knowledge of the cross-linking reaction.[36] These deviations may be due to assumptions made in deriving and simplifying the relationships between swelling and cross-link density. Indeed, the presence of long dangling chains ends, the distribution of cross-link sites, and chain entanglements (which can act like cross-link sites) will lead to deviations.

Swelling Measurements

The swelling ratio is the ratio of the solvent volume in the gel to the initial sample volume and is expressed mathematically in eq 10:

$$\text{Swelling ratio} = \frac{\text{solvent volume in gel}}{\text{volume initial sample}}$$
$$= \frac{V_{gel} - V_0}{V_0} \quad (10)$$

where V_{gel} is the volume of the gel, and V_0 is the initial sample volume.[37]

Measuring volumes, especially for a gel, can be cumbersome. It is easier to measure weights and calculate the volume via division by the density of the material.[37] Hence, the swelling ratio can be expressed as eq 11:

$$\text{Swelling ratio} = \frac{\text{wt solvent in gel}/d_s}{\text{wt initial sample}/d_0}$$
$$= \frac{(W_{gel} - W_0)/d_s}{W_0/d_0} \quad (11)$$

where W_{gel} is the weight of the swollen gel, W_0 is the weight of the initial sample, d_s is the density of the solvent, and d_0 is the density of the initial sample. Values for W_{gel}, W_0, d_s, and d_0 are determined experimentally, and then the swelling ratio is calculated from eq 11.

This procedure is based on the assumption that all the polymer chains are interconnected via cross-linking. However, un–cross-linked polymer chains may also be present. In a good solvent, these un–cross-linked chains can dissolve in the solvent and be extracted into excess solvent. Hence, the swelling would be underestimated because the polymer making up the gel is less than W_0. If the amount of extracted polymer is large, it needs to be measured and its removal from the gel needs to be included in the determination of swelling.[38]

For a cross-linked sample with a significant fraction of extractable polymer, the percent extractables and swelling ratio are derived from eqs 11 and 13, respectively.

$$\% \text{ Extractables} = \frac{W_x}{W_0} \times 100 \quad (12)$$

$$\text{Swelling ratio} = \frac{V_s}{V_p} \quad (13)$$

In eqs 12 and 13, W_x is the weight of extracted polymer (obtained by evaporation of the excess solvent), V_s is the volume of solvent in gel $= (W_{gel} - W_{dg})/d_s$, V_p is the volume of polymer in the gel $= W_{dg}/d_0$, and W_{dg} is the weight of the dried gel.

Equation 11 can be written as eq 14:

$$\text{Swelling ratio} = \frac{(W_{gel} - W_{dg})/d_s}{W_{dg}/d_0} \quad (14)$$

Values are determined experimentally for W_{gel}, W_0, W_x, d_s, d_0, and W_{dg}. Because $W_0 = W_{dg} + W_x$, which rearranges to $W_{dg} = W_0 - W_x$, then only two of these three values need be determined experimentally and the third can be calculated. The swelling ratio is calculated from eq 14.

It is common practice to add fillers to cross-linked polymers and rubbers. The presence of fillers needs to be taken into account when calculating the swelling ratio.[38] The actual weight of the polymer in the filled specimen needs to be used in calculations and is expressed by eq 15:

$$\text{Weight of polymer in sample} = W_0 f \quad (15)$$

where

$$f = \text{polymer fraction} = \frac{\text{weight of polymer in formulation}}{\text{total weight of formulation}}$$

Hence, eq 9 is modified to eq 16:

$$\text{Swelling ratio} = \frac{(W_{gel} - W_0 f)/d_s}{(W_0 f)d_0} \quad (16)$$

High swelling ratios indicate a low degree of cross-linking and that the molecular weight of the polymer between cross-link sites is large. A low swelling index indicates a higher cross-linked structure in which the molecular weight between cross-links is small. Low levels of extractables would suggest a high degree of cross-linking.

The use of swelling ratio is exemplified in the development of high-temperature, solvent-resistant poly(dodecacarborane-siloxane) elastomers. In general, the incorporation of carbon-fluorine bonds in polymers enhances the solvent resistance. Hence, a series of poly(dodecacarborane-siloxane)s was prepared with 0, 1, 2, and 3 trifluoropropyl groups per repeat unit.[39,40] The polymers were mixed with 30 parts per hundred resin (phr) fumed silica and 2.5 phr dicumyl peroxide. These materials were cured to produce elastomeric plaques from which test specimens were cut. These specimens were exposed to toluene and the swelling ratio was determined after 7 days exposure. The results appear in Table 20.8.

Fractionation

Synthetic polymers are not homogeneous. The heterogeneity of polymers can be classified according to molecular weight, chemical composition, molecular configuration, and structure. Such sources of polydispersity have a strong influence on some thermodynamic properties.

The molecular weight distribution is a general feature for most synthetic polymers. Indeed, almost all polymers are mixtures of homologues differing in molecular weight. It is very important to evaluate the molecular weight and molecular weight distribution of a polymer to establish structure-properties-processability relationships and to examine the polymerization mechanism by which the polymer was prepared. Moreover, changes in polydispersity may occur during degradation during processing, improper handling, or routine use, and may contribute substantially to an increase in the molecular weight distribution of the polymer.

The existence of molecular weight heterogeneity in synthetic polymers is quite general. It is one of their fundamental properties and is directly responsible for the need to use several molecular weight averages for their description. It also exerts an influence on all of the properties of the substance, both in solution and in the solid state.

Differences in chemical composition in polymers originate from synthetic reactions that can result in differences in substitution along the polymer backbone in random, alternating, block, and graft copolymers. Moreover, graft copolymers can contain varying amounts of homo-polymer. The polymer chains in copolymers can have varying ratios of repeat units or different sequence distributions (see chapter 20, this volume).

The third kind of heterogeneity mentioned above refers to differences in the physical configuration of the macromolecules, such as those between linear and branched polymers, that is, number, position, and length of side chains. In addition, difference in tacticity of the molecular chains present in the mixture may result in varying amounts of amorphous and crystalline segments in the polymer (see chapter 20, this volume).

The fractionation of a polymeric substance entails the orderly separation of a polymer into a number of fractions of different molecular weights, copolymer content, or structural isomerism. Polymer fractionation has been studied since the 1920s; the history of fractionation is almost as old as polymer science itself. There are several reviews on the theory and technique of fractionation.[41–45] The principles and theory of fractionation by solubility have been reviewed by several authors on the basis of the Flory-Huggins statistical thermodynamic treatment.[46–48] Early experimental techniques focused on fractionation according to molecular weight. Differences in chemical composition and physical structure are handled by modifications of techniques for molecular weight. Fractionation methods include fractionation by solubility, chromatography, sedimentation, diffusion, and ultrafiltration.[42,44,45,49]

Gel permeation chromatography (GPC) is considered the method of choice for fractionation of polymers, particularly when the instrument is equipped with an automated viscometer or multiple detectors (refractive index, ultraviolet, infrared) (see chapter 5, this volume).

Table 20.8 Swelling ratio of poly(dodecacarborane-siloxanes)

$f = 0.7/1.00 = 0.70$
d_s for toluene $= 0.866$ g/cm^3
% Extractables were negligible

$$-\left[-O-\underset{\underset{CH_3}{|}}{\overset{\overset{R_1}{|}}{Si}}-O-\underset{\underset{CH_3}{|}}{\overset{\overset{R_2}{|}}{Si}}CB_{10}H_{10}C\underset{\underset{CH_3}{|}}{\overset{\overset{R_3}{|}}{Si}}-\right]-$$

R_1	R_2, R_3	d_0 (g/cm^3)	W_0 (g)	$W_0 \times f$ (g)	W_{gel} (g)	Swelling ratio
Phenyl	CH$_3$	1.074	0.54	0.38	0.80	1.37
CH$_2$CH$_2$CF$_3$	CH$_3$	1.087	0.45	0.32	0.57	0.98
CH$_3$	CH$_2$CH$_2$CF$_3$	1.151	0.48	0.34	0.53	0.63
CH$_2$CH$_2$CF$_3$	CH$_2$CH$_2$CF$_3$	1.208	0.52	0.36	0.42	0.23

For R_1 = phenyl; R_2 and R_3 = CH$_3$
$W_0 \times f = 0.54 \times 0.70 = 0.38$

$$\text{Swelling ratio} = \frac{(0.86 - 0.38)/0.866}{0.38/1.074} = 1.37$$

Fractionation by solubility is often carried out on the basis of the decreasing solubility of a polydisperse sample with increasing molecular weight, varying levels of copolymer, or different levels of structural isomerism. For a homopolymer, the fractionation is based on lower molecular weight molecules being more soluble than higher molecular weight molecules. For copolymers or polymers with structural isomerism, fractionation is based on any heterogeneity in composition or structural differences that affect the solubility of the molecules. Polymers can be fractionated via selective precipitation from solution or selective extraction-dissolution.

Precipitation

Fractionation by precipitation is based on the dependence of solubility on molecular size and is carried out in steps with subsequent decreasing solvent power.[43,44,49] Precipitation is obtained by addition of nonsolvent, elimination of solvent by evaporation, or by lowering the solution temperature to achieve precipitation.

For example, consider a polymer in a good solvent, such as polystyrene ($\delta = 17-22$) in ethyl benzene ($\delta = 18.0$). At constant temperature a nonsolvent such as methanol with a δ value of 29.7 is added slowly to the stirred polymer solution. Gradual addition of the methanol is continued, making the ethyl benzene-methanol solvent system poorer. Eventually the polymer teeters on the brink of solubility—that is, $\Delta G_m = 0$ and $\Delta H_m = T\Delta S_m$. Ultimately, a point will be reached where the ethyl benzene-methanol solvent ratio becomes too poor to sustain solution, and the polymer starts to precipitate out; this is the onset of turbidity. This indicates that the larger polymer molecules are not soluble in the existing solvent-nonsolvent ratio. Thus, the attractive forces between the larger polymer segments become much greater than that between polymer and solvent. The methanol addition is halted and the solution is aged for at least 2 to 4 h, during which time the system equilibrates. Then slow, drop-wise addition of methanol is resumed. When the solution turns milky, the addition of nonsolvent is stopped. More polymer molecules with limited solubility in the solvent-nonsolvent will precipitate. The solution is then warmed until it becomes transparent, indicating that the polymer has redissolved. The solution is now cooled slowly to the original temperature with stirring. The solution is kept at this temperature for several hours to let the molecules precipitate and settle. The precipitate is then decanted, filtered, and dried. The supernatant liquid is further titrated with nonsolvent to obtain subsequent fractions of successively smaller molecules using the procedure outlined above. Adjusting either the temperature or the polymer-solvent system allows fractionation of the polymer according to molecular weight, as successively smaller molecules precipitate when the temperature is lowered or a poorer solvent is used.

Extraction

In extraction or column fractionation methods, the polymer is distributed on an inert support.[50] The coated support is placed in a column and successive elution takes place with solvents with increasing solvent power. The more soluble molecules are extracted first.

Initially, the process starts outside the column where the coated support for extraction is prepared.[50-54] The polymer may be deposited on glass beads, diatomaceous earth, sand, or another substrate.[55] The polymer coating on the support may be prepared by evaporation or by cooling a solution of the polymer in the presence of the support until the polymer is all precipitated on the support.[50,53,56]

Care in coating the support can improve fractionation. Ideally, each particle of support will be coated with a thin, uniform layer of polymer. This polymer layer varies in molecular weight, with the higher molecular weight polymer chains being closest to the support because they would be the least soluble and precipitate first. The lowest molecular weight polymer would be most soluble, and would likely precipitate last and be on the surface. A portion of this coated support is placed at the solvent inlet of a column packed with the same inert support. The coated material is then contacted-extracted by a solvent gradient (progressively more powerful solvents), which initially washes off the most soluble fraction and ends with the least soluble material in the last fraction.

Copolymers

Copolymers are important from both theoretical and practical viewpoints because numerous industrial polymers are copolymers. The fractionation of a copolymer consists of the separation into different molecular species, using suitable experimental techniques, to obtain fractions that are more homogeneous in composition and structure. The fractionation techniques for homopolymers are applied to the fractionation of copolymers, which have not only heterogeneity in molecular weight and structure but also heterogeneity in composition.[57-59] The choice of the fractionation conditions such as temperature, solvent, and nonsolvent are very much dependent on experimental work, and on a knowledge of the solubility parameters of the copolymer and of the homopolymers of each monomer from which the polymer is made.

Solvent Resistance Tests

The solvent or chemical resistance of a polymer is a measure of a polymer's ability to withstand chemical attack with minimal change in appearance, dimensions, mechanical properties, and weight over a period of time.[60] Polymers can absorb, dissolve, react chemically, become plasticized, or be stress cracked when exposed to chemical environments.[61] In general, a polymer's resistance to a solvent or chemical is enhanced by increased molecular weight of the polymer. Moreover, any detrimental effect of the chemical on the polymer increases as the temperature increases. If the chemical exposure is limited or infrequent, the service life of the polymeric part may still be long, but some change in properties usually occurs. Mechanical properties can be reduced and dimensions changed, but functionality may remain adequate. The final classification as chemically resistant depends on the application.

The effect of solvents and chemicals on polymers can be classified as absorption, dissolution, chemical attack, or stress cracking.

If a chemical is miscible with a polymer, absorption and plasticization may result. As the polymer absorbs the solvent, the liquid will behave like a plasticizer and will weaken cohesive forces and reduce mechanical properties. The polymer tends to swell, warpage may occur because of relaxation of molded-in stresses, and properties change. The strength, rigidity, creep resistance, and glass transition temperature will decrease, and ductility will increase.

Dissolution may result in loss of weight and mechanical strength. Most thermoplastics are soluble in some solvents. Solubility is favored by chemical similarity of polymer and solvent, that is, similar solubility parameters. However, high molecular weight plastics usually dissolve very slowly. Therefore, the solvation process can appear similar to a chemical reaction. In general, weight and dimensional changes and swelling are present along with a loss of properties.

The chemical can attack the polymer chain directly and lead to the breaking of covalent bonds along the polymer chain. This decrease of the molecular weight of the polymer will result in a decrease in mechanical properties.

The stress level is a very significant factor affecting the performance of a polymer. An unstressed plastic may appear to be unaffected by exposure to a solvent. The same solvent may cause catastrophic failure when the plastic is stressed. This phenomenon is called environmental stress cracking. As the stress level increases, resistance to a chemical environment decreases. In addition to the obvious applied loads, stresses can result from injection molding, forming, and assembly operations.

Solvent resistance of polymers is a complex subject.[61] The type of bonds in the polymer, molecular weight of the polymer, its amorphous or semicrystalline (degree of crystallinity) content, and branching are important factors in solvent resistance. Solvent test conditions include the length of exposure, concentration, temperature, and internal stress in the polymer. The type of exposure can include immersion in liquid chemicals or exposure to saturated vapors. Of prime importance is whether the chemical resistance of a polymer is reported from simple immersion tests with unstrained samples or as a function of strain in the polymer.

The type of stress on the polymer (tensile, compression, or torsional stress) during testing is critical. Another important factor is the effect of additives in the polymer such as plasticizers, fillers, stabilizers, and colorants. The chemical resistance data published by material suppliers is a source of information.

The design engineer needs to carefully consider the end-use environment when specifying a polymer for an application. Important aspects of solvent resistance that are sometimes overlooked are the processing, assembly, finishing, and cleaning operations to which the polymer will be exposed even before it reaches the end user.

Chemical resistance tests are conducted using various methods:

1. Immersion in the fluid, or exposure to the vapor, or both
2. Solvent-environmental stress-cracking resistance
   a. Tensile, compressive, or torsional stress for critical stress
   b. Constant strain jigs
   c. Elliptical strain jigs.

Sample preparation is very important. The specimen thickness influences the percentage change in dimensions and percentage change in mechanical properties. In addition, the surface area can affect the weight change caused by immersion in chemical agents. Moreover, injection-molded specimens may not exhibit the same performance as specimens cut from molded or otherwise formed sheets of a given material. Consequently, comparison of materials should be made only on the basis of results obtained from specimens of identical dimensions and like methods of preparation.

Molding conditions can affect the resistance of plastics to solvents and chemical reagents. Compression moldings should be carried out in a manner that will result in complete fusion of the plastic particles and provide a uniform specimen. Injection molding should be prepared in a manner that results in a minimum of molecular orientation and thermal stress.

Immersion Test

The immersion of plastic specimens in solvents is a standard procedure used throughout the plastics industry to measure the resistance of polymers to chemicals.[61-63] The method can be used to compare relative resistance of various polymers to typical solvents and chemical reagents. Because the test does not consider the effect of stress, the results do not necessarily provide a direct indication of suitability of a particular polymer for end-use application in certain chemical environments. The key test variables include duration of immersion, temperature of the test, and concentration of reagents. If the end-use application involves continuous exposure to a chemical, then the data obtained from short-time immersion tests are only useful in screening out unsuitable polymers before long-term immersion testing.

The test equipment for immersion testing consists of a precision scale, micrometer, vernier caliper, immersion containers (wide-mouth jars or other containers for holding the solvent), an oven or a constant-temperature bath, and the appropriate test equipment for measuring physical properties. The dimensions and type of test specimens are dependent on the form of the polymer and test to be performed. At least three test specimens are used for each material being tested and each reagent involved. Standard specimens are immersed in reagents for 168 h (7 days) at elevated or room temperature, and changes in weight, dimensions, appearance, and mechanical properties are determined.

The immersion test procedure involves the following steps:

1. The initial weight and dimensions of each specimen are measured and recorded.
2. The specimens are totally immersed in the solvent in such a way that no contact is made with the walls of the container.
3. After 7 days, the specimens are removed from the container and immediately examined and tested to prevent any loss (evaporation) of an absorbed solvent.
4. The appearance of the specimens is observed for loss of gloss, softening, crazing, warping, optical clarity, discoloration, and other features.
5. The specimens are weighed and the dimensions are remeasured and recorded.
6. Mechanical properties are determined in accordance with standard test methods on nonimmersed and immersed specimens.
7. Changes in weight, length, width, thickness, and physical properties are calculated:

Percent change in weight

= (weight after immersion − initial weight)

× 100/initial weight

Percent change in property

= (property after immersion − initial property)

× 100/initial property

Percent change in dimension

= (dimension after immersion − initial dimension)

× 100/initial dimension

8. Changes in weight, dimensions, and physical properties, and observations such as loss of gloss, swelling, clouding, rubberiness, tackiness, crazing, cracking, and bubbling are recorded.

An example of the use of the immersion test is in the chemical resistance of poly(hexamethylene sebacamide).[64] Tensile specimens were injection molded and stored in a dry environment at ambient temperatures for 24 h. The weight and length of each specimen was measured and recorded. The specimens were placed in jars, covered completely with the solvent, and stored at 23 °C. Specimens were removed after 7 and 30 days, wiped off, weighed, and the length was remeasured and recorded. The specimens were then immediately tested for tensile strength along with a control. The percent change in weight, dimension, and strength were calculated and appear in Table 20.9.

A guideline to the significance of changes in weight, dimensions, properties, and appearance appears in Table 20.10. Note that if a material is found to be incompatible in a short-term test, it will usually be found to be incompatible in a similar end-use situation. The converse, however, is not always true. Favorable performance in a short-term test is no guarantee that actual long-term, end-use conditions would be good. Indeed, a short-term test may not have duplicated actual end-use conditions. Therefore, the results of a short-term test should be used as a guide only, and it is recommended that the user test production parts under true end-use conditions.

When specimens are exposed to chemicals at elevated temperature, unless they are to be tested at the elevated temperature, they need to be placed in another container of the same reagent at the standard laboratory temperature for approximately 1 h to effect cooling before testing.

Solvent Stress-Cracking Test

Most polymers under high stress will undergo stress cracking when exposed to certain chemical environments.[61] Such cracking will occur even though some chemicals have no effect on unstressed parts; therefore, simple immersion of test specimens is an inadequate measure of chemical resistance of stressed polymers.[61] Stress cracking occurs when the polymer-to-polymer interaction is replaced by a polymer-solvent interaction, which lowers the cohesive interaction energies of the surface layers of the polymer. These new polymer-solvent interactions cannot contribute to the overall strength of the material. If the stresses present exceed the cohesive strength of the weakened polymer, then the polymer can develop microcracks and eventually rupture. As the solvent penetrates deeper with time, cracks become more extensive.[65]

Hence, the presence of stress, internal or external, is an important consideration in solvent resistance. The internal or molded-in stresses are a big problem because complete removal of such stresses can be difficult. However, the internal stresses can be minimized through proper design, optimized processing conditions, and annealing of the parts after fabrication.

Thus, when a polymer is simultaneously exposed to a solvent and a stress, it can be characterized as exhibiting a critical stress. Critical stress is defined as the stress at which the onset of crazing is observed when a specimen is exposed

Table 20.9 Chemical resistance via solvent immersion of poly(hexamethylene sebacamide)

Solvent	Immersion time (days)	Weight (g)	% Change in weight	Length (mm)	% Change in length	Tensile strength (MPa)	% Change in tensile strength
None	0	12.050	—	21.59	—	58.65	—
Toluene	7	12.086	0.3	21.60	0.06	58.31	0.5
	30	12.134	0.7	21.61	0.08	56.58	2.6
Ethyl acetate	7	12.062	0.1	21.59	0	58.31	0.5
	30	12.086	0.3	21.60	0.04	57.75	1.2
Ethanol, 50%	7	12.448	3.3	21.65	0.27	56.58	2.6
	30	12.797	6.2	21.85	1.2	52.79	7.3

to a chemical environment. Below the critical stress the solvent has no apparent effect. Above the critical stress, there is crazing or cracking. Different tests have been developed to determine critical stress.

Tensile Stress Method for Critical Stress

In the tensile stress method for determining the critical stress for craze initiation, a tensile testing machine along with a standard tensile bar are employed. This test is sometimes referred to as the calibrated solvent test. The test is carried out by stressing the tensile bar specimen to a known stress level and immediately exposing it to the chemical environment by either spraying the chemical onto the specimen or continuously wetting the specimen using a wick. The specimen is exposed to the solvent (typically for 1 min). Then the specimen is examined visually for any sign of crazing. If there is no crazing, the experiment is repeated using a new specimen each time at increasingly higher levels of stress until crazing is observed. The stress level at which initial crazing is observed is the critical stress level for crazing in that particular solvent. If no crazing is observed at the yield point of the polymer, then the material is considered safe for use in that particular chemical environment.

One of the disadvantages of the calibrated solvent test is that it requires a large number of specimens to determine the critical stress level. Moreover, the time the specimen is exposed to a solvent can be a key factor. It is possible that

the solvent can attack the polymer after exposure for a long period. However, it is not practical to expend a long time for such visual testing. Therefore, an accelerated method of testing can be used where the testing is carried out at elevated temperature and high stress.

Constant Strain Jigs

Constant strain jigs can apply a known strain to a plastic test specimen. The bar is bent on a jig that has a constant radius of curvature, thus the plastic sample has a known amount of stress. The jigs and stressed sample can then be exposed to various chemical agents.

A diagram of a constant strain jig is shown in Figure 20.8, where R is the radius of curvature. Strain jigs that can accommodate multiple samples are preferred because multiple samples are typically tested as the same time. On the left side of Figure 20.9 are strain jigs for 215-mm-long (8.5 in.) tensile specimens with 0.25%, 0.75%, 1.25%, and 1.75% strain. Each jig can hold up to three specimens. Because it is important that the specimen conform to the shape of the jig, there are two extra clamping devices in the middle of the jig. When long specimens are used, these additional clamps ensure the proper contact with the curved surface. On the right side of Figure 20.9 is a 1.5% strain jig that can hold multiple 63-mm (2.5 in.) specimens.

Table 20.10 Key to classification of test results

	Change, in percent			
Rating	Weight	Dimension	Tensile and flexural strengths	Appearance
Excellent	<0.5	<0.2	<10	Slightly discolored
Fair	>1.0	>0.5	>20	Softened, crazed, warped, distorted optically
Poor	0.5–1.0	<0.2	10–20	Discolored

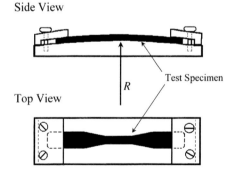

Side View

Top View

R   Test Specimen

Figure 20.8 Diagram of a constant strain jig. Key: R = radius of curvature.

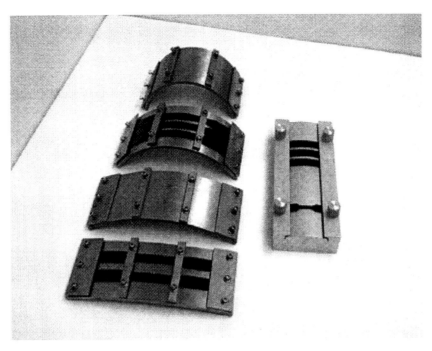

Figure 20.9 Constant strain jigs.

The strain induced in the test specimen is a function of the thickness of the specimen and the radius of curvature of the jig.[62] The radius of curvature can be determined for a specified strain level and specimen thickness using eq 17:

$$R = \frac{t}{2}\left(\frac{100}{\varepsilon} - 1\right) \qquad (17)$$

where R is the radius of curvature, t is the specimen thickness, and ε is the strain (in percent).

The jigs can be fabricated out of stainless steel by most machine shops. For convenience in carrying out exposure of multiple specimens, the strain jig can be made wider to accommodate more than one test bar. The length of the jig should be such that it can accommodate standard molded test specimens (American Society for Testing and Materials [ASTM] or International Organization for Standardization [ISO]).

Table 20.11 shows values for radii of jigs calculated from eq 17 for commonly used strains and standard thickness test bars.

The test using constant strain jigs consists of exposing standard ASTM or ISO test specimens to the chemicals being evaluated under standardized conditions of applied strain.[62,63] After exposure, the specimens are visually evaluated and mechanically tested to determine the effects of reagents on the stressed chemically exposed plastics. The test equipment consists of a micrometer, strain jigs, immersion containers, and appropriate test equipment for measuring physical properties. The constant strain test involves the following steps:

1. For tensile specimens, measure the cross-sectional area using a micrometer and record the result.
2. Mount the appropriate test specimens, as specified by the mechanical test being performed, onto strain jigs. Ensure that intimate contact of the specimens and the fixtures is maintained along the entire length of the gage area or specimen area to be tested.
3. Maintain one set of test specimens strained identically to the specimens being exposed (including 0.0%), but with no chemical exposure, to act as a control. If exposure includes elevated temperatures, a set of unexposed controls needs to be conditioned at the same temperatures and times as the exposed specimens.

Table 20.11 Radii required to produce strains for various specimen thicknesses

Strain (%)	Sample thickness in inches and (mm)			
	0.062 (1.57)	0.10 (2.54)	0.125 (3.18)	0.25 (6.35)
	Radius of curvature, R (mm)			
0.25	314.2	506.7	633.4	1266.8
0.50	156.7	252.7	315.9	631.8
0.75	104.2	168.1	210.1	420.2
1.00	78.0	125.7	157.2	314.3
1.50	51.7	83.4	104.2	208.5
2.00	38.6	62.2	77.8	155.6
2.50	30.7	49.5	61.9	123.8

4. Expose the strained test specimen to the chemical being evaluated, along with one set of 0.0% strain (unstrained) specimens. The test specimens can be immersed in chemicals. Alternatively, a wet patch method can be used that involves applying a cotton patch over the test specimens and saturating the patch with liquid. For volatile solvents, the liquid should be reapplied routinely to provide continuous saturation.

5. Exposure time is typically 168 h (7 days), but can be shorter or longer depending on the scope of the evaluation.

6. Remove the specimens from the jigs and determine the mechanical properties in accordance with standard test methods on exposed and unexposed specimens.

7. Calculate the percent change in physical properties.

Effect of strain is determined on unexposed specimens strained and unstrained:

Percent property change

= (property strained – property unstrained)

× 100/property unstrained

Effect of solvent is determined on unstrained specimens, exposed and unexposed:

Percent property change

= (property exposed – property unexposed)

× 100/property unexposed

Effect of solvent and strain (solvent stress cracking) is determined on strained specimens, exposed and unexposed:

Percent property change

= (property exposed – property unexposed)

× 100/property unexposed.

The use of constant strain jigs is exemplified in the environmental stress crack resistance of poly[4,4'-(1-phenylethylidene)diphenol]carbonate.[66] Tensile bars were injection molded and stored in a dry environment at ambient temperatures for 24 h. The test specimens were clamped into 0.5% and 1.0% strain jigs and totally immersed in the solvent for 168 h (7 days) at 23 °C. For 0% strain, jigs were not used and the molded samples were immersed in the test media under the same conditions. Control samples were clamped into 0.5% and 1.0% strain jigs without immersion in any solvent, and also were stored for 168 h at 23 °C. After 168 h, the specimens were removed from the solvent and jigs, wiped off, and tested for tensile properties. The percent change in properties is calculated as described above. The results appear in Table 20.12.

A guideline for the significance of the property changes appears in Table 20.10. The stress-cracking test is a useful screening tool. If a material is found to be incompatible in a short-term test, it will usually be found to be incompatible in a similar end-use environment. The converse, however, is not always true. Favorable performance in a short-term test is no guarantee of actual long-term, end-use conditions. The amount of molded-in stress found in any particular part will have a pronounced effect on the relative chemi-

Table 20.12 Chemical resistance via constant strain jigs of

poly[4,4-(1-phenylethylene)diphenol]carbonate

Solvent	Immersion time (days)	Temperature (°C)	Strain (%)	Tensile strength (MPa)	% Retention in tensile strength
None	7	23	0	86.5	100
	7	23	0.5	85.6	99
	7	23	1.0	84.8	98
Hexane	7	23	0	81.3	94
	7	23	0.5	79.6	92
	7	23	1.0	69.2	80
1-Butanol	7	23	0	80.4	93
	7	23	0.5	78.7	91
	7	23	1.0	67.5	78
Toluene	7	23	0	30.3	35
	7	23	0.5	27.7	32
	7	23	1.0	24.2	28

Table 20.13 Elliptical strain jigs: x and B values for calculating critical strain values

x (mm)	B	x (mm)	B	x (mm)	B
127.0	0.043745	99.0	0.003948	62.0	0.001704
126.5	0.038936	98.0	0.003815	61.0	0.001681
126.0	0.034962	97.0	0.003679	60.0	0.001661
125.5	0.031634	96.0	0.003553	59.0	0.001640
125.0	0.028813	95.0	0.003434	58.0	0.001620
124.5	0.026396	94.0	0.003324	57.0	0.001599
124.0	0.024308	93.0	0.003220	56.0	0.001582
123.5	0.022781	92.0	0.003131	55.0	0.001564
123.0	0.021148	91.0	0.003039	54.0	0.001546
122.5	0.019707	90.0	0.002952	53.0	0.001530
122.0	0.018428	89.0	0.002870	52.0	0.001514
121.5	0.017286	88.0	0.002797	51.0	0.001498
121.0	0.016230	87.0	0.002723	50.0	0.001484
120.5	0.015337	86.0	0.002654	49.0	0.001470
120.0	0.014580	85.0	0.002591	48.0	0.001456
119.5	0.013802	84.0	0.002529	47.0	0.001442
119.0	0.013104	83.0	0.002469	46.0	0.001429
118.5	0.012466	82.0	0.002418	45.0	0.001416
118.0	0.011880	81.0	0.002364	44.0	0.001405
117.0	0.010882	80.0	0.002313	43.0	0.001393
116.0	0.009990	79.0	0.002265	42.0	0.001382
115.0	0.009220	78.0	0.002218	41.0	0.001372
114.0	0.008601	77.0	0.002174	40.0	0.001361
113.0	0.008006	76.0	0.002135	39.0	0.001351
112.0	0.007482	75.0	0.002094	38.0	0.001341
111.0	0.007016	74.0	0.002056	37.0	0.001332
110.0	0.006601	73.0	0.002019	36.0	0.001323
109.0	0.006228	72.0	0.001985	35.0	0.001314
108.0	0.005892	71.0	0.001951	34.0	0.001307
107.0	0.005612	70.0	0.001918	32.0	0.001291
106.0	0.005333	69.0	0.001888	30.0	0.001277
105.0	0.005079	68.0	0.001859	25.0	0.001247
104.0	0.004857	67.0	0.001830	20.0	0.001227
103.0	0.004644	66.0	0.001804	15.0	0.001204
102.0	0.004447	65.0	0.001778	10.0	0.001191
101.0	0.004272	64.0	0.001752	5.0	0.001184
100.0	0.004104	63.0	0.001727	1.0	0.001181

cal compatibility of a polymer. The acceptable chemical compatibility of a polymer in an application can only be determined on the specific molded part.

Elliptical Strain Jigs

An alternative method for measuring solvent stress cracking uses a specimen strapped to an elliptical jig.[61,67] Instead of testing a large number of samples of the same material in the same reagent at different stress levels, elliptical jigs that have a varying radius of curvature are used. This method is useful for quickly screening the effects of a large number of chemicals or when the amount of polymer is limited. A sample clamped to an elliptical jig will have a continuum of mechanical strains. Because of the elliptical design of the jig, the stress at the thick end of the jig is extremely low. Conversely, the stress at the thin end of the jig is extremely high. The level of stress in the specimen at

different points on the jig can be calculated. The elliptical strain jigs have been called Bergin strain jigs. After exposure to a solvent, heavy crazing or cracking usually occurs in the areas under higher strain, but becomes sparser with distance along the bar and does not occur in areas of low strain. The point at which the crazing stops is considered the critical stress for craze initiation, ε_c, under the test conditions. Each specimen thus yields a value of critical strain, so that only two or three specimens are needed to determine a reasonably precise ε_c value.

The jigs comprise a quarter section of an ellipse and can be fabricated by most good machine shops from 25-mm-wide (1 in.) stainless steel blocks. The ellipse is described mathematically by eq 18:

$$(y/38.1)^2 + (x/127)^2 = 1 \qquad (18)$$

where the constants representing the semi-axes are given in millimeters, that is, the quarter section of an ellipse has a height of 38.1 mm (1.5 in.) and a length of 127 mm (5 in.). The jigs are designed to accept a specimen that is approximately 0.75 mm thick, 25 mm wide, and 75 to 127 mm long.

On a flexed sample, the strain (ε) at any point is equal to the thickness of the sample divided by twice the radius of curvature at that point. Because the jig is cut to a mathematically defined curve, the radius of curvature at any point can be calculated as a function of x, the distance from that point to the semiminor axis. A straightforward calculation yields eq 19, in which ε_c is defined as a function of x and the sample thickness, t:

$$\varepsilon_c = [0.00118(1 - 0.00000564x^2)^{-3/2}]t \qquad (19)$$

where x and t are measured in millimeters and ε_c is measured in mm/mm.

Equation 19 can be written as eq 20:

$$\varepsilon_c = Bt \qquad (20)$$

where B is defined in eq 21:

Top View

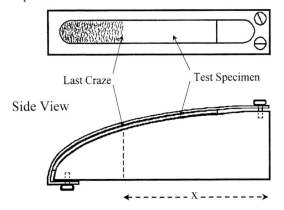

Side View

Figure 20.10 Diagram of an elliptical strain jig.

$$B = 0.00118(1 - 0.00000564x^2)^{-3/2} \qquad (21)$$

Using eq 21, B values were calculated as a function of x and appear in Table 20.13. Hence, ε can be calculated using eq 20 by determining x experimentally, obtaining B from Table 20.13, and multiplying B and t.

Thus, ε has a maximum value of $0.0437t$ at $x = 127$ mm and a minimum value of $0.00118t$ at $x = 0$ mm. For a sample 0.75 mm thick, ε would be maximum and minimum of 0.032 to 0.00089 mm/mm or 3.2% to 0.089%, respectively.

Figure 20.10 shows a diagram of an elliptical strain jig, the last crazed area, and the measurement of x for use in Table 20.13 for determining ε_C. A more sophisticated elliptical jig appears in Figure 20.11, in which there is a device at the end of the jig to pull the specimen strap tight against the jig to ensure that the specimen conforms to the shape of the jig.

The test equipment for this procedure consists of a micrometer, vernier caliper, and elliptical strain jigs. The typical procedure involves compression molding sheets of the polymer approximately 0.75 mm thick to avoid thermal stresses. Test specimens for crazing are cut in 25×100 mm strips from these sheets. The samples for testing were 25×100 mm strips approximately 0.75 mm thick. The edges of the specimens are milled smooth to reduce any tendency for defects in the cut edge to erroneously initiate cracking. The specimens are washed in soapy water, rinsed, dried with paper towels, and then handled only with gloves or tongs prior to being strained. The samples are bent around the elliptical jigs and strapped in place over the entire length by metal bands. The jig and sample are exposed to the reagent and cracks will start appearing. The cracking usually will stop completely after 1 h. The distance, x, to the point at which cracking had stopped is measured to ±0.025 mm with a vernier caliper and the thickness is measured at that point to ±0.1 mm with a micrometer. From the value of x, B is obtained from Table 20.13 and the critical strain for craze initiation, ε_C, is calculated using eq 20.

A difficulty with the use of jigs onto which specimens are strapped before immersion in a test fluid arises from edge effects.[68] Specimens cut from sheets often exhibit substantially lower resistance to crack growth from the cut edges than to crack initiation on the molded faces. This problem can be avoided by the use of a narrow wick of cloth or filter paper. The wick is laid along the center of the specimen and saturated with the test fluid, which is applied with an eyedropper. A qualitative summary of the significance of percent ε_C is summarized in Table 20.14. However, additional tests such as solvent resistance using a constant strain jig would be warranted to corroborate the performance and quantify the physical property retention.

An example of the use of elliptical strain jigs to determine the critical strain for craze initiation is the testing of polysulfone in various solvents.[68] Polysulfone pellets were dried for 4 h at 125 °C in an air-circulating oven to remove any moisture in the pellets. The pellets were compression molded into a 0.9-mm-thick plaque from which 40×100 mm strips were cut. The thickness of the strips was measured using a micrometer and recorded. The edges were milled smooth. The strips were washed in soapy water, rinsed, dried, and then handled with cotton gloves. The strips were mounted on an elliptical strain jig and a strip of filter paper placed on top of the exposed surface. Solvent was added

Figure 20.11 Elliptical strain jigs.

Table 20.14 A qualitative summary
of the significance of percent ε_C

% ε_C	Solvent resistance
<0.3	Poor
0.3–0.8	Fair to good
>0.8	Excellent

with an eyedropper and the paper was kept wet for 60 min. Then the filter paper was removed and the exposed specimen examined for crazing. A magnifying glass is useful in examining small microcracks. The last craze was noted and the distance x from the semiminor axes was measured using a vernier caliper and recorded. The x value was determined on two specimens and the average of the two x values was used to obtain a B value from Table 20.13 and then to calculate ε_C. The results for polysulfone are summarized in Table 20.15, in which the solvent, solubility parameters of the solvent, the average x value from two determinations, the ε_C and percent ε_C, are shown.

Conclusion

A polymer is often soluble in a low molecular weight liquid if the polymer and solvent have similar solubility parameters. The solubility increases if (1) the molecular weight of the polymer is low, (2) the bulk polymer is not crystalline, and (3) the temperature is elevated (except in systems with LCSTs). Hence, the method of determining solubility parameters can be useful for identifying potential solvents for a polymer. Solubility parameters can be calculated for polymers using molar attraction constants.

Cross-linked polymers cannot dissolve. However, their swelling ratio, which is related to the cross-link density and the nature of the polymer and solvent, can be determined.

Table 20.15 Critical strain for craze initiation in polysulfone

Solvent	δ	x (mm)	ε_C (mm/mm)	% ε_C
Ethanol	12.78	117	0.0098	0.98
Acetonitrile	12.11	102	0.0040	0.40
2-Propanol	11.44	114	0.0077	0.77
Cycolopentanone	10.53	78	0.0020	0.20
Acetone	9.62	54	0.0014	0.14
n-Butyl acetate	8.68	69	0.0017	0.17
Cyclohexane	8.19	118	0.0107	1.07
Isoamyl butyrate	7.8	104	0.0044	0.44
n-Hexane	7.27	116	0.0090	0.90

Date adapted from reference 68.

Solvents and chemicals can affect the performance of polymers, and immersion tests can quantify any effect on properties. In addition, solvents, which have no effect on an unstressed polymer, can cause crazing and cracking when the polymer is stressed and exposed to the solvent or chemical. The use of strain jigs helps to define and quantify solvent stress cracking in polymers.

When the solubility parameter of the solvent is similar to that of the polymer, the polymer may dissolve in the solvent. Moreover, a polymer and solvent that have similar solubility parameters suggest that there is increased interaction between the polymer and solvent where the polymer may go into solution, craze, crack, or swell. Indeed, environmental stress crazing and cracking studies of several amorphous polymers has shown a qualitative correlation between δ_p and ε_C.[68–70] Moreover, solvents that rapidly swell the polymer offer a competing mode of stress relaxation. Thus, the solvent swelling softens the polymer, reduces the polymer T_g and concurrently reduces the tendency of the polymer to stress craze.

References

1. Carpenter, D. K. In *Encyclopedia of Polymer Science and Engineering*; Kroschwitz, J. L., Ed.; Wiley-Interscience: New York, 1989; Vol. 15; pp 419–481.
2. Orwoll, R. A. In *Encyclopedia of Polymer Science and Engineering*; Kroschwitz, J. L., Ed.; Wiley-Interscience: New York, 1989; Vol. 15; pp 380–402.
3. Rätzsch, M. T. In *Encyclopedia of Polymer Science and Engineering*; Kroschwitz, J. L., Ed., Wiley-Interscience: New York, 1989; Vol. 15; pp 481–491.
4. Kwei, T. K. In *Macromolecules an Introduction to Polymer Science*; Bovey, F. A., Winslow, F. H., Eds.; Academic Press: New York, 1979.
5. Billmeyer, F. *Textbook of Polymer Science*, 3rd ed.; Wiley & Sons: New York, 1984.
6. Grulke, E. A. In *Polymer Handbook*, 3rd ed.; Brandrup, J., Immergut, E. H., Eds.; Wiley & Sons: New York, 1989; pp VII/519–VII/559.
7. Barton, A. F. M. *Handbook of Solubility Parameters and Other Cohesion Parameters*; CRC Press, Boca Raton, FL, 1991.
8. Dayantis, J. *Plast. Mod. Elastomers* **1977**, *29*, 58.
9. Rowlinson, J. S.; Swinton, F. L. *Liquids and Liquid Mixtures*, 3rd ed.; Butterworth Publications: London, 1982.
10. Freeman, P. I.; Rowlinson, J. S. *Polymer* **1960**, *1*, 20.
11. Saeki, S.; Kawahara, N.; Konno, S.; Kaneko, M. *Macromolecules* **1973**, *6*, 589.
12. Patterson, D. *Macromolecules* **1969**, *2*, 672.
13. Hildebrand, J.; Scott, R. *The Solubility of Nonelectrolytes*, 3rd Ed.; Reinhold Publishing: NewYork, 1949.
14. Scatchard, G. *Chem. Rev.* **1949**, *44*, 7.
15. Bagley, E. *J. Paint Technol.* **1971**, *43*, 35.
16. Hansen, C. M. *Solubility Parameters*; CRC Press, Boca Raton, FL, 1999.

17. Hoy, K. J. *J. Paint Technol.* **1970**, *42*, 76.
18. Karger, B. I.; Snyder, L. R.; Eon, C. *J. Chromatogr.* **1976**, *125*, 71.
19. Hilderbrand, J. H. *Chem. Rev.* **1949**, *44*, 37.
20. Hilderbrand, J. H. *Science* **1965**, *150*, 441.
21. Hilderbrand, J. H. *Proc. Nat. Acad. Sci. U.S.A.* **1979**, *76*, 6040.
22. Koenhen, D. M.; Smolders, C. A. *J. Appl. Polym. Sci.* **1975**, *19*, 1163.
23. Smidsrod, C.; Guillet, J. E. *Macromolecules* **1968**, *2*, 272.
24. Patterson, D. *Macromolecules* **1971**, *4*, 30.
25. Gee, G. *Trans. Inst. Rubber Ind.* **1943**, *18*, 266.
26. Stockmeyer, W. H., Fixman, M. *J. Polym. Sci., Part C* **1963**, *1*, 137.
27. Van Krevelen, D. W.; Hoftyzer, P. J. *Properties of Polymers. Correlations with Chemical Structure*; Elsevier: New York, 1972.
28. Fedors, R. F. *Polym. Eng. Sci.* **1974**, *14*, 147.
29. Small, P. A., *J. Appl. Chem.* **1953**, *3*, 71.
30. Krause, S. *J. Macromol. Sci., Macromol. Rev.* **1972** C7, 251.
31. Schröder, E.; Müller, G.; Arndt, K. F. *Polymer Characterization*; Hanser Gardner: Cincinnati, OH, 1989.
32. Hansen, C. M. *J. Paint Technol.* **1967**, *39*, 511.
33. *Handbook of Thermoset Plastics*, 2nd ed.; Goodman, S. H., Ed.; Plastics Design Library: Brookfield, CT, 1999.
34. Freitag, W.; Stoye, D. *Paints, Coatings and Solvents*, 2nd ed.; Wiley & Sons: New York, 1998.
35. Labana, S. S. In *Encyclopedia of Polymer Science and Engineering*; Kroschwitz, J. L., Ed.; Wiley-Interscience: New York, 1986; Vol. 4, pp 350–395.
36. Priss, L. S. *J. Polym. Sci., Part C* **1975**, *53*, 195.
37. *ASTM D3616*; American Society for Testing and Materials: Philadelphia, PA, 1988.
38. *ASTM D2765*; American Society for Testing and Materials: Philadelphia, PA, 1995.
39. Peters, E. N.; Stewart, D. D.; Bohan, J. J.; Moffitt, R.; Beard, C. D.; Kwaitkowski, G. T.; Hedaya, E. *J. Polym. Sci., Polym. Chem. Ed.* **1977**, *15*, 973.
40. Peters, E. N.; Arisman, R. K. In *Polymer Data Handbook*; Mark, J. E., Ed.; Oxford University Press: New York, 1999; pp 30–33.
41. *Fractionation of Synthetic Polymers. Principals and Practices*; Tung, L. H., Ed.; Marcel Dekker: New York, 1977.
42. *Polymer Fractionation*; Cantow, M. J. R., Ed.; Academic Press: New York, 1966.
43. Johnson, J. F.; Cantow, M. J. R.; Porter, R. S. In *Characterization of Polymers*; Bikales, N. M., Ed.; Wiley-Interscience: New York, 1971.
44. Tung, L. H. In *Encyclopedia of Polymer Science and Engineering*; Kroschwitz, J. L., Ed., Wiley-Interscience: New York, 1987; Vol. 7; pp 298–327.
45. Bello, A.; Barrales-Rienda, J. M.; Guzmán, G. M. In *Polymer Handbook*, 3rd ed.; Brandrup, J., Immergut, E. H., Eds.; Wiley & Sons: New York, 1989; pp VII/233–VII/317.
46. Huggins, M. L. *J. Polym. Sci.* **1963**, C4, 445.
47. Koningsveld, R. *Adv. Polym. Sci.* **1970**, *7*, 1.
48. Huggins, M. L.; Okamoto, H. In *Polymer Fractionation*; Cantrow, M. J. R., Ed.; Academic Press: New York, 1967; pp 1–42.
49. Kamide, K. In *Fractionation of Synthetic Polymers. Principals and Practices*; Tung, L. H., Ed.; Marcel Dekker: New York, 1977; pp 103–265.
50. Barrall, E. M.; Johnson, J. F.; Cooper, A. R. In *Fractionation of Synthetic Polymers. Principals and Practices*; Tung, L. H., Ed.; Marcel Dekker: New York, 1977; pp 267–344.
51. Baker, C. A.; Williams, R. J. P. *J. Chem. Soc.* **1956**, 2352.
52. Jungnickel, J. L.; Weiss, F. J. *J. Polym. Sci.* **1961**, *49*, 437.
53. Cooper, W.; Vaughan, G.; Yardley, J. *J. Polym. Sci.* **1962**, *59*, S2.
54. Schneider, N. S.; Loconti, J. D.; Holmes, L. G. *J. Appl. Polym. Sci.* **1961**, *5*, 354.
55. Henry, P. M. *J. Polym. Sci.* **1959**, *36*, 3.
56. Guillet, J. E.; Combs, R. L.; Slonaker, D. F.; Summers, J. T.; Coover, H. W. *SPE Trans.* **1964**, *2*, 164.
57. Riess, G.; Callot, P. In *Fractionation of Synthetic Polymers. Principals and Practices*; Tung, L. H., Ed.; Marcel Dekker: New York, 1977; pp 445–544.
58. Fuchs, O.; Schmieder, W. In *Polymer Fractionation*; Cantrow, M. J. R., Ed.; Academic Press: New York, 1967; pp 341–378.
59. Guzman, G. M. In *Progress in High Polymers*; Robb, J. C., Peaker, F. W., Eds.; Academic Press: New York, 1961.
60. McCarthy, R. A. In *Encyclopedia of Polymer Science and Engineering*; Kroschwitz, J. L., Ed.; Wiley-Interscience: New York, 1985; Vol. 7; pp 421–430.
61. Shah, V. *Handbook of Plastics Testing Technology*, 2nd ed.; Wiley-Interscience: New York 1998.
62. *ASTM D 543*; American Society for Testing and Materials: Philadelphia, PA, 1995 (www.astm.org).
63. *ISO 175*; International Organization for Standardization: Geneva, Switzerland, 1981 (www.iso.ch).
64. Bonner, R. M.; Kohan, M. I.; Lacey, E. M.; Richardson, P. N.; Roder, T. M.; Sherwood, L. T. In *Nylon Plastics*; Kohan, M. I., Ed.; Wiley-Interscience: New York, 1973; pp 327–407.
65. Baer, E. *Engineering Design for Plastics*; Reinhold: New York, 1964.
66. Schnell, H. *Chemistry and Physics of Polycarbonates*; Wiley-Interscience: New York, 1964.
67. Bergin, R. L. *SPE J.* **1962**, 667.
68. Kambour, R. P.; Romagosa, E. E.; Gruner, C. L. *Macromolecules* **1972**, *5*, 335.
69. Kambour, R. P.; Gruner, C. L.; Romagosa, E. E. *Macromolecules* **1974**, *7*, 248.
70. Kambour, R. P. *J. Polymer Sci., Macromol. Rev.* **1973**, *5*, 1.

21

Particle Size and Particle Size Distribution

DAVID F. ALLIET

LAURIE L. SWITZER

How do we define a particle? According to the *Random House College Dictionary*, a particle is a minute portion, piece, or amount; a very small bit. A particle can be as small as an electron or as large as the sun. A range of various particles of widely varying dimension is summarized in Figure 21.1. In the vast universe in which we live, even our own sun is a small particle! Thus, the area of expertise of someone studying the behavior of particles can range from astrophysics to nuclear physics. Kaye[1] defined particles as a "fineparticle." This includes all particles other than those particles that make up the structural component of the atom. He defines fineparticle science as that concerned with particle systems that are intermediate between colloidal and classical physics. In general, this would include particles that have dimensions in the range of 0.1 to 2000 µm. Most powders of interest to industrial scientists and engineers fall into this particle size range.

Figure 21.2 is a summary of some common industrial powders for which one needs to know the particle size and particle size distribution to control the behavior of the material. Industrial powders are typically small pieces of bulk material, such as coal dust, paint pigments, polymer latex, pharmaceutical powders, toners, or even nondairy creamers.

Characterization of Particle Systems

Particle size distribution is a critical parameter for a wide variety of operations in the chemical process industries.[2] Examples are the flow characteristics of granular materials, sintering behavior of metallurgical powders, and the combustion efficiency of powdered coal. The accurate measurement of size distribution is critical to the quality and performance of most industrial powders.

Today, the manufacturers of plastics, paints, and pigments are producing powders that have size distributions in the submicrometer range, which require analytical methods that are capable of accurately measuring these very fine particles. Over the last decade, there has been a rapid evolution of modern methods developed for the measurement of particle size distribution.[3,4] Each of these modern methods is unique to particular applications and each has errors and uncertainties which need to be understood.

Before modern methods of particle characterization were available, the only evaluation methods available were physical separation methods, such as sieving, which for practical purposes can only be used for dry particles larger than 25 µm. Heywood[5] referred to dry sieving as the "Cinderella" of particle size methods; it does most of the hard work and gets little consideration. Leschonski[6] stated that sieve analysis in general is the most simple, reproducible, and inexpensive method of size analysis in the 20-µm to 125-mm range. However, he showed that the results of sieve analysis may easily be misleading and highly erroneous when careful procedures are not carried out.

The modern analytical methods used in particle size distribution analysis are classified into ensemble and nonensemble methods. All nonensemble methods are composed of a separation or fractionation mechanism that allows detection of only one particle at a time. The most well-known nonensemble methods used to measure one particle at a time are the optical particle counters and the Coulter principle counter. Sedimentation, size exclusion chromatography, and field-flow fractionation are methods of fractionation that require additional detectors to complete the size measurement. More recently, a new nonensemble method has emerged that is based on the principle of particle lag in an accelerating flow field.[7,8] This is a laser velocimetry technique used to measure the time-of-flight of individual particles between two laser beams. Dahneke[9] first introduced this technology as time-of-flight aerosol beam spectrometry (TOF-ABS). The nonensemble methods have the advantage of high resolution over the other methods.

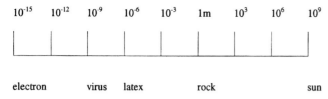

Figure 21.1 Particle size of materials (in meters).

Ensemble methods for particle size determination have evolved rapidly in the last decade. Light-scattering methods that are based on laser detection and computer-aided analysis of data have played a dominant role in the evolution of these newer methods.[10] The advantages of the ensemble methods are broad dynamic size range and fast analysis time. However, many of the newer ensemble methods have the disadvantage of low resolution compared to the nonensemble methods. There are advantages and disadvantages to both ensemble and nonensemble methods and they usually complement each other. This focuses the choice of proper analytical methods for particle size characterization on the requirements that are defined for the particular application. Users of these various modern methods often have to make compromises when they choose the best method for analysis of their materials.

Sampling of Particle Systems

The sampling of a powder before analysis of the particle size distribution is critical to obtaining an accurate measurement of the bulk material. Allen[11] has explored the various sampling methods extensively and reported on the methods used along with statistical analysis of the data obtained. He has stated two golden rules for sampling:

Rule 1: A powder should be sampled when in motion.
Rule 2: The whole of the stream of powder should be taken for many short increments of time in preference to part of the stream being taken for the whole of the time.

Allen and Khan[12] reported on the statistical reliability of five common sampling procedures, with the spinning riffling method being superior to all other methods of sampling powders. More recently, Kaye[13] developed a new sampler for very fine powders. He claims that a spinning riffler has the disadvantage of blowing the fines away because of the rotary action of the sampling device.

It is much easier to obtain a representative sample from a liquid suspension process stream. However, it is very important to properly mix and dilute the sample carefully before submitting it for particle size measurement.

Calibration Materials

Many of the manufacturers of particle size instruments claim that their equipment does not require calibration because they use absolute techniques. One who is new to this field should be skeptical of these claims and should ensure that the instrument is calibrated or the accuracy of such claims is verified.

Polymer latex microspheres[14,15] are supplied through national and international agencies. These particle standards are certified through a statistical certification procedure. Duke Scientific[16] supplies polymer latex standards and other standard materials that have third-party traceability of calibration procedures to the above agencies. Bangs Laboratories[17] is another a source of uniform polymer latex particles for calibration of particle size instruments.

Alliet[18] published analytical procedures used to standardize calibration latex particles for electrical sensing zone methods. He also discouraged the use of pollens[19] such as ragweed, which are spherical in nature. Scanning electron microscope (SEM) photos (Figure 21.3a) show the porous nature of the pollens which swell in aqueous media and lead to erroneous calibration of the nonensemble methods. Polystyrene latex particles (see Figure 21.3b) are very spherical and nonporous, which allows for more accurate calibration of both the ensemble and nonensemble methods. Some of the ensemble methods require a dry powder

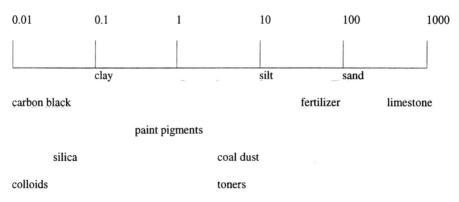

Figure 21.2 Particle size of industrial powders (in micrometers).

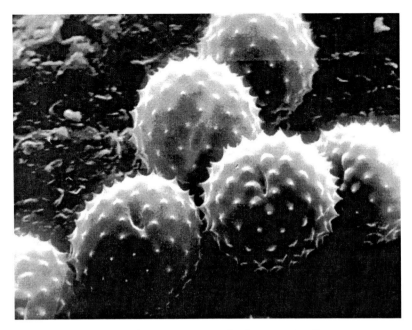

Figure 21.3a Scanning electron micrograph of ragweed pollen.

for calibration that eliminates polymer latex standards because they are usually provided as aqueous particle suspensions. Most of the suppliers also provide spherical glass powders for calibration of particle measurement devices. These standards meet the needs when a dry calibration material is required.

Measurement of Particle Size and Distribution

Particle Size

The meaning of particle size presents little difficulty when the particles have a regular shape. However, with irregularly shaped particles, the size that one assigns depends almost entirely on the method of measurement. Practitioners generally agree that some average linear measurement of many particles in a fixed direction is usually statistically adequate for describing the size distribution of particles. Thus, the size of a particle defined in a preferred manner is usually meaningless for an individual particle, but it gains value when one applies it to a large number of particles in random orientation. These measurements are called statistical diameters.

Allen[20] defines the common particle size diameters used by the various methods of particle size measurement. Some definitions of the common particle sizes are given in the Table 21.1. These diameters are usually applied to a distribution of particles and are used with a particular sizing method.

Particle Shape

Davies[21] published an extensive review article in 1975 that reveals a significant number of references dealing with particle shape. Many shape factors or correlation factors have been developed that attempt to explain the difference in particle size measurement obtained by the various methods of analysis. Kaye[22] also published extensively in an effort to quantify the shape relationships of fine particles and to better explain the significant differences one observes from data generated by the many particle sizing tools currently available. In his book on characterization of powders and aerosols, Kaye[23] develops several quantitative methods for measuring and classifying the shapes of fineparticles. His development of the chunkiness concept of particle shape distribution is quite interesting and helpful to the new powder technologist who needs a simple method for defining the shapes of particles in a sample.

Average Diameters

The term average diameter is used extensively in particle size measurement terminology. It is a statistical term that represents a group of individual values in a simple and concise manner to obtain an understanding of the group. An average is a measure of the central tendency of a group that is to be represented. The average is not affected by the relatively few extreme values that exist at the ends of the size distribution.

Because particles are three dimensional, their totality is represented by characteristics other than total number.

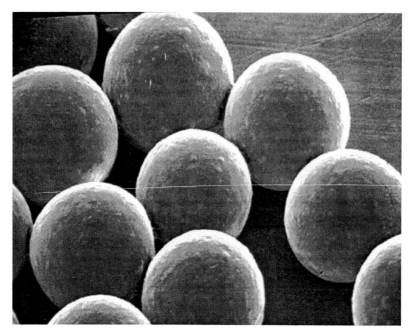

Figure 21.3b Scanning electron micrograph of polymer latex spheres.

Characteristics such as total particle volume, weight, or surface area are equally important parameters. Identification of these characteristics for a size distribution of particles is one of the most confusing areas for new scientists or engineers to understand when they attempt to use the results.

When particle size measurement data are divided into groups of equal size range called cells, each of the cells contains a certain percentage of the size measurement. A differential size frequency histogram results when this percentage is plotted against the corresponding mean of each size range (Figure 21.4). The central tendency of such a curve is described by the following terms.

Mode–particle diameter: represents the size interval of maximum frequency.

Number median diameter: the particle diameter that represents half of the particles counted with a lesser diameter and half with a greater diameter.

Volume or weight median diameter: If the data represent volume (or weight) instead of frequency of occurrence, this diameter is such that half the total particle volume (or weight) is represented by a lesser diameter and half is represented by a greater diameter.

Means or averages of a particle size distribution are additional indications of central tendency. The mean equals the median only when the distribution is symmetrical about the mean parameter. The important mean diameters are summarized here.

Number, length mean diameter, d_{NL}:

$$d_{NL} = \frac{L}{N} = \frac{\Sigma n_i d_i}{N} \tag{1}$$

where d_i is the midpoint diameter in each particle class; L is the sum total length of all particles; n_i is the number of particles in each size interval; and N is the total number of particles in all size intervals.

Length, surface mean diameter, d_{LS}:

$$d_{LS} = \frac{S}{L} = \frac{\Sigma n_i d_i^2}{\Sigma n_i d_i} \tag{2}$$

This diameter is also called the surface diameter because it represents the total surface area of the particles divided by the length of all particles in the population.

Surface, volume mean diameter, d_{VS}:

$$d_{VS} = \frac{V}{S} = \frac{\Sigma n_i d_i^3}{\Sigma n_i d_i^2} \tag{3}$$

The volume–surface mean diameter (d_{vs}) is useful in surface-area determinations for powders because its reciprocal is proportional to the specific surface.

Weight, moment mean diameter, d_{WV}:

$$d_{WV} = \frac{W}{V} = \frac{\Sigma n_i d_i^4}{\Sigma n_i d_i^3} \tag{4}$$

The weight mean diameter, also called the arithmetic mean size by weight, represents the average computed from the weights of particles, W, and volume, V. It is slightly larger than the median weight diameter.

Table 21.1 Definitions of particle size

Symbol	Name	Definition
d_V	Volume diameter	Diameter of a sphere having the same volume as the particle
d_S	Surface diameter	Diameter of a sphere having the same surface as the particle
d_{VS}	Surface volume diameter	Diameter of a sphere having the same external surface-to-volume ratio as the sphere
d_C	Perimeter diameter	Diameter of a circle having the same perimeter as the projected outline of the particle
d_{FF}	Free-falling diameter	Diameter of a sphere having the same density and the same free-falling speed as the particle in a fluid of the same density and viscosity
d_{ST}	Stokes diameter	Diameter of a sphere having the same density as the particle and settling at the same rate in a laminar flow region
d_A	Sieve diameter	The width of the minimum square aperature through which the particle will pass
d_P	Projected diameter	Diameter of a circle having area equivalent to that of the particle

Calculating Average Diameters

To illustrate the calculation of the selected average diameter for a particle size distribution of fine polymer powder, refer to Tables 21.2 and 21.3. These raw data represent an approximately log-normal distribution of particle size. Particle size distributions occurring in nature and in operating fluid systems are not likely to exhibit a symmetrical or normal distribution of particle sizes. Why naturally occurring particle size distributions are skewed has not been completely explained in the literature. However, it is reasonable to predict that because particle reduction processes are constantly occurring in systems, more small sized particles than larger sized particles are being generated.

From the data summarized in Table 21.3, one can calculate the various mean diameters as follows:

$$d_{NL} = \frac{2235}{651} = 3.43 \ \mu m \tag{5}$$

$$d_{LS} = \frac{9486}{2235} = 4.24 \ \mu m \tag{6}$$

$$d_{VS} = \frac{48,743}{9486} = 5.14 \ \mu m \tag{7}$$

$$d_{WV} = \frac{294,667}{48,743} = 6.05 \ \mu m \tag{8}$$

The practitioner should note that the average particle size increases as one emphasizes the larger particles found in a particular size distribution.

Graphical Data Analysis

Cumulative plots of particle size data can be described as those that involve plotting the percent of particles greater than or less than a given particle size against the particle size. Hence, the frequency percent in each successive size interval is cumulatively added, starting with the smallest size, and the result is plotted against the upper size in each interval. The cumulative data for the volume particle size for the fine polymer beads is summarized in Table 21.4 and plotted in Figure 21.5. The ordinate can represent total surface, external surface, weight, number of particles, or any other basis. It should be noted that this method of plotting does not differentiate between log-normal and other distributions. However, it is a simple method for representing a particle size distribution of a material.

The median or 50% size on the cumulative curve can be easily read and used in a sample-to-sample comparison. Some analysts also use other percentiles from the cumulative size distribution to define how broad or narrow a size distribution is represented. The ratio of the 90th to the 50th percentile and the ratio of the 50th to the 10th percentile are commonly used to describe the disparity of the particle size distribution.

Log probability plots of the particle size distribution can also be made for many fine powders that follow the log-normal law. The geometric mean particle diameter can be read directly from the plot. In this case, the mean is equal to the median or 50% particle size. Similarly, the geometric standard deviation can also be found from the following ratio:

$$\delta g = \frac{84.13\% \text{ size}}{50\% \text{ size}} = \frac{50\% \text{ size}}{15.87\% \text{ size}} \tag{9}$$

Particle size distributions that obey the log-normal law make possible simple transformations from weight distribution to number and surface distribution and vice versa. In this unique case, all three distributions have the same standard deviation, and the curves are parallel. Thus, a single point (e.g., 50% size or mean diameter) on each of the other distributions is all that is needed.

Light-Based Methods for Size Analysis

Optical Microscopy

Microscopy is really the only absolute method of particle size analysis. It dates back to the late 1600s when the compound microscope was invented. It is also the only method for which individual particles are observed and measured.

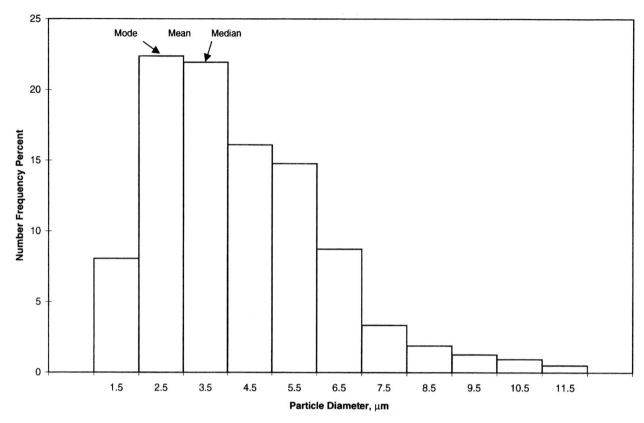

Figure 21.4 Number frequency particle size distribution for fine polystyrene powder.

A preliminary observation is highly recommended to determine the degree of dispersion and the size range of the powder under study. It is particularly critical here to ensure that a representative sample is obtained because a very minute quantity is used for this analysis.

Optical microscopy covers the size range of about 0.8 to 150 μm. It is, however, difficult to control the depth of focus if the powder or dispersion covers a wide particle size range. For this method, the more particles that are examined, the greater the accuracy of the number size distribution results. Manual microscopy is relatively inexpensive and examines specific particles as well as two-dimensional particle shape. The major disadvantages are very low throughput and high operator fatigue.

In the last two decades, image analysis microscopes have become the preferred method for two-dimensional size analysis because of the faster speed and reduced operator fatigue. The challenge becomes obtaining a particle dispersion on a glass slide or other means of sample presentation suitable for microscopic analysis. Kaye[24] has done extensive work with automated image analysis systems to measure the size distribution and has correlated this technique with other methods of size analysis. Allen[25] provides a summary of all the commercial image analysis systems available for automated size analysis by microscopy. When they were first introduced, image analysis systems were quite expensive. However, in recent years, the cost has become less prohibitive because of the lower cost of computer hardware and software.

Electron Microscopy

Transmission electron microscopy (TEM) has the advantage of imaging submicron-size particles. TEM covers a size range of 0.001 to 20 μm, so the very small size particles found in paint pigments, carbon blacks, polymer lattices, and colloids can be seen and measured. The technique is valuable for characterizing polymer latex standards used to calibrate other particle sizing methods. Figure 21.6 shows a TEM electron micrograph of a polymer latex. The particle size of the latex is measured using a Zeiss-Endter particle size instrument.[26] Measurements are made from the electron micrographs by matching the image of a variable orifice to the particle outline. Readings for about 500 to 1000 particle features are recorded and the arithmetic mean (d_{NL}) is then calculated. In more recent years, image analysis systems have been interfaced to TEM microscopes so that a larger number of particle features can be counted to improve the accuracy and precision of the size statistics.

Table 21.2 Log-normal particle size data for fine polystyrene powder

Diameter d, interval (μm)	Number of particles, n_i	Midpoint diameter, d_i	d_i^2	d_i^3	d_i^4
1.00–2.00	120	1.5	2.25	3.375	5.0625
2.00–3.00	200	2.5	6.25	15.625	39.0625
3.00–4.00	140	3.5	12.25	42.875	150.0625
4.00–5.00	80	4.5	20.25	91.125	410.0625
5.00–6.00	60	5.5	30.25	166.375	915.0625
6.00–7.00	30	6.5	42.25	274.625	1785.063
7.00–8.00	10	7.5	56.25	421.875	3164.063
8.00–9.00	5	8.5	72.25	614.125	5220.063
9.00–10.00	3	9.5	90.25	857.375	8145.063
10.00–11.00	2	10.5	110.25	1157.625	12155.06
11.00–12.00	1	11.5	132.25	1520.875	17490.06

SEM is considerably faster and gives the analyst more three-dimensional details than the TEM method. Because of its much greater depth of focus, SEM provides considerably more information about the form of a particle and its surface morphology. Earlier reference to latex and pollen particles (see Figure 21.3a,b) clearly shows the surface morphology characteristics of these particles. The SEM depth of focus is approximately 300 times greater than optical microscopy. Images from the SEM are generated by a linescan procedure and therefore the signals can be fed directly into a computer for analysis of particle characteristics via linescan logic manipulation methods.[27]

In conclusion, electron microscopy offers invaluable techniques to the particle size analyst. The optical microscope is the preferred technique for fineparticles greater than 1 μm in size, whereas the electron microscopes are preferred for submicrometer sizes. Number counts by projected area measurements can be carried out quickly and accurately, but one cannot expect to obtain accurate data for the higher moments of the distribution, such as weight or volume. This is due primarily to the lower particles counts obtained for broader particle size distributions. The narrowly distributed latex size distributions may be the exception. The SEM method is definitely preferred for studies of surface topography.

Photon Correlation Spectroscopy

In the last two decades, several instrument manufacturers have developed a popular technique for measuring submicron-size particles based on the Brownian motion of the particles in a fluid. Weiner[28] develops the theory and mathematics for this complex method called photon correlation spectroscopy (PCS). The author also lists the advantages of the technique as speed, small sample quantity, absolute sizing, and small size limit (2 nm). Several books on PCS theory and practice have been published.[29,30]

The measurement of nanometer-size particles has also been referred to as dynamic light scattering (DLS). Plantz[31] refers to the technique as DLS because it involves the shining of laser light into a liquid particle suspension (aqueous or nonaqueous), which then collects the light scattered by a photomultiplier tube. Brownian motion thereby causes each particle to change position and the scattered light wave presents a changing phase relative to the scattered light from other particles in the suspension. The Doppler shift for the frequency of light scattered at a high angle (usually 90°) is related to the velocity of the particle, which in turn is inversely related to size. By measuring the particles migration, one can determine the particle's diffusion coefficient, D. The equivalent spherical diameter, d, can then be calculated from the Stokes-Einstein equation:

$$D = kT/3\pi \mathfrak{z} d \qquad (10)$$

where k is the Boltzmann constant, T is the absolute temperature (K) and $\mathfrak{z}$ is the viscosity of the fluid.

Several instruments based on this light scattering technology have been developed to measure the particle size of many colloidal materials in the 2-nm to 5-μm size range.

Table 21.3 Calculations for various size distribution moments for fine polystyrene powder

Diameter d, interval (μm)	Number particles, n_i	$n_i d_i$	$n_i d_i^2$	$n_i d_i^3$	$n_i d_i^4$
1.00–2.00	120	180	270	405	608
2.00–3.00	200	500	1250	3125	7813
3.00–4.00	140	490	1715	6003	21009
4.00–5.00	80	360	1620	7290	32805
5.00–6.00	60	330	1815	9983	54904
6.00–7.00	30	195	1268	8239	53552
7.00–8.00	10	75	563	4219	31641
8.00–9.00	5	43	361	3071	26100
9.00–10.00	3	29	271	2572	24435
10.00–11.00	2	21	221	2315	24310
11.00–12.00	1	12	132	1521	17490
Σ	651	2235	9486	48,743	294,667

Table 21.4 Cumulative volume calculations for fine polystyrene powder

Diameter d, interval (μm)	$n_i d_i^3$	Cumulative volume $\Sigma n_i d_i^3$	Cumulative volume %
1.00–2.00	405	405	0.83
2.00–3.00	3125	3530	7.24
3.00–4.00	6003	9533	19.56
4.00–5.00	7290	16823	34.51
5.00–6.00	9983	26806	54.99
6.00–7.00	8239	35045	71.90
7.00–8.00	4219	39264	80.55
8.00–9.00	3071	42335	86.85
9.00–10.00	2572	44907	92.13
10.00–11.00	2315	47222	96.88
11.00–12.00	1521	48743	100.00

Table 21.5 lists some of the commercial instrument manufacturers for this technique. Most of the manufacturers' Web sites provide details on the operation of their instruments that are very helpful to new practitioners.

Light Scattering and Diffraction

The laser diffraction technique obtains the mean particle size and particle size distribution through a matrix conversion of scattered intensity measurements as a function of scattering angle and wavelength of light based on Mie theory. The various Web sites for manufacturers of laser diffractometers show the components of a laser diffraction particle sizer (Table 21.6). It is an absolute method without need of calibration because the measurement principles are firmly based in physics and mathematics. For particles that have a diameter larger than the wavelength of light, Fraunhofer diffraction[32] is used because the intensity of light scattered by particles is proportional to particle size, and the size of the diffraction pattern (scattering angle) is inversely proportional to size. Smaller particles scatter a small but definite amount of light at a larger angle, whereas larger particles scatter a greater amount of light at a smaller angle.

Laser diffractometers are made by many commercial manufacturers. A list of these manufacturers is presented in Table 21.6; the Web sites contain a host of information on these techniques. These ensemble methods cover a broad range of particle size from about 0.1 to 3500 μm. This makes these methods applicable to a wide variety of materials processed by methods such as grinding, milling, emulsification, and polymerization. This information becomes invaluable in the production of particles of a specific particle size distribution to control process efficiency and quality.

Bott[33] of Beckman Coulter (Miami, FL) developed an alternative technology for sizing submicron-size particles using

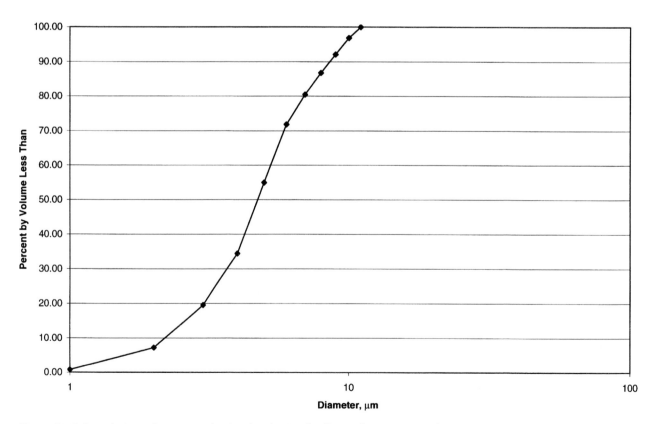

Figure 21.5 Cumulative volume particle size distribution for fine polystyrene powder.

Figure 21.6 Transmission electron micrograph of polymer latex.

polarization intensity differential scattering (PIDS). The patented technology is based on a particular property of the scattered light patterns for particles smaller than the wavelength of incident light, and extends the useful sizing range for diffraction instruments down to 0.1 µm. Polymer latex particles as small as 0.137 µm have been successfully measured by this method.

Freud et al.[34] developed a TRI-LASER system for the MICROTRAC instruments (Honeywell, Phoenix, AZ), which also allows light scattering measurements to be extended to about 0.1 µm size range for well-behaved, relatively wide, and single-mode particle systems such as polymer lattices.

All of the commercial diffractometers on the market vary in optical configuration and in the software used to transform the diffraction pattern data into a size distribution. The deconvolution procedure, which is built into the logic of these instruments, represents a major investment by the manufacturers of these instruments. The cost of these instruments can vary from about $50,000 to $100,000 de-

pending upon how much capability is needed to help solve a particle sizing problem.

Many studies have compared data from other particle sizing methods with that produced by laser diffractometers. Kaye et al.[35] correlated size distribution data from sieving, image analysis, and diffractometer methods. Significant differences in particle size distribution among the three methods were found for a series of irregular-shaped iron powders in the 30- to 300-µm size range. The most spherical powder showed much better agreement among methods, indicating that particle shape plays a role in the results being generated by the various particle size techniques. Professor Kaye[36] continues to pursue particle shape analysis and its effect on results generated by various particle size measurement equipment. Discrepancies have been reported in the literature by Merkus et al.[37] in a comparison of results from several laboratories that ran samples of a Community Bureau of Reference quartz material by laser diffractometry. The accuracy of laser diffractometer results can be improved if carefully developed mathematical models are determined on the basis of knowledge of the real and imaginary refractive index of the material in question. Mie theory models then become the better choice to use because the Fraunhofer models generally compare results to an imaginary refractive index of about 0.10. The imaginary component (absorptivity) is the primary driver in determining accurate size results.

In conclusion, the laser diffraction instruments, although not perfect for all particle size measurement applications, are helpful tools that can generate a lot of data very quickly and help to define the particle size distribution for a wide variety of materials over a very wide particle size range.

Stream Counting Methods for Size Analysis

Electrozone Sensing Counters

The electrozone sensing counters, often referred to as the Coulter principle instruments, are available commercially from two major U.S. manufacturers. They are listed in Table 21.7 along with the particle size ranges and Web site

Table 21.5 Commercial manufacturers of photon correlation spectroscopy (PCS) light-scattering instruments

Instrument name	Size range (µm)	Manufacturer	Web site
90 Plus	0.003–3.0	Brookhaven, Holtsville, NY	www.bic.com
FOQELS	0.002–5.0		
N-4	0.04–3.0	Beckman Coulter, Miami, FL	www.coulter.com
UPA 150/250	0.003–6.5	Honeywell, Phoenix, AZ	www.iac.honeywell.com
Nicomp TC100		Particle Sizing System, Santa Barbara, CA	www.pssnicomp.com

Table 21.6 Commercial manufacturers of laser diffraction instruments

Instrument name	Size range (μm)	Manufacturer	Web site
LS100Q, LS200, and LS230	0.40–1000	Beckman Coulter, Miami, FL	www.coulter.com
LA900 and LA700	0.2–900	Horiba Instruments, Irvine, CA	www.horiba.com
Microtrac X-100	0.04–700	Honeywell, Phoenix, AZ	www.iac.honeywell.com
Microtrac SRA200	3.2–2000		
Mastersizer 2000	0.02–2000	Malvern, Southborough, MA	www.malvern.co.uk
Mastersizer S	0.05–3500		

addresses for more detailed information. The technique requires an electrolyte solution (conductive liquid) and operates on the scientific principles of Ohm's law. The particles are suspended in an electrolyte solution, and a controlled vacuum draws them through a very small (e.g., 20–560 μm) sapphire orifice. (See the Beckman Coulter Web site [Table 21.7] for details on the instrument operation.) The resistance across the orifice is monitored by means of immersed electrodes on either side. As each particle passes through the small orifice, the passing of the particle changes the resistance and thereby generates a voltage pulse that is amplified, sized, and counted. These counters are calibrated using standard latex particles referred to in the Calibration Materials section of this chapter. Alliet et al.[38] showed the effects of coincidence for a model C Coulter Counter using spherical polystyrene polymer powder in a 1- to 10-μm size range. These results clearly show the significant size errors generated when the particle suspension concentration exceeds 10 mg per 250 mL of an aqueous electrolyte solution.

This technique has many applications in industry for materials such as emulsions, polymer latex, silica gel, toners, and biological cells. The beauty of the method is generation of raw number distribution data that can then be mathematically transformed to the other moments of particle size distribution. Apertures having diameters ranging from about 20 to 2000 μm are available. Each aperture covers a particle size range of about 2% to 60% of the aperture diameter, giving an overall range of 0.4 to 1200 μm. For most applications, a single aperture is sufficient. The instruments response is unaffected by particle shape, composition, or refractive index. For irregularly shaped particles, the size reported is equivalent spherical diameter. Because this non-ensemble method of particle size analysis has been available for about 45 years, hundreds of research and applications articles have been published. One can obtain summaries of this literature by contacting the manufacturers of these instruments.[39]

Alliet et al.[40] characterized the particle size distribution of multicomponent Rayleigh toners and compared electrical sensing zone counter results with light and electron microscopy. They found excellent agreement (<5% difference) between the microscope measurements and electrical sensing zone methods. This was due in part to the improved resolution of these instruments (i.e., 256 channels). Davies et al.[41] did extensive research on the investigation of flow direction and angle of particle entry through the electrical sensing zone orifice. This early work at Illinois Institute of Technology Research Institute (IITRI) clearly showed that accuracy could be improved through modifications to the sensing orifice and hydrodynamic flow.

Optical Single-Particle Counter

The light blockage (or obscuration) method of sizing particles is now referred to as the single-particle optical sensing (SPOS) method. This method counts and sizes particles as they pass individually through a small detecting zone at a rate of several thousand particles per second. These techniques are similar to the Coulter Counter in that they provide number-based raw data. However, the technique has the advantage that it does not require an electrolyte for sensing the particles in a dilute solution. It is very important to keep the particle suspensions very dilute to minimize coincidence errors during the particle size analysis.

Table 21.7 Commercial manufacturers of electrozone counter instruments

Instrument name	Size range (μm)	Manufacturer	Web site
Multisizer IIE	0.40–2000	Beckman Coulter, Miami, FL	www.coulter.com
Elzone 5380 and Elzone 5382	0.4–2000	Micrometrics, Norcross, GA	www.micrometrics.com
HELOS	0.1–3500	Sympatec, Princeton, NJ	
CILAS 920	0.3–400	Quantachrome, Boynton Beach, FL	www.quantachrome.com
CILAS 940	0.5–1500		
CILAS 1064	0.04–500		

Table 21.8 Commercial manufacturers of optical single-particle counter instruments

Instrument name	Size range (μm)	Manufacturer	Web site
Nicomp 380	0.003–5.0	Particle Sizing Systems, Santa Barbara, CA	www.pssnicomp.com
Accusizer 780	0.5–1,000		
Climet CI-100	0.65–1.0	Climet Instruments, Redlands, CA	www.climet.com
Climet CI-200	0.20–10		
Climet CI-500	0.30–25		
HIAC model 9703	1.0–600	Royco Instruments Co., Menlo Park, CA	www.HIAC.com
	0.20–10		

The early light blockage stream counters were developed by Royco Instruments Co. (Menlo Park, CA) and were called high-accuracy counters (HIAC). For details on the instrument operation, see the HIAC Web site (see Table 21.8). A relationship between the voltage output E of the photo divide and the particle cross-sectional area, a, is given as

$$E = \frac{aE_b}{A} \qquad (11)$$

where A is the cross-sectional area of the sensor in the direction of the incident beam, and E_b is the base voltage of the photo detector. This output voltage is then the size characteristic measured on the basis of the maximum cross-sectional area. It is important to note with this technique that for transparent particles, the refractive index of the suspending fluid must be different than that of the particle. A more recent higher resolution optical counter has become available from Particle Sizing Systems (Santa Barbara, CA).[42] The Accusizer 780/SPOS is compatible with any liquid or gas, and has a wide size range and a high count rate. Like all stream counting methods, this technique required concentrated suspensions to be diluted to minimize particle coincidence in the sensor. The Accusizer uses a patented method called Auto dilution, which automatically dilutes the starting sample to the optimum concentration, resulting in accurate results. This eliminates errors caused by operators either over- or under-diluting the particle suspension before analysis. Excessive dilution results in a low count rate and poor statistics, whereas overly concentrated suspensions produce coarse size results.

In summary, the optical single-particle counters are listed in Table 21.8. The Climet Instruments (Redlands, CA) and Royco instruments were developed primarily to measure gas-borne particles. These instruments measure large volume air filter systems that have high collection efficiency of 99.999% or greater for particles greater than 0.1 μm in diameter. The liquid-borne particle measuring instruments are used to ensure cleanliness of process and product liquids.

Time-of-Flight Counters

TSI Inc. (St. Paul, MN) introduced the first aerodynamic particle sizer (APS) in 1982. The APS is based on the principle of particle lag in an accelerating flow field. Laser velocimetry is a technique used to measure the time-of-flight of individual particles between two laser beams. A microprocessor then converts the time-of-flight data into a particle size distribution and a minicomputer displays the data. Blackford et al.[43] described an improved APS instrument to extend the upper size limit to 30 μm for industrial powders. The performance of the APS instruments has also been described in other literature.[44,45]

Recently, TSI Inc. acquired Amherst Process Instruments (API; Amherst, MA) to provide their customers greater advancement in this new particle size measurement technology. Time-of-flight counting instruments are sometimes referred to as aerosol beam spectrometers or TOF-ABS. Dahneke[46] first introduced the term and performed the original studies to develop the Aerosizer, which is currently manufactured by TSI Amherst. The Aerosizer measures particle size by expanding the air-particle suspension through a nozzle into a partial vacuum. A detailed description of the instrument operation can be found at the Aerosizer Web site (see Table 21.9). The air leaves the nozzle at near-sonic velocity and accelerates through the measurement region. Particles are in turn accelerated by the drag force, which is

Table 21.9 Commercial manufacturers of time-of-flight particle counter instruments

Instrument name	Size range (μm)	Manufacturer	Web site
APS 3320	0.5–20.0	TSI Inc., Particle Instrument Div., St. Paul, MN	www.TSI.com
Aerosizer 3220	0.2–200		
Aerosizer LD	0.2–700	TSI Inc., Particle Instrument/Amherst, MA	www.aerosizer.com

Table 21.10 Commercial manufacturers of sedimentation instruments

Instrument name	Size range (μm)	Manufacturer	Web site
Sedigraph 5000D	0.1–300	Micromeritics, Norcross, GA	www.micromeritics.com
Horiba CAPA-700	0.01–300	Horiba Instruments, Irvine, CA	www.horiba.com
BI-XDC	0.01–100	Brookhaven Instruments, Holtsville, NY	www.bic.com
BI-DCP Disc Centrifuge	0.01–100		

generated by the accelerating air stream. Smaller particles are accelerated to near the air velocity by the drag forces, whereas the larger particles experience lower acceleration because of their greater mass. The relationship between the particle size and the time-of-flight depends upon the density of the particle. A calibration curve is generated from particles of known size and density, and the curve is plotted against the time-of-flight. For spherical particles, converting the aerodynamic diameter measured by the instrument using the density value input generates the geometric diameter.

Niven[47] discusses the performance of the Aerosizer and its application to pharmaceutical analysis. He states that the instrument is simple to use and has proven to be reliable over time. Development of the breather unit by TSI Amherst has extended the time-of-flight technology to evaluate the performance of inhaler devices.

Allen[48] found the Aerosizer to be an exciting breakthrough for the analysis of very fine industrial powders such as titanium dioxide. The principal advantage of this instrument over other instruments is its ability to use dry samples rather than wet slurries, which also significantly reduces analysis time. He also found the accuracy to be an order of magnitude better than that of the light scattering method.

Sedimentation Methods for Size Analysis

Gravitational Sedimentation

The gravity sedimentation particle sizing technique has been available for a long time. The interpretation of particle size depends on the shape and density of the particle as well as the density and viscosity of the medium used in the analysis. If the particles are spherical, the sedimentation coefficient, s, can be represented by the Stokes' law equation:

Table 21.11 Range of field-flow fractionation applications

Polymer/lattices	Emulsions/dispersions
Polystyrene	Poly(tetrafluoroethylene) (PTFE)
Poly(methyl methacrylate)	Soft drink colorants/flavorants
Pigments/inorganics	Biological/life sciences
Carbon black	Liposomes
TiO$_2$	Viruses
Silica sols	Albumin

$$S = \frac{(p - p_0)d_{ST}^2}{18n_0} \tag{12}$$

where p and p_0 are the densities of the particle and medium, respectively; n_0 is the medium viscosity; and d_{ST} is the Stokes' particle diameter.

The most widely used instrument for particle size analysis on the basis of gravity sedimentation is the Sedigraph 5000-D (Micromeritics, Norcross, GA; see Table 21.10). This instrument uses soft X-rays to monitor the particle concentration. The sample cell is moved downward through an X-ray beam at a programmed rate so that the time and depth will define a given particle size. The results of analysis are reported as a graph of cumulative mass percent undersize as a function of particle diameter. An attractive feature of this instrument is that the X-ray monitor yields a response that is independent of the particle sizes in the disperse phase. However, for organic-based disperse phases, the X-ray absorption is too low for particle size analysis and cannot be used for these materials. Micromeritics offers another instrument, the Sedigraph L, which uses turbidity (white light) as a monitor to handle organic materials.

Centrifugal Sedimentation

Because gravity sedimentation is a relatively slow analysis, one needs to consider using a centrifugal field where a lower particle size limit can be achieved with a considerable decrease in analysis time. There are several manufacturers of commercial instruments (see Table 21.10). Brookhaven Instruments (Holtsville, NY) offers two instruments that cover a wide particle size range. The BI-XDC model uses an X-ray detector that works well for inorganic materials such as metals, ceramics, and ores over a size range of 0.01 to 100 μm. A second instrument, model BI-DCP, uses a light-scattering detector that can be used for polymer colloids, emulsions, and organic pigments over a particle range of 0.01 to 3.0 μm.

Chromatographic Methods for Size Analysis

Hydrodynamic Chromatography

Hydrodynamic chromatography (HDC) is a technique for obtaining particle size information on colloidally suspended

Table 21.12 Particle size distribution data for a toner by Coulter Multisizer IIE

Channel (bin)	Diameter range (µm)	Diameter midpoint, d_i	Frequency, N_i	$(N_i d_i)$	Frequency, number percent	Cumulative percent less than upper diameter	Diameter midpoint, d_i^3	$(N_i d_i^3)$	Frequency volume percent	Cumulative percent less than upper diameter
1	1.13–1.19	1.16	0	0	0.00	0.00	1.560896	0	0.00	0.00
2	1.19–1.26	1.23	0	0	0.00	0.00	1.860867	0	0.00	0.00
3	1.26–1.33	1.3	39	50.7	0.41	0.41	2.197	85.683	0.00	0.00
4	1.33–1.41	1.37	19	26.03	0.20	0.61	2.571353	48.855707	0.00	0.00
5	1.41–1.49	1.45	16	23.2	0.17	0.77	3.048625	48.778	0.00	0.00
6	1.49–1.57	1.53	10	15.3	0.10	0.88	3.581577	35.81577	0.00	0.00
7	1.57–1.66	1.62	15	24.3	0.16	1.03	4.251528	63.77292	0.00	0.00
8	1.66–1.76	1.71	18	30.78	0.19	1.22	5.000211	90.003798	0.00	0.00
9	1.76–1.86	1.81	2	3.62	0.02	1.24	5.929741	11.859482	0.00	0.00
10	1.86–1.96	1.91	7	13.37	0.07	1.32	6.967871	48.775097	0.00	0.00
11	1.96–2.08	2.02	11	22.22	0.11	1.43	8.242408	90.666488	0.00	0.01
12	2.08–2.19	2.14	7	14.98	0.07	1.50	9.800344	68.602408	0.00	0.01
13	2.19–2.32	2.26	7	15.82	0.07	1.58	11.54318	80.802232	0.00	0.01
14	2.32–2.45	2.39	9	21.51	0.09	1.67	13.65192	122.867271	0.00	0.01
15	2.45–2.59	2.52	8	20.16	0.08	1.76	16.00301	128.024064	0.00	0.01
16	2.59–2.74	2.67	3	8.01	0.03	1.79	19.03416	57.102489	0.00	0.01
17	2.74–2.9	2.82	14	39.48	0.15	1.93	22.42577	313.960752	0.00	0.01
18	2.90–3.06	2.98	4	11.92	0.04	1.98	26.46359	105.854368	0.00	0.02
19	3.06–3.23	3.15	11	34.65	0.11	2.09	31.25588	343.814625	0.00	0.02
20	3.23–3.42	3.33	10	33.3	0.10	2.19	36.92604	369.26037	0.00	0.02
21	3.42–3.61	3.52	10	35.2	0.10	2.30	43.61421	436.14208	0.00	0.03
22	3.61–3.82	3.72	12	44.64	0.13	2.42	51.47885	617.746176	0.01	0.04
23	3.82–4.04	3.93	14	55.02	0.15	2.57	60.69846	849.778398	0.01	0.05
24	4.04–4.27	4.15	11	45.65	0.11	2.69	71.47338	786.207125	0.01	0.05
25	4.27–4.51	4.39	19	83.41	0.20	2.88	84.60452	1607.485861	0.02	0.07
26	4.51–4.77	4.64	29	134.56	0.30	3.19	99.89734	2897.022976	0.03	0.10
27	4.77–5.04	4.91	65	319.15	0.68	3.87	118.3708	7694.100115	0.09	0.19
28	5.04–5.33	5.18	126	652.68	1.32	5.18	138.9918	17512.97083	0.20	0.39
29	5.33–5.63	5.48	231	1265.88	2.41	7.60	164.5666	38014.88275	0.43	0.82
30	5.63–5.95	5.79	306	1771.74	3.20	10.80	194.1045	59395.98893	0.67	1.48
31	5.95–6.29	6.12	432	2643.84	4.51	15.31	229.2209	99023.4409	1.11	2.60
32	6.29–6.65	6.47	484	3131.48	5.06	20.37	270.84	131086.5711	1.48	4.07
33	6.65–7.03	6.84	514	3515.76	5.37	25.74	320.0135	164486.9411	1.85	5.93
34	7.03–7.43	7.23	569	4113.87	5.95	31.69	377.9331	215043.9151	2.42	8.35
35	7.43–7.86	7.64	585	4469.4	6.11	37.80	445.9437	260877.0902	2.94	11.28
36	7.86–8.3	8.08	592	4783.36	6.19	43.99	527.5141	312288.3543	3.51	14.80
37	8.30–8.78	8.54	613	5235.02	6.41	50.39	622.8359	381798.3846	4.30	19.09
38	8.78–9.28	9.03	599	5408.97	6.26	56.65	736.3143	441052.2819	4.96	24.06
39	9.28–9.81	9.54	653	6229.62	6.82	63.48	868.2507	566967.6836	6.38	30.44
40	9.81–10.37	10.09	622	6275.98	6.50	69.98	1027.244	638945.5994	7.19	37.63
41	10.37–10.96	10.66	625	6662.5	6.53	76.51	1211.355	757097.185	8.52	46.15
42	10.96–11.58	11.27	572	6446.44	5.98	82.49	1431.435	818781.0391	9.22	55.37
43	11.58–12.24	11.91	502	5978.82	5.25	87.73	1689.411	848084.2572	9.55	64.91
44	12.24–12.94	12.59	420	5287.8	4.39	92.12	1995.617	838159.1312	9.43	74.35
45	12.94–13.68	13.31	299	3979.69	3.12	95.25	2357.948	705026.3596	7.93	82.28
46	13.68–14.46	14.07	198	2785.86	2.07	97.31	2785.366	551502.4963	6.21	88.49
47	14.46–15.28	14.87	130	1933.1	1.36	98.67	3288.008	427441.0794	4.81	93.30
48	15.28–16.15	15.72	69	1084.68	0.72	99.39	3884.701	268044.3861	3.02	96.31
49	16.15–17.07	16.61	38	631.18	0.40	99.79	4582.568	174137.5757	1.96	98.27
50	17.07–18.05	17.56	12	210.72	0.13	99.92	5414.689	64976.27059	0.73	99.01
51	18.05–19.08	18.56	5	92.8	0.05	99.97	6393.43	31967.15008	0.36	99.37
52	19.08–20.16	19.62	1	19.62	0.01	99.98	7552.609	7552.609128	0.09	99.45
53	20.16–21.31	20.74	1	20.74	0.01	99.99	8921.261	8921.261224	0.10	99.55
54	21.31–22.53	21.92	0	0	0.00	99.99	10532.26	0	0.00	99.55
55	22.53–23.81	23.17	0	0	0.00	99.99	12438.79	0	0.00	99.55
56	23.81–25.17	24.49	0	0	0.00	99.99	14688.12	0	0.00	99.55
57	25.71–26.61	25.89	0	0	0.00	99.99	17353.86	0	0.00	99.55
58	26.61–28.12	27.37	0	0	0.00	99.99	20503.33	0	0.00	99.55

(continued)

Table 21.12 Continued

Channel (bin)	Diameter range (µm)	Diameter midpoint, d_i	Frequency, N_i	$(N_i d_i)$	Frequency, number percent	Cumulative percent less than upper diameter	Diameter midpoint, d_i^3	$(N_i d_i^3)$	Frequency volume percent	Cumulative percent less than upper diameter
59	28.12–29.73	28.93	0	0	0.00	99.99	24212.82	0	0.00	99.55
60	29.73–31.42	30.58	0	0	0.00	99.99	28596.47	0	0.00	99.55
61	31.42–33.21	32.32	0	0	0.00	99.99	33760.9	0	0.00	99.55
62	33.21–35.11	34.16	1	34.16	0.01	100.00	39861.5	39861.4953	0.45	100.00
63	35.11–37.11	36.11	0	0	0.00	100.00	47084.99	0	0.00	100.00
64	37.11–39.22	38.17	0	0	0.00	100.00	55611.74	0	0.00	100.00
		Σ	9569	85792.69	100.00			8885049.38	100.00	
				$d_{10} = \Sigma nd/\Sigma n =$		8.97		$d_{30} = \sqrt[3]{\Sigma nd^3/\Sigma n} =$		9.76

particles in about the same time one obtains size analysis of molecules in solution using chromatographic methods. Small[49] made two important observations about the effluent: (1) larger particles elute faster than small ones; and (2) the smaller the packing diameter, the better the separation. The technique is limited to a size range of about 0.01 to 1.0 µm. The method is calibrated against TEM using known particle size polymer latex particles (discussed in the Calibration Materials section of this chapter). HDC has been most effec-

tive for separation of latex particles. This is primarily because latex particles are rigid spheres, whereas polymer molecules are usually deformable and nonspherical.

HDC techniques have shown considerable promise for analysis of ultra-high molecular weight macromolecules and can be applied to a variety of water-soluble polymers. The upper particle size limit of about 1 µm allows the technique to encompass even the largest water-soluble polymers. Because HDC has very low resolving power, the de-

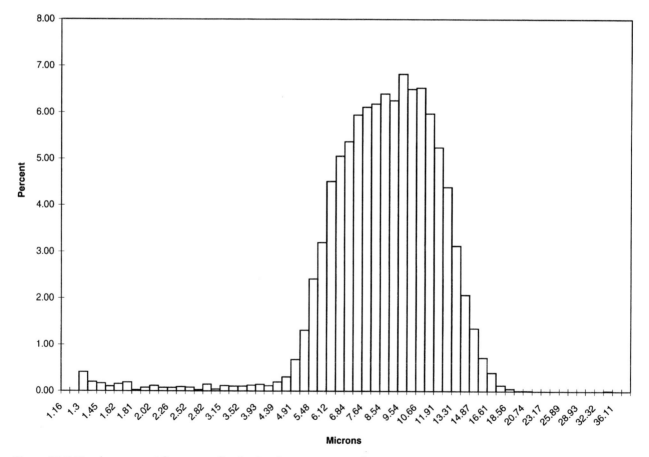

Figure 21.7 Number percent frequency distribution for a toner sample.

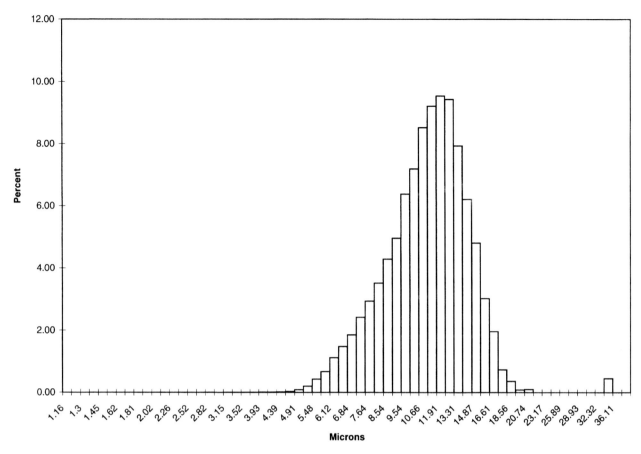

Figure 21.8 Volume percent frequency distribution for a toner sample.

termination of the chromatogram requires deconvolution of the chromatogram to make the correction for axial dispersion. McGowan and Langhorst[50] demonstrated the success of deconvolution for the separation of latex spheres by HDC.

Although the basic HDC concept has been proven through experimentation, a commercial instrument has not been developed for the particle technologist.

Field-Flow Fractionation

Sedimentation field-flow fractionation (SFFF) allows a mobile liquid phase to be delivered at a constant flow rate through a long narrow channel. A particle suspension is introduced into the liquid phase and, when it reaches the channel, the flow is momentarily stopped. Under an applied centrifugal field, the larger particles accumulate into thinner, more compact layers at the wall, whereas the small particles form thicker and more diffuse layers. An interaction with the streamlines of fluid flowing through the channel separates the various size particles, eluting the smallest ones first. Dupont Research Center (Kennett Square, PA) developed a commercial instrument around this concept called the Model S101. The instrument operated in the centrifugal mode over a size range of 0.05 to 2 μm and in the steric mode from 1 to 100 μm.

The applied field in FFF techniques can be centrifugal, gravitational, thermal, electrical, or magnetic. SFFF is commercially available. SFFF, because of open channel separation, has much greater potential than HDC. Giddings[51] invented and pioneered much of the research and applications for the FFF technique. He has applied the new particle sizing technology to a wide variety of materials including polymer lattices, emulsions, environmental particles, and biological cells (Table 21.11). For most of these applications, FFF has the advantage of high resolution and the ability to collect particle fractions that can then be examined by other techniques such as electron microscopy.

Colloidal aggregation is very common in many industrial, biological, and environmental materials. As a consequence of this physical state, the particle size distribution is altered and thus affects the quality and performance of the material. Some of the modern methods of particle size analysis cannot closely resolve polymer latex materials that are too close in particle size. This sometimes leads an analyst to gloss over significant details that may be required and can have an important bearing on product quality. Giddings et al.[52] demonstrated the very high resolution of SFFF analysis for separation of polystyrene latex samples in the submicron-size region.

Field-flow fractionation is discussed in detail in chapter 8 of this volume.

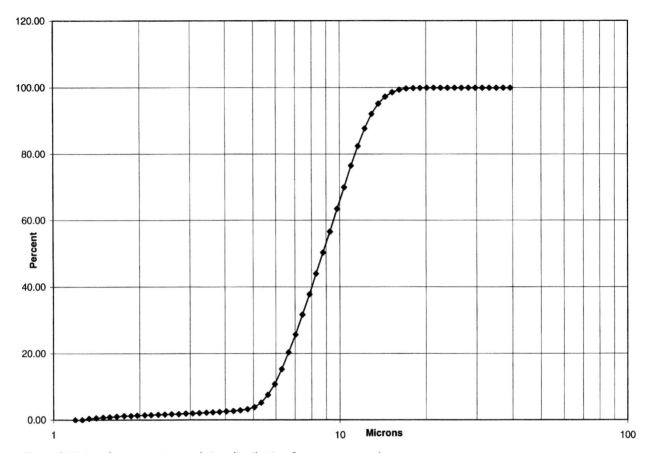

Figure 21.9 Number percent cumulative distribution for a toner sample.

On-Line and In-Line Particle Size Analysis

On-Line Methods

On-line particle size analysis requires a sample to be taken from the product stream and measured in true time, without any human intervention. This type of analysis may still require sampling but it is automatic. On-line sizing offers faster measurements and better production control of the product being processed. It is not influenced by operator errors and provides better knowledge about the influence of process parameters. Some of the disadvantages are expense, complexity, and generation of large amounts of data to interpret. However, the advantages may outweigh the disadvantages by providing enhanced product quality.

The stream counting methods were among the first techniques to be modernized for on-line particle size analysis. Coulter developed an on-line monitor[53] that operates in an identical fashion to the off-line particle sizer and can take samples continuously from a suitable outlet in a production plant or filtration system. Particle Sizing Systems recently developed on-line equipment that provides particle size distribution using the model 780/SPOS Accusizer particle counter. This equipment is still being evaluated, but holds

great promise for liquid process streams that can be diluted prior to analysis. The stream counting methods are widely used in various industries and the need for on-line particle size measurements will become increasingly important in the 21st century.

Laser light scattering methods of particle size analysis have also been automated to perform on-line particle size measurements. Hart[54] used an on-line Microtrac monitor and a specialized sample system to control a cyclone classifier for glass bead feed material. He showed the ability to control product particle size distributions within 2% at a setpoint of 13% passing through 11 μm.

Insitec Measurement Systems (San Ramon, CA, recently acquired by Malvern Industries) has been developing on-line particle size instruments for many years. The particle sizer combines laser light scattering and sophisticated information processing to determine size and velocity of particles as they pass through a laser beam. The method has been used on-line for a variety of materials including slurries, mists and liquid sprays, and solid particles. The operating particle size range is 0.2 to 300 μm at concentrations up to $10^{13}/m^3$.

The Lab-tec100 (Lasentec, Bellevue, WA) instrument uses a laser beam to rapidly scan over a small viewing area.

The beam is highly focused and illuminates individual particles in its path. Because the beam is scanned at a constant rate, the pulse width is a measure of chord size. This technique can be used on-line by focusing the beam to approximately 1 mm inside the suspension. The device covers a size range of 0.5 to 200 μm, with additional ranges up to 1.5 mm.

In-Line Methods

In-line or in situ particle size analysis removes the sampling process and thereby makes the system less complex. The measurement itself, however, is often more difficult to achieve. There is much higher demand on the correct placement of the measurement probe and protection against particle adhesion so that the measuring zone is not covered by sticky particles. Malvern/Insitec manufactures an ensemble sensor called ensemble particle concentration and sizing (EPCS), which is designed for in-line measurement of particle size over a range from 0.5 to 1000 μm in particle flows from static to supersonic.

The EPCS system is a true in-line system capable of automatically sizing particles directly from the process flow. It has been applied to powder coatings, toners, cement, and pharmaceuticals.

In Situ Methods

The in situ method of particle size analysis requires measurement in the line reactor or process equipment without disturbance. This is a nonintensive measurement technique. Weiner[55] applied the in situ method to submicrometer-size particle analysis of concentrated pigment dispersions. They used PCS and a remote fiberoptic sensor that was 40 m removed from the production process area. Haley[56] introduced the use of laser velocimetry to measure aerosols by in situ measurements over a range of 0.01 to 100 μm.

In situ methods remove the sampling process and therefore make the system less complex. The measurement itself, however, is a greater challenge to obtain. For some particle systems, this is the ideal particle size method because data can be generated as the material is being processed and there is no problem obtaining a representative sample from the product stream. This is an area of future research in particle size technology.

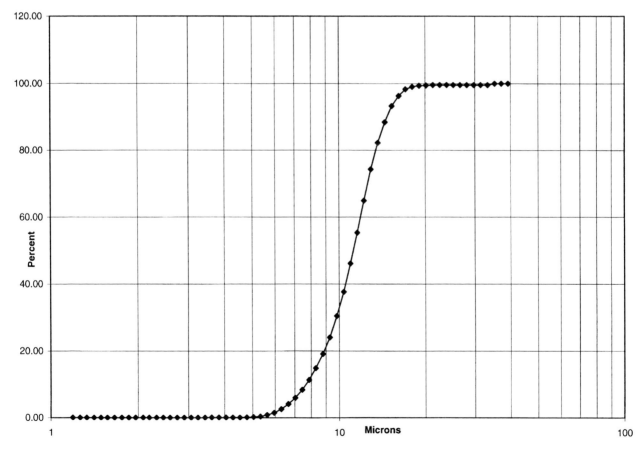

Figure 21.10 Volume percent cumulative distribution for a toner sample.

Particle Size Measurement Applications

Analytical Method

Numerous practical applications exist for the various particle size measurement techniques described in the previous sections. We have chosen to present a real-life example of the particle size measurement of xerographic toner as measured by the stream scanning method of size analysis. A standard test method for dry xerographic toner was first published by the American Society for Testing and Materials (ASTM)[57] on the basis of electrical sensing zone measurements. The sample of toner used in this analysis was pulled from a moving process stream and is representative of the bulk material. Prior to analysis, the toner sample is dispersed with Triton X-100 surfactant in an aqueous saline electrolyte. The diluted suspension of particles is then placed in the sensor compartment of a Beckman Coulter Multisizer IIE instrument equipped with a 70-μm orifice.

Data Analysis

A summary of the high-resolution Multisizer IIE data for xerographic toner is shown in Table 21.12. This table also shows the conversion of number frequency data to volume frequency and cumulative number and volume percentages. The number and volume frequency histograms in Figures 21.7 and 21.8 clearly show the shift in particle size distribution caused by conversion of the data from a number to volume basis. The cumulative frequency curves shown in Figures 21.9 and 21.10 show the same shift in particle size distribution, but this shift is usually more difficult to detect for those practitioners who have not used cumulative frequency curves for data analysis. Both the number and volume distributions are important when one needs to understand the physical behavior of a new material. Basically, the number size emphasizes the finer sized particles and the volume conversion lends more weight to the larger size particles.

Toner Performance

The optical density per unit mass, image resolution, and graininess of a xerographic print are primarily influenced by the particle size distribution of the toner. The finer size particles provide for higher resolution xerographic print qualities through better fill in of the image characters. However, too high a concentration of finer sized particles can lead to machine dirt and other copy quality problems. It is therefore necessary to carefully monitor the materials processing steps to yield optimum machine performance and print quality.

References

1. Kaye, B. H. In *Direct Characterization of Fine Particles*; Elving, P. J., Winefordner, J. D., Eds.; Monographs on Analytical Chemistry 61; Wiley & Sons: New York, 1981; pp 1–10.
2. Kaye, B. H.; Trottier, R. *Chem. Eng.* **1995**, *April*, 78.
3. Allen, T. *Particle Size Measurement*, 2nd ed.; Chapman & Hall: New York, 1992; pp 192–537.
4. Bunville, L. G. *Modern Methods of Particle Size Analyisis*; Elving, P. J., Winefordner, J. D., Eds.; Monographs on Analytical Chemistry 73; John Wiley & Sons: New York, 1984; pp 1–41.
5. Heywood, H. *Proceedings of the Particle Size Analysis Conference*; Bradford University; Society for Analytical Chemistry: London, 1970; pp 1–18.
6. Leschonski, K. *Powder Technol.* **1979**, 24, 115–124.
7. Blackford, D.; Pui, K. P. 2nd Annual Meeting of the Aerosol Society, Bournemouth, UK, March 22–24, 1988.
8. Niven, R. W. *Pharm. Technol.* **1993**, *January*, 72–78.
9. Dahneke, B. E. *Nat. Phys. Sci.* **1973**, *244*, 54–55.
10. *Particle Size Distribution II*; Provder, T., Ed.; ACS Symposium Series 472; American Chemical Society: Washington, DC, 1991.
11. Allen, T. *Particle Size Measurement*, 4th ed.; Chapman & Hall: New York, 1992, pp 1–40.
12. Allen, T.; Khan, A. A. *Chem. Eng.* **1979**, *238*, pp 108–112.
13. *The Aero Kaye Sampler*; Amherst Process Instruments, Amherst, MA, 1998.
14. Community Bureau of Reference, CEC, Brussels, Belgium.
15. National Institute of Standards and Technology (NIST), Gathersburg, MD 20899.
16. Duke Scientific Corp., 2463 Faber Place, Palo Alto, CA 94303.
17. Bangs Laboratories, 979 Keystone Way, Carmel, IN 46032.
18. Alliet, D. F. *Powder Technol.* **1976**, *13*, 3–7.
19. Alliet, D. F.; Behringer, A. J. *Particle Size Analysis 1970*; Groves, M. J. Ed.; The Society for Analytical Chemistry: London, 1970; pp 353–365.
20. Allen, T. *Particle Size Measurement*, 4th ed.; Chapman & Hall: London, 1992; pp 124–127.
21. Davies, R. *Powder Technol.* **1975**, *12*, 111–124.
22. Kaye, B. H. *Direct Characterization of Fine Particles*; Elving, P. J., Winefordner, J. D., Eds.; Chemical Analysis; Monographs on Analytical Chemistry 61; Wiley & Sons: New York, 1981; pp 338–375.
23. Kaye, B. H. *Characterization of Powders and Aerosols*; Wiley-VCH: New York, 1999; pp 21–54.
24. Kaye, B. H.; Alliet, D.; Switzer, L.; Turbitt, C. *Part. Syst. Charact.* **1998**, *15*, 180–190.
25. Allen, T. *Particle Size Measurement*, 4th ed.; Chapman & Hall: London, 1992; pp 241–245.
26. Endter, F.; Gebauer, H. *Optik* **1956**, *13*, 97.
27. Johari, O.; Bhattacharyya, S. *Powder Technol.* **1968–69**, *2*, 335–348.
28. Weiner, B. B. *Modern Methods of Particle Size Analysis*; Elving, P. J., Windfordner, J. D., Eds.; Monographs on Analytical Chemistry 73; Wiley & Sons: New York, 1984; pp 93–116.
29. Bern, B. J.; Pecora, R. *Dynamic Light Scattering with Applications to Chemistry, Biology and Physics*; Wiley-Interscience: New York, 1976.

30. *Photon Correlation Spectroscopy and Velocimetry*; Cummis, H. Z., Pike, E. R., Eds.; Plenum Press: New York, 1977.

31. Plantz, P. E.; Freud, P. J. *Powder Bulk Eng.* **1995**, *9*, 36–39.

32. Plantz, P. E. *Modern Methods of Particle Size Analysis*; Elving, P. J., Winefordner, J. D., Eds.; Monographs on Analytical Chemistry 73; Wiley & Sons: New York, 1984; pp 173–181.

33. Bott, S.; Hart, H. *Particle Size Distribution II Assessment and Characterization*; T. Provder, Ed.; ACS Symposium Series 472; American Chemical Society: Washington, DC, 1991; pp 106–122.

34. Freud, P. J.; Trainer, M. N.; Clark, A. H. Unified Scatter Technique For Full-Range Particle Size Measurement. Pittsburgh Conference, Atlanta, GA, March 1993.

35. Kaye, B. H.; Alliet, D. F.; Switzer, L.; Turbitt-Daoust, C. *Part. Syst. Charact.* **1997**, *14*, 219–225.

36. Kaye, B. H. Fineparticles Research Group, Sudbury, Ontario, Canada. Personal communication, 1998.

37. Merkus, H. G.; Bischof, O.; Drescher, S.; Scarlett, B. In *PARTEC 95, 6th European Symposium on Particle Characterization*, Nürnberg Messer GmbH, Nürnberg, Germany, March 1993.

38. Alliet, D. F.; Behringer, A. J. In *Particle Size Analysis 1970*; Groves, M. J., Ed.; The Society of Analytical Chemistry: London, 1970; pp 353–363.

39. Coulter Counter Industrial and Pharmaceutical References, Coulter Electronics Ltd., Bedfordshire, UK.

40. Alliet, D. F.; Tietjen, T. A.; Wood, D. H. Particle Size Analysis of Multicomponent Toner Systems. Presented at 3rd Chemical Society, Particle Size Analysis Conference, University of Salford, Manchester, UK, September 1977. In *Particle Size Analysis*; Groves, M. J., Ed.; Heyden, 1978; pp 31–43.

41. Davies, R.; Karuhn, R. Graf. *J. Powder Technol.* **1995**, *12*, 157–166.

42. Nicoli, D. F.; Chang, Y. J.; Wu, J. S.; Hasajidis, K. *Int. Labmate*, **1995**, *XVII*.

43. Blackford, D. *Proceedings of Ontario Research Foundation (ORF), Fine Particles and Filler*, Toronto, Ontario, Canada, May 1986; pp 89–115.

44. Chen, B. T.; Cheng, Y. S.; Yeh, H. C. *Aerosol Sci. Technol.* **1985**, *4*, 89–97.

45. Baron, P. A. *Aerosol Sci. Technol.* **1986**, *5*, 55–67.

46. Dahneke, B. E. *Nat. Phys. Sci.* **1993**, *244*, 54–55.

47. Niven, R. W. *Pharm. Technol.* **1993**, *17*, 72–78.

48. Allen, T.; Khalili, M. *Control Research and Development*; E 304/C 231; Dupont Co.: Wilmington, DE.

49. Small, H. J. *Colloid Interface Sci.* **1947**, *48*, 147.

50. McGowan, G.; Langhorst, M. J. *Colloid Interface Sci.* **1982**, *89*, 94.

51. Giddings, J. C. *Chemical & Engineering News*, October 10, 1988, 34–45.

52. Giddings, J. C.; Meyers, M. N.; Moon, M. H.; Baroon, B. N. *Particle Size Distribution II*; Provder, T., Ed.; ACS Symposium Series 472; American Chemical Society: Washington, DC, 1991; pp 198–216.

53. Barnett, M. I.; Sims, E.; Lines, R. W. *Powder Technol.* **1976**, *14*, 125–130.

54. Hart, W. H., Description and Performance of a Dry Powder Pilot Scale Process Control System. Presented at Powder and Bulk Solids Conference, Chicago, IL, May 1981.

55. Weiner, B.; Tscharnuter, W. *Proceedings of the Ontario Research Foundation, On-Line Particle Size Analysis*; Ontario Research Foundation: Toronto, Ontario, Canada, 1986; pp 115–119.

56. Haley, M. P. and Holve, D. J. *Proceedings of Ontario Research Foundation, On-Line Particle Size Analysis*; Ontario Research Foundation: Toronto, Ontario, Canada, 1986; pp 145–149.

57. ASTM F 577–583. Standard Test Method for Particle Size Measurements of Dry Toners; ASTM: West Conshohocken, PA, 1987.

22

Polymer Rheology: Principles, Techniques, and Applications

GUY C. BERRY

Introduction

The definition of rheology provided by the Society of Rheology (United States) describes it as the science of the deformation and flow of matter (e.g., see the Web site for the Society[1]). The importance and vitality of the subject are demonstrated by the many reviews and monographs available,[2-27] as well as the original literature from laboratories throughout the world devoted to scientific and technological issues in rheology. Clearly, the topics covered in a single chapter must be circumscribed, and cannot present the entire field. The subjects selected here include the following:

1. The general principles of rheometry, focusing on requirements for instrumentation.
2. The principal methods of rheometry, with emphasis on the functions determined.
3. The elements of linear viscoelasticity, including creep, recovery, stress growth and relaxation, and dynamic properties.
4. Approximations used in linear viscoelasticity, including time-temperature superposition.
5. The viscosity of dilute solutions of polymers and colloids.
6. Linear viscoelastic behavior of concentrated and undiluted polymeric fluids.
7. Nonlinear viscoelastic behavior of concentrated and undiluted polymeric fluids.
8. Strain-induced birefringence for concentrated and undiluted polymeric fluids.
9. Linear and nonlinear viscoelastic behavior of colloidal dispersions.

Although the general expressions for the linear viscoelastic behavior are presented, most of the discussion incorporates so-called incompressible behavior, for which the volume change is negligible. Many of the formulae presented in this chapter are also presented in a succinct form elsewhere.[9]

The separation into linear and nonlinear viscoelastic behavior, although very useful and employed in the following discussion, is somewhat arbitrary. A material that approximates linear viscoelastic behavior under the condition of a recently small deformation may exhibit a nonlinear response if that condition is not met.[17a,18,22a,28] In general, the constitutive equation describing the nonlinear behavior should reduce to the linear response under appropriate conditions, which are often met if the deformation has been recently small. In most of the examples discussed here, the material response may be cleanly separated into two regimes: fluid and solid behavior. These may be distinguished by their response at long time (or small frequency), with a solid exhibiting an equilibrium modulus, and a fluid presenting steady-state flow. An interesting material that may not be cleanly incorporated into these two limits is the incipient or critical gel. Materials that show solid behavior if the deformation is small, but fluid behavior if the deformation exceeds some critical value, and which exhibit yield at that deformation, are also discussed.

In studying these materials, the investigator may impose forces (or torques) and follow the resultant deformation, or vice versa. Under suitable conditions, analytical expressions that are dependent on the test geometry may be used to express the forces as stresses, and the deformation can be expressed as strains. A central tenet of a linear response is that these may be used to define compliances (given by a ratio of strain divided by stress) or moduli (given by a ratio of stress divided by strain). For a viscoelastic material, the functions so defined will depend on the time, t, for which the stress or strain is imposed. Functions appearing in linear viscoelasticity are given in Table 22.1; some corresponding functions for nonlinear behavior are also discussed. It is frequently useful to express two of these, $J(t)$ and $G(t)$, in terms of related functions and parameters:[2,9,10,11a,16a,18,23a]

Table 22.1 Linear viscoelastic functions

Shear compliance	$J(t)$
Shear modulus	$G(t)$
Bulk compliance	$B(t)$
Bulk modulus	$K(t)$
Tensile compliance	$D(t) = J(t)/3 + B(t)/9$
Tensile modulus[a]	$1/\hat{E}(s) = 1/3\hat{G}(s) + 1/9\hat{K}(s)$

a. The circumflex symbol over a variable denotes a Laplace transform.

$$R(t) = J(t) - t/\eta = J_\infty - [J_\infty - J_0]\hat{\alpha}(t) \qquad (1)$$

$$G(t) = G_e + [G_0 - G_e]\hat{\phi}(t) \qquad (2)$$

where the recoverable compliance $R(t)$ is the recoverable part of the creep compliance $J(t)$, η is the (linear) viscosity, with $1/\eta = 0$ for a solid, G_e is the equilibrium modulus, with $G_e = 0$ for a fluid, G_0 is the instantaneous modulus, with $J_0G_0 = 1$, and J_∞ is the limit of $R(t)$ for large t, with values discussed below equal to $J_e = 1/G_e$ for a solid or the steady-state recoverable compliance J_s for a fluid.

For both a linear elastic solid ($1/\eta = 0$ and $J_\infty = J_e = 1/G_e$) and a linear viscous fluid ($1/\eta > 0$ and $G_e = 0$), $\hat{\alpha}(t) = \hat{\phi}(t) = \delta(t)$, where $\delta(t)$ is the Dirac delta function (e.g., see references 13a and 23b). Consequently, $G(t) = G_0$ and $J(t) = J_0$, as expected for a linear elastic solid, or $J(t) = t/\eta$ and $G(t) = 0$ for $t > 0$ for a linear viscous fluid. For a more general linear viscoelastic material, $\hat{\alpha}(t)$ and $\hat{\phi}(t)$ are each unity for $t = 0$ and zero for large t, with $\hat{\alpha}(t) < \hat{\phi}(t)$ for $0 < t < \infty$, and in this case, $J_\infty \geq J_0$ is equal to the equilibrium compliance $J_e = 1/G_e$ for a solid or to the steady-state recoverable compliance J_s for a fluid (J_s was formerly denoted J_e^{0} [11a]). The evaluation of η and J_s from $G(t)$ for a fluid is discussed below.

The significance of the parameter J_0 (or $G_0 = 1/J_0$) in the preceding is a bit murky in practice, and in fact is operationally defined in terms of the time scale of interest as a practical matter. Effectively, it can be considered that J_0 is the value of $J(t)$ for t such that $\hat{\phi}(t) \approx 1$ on the experimental time scale, or equivalently, G_0 is the value of $G(t)$ for t such that $\hat{\phi}(t) \approx 1$ on the experimental time scale. Because these conditions are frequently reached as T approaches the glass transition temperature T_g for noncrystalline polymers, J_0 and G_0 are sometimes denoted J_g and G_g, respectively.[11a,b] Note that in fact, small contributions to $J(t)$ and $G(t)$ obtain for $T < T_g$, may be directly the subject of study, and are important in some applications.[5,12] Occasionally, the expression for $G(t)$ is characterized by a response at very small t, followed by a range of time for which time for which $\hat{\phi}(t)$ is essentially constant, before decreasing again for larger time. In such cases, the $G(t)$ may be represented by limiting the function $\hat{\phi}(t)$ to the behavior for the regime of larger t, and adding a term $\delta(t)(G_0 - G_1)$ to $G(t)$, and normalizing $\hat{\phi}(t)$ to reflect the remaining contributions $G_1 - G_e$ that relax on the longer time scale; an example of this is given in the

section Linear and Nonlinear Viscoelastic Behavior of Colloidal Dispersions.

Expressions similar to those for $J(t)$ and $B(t)$ may be introduced for the other functions in Table 22.1. For example, for the bulk properties,[11c,23a]

$$B(t) = B_e - [B_e - B_0]\hat{\beta}(t) \qquad (3)$$

$$K(t) = K_e + [K_0 - K_e]\hat{\kappa}(t) \qquad (4)$$

where $\hat{\beta}(t)$ and $\hat{\kappa}(t)$ correspond to the functions $\hat{\alpha}(t)$ and $\hat{\phi}(t)$, respectively, in the shear functions, with the same general characteristics.

The often-assumed limit of incompressibility implies that volume changes in the deformation are negligible. For a linear elastic solid, this condition is met if $K \gg G \approx E/3$ or, equivalently, $B \ll J \approx 3D$. The general relations among these functions for linear viscoelastic behavior are discussed later in this chapter. The time-dependent functions may be used to compute dynamic moduli or compliances for the response under a steady sinusoidal deformation at constant a frequency.

Finally, a number of the symbols used and the units for the quantities represented are given in Table 22.2 for easy reference; a complete list of symbols is also provided at the end of the chapter.

Principles of Rheometry

The experimenter cannot measure stresses or strains, but merely forces (torques) and deformations that may be reduced to stresses and strains under appropriate circumstances. The discipline of rheometry seeks to define practical instrumentation to provide experimental data under this constraint. Thus, some general requirements include the ability to[10,11d,e,f,14a,21]

1. Measure the torque (force) as a function of time, and have a method to compute the stress from the measured values.

Table 22.2 Functions and parameters used

Function/parameter	Symbol	Units
Time	t	T
Frequency	ω	T^{-1}
Strain component	ε_{ij}	—
Elongational strain	ε	—
Shear strain	γ	—
Rate of shear	$\dot{\gamma}, \dot{\varepsilon}$	T^{-1}
Stress component	S_{ij}	$ML^{-1}T^{-2}$
Shear stress	σ	$ML^{-1}T^{-2}$
Modulus	G, K, E	$ML^{-1}T^{-2}$
Compiance	J, B, D	$M^{-1}LT^2$
Viscosity	η	$ML^{-1}T^{-1}$

2. Measure the deformation as a function of time, and have a method to compute the strain from the measured values.

3. Precisely define the sample geometry (e.g., sample radius and height for parallel plates, sample radius and cone angle for cone and plate, and sample radii and height for concentric cylinders).

4. Minimize the influence of extraneous torques (forces) on the sample (e.g., ball bearing friction and gravitational forces).

5. Provide for measurement over as wide a span in time (frequency) as possible that is consistent with the sample behavior.

6. Control the temperature and, if necessary, the sample environment; sample geometry may require adjustment as the temperature is changed owing to sample volumetric expansion (contraction).

Increasingly, computer-aided data acquisition and instrumentation control is used in rheometry. It is indispensable in some cases, especially in experiments involving a steady sinusoidal deformation. The general features of the interactions that may be monitored by a computer system are illustrated in Figure 22.1, and simplified generalized examples of some of the instrumentation used in rheometry to study behavior in elongation or in shear appear in Figures 22.2–22.4. The simplest of these, shown schematically in Figure 22.2, is clearly limited in application to solids (or fluids with a very large viscosity), and may exhibit nonuniformity of the deformation, limiting the ability to convert forces and deformations to stresses and strains, respectively. Furthermore, even if applicable, the method has inherent limitations in the magnitude of the deformation that may be realized. Deformation of a sample confined between parallel plates (not illustrated) carries the same limitation of deformation magnitude. Oscillatory deformations with a small deformation amplitude are frequently used to study the rheological properties of solids and fluids in shear, in which case the limitation to small deformation usually is not a problem. Similar methods may be applied to the elongational deformation of a solid.[29] The expressions in Table 22.3 give the shear stress σ in terms of the force F, and the shear strain γ in terms of the displacement D for two geometries;[8a,10,11g] the displacement of coaxial cylinders usually is limited to studies of the viscosity of a fluid or the modulus of a confined solid.

Torsional deformation, using instrumentation shown schematically in Figures 22.3 and 22.4 has widespread use precisely because it is not afflicted by some of the limitations encountered in planar deformations, especially the limita-

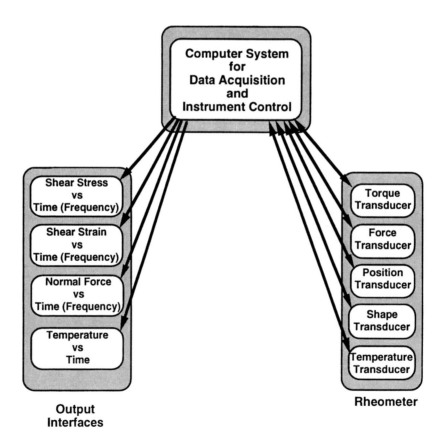

Figure 22.1 Schematic diagram of some of the principal components of a rheometer under the control of a computer and the output signals.

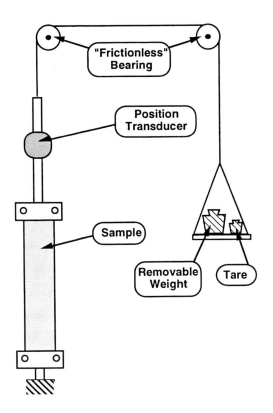

Device	Input	Output
Removable weight	Controlled weight	Controlled force
Position transducer	Measure of shaft position	Voltage (current)

Figure 22.2 Schematic drawing of a device for elongational (tensile) creep and recovery on a strip or fiber. The table at the bottom gives the input and output for the instrument.

tion to a small deformation, although torsional geometries may introduce other limitations. The torsional geometries illustrated in Figure 22.5 lend themselves to analysis under idealized conditions,[4a,8a,10,11d,14b,19a] including the assumptions of perfect alignment of the fixtures, the suppression of end effects, uniform temperature in the sample, and the condition of no slip of the sample at the fixture surfaces. The expressions in Table 22.3 give the shear stress σ in terms of the torque M, and the shear strain γ in terms of the angular deformation φ for several torsional geometries.[8a,10,11g,14b] Similar expressions are available for other arrangements, and expressions are available to determine the first normal stress difference N_1 from the total normal force required to maintain a set distance between the cone and plate fixtures.[4b,8b,14c]

Application of the scheme in Figure 22.2 to determine the elongational properties of solids may suffer from non-uniformity of the deformation along the length of the sample in applications to solids, but it is even more problematic when applied to fluids, unless the viscosity of the fluid is extremely high. Although such measurements would be of

interest, particularly with reference to nonlinear rheological behavior (see the section on nonlinear viscoelastic behavior of concentrated and undiluted polymeric fluids), methods to study the rheological properties of fluids in elongational deformations have been difficult to implement and are not discussed here because of limited space and the limitations of the available specialized methods that have been used.[14d,29,30]

The use of a controlled angular deformation, as in Figure 22.4, was historically the preferred mode for a long period, because the resultant torque could be determined, for example, using the deflection of a calibrated torsion bar supporting the fixed platen opposite that being driven by the controlled deformation. The ability to provide a controlled deformation was advanced by the use of motors under the control of a computer to impose a controlled angular rotation (e.g., a step, a ramp, or an oscillatory rotation) on one of the fixtures in the apparatus, and the sophisticated use of a beam-supported structure for the other fixture, to allow determination of the torque and normal force imposed on that fixture by analysis of the (small) deflection of those

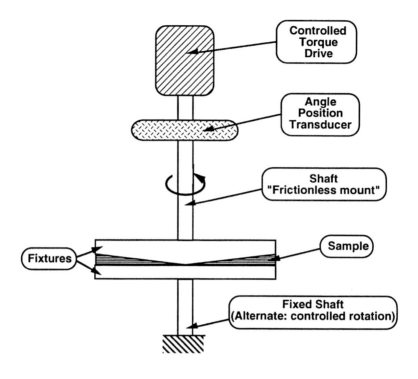

Device	Input	Output
Controlled torque drive	Controlled voltage	Controlled torque
Angle position transducer	Measure of shaft angle	Voltage (current)

Figure 22.3 Schematic drawing of a device for a torsional shear rheometer with a controlled torque input. The table at the bottom gives the input and output for the instrument.

beams in response to the deformation imposed on the sample.[31] An important consideration is the design of a transducer to cover the wide range of torques of interest experimentally, limiting, for example, the smallest torque that could be determined in a stress relaxation experiment, in which the torque is studied after a step in the strain. Because this method may be used to determine $G(t)$ (see the next section), inaccuracy in the measurement of a small torque will compromise the estimate of $G(t)$ for large t. The use of a controlled torque, as in Figure 22.3, was advanced by the ability to apply a controlled torque (e.g., a step, a ramp, or an oscillatory torque) over an arbitrarily large angular deformation by the use of an eddy-current torque motor, with a suitable position transducer to detect the angular deformation of that fixture.[32,33] In most cases, the second fixture is held fixed in the use of a controlled torque, but in some cases, it is rotated at a controlled (usually steady) rate. A major advantage of this design is the ability to determine the recoil after cessation of steady flow, from which $R(t)$ may be computed (see the next section). Any extraneous torques will seriously compromise the determination of

$R(t)$ at large t. For example, the use of ball bearings in the suspension of the rotating fixture is unacceptable, owing to unavoidable friction at some level. The use of a gas bearing reduces friction effects, but introduces a turbine torque, which may be largely suppressed by imposition of a counter torque from the eddy-current torque motor.[33] Additional reduction of the residual torque may be achieved with a magnetic suspension of the rotating member, reducing the residual torque to a contribution related to geometric asymmetry in the suspended member. The effects of this may be suppressed by aligning the instrument so that the center of gravity of the suspended member is on the axis of rotation.[32]

Inspection of the expressions in Table 22.3 shows that the shear strain is independent of position only in the cone and plate geometry, although it becomes nearly so in the concentric cylinder geometry if $\Delta/R \ll 1$. The uniform shear strain is important in studies of nonlinear rheological behavior, making these two geometries especially useful in that regard. However, corrections may be applied to the steady-state data on the torque $\boldsymbol{M}$ as a function of the rota-

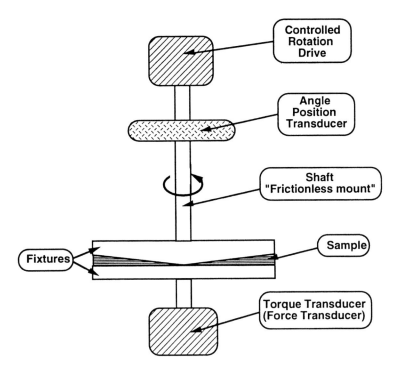

Device	Input	Output
Controlled deformation drive	Controlled voltage	Controlled shaft rotation
Torque transducer	Torque on shaft	Voltage
Angle position transducer	Measure of shaft angle	Voltage (current)

Figure 22.4 Schematic drawing of a device for a torsional shear rheometer with a controlled deformation input. The table at the bottom gives the input and output for the instrument.

Table 22.3 Geometric factors in rheometry

Geometry	Measured	Calculated[a]
Translational geometries		
Parallel plate: width, w; breadth, b; separation, h	Force, F	Stress, $\sigma = F/wb$
	Displacement, D	Strain, $\gamma = D/h$
Concentric cylinders: inner radius, R; gap, Δ; height, h	Force, F	Stress, $\sigma = F/2\pi Rh$
	Displacement, D	Strain, $\gamma = D/R\ln(1 + \Delta/R)$
Rotational geometries:		
Parallel plate: ourter radius, R; separation, H	Torque, M	Stress, $\sigma = (2r/R)M/\pi R^3$
	Rotation, Ω	Strain, $\gamma(r) = (r/h)\Omega$
Cone and plate: outer radius, R; cone angle, $\pi - \alpha$	Torque, M	Stress, $\sigma = (3/2)M/\pi R^3$
	Rotation, Ω	Strain, $\gamma = (1/\alpha)\Omega$
Concentric cylinders: inner radius, R; gap, Δ; height, h	Torque, M	Stress, $\sigma = (R/2h)M/\pi R^3$
	Rotation, Ω	Strain, $\gamma(r) = (R/\Delta R)\Omega f(R,r)$
		$f(R,r) = (R/r)^2 \dfrac{1 + \Delta/R}{1 + \Delta/2R}$

a. σ and γ are the shear stress and strain, respectively.

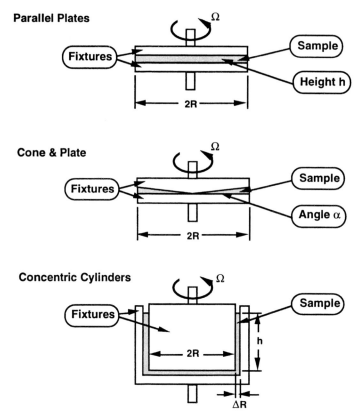

Figure 22.5 Schematic drawings of several fixtures used with torsional shear rheometers.

tional velocity Ω to give the shear stress σ_R at the periphery of the parallel plates (radius R):[4c,8c,19a]

$$\sigma_R = \frac{M}{2\pi R^3}\left(3 + \frac{\partial\ln M}{\partial\ln \Omega}\right) \tag{5}$$

Thus, the steady-state viscosity $\eta(\dot\gamma_R)$ at rate of strain $\dot\gamma_R = R\Omega/h$ at the same position may be calculated as $\eta(\dot\gamma_R) = \sigma_R/\dot\gamma_R$. For a linear response, $\partial\ln M/\partial\ln \Omega = 1$, and the relation in Table 22.3 is recovered.

A similar situation arises in capillary rheometry, in which the shear rate in steady flow is a strong function of the radius.[8d] Thus, for the flow of a linear viscous fluid through a cylindrical tube of radius R, the shear stress and shear rate $\dot\gamma_R$ at the capillary wall are given by[4d,8d,10,14e]

$$\sigma_R = -(R/2)(\Delta P/L_{cap}) \tag{6a}$$

$$\dot\gamma_R = 4Q_{vol}/\pi R^3 \tag{6b}$$

where $\Delta P/L_{cap}$ is the pressure decrease per unit capillary length, and Q_{vol} is the volume flow rate. In this case, the viscosity is simply calculated as $\eta = \sigma_R/\dot\gamma_R$. Similar expressions are available for flow in tubes with other cross-sectional geometries.[8e] It is well known that the velocity profile is parabolic in capillary flow with a linear viscous fluid, but this

profile is altered in nonlinear flow, requiring a correction to the expression for $\dot\gamma_R$ to permit calculation of η:[4d,8f,10,19a,34]

$$\dot\gamma_R = \frac{Q_{vol}}{\pi R^3}\left(3 + \frac{\partial\ln Q_{vol}}{\partial\ln \sigma_R}\right) \tag{7}$$

Thus, the steady-state viscosity $\eta(\dot\gamma_R)$ at strain rate $\dot\gamma_R$ may be calculated as $\eta(\dot\gamma_R) = \sigma_R/\dot\gamma_R$. Additional important issues involving end-effects are not discussed here, although they are important and may provide additional rheological information.[8d,10,14e] A variety of other methods to determine the viscosity are available, many of which are discussed in reference 35.

The control of temperature is almost always important in rheological studies, and can be a critical consideration under some conditions, for example, as the temperature approaches the glass transition temperature, or some equilibrium phase transition. Because the deformation should be laminar for suitable analysis, effective heat transfer to maintain isothermal conditions in the sample may be an issue, especially at high deformation rates, or with nonmetallic surfaces and large, thick samples.[14f]

Measurement of the bulk properties tends to involve customized equipment, often involving an arrangement in which an inert fluid is displaced by the volume change of the sample, with the displacement monitored by the level of the inert fluid in an attached capillary tube.[11h]

Linear Viscoelastic Phenomenology

Although limited to deformations that have been recently small, the linear viscoelastic constitutive relation provides the basis for unambiguous characterization of most materials within that constraint.[11i,13b,23a] In addition to their use in material characterization, the functions and parameters determined in the linear response provide the basis for some constitutive relations designed to describe nonlinear rheological behavior, which might often be of interest in material processing. Linear viscoelastic behavior may be approached phenomenologically, on the basis of a linear response theory with analogs in many areas of science (such as optical dispersion, dielectric dispersion, and electric circuit theory).[23c,35] The subject may also be approached using statistical mechanics with renormalized models of the real chain.[11j,k,17a,22b] The phenomenological treatment is the subject of this section; several experimentally important deformation histories are considered using the phenomenological constitutive equation. Limited aspects of theoretical treatments are discussed in a subsequent section.

The elements of the principal assumptions in linear viscoelasticity of isotropic materials can be stated in terms of experiments under a controlled strain or a controlled stress history. Thus, for a shear deformation in the linear viscoelastic regime, these may be expressed in the forms, respectively:[2,9,11m,13b,18,23d]

$$\sigma(t) = \sum_{i=1}^{N} G(t - t_i)\,\Delta\gamma_i = \int_0^{\gamma(t)} d[\gamma(u)]G(t - u) \quad (8)$$

for the response in the shear stress $\sigma(t)$ to a series of step shear strains $\Delta\gamma_i$ at times $t_1, t_2, \ldots, t_i, \ldots, t_N$, and

$$\gamma(t) = \sum_{i=1}^{N} J(t - t_i)\,\Delta\sigma_i = \int_0^{\sigma(t)} d[\sigma(u)]\,J(t - u) \quad (9)$$

for the response in the shear strain $\gamma(t)$ to a series of step shear stresses $\Delta\sigma_i$ at times $t_1, t_2, \ldots, t_i, \ldots, t_N$. The additive processes in these expressions are illustrated in Figure 22.6. The Stieltjes integrals in these expressions may be converted to give the generally more useful forms

$$\sigma(t) = \int_{-\infty}^{t} du\, G(t - u)\,\frac{\partial\gamma(u)}{\partial u}$$

$$= G_0\gamma(t) + \int_0^{\infty} du\,\gamma(t - u)\,\frac{\partial G(u)}{\partial u} \quad (10)$$

$$\gamma(t) = \int_{-\infty}^{t} du\, J(t - u)\,\frac{\partial\sigma(u)}{\partial u}$$

$$= J_0\sigma(t) + \int_0^{\infty} du\,\sigma(t - u)\,\frac{\partial J(u)}{\partial u} \quad (11)$$

where in each case, the second form follows from the first by integration by parts. Steps in the imposed strain or stress may be incorporated by use of a Dirac delta function.[13a,23b] Thus, if $\gamma(t)$ is a step shear strain γ_0 imposed at time t_1, then $\partial\gamma(t)/\partial t = \gamma_0\delta(t - t_1)$ for this piece of the strain history (or an equivalent representation using the Heaviside unit function), with a contribution $\gamma_0 G(t - t_1)$ to the total shear stress for times $t > t_1$. A similar expression applies for a step shear stress. For a linear elastic material, $\partial G(t)/\partial t = \partial J(t)/\partial t = 0$, so that $\gamma(t) = J_0\sigma(t) = (1/G_0)\sigma(t)$, as expected in this case. Both $J(t)$ and $G(t)$ are found to be non-negative for all t, and furthermore, it is usually found that $\partial J(t)/\partial t \geq 0$ and $\partial G(t)/\partial t \leq 0$. Given the functions $J(t)$ or $G(t)$, these expressions are sufficient to define the response to arbitrarily complex strain or stress histories in shear in the linear viscoelastic response approximation.

Shear deformation has been emphasized in the preceding discussion, with the stress arising from a force parallel to the surface of a deformed volume element. Normal deformations arising from a stress orthogonal to the surface of a deformed volume element are of equal interest. A similar set of expressions to those given above may be obtained for normal forces and the corresponding deformations. Although tensor analysis is not used in this presentation, it may still be useful to present the components of the strain and stress tensors as a point of reference, and to reveal the relation among the linear viscoelastic functions in Table 22.1. Furthermore, these tensors are useful in the consideration of the birefringence. Thus, the combination of normal and shear deformations may be expressed in terms of the strain and stress tensors, with components $\varepsilon_{ij}(t)$ and $S_{ij}(t)$, respectively, in Cartesian coordinates $(i,j = 1-3)$. The Einstein notation is used with its implied summation over indices appearing as a pair, here expressed as indices in Greek symbols:[2,9,11n,13b,16b,18,23a]

$$\varepsilon_{ij} = \frac{1}{2}\left(\frac{\partial u_i}{\partial x_j} + \frac{\partial u_j}{\partial x_i}\right) \quad (12)$$

$$2\varepsilon_{ij}(t) = \int_{-\infty}^{t} ds\left\{J(t - s)\left[\frac{\partial S_{ij}(s)}{\partial s} - \frac{1}{3}\delta_{ij}\frac{\partial S_{\alpha\alpha}(s)}{\partial s}\right]\right.$$

$$\left. + (2/9)\delta_{ij}\,B(t - s)\frac{\partial S_{\alpha\alpha}(s)}{\partial s}\right\} \quad (13)$$

$$S_{ij}(t) = \int_{-\infty}^{t} ds\left\{2G(t - s)\left[\frac{\partial\varepsilon_{ij}(s)}{\partial s} - \frac{1}{3}\delta_{ij}\frac{\partial\varepsilon_{\alpha\alpha}(s)}{\partial s}\right]\right.$$

$$\left. + \delta_{ij}\,K(t - s)\frac{\partial\varepsilon_{\alpha\alpha}(s)}{\partial s}\right\} \quad (14)$$

Here, **u** is the displacement vector, computed from the deformation experienced by the material. In terms of the preceding discussion, $\gamma(t) = 2\varepsilon_{12}(t)$ and $\sigma(t) = S_{12}(t)$ for shear deformation with a shear strain gradient along x_2 and de-

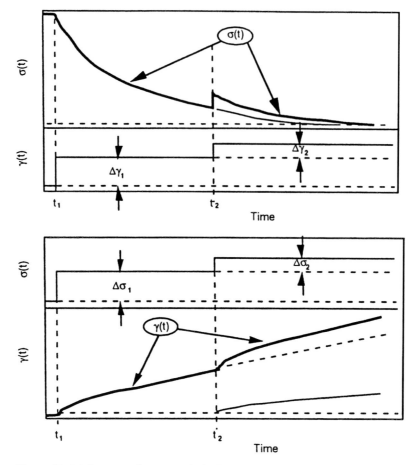

Figure 22.6 Schematic drawings of idealized rheological experiments: (*top*) the shear stress $\sigma(t)$ resulting from successive jumps in the strain $\gamma(t)$; (*bottom*) the shear strain $\gamma(t)$ resulting from successive jumps in the stress $\sigma(t)$.

formation along x_1. Note that some authors use an alternative definition of the strain tensor, differing by a factor of 2.[11p] The first and second terms in the stress and strain tensors are sometimes referred to as the deviatoric and volume components, respectively. The term involving $K(t)$ reduces to $-\delta_{ij}P$, with P representing a pressure, in the limit of incompressible behavior (negligible change in the volume); the term involving $B(t)$ may be neglected in that same limit. A similar but more complex set of equations may be obtained in the more general case with the assumption of material isotropy relaxed.[13c,36]

It can be shown that the functions $J(t)$ and $G(t)$ are related through a convolution integral:[2,9,11q,13d,16a,18,23d]

$$\frac{1}{t}\int_0^t du\, G(t-u)\,J(u) = 1 \tag{15}$$

with Laplace transform:

$$s^2 \hat{G}(s)\hat{J}(s) = 1 \tag{16}$$

Similar relations are obtained between $B(t)$ and $K(t)$, as well as $D(t)$ and $E(t)$, the latter two, being related to the other four (see Table 22.1). Although the Laplace transform appears simple, its analytic use is usually not possible (however, see Table 22.1 and reference 23e). Because these relations may be used to compute one function from its paired member [e.g., $J(t)$ given $G(t)$], in principle, an experimenter need only arrange to determine two functions, for example, $J(t)$ and $B(t)$, to fully characterize the material. Nevertheless, for various reasons (some are elaborated in the following discussion), it is frequently desirable to carry out more than one type of experiment.[10,11r]

The assumption of material isotropy may be relaxed in the linear viscoelastic response regime, with a corresponding increase in the complexity of the expressions.[13c,36] For example, for a nematic fluid exhibiting only Frank curvature elasticity and Leslie viscosity coefficients, with no true viscoelastic behavior, the minimal representation contains nine coefficients.[17b,22c,37] A discussion of the tensor character of the response function for materials with various symmetries is given in reference 36.

It should be emphasized that the general theory of linear viscoelasticity presented above does not stipulate the dependence of the functions such as $J(t)$ on temperature, important as that may be in the response obtained. Methods to incorporate the effect of temperature in the analysis of experimental data are discussed below.

The preceding discussion simplifies considerably if the functions $G(t)$, $K(t)$, and the like reduce to constants G, K, and the like, as for a linear elastic material.[2,9,23f] In this limit, the compliances and moduli are simply related: $JG = BK = DE = 1$, with relations among the moduli given in Table 22.4. The additional constant Poisson's ratio ν introduced in Table 22.4 is particularly useful in considering the volume change ΔV of a linear elastic material under an elongational deformation. For example, for a uniaxial elongation with strain ε, $\Delta V/V = \varepsilon(1 - 2\nu)$, showing that ΔV tends to zero as ν goes to $1/2$. Table 22.4 shows that ν tends to $1/2$ for $K \gg G$. In addition, the components of the strain for a deformation with normal stresses may be expressed as $\varepsilon_{11} = D[S_{11} - \nu(S_{22} + S_{33})]$. The relations given above with time-dependent moduli and compliances may be recast in terms of a time-dependent Poisson ratio $\nu(t)$.[23g]

Examples of the application of the linear viscoelastic response are given in the following sections for illustrative purposes and to develop certain important expressions for frequently used histories creep and recovery with a step shear stress; stress relaxation with a step shear strain; stress relaxation after a ramp shear strain; recovery after a ramp shear strain; oscillation with a sinusoid shear stress; oscillation with a sinusoid shear strain; and volume and elongational deformations.

Creep and Recovery with a Step Shear Stress

(See Figure 22.7a.) Stress history:

$$\sigma(t) = 0 \qquad t < 0$$
$$\sigma(t) = \sigma_0 \qquad 0 \leq t \leq T_e$$
$$\sigma(t) = 0 \qquad t > T_e$$

The strain in creep for $t \leq T_e$ is given by

$$\gamma(t) = \sigma_0 \int_0^t du\, J(t - u)\, \delta(u - 0) \qquad (17a)$$
$$= \sigma_0 J(t) = \sigma_0[R(t) + t/\eta] \qquad (17b)$$

Thus, for a fluid, a bilogarithmic plot of $\gamma(t)$ versus t approaches unit slope if T_e is large enough that $R(T_e) \approx J_s$; inspection of linear plots of $\gamma(t)$ versus t is not recommended for this analysis because deviation from linearity of $J(t)$ with t may be difficult to discern in such a representation. In principle, $R(t)$ may be computed from this response as $R(t) = \gamma(t)/\sigma_0 - t/\eta$ for a fluid, but this method will generally not give accurate results unless $t \ll \hat{\tau}_c = J_s\eta$; otherwise the method requires the difference between two large numbers for large t.[10,11s]

The strain for $t > T_e$ is given as a function of the time $\vartheta = t - T_e$ in recovery by

$$\gamma(t) = \sigma_0 \int_0^{T_e} du\, J(t - u)\, \delta(u - 0)$$
$$\qquad - \sigma_0 \int_{T_e}^t du\, J(t - u)\, \delta(u - T_e) \qquad (18a)$$
$$\gamma(\vartheta) = \sigma_0[J(\vartheta + T_e) - J(\vartheta)] = \sigma_0[R(\vartheta + T_e)$$
$$\qquad - R(\vartheta) + T_e/\eta] \qquad (18b)$$

Consequently, the permanent set or nonrecoverable strain is given by $\gamma_{NR} = \gamma(\infty) = \sigma_0 T_e/\eta$, providing a means to estimate η for a fluid, even if T_e is not large enough to reach steady-state flow in creep. Accurate measurement of $\gamma(t)$ in recovery requires the elimination of extraneous torques, such as from ball bearing suspensions.[10]

The recoverable strain $\gamma_R(\vartheta) = \gamma(T_e) - \gamma(t)$ for $t > T_e$ is given by

$$\gamma_R(\vartheta) = \sigma_0\{J(T_e) - [J(\vartheta + T_e) - J(\vartheta)]\} = \sigma_0\{R(\vartheta)$$
$$\qquad + R(T_e) - R(\vartheta + T_e)\} \qquad (19)$$

for either a solid or a fluid, with $R(\infty) = J_\infty$. Inspection shows that that $\gamma_R(\vartheta) = \sigma_0 R(\vartheta)$ for $\vartheta \ll T_e$, $\gamma_R(\vartheta) = \sigma_0 R(T_e)$ if $\vartheta \gg T_e$, and $\gamma_R(\vartheta) = \sigma_0 R(\vartheta)$ for any ϑ if T_e is large enough that $R(T_e) \approx J_\infty$. In this last case, the function $R(t)$ may be determined directly from the recoil. A priori estimation of

Table 22.4 Relations among elastic constants

	K, G	E, G	K, E	K, ν	E, ν	G, ν
K	K	$\dfrac{EG}{3[3G - E]}$	K	K	$\dfrac{E}{3[1 - 2\nu]}$	$\dfrac{2G[1 + \nu]}{3[1 - 2\nu]}$
E	$\dfrac{9KG}{3K + G}$	E	E	$3K(1 - 2\nu)$	E	$2G(1 + \nu)$
G	G	G	$\dfrac{3KE}{9K - E}$	$\dfrac{3K[1 - 2\nu]}{2[1 + \nu]}$	$\dfrac{E}{2[1 + \nu]}$	G
ν	$\dfrac{3K - 2G}{6K + 2G}$	$\dfrac{E}{2G} - 1$	$\dfrac{3K - E}{6K}$	ν	ν	ν

$J = 1/G$, $B = 1/K$, $D = 1/E$.

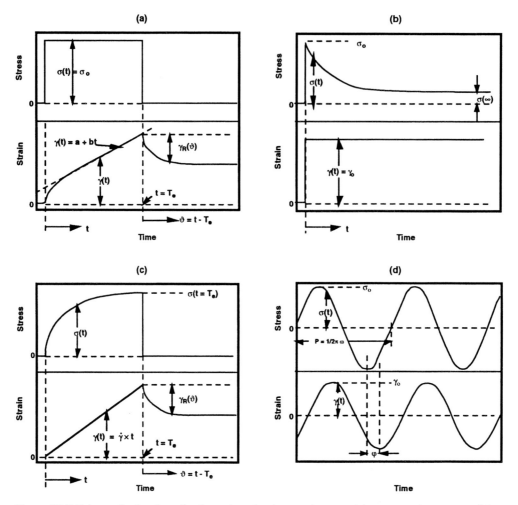

Figure 22.7 Schematic drawings for four viscoelastic experiments: (a) creep and recovery, (b) stress relaxation, (c) ramp deformation and recovery, and (d) sinusoid deformation.

T_e large enough that $\gamma_R(\vartheta) = \sigma_0 R(\vartheta)$ for any ϑ for an unknown fluid sample is not generally possible, but as a guide, T_e should be several-fold larger (e.g., 10-fold) than the time required to obtain $\partial \ln J(t)/\partial \ln t \approx 1$ during creep; an apparent constant slope $\partial J(t)/\partial t$ is usually not a sufficiently critical criterion. The function $\gamma_R(\vartheta)$ defined above should more precisely be called the constrained recoverable strain, because the experiment envisioned does not provide for relaxation of any stress normal to the shear stress. The distinction is moot for the linear viscoelastic response because normal stresses are nil in that model, but can become important in nonlinear behavior discussed below.

Note that with use of the preceding relation for the strain $\gamma(t)$ in recovery, the running sum of N successive strains for $t = nT_e$ $(n = 1, 2, \ldots, N)$ is given by

$$\sum_{n=1}^{N} \gamma(nT_e) = \sigma_0 J(NT_e) = \sigma_0 R(NT_e) + N\gamma_{NR} \qquad (20)$$

Because γ_{NR} is known, in principle this expression permits

assessment of $R(t)$ for $t = NT_e$, and that method has been proposed for use with T_e small enough that $R(T_e) < J_\infty$, to avoid the necessity of long times in creep.[38] In practice, the inaccuracies in determining the difference between the sum of successive strains and $N\gamma_{NR}$ will render this estimate inaccurate with increasing N, similar to the inaccuracy that develops in attempts to estimate $R(t)$ as $J(t) - t/\eta$ from data in creep.

Note that for recovery initiated at time T_e following an arbitrary stress history up to T_e, the recoverable strain is given by

$$\gamma_R(\vartheta) = \sigma(T_e)R(\vartheta) - \int_0^{T_e} du\, [R(\vartheta + T_e - u)$$

$$- R(T_e - u)]\frac{\partial \sigma(u)}{\partial u} \qquad (21)$$

With this expression, the total recoverable strain $\hat{\gamma}_R(T_e)$ obtained as ϑ becomes very large after the stress is reduced to zero at time T_e after an arbitrary strain history up to T_e is given by

$$\hat{\gamma}_R(T_e) = \int_0^{T_e} du \, R(T_e - u) \frac{\partial \sigma(u)}{\partial u} \qquad (22)$$

Inspection will show that eqs 21 and 22 each reduce to eq 19 for the special stress history shown in Figure 22.7a; the expression for $\hat{\gamma}_R(T_e)$ finds use in the discussion of strain-induced birefringence.

Stress Relaxation after a Step Shear Strain

(See Figure 22.7b.) Strain history:

$$\gamma(t) = 0 \qquad t < 0$$
$$\gamma(t) = \gamma_0 \qquad t \geq 0$$

The stress response for $t > 0$ is given by

$$\sigma(t) = \gamma_0 \int_0^t du \, G(t - u) \, \delta(u - 0) = \gamma_0 G(t) \qquad (23)$$

Consequently, this experiment provides direct information on $G(t)$, limited by the sensitivity of the transducer to determine the torque as it decreases to its value at long time.[10]

The integral over $\hat{\phi}(t)$ is bounded for either a fluid or a solid, and for a fluid the value must equal the steady-state viscosity as t becomes large, because in that limit of steady-state flow $\sigma = \dot{\gamma}\eta$, so that

$$\eta = G_0 \int_0^{\infty} ds \, \hat{\phi}(s) \qquad (24)$$

providing a means to determine η from $G(t)$ for a fluid. In addition, as illustrated in an example given below, for a fluid

$$J_s = G_0 \int_0^{\infty} ds \, s\hat{\phi}(s)/\eta^2 \qquad (25)$$

Although not included in the deformation history specified above, it may be noted that the total recoverable strain $\hat{\gamma}_R(T_e)$ on reduction of the stress to zero at time T_e during the stress relaxation may be computed with the expression in the preceding section for a stress history comprising a jump $\gamma_0 G_0$ followed by $\sigma(t) = \gamma_0 G(t)$ for $0 < t < T_e$, to give $\hat{\gamma}_R(T_e) = \gamma_0 \left[1 - \eta^{-1} \int_0^{T_e} du \, G(u) \right]$; this expression is used in the discussion on strain-induced birefringence.

Stress Relaxation after a Ramp Shear Strain

Strain history:

$$\gamma(t) = 0 \qquad t < 0$$
$$\gamma(t) = \dot{\gamma}t \qquad 0 \leq t \leq T_e$$
$$\gamma(t) = \dot{\gamma}T_e \qquad t > T_e$$

The stress response for $t \leq T_e$ is given by

$$\sigma(t) = \dot{\gamma} \int_0^t du \, G(t - u) = \dot{\gamma}[G_e t + (G_0 - G_e) \int_0^T ds \, \hat{\phi}(s)] \qquad (26)$$

With this expression, $\partial\sigma(t)/\partial t = \dot{\gamma}G(t)$ for a fluid, providing

an alternative (although usually less precise) means to estimate $G(t)$ than that from a step-strain. For this strain-defined history with $\gamma(t) = \dot{\gamma}t$, use of the expression

$$\gamma(t) = \int_{-\infty}^t du \, J(t - u) \frac{\partial \sigma(u)}{\partial u} \qquad (27)$$

provides an illustration of the convolution integral cited above relating $G(t)$ and $J(t)$. Thus, from above, $\partial\sigma(t)/\partial t = \dot{\gamma}G(t)$, giving

$$\gamma(t) = \dot{\gamma} \int_0^t du \, J(t - u) \, G(u) = \dot{\gamma}t \qquad (28)$$

reproducing the convolution integral. For a fluid, the strain for $t > T_e$ is given by

$$\sigma(t) = \dot{\gamma} \int_0^{T_e} du \, G(t - u) = \dot{\gamma} \int_\vartheta^{\vartheta + T_e} ds \, G(s) \qquad (29)$$

with $\vartheta = t - T_e$. Consequently, because in steady flow T_e is large compared with the time to give $G(t) \approx 0$, the final integral yields $\partial\sigma(\vartheta)/\partial\vartheta = -\dot{\gamma}G(\vartheta)$ during stress relaxation after steady flow.

Recovery after a Ramp Shear Strain

(See Figure 22.7c.) Strain history:

$$\gamma(t) = 0 \qquad t < 0$$
$$\gamma(t) = \dot{\gamma}t \qquad 0 \leq t \leq T_e$$
Stress history: $\qquad \sigma(t) = 0 \qquad t > T_e$

The stress response for $t \leq T_e$ follows that in the preceding example. The strain for $T > T_e$ is given by

$$\sigma(t) = 0 = \dot{\gamma} \int_0^{T_e} du \, G(t - u) + \int_{T_e}^t du \, G(t - u) \frac{\partial\gamma(u)}{\partial u} \qquad (30)$$

In general, a numerical iteration would be required to obtain $\gamma(t)$ in recovery, and hence the recoverable strain $\hat{\gamma}_R(t)$, but the total recoverable strain $\hat{\gamma}_R$ following steady-state flow (large T_e) for a fluid may be calculated in terms of integrals involving $G(t)$, permitting calculation of $\hat{\gamma}_R/\dot{\gamma}\eta = J_s$ for a fluid from integrals over $G(t)$:[19b]

$$\hat{\tau}_c = \eta J_s = \int_0^{\infty} ds \, s\hat{\phi}(s) / \int_0^{\infty} ds \, \hat{\phi}(s) \qquad (31)$$

The parameter $\hat{\tau}_c$ is seen to be a certain average time constant of the relaxation modulus. Note that whereas the parameters J_s and η appear directly in $J(t)$ for a fluid, their determination from $G(t)$ requires the evaluation of integrals over $G(t)$. Conversely, complete knowledge of $R(t)$ does not provide an estimate for η.

Oscillation with a Sinusoid Shear Stress

(See Figure 22.7d.) Stress history:

$$\sigma(t) = 0 \qquad t < 0$$
$$\sigma(t) = \sigma_0 \sin(\omega t) \qquad t \geq 0$$

The strain response for $t > 0$ is given by

$$\gamma(t) = \omega\sigma_0 \int_0^t du\, J(t-u)\cos(\omega u) \qquad (32)$$

After a transformation of variable with $s = t - u$, the use of trigonometric identities, and passage to the steady-state limit with large t,

$$\gamma(t) = \sigma_0\{J'(\omega)\sin(\omega t) - J''(\omega)\cos(\omega t)\} \qquad (33)$$

$$J'(\omega) = J_\infty - \omega[J_\infty - J_0]\int_0^\infty ds\,\hat{\alpha}(s)\sin(\omega s) \qquad (34)$$

$$J''(\omega) = (1/\omega\eta) + \omega[J_\infty - J_0]\int_0^\infty ds\,\hat{\alpha}(s)\cos(\omega s) \qquad (35)$$

where $J'(\omega)$ and $J''(\omega)$ are the in-phase (or real or storage) and out-of-phase (or imaginary or loss) dynamic compliances, respectively, and $\hat{\alpha}(t)$ is defined above. Inspection of these expressions shows that $J'(\omega) \approx J_\infty$, $J''(\omega) \approx 1/\omega\eta$, and $J''(\omega) - 1/\omega\eta \approx \omega$ at low frequency. In an alternative representation of the response,

$$\gamma(t) = \sigma_0|J^*(\omega)|\sin[\omega t - \delta(\omega)] \qquad (36)$$

where $|J^*(\omega)|^2 = [J'(\omega)]^2 + [J''(\omega)]^2$ and the phase angle $\delta(\omega)$ is given by $\tan\,\delta(\omega) = J''(\omega)/J'(\omega)$. For a linear viscoelastic material, the relation between the stress and the strain may be expressed in compact form in complex notation: $\gamma^* = \sigma^* J^*$; for example, $\sigma(t) = \mathrm{Im}\{\sigma^*\} = \mathrm{Im}\{\sigma_0\exp(i\omega t)\}$, $\gamma(t) = \mathrm{Im}\{\gamma^*\} = \mathrm{Im}\{\gamma_0\exp[i(\omega t - \delta)]\}$, and $J^* = |J^*|\exp(-i\delta)$, with $J^* = J' - iJ''$ equal to the complex compliance. Given their status as sine or cosine Fourier transforms, the functions given above may be used to obtain the following relations that permit conversion of the dynamic functions to the time-dependent compliances:[2,9,11q,23h]

$$R(t) = J(t) - t/\eta = J_0 + \frac{2}{\pi}\int_0^\infty d\omega\,\frac{J'(\omega) - J_0}{\omega}\sin(\omega t) \qquad (37)$$

$$R(t) = J(t) - t/\eta = J_0 + \frac{2}{\pi}\int_0^\infty d\omega\,\frac{J''(\omega) - (\omega\eta)^{-1}}{\omega}$$
$$\times[1 - \cos(\omega t)] \qquad (38)$$

As would be anticipated, Kramers-Kronig relations may be used to relate $J'(\omega)$ and $J''(\omega)$:[2,9,11t,23h]

$$J'(\omega) - J_0 = \frac{2}{\pi}\int_0^\infty du\,u\,\frac{J''(u) - (u\eta)^{-1}}{u^2 - \omega^2} \qquad (39)$$

$$J''(\omega) - (\omega\eta)^{-1} = \omega\frac{2}{\pi}\int_0^\infty du\,\frac{J'(u) - J_0}{\omega^2 - u^2} \qquad (40)$$

The Cauchy principal value of the integral is taken to resolve the apparent singularity when $u = \omega$; the calculations required are readily implemented using a desktop computer.

Oscillation with a Sinusoid Shear Strain

(See Figure 22.7d.) Strain history:

$$\gamma(t) = 0 \qquad\qquad t < 0$$
$$\gamma(t) = \gamma_0\sin(\omega t) \qquad t \geq 0$$

The stress response for $t > 0$ is given by

$$\sigma(t) = \omega\gamma_0\int_0^t du\, G(t-u)\cos(\omega u) \qquad (41)$$

Again, after a transformation of variable with $s = t - u$, the use of trigonometric identities, and passage to the steady-state limit with large t,

$$\sigma(t) = \gamma_0\{G'(\omega)\sin(\omega t) + G''(\omega)\cos(\omega t)\} \qquad (42)$$

$$G'(\omega) = G_e + \omega[G_0 - G_e]\int_0^\infty ds\,\hat{\phi}(s)\sin(\omega s) \qquad (43)$$

$$G''(\omega) = \omega[G_0 - G_e]\int_0^\infty ds\,\hat{\phi}(s)\cos(\omega s) \qquad (44)$$

where $G'(\omega)$ and $G''(\omega)$ are the in-phase (or real or storage) and out-of-phase (or imaginary or loss) dynamic compliances, respectively. In an alternative representation of the response,

$$\sigma(t) = \gamma_0|G^*(\omega)|\sin[\omega t + \delta(\omega)] \qquad (45)$$

where $|G^*(\omega)|^2 = [G'(\omega)]^2 + [G''(\omega)]^2$ and the phase angle $\delta(\omega)$ is given by $\tan\,\delta(\omega) = G''(\omega)/G'(\omega)$. In complex notation, $\sigma^* = \gamma^*G^*$ with $\gamma(t) = \mathrm{Im}\{\gamma^*\} = \mathrm{Im}\{\gamma_0\exp(i\omega t)\}$, $\sigma(t) = \mathrm{Im}\{\sigma^*\} = \mathrm{Im}\{\sigma_0\exp[i(\omega t + \delta)]\}$, and $G^* = |G^*|\exp(i\delta)$, with $G^* = G' + iG''$ equal to the complex modulus. Comparison with the expression given in the previous example shows that $G^*J^* = 1$, which provides the very useful algebraic relation $|G^*(\omega)||J^*(\omega)| = 1$, and expressions that permit algebraic conversion of the dynamic functions to the time-dependent compliances or moduli:[2,9,11u,16,23h]

$$J'(\omega) = G'(\omega)/|G^*(\omega)|^2 \qquad (46)$$

$$J''(\omega) = G''(\omega)/|G^*(\omega)|^2 \qquad (47)$$

$$G'(\omega) = J'(\omega)/|J^*(\omega)|^2 \qquad (48)$$

$$G''(\omega) = J''(\omega)/|J^*(\omega)|^2 \qquad (49)$$

$$\tan\,\delta(\omega) = J''(\omega)/J'(\omega) = G''(\omega)/G'(\omega) \qquad (50)$$

These algebraic expressions contrast with the convolution integral relating $J(t)$ and $G(t)$; note, however, that the Laplace transforms of $J(t)$ and $G(t)$ are algebraically related,[23i] just as are these Fourier transforms of $J(t)$ and $G(t)$. As shown above, the properties of sine and cosine Fourier transforms lead to useful relations:[2,9,11u,16d,23j]

$$G(t) = G_e + \frac{2}{\pi}\int_0^\infty d\omega\,\frac{G'(\omega) - G_e}{\omega}\sin(\omega t) \qquad (51)$$

$$G(t) = G_e + \frac{2}{\pi}\int_0^\infty d\omega\,\frac{G''(\omega)}{\omega}\cos(\omega t) \qquad (52)$$

It is often useful to define a complex viscosity $\eta^* = \sigma^*/\dot{\gamma}^* = (1/i\omega)\sigma^*/\gamma^* = G^*/i\omega$; the $\pi/2$ phase shift represents the dependence of the viscosity on deformation rate. Thus, $\eta^*(\omega) = \eta'(\omega) - i\eta''(\omega)$, where $\eta'(\omega) = G''(\omega)/\omega$ and $\eta''(\omega) = G'(\omega)/\omega$ are the components $\eta^*(\omega)$ that are in phase and out of phase with the shear rate, respectively. Inspection of these expressions shows that at low frequency $G'(\omega) \approx G_e$ for a solid or $(\omega\hat{\tau}_c)^2/J_s$ for a fluid, and $G''(\omega) \approx \omega\eta'(0)$ for either solid or fluid. The low-frequency limiting value $\eta'(0)$

of $\eta'(\omega)$ is given by $\eta'(0) = [G_0 - G_e]\int ds\, \hat{\phi}(s)$, so that $\eta'(0)$ is equal to the viscosity η for a fluid ($G_e = 0$) or a constant for a solid; $\eta''(\omega) = J_s\eta^2\omega$ for a fluid at low ω. For high frequencies, $\eta'(\omega)$ and $\eta''(\omega)$ are proportional to ω^{-2} and ω^{-1}, respectively. In some cases, the decline of $\eta'(\omega)$ to zero is interrupted by a high-frequency response, causing $\eta'(\omega)$ to exhibit a plateau.

Analysis of the relations for the following shows that $J'(\omega) \leq [R(t)]_{\omega t=1} < 1/G'(\omega) < [1/G(t)]_{\omega t=1}$. As would be anticipated, Kramers-Kronig relations may be used to relate $G'(\omega)$ and $G''(\omega)$:[2.9,16d,19c,23k]

$$G'(\omega) - G_e = \omega^2 \frac{2}{\pi} \int_0^\infty du\, \frac{G''(u)}{\alpha[\omega^2 - u^2]} \qquad (53)$$

$$G''(\omega) = \omega \frac{2}{\pi} \int_0^\infty du\, \frac{G'(u) - G_e}{u^2 - \omega^2} \qquad (54)$$

The Cauchy principal value of the integral is taken to resolve the apparent singularity when $u = \omega$.

Volume and Elongational Deformations

Although the preceding has been cast in terms of shear deformation, similar expressions apply in terms of the other functions, with reference to the stress and strain tensors described. For example, the relations in the first example apply to elongation in response to a step by redefinition of symbols. Thus, for a force along x_1, $\gamma(t)$ is replaced by the elongational strain $\varepsilon(t) = \varepsilon_{11}(t)$, σ_0 is replaced by the tensile stress component $S_{11}(t)$, and $J(t)$ is replaced by $D(t)$. For material behavior in the incompressible approximation (negligible volume change), $D(t) \approx J(t)/3$, giving, for example, an elongational viscosity $\eta_{elg} = 3\eta$, with η the viscosity in shear. The relations may be applied to bulk deformation under an applied pressure P, with $\gamma(t)$ replaced by the volume change $\Delta V(t)$, equal to $\varepsilon_{11}(t) + \varepsilon_{22}(t) + \varepsilon_{33}(t)$, σ_0 replaced by the $-P$, and $J(t)$ replaced by $B(t)$. Similarly, for the dynamic functions $J^*G^* = B^*K^* = D^*E^* = 1$ and the functions $B'(\omega)$ (and related functions) may be expressed in obvious ways. Note that for an isotropic linear viscoelastic material, $D^*(\omega) = J^*(\omega)/3 + B^*(\omega)/9$, analogous to the expression for $D(t)$ in Table 22.1; however, the expression for $E^*(\omega)$ becomes equally simple, in contrast to the relation for $E(t)$.[2.9] A principal difference between the shear and volume deformations is the ratio of parameters such as J_∞/J_0 and B_e/B_0. Thus, whereas these ratios may easily reach 10^6 or higher for shear deformation, they are seldom larger than 2 to 3 for a volume deformation.

Approximations Used in Linear Viscoelasticity

The information in $J(t)$ and $G(t)$ involves both parameters and functions, for example, η, J_∞, J_0, and $\hat{\alpha}(t)$ in $J(t)$, or G_0,

G_e, and $\hat{\phi}(t)$ in $G(t)$. Although it is not a necessary attribute for linear viscoelastic behavior, the functions $\hat{\alpha}(t)$ and $\hat{\phi}(t)$ are sometimes represented in terms of weighted distributions of exponential functions involving retardation times $\hat{\lambda}$ or relaxation times $\hat{\tau}$, respectively:[11v,18,23m]

$$\hat{\alpha}(t) = \frac{1}{J_\infty - J_0} \int_{-\infty}^\infty d(\ln \lambda)\, L(\lambda)\exp(-t/\lambda)$$
$$\approx \sum_m^{N-1} \alpha_i \exp(-t/\lambda_i) \qquad (55)$$

$$\hat{\phi}(t) = \frac{1}{G_0 - G_e} \int_{-\infty}^\infty d(\ln \tau)\, H(\tau)\exp(-t/\tau)$$
$$\approx \sum_1^N \hat{\phi}_i \exp(-t/\hat{\tau}_i) \qquad (56)$$

where $\Sigma\hat{\alpha}_i = \Sigma\hat{\phi}_i = 1$, and m is equal to 0 or 1 for a solid and fluid, respectively [the nomenclature $L(\hat{\lambda})$ and $H(\hat{\tau})$ preserves that commonly used[11v]]. The discrete forms, which are frequently sufficient to represent data within experimental error, may be obtained from the continuous functions, for example, $\hat{\lambda}^{-1}L(\hat{\lambda})/[J_\infty - J_0] \approx \Sigma\hat{\alpha}_i\delta(\hat{\lambda} - \hat{\lambda}_i)$. With the discrete distribution, for example, $\eta = G_0\Sigma\hat{\phi}_i\hat{\tau}_i$, $\eta^2 J_s = G_0\Sigma\hat{\phi}_i\hat{\tau}_i^2$ and $\hat{\tau}_c = \Sigma\hat{\phi}_i\hat{\tau}_i^2/\Sigma\hat{\phi}_i\hat{\tau}_i$ for a fluid, the latter revealing $\hat{\tau}_c$ as a certain average relaxation time. For the discrete representation the times alternate regularly in magnitude (with $\hat{\lambda}_0$ absent for a fluid): $\hat{\lambda}_0 > \hat{\tau}_1 > \hat{\lambda}_1 > \ldots > \hat{\lambda}_i > \hat{\tau}_i > \hat{\lambda}_{i+1} > \ldots > \hat{\lambda}_{N-1} > \hat{\tau}_N$.

The determination of $L(\hat{\lambda})$ or $H(\hat{\tau})$, or the corresponding $\hat{\alpha}_i : \hat{\lambda}_i$ or $\hat{\phi}_i : \hat{\tau}_i$ sets, from experimental data requires an iterative process. In some methods this is assisted by the use of approximations to provide the initial estimate of the desired function. A number of these have been developed,[23m] including one found to be particularly useful for computation of $L(\hat{\lambda})$:[11w]

$$L_2(\hat{\lambda}) \approx \{[\partial J(t)/\partial\ln t] - [\partial^2 J(t)/\partial(\ln t)^2]\}_{\hat{\lambda}=t/2} \qquad (57)$$

Because the differentiation removes the influence of the term t/η, the same relation applies with $J(t)$ replaced by $R(t)$. In a scheme used to estimate $L(\hat{\lambda})$, the residuals between the measured and calculated $J(t)$ using the estimate for $L(\hat{\lambda})$ begins with $L(\hat{\lambda}) = L_2(\hat{\lambda})$, with sequential improvements in the estimate for $L(\hat{\lambda})$ to minimize the residuals between the calculated and observed $R(t)$ to the level of experimental uncertainty; $L(\hat{\lambda})$ leading to nonphysical behavior, such as oscillations in $R(t)$, are rejected in this process. Other methods use inverse transforms in iterative calculations, such as the well-known CONTIN often applied to dynamic light scattering analysis,[39] or methods devised especially for use with rheological data.[40] Some commercial rheometers provide software for this purpose. Typically, these methods are designed to provide discrete weighting functions for a set of times selected to span the experimental range, with the number of times chosen to provide an acceptable representation of the data within experimental error, without introducing fictitious oscillations in the transform. Examples of $L(\hat{\lambda})$ are given in subsequent sections.

Although analytical computation of $\hat{\varphi}_i : \hat{\tau}_i$ sets from $\hat{\alpha}_i : \hat{\lambda}_i$ sets, and vice versa, is possible for small N, iterative methods are required for N normally of interest.[23n,41,42] Similarly, iterative methods are required to compute $H(\hat{\tau})$ from $L(\hat{\lambda})$, and vice versa.[11x,23m] Thus, for example, the so-called Maxwell model with $N = 1$ for a fluid is characterized by a single relaxation time $\hat{\tau}$, and no retardation times [$\hat{\alpha}(t) = \delta(t)$ and $\hat{\varphi}(t) = \exp(-t/\hat{\tau})$]. For a Maxwell solid with $N = 1$, $\hat{\varphi}(t)$ is unchanged, but $\hat{\alpha}(t) = \exp(-t/\hat{\lambda})$, with $\hat{\lambda} = (G_0/G_e)\hat{\tau}$. Although algebraic methods may be conveniently applied for N up to about 3, in general, iterative calculations are necessary to interconvert $\hat{\varphi}_i : \hat{\tau}_i$ and $\hat{\alpha}_i : \hat{\lambda}_i$ sets for larger N; these last are readily implemented on desktop computers.[42]

The so-called stretch exponential form $\exp[-(at)^n]$ has been used to represent $\hat{\alpha}(t)$ or $\hat{\varphi}(t)$.[43] The relation between $\hat{\alpha}(t)$ and $\hat{\varphi}(t)$ when one or the other is represented by a stretched exponential function has been considered in detail, showing that both cannot be stretched exponential functions for a given material.[43]

The dynamic mechanical functions take on a simple form for the special case with $\hat{\alpha}(t)$ and $\hat{\varphi}(t)$ expressed in terms of retardation or relaxation times:[11v,18,23p]

$$J_\infty - J'(\omega) = \int_{-\infty}^{\infty} d(\ln\hat{\lambda})\, L(\hat{\lambda})\, \frac{(\omega\hat{\lambda})^2}{1 + (\omega\hat{\lambda})^2}$$

$$= [J_\infty - J_0] \sum_m^{N-1} \hat{\alpha}_i \frac{(\omega\hat{\lambda}_i)^2}{1 + (\omega\hat{\lambda}_i)^2} \qquad (58)$$

$$J''(\omega) - (1/\omega\eta) = \int_{-\infty}^{\infty} d(\ln\hat{\lambda})\, L(\hat{\lambda})\, \frac{\omega\hat{\lambda}}{1 + (\omega\hat{\lambda})^2}$$

$$= [J_\infty - J_0] \sum_m^{N-1} \hat{\alpha}_i \frac{\omega\hat{\lambda}_i}{1 + (\omega\hat{\lambda}_i)^2} \qquad (59)$$

$$G'(\omega) - G_e = \int_{-\infty}^{\infty} d(\ln\hat{\tau})\, H(\hat{\tau})\, \frac{(\omega\hat{\tau})^2}{1 + (\omega\hat{\tau})^2}$$

$$= [G_0 - G_e] \sum_1^N \hat{\varphi}_i \frac{(\omega\hat{\tau}_i)^2}{1 + (\omega\hat{\tau}_i)^2} \qquad (60)$$

$$G''(\omega) = \int_{-\infty}^{\infty} d(\ln\hat{\tau})\, H(\hat{\tau})\, \frac{\omega\hat{\tau}}{1 + (\omega\hat{\tau})^2}$$

$$= [G_0 - G_e] \sum_1^N \hat{\varphi}_i \frac{\omega\hat{\tau}_i}{1 + (\omega\hat{\tau}_i)^2} \qquad (61)$$

These expressions give the expected limits, for example, $G''(\omega) = \omega\eta'(0)$ for small ω. For the special case of a fluid with only one relaxation time $\hat{\tau}_1$, it is seen that the maximum in $G''(\omega)$ occurs for $\omega\hat{\tau}_1 = 1$, at which point $G'(\omega) = G''(\omega) = G_0/2$. This feature is sometimes generalized to assert that $\omega\hat{\tau}_1 = 1$ for $G'(\omega) = G''(\omega)$ and the maximum in $G''(\omega)$ for the more general case with a number of relaxation times. Although that will not usually be correct, with the maximum in $G''(\omega)$ occurring for a larger value ω_{max} than the value of ω_x, for which $G'(\omega_x) = G''(\omega_x)$. Thus, for a fluid,

$$\omega_a = (1/\hat{\tau}_c) \frac{\int_0^\infty ds\, \cos(\omega_a s)\hat{\varphi}(s)/\int_0^\infty ds\, \hat{\varphi}(s)}{\int_0^\infty ds\, [\sin(\omega_a s)/(\omega_a s)^a]s\hat{\varphi}(s)/\int_0^\infty ds\, s(\hat{\varphi}(s)} \qquad (62)$$

with $a = 0$ or 1 for ω_a equal to for ω_{max} or ω_x, respectively. Consequently, $\omega_x \leq \omega_{max} \leq 1/\hat{\tau}_c$ in general, with equality only if $\hat{\varphi}(t)$ has decreased substantially for t such that the functions $\sin(\omega_a t)/(\omega_a t)^a$, and $\cos(\omega_a t)$ begin to deviate from their limiting values of unity for small $\omega_a t$.

As will be seen in a subsequent section, the time spanned by the functions $\hat{\alpha}(t)$ and $\hat{\varphi}(t)$ as they decay from unity to zero may easily encompass 12 to 15 decades, which is almost always too much to be covered by a single instrument at a given temperature (and pressure). In some cases, it is possible to extend the time by combination of steady-state dynamic mechanical experiments with time-dependent measurements. This is useful because dynamic experiments may be carried out for a range of ω corresponding to shorter $t(\approx 1/\omega)$ than may be convenient in the time domain, and vice versa, permitting an effective extension of the time span probed. For example, the relation

$$R(t) \approx \{[J'(\omega)]^2 + [J''(\omega) - (\omega\eta)^{-1}]^2\}_{\omega t=1}^{1/2} \qquad (63)$$

provides a useful approximation, which could be improved by the use of the appropriate exact integral relations given above; this expression is a variant of the approximation $J(t) \approx |J^*(\omega)|_{\omega t=1}$.[44]

Conversely, the relations

$$J'(\omega) \approx \{[1 - m(2t)]^{0.8}R(t)\}_{\omega t=1} \qquad (64a)$$

$$J''(\omega) - (\omega\eta)^{-1} \approx \{[m(2t/3)]^{0.8}R(t)\}_{\omega t=1} \qquad (64b)$$

have been found to be accurate representations, where $m(t) = \partial\ln R(t)/\partial\ln t$ (note typographical errors with $2t/3$ given as $2/3t$ in some references.[11y,23q,45]

A less rigorous attempt to expand the effective time (or frequency) range for which linear viscoelastic functions may be estimated beyond that available experimentally relies on the so-called time–temperature equivalence approximation that employs reduced variables. In essence, the approximation assumes that properly reduced compliances or moduli as functions of reduced time or frequency could be independent of temperature. Consideration of the behavior at intermediate and long times shows that if it obtains, such invariance should be based on the use of the dimensionless time $t/\hat{\tau}_c$ and frequency $\omega\hat{\tau}_c$, with $\hat{\tau}_c$ generalized to read $\hat{\tau}_c = \eta'(0)J_\infty$; dimensionless shear compliances by division of $J(t)$, $J'(\omega)$, or $J''(\omega)$ by J_∞; and dimensionless shear moduli by multiplication of $G(t)$, $G'(\omega)$, or $G''(\omega)$ by J_∞.[46,47] Similar expressions may be written for the other compliances and moduli. For example, for a fluid,

$$J(t/\hat{\tau}_c)/J_s = 1 - [1 - J_0/J_s]\hat{\alpha}(t/\hat{\tau}_c) + t/\hat{\tau}_c \qquad (65)$$

Because normally (except for small molecules), $J_0/J_s \ll 1$, it can be seen that the time–temperature approximation in which $J(t/\hat{\tau}_c)/J_s$ is considered to be a function of $t/\hat{\tau}_c$, and not otherwise dependent on temperature, will hold if $\hat{\alpha}(t/\hat{\tau}_c)$ proves to be independent of temperature. Examples of other reduced functions are given in Table 22.5.

Table 22.5 Time–temperature superposition approximation

Temperature-dependent parameters

Fluid	Solid
$\eta'(0) = \eta$	$\eta'(0) = cst$
$J_\infty = J_s$	$J_\infty = J_e = 1/G_e$
$\tau_c = \eta'(0)J_\infty = \eta J_s$	$\tau_c = \eta'(0)J_\infty = \eta'(0)J_s$

Key: cst = a constant number.

Functions approximately independent of temperature

Compliances	Moduli	
$J(t/\tau_c)/J_\infty$	$J_\infty G(t/\tau_c)$	
$J'(\omega\tau_c)/J_\infty$	$J_\infty G'(\omega\tau_c)$	$\eta'(\omega\tau_c)/\eta'(0)$
$J''(\omega\tau_c)/J_\infty$	$J_\infty G''(\omega\tau_c)$	$\eta''(\omega\tau_c)/\eta'(0)$

Relative shift factors at temperature T_{REF}

$$b_T = b(T, T_{REF}) = J_\infty(T)/J_\infty(T_{REF})$$
$$h_T = h(T, T_{REF}) = \eta(T)/\eta(T_{REF})$$
$$a_T = a(T, T_{REF}) = \tau_c(T)/\tau_c(T_{REF}) = h_T b_T$$

Use of relative shift factors to produce a master curve

Compliances	Moduli
$J(t/a_T;T)/b_T \approx J(t;T_{REF})$	$b_T G(t/a_T;T) \approx G(t;T_{REF})$
$J'(\omega a_T;T)/b_T \approx J'(\omega;T_{REF})$	$b_T G'(\omega a_T;T) \approx G'(\omega;T_{REF})$
$J''(\omega a_T;T)b_T \approx J''(\omega;T_{REF})$	$b_T G''(\omega a_T;T) \approx G''(\omega;T_{REF})$
	$\eta'(\omega a_T;T)/h_T \approx \eta'(\omega;T_{REF})$
	$\eta''(\omega a_T;T)/h_T \approx \eta''(\omega;T_{REF})$

Although J_∞ is frequently only slightly dependent on temperature,[18] the viscosity will generally depend markedly on temperature, with this dependence increasing markedly for T near the glass transition temperature T_g.[11z,18] In the latter case, it is useful to correlate the temperature dependence of η [or $\eta'(0)$] with the simple relation[11z,48,49]

$$\eta(T)/\eta(T_{REF}) = \exp[C/(T - T_0) - C/(T_{REF} - T_0)] \quad (66a)$$

$$= \exp\left(-\frac{C(T - T_{REF})}{\Delta_{REF}(T - T_{REF} + \Delta_{REF})}\right) \quad (66b)$$

with C and T_0 being constants, and $\Delta_{REF} = T_{REF} - T_0$. Thus, with this representation, $\eta(T)/\eta(T_{REF})$ becomes a function of $T - T_{REF}$, with a constant Δ_{REF} that depends on the arbitrarily chosen reference temperature T_{REF}. Fitting this relation to experimental data permits estimation of C and T_0.[11z,18,48] If T_{REF} is put equal to T_g then

$$\eta(T)/\eta(T_g) = \exp\left(\frac{K(T - T_g)}{T - T_g + \Delta}\right) \quad (67)$$

where $\Delta = T_g - T_0$ and $K = C/\Delta$. This form emphasizes the role of T_g in the temperature dependence, and is especially useful because the parameters K and Δ are often found to be nearly equal to universal values of $K = 2300$ K and $\Delta = 57.5$ K for polymeric materials.[11z,18,48] This form takes account of the anticipated dependence of T_g on the number-average molecular weight M_n for low molecular weight polymers,[11a',48] as well as the dependence of T_g on the volume fraction of polymer in a solution. Because $T_m/T_g \approx 1.4$ to 2[50] for crystallizable polymers, then T is far enough above T_0 to permit simplification of this expression to the Arrhenius relation $\eta(t) \propto \exp(W/T)$, in which W is a constant; similarly, the Arrhenius relation applies if T is greater than about $(1.4$ to $2)T_g$ for noncrystallizable polymers.

Because J_s and η may not be known as functions of T, the time-temperature approximation is usually implemented in terms of reference variables $b(T,T_{REF}) = J_\infty(T)/J_\infty(T_{REF})$, $h(T,T_{REF}) = \eta(T)/\eta(T_{REF})$, and $a(T,T_{REF}) = b(T,T_{REF})h(T,T_{REF})$, which are called shift factors (see Table 22.5).[11b',47] The notation for T_{REF} is usually suppressed, that is $a(T,T_{REF})$ is usually written a_T or occasionally $a(T)$.[11b'] Implementation of this approximation is usually accomplished by finding the values of the reference variables that will superpose bilogarithmic plots of data at temperature T as a function of time or frequency onto data at some convenient T_{REF}. For example, as illustrated in Figure 22.8, data on $\log[J(t) - J_0]$ as a function of $\log(t)$ at temperature T are shifted horizontally to coincide with data at T_{REF} to estimate $\log[a(T,T_{REF})]$; $b(T,T_{REF})$ is often small or negligible, and is taken to be unity for these data, but otherwise, a vertical shift needed to superpose the data would give $\log[b(T,T_{REF})]$. The shift factors for other functions are given in Table 22.5. Implementation of this procedure with dynamic mechanical data is facilitated by examination of $J'(\omega)$ and $\eta'(\omega)$ because in each case, these approach a constant value at low ω, making it simple to assess the shift factors $b(T,T_{REF})$ and $h(T,T_{REF})$, respectively, from the vertical shift for those portions of the data for which ω is small enough to approximate constancy in $J'(\omega)$ and $\eta'(\omega)$. Then, a degree of consistency is imposed by the requirement that $a(T,T_{REF}) = b(T,T_{REF})h(T,T_{REF})$ for the shift factor to be applied to the frequency.

In some cases experimenters carry out dynamic experiments as a function of temperature at a fixed frequency (so-called isochronal experiments). This practice arose in an age when it was not easy to do measurements as a function of ω, and has become embedded in certain fields. Insofar as the time-temperature equivalence behavior is a reasonable approximation, there is a simple mapping between the time and frequency domains. An example based on the use of the empirical temperature dependence discussed below is given in Figure 22.9 for a viscoelastic solid.

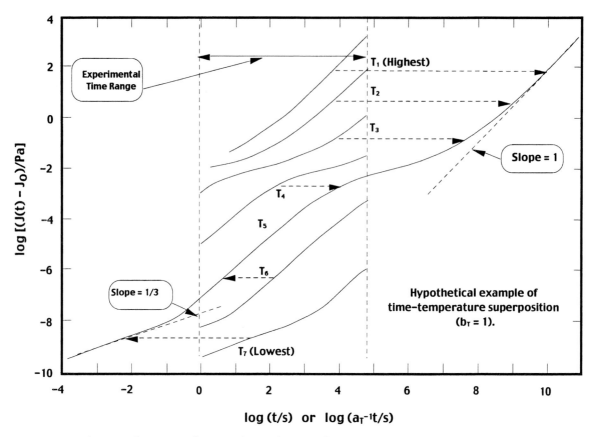

Figure 22.8 Schematic drawing to illustrate the application of time-temperature superposition for data on the shear creep compliance (assuming that $b_T = 1$ for simplicity). The shifts of the data at each temperature to the reference plot at temperature $T_{REF} = T_5$ are shown by the dashed lines. The indicated experimental range is given as a guideline of actual practice, but could be extended to shorter or longer times on occasion.

The Viscosity of Dilute Polymer Solutions and Colloidal Dispersions

Although viscoelastic measurements can be carried out on dilute solutions of polymers, specialized equipment is required and the result is not normally of interest; however, a brief discussion of such measurements is provided. The viscosity, however, is widely used to characterize macromolecules or colloidal particles in dilute solution, and capillary viscometry is by far the most widely used method for that purpose. The parameters of usual interest arise in a series expansion of the viscosity in terms of the polymer concentration c(wt/vol):

$$\eta = \eta_{LOC}^{(c)}\{1 + [\eta]c + k'([\eta]c)^2 + k''([\eta]c)^3 + \ldots\} \quad (68)$$

with the local viscosity $\eta_{LOC}^{(c)}$ usually assumed to be approximated by the solvent viscosity η_{solv} in a dilute solution, and k' is a parameter of order unity.[15a] Although the use of this virial expansion is usually appropriate for a dilute solution ($[\eta]c < 1$), it may not apply to all systems, with solutions of charged species in a solvent of low dielectric strength providing one example where it may fail.[15b] The expression for the relative viscosity $\eta_{rel} = \eta/\eta_{solv}$ may be transformed into

several useful forms for analysis, assuming that $\eta_{LOC}^{(c)} \approx \eta_{solv}$ to determine $[\eta]$ and k' (possible exceptions to this widely applicable assumption are discussed at the end of this section), including

$$\eta_{sp}/c = (\eta_{rel} - 1)/c = [\eta] + k'[\eta]^2c + k''[\eta]^3c^2 + \ldots \quad (69a)$$

$$\ln(\eta_{rel})/c = [\eta] - (1/2 - k')[\eta]^2c$$
$$+ (1/3 - k' + k'')[\eta]^3c^2 + \ldots \quad (69b)$$

$$\{2[\eta_{sp} - \ln(\eta_{rel})]\}^{1/2}/c = [\eta] - (1/3 - k')[\eta]^2c + (7/36$$
$$- 2k'/3 + k'')[\eta]^3c^2 + \ldots \quad (69c)$$

all to the same order in c, with η_{sp} representing the specific viscosity. As shown by these expressions, $[\eta] = (\partial\eta_{sp}/\partial c)_{c=0} = [\partial\ln(\eta_{rel})/\partial c]_{c=0} = 2(\partial\{[\eta_{sp} - \ln(\eta_{rel})]\}^{1/2}/\partial c)_{c=0}$ (under the assumption $\eta_{LOC}^{(c)} \approx \eta_{solv}$). Additional transformations of the basic series expansion have sometimes been used to aid analysis to determine $[\eta]$ and k',[15a] but these are the forms most commonly encountered. Simultaneous analysis with these three forms should yield common values for $[\eta]$ and k', and failure to do so should be taken as evidence of inaccuracy in the data, or data confined to c too large to permit a reliable extrapolation to infinite dilution. In particular, if

$k' \approx 1/2$, as with flexible chain polymers in solutions with $A_2 \approx 0$ in a Flory Theta solvent, then the curvature in η_{sp}/c will be enhanced, whereas the initial tangent for $\ln(\eta_{rel})/c$ will be zero, making it easy to underestimate $[\eta]$ and overestimate k' by assuming linearity in both of these functions over the measured range of c.[51] The third form is particularly useful if $k' \approx 1/3$, as is often the case for flexible chain polymers in so-called good solvents,[15a] and can be recommended for use in the analysis of viscometric data obtained for the effluent in a chromatographic analysis.

Capillary viscometry using a suspended-level Ubbelohde capillary viscometer to permit successive dilutions in the viscometer is the ususal method to obtain η_{rel} as a function of c for analysis with the preceding relations. Suspended-level Ubbelohde capillary viscometers are commercially available that suppress end-effects, so that the efflux time for a calibrated volume is accurately proportional to the viscosity, permitting evaluation of η_{rel} as the ratio of the efflux times for solution and solvent; consideration of the analysis to determine $[\eta]$ and k' will show that η_{rel} should be determined to ± 0.0005 to obtain the desired precision in $[\eta]c$. A viscometer with an efflux time of about 100 s should normally be used, with means to determine the efflux time to ± 0.03 s applied to obtain the necessary precision in the analysis. The viscometer must be held to ± 0.02 °C to obtain this precision in the efflux times. It is essential that solution and solvent be filtered to remove extraneous matter that might lodge in the capillary or otherwise alter the efflux time, and that η_{rel} not exceed about 1.8 for the most concentrated solution, and preferably be less than 1.5 if $k' \approx 1/2$; typically, the solution is diluted until $\eta_{rel} \approx 1.1$. Application of a vacuum to force the solution from the dilution bulb of the viscometer into the capillary should be avoided, especially with volatile organic reagents; the application of

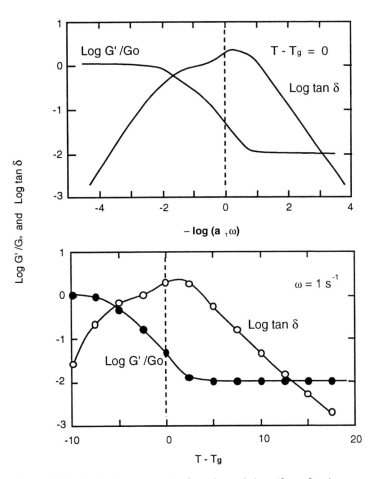

Figure 22.9 Illustrative example of isochronal data (for a fixed frequency) corresponding to isothermal data (reduced to the glass transition temperature) for a linear viscoelastic solid. The isochronal data were computed from the isothermal data by the use of the universal expression for the temperature dependence of the viscosity, with $b_T = 1$ for simplicity. The vertical dashed lines mark positions of frequency and temperature common to the two representations.

a pressure is preferred. Specialized sealed capillary viscometers are available for work above the normal boiling temperature of the solvent. Studies on very high molecular weight polymers can introduce two special problems: (1) degradation of the polymer in solution in the filtration process or in the capillary in flow, and (2) nonlinear behavior such that $[\eta]$ becomes a function of the rate of shear in the capillary flow, for example, the rate of strain at the wall of a capillary viscometer will typically be in the range 10^2 to $10^3 \ s^{-1}$. Both of these difficulties may be suppressed by use of specialized concentric cylinder viscometers designed to operate at a low rate of strain ($<1 \ s^{-1}$), with a gap between the cylinders large enough to suppress the effects of a few dust particles, and sufficiently precise to give the necessary precision in η_{rel}.[51]

Consideration of the preceding shows that for a heterodisperse sample, the observed $[\eta]$ is an average over the $[\eta]_i$ for the components with weight fraction $w_i = c_i/c$ given by

$$[\eta] = \Sigma[\eta]_i w_i \qquad (70)$$

Thus, if $[\eta] = KM^{\mu}$ for each of the components, the average molecular weight corresponding to the observed $[\eta]$ is given by $M_{(\mu)} = (\Sigma M_i^{\mu} w_i)^{1/\mu}$. For example, a Schulz-Zimm (two-parameter exponential) distribution of molecular weight with $Z^{-1} = (M_w/M_n - 1)$ gives $M_{(\mu)} = [M_w/(Z+1)]\{\Gamma(Z+1+\mu)/\Gamma(Z+1)\}^{1/\mu}$ so that $M_{(\mu)}$ lies intermediate to M_n and M_w for μ in the range 0.5 to 1.0;[52] (see reference 53 for additional examples).

The intrinsic viscosity $[\eta]$ has a simple interpretation for rigid spherical particles of molecular weight M and radius R because the product $[\eta]c = 5\varphi/2$,[53a] with $\varphi = cN_A(4\pi R^3/3)/M$ the volume fraction of the spheres: $[\eta] = 5\varphi/2c = 10\pi N_A R^3/3M$. More generally, it is convenient to express $[\eta]$ in the form[54]

$$[\eta] = \pi N_A K_{\eta} R_G^2 R_H/M \qquad (71)$$

where R_G is the root-mean-square radius of gyration; R_H is the hydrodynamic radius, equal to $kT/6\pi\eta_{solv}D_T$, with D_T the translational diffusion constant; and the dimensionless parameter K_{η} depends on the solute shape and other characteristics, with K_{η} in the range 1 to 10/3 for polymeric solute and ranging up to 50/9 for spherical particles (for which $R_G^2 R_H = 3R^3/5$).

For high molecular weight flexible chain polymers, $R_H \approx 2R_G/3$ and $K_{\eta} \approx 10/3$ in the so-called nondraining hydrodynamic limit,[54,55a] leading to the well-known Flory-Fox relation $[\eta]M = \Phi' R_G^3$, with $\Phi' = \pi N_A K_{\eta} R_H/R_G$, or $\Phi' \approx 20\pi N_A/9$ for linear flexible chain polymers. The proportionality $[\eta]M \propto R_G^3$ was originally suggested by analogy with the behavior of hard spheres.[56,57] In this form, the molecular weight dependence of $[\eta]M$ is attributed to that of R_G^3, which in turn is given by the relation $R_G^2 = \hat{a}L\alpha^2/3 \propto M^{\varepsilon}$ for high molecular weight polymers, where α is the excluded

volume expansion factor, $\hat{a}$ is the persistence length of the chain, and L its contour length ($L = M/M_L$ with M_L the mass per unit length of the chain).[58] Under Flory Theta conditions, defined by the condition that the second virial coefficient A_2 is zero, $\alpha = 1$ and $R_G^2 = \hat{a}L/3 \propto M$. In the opposite extreme, repulsive interactions among the chain segments saturate to give $A_2 M^2 \propto R_G^3 \propto [\eta]$ and $\alpha \propto z^{1/5}$, where the interaction parameter $z \propto (L/\hat{a})^{1/2}$, so that ε varies from unity at the Theta temperature to 6/5 in a so-called good solvent. Correspondingly, $\mu = \partial \ln [\eta]/\partial \ln M = (3\varepsilon/2) - 1$ varies from 1/2 to 4/5. Values of the so-called Mark-Houwink-Sakarada parameters μ and $K = [\eta]/M^{\mu}$ are listed in handbooks for the range of M for which these are essentially constant for a given polymer-solvent pair.[59]

The presence of branching can alter the analysis of $[\eta]$. The effects of short branches placed randomly, or uniformly along a much longer backbone of an otherwise linear chain, are readily accommodated by assuming that K_{η} and R_G/R_H will be unaltered, and that R_G^2 will be essentially the value for the linear backbone in the absence of the short branches.[60] In this case, the ratio $g' = [\eta]/[\eta]_{LIN}$ of the intrinsic viscosities of chains with common M is given by $g' \approx g^{3/2}$, with $g = R_G^2/(R_G^2)_{LIN}$ the ratio of the mean-square-radii of gyration of chains with common M. The introduction of long-chain branching, as with randomly branched or comb-shaped branched chains or star-branched chains, introduces additional complications because then both K_{η} and R_G/R_H may be altered. Thus, the ratio $h = R_H/(R_H)_{LIN}$ of the hydrodynamic radii of chains with common M is in the range $1 \leq hg^{-1/2} \leq 1.4$ for Gaussian comb- or star-branched chains, and for those same models, K_{η} increases with decreasing g for the same model.[54] These effects can be expressed in the form $g' \approx g^{m(\lambda)}$, with λ the fraction of the repeating units in the longest linear segment of the branched structure; $m(\lambda) \approx 0.44 + \lambda^{10/3}$ and $g \approx \lambda + [(3p-2)/p^2](1 - \lambda)^{7/3}$ comb- or star-branched molecules with p branches per molecule.[54] Consequently, $m(\lambda)$ varies from about 1/2 for regular star-branched chains to 3/2 for comb-branched chains with short branches (as in the preceding discussion). This same form may approximate g' for a randomly branched chain based on the statistically longest linear segment in those structures.

With decreasing $L/\hat{a}$ for flexible chain polymers, α tends to unity, even in systems with large A_2, and R_H/L and K_{η} tend to unity, giving $[\eta] \approx \pi N_A R_G^2/M_L \propto M$ in the free-draining hydrodynamic limit.[54,55a] This regime is not usually encountered except for low M, but may be observed with semiflexible chains for which $\hat{a}$ is large enough that $L/\hat{a}$ is in the range for this behavior even for chains with relatively high molecular weight, for example, with cellulosic polymers and certain other chains with relatively large $\hat{a}$.[54] Expressions are available for a variety of other solute shapes, such as rods, semiflexible wormlike chains, and ellipsoids of revolution. For rods with diameter $d \ll L$, $K_{\eta} = 1$, $R_G^2 = L^2/12$, and $R_H \approx L/2\ln(3L/2d)$. Relations for the wormlike chain and prolate and oblate ellipsoids of revolution are given elsewhere.[54]

For very small M, the observed $(\partial\ln\eta/\partial c)_{c=0}$, interpreted as $[\eta]$ by the preceding equations, may be negative. This effect is attributed to a failure of the assumption $\eta_{LOC}^{(c)} = \eta_{solv}$ for nonzero c if η is to be interpreted by the preceding relations or more sophisticated variations of them.[61] For example, in a dilute solution, $\eta_{LOC}^{(c)}$ might be expressed as a Taylor series in c with $(\eta_{LOC}^{(c)})_{c=0} = \eta_{solv}$, but with $|\partial\ln\eta_{LOC}^{(c)}/\partial c|_{c=0}$ not negligible in comparison with $[\eta]$, as is usually assumed for dilute solutions. In this case, the series expansion for η given above must be modified, so that $(\partial\ln\eta/\partial c)_{c=0} = [\eta] + (\partial\ln\eta_{LOC}^{(c)}/\partial c)_{c=0}$, requiring an estimate of $(\partial\ln\eta_{LOC}^{(c)}/\partial c)_{c=0}$ to evaluate $[\eta]$ from $(\partial\ln\eta/\partial c)_{c=0}$. This correction is expected to be negligible for high molecular weight solute, because then $[\eta] \gg |\partial\ln\eta_{LOC}^{(c)}/\partial c|_{c=0}$, but could become important for oligomeric polymers, for which $[\eta]$ is smaller. In some such cases it appears that $(\partial\ln\eta_{LOC}^{(c)}/\partial c)_{c=0} < 0$ may be larger than $[\eta]$, leading to negative $(\partial\ln\eta/\partial c)_{c=0}$.[61-65] Nonzero $(\partial\ln\eta_{LOC}^{(c)}/\partial c)_{c=0}$ for a dilute solution may reflect a number of effects, including the situation in which a solvent with a high glass temperature (T_g) is mixed with a polymer with a low T_g, so that $\partial\ln\eta_{LOC}^{(c)}/\partial c < 0$, although this mechanism has been challenged.[66]

The Huggins constant k' appearing in the virial expansion is expected to depend on both hydrodynamic and thermodynamic effects, and for flexible chain polymers can usually be approximated by the relation

$$k' = c_1 + c_2 A_2 M/[\eta] \tag{72}$$

For example, for linear flexible chain polymers, $c_1 \approx 1/2$ and $c_2 \approx -1/6$,[15c,55b,67] in reasonable agreement with the experimental observations that $k' \approx 1/2$ under Flory Theta conditions $(A_2 = 0)$ and $k' \approx 1/3$ as $A_2M/[\eta]$ approaches its asymptotic value of approximately unity in a so-called good solvent. For rigid spheres interacting through a hard-core potential, the dependence on $A_2M/[\eta]$ is moot because $A_2M/[\eta] = 8/5$. Numerous estimates of k' for rigid spheres have appeared, but as discussed in the final section, the accepted theoretical and experimental estimates give $k' \approx 1.0$.[53a,68,69]

Most of the linear viscoelastic studies on dilute polymer solutions have involved either dynamic mechanical or dynamic birefringence measurements. The stress-optic approximation discussed in the final section has been invoked in the latter method. These studies often have the objective of comparison of the reduced responses $[\eta]'(\omega) = \{\eta'(\omega)/\eta_{LOC}^{(c)}c\}_{c=0}$ and $[\eta]''(\omega) = \{[\eta''(\omega) - \eta_{LOC}^{(c)}]/\eta_{LOC}^{(c)}c\}_{c=0}$ with theoretical estimates of the same functions, where it is usual to assume that $\eta_{LOC}^{(c)} \approx \eta_{solv}$.[11j,63] An anomaly at high frequency suggests that $\eta_{LOC}^{(c)} \neq \eta_{solv}$, similar to the behavior discussed above in which $(\partial\ln\eta/\partial c)_{c=0}$ may be negative for solutions of oligomeric polymers, and understood in the same general way, although care must be taken to account for any viscoelastic character of the solvent, which could make $\eta'(\omega) \neq 0$ and $\eta''(\omega) < \eta_{solv}$ for the solvent even if $\eta_{LOC}^{(c)} \approx \eta_{solv}$.[61,64,65]

With increasing c, in the range of moderately concentrated (or semidilute) solutions, the virial expansion of η given above fails. This regime is marked approximately by the condition that R_g is about equal to the mean separation $\Lambda = (M/cN_A)^{1/3}$ of molecular centers. The behavior in this regime is conveniently represented by the expression

$$\eta = \eta_{LOC}^{(c)}\{1 + [\eta]^{(c)}c\} \tag{73}$$

For dilute solutions, $[\eta]^{(c)} \approx [\eta]\{1 + [\eta]c\}^{k'}$ to approximate the virial expression to order c^2. With increasing c, both $K_\eta R_H$ and α reduce to unity through screening of hydrodynamic and thermodynamic interactions, respectively, and $[\eta]^{(c)}c \approx H^{(c)}\bar{X}$, with $\bar{X} = [\pi N_A \hat{a}(\alpha^{(c)})^2/3M_L^2]cM$, provided cM is small enough to suppress intermolecular entanglement interactions discussed in the next section.[48,49] Here, $H^{(c)}$ is a function to account for the dependence of $K_\eta R_H$ on c, and $\alpha^{(c)}$ is the expansion factor for the root-mean-square radius of gyration of the chain at polymer concentration c, and other parameters are introduced above. A semi-empirical cross-over expression to span these two extremes is presented in reference 49; a schematic representation of the range of the concentration for the successive screening of thermodynamic and hydrodynamic interactions and the development of entanglement effects is given in Figure 22.10. Although the dependence of $H^{(c)}$ and $\alpha^{(c)}$ may each be represented as power laws in $[\eta]c$ over appreciable ranges of concentration, care should be used in the application of these, especially in poor solvents.[48,49] The proportionality $[\eta]M \propto R_G^3$ for high molecular weight linear polymer (or hard spheres) is suggestive of an overlap condition for a flexible chain polymer, in which the concentration has increased to a level c^* such that $\Lambda \approx R_G$. Thus, for a flexible chain polymer, $c^* \approx M/[(4/3)\pi N_A R_G^3] \approx 5/(3[\eta])$. Discussions of the concentration dependence of $\eta/\eta_{LOC}^{(c)}$ are sometimes presented in terms of c/c^* in lieu of $[\eta]c$.[70] The dependence of $\eta_{LOC}^{(c)}$ on c may be important in this regime, even if $\eta_{LOC}^{(c)} \approx \eta_{solv}$ for dilute solutions. In some cases, $\eta_{LOC}^{(c)} \approx \eta_{solv}\exp(b_T\varphi)$, with b_T a temperature-dependent parameter, before reaching the behavior described in the preceding, with $\eta_{LOC}^{(c)}$ a function of the glass transition temperature.[71]

Linear Viscoelastic Behavior of Concentrated and Undiluted Polymeric Fluids

According to the preceding discussion, linear viscoelastic behavior may be represented by several parameters and functions. For example, in shear deformation a set would

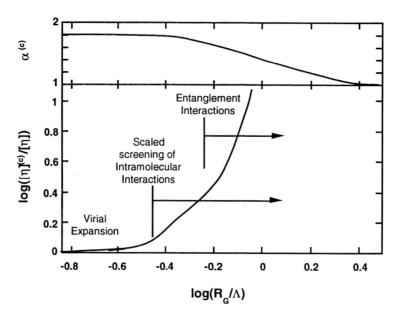

Figure 22.10 Schematic drawing of the viscosity (*bottom*) and the concentration-dependent excluded volume factor (*top*) versus R_G/Λ based on a relations discussed in the text to illustrate regimes over which various approximations apply; $\Lambda = (M/N_{AC})^{1/3}$ is equal to the mean separation of molecular centers. (Reproduced with permission from reference 49. Copyright 1996 by the Society of Rheology.)

comprise the parameters $J_0 = 1/G_0$, J_∞ (equal either to $J_e = 1/G_e$ for a solid or J_s for a fluid), η for a fluid, and the functions $\hat{\alpha}(t)$ or $\hat{\phi}(t)$, with the latter related through a convolution integral. Similar representations may be written for the volume functions $B(t)$ and $K(t)$, and for the elongational functions $D(t)$ and $E(t)$, with the latter approximated by $D(t) \approx J(t)/3$ and $E(t) \approx 3G(t)$ in the incompressible approximation (negligible volume change).

It is convenient to discuss η before turning to J_s and the time-dependent functions $G(t)$ and $J(t)$. The viscosity of polymers and their concentrated solutions has received considerable attention.[48,49,72,73] Empirically, the viscosity of polymers and their concentrated solutions may be represented for a linear polymer by the simple expressions[49]

$$\eta = \eta_{LOC}^{(c)} H^{(c)} \tilde{X} [1 + (\tilde{X}/\tilde{X}_c)^{4.8}]^{1/2} \tag{74a}$$

$$\tilde{X} = [\pi N_A \rho \hat{a}(\alpha^{(c)})^2 / 3 M_L^2] \varphi M \tag{74b}$$

where $\hat{a}$ is the persistence length, $\alpha^{(c)}$ is expansion factor at polymer volume fraction φ, ρ is the polymer density, $M_L = M/L$ with L the contour length of the chain, N_A is the Avogadro number, $\eta_{LOC}^{(c)}$ is a local viscosity, $H^{(c)} \approx \varphi^{1/2}$, and $\tilde{X}_c \approx 100$ for a variety of chains. The assumed form with factor-

ization of the local frictional effects included in $\eta_{LOC}^{(c)}$ is consistent with a study that showed $\eta/\eta_{LOC}^{(c)}$ to be the same at the upper and lower Flory Theta temperatures, some 100 °C apart, for moderately concentrated solutions over a range of c and M.[73] For Gaussian chain statistics, as would occur in the absence of excluded volume effects, the mean-square-radius of gyration $R_G^2 = \hat{a}L/3 \propto M$.[58] Because $\alpha^{(c)} \approx 1$ for concentrated solutions and undiluted polymer,[48,49,58,70] $\tilde{X} \propto \varphi M$ for such, in which case $\eta \propto \varphi M$ for $\tilde{X} \leq \tilde{X}_c$ or $\eta \propto (\varphi M)^{3.4}$ for $\tilde{X} > \tilde{X}_c$. The molecular weight M_c corresponding to this condition for undiluted polymer is seen to be given by $M_c \approx 100/[\pi N_A \rho \hat{a}/3 M_L^2]$ insofar as $\tilde{X}_c \approx 100$. An illustration of this behavior is given in Figure 22.11, which shows $\eta/\eta_{LOC}^{(c)}$ versus a parameter $X = \tilde{X}/\pi N_A m_a/M_L$, with m_a the molar mass per repeat atom of the chain;[48] m_a/M_L is essentially a constant for the examples shown. The division by $\eta_{LOC}^{(c)}$ corrects for the variation of $\eta_{LOC}^{(c)}$ with M_n at low molecular weight owing to the variation of T_g with M_n. The general features of this behavior may be understood in terms of models based on Rouse-like dynamics for $\tilde{X} \leq \tilde{X}_c$, modified by entanglement effects, perhaps represented by some model involving reptation concepts, for $\tilde{X} > \tilde{X}_c$.[17c]

The local viscosity $\eta_{LOC}^{(c)}$ is not precisely defined, but $\eta_{LOC}^{(c)}$

FACING PAGE

Figure 22.11 Bilogarithmic plot of the reduced viscosity $\eta/\eta_{LOC}^{(c)}$ versus $\tilde{X}$ for various polymers. The arbitrary constant provides vertical separation of the data. (Adapted from reference 48.)

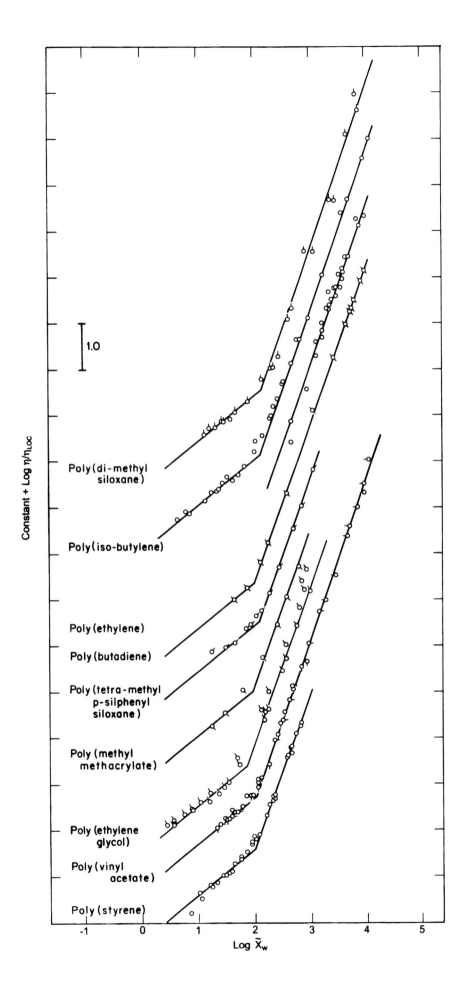

Poly(di-methyl
siloxane)

Poly(iso-butylene)

Poly(ethylene)

Poly(butadiene)

Poly(tetra-methyl
p-silphenyl
siloxane)

Poly(methyl
methacrylate)

Poly(ethylene
glycol)

Poly(vinyl
acetate)

Poly(styrene)

Constant + Log η/η_{LOC}

1.0

Log $\tilde{X}_w$

$(T)/\eta_{LOC}^{(c)}(T_g)$ can be represented by the expression used above for $\eta(T)/\eta(T_g)$. An additional, usually very much smaller, temperature dependence to the viscosity is possible through the dependence of the paramaters ρ, $\hat{a}$, and $\alpha^{(c)}$ on temperature.[49] Without further information, the approximation $\log\{\eta_{LOC}^{(c)}(T_g)/Pa \cdot s\} \approx 10$ to 12 sometimes provides an adequate approximation.[74,75] For more precise work, $\eta_{LOC}^{(c)}(T_g)$ should be evaluated from experimental data on the system of interest. In this representation, the dependence of $\eta_{LOC}^{(c)}$ on polymer concentration through the dependence of T_g on concentration may be much larger than that from the variation of $\tilde{X}$ with concentration.[49] In most cases, the effect of molecular weight heterodispersity on the viscosity may be represented by the use of the weight average molecular weight, M_w, in the calculation of $\hat{X}$, and the use of the number average molecular weight to obtain T_g appearing in $\eta_{LOC}^{(c)}$;[48] the simple dependence on M_w may be inadequate for especially broad or strongly binodal distributions.[48,76] Some of these effects are illustrated in Figure 22.12 for data on the viscosity of fractions of a polymer as a function of φM_w for five series at fixed φ over a range of M_w.[77] The deviation of the data from the dashed lines at lower φM_w for the samples with larger φ is caused by the variation of T_g with molecular weight, and the corresponding effect on $\eta_{LOC}^{(c)}$.[48,49] Other than this effect, $\eta_{LOC}^{(c)}$ is constant at fixed φ, and the data for $\log(\eta)$ versus $\log(\varphi M_w)$ are parallel but offset by the variation of $\eta_{LOC}^{(c)}$ with φ owing to the variation of T_g with $\varphi^{48,49}$ because $\alpha^{(c)} \approx 1$ for the range of φ encompassed by the data.[77] In some cases, attempts are made to estimate the critical value of φM_w for the onset of entanglement effects without consideration of the variation of $\eta_{LOC}^{(c)}$ with φ, often leading to erroneous results.

Schematic illustrations of the behavior for the shear compliances and moduli, along with the corresponding distribution of retardation or relaxation times for glass-forming polymeric fluids over a range of molecular weights are shown in Figure 22.13. The schematic illustrations in Figure 22.13 show essential features observed for fluids as the molecular weight increases from the oligomeric range through values with $M < M_e$ and $M > M_e$, where M_e is the mean molecular weight between entanglement loci; typically, $M_e \approx M_c/2$. The gradual expansion of the time scale required for $G(t)$ to relax to zero and $J(t)$ to approach linearity in t, is seen to increase with increasing M, with corresponding effects on $H(\hat{\tau})$ and $L(\hat{\lambda})$, respectively. The figures for with $M < M_e$ and $M > M_e$ illustrate behavior for Rouse-like response and a fluid with entanglements, respectively. A more detailed example for a high molecular weight polymer is given in Figure 22.14, which shows a comparison of various functions for the same material. As may be seen, $J(t)$ differs from $R(t)$ only for larger t, wherein the term t/η becomes dominant. As expected, $1/G(t)$ is generally larger than $J(t)$, with the two being about equal in regions of intermediate t for which $\partial J(t)/\partial t \approx 0$, for example, in the first

plateau, for which $J(t) \approx R(t) \approx J_N$ and $G(t) \approx G_N = 1/J_N$. Furthermore, $J'(\omega) \leq [R(t)]_{\omega t=1}$, with equality occurring only in the plateau regions, and $G'(\omega) \geq [G(t)]_{\omega t=1}$, with the latter pair being very close except for large t, for which $G'(\omega) \propto (\omega\hat{\tau}_c)^2/J_s$, but $G(t)$ decreases exponentially with t. These functions demonstrate the inequalities $J'(\omega) \leq [R(t)]_{\omega t=1} < 1/G'(\omega) < [1/G(t)]_{\omega t=1}$ expected for any linear viscoelastic material. As mentioned above, a close approximation to $[R(t)]_{\omega t=1}$ is given by $\{[J'(\omega)]^2 + [J''(\omega) - (\omega\eta)^{-1}]^2\}^{1/2}$ over a wide range of ω.

A schematic illustration for $R(t)$ and $L(\hat{\lambda})$ for a high molecular weight monodisperse polymeric fluid is given in Figure 22.15.[9,18] As may be seen, the functions $R(t)$ and $L(\hat{\lambda})$ reveal several distinctive and characteristic features for a monodisperse high molecular weight polymer in the entanglement regime, arbitrarily arranged here in order of their importance with increasing t:[11c',18]

1. A response at small t with $R(t)$ linear in $t^{1/3}$, following the initial instantaneous J_0
2. A range of t with $R(t)$ increasing more sharply, but with $\partial R(t)/\partial t < 1$, to reach a plateau for which $\partial R(t)/\partial t \approx 0$ and $R(t) \approx J_N$
3. A second increase in $R(t)$ to a final plateau, for which $R(t) = J_s$.

As may be seen, $J(t)$ differs from $R(t)$ only for larger t, wherein the term t/η becomes dominant. As expected, $1/G(t)$ is generally larger than $J(t)$, with the two being about equal in regions of intermediate t for which $\partial J(t)/\partial t \approx 0$, for example, in the first plateau, for which $J(t) \approx R(t) \approx J_N$.

The regimes described above in $R(t)$ are mirrored in the function $L(\hat{\lambda})$ by several peaks that appear successively with increasing $\hat{\lambda}$, so that, respectively:[11c',18]

1. Peak 1 with $L(\hat{\lambda})$ linear in $\hat{\lambda}^{1/3}$ before the peak decreases sharply to zero (the decrease is obscured by overlap with peak 2 in the example);
2. Peak 2 that increases in peak area with increasing M until reaching a certain level, beyond which the peak is invariant with increasing M both in area and position in $\hat{\lambda}$;
3. Peak 3 that develops as peak 2 area ceases to increase with increasing φM, with peak 3 developing an area invariant with φM, and a maximum at $\hat{\lambda}_{MAX}$ that moves to larger $\hat{\lambda}$ as $\hat{\lambda}_{MAX} \propto (\varphi M/M_c)^{3.4}$ for $\varphi M > M_c$.

Peaks 2 and 3 lead to the first and second plateaus in $R(t)$ seen in Figure 22.15. These effects are also reflected in $H(\hat{\tau})$, but the dominant role of the longer relaxation times on $H(\hat{\tau})$ for larger $\hat{\tau}$, largely reflecting viscous deformation, tends to obscure the additive features seen in $L(\hat{\lambda})$. Peak 3 does not develop for chains with $\varphi M < M_c$.

The behavior ascribed to peak 1, first reported by da Andrade,[78] is seen in a variety of materials, such as metals, ceramics, crystalline and glassy polymers, and small organic molecules;[79] the decrease of $L(\lambda)$ to zero is evident in

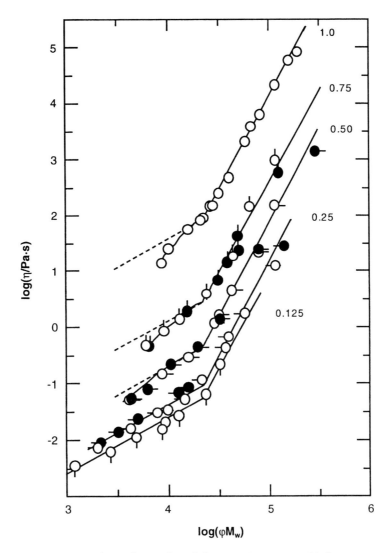

Figure 22.12 Bilogarithmic plot of the viscosity versus φM_w for fractions of poly(vinyl acetate) and its concentrated solutions. Each data set is for the fixed volume fraction φ indicated. The dashed lines indicate the curve that would be obtained if data would be plotted after reduction for the change in T_g with molecular weight and φ. The downward displacement of the data sets for $\varphi < 1$ reflects the suppression of $\eta_{LOC}^{(c)}$ as T_g decreases with decreasing T. The filled points are for data under Flory Theta conditions. (Adapted from an example in reference 77.)

examples of the latter.[80] The area under peak 1 provides the contribution $J_A - J_0$ to the total recoverable compliance J_s. It seems likely that the mechanism giving rise to peak 1 may be distinctly different from the largely entropic origins of peaks 2 and 3 described in the following discussion.

Peak 2 is ascribed to Rouse-like modes of motion, either fluidlike for low molecular weight in the range for which the area increases with M, or pseudo-solid-like (on the relevant time scale) in the range of M after peak 3 develops.[11c',17d,18,22d] For low molecular weight, the Rouse model gives the area of peak 2 as $J_s - (J_A + J_0) = (2M/5\varphi \underline{R}T)$. For

the pseudo-solid-like behavior, obtained when peak 3 has developed, reflecting the effects of intermolecular entanglements with molecular weight M_e between the entanglement loci, the area of peak 2 becomes $J_N - (J_A + J_0) = (M_e/\varphi RT)$, invariant with M. The area under peak 3, also invariant with M, ascribed to the effects of chain entanglements,[17e,18] is given by $J_s - (J_N + J_A + J_0) = (kM_e/\varphi^{2+s}\underline{R}T)$, where k is in the range 6 to 8 in most cases and $s \approx 2(\varepsilon - 1)/(3\varepsilon - 2)$, with $\varepsilon = \partial \ln R_G^2/\partial \ln M$.[18,76] Overall,

$$J_s - (J_A + J_0) = (2M/5\varphi \underline{R}T)[1 + (\varphi^{1+s}M/kM_e)^\varepsilon]^{-1/\varepsilon} \quad (75)$$

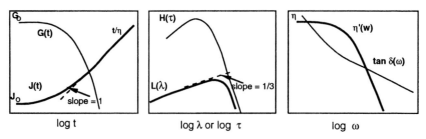

Low Molecular Weight Glass Former

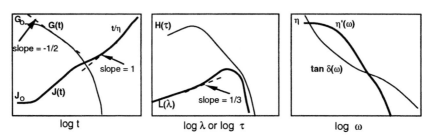

Polymeric Fluid with M < Mₑ

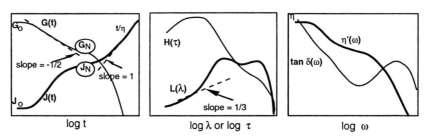

Polymeric Fluid with M >> Mₑ

Figure 22.13 Schematic drawings to illustrate various linear viscoelastic functions for an oligomer with a low *M* (*top*), a polymer with *M* less than that for entanglement effects (*middle*), and a polymer with *M* large enough to exhibit entanglement effects (*bottom*).

where the sharpness of the "transition" may be represented by adjustment of ε.[18] The term in square brackets accounts for effects caused by entanglements, making J_s independent of M for φM greater than kM_c, with $k \approx 2$ to 4.[18,70,76] A reptation model, developed to give $G(t)$ as a discrete sum of weighted exponentials for $\hat{\phi}(t)$, may be used to compute both η and $R(t)$ as a discrete sum of weighted exponentials for $\hat{\alpha}(t)$.[81]

Increasing molecular weight heterodispersity affects $R(t)$ in various ways, conveniently discussed in terms of the effects on $L(\hat{\lambda})$. The effects of molecular weight heterogeneity are marked for J_s in both the Rouse and entanglement regimes.[11d,18] Calculation with the Rouse model gives the average $M_z M_{z+1}/M_w$ for the molecular weight average in that regime, reflecting a broadening of peak 2.[11e'] Experience in the entanglement regime reveals an equally strong dependence of J_s on the molecular weight distribution, reflecting

a broadening of peak 3, with effects on peak 2 being nil if the molecular weights of all components exceeds M_e.[82-89] Thus, the dominant effect for high molecular weight polymer in the entanglement regime is to broaden the final peak 3, causing it to encompass a larger range in $\hat{\lambda}$, as shown schematically in Figure 22.15, reflecting enhancement of J_s, with little influence on J_N, for example, with $J_s \propto (M_z/M_w)^{2.5}$ in some experiments for all $M > M_c$.[11f,82-89] For a distribution containing both low ($M_1 < M_c$) and high ($M_2 > M_c$) molecular weight components, one can expect effects on both peaks 2 and 3, as well as η.[80,83,88] The short chains essentially act as a diluent for the longer chains, such that J_s increases approximately in proportion to the inverse of the square of the volume fraction φ_2 of the longer chains. A single peak 1 is observed, but two peak 2s are observed, one for each component, with the separation between peak 2 for the higher molecular weight component and peak 3 given by $M_2\varphi_2$.[80] Thus, the effect of molecular weight heterodispersity can be simply represented in terms of $L(\hat{\lambda})$ [or $R(t)$]:

1. Peak 1 is essentially unaffected by molecular weight dispersion.

2. Peak 2 gives an area proportional to $\varphi_L M_z M_{z+1}/M_w$, with the averages calculated for chains at volume fraction φ_L with $M < M_c$, and a separate peak for chains with area proportional to $(1 - \varphi_L)M_c$ for chains at volume fraction $1 - \varphi_L$ with $M > M_c$.

3. Peak 3 with an area that varies approximately in proportion to $(1 - \varphi_L)^{-2}(M_z/M_w)^{2.5}$, with the averages calculated for chains with $M > M_c$; the maxima for

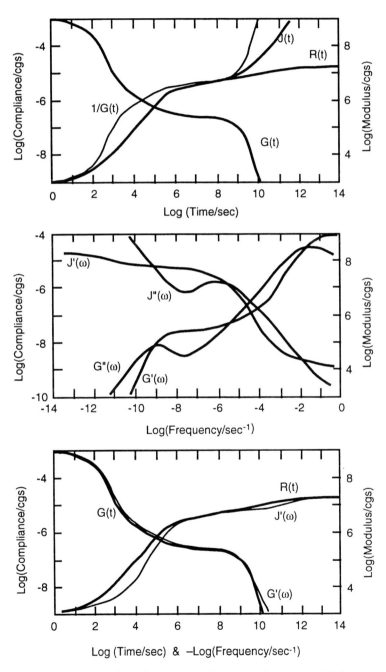

Figure 22.14 Examples of creep compliances and shear moduli for a high molecular weight polymer; the data exhibit entanglement effects for large t (or small frequency). (Adapted from an example in reference 18.)

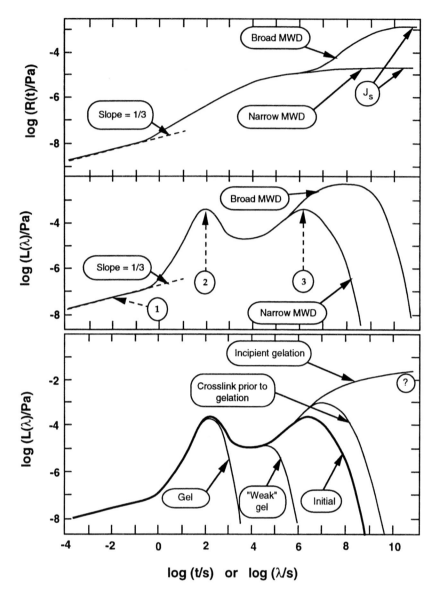

Figure 22.15 Schematic drawings to illustrate the linear recoverable compliance and the associated retardation spectrum for several cases: (*top*) the recoverable compliance and the associated retardation spectra for polymers with narrow and broad molecular weight distributions and with M large enough to exhibit entanglement effects; (*bottom*) the effect of random cross-linking on the retardation spectrum for a polymer initially with a narrow molecular weight distribution, and with M large enough to exhibit entanglement effects.

peaks 2 and 3 separate in $\hat{\lambda}$ as $(1 - \varphi_L)M_w$; no direct theoretical expression for $L(\hat{\lambda})$ in terms of the distribution in M in this range of $\hat{\lambda}$ exists.

In addition to these effects, the reduced parameters $\hat{\lambda}/\hat{\tau}_c$ and $t/\hat{\tau}_c$ will both depend on M_n for low M_n through the variation of T_g with M_n, and the consequent effect on $\eta_{LOC}^{(c)}(T)/\eta_{LOC}^{(c)}(T_g)$. The complex behavior characterized here would be difficult to represent by the expression given above for

$G(t)$ because it must include the (approximate) correlation of η on M_w, as well as the dependence of J_s on molecular weight heterogeneity, along with the other features noted above. Nevertheless, available theoretical treatments focus on consideration of $G(t)$, and a number of treatments of this type that have been proposed for fluids may be cast in the form[11g,82–89]

$$G(t) = \left[\sum_i w_i G_i(t)^\nu \right]^{1/\nu} \tag{76}$$

A recent treatment for mixtures of chains with all components having molecular weights in excess of M_c gives a result of this form with $v = 1/2$, and an expression for $G(t)$;[86,87] by comparison, in the original reptation model, $v = 1$.[17f]

The effect of a distribution of species shows up especially strongly in a material undergoing cross-linking to form a network. For example, for a linear polymer undergoing random cross-linking, for example by radiation chemistry, $L(\hat{\lambda})$ first extends to longer $\hat{\lambda}$, as with heterogeneity in molecular weight, corresponding to increased J_s for the prenetwork fluid.[11h',90] At a certain level of cross-link density at incipient network formation, $R(t)$ extends to such long times that the long time-limiting behavior cannot be experimentally assessed. In the postnetwork solid the contributions with longer $\hat{\lambda}$ are successively suppressed as the cross-link density increases, until the retardation exhibits only peaks 1 and 2 for networks with a molecular weight M_{XL} between crosslinks less than M_c, with the area of peak 2 proportional to M_{XL}. These effects are shown schematically in Figure 22.15. Similar behavior is observed on forming a network through the polymerization of a mixture of difunctional monomer containing some trifunctional monomer. The behavior for the incipient network formation frequently approximates a power law, with $R(t) \approx J(t) = J_0[1 + (t/\hat{\lambda}_p)^\alpha]$, where $\hat{\lambda}_p$ and α are constants. In this power-law regime, $\tan \delta(\omega) \approx \tan[\sin(\alpha\pi/2)]$, so that over a wide range of ω, $J'(\omega) \propto J''(\omega)$ and $G'(\omega) \propto G''(\omega)$, and for the special case $\alpha = 0.5$, $J'(\omega) \approx J''(\omega)$ and $G'(\omega) \approx G''(\omega)$.[91] Note that behavior with $\alpha \approx 0.5$ would obtain for a wide range in ω according to the Rouse model with a proliferation of relaxation times, particularly long relaxation times corresponding to larger species, as might result from the wide range of molecular structures present at incipient network formation. In such a case, the anticipated behavior at low ω for either a fluid or a solid might be suppressed to such low ω as to be experimentally unattainable.

As might be anticipated, the time-temperature equivalence approximation is not truly adequate over the full range of material response embodied in $J(t)$ or $G(t)$.[92] For example, the Andrade creep reflected in peak 1 in $L(\hat{\lambda})$ appears to have a different dependence on temperature than that for other contributions to $L(\hat{\lambda})$, as might be expected given its rather different origin.[75,92–94] Additional departures from the approximation have also been reported for behavior in connection with peaks 2 and 3.[92]

The behavior for the linear viscoelastic volume properties $B(t)$ or $K(t)$, or their dynamic counterparts, has not been explored in as much detail as that for the shear functions such as $J(t)$.[75,95] It has been reported that the distribution of retardation times $\hat{\lambda}_{vol}$ found in $\hat{\beta}(t)$ is similar to that of the retardation times $\hat{\lambda}$ in $\hat{\alpha}(t)$ for shear deformation, except that they terminate at much smaller $\hat{\lambda}_{vol}$, corresponding to the initial parts of peak 1.[96]

Some of the early development of linear viscoelasticity was cast in terms of mechanical models of arrays of springs and dashpots, by virtue of the mathematical homology of the linear response theory for familiar models with that for the retarded response and relaxation behavior of linear viscoelasticity. A thorough discussion of these models and their relation to linear viscoelastic behavior appears in reference 23c.

Nonlinear Viscoelastic Behavior of Concentrated and Undiluted Polymeric Fluids

All polymers and their solutions will exhibit viscoelastic behavior if the strain is not recently small, where the specifics of that criterion depend on the linear viscoelastic functions, for example, $G(t)$ or $J(t)$ in shear deformation. Nonlinear behavior can take many forms, and it is possible to consider only a very limited range of those forms here. The well-studied behavior of nonlinear elastic materials provides the motivation for a constitutive equation often used to approximate the properties of nonlinear viscoelastic materials, in the incompressible limit (negligible volume change). For example, the Mooney-Rivlin theory gives a widely used relation for an incompressible nonlinear elastic solid[3a,9]

$$\tau(\lambda) = 2(\lambda^2 - \lambda^{-1})C_1 + C_2\lambda^{-1}) \qquad (77)$$

where $\tau = S_{11} - S_{22}$ is a tensile stress, $\lambda = 1 + \varepsilon_{11}$, for an elongational deformation along x_1, and the parameters C_1 and C_2 are called the Mooney-Rivlin constants; this expression is also seen in the form using the apparent stress $\tau_{APP}(\lambda) = \tau(\lambda)/\lambda$ on the basis of the original cross-sectional area. Inspection for small strain shows that $2(C_1 + C_2) = G_e$. The expression for an ideal network results with $C_2 = 0$.[3b] The Mooney-Rivlin model also gives the shear stress $\sigma = S_{12}$, and first and second normal stresses $N_1 = S_{11} - S_{22}$ and $N_2 = S_{22} - S_{33}$, respectively, for a shear deformation with shear strain γ along x_1, and gradient along x_2:[3c,9]

$$\sigma(\gamma) = G_e\gamma \qquad (78a)$$

$$N_1(\gamma) = G_e\gamma^2 \qquad (78b)$$

$$N_2(\gamma) = -2C_2\gamma^2 \qquad (78c)$$

The third normal stress difference is given by $N_3 = N_1 + N_2 = S_{11} - S_{33}$, or $N_3 = 2C_1\gamma^2$ for this model; the result, $N_1(\gamma)/\sigma(\gamma) = \gamma$, obtained for this model also will be applied for fluids in the following discussion. These results are used in the birefringence section.

Some of these features are carried over in theories of the nonlinear deformation of viscoelastic materials. Thus, a number of models constructed to account for nonlinear dependence on the strain for an incompressible fluid can be expressed in the form[17g]

$$S_{ij}(t) = -\int_0^\infty du\, Q_{ij}(t,u)\frac{\partial G(t-u)}{\partial u} - \delta_{ij}P \tag{79a}$$

$$\mathbf{Q} = \mathbf{Q}^o \cdot \mathbf{F} \tag{79b}$$

where the tensor $\mathbf{Q}^o$ represents the response under a certain deformation range discussed below for which $\mathbf{F}$ may be approximated by the unitary tensor, and the tensor $\mathbf{F}$ accounts for nonlinear departures from that response. Not all models are cast in the form of integral expressions; a number of forms are discussed in reference 22e, many of these having the form of differential equations. In Cartesian coordinates for shear deformation along x_1, shear gradient along x_2, and uniaxial elongational deformation with deformation along x_1:

$$\mathbf{Q_{sh}}^o[\gamma(t,u)] = \begin{pmatrix} 1+\gamma(t,u)^2 & \gamma(t,u) & 0 \\ \gamma(t,u) & 1 & 0 \\ 0 & 0 & 1 \end{pmatrix};$$

$$\gamma(t,u) = \int_u^t ds\, \dot{\gamma}(s) \tag{80a}$$

$$\mathbf{Q_{el}}^o[\lambda(t,u)] = \begin{pmatrix} \lambda(t,u)^2 & 0 & 0 \\ 0 & \lambda(t,u)^{-1} & 0 \\ 0 & 0 & \lambda(t,u)^{-1} \end{pmatrix};$$

$$\ln[\lambda(t,u)] = \int_u^t ds\, \dot{\gamma}(s) \tag{80b}$$

Several applications employing $\mathbf{Q}^o$ to compute the shear stress $\sigma(t) = S_{12}(t)$ and the first normal stress difference $N_1(t) = S_{11}(t) - S_{22}(t)$ are of interest; these represent the second-order approximation, with the first appearance of normal stresses in shear deformation of an incompressible fluid. The stress relaxation of steady-state shearing flow after a small step strain of γ is found to be[97,98]

$$N_1(t,\gamma) \approx \gamma\sigma(t,\gamma) = \gamma^2 G(t) \tag{81}$$

similar to the result noted above for an elastic solid. This result is discussed below in the birefringence section. The stresses following imposition of a constant shear rate $\dot{\gamma}$ are given by[98]

$$N_1(t,\dot{\gamma}) \approx 2\dot{\gamma}^2\int_0^t du\, uG(u) \tag{82a}$$

$$\sigma(t,\dot{\gamma}) \approx \dot{\gamma}\int_0^t du\, G(u) \tag{82b}$$

whereas during relaxation following steady-state flow,[98]

$$\partial N_1(t,\dot{\gamma})/\partial t \approx -2\dot{\gamma}\sigma(t,\dot{\gamma}) = -2\dot{\gamma}\int_0^t du\, G(u) \tag{83}$$

For steady shear flow, eqs 82a and 82b reduce to the important result[99]

$$N_1(\dot{\gamma})/2\sigma(\dot{\gamma})^2 = J_s \tag{84}$$

which is used in the following discussion. Finally, for a steady-state oscillation induced by a strain $\gamma(t) = \gamma_0\sin(\omega t)$, with a small strain amplitude γ_0, use of the strain tensor $\mathbf{Q}^o$ gives[98,100]

$$N_1(t) = S_{11}(t) - S_{22}(t) = \gamma_0^2\{G'(\omega) + [G''(\omega) - G''(2\omega)/2]$$
$$\times \sin(2\omega t) - [G'(\omega) - G'(2\omega)/2]\cos(2\omega t)\} \tag{85a}$$

$$N_3(t) = S_{11}(t) - S_{33}(t) = (1-\beta)[S_{11}(t) - S_{22}(t)] \tag{85b}$$

in the Cartesian coordinate system introduced above, where $\beta = -N_2/N_1$, with $N_2 = N_3 - N_1$. Theoretical and experimental estimates give $\beta \approx 0.1$ to 0.3.[101] The normal stresses oscillate at frequency 2ω for the deformation at frequency ω, reflecting symmetry requirements; the response is offset by a time-independent term involving $G'(\omega)$, so that $S_{11}(t) - S_{22}(t) > 0$ in this approximation.

These approximations fail for larger deformations that have not been recently small, and a form for the nonlinear response tensor $\mathbf{F}$ is required. A particular formulation for $\mathbf{F}$ given by theoretical considerations for entangled flexible chain polymers gives the following forms for shear and elongation, respectively:[17g]

$$\mathbf{F_{sh}}[\gamma(t,u)] = \begin{pmatrix} F_1[\gamma(t,u)] & 0 & 0 \\ 0 & F_1[\gamma(t,u)] & 0 \\ 0 & 0 & F_2[\gamma(t,u)] \end{pmatrix} \tag{86a}$$

$$\mathbf{F_{el}}[\lambda(t,u)] = \begin{pmatrix} F_3[\lambda(t,u)] & 0 & 0 \\ 0 & F_3[\lambda(t,u)] & 0 \\ 0 & 0 & F_3[\lambda(t,u)] \end{pmatrix} \tag{86b}$$

A model based on reptational diffusion in the entanglement regime gives results that may be expressed in the forms

$$F_1[\gamma] \approx [1 + (|\gamma|/\gamma'')^2]^{-1} \tag{87a}$$

$$F_2[\gamma] - F_1[\gamma] \approx (2/7\gamma^2)[1 + k(|\gamma|/\gamma'')^2]^{-3/2} \tag{87b}$$

$$F_3[\lambda] \approx \left[1 + \left(\frac{\lambda^3-1}{\lambda''^3-1}\right)^{2/3}\right]^{-1} \tag{87c}$$

where k, γ'', and λ'' are expected to be universal constants.[17g] Thus, for the shear deformation,

$$\sigma(t) = \int_0^t du\, \frac{\partial G(u)}{\partial u}\gamma(t-u)F_1[\gamma(t-u)] \tag{88a}$$

$$N_1(t) = \int_0^t du\, \frac{\partial G(u)}{\partial u}\gamma(t-u)^2 F_1[\gamma(t-u)] \tag{88b}$$

with $F_1[\gamma(t)] \approx 1$ in the limit discussed above. A principal feature of this form is the factorization of effects due to time and strain, a form often referred to as a Bernstein-Kearsley-Zapas (BKZ) form after investigators who exploited it on a phenomenological basis.[102] For a strain-defined history with a step γ_0 in the strain at time zero, the stress relaxation is found to be simply $\sigma(t,\gamma_0) = \gamma_0 F_1(\gamma_0)G(t)$ and $N_1(t,\gamma_0) = \gamma_0^2 F_1(\gamma_0)G(t)$, so that $N_1(t,\gamma_0)/\sigma(t,\gamma_0) = \gamma_0$, independent of t, in reasonable accord with data in the so-called terminal time response of interest here;[17g] this result gives the same behavior $N_1(t)/\sigma(t) \approx \gamma$ noted above for smaller γ.

Restricting attention to deformation at constant strain rate $[\gamma(t) = \dot{\gamma}t$ in shear, or $\lambda(t) = \dot{\lambda}t = 1 + \dot{\epsilon}t$ in elongation], nonlinear models of this type have been developed for a number of deformations. These include the shear stress $\sigma(t,\dot{\gamma}) = S_{12}(t,\dot{\gamma})$ and the first-normal stress difference $N_1(t,\dot{\gamma}) = S_{11}(t,\dot{\gamma}) - S_{22}(t,\dot{\gamma})$ as functions of the time and shear strain γ in shear deformations at constant rate $\dot{\gamma}$, and the tensile stress $\tau(t,\dot{\varepsilon}) = S_{11}(t,\dot{\varepsilon}) - S_{22}(t,\dot{\varepsilon})$ as a function of the time and elongational strain ε in uniaxial elongation at constant rate $\dot{\varepsilon}$:

$$\sigma(t,\dot{\gamma}) = -\int_0^t du\, \Delta\gamma(t,u) F_1[\Delta\gamma(t,u)] \frac{\partial G(u)}{\partial u} \quad (89)$$

$$N_1(t,\dot{\gamma}) = -\int_0^t du\, [\Delta\gamma(t,u)]^2 F_1[\Delta\gamma(t,u)] \frac{\partial G(u)}{\partial u} \quad (90)$$

$$\tau(t,\dot{\varepsilon}) = -\int_0^t du\, \Delta\varepsilon(t,u) F_3[\Delta\varepsilon(t,u)] \frac{\partial E(u)}{\partial u} \quad (91)$$

for incompressible fluids, where $\Delta\gamma(t,u) = (t - u)\dot{\gamma}$, and so on. Although the theoretical expressions given above for F_1 and F_3 may be used, it is convenient to adopt simpler forms more amenable to integration for some purposes. For example, the simple expression

$$F_1(\gamma) = 1 \qquad \text{for } |\gamma| \le \gamma' \quad (92a)$$

$$\qquad = \exp(-[\,|\gamma| - \gamma'\,]/\gamma'') \qquad \text{for } |\gamma| > \gamma' \quad (92b)$$

on the basis of the proposition that a linear response obtain for $|\gamma| \le \gamma'$ for any deformation rate provides a useful approximation to both the theoretical expression given above and experimental data, but facilitates analytical integration if $G(t)$ is expressed as a sum of exponential functions of time.[18,42] More complex forms involving sums of exponential functions have also been used.[103,104] This simple single-exponential form gives a maxima in $\sigma(t,\dot{\gamma})$ and $N_1(t,\dot{\gamma})$ for $\dot{\gamma}t$ equal to γ'' and $2\gamma''$, respectively, in qualitative accord with experiment.[17,22] The simple form may also be used to obtain observable shear-rate-dependent functions in steady-state flow given information on the linear viscoelastic shear relaxation modulus $G(t)$:[18,42]

$$\eta(\dot{\gamma}) = \eta H_\eta(\tilde{\beta}\hat{\tau}_c\dot{\gamma}) \quad (93)$$

$$S^{(1)}(\dot{\gamma}) = J_s S(\tilde{\beta}\hat{\tau}_c\dot{\gamma}) \quad (94)$$

$$J_s(\dot{\gamma}) = J_s P(\tilde{\beta}\hat{\tau}_c\dot{\gamma}) \quad (95)$$

where η, J_s, and $\hat{\tau}_c$ are the linear viscoelastic parameters, $\tilde{\beta}$ is a weak function of γ'/γ'', often about 0.3 to 0.5, and $H_\eta(0) = S(0) = P(0) = 1$. Here, $S^{(1)}(\dot{\gamma}) = N_1(\dot{\gamma})/2[\dot{\gamma}\eta(\dot{\gamma})]^2$, and $J_s(\dot{\gamma}) = \hat{\gamma}_R(\dot{\gamma})/\dot{\gamma}\eta(\dot{\gamma})$, with $\hat{\gamma}_R(\dot{\gamma})$ the total constrained recoverable strain following steady-state flow at shear rate $\dot{\gamma}$; as noted above, the equality $S^{(1)}(\dot{\gamma}) = J_s$ for small $\hat{\tau}_c\dot{\gamma}$ is found in a more general continuum mechanical treatment.[99] It should be remembered that the constrained recoverable strain is determined after reduction of the shear stress to zero, but that the normal stress is allowed to relax during the recovery process. Although the parameters γ' and γ'' are expected to be universal constants according to a theoretical model in which $F_{sh}(\gamma)$ and $F_{el}(\lambda)$ are universal functions,[17g] in practice they both depend weakly on the molec-

ular weight distribution and on the polymer concentration in polymer solutions.[18,42] The functions $H_\eta(\tilde{\beta}\hat{\tau}_c\dot{\gamma})$, $S(\tilde{\beta}\hat{\tau}_c\dot{\gamma})$, and $P(\tilde{\beta}\hat{\tau}_c\dot{\gamma})$ all depend markedly on the molecular weight distribution through its effect on $G(t)$. A comparison of experimental data on $\eta(\dot{\gamma})/\eta$, $J_s(\dot{\gamma})/J_s$, and $S^{(1)}(\dot{\gamma})/J_s$ versus $\hat{\tau}_c\dot{\gamma}$ for a high molecular weight polyethylene[103,104] includes calculated curves determined with $\hat{\varphi}(t)$ represented by the discrete form $\hat{\varphi}(t) = \Sigma\hat{\varphi}_i\exp(-t/\hat{\tau}_i)$ to reproduce the linear viscoelastic response. $F_1(\gamma)$ given above is shown in Figure 22.16;[42] estimation of $J_s(\dot{\gamma})$ requires an iterative calculation. Additional examples for a range of polymers are given in reference 42. As demonstrated in this example, the function $H_\eta(\tilde{\beta}\hat{\tau}_c\dot{\gamma})$ is essentially unity for $\tilde{\beta}\hat{\tau}_c\dot{\gamma} < 1$, and deceases for larger $\tilde{\beta}\hat{\tau}_c\dot{\gamma}$, perhaps reaching a constant value again for $\hat{\tau}_c\dot{\gamma}$ too large for the behavior to be represented by this simple model. Similarly, for a sample with a broad distribution of molecular weight, both $S(\tilde{\beta}\hat{\tau}_c\dot{\gamma})$ and $P(\tilde{\beta}\hat{\tau}_c\dot{\gamma})$ are predicted to decrease with increasing $\dot{\gamma}$ for polymers with $P(\tilde{\beta}\hat{\tau}_c\dot{\gamma}) \ge S(\tilde{\beta}\hat{\tau}_c\dot{\gamma})$, in accordance with experiment. In contrast, both $S(\tilde{\beta}\hat{\tau}_c\dot{\gamma})$ and $P(\tilde{\beta}\hat{\tau}_c\dot{\gamma})$ are predicted to increase with increasing $\dot{\gamma}$ for polymers with a narrow distribution of relaxation times, as would be expected with a narrow distribution of molecular weight. An example of data on $\eta(\dot{\gamma})/\eta$ and $S^{(1)}(\dot{\gamma})/J_s$ versus $\hat{\tau}_c\dot{\gamma}$, along with the behavior of $\eta'(\omega)/\eta$ and $J'(\omega)/J_s$ versus $\hat{\tau}_c\omega$ for solutions of an anionically synthesized polystyrene, are shown in Figure 22.17,[105] showing the increase in $S^{(1)}(\dot{\gamma})$ with increasing $\hat{\tau}_c\dot{\gamma}$, in contrast to the behavior for a sample with a broad distribution of molecular weight; examples for $P(\tilde{\beta}\hat{\tau}_c\dot{\gamma})$ show qualitatively similar behavior.[42] Note that the general form of these relations is the same for polymers and their concentrated solutions, although the value of $\hat{\tau}_c$ and to a lesser extent $\tilde{\beta}$ will depend on the polymer concentration. In some cases, for polymer solutions of polymer with a narrow molecular weight distribution, $\hat{\tau}_c$ may be small enough that $H_\eta(\tilde{\beta}\hat{\tau}_c\dot{\gamma})$, $P(\tilde{\beta}\hat{\tau}_c\dot{\gamma})$, and $S(\tilde{\beta}\hat{\tau}_c\dot{\gamma})$ are all essentially unity, even though the sample exhibits measurable $N_1(\dot{\gamma}) \propto \dot{\gamma}^2$, with $S^{(1)}(\dot{\gamma}) \approx J_s(\dot{\gamma})$, both independent of $\dot{\gamma}$ over a substantial range of $\dot{\gamma} < 1/\hat{\tau}_c$. Such systems are sometimes called Boger fluids.[22f]

With $\hat{\varphi}(t)$ expressed in terms of a distribution of relaxation times, a close approximation to $\eta(\dot{\gamma})$ with the preceding is given by[18,42]

$$\eta(\dot{\gamma}) = G_0 \sum_1^N \hat{\varphi}_i\hat{\tau}_i \frac{1}{[1 + (\tilde{\beta}\dot{\gamma}\hat{\tau}_i)^\varepsilon]^{2/\varepsilon}} \quad (96)$$

with $\varepsilon \approx 6/5$. In comparison, for a linear viscoelastic response for a fluid, as seen above

$$\eta'(\omega) = G_0 \sum_1^N \hat{\varphi}_i\hat{\tau}_i \frac{1}{1 + (\omega\hat{\tau}_i)^2} \quad (97)$$

In effect, the factors in the denominators act as a filter in each case, serving to delete the contributions of successively smaller relaxation times $\hat{\tau}_i$ as ω or $\dot{\gamma}$ increase, and they do this in more or less equivalent ways, leading to the Cox-Merz

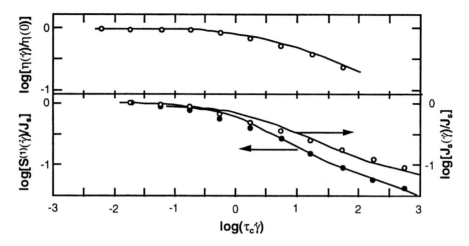

Figure 22.16 Bilogarithmic plots of the several reduced functions for a high molecular weight polyethylene with a broad molecular weight distribution: the viscosity $\eta(\dot{\gamma})/\eta(0)$, the total recoverable compliance $J_s(\dot{\gamma})/J_s$, and the first-normal stress difference $S^{(1)}(\dot{\gamma})/J_s$ as functions of $\hat{\tau}_c\dot{\gamma} = J_s\eta(0)\dot{\gamma}$. The curves are calculated as described in the text. (Adapted from an example in reference 42 based on data in reference 103.)

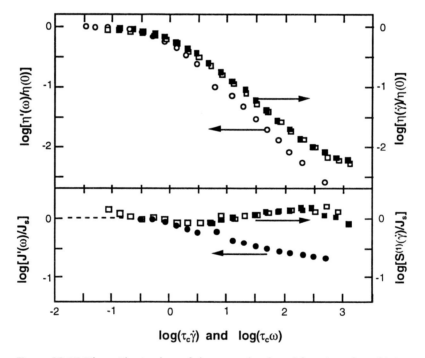

Figure 22.17 Bilogarithmic plots of the several reduced functions for a high molecular weight polystyrene with a narrow molecular weight distribution: the viscosity $\eta(\dot{\gamma})/\eta(0)$, the dynamic viscosity $\eta'(\omega)/\eta(0)$, the total recoverable compliance $J_s(\dot{\gamma})/J_s$, the dynamic storage compliance $J'(\omega)/J_s$, and the first-normal stress difference $S^{(1)}(\dot{\gamma})/J_s$ as functions of $\hat{\tau}_c\dot{\gamma} = J_s\eta(0)\dot{\gamma}$. (Adapted from an example in reference 105.)

approximation $\eta(\dot{\gamma}) \approx \eta'(\omega = \dot{\gamma})$ [or perhaps better $\eta(\dot{\gamma}) \approx |\eta^*(\omega = \dot{\gamma})|$] proposed long ago.[106] This approximation is likely to improve as the distribution of relaxation time broadens. Similarly, $J_s(\dot{\gamma}) \approx R(t = \dot{\gamma}^{-1})$ for samples with a reasonably broad distribution of relaxation times, and $S^{(1)}(\dot{\gamma})/J_s(\dot{\gamma}) \leq 1$, tending toward a ratio of about 0.5 with increasing $\hat{\tau}_c\dot{\gamma}$, similar to the m value noted above for a linear elastic solid.[18,42]

Nonlinear behavior may also be exhibited in transient behavior. For example, maxima are observed in the shear stress $\sigma(t; \dot{\gamma}) = \eta(t; \dot{\gamma})\dot{\gamma}$ and the first-normal stress difference during stress growth $N_1(t; \dot{\gamma})$ at constant shear rate $\dot{\gamma}$,[22g,103,105,107–111] and the recoverable compliance $R(t; \sigma)$ observed after cessation of nonlinear steady flow depends on the steady state stress $\sigma(\dot{\gamma}) = \dot{\gamma}\eta(\dot{\gamma})$.[42,104] The maxima in the stress growth tend to occur at a given strain $\gamma_c = \dot{\gamma}t_c$, independent of $\dot{\gamma}$, with the γ_c for the normal stress about twice that for the shear stress difference, for example, $\gamma_c \approx 2\gamma''$ and $\gamma_c \approx \gamma''$ for normal and shear stresses, respectively.[109] The function $R(t; \sigma)$ is found to be equal to the linear response $R(t)$ for small t, but tends to a steady state limit $J_s(\dot{\gamma})$ that depends on the steady state shear rate $\dot{\gamma}$.[42,104] These features may be represented reasonably well by the simple function for $F_1(\gamma)$ given above, or by other similar expressions.[103] Some other stress histories are less satisfactorily represented by this simple expression, especially those with multiple reversing steps in the stress.[22g]

Nonlinear behavior with an entirely different origin occurs in some systems exhibiting a yield stress behavior.[24a,26a] In these systems, the material behaves as a linear viscoelastic solid, provided the strain does not exceed some critical value γ^0, but exhibits the nonrecoverable deformation of a fluid if the strain exceeds γ^0. It is presumed that the intermolecular (or interparticle) interactions responsible for the solidlike behavior rupture under deformation. The most easily interpreted but most time consuming means to determine γ^0 is to investigate successive creep and recovery cycles with increasing applied stress σ_0 until a nonrecoverable strain is observed. In some cases, the yield will occur with a clear enhancement of the compliance for $\gamma(t) = \gamma^0$ during creep in comparison with the behavior for the linear response for smaller strain. An example of this is given if Figure 22.18 for a thermally reversible gel of a rodlike polymer.[112] In this case, the deformation of the gel for $\gamma(t) < \gamma^0$ follows Andrade creep. Moreover, after deformation with $\gamma(t) > \gamma^0$, the recoverable strain is given by the formula for a linear viscoelastic deformation. The latter behavior may not be unusual, providing an example of a quasilinear behavior with an overall nonlinear response. In many cases, γ^0 is essentially independent of the applied stress, such that the stress σ^0 at yield may vary with the rate of deformation, even though $\hat{\gamma}^0$ is essentially invariant. For example, under a shear deformation at constant shear rate $\dot{\gamma}$, $\sigma(t) = \dot{\gamma}\int_0^t du\, G(u)$ provided $\dot{\gamma}t < \gamma^0$, but $\sigma(t)$ will deviate markedly from this response for larger strain. Clearly, in a steady-state deformation, the strain will always exceed γ^0 if t is large enough, making it difficult or impossible to obtain a reliable estimate for γ^0 from such experiments. In some cases, the yielding gives an approximately constant stress in steady-state flow over a range of shear rate, providing an approximate estimate for σ^0.[24b,26b] Moreover, the apparent viscosity determined in such a case will differ markedly from $\eta'(\omega = \dot{\gamma})$ obtained with a strain amplitude less than γ^0. A plot of the steady-state stress $[\sigma_{ss}(\dot{\gamma})]^{1/2}$ versus $\dot{\gamma}^{1/2}$ is sometimes used to estimate a yield stress as the intercept of a straight line fitted to the data in a so-called Casson plot.[113] However, evaluation of viscoelastic data at still smaller stress usually will show that such an estimated yield stress is too large, and may even be entirely fallacious, with the material behaving as a viscoelastic liquid to arbitrarily small stress.

A nonlinear behavior with a different origin occurs in measurements of volume properties. The general principles of linear viscoelasticity may be applied to deduce exact relations among the relative rates of creep and relaxation in shear, elongation, and volume. Thus,[114]

$$\frac{\partial \ln J(t)/\partial \ln t}{\partial \ln D(t)/\partial \ln t} = 1 + \frac{B(t)}{3J(t)}\left(1 - \frac{\partial \ln B(t)/\partial \ln t}{\partial \ln J(t)/\partial \ln t}\right) \quad (98)$$

with a similar expression for the moduli with J, D, and B replaced by G, E, and K, respectively. Consequently, if $0 < [\partial \ln B(t)\partial \ln t]/[\partial \ln J(t)/\partial \ln t] < 1$, the shear compliance changes faster than the elongational compliance. Opposite behavior has been cited in the literature, suggesting a nonlinear behavior, even though the deformation would appear to be small, even recently small.[75,95] The behavior, which manifests itself in the so-called aging behavior of glasses, is attributed to a nonlinear behavior reflecting a change in the underlying retardation or relaxation times with the volume of the sample.[11i,75,95,115] Schemes to represent this behavior in approximate ways have been considered by a number of authors.[11i,25,74,75,95] In one approach, it is assumed that the expressions for the linear viscoelastic constitutive equation may be adopted if the shift factor a_V is allowed to vary as the volume of the sample changes, and a time-averaged shift factor is adopted in describing the behavior. Thus, the function $B(t_2 - t_1)$ appearing in expressions for $\Delta V/V$ in terms of the stress history would become[75,115]

$$B[(t_2 - t_1)\langle a_V^{-1}(t_2,t_1)\rangle] = B_e - (B_e - B_0)\hat{\beta}[(t_2 - t_1)$$
$$\langle a_V^{-1}(t_2,t_1)\rangle] \quad (99a)$$

$$\langle a_V^{-1}(t_2,t_1)\rangle = \frac{1}{t_2 - t_1}\int_{t_1}^{t_2} du\, a_V^{-1}(u) \quad (99b)$$

The linear behavior is recovered if a_V is a constant. Typically, as stated above, $\hat{\beta}(t)$ would exhibit Andrade creep behavior, with $\hat{\beta}(t)$ linear in $t^{1/3}$.[75] This nonlinear form is able to represent many, but not all, aspects of the behavior.[11i,75]

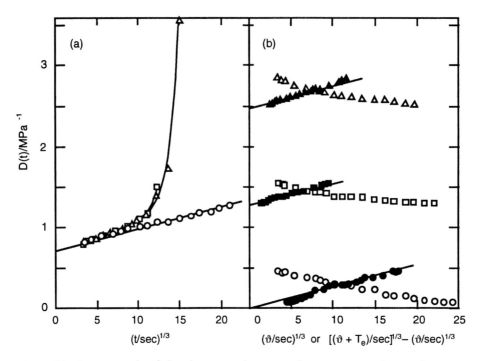

Figure 22.18 An example of the elongational creep and recovery compliances for a material exhibiting a yield strain behavior. The unfilled and filled symbols in (b) show the strain versus $\vartheta^{1/3}$ and the recoverable strain versus the function $(\vartheta + T_e)^{1/3} - \vartheta^{1/3}$ computed for a linear viscoelastic response; the data exhibit quasilinear response in recovery following creep beyond the yield strain. (Adapted from an example in reference 112.)

Strain-Induced Birefringence for Concentrated and Undiluted Polymeric Fluids

Rheo-optical measurements have long been used in conjunction with rheological measurements. Examples include strain-induced birefringence, optical dichroism, and light scattering.[16,17h,24c,27,98,100,116–121] Only birefringence measurements are discussed in this chapter. The increasing availability of commercial rheo-optical equipment affords enhanced opportunities for the use of birefringence techniques. For an isotropic material, the principal components (proper vectors) n_1, n_2, and n_3 of the refractive index tensor are equal, but this symmetry is usually lost under deformation. Thus, in the shear and elongational deformations discussed above, with deformation in the 1,3-plane (in Cartesian coordinates), the birefringence $\Delta n = n_1 - n_2$ measured by light propagating along x_3 will generally not be zero, and the angle $\chi_n \leq \pi/4$ between n_1 and x_1 will depend on the deformation. The angle χ_n may be visualized by the orientation of the extinction cross of isocline viewed between crossed polars under appropriate optical arrangements. Several effects may contribute to the strain-induced optical anisotropy, including (1) orientation of optically anisotropic chain elements and (2) an internal field effect called form birefringence. Form birefringence arises from spatial varia-

tions of the refractive index, as with the anisotropy in a deformed single chain in a dilute solution, or phase-separated regimes in a blend or a semicrystalline polymer.[17i,27a] Here, consideration is limited to homogeneous amorphous polymers or their concentrated solutions, for which the form birefringence may be neglected. In the simplest case, of interest here, the optical anisotropy of the chain elements is assumed to have ellipsoidal symmetry, with a major axis along the chain axis, so that the optical anisotropy is directly associated with anisotropy of the chain conformation.[16e,27b,121] Clearly, this is a major approximation that may not always be valid, and deviations from the behavior anticipated with the simplification may provide insight on the material.[94,122,123]

Optical arrangements for the measurement of Δn and χ_n are amply discussed in the literature[16f,27c,118,120,121] and will not be elaborated on here. Discussion here is limited to planar samples, with the sample in the 1,3-plane, and deformation along x_1; in shear, the gradient is along x_2, all in Cartesian coordinates. Measurements may be made with a light beam propagating in the 2,3-plane: a birefringence $\Delta n_{1,2}$ and the extinction angle χ_n are determined with a beam propagating along x_3 (i.e., in the sample plane); a birefringence $\Delta n_{1,3}$, but not χ_n, is determined with a light beam propagating along x_2 (i.e., orthogonal to the sample plane); or components of $\Delta n_{1,2}$, $\Delta n_{1,3}$, and χ_n may be determined with a light beam oriented between x_2 and x_3.[124,125] In another arrangement, measurements with a light beam along x_1, as along the axis of flow in a tube, give a birefringence $\Delta n_{2,3}$. In some arrangements an average birefringence is obtained, and additional analysis is required (as with a beam along x_3 in flow in a rectangular channel or a beam along x_3 in rotating parallel plates), but that complication is not considered here.[16f,120] It is sometimes convenient to define a ratio $\hat{\psi} = -[n_{22} - n_{33}]/[n_{11} - n_{22}]$ so that for light propagating along x_2, $\Delta n_{1,3} \approx (1 - \hat{\psi})\Delta n_{1,2}$, with $\Delta n_{1,2} = n_{11} - n_{22}$ and $\Delta n_{1,3} = n_{11} - n_{33}$. For shearing flow with velocity along x_1, gradient along x_2, and light propagating along x_3, the optical extinction angle $\chi_n(t)$ and birefringence $\Delta n_{1,2}(t)$ are given by[16f,27d,98,100,120]

$$\Delta n_{1,2}(t) = [n_{11}(t) - n_{22}(t)]\cos[2\chi_n(t)]$$
$$+ 2n_{12}(t)\sin[2\chi_n(t)] \qquad (100)$$

in terms of the components of the refraction tensor, with $\pi/4 > \chi_n(t) > 0$. Consequently,

$$\Delta n_{1,2}(t)\sin[2\chi_n(t)] = 2n_{12}(t) \qquad (101a)$$
$$\cot[2\chi_n(t)] = [n_{11}(t) - n_{22}(t)]/2n_{12}(t) \qquad (101b)$$

for this case. For the same deformation, with light propagating along x_2, $\Delta n_{1,3}(t) = n_{11}(t) - n_{33}(t)$, with no dependence on $\chi_n(t)$. The same result holds for a uniaxial extension along x_1 with light propagating along x_2.

It has been shown by a continuum mechanical argument that for recently small deformations, the components n_{ij} of the refractive index tensor in Cartesian coordinates in the linear viscoelastic regime may be expressed by relations similar to the expressions given above, with mechanical compliances and moduli replaced by similarly defined optical compliances and moduli, respectively. These are denoted by the subscript n:[100,119,126]

$$n_{ij}(t) = \int_{-\infty}^{t} ds \left\{ J_n(t-s) \left[\frac{\partial S_{ij}(s)}{\partial s} - \frac{1}{3}\delta_{ij}\frac{\partial S_{\alpha\alpha}(s)}{\partial s} \right] \right.$$
$$\left. + \delta_{ij} \left[n_0 + (2/9)B_n(t-s)\frac{\partial S_{\alpha\alpha}(s)}{\partial s} \right] \right\} \qquad (102)$$

$$n_{ij}(t) = \int_{-\infty}^{t} ds \left\{ 2G_n(t-s) \left[\frac{\partial \varepsilon_{ij}(s)}{\partial s} - \frac{1}{3}\Delta_{ij}\frac{\partial \varepsilon_{\alpha\alpha}(s)}{\partial s} \right] \right.$$
$$\left. + \delta_{ij} \left[n_0 + K_n(t-s)\frac{\partial \varepsilon_{\alpha\alpha}(s)}{\partial s} \right] \right\} \qquad (103)$$

for stress- and strain-defined histories, respectively, where n_0 is the refractive index of the undistorted isotropic material, and $J_n(t)$ is defined by

$$J_n(t) = G_n(0)J(t) + \int_0^t du\, J(t-u)\frac{\partial G_n(u)}{\partial u} \qquad (104)$$

with a similar expression for $B_n(t)$ in terms of $B(t)$ and $K_n(t)$. The second form is obtained from the first using the convolution integrals for $J(t)$ in terms of $G(t)$, and $B(t)$ in terms

of $K(t)$ given above. That integral may be transformed by differentiation with respect to t to facilitate comparison with the expression for $J_n(t)$:

$$1 = G(0)J(t) + \int_0^t du\, J(t-u)\frac{\partial G(u)}{\partial u} \qquad (105)$$

Comparison of these two relations shows that if $G_n(t)/G(t)$ is independent of t, $J_n(t)$ is a constant, with consequences discussed below. Through an integration by parts, the expressions for the deviatoric components of $n_{ij}(t)$ may be written in the forms

$$\{n_{12}(t)\}_{\text{Deviatoric}} = G_n(0)\varepsilon_{12}(t)$$
$$+ \int_0^\infty du\, \varepsilon_{12}(t-u)\frac{\partial G_n(u)}{\partial u} \qquad (106a)$$

$$\{n_{11}(t) - n_{22}(t)\}_{\text{Deviatoric}} = G_n(0)[\varepsilon_{11}(t) - \varepsilon_{22}(t)] + \int_0^\infty du\, [\varepsilon_{11}(t-u)$$
$$- \varepsilon_{22}(t-u)]\frac{\partial G_n(u)}{\partial u} \qquad (106b)$$

for a strain-defined history, and

$$\{n_{12}(t)\}_{\text{Deviatoric}} = J_n(0)S_{12}(t)$$
$$+ \int_0^\infty du\, S_{12}(t-u)\frac{\partial J_n(u)}{\partial u} \qquad (107a)$$

$$\{n_{11}(t) - n_{22}(t)\}_{\text{Deviatoric}} = J_n(0)[S_{11}(t) - S_{22}(t)]$$
$$+ \int_0^\infty du\, [S_{11}(t-u)$$
$$- S_{22}(t-u)]\frac{\partial J_n(u)}{\partial u} \qquad (107b)$$

for a stress-defined history. The normal stresses are zero in a shear deformation in this linear approximation, but modifications to this for nonlinear behavior are included below (e.g., see eq 119 ff). With these results, it can be seen that validity of the stress-optic approximation $G_n(t)/G(t) = C_{\text{optic}}$ for all t, which requires that $\partial J_n(t)/\partial t = 0$ for all t, so that the retardation behavior embodied in the integrals involving $\partial J_n(t)/\partial t$ in eq 107 vanishes, with consequences elaborated below.

In general, the continuum mechanical model cannot provide the relation between the optical functions and their mechanical analogs; for example, it cannot specify the nature of $G_n(t)/G(t)$, and other values. However, for the special case of a linear elastic solid this relation is not necessary to obtain a useful result because $\partial G_n(t)/\partial t = \partial J_n(t)/\partial t = 0$ in this case. Consequently, $n_{12}(t) = G_n(0)\varepsilon_{12}(t)$ for the strain-defined history and $n_{12}(t) = J_n(0)S_{12}(t) = G_n(0)J(t)S_{12}(t) = G_n(0)\varepsilon_{12}(t)$ for the stress-defined history, showing the representations to be equivalent for small strain of an elastic solid, as expected. Similar expressions may be written for $n_{11}(t) - n_{22}(t)$. For the more general linear viscoelastic material, if $G_n(t)/G(t) = C_{\text{optic}}$ for all t, as presumed in the stress-optic approximation, then the deviatoric components may be readily evaluated in the limit of an incompressible material for a sample relaxed for $t < 0$. Thus, for an elongational deformation ($\chi_n = 0$)

of an incompressible linear viscoelastic material under a strain-defined history resulting in a transient stress $\tau(t) = S_{11}(t) - S_{22}(t)$ and transient strain $\varepsilon(t) = \varepsilon_{11}(t)$ for a uniaxial deformation along x_1, $\Delta n(t) = n_{11}(t) - n_{22}(t)$ for a light beam propagating along x_3 is given by

$$\Delta n(t)/C_{\text{optic}} = 3 \int_0^t du\, G(t-u) \frac{\partial \varepsilon(u)}{\partial u} = \tau(t) \qquad (108)$$

For the special case of a linear elastic solid, $G(t) = G_e$ and $\Delta n(t)/C_{\text{optic}} = 3G_e\varepsilon(t)$. Similarly, for a recently small shear deformation to obtain a linear viscoelastic response, $\Delta n(t) = 2n_{12}(t)\sin[2\chi_n(t)]$, with $\sin[2\chi_n(t)] \approx 1$,

$$\Delta n(t)/2C_{\text{optic}} = \int_0^t du\, G(t-u) \frac{\partial \gamma(u)}{\partial u} = \sigma(t) \qquad (109)$$

for a strain-defined history. Thus, for stress relaxation following a small step-strain γ_0, or a slow shear flow at shear rate $\dot{\gamma}_{ss}$, such that $\hat{\tau}_c\dot{\gamma}_{ss} \ll 1$,

$$\Delta n(t)/2C_{\text{optic}} = \gamma_0 G(t) \qquad (110)$$

$$\Delta n(t)/2C_{\text{optic}} = \dot{\gamma}_{ss} \int_0^t du\, G(u) \qquad (111)$$

respectively. In the latter case, the steady-state flow birefringence Δn is given by

$$\Delta n/2C_{\text{optic}} = \dot{\gamma}_{ss} \int_0^\infty du\, G(u) = \eta\dot{\gamma}_{ss} = (1/J_s)\hat{\gamma}_R(\dot{\gamma}_{ss}) \qquad (112)$$

as expected, where $\hat{\gamma}_R(\dot{\gamma}_{ss})$ is the total strain recoil on cessation of steady flow at shear rate $\dot{\gamma}_{ss}$. Experiments over a reasonable time span appear to be in accordance with these predictions.[16,27,118,120] Note that the total recoverable strain $\hat{\gamma}_R(T_e, \gamma_0)$ that would be measured if the stress were dropped to zero at time T_e during stress relaxation of a fluid following a jump γ_0 in the strain is given by

$$\hat{\gamma}_R(T_e, \gamma_0) = \gamma_0[1 - \eta^{-1} \int_0^{T_e} ds\, G(s)] \qquad (113)$$

Consequently, $(1/J_s)\hat{\gamma}_R(t,\gamma_0)$ is not proportional to $\Delta n(t)/2C_{\text{optic}} \approx G(t)$ given above using the stress-optic approximation, unless $G(t)$ is given by an exponential function, in which case $R(t)$ is a constant, as would be consistent with the stress-optic approximation. Because $J_n(t) = C_{\text{optic}}$ for this model, the birefringence may also be obtained with a stress-defined deformation under the same conditions. For example, for a uniaxial deformation along x_1, $\Delta n(t) = n_{11}(t) - n_{22}(t)$ is given by

$$\Delta n(t)/2C_{\text{optic}} = [S_{11}(t) - S_{22}(t)]/3 \qquad (114)$$

Similarly, for a recently small shear deformation to obtain a linear viscoelastic response, $\Delta n(t) = 2n_{12}(t)\sin[2\chi_n(t)]$, with $\sin[2\chi_n(t)] \approx 1$,

$$\Delta n(t)/2C_{\text{optic}} = S_{12}(t) \qquad (115)$$

According to this result, in a creep experiment $\Delta n(t)/2C_{\text{optic}}$ must be equal to its value $(1/J_s)\hat{\gamma}_R(\dot{\gamma}_{ss})$ in steady flow for all t, because that must be its value for large t in steady flow.

Similarly, $\Delta n(t)$ must be zero for all t during recovery on cessation of creep. No assessment of these peculiar results appears to be available.

Application of the preceding to steady-state response to a small-amplitude oscillatory shear deformation may also be expressed in forms similar to those given above for the mechanical response:[100,126]

$$n_{12}(t) = \gamma_0\{G'_n(\omega)\sin(\omega t) + G''_n(\omega)\cos(\omega t)\} \qquad (116a)$$

$$n_{12}(t) = \sigma_0\{J'_n(\omega)\sin(\omega t) - J''_n(\omega)\cos(\omega t)\} \qquad (116b)$$

for the strain- and stress-defined experiments, respectively, where $G'_n(\omega)$ is defined in terms of $G_n(t)$ in a way analogous to the relation between $G'(\omega)$ and $G(t)$. Use of the approximation $G_n(t)/G(t) = C_{\text{optic}}$ for all t gives $G'_n(\omega) = CG'(\omega)$ and $G''_n(\omega) = CG''(\omega)$, so that the birefringence under a sinusoidal strain history is proportional to the stress. This appears to agree with experiment over a reasonable range in ω.[65,101,127–129] Nevertheless, the stress-optic approximation is known to be inadequate in some situations, especially for the response at a temperature close to the glass temperature. For example, in the latter case, the birefringence has been observed to change sign, inconsistent with the approximation $G_n(t)/G(t) = C_{\text{optic}}$ for all t.[94,130] This behavior may be attributed to the substantially different mechanisms of deformation in the Andrade behavior sampled at higher $\omega\hat{\tau}_c$, and the Rouse or entanglement behavior that may dominate at lower $\omega\hat{\tau}_c$.[130,131]

Because $J^*_n(\omega) = G^*_n(\omega)J^*(\omega)$ according to the relations given above,[100,132] it follows that

$$J'_n(\omega) = G'_n(\omega)J'(\omega) + G''_n(\omega)J''(\omega) \qquad (117a)$$

$$J''_n(\omega) = G'_n(\omega)J''(\omega) - G''_n(\omega)J'(\omega) \qquad (117b)$$

Consequently, if $G_n(t)/G(t) = C$ for all t, then use of the expressions relating $G'(\omega)$ and $G''(\omega)$ to $J'(\omega)$ and $J''(\omega)$ given above yields $J'_n(\omega) = C_{\text{optic}}$ and $J''_n(\omega)$ for all ω, with the result that the dynamic birefringence is in phase with the applied stress in a steady-state oscillatory deformation. This appears to be in accord with experiment over a wide range in ω.[65,101]

An important application of strain-induced birefringence measurements is in the estimation of the first-normal stress difference $N_1(t)$ in a transient response in a nonlinear response. Neither $N_1(t; \gamma_0)$ in relaxation following a step-strain γ_0 nor $N_1(t; \dot{\gamma})$ in shear deformation in nonlinear flows at a constant shear rate are trivial to measure mechanically. For example, although $N_1(t; \dot{\gamma})$ may be determined from the force required to prevent separation of a cone and plate during a shear deformation, the slight movement that is unavoidable in the feedback mechanism to produce a measure of the required force may complicate the interpretation of the transient response.[133] The error should be less critical in a steady flow.

Use of the approximation $G_n(0) = C_{\text{optic}}G_e = C_{\text{optic}}/J_e$ for a linear elastic solid, along with the generalized constitutive relation incorporating the strain tensor $\mathbf{Q}$ to encompass nonlinear behavior, provides predictions for the birefringence in a nonlinear deformation. Thus, with some expressions for $\mathbf{Q}$ for an incompressible material,[16g,117]

$$\Delta n/C_{\text{optic}} = \tau(\lambda) = (1/D_e)[D_e/D_e(\lambda)]\hat{\varepsilon}_R(\lambda) \qquad (118)$$

for a uniaxial elongation with stretch $\lambda = 1 + \varepsilon_{11}$ along x_1 ($\chi = 0$) under a tensile stress $\tau = S_{11} - S_{22}$; this reduces to the linear approximation given above for small λ. Here $\hat{\varepsilon}_R(\lambda) = D_e(\lambda)\tau(\lambda)$ is the total recoverable strain on removal of the stress, $D_e = 1/3G_e$ is the linear equilibrium tensile compliance, and $D_e(\lambda)$ is the corresponding nonlinear compliance at elongation λ. Similarly, under a shear strain γ leading to a shear stress $\sigma(\gamma)$,

$$\Delta n \sin[2\chi_n(\gamma)]/C_{\text{optic}} = 2\sigma(\gamma) = 2G_e[J_e/J_e(\gamma)]\hat{\gamma}_R(\gamma) \qquad (119a)$$

$$\Delta n \cos[2\chi_n(\gamma)]/C_{\text{optic}} = N_1(\gamma) = mG_e[J_e/J_e(\gamma)]\hat{\gamma}_R^2(\gamma) \qquad (119b)$$

$$\cot[2\chi_n(\gamma)] = N_1(\gamma)/2\sigma(\gamma) = m\hat{\gamma}_R(\gamma)/2 \qquad (119c)$$

for a shear strain $\gamma = 2\varepsilon_{12}$, with light beam propagating along x_3, deformation along x_1, and gradient along x_2, giving rise to a shear stress $\sigma = S_{12}$ and a first-normal stress $N_1 = S_{11} - S_{22}$. Here $\hat{\gamma}_R(\gamma) = J_e(\gamma)\sigma(\gamma)$ is the strain that would be recovered on removal of the stress, and m, $D_e(\gamma)/D_e$, and $J_e(\gamma)/J_e$ depend on the constitutive equation for the elastic solid. Because $J_e(\gamma)/J_e = D_e(\gamma)/D_e = 1$ and $\sin(2\chi_n) \approx 1$ for the linear elastic response for a small strain, these expressions reduce to the relations for a linear elastic solid for small strain. In these expressions, both $J_e(\gamma)/J_e$ and m are unity for the Mooney-Rivlin network model or for an ideal rubber network for which $C_2 = 0$,[3b] showing that the linear elastic expressions may be applied with equivalent results for either a stress- or strain-defined shear deformation for this model. In contrast, because $D_e(\lambda)/D_e = \tau(\lambda)/(\lambda - 1)$ depends on λ for the Mooney-Rivlin model, the linear elastic expression in terms of $\hat{\varepsilon}_R$ would not apply to the nonlinear response using the approximation $D_e(\lambda)/D_e = 1$ obtaining in the linear response. These relations have been applied to interpret the birefringence under nonlinear deformations.[16g]

Although normal stresses are absent in shear deformation with the linear viscoelastic fluid, as noted above, a first-normal stress N_1 does appear with use of the strain tensor $\mathbf{Q_o}$, with several predicted results given following eq 79. Assumption of a similar form may be adopted to introduce normal stresses in the refractive index tensor. For example, during stress relaxation following a step strain γ_0,

$$\Delta n \sin[2\chi_n(t, \gamma_0)]/C_{\text{optic}} = 2\sigma(t; \gamma_0) = 2G(t)\gamma_0 \qquad (120a)$$

$$\Delta n \cos[2\chi_n(t, \gamma_0)]/C_{\text{optic}} = N_1(t; \gamma_0) = G(t)\gamma_0^2 \qquad (120b)$$

$$\cot[2\chi_n(t; \gamma_0)] = N_1(t, \gamma_0)/2\sigma(t; \gamma_0) = \gamma_0/2 \qquad (120c)$$

for the 1,2-plane; the similarity with the results given in the preceding paragraph for the Mooney-Rivlin model for a linear elastic solid is evident. Similarly, in the transient response for deformation at constant shear rate $\dot{\gamma}_{ss}$,

$$\Delta n \sin[2\chi_n(t, \dot{\gamma}_{ss})]/C_{\text{optic}} = 2\sigma(t, \dot{\gamma}_{ss}) = 2\dot{\gamma}_{ss}\int_0^t du\, G(u) \qquad (121a)$$

$$\Delta n \cos[2\chi_n(t, \dot{\gamma}_{ss})]/C_{\text{optic}} = N_1(t, \dot{\gamma}_{ss}) = \dot{\gamma}_{ss}^2\int_0^t du\, uG(u) \qquad (121b)$$

$$\cot[2\chi_n(t, \dot{\gamma}_{ss})] = N_1(t, \dot{\gamma}_{ss})/2\sigma(t; \dot{\gamma}_{ss})$$

$$= \dot{\gamma}_{ss}\int_0^t du\, uG(u)/2\int_0^t du\, G(u) \qquad (121c)$$

In steady-state flow at shear rate $\dot{\gamma}_{ss}$ and shear stress $\sigma(\dot{\gamma}_{ss}) = \dot{\gamma}_{ss}\eta(\dot{\gamma}_{ss})$:

$$\Delta n \sin[2\chi_n(\dot{\gamma}_{ss})]/C_{\text{optic}} = 2\sigma(\dot{\gamma}_{ss}) = (2/J_s)\hat{\gamma}_R(\dot{\gamma}_{ss}) \qquad (122a)$$

$$\Delta n \cos[2\chi_n(\dot{\gamma}_{ss})]/C_{\text{optic}} = N_1(\dot{\gamma}_{ss}) = S^{(1)}(\dot{\gamma})2\sigma(\dot{\gamma}_{ss})^2$$

$$= (2/J_s)\hat{\gamma}_R(\dot{\gamma}_{ss})^2 \qquad (122b)$$

$$\cot[2\chi_n(\dot{\gamma}_{ss})] = S^{(1)}(\dot{\gamma}_{ss})\sigma(\dot{\gamma}_{ss}) = \hat{\gamma}_R(\dot{\gamma}_{ss}) \qquad (122c)$$

for the 1,2-plane, where in this regime, $S^{(1)}(\dot{\gamma}_{ss}) = N_1(\dot{\gamma}_{ss})/2\sigma(\dot{\gamma}_{ss})^2 \approx J_s(\dot{\gamma}_{ss}) \approx J_s$, and the total recoverable strain $\hat{\gamma}_R(\dot{\gamma}_{ss})$ following cessation of the slow steady shear flow is given by $\hat{\gamma}_R(\dot{\gamma}_{ss}) \approx J_s\sigma(\dot{\gamma}_{ss})$. On the basis of the preceding discussion, the birefringence in steady-state slow flow at shear rate $\dot{\gamma}$, for which $\sin[2\chi_n(\dot{\gamma}_{ss})] \approx 1$, is given by $\Delta n_{1,3}(\dot{\gamma}_{ss})/C_{\text{optic}} = (1 - \hat{\psi})N_1(\dot{\gamma}_{ss}) \approx (2/J_s)(1 - \hat{\beta})(\hat{\tau}_c\dot{\gamma}_{ss})^2$ for shear in the 1,3-flow plane, or $\cot[2\chi_n(\dot{\gamma}_{ss})] \approx \hat{\tau}_c\dot{\gamma}_{ss} = \hat{\gamma}_R(\dot{\gamma}_{ss})$ in the 1,2-flow plane. These expressions, which appear to be in accord with experiment in this response range,[16,27e,120] make use of the relation $J_s = N_1/2\sigma^2$, obtained under rather general considerations for in a steady slow flow,[99] and further emphasize the direct connection between the birefringence behavior and the total recoverable strain $\hat{\gamma}_R(\dot{\gamma})$.

Dynamic birefringence measurements of $\Delta n_{1,3}(t)$ in the 1,3-plane for a range of γ_0, for which $\Delta n_{1,3}(t) = C_{\text{optic}}(1 - \hat{\psi})N_1(t)$, with $N_1(t)$ given by the expression for nonlinear normal stress in oscillatory shear deformation using the strain tensor $\mathbf{Q^o}$, have been used to estimate $\hat{\beta}$ on the basis that $\hat{\psi} = \hat{\beta}$, as expected with the stress-optic approximation.[101]

Birefringence measurements are often of interest in steady-state flow for a nonlinear response for which the use of $\mathbf{Q_o}$ is inadequate. For example, the stress-optic relation given has been invoked to estimate $N_1(\dot{\gamma})$ in nonlinear steady-state shear flow ($\hat{\tau}_c\dot{\gamma} > 1$) with apparent success.[16,27,110,111] Following the examples given, the use of the strain tensor $\mathbf{Q}$ to encompass nonlinear behavior in the response to a step-strain γ_0 in the 1,2-plane gives

$$\Delta n \sin[2\chi_n(t; \gamma_0)]/C_{\text{optic}} = 2\sigma(t; \gamma_0) = 2\gamma_0 F(\gamma_0)G(t) \qquad (123a)$$

$$\Delta n \cos[2\chi_n(t; \gamma_0)]/C_{\text{optic}} = N_1(t; \gamma_0) = \gamma_0^2 F(\gamma_0)G(t) \qquad (123b)$$

$$\cot[2\chi_n(t; \gamma_0)] = N_1(t; \gamma_0)/2\sigma(t; \gamma_0) = \gamma_0/2 \qquad (123c)$$

for the 1,2-plane, in accord with data on some polymeric fluids.[107,134] Similarly, the expressions given for a shear deformation at constant shear rate $\dot{\gamma}_{ss}$ may be used, with $\sigma(t; \dot{\gamma}_{ss})$ and $N_1(t; \dot{\gamma}_{ss})$ computed as discussed in the section on nonlinear viscoelasticity in place of the expressions for a linear viscoelastic fluid. In this case, $\chi_n(t; \dot{\gamma}) \leq \pi/4$ decreases with increasing $\dot{\gamma}_{ss}$, and $S^{(1)}(t;\ \dot{\gamma}_{ss}) = N_1(t;\ \dot{\gamma}_{ss}/2[\sigma(t;\ \dot{\gamma}_{ss})]^2)$. Similarly, $\Delta n_{1,3}(t;\ \dot{\gamma}_{ss}) = C(1 - \hat{\psi})N_1(t;\ \dot{\gamma}_{ss})$ in the 1,3-flow plane in this approximation if it is assumed that $\hat{\psi}$ is independent of t and $\dot{\gamma}_{ss}$. At steady-state flow, these relations become

$$\Delta n \sin[2\chi_n(\dot{\gamma}_{ss})]/C_{\text{optic}} = 2\sigma(\dot{\gamma}_{ss}) = (2/J_s)[J_s/J_s(\dot{\gamma}_{ss})]\tilde{\gamma}_R(\dot{\gamma}_{ss}) \quad (124a)$$

$$\Delta n \cos[2\chi_n(\dot{\gamma}_{ss})]/C_{\text{optic}} = N_1(\dot{\gamma}_{ss}) = (2/J_s)[S^{(1)}(\dot{\gamma}_{ss})/J_s(\dot{\gamma}_{ss})]$$
$$\times [J_s/J_s(\dot{\gamma}_{ss})]\tilde{\gamma}_R(\dot{\gamma}_{ss})^2 \quad (124b)$$

$$\cot[2\chi_n(\dot{\gamma}_{ss})] = S^{(1)}(\dot{\gamma}_{ss})\sigma(\dot{\gamma}_{ss})$$
$$= [S^{(1)}(\dot{\gamma}_{ss})/J_s(\dot{\gamma})]\tilde{\gamma}_R(\dot{\gamma}_{ss}) \quad (124c)$$

where $\sigma(\dot{\gamma}_{ss}) = \dot{\gamma}_{ss}\eta(\dot{\gamma}_{ss})$, and $\tilde{\gamma}_R(\dot{\gamma}_{ss}) = J_s(\dot{\gamma}_{ss})\dot{\gamma}_{ss}\eta(\dot{\gamma}_{ss})$ is the total recoverable strain following cessation of steady-state flow. Similarly, $\Delta n_{1,3}(\dot{\gamma}_{ss}) = C_{\text{optic}}(1 - \hat{\psi})N_1(\dot{\gamma}_{ss})$ in the 1,3-flow plane if it is assumed that $\hat{\psi}$ is independent of $\dot{\gamma}_{ss}$. The final forms emphasize the relation of the birefringence to the constrained recoil $\tilde{\gamma}_R(\dot{\gamma}_{ss})$ on cessation of flow, and are analogous to the relations given for an elastic solid (i.e., replace $\dot{\gamma}_{ss}$ by γ, J_s by J_e, and $S^{(1)}(\dot{\gamma}_{ss})/J_s(\dot{\gamma}_{ss})$ by $m/2$). As discussed in the preceding section, $S^{(1)}(\dot{\gamma})/J_s(\dot{\gamma})$ depends on the distribution of components (e.g., the molecular weight distribution), where $S^{(1)}(\dot{\gamma})/J_s(\dot{\gamma}) \leq 1$, with the equality occurring for flows of monodisperse polymer with $\hat{\tau}_c\dot{\gamma} < 1$. The analogy with the behavior for a linear elastic solid is strengthened by noting that $S^{(1)}(\dot{\gamma})/J_s(\dot{\gamma})$ tends toward 0.5 with increasing $\hat{\tau}_c\dot{\gamma}$, similar to the value of m noted for a linear elastic solid. Estimates of $N_1(t;\ \dot{\gamma})$ made in this way appear to be similar to those obtained by direct mechanical measurements.[107,108,110,111] This correspondence could reflect the marked weighting of the response toward the longer time response in $G_n(t)$, and may not provide definitive evidence for constant $G_n(t)/G(t)$ for all t. In part, the use of the stress-optic approximation here is motivated by statistical mechanical theories that suggest that Δn and ΔS depend on similar averages over chain conformations for flexible chain polymers.[16h,22b,27b,121] However, the similarity with the expression for an elastic solid given above is evident, and is not accidental because the theories in both cases are based on additivity of incremental stresses attributed to molecular deformations, similar to the treatment of a collection of macroscopic beads and dashpots.[22h]

Relations comparable to the preceding do not appear to be available for a stress-defined history. Moreover, there are few data to evaluate the nature of $J_n(t)$, a situation that may change with the availability of commercial rheo-optical instrumentation to permit the implementation of stress-defined histories. The lack of experimental and theoretical attention may reflect both the lack of commercially available equipment to study the transient response with a stress-defined shear deformation and the perception that $J_n(t)$ is a constant, as required by the stress optic approximation with constant $G_n(t)/G(t)$, in which case birefringence measurements for a stress-defined history would be relatively uninteresting. However, as noted in the preceding

discussion, the evidence that $G_n(t)/G(t)$ must be considered to be independent of t is not definitive. For example, $G_n(t)$ could relax more slowly than $G(t)$ for large t, but still decay fast enough to give the response $G'_n(\omega) \propto \omega^2$ and $G''_n(\omega) \propto \omega$ reported experimentally for small ω.[101] In addition, as may be seen in the expression for $J_n(t)$, deviation of $G_n(t)/G(t)$ from a constant for small t will have an impact on $J_n(t)$ at large t. As mentioned above, birefringence measurements are sometimes made on amorphous materials near T_g, with the finding that a single constant C_{optic} cannot be applied over the entire range of ω at a given T.[130,131] This result is readily accommodated in terms of a distribution of retardation times to represent $\hat{\alpha}(t)$, with the response corresponding to an Andrade creep region having a different value of C_{optic} than that associated with the portion of the retardation times associated with the Rouse-like or entanglement responses.

In summary, it appears that the stress-optic approximation embodied in the expression $G_n(t)/G(t) \approx C_{\text{optic}}$ for all t may sometimes be used for strain-defined deformations to estimate the first-normal stress and the shear stress in a shear deformation and the tensile stress in an elongational deformation, although it is clear that the approximation is not always valid, especially for deformation involving the short-time features of the response, for example, the transition from the Andrade creep to the Rouse-like response in the retardation spectrum. The experimental situation is not clear for transient stress-defined deformations, with strict compliance with the relation $G_n(t)/G(t) \approx C_{\text{optic}}$ requiring that $J_n(t)$ be independent of time.

Linear and Nonlinear Viscoelastic Behavior of Colloidal Dispersions

The viscoelastic behavior of colloidal dispersions is an enormous topic, the subject of numerous reviews and monographs,[53,69,135] and one that can receive only incomplete discussion here. The interactions among colloidally dispersed particles can involve diverse phenomena, including dispersion forces, electrostatic interactions, interactions with dissolved polymeric components, and interactions mediated by adsorbed surfactants.[53b,c,d,69] In many, perhaps most, cases there is a tendency for cluster formation among the colloidal particles, leading to complex behavior, including yield phenomena in which the material may be considered to be a solid provided the strain is small, but will flow if the strain is increased beyond some value γ^0, similar to the behavior discussed for certain polymeric systems.

The simplest behavior, that of dilute dispersion of spheres interacting only through a hard-core potential, is discussed above in terms of a virial expansion for the viscosity, rewritten here in a form to emphasize the dependence on the volume fraction φ of spheres:

$$\eta = \eta_{LOC}^{(c)}\{1 + (5/2)\varphi + k'(5/2)^2\varphi^2 + \ldots\} \quad (125)$$

where $(5/2)\varphi = [\eta]c$, and it is usually assumed that $\eta_{LOC}^{(c)} \approx \eta_{solv}$.[15a,53a,69] As noted above, $k' \approx 1.0$ for this model, and as usual, the virial expansion is inadequate with increasing concentration. A number of expressions has been developed to approximate η as φ is increased toward the maximum value φ_{max} possible with hard-core spheres. Several general approximate forms are designed to approximate the behavior exactly for dilute suspensions and approximate the rapid increase in η as φ approaches φ_{max}.[15c,53e,69,135] The simple expressions

$$\eta \approx \eta_{LOC}^{(c)}\{1 - \varphi/n_1\}^{-5n_1/2} \quad (126)$$

and

$$\eta \approx \eta_{LOC}^{(c)}\{1 - (5/2)\varphi[1 - \varphi/n_2]^{-5n_2k'/2}\} \quad (127)$$

are designed to force agreement with the virial expansion at least to order φ and φ^2, respectively, where n_1 and n_2 are constants, and again it is usual to assume that $\eta_{LOC}^{(c)} \approx \eta_{solv}$. The second expression will be recognized to involve a variation of the relation $[\eta]^{(c)} \approx [\eta]\{1 + \eta]c\}^{k'}$ introduced above in the discussion of the viscosity of dilute solutions of polymers. An application of the first expression puts $n_1 = \varphi_{max}$;[69,135] the value $n_1 = 5/8$ would be required to give $k' \approx 1.0$, which is close to the estimate $\varphi \approx 0.64$ for spontaneous ordering of the spheres in this model.[136] It is noteworthy that η does not depend on the sphere radius R in these expressions, but only on the volume fraction of spheres.

Certain theoretical treatments may be put in the (closed) form

$$\eta = \eta_{LOC}^{(c)}\{1 + (5/2)\varphi + k'[\psi_1(\varphi) + \psi_2(\varphi)](5/2)^2\varphi^2\} \quad (128)$$

where it is assumed that $\eta_{LOC}^{(c)} \approx \eta_{solv}$. The functions $\psi_1(\varphi)$ and $\psi_2(\varphi)$, reflecting hydrodynamic and thermodynamic effects, respectively, depend on the form of the particle interaction potential; $\psi_1(0) + \psi_2(0) = 1$, and both $\psi_1(\varphi)$ and $\psi_2(\varphi)$ increase with increasing φ.[53f,69,136-141] Experimental data for η from slow steady flow and $\eta = \eta'(0)$ from small strain amplitude dynamic measurements for dispersions of spheres over a range of R from different investigators in Figure 22.19 for systems artfully designed to behave as hard spheres demonstrate the dependence of η on φ.[137,139,142] The solid curve is calculated with the semi-empirical expressions $\psi_1(\varphi) \approx (4/5)(1 - \varphi/\varphi_{max})$ and $\psi_2(\varphi) \approx (1/5)(1 - \varphi/\varphi_{max})^2$, which are based on more complex theoretical expressions.[136,138,140,141,143,144] The semi-empirical expression provides a reasonable representation and is given over the entire range of φ by the empirical relations, as would the second of the empirical relation above with $n_2 = 0.605$. The situation is more complex with nonspherical particles, with the possibility of producing ordered phases with plate or rod shaped particles, even with the simple hard-core interaction.[145] In any case, η will involve a measure of the particle

shape, such as the L/d ratio for a rod, or the ratio of the principal axes for spheroids of revolution.[67]

Linear viscoelastic behavior has been predicted and observed for dispersions of rigid spheres.[53f,136-141,144] Experimental data on the designed dispersions exhibit the behavior of hard spheres (Figure 22.20);[139] similar data are available over a range of particle size,[139] as well as from other investigators.[137,146] The data for several temperatures have been superposed, assuming that $b(T, T_{REF}) = 1$. The data on $\eta'(\omega)$ exhibit a limiting value $\eta'(0)$ for small ω, as expected, but also show a plateau $\eta'(\omega_L)$ for a regime at an intermediate range of $\omega \approx \omega_L$, before decreasing to zero with increasing ω, beyond the range of ω usually accessible. Because only the frequency range $\omega \leq \omega_L$ is of usual interest, the models lump the effects at high ω (e.g., $\omega > 1\hat{\tau}_1$) into an additive term $(G_0 - G_1)\delta(t)$ to $G(t)$, as discussed in the first section of this chapter, so that $\eta'(\omega_L) \approx (G_0 - G_1)\hat{\tau}_1$. Because $J'(\omega)$ would tend to a constant for small ω, but would decrease with increasing ω for larger ω, reflecting the term $(G_0 - G_1)\delta(t)\hat{\tau}_1$ in $G(t)$, it is convenient to remove the latter by use of $G''_{EFF}(\omega) = \omega[\eta'(\omega) - \eta'(\omega_L)]$ instead of $G''(\omega)$ to compute a $J'_{EFF}(\omega)$ from the reported moduli to suppress the response for $\omega > \omega_L$. The data in Figure 22.20 show that $J'_{EFF}(\omega)$ decreases with increasing ω, to reach a plateau $J'_{EFF}(\omega_L) \approx 1/G'(\omega_L) \approx 1/G_1$ for ω in the regime for which $\eta'(\omega) \approx \eta'(\omega_L)$. The theoretical expression given above for η is taken to estimate $\eta'(0) = \eta$, and it is assumed that $\eta'(\omega_L)$ is given by the theoretical relation for η, with $\psi_2(\varphi) = 0$, reflecting the suppression of thermodynamic interactions at high ω. Comparison with experimental data in Figure 22.19 shows this to be reasonably accurate. For most models, $J'_{EFF}(\omega_L) \approx 1/G'(\omega_L) \approx 1/G_1$ is expected to vary markedly with φ, with $G_1R^3/kT\varphi^2 \approx \psi_0(\varphi)$ for spheres of radius R. The expression $\psi_0(\varphi) \approx 0.78\ (\eta'(\omega_L)/\eta_{solv})g(2, \varphi)$ given by an approximate model[140] is in reasonable accord with numerical results from a more rigorous treatment, where $g(2, \varphi)$ is the value of the radial distribution $g(r/R, \varphi)$ at the contact condition $r/R = 2$; $g(2, \varphi) = (1 - \varphi/2)^2/(1 - \varphi)^3$ for $\varphi < 0.5$, and $g(2, \varphi) = (6/5)(1 - \varphi/\varphi_{max})$ for $\varphi \geq 0.5$.[138-140] These expressions are compared with experimental data on dispersions of spherical particles in Figure 22.19, revealing reasonable agreement with the theoretical model. Theoretical models predict a region of the response, which would have $J'_{EFF}(\omega) \propto \omega^{-1/2}$;[138,140] as shown in Figure 22.20, experimental data are consistent with this, even though the agreement is not definitive.

The nonlinear steady-state viscosity of dispersions of spherical particles is frequently characterized by a plateau $\eta_{plateau}(\dot{\gamma})$ over a range of shear rate with $\dot{\gamma} \approx \omega_L$.[53e,69,135] It has been suggested that $\eta_{plateau}(\dot{\gamma}) \approx \eta'(\omega_L)$, reflecting suppression of the same contributions to $\eta(\dot{\gamma})$ that are lost in $\eta'(\omega)$ for $\omega \approx \omega_L$.[53f,136] Comparison of $\eta(\dot{\gamma}) \approx \eta'(\omega = \dot{\gamma})$ shows that the approximation is qualitatively useful, but not fully accurate.[146] The behavior is similar to the Cox-Merz approximation $\eta(\dot{\gamma}) \approx \eta'(\omega = \dot{\gamma})$, but the paucity of relaxation times

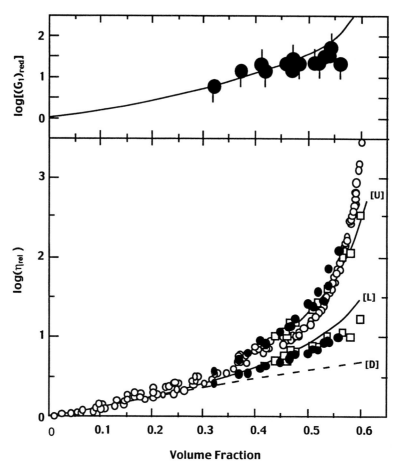

Figure 22.19 Semilogarithmic plots of the relative viscosity $\eta_{rel} = \eta/\eta_{solv}$ and a reduced modulus $(G_1)_{red}$ versus volume fraction for hard spheres over a range of sphere radius R; $(G_1)_{red} = G_1R^3/kT$, with $G_1 \approx G'(\omega_L)$ in a certain range of frequency for which $G'(\omega_L)$ is about constant (see text). The curves in the lower figure labeled U and L correspond to the formulae discussed in the text for $\eta'(0)$ and $\eta'(\omega_L)$, respectively, and the curved labeled D corresponds to the dilute solution virial series truncated at the term in φ^2, with $k' = 1.0$. The curve in the upper figure is calculated as described in the text. The data for η_{rel} are from references 136, 138, and 141 for unfilled circles, filled circles, and squares, respectively; the data from reference 141 represent data from several sources. The data for $(G_1)_{red}$ are from reference 138, for spheres with $R/\mu m$ equal to 0.060 (pip down), 0.125 (no pip), and 0.225 (pip up).

may play a role in rendering the approximation less accurate than sometimes found with polymers, which display a relative abundance of relaxation times.

The effect of more specific or long-range interparticle interactions can substantially modify the behavior described above. For example, charged spheres dispersed in a solvent with a very low ionic strength may form an ordered phase, with the spheres tending to lie on a lattice through the effects of electrostatic repulsion among the spheres, forming a viscoelastic solid with an equilibrium modulus if the deformation is small and the sphere concentration is large enough.[53g] Alternatively, with a lower concentration of spheres, the spheres may phase separate into clusters of ordered arrays, in equilibrium with essentially pure solvent.[147-149] Similar clusters may be formed by the addition of a polymeric solute that does not adsorb on the particles, with the particle flocculation stabilized by the loss of entropy associated with a polymer chain inserted between two closely spaced spheres.[53g,150] In either case, the clusters may be disrupted by an imposed strain, often leading to a behavior in which $\eta(\dot{\gamma})$ appears to be nearly proportional to $\dot{\gamma}^{-1}$, so that the stress $\dot{\gamma}\eta(\dot{\gamma})$ is essentially constant in flow.[53g] For example, studies on a dispersion of polystyrene beads in a polystyrene solution ($w = 0.15$ for the polymer, with 170-nm beads of weight fractions w_B of 0.05, 0.10, 0.15, and 0.20 dispersed in polystyrene solutions in either tricresyl phosphate [TCP] or 1,2-di(2-ethyl hexyl)phthalate [DOP], a Flory Theta solvent at 22 °C) revealed interconnected strings of close-packed beads coexisting with regions essentially free of beads,[150] similar to structures reported for charged beads in a low ionic strength solvent.[148] The filled polymer solution exhibited a weak yield stress determined as the maximum stress permitting fully recoverable strain in creep, with the yield stress in good accord with that expected from the osmotic pressure of the polymer solution.[150] Data on the steady-state viscosity and the recoverable compliance following steady flow are shown in Figure 22.21 for the filled polymeric system. The data reveal a plateau viscosity η_P with $\eta(\dot{\gamma})$ essentially independent of $\dot{\gamma}$ at intermediate $\dot{\gamma}$, and a limiting value J_s of $J(\dot{\gamma})$ at low $\dot{\gamma}$. Except for the increase of $\eta(\dot{\gamma})$ with decreasing $\dot{\gamma}$ at low $\dot{\gamma}$, with $\partial\ln(\eta(\dot{\gamma}))/\partial\ln\dot{\gamma} < -1$, the behavior is similar to that observed with polymers and their solutions, with remarkable reduction over a range of bead concentration and solution temperature in two solvents. The data on $\eta(\dot{\gamma})$ at low $\dot{\gamma}$ for the dispersion with $w_B = 0.05$ exhibit largest deviation. The data on η_P are seen to exceed η for the bead-free polymer solution by about the amount expected for dispersed beads, but the data on the recoverable compliance show a much enhanced J_s, attributed to the elasticity from distortion of the strings of close-packed beads in steady-flow. These rheological features will have a substantial impact on the processing characteristics of such filled suspensions.

As indicated above, the subject of this section is very broad, and many additional examples could be discussed, including electrostatic interactions among particles and dissolved polymer, the effects with polymer adsorbed on a particle, interactions of polymeric solute with micelle structures, and the effects of nonspherical shapes of the dispersed particles (see reference 69).

Appendix

Example Problem

A research laboratory has generated data on the dynamic moduli $G'(\omega)$ and $G''(\omega)$ over a range of temperatures for an aqueous solution of a certain polymer, with the volume fraction of polymer $\varphi \approx 0.05$. Data on the dependence of the viscosity $\eta(\dot{\gamma})$ as a function of the shear rate $\dot{\gamma}$ are also available at a single temperature. It is suspected that the

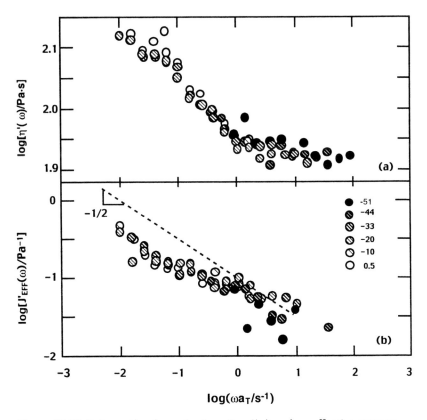

Figure 22.20 Data on the dynamic viscosity $\eta'(\omega)$ and an effective storage creep compliance $J'_{EFF}(\omega)$ versus reduced frequency ωa_T for hard spheres ($R = 0.060\ \mu m$, $\varphi = 0.37$) over a range of temperatures as indicated. The dashed line has slope $-1/3$, as expected in some treatments discussed in the text. (From data given in reference 138.)

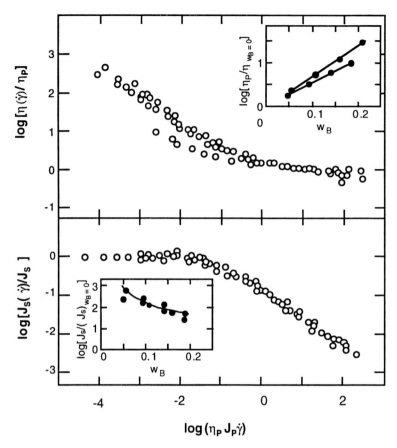

Figure 22.21 Bilogarithmic plots of the reduced functions for a solution of a high molecular weight polystyrene with a narrow molecular weight distribution filled with cross-linked polystyrene spheres: the viscosity $\eta(\dot{\gamma})/\eta_P$ and the total recoverable compliance $J_s(\dot{\gamma})/J_s$ as functions of $\hat{\tau}_c\dot{\gamma} = J_s\eta_P\dot{\gamma}$, where η_P is the value of $\eta(\dot{\gamma})$ over an intermediate range of $\dot{\gamma}$ for which $\eta(\dot{\gamma})$ is essentially a constant. The weight fraction w_B of beads covers the range shown in the insets; the data for $\eta(\dot{\gamma})/\eta_P$ that tend to lie below the bulk of the data are for $w_B = 0.05$. One of the two solvents used corresponded Flory Theta conditions for polystyrene at the experimental temperature. (Adpated from an example in reference 150.)

polymer might associate in solution, and the question is whether the data can be used to assess that possibility.

The dynamic data obtained at temperatures of 25, 35, 50, and 60 °C are shown in Figure 22.A1 in a reduced plot, with a reference temperature of 25 °C, prepared under the assumption that the vertical shift factor for the modulus $b(T) = 1$. The various shift factors used in this example are defined in the text and Table 22.5. The values of the horizontal shift factor for the frequency $\log[a(T)]$ were found to be -0.32, -0.60, and -0.80 for T equal to 35, 50, and 60 °C, respectively. It would have been appropriate to subtract $\omega\eta_{solv}$ from $G''(\omega)$, but the difference would have been too small to be important in the conclusions reached at this point. In the analyses in this example, the use of a spreadsheet with calculation and graphical capabilities provided a convenient method to estimate the parameters $a(T)$, $b(T)$, and $h(T)$. The sheet can be set up so that changes in these

parameters are affected globally, with results shown in bilogarithmic plots of the data. Inspection of Figure 22.A1 shows that the reduction is imperfect, with some suggestion of systematic deviations in the fit. Because it is assumed that $b(T) = 1$, the frequency shift factor $a(T)$ is equal to the viscosity shift factor $h(T)$, and for a solution as dilute as this, one might expect $h(T)$ to be approximately equal to $h_{solv}(T)$, the ratio of the solvent viscosity at temperature T to that at the reference temperature of 25 °C, given by -0.09, -0.21, and -0.28 for T equal to 35, 50, and 60 °C, respectively. The substantial deviation between $h(T)$ and $h_{solv}(T)$, and the apparent systematic deviations in the reduced plot both suggest a further analysis.

Following the suggestions in the text regarding the implementation of time-temperature reduction, the data on $G'(\omega)$ and $G''(\omega)$ were transformed to give $\eta'(\omega) = G''(\omega)/\omega$ and $J(\omega) = G'(\omega)/[G'(\omega)^2 + G''(\omega)^2]^{1/2}$. Reduced bilogarithmic plots

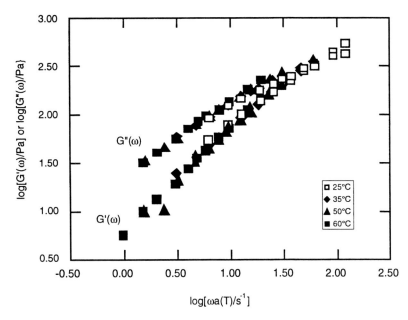

Figure 22.A1 $G'(\omega)$ and $G''(\omega)$ as functions of $\omega a(T)$ for a polymer solution at the indicated temperatures, with $a(T)$ chosen to give the best superposition possible; $T_{REF} = 25\ °C$.

of $[\eta'(\omega) - \eta_{solv}]/a(T)$ and $J'(\omega)$ versus $\omega a(T)$ prepared with the assumed $b(T) = 1$, and consequently $a(T) = h(T)$, are shown in Figure 22.A2, along with a data on $[\eta(\dot{\gamma}) - \eta_{solv}]$ versus $\dot{\gamma}$. Though subtraction of η_{solv} has little effect in this case owing to the magnitude of $\eta'(\omega)$ and $\eta(\dot{\gamma})$, it is included for completeness; with this inclusion, $h(T)$ is the ratio of $\eta - \eta_{solv}$ at temperatures T and T_{REF}. It is clear that systematic deviations appear in both plots of $[\eta'(\omega) - \eta_{solv}]/a(T)$ and $J'(\omega)$ versus $\omega a(T)$, suggesting that the assumption $b(T) = 1$ is inadequate.

Removal of the constraint on $b(T)$ results in the superposed plots shown in Figure 22.A3. In this case, the superposition of $[\eta'(\omega) - \eta_{solv}]/h(T)$ and $J'(\omega)/b(T)$ versus $\omega a(T) = \omega h(T)b(T)$ is much improved, and the data on $[\eta'(\omega) - \eta_{solv}]/h(T)$ versus $\omega a(T)$ are seen to correspond closely to those on $[\eta(\dot{\gamma}) - \eta_{solv}]$ versus $\dot{\gamma}$ at the reference temperature of 25 °C, as expected with the Cox-Merz relation discussed in the text. The values of $h(T)$ and $b(T)$ obtained in this way are given in Figure 22.A4, along with $h_{solv}(T)$. Inspection of Figure 22.A4 shows that both $b(T)$ and $h(T)$ increase with decreasing temperature, with the change in $h(T)$ far greater than that for $h_{solv}(T)$. Both of these results suggest that association does occur with these aqueous solutions, with the degree of association increasing with decreasing temperature. In addition to an enhanced viscosity, the association increases the distribution of species present, with the consequence that J_s and hence $b(T)$ increases markedly with decreasing temperature.

The presumption that superposition should obtain for data over a range of T with a changing extent of association with T is certainly open to question. Such a superposition

implies that the functional dependence of $[\eta'(\omega) - \eta_{solv}]/[\eta - \eta_{solv}]$ and $J'(\omega)/J_s$ on $\omega\eta J_s$ is not significantly altered by the formation of associated species, even though η and J_s might individually be strongly altered. The observed reasonable applicability of the Cox-Merz relation supports this assumption, but does not provide definitive proof of its validity. Further assessment would have been possible if data on the recoverable compliance $R(t)$ had been obtained at 25 °C for comparison with the corresponding estimate based on the approximation $R(t) \approx \{[J'(\omega)]^2 + [J''(\omega) - (\omega\eta)^{-1}]^2\}^{1/2}$ discussed in the text. If accepted as meaningful, further interpretation of the imputed changes in η and J_s based on $h(T)$ and $b(T)$ would be required in a model for the association process. For example, it might be assumed that the association creates a randomly branched structure, with the modified Fox parameter $\tilde{X}$ discussed in the text proportional to the ratio g of the mean-square-radii of branched and linear chains of the same molecular weight. Then, if it can be assumed that $\eta_{LOC}^{(c)} \approx \eta_{solv}$, the ratio $h(T)/h_{solv}(T)$ would be about proportional to $(gM)^a$ with $a \approx 3.4$ if $\tilde{X} > \tilde{X}_c$, or $a \approx 1$ otherwise. Use of g for a randomly branched polymer could then give the change in the association with T.

List of Symbols

$\tilde{a}(y_1, y_2, \ldots)$ A notation used in general to indicate that a function $\tilde{a}$ depends the variables $y_1, y_2, \ldots$ (e.g., $\sigma(t, \dot{\gamma})$ is the shear stress at time t and shear rate $\dot{\gamma}$)

$\hat{a}$ The persistence length, given by $3R_G^2/L$ for a random-flight chain model

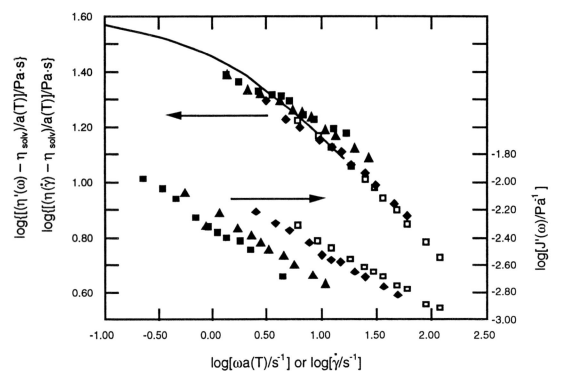

Figure 22.A2 $[\eta'(\omega) - \eta_{solv}]/a(T)$ and $J'(\omega)$ as functions of $\omega a(T)$ for a polymer solution at the indicated temperatures (see Figure 22.A1), and $[\eta(\dot\gamma) - \eta_{solv}]$ versus $\dot\gamma$ at 25 °C (solid line). Values of $a(T)$ were chosen to give the best superposition possible; $T_{REF} = 25$ °C.

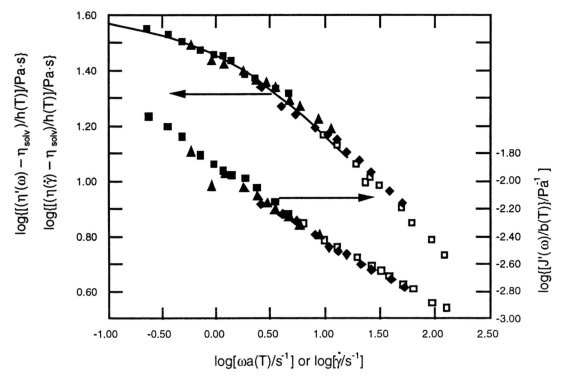

Figure 22.A3 $[\eta'(\omega) - \eta_{solv}]/h(T)$ and $J'(\omega)/b(T)$ as functions of $\omega a(T) = \omega h(T)b(T)$ for a polymer solution at the indicated temperatures (see Figure 22.A1), and $[\eta(\dot\gamma) - \eta_{solv}]$ versus $\dot\gamma$ at 25 °C (solid line). Values of $h(T)$ and $b(T)$ chosen to give the best superposition possible; $T_{REF} = 25$ °C.

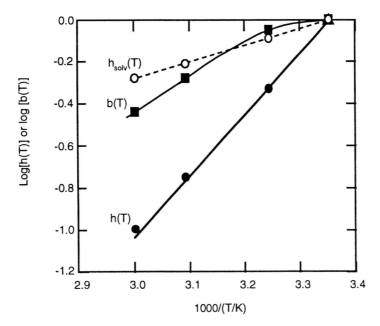

Figure 22.A4 The functions $h(T)$ and $b(T)$ from the analysis in Figure 22.A3 as functions of temperature, and $h_{solv}(T)$ for the solvent (water).

$a(T,T_{REF})$ The ratio $\eta(T)J_\infty(T)/\eta(T_{REF})J_\infty(T_{REF})$, equal to $b(T, T_{REF})h(T,T_{REF})$

$a_V(t)$ A shift factor at time t in a bulk deformation (see eq 99)

$\langle a_V^{-1}(t_2,t_1)\rangle$ An average shift factor averaged between times t_1 and t_2 in a bulk deformation (see eq 99)

$b(T,T_{REF})$ The ratio $J_\infty(T)/J_\infty(T_{REF})$ of the linear function J_∞ at temperatures T and T_{REF}

B_e The limiting value of $B(t)$ for large t

B_0 The limiting value of $B(t)$ for small t

$B(t)$ The bulk compliance

$B_n(t)$ A function appearing in an expression for $n_{ij}(t)$ (see eq 102)

c The solute concentration (weight/volume)

$c*$ The overlap concentration, defined by $M/[(4/3)\pi N_A R_G^3]$

C A constant in eq 66

C_{optic} The stress-optic coefficient

$D(t)$ The tensile compliance (see Table 22.1)

$D_e(\lambda)$ The equilibrium recoverable compliance for nonlinear viscoelastic behavior of a solid under an elongational deformation λ

$\mathbf{F}$ A strain tensor in a nonlinear constitutive equation (see eq 79)

$\mathbf{F}_{el}$ A form of $\mathbf{F}$ for an elongational deformation (see eq 86b)

$\mathbf{F}_{sh}$ A form of $\mathbf{F}$ for a shear deformation (see eq 86a)

$F_1[\gamma]$, $F_2[\gamma]$ Functions of the shear strain appearing in $\mathbf{F}_{sh}$ (see eqs 87a,b)

$F_3[\gamma]$ Functions of the elongational strain appearing in $\mathbf{F}_{el}$ (see eq 87c)

g The ratio $R_G^2/(R_G^2)_{LIN}$ of the mean-square-radii of gyration of branched and linear chains with common M

g' The ratio $[\eta]/[\eta]_{LIN}$ of the intrinsic viscosities of branched and linear chains with common M

G_e The equilibrium modulus

G_N The entanglement (or pseudo-network) modulus, the reciprocal of J_N

G_0 The instantaneous shear modulus, given by the limiting value of $G(t)$ for small t

$G(t)$ The linear shear stress modulus

$G'(\omega)$ The linear in-phase (real or storage) dynamic modulus

$G''(\omega)$ The linear out-of-phase (imaginary or loss) dynamic modulus

$G^*(\omega)$ The complex dynamic modulus, given by $G'(\omega) + iG''(\omega)$

$G_n(t)$ A function appearing in an expression for $\eta_{ij}(t)$ (see eq 103)

$G'_n(\omega)$ An in-phase function appearing in the response of $n_{ij}(t)$ to an imposed oscillatory strain (see eq 116a)

$G''_n(\omega)$ An out-of-phase function appearing in the response of $n_{ij}(t)$ to an imposed oscillatory strain (see eq 116a)

h The ratio $R_H/(R_H)_{LIN}$ of the hydrodynamic radii of branched and linear chains with common M

$h(T,T_{REF})$ The ratio $\eta(T)/\eta(T_{REF})$ of the linear steady-state viscosities at temperatures T and T_{REF}

$H(\hat{\tau})$ The relaxation spectrum (see eq 56)

$H_\eta(\hat{\beta}\hat{\tau}_c\dot{\gamma})$ The function $\eta(\hat{\tau}_c\dot{\gamma})/\eta(0)$ for the nonlinear steady-state viscosity

J_e The equilibrium recoverable compliance for linear viscoelastic behavior of a solid, that is, J_∞ for a solid

J_N The entanglement (or pseudo-network) compliance, the reciprocal of G_N

J_s The steady-state recoverable compliance for linear viscoelastic behavior of a fluid, that is, J_∞ for a fluid

J_0 The instantaneous shear compliance, given by the limiting value of $J(t)$ for small t

J_∞ The limiting value of $R(t)$ for large t for linear viscoelastic behavior

$J(t)$ The linear shear creep compliance

$J'(\omega)$ The linear in-phase (real or storage) dynamic compliance

$J''(\omega)$ The linear out-of-phase (imaginary or loss) dynamic compliance

$J^*(\omega)$ The complex dynamic compliance, given by $J'(\omega) - iJ''(\omega)$

$J_n(t)$ A function appearing in an expression for $n_{ij}(t)$ (see eq 102)

$J'_n(\omega)$ An in-phase function appearing in the response of $n_{ij}(t)$ to an imposed oscillatory stress (see eq 116b)

$J''_n(\omega)$ An out-of-phase function appearing in the response of $n_{ij}(t)$ to an imposed oscillatory stress (see eq 116b)

$J_e(\gamma)$ The equilibrium recoverable compliance for nonlinear viscoelastic behavior of a solid under a shear strain γ

$J_s(\dot{\gamma})$ The steady-state recoverable compliance for nonlinear viscoelastic behavior of a fluid on cessation of steady-state flow at shear rate $\dot{\gamma}$

k' The coefficient of $[\eta]c$ in a Taylor series expansion of η/η_{solv}

k'' The coefficient of $([\eta]c)^2$ in a Taylor series expansion of η/η_{solv}

K A parameter given by $[\eta]/M^\mu$, about constant for a given material

$\boldsymbol{K}$ A constant in eq 67

K_e The limiting value of $K(t)$ for large t

K_0 The instantaneous bulk modulus, given by the limiting value of $K(t)$ for small t

K_η The function $[\eta]M/\pi N_A R_G^2 R_H$

$K(t)$ The linear bulk modulus

$K_n(t)$ A function appearing in an expression for $n_{ij}(t)$ (see eq 103)

L_{cap} The length of a capillary

L The contour length of a linear polymer

$L(\hat{\lambda})$ The retardation spectrum (see eq 55)

M The molecular weight

$\boldsymbol{M}$ The torque

M_e The molecular weight between entanglement loci, calculated as $J_N/\varphi RT$

M_L The mass per unit length of a polymer chain, given by M/L

$M_{(\mu)}$ A generalized molecular weight average given by $(\Sigma M_i^\mu w_i)^{1/\mu}$ (see eq 70)

M_n, M_w, M_z, M_{z+1} The number, weight, z and, $z + 1$ aver- age molecular weights, given by $M_{(\mu+1)}^{\mu+1}/M_{(\mu)}^\mu$ with μ equal to -1, 0, 1, and 2, respectively.

$n_{ij}(t)$ The components of the linear refractive index tensor (see eqs 102 and 103

$\{n_{11}(t) - n_{22}(t)\}_{Deviatoric}$ Deviatoric part of the normal components in $n_{ij}(t)$

$\{n_{12}(t)\}_{Deviatoric}$ Deviatoric part of the shear component in $n_{ij}(t)$

N_A The Avogadro number

N_1 The first normal stress difference $S_{11} - S_{22}$ for a shear deformation

N_2 The second normal stress difference $S_{22} - S_{33}$ for a shear deformation

$N_1(t, \dot{\gamma})$ The first normal stress difference for a fluid at time t after onset of shear flow at shear rate $\dot{\gamma}$

$N_1(\dot{\gamma}_{ss})$ The first normal stress difference for a fluid in steady flow at shear rate $\dot{\gamma}_{ss}$

$N_1(\gamma)$ The first normal stress difference for a solid under a shear strain γ

ΔP The pressure drop in a capillary flow

$P(\tilde{\beta}\hat{\tau}_c\dot{\gamma})$ The function $J_s(\hat{\tau}_c\dot{\gamma})/J_s$ for the nonlinear steady-state recoverable compliance

Q_{vol} The volumetric flow rate in capillary flow

$\boldsymbol{Q}$ A strain tensor in a nonlinear constitutive equation (see equation 79) equal to $\boldsymbol{Q}^o \cdot \boldsymbol{F}$

$\boldsymbol{Q}^o$ A strain tensor in a nonlinear constitutive equation (see eq 79)

$\boldsymbol{Q}_{el}^o$ A form of $\boldsymbol{Q}^o$ for an elongational deformation (see eq 80b)

$\boldsymbol{Q}_{sh}^o$ A form of $\boldsymbol{Q}^o$ for a shear deformation (see eq 80a)

R The gas constant

R The radius of a sphere, a capillary, or a disc

$R(t)$ The linear recoverable compliance (see eq 1)

R_G The root-mean-square radius of gyration

R_H The hydrodynamic radius, given by $k_B T/6\pi\eta D_T$, with D_T the transactional diffusion constant

$S_{ij}(t)$ The components (in Cartesian coordinates) of the stress tensor ($i,j = 1$ to 3) (see eq 14)

$S(\tilde{\beta}\hat{\tau}_c\dot{\gamma})$ The function $S^{(1)}(\hat{\tau}_c\dot{\gamma})/J_s$ for the nonlinear steady-state first-normal stress difference

$S^{(1)}(\dot{\gamma})$ The function $N_1(\dot{\gamma})/2[\dot{\gamma}\eta(\dot{\gamma})]$ for the nonlinear first-normal stress difference for a fluid in steady flow

t time

T_g The glass transition temperature

T_0 A constant in eq 66

$\boldsymbol{u}$ The displacement vector, with components $u_i(i = 1$ to 3 in Cartesian coordinates)

$\bar{X}$ The modified Fox parameter, given by $[\pi N_A \hat{a}(\alpha^{(c)})^2/3M_L^2]cM$

$\bar{X}_c$ A critical value of $\bar{X}$ (see eq 74)

$\hat{\alpha}(t)$ The shear retardation function (see eq 1)

$\hat{\alpha}_i$ Weight factors in a discrete representation of $\hat{\alpha}(t)$ (see eq 55)

$\alpha^{(c)}$ The expansion factor for R_G in a solution at concentration c, with value α at infinite dilution

β The ratio $-N_2/N_1$

$\hat{\beta}(t)$ The bulk retardation function (see eq 3)

$\gamma(t)$ The shear strain, equal to $2\varepsilon_{12}(t)$

$\gamma_R(\vartheta)$ The (constrained) recoverable strain at time $\vartheta = t - T_e$, calculated from the strain as $\gamma(T_e) - \gamma(t)$

$\hat{\gamma}_R(\gamma)$ The total recoverable strain on cessation of equilibrium shear strain γ

$\hat{\gamma}_R(T_e)$ The total recoverable strain after a shear deformation to terminate at time T_e

$\hat{\gamma}_R(T_e, \gamma_0)$ The total recoverable strain after a stress relaxation under a strain γ_0 for duration T_e

$\hat{\gamma}_R(\dot{\gamma}_{ss})$ The total recoverable strain on cessation of steady-state shear at shear rate $\dot{\gamma}_{ss}$

$\dot{\gamma}_R$ The shear rate at the wall in capillary flow, or at the outer radius in shear between parallel disks

$\dot{\gamma}_{ss}$ The shear rate in steady-state shear flow

δ_{ij} The components (in Cartesian coordinates) of the unitary tensor $(i,j = 1$ to $3)$, with $\delta_{ij} = 1$ if $i = j$, and $\delta_{ij} = 0$ otherwise

$\delta(\omega)$ The phase angle, given by $\arctan[J''(\omega)/J'(\omega)] = \arctan[G''(\omega)/G'(\omega)]$

Δ The difference $T_g - T_0$

ε_{ij} The components (in Cartesian coordinates) of the strain tensor $(i,j = 1$ to 3 (see eq 13)

$\hat{\varepsilon}_R(\lambda)$ The total recoverable strain on cessation of equilibrium tensile elongation λ

η The steady-state viscosity in a linear response

$\eta(\dot{\gamma})$ The steady-state viscosity at shear rate $\dot{\gamma}$, with limiting value $\eta = \eta(0)$ at small $\dot{\gamma}$ if that limit exits

$\eta(T)$ The linear steady-state viscosity at temperature T

η_{solv} The viscosity of the solvent in a solution (or dispersion)

$\eta_{LOC}^{(c)}$ The local viscosity (see eq 74)

$[\eta]$ The intrinsic viscosity, given by the limiting value of $\{(\eta/\eta_{solv}) - 1\}/c$ at infinite dilution

$[\eta]^{(c)}$ $\{(\eta/\eta_{LOC}^{(c)}) - 1\}/c$ (see eq 73)

$\hat{\kappa}(t)$ The bulk relaxation function (see eq 4)

λ $1 + \varepsilon_{11}$, an elongational deformation

$\hat{\lambda}$ The retardation time in the function $L(\hat{\lambda})$

$\hat{\lambda}_i$ Time constants in a discrete representation of $\hat{\alpha}(t)$ (see eq 55)

Λ The mean separation of molecular (particle) centers of gravity, given by $(M/cN_A)^{1/3}$ for an isotropic solution

μ $\partial \ln [\eta]/\partial \ln M$

$\sigma(t)$ The shear stress, equal to the component $S_{12}(t)$ of the stress tensor

σ_R The shear stress at the wall in capillary flow, or at the outer radius in shear between parallel disks

$\sigma(t,\dot{\gamma})$ The shear stress at time t in a deformation at constant shear rate $\dot{\gamma}$

$\sigma(\dot{\gamma}_{ss})$ The shear stress in a steady-state deformation at shear rate $\dot{\gamma}_{ss}$

τ A tensile stress given by $S_{11} - S_{22}$ for a deformation in simple elongation

$\hat{\tau}$ The relaxation time in the function $H(\hat{\tau})$

$\hat{\tau}_c$ An average relaxation time, given by ηJ_s

$\hat{\tau}_i$ Time constants in a discrete representation of $\hat{\phi}(t)$ (see eq 56)

$\tau(t)$ The tensile stress at time t in an elongational deformation

$\tau(t\dot{\varepsilon})$ The tensile stress at time t in an elongational deformation at constant strain rate $\dot{\varepsilon}$

$\tau(\lambda)$ The tensile stress under equilibrium shear strain γ

φ The volume fraction of solute (or dispersed particles)

φ_L Volume fraction of polymer in a heterodisperse linear polymer with $M < M_c$

$\hat{\varphi}_i$ Weight factors in a discrete representation of $\hat{\phi}(t)$ (see eq 56)

$\varphi(t)$ The shear relaxation function (see eq 2)

Φ' The Flory-Fox parameter, given by $[\eta]M/R_G^3$

$\chi_n(t)$ The extinction angle of the birefringence at time t in a deformation

$\chi_n(\gamma)$ The extinction angle of the birefringence under equilibrium shear strain γ

$\chi_n(\dot{\gamma}_{ss})$ The extinction angle of the birefringence under steady-state shear at shear rate $\dot{\gamma}_{ss}$

$\psi_1(\varphi), \psi_2(\varphi)$ Functions in an expression for the viscosity of dispersions of spheres (see eq 128)

ω The frequency in a steady-state oscillatory deformation

Ω The rotational rate in a torsional rheometer

References

1. Society of Rheology. http://www.rheology.org/sor/default.htm (accessed Jan. 2002).

2. Staverman, A. J.; Schwarzl, F. Linear Deformation Behavior of High Polymers. In *Die Physik der Hochpolymeren*; Stuart, H. A., Ed.; Springer-Verlag: Berlin, 1956; Vol. IV, Chapter 1.

3. Treloar, L. R. G. *The Physics of Rubber Elasticity*, 2nd ed.; Oxford University Press: London, 1958; pp. (a) 156, (b) 64, (c) 177.

4. Coleman, B. D.; Markovitz, H.; Noll, W. *Viscometric Flows of Non-Newtonian Fluids*; Springer-Verlag: New York, 1966; pp. (a) 34, (b) 55, (c) 54, (d) 46.

5. McCrum, N. G.; Read, B. E.; Williams, G. *Anelastic and Dielectric Effects in Polymeric Solids*; Wiley: New York, 1967.

6. Aklonis, J. J.; MacKnight, W. J.; Shen, M. C. *Introduction to Polymer Viscoelasticity*; Wiley-Interscience: New York, 1972.

7. Bartenev, G. M.; Zelenev, Y. V.; Eds. *Relaxation Phenomena in Polymers*; Wiley & Sons (Halsted Press): New York, 1974.

8. Walters, K. *Rheometry*; Wiley & Sons: New York, 1975, (a) 44, (b) 48, (c) 52, (d) 93, (e) 102, (f) 100.

9. Markovitz, H. Rheology. In *Physics Vade Mecum*; Anderson, H. L., Ed.; American Institute of Physics: New York, 1981; pp 274–286.

10. Plazek, D. J. Viscoelastic and Steady-State Rheological

Response. In *Methods of Experimental Physics*; Fava, R. A., Ed., Academic Press: New York, 1980; Vol. 16C, Chapter 11.

11. Ferry, J. D. *Viscoelastic Properties of Polymers*, 3rd ed.; Wiley & Sons: New York, 1980; pp (a) 37, (b) 40, (c) 48, (d) 96, (e) 132, (f) 154, (g) 603, (h) 168, (i) 1, (j) 177, (k) 224, (m) 17, (n) 3, (p) 5, (q) 68, (r) 33, (s) 103, (t) 70, (u) 69, (v) 60, (w) 83, (x) 63, (y) 91, (z) 287, (a') 227, (b') 266, (c') 366, (d') 229, (e') 232, (f') 390, (g') 388, (h') 411, (i') 545.

12. Bailey, R. T.; North, A. M.; Pethrick, R. A. *Molecular Motion in High Polymers*; Baldwin, J. E., Goodenough, J. B., Halpern, J., Rowlinson, J. S., Eds.; The International Series of Monographs on Chemistry; Clarendon Press: Oxford, 1981.

13. Christensen, R. M. *Theory of Viscoelasticity*, 2nd ed.; Academic Press: New York, 1982; pp (a) 353, (b) 1, (c) 5, (d) 7.

14. Dealy, J. M. *Rheometers for Molten Plastics*; Van Nostrand Reinhold Co.: New York, 1982; pp (a) 1, (b) 116, (c) 126, (d) 148, (e) 75, (f) 61.

15. Bohdanecký, M.; Kovář, J. *Viscosity of Polymer Solutions*; Jenkins, A. D., ed.; Polymer Science Library; Elsevier: Amsterdam, 1982; Vol. 2, pp (a) 167, (b) 175, (c) 177.

16. Janeschitz-Kriegl, H. *Polymer Melt Rheology and Flow Birefringence*; Springer-Verlag: Berlin, 1983; pp (a) 455, (b) 4, (c) 467, (d) 464, (e) 355, (f) 60, (g) 177, (h) 217.

17. Doi, M.; Edwards, S. F. *The Theory of Polymer Dynamics*; Clarendon Press: Oxford, UK, 1986; pp (a) 222, (b) 366, (c) 218, (d) 91, (e) 226, (f) 281, (g) 255, (h) 121, (i) 166.

18. Berry, G. C.; Plazek, D. J. Rheology of Polymeric Fluids. In *Glass: Science and Technology*; Uhlmann, D. R., Kreidl, N. J., Eds.; Academic Press: New York, 1986; Vol. 3, pp 319–362.

19. Bird, R. B.; Armstrong, R. C.; Hassager, O. *Dynamics of Polymeric Liquids. Vol. 1. Fluid Mechanics*, 2nd ed.; Wiley & Sons: New York, 1987; pp (1) 158, (b) 269, (c) 290.

20. Bird, R. B.; Curtiss, C. C.; Armstrong, R. C.; Hassager, O. *Dynamics of Polymeric Liquids. Vol. 2. Kinetic Theory*, 2nd ed.; Wiley & Sons: New York, 1987.

21. Collyer, A. A., Clegg, D. W.; Eds. *Rheological Measurement*; Elsevier Science: New York, 1988.

22. Larson, R. G. *Constitutive Equations for Polymer Melts and Solutions*; Butterworths: Boston, 1988; pp (a) 49, (b) 93, (c) 318, (d) 118, (e) 120, (f) 233, (g) 120, (h) 95.

23. Tschoegl, N. W. *The Phenomenological Theory of Linear Viscoelastic Behavior*; Springer-Verlag: Berlin, 1989; pp (a) 513, (b) 551, (c) 69, (d) 47, (e) 560, (f) 21, (g) 508, (h) 55, (i) 413, (j) 401, (k) 406, (m) 157, (n) 409, (p) 157, (q) 425.

24. White, J. L. *Principles of Polymer Engineering Rheology*; Wiley & Sons: New York, 1990; pp (a) 83, (b) 163, (c) 174.

25. Matsuoka, S. *Relaxation Phenomena in Polymers*; Hanser: Munich, 1992.

26. Yanovsky, Y. G. *Polymer Rheology: Theory and Practice*; Chapman & Hall: New York, 1993; pp (a) 200, (b) 235.

27. Fuller, G. G. *Optical Rheometry of Complex Fluids*; Gubbins, K. E., Ed.; Topics in Chemical Engineering; Oxford University Press; New York, 1995; pp (a) 117, (b) 109, (c) 149, (d) 167, (e) 193.

28. Coleman, B. D.; Noll, W. Foundations of linear viscoelasticity. *Rev. Mod. Phys.* **1961**, *33*, 239–249.

29. Petrie, C. J. S. *Elongational Flows: Aspects of the Behaviour of Model Elasticoviscous Fluids*; Pitman: London, 1979.

30. Meissner, J.; Hostettler, J. A new elongational rheometer for polymer melts and other highly viscoelastic liquids. *Rheol. Acta* **1994**, *33*, 1–21.

31. Macosko, C.; Starita, J. M. New rheometer is put to the test. *Soc. Plast. Eng. J.* **1971**, *27*(11), 38–42.

32. Plazek, D. J. Magnetic bearing torsional creep apparatus. *J. Polym. Sci., Part A-2* **1968**, *6*, 621–638.

33. Berry, G. C.; Birnboim, M. H.; Park, J. O.; Meitz, D. W.; Plazek, D. J. A rotational rheometer for rheological studies with prescribed strain or stress history. *J. Polym. Sci., Part B: Polym. Phys.* **1989**, *27*, 273–296.

34. Rabinowitsch, B. Über die Viskosität und Elastizität von Solen. *Z. Phys. Chem., Abt. A* **1929**, *145*, 1–26.

35. Barr, G. A. *Monograph of Viscometry*; Oxford University Press (Authorized facsimile by University Microfilms, Ann Arbor, MI, 1967); London, 1931.

36. Shermergor, T. D. Description of Relaxation Phenomena in Structurally Nonhomogeneous Polymers by Correlation Functions. In *Relaxation Phenomena in Polymers*; Bartenev, G. M., Zelenev, Yu. V., Eds.; Wiley & Sons (Halsted Press): New York, 1974.

37. Berry, G. C. Rheological and rheo-optical studies on nematic solutions of a rodlike polymer: Bingham Award Lecture. *J. Rheol.* **1991**, *35*, 943–983.

38. Meissner, J. Experimental problems and recent results in polymer melt rheometry. *J. Macromol. Chem., Macromol. Symp.* **1992**, *56*, 25–42.

39. Ostrowsky, N.; Sornette, D.; Parker, R.; Pike, E. R. Exponential sampling method for Light scattering polydispersity analysis. *Opt. Acta* **1981**, *28*, 1059–1070.

40. Baumgaertel, M.; Winter, H. H. Determination of discrete relaxation and retardation time spectra from dynamic mechanical data. *Rheol. Acta* **1989**, *28*, 511–519.

41. Sips, R. General theory of deformation of viscoelastic substances. *J. Polym. Sci.* **1951**, *7*, 191–205.

42. Nakamura, K.; Wong, C. P.; Berry, G. C. Strain criterion in nonlinear creep and recovery in concentrated polymer solutions. *J. Polym. Sci., Polym. Phys. Ed.* **1984**, *22*, 1119–11148.

43. Berry, G. C.; Plazek, D. J. On the use of stretched-exponential functions for both linear viscoelastic creep and stress relaxation. *Rheol. Acta* **1997**, *36*, 320–329.

44. Riande, E.; Markovitz, H. Approximate relations among compliance functions of linear viscoelasticity for amorphous polymers. *J. Polym. Sci., Polym. Phys. Ed.* **1975**, *13*, 947–951.

45. Plazek, D. J.; Raghupathi, N.; Orbon, S. J. Determination of dynamic storage and loss compliances from creep data. *J. Rheol.* **1979**, *23*, 477–488.

46. Markovitz, H. The reduction principle in linear viscoelasticity. *J. Phys. Chem.* **1965**, *69*, 671.

47. Markovitz, H. Superposition in rheology. *J. Polym. Sci., Polym. Symp.* **1975**, *50*, 431–456.

48. Berry, G. C.; Fox, T. G The viscosity of polymers and their concentrated solutions. *Adv. Polym. Sci.* **1968**, *5*, 261–357.

49. Berry, G. C. Crossover behavior in the viscosity of semiflexible polymers: Intermolecular interactions as a function of concentration and molecular weight. *J. Rheol.* **1996**, *40*, 1129–1154.

50. Boyer, R. F. The relation of transition temperatures to chemical structure in high polymers. *Rubber Rev.* **1963**, *36*, 1303–1421.

51. Berry, G. C. Thermodynamic and conformational studies of polystyrene. II. Intrinsic viscosity studies on dilute solutions of linear polystyrenes. *J. Chem. Phys.* **1967**, *46*, 1338–1352.

52. Berry, G. C. Molecular Weight Distribution. In *Encyclopedia of Materials Science and Engineering*; Bever, M. B., Ed.; Pergamon Press: Oxford, 1986; pp 3759–3767.

53. Russel, W. B.; Saville, D. A.; Schowalter, W. R. *Colloidal Dispersions*; Cambridge University Press: Cambridge, UK, 1989; pp (a) 498, (b) 88, (c) 129, (d) 162, (e) 466, (f) 469, (g) 474.

54. Berry, G. C. Remarks on a relation among the intrinsic viscosity, the radius of gyration, and the translational friction coefficient. *J. Polym. Sci., Part B: Polym. Phys.* **1988**, *26*, 1137–1142.

55. Yamakawa, H. *Modern Theory of Polymer Solutions*; Harper and Row: New York, 1971; pp (a) 266, (b) 314.

56. Flory, P. J.; Fox, T. G Molecular configuration and thermodynamic properties from intrinsic viscosities. *J. Polym. Sci.* **1950**, *5*, 745–747.

57. Casassa, E. F., Berry, G. C. Reflections and comments on Molecular configuration and thermodynamic properties from intrinsic viscosities by Paul J. Flory and Thomas G Fox. *J. Polym. Sci., Part B: Polym. Phys.* **1996**, *34*, 203–206.

58. Casassa, E. F.; Berry, G. C. Polymer Solutions. In *Comprehensive Polymer Science*; Allen, G., Ed.; Pergamon Press: New York, 1988; Vol. 2, Chapter 3.

59. Brandrup, J.; Immergut, E. H., Eds., *Polymer Handbook*, 3rd ed.; Wiley: New York, 1989.

60. Casassa, E. F.; Berry, G. C. Angular distribution of intensity of Rayleigh scattering from comblike branched molecules. *J. Polym. Sci., Part A-2* **1966**, *4*, 881–897.

61. Lodge, T. P. Solvent dynamics, local friction, and the viscoelastic properties of polymer solutions. *J. Phys. Chem.* **1993**, *97*, 1480–1487.

62. Harrison, G.; Lamb, J.; Matheson, A. J. The viscoelastic properties of dilute solutions of polystyrene in toluene. *J. Phys. Chem.* **1966**, *68*, 1072–1078.

63. Osaki, K. Viscoelastic properties of dilute polymer solutions. *Adv. Polym. Sci.* **1973**, *12*, 1–64.

64. Birnboim, M. H. The viscoelastic properties of low molecular weight polystyrene solutions in high frequency regime: Polymer-solvent interaction. *Proc. IUPAC Macro 82* **1982**, *July 12–16*, 872.

65. Schrag, J. L.; Stokich, T. M.; Strand, D. A.; Merchak, P. A.; Landry, C. J. T.; Radtke, D. R.; Man, V. F.; Lodge, T. P.; Morris, R. L.; Hermann, K. C.; Amelar, S.; Eastman, C. E.; Smeltzly, M. A. Local modification of solvent dynamics by polymeric solutes. *J. Non-Cryst. Solids* **1991**, *131*, 537–543.

66. Yoshizaki, T.; Takaeda, Y.; Yamakawa, H. On the correlation between the negative intrinsic viscosity and the rotatory relaxation time of solvent molecules in dilute polymer solutions. *Macromolecules* **1993**, *26*, 6891–6896.

67. Yamakawa, H. Concentration dependence of polymer chain configurations in solution. *J. Chem. Phys.* **1961**, *34*, 1360–1372.

68. Batchelor, G. K. The effect of Brownian motion on the bulk stress in a suspension of spherical particles. *J. Fluid Mech.* **1977**, *83*, 97–117.

69. Pal, R. Rheology of Emulsions Containing Polymeric Liquids. In *Encyclopedia of Emulsion Technology*; Becher, P., Ed.; Marcel Dekker: New York, 1996; Vol. 4, pp 93–263.

70. de Gennes, P.-G. *Scaling Concepts in Polymer Physics*: Cornell University Press: Ithaca, NY, 1979.

71. Park, J. O.; Berry, G. C. Moderately concentrated solutions of polystyrene. 3. Viscoelastic measurements at the Flory Theta temperature. *Macromolecules* **1989**, *22*, 3022–3029.

72. Onogi, S.; Masuda, T.; Miyanaga, N.; Kimura, Y. Dependence of viscosity of concentrated polymer solutions upon molecular weight and concentration. *J. Polym. Sci., Part A-2* **1967**, *5*, 899–913.

73. Hager, B. L.; Berry, G. C. Moderately concentrated solutions of polystyrene—1. Viscosity as a function of concentration, temperature, and molecular weight. *J. Polym. Sci., Polym. Phys. Ed.* **1982**, *20*, 911–928.

74. Narayanaswamy, O. S. Annealing of Glass. In *Glass: Science and Technology*; Uhlmann, D. R., Kreidl, N. J., Ed.; Academic Press: New York, 1986; Vol. 3, pp 275–318.

75. Plazek, D. J.; Berry, G. C. Physical Aging of Polymer Glasses. In *Glass: Science and Technology*; Uhlmann, D. R., Kreidl, N. J., Ed.; Academic Press: New York, 1986; Vol. 3, pp 363–399.

76. Graessley, W. W. The entanglement concept in polymer rheology. *Adv. Polym. Sci.* **1974**, *16*, 1–179.

77. Berry, G. C.; Nakayasu, H.; Fox, T. G Viscosity of poly (vinyl acetate) and its concentrated solutions. *J. Polym. Sci., Polym. Phys. Ed.* **1979**, *17*, 1825–1844.

78. da Andrade, E. N. *Viscosity and Plasticity*; Chemical Publishing Co.: New York, 1951.

79. Berry, G. C. The stress-strain behavior of materials exhibiting Andrade creep. *Polym. Eng. Sci.* **1976**, *16*, 777–781.

80. Orbon, S. J.; Plazek, D. J. The recoverable compliance of a series of bimodal molecular weight blends of polystyrene. *J. Polym. Sci., Polym. Phys. Ed.* **1979**, *17*, 1871–1890.

81. Berry, G. C. Terminal retardation times and weights for the Rouse model for a crosslinked network. *J. Polym. Sci., Part B: Polym. Phys.* **1987**, *25*, 2203–2205.

82. Masuda, T.; Takahashi, M.; Onogi, S. Steady-state compliance of polymer blends. *Appl. Polym. Symp.* **1973**, *20*, 49–60.

83. Kurata, M. Effect of molecular weight distribution on vis-

coelastic properties of polymers. 2. Terminal relaxation time and steady-state compliance. *Macromolecules* **1984**, *17*, 895–898.

84. Fujita, H.; Einaga, Y. Self diffusion and viscoelasticity in entangled systems. II. Steady-state viscosity and compliance of binary blends. *Polym. J.* **1985**, *17*, 1189–1195.

85. Montfort, J. P.; Marin, G.; Monge, P. Molecular weight distribution dependence of the viscoelastic properties of linear polymers: The coupling of reptation and tube-renewal effects. *Macromolecules* **1986**, *19*, 1979–1988.

86. des Cloizeaux, J. Relaxation of entangled polymeric melts. *Macromolecules* **1990**, *23*, 3992–4006.

87. Tsenoglou, C. Molecular weight polydispersity effects on the viscoelasticity of entangled linear polymers. *Macromolecules* **1991**, *24*, 1762–1767.

88. Berry, G. C. Rheological properties of blends of rodlike chains with flexible or semiflexible chains. *Trends Polym. Sci.* **1993**, *1*, 309–314.

89. Berry, G. C. Rheology of blends of liquid crystalline and flexible chain polymers. *Trends Polym. Sci.* **1996**, *4*, 289–292.

90. Plazek, D. J.; Chay, I. C. The evolution of the viscoelastic retardation spectrum during the development of an epoxy resin network. *J. Polym. Sci., Part B: Polym. Phys.* **1991**, *29*, 17–29.

91. Winter, H. H.; Mours, M. Rheology of polymers near liquid-solid transitions. *Adv. Polym. Sci.* **1997**, *134*, 167–234.

92. Plazek, D. J. Oh, thermorheological simplicity, wherefore art thou? *J. Rheol.* **1996**, *40*, 987–1014.

93. Plazek, D. J.; Chelko, A. J., Jr. Temperature dependence of the steady state recoverable compliance of amorphous polymers. *Polymer* **1977**, *18*, 15–18.

94. Inoue, T.; Mizukami, Y.; Okamoto, H.; Matsui, H.; Watanabe, H.; Kanaya, T.; Osaki, K. Dynamic birefringence of vinyl polymers. *Macromolecules* **1996**, *29*, 6240–6245.

95. Struik, L. C. E. *Physical Aging in Amorphous Polymers and Other Materials*; Elsevier: Amsterdam, 1978.

96. Bero, C. A.; Plazek, D. J. Volume-dependent rate processes in an epoxy resin. *J. Polym. Sci., Part B: Polym. Phys.* **1991**, *29*, 39–47.

97. Lodge, A. S.; Meissner, J. On the use of instantaneous strains, superposed on shear and elongational flows of polymeric liquids to test the Gaussian network hypothesis and to estimate the segment concentration and its variation during flow. *Rheol. Acta* **1971**, *11*, 351–352.

98. Coleman, B. D.; Markovitz, H. Asymptotic relations between shear stresses and normal stresses in general incompressible fluids. *J. Polym. Sci., Polym. Phys. Ed.* **1974**, *12*, 2195–2207.

99. Coleman, B. D.; Markovitz, H. Normal stress effects in second-order fluids. *J. Appl. Phys.* **1964**, *35*, 1–9.

100. Coleman, B. D.; Dill, E. H.; Toupin, R. A. A phenomenological theory of streaming birefringence. *Arch. Ration. Mech. Anal.* **1970**, *39*, 358–399.

101. Kannan, R. M.; Kornfield, J. A. The third-normal stress difference in entangled melts: Quantitative stress-optical measurements in oscillatory shear. *Rheol. Acta* **1992**, *31*, 535–544.

102. Bernstein, B.; Kearsley, E. A.; Zapas, L. J. A study of

stress relaxation with finite strain. *Trans. Soc. Rheol.* **1963**, *7*, 391–410.

103. Wagner, M. H.; Laun, H. M. Nonlinear shear creep and constrained recovery of a LDPE melt. *Rheol. Acta* **1978**, *17*, 138–148.

104. Wagner, M. H.; Stephenson, S. E. The irreversibility assumption of network disentanglement in flowing polymer melts and its effects on elastic recoil predictions. *J. Rheol.* **1979**, *23*, 489–504.

105. Graessley, W. W.; Park, W. S.; Crawley, R. L. Experimental tests of constitutive relations for polymers undergoing uniaxial shear flows. *Rheol. Acta* **1977**, *16*, 291–301.

106. Cox, W. P.; Merz, B. H. Correlation of dynamic and steady flow viscosities. *J. Polym. Sci.* **1958**, *28*, 619–622.

107. Osaki, K.; Bessho, N.; Kojimoto, T.; Kurata, M. Flow birefringence of polymer solutions in time-dependent field. Relation between normal and shear stresses on application of step-shear strain. *J. Rheol.* **1979**, *23*, 617–624.

108. Takahashi, M.; Masuda, T.; Bessho, N.; Osaki, K. Stress measurements at the start of shear flow: Comparison of data from a modified Weissenberg Rheogoniometer and from flow birefringence. *J. Rheol.* **1980**, *24*, 516–520.

109. Menezes, E. V.; Graessley, W. W. Nonlinear rheological behavior of polymer systems for several shear-flow histories. *J. Polym. Sci., Polym. Phys. Ed.* **1982**, *20*, 1817–1833.

110. Pearson, D. S.; Kiss, A. D.; Fetters, L. J.; Doi, M. Flow-induced birefringence of concentrated polyisoprene solutions. *J. Rheol.* **1989**, *33*, 517–535.

111. Rochefort, W. E.; Heffner, G. W.; Pearson, D. S.; Miller, R. D.; Cotts, P. M. Rheological and rheooptical studies of poly(alkylsilanes). *Macromolecules* **1991**, *24*, 4861–4864.

112. Wong, C. P., Berry, G. C. Rheological studies on concentrated solutions of heterocyclic polymers. *Polymer* **1979**, *20*, 229–240.

113. Casson, N. A Flow Equation for Pigment-Oil Suspensions of the Printing Ink Type. In *Rheology of Disperse Systems*; Mill, C. C., Ed.; Pergamon Press: London, 1959; pp 84–104.

114. Markovitz, H. Relative rates of creep and relaxation in shear, elongation and isotropic compression. *J. Polym. Sci., Polym. Phys. Ed.* **1973**, *11*, 1769–1777.

115. Kovacs, A. J.; Aklonis, J. J.; Hutchinson, J. M.; Ramos, A. R. Isobaric volume and enthalpy recovery of glasses. II. A transparent multiparameter theory. *J. Polym. Sci., Polym. Phys. Ed.* **1979**, *17*, 1097–1162.

116. Philippoff, W. Flow birefringence and stress. *J. Appl. Phys.* **1956**, *27*, 984–989.

117. Philippoff, W. Elastic stresses and birefringence in flow. *Trans. Soc. Rheol.* **1961**, *5*, 163–191.

118. Janeschitz-Kriegl, H. Flow birefringence of elastico-viscous polymer systems. *Adv. Polym. Sci.* **1969**, *6*, 170–318.

119. Coleman, B. D.; Dill, E. H. Photoviscoelasticity: Theory and Practice. In *The Photoelastic Effect and Its Applications*; Kestens, J., Ed.; Springer-Verlag: Berlin, 1975; pp 455–505.

120. Wales, J. L. S. *The Application of Flow Birefringence to*

Rheological Studies of Polymer Melts; Delft University Press: Delft, The Netherlands, 1976.

121. Riande, E.; Saiz, E. *Dipole Moments and Birefringence of Polymers*; Prentice Hall: Englewood Cliffs, NJ, 1992.

122. Read, B. E. Viscoelastic behavior of amorphous polymers in the glass-rubber transition region: Birefringence studies. *Polym. Eng. Sci.* **1983**, *23*, 835–843.

123. Kornfield, J. A.; Fuller, G. G.; Pearson, D. S. Third normal stress difference and component relaxation spectra for bidisperse melts under oscillatory shear. *Macromolecules* **1991**, *24*, 5429–5441.

124. Takahashi, T. Measurement of first and second normal stress difference of a polystyrene solution using a simultaneous mechanical and optical measurement technique. In *Fluid Measurement and Instrumentation: 1995*; Morris, G. L., Nishi, M., Morrow, T. B., Gore, R. A., Eds.; American Society of Mechanical Engineers: New York, 1995; FED-211, pp 31–34.

125. Olson, D. I.; Brown, E. F.; Burghardt, W. R. Second normal stress difference relaxation in a linear polymer melt following step-strain. *J. Polym. Sci., Part B: Polym. Phys.* **1998**, *36*, 2671–2675.

126. Coleman, B. D.; Dill, E. H. Theory of induced birefringence in materials with memory. *J. Mech. Phys. Solids* **1971**, *19*, 215–243.

127. Lodge, T. P.; Schrag, J. L. Oscillatory flow birefringence properties of polymer solutions at high effective frequencies. *Macromolecules* **1984**, *17*, 352–360.

128. Chirinos, M. L.; Crain, P.; Lodge, A. S.; Schrag, J. L.; Yaritz, J. Measurements of N_1-N_2 and eta in steady shear flow and eta', eta", and birefringence in small-strain oscillatory shear for the polyisobutylene solution M1. *J. Non-Newtonian Fluid Mech.* **1990**, *35*, 105–119.

129. Ahn, K. H.; Schrag, J. L.; Lee, S. J. Bead-spring chain model for the dynamics of dilute polymer solutions Part 2. Comparisons with experimental data. *J. Non-Newtonian Fluid Mech.* **1993**, *50*, 349–373.

130. Osaki, K.; Okamoto, H.; Inoue, T.; Hwang, E.-J. Molecular interpretation of dynamic birefringence and viscoelasticity of amorphous polymers. *Macromolecules* **1995**, *28*, 3625–3630.

131. Mott, P. H.; Roland, C. M. Birefringence of polymers in the softening zone. *Macromolecules* **1998**, *31*, 7095–7098.

132. Osaki, K.; Inoue, T. Some phenomenological relations for strain-induced birefringence of amorphous polymers. *Nihon Reoroji Gakkaishi* **1991**, *19*, 130–132 (in English).

133. Lodge, A. S. Stress relaxation after a sudden shear strain. *Rheol. Acta* **1975**, *14*, 664–665.

134. Osaki, K.; Kimura, S.; Kurata, M. Relaxation of shear and normal stresses in step-shear deformation of a polystyrene solution. Comparison with the predictions of the Doi-Edwards theory. *J. Polym. Sci., Polym. Phys. Ed.* **1981**, *19*, 517–527.

135. Krieger, I. M. Rheology of monodisperse latices. *Adv. Colloid Interface Sci.* **1972**, *3*, 111–136.

136. Russel, W. B.; Gast, A. P. Nonequilibrium statistical mechanics of concentrated colloidal dispersions: Hard spheres in weak flows. *J. Chem. Phys.* **1986**, *84*, 1815–1826.

137. van der Werff, J. C.; de Kruif, C. G.; Blom, C.; Mellema, J. Linear viscoelastic behavior of dense hard-sphere dispersions. *Phys. Rev. A* **1989**, *39*, 795–807.

138. Brady, J. F. The rheological behavior of concentrated colloidal dispersions. *J. Chem. Phys.* **1993**, *99*, 567–581.

139. Shikata, T.; Pearson, D. S. Viscoelastic behavior of concentrated spherical suspensions. *J. Rheol.* **1994**, *38*, 601–616.

140. Lionberger, R. A.; Russel, W. B. High frequency modulus of hard sphere colloids. *J. Rheol.* **1994**, *38*, 1885–1908.

141. Lionberger, R. A.; Russel, W. B. Effectiveness of nonequilibrium closures for the many body forces in concentrated colloidal dispersions. *J. Chem. Phys.* **1997**, *106*, 402–416.

142. Chong, J. S.; Christiansen, E. B.; Baer, A. D. Rheology of concentrated suspensions. *J. Appl. Polym. Sci.* **1971**, *15*, 2007–2021.

143. Beenakker, C. W. J. The effective viscosity of a concentrated suspension of spheres (and its relation to diffusion). *Physica A* **1984**, *128A*, 48–81.

144. Beenakker, C. W. J.; Mazur, P. Diffusion of spheres in a concentrated suspension. II. *Physica A* **1984**, *126A*, 349–370.

145. de Gennes, P. G.; Prost, J. *The Physics of Liquid Crystals*; Oxford University Press: New York, 1993.

146. Mellema, J.; de Kurif, C. G.; Blom, C.; Vrij, A. Hard sphere colloidal dispersions: Mechanical relaxation pertaining to thermodynamic forces. *Rheol. Acta* **1987**, *26*, 40–44.

147. Arora, A. K.; Tata, B. V. R.; Eds. *Ordering and Phase Transitions in Charged Colloids*; VCH Publishers: New York, 1996.

148. Yoshida, J.; Yamanaka, J.; Koka, T.; Ise, N.; Hashimoto, T. Novel crystallization process in dilute ionic colloids. *Langmuir* **1998**, *14*, 569–574.

149. Weiss, J. A.; Larsen, A. E.; Grier, D. G. Interactions, dynamics, and elasticity in charge-stabilized colloidal crystals. *J. Chem. Phys.* **1998**, *109*, 8659–8666.

150. Meitz, D. W.; Yen, L.; Berry, G. C.; Markovitz, H. Rheological studies of dispersions of spherical particles in a polymer solution. *J. Rheol.* **1988**, *32*, 309–351.

23

Mechanical Properties

GARTH L. WILKES

The mechanical response of polymeric materials is one of the most important subjects within the field of polymer science. Although there are many other properties that are of equivalent or greater importance in specific applications, the utility of polymeric systems for structural applications is certainly one of the reasons for their extensive growth and size of their commercial markets. Why is this so? Structural materials, in general, fall into the categories of metallics, inorganics (ceramics), or polymers. Although some crossover occurs among these fields, particularly in the area of composite materials, polymers tend to stand apart from the other two for distinct reasons. One reason is that polymers are generally of considerably lower density than the other two classes of materials. Polymers tend to have densities of the order of 1 g/cm^3, although poly(tetrafluoroethylene), a high-density polymer, is about 2.2 g/cm^3. In contrast, metallics such as mild steel have a density of more than 7 g/cm^3, whereas many refractory or ceramic materials are often 3 g/cm^3 or above. Porosity, of course, can be induced in any of these materials, which thereby lowers the overall bulk density, a feature of importance with regard to formation of cellular materials or foams—a topic that is discussed near the end of this chapter.

Often, one wishes to produce a material with some set of desired properties but with minimal mass (such as for applications in the transportation industry). Thus, the nature of specific properties becomes extremely important. Specific mechanical properties are defined by dividing a given important parameter (such as stiffness, tensile strength, or other variables that are addressed later in the chapter) by the density of the material. This is one of the major features that promotes polymers for their present applications, that is, they have high specific properties in many arenas of mechanical behavior.

Another important issue is the processability of polymers. Polymers require relatively low processing energy relative to metals or ceramics, and therefore the economics of their production is often much more favorable. Because of the tremendous range of chemical compositions and structures for polymeric materials, some will display high chemical-solvent resistance in some cases, whereas others may become soluble or degrade in the same media. Hence, recognition of the environmental conditions that may occur for a given structural application is required.

It is not the principal goal of the author to address the specifics of macromolecular chemistry and structural details, although they are used where necessary, in this chapter to demonstrate how mechanical response can be altered by some of the more important molecular parameters. As a result, the next section addresses some useful concepts that are helpful in promoting an understanding of the mechanical response of polymeric systems when measured under a given set of conditions and environment. Specifically, the intent is to promote a molecular and qualitative approach with minimal use of mathematics to facilitate an understanding of mechanical response. There are many references dealing with a higher level presentation of both phenomenological and molecularly based theoretical aspects of mechanical behavior,[1-8] and the interested reader may wish to pursue these more detailed sources.

Useful Concepts Concerning Polymers and Their Importance with Respect to Mechanical Behavior

The majority of older polymer texts, as well as some newer ones, often state that polymers are like spaghetti because of their chainlike structure. This concept is one that the author rejects for two principal reasons. The first, and possibly the lesser important of the two, is that if one calculates the aspect ratio for a polymer chain (the ratio of the extended chain length to its diameter) that ratio is typically far greater than that of a strand of spaghetti, even the thinnest angel hair pasta! Certainly the aspect ratio for a given chemical structure and therefore a given molecular diameter will scale directly with molecular weight because doubling the molecular weight will also double the length of the extended molecule. In addition, more bulky polymer chains containing a side group on the repeat unit will have a larger molecular diameter and therefore a lower aspect ratio for a given extended length. However, in general, for most polymeric systems that are applicable for structural use, the molecular weights are such that the corresponding aspect ratios, irrelevant of chemistry, are typically well in excess of the aspect ratios of spaghetti. Generally, it is only the low molecular weight or oligomeric forms of polymers that may be rightfully compared with the pasta analogy. To further emphasize this point, consider that the calculation for linear polyethylene would show that for a molecular weight of 2.8 million g/mol, which is a particularly high molecular weight polyethylene but still commonly a part of many polyethylene materials, the aspect ratio is of the order of 50,000 and the molecular extended length is of the order of 250,000 Å (25 μm). This latter figure is four times the thickness of a quarter mil film. At the lower end, where polyethylene is not really usable for structural applications (e.g., 10,000 g/mol), the aspect ratio is of the order of 200, which is still well in excess of a strand of spaghetti. As stated above, polymers containing bulky side groups such as polystyrene will have lower relative aspect ratios for comparable molecular weights, but the general trends will be the same.

The second and more important feature that addresses the lack of reality in the spaghetti comparison is the fact that polymer molecules vary in terms of their molecular mobility, particularly depending on their state of thermal energy. Spaghetti, however, never undergoes significant thermally induced motions. As is discussed later, if a polymer is below its glass transition temperature, T_g, or it is within a crystal phase, multisegmental cooperative backbone motion is typically limited, and one will find that the chain backbone is relatively rigid or stiff. However, at a higher temperature above T_g of amorphous materials or above the melt temperature for a crystal phase, the molecular backbone will typically display high cooperative segmental mobility, which will be either increased or decreased as the thermal energy is removed or added to the system accordingly. Certainly there are many molecular parameters that can also influence mobility, including those of cross-linking, plasticizer effects, and crystallinity. However, the important issue here, and one that is discussed many times in this chapter, is that polymers are much more analogous to the appropriate aspect ratio of a worm in contrast to a strand of spaghetti, which does not move. The analogy of polymers being similar to worms is in line with the well-founded theories of polymer reptation, where the term reptation refers to snakelike or wormlike motion caused by thermally induced Brownian mobility as discussed by, for example, Nobel laureate Pierre DeGennes, who helped to formulate some of the general theories of reptation[9,10] in conjunction with Doi and Edwards.[11] The concept of thermally induced molecular mobility and its influence on the wormlike motion of polymer chains are used considerably within this chapter, and cannot be overstated.

The level of mobility for a given chemistry will depend on the amount of thermal energy in a system, which can be expressed by the Boltzmann constant, k, multiplied by the absolute temperature, the product of which carries the dimensions of energy. The equivalent on a molar basis would be the product of the molar gas constant, R, with the absolute temperature, so the only difference between these two expressions of energy is the factor of Avogadro's number. Taking the analogy of worms further, if one were to place an entangled group of these creatures within liquid nitrogen, this would minimize their backbone motion and would provide a material that would become glasslike, whereas in contrast, placing these same species on a hot stove would provide a very rapid motion (at least for a short time) relative to a lower temperature. The description may seem overly simplistic, but its utility has much merit in addressing many of the features of the mechanical response of polymeric systems when measured at different temperatures and/or rates of deformation, as is apparent within this chapter.

Another important relevant concept regarding polymer systems is the issue of order. If the system displays only random packing of the chains, it is said to be amorphous and therefore of zero crystallinity. If molecules in a system are packed and ordered in three dimensions, the system is said to be crystallizable and will generally transform at the melting point, T_m, to a liquid. Amorphous materials, therefore, do not display a melting point but do display the principal transition, T_g, below which (for a given loading rate experiment) cooperative segmental backbone motion no longer occurs, thereby causing the material to appear rigid and stiff. This does not mean that below T_g there are no local-scale motions within the molecular system, but only that the backbone of the macromolecular chain is limited in cooperative segmental mobility. It is also apparent from the discussion that T_g is not a single fixed number, but is kinetically influenced, that is, it changes with the rate (frequency) of loading. This is addressed in the discussions of dynamic mechanical properties. The glass transition also changes with rate of heating or cooling in a calorimetric (differential scanning calorimetry [DSC]) measurement, but this subject is outside the scope of this chapter.

Certainly there can be intermediate order between that of the amorphous and fully three-dimensionally crystallized regions. Some materials fall into the category of liquid crystalline order, and may display one- or two-dimensionally ordered regions, such as those illustrated in Figure 23.1. Those molecules (small molecules as well as some polymers) displaying liquid crystalline textures are often com-

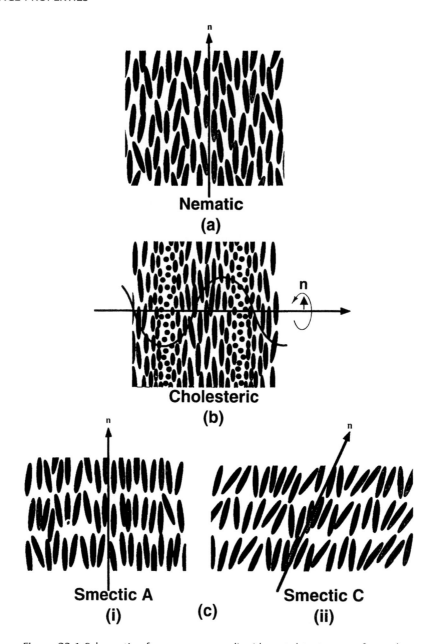

Figure 23.1 Schematic of some common liquid crystal textures as formed from rodlike molecules: (a) nematic, (b) cholesteric or twisted nematic, and (c) (i) smectic A and (ii) smectic C. n refers to the general orientation of the molecular axis and is commonly called the order parameter. (Adapted with permission from reference 12.)

posed of rodlike or platelike molecular architectures or they may contain such elements within their structure, as in the case of polymers, which in turn tend to favor local ordering such as do logs floating down a river. Because of space limitations, all facets of morphological characteristics of polymer materials are not discussed in great detail, but it should be recognized that these features can play a crucial role in determining the mechanical response of a system. For polymeric materials, the reader is likely familiar with the fact that those that have molecular symmetry may display

crystallinity if the kinetics are appropriate, and will develop a semicrystalline morphology with the presence of both crystalline and amorphous regions in the same solid. Typically, semicrystalline polymers are made up of folded chain lamellae often interconnected with amorphous regions. The lamellae are often the order of 100 Å in thickness but their lateral dimensions can vary considerably depending on crystallization conditions and chain perfection.[13] The nature of connectivity of the crystals with either themselves or through the amorphous regions is of great importance in determining mechanical response—particularly when one is above the glass transition temperature of the amorphous phase. The level of crystallinity will also play a very influential role, as can the type of crystalline morphology for a given level of crystallinity.

Another key parameter affecting mechanical behavior is the level of molecular orientation often induced by a specific processing step, such as fiber spinning or film drawing. Because of the wormlike or chainlike nature of these systems, the chain backbone can often be promoted to partially or highly orient along a specific axis or within a given plane, thereby leading to high anisotropy in physical properties where the term anisotropy refers to directional dependence. This topic is also discussed late in the chapter because of its importance in the practical usage of polymers.

A critical concept for this chapter is that of molecular entanglements between chains. Polymers, in general, do not display entanglements at low molecular weights, but once they reach a critical mass, which varies for different chemistry, entanglements and their important effects are noted. These entanglements play an enormous role in influencing the mechanical response of the system—particularly ultimate properties. The molecular concept of an entanglement is illustrated in Figure 23.2a–c, where some additional caveats are provided. First, the entanglements promoted by neighboring chains will distinctly begin to place restrictions on the direction in which a given backbone can move or diffuse. As noted in Figure 23.2a, where no entanglements exist, the molecular structure should be able to diffuse, in general, along any particular direction in a similar way that Brownian motion occurs for lower molecular weight molecules that do not possess entanglements. However, once the molecular weight is sufficiently high to promote entanglements for a linear system as shown in Figure 23.2b, the entanglement points serve to help define a tubelike structure as illustrated with the dotted lines that limit the direction that the overall backbone can diffuse through by Brownian motion if sufficient thermal energy and time exist. This does not indicate that no motion or "wiggling" occurs between entanglement points, but only that the full movement of the molecule would be limited more to the tubelike structure. Finally, in Figure

23.2c, a long-chain branch (LCB) is shown, where LCB denotes a branch that is long enough to entangle. The presence of a single LCB will make it considerably difficult for that molecule to move through the entanglement "hoops" of its neighbors, thereby greatly increasing its molecular relaxation time for flow accordingly.

What are the values for the molecular weight between entanglements (as often determined by melt rheological investigations)? The value determined by melt rheology is typically denoted by M_c, and is often taken to be about twice that of the true molecular weight between entanglements, M_e. Table 23.1 is provided as adapted from the work of Zhang and Carreau.[14] It is noted that many chemistries have relatively low M_c values, for example, polyethylene, which possesses a value of about 3800 g/mol and repre-

Table 23.1 Critical weight average molecular weight between entanglements $(M_{w_c})^*$

Polymer	M_{w_c}
Polyethylene	3800
Polypropylene	7000
Polystyrene	35,000
Poly(vinyl chloride)	6250
Poly(vinyl acetate)	24,500
Poly(vinyl alcohol)	5300
Polyacrylamide	9100
Poly(a-methyl styrene)	40,800
Polyisobutylene	15,200
Poly(methyl acrylate)	24,100
Poly(ethyl acrylate)	31,300
Poly(methyl methacrylate)	31,000
Poly(n-butyl methacrylate)	60,400
Poly(n-hexyl methacrylate)	91,900
Poly(n-octyl methacrylate)	114,000
Poly(2-ethylbutyl methacrylate)	42,800
Poly(dimethyl siloxane)	24,500
Poly(ethylene oxide)	4400
Poly(propylene oxide)	7700
Poly(tetramethylene oxide)	2500
Cis-polyisoprene	7700
Hydrogenated polyisoprene	4000
Cis, trans, vinyl-polybutadiene	4500
Cis-polybutadiene	5900
1,2-Polybutadiene	12,700
Hydrogenated 1,2-polybutadiene	26,700
Poly(ε-caprolactam) nylon 6	5000
Poly(hexamethylene adipamide) nylon 66	4700
Poly(decamethylene succinate)	4600
Poly(decamethylene adipate)	4400
Poly(decamethylene sebacate)	4500
Poly(diethylene adipate)	4800
Poly(ethylene terephthalate)	3300
Poly(carbonate of bisphenol A)	4900
Poly(ester carbonate of 1-bisphenol A and 2-terephthalic acid)	4800
Poly(ester of bisphenol A and diphenyl sulfone)	7100

*Generally determined from melt rheological measurements.

Adapted from reference 14.

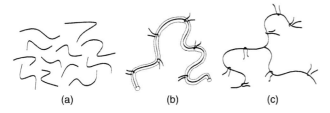

(a)   (b)   (c)

Figure 23.2 Schematic illustrating (a) short linear unentangled polymer, (b) a longer linear entangled polymer, and (c) an entangled polymer having a long chain branch.

sents a degree of polymerization of 135. However, atactic polystyrene (which has the same tetrahedrally bonded carbon-carbon backbone) has an M_c of about 35,000 g/mol, which represents a degree of polymerization of 346! A general trend is that as the chain stiffness tends to increase, the molecular weight between entanglements also tends to increase.[15] Another correlation is that as the molecular diameter becomes greater, so does the molecular weight between entanglements.[16] This last statement is relatively easy to understand if the reader considers the ease of entangling three types of flexible materials that have considerably different diameters. A simple analogy often used by the author in the discussion of molecular weight between entanglements is to consider that the strand or mass length between entanglements in yarn versus a garden hose versus a large flexible irrigation pipe would be considerably different—the lowest for the smallest diameter unit. Simplistic as this seems, the concept is worthwhile when considering polymers of different chemistries.

Chain configurational features can also alter these critical molecular weights (such as through the influence of tacticity [stereoregularity]) so that flow occurs in the presence of entanglements. A polymer chain will either need to slither through the entanglements, or the entanglements will have to disappear (loss of chain constraint), which may happen when its neighbors undertake the same slithering motion. This raises the concept of molecular relaxation times, that is, the time for a molecule or some portion of the molecule to undertake some particular type of motion—one example being the time to flow or reptate through the entanglement tube. Hence, there is a spectrum of relaxation times that occurs in a given system because of the nature of different arrangements of the polymers, their molecular weight variation and molecular weight distribution, and their general chain architecture or chain topology (recall Figure 23.3). Thus, this concept of relaxation behavior that may or may not occur in a given timescale at a given temperature will very much influence the mechanical response for a given set of polymeric "worms" of suitable aspect ratio. With these simplistic yet important concepts in mind, we now begin a more direct discussion of mechanical features of polymeric systems and the types of deformation that they may undergo.

Types of Deformation

A material will undergo mechanical deformation by one of three principal modes or some combination thereof, depending on how the force (or stress) is applied. These deformation types are (1) tensile or extensional deformation, (2) shear deformation, and (3) bulk or hydrostatic deformation. In many tests, the loading profile may be a combination of more than one of the above types, but for simplicity in this chapter, we discuss each of these separately. Emphasis is placed on uniaxial deformation because this is a particularly common procedure for testing the mechanical response of a solid polymer. Interestingly, when probing fluid or melt response, the more typical mode of deformation used in common melt rheological investigations of polymer systems is shear.[18,19]

The three types of deformation are schematically outlined in Figure 23.4a–c. Only a uniaxial deformation scheme is illustrated for the extensional deformation, but biaxial or multiaxial tensile (or compression) behavior certainly occurs in a number of polymer processes, such as in the fabrication of biaxially oriented film. Furthermore, the shear deformation mode shown is that for simple shear, in which a rotational component of the deformation exists. This component is obvious when one shears a ball between his or her hands and rotation occurs. Hence, shear flows are typically referred to as rotational flows, whereas no rotation occurs in the case of a tensile deformation of the same ball. Rather, the deformation promotes only a change in shape of the ball to that of an ellipsoid if uniaxial deformation is applied—a quite different phenomenon from that observed for simple shear. Hence, extensional flow is referred to as an irrotational flow or deformation. This is the first distinct indication that there are likely to be differences in the type of mechanical response that a material will display when placed under a tensile mode of deformation in contrast to shear. Furthermore, from the diagram in Figure 23.4c, bulk or hydrostatic deformation also carries no rotational component. This last type of deformation is much less common but is important in some applications.

Some terminology that will be useful in describing mechanical response needs to be defined (see Figure 23.5a). Regarding uniaxial tensile deformation, if one deforms a rubber band in this mode, there will be a positive force, F, generated along the deformation axis that reflects the molecular resistance to further deformation. If the cross-sectional area of the rubber band is increased, the force will also increase for the same elongation. This suggests the need for normalization of sample size because of the variation in specimen cross-sectional area so that the results from different size specimens can be compared. This is accomplished by one of two approaches. The first and most common is to divide the measured force by the initial cross-sectional area of the test specimen. If the deformation were along the z axis, this would be represented by dividing F by the product $\ell_{0x}\ell_{0y}$. This product is known as the initial cross-sectional area, which we denote by A_0, and which is not a function of deformation. The ratio of F/A_0 provides what is commonly known as the engineering stress σ_0. It is also referred to as the nominal stress. The use of σ_0 rather than the values of force should therefore lead to the same result for the stress-elongation behavior of two identical materials regardless of their cross-sectional area (Figure 23.5b).

An alternate means of defining stress is to use the true stress, denoted by σ_t. This parameter is given as the force

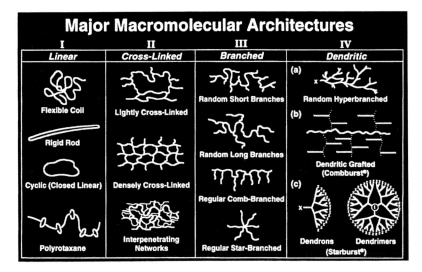

Figure 23.3 Some examples of various molecular architectures that polymer systems may display. (Reproduced with permission from reference 17.)

divided by the actual cross-sectional area, A, that exists at the time the force is determined. Generally, A is a decreasing function of the degree of deformation, and this makes it necessary to determine the cross-sectional area over the desired deformational range if the true stress is the required parameter. As indicated in Figure 23.6, for any material that undergoes considerable deformation, such as a rubber band, there is an ever increasing difference between the true stress and the actual stress as deformation proceeds because A becomes smaller with respect to A_0. The result is that if one were to express the true tensile strength (stress at the break point σ_B), there would be a considerable difference between the two values depending on which choice of stress were used; clearly, the realistic value would be that based on the true stress. However, for the sake of convenience, it is much more common for the results to be reported using the engineering stress versus that of the true stress, thereby underestimating the actual stress at the point of failure for materials that show significant deformation. For specimens that undergo low deformation before fracture (e.g., many brittle glassy polymers), the difference between these two stress parameters is not great, and becomes zero in the limit of no deformation. The important issue is how the results will be provided in describing the mechanical response of the deformed system. It would be misleading to report stress arbitrarily without specifying whether it is engineering stress or true stress.

In the case of shear deformation (Figure 23.4b), if simple shear is imposed, one notes that there is no change in the cross-sectional area and hence only a single stress value (F/A_0) must be reported. This is conventionally denoted by τ in contrast to σ for purposes of separating shear from extensional deformation.

To specify the magnitude of the deformation level, four parameters are introduced. The first is denoted by ε_i and is called the engineering strain (it is also called the Cauchy strain). Using the tensile mode of deformation, the strain along any principal axis is given as

$$\varepsilon_i = (\ell_i \ell_{0i})/\ell_{0i} \tag{1}$$

where ℓ_i represents the new length along the ith axis (x, y, or z), and ℓ_{0i} represents its initial dimension before deformation. Similar values for the strain along the other two principal axes can also be specified in a comparable way. Generally, the strain value of interest in a given deformation is that along the principal axis of deformation. For the uniax-

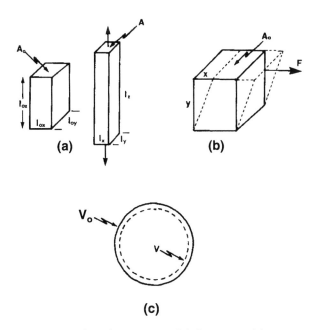

Figure 23.4 Three basic types of deformation: (a) uniaxial (an extensional deformation); (b) simple shear; and (c) hydrostatic or bulk compression.

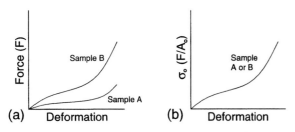

Figure 23.5 (a) Force-extensional deformation sketches for two samples, A and B, of the same material, but where A and B have different cross-sectional area (A is less than B). (b) Engineering stress versus deformation for the same two samples A and B.

ial deformation of an elastomer, this value of strain would increase from zero, whereas the two orthogonal strain values along the respective directions of thickness and width would decrease to negative values as the deformation occurs.

In the case of simple shear (see Figure 23.5b), the shear strain, γ, is expressed for small deformations as x/y, which again is a dimensionless number, as is ε_i. In the case of shear strain, however, the conventional symbol is γ rather than ε. For shear deformation, for the same degree of movement in the shear direction (i.e., along the x direction in Figure 23.5b), the thickness of the element undergoing deformation influences the level of shear strain in a reciprocal manner. Thus, for a thin adhesive bond line between two substrates, a large shear strain may be promoted with relatively small sliding movement of one of the substrates.

Returning to extensional deformation, the principal strain or deformation direction is correlated to another parameter that is often used to express the magnitude of deformation. This is percent elongation and is written as

$$(\% \text{ elongation})_i = \varepsilon_i \times 100 \qquad (2)$$

A third means of expressing deformation is to use extensional or draw ratios, which are denoted by the symbol λ_i. Three such values exist, one correlated to each of the principal axes. These are defined as

$$\lambda_x = \ell_x/\ell_{0x} \qquad (3a)$$
$$\lambda_y = \ell_y/\ell_{0y} \qquad (3b)$$
$$\lambda_z = \ell_z/\ell_{0z} \qquad (3c)$$

In the limit of zero deformation, each respective value of λ becomes unity, in contrast to zero for the respective values of ε_i. Hence, there is a factor of unity that relates λ_i to ε_i, that is,

$$\lambda_i = \varepsilon_i + 1 \qquad (4)$$

There are certain relationships between the respective values of strain, percent elongation, or λ that are useful with regard to determining how the dimensions of a specimen change in a given deformation. In particular, for materials

that undergo what is considered to be a constant volume deformation, and addressing only the values of λ, the product of $\lambda_x \lambda_y \lambda_z$ is unity. In fact, good elastomers closely follow this behavior, and because they generally deform in a uniform fashion, it can be shown that for uniaxial deformation along z

$$\lambda_x = \lambda_y = \lambda_z^{-1/2} \qquad (5)$$

The same approximation can be used to define the deformation of many other materials, although some error is introduced, as is discussed later.

A final parameter describing the degree of deformation for extensional behavior is that of true strain ε_t. Through the use of calculus one can show that this definition is given as

$$\varepsilon_{ti} = \ln(\ell_i/\ell_{0i}) = \ln \lambda_i \qquad (6)$$

One immediately notes that ε_{ti} (also known as the Hencky strain) is also given by the natural logarithm of the respective extension ratio for the same axis (see eq 6 and eq 3). Generally, in the deformation behavior of solid polymeric systems, the true strain is less commonly utilized. However, some data exist in the literature in this form and therefore its definition must be recognized. It is also common to use this variable for extensional flows of polymeric fluids.

Sample Preparation Considerations

The properties that one will determine from a polymeric specimen may dependent on the sample history as well as the conditions under which the measurements are made. It is critical that when the mechanical properties are to be compared between specimens, the thermal history conditions are also recognized as possible contributors to any differences in behavior. For example, many samples may be prepared by a melt process of injection molding, extrusion, or compression molding. The latter process will minimize and, it is hoped, eliminate any molecular orientation effects

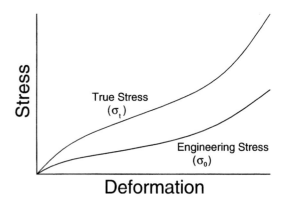

Figure 23.6 General plot of true (σ_t) and engineering stress (σ_0) versus deformation.

within the sample, whereas the other two processes will often promote residual orientation that will generate anisotropy in mechanical as well as other properties. However, in many cases, it may be the effect of orientation that one wishes to denote with respect to its influence on mechanical behavior, as for oriented fibers and films, for example, tear anisotropy in the case of films.

Sample dimensions can also play an important role as, for example, in injection molded samples. Specifically, for a given melt temperature, mold temperature, fill rate, and a particular polymer, a thicker molded part may lead to less orientation along the flow direction because of the longer cooling time that will be needed to solidify the thicker part. This provides additional time for any oriented molecules flowing into the mold to relax to an unoriented isotropic state prior to solidification, be it by vitrification (glass formation) or crystallization. For a rubbery material, relaxation will have time to occur in general because it is above the glass transition. An exception to this would be locking in orientation through a rapid cross-linking reaction if that were part of the fabrication process.

In some instances, solvent casting may be used to prepare a particular sample, such as by film casting, and the film may then be cut into appropriate test specimens. It is often thought that solvent casting will lead to an isotropic sample, but many times this is not the case. In particular, if the polymer has a particular affinity to adhere to the casting substrate as the solvent is lost through evaporation, the polymer film will often display planar orientation. That is, the molecules tend to be somewhat stretched out and lie randomly in the plane of the film, producing the planar structure. Thus, although measurement of a given mechanical property parallel to the plane of the film may not depend on where one cuts the specimen, it still is a property that is not truly representative of the isotropic material where such planar orientation would not exist. Planar orientation can often be recognized in amorphous specimens by noting the specimen dimensions and then raising the temperature above the glass transition temperature, which will tend to promote some level of dimensional change leading to a thickening of the specimen and reduction of the dimensions in the other two directions, as promoted by the retraction of the planar oriented chains.

The considerable variation in properties that sample preparation can promote is one of the major reasons that the American Society for Testing Materials (ASTM)[20] and the International Organization for Standardization (ISO)[21] have formulated specific protocols for specimen preparation and types of tests. Such protocols empower investigators from different laboratories or industries to fairly compare the results of a given test with their peers or competitors because there will be a common thermal and processing history for each material. Any reader who is unfamiliar with the ASTM or ISO protocols should take the time to become familiar with these guidelines.

Testing Conditions

Clearly, the external (nonsystem) variables of temperature and rate of deformation (or frequency of loading) can play significant roles on the properties of polymeric materials. In fact, these two parameters are likely the two most important external influences on the behavior of a given sample. Hence, there is an emphasis on illustrating the impact of time and temperature throughout this chapter. As an example, on the basis of our prior discussion, if a measurement is made under conditions in which the material is glassy versus rubbery, a major difference in behavior will be observed.

Effect of atmosphere can also have an important influence on the state of the material and the mechanical results one obtains. For example, relative humidity can play a major role in the response of hydrophilic polymers such as the polyamides, polyurethanes, and other materials containing many polar groups that possess an affinity for water. Detergent species can also have great influence on the stress cracking behavior of polymeric materials, and hence tests have been developed to investigate the influence of such fugitive species.[20] For example, if the polymer is to be used in the presence of an organic environment, it therefore may be useful to carry out tests on samples that either have been or are being soaked in an organic solvent at the time of the test. This is particularly true for applications of aerospace structural adhesives and related polymeric materials, for which the presence of jet fuel, brake fluids, and other organic liquids may interact and promote changes in property response.[20,21]

In special cases in which a sample is under high pressure such as that promoted by carbon dioxide or some other gas, should that gas serve as a plasticizer for the material, the properties again can be greatly altered.[22] Hence, recognition of the atmospheric conditions surrounding a sample at the time of its mechanical measurement is also a crucial issue in terms of reporting the final results. A final comment regarding pressure is that if the pressure is promoted by a noninteracting gas, the effect of this variable is to squeeze out excess space or free volume within the molecular structure, which therefore places restrictions on molecular mobility and in turn promotes a more brittle-like response of the system. Pressure often is not a principal variable in common polymer testing, but in some applications it is of critical importance as, for example, in transatlantic cable applications in which a polymer is under major pressures at great depths.

Deformation Behavior and Characteristic Parameters Thereof

The reader is reminded that there are three particular types of deformation or combinations thereof: tensile, shear, and hydrostatic deformation. To bring out many of the specific types of deformation behavior, simple tensile deformation will be discussed in some detail. However, the reader should also consider the loading profile for other situations and the corresponding stress-strain behavior. With this preface, several of the important mechanical parameters obtained in tensile loading are now discussed

Generally, the tensile sample, if molded or prepared by a machining operation, will be shaped in the form of a dog bone or dumbbell, as illustrated in Figure 23.7a. Briefly, mechanical jaws of the testing machine must be able to grip the material, which will place a complex loading pattern on the system. To minimize the influence of the mechanical jaws on the test, the specimen is made with a tab that has a larger cross-sectional area at each end. During the test, the stress will be higher along the gauge length, and therefore, the mechanical nature of this lower cross-sectional area will behave independently of the nature of the tab. Note that for many tensile tests, the sample may not be able to be placed into the form of a dog bone shape, for example, in the testing of a fiber. In such cases, one has to try to minimize any complex stress field at the ends of the specimen caused by holding of the sample during the test. For fibers, one such method is to wrap a portion of the fiber at each end around high-friction surface rolls such that when the material is extended, the frictional force will serve as a means by which the fiber will be held in place at the roll surface.

The stress-strain or extensional behavior of polymeric materials can be extremely varied for different systems, as well as for a given system, depending on test conditions. Figure 23.8 illustrates one example of stress-strain behavior in which the curve shows two somewhat different final paths in loading behavior that will be used to develop some of the important properties that would be valuable in reporting. The data are presented as engineering stress-strain

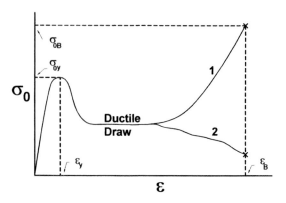

Figure 23.8 Generalized engineering stress-engineering strain plots for two different specimens, 1 and 2.

behavior and this general trend will continue within this chapter. The reader is reminded, however, that there are times where true stress, as well as use of other deformation or extension ratio parameters, are employed in reporting data, and the interconversion between these parameters can be made, as described in the discussion above.

In path 1 of Figure 23.8, if one wishes to specify the degree of deformation at failure of the material (as indicated by an X in Figure 23.8), strain to break, ε_B, is clearly that parameter. In addition, the stress at the break point denoted by σ_{0B} provides an index of the tensile strength. However, in pathway 2 of this figure, although the percent elongation to break is equivalent to that obtained by path 1, the tensile strength is more commonly reported as the high point on the curve (tensile strength at yield), and in this case it will also represent the yield strength, σ_{0y}. Just why are paths 1 and 2 so different, and more specifically, why is path 2 tapering downward at the end of the deformation? This often arises when a tear develops in the side of a sample. As further deformation occurs, the tear proceeds laterally across the material, thereby leaving less material to undergo further deformation, and hence, a decrease in the stress (recall that the stress is based on the observed force divided by the original cross-sectional area). Because a defect such as a tear detracts from the true properties of the material, and to enhance engineering applications of materials, it is usually more important to specify the engineering stress at the high point of the stress-strain curve. In some tests, however, the actual stress at break is desired; in this case, the stress associated with the end point of the B pathway would be reported, but it is less common in practice to utilize this particular parameter.

While ultimate properties are addressed (i.e., properties that relate to the failure conditions of the sample), the reader should recognize that typically any ultimate property is dependent on defects or flaws in the material that can greatly limit the inherent properties of the system. It is also for this reason that a single specimen test value may

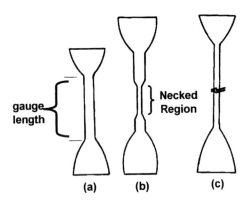

Figure 23.7 Sketch illustrating (a) general "dog bone" or dumbbell shaped tensile specimen with gauge length defined, (b) the same sample, which has undergone yielding and promoted a "neck", (c) fully necked sample.

not necessarily be indicative of the material's behavior. Thus, when measuring ultimate properties, it is necessary to measure many samples and then average these properties accordingly to obtain a meaningful value. In some cases, one may wish to remove major outlier data points, but this will depend on the specifications of a given test, such as might be included or defined within the ASTM procedures.

In Figure 23.8, curve 1, the yield behavior is denoted by σ_{0y} and ε_y, which refer to the yield stress and yield strain, respectively. Up until the yield point, as specified by ε_y, the material, if unloaded, will often return to its original dimensions and therefore show elastic behavior. However, crossing the yield point generally promotes a decrease in the engineering stress and also causes the sample to yield, generally leading to irrecoverable strain at least under the current test conditions, including temperature and other parameters. The occurrence of yield is a particularly significant point with regard to engineering applications because it can occur at relatively small strains or elongations—often much less than 15% elongation. This yield behavior arises in some local region within the sample that undergoes localized drawing or orientation and promotes local "draw down" into distinctly smaller dimensions than the rest of the sample; this is often easily observed by the bottleneck or simply the neck that develops at the localized yield zone (see Figure 23.7b). This yield behavior then promotes a sample that has very different cross-sectional areas in different regions of the sample, which in turn would make specification of the true stress more complex. It is also one of the reasons why engineering stress is a more simplified approach and more commonly used because only the cross-sectional area of the original material is used to determine the engineering stress values.

As the stress decreases from the yield point down to the flatter region of the curve that exists at higher elongations, it may appear at first to represent strain softening, that is, higher levels of strain lead to lower stress. This is somewhat misleading because this decrease in stress is often due to the "neck down" of the sample, which promotes a smaller cross-sectional area. Often, if the area of the thinner region of the sample were utilized to divide into the force to generate the stress instead of the original cross-sectional area, the "turn down" region just past the yield point might well disappear and the yield point would simply become a knee in the curve in contrast to that shown in Figure 23.8. Hence, often the level of stress drop following a yield point simply relates to the level of draw down that a sample undergoes locally, and it can be used to determine what is often referred to as a natural draw ratio for that set of testing conditions. It is important to note, however, that even after one accounts for the differences in cross-sectional area, it is possible that there still may be a decrease in the stress following the yield point, which has arisen from viscous heating of the sample. Specifically, because a material will undergo major local viscous dissipation (loss) of energy caused by molecular flow in a yield zone, and because polymers are poor heat conductors, the portion of mechanical energy converted into heat will lead to a localized softening of the polymeric matrix. This, in turn, can cause a true strain softening to occur depending on sample thickness, rate of deformation, and other parameters. Hence, the yield behavior and what immediately follows is rather complex and the actual behavior is often a combination of both true strain softening and natural draw down.

Following the yield point, one notes that the stress-strain curve remains relatively flat, although real samples may show some undulation in stress behavior along the horizontal region labeled ductile or cold flow (see Figure 23.8). It is this region of the curve that denotes transferring of the neck boundary along the sample length as region by region undergoes yielding and local drawing. Once the neck has been transmitted throughout the sample, unless failure occurs otherwise, there will then typically be an upturn in the stress with further deformation, and this can be denoted as strain hardening (curve 1 in Figure 23.8). This results from additional promotion of molecular orientation throughout the material until failure occurs. At the point of failure, however, full orientation does not exist. Generally, the chainlike polymer molecules are often far removed from full orientation at the time of failure, particularly for amorphous systems. For crystalline polymers, the crystalline regions can indeed develop very high orientation of the chain axis along the deformation direction. (Additional comments about orientation are provided toward the end of this chapter.) Hence, mechanical failure may occur as a result of the inability of the chains to provide further slippage because of entanglements or the associated long relaxation times of the material. Another cause may arise from stress concentrators such as scratches, nicks, or voids in the bulk or surface of the sample.

Figure 23.9 shows another stress-strain curve. This curve is shown without yield, as is often the behavior in amorphous materials when measurements are made above

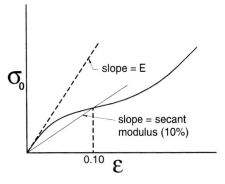

Figure 23.9 Engineering stress-strain plot showing how Young's modulus E and the 10% secant modulus are determined.

the glass transition temperature (e.g., for measurements of a rubber band). In this curve or those discussed previously, a particularly important parameter is the tensile modulus of a material. This is designated as the slope of the initial linear region promoted during deformation. In some cases, the linear region is not necessarily observed at the instant that deformation occurs because of poor mounting of the sample, which may lead to slack and thus to a "toe" being developed in the stress-strain curve ahead of the linear region. In this case, the modulus would be determined by the slope of the curve in the initial linear portion of the stress deformation behavior. It is this parameter that signifies the initial resistance to deformation of the material and also provides a mechanical fingerprint of the initial structure. Definition of this parameter denoted as Young's modulus, modulus of elasticity, or tensile modulus is given as

$$E = \lim_{\Delta\varepsilon\to 0} \frac{\Delta\sigma_0}{\Delta\varepsilon} \qquad (7)$$

where lim represents "in the limit" as the deformation goes to zero; this parameter assumes that no toe in the curve occurs. If the toe exists, extrapolation of the linear region to zero stress will serve as the onset of strain. Note that on the upper side of the initial linear region, where deviation occurs from linear behavior, the associated strain value is often referred to as the proportional limit. This term simply indicates the strain at which, above this value, linear stress-strain response is no longer observed.

Similar definitions for the shear modulus, G, and the bulk modulus, B, are given in eq 8 and eq 9 and relate to the Young's modulus, in that they are also tangent moduli and represent the stiffness of the material under these two different types of loading procedures, that is, shear and bulk compression in contrast to uniaxial extension.

$$G = \lim_{\Delta\gamma\to 0} \frac{\Delta\tau}{\Delta\gamma} \qquad (8)$$

$$B = \lim_{\Delta p\to 0} \left(\frac{\Delta P}{\Delta V/V_{initial}}\right) \qquad (9)$$

where $\Delta V = V_{initial} - V_{new}$, and V and P are volume and pressure, respectively.

As is discussed below, because the modulus provides a signature of initial structure, it is often used as a major parameter in reporting property changes with different test conditions (such as temperature or rate of deformation) or sample parameters (such as plasticizer or filler content). Certainly, there are other higher deformation properties that are also important with regard to their dependence on similar test variables, and one may wish also to report these as a function of such variables as temperature, rate, and environment. Because of limited space within this chapter, however, principal reliance is placed on the tensile-strain parameters when addressing many of the test condition variables in discussions that follow.

Secant modulus is another modulus term that is sometimes used. The formerly defined Young's modulus is a tangent modulus (i.e., slope of a curve), whereas the secant is truly a secant extending from the zero point of strain to the point on the stress deformation curve whereby the secant modulus is to be determined. For example, a 10% secant modulus would represent the slope of the line extending from zero deformation through the 10% deformation point on the loading curve (see Figure 23.9). Hence, a secant modulus must be reported along with the level of deformation with which it is associated, in contrast to the tensile modulus, which represents the slope of the initial linear region of the curve.

What is the relationship between the stiffness parameters such as Young's modulus, E, the shear modulus, G, and the bulk modulus, B, for a given isotropic material? Specifically, will a material be of the same stiffness in tension as it is in shear or bulk deformation? For isotropic materials, a fundamental law exists that shows an interrelationship among these three fundamental parameters with one additional parameter that needs to be specified: Poisson's ratio, μ. This is defined for very small strain levels as

$$\mu = -\frac{\text{transverse strain}}{\text{longitudinal strain}} \qquad (10)$$

Here the transverse strain relates to an orthogonal strain relative to the longitudinal or principal stretch direction in a low-strain tensile experiment. Note that because the typical orthogonal strain value will be negative for most materials (i.e., thinning of the sample will occur orthogonal to the stretch direction), the result of the negative sign in eq 10 will lead to Poisson's ratio being a positive quantity, except for some very unusual materials that will not be discussed here. What does the molecular significance of Poissons ratio convey? Briefly, it is a measure of the change in volume that the material will display at small strains in a tensile experiment. For materials that display a constant volume deformation, it can be shown that μ equals 0.5, which applies to liquids in general and also good elastomers where μ values equal to 0.498 have been noted. In contrast, glassy materials actually undergo local dilatation or expansion typically giving μ values of about 0.33. It might be mentioned that metallics and ceramics can often display rather similar values as well (i.e., approximately 0.33). Conversely, the μ value for semicrystalline polymers cannot be easily specified because such a wide range of levels of crystallinity is possible, and, in addition, when the experiment is done, the amorphous phase of the material may be above rather than below its glass transition temperature. Briefly, if the latter case is true, irrelevant of the degree of crystallinity, Poisson's ratio will be closer to 0.33, but as the level of crystallinity decreases and if the measurements are made above T_g, where the amorphous phase is more rubber-like, then μ will approach the value of 0.5. The reader is reminded that the value of Poissons ratio represents the initial change in volume beginning from an isotropic state. Hence, it should not arbitrarily be applied over a range of deformation because the structure will generally change with degree of deformation, and phase changes may occur as well, such as through strain-induced crystallization for some materials.

With Poisson's ratio now defined, the following equation shows the fundamental relationship between the moduli for isotropic materials:

$$E = 2G(1 + \mu) = 3B(1 - 2\mu) \qquad (11)$$

Note that for the case of constant volume deformation ($\mu = 0.5$), the value of the tensile modulus would be three times the shear modulus and the bulk modulus would go to infinity in the limit as μ approaches 0.5. This would be required to maintain the equality in eq 11. For μ values typical of glasses, (i.e., μ equals 0.33), one still notes that E is greater than two times the shear modulus. This same condition would also show that the tensile modulus is equal to the bulk modulus, as can be seen from eq 11. As is shown later in the chapter, because there is a very wide range of moduli values associated with glassy or highly semicrystalline materials compared to the same material in a rubbery state, moduli are often expressed on a logarithmic scale when plotted against a number of important parameters including temperature and rate of deformation. Thus, the approximation that E is often about equal to $3G$ is not unreasonable in many instances for isotropic polymeric materials.

Table 23.2 highlights the importance of the difference in moduli values and enforces the earlier comment about specific properties (i.e., a mechanical property normalized on the materials density). The table not only includes entries for polymeric materials but also for metallics and ceramics to allow a broader comparison. Note that although mild steel may display a much higher tensile modulus than that of aluminum, silicate glass, or polystyrene, once the moduli values are normalized by the respective densities, there is a much closer similarity among all materials, although the stiffness of the organic glassy polymer is still below that of the metallics or inorganic glass. However, the reader should keep in mind that polymers can undergo significant chain orientation, leading to much higher stiffness along the stretch direction, thereby greatly enhancing the moduli characteristics. When high orientation exists, the properties of such polymeric materials can often well exceed the specific properties of metallics or inorganics. This is discussed later in the chapter.

Mechanical energy or work is involved in ultimately promoting the failure of a sample, and it is not surprising that one important parameter often of interest is the energy required to induce failure at a given temperature and loading rate. This is also denoted as the toughness of a system. One can show that the toughness is the area under an engineering stress-strain curve (shown in Figure 23.10), which represents the energy per unit initial volume of the material. For a linear material, the stress of which increases linearly with strain, it can also be shown that the toughness or area under the curve will increase with the square of the strain, thereby indicating that a small incremental increase in higher level deformation can make more major contributions to the toughness than deformations at lower strains. This simply says that there will be more area under a seg-

Table 23.2 Comparison of the moduli for several different types of materials. Poisson's ratio, density, and the specific Young's and shear moduli

Material	Poisson's ratio (μ)	Young's modulus (GPa)	Shear modulus (G) (GPa)	Bulk modulus (B) (GPa)	Density ρ (g cm^3)	Specific properties	
						$10^6 E/\rho$ (m^2/s^2)	$10^6 G/\rho$ (m^2/s^2)
Metals							
Cast iron	0.27	90	35	66	7.5	12.0	4.7
Steel (mild)	0.28	220	86	166	7.8	28.0	11.0
Aluminum	0.33	70	26	70	2.7	26.0	9.6
Copper	0.35	120	45	134	8.9	13.5	4.5
Lead	0.43	15	5.3	36	11.0	13.6	4.8
Inorganics							
Quartz	0.07	100	47	39	2.65	38.0	17.8
Vitreous silica	0.14	70	31	33	2.20	32.0	14.0
Glass	0.23	60	25	37	2.5	24.0	9.8
Polymers							
Polystyrene	0.33	3.2	1.2	3.0	1.05	3.05	1.15
Poly(methyl methacrylate)	0.33	4.15	1.55	4.1	1.17	3.55	1.33
Nylon-6,6	0.33	2.35	0.85	3.3	1.08	2.21	0.79
Polyethylene (low density)	0.45	1.0	0.35	3.33	0.91	1.1	0.385
Ebonite	0.39	2.7	0.97	4.1	1.15	2.35	0.86
Rubber	0.49	0.002	0.0007	0.033	0.91	0.002	0.00075

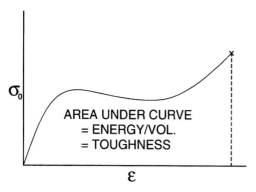

Figure 23.10 Generalized engineering stress-strain plot illustrating that the area under the curve is equal to the toughness or energy-volume of the material tested.

ment of the stress-strain curve for an incremental strain at higher levels of elongation than for lower levels. Most materials are nonlinear, and therefore, this last comment should not be applied irrationally. Because toughness is an ultimate property, it too can vary considerably because of differences in defect content in a group of specimens. Therefore, averaging of the data from several specimens is generally desirable.

In light of the definitions given so far, Figure 23.11a–d illustrates four deformation curves in which some relative comparisons of some common terms are made. If one assumes that the axes are equal in terms of their scales, note that Figure 23.11a indicates that a hard-brittle material refers to a high modulus-low strain to break (low toughness),

whereas the hard-tough material (Figure 23.11b) refers to a high modulus-high toughness material as denoted by a considerably increased area under the deformation curve. In contrast, a soft-weak material relates to a lower modulus and low strain to break material (Figure 23.11c), whereas Figure 23.11d illustrates a soft (low modulus) and tough material, which might be denoted from an elastomeric material that can undergo considerable deformation by strain hardening prior to failure.

It is likely almost a daily ritual for the reader to promote a uniaxial deformation of a rubber band for one application or another. In doing so, at the end of the loading profile, the stress is released and the material can return to its original zero point deformation. How does the unloading profile compare to the loading profile? Figure 23.12 provides some sketches of where, following a given loading profile, a sample has been considered to be allowed to return to its zero point deformation, as indicated by curves A, B, and C. For path B, both the loading and unloading profiles are identical. Which of these unloading profiles is realistic in terms of material behavior? Clearly, unloading profile A is unrealistic and violates the laws of thermodynamics. The area under a loading curve represents the energy input per unit volume of the material. Hence, if there is a greater area under the unloading curve than that of the loading profile, energy would have been created; this is prohibited by the first law of thermodynamics, which requires energy to be conserved.

Unloading profile B indicates that all energy placed into the sample was stored and could be returned to carry out

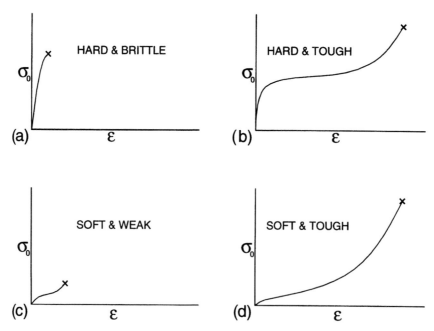

Figure 23.11 Four engineering stress-strain plots (assume equal scales) showing the general differences between the labeled material behaviors for the four systems.

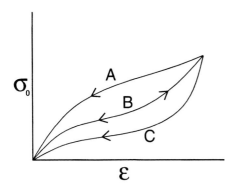

Figure 23.12 An engineering stress-strain plot showing a loading profile (curve B) as well as three unloading profiles A, B, and C. See text for details regarding hysteresis.

mechanical work. Because the curve is nonlinear, this would represent a perfectly nonlinear elastic material and would be in contrast to linear elastic behavior such as denoted by Hooke's law in which the stress is directly proportional to the strain, regardless of whether one is loading or unloading a specimen. Unloading path C is more realistic for viscoelastic materials such as polymers (Figure 23.12), and is typical of what might be expected. Here, the area under the unloading curve is less than that of the loading profile, which implies that a portion of the energy has been lost in this cycle of deformation. This leads to the important concept known as mechanical hysteresis, which represents the difference in loading and unloading pathways. Because the energy lost represents the difference between the loading and unloading profiles, a percent mechanical hysteresis, MH, is defined as

$$\text{percent MH} = \frac{\begin{array}{c}\text{area under loading profile}\\ - \text{ area under unloading profile}\end{array}}{\text{area under loading profile}} \times 100 \quad (12)$$

This relationship is important because it allows calculation of the energy loss per loading cycle. It is particularly significant in dynamic loading applications of polymeric materials. The importance of this is emphasized by a simple application such as cyclic loading of automobile tires during rotation. Any hysteresis energy observed clearly must come from energy taken from the gas tank, which leads to lower mileage. The energy is, of course, dissipated generally in the form of heat caused by molecular friction, and is important in terms of heat buildup, particularly in the cyclic loading of thick samples. Because the localized heat generated by molecular friction will not readily be dissipated in polymers due to their poor heat conductivity, cyclic loading can even promote such a rapid rise in temperature that thermally induced chain degradation with volatile by-products can occur. In fact, one can even promote a local explosion of a product caused by the buildup of internal volatiles caused by thermal degradation.[23]

Figure 23.13 shows an example of the cyclic loading of a thin strip of natural rubber at ambient conditions. Hysteresis behavior illustrates the first, second, and eighth cycle of deformation for a thin strip of natural rubber as deformed at ambient conditions. Note that the first cycle achieves a considerably higher stress level at an extension ratio of six (500% elongation) than it does on the second and additional cycles, where only the second and eighth are shown. In fact, the upper stress level on the second cycle and above is only about 50% of that observed on the first cycle. In addition, the percent mechanical hysteresis is also higher in the first cycle than calculations reveal for the second or higher cycles, thereby indicating that the energy loss per cycle is high in cycle one but then immediately decreases to a lower level and continues in that fashion as long as the dynamic test continues.

Why is the mechanical hysteresis and general stress-strain behavior so different in the first cycle in contrast to the remaining cycles? This arises from the fact that in addition to the cross-linked network of the natural rubber, there are also a number of dangling chains from the network itself, which can entangle with the network strands, but in which the end of the dangling chain is free to move. In addition, some entangled chains may not be a part of the covalent network at all and therefore would represent an extractable or sol component because both ends are free.

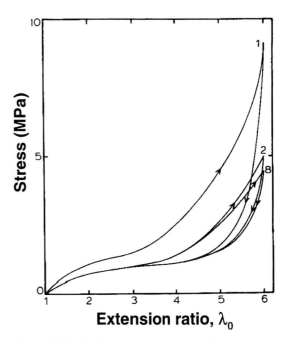

Figure 23.13 Engineering stress-extension ratio plots for a strip of cross-linked natural rubber. The numbers 1, 2, and 8 represent the first, second, and eighth loading (and unloading) profiles. (Reproduced with permission from reference 24.)

Upon deformation, the dangling ends and loose molecules can start to unentangle from the network as deformation proceeds. In addition, when the number of entanglements is reasonably high, as it is in the first cycle, natural rubber can also undergo strain-induced crystallization as the orientation builds, and this can also give rise to an enhanced tensile stress or strain hardening. However, when release is allowed from the initial deformation, not only do the crystals melt out again, but there are fewer entanglements in the system, so that during the second cycle the stress level will not be as great because the lost entanglements will not have had sufficient time to redevelop. Hence, the orientation behavior will be different and less or no strain-induced crystallization will occur in the same deformation range under the same loading conditions. The result is that because of the loss of entanglements in the first deformation and also the reduced likelihood of strain-induced crystallization, the second deformation cycle will be very different. Specifically, the first cycle would not represent the expected hysteresis in dynamic behavior, and therefore measurements of this parameter for real applications must recognize this consideration.

As might be expected, the rate of loading and the temperature at which the material is deformed (let alone the molecular morphology of the system) will have major ramifications on the hysteresis observed whether it involves the first or later cycles. In addition, higher levels of cross-linking will minimize chain entanglement loss, thereby lowering hysteresis. One would anticipate that more highly cross-linked systems may not be able to achieve high elongation, and that if the system has molecular symmetry to crystallize, the presence of cross-link points interferes with crystallization, thereby minimizing this solidification process as the cross-link density increases.

There are many cases in which removing the load from a deformed material does not lead to the recovery of the sample to its original dimensions, in contrast to the case of a rubber band or high-quality elastomer. Rather, there might be no or only partial recovery. For example, in deforming a piece of bubble gum, if the material is held for a moment following the extension, generally release of that drawn material will show no change in dimensions. (In passing, it might be noted that bubble gum is comprised of polymeric materials as well as sugar and other components.) Often, however, nonelastomeric materials may show some partial recovery, but whatever remaining unrecovered strain exists is often referred to as permanent set or irrecoverable flow. However, because of the viscoelastic nature of polymeric materials, the magnitude of the elastic component often depends on time as well as temperature, and therefore the parameter of permanent set may actually vary with time at a given set of conditions. This is because there are some extremely long relaxation time characteristics associated with deformed macromolecules. This has been alluded to before and arises from the high level of entanglements for the higher molecular weight species. Similarly, when oriented amorphous chains or sections thereof are above their glass transition but are more confined in their molecular mobility because of the presence of crystallinity or fillers, this also will impede their thermal Brownian motion, thereby lengthening the relaxation process. When permanent set is reported, the exact conditions under which it was determined must be indicated.

It also may be useful to make the measurement of permanent set or so-called irrecoverable flow as a function of time at a given temperature to determine if there is a transient behavior to this parameter. In cases where no such time dependence is observed, one may have the initial feeling that no recovery is possible, but often this is only because of existing restrictions on the mobility of oriented segments or chains. If those restrictions are removed, however, such as by increasing the temperature, which promotes greater "molecular kicking" or thermal motion of the wormlike segments, then further recovery may well occur. Current packaging that makes much use of heat-shrink film is familiar to most readers. Such polymeric films are typically oriented at higher temperatures and then quenched, leading to either vitrification or, in some cases, to the development of a sufficient amount of crystalline content to restrict the mobility of the remaining amorphous oriented material. The article in question is wrapped in the film and then heat is applied, thereby either increasing the mobility of the amorphous state above the glass transition temperature, which promotes a "kicking of the molecules" to revert to a more coiled conformation, leading to shrinkage or, in the case of semicrystalline materials, partially melting out some of the crystals thereby again releasing the restrictions on the normally thermally active oriented amorphous material. In many such instances of heat-shrink materials, the initial system prior to elongation may have been partially cross-linked, such as through the use of electron beam irradiation (e.g., polyethylene or related copolymers), which cross-links many of the molecules together into a covalent network (gel) structure that has the characteristics of a typical rubber band when stretched above the glass or melt transition temperature.

The associated retractive force or stress associated with the shrinkage is a significant point of interest to such packaging materials because if the force is too great, the packaging material can crush the contents or distort the desired package container, and perhaps ruin the product inside. Similar retractive force, stress, or sometimes retractive power, is important in other applications such as in elastic polymeric filaments in garments (i.e., Lycra spandex). Certainly if the stress is too great, circulation of the garment bearer is unpleasantly affected; if the stress is not sufficient, the garment may slip from its desired placement. In summary, it is obvious that the level of set and the associated retractive forces or stresses will be dependent upon the morphological conditions of the system, its temperature, and the level of oriented amorphous material.

Temperature Effects on Stress-Strain Behavior and Its Promotion of the Thermal Mechanical Spectrum

The importance of temperature with regard to its influence on mechanical response was alluded to previously. The reader certainly is also familiar with the deformation of a rubber band and its low modulus and high extensibility as well as its recovery. However, when this material is placed into a very cold environment, such as liquid nitrogen, the material now shows a high modulus, low strain to break, and low toughness, and would be defined as a brittle material. In fact, most any cross-linked or uncross-linked high molecular weight amorphous polymer could sweep out a considerable range of stress-strain responses at a given rate of deformation as the variable of temperature is changed and this is illustrated in Figure 23.14a. It therefore becomes obvious that any mechanical parameter determined must typically have associated with it the temperature at which that measurement was made.

With such a wide range of behavior that is influenced by temperature, there is often the desire to represent the thermal dependence of a given property by plotting that property against temperature. In many cases, the variable of importance may be a high deformation parameter, such as the tensile strength, elongation to break, energy to fail, or the like. The low deformation parameter modulus, however, provides a fingerprint of the original structure, whether it is in shear or tensile deformation. Thus, as shown in Figure 23.14a, if the initial slopes of the respective curves or moduli are plotted as a function of temperature for an amorphous, uncross-linked, high molecular weight polymer, a nearly universal plot develops, which is known as the thermal mechanical spectrum (TMS) (Figure 23.14b).

In this spectrum, one notes four distinct regions. The first represents the modulus well below the glass transition temperature. This value is in the range of 10^9 to 10^{10} Pa (10^9 Pa = 1 GPa = 145×10^3 psi). This behavior is representative of unoriented organic glasses and even nonpolymeric glasses will fall into this category if they are organically based. Hence, the level of modulus deep within the glassy state is approximately 10^9 to 10^{10} Pa. The level of modulus will be at the higher end of this range for those systems showing strong secondary bonding (e.g., polyimides and polyureas), as opposed to those showing weak secondary bonding, which will be at the lower end (atactic polypropylene). When the temperature is raised and additional thermal energy is provided to the system, once the energy is great enough to permit cooperative segmental backbone

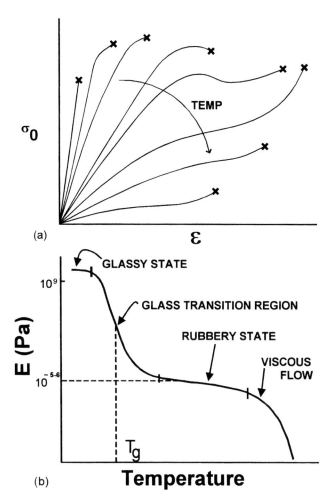

Figure 23.14 (a) Generalized engineering stress-strain plots for a material that shows variation in properties with increasing temperature, that is, a softening response. (b) Typical thermomechanical spectrum plot of E versus temperature for a high molecular weight linear amorphous polymer.

motion in the time scale that the measurement is made, the material will begin to soften; this is the second region, known as the glass transition. The inflection point associated with the TMS as indicated in Figure 23.14b illustrates one index of expressing T_g. For a homogeneous, unoriented, amorphous homopolymer that is linear, the glass transition will generally extend over a range of 15 to 30 °C although it will vary depending on the method of measurement. During that transition, the modulus will generally decrease about 10^3 to 10^4 Pa, with the former being the more representative value. This level of change is enormous, and indicates why the glass transition is so important to the mechanical properties of materials.

Once the glass transition region is passed, one enters a second region where the modulus appears again to be relatively independent of temperature, at least on a log scale, and this is the well-known rubbery region or state. This plateau is due to the occurrence of an entangled network of the chains that do not have sufficient time to completely relax during the time of measurement. However, upon further promotion of thermal energy into the uncross-linked system, molecular motion is even more greatly enhanced and the relaxation times become shorter. This enhanced mobility will allow the molecule to move more freely and escape from neighboring entanglements, and therefore to flow more readily at higher temperatures, which then promotes the so-called viscous flow state—the last region in Figure 23.14b. It is within this region that melt processing of polymers typically occurs, although there are some forming operations that clearly do not require as high a temperature. Because of the near-universal nature of the TMS for linear amorphous polymers, the TMS is an absolutely critical concept to understand in light of mechanical response.

Not only will temperature promote the change in a material's characteristic behavior from that of glassy to one of the other states or regions, but another common parameter, rate of deformation or frequency, can also achieve a similar result at an isothermal condition. The reader should recognize this general universal TMS behavior for amorphous, uncross-linked, higher molecular weight polymers, even though the chemical structure of the repeat units can vary widely. For example, polystyrene that is a glassy polymer at ambient conditions will show a TMS similar to that illustrated in Figure 23.14b, and its general glass transition will usually fall in the range of about 100 °C. Amorphous *cis/trans*-polybutadiene rubber will also have a very similar TMS, but its glass transition will be of the order of −85 °C and therefore, at ambient conditions, we typically experience its performance in the rubbery region in contrast to the glassy region for polystyrene.

What systems or molecular variables influence the TMS? The list of parameters is extremely long. A few of these parameters are briefly addressed. As the molecular weight increases in a systematic fashion for a narrow molecular distribution, homogeneous, amorphous linear polymer, the general behavior noted for the TMS response would be as shown in Figure 23.15 (curves A–E). Below a critical molecular weight, the molecular weight for entanglements, no sign of a rubbery plateau is observed because an entangled network does not exist and rubbery behavior is not observed (Figure 23.15, curve A). (At a very low molecular weight, where no molecular entanglements occur, the glass transition temperature will also be typically lowered because of a higher concentration of "chain ends", each of which possesses a greater molecular mobility or free volume than internal chain segments. This effect would depress T_g somewhat, but this is not illustrated in Figure 23.15 for brevity.) Note that there is no major effect, however, on the glassy modulus, and this is typical because the high rigidity in the glassy state depends little on how large

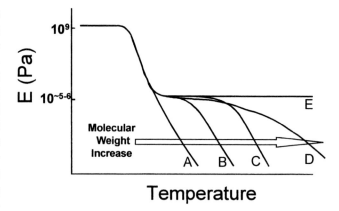

Figure 23.15 Effect of increasing the molecular weight on the thermomechanical spectrum for an amorphous linear polymer, but for which the molecular weight distribution is narrow (curves A, B, and C). Curve D illustrates the consequence of a broadened molecular weight distribution, whereas curve E indicates that the gel point has been reached, thereby eliminating the viscous flow region.

the molecule is, but principally only on limited local-scale motions.

Note that the rubbery plateau extends systematically to higher temperatures as the molecular weight increases because of the enhancement in the number of entanglements per chain, which therefore lengthens their relaxation time accordingly. Hence, to achieve more viscous-like flow behavior, a higher temperature will be required for the same time scale of loading (rate of deformation). Figure 23.15 also shows curve D, which is more representative of a system having a broad molecular weight distribution, that is, there are a range of chain lengths present. The shorter chain lengths have shorter relaxation times than the longer ones and this influences the overall rubbery plateau by generally showing a somewhat more rapid decrease in its behavior into the viscous flow state. This behavior is in contrast to a narrow distribution polymer, which tends to simply extend the plateau in a more nearly flat manner to higher temperatures before entering the viscous state. The presence of some long-chain branching would also influence the general state of the rubbery plateau and its shape dependence in a similar way. The important point, however, is that without an entangled network, the rubbery plateau would not exist for polymers that posses only van der Waals interactions.

As mentioned, branching can alter the shape of the rubbery plateau, but let us further consider what would happen to the TMS if cross-linking reactions were permitted (which can promote branching and network development) within an initial linear polymer. Through cross-linking and branching reactions, the molecular weight will increase and the relaxation times will become longer, thereby generally extending the rubbery plateau, but there will be little influence on the glassy state and glass transition behavior. However, a critical level of cross-linking will be reached whereby a three-dimensional network is formed. This critical level of cross-linking is known as the gel point. Above this point, no longer will steady state flow be permitted because of the "tying together" or cross-linking of the molecules into a three-dimensional "fishnet," and hence, the upper region of the TMS, viscous flow, will be eliminated (Figure 23.15, curve E). This process is essential to recognize because if a cross-linking reaction is developed during the midst of processing a material, steady-state flow will cease to occur above the gel point and the ability to remove the polymer from a reactor, extruder, or other instrument will become a major task. Furthermore, the TMS rubbery region would simply extend to higher temperatures until the polymer finally degrades either by chain scission or cross-linking. The presence of a gel point as obtained through inducement of cross-linking reactions is very important in systems such as epoxies, or any related vulcanized materials that are elastomeric or semicrystalline.

One must also be aware that a given material that is being tested as a function of temperature may begin to promote such reactions as the temperature is increased. When this occurs, the mechanical response that is measured is no longer characteristic of the original structure, but is influenced by the chemical changes that are occurring within the testing environment. This suggests that in materials for which there is some concern that such reactions may occur, retesting the material again at lower temperatures may be useful to determine if the original structure or response is still obtained after it has passed through a higher temperature cycle. As the level of cross-linking increases above the gel point, it will be the rubbery region that is enhanced in terms of modulus. A broadening and increase in the inflection point associated with the glass transition behavior will also occur as a result of the restrictions placed on the chain segments that reside near a cross-link, the latter of which limits thermal Brownian motion (Figure 23.16). Briefly, rubber elasticity theory shows that the modulus will scale proportionally to the number of network chains per unit volume (i.e., a higher cross-link density).[25]

Other variables also play important roles in altering the TMS of an amorphous polymer. These include rigid fillers, often inorganic in nature, which will begin to enhance the glassy modulus. A broadening of the glass transition will also occur and an enhanced rubbery modulus may be noted. Viscous flow will also occur but often it will be somewhat shifted to higher temperatures because of flow restrictions caused by the presence of the filler particles; this effect is particularly prominent if the level of polymer interaction with the filler is high. For example, if the polymer strongly absorbs on the dispersed filler and if there is high specific surface and content of the filler, then limitations on molecular mobility will affect the TMS in the expected way. One

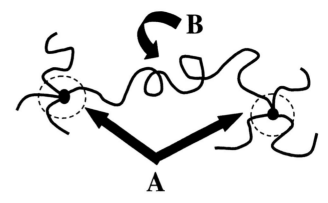

Figure 23.16 Schematic attempting to show regions of the chain near cross-link points that may have more restriction on mobility (regions A) as compared to chain segments far away from cross-links (region B).

example of this last effect would be when carbon black is used as a common reinforcing particle in rubber materials.[26]

In contrast to the presence of rigid fillers, the addition of a plasticizer, which is often a low molecular weight mobile component or a lower T_g miscible polymer, will in turn depress the glass transition temperature in proportion to its added amount but with a dependence on the relative T_g of the starting matrix polymer and the T_g of the added plasticizer. This concept is also cleverly used in developing copolymers to tailor the TMS glass transition. For example, in Figure 23.17 the modulus-temperature behavior is shown for a series of styrene-butadiene random copolymers of varying composition; as the composition of the higher T_g comonomer (homopolymer of a given comonomer) is used, the glass transition temperature systematically is increased. Of-

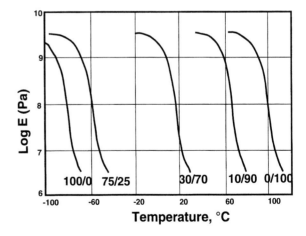

Figure 23.17 Thermomechanical spectrum profiles for random butadiene-styrene copolymers as well as the homopolymers of butadiene (100/0) and styrene (0/100). (Adapted with permission from reference 27.)

ten, random copolymers allow one to roughly predict where the glass transition temperature will be for a given copolymer composition, T_{gc}, by using the well-known Fox equation:

$$\frac{1}{T_{gc}} = \frac{W_1}{T_{g1}} + \frac{W_2}{T_{g2}} \qquad (13)$$

where T_{g1} represents the T_g of the homopolymer containing comonomer 1, and likewise for T_{g2}. The values of W_1 and W_2 refer to the weight fractions of the corresponding comonomer species. Note that application of this relationship to a given TMS would also require that the glass transitions T_{g1} and T_{g2} be determined by the same procedure and conditions as used in determining the T_{gc} of the copolymer.

We have focused on the TMS concept, which is extremely critical to understand when dealing with the mechanical response of polymer systems. It is worth restating that although only modulus has been addressed for the majority of this section, the other parameters discussed earlier that deal with different facets of the stress-strain curve could also be considered as a function of temperature. However, many of these parameters would not be representative of the original structure of the material when plotted in a similar way as a function of temperature. Specifically, once deformation begins, the morphological texture may be altered, and even phase changes such as crystallization may occur with further orientation and deformation. Hence, it is the parameter of modulus that often serves as the principal fingerprint of a given structure in contrast to many of the other variables discussed, even though other variables may be critical for a given application of some particular material in practice.

We have addressed the general TMS behavior for amorphous systems, some of the factors that can influence the shape of this curve, and how the glass transition may depend upon various molecular factors. However, we mentioned that if the molecular structure of the polymer is symmetric, a portion of it may undergo crystallization, leading thereby to a rigid phase. The crystalline content promotes an important influence on stiffness, and a new transition becomes apparent—the melting point. Polymers typically display a crystalline structure that is of a lamellar nature, although certainly there are variants of this. A simplified sketch of this is shown in Figure 23.18a, which attempts to not only illustrate the folded-chain lamellar texture but also larger scale superstructure denoted as spherulitic textures that generally arise when a polymer is crystallized from its unoriented melt and which will lead to a globally isotropic sample. (Exceptions to this type of structure distinctly appear when the material is crystallized from an oriented melt, leading to a strain-induced crystallized texture.)

Figure 23.18b is an optical micrograph taken between cross-polarizers showing a partially spherulitic texture where the distinct multimicron spherulitic structures can be seen.

For further detail concerning the morphological features of crystalline polymers, the reader should consult reference 13. The impact of crystalline structure on the thermal mechanical spectrum is outlined in Figure 23.19. It is clear that as crystallinity increases, the modulus characteristics are particularly enhanced above the glass transition temperature, but there is relatively little change below T_g. The glass transition temperature tends to be broadened and increased if T_g is taken as an inflection point as defined earlier. (Note some strong similarities between the influence of crystallinity and cross-linking, at least in the T_g and rubbery regions of the TMS.) The melting regime leads to the final softening of the material as it enters into the viscous flow state. (An exception to this would be where the crystalline phase transforms to a liquid crystalline [mesophase] morphology. The reader interested in this latter, more specialized structure should consult references 12 and 30.) Because the melting regime in a semicrystalline polymer is generally not sharp, and in fact can occur over tens of degrees because of variation in crystal size and perfection, the distinct nature of the melting transition is not always as sharp as illustrated in Figure 23.19.

Influence of Deformation Rate (Frequency) on TMS Behavior

As described previously, the parameter of rate of deformation (or frequency of loading in a dynamic experiment), which carries reciprocal time as its dimensions, can play an important role in the observed mechanical response at a given temperature. From the author's viewpoint, this is the second most important external parameter to influence general mechanical response; the variable of temperature is the most important parameter.

To further illustrate the importance of the rate of deformation, there is a well-known hobby shop material called Silly Putty that is generally a filled poly(dimethylsiloxane) polymer. The reader is probably familiar with this material because it is a common childhood toy that has interesting mechanical response. It can be used to discuss the issue at hand, that is, the rate of deformation and its influence on mechanical behavior. If the Silly Putty system is drawn slowly (low rate of deformation) at ambient conditions, there is very little force noted and it behaves somewhat like a piece of warm bubble gum, that is, it draws well, displaying low modulus, high extensibility, and little to no recovery. At the other extreme, but at the same temperature, if the material is rapidly deformed (or at least as rapidly as one can stretch the material), the system will typically display near-brittle fracture with much higher modulus characteristics, low energy to fail (on a relative basis), and other characteristics of a near-glasslike system. If one uses an intermediate loading rate, which typically amounts to reshaping

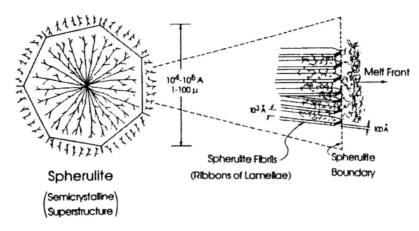

Figure 23.18a General schematic of the morphological texture of a spherulite. (Reproduced with permission from reference 28.)

the Silly Putty into the form of a round ball and dropping it or tossing it against the wall, the material will distinctly behave like a rubbery system with good elasticity and recoverability in that loading time scale. Hence, one can use this single material to demonstrate essentially all of the regions of the TMS and yet the temperature is held constant.

From the discussion above addressing Silly Putty, we note the rate of deformation or loading that determines the properties at a given temperature. In essence, this means that a series of stress-strain curves for most materials can, in principle, display the same general behavior that was shown in Figure 23.14a, but the arrow representing temperature will be reversed if the variable is rate of deformation (or frequency of loading in a dynamic experiment). At first, this strong similarity in the stress-strain response of a material with increasing temperature and its similarity with decreasing rate of deformation seems possibly foreign in that the two variables, temperature and rate, are extremely different and possess different dimensional units. Why is it that these two parameters essentially can induce very similar behavior?

The analogy discussed earlier about the similarity between worms and polymers (versus spaghetti and poly-

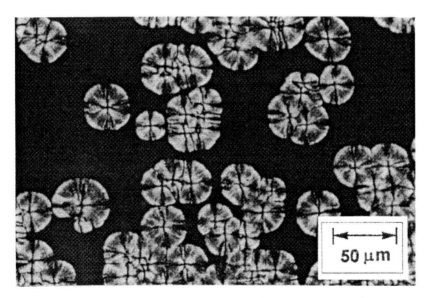

Figure 23.18b An optical micrograph of a partially spherulite film of a semicrystalline polyimide taken between crossed polarizers. (Reproduced with permission from reference 29.)

mers) is again extremely important. As recalled from that initial simplistic discussion, if one has a molecular system at a given thermal energy, there is a certain level of molecular mobility permitted, depending on the nature of the chemical structures involved. If a short time window (high rate of loading or frequency) is used for that system, there will be less time for the molecular system to respond and therefore the material will be more apt to appear rigid or glass-like. However, for exactly the same temperature, opening the time window for molecular motion (by going to a slower rate of deformation) will allow more molecular response to occur (if it can), and therefore a softer material would result. This is analogous to the earlier discussion about carrying out a stress-strain experiment on a group of entangled live worms, that is, a high rate of deformation leads more to brittle fracture, whereas a slow rate of deformation provides a relatively lower modulus and higher extensibility. In many respects, the same analogy applies to our polymeric "worms" as well. In contrast to plotting a specific mechanical property as a function of rate of deformation, as was done with temperature, it is better to plot that property as a function of the logarithm of the rate of deformation (or logarithm of loading frequency) because an extremely wide range of rates can be covered. When this is done for the modulus at constant temperature, it will basically promote a mirror image effect of our former TMS. This occurs because increasing the rate of deformation is similar to decreasing the temperature since the time window for mobility is reduced. Thus, rate data are often plotted against the logarithm of the inverse loading rate or frequency to invert the mirrored image TMS curve to what appears in Figure 23.14b. Because of the importance of this point, Figure 23.20a,b provides an illustrated form of the original mirror image and inverted mechanical spectra, in which the logarithm of the rate of deformation and its inverse are the independent variables, respectively.

The influence of the molecular variables that were used earlier for discussion of the TMS in Figure 23.14b (such as

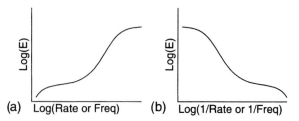

Figure 23.20 General comparisons between (a) a plot of log E vs log (rate or frequency of loading) with that of (b) log E vs log (1/rate or 1/frequency). Note the mirror image effect.

molecular weight, branching, and filler effects) would also apply to the influence of the rate of deformation. That is, the important consideration is the time window for mobility to be noted at a given temperature versus the time for mobility when temperature is varied for a given rate of deformation. The importance of this interrelationship between time and temperature cannot be overstated, and it serves as the basis for the development of a very important feature of mechanical properties of polymers called time-temperature superposition (TTS).

Introduction to the Deborah Number (D_e) and TTS

In view of the discussion showing the interrelationship between time and temperature, we have reached the conclusion that the general origin of this inter-relationship applies to how much molecular motion or mobility can be displayed in a given observation time as denoted by the experimental time frame. This is in turn inversely related, for example, to the rate of deformation, that is, a faster rate of deformation provides a shorter observation time. In essence, we wish to determine what is the relaxation characteristics or time frame of the molecular system relative to the experimental time frame as dictated by deformation rate. This can be simply expressed as a dimensionless number and is known by engineers as the Deborah number, D_e. This can be written as

$$D_e = \frac{\text{characteristic system or molecular relaxation time}}{\text{observation time}} \qquad (14)$$

The denominator will be a function of the rate of deformation. Note that the observation time is not equal, however, to the rate of deformation because one parameter (observation time) carries time to the first power, whereas the second (rate of deformation) carries the dimensions of reciprocal time as would strain rate, frequency, or shear rate. Hence, if one wishes to utilize directly the Deborah number as formulated by a characteristic relaxation time and a deforma-

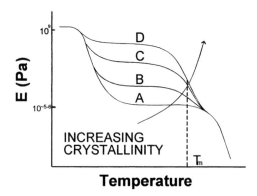

Figure 23.19 Generalized thermomechanical spectrum for an uncross-linked polymer having different levels of crystallinity.

tion rate, it would be the product of these two parameters that would provide a dimensionless number.

Let us attempt to apply this concept. Suppose one were to sit and watch, at ambient conditions, the flow of a glassy polystyrene window material out of its window frame structure. What would they observe? Even if one watched for considerably long periods of time, the Deborah number would be far in excess of unity because the relaxation or flow time of the polystyrene (at ambient temperature) would be excessive and far overpower any reasonable observation time. In fact, because the Deborah number is considerably greater than 1, under these conditions the observer would view the substrate material as a rigid solid (glassy in this case) because no flow was observed during any reasonable experimental time frame. At the other extreme, if one tipped over a vessel of a low viscosity polymer solution, it would yield a Deborah number that was very much less than unity because flow would be noted as fast as one could visually record the event. As a result, one would classify this system (i.e., very low Deborah number) as a low-viscosity fluid due to its liquid-like response. Under intermediate conditions for which the Deborah number is of the order unity (i.e., one observes the flow over the time scale of the observation and thus experiences the transitional nature), the response might be classified as that of the glass transition (for amorphous systems). Although the nature of the experiment would determine whether the material is more rubber-like or glasslike, it would be clear that a unity value of the Deborah number would represent transitional behavior between rigid elastic response and that of fluid or viscous flow.

Before continuing to the next point of discussion, it is important to reinforce the concept of the Deborah number by reminding the reader that the origin of this number relates to biblical times and extends from the book of Judges, Chapter 5, verse 5. It is there, depending on which translation the reader inspects, they will note that Deborah, one of the Judges, states, "the mountains flowed before the eyes of the Lord." The basic concept is that His time scale is likely longer than that of the readers, and therefore He may well be able to see flow occur on the surface of the earth given a sufficient time scale. The reader, on the other hand, will not likely observe this occurrence unless he or she lives in a zone where earthquakes or volcanoes might occur! In summary, the Deborah number concept is extremely important, and there are two ways to influence it. One is through the internal or system variables, which affect the molecular relaxation time. Some examples of these variables include molecular weight, its distribution, molecular architecture, fillers, plasticizers, and crystallinity. However, one of the most important variables that influences relaxation time is the external variable of thermal energy as expressed by kT. The other approach is to vary the experimental time frame (the denominator in eq 13), and this is influenced by, or

coupled to, the rate of deformation in an inverse matter, as discussed earlier.

Now that the interrelationship between time and temperature is clear, one can utilize this concept to predict the properties in regimes of either time (frequency domain) or temperature, neither of which is necessarily easily accessible to the investigator. A method that uses temperature variation as a means of extending the time scale is now described.

Consider an amorphous polymeric material that is placed under a stress at a fixed strain at a given temperature (i.e., a stress relaxation experiment). Suppose one observes that the stress relaxes with time in that system because of molecular flow (note curve A in Figure 23.21a). Let us assume that these data are obtained at some low temperature (e.g., somewhat above T_g), and because one is near T_g, in this selected observation time frame, only a relatively small amount of stress relaxation is observed because of the more rigid

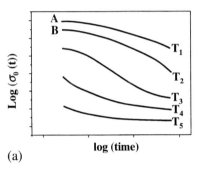

(a)

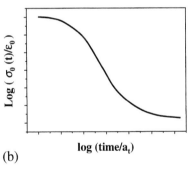

(b)

Figure 23.21 (a) A sketch of log σ_0 (*t*) vs log time for an amorphous polymer determined in the time range shown but at different temperatures where $T_1 < T_2 < T_3 < T_4 < T_5$. (b) Master curve of log $\sigma_0(t)/\varepsilon_0 = \log E_r(t)$ vs log t/a_t as might result in the time-temperature superposition of the data in (a). Note the expanded time scale obtained by this method (i.e., each tic mark on the horizontal axes of Figure 23.21a,b represents one decade).

(less mobile) nature of the system. However, if the same time frame is used and the experiment is carried out again, but at a higher temperature, one would anticipate that the flow will occur more readily and therefore over the same time period the stress values will be respectively lower (see curve B in Figure 23.21a). One can easily continue this increase of temperature for similar tests carried out within the same time frame, thereby generating the series of curves T_1 to T_5 that might be noted in Figure 23.21a.

Suppose one wishes to determine the stress relaxation that would have occurred in a much longer time frame, for example, at temperature T_4. By noting the shapes of the curves given in Figure 23.21a, one can see that the upper curves could be slid to the left (lower time frames) to match or superpose the left-hand terminal region of curve T_3, whereas the lower curves could be slid to the right with the anticipation that these will superpose with those of the longer time portion of T_3 (see Figure 23.21b). The corresponding shift factor or log time factor that is involved in providing the level of shift for a given temperature segment is denoted as a_t.

In the final superposed master curve (Figure 23.21b), one sees that the original data obtained in a relatively short time frame (possibly a few hours) is now extended both to much longer and shorter times, and has been obtained by using temperature as the variable. Because this test considered that the stress was always carried out with the same fixed strain, the relaxation master curve could also be referred to as a relaxation modulus if the variable plotted against log time is the ratio of the time-dependent stress divided by the fixed strain. This procedure helps one to visualize the extremely short or long time frames that can be involved with the relaxation behavior of a polymeric material placed under a given strain.

One may ask if TTS always works, and the answer is clearly no. It will work when the relaxation mechanism involved with a given molecular process is only changed in time scale as temperature is altered, but no new molecular mechanisms occur, that is, no changes in structure occur. One possible example where TTS would not be applicable would be when crystallization might occur at higher temperatures (when coming from the glassy state), thereby altering the morphology of the system, leading to a lengthening of the relaxation response of the now more rigid material. A second exception would be if cross-linking takes place at higher temperatures; this also would lead to branching and potential gel formation, thereby lengthening the relaxation rate and inhibiting mobility. Thus, to use TTS one must ask whether changing the temperature will only lead to a change in the time scale of the molecular process but not alter the general relaxation mechanism responsible for the relaxation process. If this can be answered affirmatively, then TTS is likely to be a useful tool in gathering information for longer or shorter times that may be difficult to obtain otherwise.

Note also that although the concept of TTS discussed above considers only stress relaxation (an important mechanical process), the same general scheme will apply to many other mechanical parameters as well. The reader is referred to other more in-depth sources that discuss TTS and its applicability.[2,3,8,31,32] Briefly, it can be shown that when the shift factor value, a_t (determined by shifting the various experimental data to superpose and generate a master curve), is plotted against temperature, the plot can later be used to develop a master curve for any other selected temperature between the extremes used in collecting the data. It is not simple, however, to predict behavior in the glassy state by TTS for reasons that will become clear when the topic of physical aging is discussed.

Time-Dependent Dimensional Changes Under Load-Creep Behavior

In the preceding section, reference is made to the use of stress relaxation to follow molecular flow; that is, a fixed strain is placed on a material and then one follows the decrease in stress with time if molecular flow occurs. Transient behavior, if observed, is certainly indicative of the loss of mechanical energy that has been placed into the material and therefore dissipation of the energy in the form of heat by molecular friction represents the viscous component of the viscoelastic behavior of the material. A second important transient mechanical parameter is that of creep. The fundamentals of creep are nearly the reverse of the stress relaxation experiment. This is expressed in Figure 23.22, which shows that at zero time, a constant stress, σ_0, is applied to the sample and it responds quickly by displaying a step strain, ε, but with time, the strain may then either increase or stay steady, depending on whether the material displays flow. Note that in this test, the stress stays as a

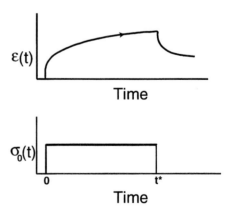

Figure 23.22 General creep and creep recovery behavior for a high molecular weight polymer. Note that the load was removed at $t = t^*$ and that some strain recovery occurred.

constant, as did the strain in the stress relaxation experiment. In systems that would display molecular flow, the strain will increase with time but may reach an equilibrium value if, for example, the system is sufficiently cross-linked such that finally the true network (gel) reaches an equilibrium strain value as promoted by the retraction (molecular kicking) forces of the interconnected chains. In fact, upon removal of the load, this latter system may lead to full recovery of the material with time, although it will not be instantaneous, as shown in Figure 23.22. This experiment could be carried out in tensile-compression deformation or in shear, and could in principle be performed in hydrostatic compression, although the latter scheme is certainly much less common. The important point to recognize is that both stress relaxation and creep are a way of denoting the transient or viscous characteristics of a material for a given time scale and which originate from molecular motion. At the same time, at least in the case of our recent discussion of creep recovery, this transient response can also illustrate the time-dependent elastic (recovery) character of a material as well. It is these two components together (viscous and elastic) that designate the level of viscoelastic character of a polymeric material for a given time and temperature. It is important to be able to delineate, for a given time and temperature, what is the magnitude of the viscous versus the elastic component of a viscoelastic material because it helps dictate what will be the nature of hysteresis, elasticity, and other parameters.

One key method allows this separation of the viscous and elastic contributions for a given loading scheme and the principal approach or technique is dynamic mechanical analysis (DMA). Typically, the application of a small dynamic strain is applied to a material at a given temperature either in tension or in shear. One then monitors the corresponding stress state of the material during the applied sinusoidal strain function. With this information, one has the ability to separate the relative magnitudes of the viscous and elastic character. What is the formalism behind this technique and how does it work? Because this technique is one of the principal tools used in polymer characterization throughout all areas of polymer science applied to mechanical response, further discussion of DMA follows.

DMA: Basic Theory and Application

DMA is one of the most common methods for characterizing the TMS behavior of polymeric materials. However, it provides more than just the thermal mechanical spectrum because it can separate the relative components of viscous versus elastic behavior, and it can do this while changing temperature as well as rate of loading (frequency). Furthermore, a single sample specimen is all that is necessary in contrast to the requirement for many specimens in some of the other tests described so far. A restriction that should be noted, however, is that the dynamic measurements are

made at small strains where one is in the initial linear region of the stress-strain response (Figure 23.9). Therefore, although a tremendous amount of information is provided from this test, the data do not necessarily describe the characteristics of the structure at high levels of deformation, although correlations may be made in many cases.

Recognizing that polymers are viscoelastic, suppose we first address how a so-called linear elastic versus a linear viscous body would behave if a dynamic strain were applied to either of these systems. For a linear elastic system, the analogy would be a Hookean spring (Figure 23.23a), in which the stress would be directly proportional to strain through a single constant (the modulus) because it represents the slope of the linear stress-strain relationship. Let us apply a sinusoidal strain to the Hookean spring where the degree of oscillation will be within the linear region of the stress-strain curve. This can be expressed as

$$\varepsilon(t) = \varepsilon_0 \sin(\omega t) \qquad (15)$$

where ε_0 represents the maximum amplitude of strain promoted by the sinusoidal variation and ω represents $2\pi f$, where f is the frequency of loading. Hookes law in tensile form states that

$$\sigma(t) = E\,\varepsilon(t) \qquad (16)$$

Combining eq 15 and eq 16 indicates that the dynamic stress response of a Hookean spring undergoing a sinusoidal strain deformation will be given as

$$\sigma(t) = E\varepsilon_0 \sin(\omega t) \qquad (17)$$

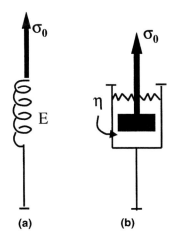

Figure 23.23 Schematic of (a) a Hookean spring of modulus E, which can be used to represent or model a linear elastic response, and (b) a dashpot having a Newtonian viscosity η, which can be used to represent a linear viscous response.

Equation 17 shows that the stress scales in proportion to the modulus, as would be anticipated from Hooke's law, but a particularly critical feature is that the stress and the strain are identically in phase; that is, both have a sine function temporal dependence of identical form.

In contrast, we now consider the same sinusoidal strain applied to a linear viscous body, which can often be represented by a so-called dashpot (Figure 23.23b). In this figure, the piston can be viewed as being surrounded by a Newtonian viscous fluid, the viscosity of which is independent of the rate of deformation. When one pulls on the piston, the fluid must flow around the edges of the piston and in between the cylinder, thereby promoting resistance as the piston moves through the cylinder. Certainly, work is being done as the piston is pulled or pushed through the viscous fluid, but all of that work will be lost or dissipated when the stress is removed because there is no elastic character to this body as there was in the case of a Hookean spring. That is, the Hookean spring retraces its original loading path as a result of its perfect elasticity. Thus, we see a major difference between these two bodies; one stores the energy (Hookean spring), whereas the other dissipates all energy (loss), and therefore the latter is viewed as purely viscous or dissipative in nature. To determine the time-dependent stress of the linear viscous dashpot, we recall Newton's law of viscosity, which holds for linear viscous materials. This law is given in tensile form as

$$\sigma(t) = \eta \dot{\varepsilon}(t) \tag{18}$$

where the variable $\dot{\varepsilon}$ represents the strain rate. This well-known relationship, written here in its simplest form, shows that the time-dependent stress does not scale with the strain function as was the case for the Hookean behavior, but rather it is proportional to the strain rate (the derivative of the strain with respect to time). Taking the derivative of the strain function given in eq 15 provides the relationship

$$\sigma(t) = \eta \omega \cos(\omega t) \tag{19}$$

It is significant that the stress now scales directly in proportion to the angular frequency, ω, but there is also a 90° phase shift because the stress function now depends on $\cos(\omega t)$, whereas the strain function is expressed as $\sin(\omega t)$. Figure 23.24a schematically shows that the stress and strain functions are 90° out of phase and also shows the earlier Hookean spring result (Figure 23.24b). Equation 19 shows that doubling the rate of frequency will double the stress level for a purely viscous material. This is expected from simple experiments such as doubling the rate of deformation on shearing a Newtonian fluid (such as water), for which the corresponding stress will also double.

Because a perfectly linear elastic material displays no phase shift with the applied strain function, whereas a 90° phase shift occurs for the purely viscous material, it is obvious that for a linear viscoelastic material, the observed

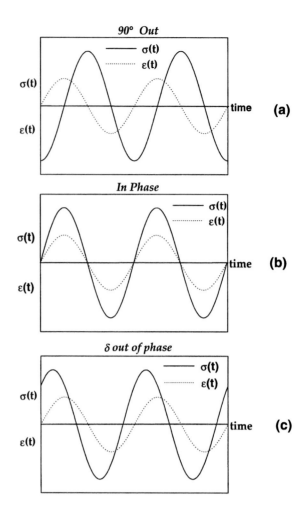

Figure 23.24 Plots of the time-dependent stress and strain vs time in a dynamic mechanical experiment: (a) for a linear viscous body, $\delta = 90°$; (b) for a linear elastic body, $\delta = 0°$; and (c) for a viscoelastic body, $0° < \delta < 90°$.

phase angle will lie between zero and 90° and can be easily determined (Figure 23.24c). This measurement is easy to make: If a small sample is placed in an appropriate thermally insulated chamber and a sinusoidal strain function is applied at one end with a strain transducer, the other end of the sample can be attached to a second transducer. The latter will record the force, and with the cross-sectional area of the sample, the engineering stress and its time dependence can be determined as the oscillatory strain is applied. These two electrical signals from the two transducers, when appropriately analyzed, directly lead to the determination of the phase angle, which is commonly denoted as δ. Measurement of δ must somehow relate to the level of viscous versus elastic behavior for a given cycle of deformation, but how can these two components be deconvoluted from that variable? Because the experiment is dynamic and at small strains, then the stress divided by the strain provides a dynamic modulus (often denoted as E^*), but it is recognized that this variable is made up of both a viscous and an elastic component. The relative magnitudes of these two components differ depending on the level of elastic versus viscous behavior in the system.

As illustrated in Figure 23.25, the dynamic stress sensed in the material can be represented by two sinusoidal functions added together, where one of these functions is directly in phase with the applied strain and the other is 90° out of phase with that function. If the two functions just described are of appropriate amplitude and are added together, they will always achieve the value of the true or dynamic stress in the system at any given time. One can therefore represent the dynamic modulus, which extends from the measurement of the dynamic stress divided by the dynamic strain, by considering $E*$ to be made up of two vectors that are orthogonal to one another and which when added together provide the real behavior, $E*$, as represented in Figure 23.26. This is often written conventionally as

$$E* = E' + iE'' \qquad (20)$$

The symbol i may be unfamiliar to some readers: It represents the square root of -1 and arises because of the considerations of angular frequency in conventional complex notation, but this aspect is not really needed for our discussion here. Figure 23.26 illustrates how the relative (vectorial) components shown along the $X(E')$ and $Y(E'')$ axes comprise the dynamic modulus vector behavior, $E*$. To help clarify this point, suppose one were to undertake a dynamic experiment only on a linear elastic material. Then, $E*$ has

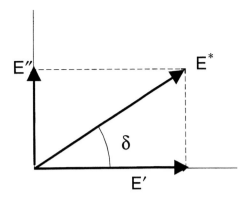

Figure 23.26 Vector diagram of $E*$, E', E'', and their associated phase angle δ.

to be composed only of an in-phase component ($\delta = 0$) denoted by E'; this would require that the vector $E*$ lie identically on the E' axis and there would be no component arising from E''. In a similar way, a dynamic experiment of a perfectly linear viscous body would require $E*$ to lie parallel to the E'' axis ($\delta = 90°$) with no component from E'. Thus, as $E*$ rotates further away from the E' axis, the phase angle δ in Figure 23.26 is directly a measure of the phase difference between the real behavior and that of a perfectly elastic body where δ is zero. Applying trigonometry in light of Figure 23.26 leads to the very important relationship:

$$\tan \delta = \frac{E''}{E'} = \frac{\text{loss}}{\text{storage}} \qquad (21)$$

Note that the parameter $\tan \delta$ is a direct measure of the ratio of loss (viscous dissipation) to storage (elastic). For example, when $\tan \delta$ equals unity, this represents a case where there is an equal amount of energy dissipated as well as stored during a deformation cycle, that is, half of the energy is lost. It also follows that as the material behaves increasingly elastic, $\tan \delta$ will become smaller and in the limit approach zero, whereas for a perfectly viscous body, $\tan \delta$ would increase to infinity. For viscoelastic materials such as polymer systems, $\tan \delta$ typically will range from near zero to values generally less than 10, at least for the general frequency and thermal ranges used for the common DMA instruments.

Equation 21 clearly illustrates that there are only two independent variables that one needs to report; the third can be determined. Generally, the values of interest are either E' and $\tan \delta$, or E' and E''. It is also important to note that E' (commonly called the storage modulus, real part of the modulus, or in-phase modulus) is essentially equivalent to the slope of a typical stress-strain response that we addressed previously in this chapter, but for which the temperature and rate of deformation appropriately match the temperature and the frequency used in the DMA experiment. Hence, E', when plotted against temperature, provides the TMS that we referred to previously in this chapter

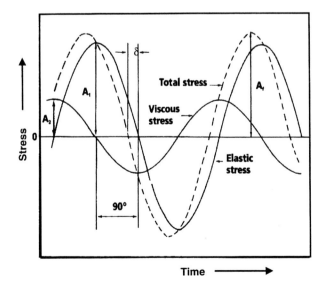

Figure 23.25. The dynamic total (observed) stress, the associated elastic (in phase) stress, and the viscous (90° out of phase) stress components all plotted vs time. Note that the addition of the elastic and viscous stress values will regenerate the total or observed stress. (Reproduced with permission from reference 23.)

(Figure 23.4b). What do the other functions tan δ or E'' look like when plotted against temperature? Even though frequency will be an important variable as well, let us consider a constant frequency test and the general temperature behavior of the dynamic parameters.

Figure 23.27a,b,c shows DMA examples of the behavior of poly(vinyl chloride) (PVC), where (in Figure 23.27a) not only is the temperature dependence of E' and tan δ displayed, but also four different frequencies were used for this test. For this test, it is clear that measurements extend from ambient temperature, which is below T_g, to somewhat beyond that transition. Of particular significance is that the low temperature modulus for all four frequencies is in the range of 10^9 to 10^{10} Pa, as was discussed for isotropic glasses. Conventional PVC generally possesses a very small amount of crystallinity, which therefore plays relatively little role in influencing the data shown. The decrease in glass transition in this material is approximately three orders of magnitude and the rubbery plateau was not completely determined because the tests were terminated in the temperature range of approximately 140 °C. It is significant that as the frequency is increased, the curves for the modulus-temperature behavior are shifted to higher temperatures but the same general shape is maintained. In addition, the glass transition associated with the tan δ peak (another index of T_g) for each frequency is also shifted to higher temperatures. Thus, whether T_g is determined by the inflection point associated with E' or the somewhat higher value taken from the peak of the corresponding tan δ plot, it is noted that as frequency increases, so does T_g. This is our first example to illustrate that T_g is not a fixed number, but rather it is dependent on the observation time discussed previously (the rate of loading or frequency). This kinetic or time-dependent effect results from the fact that if we are looking for a softening or glass transition point (when using a higher frequency or loading rate), then to observe the same level of softening relative to a lower frequency where the window of time is longer, molecular motion must be enhanced by a higher thermal energy or temperature. A general rule for the upward shift in T_g as frequency increases is 3 to 8 °C per decade.

Briefly, if one were to determine a glass transition at 1 Hz to be 100 °C, such as might be the case for common atactic polystyrene, that same transition would be found

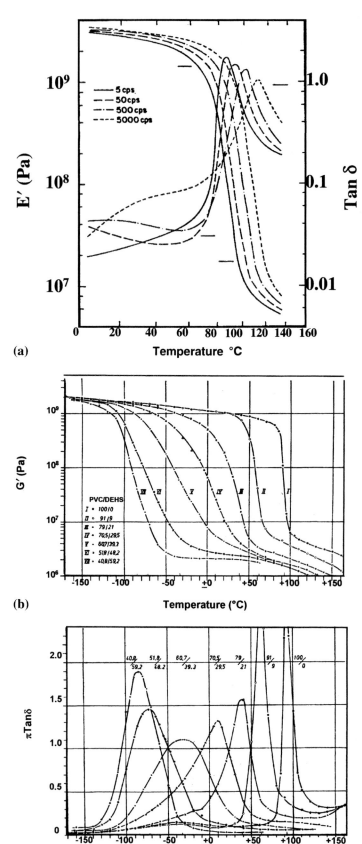

(a)

(b)

(c)

Figure 23.27 (a) E' and tan δ vs temperature for a polyvinyl chloride (PVC) specimen as determined at four different frequencies. Note that the temperature axis is nonlinear. (Adapted with permission from reference 34.) (b,c) G' and damping (π tan δ) plotted against temperature for a PVC material containing different levels of a low molecular weight plasticizer (diethyl hexyl succinate). Measurement frequency ~ 1 Hz. Note: When tan δ is multiplied by the constant π, it is known as the "log decrement". (Adapted with permission from reference 35.)

generally between 109 and 124 °C at a frequency of 1 kHz. One can often use a range of frequencies to allow predictions of the shift of not only T_g, but other loss peaks described by plotting the log of the frequency versus reciprocal temperature, where the peak in the tan δ transition occurs. Generally, this plot will be linear (Arrhenius) or will show near-linear behavior, particularly for sub-T_g loss peaks. By extrapolating these data to the frequency that may not be easy to measure within the laboratory, one can often nearly predict where the transition will occur. In fact, from the slope of these lines, if linear, one can calculate an activation energy, which indicates how that particular transition depends on temperature. Additional details about activation energies can be found elsewhere.[36]

A very significant point regarding Figure 23.27a is that there is also a sub-T_g loss peak. It is easiest to see this peak, which is much smaller than the magnitude of tan δ increase associated with the glass transition, at the highest frequency where the lower temperature shoulder is quite distinct. The other three frequencies do not display this loss peak as well, particularly for the lowest frequency. However, it is clear that there is a secondary or sub-T_g loss peak that is associated with a local-scale motion in the PVC chain, and that has a minor but real effect on the softening characteristics in the glassy modulus area. The ability to determine the onset of local-scale motions, whether in the glassy state or potentially in the crystalline state, is one of the outstanding features of the DMA technique. What is the significance of sub-T_g loss peaks? Certainly, if a local-scale motion is occurring, there is a dissipation of energy on a local scale and therefore a means of dissipating the mechanical energy that has been transferred into the material. This can play an important role in many systems by improving the impact properties below T_g. That is, if indeed the system shows little sign of molecular mobility in the glassy state, the glassy material will often behave more brittle-like because any mechanical energy put into the system in that state will often lead to fracture. However, if indeed a local-scale motion can occur, particularly in the backbone versus a side group, then that local-scale motion can serve to dissipate mechanical energy being transferred through the material, and thereby transform that energy into the form of local heating. Hence, sub-T_g mobility may play an extremely important role in fracture toughness of glasses. These same arguments regarding molecular motion are also important in considering the dielectric properties of polymers, and are addressed in a somewhat similar way by applying an oscillating electrical field across a sample and noting the dissipation of that electrical energy through dipole motion. (See reference 37 for more details about the technique of dielectric spectroscopy.)

We now describe a second example using PVC as the polymer, but for which the effect of plasticization by a low molecular weight molecule is considered. In Figure 23.27b,c, the storage modulus as well as tan δ are illustrated for a constant frequency test as a function of temperature, but the variable of plasticizer content is changed. Curve 1 represents the unplasticized material and is similar in nature to Figure 23.27a except that the measurements were made in dynamic shear versus extension. In addition, because the PVC materials may be somewhat different in crystallinity, direct comparison of the data in Figure 23.27b with the data in Figure 23.27a may not be exact. Note that as one adds plasticizer, the glass transition temperature is systematically decreased and broadened relative to the unplasticized sample. One also sees the presence of the lower sub-T_g peak that was mentioned earlier in the materials containing high or low amounts of plasticizer. The glass transition can be altered systematically with plasticizer, but particularly the damping characteristics can be controlled to provide differences in behavior that might be desired in specific applications. For example, an appropriately plasticized material in which tan δ maximizes near ambient conditions will provide a very high mechanical damping material that can be useful in applications for floor coverings; the damping characteristics provide more comfort for those who walk (a lower frequency loading) on such a floor for long periods of time. Figure 23.27 shows that with plasticizer addition, a broadening of the glass transition temperature occurs, as noted by the tan δ response, but also it promotes a lower magnitude for the corresponding tan δ peak. There are, however, exceptions to this behavior: the addition of some plasticizers tends to simply systematically shift the G' or tan δ curve to lower temperatures relative to values for the unplasticized material.

We now address the variable of cross-linking and its influence on the DMA curve. Figure 23.28 provides an example of a phenol formaldehyde resin that contains three different levels of the cross-linking agent hexamethylenetetramine. Figure 23.28a shows the behavior of G' with temperature and Figure 23.28b illustrates damping (loss) behavior. Note that as the cross-linking increases, the glass transition region is shifted upward and broadened greatly. At the highest level of cross-linking additive, it becomes difficult to define a T_g region at all. This is due to the limitations placed on cooperative segmental mobility caused by the extensive level of cross-linking. Hence, highly cross-linked materials, when measured by DMA, may not easily display a recognizable glass transition. In many cases, the transition occurs over a very wide range of temperature in nonhomogeneously cross-linked materials because chains are present that are not near a cross-linking point, and therefore undergo the transition as if it were uncross-linked. Many other segmental regions are highly restricted because of the presence of cross-links and thus require a higher thermal mobility to undergo cooperative segmental motion (see Figure 23.16). Hence, broadening of the glass transition is often observed in cross-linked systems.

The DMA behavior for a crystalline polymer for which the level of crystallinity is varied is also considered. In this

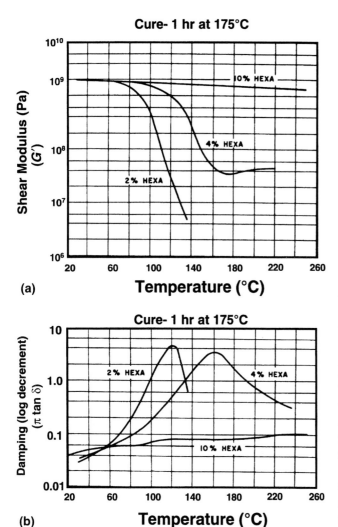

Figure 23.28 Plots of G' (a) and damping (π tan δ) (b) for phenol formaldehyde resin (novolac) cross-linked with various levels of hexamethylene-tetramine. (Reproduced with permission from reference 37.)

case, the example is linear polyethylene, in which the molecular weight varies but the thermal history has been maintained constant, that is, the samples were all slowly cooled (Figure 23.29a,b). At least three distinct transitions occur at approximately -120, -40 °C, and a third beginning above 0 °C, which is a broad transition that is really a composite of more than a single transition. The lowest temperature peak, usually called the γ peak, is associated with cooperative segmental motion of a very few chain segments (rotation of multiple methylene units). The -40 °C transition (β peak) is influenced by short chain branching or amorphous content and is believed to be the glass transition, although there is some controversy about this assignment.[39] The upper transition is associated with motions related to the crystalline state and is often called the α_c transition. The α_c transitions are not observed in all semi-crystalline polymers, but if sufficient mobility can be gener-

ated within the crystal prior to melting, then this motion can be identified by DMA techniques and is important in thermal processing considerations. Generally, the presence of an α_c transition permits an annealing step to allow lamellar thickening of the initial folded chain lamellae, and also permits solid-state forming of the material with greater ease.

One of the particularly strong features of DMA is that often both frequency and temperature can be varied, thereby providing a dual mapping of these two important external variables on the dynamic modulus and its respective components. This allows one to construct (sometimes in combination with a time-temperature-superposition technique discussed earlier) the response surface of a particular mechanical parameter. An example of this is illustrated in Figure 23.30a,b, in which the storage modulus and tan δ are each plotted as a function of both temperature and log frequency associated with the dynamic test. These data reinforce earlier findings that raising temperature at a fixed frequency will lead to softening of the material and a decrease in modulus, or lowering the frequency at a constant temperature will achieve the same result. A response surface can also be used for many of the other variables that have been discussed in this chapter, such as yield stress and toughness. The concept of a response surface for practical applications of a material when exposed to variation in different external parameters (such as temperature or loading rate) is very significant.

Although the examples presented so far have demonstrated only the behavior of E' and tan δ, note that the general behavior of E'' is somewhat similar to that of tan δ except that it carries dimensions of stress. In addition, if one is making such a plot at a constant frequency, the peak temperatures in E'' will typically be somewhat lower than that determined in the tan δ plot, particularly if the material is softening with temperature. Hence, if a given tan δ peak temperature is taken as an index of T_g or some other transition, and if another researcher reported the same values using the respective peaks in E'', those values would be lower for the latter researcher. Thus, the reader should be cautious about noting exactly what transitions are specified and how they were determined if gathered from DMA analysis.

In summary, the DMA technique is a powerful tool, and is commonly used to characterize the mechanical characteristics of polymers and other materials. The measurements, however, are made in the linear region at small strains. Therefore, the results do not reflect high deformation characteristics of the system, but at times provide some indication of the expected behavior at given temperatures or frequencies by informing the investigator about the transitional characteristics associated with sub-T_g motions, the glass transition, and even local-scale motions in crystals (α_c). The DMA method can also be used to characterize the melt and solution viscoelastic properties of polymers.[18,19]

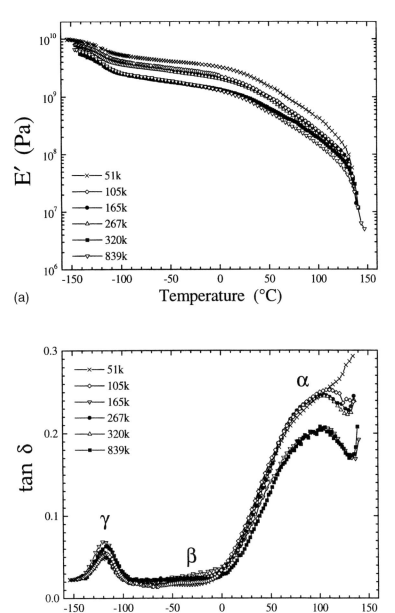

Figure 23.29 Dynamic mechanical response for several metallocene linear polyethylenes of different average molecular weight as labeled; data obtained at 1 Hz: (a) E' vs temperature, (b) tan δ vs temperature. (Reproduced with permission from reference 38.)

These topics are outside the scope of this chapter, although the same principles would apply.

Some Mechanical Considerations of Multiphase Materials

Although mechanical considerations of multiphase materials is a major topic in the study of the mechanical behavior of polymers, space limitations restrict our discussion. The number of multiphase polymeric-based systems is enormous. Any semicrystalline polymer will contain both crystalline and amorphous regions, but incompatible polymer blends (of which there are many commercially) are also made up of different phases represented by the individual polymeric ingredients. In addition, many polymers are used in the form of filled systems in which either a soft or rigid particulate has been added to influence the properties of the material. Such fillers may be inorganic, such as talc or glass, and may be in fiber form as well as spherical parti-

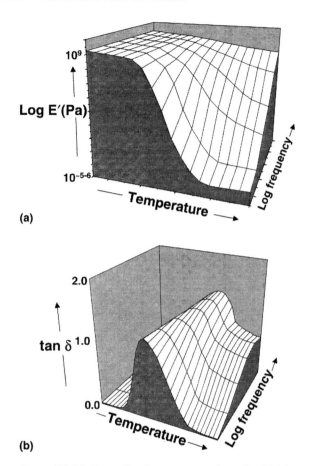

(a)

(b)

Figure 23.30 Generalized response surfaces for E' (a) and tan δ (b) as a function of temperature and log frequency. Note the dependence of the transition, T_g, on each independent variable.

cles; thus, the aspect ratio of the particulate phase can vary considerably. Rubber-based particulate phases are often used to promote enhanced toughness and strain to break in some cases. One example is in impact-modified polystyrene in which butadiene serves as a phase-separated component in that system, although there is some actual grafting of the butadiene with the polystyrene to help minimize massive macrophase separation (see Figure 23.31). Although we cannot discuss all of these parameters, some of the important aspects of these systems that relate to the types of properties that one will observe mechanically are level of crystallinity, measurement temperature (above or below the glass transition temperature of whatever components are present), presence of orientation of any geometrically anisotropic dispersed phase, and the degree of difference in elastic characteristics of the dispersed phase relative to the matrix phase, which in turn promotes major differences in local mechanical response. A brief sampling of some of the topics associated with multiphase systems now follows.

As an example of crystallinity, Figure 23.32a,b illustrates the stress-strain behavior measured on a series of narrow molecular weight distribution linear polyethylenes with varying molecular weights. In Figure 23.32a, these unoriented samples were prepared by compression molding and rapidly quenched from the melt to produce a lower crystallinity sample than those in Figure 23.32b, in which the cooling rate from the melt was slow, thereby allowing considerably more time for crystallization to occur and allowing the crystals to become more perfected. The testing temperature was at ambient and a constant sample size and crosshead speed (initial strain rate) were used. The general behavior of the stress-strain response is somewhat the same for the two sets of data with some distinct exceptions. For example, the lowest molecular weight ($\overline{M_w}$) sample shown (51,000 molecular weight) is a ductile high-deformation material when measured in the quenched form, but displays very brittle behavior in the slow-cooled, that is, the higher crystalline form. This results from the differences in the crystallinity and morphological features promoted by the different thermal histories.

Note also that as the molecular weight increases, even though a ductile response tends to be observed in most of the materials after yielding, the degree of strain hardening becomes more enhanced earlier with higher molecular weight materials. This is because higher molecular weight species will tie together more folded-chain lamellar crystals, and this will lead to longer relaxation times and therefore strain hardening occurs earlier in the deformation.

Figure 23.33a–c shows the parameters of modulus, yield stress, and yield strain plotted against the percent crystallinity observed for the respective samples. As might be anticipated, because the measurements in this case were made well above the glass transition temperature of the amorphous phase, the addition of crystallinity will enhance the stiffness or modulus of a material. This is distinctly shown in Figure 23.33a, in which the modulus systematically increases in a near-linear fashion with level of crystallinity. As anticipated, yield stress increases with crystallinity,

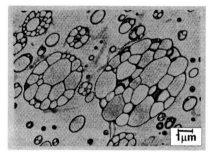

Figure 23.31 Transmission electron micrograph of a high-impact polystyrene specimen. The black regions are butadiene as stained with osmium tetraoxide, whereas the light regions are polystyrene. (Reproduced with permission from reference 41.)

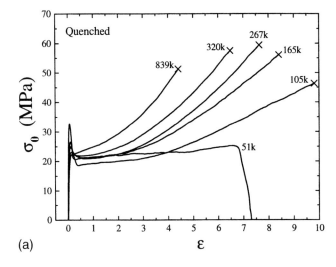

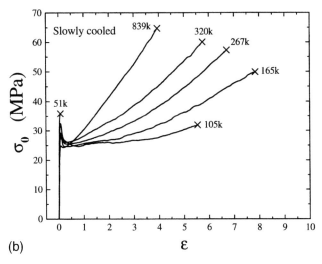

Figure 23.32 Engineering stress-strain plots for several compression molded metallocene linear polyethylenes of different average molecular weight as designated; sample gauge length = 0.7 cm and crosshead speed = 25.4 mm/min: (a) rapidly quenched from the melt, (b) slowly cooled from the melt. (Reproduced with permission from reference 38.)

however, less zero-slope behavior occurs, indicating that other variables in addition to crystallinity are important. Other types of mechanical analysis could also be discussed, but these examples illustrate the general nature of how crystallinity and changes in molecular weight affect a given polymer.

The addition of a rigid nonpolymeric filler (at least relative to the matrix polymer) will also enhance modulus, and this is often done to further stiffen a system above that of even glassy polymers because inorganic fillers will increase the modulus above what can be obtained by typical isotropic organic systems. However, as rigid fillers are added to the material, typically the strain to break tends to decrease and often toughness decreases because the rigid filler cannot absorb energy by deformation or flow. In some cases, the rigid filler may be ductile in nature (some rigid polymeric fillers), and if these materials undergo yielding, the dissipation of energy can be increased yet the modulus is not lowered as it is by addition of a soft filler.

When there is a dispersed phase with different stiffness characteristics than the matrix material about that phase,

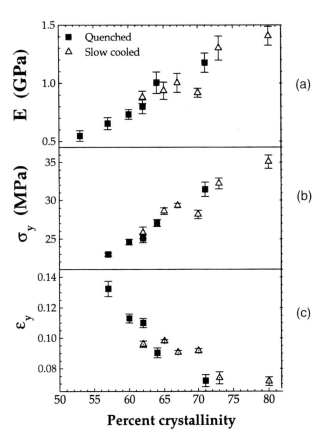

Figure 23.33 Plots of tensile modulus (*E*), yield stress (σ_y), and yield strain (ε_y), vs percent crystallinity for the same linear polyethylene materials in Figures 23.29 and 23.32. Data include results from both quenched and slow cooled specimens. (Reproduced with permission from reference 38.)

whereas yield strain decreases because a smaller amount of the more compliant amorphous material is present. It might be expected that the yield stress increases with crystalline content and yield strain decreases with crystalline content if measurements are made above the T_g.

If the data above are replotted (not shown) in a normalized form (yield stress or yield strain are, respectively, divided or multiplied by the initial crystallinity and again normalized by the highest value of the series), for σ_y, this normalized form of the yield stress becomes nearly independent of crystalline content, indicating that crystallinity highly dominates that particular mechanical response in these isotropic samples. For the normalized yield strain, once the material is loaded, local points of stress concentration develop. For example, Figure 23.34 shows a schematic of a soft phase, which is represented as a hole (a very soft phase) punched into a matrix material that is being loaded along the vertical direction. Without details, it can be shown that the tangential stress, σ_{tan}, at the edge of the hole will be a function of the applied stress on the sample, σ_0, and the angle θ shown in Figure 23.34. The value can be derived from mechanics and the tangential stress is given by

$$\sigma_{tan} = \sigma_0[1 - 2\cos(2\theta)] \qquad (22)$$

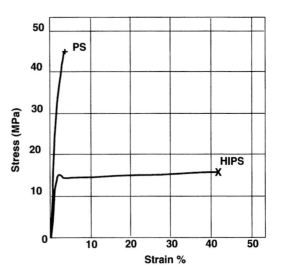

Figure 23.35 Engineering stress-strain plots for a polystyrene specimen as well as a high-impact polystyrene material. (Adapted with permission from reference 3.)

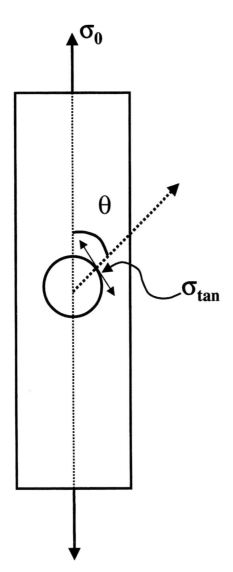

Figure 23.34 Schematic showing a rectangular sample possessing a hole (soft phase) that is loaded in the vertical direction with a stress σ_0. The tangential stress, σ_{tan}, at the edge of the hole is a function of θ.

This equation shows that when θ is equal to zero or 180° (i.e., the tangential stress at either the north or south pole of the sample in Figure 23.34), σ_{tan} would actually be equal to the negative of the applied stress. Therefore, if one were sitting at the edge of the hole at either the north or south pole position, they would be in compression. On the other hand, when the value of θ is equal to 90° (or 270°), which places the tangential stress at the equatorial regions of the matrix material, this leads to an enhanced or concentrated stress by a factor of 3 at the edge of the hole. (For a three-dimensional soft phase as a void, it can be shown that the stress concentration factor decreases to 2.) This simplest case demonstrates what is meant by stress concentration arising by induction of a soft phase into a hard matrix material.

At first consideration, promoting stress concentration would seem to be undesirable because this may promote failure under a lower load than would have occurred without the stress concentration point. However, in many materials that are brittle, such as atactic polystyrene at ambient conditions, the presence of a soft, rubbery, spherical phase in its midst can induce the material to become ductile by local yielding, in contrast to brittle behavior. This behavior greatly enhances the ability of the sample to absorb energy and the toughness is also greatly increased. This change in behavior, however, leads to a loss in modulus caused by the addition of a soft phase (note that a rubber phase is roughly 10^3 lower in modulus than that of an organic glass), and often the tensile strength is also decreased. To illustrate this point, Figure 23.35 shows the stress-strain behavior of a rubber-modified impact polystyrene with that of an unmodified styrene behavior. Note that the tensile strengths are greatly different, as are the toughness values (areas under the curve), and that the modulus of the rubber-modified material is a bit lower, as expected. Hence, when one attempts to modify a particular polymer to enhance a given mechanical parameter such as toughness, there is often a trade-off: other mechanical properties such as tensile strength might be unfavorably affected. However, the exact character displayed by the different parameters will depend on many factors, such as particle size, interaction of the particle with the matrix, and test temperature. That is, both system and external variables will play important roles.[41]

What happens to the local mechanical response if the particulate phase is rigid relative to the matrix? An example of this is illustrated in Figure 23.36a, in which a glass sphere is placed into a soft silicone rubber and then deformed along the vertical axis. In this case, pulling in the vertical axis would desirably extend the rubbery phase in that direction and therefore place a hoop stress on the equatorial region of the particle as the rubber phase tries to decrease its lateral dimensions. This is in great contrast to the matrix and dispersed phase properties that were reversed relative to our previous example. In the present case, it is the polar regions that become important because the rubber phase would tend to extend along the vertical direction, whereas the hard glassy particle would tend to stay undeformed because of its much higher modulus.

This often leads to two associated phenomena. One phenomenon is caused by the complex stress field at the poles, which can promote a local cavitation in the matrix material, as is illustrated in Figure 23.36a, frame 2. With further deformation, unless there is good coupling of the matrix phase with the particle surface, delamination of that matrix phase can occur, thereby leading to a major void or cavity being induced between particle and matrix (this is often called dewetting). This is also illustrated in Figure 23.36a, frame 3, where not only did cavitation occur in the matrix at lower strain, but this was then followed by further deformation and formation of a large void near the south pole. Dewetting can sometimes be prevented by the development of an appropriate chemical coupling between the matrix and particulate phase (see Figure 23.36b). Through the use of a silane-coupling agent that bonds the glass to the rubber, even higher levels of deformation were obtained without dewetting, although cavitation in the matrix still occurred. The result is an influence on the stress strain properties of these different materials with and without the presence of a localized coupling agent. This example strongly indicates how the mechanical properties can be

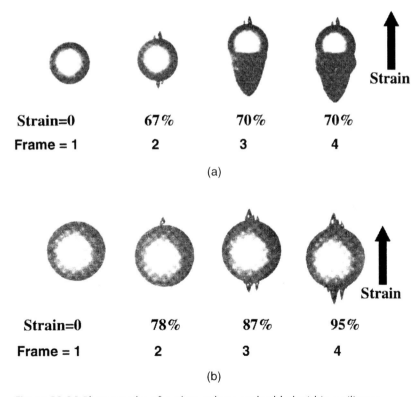

Figure 23.36 Photographs of a glass sphere embedded within a silicone rubber matrix: (a) the different frames show the material for different levels of deformation along the vertical axis, (b) a silane coupling agent has been used to promote better adhesion of the glass and silicone. (Reproduced with permission from reference 42.)

influenced by a very small amount of a chemical compound that can better bridge or chemically attach a matrix to a dispersed phase.

Before leaving the topic of stress concentrators, it is very pertinent to point out that scratches, nicks, or defects in the surface of any polymer sample will often become points of stress concentration and may greatly limit the ultimate properties of a material. Figure 23.37 shows a sketch of a specimen with a crack or notch that is orthogonal to the loading axis. As the radius of curvature of that notch becomes smaller (i.e., the crack becomes sharper), the stress that exists in the matrix at the tip of the crack when the sample is deformed vertically will be greatly enhanced. This relationship can be given by

$$\sigma_{\text{tip}} = \sigma_0 \left[1 + 2\left(\frac{a}{r}\right)^{1/2} \right] \tag{23}$$

where σ_{tip} represents the stress at the tip of the crack, σ_0 is the applied stress, a is the length of the crack, and r is the radius of curvature at the crack tip. Note that for a given crack length a, as the radius of curvature becomes smaller, the stress concentration factor will increase. It is therefore important to learn how different polymers respond mechanically, either in impact or through some other test, because defects are commonly found or induced in materials during use. So-called notched versus unnotched impact tests have been established, for example, by ASTM.[20] These tests allow one to determine the sensitivity of a material to such notches and the associated radius of curvature. For example, bisphenol-A polycarbonate is a well-known high-impact glass, but it is particularly susceptible to significantly reduced toughness if any sharp notch is placed within the material. Some other polymers, however, are much less dependent on the presence of a notch or a sharp-tipped crack.[3,41] Thus, we see again that different chemical structures and morphological features of polymers can significantly alter the actual results of a given stress concentration test.

Influence of Molecular Orientation on Mechanical Behavior

Because of the chainlike nature of polymeric materials, when the materials have been processed into a particular article of commerce, they commonly contain preferential orientation in a specific direction, such as the machine direction (MD) or the direction in which the process line moves. Many products (such as drawn fibers and films, and stretched blow-molded bottles) are intentionally oriented to provide enhanced properties in particular directions. In fact, it is likely that the majority of commercial polymer articles have some level of preferential orientation if they have been melt processed. Even for polymers such as cellulose that are produced naturally in the morphological texture of the tree, preferential orientation is critical for providing the necessary rigidity and strength to maintain the endurance of a tree under the stress of a high wind. Hence, it is important to recognize that mechanical anisotropy is a major factor in oriented materials, and therefore, their directionality dependence must be considered when carrying out mechanical property tests.

To provide some indication of these dependencies, we first address the issue of orientation itself. This variable can be difficult to understand and it is often confused with some of the deformation parameters such as elongation, strain, and extension ratio. Although there may be a correlation of the degree of orientation with these parameters, polymer molecules will often undergo relaxation and loss of orientation at the same time a sample is being extended. Hence, if only draw ratio or extrusion ratio are stated, little insight about the level of final true molecular orientation is provided. Figure 23.38 more distinctly defines orientation. A single chain is shown preferentially oriented at an angle θ with respect to some reference axis of interest, which is often the

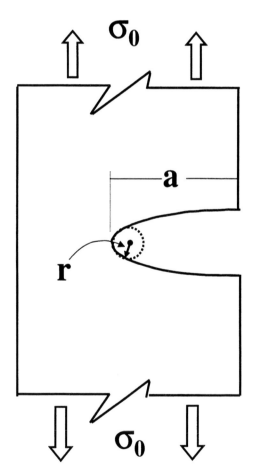

Figure 23.37 Schematic showing a horizontal crack in a specimen that is loaded with a stress, σ_0, along the vertical. The crack has length a and radius of curvature r at the tip.

Reference Axis

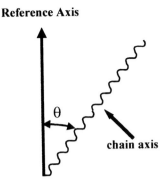

Figure 23.38 Schematic showing a chain oriented at an angle θ with respect to the vertical (stretch axis).

deformation axis. If the orientation is uniaxial, it is symmetric about the draw axis. This commonly occurs in a fiber where a single draw axis is used.

To specify the state of orientation of the chain or molecular axis, the Hermans orientation function, f_H, is used. This function is identical to what was referred to previously as the order parameter (see Figure 23.1). The Hermans function is defined as

$$f_H = \frac{3<\cos^2\theta> - 1}{2} \qquad (24)$$

where the term $<\cos^2\theta>$ represents the average over all of the chains in the system. If θ were equal to zero, the system would be perfectly oriented and f_H would equal 1. On the other hand, for perpendicular orientation, θ is 90° and $f_H = -0.5$, thereby giving the second extreme. For random orientation, the $<\cos^2\theta>$ term is equal to 0.33, and hence f_H equals zero. Thus, a better measure of the state of orientation is to specify the actual determination of chain orientation through such methods (or their combination) of birefringence, linear dichroism, X-ray scattering, polarized Raman scattering, and some selected, other less common methods. The details of these methods are not addressed in this chapter, but the interested reader can consult the appropriate references.[43,44]

The reader is reminded that although warm bubble gum (or even Silly Putty) can be highly drawn, because of the rapid relaxation of the partially oriented polymeric molecules making up this system, no residual orientation occurs, and specification of draw ratio alone is not a sufficient index of orientation. With variation in molecular weight, molecular weight distribution, draw rate, and temperature, the reader can easily understand why orientation cannot quantitatively be coupled to any deformation variable alone. However, for a given molecular weight, temperature, draw rate, or some set of conditions for which only one variable is altered, a correlation with level of deformation will often be present if it is quantitatively measured. Unfor-

tunately, quantitative orientation is often not determined and therefore the directionality or draw dependence of some specific property is simply noted by plotting that property versus the level of deformation.

Figure 23.39a illustrates one example of this orientation that also considers morphological structure. The general morphological structure of this extruded oriented linear polyethylene displays a stacked, folded-chain, lamellar structure that is schematically illustrated in Figure 23.39b. The general chain orientation in the lamellae is preferentially along

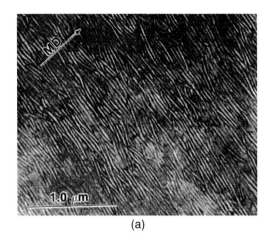

(a)

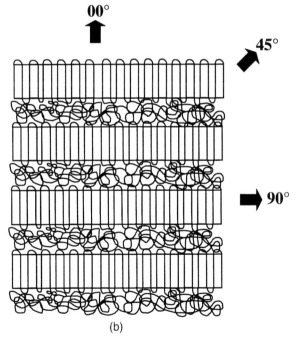

(b)

Figure 23.39 (a) Transmission electron micrograph of a stacked lamellar morphology for an extruded linear polyethylene. The extrusion axis is labeled as machine direction (MD). (Reproduced with permission from reference 45.) (b) Schematic of molecular structure associated with (a).

the extrusion or MD, whereas the lamellar platelets are oriented more nearly perpendicular to the MD. The interlying amorphous layers between lamellae are essentially unoriented, as shown by appropriate methods.[46] The usual belief would be that as the chains become oriented along a specific direction (in this case the MD), higher stiffness would occur in that direction and lower modulus would be perpendicular to the MD. Figure 23.40 shows the stress-strain behavior of this material when measured parallel, perpendicular, and 45° to the original extrusion axis. Note that the actual modulus is higher in the orthogonal direction (this rarely occurs). Furthermore, the degree of extension to fail in this direction essentially exceeded the instrument's ability to determine it; the material extended well over 1000% and the crosshead of the stretching apparatus was limited in its degree of extension, thereby requiring the experiment to be discontinued. On the other hand, the degree of extension in the other two directions (i.e., the 0° [MD] and 45° directions) is quite different in behavior.

Detailed reasons for this difference in directionality have been given elsewhere,[45-47] but basically the differences relate to the morphological texture of the material. Because the lamellae are more rigid than the interlayer amorphous material that is well above its glass transition temperature, the stacking of the lamellae promotes what is often referred to as a series model of the two components: a soft amorphous component and a more rigid crystalline component, as indicated in Figure 23.39b. The interlying amorphous material, which is soft, will more easily deform than the crystalline regions, and is the more compliant of the two components. This is why the modulus is lower in the MD than in the orthogonal direction, where the two phases (amorphous and crystalline) act more in parallel. When this type of structure is stretched at 45°, a complex shear deformation of the lamellae occurs that provides a different modulus than either of the two orthogonal directions. At 90°, the degree of high extension arises because the chain axis is essentially unfolding from the folded-chain crystalline lamellae, thereby giving rise to extremely high extensibility. The important point is that the nature of the morphology drastically alters the directionality of the mechanical properties, and this must be recognized when mechanical results are reported on any sample that contains orientation.

Another common example of orientation effects is illustrated in Figure 23.41, in which the stress-strain properties are shown for a semicrystalline thermoplastic fiber poly(ethylene terephthalate) (PET) that was spun from a melt extrusion process at different speeds, but all other conditions were held as constant as possible. As a result of higher spinning rates, and less time for molecular relaxation to occur prior to quenching and crystallization, the material displays a higher tenacity (tensile strength) but correspondingly has a lower elongation to break.

Biaxial orientation is also common and is more complex than the uniaxial process. Biaxial orientation generally exists in blown films used in packaging, but also is induced in the production of stretch blow-molded bottles such as the common PET soda bottle. In these processes there is often a simultaneous drawing of the chains in two directions (the plane of the film) in contrast to some other biaxial processes in which there is an initial forward draw along the machine

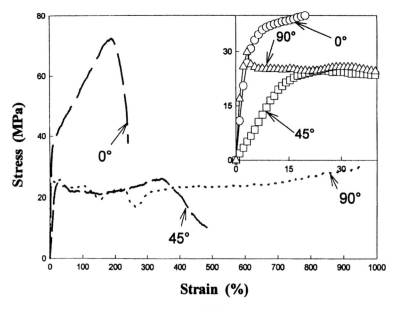

Figure 23.40 Engineering stress-strain behavior for the specimen shown in Figure 23.39a. The three different plots are for deformation parallel (0°), perpendicular (90°), and at 45° to the extrusion (machine direction) axis. (Reproduced with permission from reference 47.)

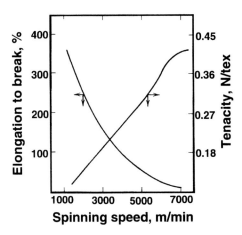

Figure 23.41 General expected behavior for elongation to break (%) and tensile strength (tenacity) vs melt specimen speed for a material such as poly(ethylene terephthalate). (Reproduced with permission from reference 48.)

direction followed by a later transverse or orthogonal draw. In some cases, a third step may also lead to another forward draw as well.[49] Materials prepared by this sequential process often are biaxially oriented PET films used in packaging and tape applications. Biaxially oriented polypropylene is also prepared by this process, which is referred to as tentering.[49,50] Specification of the orientation is more complex than that expressed by eq 24 because the degree of orientation along two orthogonal axes must be specified.

Quantitative methods of specifying biaxial orientation are available, as are means of specifying the appropriate biaxial state of orientation through what is known as the biaxial orientation functions.[51] To illustrate an example of impact strength and its dependence on biaxial orientation, Figure 23.42a,b shows low-density polyethylene that has undergone biaxial orientation in the plane of the film but to different extents. The draw ratios for the two different axes are also varied. Note that for a given draw ratio in one direction, balanced orientation (i.e., equal draw in the two orthogonal directions) tends to lead to better impact properties in contrast to unbalanced orientation. This is typically true because unbalanced orientation often promotes ease in fracture caused by weakness that extends orthogonal to the chain direction; that is, the material will often tear or split along the molecular axis because the secondary forces between molecules will be more easily cleaved than fracture across the chain axis. One of the principal properties for packaging films that have been biaxially oriented are their tear properties and the anisotropy ratio of that property. Toughness or the ability to absorb energy by an impact normal to the plane of the film is also very important in these applications.

Specific properties were discussed previously when moduli characteristics were defined (see Table 23.2). Specific properties were also addressed through the normalization of that parameter on the basis of the materials density. It was also indicated that the specific properties of polymers could also exceed those of metallics if molecular orientation were taken into account, at least for some materials. Figure 23.43 provides reinforcement and evidence that this occurs; that is, the specific modulus and specific tensile strength for polyethylene, as well as the liquid crystalline aromatic polyamide Kevlar, easily overpower the same parameters of the inorganic glass and metallic fibers shown because of the high increase in stiffness and tensile strength along the drawn axis of the respective polymer fibers. The orthogonal properties would be decreased considerably, but from the viewpoint of application of the fiber, these data clearly illustrate the high importance of orientation in polymeric materials and the major effect it can have on mechanical response.

Comments on the Mechanical Properties of Cellular (Foam) Materials

Because approximately 10% of the world's structural materials are cellular in nature, it seems appropriate to provide a few brief comments on these material systems. First, very few textbooks address the properties of cellular systems such as the common open-cell urethane cushioning foams and the closed-cell urethane or polystyrene and related insulation foam materials. Wood is a cellular system, as are bone and several other natural materials. It should be realized that the presence of cells in the system, particularly if cells are viewed as pockets of air or gas, will lower the density of the material and therefore provide potentially very different specific properties depending on the overall bulk properties displayed (recall that a specific property is exhibited when the bulk property has been normalized on the density of that material). Gibson and Ashby[53] provide a highly recommended text that addresses the mechanical features of cellular materials.

A feature that arises in the development of some foam materials, such as the common slabstock polyurethanes used in many applications of furniture and other cushioning needs, is the induced anisotropy caused by the chemical reactions involved when gas is produced to generate the cells. These cells or bubbles tend to rise (as they do when a cake is baked), thereby producing a geometric anisotropy of the cell structure. As shown in Figure 23.44a,b, which illustrates a typical slabstock urethane foam texture, if the cell structure is observed normal to the rise direction of the foam, a near circular symmetry is noted. An orthogonal view, however, shows that the cellular structure is more elliptical in nature, which is caused by the rise of the bubble that later undergoes cell opening, leading to the observed

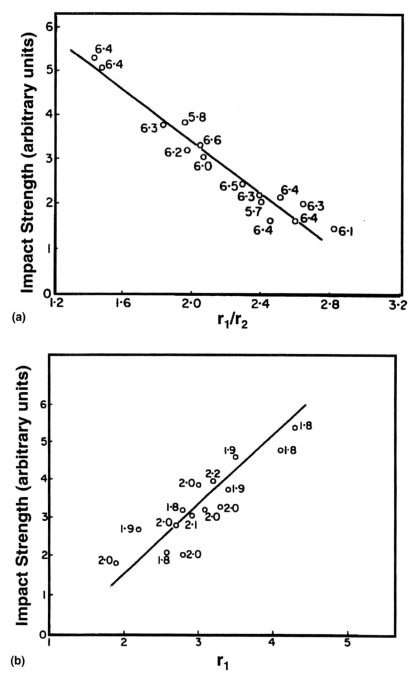

Figure 23.42 (a) Plot of impact strength vs the ratio of the draw ratios along the machine (r_1) and transverse axis (r_2) for a low-density polyethylene film, but where the draw ratio along the machine axis is given by the data points. (b) Plot of impact strength vs the extension ratio along the machine axis for a low-density polyethylene film. The numbers by the data points were obtained by dividing the extension ratio for the machine direction (MD) by the extension ratio for the transverse axis (TD). (Adapted with permission from reference 52.)

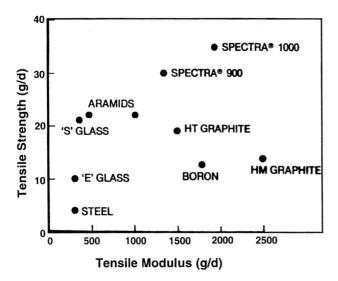

Figure 23.43 Plot of specific tensile strength vs specific modulus for several materials, including steel, glass-oriented polyethylene (Spectra), liquid crystalline aromatic polyamides (Kevlar), and others.

strut development. The open windows are required for airflow, which is necessary for comfort in cushioning applications. The important point about this texture is that the mechanical properties will in turn be different in the two orthogonal directions that relate to the observation directions in Figure 23.44a,b. The material is stiffer when compressed along the rise direction than it is when compressed in the other orthogonal direction. Recognition of the poten-

tial of geometric cell anisotropy becomes important when a given directional test on cellular materials is evaluated.

Other Related Topics

This chapter has focused so far on topics dealing with mechanical property behavior, but has also tried to assign some molecular significance to the origin of the mechanical response. Many areas, such as specific types of tests, procedures, or types of properties determined, have not been described. For example, crack growth and its coupling to fatigue behavior at high strains (hinge life), adhesive testing, such as peel strength and tack, and abrasion resistance have not been described in any detail. All of these procedures are important to many areas of mechanical testing of polymers. However, the majority of these procedures pose questions similar to those that have been raised in this chapter. For example, under a given load, time scale, and temperature, in what general state is the polymeric material (glassy, glass transition region, rubbery, viscous flow)? Is it multiphase in nature, as would be caused by the presence of crystals or fillers? What is its molecular weight? Is it cross-linked? A variety of other molecular-related questions can be asked. In some cases (such as adhesive behavior, for which the polymer material may be the substrate, the adhesive, or both), the wettability of one particular species compared to that of another should also be determined. Where compatibility exists and leads to wettability, is there the possibility that the adhesive and substrate have interpenetrated by molecular diffusion, thereby minimizing a sharp interface between them?[55]

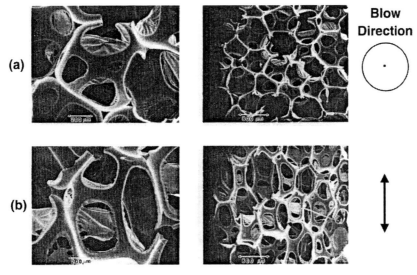

Figure 23.44 Scanning electron micrograph of a common open-cell polyurethane slabstock foam: (a) view down the rise or blow axis, (b) view orthogonal to the rise axis. Note the geometry anisotropy in the latter case. (Reproduced with permission from reference 54.)

Another issue that has not been discussed relates to the complex systems of fiber-reinforced composites, in which polymeric materials often serve as the matrix system (either a thermoplastic or a thermoset form) and high-modulus, high-strength fibers, such as those based on carbon or other materials, serve to reinforce the system. Depending on the way that the composite is constructed, such as through multi-ply lay-up and varied orientation of those plies, the properties of the final composite can vary tremendously for a given composition. The interested reader wishing to further their knowledge on the mechanical properties of composite systems should consult references 56–58.

There is one final area that deserves mention because the topic frequently arises in the mechanical testing of polymers containing a glassy matrix. This topic is physical aging. Physical aging refers to the transient behavior of the glassy phase once an amorphous material is quenched below its glass transition. The cooling process typically provides a nonequilibrium state and, if local-scale molecular motion can occur (as it often does in many polymer glasses), then local changes can lead to an increase in the packing density of the molecules. This in turn will alter the mechanical response of the system and other properties as well. Why does physical aging occur? Briefly, when the material is above its T_g, it is essentially a liquid at equilibrium, and therefore there is considerable molecular motion and also considerable free volume or unoccupied space between molecules because of their kicking nature. However, for a given observation time, as the system is cooled and the material thermally contracts according to its general thermal expansion coefficient, the viscosity will begin to increase as the glass transition region is approached, thereby greatly enhancing the relaxation times of the molecules. This limits their ability to locally pack into as dense a state as they would prefer (this does not imply crystallization, i.e., amorphous materials are being discussed). This in turn leads to essentially a molecular "log jam" as the glassy state is approached and the system is left in a nonequilibrium state (i.e., lower density is present because of poorly packed molecular segments caused by their inability to move within the time frame imposed on them by the quenching experiment). This leaves a higher state of free volume in the material than what would be expected at true equilibrium behavior. Thus, there is more freedom because of the excess free volume in the system, so that local-scale motions within the backbone or in side groups can occur more freely than would have been the case in the equilibrium state at the same temperature.

From a thermodynamic viewpoint, this driving force to approach the lower energy form or equilibrium state leads, with time (often logarithmic in behavior), to a more densified material, which means that free volume is lost with time. This loss is a measurable quantity and often relates to changes in the third decimal place of density.[59] The impor-

tant aspect, however, is that a small difference in density plays a tremendous role on many mechanical and other properties, that is, permeability is affected because if there is less free volume and mobility, the system will appear more rigid and less porous. As one example of physical aging, Figure 23.45 illustrates amorphous PET ($T_g \sim 65\ °C$) and a series of stress-strain curves determined on that material under exactly the same conditions; only the aging time in the glass varied for the sample. Note that as the aging time at the fixed sub-T_g storage temperature (23 °C) increases, there is a very pronounced enhancement in the yield stress, which is a very important engineering property. There is a slight increase in the modulus but it is hard to see in Figure 23.45 because the modulus enhancement only represents a few percentage points. Often, physically aged materials will show less strain to break, and therefore will become more brittle-like as aging occurs. Glassy materials often embrittle and lose toughness with time of aging in the glassy state.[61,62] Therefore, it is necessary to be aware of the measurement conditions for a given test, but it is also critical to reflect on the thermal history of the material.

Note that the former history caused by physical aging can be easily erased by raising the temperature above the glass transition temperature, which will rejuvenate the system by returning it to the equilibrium liquid state. After the material is cooled again into the glassy state, if the quench rate is identical to the earlier quench, the same time-dependent behavior in terms of the physical aging property response will be observed. This distinctly identifies the process as being physical in nature and not chemical, that is, it is not chemical degradation that influences the properties, but a morphological or local structural rearrangement process. This very

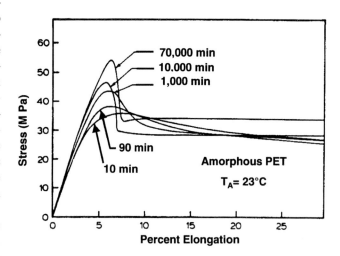

Figure 23.45 Engineering stress-strain plots for glassy poly(ethylene terephthalate) determined at different times following physical aging at 23 °C. (Reproduced with permission from reference 60.)

important topic is seldom discussed in texts or mechanical properties references, yet it plays an extremely important role in many mechanical and other properties of glassy systems.

Concluding Comments

An attempt has been made to provide an overview of many of the fundamental parameters and particularly the associated molecular considerations behind the mechanical response of polymeric materials. Many topics have not been fully developed. It is hoped that the reader, particularly one who is newer to the polymer area, will have learned a few important points that, although they may seem very simplistic, are essential to a fuller understanding of the molecular origins of measured mechanical properties. Figure 23.46 is designed to help the reader locate appropriate references for specific topics associated with the mechanical testing and behavior of polymeric materials. In closing, the reader should recall that polymers are not analogous to spaghetti, but are better represented by the analogy with the appropriate aspect ratio worm, which responds and relaxes at rates that depend on the thermal energy of the system. The features of entanglements, orientation behavior, crystallinity, type of morphology, and the state of matter the system at a given set of time-temperature conditions are also critical. Molecular weight and its distribution, and the potential of cross-linking into the gel state need to be understood as well. It is hoped that the reader has gained a knowledge base for understanding the mechanical properties of these diverse and highly important commercial systems.

Glossary

anisotropy   Implies a directional dependence to some property, such as mechanical or electrical.

bulk modulus, B   In the limit of small pressures, it represents the resistance to a change in volume with hydrostatic pressure when a material is placed under hydrostatic loading. This parameter is inversely related to the compressibility of a material.

chain aspect ratio   This ratio is formed by taking the extended end-to-end length of a given chain and dividing by the diameter of that chain, that is, the diameter of a tube that would fit around the extended chain.

chain configuration   The way the monomer units have been coupled together during polymerization. It takes into account such features as stereoregularity, geometric isomerism, and regiospecifity.

chain conformation   The spatial arrangement of the chain. Examples are a coil, rod, or helical conformation.

Deborah number, D_e   A dimensionless parameter related to the ratio of the molecular relaxation time of a material to that of the time frame over which the material is observed; that is, the experimental time window.

engineering or nominal strain, ε_i   A measure of deformation expressed as $(\ell_i - \ell_{0i})/\ell_{0i}$, where ℓ_{0i} and ℓ_i refer to the initial and deformed dimensions, respectively, along the *i*th axis.

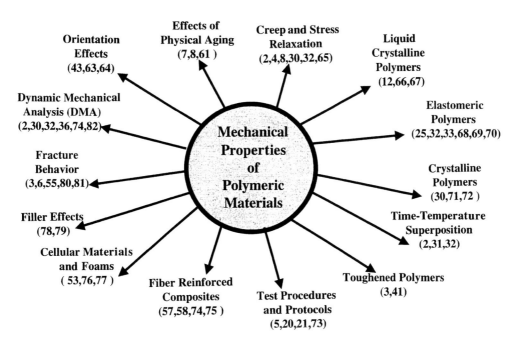

Figure 23.46 General schematic with locating resources for specific areas of mechanical behavior of polymeric materials. The numbers in parentheses denote the reference numbers.

engineering stress, σ_0 Force obtained in extension divided by the samples original cross-sectional area.

entanglement   The intertwining of two different polymer molecules; more specifically, there is usually interest in the polymer-dependent average molecular weight between entanglement points.

extension or draw ratio, λ_i A variable expressing how a given sample dimension is changing with deformation. It is defined as the deformed length divided by the undeformed length in the same direction.

gel point   The point at which cross-linking between different molecules has led to the development of an interconnected three-dimensional network structure. Generally, a sol fraction is still contained within the material, but an insoluble gel also exists. At and above the gel point, steady-state flow no longer occurs.

glass transition temperature, T_g This important parameter is related to the temperature at which the onset of cooperative segmental backbone motion occurs for a polymer within the time frame of the experiment. It applies only to amorphous regions of a material.

long-chain branching (LCB)   A branch extending from the general main backbone of the chain whereby the mass of the branch exceeds the molecular weight between entanglements.

loss modulus   A parameter related to the viscous dissipation of a material undergoing small cyclic deformations; generally obtained by dynamic mechanical spectroscopy techniques.

mechanical hysteresis   An important parameter that is related to the amount of energy dissipated during cyclic deformation. No hysteresis would be obtained if the unloading stress-strain profile were identical to the loading profile.

permanent set   A parameter related to the degree of irrecoverable flow that exists after a material has been deformed in either shear or tension.

Poisson's ratio, μ A parameter indicating whether a material undergoes dilation during the initial stages of extensional deformation; usually denoted by μ.

reptation   Refers to the process of snakelike or reptile motion that describes the slithering of a chain in the midst of its entanglement constraints.

secant modulus   Slope of the line that extends from the origin of an engineering stress-strain curve and intersects that curve at a given elongation. The secant modulus must be specified with respect to the degree of deformation where the secant crosses the stress-strain curve.

shear modulus, G Initial slope of a shear stress-shear strain deformation curve; an indication of the resistance to deformation by shear.

shear strain, γ A measure of the degree of deformation in a shear experiment.

shear stress, τ The force per unit area in shear deformation.

storage modulus   A parameter that is related to the elastic behavior of a material when undergoing small cyclic deformations; generally obtained by dynamic mechanical spectroscopy.

strain   Variable expressing how the dimensions of a material change with deformation.

tensile strength   Generally, the peak or highest value of the engineering stress with strain.

true strain, ε A measure of deformation given by $\ln(\ell_i/\ell_{0i}) = \ln(\lambda_i)$, where ℓ_i, ℓ_{0i} and λ_i refer, respectively, to the deformed length, the undeformed length, and the extension ratio along the ith axis.

true stress, σ_t Force of deformation divided by the cross-sectional area that exists at the elongation at which the stress is determined.

ultimate property   Any property that is determined at the point of failure, for example, strain to break.

Young's modulus, E (tensile modulus or modulus of elasticity)   Initial slope of a tensile stress-strain deformation curve; an indication of the resistance to a tensile deformation.

References

1. McCrum, N. G.; Buckley, C. P.; Bucknall, C. B. *Principles of Polymer Engineering*; Oxford University Press: New York, 1988.
2. Ward, I. M. *Mechanical Properties of Polymers*; Wiley & Sons: London, 1971.
3. Kinloch, A. J.; Young, R. J. *Fracture Behaviour of Polymers*; Applied Science Publishers: London, 1983.
4. Powell, P. C. *Engineering with Polymers*; Chapman & Hall: New York, 1983.
5. Brown, R. R., Ed. *Handbook of Polymer Testing*; Marcel Dekker: New York, 1998.
6. Brostow, W.; Corneliussen, R. D., Eds. *Failure of Plastics*; Hanser Gardner: Cincinnati, OH, 1986.
7. Struik, L. C. E. *Internal Stresses, Dimensional Inabilities and Molecular Orientation in Plastics*; Wiley & Sons; New York, 1990.
8. Matsuoka, S., Ed. *Relaxation Phenomena in Polymers*; Oxford University Press: New York, 1992.
9. deGennes, P. G. *J. Chem. Phys.* **1971**, 55, 572–579.
10. deGennes, P. *Scaling Concepts in Polymer Science*; Cornell University Press: Ithaca, NY, 1979.
11. Doi, M.; Edwards, S. F. *The Theory of Polymer Dynamics*; Clarendon Press: Oxford, UK, 1986.
12. Donald, A. M.; Windle, A. H. *Liquid Crystalline Polymers*; Cambridge University Press: New York, 1992.
13. Wunderlich, B. *Macromolecular Physics, Vol. 1—Crystal Structure, Morphology, Defects*; Academic Press: New York, 1973; *Macromolecular Physics, Vol. 2—Crystal Nucleation Growth, Annealing*; Academic Press: New York, 1976; *Macromolecular Physics, Vol. 3—Crystal Melting*; Academic Press: New York, 1980.
14. Zhang, Y. H.; Carreau, P. J. *J. Appl. Polym. Sci.* **1991**, 42, 1965.

15. Wu, S. *J. Polym. Sci., Part B: Polym. Phys.* **1989**, *27*, 723.

16. Miller, R.; Boyer, R. *Polym. News* **1978**, *4*, 255.

17. Tomalia, D. A.; Brothers, H. M. II; Piehler, L. T.; Hsu, Y. *PMSE Prepr.* **1995**, *73*, 75.

18. Dealy, J. M.; Wissbrun, K. F. *Melt Rheology and Its Role in Plastics Processing Theory and Applications*; Van Norstrand Reinhold: New York, 1990.

19. Dealy, J. M. *Rheometers for Molten Plastics*; Van Nostrand Reinhold: New York, 1982.

20. ASTM D 256–D 2343. *Annu. Book ASTM Stand.* **1999**, Sect. 8 (1).

21. International Organization for Standardization, Geneva, Switzerland. www.iso.ch.

22. Risch, B. G.; Wilkes, G. L. *J. Appl. Polym. Sci.* **1995**, *56*, 1511–1517.

23. Reed, T. F. *Elastomerics* **1989**, Nov. 15.

24. Harwood, J. A. C.; Payne, A. R. *J. Appl. Polym. Sci.* **1996**, *10*, 1203.

25. Treloar, L.R.G. *The Physics of Rubber Elasticity*; Clarendon Press: Oxford, UK, 1958.

26. Wang, M. *Rubber Chem. Tech. Rev.* **1998**, *71* (3), 520.

27. Tobolsky, A. V. *Properties and Structure of Polymers*; Wiley & Sons: New York, 1960.

28. DePorter, J. K.; Baird, D. G.; Wilkes, G. L. *J. Macromol. Sci., Part C* **1993**, *C33* (1), 1.

29. Ratta, V.; Stancik, E. J.; Ayambem, A.; Parvatareddy, H.; McGrath, J. E.; Wilkes, G. L. *Polymer* **1999**, *40*, 1889–1902.

30. Van Krevelen, D.W. *Properties of Polymers*; Elsevier: New York, 1990.

31. Ferry, J. *Viscoelastic Properties of Polymers*, 3rd ed.; Wiley & Sons: New York, 1981.

32. Aklonis, J.; MacKnight, W. *Introduction to Polymer Viscoelasticity*; Wiley-Interscience: New York, 1972.

33. Gent, A. N., Ed. *Engineering with Rubber—How to Design Rubber Components*; Hanser Gardner: Cincinnati, OH, 1992.

34. Becker, G. W. *Kolloid Z.* **1955**, *140*, 1.

35. Schmieder, K.; Wolf, K. *Kolloid Z.* **1952**, *127*, 65.

36. McCrum, N. G.; Read, B. F.; Williams, G. *Anelastic and Dielectric Effects in Polymeric Solids*; Wiley & Sons: New York, 1967.

37. Murayama, T.; Bell, J. P. *J. Polym. Sci.* **1970**, *A2* (8), 437.

38. Jordens, K.; Wilkes, G. L.; Janzen, J. D.; Rohlfing, C.; Welch, B. *Polymer*, **2000**, *41*, 7175–7192.

39. McCrum, N. G. In *Molecular Basis of Transitions and Relaxations*; Meier, D. J., Ed.; Gordon & Breach: Langhorne, PA, 1978; pp 167; Keating, M. Y.; Lee, I. *J. Macromol. Sci., Part B* **1959**, *B38* (4), 379.

40. Wagner, E. R.; Robeson, L. *Rubber Chem. Technol.* **1970**, *43*, 1129.

41. Bucknall, C. B. *Toughened Plastics*; Applied Science: London, 1977.

42. Gent, A. N.; Park, B. *J. Mater. Sci.* **1984**, *19*, 1947–1956.

43. Wilkes, G. L. In *Structure and Properties of Oriented Polymers*, 2nd ed.; Ward, I. M., Ed; Chapman and Hall: New York, 1997; p 44.

44. Wilkes, G. L. In *Encyclopedia of Polymer Science and Engineering*; Kroschwitz, J., Ed.; Wiley & Sons, New York, 1988; Vol. 14, p 542

45. Zhou, H.; Wilkes, G. L. *J. Mater. Sci.* **1998**, *33*, 287.

46. Yu, T.-H.; Wilkes, G. L. *Polymer* **1996**, *37* (21), 4675.

47. Zhou H.; Wilkes, G. L. *Polymer* **1998**, *39* (16), 3597.

48. Davis, G. W.; Talbot, J. R. In *Encyclopedia of Polymer Science and Engineering*; Kroschwitz, J., Ed.; Wiley & Sons: New York, 1988; Vol. 12, p 118.

49. Werner, E.; Janocha, S.; Hopper, M. J.; MacKenzie, K. J. In *Encyclopedia of Polymer Science and Engineering*; Kroschwitz, J., Ed.; Wiley & Sons: New York, 1988; Vol. 12, p 193.

50. Benning, C. J. *Plastic Films for Packaging*; Technomic: Lancaster, PA, 1983.

51. White, J. L.; Spruiell, J. E. *Polym. Eng. Sci.* **1981**, *21*, 859.

52. Duckett, R. A. In *Structure and Properties of Oriented Polymers*; Ward, I. M., Ed.; Applied Science: London, 1975; p 366.

53. Gibson, L. J.; Ashby, M. E. *Cellular Solids: Structure and Properties*; Pergamon Press: New York, 1988.

54. Armistead, P. M. S. Thesis, Virginia Polytechnic Institute and State University, Blacksburg, VA, 1985.

55. Wool, R. P. *Polymer Interfaces, Structure and Strength*; Hanser Gardner: Cincinnati, OH, 1995.

56. Agarwal, B. D.; Broutman, L. J. *Analysis and Performance of Fiber Composites*; Wiley & Sons: New York, 1980.

57. Piggot, M. R. *Load Bearing Fiber Composites*; Pergaman Press: New York, 1980.

58. Carlsson, L. A.; Pipes, R. B. *Experimental Characterization of Advanced Composite Materials*, 2nd ed.; Technomic: Lancaster, PA, 1996.

59. Shelby, M. D.; Wilkes, G. L. *Polymer* **1998**, *39*, 6767.

60. Tant, M. R.; Wilkes, G. L. *J. Appl. Polym. Sci.* **1981**, *26*, 2813.

61. Struik, L. C. E. *Physical Aging in Amorphous Polymers and Other Materials*; Elsevier Science: New York, 1978.

62. Tant, M. R.; Wilkes, G. L. *Polym. Eng. Sci.* **1981**, *21* (14), 874.

63. Samuels, R. *Structured Polymer Properties*; Wiley & Sons: New York, 1974.

64. *Ultra High Modulus Polymers*; Ciferri, A., Ward, I. M., Eds.; Applied Science: London, 1977.

65. Riande, E.; Diaz-Calleja, R.; Prolongo, M. G.; Masegosa, R. M.; Salom, C. *Polymer Viscoelasticity—Stress and Strain in Practice*; Plastics Engineering Series; Marcel Dekker: New York, 1999.

66. *Liquid Crystal Order in Polymers*; Blumstein, A., Ed.; Academic Press: New York, 1976.

67. *Polymer Liquid Crystals*; Ciferri, A., Krigbaum, W. R., Meyer, R. B., Eds.; Academic Press: New York, 1982.

68. Mark, J. E.; Erman, B. *Rubberlike Elasticity: A Molecular Primer*; Wiley & Sons: New York, 1988.

69. *Thermoplastic Elastomers—A Comprehensive Review*, 2nd ed.; Holden, G., Legge, N. R., Quirk, R., Schroeder, H. E., Eds.; Hanser Gardner: Cincinnati, OH, 1996.

70. *Engineering with Rubber—How to Design Rubber Components*; Gent, A. N., Ed.; Hanser Gardner: Cincinnati, OH, 1992.

71. *Nylon Plastics Handbook*; Kohan, M. I., Ed.; Hanser Gardner: Cincinnati, OH, 1995.

72. Moore, E. P., Jr. *Polypropylene Handbook: Polymerization, Characterization, Properties, Processing, Applications*; Hanser Gardner: Cincinnati, OH, 1996.

73. *Handbook of Plastics Test Methods*, 2nd ed.; Brown, R. P., Ed.; George Goodwin Ltd., in association with Plastics & Rubber Institute: London, 1981.

74. Read, B.; Dean, G. D. *The Determination of Dynamic Properties of Polymers and Composites*; Wiley & Sons: New York, 1978.

75. Jones, R. F. *Guide to Short Fiber Reinforced Plastics*; Hanser Gardner: Cincinnati, OH, 1998.

76. *Polymeric Foams*; Klempner, D., Frisch, K. C., Eds.; Hanser Gardner: Cincinnati, OH, 1991.

77. *Dow Polyurethanes, Flexible Foams*, 2nd ed.; Herrington, R., Hock, K., Eds.; The Dow Chemical Company: Midland, MI, 1997.

78. Titow, W. V.; Lanham, B. J. *Reinforced Thermoplastics*; Wiley & Sons: New York, 1975.

79. Wypych, G. *Handbook of Fillers*, 2nd ed.; ChemTec Publishing: Brookfield, CT, 2000.

80. Williams, J. G. *Fracture Mechanics of Polymers*; Ellis Horwood Series in Engineering Science; Halsted Press, a Division of Wiley & Sons: New York, 1987.

81. Ezrin, M. *Plastics Failure Guide*; Hanser Gardner: Cincinnati, OH, 1996.

82. Menard, K. P. *Dynamic Mechanical Analysis: A Practical Introduction*; CRC Press: Boca Raton, FL, 1999.

24

Dielectric Properties

WILLIAM STEPHEN TAIT

Polymers are dielectric materials, the dielectric properties of which change when water or electrolyte diffuses into the polymer matrix. Electronic impedance spectroscopy (EIS) is used to measure polymer dielectric properties and changes in those properties with exposure time.[1,2] This chapter discusses the fundamentals of EIS, the use of EIS to measure dielectric properties of polymer coatings and films, and ways to characterize and predict long-term performance of polymer coatings and films. Additional information on EIS can be found in the references at the end of this chapter.

Polymer coatings are commonly used to decorate metals and protect them from corrosion, and polymer films are used as packaging to protect a wide variety of items, such as foods and electronic circuit boards, from environmental degradation. There are several theories about how polymer coatings protect metals from corrosion; however, it is generally agreed that metallic corrosion occurs when water accumulates at the polymer-metal interface.[3-5] Polymer films are used as barriers against the environment to prevent spoilage or degradation from atmospheric water and pollutants.

Polymer molecules consist of relatively short side chains attached to a longer backbone chain. Polymer coatings and films are formed when the numerous molecules bond together via side chain cross-linking or van der Waals forces into a seemingly impenetrable matrix of intertwined polymer molecules.[6] However, water molecules are small enough that they can diffuse through the dense tangle of polymer molecules.

One can visualize water's journey through a polymer coating or film as being analogous to a large group of people walking through a forest. The forest appears to be impenetrable from a distance; however, gaps in the trees (polymer backbones) and overlapping branches (polymer side chains) are clearly seen when one stands at the edge of the forest. The gaps provide paths into and through the forest, and overlapping branches can be moved or bent to allow passage through the forest. A large group of people moving through the forest will alter the forest's morphology as branches are permanently bent or broken, and underlying brush is trampled while people pass through the forest. Passage through a forest is more difficult when the forest is very dense, much like a highly cross-linked polymer is much more chemically resistant than a water-base latex paint.

Water diminishes attractive forces between polymer molecules, often causing the polymer to swell, and thus altering its morphology.[7] Water-soluble salts also can diffuse with their water shells into and through a polymer matrix, thereby changing polymer morphology.[8,9] Figure 24.1 illustrates how different salt solutions (electrolytes) affect coated metal failure.[10] The Y axis shows the cumulative percent failure of a coated metal container population, and the X axis gives the exposure time prior to failure. Failure is defined as perforation of the coated metal with extensive delamination, or extensive coating delamination. Figure 24.1 illustrates the following:

1. Cations enhance coating damage and subsequent coated metal failure.
2. Cations can be ranked according to the magnitude of their effect on failure.
3. Electrolytes are more effective at damaging polymer coatings than purified water.
4. Purified water also will eventually cause coating and coated metal failure.

EIS is used to measure polymer dielectric properties and changes in those properties with exposure time.[1,2]

This chapter discusses how to use EIS to measure polymer coating and film dielectric properties, and how to use EIS measurements to characterize and predict long-term performance of polymer coatings and films.

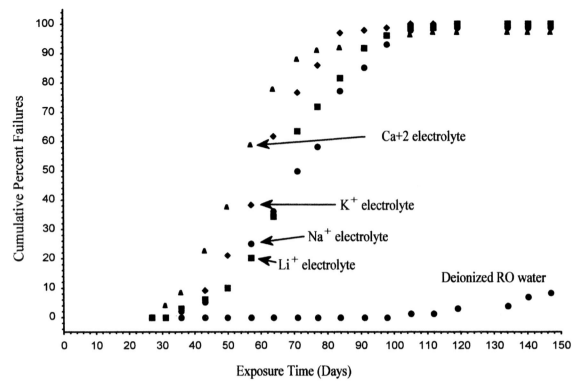

Figure 24.1 Illustration of how different cations and purified water affect the lifetime of coated metal containers.

Coating Dielectric Properties

A few definitions will provide a common framework for the discussion on measurement of polymer dielectric properties with EIS.

Current Current is the movement of electrical charge through matter.[11] Ionic current occurs when ions move in water in response to an electrical potential gradient between two electrodes. Electrical current in a solid material is the movement of electrons and positive holes in response to an electrical potential gradient between two points in the solid material.[12] EIS measurements require a balance of electrical current in the solid electrodes (used to make measurements) and ionic current between the counter and test electrodes in the electrolyte.

Dielectric A dielectric material is one that does not conduct electricity. Dielectric materials impede the flow of current between two oppositely charged plates, with the result that more voltage is required to bring the potential to the same magnitude that is possible without the dielectric between the two plates.

Electrolyte An electrolyte is an aqueous solution of water-soluble salts, gases, or molecules.

Electrode An electrode is a metal or coated metal that is submerged in an electrolyte. Three types of electrodes are used in electrochemical testing of coated metals.

1. The *counter electrode* is a metal that is used to provide electrical balance when a test electrode is polarized during an EIS measurement (see the definition of current).

2. The *reference electrode* is a metal, or glass tube containing a metal, that is used to measure the electrical potential of the test electrode.

3. A *test electrode* is a coated metal sample that is used for EIS measurements.

Capacitance Capacitance is the amount of charge that can be stored between two parallel plates at a given potential difference.[12]

The capacitance of a dielectric material is a function of its dielectric constant, area, and thickness as shown by the following equation:[13]

$$C = \frac{\varepsilon \varepsilon^0 A}{d} \qquad (1)$$

where d is the material thickness, A represents the geometric area, ε^0 is the dielectric constant for a vacuum, 8.85×10^{-14} Farads/cm, and ε is the dielectric constant of the material (i.e., polymer coating or film). Polymers typically have dielectric constants ranging from 2 to 7.[14]

Equation 1 also illustrates that altering thickness, area, or the dielectric constant of the polymer can change the capacitance of the polymer. Remember that water and electrolyte can diffuse into polymer coatings and films, and thus change the coating or film capacitance.

Brasher and Kinsbury[15] developed an empirical relationship between capacitance and percent water uptake in a polymer:

$$X_V = \frac{\log (C_t/C_0) \ 100\%}{\log (\varepsilon_{H_2O})} \qquad (2)$$

where X_V is the percent of absorbed water in the polymer, C_t is the polymer capacitance at a given time after initial exposure to an environment containing water or electrolyte, and C_0 is the polymer capacitance prior to exposure to water or an electrolyte. ε_{H_2O} is the dielectric constant of water (80 at 20 °C). Equation 2 can be made more general by substituting electrolyte dielectric constant, $\varepsilon_{electrolyte}$, for ε_{H_2O}. Equation 2 illustrates that the dielectric constant of a polymer coating or film approaches that for water or the electrolyte as the coating becomes saturated with water or electrolyte.

The interfacial area between a polymer coating and a metal surface submerged in water or electrolyte contains water molecules and ions from both the metal and the electrolyte.[10,16] The chemical composition of this interfacial area is significantly different from that of the bulk electrolyte in which the metal is submerged. Metal surfaces are essentially negatively charged because of the valence electrons that form metallic bonds. The metal surface and the adjacent electrolyte layer are collectively referred to as the electrical double layer.[17] Addition of a polymer coating to the metal surface produces a composite interface consisting of the electrical double layer and polymer coating with electrolyte or water dispersed throughout the coating. Figure 24.2 contains a model of this composite interface.[18] Diffusion of electrolyte or water will cause a concentration gradient through the thickness of the coating. This gradient will cause charge separation in the coating, and the charge separation produces capacitance. Polymer films will also have a concentration of water and electrolyte through their thickness, and thus will also have a charge separation that produces capacitance.

The composite interface in Figure 24.2 has two charge transfer processes: one for transfer of electrons from the metal to electrochemically active species, and another for movement of charge through the polymer coating. Capacitance also arises in the interface because of charge separation in the electrical double layer and polymer coating. Consequently, the interface has both capacitance and resis-

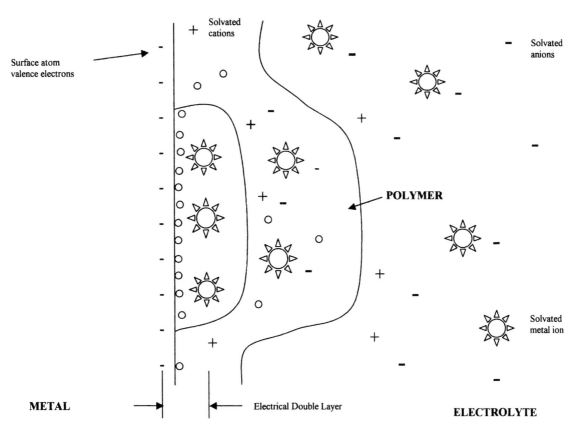

Figure 24.2 A model for the electrical double layer associated with a corroding coated metal.

tance properties, and the interface will react to an alternating current (ac) voltage much like a simple electronic circuit made from resistors and capacitors.

EIS can be used to measure polymer capacitance, dielectric constant, and resistance, and these properties can be used to quantify polymer performance. For example, this author and other researchers have demonstrated that there is a relationship between coating capacitance and the amount of area of delaminated coating.[1,2,19] EIS has also been used to measure substrate metal corrosion, and is thus a very useful method for studying corrosion protection of metals by polymer coatings.[2,20]

What Is EIS?

EIS uses ac voltage with a small amplitude range of frequencies. The ac voltage is applied between the polymer film or coated metal and a counter (or auxiliary) electrode. Figure 24.3 contains an example of a test cell that is used for EIS measurement on coated metal samples. A sample of the coated metal is clamped between the glass body of the cell and the cell bottom. The sample is typically a circular shape with a small tab for clamping under the rectangular bar that is used to electrically connect the sample to measurement equipment. Electrical connections for the counter

Figure 24.3 An example of an electrochemical test cell that can be used to measure dielectric properties of polymer coated metals.
(Reproduced with permission of Pair O Docs Professionals, LLC.)

and reference electrodes are located on the top of the cell. EIS measurements can also be made on polymer films by placing the counter and reference on one side of the polymer film, or clamping the film between two electrodes.

A spectrum of ac-voltage frequencies is applied to a polymer film or coated metal sample to generate EIS data. A typical spectrum can include frequencies from 100,000 to 0.005 Hz. A 0.005-V amplitude is often used for the entire frequency range.[21] Current and phase angle difference between voltage and current are measured and recorded at each ac-voltage frequency.

Voltage, ac, and impedance are sinusoidal functions of time, and ac voltage can lag its corresponding current by a phase angle Φ.[22] One can envision phase angle by plotting the voltage and current magnitudes as a function of angle, using the equations that describe ac voltage, current, and impedance:

$$V = V_0 \left[\sin \left(2\pi t + \Phi \right) \right] \tag{3}$$

$$I = I_0 [\sin (2\pi t)] \tag{4}$$

$$Z = \frac{V_0 [\sin (2\pi t + \Phi)]}{I_0 [\sin (2\pi t)]} = \frac{Z_0 [\sin (2\pi t + \Phi)]}{\sin (2\pi t)} \tag{5}$$

where V_0 is the amplitude of the applied alternating current voltage, I_0 is the maximum amplitude of current corresponding to V_0, t is time, Z is the total impedance, and Φ is the phase angle difference between the ac voltage and its corresponding current.

Equation 5 is the ac analogue of Ohms law, and impedance is analogous to direct current (dc) resistance; ac current continues to move through a capacitor, unlike dc current, which stops moving after the capacitor is fully charged.

The equipment used for EIS measurements includes a potentiostat to apply ac voltage to samples, and an amplifier to measure ac current and phase angle difference between current and voltage. A potentiostat is typically an operational amplifier that supplies the amount of electrical current needed to maintain the desired electrical potential for the test electrode. The amplifier is used to measure the current and phase angle difference between the voltage and current.

EIS spectra can be represented graphically in several different ways. Figures 24.4, 24.5, and 24.6 contain three common ways to graph EIS spectra. Impedance (eq 5) is plotted as a log/log function of frequency in Figure 24.4 to produce a Bode magnitude plot. Horizontal areas occur in the plot when the phase angle difference between the current and voltage is zero, corresponding to resistance to charge transfer of electrons from the metal, or ion movement in the polymer coating or film. Sloped portions of the curve arise from the capacitance resulting from charge separation in the electrical double layer, or the polymer coating or film.

The phase angle difference between the voltage and current is plotted as a function of frequency in Figure 24.5 to

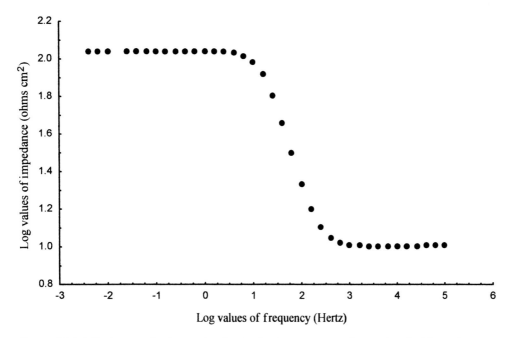

Figure 24.4 A Bode magnitude graph of a single time constant electrochemical impedance spectroscopy spectrum. (Reproduced with permission from reference 19. Copyright 1994 Pair O Docs Professionals, LLC.)

produce a Bode phase plot. The minimum in Figure 24.5 corresponds to the midpoint of the sloped line between resistance areas of the Bode magnitude plot in Figure 24.4.

Impedance is a vector because it has magnitude and direction, and the vector can be separated into its X and Y components. The mathematical equation for the magnitude (tensor) of the two components is given in eq 6.[23]

$$Z = R_1 + \frac{R_2}{1 + (\omega R_2^2 C)^2} + \frac{j(-\omega R_2^2 C)}{1 + (\omega R_2^2 C)^2} \qquad (6)$$

R_1 is the resistance to ion movement in solution, R_2 is the corrosion or polymer resistance, C is the capacitance associated with R_2, and ω is 2π times the ac-voltage frequency.

The term in eq 6 that is multiplied by the imaginary number j (which is $\sqrt{-1}$) is referred to as the imaginary component of impedance, and the other term containing R_2 is the real component of impedance. A graph of the absolute value of the imaginary component as a function of the real component is shown in Figure 24.6, and is known as a complex plane plot, or a Nyquist plot.

How Are Data Analyzed?

EIS spectra can have one or more responses depending on how many processes are present in a polymer film or coated metal. Processes can be metallic corrosion, water absorption, or ion movement in the polymer. For example, a corroding coated metal can have two responses in the spectrum, such as that shown by the complex plane plot in Figure

24.7. The resistance and capacitance associated with each response are used to evaluate or predict long-term behavior such as coating delamination or water absorption. Resistance multiplied by capacitance has units of time, so each response in a spectrum is referred to as the time constant for the individual process. Multiple time constants can be observed when they are separated by a ratio of approximately 20.[24]

There are several ways that time constants can be extracted from EIS spectra.

1. Parameters for each process can be extracted by fitting experimental spectra to a mathematical model that is developed from reaction or diffusion kinetics.
2. An equivalent electrical circuit model can be used to extract parameters from the spectrum.
3. Parameters can be extracted from spectra using the measurement model method.

Using Diffusion and Chemical Reaction Kinetics to Model EIS Spectra

Diffusion and chemical reaction rate constants are functions of time, so mathematical relationships can be developed from diffusion and kinetics.[25] These mathematical relationships are regressed around experimental spectra to extract polymer dielectric parameters from the spectra. Unfortunately, the mathematics for diffusion and kinetics are rarely known for industrial situations, and often cannot be determined because of the complexity of the process. The

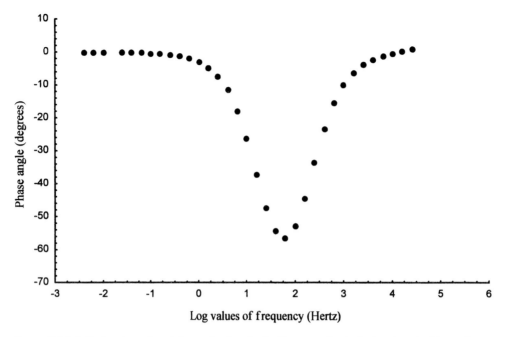

Figure 24.5 A Bode phase (angle) graph of a single time constant electrochemical impedance spectroscopy spectrum. The data for this figure are the same data used for Figure 24.4. (Reproduced with permission from reference 19. Copyright 1994 Pair O Docs Professionals, LLC.)

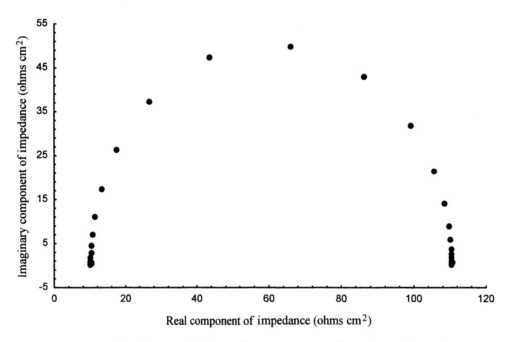

Figure 24.6 A complex plane graph of a single time constant electrochemical impedance spectroscopy spectrum. The data for this figure are the same data used for Figures 24.4 and 24.5. (Reproduced with permission from reference 19. Copyright 1994 Pair O Docs Professionals, LLC.)

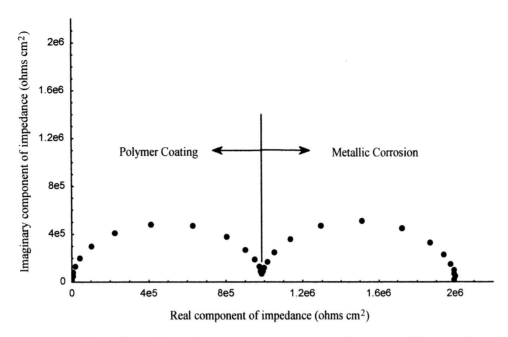

Figure 24.7 Electrochemical impedance spectroscopy spectrum from a corroding coated metal. Two time constants, one for corrosion and one for the coating, are noted in this spectrum. (Reproduced with permission from reference 19. Copyright 1994 Pair O Docs Professionals, LLC.)

reference listed for this method[25] contains a more detailed discussion on how to build models for EIS spectra from diffusion and reaction kinetics

Using Equivalent Electrical Circuit Models to Extract Parameters from EIS Spectra

The equivalent electrical circuit model (EECM) method is the most common method for extracting parameters from EIS spectra. EECMs are combinations of capacitors and resistors such as that depicted in Figure 24.8. The EECM in Figure 24.8 is often referred to as the Randles circuit. The capacitor represents the double-layer capacitance of an un-

coated metal, polymer coating, or film. Resistor R_2 represents the electrical resistance (corrosion resistance) to transferring electrons from the metal, or resistance of ion motion through the polymer. Resistor R_1 represents a correction factor that accounts for resistance to motion of ions in water (ionic current).

The mathematical expression in eq 6 models how a Randles circuit responds to an ac voltage at different frequen-

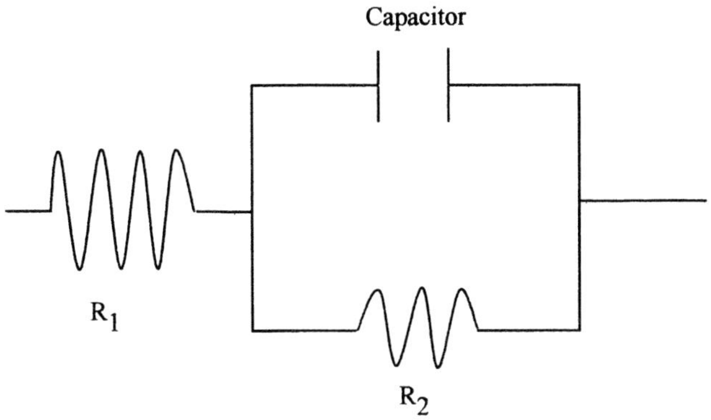

Figure 24.8 An example of a simple equivalent electrical circuit model that is often referred to as a Randles circuit.

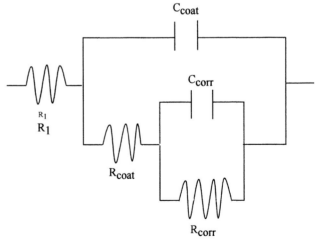

Figure 24.9 An example of an equivalent electrical circuit model that is often used to model electrochemical impedance spectroscopy spectra that contain two time constants.

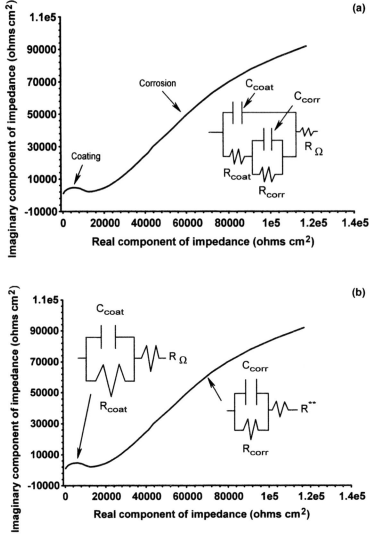

Figure 24.10 Two methods for modeling electrochemical impedance spectroscopy spectra that contain more than one time constant: (a) models the entire spectrum with one EECM; (b) models the spectrum in sections using a Randles circuit for each section. (Reproduced with permission from reference 19. Copyright 1994 Pair O Docs Professionals, LLC.)

cies. Equation 6 can be fitted (regressed) to an experimental spectrum in an iterative process by adjusting the magnitude of the resistors and capacitor. Iteration continues until the difference between the spectrum developed from the equation and the experimental spectrum are the same. However, spectra generated with eq 6 rarely match experimental spectra exactly, because an electrochemical interface rarely behaves like an electronic capacitor. Consequently, making the exponent in the denominator a variable parameter is one way to modify eq 6 so that the model more closely fits the experimental spectra.[26]

$$Z = R_1 + \frac{R_2}{1 + (\omega R_2^2 C)^\phi} + \frac{j(-\omega R_2^2 C)}{1 + (\omega R_2^2 C)^\phi} \qquad (7)$$

The exponent [ϕ] is an adjustable parameter, and the capacitor is often referred to as a constant phase element (CPE) to infer that the electrochemical interface does not respond to an ac voltage like an ideal electronic capacitor.

Figure 24.9 illustrates how multiple Randles circuits can be nested to form more complex circuits when more than one time constant is present in a spectrum. This EECM is commonly used to model EIS spectra from corroding coated metals. The outer circuit (C_{coat}, R_{coat}) is for the polymer coating, and the inner (nested) circuit (C_{corr}, R_{corr}) is for metal corrosion under the polymer coating.

The equation for an EECM can be fitted to an entire spectrum, as illustrated in Figure 24.10a, or individual time constants can be modeled separately using a Randles circuit for each, as illustrated in Figure 24.10b.[27] Parameters obtained from separately modeling time constants may not be as accurate as those obtained from modeling the entire spectrum. However, estimated parameters obtained from separately modeling time constants are often sufficient for industrial corrosion and service lifetime estimations.[28]

Using the Measurement Model Method to Extract Parameters from EIS Spectra

Statistical weighting is often used with EECMs in an effort to make the model more closely fit the experimental spectrum. Several different weighting schemes have been devised, but most make assumptions about the error structure of the data that may not account for electrical noise and nonsteady-state electrochemical processes.[29]

The measurement model (MM) method is widely used for identification of error structures of spectroscopic measurements.[30] The MM method uses estimates of the standard deviation of the experimental spectra to develop weighting factors for each frequency. Thus, the weighting strategy is based on the experimentally determined variance of the measurement, avoiding some of the limitations of other

weighting procedures. The MM can also help determine whether the lack of complete agreement between the spectrum and the model is attributable to an inadequate model, to nonsteady-state processes, or to instrument electrical noise. Data that are due to nonsteady-state processes or instrumental artifacts can be deleted from the modeling process, and the MM fitted (regressed) to only that part of the spectrum that is consistent with statistical tests for model correctness, such as the Kramers-Kronig relations.[31] Impedances calculated from the MM have been in good agreement with results obtained from extrapolation of process models.[32,33]

How Are EIS Data Used to Predict Long-Term Behavior?

EIS spectra are often collected over a period of time, and the parameters from the spectra are plotted as a function of time so that the data can be extrapolated. Figure 24.11 contains several Bode magnitude plots at different exposure times from a 0.002-in.-thick poly(ethylene terephthalate) (PET) film exposed to liquid water on one side of the film. These measurements were conducted to determine how fast water permeates PET bottles. The test cell used for the measurements was like that in Figure 24.3, but the cell was modified so the counter and reference electrodes could be positioned on top of the polymer film.

The top spectrum indicates that little or no water was absorbed into the PET film after 2 h. The 4- and 6-h EIS

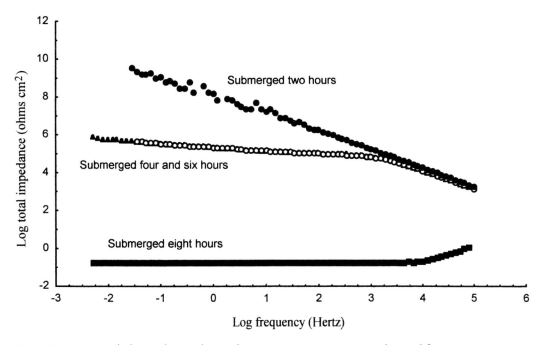

Figure 24.11 Several electrochemical impedance spectroscopy spectra obtained from a 0.002-in.-thick poly(ethylene terephthalate) film that is in contact with water on one side.

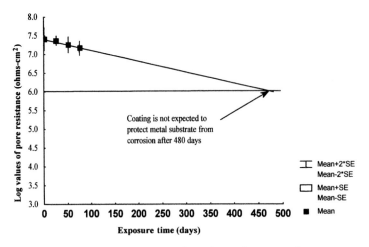

Figure 24.12 Coating resistance plotted as a function of time in the box and whisker format. The extrapolation of the data indicates that the coating will cease providing corrosion protection at approximately 480 days.

measurements indicate that the film absorbed water after being submerged for 4 h and the amount of absorbed water did not change between 4 and 6 h of submersion. The bottom spectrum indicates that the polymer film is most likely saturated with water after 8 h.

EIS parameters such as coating resistance can also be graphed as a box and whisker plot versus time, as illustrated in Figure 24.12. Each data point in the box and whisker plot consists of the mean value for all replicates, the mean plus one standard error (the box), and the mean plus two standard errors (the vertical line for each box, or whisker). Standard error is calculated from the standard deviation of a set of replicate measurements and the number of replicates.[34] Data in Figure 24.12 are extrapolated to estimate the time when polymer-coating resistance reaches a critical value. Several researchers have determined that polymer coatings will provide good protection for the substrate metal when the resistance is above 10^9 ohms cm^2, and that polymers do not provide corrosion protection when the resistance is below 10^6 ohms cm^2.[35-37] The extrapolation in Figure 24.12 illustrates that the coating is expected to cease protecting the metal from corrosion after 480 days of exposure.

Statistical Considerations

It has been shown that EIS data on polymer-coated metals can have a significant amount of variability.[38-40] Appropriate methods for dealing with variability are:[41-44]

Use a large enough number of replicate test samples for each test variable so that the sensitivity of statistical error to variability is minimal. Twelve or more replicates per test variable are recommended by this author.

Use a large enough number of replicate test samples for each test variable so that the statistical confidence is at the desired level (e.g., 25 replicates provide a 95% statistical confidence).

Conduct a series of measurements on each test variable over a sufficiently long exposure time so that data can be extrapolated to longer times.

References

1. Tait, W. S. *J. Coat. Technol.* **1989**, *61* (768), 57–61.
2. Tait, W. S. *J. Coat. Technol.* **1990**, *62* (781), 41–44.
3. Perez, C.; Collazo, A.; Izquierdo, M.; Merino, P.; Novoa, X. R. *Prog. Org. Coat.* **1999**, *36*, 102–108.
4. Madani, M. M.; Miron, R. R.; Granata, R. D. *J. Coat. Technol.* **1997**, *69* (872), 45–54.
5. Bierwagen, G. P. *Prog. Org. Coat.* **1996**, *28*, 43–48.
6. Billmeyer, F. W. *Textbook of Polymer Science*; Wiley-Interscience: New York, 1971; pp 141–154.
7. Negele, O.; Funke, W. *Prog. Org. Coat.* **1996**, *28*, 285–289.
8. Leidheiser H., Jr.; Wang, W. In *Corrosion Control by Organic Coatings*; Leidheiser, H., Jr., Ed.; NACE: Houston, TX, 1981.
9. Leidheiser, H., Jr.; Granata, R. D.; Turoscy, R. *Corrosion* **1987**, *43* (5), 296.
10. Tait, W. S.; Handrich, K. A. *Corrosion* **1994**, *50* (5), 373–378.
11. Sternheim, M. M.; Kane, J.W. *General Physics*; Wiley and Sons: New York, 1986; p 361.
12. Bockris, J. O'M.; Reddy, A. K. N. *Modern Electrochemistry*; Plenum Press: New York, 1977; Vol. 1, pp 132–134.
13. Murray, J. N. *Prog. Org. Coat.* **1997**, *31*, 375–391.
14. *Handbook of Chemistry and Physics*, 50th ed.; Weast, R. C., Ed.; The Chemical Rubber Company: Cleveland, OH, 1970.

15. Brasher, D. M.; Kinsbury, A. H. *J. Appl. Chem.* **1954**, *4*, 62.
16. Leidheiser, H., Jr.; Vertes, A.; Czako-Nagy, I. *J. Electrochem. Soc.* **1987**, *134* (6), 1470.
17. Tait, W. S. *An Introduction to Electrochemical Corrosion Testing for Practicing Engineers and Scientists*; Pair O Docs Professionals: Madison, WI, 1994; pp 5–7.
18. Tait, W. S. *An Introduction to Electrochemical Corrosion Testing for Practicing Engineers and Scientists*; Pair O Docs Professionals: Madison, WI, 1994; p 86.
19. Kellner, J. D. *Corrosion/85* **1985**, *March*, paper no. 73.
20. Mansfeld, F.; Kendig, M. W.; Tsai, S. *Corrosion* **1982**, *38* (9), 478.
21. Tait, W. S.; *An Introduction to Electrochemical Corrosion Testing for Practicing Engineers and Scientists*; Pair O Docs Professionals: Madison, WI, 1994; p 22.
22. Brophy, J. J. *Basic Electronics for Scientists*; McGraw-Hill: New York, 1972; p 102.
23. Tait, W. S.; *An Introduction to Electrochemical Corrosion Testing for Practicing Engineers and Scientists*; Pair O Docs Professionals: Madison, WI, 1994; p 83.
24. Walters, G. *Corrosion Sci.* **1986**, *26* (9), 681–703.
25. Armstrong, R. D.; Bell, M. F.; Metcalfe, A. A. *Electrochemistry* **1978**, *6*, 98–127.
26. Cole, K. S.; Cole, R. H. *J. Chem. An Introduction to Electrochemical Corrosion Testing for Practicing Engineers and Scientists*; Pair O Docs Professionals: Madison, WI, 1994; p 110.
27. Tait, W. S.; Handrich, K. A.; Tait, S. W.; Martin, J. W. In *Electrochemical Impedance: Analysis and Interpretation*; Scully, J., Silverman, D., Kendig, M., Eds.; ASTM STP 1188; American Society for Testing and Materials: Philadelphia, PA, 1993; pp 428–437.
28. Agarwal, P.; Orazem, M. E.; García-Rubio, L. H. *J. Electrochem. Soc.* **1992**, *139*, pp. 1917–1927.
29. Agarwal, P.; Orazem, M. E.; García-Rubio, L. H. In *Electrochemical Impedance: Analysis and Interpretation*; Scully, J., Silverman, D., Kendig, M., Eds.; ASTM STP 1188; American Society for Testing and Materials: Philadelphia, PA, 1993; pp 115–139.
30. Orazem, M. E.; El Moustafid, T.; Deslouis, C.; Tribollet, B. *J. Electrochem. Soc.* **1996**, *143*, 3880.
31. Wojcik, P. T.; Orazem, M. E. *Corrosion* **1998**, *54*, 289–298.
32. Frateur, M.; Deslouis, C.; Orazem, M. E.; Tribollet, B. *Electrochim. Acta* **1999**, *44*, 4345.
33. Miller, I.; Freund, J. E. *Probability and Statistics for Engineers*; Prentice-Hall: Englewood Cliffs, NJ, 1997; p. 297.
34. Leidheiser, H., Jr. *Prog. Org. Coat.* **1979**, *7*, 79.
35. Mayne, J. E. O.; Mills, D. J. *Oil Colour Chem. Assoc.* **1975**, *58*, 155.
36. Bacon, R. C.; Smith, J. J.; Rugg, F. M. *Ind. Eng. Chem.* **1948**, *40*, 161.
37. Murry, J. N.; Hack, H. P. *Corrosion/91*, Cincinnati, OH, 1991, paper 131; National Association of Corrosion Engineers: Houston, TX.
38. Feliu, S.; Morcillo, M.; Galvan, J. C. In *Advances in Corrosion Protection by Organic Coatings*; Scantlebury, D., Kendig, M. W., Leidheiser, H., Jr., Eds.; The Electrochemical Society: Pennington, NJ, 1987; Vol. 87-2, p 280.
39. Martin, J. W.; McKnight, M. E.; Nguyen, T; Embree, E. *J. Coat. Technol.* **1989**, *61* (772), 39–48.
40. Tait, W. S. *J. Coat Technol.* **1994**, *66* (834), 59–61.
41. Tait, W. S. *Mater. Perform.* **2001**, *40*, 58.
42. Tait, W. S. *Spray Technol.* **2001**, *11*, 28.
43. Tait, W. S. *Spray Technol.* **2001**, *11*, 24.

25

Transport and Barrier Properties

WILLIAM J. KOROS

CATHERINE M. ZIMMERMAN

Mass transport properties of polymers depend upon a diverse array of factors, including sample preparation and the measurement technique used for assessment. Traditional barrier packaging,[1] controlled-atmosphere packaging,[2] membrane electrodes,[3] membrane separators,[4] and controlled-release devices[5] all share a need for high-quality characterization of transport properties. Multicomponent effects caused by plasticization[6] and more subtle bulk flow artifacts caused by so-called frame of reference contributions[7] also can alter the interpretation of measurements. Finally, perplexing history-dependent phenomena that defy conventional wisdom for transport through equilibrium materials are often encountered in glassy polymer samples. These effects require an understanding of glassy-state characteristics to avoid mistakes in interpretation of transport and barrier observations.[8,9]

This chapter highlights key aspects of the above-described issues with adequate references for the reader to follow up in considerable depth. In addition to monolithic polymer layers, complex higher order structures such as laminates and dispersed phases are often encountered in practical applications. To better illustrate such a case, a short example is discussed toward the end of this chapter. This example extends the discussion beyond the measurements and interpretations for monolithic layers to include a multilayer laminate. Our overall goal for this chapter is to enable readers to connect with and understand the large network of information related to the subject of transport and barrier properties.

Polymer Classifications

To understand transport properties of polymers, it is useful to group materials according to whether they are (1) amorphous versus semicrystalline on a supersegmental level, and (2) rubbery versus glassy in nature on a segmental level (Figure 25.1). Combinations of these two categories provide subgroups of (1) amorphous rubbery, (2) amorphous glassy, (3) semicrystalline rubbery, and (4) semicrystalline glassy types of samples. These latter two categories indicate the degree of segmental motion in the noncrystalline regions of a semicrystalline sample. The crystalline regions are generally impermeable to all penetrants.[9]

The characteristic physical properties of materials in these four subcategories affect the ability of small penetrants to sorb, diffuse, and permeate in response to a chemical potential driving force. These characteristic differences between the subcategories reflect morphological factors that are both segmental level and larger than segmental level.[10] Imposed orientation at both the segmental and supersegmental levels can cause further differentiation in the properties of chemically identical samples within these four subcategories. Efficient orientation of impermeable crystalline domains in a sample increases the effective diffusion path of a penetrant within the sample (Figure 25.2). The added pathlength observed compared to that for a sample without the impediments can be interpreted as an increased tortuosity. This added pathlength effectively decreases the ability of penetrants to cross through a film composed of the material, thereby making it a better barrier.

In addition to tortuosity, a more subtle chain restriction factor can also suppress permeation in semicrystalline media. Especially for flexible matrices, amorphous segments between ordered crystalline domains can experience suppressed mobility. This suppressed segmental mobility inhibits diffusion compared to the case for a purely amorphous material of the same chemical composition. These effects become increasingly apparent for large penetrants. Thus, crystallinity has a more potent ability to suppress diffusion coefficients for relatively large penetrants (e.g., C_3H_8) compared to smaller penetrants (e.g., He).[9,11]

Even for totally amorphous materials, the much larger amplitude segmental rotations and vibrations in rubbery

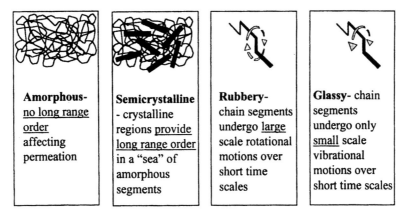

Figure 25.1 Useful categories of polymers for discussion of transport properties.

materials compared to glassy materials tend to cause much higher permeability in the rubbery materials (see Figure 25.1). In addition to intrinsic chain rigidity, intersegmental attractions can affect the segmental level of motion involved in diffusion. For example, within a given family of materials (e.g., substituted polyolefins), adding polar substituents increases intersegmental attractions between neighboring segments, thereby hindering segmental motions and transport through the medium.[12-15]

Fundamentals

Systematic transport studies and analyses date back well over 150 years. In 1831, Mitchell observed that natural rubber balloons exposed to different gas atmospheres deflated at different rates.[16] More than two decades later, Fick's law was formulated to describe the process of diffu-

sion in liquids by analogy to heat transport.[17] In 1866, Graham quantitatively measured gas permeation through a membrane and proposed the qualitative solution-diffusion mechanism to describe his results.[18] In 1879, von Wroblewski[19] defined the permeability coefficient as the penetrant flux times the membrane thickness divided by the pressure difference acting across the film.

$$\text{Flux of component } i = P_i \left(\frac{\Delta p_i}{\ell} \right) \tag{1}$$

The external upstream and downstream partial pressures and concentrations of components i, p_{i2}, and p_{i1}, and C_{i2} and C_{i1}, respectively, acting across the barrier layer provide the driving force for permeation, as shown in Figure 25.3. Effective barrier materials must have a low permeability for undesirable penetrants (e.g., O_2) to be economically viable as a single layer. Diffusion of most penetrants through a polymer follows Fick's laws of diffusion and hence after rearrangement, the permeability coefficient can be represented by the following equation:

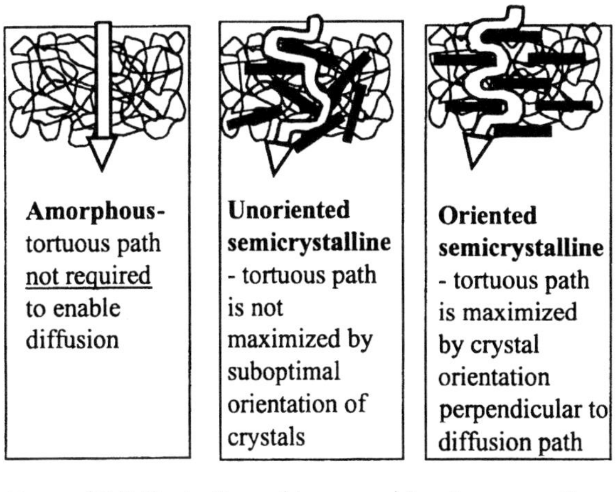

Figure 25.2 Illustration of improved barrier properties achievable when impermeable crystalline phase is oriented with largest dimensions perpendicular to the direction of permeation.

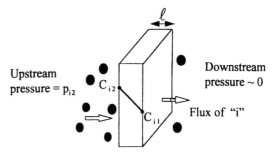

Figure 25.3 Schematic representation of relationship between film thickness, partial pressure, and concentration driving forces responsible for permeation through barrier films.

$$P_i = -D_i \left[\frac{\partial C_i}{\partial x}\right] \frac{\ell}{(p_{i2} - p_{i1})} \qquad (2)$$

where C_i is the concentration of the permeating component in the barrier material. In ideal barrier and membrane applications, the partial pressure of a component is negligible on the downstream side as compared to that on the upstream side. For such a case, the permeability can be expressed simply as a product of a kinetic factor, D_i, which is an effective diffusion coefficient, and a thermodynamic factor, S_i, which is a solubility or sorption coefficient. In this case, the sorption coefficient equals the equilibrium-sorbed concentration of the penetrant in the polymer at the upstream partial pressure (i.e., $S_i = C_{i2}/p_{i2}$). Dropping the subscript i for convenience, we see:

$$P = \int_0^\ell \frac{P}{\ell}\,dx = \int_0^{C_2} \left(\frac{D \cdot dC}{C}\right)\left(\frac{C_2}{p_2}\right) = D \cdot S \qquad (3)$$

Consistent with Graham's early qualitative description noted above, penetrant molecules sorb into the polymer on the high partial pressure side, diffuse through the matrix, and desorb from the low partial pressure side. To design a product for a specific application, knowledge of permeability coefficients and their dependence on temperature, pressure, and penetrant composition is adequate. More detailed understanding of the transport mechanism of penetrants in polymers allows one to tailor molecular structures to optimize properties.[20] For this purpose, it is essential to study the effects of these factors on diffusivities and solubilities as well. Characterization of the individual sorption and diffusion coefficients comprising the permeability coefficient was first addressed by using transient as well as steady-state permeation. Daynes[21] introduced, and Barrer[22] later refined the so-called time lag method for estimation of the diffusion coefficient from the transient portion of the permeation process.

A typical gas permeability measurement apparatus is shown in Figure 25.4.[23] This unit operates with a constant downstream volume connected to an accurate pressure transducer. A pressure release mechanism is required in the downstream section to prevent excessive pressure buildup

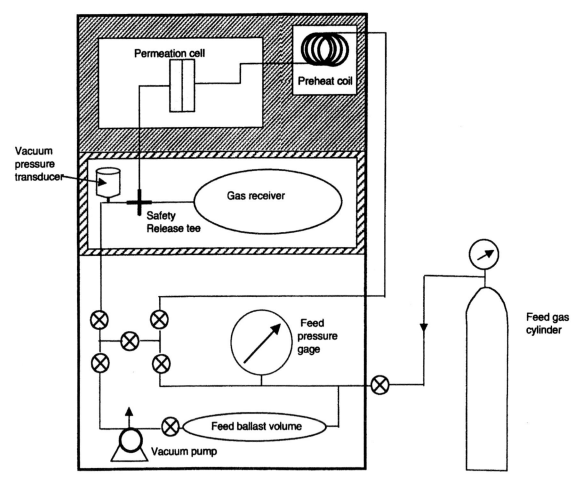

Figure 25.4 Typical constant receiver volume permeation cell. The mid-section is maintained at a fixed well-defined temperature, whereas the top section with the cell holding the test film can be varied over a wide range of temperatures without thermally cycling fittings and the sensitive pressure transducer.

in the case of failed membranes. Caution should be used; and because permeation experiments can involve flammable gases, the work environment should be periodically reviewed for possible ignition sources. When venting the system for any reason, gas must be directed into a fume hood to avoid inhalation or its accumulation in the work environment.

Tracking the steady-state pressure versus time in the fixed downstream volume under vacuum conditions allows a straightforward determination of the moles of total gas permeated. This permeation rate can easily be converted to a molar flux using the known area of the film. Normalizing this value by the transmembrane pressure and the membrane thickness according to eq 1, therefore, yields the permeability coefficient for the penetrant. The steady-state permeation rate should be determined by re-evacuation of the downstream receiver and redetermination of the apparent steady-state slope to verify that a true steady state has been achieved. Units for the permeability vary considerably among the various communities measuring data, but the most common units and their conversions for gas penetrants are given in Table 25.1.

Steady-state permeabilities can also be measured using a variable-volume cell, which is best suited for high steady-state permeation rates (Figure 25.5). Permeation cells like those used in the preceding constant-volume method can be used for the variable method as well.[24] To run an experiment, gas pressure is applied to the upstream membrane face, and gas permeates through the film into the downstream section as it does in the constant-volume experiment. For the variable-volume system, the downstream is at constant pressure, usually atmospheric. This variable downstream volume increases as gas permeates through the film. The displacement of a mercury or silicone oil plug or a soap bubble in a capillary column measures the rate of downstream volume increase. By measuring the volume increase with time, the permeate flux can be calculated using the gas law and the area of the film open to permeation.

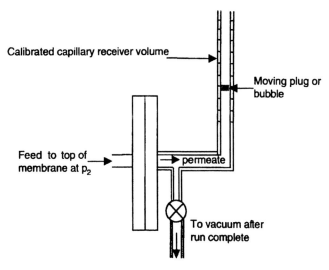

Figure 25.5 Schematic of simplified variable volume permeation cell using a calibrated capillary to monitor permeation. This type of cell is very useful for high-flux asymmetric membranes, but for barrier polymers, it is not very satisfactory due to low sensitivity.

The technique is particularly useful for measuring flux through thin-skinned asymmetric membranes.[25] Although variable-volume permeation measurements are easier to perform than variable-pressure measurements, they are often less sensitive and require a precisely measured downstream capillary volume.

Both the constant-volume and constant-pressure permeation systems can be modified to include multicomponent measurement capability by including a gas chromatograph or mass spectrometer.[26] Similarly, a carrier gas system with separate flowing streams on the top and bottom of the membrane can also be used in multicomponent as well as in single-permeate studies.[27] The extra complexity associated with multicomponent characterization is justified in cases where one of the components is strongly sorbing and can affect the steady-state permeation of other components. The CO_2/CH_4 gas pair typifies this situation, because CO_2 can increase the permeation rate of CH_4 greatly compared to the pure component CH_4 situation. Thus, in a CO_2/CH_4 membrane separator application, high-pressure, mixed-gas permeation rates should be characterized with a multicomponent testing system (Figures 25.6 and 25.7), rather than relying only on individual pure component tests.

Specialty Sensors for High-Barrier Permeation Characterization

The low permeabilities of many gases and vapors in barrier materials present challenges to characterization because fluxes are very low. A reliable fuel cell detector that is based on the reaction of permeating O_2 with H_2 provided from

Table 25.1 Permeability units and their conversions for gas penetrants

Unit	To convert to Barrers, multiply by	To convert to $\dfrac{cm^3 \text{ (STP) mil}}{100 \text{ in.}^2 \text{ day atm}}$ multiply by
Barrers[a]	1	167
$\dfrac{cm^3 \text{ (STP) mil}}{100 \text{ in.}^2 \text{ day atm}}$	6×10^{-3}	1

a. 1 Barrer $= \dfrac{1 \times 10^{-10} \ cm^3 \text{ (STP) cm}}{cm^2 \ sec \ cm \ Hg}$.

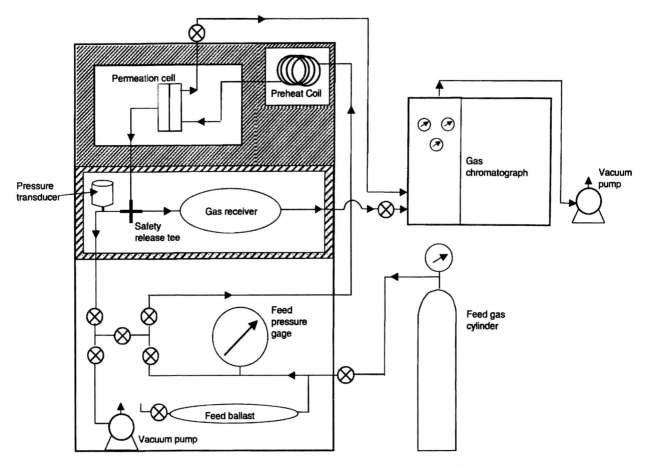

Figure 25.6 Schematic of upgraded version of constant volume cell in Figure 25.5 to allow measurement of multicomponent feeds.

a cylinder provides improved sensitivity compared to the pressure or volume increase measurements or a simple gas chromatograph described above. This is particularly useful for the ultra-low permeability materials needed for the most effective O_2 barrier applications. In such a sensor, four electrons are released electrochemically for every O_2 molecule that passes through it to react with H_2.[28] The resulting electrons form a current, which passes through a resistor, thereby creating a voltage. This voltage is then calibrated to indicate the amount of O_2 permeating through the barrier material by comparison to a known standard film. MOCON, Inc., Minnetonka, MN, produces a widely used commercial device, the OX-TRAN, for measuring O_2 transmission rates in flat films and finished package configurations such as bottles, bags, and glove boxes. For flat film measurements, the samples are positioned in a diffusion cell similar to those described previously. On the upstream side of the film, a steady flow of O_2 or air establishes a fixed upstream driving force. The option of maintaining a specified relative humidity also exists. Moist N_2 at a specified relative humidity (RH) passes across the downstream face of the film and is then sent to the sensor for O_2 analysis. Gas permeation through

barrier materials is often expressed in units of cubic centimeters (standard temperature and pressure [STP]) mil/100 in.^{2} day atm (see Table 25.1). Many polymers show significant dependence of O_2 permeability on the percent RH of the external gas. Therefore, when characterizing the O_2 barrier properties of a film, it is advisable to perform the permeation experiment under the same conditions of relative humidity and temperature as those the final package will be exposed to. Both temperature and relative humidity must be controlled well ($+1 - 2$ °C and $+1 - 2$% RH) to avoid scattered and poorly reproducible data. The O_2 permeation rate in Nylon 6 increases by over a factor of 3 in the range of 0% to 90% RH, and even larger changes can be expected for a more hydrophilic material such as poly-(vinyl alcohol) or ethylene-vinyl alcohol copolymers.[29]

The water vapor transmission rate (WVTR) describes the rate at which water molecules permeate through a film exposed to a humid environment on one side (upstream) and a dry environment on the other (downstream) side. The WVTR parameter is another important quantity to characterize barrier properties of materials.[30-32] The common units for WVTR, (g mil)/(100 in.^{2} day), differ from those of true

Simplified schematic for sweep gas permeation system

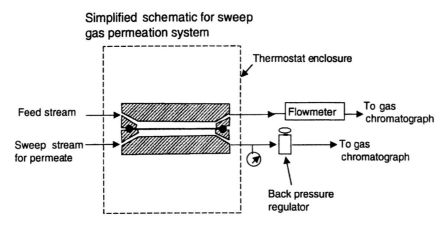

Figure 25.7 Schematic of a carrier gas permeation cell where the concentration of permeate in the carrier gas must be measured to determine the permeation flux. Accurate flow control of the sweep gas is very important in this type of cell.

permeability in eq 1, because they lack a pressure in the denominator to normalize the flux. As a result, a complete WVTR report must always include the experimental water vapor pressure used in the study. Two common methods for WVTR measurements exist: American Society for Testing and Materials (ASTM) E96 and ASTM F372. In the former method, a flat film is exposed to the penetrant vapor on one side while the other side has a chamber that contains desiccant. As the moisture permeates through the polymer, the increased water concentration in the desiccant is monitored. In ASTM F372, a film separates wet and dry chambers. An infrared (IR) detector monitors permeation of water molecules into the dry chamber with time. A commercial instrument, PERMATRAN-W (MOCON, Inc.), uses IR in a similar manner to monitor the water concentration in a flowing dry air stream that passes along one face of the film.[28]

Carbon dioxide permeation measurements are of special interest when developing polymers for carbonated beverage packaging applications. Besides the constant volume and constant pressure measurement systems, a simple method to measure CO_2 permeability is also offered by MOCON, Inc., using the PERMATRAN-C unit.[28] In this device, CO_2 is applied to one side of a film or container. Air is circulated past the other surface of the test film or container in a closed sample loop to monitor the increase in CO_2 concentration with time using an IR detector to allow determination of the CO_2 permeability.

The techniques above are typically used to determine the steady-state permeation properties of a barrier material. As noted, it is sometimes important to characterize the individual components of diffusion and sorption coefficients comprising the overall permeability coefficient. The so-called isothermal time lag transient permeation experiment developed by Daynes[20] and Barrer[21] is commonly used for this purpose. To perform this experiment, complete evacuation of penetrant from a film of thickness, ℓ, is followed by im-

posing a constant upstream pressure or concentration on the film. The evacuated downstream receiver is then monitored as it approaches steady-state permeation to yield a response like that in Figure 25.8. The time lag, θ, is determined by extrapolating the linear pseudosteady-state region back to the time axis. To run such transient experiments accurately, it is important that the apparent linear region from the beginning to the end of the run be sufficiently long. A useful rule of thumb is that the time period over which extrapolation is performed should comprise at least three time lags (e.g., 3θ), so the minimum valid run should comprise a time of 4θ to avoid underestimation of the true time lag.[33] More conservatively, it is wise to use a run time of five to six time lags and then evacuate and verify that the apparent steady-state slope after six time lags is truly consistent as shown in Figure 25.8. The most common mistake made in such runs is due to impatience. For a constant diffusion coefficient, the experimental time lag, θ, is related simply to the effective diffusion coefficient, D_{eff}, by eq 4[34].

$$D_{\text{eff}} = \frac{\ell^2}{6\theta} \qquad (4)$$

On the basis of eq 3, because $P = DS$, determination of D_{eff} from eq 4 and P from eq 1 also allows determination of $S = C/p$ for this ideal case of constant coefficients. This time-lag analysis is rigorously valid only for the case of a constant diffusion coefficient with a constant sorption coefficient (i.e., Henry's law). As a rule of thumb, rubbery polymers typically satisfy this requirement for penetrants that do not sorb at an equilibrium level much above 2% to 3% by weight.

For cases in which the coefficients show considerable dependency on external penetrant activity (e.g., partial pressure or concentration dependence), more complex interpretations are involved in rigorous analysis of the data. Nevertheless, as long as Fick's law applies with only a concentration-dependent diffusion coefficient, even with non-

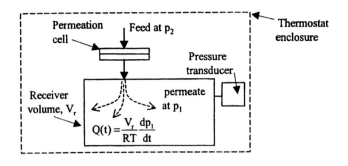

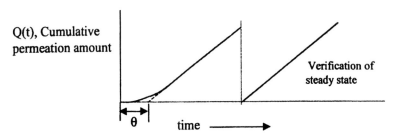

Figure 25.8 Illustration of key variables in a traditional time-lag experiment.

linearity in the sorption isotherms, the diffusivity, sorption isotherm, and steady-state permeation can be determined in a straightforward manner. In practice, it is common to also use the above technique to estimate an effective value for D using the time lag, even when the effective coefficient is a function of the upstream driving pressure. In this case, the diffusion coefficient calculated from eq 4 is an estimate of the local diffusion coefficient averaged between the upstream concentration and downstream (effectively zero) concentrations. More rigorous (and tedious) analyses of the pressure and concentration dependence in such cases have been reported by several authors;[33-39] however, for many cases, this approximate analysis is adequate. By using the ratio of the steady-state permeability with such an average effective time lag D, one can even obtain a reasonable measure of the secant slope solubility coefficient, $S = C_2/p_2$, evaluated at the upstream conditions of C_2 and p_2. Fortunately, even when the sorption isotherm is more complex than a simple Henry's law relation, this approach still allows reasonable estimates of the sorption, permeation, and diffusion coefficient's dependency on concentration without resorting to actual sorption studies.

An example of such a case is given by gas permeation through barrier polymers such as poly(ethylene terephthalate) (PET). The actual detailed time lag, permeation, and sorption isotherms for CO_2 in 2-mil-thick PET film at 35 °C are shown in Figures 25.9a,b and 25.10.[40,41] The peculiar concave shape of the sorption isotherm reflects the gradual saturation of unrelaxed segmental packing defects trapped in the low-mobility glassy matrix. Equipment to enable

such sorption measurements is discussed later in the chapter. One can see that at an upstream pressure of 20 atm of CO_2, the time lag is approximately 44 min (i.e., 2640 s), the permeability is approximately 0.14 Barrers, and the true secant slope at 20 atm is 0.60 cm^3 (STP)/(cm^3 polymer atm). An effective average diffusion coefficient equal to 1.6 $\times 10^{-9}$ cm^2/s can be estimated from eq 4 using the time lag at 20 atm. Dividing this coefficient into the observed permeability at 20 atm gives an estimated value of $S = 0.0864$ cm^3 (STP)/(cm^3 cm Hg) = 0.66 cm^3 (STP)/(cm^3 polymer atm). This estimated value of S is reasonably close to the true secant slope at 20 atm estimated from Figure 25.10 as 0.60 cm^3 (STP)/(cm^3 polymer atm). This convenient approximate method, therefore, is quite useful for estimation purposes.

Actual independent sorption equilibria measurements such as those shown in Figure 25.10 for gases and vapors in polymers typically involve gravimetric and pressure decay methods.[23,42] Equipment for characterization of sorption isotherms in such cases is shown in Figures 25.11 and 25.12. Before operating at elevated pressures, equipment must be adequately constructed and pressure tested. Even the simple quartz spring microbalance, which was developed more than 70 years ago, continues to have applications for the diverse systems that must be studied.[43] The McBain type quartz spring microbalance has evolved somewhat over the years, with the introduction of accurate pressure transducers and extremely accurate optical readers to detect spring deflections. The main advancement in this area of gravimetric sorption studies involves replacement of

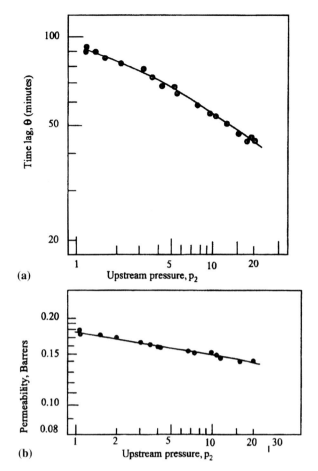

(a)

(b)

Figure 25.9 Time lag (a) and permeation (b) data for carbon dioxide in poly(ethylene terephthalate) at 35 °C. (Reproduced with permission from reference 40. Copyright 1978 Wiley and Sons.)

the optical reader with a photometric detector that senses deviations from standard position of a lever attached to the sample. Measurement of the electrical current required to return the beam to its neutral position provides a direct measure of the sorption uptake that must have occurred to cause the original deviation in beam position.[44] This electrobalance allows automated recording of data for kinetics and equilibrium studies. The high-pressure, dual-volume sorption cell shown in Figure 25.11 also allows accurate sorption kinetics and equilibria to be measured by monitoring pressure decay in the known void volume of the cells. This technique allows easy data recording, but the gas phase boundary condition is not constant, and special analyses to account for the changing pressure boundary condition are needed.[45]

For liquid systems, simple immersion and weighing measurements are often adequate for sorption characterization.[46] In this case, the steady-state barrier properties coupled with this equilibrium sorption experiment provide an estimate of the diffusion coefficient averaged between the conditions of saturated liquid and infinite dilution sorption.

Note that in this case, one could use the reverse procedure to that illustrated for the estimated secant slope described earlier. One would estimate an average diffusion coefficient by taking the ratio of the steady-state permeability and the experimental secant slope, C_{sat}/p_{sat}, where the p_{sat} is simply the saturated vapor pressure of the liquid at the temperature of the measurement, and C_{sat} is the measured uptake of the polymer at sorption equilibrium in the saturated (pure) liquid. This approximation is usually reasonably satisfactory to estimate the average diffusion coefficient in the entire range between infinite dilution and saturation condition in most cases if the liquid uptake is less than 2% to 3%. If swelling exceeds 5% to 6%, stronger concentration dependency is expected and such an estimated diffusion coefficient may considerably underestimate the true average penetrant mobility under conditions in which the downstream penetrant activity differs significantly from zero. Nevertheless, as a first approximation, the procedure is a useful estimator.

Transient Sorption Measurements

Sorption-desorption kinetics accompanying the measurement of equilibrium sorption uptake data can also be analyzed to estimate diffusion coefficients for systems that obey Fick's law. Non-Fickian effects reflected by time-dependent diffusion coefficients or nonuniform material properties greatly complicate the transient permeation and kinetic sorption analyses. Probably the most reliable way to determine if non-Fickian effects are involved is to run kinetic sorption (or desorption) experiments with films of different characteristic thicknesses. The various uptake or desorption kinetics should be placed on a single plot as a function of the square root of time divided by the thickness of the samples. If (1) the short time response (less than 50% uptake or desorption completed) for all of the different samples superimpose, and (2) the short time uptake is essentially linear, this is a strong indication that Fickian transport is occurring. This test is valid even if a very strong concentration dependence is present for the diffusion coefficient and the sorption isotherm is nonlinear.[3] An important special case for which position-dependent properties can be handled relatively simply includes laminates and the related problem with disperse impermeable flakes in a continuous medium. These cases are considered briefly later in this chapter.

For the case in which Fickian uptake kinetics apply with a constant diffusion coefficient, Fick's second law can be written for one-dimensional transport such as that which controls diffusion in a film:

$$\frac{\partial C}{\partial t} = D \frac{\partial^2 C}{\partial x^2} \qquad (5)$$

where C is the concentration of the diffusing penetrant in the sample and t is the time. Equation 5 can be solved using

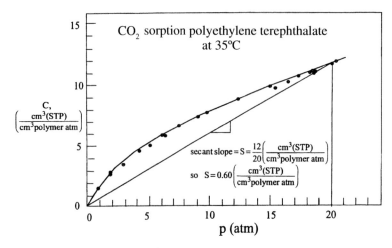

Figure 25.10 Sorption data for carbon dioxide in poly(ethylene terephthalate). (Reproduced with permission from reference 41. Copyright 1978 Wiley and Sons.)

Laplace transforms and related to M_t and M_∞, the total mass uptake at time t and at equilibrium, respectively. For desorption, M_t and M_∞ simply refer to the mass desorbed at time t and at equilibrium in the experiment. Many diffusion coefficient measurements employ a large surrounding penetrant volume, which is assumed to be infinite, so the penetrant concentration is essentially constant over the entire experiment. The solution for a plane sheet with two-sided symmetrical penetrant diffusion can be reduced to a simplified form that is valid for short times ($M_t/M_\infty \leq 0.5$) for the case of a constant diffusion coefficient:

$$\frac{M_t}{M_\infty} = \frac{2}{\sqrt{\pi}} \left[\frac{4Dt}{\ell^2} \right]^{\frac{1}{2}} \quad (6)$$

where ℓ is the total film thickness. The diffusion coefficient can be easily calculated by plotting M_t/M_∞ versus $\sqrt{t}$ and considering the initial linear slope of the response such as that shown in Figure 25.13a. Likewise, the solution of eq 5 for a plane sheet with two-sided penetrant diffusion and a constant diffusion coefficient at long times ($M_t/M_\infty \geq 0.6$) can be approximated as:[47]

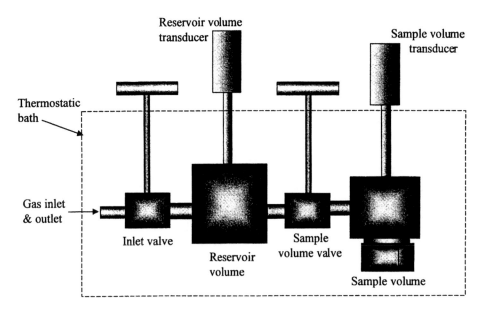

Figure 25.11 Schematic representation of a two-volume "pressure decay" sorption cell. (Reproduced with permission from reference 42. Copyright 1976 Wiley and Sons.)

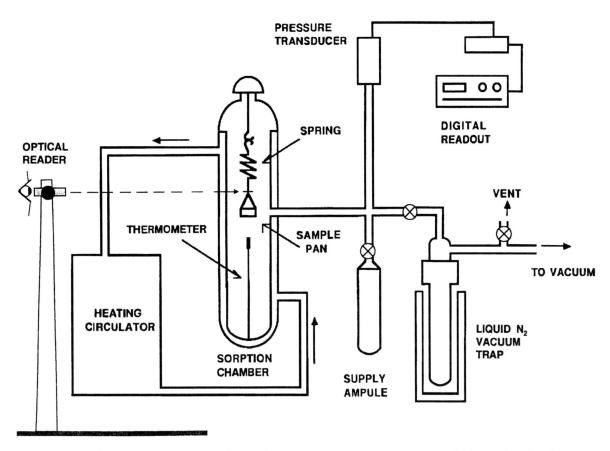

Figure 25.12 Schematic representation of a modern McBain quartz spring sorption cell. (Reproduced with permission from reference 43. Copyright 1982 Wiley and Sons.)

$$\frac{M_t}{M_\infty} = 1 - \frac{8}{\pi^2} \exp\left[-\frac{\pi^2 D t}{\ell^2}\right] \qquad (7)$$

$$D_{\text{eff}} = \frac{\int_0^{C_2} (D \cdot dC)}{\int_0^{C_2} (dC)} = \int_0^{C_2} \frac{D \cdot dC}{C_2} \qquad (8)$$

Clearly, a plot of $\ln M_t/M_\infty$ versus time enables the diffusion coefficient to be determined from the slope, as shown in Figure 25.13b. Crank[47] also presents similar solutions for other geometries such as cylinders and spheres as well as solutions in which sorption or desorption occurs in a finite, rather than an ideal infinite volume.

The above expressions apply rigorously only for the case of a constant-diffusion coefficient. In practice, as is discussed for the case of transient permeation, these simple formulas can still be used approximately, despite concentration-dependent diffusion coefficients and nonlinear sorption isotherms. In the case of sorption into a totally evacuated film from an infinite bath in which the equilibrium sorption uptake will be C_2 (corresponding to $M_\infty = V_{\text{sample}} \cdot C_2$), the following approximate expression describes the meaning of the effective diffusion coefficient that will result from a sorption experiment[9]:

Although diffusion coefficient calculations from kinetic sorption experiments are straightforward, care is needed to ensure the minimum dimension is at least 10-fold smaller than the next smallest dimension to ensure the validity of the assumption of one-dimensional diffusion. In addition, uniformity of sample dimensions is important. Berens[48] has shown that sorption in a distribution of sphere sizes results in an unexpected uptake curve that complicates analysis. The time scale of the experiment must also be considered. If the time needed to reach $M_t/M_\infty = 0.5$ is relatively short (e.g., a few seconds) the short time solution estimate of the diffusion coefficient may be inaccurate because of difficulties in valve closures and stabilizations of pressures or solution concentrations. In this case, the long time analysis would be preferable. It is also important to determine M_∞ accurately. Premature identification of M_∞ causes errors in both

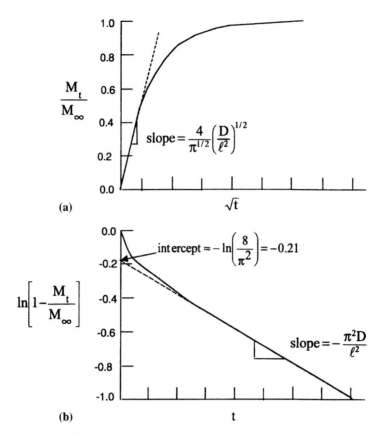

$$\text{slope} = \frac{4}{\pi^{1/2}}\left(\frac{D}{\ell^2}\right)^{1/2}$$

$$\text{intercept} \approx -\ln\left(\frac{8}{\pi^2}\right) = -0.21$$

$$\text{slope} = -\frac{\pi^2 D}{\ell^2}$$

Figure 25.13 Schematic representation of "short time" (a) and "long time" (b) sorption kinetic responses for a constant diffusion coefficient case.

the long time and short time solutions. As with the time lag work, impatience can lead to unnecessary errors. A good policy is to repeat a sorption run and wait at least five times as long as is initially estimated to be required to achieve 50% uptake to demonstrate that the apparent value estimated for M_∞ has been accurately determined.

Transient sorption measurements of gases, vapors, and liquids can be monitored using a piezoelectric quartz crystal microbalance (QCM).[49-52] The diffusion coefficient analysis follows the same procedure as described above for M_t/M_∞, as described in either eqs 6 or 7. A schematic diagram of the QCM apparatus is shown in Figure 25.14.[49] In this technique, a quartz crystal with metal electrodes is coated with a polymeric film that is a few micrometers thick. The coated film consists of only milligram polymer quantities. After the coated crystal is connected to an external driving oscillating circuit, the crystal oscillates at its resonant frequency. The crystal is then exposed to the gaseous or vapor penetrant, which sorbs into the thin polymer film. As the film mass increases with penetrant sorption, a frequency shift, Δf, is observed, which is proportional to the change in mass, Δm, according to the following equation:[50]

$$\Delta m = -C\Delta f \qquad (9)$$

where C is a constant defined by the intrinsic quartz slab properties and the quartz thickness.

Additional Techniques

Recent liquid and vapor diffusion coefficient studies using Fourier transform infrared attenuated total reflectance (FTIR-ATR) spectroscopy demonstrate good agreement with data obtained from other techniques. The FTIR-ATR technique involves inserting a film-coated ATR crystal in an ATR cell where the free side of the film is contacted with the penetrant as shown in Figure 25.15. Details of the FTIR-ATR system construction are described elsewhere.[53,54] Penetrant diffusion is detected by the IR light reflected in the crystal, which is directed to a detector. The resulting spectrum, possessing one or more significant penetrant absorbance peaks, is then related to the penetrant concentration. This technique offers the opportunity to explore specific polymer-penetrant interactions, such as hydrogen bonding, and their effect on the diffusion coefficient that simple gravimetric techniques do not allow.[55] In addition, the technique enables multicomponent diffusion behavior to be analyzed.[53] For some gas pairs such as CO_2/CH_4, pure

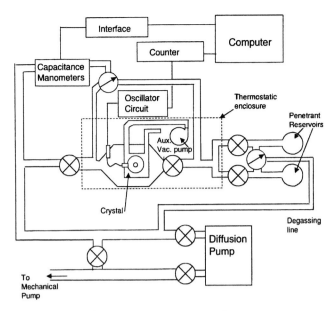

Figure 25.14 Schematic representation of a resonant quartz crystal microbalance. (Reproduced with permission from reference 49. Copyright 1991 Wiley and Sons.).

gas measurements often fail to describe accurately the gas transport properties of a mixture. Although mixed gas permeation experiments are relatively simple to complete, separation of mixed gas permeability coefficients into their respective diffusion and sorption coefficient terms can be complex. Marand and Dhingra[56] propose a method for estimating these mixed gas diffusion and sorption coefficients that uses the permeate flux profile and the permeate gas concentrations under steady-state conditions. The neces-

sary equipment and its construction are nearly identical to that described previously for typical permeation measurements.[56] This analysis assumes that gas sorption obeys Henry's law, which is a good assumption for glassy polymers at low penetrant concentrations (pressure < 2 atm) and rubbery polymers over a wide pressure range.

Inverse gas chromatography (IGC) is a technique for measuring diffusion coefficients in the range of 10^{-6} to 10^{-11} cm^2/s[57] that has been developed over the last decade. Initial IGC diffusion coefficients measurements were limited to infinite dilution conditions, but recent work using the system shown in Figure 25.16 enable diffusion coefficients at finite concentrations to be measured.[57] Initial tests were limited by packed columns, which often had an irregular distribution of polymer particles. Typical applications currently use capillary columns, which possess a stationary, thin, uniform polymeric coating. A carrier gas, which flows through the column, is either saturated or injected with the solute. The solute retention time and the shape of the elution profile can be analyzed to determine the solute solubility and diffusion coefficient in the polymer. Diffusion coefficient experiments using IGC show reasonable agreement with gravimetric vapor sorption data.[57-60] This technique is especially useful at low solute concentrations as well as situations for which gravimetric vapor sorption experiments take an extended time to complete. Some limitations exist for polymeric materials that have difficulty coating the column uniformly.

Nuclear magnetic resonance (NMR) investigations have applied the spin-echo method with pulsed magnetic field gradient to measure penetrant diffusion in polymers as well as self-diffusion coefficients.[61-65] The spin-echo method for

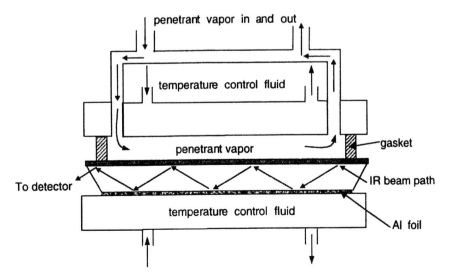

Figure 25.15 Schematic representation of an attenuated total reflectance infrared sorption cell. (Reproduced with permission from reference 54. Copyright 1993 Butterworth-Heinemann.)

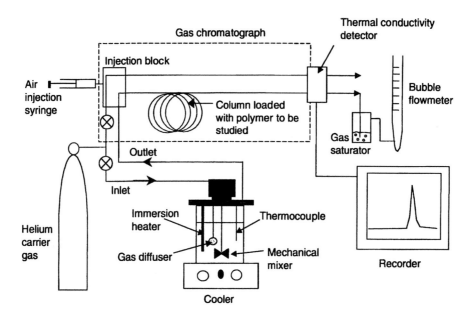

Figure 25.16 Schematic representation of a finite concentration gas chromatography unit for evaluation of concentration dependent diffusion coefficients. (Reproduced with permission from reference 57. Copyright 1997 Wiley and Sons.)

determination of the diffusion coefficient is based on the difference between the amplitude of the echo with and without a magnetic field gradient applied for a specified period.[63] Both solvent and low-concentration tracer diffusivities can be determined for several polymer-solvent systems using the pulsed-gradient spin-echo NMR (PGSE-NMR) method.[64,65] The dynamics of sorbed gas, such as $^{13}CO_2$ in glassy and rubbery polymers has also been investigated using NMR.[66,67] The NMR technique has been commonly used to characterize complex microporous and molecular sieving media such as zeolites and molecular sieving carbons. The technique is also attractive for use with highly rigid and brittle polymers. In many cases, such media cannot be characterized by actual steady-state permeation techniques via the classical methods described above. Accumulation of more experience with the use of nonclassical chromatographic and spectroscopic methods should eventually provide confidence that the classical sorption and permeation based techniques and these more advanced instrumental techniques yield essentially equivalent results. For the time being, results from the classical techniques are recommended for use in practical barrier application calculations when they can be applied.

Barrier Structures Comprising Supermolecular Domains

Semicrystalline polymers represent the simplest example of structures that can produce improved barrier efficacy. Barrier performance is greatly increased when orientation places the largest dimension of the domains perpendicular to the permeation direction. Figure 25.17 shows a relatively complete representation of the range of barrier structures usually considered.[68] The discussion that follows summarizes important issues related to these barrier structures.

Chemical modification of the surface of polymers is an attractive method of improving barrier characteristics of polymers.[69] Practically speaking, this treatment corresponds to a reactively formed composite structure. Other relatively recent developments in barrier structures include inorganic glass-coated polymers exhibiting excellent gas barrier and retort properties and bending resistance.[70] Deposition of a 300- to 1000-Å layer of Si-SiO_2 on PET is reported to result in an excellent gas barrier film.[71]

Most applications require both gas-barrier and water-barrier properties. Many high-barrier polymers for oxygen are poor water barriers or have oxygen-barrier properties that are sensitive to water vapor. Multilayer laminates of barrier polymers with good O_2-barrier properties, such as ethylene vinyl alcohol copolymer, sandwiched between hydrophobic layers, such as polypropylene, are commonly used.[30,68,72–74] Laminates of from five to seven layers are formed by melt extrusion to provide multilayer packages.

Even in the laminate structure, the water-sensitive polymers can lose barrier properties under sterilization or retort conditions, thereby leading to complex history-dependent behavior. During such processes, the package is exposed to steam or pressurized water at 100 to 135 °C for 1 to 2 h. Under these conditions, water permeates through the hydrophobic polymers and reaches the water-sensitive oxygen-barrier polymer [typically poly(ethylene vinyl alcohol)] sandwiched between the hydrophobic layers, thereby significantly reducing its oxygen-barrier properties. The package is then stored with the inside wall contacting food at conditions as high as 100% RH and various external RH values. This history-dependent loss of poly(ethylene vinyl alcohol) barrier properties as a result of the retort process is complex.[8,30]

Addition of desiccants into the multilayer structure overcomes the effects of the retort process, resulting in substantially lower oxygen permeation after the retort shock. The desiccant must have a high capacity for holding water (one to two times its own weight) at high RH values. It should also be processable and meet the U.S. Food and Drug Administration (FDA) regulations.[8]

Interest in use of single-layer packaging materials capable of controlling the package atmosphere over fresh fruits and vegetables has stimulated new processing approaches in some cases.[2] Melt blending of a polyolefin film with inert fillers, a processing aid such as calcium stearate, and a stabilizer is reported to maintain the moisture level below 700 ppm in the melt blend. The polyolefin film cast from this blend can be uniaxially oriented until it has a sufficient number of elongated, narrow-shaped, microporous voids that allow permeation of oxygen and carbon dioxide in the range of about 5,000 to 10,000 (cm^3/[100 in.^{2} atm day]).[75] Unlike the previously discussed cases in which the layer provides a molecularly selective resistance, this type of barrier would simply serve to keep particulates and microbes out of the package.

Other examples of mixed-matrix layers cover a wide spectrum of properties when the matrix and flake are well-bonded. In this well-bonded case, a mixed-matrix barrier containing either inorganic flakes such as mica or semipermeable polymer flakes also depends strongly on the so-called aspect ratio of the dispersed-phase particles used in the formulation. The aspect ratio of a typical square platelet filler is defined as the lateral dimension divided by the thickness dimension. As the aspect ratio for an impermeable flake increases, a penetrant experiences a huge increase in effective pathlength.[76] This added tortuosity (see Figure 25.2) greatly reduces the rate of transport across the barrier layer. The use of nano-sized clay particles with enormous aspect ratios in polymer matrices allows one to incorporate relatively small amounts of the material.[77-79] Barrier enhancements caused by incorporation of less than 10% by weight of nano-clay particles has been shown to yield enormous reductions in barrier properties; however, processing challenges must be overcome.[79] As with the crystalline domains in Figure 25.2, uniform distribution and orientation of the impermeable phase with the largest dimensions oriented perpendicular to the transport direction provides the largest improvement in barrier properties.

A simpler totally polymeric mixed-matrix barrier has been shown to improve toluene[80,81] barrier properties by incorporating low-permeability nylon flakes with large aspect ratios into a polyethylene layer. Larger aspect ratio flakes can be incorporated using this method with the smeared-out polymer flakes as compared to that which is practical with brittle inorganic flakes such as mica.[29] The aspect ratios are not as large as those believed to be achievable with the nano-clay composites, but impressive achievements are still apparent. The large increase in barrier properties shown as the ordinate in Figure 25.18 is very impressive. For instance, a 0.1-in. mixed-matrix wall with only 40% nylon provides a barrier to toluene permeation equivalent

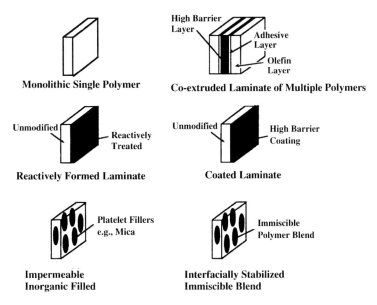

Figure 25.17 Illustration of the diverse range of barrier structures currently in use. (Reproduced with permission from reference 68. Copyright 1990 American Chemical Society.)

to a 4.5-in.-thick slab of pure polyethylene. This barrier enhancement factor of 45 at 40% loading is improved even more at higher loadings, but to maintain flexibility and high-speed processability, the 20% to 50% loading range is desirable.[23] In principle, for slightly permeable flakes, this complex morphology is difficult to model, but as these authors note, for satisfactory interfacial adhesion between matrix, the barrier performance of such a layer for component i (toluene in this case) is ideally bounded by the two limiting cases given by eqs 10 and 11 and illustrated in Figure 25.18 by the two theoretical lines:[23]

(a) a laminated structure of components 1 and 2 with thickness or volume fraction

$$(P_i)_{\text{eff}} \frac{\ell_{\text{tot}}}{\dfrac{\ell_1}{P_{i1}} + \dfrac{\ell_2}{P_{i2}}} = \frac{1}{\dfrac{\ell_1/\ell_{\text{tot}}}{P_{i1}} + \dfrac{\ell_2/\ell_{\text{tot}}}{P_{i2}}} = \frac{1}{\dfrac{\Phi_1}{P_{i1}} + \dfrac{\Phi_2}{P_{i2}}} \qquad (10)$$

(b) uniformly dispersed spheres with volume fraction, Φ_2

$$(P_i)_{\text{eff}} = P_{i1} \left[\frac{P_{i2} + 2P_{i1} - \Phi_2(P_{i1} - P_{i2})}{P_{i2} + 2P_{i1} + \Phi_2(P_{i1} - P_{i2})} \right] \qquad (11)$$

In eq 10, the thickness fraction, ℓ_i/ℓ_{tot}, can generally be related to a simple volume fraction, Φ_i, for the laminated structure. The data points in Figure 25.18 clearly illustrate that large aspect ratio platelet fillers are greatly superior to spherical platelet fillers (eq 11) for adding barrier efficacy caused by tortuosity. Moreover, for the large aspect ratio flakes, it is apparent that within experimental error, the actual data points agree with eq 10, indicating essentially series resistance contributions from both the dispersed and continuous phases. This fortunate coincidence is useful for estimating permeation resistance in the mixed-matrix formulation for semipermeable flakes.

Example of Barrier Laminate Application

Generalization of the expression in eq 10 for a two-component laminate gives an expression for the overall permeability of a multilayer laminate structure in the form of slabs of thickness ℓ_i. The overall permeability, P_{total}, is determined by the series contribution of the individual layers as shown for the five-layer composite in Figure 25.17.

$$P_{\text{total}} = \frac{\ell_{\text{total}}}{\left[\sum_i^N \left(\ell_i / P_i \right) \right]} \qquad (12)$$

In a typical barrier laminate, the five total layers are comprised of a barrier core, two adhesive layers, and two outer hydrophobic layers. For a 16-oz (454 g) container with a maximum oxygen uptake allowed of 1 ppm (4.54×10^{-4} g $= 1.42 \times 10^{-5}$ mol $= 0.318$ cm^3 [STP]), one can estimate

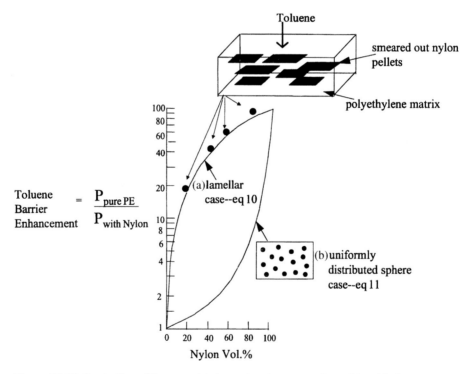

Toluene Barrier Enhancement $= \dfrac{P_{\text{pure PE}}}{P_{\text{with Nylon}}}$

Figure 25.18 Illustration of improved toluene barrier properties achievable by incorporating high aspect ratio nylon flake into polyethylene bottle walls.
(Reproduced with permission from reference 80. Copyright 1985 Society of Plastics Engineers.)

the container shelf life at 25 °C assuming no leaks occur through seams. To use the data in Table 25.2, assume dry storage conditions (0% RH). The external surface area of such a container can be estimated to be roughly 240 cm^2 (or 37.2 in.^{2}). For simplicity, assume two identical combined hydrophobic and adhesive layers of 15 mil equivalent to polypropylene surrounding a 1-mil layer of 70% vinyl

alcohol-30% ethylene composition as the core. Accepted values for the oxygen and water vapor permeabilities for various barrier materials are given in Table 25.2. Additional values can be found in a number of sources.[15,20,82,83] For the current example, one can use the value for polypropylene and 30% ethylene-70% vinyl alcohol at 25 °C given in the table:

Table 25.2 Summary of permeability data for various common polymers near 25 °C and 1 atm upstream pressure

Polymer	23–25 °C O$_2$ permeability (Barrers)	23–25 °C CO$_2$ permeability (Barrers)	23–25 °C H$_2$O permeability (Barrers)	Reference no.
1. Poly(vinyl alcohol)	0.00006 (0% RH)	—	19 (40% RH)	13
Poly(vinyl alcohol)	0.11 (90% RH)	1.3 (90% RH)	—	83
2. Poly(ethylene-vinyl alcohol), 70% VOH	0.00010 (0% RH)	—	—	13
3. Polyacrylonitrile	0.00015–0.000024	0.00060	230	13,82
4. Vectra liquid crystal polyester	0.00047	—	—	13
5. Poly(vinylidene chloride)	0.00060	—	1.5–4.5	13
6. Cellophane	0.0010 (0% RH)	0.00353	8100–18,900 (100% RH)	13,82[a]
7. Poly(ethylene-vinyl alcohol), 60% VOH	0.0010 (0% RH)	—	—	13
8. Poly(vinylidene chloride-vinyl chloride) copolymer, 80% vinylidene	0.0015	—	—	13
9. Saran	0.003	—	0.001	13
10. Poly(acrylonitrile-styrene) copolymer (86% ACN)	0.0032	0.011	640	82
11. EK 80 liquid crystal polyester	0.005	—	—	13
12. Poly(acrylonitrile-styrene) copolymer (66% ACN)	0.036	0.160	1500	13
13. Nylon 6	0.009 (0% RH)	—	400	13
14. Nylon 66	0.015 (0% RH)	0.052–0.071	433	13,82
15. Epoxy, thermoset (Bis A/amine)	0.018	—	—	13
16. Poly(ethylene terephthalate) (semicrystalline), dependent on crystallinity	0.018–0.035	0.12	7.5–98	13,82
17. Poly(chlorotrifluoroethylene)	0.018	—	—	13
18. Poly(vinylidene fluoride)	0.027	—	—	13
19. Carilon polyketone	0.027	—	0.07	13
20. Nylon 6,10	0.04	—	—	13
21. Polybisphenol-epichlorohydrin	0.042	0.086	—	13[a]
22. Poly(vinyl chloride)	0.05–0.26	1.3	37–562	13[a]
23. Polyethylene terephthalate (amorph)	0.06	0.227	175	82
24. Polyacetal	0.06 (0% RH)	1.28	422	13,82
25. Poly(vinyl fluoride)	0.09	—	30	13
26. Poly(methyl methacrylate)	0.10–0.15	—	480–2500	13,82
27. Nylon 11	0.16	—	—	13
28. Poly(acrylonitrile-styrene) copolymer (39% ACN)	0.35	1.0	1900	82
29. Polychloroprene, Neoprene W	4.0	22.0	—	83
30. Ethyl cellulose	11.0	84.5	6700	82
31. Polyethylene (0.96 g/cm^3)	0.3–0.7	—	9–12	13,82
32. Polyurethane elastomer	0.81 (0% RH)	—	—	13
33. Polypropylene	0.9–1.8	4.1–6.9	15–58	13,82
34. Polysulfone	1.18	—	2149	13
35. Bisphenol-A polycarbonate	1.1–1.35	6	1050–1274	13,82
36. Polybutene	1.98	—	—	13
37. Polystyrene	1.9–2.5	7.9	840–1350	13,82
38. Polyethylene (0.914 g/cm^3)	2.2	9.5	68	13,82
39. Teflon polytetrafluoroethylene	3–4.2	11.7	—	13[a]
40. SBR Rubber	9	—	—	82
41. Polybutadiene	15	—	5070	13,82
42. Natural rubber	24	131	—	83
43. Silicone rubber dimethyl siloxane	540–600	3230	—	13,83

Key: VOH = vinyl alcohol; ACN = acrylonitrile.

a. Hwang, S. T.; Choi, C. K.; Kammermeyer, K. *Sep. Sci.* **1974**, *9* (6), 461.

$$P_{PP} = 150 \frac{cm^3(STP) \, mil}{100 \, in.^2 \, day \, atm} \text{ for polypropylene}$$

$$P_{EVOH} = 0.017 \frac{cm^3 \, (STP) \, mil}{100 \, in.^2 \, day \, atm}$$

$$\text{for 30\% ethylene-70\% vinyl alcohol} \quad (12a)$$

If one wished to consider different permeabilities for the adhesive and polypropylene layers, five terms will be needed. The core layer is really the main barrier, as seen below. Therefore, for this illustrative example, this added complexity is questionable and two effective layers of 15 mils each, with permeability equal to that of polypropylene, is a reasonable approximation.

$$P_{total} = \frac{\ell_{total}}{\left[\sum_{i}^{N} (\ell_i/P_i)\right]} = \frac{31}{\frac{15}{150} + \frac{1}{0.017} = \frac{15}{150}}$$

$$\text{thus, } P_{total} = 0.525 \frac{cm^3 \, (STP) \, mil}{100 \, in.^2 \, day \, atm} \quad (12b)$$

Neglecting the transient period (Figure 25.8) of permeation is reasonable and provides a small safety margin, before oxygen truly achieves pseudosteady-state permeation into the package. It is assumed that the oxygen that permeates into the package is consumed immediately to produce an off-specification compound. This means that effectively, 0.21 atm of oxygen partial pressure applies in eq 1, for the cumulative allowed permeation weight of oxygen, 0.318 cm^3 (STP), into the package over a period of time t days:

$$0.318 \, cm^3 \, (STP) =$$

$$\frac{0.525 \, cm^3 \, (STP) \, mil}{100 \, in.^2 \, day \, atm} \left| \frac{0.21 \, atm}{31 \, mil} \right| 37.2 \, in^2 \left| \frac{t \, day}{} \right. \quad (12c)$$

Thus, $t = 240$ days $= 8.0$ months shelf life without seam or closure leaks.

If the package were stored in 80% RH conditions, the polypropylene would not be affected, but the ethylene-vinyl alcohol permeability would increase to 0.11 (cm^3 [STP] mil)/(100 in.^{2} day atm).[29] Refiguring the effective total permeability for the composite gives a value of 3.34 (cm^3 [STP] mil)/(100 in.^{2} day atm), and with this permeability, the shelf life decreases to 1.26 months. Clearly, the conditions of storage affect shelf life more than would be found for a metal container. Nevertheless, adequate shelf lives are still accessible under proper storage conditions, or by increasing the core barrier thickness.

Temperature Dependence

Permeability and diffusivity are exponential functions of temperature as described by the Arrhenius equation for rubbery and glassy polymers, as shown in eq 13 for permeation coefficients:[15]

$$P = P_0 \exp\left[\frac{-E_p}{RT}\right] \quad (13)$$

where E_p is the activation energy for permeation, T is the absolute temperature, and P_0 is a temperature-independent pre-exponential factor. The thermodynamic solubility coefficient can be described by a van't Hoff relationship as in eq 13, where H_s is the exothermic heat of sorption of the penetrant in the polymer. The activation energies of permeation can then be defined as the sum of the heat of sorption and the activation energy for diffusion of the penetrant:

$$S = S_0 \exp\left[\frac{-H_s}{RT}\right] \quad (14)$$

$$E_p = E_d + H_s \quad (15)$$

For the more complex barrier structures discussed in the preceding section, the temperature dependence of the various components in the structure must be used in a weighted average to obtain the effective properties for the overall structure. This averaging can be quite complex and is not a well-studied topic except for very simple cases, such as the series resistance model for laminates discussed in the earlier example.

Complementary Physical Characterizations

To assist in understanding permeation, diffusion, and sorption measurements, additional techniques can help explain causes for differences in the transport properties between various samples. As discussed, the degree of polymer chain packing heavily influences the rate of penetrant diffusion through a polymeric material as well as material selectivity to penetrants of varying sizes. When the polymeric transport properties are studied, density and X-ray diffraction are sometimes used to characterize polymeric physical properties. Although these methods are likely discussed in greater detail elsewhere in this volume, it is useful to identify and briefly discuss these measurements in the context of correlating polymer structures with their respective transport properties.

A simple density measurement can often provide an indication of the relative tightness of chain packing. In general, a more dense polymer would have less free volume to enable penetrants to permeate through the polymer. In contrast, a low-density polymer potentially has a high degree of free volume between polymer chains, resulting in a high permeability. Density measurements are commonly performed using an aqueous gradient column, but it must be verified that negligible polymer-gradient column fluid interactions exist and that the polymer sorbs a negligible amount (<0.25-0.5 wt %) of the fluid.

The fractional free volume (FFV) is another quantity that indicates the relative degree of packing between polymer chain segments. Although density measurements provide a similar analysis in certain systems, heavy atoms such as fluorine (relative to carbon, hydrogen, and nitrogen) often distort densities with respect to overall chain packing. Hence, the relative comparison of chain packing using FFV is, in many cases, considered superior to that of density. The FFV is the ratio of the specific free volume to the specific volume as estimated by the method of Lee.[84] The specific free-volume calculation, which uses the group contribution method of van Krevelen,[85] is determined by the difference between the experimentally measured specific volume and the specific occupied van der Waals volume given by Bondi.[86] Although the FFV alone can often be correlated to material permeability and permselectivity, it has been suggested that the FFV distribution also plays a key role in the gas transport properties.[87–89]

Wide angle X-ray diffraction (WAXD), which is sensitive to the electron density in a material, is an analytical method that provides information about the polymeric intersegmental packing. Crystalline materials, possessing a high degree of order, exhibit very sharp peaks indicative of very precise distances between polymer chains. In contrast, amorphous polymers exhibit broad peaks because of a lack of long-range order. Nevertheless, this technique, like the FFV estimation technique, provides a general indication of the degree of packing within the polymer matrix when intermolecular contributions dominate.[90] From this technique, the d-spacing, which measures the average distance between the center of polymer chain axes, is determined from the peak maxima. Hence, as the d-spacing and FFV increase, intersegmental packing decreases. The d-spacing is determined from Bragg's law

$$n\lambda = 2d \sin \theta \qquad (16)$$

where n is the order of reflection, λ is the X-ray wavelength, d is the d-spacing, and θ is the angle of incidence. When intramolecular contributions dominate, the WAXD spectrum is essentially independent of polymer chain packing.

Positron annihilation lifetime (PAL) spectroscopy has been used to study the free-volume size, fraction, and distribution in polycarbonate,[91] poly(aryl ether)-ether-ketone,[92] polyimides,[93] and polysulfone.[93] Often complementing gas transport studies, this technique provides an additional method to probe the segmental-scale polymer morphology. Positrons, which are positively charged electrons, are not abundant in the natural world because of their rapid reaction with electrons. One positron and one electron react to form two photons or a high-energy light ray, which results in their mutual annihilation.[94] Positrons are generated by radioisotope decay or other nuclear reactions. Lifetimes of positrons and positroniums (Ps), a bound state between a positron and an electron, are experimentally measured using a conventional fast-fast coincidence method that monitors the positron decay signal.[95]

The lifetimes measured can be correlated to lifetimes of specific species and physical properties.[91] As for hydrogen, Ps exhibit two spin states, ortho and para. The o-Ps lifetime, t3 (~1–3 ns), is the most useful for determining polymeric free-volume properties because it is the longest and hence most accurately measurable. The lifetime of the o-Ps is considered to correlate directly to the hole size, whereas the intensity measures the free-volume hole fraction and gives a measure of relative free volume and its distribution.[91]

References

1. DeLassus, P. Barrier Polymers. In *Kirk-Othmer Concise Encyclopedia of Chemical Technology*, 4th ed.; Grant, M. H., Ed.; Wiley & Sons, New York, 1999; pp 213–216.

2. Wilkinson, S. *Chem. Eng. News* **1998**, June 15, 26.

3. Lee, E. K. In *Encyclopedia of Physical Science and Technology*; Academic Press: New York, 1987; Vol. 8, pp 21–55.

4. Zolandz, R. R.; Fleming, G. K. In Gas Permeation. *Membrane Handbook*; Ho, W. S., Sirkar, K. K., Eds.; Van Nostrand Reinhold, New York, 1992; Section II.

5. Edgren, D.; Leeper, H.; Nichols, K.; Wright, J. In *Kirk-Othmer Concise Encyclopedia of Chemical Technology*, 4th ed.; Grant, M. H., Ed.; Wiley & Sons, New York, 1999; p 533.

6. Koros, W. J.; Hellums, M. W. *Fluid Phase Equilib.* **1989**, 53, 339.

7. Kamaruddin, H. D.; Koros, W. J. *J. Membr. Sci.* **1997**, 135, 147.

8. Tsai, B. C.; Wachtel, J. A. In *Barrier Polymers and Structures*; ACS Symposium Series 423; American Chemical Society: Washington, DC, 1990; Chapter 9.

9. Koros, W. J.; Hellums, M. W. In *Encyclopedia of Polymer Science*, 2nd ed.; Siegel, P., Ed.; Wiley-Interscience, New York, 1989; Suppl. Vol., p 724.

10. Barrer, R. M.; Petropoulos, J. H. *Br. J. Appl. Phys.* **1961**, 12, 691.

11. Michaels, A. S.; Vieth, W. R.; Barrie, J. A. *J. Appl. Phys.* **1963**, 34, 13.

12. Brennan, D. J.; White, J. E.; Haag, A. P.; Kram, S. L.; Mang, M. N.; Pikulin, S.; Brown, C. *Macromolecules* **1996**, 29, 3707.

13. Koros, W. J.; Moaddeb, M. In *Polymeric Materials Encyclopedia*; Salamone, J. C., Ed.; CRC Press: New York, 1996; Vol. 4, p 2697.

14. Meares, P. *J. Am. Chem. Soc.* **1954**, 76, 3415.

15. Van Amerongen, G. J. *Rubber Chem. Technol.* **1951**, 24, 1951.

16. Mitchell, J. K. *J. Med. Sci.* **1831**, 13, 36.

17. Fick, A. *Ann. Phys. Chem.* **1855**, 94, 59.

18. Graham, T. *Phil. Mag.* **1866**, 32, 401 and 503.

19. Von Wroblewski, S. *Ann. Phys. Chem.* **1879**, 8, 29.

20. Koros, W. J.; Coleman, M. R.; Walker, D. R. B. Controlled Permeability Polymer Membranes. *Annu. Rev. Mater. Sci.* **1992**, 22, 47.

21. Daynes, H. *Proc. R. Soc.* **1920**, *97A*, 286.
22. Barrer, R. M. *Diffusion in and Through Solids*; Cambridge University Press: London, 1951.
23. Felder, R. M.; Huvard, G. S. *Methods Exp. Phys.* **1980**, *16C*, 315.
24. Stern, S. A.; Gareis, P. J.; Sinclair, T. F.; Mohr, P. H. *J. Appl. Polym. Sci.* **1963**, *7*, 2035.
25. Koros, W. J.; Chern, R. T. Separation of Gaseous Mixtures Using Polymer Membranes. In *Handbook of Separation Process Technology*; Rousseau, R. W., Ed.; Wiley & Sons, New York, 1987; Chapter 20.
26. O'Brien, K. C.; Koros, W. J.; Barbari, T. A.; Sanders, E. S. A new technique for the measurement of multicomponent gas transport through polymer films. *J. Membr. Sci.* **1986**, *29*, 229.
27. Pye, D. G.; Hoehn, H. H.; Panar, M. *J. Appl. Polym. Sci.* **1976**, *20*, 1921.
28. OX-TRAN, PERMATRAN-W, and PERMATRAN-C. Product Information, MOCON, Minneapolis, MN, 1999.
29. Bissot, T. C. In *Barrier Polymers and Structures*; ACS Symposium Series 423; American Chemical Society: Washington, DC, 1990; Chapter 11.
30. Alger, M. M.; Stanley, T. J.; Day, J. In *Barrier Polymers and Structures*; ACS Symposium Series 423; American Chemical Society: Washington, DC, 1990; Chapter 10.
31. Gerlowski, L. *Barrier Polymers and Structures*; ACS Symposium Series 423; American Chemical Society: Washington, DC, 1990; Chapter 8.
32. Salame, M. *Plast. Packag.* **1988**, *1*, 28.
33. Crank, J. *The Mathematics of Diffusion*, 2nd ed.; Clarendon Press: Oxford, UK, 1975; Chapter 4.
34. Stannett, V. T. Simple Gases. In *Diffusion in Polymers*; Crank, J., Park, G. S., Eds; Academic Press: London, 1968; Chapter 2.
35. Stern, S. A. *Barrier Polymers and Structures*; ACS Symposium Series 423; American Chemical Society: Washington, DC, 1990; Chapter 2.
36. Paul, D. R.; Koros, W. J. *J. Polym. Sci., Phys. Ed.* **1976**, *14*, 675.
37. Petropoulos, J. H. *J. Polym. Sci., Part A: Polym. Chem.* **1970**, *8*, 1797.
38. Karger, J.; Ruthven, D. M. *Diffusion in Zeolites and Other Microporous Solids*; Wiley & Sons: New York, 1992; Chapter 1.
39. Crank, J. *The Mathematics of Diffusion*, 2nd ed.; Clarendon Press: Oxford, UK, 1975; Chap 9.
40. Koros, W. J.; Paul, D. R. *J. Polym. Sci., Phys. Ed.* **1978**, *16*, 2171.
41. Koros, W. J.; Paul, D. R. *J. Polym. Sci., Phys. Ed.* **1978**, *16*, 1947.
42. Koros, W. J.; Paul, D. R. *J. Polym. Sci., Phys. Ed.* **1976**, *14*, 1903.
43. Iler, L. R.; Laundon, R. C.; Koros, W. J. *J. Appl. Polym. Sci.* **1982**, *27*, 1163.
44. Connelly, R.; McCoy, N.R.; Hopfenberg, H. B.; Stewart, M. E.; Koros, W. J. *J. Appl. Polym. Sci.* **1987**, *34*, 703.
45. Zimmerman, C. M. A.; Singh, A.; Koros,W. J. *J. Polym. Sci., Polym. Phys. Ed.* **1998**, *36*, 1747.
46. Al-Hussaini, H. S. A. Penetrant permeation through fluorinated and untreated polyethylene films. M. S. Thesis, North Carolina State University, Raleigh, NC, 1983.
47. Crank, J. *The Mathematics of Diffusion*, 2nd ed.; Clarendon Press: Oxford, 1975; Chapters 4–6.
48. Berens, A. R. *J. Membr. Sci.* **1978**, *3*, 247.
49. Moylan, C. R.; Best, M. E.; Ree, M. *J. Polym. Sci., Polym. Phys. Ed.* **1991**, *29*, 87.
50. Rodahl, M.; Höök, F.; Krozer, A.; Brzezinski, P.; Kasemo, B. *Rev. Sci. Instrum.* **1995**, *66*, 3924.
51. Czanderna, A. W.; Thomas, T. M. *J. Vac. Sci. Technol.* **1987**, *A5*, 2412.
52. Best, M. E.; Moylan, C. R. *J. Appl. Polym. Sci.* **1992**, *45*, 17.
53. Hong, S. U.; Barbari, T. A.; Sloan, J. M. *J. Polym. Sci., Polym. Phys. Ed.* **1998**, *36*, 337.
54. Fieldson G. T.; Barbari, T. A. *Polymer* **1993**, *34*, 1146.
55. Fieldson G. T.; Barbari, T. A. *AIChE J.* **1995**, *41*, 795.
56. Dhingra, S. S.; Marand, E. *J. Membr. Sci.* **1998**, *141*, 45.
57. Tihminlioglu, F.; Surana, R. K.; Danner R. P.; Duda, J. L. *J. Polym. Sci., Polym. Phys. Ed.* **1997**, *35*, 1279.
58. Surana, R. K.; Danner, R. P.; Tihminlioglu, F.; Duda, J. L. *J. Polym. Sci., Polym. Phys. Ed.* **1997**, *35*, 1233.
59. Xie, L. Q. *Polymer* **1993**, *34*, 4579.
60. Pawlisch, C. A.; Macris, A.; Laurence, R. L. *Macromolecules* **1987**, *20*, 1564.
61. Meresi, G. H.; Tao, A.; Gong, X.; Wen, W. Y.; Inglefield, P. T.; Jones, A. A. *Polym. Prepr.* **1998**, *39*, 886.
62. Fleischer, G.; Fujara, F. *Macromolecules* **1992**, *25*, 4210.
63. Zupancic, I.; Lahajnar, G.; Blinc, R.; Reneker, D. H.; Peterlin, A. *J. Polym. Sci., Polym. Phys. Ed.* **1978**, *16*, 1399.
64. Waggoner, R. A.; Blum, F. D.; McElroy, J. M. *Macromolecules* **1993**, *26*, 6841.
65. Petit, J. M.; Zhu, X. S.; McDonald, P. M. *Macromolecules* **1996**, *29*, 70.
66. Wen, W.-Y.; Cain, E.; Ingelfield, P. T.; Jones, A. A. *Polym. Prepr.* **1987**, *28*, 225.
67. Dong, Z. P.; Cauley, B. J.; Bandis, A.; Mou, C. W.; Inglefield, C. E.; Jones, A. A.; Inglefield, P. T.; Wen, W. Y. *J. Polym. Sci., Polym. Phys. Ed.* **1993**, *31*, 1213.
68. Koros, W. J. In *Barrier Polymers and Structures*; ACS Symposium Series 423; American Chemical Society: Washington, DC, 1990; Chapter 1.
69. Gentilcore, J. F.; Trialo, M. A.; Woytek, A. *J. Plast. Eng.* **1978**, *34*, 40.
70. Deak, G. I.; Jackson, S. C. *Annu. Tech. Conf. Proc. Soc. Vac. Coaters* **1993**, *36*, 318.
71. Shusei, M.; Toshio, U.; Kiyoshi, I.; Toru, K.; Toshiyuki, O.; Yozo, Y.; Teizo, H.; Hideomi, K.; Yoshiharu, M. European Patent Application 550,039 A2 930707, 1993.
72. Foster, R. In *The Wiley Encyclopedia of Packaging Technology*; Bakker, M., Ed.; Wiley & Sons: New York, 1986; p 270.
73. Wachtel, J. A.; Tsai, B. C.; Farrell, C. *J. Plast. Eng.* **1985**, *41* (2), 41.
74. Blackwell, A. L. In *Plastic Film Technology*; Finlayson, K. M., Ed.; Technomic Publishing: Lancaster, PA, 1989; Vol. 1, p 41.
75. Antoon, M. K., Jr. U. S. Patent 4,879,078, 1989.
76. Sen, L.E. *J. Macromol. Sci., Chem.* **1967**, *A1* (5), 929.
77. Lan, T.; Kaviratna, P. D.; Pinnavaia, T. *J. Chem. Mater.* **1994**, *6*, 573.

78. Miller, B. *Plast. Formul. Compd.* **1997**, *May/June*, 30.

79. Sherman, L. M. *Plast. Technol.* **1999**, *June*, 1–5.

80. Subramanian, P. M. *Polym. Engr. Sci.* **1985**, *25*, 483.

81. Subramanian, P. M. U. S. Patent 4,444,817, 1984.

82. Pauly, S. In *Polymer Handbook*, 3rd ed.; Brandrup, J., Immergut, E. H., Eds.; Wiley & Sons: New York, 1989; Chapter VI, pp. 545–567.

83. Bixler, H. J.; Sweeting O. J. In *The Science and Technology of Polymer Films*; Sweeting, O. J., Ed.; Wiley & Sons: New York, 1971; Vol. II, p 2.

84. Lee, W. M. *Polym. Eng. Sci.* **1980**, *20*, 65.

85. van Krevelen, D. W. *Properties of Polymers: Their Structure and Correlation with Chemical Structure*; Elsevier: New York, 1976.

86. Bondi, A. A. *Physical Properties of Molecular Crystals, Liquids, and Glasses*; Wiley & Sons: New York, 1968.

87. Koros, W. J.; Walker, D. R. B. *Polym. J.* **1991**, *23*, 481.

88. Park, J. Y.; Paul, D. R. *J. Membr. Sci.* **1997**, *125*, 23.

89. Koros, W. J.; Fleming, G. K. *J. Membr. Sci.* **1993**, *83*, 1.

90. Koros, W. J.; Fleming, G. K.; Jordan, S. M.; Kim T. H.; Hoehn, H. H. *Prog. Polym. Sci.* **1988**, *13*, 339.

91. Hong, X.; Jean, Y. C.; Yang, H.; Jordan, S. S.; Koros, W. J. *Macromol.* **1996**, *29*, 7859.

92. Nakanishi, H.; Jean, Y. C. *J. Polym. Sci., Polym. Phys. Ed.* **1989**, *27*, 1419.

93. Tanaka, K.; Okamoto, K.; Kita, H.; Ito, Y. *Polym. J.* **1993**, *25*, 577.

94. Latimer, W. M.; Hildebrand, J. H. *Reference Book of Inorganic Chemistry*, 3rd ed.; Macmillan Publishing: New York, 1964.

95. Schrader, D. M.; Jean, Y. C. Experimental Techniques in Positron and Positronium Chemistry. In *Positron and Positronium Chemistry*; Jean, Y. C., Schrader, D. M., Eds.; Elsevier: New York, 1988.

26

Flammability and Fire Performance of Polymers

MARCELO M. HIRSCHLER

Introduction

Organic polymers are combustible: when sufficient heat is supplied to an organic polymer it will thermally decompose, and its thermal decomposition products will burn. There are, however, vast differences in both the rates and the mechanisms of polymer breakdown and in aspects of the subsequent combustion. This chapter analyzes the similarities and differences.

Application of excessive heat to polymeric materials results in both physical and chemical changes, often with undesirable changes to the material properties. A clear distinction must be made between thermal decomposition and thermal degradation. The American Society for Testing and Materials (ASTM) generated definitions that provide helpful guidelines. Thermal decomposition is defined as "a process of extensive chemical species change caused by heat". Thermal degradation, on the other hand, is defined as "a process whereby the action of heat or elevated temperature on a material, product, or assembly causes a loss of physical, mechanical, or electrical properties."[1] In terms of fire, the important change is thermal decomposition, whereby the chemical decomposition of a solid material generates gaseous fuel vapors, which can burn above the solid material.

For a combustion process to be self-sustaining, burning gases resulting from thermal decomposition must feed back sufficient heat to the material to continue the production of gaseous fuel vapors or volatiles. As such, the process can be a continuous feedback loop if the material continues burning. In that case, heat transferred to the polymer causes the generation of flammable volatiles (thermal decomposition products); these decomposition products then react with the oxygen in the air (or, more correctly, with OH· or O$_2$H· free radicals, in an exothermic chain reaction) surrounding the polymer condensed phase to generate heat, and a part of this heat is transferred back to the polymer to continue the combustion process. Flames (usually accompanied by light and soot, and often by sound) are a common part of this process.

$$\text{Polymer} + \text{heat} \Rightarrow \text{thermal decomposition products}$$
$$\text{Decomposition products} + \text{oxygenated radicals}$$
$$\Rightarrow \text{combustion products} + \text{heat}$$

The chemical thermal decomposition processes are responsible for the generation of flammable volatiles, whereas physical changes, such as melting and charring, can markedly alter the decomposition and burning characteristics of a material, but will not be addressed much further.

The formation of volatile products from polymers is generally much more complex than that from flammable liquids, where gasification usually is a simple evaporation process. For polymeric materials, the condensed phase material itself is essentially involatile, and chemical breakdown (thermal decomposition) into smaller molecules that can vaporize is essential. In most cases, a solid polymer breaks down into a variety of smaller molecular fragments made up of a number of different chemical species, so that each fragment has a different equilibrium vapor pressure. The lighter fragments will vaporize immediately upon their creation, whereas other heavier molecules will remain in the condensed phase (solid or liquid) for some time. The molecules that remain in the condensed phase can undergo additional thermal decomposition to lighter fragments, which are, in turn, more easily vaporized. A few polymers break down completely so that virtually no solid residue remains. More likely, however, is that materials leave behind solid residues, which can be carbonaceous (char), inorganic (originating from heteroatoms contained in the original polymer, either within the structure or as a result of the incorporation of additives; for example, as fillers) or a combination of both. Charring materials, for which wood is the typical example, leave large fractions of the original carbon content as carbonaceous residue, often as a porous char. (It must be clarified, however, that many plastic materials can also be charring materials.) When thermal decomposition of deeper layers of such a material continues, the volatiles produced must pass through the char above them to reach the surface. During this travel, the hot char may cause secondary reactions to occur in the volatiles. Carbonaceous chars can be intumescent layers, when appropriately formed, which slow further thermal decomposition considerably. Inorganic residues, on the other hand, can form glassy layers that may then become impenetrable to volatiles and protect the underlying layers from any additional thermal breakdown. Unless such inorganic barriers form, purely carbonaceous chars can always be burned off by oxidation at higher temperatures.

This brief description of thermal decomposition shows that the processes involved are varied and complex. The rates, mechanisms, and product composition of thermal decomposition depend both on the physical properties of the original material and on its chemical composition.

Classification of Polymers

Polymers can be classified in a variety of ways:

1. By their source (or origin), which includes natural and synthetic (sometimes including also a third category of seminatural or synthetic modifications of natural polymers).
2. By their physical properties, in particular the elastic modulus and the degree of elongation. According to this criterion, polymers can be classified into elastomers, plastics, and fibers. Plastics can be further subdivided into thermoplastics (which deform reversibly at elevated temperature, above their melting point but below their chemical decomposition point) and thermosets (which change irreversibly when heated).
3. By their polymerization mechanism (synthetic polymers only). Synthetic polymers are prepared mainly by either addition polymerization or condensation polymerization, with the occasional additional category of poly-insertion.
4. By their chemical composition.

The most useful classification method for polymers is based on their chemical composition. This gives a clear indication about polymer reactivity, including the mechanism of thermal decomposition and the fire performance. In this way, organic polymers can be subdivided among those containing:

1. Carbon-carbon chains, with no heteroatoms in the chain (principally polyolefins, polydienes, and aromatic hydrocarbon polymers, such as styrenics). The main polyolefins are polyethylene (PE) and polypropylene (PP), which are two of the three most widely used synthetic polymers. Polydienes contain one double bond per repeating unit and are generally elasto-

meric, and are often copolymers. The most important aromatic hydrocarbon polymers are based on polystyrene (PS), but numerous styrenic copolymers and blends have found wide use.

2. Oxygen-containing polymers. These are principally cellulosics, polyacrylics, and polyesters, some vinyl polymers, such as poly(vinyl alcohol) or poly(vinyl acetate) (PVA), phenolic resins, polyethers, epoxy resins, and polyacetals.
3. Nitrogen-containing materials. These include polyamides (both aliphatic nylons and aromatic polyamides), polyurethanes, and polyacrylonitrile.
4. Chlorine-containing polymers. Such polymers usually are carbon-carbon chains with chlorine atoms substituting hydrogen atoms in the polymer backbone. They are typified by poly(vinyl chloride) (PVC, the third of the three most widely used synthetic polymers). PVC is unique in that it is used both as a rigid material (unplasticized) and as a flexible material (plasticized). Flexibility is achieved by incorporating plasticizers or flexibilizers together with the polymer (so that flexible PVC often contains less than 50% actual PVC). Through the additional chlorination of PVC, another member of the family of vinyl materials is made: chlorinated poly(vinyl chloride) (CPVC) with different physical and fire properties from PVC. Two other chlorinated materials of commercial interest are polychloroprene (a polydiene) and poly(vinylidene chloride).
5. Fluorine-containing polymers. Such polymers usually, just like chlorine-containing polymers, are carbon-carbon chains with fluorine atoms substituting hydrogen atoms in the polymer backbone. They are characterized by high thermal and chemical inertness and low coefficient of friction. The most important individual material in the family is poly(tetrafluoroethylene) (PTFE), while others include poly(vinylidene fluoride) (PVDF), poly(vinyl fluoride) (PVF), and various fluorinated ethylene polymers (FEP).

There is a natural tendency to assume that all materials with the same generic name, such as poly(methyl methacrylate) (PMMA), have the same properties. Because these are commercial products, the preparation methods are dictated by the required physical and chemical properties of the material for normal use. Additives, both intentional and inadvertent, may be present, and other characteristics may vary, too, including the method of polymerization, the method of incorporation of the additives, and the molecular weight of the polymer chains.

Two practical terms from the plastics industry are key for this chapter: resin and compound.

A resin is a commercially manufactured polymer, without additives, but which may still contain residues from the polymerization process.

A compound is a material that is used to fabricate a product commercially, and which contains one or more resins and additives. Usually, one resin forms the core, and all other components are described in parts per hundred resin (phr), on the basis of the core resin. The additives contained in a compound include copolymers, fillers, flame retardants, smoke suppressants, synergists, plasticizers, colorants, and processing aids; additives often have more than one function.

The distinction between the properties of different types of materials with the same generic designation is particularly important for polymeric compounds that contain a number of additives. In some polyolefins, the fraction of polymer (or resin) is less than half of the total mass of the compound because of the presence of large amounts of fillers; therefore, the material will have improved fire performance. In some compounds derived from PVC, the flexibility is introduced by means of plasticizers. If the plasticizer is fire retarded, the fire performance will be much better than if it simply adds easily combustible fuel.

Thermal Decomposition of Polymers

Techniques for Studying Thermal Decomposition of Polymers

Thermogravimetric Analysis

By far, the most commonly used thermal decomposition test is thermogravimetric analysis (TGA). In TGA experiments, the sample (in milligram size) is heated quickly to the desired temperature and the sample weight is monitored during the course of thermal decomposition. The most common process is to subject the sample to a linearly increasing temperature at a predetermined rate of temperature rise, which is critical because the thermogram (plot of weight vs temperature) depends on the heating rate chosen (Figure 26.1). Traditional equipment rarely exceeds heating rates

of 0.5 K/s, but modifications can be made to obtain rates of up to 10 K/s. The rate of thermal decomposition is a function of the temperature, as well as of the amount and nature of the decomposition process that has preceded it. A more recent associated technique worth mentioning is high-resolution TGA.

The relevance of thermogravimetric studies to fire performance is questionable, but analytical thermogravimetric studies do give important information on the decomposition process, even though caution must be exercised in their direct application to fire behavior.

Differential TGA

Differential thermogravimetry (DTG) is the same as TGA, except the mass loss versus time output is differentiated automatically to give the mass loss rate versus time. Often, both the mass loss and the mass loss rate versus time are produced automatically. This is important because the rate of thermal decomposition is proportional to the volatilization or mass loss rate, and thus DTG helps in mechanistic studies. For example, it is an excellent indicator of the temperatures at which the various stages of thermal decomposition take place and the order in which they occur.[2] The effects of a variety of additives on polymer breakdown (and fire performance) can be studied by DTG. DTG shows whether some additives are effective flame retardants and others are not: the amount of overlap between the thermal decomposition stages of polymer and additives is an indication of the effectiveness of the additive.[3,4]

Differential Thermal Analysis

In addition to the rate of decomposition, it is important to determine the heat of reaction of the decomposition process. Usually, heat must be supplied to the sample to get it to a temperature where significant thermal decomposition will occur. However, once that temperature is reached, thermal decomposition may either generate or use additional heat

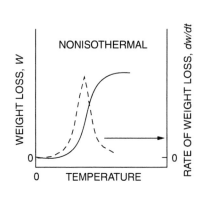
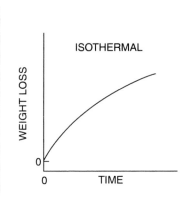

Figure 26.1 Typical thermogravimetric curves.

(i.e., it is exothermic or endothermic). In differential thermal analysis (DTA), a sample and a reference inert material with approximately the same heat capacity are both subjected to the same linear temperature program. The sample and reference material temperatures are measured and compared. If the thermal decomposition of the sample is endothermic, the temperature of the sample will lag behind the reference material temperature; if the decomposition is exothermic, the temperature of the sample will exceed the reference material temperature. Very often, the sample is held in a crucible, and an empty crucible is used as a reference (see Figure 26.2 for an example of a DTA pair of crucibles, and Figure 26.3 for an example of a single crucible). Such a test can be quite difficult to calibrate to yield quantitative heats of reaction.

Differential Scanning Calorimetry

Another method, which more easily than DTA yields quantitative results, is differential scanning calorimetry (DSC). In DSC, the sample and a reference material are kept at the same temperature during the linear temperature program, and the heat of reaction is measured as the difference in heat input required by the sample and the reference material. The system is calibrated using standard materials, such as melting salts, with well-defined melting temperatures and heats of fusion. However, because DSC experiments are normally conducted with a sample in a sealed sample holder, this technique is usually unsuitable for thermal decomposition: It is excellent for physical changes, but not for chemical processes.

Simultaneous Thermal Analysis

In view of the considerable importance of the exact process of thermal decomposition, it is advantageous to carry out simultaneously the measurements of TGA, DTG, and DTA. This can be achieved by using a simultaneous thermal analyzer (STA), which uses a dual sample and reference material system. In the majority of cases, polymeric materials are best represented by a reference material that is simply air; that is, an empty crucible.[5,6] STA equipment is often connected to a Fourier transform infrared spectrometer (FTIR) for a complete chemical identification and analysis of the gases evolved at each stage, making the technique even more powerful. Interestingly, some commercial STA instruments are, in fact, based on DSC rather than DTA techniques for obtaining the heat input. Figure 26.4 shows the schematic diagram for the balance of an early STA apparatus.

Thermal Volatilization Analysis

Another method for determining the rate of mass loss is thermal volatilization analysis (TVA).[7] In this method, a

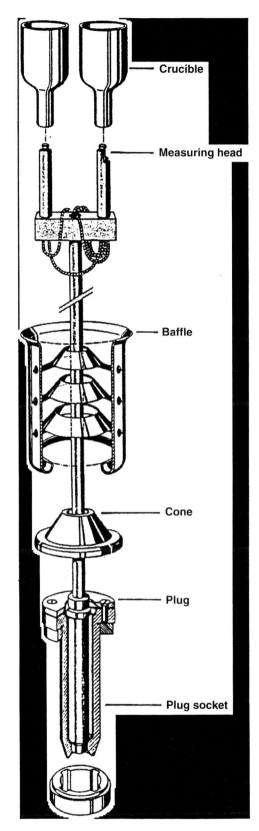

Figure 26.2 Differential thermal analyzer crucible holder (Mettler simultaneous thermoanalyzer II).

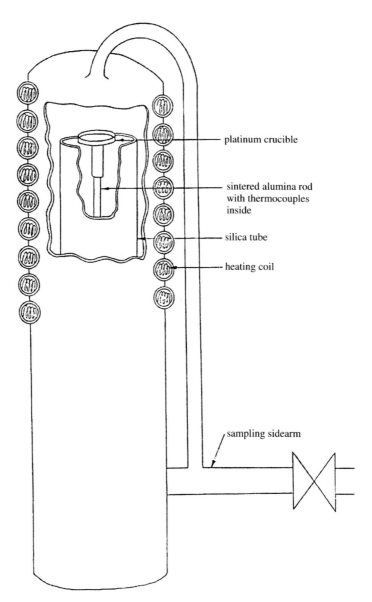

platinum crucible

sintered alumina rod
with thermocouples
inside

silica tube

heating coil

sampling sidearm

Figure 26.3 Differential thermal analyzer furnace with platinum
crucible (Mettler simultaneous thermoanalyzer II).

sample is heated in a vacuum system (0.001 Pa) equipped with a liquid nitrogen trap (77 K) between the sample and the vacuum pump. Any volatiles produced will increase the pressure in the system until they reach the liquid nitrogen and condense out. Thus, the pressure is proportional to the mass volatilization rate, and a pressure transducer, rather than a sample microbalance, is used to measure the decomposition rate.

Pyrolysis Gas Chromatography

The experimental methods discussed above have addressed the kinetics and thermodynamics of the thermal decomposi-

tion process. The decomposition products will affect smoke and toxicity. Chemical analysis of the volatiles exiting from any of the above instruments is possible. However, it is often convenient to design a special decomposition apparatus to attach directly to an existing analytical instrument. Given the vast numbers of different products that can result from the decomposition in a single experiment, separation of the products is often required. Hence, pyrolysis can be carried out in the injector of a gas chromatograph (PGC). Column packing, column temperature programming, carrier gas flow rate, sample size, and detector type all affect the response of the instrument, and must be adjusted for optimum discrimination of the decomposition products.

Gas Chromatography/Mass Spectrometry

Once the gases have been separated, any number of analytical techniques can be used for identification. The most powerful technique used to be mass spectroscopy (MS), in which the chemical species is ionized and the atomic mass of the ion is determined by the deflection of the ion in a magnetic field. Generally, the ionization process will also result in the fragmentation of the molecule, so the fingerprint of the range of fragments and their masses must be interpreted to determine the identity of the original molecule. In recent years, there has been a tendency to replace gas chromatography/mass spectrometry (GC/MS) by FTIR, but there is some indication that GC/MS is making a comeback.

Thermomechanical Analysis

Useful physical data can be obtained by thermomechanical analysis (TMA). This is a general name for the determination of a physical or mechanical property of a material subjected to high temperatures. Compressive and tensile strength, softening, shrinking, thermal expansion, glass transition, and melting can be studied using TMA.

Product Analysis

Many of these tests can be performed under vacuum, in inert atmospheres, and in oxidizing atmospheres. Each has come theoretical importance, but it is the tests conducted in air that are most relevant to fires, because the decomposition mechanism and the products of decomposition may be affected by the atmosphere used. Under vacuum, secondary reactions are minimized so the initial decomposition product may reach an analytical instrument intact. When some materials burn, the flow of combustible volatiles from the surface and the flame above the surface often effectively excludes oxygen at the material's surface (making oxidative processes unimportant). For other materials, oxidative processes are critical, and changing from nitrogen to air in thermal decomposition experiments will help to determine this. Thermal decomposition is usually monitored by sample mass loss. The partially decomposed solid residue is generally only analyzed when charring materials, such as cel-

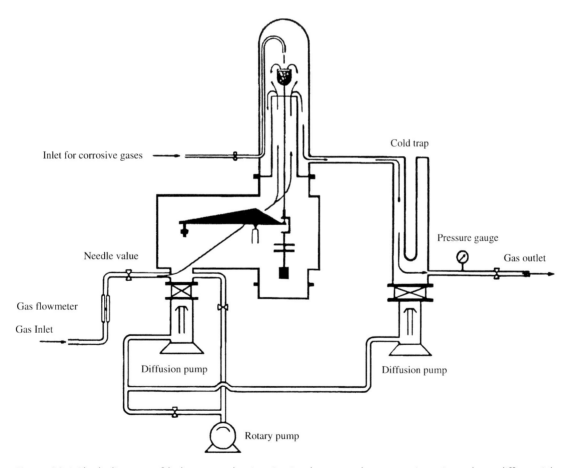

Figure 26.4 Block diagram of balance mechanism in simultaneous thermogravimetric analyzer-differential thermal analyzer (Mettler simultaneous thermoanalyzer II).

lulosics or thermosets, are tested. In such cases, the analysis usually involves a search for heteroatoms (such as phosphorus, halogens, or sulfur) arising from additives. Other techniques include the search for free radicals in this residual polymer by electron spin resonance (ESR) or electron paramagnetic resonance (EPR), the detection of bonds absent in the original polymer by infrared spectroscopy (IR) (because of double-bond formation via chain-stripping or the incorporation of oxygen into the polymer), and solid-phase structure analysis by solid-state nuclear magnetic resonance (NMR), often used on partially decomposed (and thus usually insoluble) polymers.

Physical Issues Involved in Thermal Decomposition

Many materials (cellulosics and thermosetting, as well as some thermoplastics) produce carbonaceous chars when they decompose thermally. Physical structure of char markedly affects the continued thermal decomposition process. The characteristics of the char often dictate the thermal decomposition rate of the rest of the polymer. Among the most important characteristics of char are density, continuity, coherence, adherence, oxidation-resistance, thermal insulation properties, and permeability. Low-density–high-porosity chars tend to be good thermal insulators; they can significantly inhibit heat flow from the gaseous combustion zone back to the condensed phase behind it, and thus slow thermal decomposition. This is an excellent means of decreasing polymer flammability (through additive or reactive flame retardants). As the char layer thickens, the heat flux to the virgin material decreases, and the decomposition rate is reduced. The char itself can glow when exposed to air. However, it is unlikely that both glowing combustion of the char and significant gas-phase combustion can occur simultaneously in the same zone above the surface because the flow of volatiles through the char will tend to exclude air from direct contact with the char. Therefore, in general, solid-phase char combustion tends to occur after volatilization has largely ended.

The volatile thermal decomposition products are dictated by the chemical and physical properties of both the polymer and the products of decomposition. The size of the molecular fragments must be small enough to be volatile at the decomposition temperature. This sets a practical upper limit on the molecular weight of the volatiles. If larger chain fragments are created, they tend to stay in the condensed phase and be further decomposed to smaller fragments, which can vaporize. Both physical and chemical changes can occur when a polymer is volatilized. The type of changes can include simple phase transformations, sublimation, vaporization, melting of thermoplastics, charring, and a variety of chemical mechanisms. In some cases, polymers can decompose by several mechanisms at the same time (typical examples include cellulosics). These varied thermal degra-

dation-decomposition mechanisms have clear effects on fire behavior.

Generic Chemical Breakdown Mechanisms

General

Thermal decomposition of polymers may proceed oxidatively or simply by the action of heat. In many polymers, thermal decomposition is accelerated by oxidants (such as air or oxygen); in this case, minimum decomposition temperatures are lower in the presence of an oxidant. This prediction of thermal decomposition rates as the prediction of the concentration of oxygen at the polymer surface during thermal decomposition or combustion is quite difficult. Despite its importance to fire, there have been many fewer studies of thermal decomposition processes in oxygen or air than in inert atmospheres.

It is worthwhile highlighting, however, that detailed measurements of oxygen concentrations and of effects of oxidants have been made by Stuetz et al.[8] in the 1970s and more recently by Kashiwagi et al.,[9–13] Brauman,[14] and Gijsman et al.[15] Stuetz found oxygen below the surface of PP, at distances at least 10 mm from the surface. Moreover, for both PE and PP, the oxygen present determined the thermal decomposition rates and mechanisms. Brauman[14] also suggested that the thermal decomposition of PP is affected by the presence of oxygen (a fact confirmed more recently by Gijsman,[15] whereas PMMA thermal decomposition is not. Kashiwagi[9–13] found that thermal and oxidative decomposition of thermoplastics are affected by several factors, including molecular weight, prior thermal damage, weak linkages, and primary radicals. Of particular interest is the fact that the effect of oxygen (or air) on thermal decomposition depends on the polymerization mechanism: free-radical polymerization leads to a neutralization of the effect of oxygen. Oxygen affects both reaction rate and kinetic order of reaction during the thermal decomposition of PVF.[16] Kashiwagi's work led to the development of models for the kinetics of general random-chain scission thermal decomposition[17] as well as for the thermal decomposition of cellulosics[18] and thermoplastics.[19]

A number of general classes of chemical mechanisms are important in the thermal decomposition of polymers: (1) random-chain scission, in which chain scissions occur at random locations in the polymer chain; (2) end-chain scission, in which individual monomer units are successively removed at the chain end; (3) chain stripping, in which atoms or groups not part of the polymer chain (or backbone) are cleaved; and (4) cross-linking, in which bonds are created between polymer chains. These are discussed in some detail in later sections. These reactions can be subdivided into those involving atoms in the main polymer chain (the chain-scission mechanisms) and those involving principally side chains or groups (chain stripping) or both (cross-linking). Although the decomposition of some polymers can be explained by one of these general mechanisms, others involve combinations of these four general mechanisms. Nonetheless, these general classes provide a conceptual framework useful for understanding and classifying polymer decomposition behavior.

Random- and End-Initiated Chain Scission

The most common thermal decomposition reaction mechanism among thermoplastic polymers involves bond breaking in the main polymer chain. These chain scissions may occur at random locations in the chain (random-initiated chain scission or random-chain scission) or at the chain end (end-initiated chain scission or end-chain scission). End-chain scissions result in the production of monomer, and the process is often known as unzipping. Random-chain scissions generally result in the generation of both monomers and oligomers as well as a variety of other chemical species. They type and distribution of volatile products depend on the relative volatility of the resulting molecules.

Chain-scission decomposition is typical for vinyl polymer decomposition. The process is a multistep radical reaction with all of the general features of such reaction mechanisms: initiation, propagation, branching, and termination steps. Initiation reactions are of two basic types: random-chain scission and end-chain scission. Both, of course, result in the production of free radicals. The random scission, as the name suggests, involves the breaking of a main chain bond at a seemingly random location; all such main chain bonds are nearly equal in strength. End-chain initiation involves the breaking off of a small unit or group at the end of the chain. This may be a monomer unit or some smaller substituent. These two types of initiation reactions may be presented by the following generalized reactions:

Random-chain scission

$$P_n \Rightarrow R_r\cdot + R_{n-r}\cdot$$

End-chain scission

$$P_n \Rightarrow R_n\cdot + R_E\cdot$$

where P_n is a polymer containing n monomer units and $R_r\cdot$ is a radical containing r monomer units. $R_E\cdot$ refers to an end group radical. The symbol $\cdot$ indicates an unpaired electron, and, hence, a radical site.

Propagation reactions in polymer decomposition are often called depropagation reactions, rather than decomposition. There are several types of reactions in this class:

Intramolecular H transfer, random-chain scission

$$R_n\cdot \Rightarrow R_{n-m}\cdot + P_m$$

Intermolecular H transfer

$$P_m + R_n\cdot \Rightarrow P_{m-j} + P_n + R_j\cdot$$

Unzipping, depropagation, depolymerization

$$R_n\cdot \Rightarrow R_{n-1}\cdot + P_1$$

The first of these reactions involves the transfer of a hydrogen atom within a single polymer chain; that is, intramolecular hydrogen atom transfer. The m value usually lies between one and four, because polymer molecules are often oriented such that the location of the nearest available H within the chain is one to four monomer units away from the radical site. The m value need not be a constant for a specific polymer because the closest available hydrogen atom in the chain may vary as a result of conformational variations. Decomposition mechanisms based on this reaction are sometimes known as random-chain scission mechanisms. The second reaction involves the transfer of a hydrogen atom between polymer chains; that is, intermolecular hydrogen atom transfer. The original radical, $R_n\cdot$, abstracts a hydrogen from the polymer, P_m. Because this makes P_m a radical with the radical site more often than not within the chain itself (i.e., not a terminal radical site), the newly formed radical breaks up into an unsaturated polymer, P_{m-j}, and a radical, $R_j\cdot$. In the final reaction, no hydrogen transfer occurs. It is essentially the reverse of the polymerization step and, hence, is called unzipping, depropagation, or depolymerization. Whether the decomposition involves principally hydrogen transfer reactions or unzipping can be determined by examining the structure of the polymer, at least for polymers with only carbon in the main chain. If hydrogen transfer is impeded, then the unzipping reaction is likely to occur.

Vinyl polymers, strictly speaking, are functionally derived from a vinyl repeating unit:

$$-[CH_2-CH_2]_n-$$

where n is the number of repeating monomers. In this structure, each of the hydrogen atoms can be substituted, leading to a repeating unit of the following form:

$$-[\overset{\displaystyle W}{\underset{\displaystyle X}{C}}-\overset{\displaystyle Y}{\underset{\displaystyle Z}{C}}]_n-$$

where W, X, Y, and Z are substituent groups, which can be hydrogen atoms, halogen atoms, methyl groups, or larger groups. (Note that X indicates any substituent group and not necessarily a halogen atom.) Consider that the C—C bond connecting monomer units is broken and that a radical site results from the scission shown as

$$-[\overset{\displaystyle W}{\underset{\displaystyle X}{C}}-\overset{\displaystyle Y}{\underset{\displaystyle Z}{C}}]_j-\overset{\displaystyle W}{\underset{\displaystyle X}{C}}-\overset{\displaystyle Y}{\underset{\displaystyle Z}{C}}\cdot$$

where the symbol · again indicates an unpaired electron. Table 26.1[20] shows the monomer yields following thermal decomposition of vinyl addition polymers. When neither W nor X are hydrogen atoms but Y and Z are, yields of monomer are very high. However, when the substituent in X is replaced by a hydrogen atom, the monomer yield decreases significantly [compare PS with poly(α-methylstyrene)]. If X is a deuterium atom, the yield of monomer still decreases but less markedly [compare poly(α-deuterostyrene)]. This emphasizes the key role played by α-hydrogen atoms in the decomposition mechanism [compare poly(β-deuterostyrene) with PS].

The first step in any chain scission mechanism is the homolytic breakdown of a C—C bond, followed by intramolec-

ular or intermolecular transfer reactions. Larger substituents at the X and W positions sterically hinder the radical in its intermolecular attack on a hydrogen atom and make the radical more likely to intramolecularly eliminate a monomer unit. On the other hand, if there is a hydrogen atom at the X or W positions (i.e., hydrogen atom on the α-carbon) the probability of monomer formation decreases considerably. If the availability of α-hydrogen atoms were the only important factor governing the decomposition behavior of these polymers, monomer yields would be expected to be either very small, as with PE and PP, or very large as with PMMA, but not moderately large as with PS and polyisobutene. Another significant factor is the reactivity of the propagating radical as a chain-transfer species for the reac-

Table 26.1 Monomer yield following thermal decomposition of vinylic addition polymers[a]

Polymer	W	X	Y	Z	Mon. yld. (wt %)	Decomposition mechanism
Poly(methyl methacrylate)	H	H	CH$_3$	CO$_2$CH$_3$	91–98	E
Polymethacrylonitrile	H	H	CH$_3$	CN	90	E
Poly(α-methyl styrene)	H	H	CH$_3$	C$_6$CH$_5$	95	E
Polyoxymethylene[b]	—	—	—	—	100	E
Polytetrafluoroethylene	F	F	F	F	95	E
Poly(methyl atropate)	H	H	C$_6$H$_5$	CO$_2$CH$_3$	>99	E
Poly(p-bromo styrene)[c]	H	H	H	C$_6$H$_4$BR	91–93	E
Poly(p-chloro styrene)[c]	H	H	H	C$_6$H$_4$Cl	82–94	E
Poly(p-methoxy styrene)[c]	H	H	H	C$_7$H$_7$O	84–97	E
Poly(p-methylstyrene)[c]	H	H	H	C$_7$H$_7$	82–94	E
Poly(α-deuterostyrene)	H	H	D	C$_6$H$_5$	70	E
Poly(α, β, β-trifluorostyrene)	F	F	F	C$_6$H$_5$	44	E/R
Polystyrene	H	H	H	C$_6$H$_5$	42–45	E/R
Poly(m-methyl styrene)	H	H	H	C$_7$H$_8$	44	E/R
Poly(β-deutero styrene)	H	D	H	C$_6$H$_5$	42	E/R
Poly(β-methyl styrene)	H	CH$_3$	H	C$_6$H$_5$		E/R
Poly(p-methoxy styrene)[d]	H	H	H	C$_7$H$_7$O	36–40	E/R
Polyisobutene	H	H	CH$_3$	CH$_3$	18–25	E/R
Poly(chlorotrifluoroethylene)	F	F	Cl	F	18	E/S
Poly(ethylene oxide)[b]	—	—	—	—	4	R/E
Poly(propylene oxide)[b]	—	—	—	—	4	R/E
Poly(4-methyl pent-1-ene)	H	H	H	C$_4$H$_9$	2	R/E
Polyethylene	H	H	H	H	0.03	R
Polypropylene	H	H	H	CH$_3$	0.17	R
Poly(methyl acrylate)	H	H	H	CO$_2$H$_3$	0.7	R
Poly(trifluoroethylene)	F	F	H	F	—	R
Polybutadiene[b]	—	—	—	—	1	R
Polyisoprene[b]	—	—	—	—	0–0.07	S
Poly(vinyl chloride)	H	H	H	Cl	—	S
Poly(vinylidene chloride)	H	H	Cl	Cl	—	S
Poly(vinylidene fluoride)	H	H	F	F	—	S
Poly(vinyl fluoride)	H	H	H	F	—	S
Poly(vinyl alcohol)	H	H	H	OH	—	S
Polyacrylonitrile	H	H	H	CN	5	C

Key: monomer yield (Mon. yld.); random-chain scission (R); end-chain scission (E); chain stripping (C); cross-linking (C).

a. Data from reference 20.

b. Not of general formula [CWX-CYZ]$_n$.

c. Cationic polymerization.

d. Free-radical polymerization reaction occurs immediately after an initiation reaction; no unzipping.

tion in question. If the polymer radical has a very low transfer rate constant, random-chain scission will be less likely than end-chain scission and the yield of monomer will be correspondingly higher. This occurs with the polystyryl radical where the benzene ring clearly has a stabilizing effect. The stability of this radical may be significantly affected by substituents in the ring, the effect depending both on the nature of the substituent and the polymerization mechanism. The introduction of an alkyl substituent in the meta position, followed by uncatalyzed thermal polymerization, has no effect on the ratio of the rates of depolymerization and of chain transfer, and consequently on the monomer yield [e.g., poly(m-methylstyrene)]. The same substituent in the para position (especially with a cationically produced polymer) will, however, significantly increase the activation energy of the transfer reaction so that hardly any transfer occurs; the polymer thus becomes more stable and the predominant product formed is monomer, as a result of end-chain scission.

Table 26.1 shows examples of the monomer yields discussed here. Polymers near the top of the table have Y and Z substituents that are generally large, with a resulting high monomer yield, characteristic of unzipping reactions, whereas those near the bottom of the table, where Y and Z are small, form negligible amounts of monomer because other mechanisms dominate.

Although chain branching is of little importance in polymer decomposition, termination reactions are required in all chain mechanisms. Several termination reactions are common.

Unimolecular termination

$$R_m \cdot \Rightarrow P_m$$

Recombination

$$R_m \cdot + R_n \cdot \Rightarrow P_{m+n}$$

Disproportionation

$$R_n \cdot + R_n \cdot \Rightarrow P_m + P_n$$

The first of these reactions, strictly speaking, is not possible, because an unpaired electron (radical) must be formed. Nonetheless, there are instances where the observed termination reaction appears to be first order (at least empirically), and thus kinetically corresponds to the chemical equation shown. It is theoretically impossible to remove the radical site from a polymer radical without adding or subtracting at least one hydrogen atom, or another radical. What probably occurs is that the termination reaction, in fact, is second order, but that the other species involved is so little depleted by the termination reaction that the termination reaction appears not to be affected by the concentration of that species. This is known as a pseudo-first-order reaction. The recombination reaction is a classical termination step that is actually just the reverse of the random-chain scission initiation reaction. Finally, the disproportion-

ation reaction involves the transfer of a hydrogen from one radical to the other. The hydrogen donor forms a double bond as a result of the hydrogen loss, and the acceptor is fully saturated. If this sort of reaction occurs, no unzipping or other propagation reaction will follow, and the polymer decomposition will be fully characterized by a random-chain scission process.

In this regard, it is interesting and important to note that polymers tend to be less stable than their oligomeric counterparts. In the production and aging of polymers, there are opportunities for the production of abnormalities in the polymer chains (including presence of groups unrelated to the repeating unit, chain branching, unsaturation, and head-to-head linkages) because of the mode of synthesis and thermal, mechanical, and radiation effects during aging. Such abnormalities can become the site for initiating thermal decomposition reactions, and can then lead to far more monomer units being generated, relative to the short-chain oligomeric analog.

The molecular weight of the polymer can markedly affect not only the thermal decomposition rates, but also their mechanisms. An example of such a change might involve chain reactions occurring at the location of impurities in the chain of a polymer which, if pure, would principally be subject to end-chain initiation. Not all polymer defects degrade polymer thermal performance. In a polymer that decomposes by unzipping, a head-to-head linkage can stop the unzipping process. Thus, for an initiation that would have led to the full polymer being decomposed, only the part between the initiation site and the head-to-head link is affected. At least one additional initiation step is required to fully decompose the chain. This has been studied in detail by Kashiwagi.[9–13]

Monomer formation is primarily the result of unzipping (end-chain scission) reactions on multisubstituted vinyl structures (see Table 26.1). It is rare for any significant fraction of monomer to be generated if the structure of the polymer is nonvinylic or if there are less than two (out of four) substituted positions on the repeating unit.

Chain Stripping

Chain stripping involves the loss of small molecules by reactions that strip side-chain substituents from the main chain and form small molecules. The main reaction types involving side chains or groups are elimination reactions and cyclization reactions. In cyclization reactions, two adjacent side groups react to form a bond between them, resulting in the production of a cyclic structure. In elimination reactions, the bonds connecting side groups of the polymer chain to the chain itself are broken, with the side groups often reacting with other eliminated side groups. The products of these reactions are generally small enough to be volatile. Elimination is also important in char formation because, as the reaction scheme shows, the residue is much richer in carbon than the original polymer, as seen, for example, for PVC or for PVDC. Chain stripping in PVC,

$$-CH_2-CHCl- \Rightarrow -CH=CH- + HCl$$

leads to a hydrogenated char, whereas in PVDC,

$$-CH_2-CCl_2- \Rightarrow -C{\equiv}C- + 2HCl$$

yields a purely carbonaceous char with an almost graphitic structure. These chars will tend to continue breaking down by chain scission, but only at high temperatures, during the carbon burn-off stage.

Cross-Linking

Cross-linking involves the creation of bonds between two adjacent polymer chains, and generally occurs only after some stripping of substituents. The process is critical in char formation, because it generates a structure that is more compact, and thus less easily volatilizable. Cross-linking also increases the melting temperature and, like chain stiffening, can render a material infusible. Cross-links created in fabrication or during heating are also important in thermoplastics. The glass transition temperature can be increased in amorphous polymers by the inclusion of cross-links during fabrication. Random-chain scissions can quickly render a material unusable by affecting its physical properties unless cross-linking occurs. Such cross-linking in thermoplastics on heating may be regarded as a form of repolymerization. The temperature at which polymerization and depolymerization reactions are equally fast is known as the ceiling temperature. Clearly, above this temperature catastrophic decomposition will occur.

Although char formation is a chemical process, the significance of char formation largely is due to its physical properties. Clearly, if material is left in the solid phase as char, less flammable gas is given off during decomposition. More importantly, the remaining char can be a low-density material and is a barrier between the source of heat and the virgin polymer material. As such, the flow of heat to the virgin material is reduced as the char layer thickens, and the rate of decomposition is reduced, depending on the properties of the char.[21] If the heat source is the combustion energy of the burning volatiles, not only will the fraction of the incident heat flux flowing into the material be reduced, but the incident heat flux as a whole will be reduced as well. Unfortunately, char formation is not always an advantageous process. The solid-phase combustion of char can cause sustained smoldering combustion. Thus, by enhancing the charring tendency of a material, flaming combustion rates may be reduced, but perhaps at the expense of creating a source of smoldering combustion that would not otherwise have existed.

Charring is enhanced by many of the same methods used to increase the melting temperature. Thermosetting materials are typically highly cross-linked and/or chain-stiffened.

However, charring is not restricted to thermosetting materials. Cross-linking may occur as a part of the decomposition process, as is the case in PVC and polyacrylonitrile (PAN).

Special Thermal Decomposition Mechanisms

The physical changes that occur on heating a material are both important in their own right and also impact the course of chemical decomposition significantly. The nature of the physical changes and their impact on decomposition vary widely with material type. This section addresses the general physical changes that occur for thermoplastic (glass transition, melting) and thermosetting (charring, water desorption) materials.

Exothermic Reactions: Self-Heating

Self-heating of polymers is caused by spontaneous exothermic reactions, that is, reactions where chemical change occurs within the bulk of the polymer without being caused by external heat sources. It is a serious problem during production and storage of some materials, typically cellulosic materials.[22–25] The most commonly encountered example involves cotton cloth impregnated with linseed oil. Recently, it has been found that the problem associated with charring polymers is much broader than cellulosic polymers, and can be found with synthetic polymers,[26] typically elastomeric polymers. Examples of polymers found to cause self-heating are styrene-butadiene copolymers or blends, ethylene-propylene dienes, polyurethane elastomers, latex gloves, and tires. This is a frequent cause of fires in agricultural silos and it has also been encountered in warehouses.

Examples Applicable to Some Individual Classes of Polymers

The discussion thus far has been general, focusing on thermal decomposition without the inevitable complications in the treatment of a particular polymer. This approach may also tend to make the concepts abstract. Through the treatment of individual polymers by polymer class, this section provides an opportunity to apply general concepts to real materials. In general, the section is restricted to polymers of commercial importance.

Polyolefins Of the polyolefins, low-density polyethylene (LDPE), high-density polyethylene (HDPE), and PP are of the greatest commercial importance because of the large amount produced each year. Upon thermal decomposition, very little monomer formation is observed for any of these polymers (although polyisobutene may give appreciable levels; see Table 26.1); they form a large number of different small molecules (up to 70), mostly hydrocarbons. Thermal stability of polyolefins is strongly affected by branching, with linear PE most stable and polymers with branching less stable. The order of stability is illustrated as follows:

$$\begin{matrix} \text{H} & \text{H} & & \text{H} & \text{CH}_3 & & \text{H} & \text{R} & & \text{H} & \text{R} \\ | & | & & | & | & & | & | & & | & | \\ -\text{C}-\text{C}- & > & \text{C}-\text{C}- & > & \text{C}-\text{C}- & > & \text{C}-\text{C}- \\ | & | & & | & | & & | & | & & | & | \\ \text{H} & \text{H} & & \text{H} & \text{H} & & \text{H} & \text{R} & & \text{X} & \text{Z} \end{matrix}$$

with R being any hydrocarbon group larger than a methyl group, and X and Z as other substituents.

PE In an inert atmosphere, PE begins to cross-link at 475 K and to decompose (reductions in molecular weight) at 565 K, although extensive weight loss is not observed below 645 K. Piloted ignition of PE caused by radiative heating has been observed at a surface temperature of 640 K. The products of decomposition include a wide range of alkanes and alkenes. Branching of PE causes enhanced intramolecular hydrogen transfer and results in lower thermal stability. The molecular weight changes at low temperature without volatilization principally are due to the scission at weak links, such as where an oxygen atom is found, incorporated into the main chain as impurities. Initiation reactions at higher temperatures involve scission of tertiary carbon bonds or ordinary carbon-carbon bonds in the beta position to tertiary carbons. The major products of decomposition are propane, propene, ethane, ethene, butene, hexene-1, and butene-1. Propene is generated by intramolecular hydrogen transfer to the second carbon and by scission of the bond beta to terminal $=\text{CH}_2$ groups.

The intramolecular transfer route is most important, with molecular coiling effects contributing to its significance. A broad range of activation energies has been reported, depending on the percent conversion, the initial molecular weight, and whether the mass or the molecular weight of the polymer remaining after decomposition was being monitored. Decomposition is strongly enhanced by the presence of oxygen, with significant effects detectable at 423 K in air.

PP In PP every other carbon atom in the main chain is a tertiary carbon, and thus both tertiary C—H bonds and all C bonds are relatively prone to attack. This lowers the stability of PP, when compared to that of PE. As with PE chain scission and chain-transfer reactions are important during decomposition. By far, radical sites on a substituted carbon are more important than primary radicals. This is shown by the major products formed (i.e., pentane [24%], 2 methyl-1-pentene [15%], and 2-4-dimethyl-1-heptene [19%]), which are more easily formed from intramolecular hydrogen transfer involving tertiary radicals. Reductions in molecular weight are first observed at 500 to 520 K and volatilization becomes significant above 575 K. Piloted ignition of PP caused by radiative heating has been observed at a surface temperature of 610 K. Oxygen drastically effects both the mechanism and rate of decomposition. The decomposition temperature is reduced by about 200 K, and the

products of oxidative decomposition mainly include ketones. Unless the polymer samples are very thin (less than 0.010–0.012 in. thick), oxidative pyrolysis can be limited by diffusion of oxygen into the material. At temperatures below the melting point, PP is more resistant to oxidative pyrolysis because oxygen diffusion into the material is inhibited by the higher density and crystallinity of PP.

Styrenics

PS PS shows no appreciable weight loss below 575 K, although there is a decrease in molecular weight caused by scission of weak links. Above this temperature, the products are principally monomer, with decreasing amounts of dimer, trimer, and tetramer. There is an initial sharp decrease in molecular weight followed by slower rates of molecular weight decrease. The mechanism is thought to be dominated by end-chain initiation, depolymerization, intramolecular hydrogen transfer, and bimolecular termination. The changes in molecular weight are principally caused by intermolecular transfer reactions, whereas volatilization is dominated by intramolecular transfer reactions. Despite the lack of steric hindrance, depropagation is prevalent and is caused by the stabilizing effect of the electron delocalization associated with the aromatic side group. The addition of an alpha methyl group to form poly(α-methylstyrene) provides additional steric hindrance such that only monomer is produced during decomposition, and the thermal stability of the polymer is lessened. Free radical polymerized PS is less stable than anionic PS, and the rate of decomposition is dependent on the end group.

Most of the other styrenics tend to be copolymers of PS with acrylonitrile (SAN), acrylonitrile and butadiene (ABS), or methyl methacrylate and butadiene (MBS), and their decomposition mechanisms are hybrids between those of the individual polymers.

Polydienes and Rubbers with No Heteroatoms

Polyisoprene Synthetic rubber or polyisoprene decomposes by random-chain scission with intramolecular hydrogen transfer. This, of course, gives small yields of monomer. Other polydienes appear to decompose similarly, although the thermal stability can be considerably different. The average size of fragments collected from isoprene decomposition are 8 to 10 monomer units long. This supports the theory that random-chain scission and intermolecular transfer reactions are dominant in the decomposition mechanism. In nitrogen, decomposition begins at 475 K. At temperatures above 675 K, increases in monomer yield are attributable to the secondary reaction of volatile products to form monomer. Between 475 and 575 K, low molecular weight material is formed, and the residual material is progressively more insoluble and intractable. Preheating at between 475 and 575 K lowers the monomer yield at higher temperatures. Decomposition at less than 575 K results in a viscous liquid and, ultimately, a dry solid. The monomer is prone to dimerize to dipentene, as it cools. There seems to be little significant difference in the decomposition of natural rubber and synthetic polyisoprene.

Polybutadiene Polybutadiene is more thermally stable than polyisoprene because of the lack of branching. Decomposition at 600 K can lead to monomer yields of up to 60% with lower conversions at higher temperatures. Some cyclization occurs in the products. Decomposition in air at 525 K leads to a dark, impermeable crust, which excludes further air. Continued heating hardens the elastomer.

Polyacrylics

PMMA PMMA is a favorite material for use in fire research because it decomposes almost solely to monomer, and burns at a very steady rate. Methyl groups effectively block intramolecular H transfer as discussed in the General Chemical Mechanisms section, leading to a high monomer yield. The method of polymerization can markedly affect the temperatures at which decomposition begins. Free radical polymerized PMMA decomposes around 545 K, with initiation occurring at double bonds at chain ends. A second peak between 625 and 675 K in dynamic TGA thermograms is the result of a second initiation reaction. At these temperatures, initiation is by both end-chain and random-chain initiation processes. Anionically produced PMMA decomposes at about 625 K; the end-chain initiation step does not occur during the polymerization because of the lack of double bonds at the chain. This may explain the range of observed piloted ignition temperatures (550 to 600 K). Decomposition of PMMA is first order, with an activation energy of 120 to 200 kJ/mol, depending on the end group. The rate of decomposition is also dependent on the tacticity of the polymer and on its molecular weight. These effects can also have a profound effect on the flame spread rate. The most important use of PMMA in terms of fire safety is as a glazing material.

Poly(methyl acrylate) Poly(methyl acrylate) (PMA) decomposes by random-chain scission rather than end-chain scission, with almost no monomer formation. This results because the methyl group blocking intramolecular hydrogen transfer in PMMA is lacking in PMA. Initiation is followed by intra- and intermolecular hydrogen transfer.

PAN PAN begins to decompose exothermically between 525 and 625 K with the evolution of small amounts of ammonia and hydrogen cyanide. These products accompany cyclization reactions involving the creation of linkages between nitrogen and carbon on adjacent side groups. The gaseous products are not the result of the cyclization itself, but arise from the splitting off of side or end groups not involved in the cyclization. The ammonia is derived principally from terminal imine groups (NH), whereas HCN results from side groups that do not participate in cyclization or other polymerization reactions. When the polymer is not isotactic, the cyclization process is terminated when hydrogen is abstracted by the nitrogen atom, and then the cyclization process is reinitiated intramolecularly. Then, the nitrile (CN) groups not involved in the cyclization are ultimately removed and appear among the products as HCN. Typically, there are between zero and five chain polymerization steps between each hydrogen abstraction. At temperatures of 625 to 975 K, hydrogen is evolved as the cyclic structures carbonize. At higher temperatures, nitrogen is evolved as the char becomes nearly pure carbon. In fact, with adequate control of the process, this method is widely used to produce carbon fibers. Oxygen stabilizes PAN, probably by reacting with initiation sites for the nitrile polymerization. The products of oxidative decomposition are highly conjugated and contain ketonic groups.

Polyhalogens

PVC The most common halogen-containing polymer is PVC. Between 500 and 550 K, hydrogen chloride gas is evolved nearly quantitatively by a chain-stripping mechanism. It is very important to point out, however, that the temperature at which hydrogen chloride starts being evolved in any measurable way is heavily dependent on the stabilization package used, and all commercial materials will be stabilized. Thus, commercial PVC compounds (as defined earlier) have been shown to start evolving hydrogen chloride at temperatures in excess of 520 K and to have a dehydrochlorination stage starting at 600 K.[27] (Unstabilized PVC polymer samples used for academic research have, in some work, released hydrogen chloride at lower temperatures, but such materials are not commercially available.) Between 700 and 750 K, hydrogen is evolved during carbonization, following cyclization of the species evolved. At higher temperatures, cross-linking between chains results in a fully carbonized residue. The rate of dehydrochlorination depends on molecular weight, crystallinity, presence of oxygen, hydrogen chloride gas, and stabilizers. The presence of oxygen accelerates the dehydrochlorination process, produces main-chain scissions, and reduces cross-linking. At temperatures above 700 K, the char (resulting from dehydrochlorination and further dehydrogenation) is oxidized, leaving no residue. The rate of dehydrochlorination increases as the molecular weight decreases. Dehydrochlorination stabilizers include zinc, cadmium, lead, calcium, barium soaps, and organotin derivatives. The stability of model compounds indicates that weak links are important in decomposition.

This is one of the polymers for which the thermal decomposition process has been most exhaustively studied. There has been considerable controversy, particularly in terms of explaining the evolution of aromatics in the second decomposition stage. Some recent evidence[27] seems to indicate a simultaneous cross-linking and intramolecular decomposition of the polyene segments resulting from the dehydrochlorination, via polyene free radicals, although more recently it appears that the mechanism is ionic (e.g., see reference 28). Earlier evidence suggested Diels-Alder cyclization (which can only be intramolecular if the double bond ends up in a cis orientation[29]). Evidence for this was given by the fact that smoke formation (an inevitable consequence of the emission of aromatic hydrocarbons) was decreased by introducing cross-linking additives into the polymer.[30] Thus, it has now become clear that formation of any aromatic hydrocarbon occurs intramolecularly. The chemical mechanism for the initiation of dehydrochlorination also was reviewed a few years ago.[28,31] More recently, a series of papers was published investigating the kinetics of chain stripping, based on PVC.[32] The work by Starnes et al. was key in solving these mechanisms and the ways in which reductive coupling cross-linking can be used to decrease smoke release from PVC.[33-35] With PVC it must be remembered that both the thermal decomposition and fire performance, and the applications of rigid (unplasticized) compounds and of flexible (plasticized) compounds are very different.

CPVC An interesting derivative of PVC is CPVC, resulting from postpolymerization chlorination of PVC. It decomposes at a much higher temperature than PVC, but by the same chain-stripping mechanism. The resulting solid is a polycetylene (for example, polyacetylene), which gives off much less smoke than PVC and is also more difficult to burn.[36] The most common application of CPVC is in sprinkler pipes.

Polychloroprene Polychloroprene, which is a diene, decomposes in a manner similar to that of PVC, with initial evolution of hydrogen chloride at around 615 K and subsequent breakdown of the residual polyene. The sequences of the polyene are typically around three (trienes), much shorter than PVC. Polychloroprene melts at around 325 K. A common application of this polymer is as a flexible foam in fire-retarded upholstery, as a partial or total replacement for polyurethane foam. Other applications include oil-resistant wire and cable materials (where it is compounded and usually exhibits more mediocre fire performance) and footwear.

PTFE PTFE is a very stable polymer owing to the strength of C−F bonds and shielding by the very electronegative fluorine atoms. Decomposition starts between 750 and 800 K. The principal product of decomposition is the monomer with small amounts of hydrogen fluoride, CF_4, and hexafluoroproprene. Decomposition is initiated by random-chain scission, followed by depolymerization. Termination is by disproportionation. It is possible that the actual product of decomposition is CF_2, which immediately forms in the gas phase. The stability of the polymer can be further enhanced by promoting chain transfer reactions, which can effectively limit the zip length. Under oxidative pyrolysis conditions, no monomer is formed. Oxygen reacts with the polymeric radical, releasing carbon monoxide, carbon

dioxide, and other products. PTFE is widely used in household and electrical applications.

FEP and PVDF Other fluorinated polymers are less stable than PTFE but are still generally more stable than their unfluorinated analogs, particularly in an oxidizing atmosphere. Hydrofluorinated polymers produce hydrogen fluoride directly by chain-stripping reactions, but the source of hydrogen fluoride from perfluorinated polymers, such as PTFE, is less clear. It is related to the reaction of the decomposition products (including tetrafluoroethylene) with atmospheric humidity. Both of these materials are used extensively for wire and cable in applications where fire performance is critical, typically in plenum cables for communications.

Other Substituted Vinyl Polymers All vinyl polymers having a single substituent (replacing hydrogen) on the basic repeating unit decompose by mechanisms similar to that of PVC. Such polymers include PVA, poly(vinyl alcohol), and poly(vinyl bromide), and result in gas evolution of acetic acid, water, and hydrogen bromide, respectively. Although chain stripping of each of these polymers occurs at different temperatures, all of them aromatize by hydrogen evolution at roughly 720 K.

Polymers Containing Oxygen in the Chain (Synthetic)

Poly(ethylene terephthalate) Poly(ethylene terephthalate) (PET) decomposition is initiated by scission of an alkyl-oxygen bond. Decomposition kinetics suggest random-chain scission. Principal gaseous products are acetaldehyde, water, carbon monoxide, carbon dioxide, and compounds with acid and anhydride end groups. The decomposition is accelerated by the presence of oxygen. Recent evidence indicates that both PET and poly(butylene terephthalate) (PBT) decompose via the formation of cyclic or open-chain oligomers with olefinic or carboxylic end groups.

Polycarbonates Polycarbonates (PCs) yield substantial amounts of char if products of decomposition can be removed (the normal situation). If volatile products are not removed, cross-linking does not occur because of competition between condensation and hydrolysis. The decomposition is initiated by scission of the weak $O-CO_2$ bond, and the major volatile product is the resulting carbon dioxide (which often represents 35% of the products released); bisphenol A and phenol are other major combustion products. Decomposition is a mixture of random-chain scission and cross-linking, initiated intramolecularly. Decomposition begins at 650 to 735 K, depending on the PC in question. PCs are used for glazing applications (as a better fire-performing alternative to PMMA), for appliance housings, and as part of blends to form engineering thermoplastics.

Engineering thermoplastic blends of PC and styrenics (such as ABS) have properties intermediate between those of the forming polymers, both in terms of physical properties (and processability) and in their modes of thermal breakdown.

Phenolic Resins Phenolic resins are based on phenol formaldehyde precursors and have very high thermal stability and low flammability.[37] Thermal decomposition begins at 575 K and is initiated by the scission of the methylene-benzene ring bond.[20] At 633 K, the major products are C_3 compounds. In continued heating (725 K and higher), char (carbonization), carbon oxides, and water are formed. Above 770 K, a range of aromatic, condensable products are evolved. Above 1075 K, ring breaking yields methane and carbon oxides. In TGA experiments at 3.3 K/min, the char yield is 50% to 60%. The weight loss at 700 K is 10%. All decomposition is oxidative in nature (oxygen is provided by the polymer itself).

Polyoxymethylene Polyoxymethylene (POM) decomposition yields formaldehyde almost quantitatively. The decomposition results from end-chain initiation followed by depolymerization. The presence of oxygen in the chain prevents intramolecular hydrogen transfer quite effectively. With hydroxyl end groups, decomposition may begin at temperatures as low as 360 K, whereas with ester end groups, decomposition may be delayed to 525 K. Piloted ignition caused by radiative heating has been observed at a surface temperature of 550 K. Acetylation of the chain end group also improves stability. Upon blocking the chain ends, decomposition is by random-chain initiation, followed by depolymerization, with the zip length less than the degree of polymerization. Some chain transfer occurs. Amorphous POM decomposes faster than crystalline POM, presumably because of the lack of stabilizing intermolecular forces associated with the crystalline state (below the melting temperature). Incorporating oxyethylene in POM improves stability, presumably as a result of hydrogen transfer reactions that stop the unzipping. Oxidative pyrolysis begins at 430 K and leads to formaldehyde, carbon monoxide, carbon dioxide, hydrogen, and water vapor.

Epoxy Resins Epoxy resins are characterized by an epoxy group (where an oxygen atom bridges between two adjacent carbon atoms) and may contain aliphatic, aromatic or heterocyclic structures. They tend to be less thermally stable than phenolic resins, PC, polyphenylene sulfide, and PTFE, and have a complex decomposition mechanism (actually a combination of different mechanisms) usually yielding mainly phenolic compounds, and often involving cross-linking. Thermal decomposition starts at the glycidyl group, often followed by dehydration of secondary alcohols and formation of double bonds. The thermal stability will be a function of the repeating-unit structure, and of the level of cross-linking.[37,38] These materials have been shown to provide high service time with good physical properties. Thus, because of the extensive use of these materials for printed circuit boards and encapsulation of semiconductors, a large focus of many studies has been the improvement in their poor fire performance, by one of three methods: (1)

incorporation of additives (typically either containing phosphorus or halogens or both or containing metal hydroxides), (2) use of reactive phosphorus-containing monomers,[39] or (3) use of thermally stable charring comonomers.[40]

Nitrogen-Containing Polymers

Nylons The principal gaseous products of decomposition of aliphatic polyamides, or nylons, are carbon dioxide and water. Nylon 6 gives small amounts of various simple hydrocarbons, whereas Nylon 6-10 produces notable amounts of hexadienes and hexene. As a class, nylons do not notably decompose below 615 K. Nylon 6-6 melts between 529 and 532 K, and decomposition begins at 615 K in air and 695 K in nitrogen. At temperatures in the range 625 to 650 K, random-chain scissions lead to oligomers. The C−N bonds are the weakest in the chain, but the CO−CH$_2$ bond is also quite weak, and both are involved in decomposition. At low temperatures, most of the decomposition products are nonvolatile, although above 660 K main-chain scissions lead to monomer and some dimer and trimer production. Nylon 6-6 is less stable than nylon 6-10 because of the ring closure tendency of the adipic acid component. At 675 K, if products are removed, gelation and discoloration begin.

Aromatic polyamides have good thermal stability and some can be stable in air to 725 K. The major gaseous decomposition products at low temperatures are water and carbon oxides. At higher temperatures, CO, benzene, hydrogen cyanide (HCN), toluene, and benzonitrile are produced. Above 825 K, hydrogen and ammonia are formed, leaving a highly cross-linked residue.

Wool On decomposition of wool, approximately 30% is left as a residue. The first step in decomposition is the loss of water. Around 435 K, some cross-linking of amino acids occurs. Between 485 and 565 K, the disulfide bond in the amino acid cystine is cleaved with carbon disulfide and carbon dioxide being evolved. Pyrolysis at higher temperatures (873 to 1198 K) yields large amounts of HCN, benzene, toluene, and carbon oxides.

Polyurethanes As a class, polyurethanes do not break down below 475 K, and air tends to slow decomposition. The production of HCN and CO increases with the pyrolysis temperature. Other toxic products formed include nitrogen oxides, nitriles, and toluene diisocyanate (TDI) and other isocyanates. A major breakdown mechanism in urethanes is the scission of the polyol-isocyanate bond formed during polymerization. The isocyanate vaporizes and recondenses as a smoke, and liquid polyol remains to further decompose. Foamed polyurethanes can be subdivided in flexible foams (often used for upholstery) and rigid foams (most often used in construction applications).

Cellulosics

Cellulose The thermal decomposition of cellulose involves at least four processes in addition to simple desorption of physically bound water. The first is the cross-linking of cellulose chains, with the evolution of water (dehydration). The second concurrent reaction is the unzipping of the cellulose chain, which forms levoglucosan from the monomer unit. The third reaction is the decomposition of the dehydrated product (dehydrocellulose) to yield char and volatile products. Finally, levoglucosan itself can further decompose to yield smaller volatile products, including tars and, eventually, carbon monoxide. Some levoglucosan may also repolymerize. Below 550 K, the dehydration reaction and the unzipping reaction proceed at comparable rates, and the basic skeletal structure of the cellulose is retained. At higher temperatures, unzipping is faster, and the original structure of the cellulose begins to disappear. The cross-linked dehydrated cellulose and the repolymerized levoglucosan begin to yield polynuclear aromatic structures, and graphite carbon structures form at around 770 K. The char yield is quite dependent on the rate of heating of the sample. At very high rates of heating, no char is formed. On the other hand, preheating the sample at 520 K will lead to 30% char yields. This is due both to the importance of the low-temperature dehydration reactions for ultimate char formation and the increased opportunity for repolymerization of levoglucosan that accompanies slower heating rates.

Wood Wood is composed of 50% cellulose, 25% hemicellulose, and 25% lignin. Yields of gaseous products and kinetic data indicate that the decomposition is probably a combination of the decomposition mechanisms of the individual constituents. On heating, hemicellulose first decomposes (475 to 535 K), followed by cellulose (525 to 625 K) and lignin (555 to 775 K). The decomposition of lignin contributes significantly to the overall char yield. Piloted ignition of woods caused by radiative heating has been observed at a surface temperature of 620 to 650 K.

Sulfur-Containing Polymers Polysulfides are generally stable to 675 K. Poly(1,4 phenylene sulfide) decomposes at 775 K. Below this temperature, the principal volatile product is hydrogen sulfide. Above 775 K, hydrogen, evolved in the course of cross-linking, is the major volatile product. In air, the gaseous products include carbon oxides and sulfur dioxide. The decomposition of polysulfones is analogous to that of polycarbonates. Below 575 K, decomposition is by heteroatom bridge cleavage, and above 575 K, sulfur dioxide is evolved from the polymer backbone.

Thermally Stable Polymers The development of thermally stable polymers is an area of extensive ongoing interest. Relative to many other materials, most polymers have fairly low use temperatures, which can reduce the utility of the product. This probable improvement in fire properties is often counterbalanced by a decrease in both processability and favorable physical properties. Of course, materials that are stable at high temperatures are likely to be better per-

formers in relation to fire properties. The high-temperature physical properties of polymers can be improved by increasing interactions between polymer chains or by chain stiffening. Chain interactions can be enhanced by several means. As noted previously, crystalline materials are more stable than their amorphous counterparts as a result of chain interactions. Of course, if a material melts before volatilization occurs, this difference will not affect chemical decomposition. Isotactic polymers are more likely to be crystalline because of increased regularity of structure. Polar side groups can also increase the interaction of polymer chains. The softening temperature can also be increased by chain stiffening. This is accomplished by the use of aromatic or heterocyclic structures in the polymer backbone, such as in poly(*p*-phenylene) or poly(*p*-xylene). Poly(*p*-phenylene) is quite thermally stable but is brittle, insoluble, and infusible. Thermal decomposition begins at 870 to 920 K, and up to 1170 K, only 20% to 30% of the original weight is lost. Introduction of the following groups

$$-O-; \quad -CO-; \quad -NH-; \quad -CH_2-; \quad -O-CO-; \quad -O-CO-O-$$

into the chain can improve workability, although at the cost of some loss of oxidative resistance. Poly(*p*-xylene) melts at 675 K and has good mechanical properties, although it is insoluble and cannot be thermoprocessed. Substitution of halogen, acetyl, alkyl, or ester groups on aromatic rings can help the solubility of these polymers at the expense of some stability. Several relatively thermostable polymers (such as PCs or polysulfones) can be formed by condensation of bisphenol A with a second reagent. The stability of such polymers can be improved if aliphatic groups are not included in the backbone, because the $-C(CH_3)_2-$ groups are weak links.

Other thermostable polymers include ladder and extensively cross-linked polymers. Cyclized PAN is an example of a ladder polymer with two chains periodically interlinked. Other polymers (e.g., rigid polyurethanes) are sufficiently cross-linked that it becomes impossible to speak of a molecular weight or definitive molecular repeating structure. As in polymers that gel or cross-link during decomposition, cross-linking of the original polymer yields a carbonized char residue upon decomposition, which can be oxidized at temperatures above 775 K.

Implications of Thermal Decomposition to Fire Properties

One of the major reasons thermal decomposition of polymers is studied is because of its importance in terms of fire performance. This issue has been studied extensively. Early on, Van Krevelen[41,42] showed that, for many polymers, the oxygen index (or limiting oxygen index [LOI], an early flammability measure) could be linearly related to char yield as measured by TGA under specified conditions. After

Van Krevelen showed how to compute char yield to a good approximation from structural parameters, LOI values became predictable, in theory; in fact they can be predicted to some extent for those pure polymers having substantial char yields. Since then, however, comparisons have been made between the minimum decomposition temperature (or the more meaningful temperature for 1% thermal decomposition) and the LOI. The conclusion was that although in general, low flammability resulted from high minimum thermal decomposition temperatures, no easy comparison could be found between the two, with some notable cases of polymers with both low thermal stability and low flammability.[3,20] This type of approach has since fallen into disrepute, particularly in view of the lack of confidence remaining today in the LOI technique.[43] There is now significant consensus that LOI values are of interest mainly as an indication of ease of extinction, and relate to very low heat inputs rather than to those inputs associated with most fires. Work by Lyon,[44] however, indicates predictability of heat release information from thermoanalytical data obtained on a new piece of equipment. Mechanisms of action of fire retardants and potential effectiveness of fire retardants can be well predicted from thermal decomposition activity,[3,4] but it is often necessary to have an added understanding of the chemical reactions involved. Camino[45,46] and Bourbigot[47] recently used thermal decomposition techniques to determine the mechanism of fire retardance and to synthesize new polymers. The newest approach is the use of the mass loss calorimeter, which uses heat fluxes and sample sizes typical of fires.[48-50]

However far science has progressed in understanding how to predict fire performance from thermal decomposition data, it is clear that this remains a critical goal, because polymers cannot burn if they do not break down.

Flames and Polymers

All flames are associated with very exothermic chemical reactions in the gaseous phase. Although the term flame conveys a widely understood meaning, some features generally associated with flames are sometimes lacking.[20] Although flames are usually connected with the emission of light, some flames such as that of hydrogen burning in dust-free air or oxygen are practically nonluminous. Furthermore, although most flames are characterized by relatively high temperatures, cool flames are observed when oxygen is mixed with many organic compounds, and these emit light at fairly low temperatures.

When gaseous fuels burn in a gaseous oxidant, flames of two basic but extreme types are produced. Flame reactions are either controlled by chemical kinetics or by physical mixing processes such as diffusion or flow. Premixed flames are kinetically controlled, with the fuel, oxidant, and product uniformly distributed in the gas phase; gas explosions are typical

examples. Diffusion flames have sizable gradients of chemical species and temperatures, with reactants diffusing into the flame zone, and products and heat diffusing out: the combustion is controlled by diffusion. In practice, however, many flames are intermediate between the two extreme types.

Premixed flames have been the subject of more study than diffusion flames because they can give information about a number of fundamental properties of the gaseous mixture, such as burning velocity, flame temperature, and the chemical reactions occurring. The burning velocity is the rate at which the flame front travels in a direction normal to its surface into the adjacent unburnt gas and relative to the rate of movement of the unburnt mixture. In practice, the position of the flame front in burners is held steady by an opposing flow of fuel and oxidant, the magnitude of which is used to measure the burning velocity.

Diffusion flames are occasionally homogeneous, when all the reactants are in the gas phase, but they are more often heterogeneous, such as solid or liquid fuels burning in oxidizing gas atmospheres. Most flames of commercial importance are diffusion flames, rather than premixed flames, for safety reasons. Despite the economic importance of diffusion flames, there have been fewer fundamental studies on them, because they are not one dimensional. The chemical reaction rate is normally much higher than the interdiffusion rate of fuel and oxidant, which thus controls flame reaction velocities, and the gas composition changes steadily as the flame is traversed. Thus, diffusion flames yield fewer important parameters, except geometric parameters or temperature gradients. The position with diffusion flames may be further complicated by thermal decomposition of the fuel prior to its contact with the oxidant.

A burning organic polymer constitutes a highly complex combustion system. The associated flames are invariably stationary flames, as opposed to propagating or exploding flames, which may be observed in purely gaseous systems. In contrast to flames of gaseous fuels, where the fuel is supplied to the flame front at an arbitrary rate, a burning polymer generates its own fuel at a rate that must satisfy the equilibrium condition for maintenance of the burning process. Because condensed phase fuel generation is closely coupled with gas phase fuel consumption, the rates of these processes must be in equilibrium to maintain a steady state of combustion. Acceleration or deceleration of either process will lead to disequilibrium and eventually to collapse or blow-off of the flame. In general, however, polymer combustion flames are much more similar to gaseous diffusion flames than to premixed flames, although with certain polymers air may be in contact with the polymer surface so that the flame is at least partly premixed.

Composition profiles of hydrocarbon flames show that chemical reaction takes place in two distinct regions and in two well-defined stages. In the primary reaction zone the initial hydrocarbon is converted largely to carbon monoxide, hydrogen, and water, although other intermediate products, such as lower alkanes and alkenes, may also be present to an extent dependent on the fuel-to-oxidant ratio. The secondary zone is always characterized by the oxidation of carbon monoxide to carbon dioxide. In other words, a hydrocarbon flame can be regarded as a chemical reaction producing a carbon monoxide-hydrogen-oxygen flame. In effect, therefore, the original hydrocarbon, carbon monoxide, and hydrogen are all being oxidized simultaneously. Direct molecule-molecule reactions are not significant within the short residence time available and free radical reactions are again involved. The most important radicals in all hydrocarbon flames are simple species such as H·, O·, and OH·, and to a smaller extent HO$_2$·, HCO·, and CH$_3$·. In flames of alkanes (other than methane, which is a special case, and of little interest to polymer combustion[20]), alkenes and aromatic hydrocarbons, and indeed most organic polymers, the larger radicals produced initially rapidly break down to give smaller species such as CH$_3$·. The latter radicals therefore tend to be involved in the flames of all hydrocarbons, with the result that the chain-propagating species are largely independent of the initial fuel. Thus, the chemistry of hydrocarbon flames is effectively the high-temperature chemistry of a small number of free radicals. In practice, therefore, the number of relevant reactions is limited to perhaps five or six times the number of fuel species rather than to the square of the number of these species as would be expected if every molecule reacted with every other molecule. However, the nature of the chemical reactions in hydrocarbon flames does depend on the fuel-to-oxidant ratio. Premixed flames can cover the whole gamut of such ratios, but diffusion flames probably represent the ultimate in fuel-rich flames. Fuel-lean alkane-oxygen flames have been most fully studied, and here OH· radicals are the main species that attack the fuel. As the mixture becomes more fuel rich, H atoms progressively replace OH radicals. When the fuel is in considerable stoichiometric excess, the principal final products are carbon monoxide, hydrogen, water, and solid carbon, some of which generally appears as soot. In addition, hydrocarbons of molecular weight higher than that of the initial fuel may be present. Polyacetylenes are formed in these very hydrocarbon-rich flames and these may be the precursors of soot. Other elements may also be present in the final products of the combustion of organic polymers, especially in fuel-rich flames. The most important of these are nitrogen, sulfur, chlorine, and bromine, and some of the products are derived from those heteroelements contained in the polymers. Most of the nitrogen appearing as nitrogen oxides is, however, formed mainly by reaction of molecular nitrogen in the atmosphere—a process called nitrogen fixation. In contrast, chlorine and bromine in organic polymers are converted during combustion almost exclusively into the corresponding hydrogen halides (although bromine-containing polymers sometimes result in volatilization of the bromine-containing additive rather than in generation of hydrogen bromide), and sulfur appears as oxides of sulfur, mainly sulfur dioxide, especially at high temperatures.

The nature of the fuel supplied to the flame when an organic polymer burns is clearly related to the structure of the original polymer. It also depends on the extent to which the polymer or its breakdown products react with the oxidant prior to entering the flame zone. Nevertheless, the chemistry of the flames of hydrocarbons and that of other organic compounds in a given oxidant are remarkably similar, involving largely the same chain-propagating radicals. Indeed, because large hydrocarbon radicals are thermally unstable and thus usually decompose to give smaller radicals such as methyl, chain propagation normally involves the same simple species. Any hydrocarbon flame can thus be envisaged as a reaction leading to a carbon monoxide-hydrogen-oxygen flame. In the same way, a flame above an organic polymer can be thought of as a chemical reaction taking place at a solid or liquid surface yielding a hydrocarbon flame. Flames above different burning organic polymers do not therefore differ greatly, from the chemical point of view, from one polymer to another, although they are highly dependent on the nature of the gaseous oxidant.

Fire Performance of Polymers

Fire hazard is the result of a combination of factors including ignitability, ease of extinction, flammability of the products generated, amount of heat released on burning, rate of heat release, flame spread, smoke obscuration, and smoke toxicity, as well as the fire scenario. Some of these concepts are simple to understand, but others are more complex; definitions are provided for the most important of them, to a large extent based on ASTM E176.[1] The following sections discuss these fire properties and test methods that can be used to assess them. Compliance with fire properties is often required by different codes or regulations (usually for products, but occasionally also for materials) and very frequently in specifications for materials or combustion methods. Advances in science and technology have resulted (and continue to do so) in the development of a better understanding of critical fire properties and of improved methods to assess them. Thus, it is often the case that traditional fire testing procedures are no longer considered to be appropriate by fire scientists, but they are still used because of the existence of commercial interests and the time needed to change requirements. The following sections address both traditional and newer tests methods, and discuss their usefulness.

Some data presented in these sections are based on tests of a series of 35 materials (all but one of them plastics) in use in the 1980s and 1990s (Table 26.2). Table 26.2 lists identification numbers (used in several subsequent figures and tables) and short descriptions of each material. Data on fire properties are presented in such a way that the better fire responses tend to be at the top.

Table 26.2 Description of materials used in cone calorimeter (and some other) fire tests

A: Nonvinyls

ABS	Cycolac CTB acrylonitrile butadiene styrene terpolymer (Borg Warner) (no. 29)
ABS FR	Cycolac KJT acrylonitrile butadiene styrene terpolymer fire retarded with bromine compounds (Borg Warner) (no. 20)
ABS FV	Polymeric system containing acrylonitrile butadiene styrene and some poly(vinyl chloride) as additive (no. 19)
ACET	Polyacetal: polyformaldehyde (Delrin, Commercial Plastics) (no. 24)
DFIR	Douglas fir wood board (no. 22)
EPDM	Copolymer of ethylene propylene diene rubber (EPDM) and styrene acrylonitrile (SAN) (Rovel 701) (no. 31)
KYD	Kydex: fire retarded acrylic paneling, blue (samples were 4 sheets at 1.5 mm thickness each, Kleerdex) (no. 15)
PCARB	Polycarbonate sheeting (Lexan 141-111, General Electric) (no. 5)
PCARB B	Commercial polycarbonate sheeting (Commercial Plastics) (no. 16)
NYLON	Nylon 6,6 compound (Zytel 103 HSL, Du Pont) (no. 28)
PBT	Poly(butylene terephthalate) sheet (Celanex™ 2000-2 polyester, Hoechst Celanese) (no. 32)
PE	Polyethylene (Marlex HXM 50100) (no. 34)
PET	Poly(ethylene terephthalate) soft drink bottle compound (no. 33)
PMMA	Poly(methyl methacrylate) (25 mm thick, lined with cardboard, standard RHR sample) (no. 26)
PP	Polypropylene (Dypro 8938) (no. 35)
PPO/PS	Blend of polyphenylene oxide and polystyrene (Noryl N190, General Electric) (no. 18)
PPO GLAS	Blend of poly(phenylene oxide) and polystyrene containing 30% fiberglass (Noryl GFN-3-70, General Electric) (no. 17)
PS	Polystyrene, Huntsman 333 (Huntsman) (no. 30)
PS FR	Fire retarded polystyrene, Huntsman 351 (Huntsman) (no. 23)
PTFE	Poly(tetrafluoroethylene) sheet (samples were two sheets at 3 mm thickness each, Du Pont) (no. 1)
PU	Polyurethane flexible foam, not fire retarded (25 mm thick, Jo-Ann Fabrics) (no. 25)
THM PU	Thermoplastic polyurethane containing fire retardants (estane, BF Goodrich) (no. 27)
XLPE	Black non-halogen flame retardant, irradiation crosslinkable, polyethylene copolymer cable jacket compound (Unigard DEQD-1388, Union Carbide) (no. 11)

B: Vinyls

Rigid vinyls

PVC EXT	Poly(vinyl chloride) rigid weatherable extrusion compound with minimal additives (BF Goodrich) (no. 13)
PVC LS	Poly(vinyl chloride) rigid experimental sheet extrusion compound with smoke-suppressant additives (BF Goodrich) (no. 10)
PVC CIM	Poly(vinyl chloride) general-purpose rigid custom injection molding compound with impact modifier additives (BF Goodrich) (no. 8)
CPVC	Chlorinated poly(vinyl chloride) sheet compound (BF Goodrich) (no. 7)

Flexible vinyls

FL PVC	Standard flexible poly(vinyl chloride) compound (noncommercial; similar to a wire and cable compound) used for various sets of testing (including Cone Calorimeter RHR ASTM round robin; it contains PVC resin 100 phr; diisodecyl phthalate 65 phr; tribasic lead sulphate 5 phr; calcium carbonate 40 phr; stearic acid 0.25 phr) (no. 21)
PVC WC	Flexible wire and cable poly(vinyl chloride) compound (not fire retarded) (BF Goodrich) (no. 14)
PVC WC SM	Flexible wire and cable poly(vinyl chloride) compound (containing minimal amounts of fire retardants) (BF Goodrich) (no. 12)
PVC WC FR	Flexible wire and cable poly(vinyl chloride) compound (containing fire retardants) (BF Goodrich) (no. 9)
VTE 1	Flexible vinyl thermoplastic elastomer alloy wire and cable jacket experimental compound, example of the first of several families of VTE alloys (no. 6)
VTE 2	Flexible vinyl thermoplastic elastomer alloy wire and cable jacket experimental compound, example of the second of several families of VTE alloys (no. 3)
VTE 3	Flexible vinyl thermoplastic elastomer alloy wire and cable jacket experimental compound, example of the third of several families of VTE alloys (no. 2)
VTE 4	Semi flexible vinyl thermoplastic elastomer alloy wire and cable jacket experimental compound, example of a family of VTE alloys containing CPVC (no. 4)

All samples are at 6 mm thickness, except as indicated.

Definitions

The following definitions have been transcribed verbatim from ASTM E176,[1] unless indicated otherwise.

Burn: to undergo combustion.
Char (*v*): to form carbonaceous residue during pyrolysis or during incomplete combustion

Char (*n*): a carbonaceous residue formed by pyrolysis or incomplete combustion.
Combustion: a chemical process of oxidation that occurs at a rate fast enough to produce temperature rise and usually light either as a glow or flame.
Ease of extinction: facility of quenching of combustion occurring with a material, product, or assembly.

Fire: destructive burning as manifested by any or all of the following: light, flame, heat, smoke.

Fire hazard: the potential for a fire to cause harm to people or property.

Discussion: A fire may pose one or more types of hazard to people, animals, or property. These hazards are associated with the environment and with a number of fire-test-response characteristics of materials, products, or assemblies including but not limited to ease of ignition, flame spread, rate of heat release, smoke generation and obscuration, toxicity of combustion products, and ease of extinguishment.

Fire performance: response of material, product, or assembly in a particular fire, other than in a fire test involving controlled conditions (different from fire-test-response characteristic).

Discussion: The ASTM Policy on Fire Standards distinguishes between the response of materials, products, or assemblies to heat and flame under controlled conditions (which is fire-test-response characteristic), and under actual fire conditions (which is fire performance). Fire performance depends on the occasion or environment and may not be measurable. In view of the limited availability of fire performance data, the response to one or more fire tests, appropriately recognized as representing end-use conditions, is generally used as a predictor of the fire performance of a material, product, or assembly.

Fire resistance: the property of a material, product, or assembly to withstand fire or give protection from it.

Discussion: As applied to elements of buildings, it is characterized by the ability to confine a fire or to continue to preform a given structural function, or both.

Fire risk: an estimation of expected fire loss that combines the potential for harm in various fire scenarios that can occur with the probabilities of occurrence of those scenarios.

Discussion: Risk may be defined as the probability of having a certain type of fire, where the type of fire may be defined in whole or in part by the degree of potential harm associated with it, or as potential for harm weighted by associated probabilities. However it is defined, no risk scale implies a single value of acceptable risk. Different individuals presented with the same risk situation may have different opinions on its acceptability.

Fire scenario: a detailed description of conditions, including environmental, of one or more of the stages from before ignition to the completion of combustion in an actual fire, or in a full scale simulation.

Discussion: The conditions describing a fire scenario, or a group of fire scenarios, are those required for the testing, analysis, or assessment that is of interest. Typically, they are those conditions that can create significant variation in the results. The degree of detail necessary will depend upon the intended use of the fire scenario. Environmental conditions may be included in a scenario definition but are not required in all cases. Fire scenarios often define conditions in the early stages of a fire while allowing analysis to calculate conditions in later stages.

Fire-test-response characteristic: a response characteristic of a material, product, or assembly, to a prescribed source of heat or flame, under controlled fire conditions; such response characteristics may include but are not limited to ease of ignition, flame spread, heat release, mass loss, smoke generation, fire endurance, and toxic potency of smoke.

Discussion: A fire-test-response characteristic can be influenced by variables of exposure such as ignition source intensity, ventilation, geometry of item or enclosure, humidity, or oxygen concentration. It is not an intrinsic property such as specific heat, thermal conductivity, or heat of combustion, where the value is independent of test variables. A fire-test-response characteristic may be described in one of several terms. Smoke generation, for example, may be described as smoke opacity, change of opacity with time, or smoke weight. No quantitative correlation need exist between values of a fire-test-response characteristic for different materials, products, or assemblies, as measured by different methods or tested under different sets of conditions for a given method.

Flame: a hot, usually luminous zone of gas that is undergoing combustion.

Discussion: The luminosity of a flame is frequently caused by the presence of glowing particulate matter suspended in the hot gases.

Flame spread: the propagation of a flame away from the source of ignition.

Flashover: the rapid transition to a state of total surface involvement in a fire of combustible materials within an enclosure.

Discussion: Flashover occurs when the surface temperatures of an enclosure and its contents rise, producing combustible gases and vapors, and the enclosure heat flux becomes sufficient to heat these gases and vapors to their ignition temperatures. This commonly occurs when the upper layer temperature reaches 600 °C or when the radiant heat flux at the floor reaches 20 kW/m^2.

Heat release rate: the heat evolved from the specimen, per unit of time.

Ignition: the initiation of combustion.

Discussion: The combustion may be evidenced by glow, flame, detonation, or explosion. The combustion may be sustained or transient.

Flash ignition temperature: the minimum temperature at which, under specified test conditions, sufficient flammable gases are emitted to ignite momentarily upon application of a small external pilot flame. (Contained in ASTM D1929.[51])

Ignition temperature: the lowest temperature at which sustained combustion of a material can be initiated under specified test conditions.

Discussion: Although the phenomenon of combustion may be transient or sustained, in fire testing practice, the ignition temperature is reached when combustion continues after the pilot source is removed.

Piloted ignition: ignition of combustible gases or vapors by a pilot source of ignition.

Pilot source of ignition: a discrete source of energy, such as a flame, spark, electrical arc, or glowing wire.

Self ignition: the minimum temperature at which the self-heating properties of the specimen lead to ignition or ignition occurs by itself, under specified test conditions, in the absence of any additional flame ignition source. (Contained in ASTM D1929.[51])

Unpiloted ignition: ignition caused by one or more sources of energy without the presence of a pilot source of ignition.

Oxygen consumption principle: the expression of the relationship between the mass of oxygen consumed during combustion and the heat released.

Smoke: the airborne solid and liquid particulates and gases evolved when a material undergoes pyrolysis or combustion.

Smoke obscuration: reduction of light transmission by smoke, as measured by light attenuation.

Smoke toxicity: the propensity of smoke to produce adverse biochemical or physiological effects.

Toxic potency: as applied to inhalation of smoke or its component gases, a quantitative expression relating concentration and exposure time to a particular degree of adverse physiological response, for example, death, on exposure of humans or animals.

Discussion: The toxic potency of the smoke from any material, product, or assembly is related to the composition of that smoke, which, in turn, is dependent on the conditions under which the smoke is generated.

Ignitability

Unless one of the materials present in a certain environment ignites, there is no fire. Therefore, low ignitability is the first line of defense in a fire. All organic polymers do ignite, but the higher the temperature a material has to reach before it ignites, the safer it is. Thus, it is possible to determine ignition temperatures using a traditional test method: ASTM D1929 (Setchkin test).[51] Figure 26.5 presents the self-ignition and flash-ignition (with a pilot flame) temperatures of many common materials.[52-54] Test Method ASTM D1929 is a test specific for ignition of plastics, and uses small pieces or pellets as specimens, exposed inside a vertical furnace tube, electrically heated to a pre-set temperature increase rate, under a slow air flow. Its results are no longer considered to be as valuable as they once were because the exposure is not truly representative of real fires. However, results from this test are frequently required in

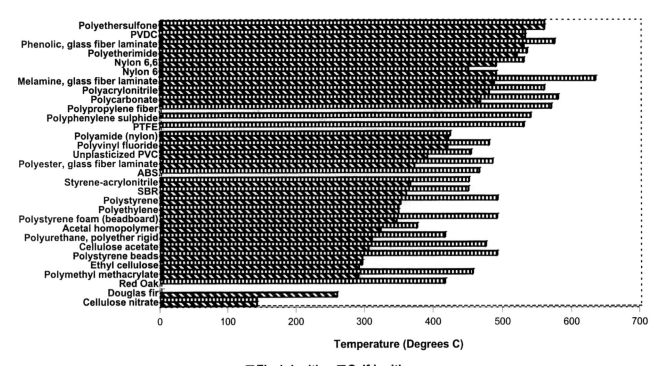

Figure 26.5 Ignition temperature in the Setchkin test (ASTM D1929); flash ignition and self-ignition.

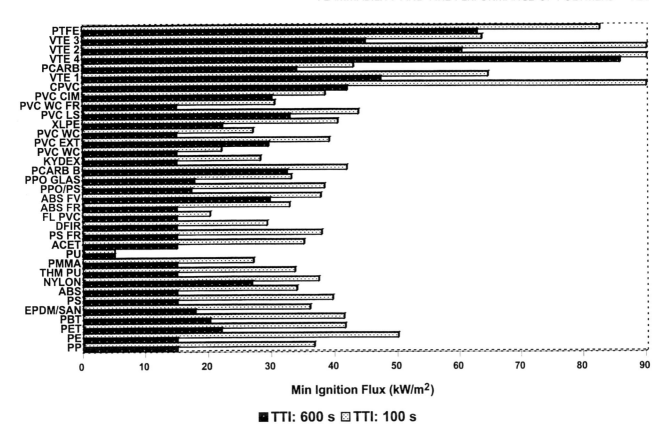

TTI: 600 s   TTI: 100 s

Figure 26.6 Minimum heat flux for ignition in the cone calorimeter (ASTM E1354).

specifications and quoted in data sheets, and the test is referenced in the three regional model building codes, National Building Code, Standard Building Code, and Uniform Building Code, as well as the new country-wide International Building Code, as a method for determining the suitability of plastic materials for use in construction. Ignitability is also assessed, in a better way, by determining time to ignition or a minimum heat input needed to ignite a material, when testing for other properties. Fire performance improves as either one of these properties becomes larger. Figure 26.6 shows minimum ignition fluxes to cause ignition within 100 s or 10 min, from a more modern standard test, ASTM E1354 (cone calorimeter),[55] for the 35 materials in Table 26.2.[56] A schematic diagram of the cone calorimeter is shown in Figure 26.7. Alternative test methods to assess ignition are ASTM E906 (Ohio State University calorimeter, primary purpose heat and smoke release[57]) and ASTM E1321 (Lateral Ignition and Flame Spread Test [LIFT], primary purpose flame spread[58]).

Ease of Extinction

Once ignited, the easier to extinguish a material is, the lower its associated fire hazard. ASTM D2863[59] is used to determine the oxygen index (or LOI), which is the minimum oxygen concentration (in a flowing mixture of oxygen and nitrogen) required to support candle-like downward flaming combustion. The test actually serves as a measure of the ease of extinction of the material. The small specimen is placed vertically inside a glass column and ignited at the top with a small gas flame. The repeatability and reproducibility of this test method are excellent, and it can generate numerical data covering a very broad range of responses. However, once more, the test method is inappropriate as a predictor of real-scale fire performance, mainly because of the low heat input and the artificiality of the high oxygen environments used. It is widely required in specifications and quoted in data sheets, and is suitable as a quantitative quality control tool during manufacturing and as a semi-qualitative indicator of the effectiveness of additives during research and development, for low incident energy situations. Figure 26.8 shows results on a number of common materials.[54]

Flame Spread

The tendency of a material to spread a flame away from the fire source is critical to the understanding of the potential fire hazard. Flame spread tests may refer to organic polymers themselves, to materials in diverse applications (such as textiles or electrical insulation sleeving), or to whole structures (such as furniture or building components). Most

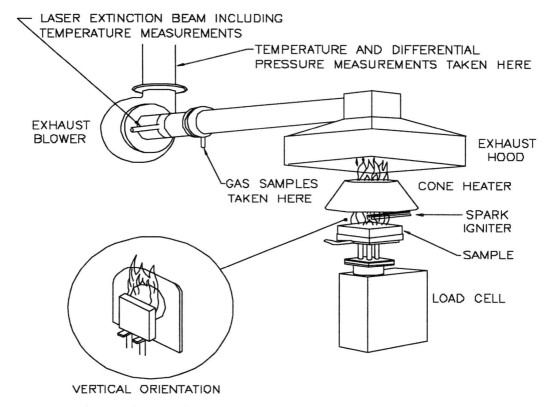

Figure 26.7 Schematic diagram of the cone calorimeter.

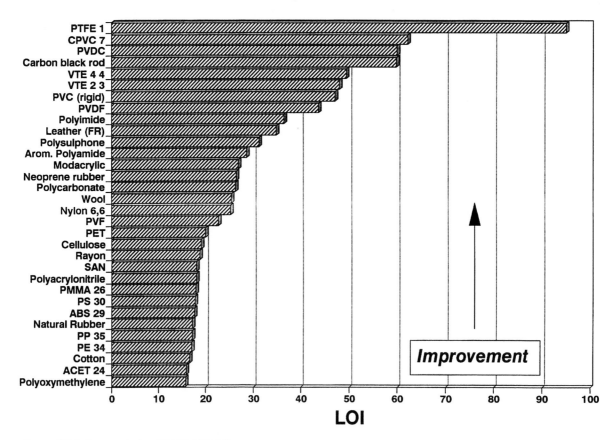

Figure 26.8 Oxygen index (ASTM D2863).

of the tests can be classified in terms of the angle between the exposed surface and the horizontal. Thus, this angle may vary between 0° for a horizontal material burning on its upper surface, through 180° for a horizontal material burning on its lower surface, and 240° for a material burning on its lower surface but rotated a further 60° from the horizontal, up to 270° for a material burning vertically downward.[20] This surface angle is closely related both to the extent to which the gaseous products of combustion heat the surface before the flame front reaches it and to the extent to which the hot gases penetrate into the polymer. Sample sizes range widely, from very small (127 mm × 13 mm, or 5 in. × 0.5 in. Underwriters Laboratories [UL] 94[60]) to quite large (7.3 m × 0.56 m, or 24 ft × 22 in. ASTM E84,[61] Steiner tunnel). These two tests are the most widely used for various specifications and for building code requirements. Two other test apparatuses are also suitable to assess flame spread of materials: those used in ASTM E162[62] (radiant panel, which is also used for cellular plastics in ASTM D3675[63]) and in ASTM E1321[58] (also known as LIFT).

The UL 94 test method applies a small Bunsen-burner-type flame to a plastic specimen. The test has vertical and horizontal versions, depending on the required degree of fire performance for the material; a vertical test is more severe than a horizontal test. The following ratings are normally obtained: HB (with the horizontal test), V-0, V-1, and V-2 (with the vertical test), and 5-V (with a test involving a larger sample exposed vertically and a flame 10 times larger). It is the most widely used fire test specification for plastic materials, especially fire retarded materials, and

forms the basis of the well-known Yellow Card used by UL to list plastic materials.

The Steiner tunnel test method (ASTM E84[61] was developed by Al Steiner for building materials[64] such as wood or gypsum board. It has been adopted by every building and fire code in the United States. It is perhaps the most widely accepted test for surface flammability in North America, and a large number of applications tests are developed from it. The most well-known test is that used for electrical cables needed for plenum use (National Fire Protection Association [NFPA] 262, also known as UL 910[65]), and there are also application tests for pneumatic tubing, sprinkler piping, and so on. The specimen in ASTM E84 is a building material (normally up to 0.15 m thick), either in one unbroken length or in separate sections joined end to end, which is mounted face downward so as to form the roof of a horizontal tunnel 305 mm high. The fire source, two gas burners, ignites the sample from below with an 89-kW fire source. When plastics started being used in construction, this test was applied to them, despite the fact that it is not always appropriate. For example, samples that cannot be retained in place above the tunnel floor, or which melt and continue burning on the tunnel floor (typical behavior for most thermoplastics), are still being tested with this equipment even though the results are misleading. The test can also produce wrong results for thin materials.[66] The normal output is a flame spread index (FSI), calculated on the basis of the area under the flame spread distance versus time curve. The FSI is a relative number based on an FSI = 0 for cement board and FSI = 100 for red oak flooring. Results

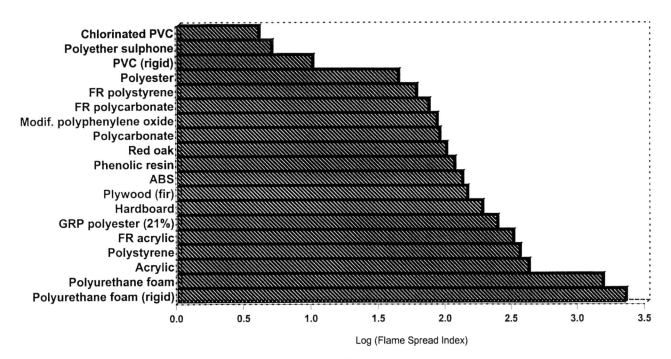

Figure 26.9 Flame spread index (ASTM E162). Improvement in flame spread is indicated by a higher vertical position.

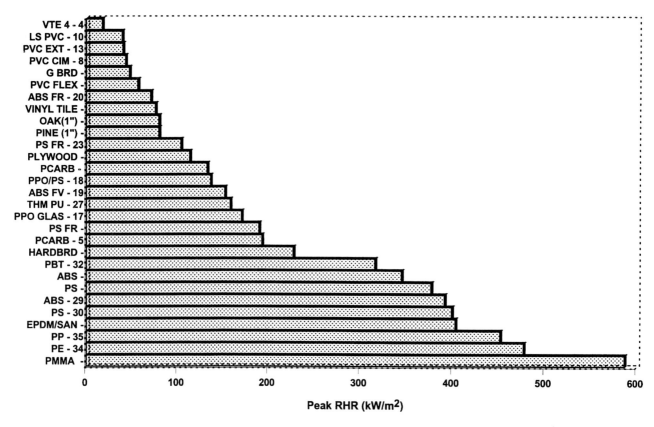

Figure 26.10 Peak rate of heat release in the Ohio State University calorimeter (ASTM E906) at 20 kW/m² flux.

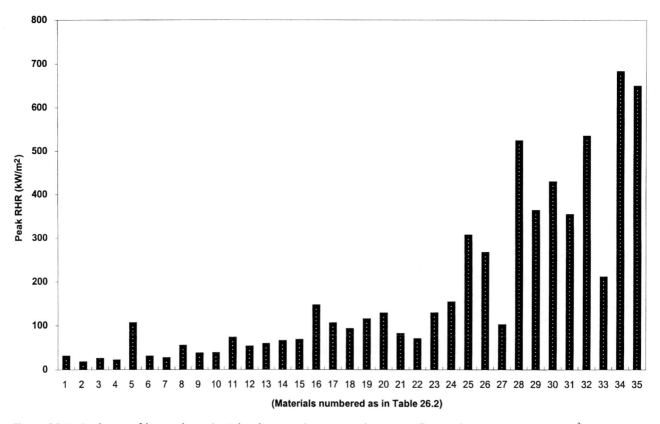

Figure 26.11 Peak rate of heat release (weighted average) in cone calorimeter (fluxes of 20, 40, and 70 kW/m²).

are used in the codes as classes, with Class A being an FSI of up to 25, Class B being an FSI of 26 to 75, and Class C being an FSI of 76 to 200. The test method also assesses smoke obscuration, and assigns a smoke developed index (SDI), on the basis of concept similar to the FSI. To be classified as Class A, B, or C, a material must have an SDI not exceeding 450.

ASTM Test Method E162[62] is also used to determine an FSI, albeit different from that assessed by the Steiner tunnel. The FSI is calculated as the product of a flame spread factor and a heat evolution factor, using techniques described in the standard. The test apparatus consists of a gas-fed radiant panel in front of which an inclined (at a 30° angle) specimen (150 × 460 mm) is exposed to a radiant flux equivalent to a black-body temperature of 670 °C (943 K), which is approximately 45 kW/m^2, in the presence of a small gas pilot flame. The maximum thickness that can be tested in the normal specimen holder is 25 mm, but alternative specimen holders can accommodate thicker specimens. The ignition is forced near the upper edge of the specimen and the flame front progresses downward. The FSI is calculated as the product of a flame spread factor, which results from the measurements of flame front position and time, and a heat evolution factor, which is proportional to the maximum temperature measured in the exhaust stack. This test has also not been shown to be an adequate predic-

tor of real-scale fire performance: if the specimen melts or causes flaming drips, this is likely to affect the flame spread in a way that is uneven; the test method simply requires that such events be reported. Moreover, if flame spread is very rapid, the flame spread is potentially lost unless recording is continuous. This apparatus is often referred to as the radiant panel, and results from this test are frequently required in regulations and detention environment specifications, and are quoted in data sheets. Results from this test are shown in Figure 26.9, on a logarithmic scale.[54]

The LIFT apparatus (ASTM E1321[58]) was developed as an improvement on the apparatus in ASTM E162.[67] The specimen size for flame spread studies is 155 × 800 mm by a maximum thickness of 50 mm (a smaller specimen is used for ignition studies). The test method determines the critical flux for flame spread, the surface temperature needed for flame spread, and the thermal inertia or thermal heating property (product of the thermal conductivity, the density, and the specific heat) of the material being tested. These properties are mainly used for assessment of fire hazard and for input into fire models. A flame spread parameter, Φ, is also determined, and this can be used as a direct way of comparing the responses of the specimens. This test method appears well suited for materials (or composites) that are nonmelting and can be ignited without raising the incident flux to potentially dangerous limits. It has been

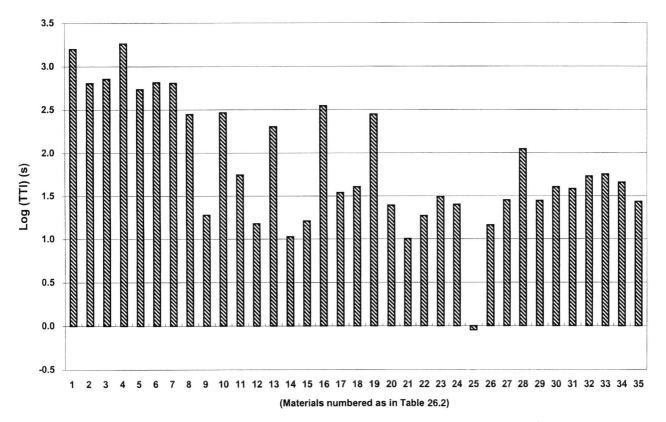

Figure 26.12 Time to ignition (weighted average) in cone calorimeter (fluxes of 20, 40, and 70 kW/m^2).

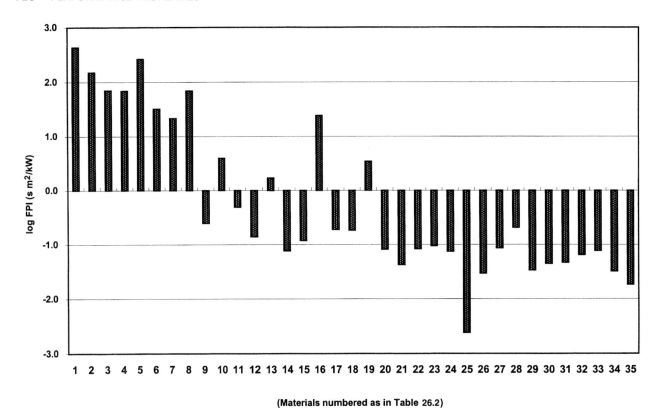

Figure 26.13 Fire performance index (weighted average) in cone calorimeter (fluxes of 20, 40, and 70 kW/m²).

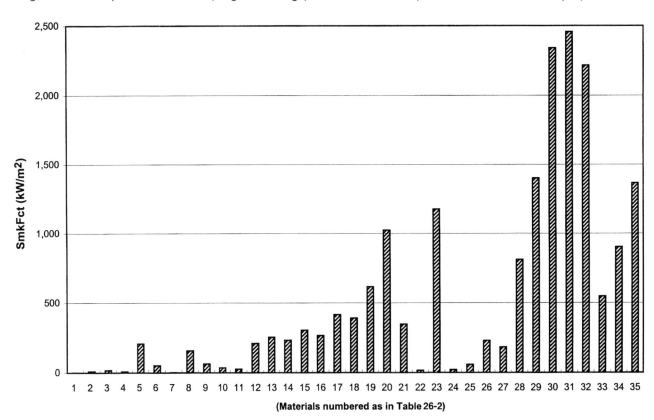

Figure 26.14 Smoke factor (weighted average) in cone calorimeter (fluxes of 20, 40, and 70 kW/m²).

used successfully for predictions of full-scale flame spread performance.[68]

Heat Release

The key question to ask in a fire is, "How big is the fire?" The fire property that answers that question is the rate of heat release. A burning product will spread a fire to nearby products only if it gives off enough heat to ignite them. Moreover, the heat has to be released fast enough not to be dissipated or lost while traveling through the cold air surrounding any product that is not on fire. Therefore, fire hazard is dominated by the rate of heat release.[69-72] In fact, the rate of heat release has been shown to be much more important than ease of ignition, smoke toxicity, or flame spread in controlling the time available for potential victims of a fire to escape.[73] In the late 1960s, Professor Edwin Smith at Ohio State University (OSU) developed the first test instrument—the OSU rate of heat release (RHR) calorimeter (ASTM E906[57])—to measure rates of heat release.[74] Figure 26.10 presents results of maximum rate of heat release for a variety of materials, at an incident heat flux of 20 kW/m^2, as measured in an OSU RHR calorimeter.[52,75] The heat flux used, 20 kW/m^2, is sufficient to ignite a sheet of newspaper from a 2.4 m (8 ft) height off the floor (a normal distance from ceiling to floor). In the early 1980s, the National Institute of Standards and Technology (NIST, then National Bureau of Standards) developed a more advanced test method to measure RHR: the cone calorimeter (ASTM E1354[55,75]). This instrument can also be used to assess other fire properties, the most important of which are the ignitability, the mass loss, and the smoke released. Moreover, results from this instrument correlate with those from full-scale fires.[76-80] To obtain the best overall understanding of the fire performance of materials, it is important to test the materials under a variety of conditions. Therefore, tests are often conducted at a variety of incident heat fluxes. Figures 26.11–26.14 show the four most important fire properties from the cone calorimeter: peak RHR, time to ignition, fire performance index (ratio of the time to ignition to the peak rate of heat release), and smoke factor (a smoke hazard parameter). All of these results are weighted averages of three fluxes: 20, 40, and 70 kW/m^2, for the set of 35 materials in Table 26.2.[56] Figure 26.6 presented ignitability data from the same test in a different way. The peak rates of heat release of the materials at each incident flux are shown in Table 26.3.

A number of modern full-scale fire test methods that rely on RHR measurements have been developed for products such as those involving testing of upholstered furniture (ASTM E1537[81]), mattresses (ASTM E1590[82]), stacking chairs (ASTM E1822[83]), electrical cables (ASTM D5424[84] and ASTM D5537[85]), plastic display stands (UL 1975[86]), or wall lining products (NFPA 265,[87] NFPA 286,[88] and International Organization for Standardization [ISO] 9705[89]). In

fact, room-corner tests (Figure 26.15) are being used in the newest editions of the codes, as alternatives to replace the Steiner tunnel test, thus generating more useful results. Figures 26.16–26.18 show, respectively, the predictability of the cone calorimeter. In Figure 26.16,[90] the cone calorimeter fire performance index (with tests conducted at 50 kW/m^2) was a good predictor of time to flashover in Federal Aviation Administration (FAA) full aircraft fires[91] and in the ISO 9705 room-corner test.[92] In Figure 26.17, the same cone calorimeter tests, but using only heat release criteria, have almost perfect predictability of ISO 9705 room-corner test rankings.[92] In Figure 26.18, cone calorimeter tests at 35 kW/m^2 identify the safe zone when compared to full-scale furniture heat release tests.[90]

Table 26.3 Peak rate of heat release in the cone calorimeter of the materials in Table 26.2

Sample no.	Material	Pk RHR at 20 kW/m^2	Pk RHR at 40 kW/m^2	Pk RHR at 70 kW/m^2
1	PTFE	3	13	161
2	VTE 3	4	43	70
3	VTE 2	9	64	100
4	VTE 4	14	87	66
5	PCARB	16	429	342
6	VTE 1	19	77	120
7	CPVC	25	84	93
8	PVC CIM	40	175	191
9	PVC WC FR	72	92	134
10	PVC LS	75	111	126
11	XLPE	88	192	268
12	PVC WC SM	90	142	186
13	PVC EXT	102	183	190
14	PVC WC	116	167	232
15	KYD	117	176	242
16	PCARB B	144	420	535
17	PPO GLAS	154	276	386
18	PPO/PS	219	265	301
19	ABS FV	224	291	409
20	ABS FR	224	402	419
21	FL PVC	233	237	252
22	DFIR	237	221	196
23	PS FR	277	334	445
24	ACET	290	360	566
25	PU	290	710	1221
26	PMMA	409	665	988
27	THM PU	424	221	319
28	NYLON	517	1313	2019
29	ABS	614	944	1311
30	PS	723	1101	1555
31	EPDM/SAN	737	956	1215
32	PBT	850	1313	1984
33	PET	881	534	616
34	PE	913	1408	2735
35	PP	1170	1509	2421

See Table 26.2 for definitions of material codes.
Key: Pk RHR at 20, Pk RHR at 40, and Pk RHR at 70: peak rate of heat release at incident heat fluxes of 20, 40, and 70 kW/m^2, respectively.

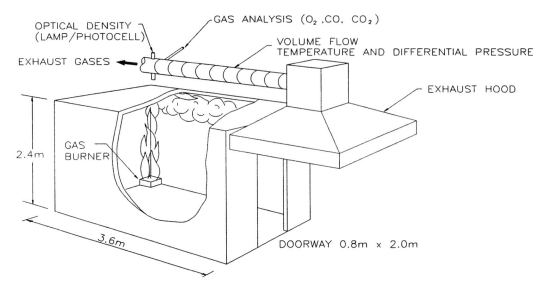

Figure 26.15 Schematic diagram of the room-corner apparatus.

The heat release of materials can be decreased by adding fire retardants. Table 26.4[93] is a compilation of a few of the data available wherein the same basic material was tested, using the same technique before and after the addition of flame retardants. The data indicate that considerable improvements in RHR are commonplace, including more than 10-fold increases. The range of materials presented in the table is also very broad, including thermoplastics, cross-linked materials, thermosets, and cellulosics. In view of the effect of heat release on fire hazard, this indicates a beneficial effect of fire retardants.

Smoke Obscuration

Obscuration is a serious concern in a fire, because a decrease in visibility reduces the light available, thus hindering both escape from the fire and rescue by safety personnel. The emission of smoke is the main way in which a fire causes decreases in visibility. This decrease in visibility is the result of a combination of two factors: how much material is burned in the real fire (which will be less if the material has better fire performance) and how much smoke is released per unit material burned. Several empirical param-

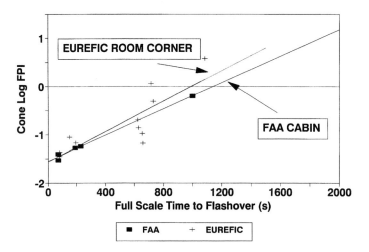

Figure 26.16 Comparison of predicted time to flashover from cone calorimeter (ASTM E1354) at 50 kW/m^2 with room-corner test results and with aircraft test results.

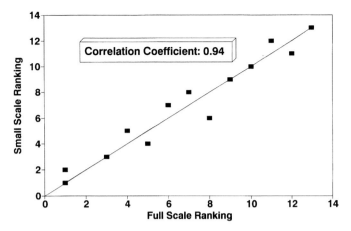

Figure 26.17 Comparison of predicted rankings from cone colorimeter (ASTM E1354) tests and room-corner burn test (ISO 9705) rankings.

eters have been proposed to make this compensation for incomplete sample consumption, including one called the smoke factor, determined in small-scale RHRs.[94] It combines the two aspects mentioned above: the light obscuration and the RHR. Results are shown in Figure 26.14.

Despite the understanding that smoke obscuration ought to be measured in a large scale, or by a method the results of which can predict large-scale smoke release, the most common small-scale test method for measuring smoke from burning products is the traditional smoke chamber in the vertical mode (ASTM E662[95]). The test results are expressed in terms of a quantity called specific optical density, which is defined in the test standard. This test has been shown to have some serious deficiencies, the most important problems being misrepresentation of the smoke obscuration

found in real fires and found for melting materials, such as thermoplastics.[96–100] When those materials that melt or drip when exposed to flame are exposed vertically in the smoke chamber test, the molten portions will have escaped the effect of the radiant heat source. This means that some of the material does not burn during the test (and does not give off smoke). In a real fire all of the molten material will burn and generate smoke. If these dripping products are exposed horizontally, the entire sample will be consumed (see the difference in some test results shown in Table 26.5[101]). Smoke chamber test results are shown in Figure 26.19. A comparison of Figures 26.14 and 26.19 shows the different rankings for similar materials, when tested in a test that can, at least to some extent, predict full-scale smoke obscuration, such as in the cone calorimeter.

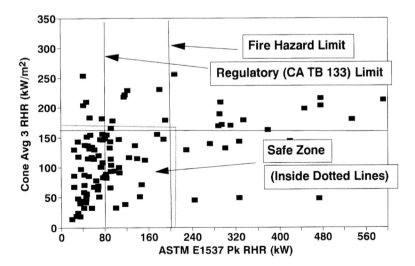

Figure 26.18 Comparison of average (3 min) heat release rates from cone colorimeter (ASTM E1354) at 35 kW/m² with full-scale chair fire test peak heat release rates (ASTM E1537).

Table 26.4 Comparison of heat release rate data for fire retarded (FR) and non-fire-retarded (NFR) materials

Material	Heat flux kW/m²	NFR Pk RHR kW/m²	FR Pk RHR kW/m²	NFR/FR ratio
ABS (+FR1)	20	614	224	2.7
ABS (+FR1)	40	944	402	2.3
ABS (+FR1)	70	1311	409	3.2
ABS (+FR2)	20	614	224	2.7
ABS (+FR2)	40	944	291	3.2
ABS (+FR2)	70	1311	419	3.1
EVA (cross-linked)	30	463	110	4.2
EVA (thermoplastic)	30	574	83	6.9
HDPE	30	1803	114	15.8
HDPE no. 2	50	1167	476	2.5
LDPE	20	913	88	10.3
LDPE	40	1408	192	7.3
LDPE	70	2735	268	10.2
PP	30	1555	174	8.9
PVC rigid	20	102	25	4.0
PVC rigid	40	183	84	2.2
PVC rigid	70	190	93	2.1
PVC rigid no. 2	30	98	42	2.3
PVC rigid no. 3	30	118	56	2.1
PVC WC	20	116	9	12.8
PVC WC	40	167	64	2.6
PVC WC	70	232	100	2.3
PVC WC no. 2	20	116	72	1.6
PVC WC no. 2	40	167	92	1.8
PVC WC no. 2	70	232	134	1.7
Particle board	25	151	66	2.3[a]
Particle board B (+FR1)	25	160	70	2.3
Particle board B (+FR1)	50	227	141	1.6
Particle board B (+FR2)	50	227	52	4.4
Plywood	25	114	43	2.7
Plywood	50	150	75	2.0
Polyester	30	186	95	2.0
PS	20	723	277	2.6
PS	40	1101	334	3.3
PS	70	1555	445	3.5

See Table 26.2 for definitions of material codes.

a. All data presented were obtained using the cone calorimeter (ASTM E1354), except for these data, obtained using ASTM E906 (Ohio State University rate of heat release calorimeter).

The majority of materials with low flame spread (or low heat release) also have low smoke. However, it has been shown in several series of room-corner test projects (with the tested material lining either the walls or the walls and the ceiling) that approximately 10% of the materials tested (8 out of 84) exhibited adequate heat release (or fire growth) characteristics, but had very high smoke release.[102,103] These materials would cause severe obscuration problems if used in buildings. A combination of this work, and the concept that a visibility of 4 m is reasonable for people familiar with their environment,[104] has led all U.S. codes to include smoke pass-fail criteria when room-corner tests are used as alternatives to the ASTM E84 Steiner tunnel test.

Smoke Toxicity

The majority of fire fatalities are a result of inhalation of smoke and combustion products, rather than as a consequence of burns. Various organizations are trying to develop test methods and guidance documents on smoke toxicity, but emotional responses arise from discussions on interpretation of results or requirements for the use of animals as test surrogates. The following is now accepted by many fire scientists,[105-111] and is critical to understanding how to assess fire hazard:

- Most fire fatalities occur in fires that become very large. In fact, U.S. statistics indicate that such fires account for more than six times more fatalities than all other fires. This is illustrated by statistics of U.S. fires in the 1986 to 1990 time period (Table 26.6[112]).
- Carbon monoxide concentrations in the atmospheres of flashover fires (the fires most likely to produce fatalities), are determined by geometric variables and oxygen availability, but are virtually unaffected by chemical composition of fuels.
- CO yields in full-scale flashover fires are approximately 0.2 g/g, which can be calculated to correspond to a toxicity of 25 mg/L.[107] This consistent yield of CO is illustrated by a set of 24 studies where such results are shown (Table 26.7).
- Toxic potency values from the most suitable small-scale smoke toxicity test (NIST radiant test, using rats as the animal model, but only for confirmatory purposes, standardized in ASTM E1678[113] and NFPA 269[114]) have been well validated with regard to toxicity in full-scale fires. However, such validation cannot be done to a better approximation than a factor of 3. Table 26.8 shows a series of toxic potencies of materials using this test method.
- The consequence of this is that any toxic potency (median lethal concentration [LC_{50}]) higher than 8 mg/L (i.e., any toxicity lower than 8 mg/L) will be subsumed within the toxicity of the atmosphere, and is of no consequence. Thus, LC_{50} values of 8 or greater should be converted to 8 mg/L for reporting purposes.

Table 26.5 Effect of orientation on the specific optical density of smoke measured by test method ASTM E662 on melting thermoplastic materials

	Horizontal orientation	Vertical orientation
Polypropylene	398	57
Polyethylene	286	35
Nylon	264	48
Paraffin wax	228	83

Data are maximum specific optical density of smoke; in the flaming mode.

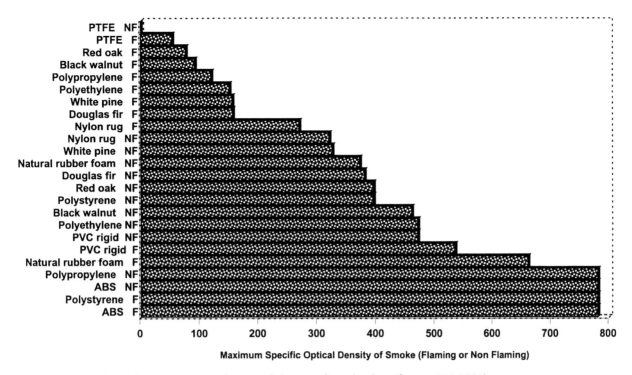

Figure 26.19 Traditional maximum specific optical density of smoke data (from ASTM E662).

Thus, as shown in Table 26.8, all common materials have virtually the same smoke toxicity; their associated fire hazard will not be a function of smoke toxic potency.

- All small-scale fire tests underpredict CO yields, but do not underpredict yields of other toxicants, such as hydrogen chloride or hydrogen cyanide. Thus, small-

scale fire tests cannot be used to predict toxic fire hazard for ventilation-controlled flashover fires, unless CO yields are recalculated by analogy with full-scale fire test results,[109] because a comprehensive study of fire (and nonfire) fatalities associated with CO showed that CO inhalation is a good statistical indicator for fire fatalities.[115]

- Fire retardants actually decrease the toxicity of the atmosphere (Table 26.9), despite some misinformation that indicates the opposite, because of the effect fire retardants have on decreasing the amount of burning that takes place.

Fire Modeling

Overall fire safety is generally achieved by deciding if materials meet certain pre-set objectives. However, generally it is necessary to combine various properties and calculate results that are based on certain fire models. Occasionally, such models are theoretically developed, and start from first principles. More frequently, work is, to a large extent, not based on a fundamental theoretical model, but on empirical correlations, which would help to determine what is needed to develop safe products. To that extent, correlation results may generate safe errors, and be preliminarily acceptable. A safe error is a case where a prediction, on the basis of a small-scale test, suggests that a material is unsafe, when actual full-scale testing indicates it to be safe. Accepting such a safe error would not cause a fire safety concern, but

Table 26.6 Statistics of fire fatalities in post-flashover structure fires in the United States, 1986 to 1990

Details	Number of fatalities	Percentages of total
Post-flashover fires		
Civilians killed by smoke inhalation only	621	78.3
Civilians killed by burns only	125	58.1
Civilians killed by combination of smoke inhalation and burns	2406	83.3
Combination of all of the above	3152	80.9
Other data, presented for comparison		
Total civilian fire fatalities in structure fires	3896	100.0
Total civilian fire fatalities in structure fires, intimate with ignition	670	17.2

Data are from reference 112.

Table 26.7 Yields of carbon monoxide in literature

Material	Yield	Reference	Organization
Plywood walls, wood fiberboard ceiling	0.35	116	HUD
FR plywood walls, wood fiberboard ceiling	0.42	116	HUD
Plywood walls, no ceiling	0.10	116	HUD
Upholstered chair, bed FR plywood walls	0.36	117	HUD
Plywood walls, bed	0.31	117	HUD
Plywood walls, wood cribs, cellulosic ceiling tile	0.29	118	NIST
Non FR chair, TV cabinets, cables, etc	0.22	105	FRCA
FR chair, TV cabinets, cables, etc	0.23	105	FRCA
PMMA walls	0.28	NIST, unpublished	NIST
Wood cribs	0.15	119	VPI
Flexible PU foam	0.25	119	VPI
PMMA	0.30	119	VPI
Hexane	0.23	119	VPI
Propane	0.23	120	Harvard
Propene	0.20	120	Harvard
Hexanes	0.20	120	Harvard
Toluene	0.11	120	Harvard
Methanol	0.24	120	Harvard
Ethanol	0.22	120	Harvard
Isopropanol	0.17	120	Harvard
Acetone	0.30	120	Harvard
Polyethylene	0.18	120	Harvard
PMMA	0.19	120	Harvard
Pine	0.14	120	Harvard
Average	0.236	Total cases: 24	Total number: 5

Key: FR = fire retarded; HUD = U.S. Department of Housing and Urban Development; NIST = National Institute of Standards and Technology; FRCA = Fire Retardant Chemicals Association; PMMA = poly(methyl methacrylate); PU = polyurethane; VPI = Virginia Polytechnic Institute.

would prevent safe materials from being used in a suitable application. On the other hand, errors that are always unacceptable would be encountered when the small-scale test predicts that a material is safe, and that result is contradicted by the full-scale test. Correlation fire models now exist (e.g., see reference 121), and new ones are being developed continually, that take fire test results developed in fire safety engineering units and predict full-scale fire outcomes. This is particularly true in the areas of wall linings, seating furniture, and electric cables, which have been assessed most frequently.

Fire Hazard Assessment

Most of the prescriptive techniques used for fire safety requirements (fire tests) were developed many years ago, and tend to have some deficiencies when applied to new materials (and in the context of the regulatory world, plastics are considered new materials in many instances). For example, techniques designed with traditional materials in mind often involve vertical or ceiling mounting, both of which can generate misleading results with melting materials, as discussed. Traditional techniques can also have additional problems in that they may not be able to generate data that can be used in fire safety engineering applications (such as fire models), and thus simply represent an

end point rather than creating a start for additional studies of fire safety.

If an inappropriate assessment technique is used for a material or product, there are three possible outcomes: (1) the technique gives the right answer (not a likely result, because the assumption is that the technique is inappropriate, but obviously the most desirable one); (2) the technique assesses the material or product as being better than it really is; or (3) the technique assesses the material or product as being worse than it really is. If the technique is too lenient on the material, an unsafe material can be introduced into use, with safety concerns as the potential outcome. If the technique is too severe on the material, a novel and suitable material may be unfairly excluded from an application where a safe alternative was developed. If the technique is too severe on a material, but still classifies it as suitable, there is no problem. If a technique is used which adequately assesses a material, but cannot generate results in the appropriate fire safety engineering units, the assessment is incomplete because it can only be used for comparisons with other materials for the same application. Thus, such an assessment technique will need to be supplemented by other techniques to ensure safety of the material under study. Most traditional fire property assessment techniques are of this kind: suitable for limited use but unable

Table 26.8 Radiant toxicity test results (ASTM E1678, NFPA 269)

Specimen	LC_{50} (mg/L)	Corr LC_{50} (mg/L)
ABS	17.8	11.8
Acrylic F + MELFM	9.6	6.9–8.2
Ceiling tile	30.5	21.9
Composite	20.0	Not given
Cork	ca. 40	Not given
Dg FIR	100–200	21–23
Dg FIR	56.0	21.0
Dg FIR (full scale)	>70	>70
FLX PU FM	52.0	18.0
MELFM	12.5	8.0
Nylon	36.7	17.0
Nylon rug (Tr)	28.5	14.2 FED 1.2
Nylon rug (Tr)	42.9	18.1 FED 2.0
Nylon rug (Un)	>41	>16
PVC CB	36.0	Not given
PVC INS	33.4	22.7
PVC INS	29.2	14.9
PVC JK	53.1	25.9
PVC Lw HCl	146.9	28.8
PVC Md HCl	86.2	26.7
PVC PRF	26.0	16.0
PVC PRF	20–30	13–17
PVC PRF (full scale)	35–45	35–45
Particle board	120–138	Not given
Rg PU FM	22.0	14.0
Rg PU FM	20–30	14–19
Rg PU FM (full scale)	30–40	30–40
Vinyl F	32.0	19.0
Vinyl F + MELFM	26.0	15.0
Vinyl FLR	82.0	Not given
VTE 1–6	18.2	10.9
VTE 2–3	45.9	16.9
VTE 3–2	35.8	15.4
Pr Full		8.0

Key: *General:* LC_{50} = toxic potency; Corr LC_{50} = toxic potency corrected for CO yields; FED = fractional effective doses. *Materials:* Acrylic F = acrylic fabric; Composite = naval composite board; Dg FIR = fire retarded Douglas fir board; FLX PU FM = flexible polyurethane foam; MELFM = melamine polyurethane foam; Nylon = nylon wire coating compound; Nylon Rug (Tr) = rug treated with PTFE coating; Nylon Rug (Un) = untreated rug; Pr Full = predicted carbon monoxide post flashover toxicity; PVC CB = PVC cable insulation; PVC INS = traditional PVC wire insulation compound; PVC JK = traditional PVC wire jacketing compound; PVC Lw HCl = PVC jacket compound and abundant amounts of acid retention filler; PVC Md HCl = PVC jacket compound and moderate amounts of acid retention filler; PVC PRF = rigid PVC profile; Rg PU FM = rigid polyurethane foam; Vinyl F = vinyl fabric; Vinyl FLR = vinyl flooring over plywood. Please see Table 26.2 for additional material designations.

to provide fire safety engineering information. Fire performance-based safety techniques involve fire hazard assessment or fire risk assessment, frequently, but not necessarily, via fire modeling.

An example of the type of results that can be obtained by applying fire hazard assessment concepts to the selection of a new (alternative) product to an existing one follows. The example is the choice of upholstered seating furniture in the patient room of a healthcare occupancy. The process and its potential outcomes include the following:

1. Choose fire scenarios.
2. Develop fire safety objectives and apply them to the scenarios chosen.
3. Use assessment techniques (or test methods) as input into calculation methods.
4. Assess the new item of upholstered seating furniture being considered for use in a certain healthcare facility, and reach one of the following five conclusions:
   a. The new chair is safer, on the basis of fire performance, than the one in established use. Then the new product would be desirable, from the view point of fire safety.
   b. The fire safety of the new and established chairs are the same. Then there would be no disadvantage in using the new product, from the view point of fire safety.
   c. The new chair is less safe, on the basis of fire performance, than the one in established use. Then the new chair would be undesirable, from the view point of fire safety.
   d. The new chair could only be used with suitable fire safety if the design of the patient room is altered, either by changing the layout or geometry, other furnishings or contents, or by adding protective measures, such as automatic sprinklers or smoke detectors, to get equivalent fire safety.
   e. The new chair offers some safety advantages and some safety disadvantages over the one in established use. This could mean more smoke obscuration with less heat release. Then a more in-depth fire hazard assessment would have to be conducted to balance the advantages and disadvantages.
5. If the patient room does not have a chair, then the same fire hazard assessment is done, but the comparison would be to see whether the new chair introduces unacceptable additional fire hazard.
6. The fire hazard assessment reaches a conclusion regarding desirability of the new chair.

The process described above has now become a national standard in the area of rail transportation vehicles.[122] Such a process, if properly conducted, ensures a free flow of ideas and the introduction into commerce of innovative designs, which can then compete on a level playing field with established solutions to a certain need. Rejection of this process freezes innovation. When a performance-based approach is introduced into a certain environment, it is essential, as an interim solution, to retain a prescriptive approach as an acceptable alternative. This would give the necessary leeway to those familiar with the existing system to continue using it.

Final Thoughts

Fire safety of polymeric materials is a combination of many aspects: thermal decomposition, ignition, flame spread, heat

Table 26.9 Effects of flame-retardant additives on overall smoke toxicity

Products	Peak temperature (K)	Smoke toxicity (kg CO)	Tenability time in room		CO yield (g/g)	Pk RHR (kW)
			Burn (s)	Target (s)		
Non-FR	>600	21	110	200	0.22	1590
Non-FR	>875	17	112	215	0.18	1540
Non-FR	>875	16	116	226	0.14	1790
FR 1B	458	2.6	NA	NA	0.22	220
FR	546	5.5	1939	NA	0.23	370
FR	558	6.1	2288	NA	0.23	350
FR	507	5.6	1140	1013	0.23	450

Key: Maximum temperature in burn room (peak temperature); level of toxicity, calculated in equivalent mass of carbon monoxide (smoke toxicity); time to reach untenable conditions in burn room or target room (via smoke toxicity or by having reached flashover, that is, temperatures over 600 °C, 873 K) (tenability time in room); mass of carbon monoxide formed per mass of fuel burned (CO yield); maximum heat release rate in room (Pk RHR); non-fire-retarded products (Non-FR); fire-retarded products (FR); fire-retarded products without an auxiliary burner (FR 1B).

release, and smoke obscuration, among others. Polymers have the added problem of being materials that still have not been fully studied and that have fire performance characteristics different from those of traditional materials they are replacing. Thus, innovative approaches must be applied to their study, including fire safety engineering techniques. It is critical to realize that safety, and especially fire safety, is not an absolute, but a relative proposition, and that each person, and society as a whole, sets its levels.

References

1. ASTM E176. Standard Terminology of Fire Standards. *Ann. Book ASTM Stand.*; ASTM: West Conshohocken, PA, Vol. 4.07.
2. Beyler, C. L., Hirschler, M. M. Thermal Decomposition of Polymers. In *SFPE Handbook of Fire Protection Engineering*, 2nd ed.; Society for Fire Protection Engineering: Boston, MA, 2002; Section 1, Chapter 7, pp 1/99–1/119.
3. Cullis, C. F.; Hirschler, M. M. The significance of thermoanalytical measurements in the assessment of polymer flammability. *Polymer* **1983**, *24*, 834–840.
4. Hirschler, M. M. Thermal analysis and flammability of polymers: Effect of halogen-metal additive systems. *Eur. Polym. J.* **1983**, *19*, 121–129.
5. Chandler, L. A.; Hirschler, M. M.; Smith, G. F. A heated tube furnace test for the emission of acid gas from PVC wire coating materials: effects of experimental procedures and mechanistic considerations. *Eur. Polym. J.* **1987**, *23*, 51–61.
6. Hirschler, M. M. Thermal decomposition (STA and DSC) of poly(vinyl chloride) compounds under a variety of atmospheres and heating rates. *Eur. Polym. J.* **1986**, *22*, 153–160.
7. McNeill, I. C. The Application of Thermal Volatilization Analysis to Studies of Polymer Degradation. In *Developments in Polymer Degradation*; Grassie, N., Ed.; Applied Science: London, 1977; Vol. 1, p 43.
8. Stuetz, D. E.; DiEdwardo, A. H.; Zitomer, F.; Barnes, B. F.; Polymer flammability. II. *J. Polym. Sci., Polym. Chem. Ed.* **1980**, *18*, 987–1009.
9. Kashiwagi, T.; Ohlemiller, T. J. A Study of Oxygen Effects on Flaming Transient Gasification of PMMA and PE during Thermal Irradiation. In *Proceedings of the Nineteenth Symposium (International) on Combustion*, The Combustion Institute: Pittsburgh, PA, 1982; pp 1647–1654.
10. Kashiwagi, T.; Hirata, T.; Brown, J. E. Thermal and oxidative degradation of poly(methyl methacrylate), molecular weight. *Macromolecules* **1985**, *18*, 131–138.
11. Hirata, T.; Kashiwagi, T.; Brown, J. E. Thermal and oxidative degradation of poly(methyl methacrylate), weight loss. *Macromolecules* **1985**, *18*, 1410–1418.
12. Kashiwagi, T.; Inabi, A.; Brown, J. E.; Hatada, K.; Kitayama, T.; Masuda, E. Effects of weak linkages on the thermal and oxidative degradation of poly(methyl methacrylates). *Macromolecules* **1986**, *19*, 2160–2168.
13. Kashiwagi, T.; Inabi, A. Behavior of primary radicals during thermal degradation of poly(methyl methacrylate). *Polym. Degrad. Stab.* **1989**, *26*, 161–184.
14. Brauman, S. K. Polymer degradation during combustion. *J. Polym. Sci., Part B: Polym. Chem.* **1988**, *26*, 1159–1171.
15. Gijsman, P.; Hennekens, J.; Vincent, J. The mechanism of the low-temperature oxidation of polypropylene. *Polym. Degrad. Stab.* **1993**, *42*, 95–105.
16. Hirschler, M. M. Effect of oxygen on the thermal decomposition of poly(vinylidene fluoride). *Eur. Polym. J.* **1982**, *18*, 463–467.
17. Inaba, A.; Kashiwagi, T. A calculation of thermal degradation initiated by random scission, unsteady radical concentration. *Eur. Polym. J.* **1987**, *23* (11), 871–881.
18. Kashiwagi, T.; Nambu, H. Global kinetic constants for thermal oxidative degradation of a cellulosic paper. *Combust. Flame.* **1992**, *88*, 345–368.

19. Steckler, K. D.; Kashiwagi, T.; Baum, H. R.; Kanemaru, K. Analytical Model for Transient Gasification of Noncharring Thermoplastic Materials. In *Fire Safety Science, Proceedings of the Third International Symposium*; Cox, G., Langford, B., Eds.; Elsevier: London, 1991.

20. Cullis, C. F.; Hirschler, M. M. *The Combustion of Organic Polymers*; Oxford University Press: Oxford, 1981.

21. Weil, E. D.; Hansen, R. N.; Patel, N. Prospective Approaches to More Efficient Flame-Retardant Systems. In *Fire and Polymers: Hazards Identification and Prevention*; Nelson, G. L., Ed.; ACS Symposium Series 425, American Chemical Society: Washington, DC, 1990; Chapter 8, pp 97–108.

22. Bowes, P. C. *Self-Heating: Evaluating and Controlling the Hazards*; Elsevier: Oxford, UK, 1984.

23. Beever, P. F. Self-Heating and Spontaneous Combustion. *SFPE Handbook of Fire Protection Engineering*, 2nd ed. Society for Fire Protection and Engineering: Boston, 1995; Section 2, Chapter 12, pp 2/180–2/189.

24. Howitt, D. G.; Zhang, E.; Sanders, B. R. The Spontaneous Combustion of Linseed Oil. In *Proceedings of the 20th International Conference on Fire Safety*, Hilado, C. J., Ed.; Jan. 9–13, 1995, San Francisco, 1995.

25. Jones, J. C. Self-heating tests to establish the shipping safety of solid materials: the case for briquetted peat. *J. Fire Sci.* **1996**, *14*, 342–345.

26. Hill, S. M.; Quintiere, J. G. Investigating Materials from Fires Using a Test Method for Spontaneous Ignition: Case Studies. In *Proceedings of the 5th International Fire and Materials Conference*; Grayson, S. J., Ed.; Interscience Communications: London, 1998; pp 171–181.

27. Montaudo, G.; Puglisi, C. Evolution of aromatics in the thermal degradation of poly(vinyl chloride): a mechanistic study, *Polym. Degrad. Stab.* **1991**, *33*, 229–262.

28. Starnes, W. H., Jr.; Girois, S. *Polymer Yearbook*; Pethrick, R. A., Ed.; Gordon & Breach: Langhorne, PA, 1995; Vol. 12, p 105.

29. Starnes, W. H., Jr.; Edelson, D. *Macromolecules* **1979** *12*, 797.

30. Edelson, D.; Lum, R. M.; Reents, W. D., Jr.; Starnes, W. H., Jr.; Westcott, L. D., Jr. New Insights into the Flame Retardance chemistry of Poly(vinyl chloride). In *Proceedings of the Nineteenth (International) symposium on Combustion*; The Combustion Institute: Pittsburgh, PA, 1982; pp 807–814.

31. Minsker, K. S.; Klesov, S. V.; Yanborisov, V. W.; Berlin, A. A.; Zaikov, G. E. The reason for the low stability of poly(vinyl chloride)—A review. *Polym. Degrad. Stab.* **1986**, *16*, 99–133.

32. Simon, P.; et al. Kinetics of polymer degradation involving the splitting off of small molecules. Parts 1–7. *Polym. Degrad. Stab.* (a) **1990**, *29*, 155, 253, 263; (b) **1992**, *35*, 157, 249; (c) **1992**, *36*, 85.

33. Jeng, J. P.; Terranova, S. A.; Bonaplata, E.; Goldsmith, K.; Williams, D. M.; Wojciechowski, B. J.; Starnes, W. H., Jr. ACS Symposium Series 599; American Chemical Society: Washington, DC, 1995; p 118.

34. Pike, R. D.; Starnes, W. H., Jr.; Jeng, J. P.; Bryant, W. S.; Kourtesis, P.; Adams, C. W.; Bunge, S. D.; Kang, Y. M.; Kim, A. S.; Kim, J. H.; Macko, J. A.; O'Brien, C. P. *Macromolecules* **1997**, *30*, 6957.

35. Starnes, W. H., Jr.; Frantz, S.; Chung, H. T. *Polym. Degrad. Stab.* **1997**, *56*, 103.

36. Chandler, L. A.; Hirschler, M. M. Further chlorination of poly(vinyl chloride): Effects on flammability and smoke production tendency. *Euro. Polym. J.* **1987**, *23*, 677–683.

37. Levchik, S. V. Thermosetting Resins. In *International Plastics Flammability Handbook*, 3rd ed. Troitszch, J., Ed.; in preparation.

38. Levchik, S. V.; Camino, G.; Luda, M. P.; Costa, L.; Costes, B.; Henry, Y.; Muller, G.; Morel, E. Mechanistic Study of Thermal Behaviour and Combustion Performance of Epoxy Resins. II TGDDM/DDS System. *Polym. Degrad. Stab.* **1995**, *48*, 359–370.

39. Levchik, S. V.; Levchik, G. F.; Antonov, A. V.; Yablokova, M. Yu.; Tuzhikov, O. I.; Tuzhikov, O. O.; Costa, L. Fire Retardant Action of Phosphorus-Based Fire Retardants in Epoxy Resins. In *Proceedings of the 10th Annual Business Communications Company Conference on Recent Advances in Flame Retardancy of Polymeric Materials* in Stamford, CT, May 24–26, 1999; Lewin, M., Ed.; Business Communications Co.: Norwalk, CT, 1999.

40. Pearce, E. M. Cross-linking and Char Formation. In *Proceedings of the 5th Annual Business Communications Company Conference on Recent Advances in Flame Retardancy of Polymeric Materials* in Stamford, CT, May 24–26, 1994; Lewin, M., Ed.; Business Communications Co.: Norwalk, CT, 1994; pp 47–55.

41. Van Krevelen, D. W. Thermal Decomposition, and Product Properties: Environmental Behavior and Failure. In *Properties of Polymers*, 3rd ed. Elsevier: Amsterdam, 1990; Chapter 21, pp 641–653 and Chapter 26, pp 725–743.

42. Van Krevelen, D. W. Some basic aspects of flame resistance of polymeric materials. *Polymer* **1975**, *16*, 615–620.

43. Weil, E. D.; Hirschler, M. M.; Patel, N. G.; Said, M. M.; Shakir, S. Oxygen index: Correlation to other tests. *Fire Mater.* **1992**, *16*, 159–167.

44. Lyon, R. E.; Walters, R. N. Microscale Heat Release Rate of Polymers. In *Proceedings of the 5th International Fire and Materials Conference* in San Antonio, TX; Grayson, S. J., Ed.; Interscience Communications, London, 1998; pp 195–203.

45. Camino, G.; Costa, L. Performance and mechanisms of fire retardants in polymers—a review. *Polym. Degrad. Stab.* **1988**, *20*, 271–294.

46. Bertelli, G.; Costa, L.; Fenza, S.; Marchetti, F. E.; Camino, G.; Locatelli, R. Thermal behaviour of bromine-metal fire retardant systems. *Polym. Degrad. Stab.* **1988**, *20*, 295–314.

47. Bourbigot, S.; LeBras, M.; Delobel, R.; Breant, P.; Tremillon, J. M. *Polym. Degrad. Stab.* **1996**, *53*, 275.

48. Horrocks, A. R.; Kandola, B. K. The Use of Intumescents to Render Flame-Retardant Cotton Fibers More Reactive. In *Proceedings of the 8th Annual Business Communications Company Conference on Recent Advances in Flame Retardancy of Polymeric Materials* in Stamford, CT, June 2–4, 1997; Lewin, M., Ed.; Business Communications Co.: Norwalk, CT, 1997.

49. Hirschler, M. M. Survey of American test methods asso-

ciated with fire performance of materials or products. *Polym. Degrad. Stab.* **1997**, *57*, 333–343.

50. ASTM E2102. Standard Test Method for Screening Purposes for Measurement of Mass Loss and Ignitability Using a Conical Radiant Heater. *Annu. Book ASTM Stand.*; ASTM: West Conshohocken, PA, Vol. 4.07.

51. ASTM D1929. Standard Test Method for Ignition Properties of Plastics. *Ann. Book of ASTM Stand.*; ASTM: West Conshohocken, PA, Vol. 8.02.

52. Hilado, C. J. *Flammability Handbook of Plastics*, 3rd ed.; Technomic: Lancaster, PA, 1982.

53. Landrock, A. H. *Handbook of Plastics Flammability and Combustion Toxicology*. Noyes: Park Ridge, NJ, 1983.

54. Hirschler, M. M. Fire hazard and toxic potency of the smoke from burning materials. *J. Fire Sci.* **1987**, *5*, 289–307.

55. ASTM E1354. Standard Test Method for Heat and Visible Smoke Release Rates for Materials and Products Using an Oxygen Consumption Calorimeter. *Annu. Book ASTM Stand.*; ASTM: West Conshohocken, PA, Vol. 4.07.

56. Hirschler, M. M. Heat Release from Plastic Materials. In *Heat Release in Fires*; Babrauskas, V., Grayson, S. J., Eds.; Elsevier: London, pp 375–422 (1992).

57. ASTM E906. Standard Test Method for Heat and Visible Smoke Release Rates for Materials and Products. *Annu. Book ASTM Stand.*; ASTM: West Conshohocken, PA, Vol. 4.07.

58. ASTM E1321. Standard Test Method for Determining Material Ignition and Flame Spread (LIFT). *Annu. Book ASTM Stand.*; ASTM: West Conshohocken, PA, Vol. 4.07.

59. ASTM D2863. Standard Test Method for Measuring the Minimum Oxygen Concentration to Support Candle-like Combustion of Plastics (Oxygen Index). *Annu. Book ASTM Stand.*; ASTM: West Conshohocken, PA, Vol. 8.02.

60. UL 94. Standard for Test for Flammability of Plastic Materials for Parts in Devices and Appliances; Underwriters Laboratories: Northbrook, IL, October 29, 1996.

61. ASTM E84. Standard Test Method for Surface Burning Characteristics of Building Materials. *Annu. Book ASTM Stand.*; ASTM: West Conshohocken, PA, Vol. 4.07.

62. ASTM E162. Standard Test Method for Surface Flammability of Materials Using a Radiant Heat Energy Source. *Annu. Book ASTM Stand.*; ASTM: West Conshohocken, PA, Vol. 4.07.

63. ASTM D3675. Standard Test Method for Surface Flammability of Flexible Cellular Materials Using a Radiant Heat Energy Source. *Annu. Book ASTM Stand.*; ASTM: West Conshohocken, PA, Vol. 8.02.

64. Steiner, A. J. Research Bulletin No. 32; Underwriters Laboratories: Northbrook, IL, 1944.

65. NFPA 262. Standard Method of Test for Flame Travel and Smoke of Wires and Cables for Use in Air-Handling Spaces. National Fire Protection Association: Quincy, MA.

66. Belles, D. W.; Fisher, F. L.; Williamson, R. B. How well does the ASTM E84 predict fire performance of textile wallcoverings? *Fire J.* **1988**, *82* (1), 24–30, 74.

67. Quintiere, J. G.; Harkleroad, M. New Concepts for Measuring Flame Spread Properties, NBSIR 84-2943, U.S. National Bureau Standards: Gaithersburg, MD, 1984.

68. Cleary, T. G.; Quintiere, J. G. A Framework for Utilizing Fire Property Tests, NISTIR 91-4619, U.S. National Institute Standards and Technology: Gaithersburg, MD, 1991.

69. Babrauskas, V. Effective Measurement Techniques for Heat, Smoke and Toxic Fire Gases. In *International Conference on FIRE: Control the Heat—Reduce the Hazard*, October 24–25, 1988; Fire Research Station: London, 1988; no. 4.

70. Babrauskas, V.; Grayson, S. J. *Heat Release in Fires*; Elsevier: London, 1992.

71. CBUF Report. Fire Safety of Upholstered Furniture—The Final Report on the CBUF Research programme; Sundstrom, B., Ed.; EUR 16477 EN, European Commission, Measurements and Testing Report; Contract No. 3478/1/0/196/11-BCR-DK(30), Interscience Communications: London, 1995.

72. Hirschler, M. M. Analysis of and potential correlations between fire tests for electrical cables, and how to use this information for fire hazard assessment. *Fire Technol.* **1997**, *33*, 291–315.

73. Babrauskas, V.; Peacock, R. D. Heat release rate: The single most important variable in fire hazard. *Fire Safety J.* **1992**, *18*, 255–272.

74. Smith, E. E. Heat Release Rate of Building Materials. In *Ignition, Heat Release and Noncombustibility of Materials*, ASTM STP 502, Robertson, A. F., Ed.; ASTM: West Conshohocken, PA, 1972, p 119.

75. Hirschler, M. M. The measurement of smoke in rate of heat release equipment in a manner related to fire hazard. *Fire Safety J.* **1991**, *17*, 239–258.

76. Babrauskas, V. *Development of the Cone Calorimeter. A Bench-Scale Heat Release Rate Apparatus Based on Oxygen Consumption.* NBSIR 82-2611; National Bureau of Standards: Gaithersburg, MD, 1982.

77. Babrauskas, V. *Bench-Scale Methods for Prediction of Full-Scale Fire Behavior of Furnishings and Wall Linings.* Technology Report 84-10; Society of Fire Protection Engineers: Boston, 1984.

78. Babrauskas, V. Upholstered furniture room fires—measurements, comparison with furniture calorimeter data, and flashover predictions. *J. Fire Sci.* **1984**, *2*, 5–19.

79. Babrauskas, V.; Krasny, J. F. Prediction of Upholstered Chair Heat Release Rates from Bench-Scale Measurements. In *Fire Safety. Science and Engineering*, ASTM STP 882; Harmathy, T. Z., Ed.; ASTM: West Conshohocken, PA, 1985; p 268.

80. Hirschler, M. M. Tools Available to Predict Full Scale Fire Performance of Furniture. In *Fire and Polymers II. Materials and Tests for Hazard Prevention*, Nelson, G. L., Ed.; ACS Symposium Series 599, American Chemical Society: Washington, DC, 1995; Chapter 36, pp 593–608.

81. ASTM E1537. Standard Test Method for Fire Testing of Upholstered Seating Furniture. *Annu. Book ASTM Stand.*; ASTM: West Conshohocken, PA, Vol. 4.07.

82. ASTM E1590. Standard Test Method for Fire Testing of Mattresses. *Annu. Book ASTM Stand.*; ASTM: West Conshohocken, PA, Vol. 4.07.

83. ASTM E1822. Standard Test Method for Fire Testing of Real Scale Stacked Chairs. *Annu. Book ASTM Stand.*; ASTM: West Conshohocken, PA, Vol. 4.07.

84. ASTM D5424. Standard Test Method for Smoke Obscuration Testing of Insulating Materials Contained in Electrical or Optical Fiber Cables When Burning in a Vertical Configuration. *Annu. Book ASTM Stand.*; ASTM: West Conshohocken, PA, Vol. 10.02.

85. ASTM D5537. Standard Test Method for Heat Release, Flame Spread and Mass Loss Testing of Insulating Materials Contained in Electrical or Optical Fiber Cables When Burning in a Vertical Cable Tray Configuration. *Annu. Book ASTM Stand.*; ASTM: West Conshohocken, PA, Vol. 10.02.

86. UL 1975. *Standard for Fire Tests for Foamed Plastics Used for Decorative Purposes*; Underwriters Laboratories: Northbrook, IL.

87. NFPA 265. *Standard Methods of Fire Test for Evaluating Room Fire Growth Contribution of Textile Wall Coverings*; National Fire Protection Association: Quincy, MA.

88. NFPA 286. *Standard Methods of Fire Test for Evaluating Room Fire Growth Contribution of Interior Finish*; National Fire Protection Association: Quincy, MA.

89. ISO 9705. *Room Fire Test in Full Scale for Surface Products*; International Organization for Standardization: Geneva, Switzerland.

90. Hirschler, M. M. Use of heat release rate to predict whether individual furnishings would cause self propagating fires. *Fire Safety J.* **1999**, *32*, 273–296.

91. Sarkos, C. P.; Hill, R. G. *Evaluation of Aircraft Interior Panels Under Full-Scale Cabin Fire Test Conditions*; AIAA-85-0393; AIAA 23rd Aerospace Sciences Meeting, Reno, NV, 1985.

92. Wickstrom, U. *Proceedings of the International EUREFIC Seminar 1991*, Sept. 11–12, 1991; Copenhagen, Denmark; Interscience Communications: London, 1991.

93. Hirschler, M. M. Fire Retardance, Smoke Toxicity and Fire Hazard. In *Proceedings of Flame Retardants '94*, British Plastics Federation, Ed.; Interscience Communications: London, 1994; pp 225–237.

94. Hirschler, M. M. Smoke and heat release and ignitability as measures of fire hazard from burning of carpet tiles. *Fire Safety J.* **1992**, *18*, 305–324.

95. ASTM E662. Standard Test Method for Specific Optical Density of Smoke Generated by Solid Materials. *Annu. Book ASTM Stand.*; ASTM: West Conshohocken, PA, Vol. 4.07.

96. Babrauskas, V. Applications of predictive smoke measurements. *J. Fire Flammability* **1981**, *12*, 51.

97. Quintiere, J. G. Smoke measurements: an assessment of correlations between laboratory and full-scale experiments. *Fire Mater.* **1982**, *6*, 145.

98. Babrauskas, V. Use of the Cone Calorimeter for Smoke Prediction Measurements. In *Society of Plastics Engineers RETEC Conference on PVC: The Issues*, Atlantic City, NJ, 1987, p 41.

99. Hirschler, M. M. Smoke in Fires: Obscuration and Toxicity. In *Proceedings of the Annual Business Communications Company Conference on Recent Advances in Flame Retardancy of Polymeric Materials* in Stamford, CT; Kirshenbaum, G. S., Lewin, M., Eds.; Business Communications Co.: Norwalk, CT, 1990; pp 70–82.

100. Hirschler, M. M. How to measure smoke obscuration in a manner relevant to fire hazard assessment: Use of heat release calorimetry test equipment. *J. Fire Sci.* **1991**, *9*, 183–222.

101. Breden, L. H.; Meisters, M. The effect of sample orientation in the smoke density chamber. *J. Fire Flammability* **1976**, *7*, 234.

102. Hirschler, M. M.; Janssens, M. L. Smoke Obscuration Measurements in the NFPA 265 Room-Corner Test. In *Proceedings of the 6th Fire and Materials Conference* in San Antonio, TX; Grayson, S. J., Ed.; Interscience Communications: London, 1999; pp 179–198.

103. Hirschler, M. M. Fire Performance of Organic Polymers, Thermal Decomposition, and Chemical Composition. In *Fire and Polymers III, Materials and Solutions for Hazard Prevention*; Nelson, G. L., Wilkie, C. A., Eds.; ACS Symposium Series 797; American Chemical Society: Washington, DC, 2001; pp 293–306.

104. Jin, T. Studies of emotional instability in smoke from fires. *J. Fire Flammability* **1981**, *12*, 130–142.

105. Babrauskas, V.; Harris, R. H.; Gann, R. G.; Levin, B. C.; Lee, B. T.; Peacock, R. D.; Paabo, M.; Twilley, W.; Yoklavich, M. F.; Clark, H. M. *Fire Hazard Comparison of Fire-Retarded and Non-Fire-Retarded Products*; NBS Special Publication 749; National Bureau of Standards: Gaithersburg, MD, 1988.

106. Mulholland, G. W. In *Executive Summary for the Workshop on Developing a Predictive Capability for CO Formation in Fires*; NISTIR 89-4093; Pitts, W. M., Ed.; National Institute of Standards and Technology: Gaithersburg, MD, 1989; p 25.

107. Babrauskas, V.; Harris, R. H.; Braun, E.; Levin, B. C.; Paabo, M.; Gann, R. G. *The Role of Bench-Scale Data in Assessing Real-Scale Fire Toxicity*; NIST Technical Note 1284; National Institute of Standards and Technology: Gaithersburg, MD, 1991.

108. Hirschler, M. M. General Principles of Fire Hazard and the Role of Smoke Toxicity. In *Fire and Polymers: Hazards Identification and Prevention*; Nelson, G. L., Ed.; ACS Symposium Series 425; American Chemical Society: Washington, DC, 1990; Chapter 28, pp 462–478.

109. Babrauskas, V.; Levin, B. C.; Gann, R. G.; Paabo, M.; Harris, R. H.; Peacock, R. D.; Yusa, S. *Toxic Potency Measurement for Fire Hazard Analysis*; NIST Special Publication 827; National Institute of Standards and Technology: Gaithersburg, MD, 1991.

110. Debanne, S. M.; Hirschler, M. M.; Nelson, G. L. The Importance of Carbon Monoxide in the Toxicity of Fire Atmospheres. In *Fire Hazard and Fire Risk Assessment*; Hirschler, M. M., Ed.; ASTM STP 1150; ASTM: West Conshohocken, PA, 1992; pp 9–23.

111. Hirschler, M. M.; Debanne, S. M.; Larsen, J. B.; Nelson, G. L. *Carbon Monoxide and Human Lethality—Fire and Non-Fire Studies*. Elsevier: London, 1993.

112. Gann, R. G.; Babrauskas, V.; Peacock, R. D.; Hall, J. R. Fire conditions for smoke toxicity measurement. *Fire Mater.* **1994**, *18*, 193–199.

113. ASTM E1678. Standard Test Method for Measuring Smoke Toxicity for Use in Fire Hazard Analysis. *Annu. Book ASTM Stand.*; ASTM: West Conshohocken, PA, Vol. 4.07.

114. NFPA 269. *Standard Test Method for Developing Toxic Po-

tency Data for Use in Fire Hazard Modeling; National Fire Protection Association: Quincy, MA.

115. Hirschler, M. M. The Role of Carbon Monoxide in the Toxicity of Fire Atmospheres. In *Proceedings of the 19th International Conference on Fire Safety*; Hilado, C. J., Ed.; Product Safety Corp.: San Francisco, 1994; pp 163–184.

116. Budnick, E. K. *Mobile Home Living Room Fire Studies.*; NBSIR 78-1530; National Bureau of Standards: Gaithersburg, MD, 1978.

117. Budnick, E. K.; Klein, D. P.; O'Laughlin, R. J. *Mobile Home Bedroom Fire Studies: The Role of Interior Finish.* NBSIR 78-1531; National Bureau of Standards: Gaithersburg, MD, 1978.

118. Levine, R. S.; Nelson, H. E. *Full Scale Simulation of a Fatal Fire and Comparison of Results with Two Multiroom Models*; NISTIR 90-4268; National Institute of Standards and Technology: Gaithersburg, MD, 1990.

119. Gottuk, D. T.; Roby, R. J.; Peatross, M. J.; Beyler, C. L. CO production in compartment fires. *J. Fire Protect. Eng.* **1992**, *4*, 133–150.

120. Beyler, C. L. Major species production by diffusion flames in a two-layer compartment fire environment. *Fire Safety J.* **1986**, *10*, 47–56.

121. Karlsson, B. Models for calculating flame spread on wall lining materials and the resulting heat release rate in a room. *Fire Safety J.* **1994**, *23*, 365–386.

122. ASTM E2061. Standard Guide for Fire Hazard Assessment of Rail Transportation Vehicles. *Annu. Book ASTM Stand.*; ASTM: West Conshohocken, PA, Vol. 4.07.

Index